MW01328691

Jürgen Sachs

Handbook of Ultra-Wideband Short-Range Sensing

Related Titles

Haykin, S. (ed.)

Adaptive Radar Signal Processing

230 pages
2006
Hardcover
ISBN: 978-0-471-73582-3

Allen, B., Dohler, M., Okon, E., Malik, W., Brown, A., Edwards, D. (eds.)

Ultra Wideband Antennas and Propagation for Communications, Radar and Imaging

508 pages
2006
E-Book
ISBN: 978-0-470-05682-0

Arslan, H., Chen, Z. N., Di Benedetto, M.-G. (eds.)

Ultra Wideband Wireless Communication

520 pages
2006
Hardcover
ISBN: 978-0-471-71521-4

Sengupta, D. L., Liepa, V. V.

Applied Electromagnetics and Electromagnetic Compatibility

486 pages
2005
Hardcover
ISBN: 978-0-471-16549-1

Jürgen Sachs

Handbook of Ultra-Wideband Short-Range Sensing

Theory, Sensors, Applications

WILEY-VCH Verlag GmbH & Co. KGaA

The Author

Dr.-Ing. Jürgen Sachs
Ilmenau University of Technology
Electrical Engineering and Information Technology
Institute for Information Technology
Electronic Measurement Research Lab
Ilmenau, Germany

Library of Congress Card No.: applied for

British Library Cataloguing-in-Publication Data
A catalogue record for this book is available from the British Library.

Bibliographic information published by the Deutsche Nationalbibliothek
The Deutsche Nationalbibliothek lists this publication in the Deutsche Nationalbibliografie; detailed bibliographic data are available on the Internet at http://dnb.d-nb.de.

Print ISBN: 978-3-527-40853-5
ePDF ISBN: 978-3-527-65184-9
ePub ISBN: 978-3-527-65183-2
mobi ISBN: 978-3-527-65182-5
oBook ISBN: 978-3-527-65181-8

Cover Design Adam-Design, Weinheim

Typesetting Thomson Digital, Noida, India

Printing and Binding Markono Print Media Pte Ltd, Singapore

Printed in Singapore
Printed on acid-free paper

Contents

Preface

Ultra-wideband (UWB) sensors exploit the manifold interactions between electromagnetic fields and matter. They aim at gaining information about a target, the environment, a technical process, or a substance by remote, non-destructive, continuous and fast measurement procedures. The applied sounding waves are of low power (typically thousand times less than the radiation of a mobile phone) and their frequencies range within the lower gigahertz domain. This enables acceptable wave penetration into optically opaque objects. Furthermore, the electromagnetic waves are harmless due to their low power and they are non-ionizing. This makes ultra-wideband sensors attractive for many short-range sensing tasks covering industrial, medical, security and rescue issues of detection, recognition, tracking, surveillance, quality control and so on.

Electromagnetic sounding is an inverse and indirect approach to gather information about the objects of interest. Inverse and indirect methods – independent of the underlying physical principles – are always prone to ambiguities with unwanted objects or events. The reduction of such cross-sensitivity needs diversity in data capturing in order to comprise preferably orthogonal interaction phenomena into data interpretation, data fusion and decision. Restricting to the determination of purely electrical properties, some ways to increase the diversity of data capturing consist of (a) measuring over a large frequency band, (b) measuring at many points in space, (c) observing the evolution of the scenario over a long time and (d) respecting the vector nature of the electromagnetic field by including polarimetric data. In addition, the measurement of non-electric properties may further reduce the ambiguities. However, we will exclude non-electric measurements from our discussions here. As is clear from the title of the book itself, it is point (a) that will be in the foreground of our interest, giving an outline of this book without losing however our other important viewpoints.

With a few exceptions, the classical electrical engineering is 'narrowband'. This has not only historical but also theoretical roots. One, essentially by theory-motivated reason, to deal with narrowband (i.e. 'gently' modulated sine wave) signals is the property of a sine wave to keep its time evolution by interacting with a (linear) object. This reduces and simplifies theoretical evaluations to (not necessarily facile) magnitude and phase calculations and accelerates numerical computations. But corresponding considerations presume steady-state conditions which often obscure the intuitive understanding of the occurring processes.

Short pulses count to the class of wideband signals that are of major interest here. They are traditionally used in ultra-wideband technique. Pulse propagation is easy to comprehend. However, such signals are subjected to a modification of their time shape if interacting with an object or propagating in a lossy medium. These time shape variations are our major source of information about the objects of interest, but they are also more demanding with respect to modelling and mathematical treatment.

The propagation of time-extended wideband signals (e.g. random or pseudo-random signals) is intuitively less comprehensive, though the technical implementation of related sensor electronics provides some advantages in favour of pulse systems. In order to retrieve the illustrative understanding of wave propagation, we will trace back such signals to impulse-like time functions which can be handled like real pulses for our scope of application.

The purpose of this book is to give an overview of theoretical, implementation and application aspects of low-power – and hence short-range – ultra-wideband sensing. Intended readers are students, engineers and researches with a background in undergraduate level of mathematics, signal and system theory, electric circuit theory and electromagnetic wave propagation. The introductory part of the book introduces the definition of the UWB term and it gives a short overview of UWB history, the radiation regulations, possible fields of application and the basic approach of information gathering by UWB sensors.

Chapter 2 defines characteristic functions and parameters and summarizes basic concepts of signal and system theory which are important for the understanding of functioning of UWB sensors and their applications. It is mainly targeted at the less experienced reader in this field. Numerous figures are inserted to illustrate the basic relations and, in addition, an annex available at Wiley homepage (http://www.wiley-vch.de, search for ISBN 978-3-527-40853-5) provides a collection of useful mathematical rules, properties of signal transformations, and some basic considerations on signals and elementary signal operations. Furthermore, the reader can also find also some colored figures here and a couple of movies which complement several figures to better illustrate three-dimensional data sets.

Chapter 3 deals with the different concepts of UWB sensing electronics and their key properties. Developments within the last decade educe new and improved sensor principles allowing manufacturing inexpensive, monolithically integrated, lightweight and small microwave devices. They are prerequisites to pave the way for UWB sensing from laboratory to the field.

Chapter 4 discusses some peculiarities of UWB radar whose physical principle is indeed the same as for the 'traditional' radar, but the large fractional bandwidth requires some extensions and specifications of the classical radar theory. They are mainly required by the fact that achievable UWB resolutions may be far better than the geometric size of involved bodies as antennas and targets. The considerations are focused on wideband aspects of wave propagation but they are restricted to a simplified model of scalar waves.

In Chapter 5, the actual vector character of the electromagnetic wave and some of its implications are briefly treated, introducing some basic time domain models of wave propagation.

In the final chapter, some selected aspects of sensor implementation and application are discussed. These topics are contributed by several co-authors who were cooperating with me in numerous UWB projects during the last years.

I take this opportunity to thank all these co-authors and ancient project collaborators, institutions and companies for their fruitful work and pleasant collaboration. These projects and consequently this book could not have been possible without the chance to work on interesting tasks of the ultra-wideband technique at Ilmenau University of Technology and without the encouragement and support of many of the colleagues from research and technical staff as well as the administration. In particular, I would like to thank my colleagues at Electronic Measurement Research Lab for their support, engagement and productive interaction. I also owe a great deal to Stephen Crabbe who promoted and managed our first projects which bred the ultra-wideband M-sequence approach. The support of our research from the German Science Foundation (DFG) through the UKoLoS Program and also from various national and European projects is gratefully acknowledged. I am indebted to Valerie Molière of Wiley-VCH Verlag GmbH for inviting me to write this book and to Nina Stadthaus for her help and patience with any delays and project modifications. Thanks are due to Vibhu Dubey from Thomson Digital for manuscript corrections and typesetting. Last but not least, I am deeply grateful to my wife and my family, who displayed such appreciation, patience, support and love while I was occupied with this book.

Even if the extent of the book largely exceeds the initial intention, the text is by no means complete and all-embracing. In order to limit the number of pages and to complete the work in a reasonable amount of time, some decisions had to make what topics could be appropriate and what subjects could be omitted or reduced in the depth of their consideration. I may only hope that the reader can identify with my decision and that the work of the many researchers in the field of ultra-wideband technique is adequately appreciated. Clearly, the blame for any errors and omissions lies with the author. Let me know if you encounter any.

Schmiedefeld, Germany *Jürgen Sachs*
August 2012

List of Contributors

Frank Bonitz
Materialforschungs- und -prüfanstalt
Weimar an der Bauhaus-Universität
Weimar (MFPA)
Coudraystraße 9
D-99423 Weimar
Germany

Frank Daschner
University of Kiel
Technical Faculty
Institute of Electrical Engineering
and Information
Engineering
Microwave Group
Kaiserstrasse 2
24143 Kiel

Matthias A. Hein
Ilmenau University of Technology
RF and Microwave Research
Laboratory
P.O. Box 100565
98684 Ilmenau
Germany

Marko Helbig
Ilmenau University of Technology
Electronic Measurement Research
Lab
P.O.Box 100565
98684 Ilmenau
Germany

Ralf Herrmann
Ilmenau University of Technology
Electronic Measurement Research
Lab
P.O.Box 100565
98684 Ilmenau
Germany

Michael Kent
University of Kiel
Technical Faculty
Institute of Electrical Engineering
and Information
Engineering
Microwave Group
Kaiserstrasse 2
24143 Kiel

Martin Kmec
Ilmenau University of Technology
Electronic Measurement Research
Lab
P.O.Box 100565
98684 Ilmenau
Germany

Reinhard Knöchel
University of Kiel
Technical Faculty
Institute of Electrical Engineering
and Information Engineering
Microwave Group
Kaiserstrasse 2
24143 Kiel

Dušan Kocur
Technická univerzita v Košiciach
Fakulta elektrotechniky a informatiky
Katedra elektroniky a multimediálnych telekomunikácií
Letná 9
041 20 Košice
Slovenská republika

Olaf Kosch
Physikalisch-Technische Bundesanstalt (PTB)
Abbestraße 2-12
10587 Berlin
Germany

Jana Rovňáková
Technická univerzita v Košiciach
Fakulta elektrotechniky a informatiky
Katedra elektroniky a multimediálnych telekomunikácií
Letná 9
041 20 Košice
Slovenská republika

Jürgen Sachs
Ilmenau University of Technology
Electronic Measurement Research Lab
P.O. Box 100565
98684 Ilmenau
Germany

Ulrich Schwarz
BMW Group
Entertainment and Mobile Devices
Max-Diamand-Straße 15–17
80937 Munich, Germany

Francesco Scotto di Clemente
Ilmenau University of Technology
RF and Microwave Research Laboratory
P.O. Box 100565
98684 Ilmenau
Germany

Frank Seifert
Physikalisch-Technische Bundesanstalt (PTB)
Abbestraße 2-12
10587 Berlin
Germany

Florian Thiel
Physikalisch-Technische Bundesanstalt (PTB)
Abbestraße 2-12
10587 Berlin
Germany

Daniel Urdzík
Technická univerzita v Košiciach
Fakulta elektrotechniky a informatiky
Katedra elektroniky a multimediálnych telekomunikácií
Letná 9
041 20 Košice
Slovenská republika

Egor Zaikov
Institute for Bioprocessing and Analytical Measurement Techniques e.V.
Rosenhof
37308 Heilbad Heiligenstadt
Germany

Rudolf Zetik
Ilmenau University of Technology
Electronic Measurement Research Lab
P.O. Box 100565
98684 Ilmenau
Germany

1
Ultra-Wideband Sensing – An Overview

1.1
Introduction

For the human beings (and most of the animals), the scattering of electromagnetic waves, for example the scattering of sunlight at trees, buildings or the face of a friend or an enemy, is the most important source to gain information on the surroundings. As known from everybody's experience, the images gained from light scattering (i.e. photos) provide a detailed geometrical structure of the surroundings since the wavelengths are very small compared to the size of the objects of interest. Furthermore, the time history of that 'scattering behaviour' gives us a deep view inside the nature of an object or process. However, there are many cases where light scattering fails and we are not able to receive the wanted information by our native sense. Therefore, technical apparatuses were created which use different parts of the electromagnetic spectrum as X-rays, infrared and Terahertz radiation or microwaves exploiting each with specific properties of wave propagation.

Ultra-wideband (UWB) sensors are dealing with microwaves occupying a very large spectral band typically located within the lower GHz range. On the one hand, such waves can penetrate most (non-metallic) materials so that hidden objects may be detected, and on the other hand, they provide object resolution in the decimetre, centimetre or even millimetre range due to their large bandwidth. Moreover, polar molecules, for example water, are showing relaxation effects within these frequency bands which give the opportunity of substance characterization and validation. In general, it can be stated that a large bandwidth of a sounding signal provides more information on the object of interest.

With the availability of network analysers and the time domain reflectometry (TDR) since the 60th of the last century, very wideband measurements have been established but they were banned to laboratory environments. New and cheaper solutions for the RF-electronics, improved numerical capabilities to extract the wanted information from the gathered data and the effected ongoing adaptation of radio regulation rules by national and international regulation authorities allow this

Handbook of Ultra-Wideband Short-Range Sensing: Theory, Sensors, Applications, First Edition. Jürgen Sachs.

sensor approach to move stepwise in practice now. Ultra-wideband sensing is an upcoming technique to gather data from complex scenarios such as nature, industrial facilities, public or private environments, for medical applications, non-destructive testing, security and surveillance, for rescue operations and many more. Currently, it is hard to estimate the full spread of future applications.

The objective of this book is to introduce the reader to some aspects of ultra-wideband sensing. Such sensors use very weak and harmless electromagnetic sounding waves to 'explore' their surroundings. Sensor principles using electromagnetic waves are not new and are in use for many years. But they are typically based on narrowband signals. In contrast, the specific of UWB sensors is to be seen in the fact that they apply sounding signals of a very large bandwidth whereas bandwidth and centre frequency[1] are of the same order.

Concerning their application there are four major consequences:

- As a generic rule of thumb one can state that increasing frequency diversity leads to more information about the scenario under test. This observation is well respected by UWB sensors due to their large bandwidth. Hence they will have better resolution, lower cross-ambiguities or better recognition capabilities than their narrowband 'brothers'.
- The spectral band occupied by UWB sensors is placed at comparatively low frequencies. Typical values are 100 MHz–10 GHz. This involves a good (reasonable) penetration of the sounding wave in many materials (except metal) which makes such sensors useful to investigate opaque objects and detect hidden targets.
- In the past, UWB techniques were largely banned to the laboratory due to the need of bulky and expensive electronic devices. But recent developments in RF-system and antenna design, RF-circuit integration and digital signal processing promote the step from the laboratory into the real world. Costs, robustness and power consumption of the sensor devices as well as reliability of the sensing method will be important aspects for the future application of UWB sensing.
- The large bandwidth of UWB devices causes inevitably interferences with other electronic devices, that is mainly with classical narrowband radio services and with other UWB devices. Simply spoken, UWB sensors increase the background noise. In order to limit this noise, the maximum power emission of UWB devices is typically restricted to an interference level which is generally accepted for unintentional radiations of all electric devices. Exceptions are high-power devices for research or military purposes [1].

In this book, we will discuss various UWB sensing approaches exclusively based on low-power emission. The most applied and considered one is probably the radar principle which is meanwhile more than 100 years old. But so far most radar devices are working with a sounding signal of comparatively narrow bandwidth. Here, we will address specific features of very wideband systems because 'Future Radar development must increase the quantity and quality of information for the user. The long-term objective is to provide radar sensing to aid human activities

1) Please note that we do not speak about carrier frequency.

with new and unique capabilities. Use of UWB radar signals appears to be the most promising future approach to building radar systems with new and better capabilities and direct applications to civil uses and environmental monitoring [2]'.

A further principle is the impedance or dielectric spectroscopy which is aimed to determine the electric material parameters $(\varepsilon, \mu, \sigma)$ as function of the frequency. These parameters allow interfering with the state or quality of various substances by a non-destructive and continuously working method. Narrowband measurements suffer from cross-ambiguities since the electric material parameters depend on many things. Observations over a larger bandwidth may possibly reduce these indeterminacies if, for example, superimposed material effects show different frequency behaviour. So far, wideband measurements of this kind were mainly band to the laboratory due to the need of expensive and bulky devices, for example network analysers requiring specifically skilled persons to operate them. New UWB device conceptions adapted to a specific task will promote the dissemination of such sensing methods for industrial purposes or in our daily life.

The previous remarks gave a guideline of possible applications for UWB sensors. As long as the sensors should be accessible for a larger community, they must be restricted to low-power emissions which entitles them to short-range sensing, that is up to about 100 m. A large bandwidth combined with high jitter immunity provides high-range resolution and accuracy down to the μm range, permitting high-resolution radar images and recording of weak movements or other target variations. Furthermore, the interaction of electromagnetic waves with matter provides the opportunity of remote material characterization via permittivity measurements. As examples, we can find applications in following arbitrarily ordered areas:

- Geology, archaeology
- Non-destructive testing
- Metrology
- Microwave imaging
- Quality control
- Inspection of buildings
- Medical engineering
- Search and rescue
- Localization and positioning
- Ranging, collision avoidance
- Law enforcement, intrusion detection, forensic science
- Assistance of handicapped people
- Labour protection[2] and others.

Due to the low emissions, ultra-wideband short-range sensing will become an interesting extension of the radar approach to daily life applications. For large-scale applications, the step from the laboratory into the real world will require further system integration as well as reduction of costs and power consumption. The

2) For example, against severe injuries of hands and arms by rotating tools or squeezing machines; admission control for hazard zones.

processing of the captured data may become a challenging issue, since the measurement principles are often indirect approaches which involve the solution of ill-conditioned inverse problems.

The book is addressed to all who are interested in sensing technology specifically in microwave sensing. Particularly it is addressed to students of electrical engineering, sensor developers, appliers and researches. Reading the book supposes only some very basic knowledge in mathematics, electromagnetic field theory as well as system and signal theory. In order to concentrate on the main aspects and features of UWB sensing, the discussions of many issues will be based on an 'engineering approach' than on rigorous solutions of mathematical or electric field problems. Some considerations within the book may possibly appear somewhat unusual since it was tried to follow a way which is a bit different than usually applied. The classical signal and system theory and the theory of electromagnetic fields too are more or less 'narrowband'. That is, the majority of publications and text books in that area deal with sinusoidal signals or waves. There are two simple reasons why: First, it was forbidden to transmit wideband signals since the second decade of the twentieth century, and narrowband devices were more efficient in the early days of radio development and they cause less interference.

The second reason is of theoretical nature. Let us consider, for example the propagation of waves and their interaction with objects. As long as this interaction is based on linear dynamic effects (Section 2.4 explains what this means), the actual type of the sounding signals does not matter. Hence, one can look for a signal which largely simplifies the solution of equations that model the test scenarios. The sine wave (the decaying sine wave[3)]) is such a signal since it always maintains its shape and frequency. As a consequence of a linear interaction, a sinusoid can only sustain a variation of its amplitude and a time delay usually expressed as phase shift. This simple signal enables to find a rigorous solution of the equations for many cases since linear differential equations can be reduced to algebraic ones. However, the resulting equations are often quite complex and less comprehensible or illustrative. They represent the steady state of superimposed components of wave fields at a single frequency which often leads to not apprehensible interference pattern.

The developer and the applier of UWB sensors need a more pristine view on the wave phenomena, which allows him to understand and interpret measurement data even from complex test scenarios. The human brain is trained since childhood to observe and analyse processes by their temporal evolution and causality. Hence it is much easier to understand propagation, reflection, diffraction and so on of short and pulse-like waves than the superposition of infinitely expanded sine waves of different frequencies. Mathematically both considerations will lead to the same result since they may be mutually transformed by the Fourier transform (the Laplace transform). But the solution of system equations with respect to pulse

3) In the literature of electrical engineering, a sine or a damped sine are usually expressed by exponential functions $\sin 2\pi f t = \Im\{e^{j2\pi f t}\}$ and $e^{\sigma t} \sin 2\pi f t = \Im\{e^{st}\}$; $s = \sigma + j2\pi f$; $\sigma \leq 0$. s is also assigned as complex frequency.

excitation (or generally signals of arbitrary shape) is much more complicated and requires bigger computational effort than for sine waves. The reason is that non-sinusoidal waves typically cannot maintain their shape by interacting with an object. As we will see later, such kind of signal deformations will be expressed by a so-called convolution.

If interaction phenomena between sounding signal and test object[4] are discussed on the base of sine wave signals, one usually talks from frequency (spectral) domain consideration. If one deals with pulse-shaped signals, then it is assigned as time domain consideration which our preferential approach will be. It should be noted that the expression 'time domain consideration' is usually not limited to the exclusive use of pulse-shaped signals. In this book, however, we will restrict ourselves to pulse signals in connection with this term since they best illustrate wave propagation phenomena. But it doesn't mean at all that we will only deal with pulses here. Quite the contrary, we will also respect wideband signals which are expanded in time as a sine wave. They are often referred to as continuous wave (CW) UWB signals. But then, we run in the same or even more critical incomprehensibility of wave propagation phenomena as for sine waves. Fortunately, we can resolve this by applying the correlation approach which transforms a wideband signal of any kind into a short pulse so that 'impulse thinking' is applicable as before.

In this book, we will follow both approaches – time and frequency domain consideration – for three reasons. Recently, most parameters or characteristic functions of electronic devices or sensors and test objects are given by frequency domain quantities. Consequently, we need some tools to translate them in time domain equivalences and to assess their usefulness for our 'time domain' purposes. A further point is that wave propagation is indeed quite comprehensible in time domain but some interaction phenomena between the electromagnetic field and matter as well as some measurement approaches are easier to understand in the frequency domain, that is with sine wave excitation. Finally, many algorithms of signal processing are running much faster in the spectral domain than in time domain.

The book is organized in six chapters. Here, in this chapter, we will introduce the UWB term, give a short historical overview about the UWB technique and consider some aspects of information gathering by UWB sensors from a general point of view.

The second chapter reviews basic aspects of signal and system theory focused on topics of wideband systems. This involves the definition of common signal and system parameters and characteristic functions in time and frequency domains, overview of important wideband signals as well as a discussion of deterministic and random errors.

4) Depending on the actual application of UWB sensors, the items of interest may be quite different. It may be an individual body, a number of bodies, a certain arrangement or a whole scenario including its temporal evolution. For these items, we will use several synonyms – object under test, object of interest, observed object and correspondingly for scenario, system, process or target.

The working principles of the various UWB sensor approaches are discussed in Chapter 3. It covers the basics on generating and gathering very wideband signals as well as the fundamental measurement circuitry. The different UWB approaches are usually designated by the type of sounding signal which is applied. Here, we will distinguish between pulse, pseudo-random code, sine wave and random noise-based sensor conceptions.

Chapter 4 discusses specific aspects of UWB radar and remote sensing. In generalized terms, it means the investigation of distributed systems and scenarios which extends the considerations in Chapter 2 on systems with a finite number of lumped ports to scenarios with a theoretically infinite number of measurement positions. Some important differences between narrowband and UWB radar are analysed. Obtaining an intuitive idea on how UWB radar works, we deal with a very elementary and simplified understanding of wave propagation. Furthermore, we discuss some issues of resolution limits and sensitivity, try to systematize the various approaches of UWB imaging and introduce characteristic values and functions in the time domain in order to quantify UWB antennas.

The fifth chapter summarizes basics on electromagnetic wave propagation. Up to this point, the vector nature of the electromagnetic field, dispersion, interaction with matter and so on were omitted. We will catch up on this in this chapter so that the reader can assess the validity of the simplified approaches considered in Chapter 4 in case of a specific application.

The final chapter describes several applications. For brevity, we only refer to a few examples but they originate from quite different sensing topics in order to give the reader an idea about possible applications of UWB sensing in the future.

The book is closed with a list of applied symbols and notations. Additionally, some appendices, colored pictures and movies are available online, which may be downloaded from the Wiley Homepage www.wiley.com by searching the book title or the ISBN number. The appendices summarize useful mathematical basics, some fundamentals of signal and system theory and selected aspects of electromagnetic field theory. A couple of gray scaled pictures of the book are reproduced there in a colored fashion and further we have inserted some movies with the aim to better illustrate time dependent processes. The availability of related online topics will be indicated at corresponding passages in the text.

About 30 years ago, H. Harmuth, one of the pioneers of the UWB technique, stated [3]: 'The relative bandwidth[5] $\eta = (f_H - f_L)/(f_H + f_L)$ can have any value in the range $1 \geq \eta \geq 0$. Our current technology is based on a theory for the limit $\eta \to 0$. Both theory and technology that apply to the whole range $1 \geq \eta \geq 0$ will have to be more general and more sophisticated than a theory and technology that apply to the limit $\eta \to 0$ only. Skolnik, M. I. Radar Handbook, McGraw-Hill, Inc. 1970.

Meanwhile we have seen some remarkable progress in theory and sensor technology. Nevertheless, 'the insufficiency of the theoretical basis for the development of UWB technique and technology as a system and as a tool for the design of individual

5) Currently, a slightly modified definition with different symbols is in use. We will introduce it in the next section (see (1.1)–(1.3)).

devices, especially antenna systems' and scattering, 'remains an obstacle to further progress' [4]. Therefore, we can only hope that electrical engineers and appliers are furthermore attracted by the potentials of UWB sensing in order to improve the theoretical basis and the sensor technique and to make them accessible to a wide audience.

In this context, we have to focus on remarkable differences in the development of classical narrowband radar and UWB sensing. Narrowband and UWB long-range radar were, and will be, mainly pushed by military needs, and are reserved to a comparatively small community of specialists. In the case of UWB high-resolution short-range sensing the situation is different. There exists an industrial and civil interest besides the military one. On one hand this will widen the field of applications with all its impacts on sensor and theory development. On the other hand, the audience involved in such developments and applications will be less homogenous than the (classical) radar community. Hence, one has to find out a reasonable way to communicate this sensing technique effectively.

1.2 Ultra-Wideband – Definition and Consequences of a Large Bandwidth

The UWB term actually implies two aspects – a large fractional bandwidth and a huge absolute bandwidth. The operational band typically occupies a certain part of the spectrum within the range of 100 MHz to 10 GHz. If UWB devices are used on a large scale, they have to be bound to low emission levels in order to avoid interference with other communication systems [5]. Therefore, high-power medium- and long-range radar systems will always be reserved for special (usually military) use. This book considers high-resolution short-range devices which deal with low radiation power (typical power < 1 mW) and may be of interest for a wider audience than military only.

Let us now take a closer look at the definition and role of a large fractional bandwidth. The term ultra-wideband relates to a normalized width of a spectral band which is occupied by a signal and in which a device is able to operate. For example, we consider Figure 1.1a. It illustrates either the power spectrum of a signal or the

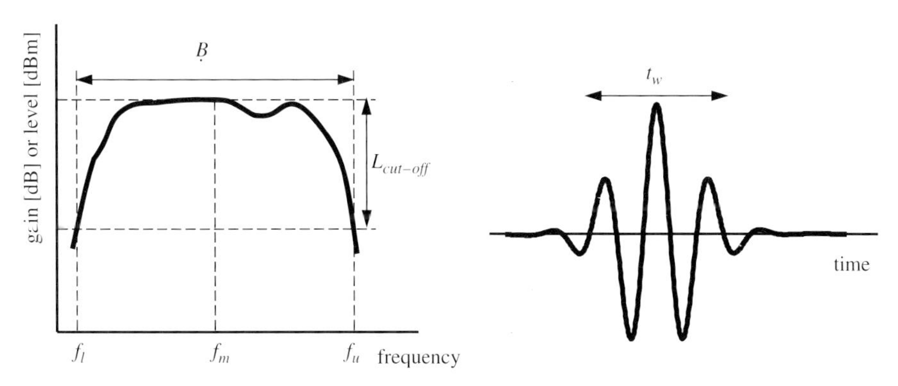

Figure 1.1 Example of power spectrum or transfer function (a) and corresponding time shape or pulse response (b) of a signal or transmission system.

gain of a transmission system. Herein, $f_{\mathrm{l}}, f_{\mathrm{u}}$ are the lower and upper cut-off frequencies referring to an (more or less) arbitrary fixed threshold level $L_{\text{cut-off}}$. The arithmetic mean between both cut-off frequencies is $f_{\mathrm{m}} = (f_{\mathrm{u}} - f_{\mathrm{l}})/2$. It represents the centre frequency. The absolute bandwidth of the signal and device is given by $B = f_{\mathrm{u}} - f_{\mathrm{l}}$ supposing that the spectral power or gain exceeds the threshold level $L_{\text{cut-off}}$ within the whole range. We have several possibilities to normalize the involved frequency quantities. Three of them are shown here:

$$\text{Frequency ratio:} \quad b_{\mathrm{fr}} = \frac{f_{\mathrm{u}}}{f_{\mathrm{l}}} \quad b_{\mathrm{fr}} \in [1, \infty) \tag{1.1}$$

$$\text{Relative bandwidth:} \quad b_{\mathrm{rb}} = \frac{f_{\mathrm{u}} - f_{\mathrm{l}}}{f_{\mathrm{u}} + f_{\mathrm{l}}} \quad b_{\mathrm{rb}} \in [0, 1] \tag{1.2}$$

$$\text{Fractional bandwidth:} \quad b_{\mathrm{f}} = \frac{B}{f_m} = 2\frac{f_{\mathrm{u}} - f_{\mathrm{l}}}{f_{\mathrm{u}} + f_{\mathrm{l}}} \quad b_{\mathrm{f}} \in [0, 2] \tag{1.3}$$

We will only deal with the last definition (1.3) by which a signal or device is called ultra-wideband if its fractional bandwidth exceeds a lower bound $b_{\mathrm{f}} \geq b_{\mathrm{f0}}$. This lower bound is typically fixed at $b_{\mathrm{f0}} = 0.2$ in connection with a cut-off level of $L_{\text{cut-off}} = -10\,\mathrm{dB}$. In order to avoid ambiguities with acoustic systems (audio devices, ultrasound or sonar sensors), one additionally requires an absolute bandwidth larger than a given value $B \geq B_0$ for UWB signals and devices: $B_0 = 50 - 500\,\mathrm{MHz}$ depending on the country.

Nevertheless, a comparison with acoustic sensing approaches reveals interesting similarities. Namely, the human ear and UWB sensors are working at comparable wavelengths (see Table 1.1). Furthermore, the human eye operates on visible light with wavelength ranging from 390 to 780 nm. Using (1.3) leads to a fractional bandwidth of $b_{\mathrm{f}} \approx 0,63$ for the visual sense. Hence, the human beings (and many animals too) have a set[6] of highly sensitive 'ultra-wideband sensors' in order to capture information about their environment. With these sensors, they capture most of the information about their environment and control their lives. It is amazing which information a human being can infer from a physical phenomenon such as reflection and diffraction captured with their 'visual sensors'. But, beside these sensors, an efficient and powerful instrument such as the brain is required to interpret the incoming sensor stimuli. The brain – as synonym for data processing, feature extraction, data mining and so on –plays an even more important role as the bandwidth of the sensors increases since more information must be decrypted from an ever-growing amount of data.

These examples underline the need and the potential of ultra-wideband sensors for future technical and industrial developments, whereas the term 'ultra-wideband' should be seen under a generic aspect[7] for any type of sensing principle and not only restricted to the frequency band of sounding waves. It also shows that signal processing and data mining will become a key point in future development of

6) Including'ultra-wideband' chemical sensors for various substances.
7) The usually applied term in this connection is 'diversity'.

Table 1.1 Wavelength versus frequency for electromagnetic and acoustic waves for propagation in air and water.

Wavelength	Electromagnetic wave		Acoustic wave	
	Air $c \approx 30$ cm/ns	Water $c \approx 3.3$ cm/ns	Air $c \approx 330$ m/s	Water $c \approx 1.5$ m/ms
3 m	100 MHz	11 MHz	110 Hz	500 Hz
30 cm	1 GHz	110 MHz	1.1 kHz	5 kHz
3 cm	10 GHz	1.1 GHz	11 kHz	50 kHz

UWB sensing in order to be able to explore and exploit the wanted information hidden in the captured data. Data processing will gain much more importance than it has in narrowband sensing.

Let us come back to our initial discussion of UWB signals and systems. What is the impact of a large bandwidth on the time evolution of the involved signals? In order to be illustrative, we restrict ourselves to short pulses first (refer to Figure 1.1b). Its duration is t_w (FDHM[8]) – full duration at half maximum) and it is composed of N oscillations. Thus, a (centre) frequency f_m can be roughly assigned to our pulse which is calculated from $N \approx f_m t_w = (f_u + f_l) t_w / 2$. Since for a band-pass pulse $t_w B = (f_u - f_l) t_w \approx 1$ (see Section 2.2.5, Eq. (2.77)), relation (1.3) can also be expressed as

$$b_f N \approx 1 \tag{1.4}$$

Hence, pulse-like wideband signals are composed of only few oscillations because b_f takes values from 0.2 to 2 in the UWB case while narrowband signals ($b_f \to 0$) have many of them. We can extend this condition to any type of wideband signal by referring to the auto-correlation function instead of the actual time shape (see Section 2.2.4).

The behaviour of an UWB system/device also underlies the condition, (1.4) that is, if they are stimulated by a short impact, they will react typically with only a few oscillations that strongly decay.

1.2.1 Basic Potentials of Ultra-Wideband Remote Sensing

In anticipation of later discussions (mainly in Chapters 4 and 6), we will shortly summarize the main consequences of a large absolute and fractional bandwidth on remote sensing applications. As long as the targets do not move too fast, all objects included in the radar channel (antennas and scatterers) may be considered as LTI (linear time invariant) systems. Hence, the electrodynamics of transmission, receiving and scattering, can be formally described by impulse response functions. The target responses can be interpreted either as the reaction of different bodies to an incident field or the reaction of distinct scattering centres of a composed target.

8) One can also find the notation: FWHM – full width at half maximum.

The scatterer response contains all information about the target accessible by radar measurement. Some sensing features are summarized below that are promoted by a large bandwidth:

Target identification [6–8]: We consider a single body of complex structure. Its total scattering response typically comprises a number of peaks (caused, e.g., by specular reflections) and damped oscillations (representing the eigenmodes of the target). In order to resolve these properties, the temporal width t_p of the sounding wave must be shorter than the time distances between adjacent peaks in the response function and the sounding bandwidth should cover at least a few eigenfrequencies of the target. The temporal structure of the specular reflections and the eigenfrequencies are distinctive parameters since they relate to characteristic body dimensions. Hence, to achieve both – separation of specular reflections and a mix of natural frequencies – a large fractional bandwidth is needed. The demands on absolute bandwidth result from the smallest dimensions to be resolved.

Detection of hidden targets and investigation of opaque structures [9]: On one hand, microwave penetration in most of the substances or randomly distributed bodies (e.g. foliage, soil) is restricted to low frequencies, but on the other hand, reasonable range resolution requires bandwidth. To bring both aspects together, the fractional bandwidth must be large. The absolute bandwidth is typically limited by the properties of the propagation medium.

Separation of stationary targets [10]: Scattered waves of two targets located nearby will overlap. They may be separated as long as the signals do not constitute too many oscillations, that is if their fractional bandwidth is large enough. Otherwise a periodic ambiguity of the separated signals will arise. SAGE and MATCHING PURSUIT are numerical techniques using such approaches.

Small target detection and localization: Backscattering from thin layers or cracks is proportional to the first temporal derivative of the sounding wave; small volume scatterers (Rayleigh scattering) cause a second derivation. That is, high frequencies promote the detection of small defects and a large bandwidth leads to their localization. If the small targets of interest are embedded in any substance, a frequency and bandwidth compromise between penetration, detection and localization precision must be found.

Moving target detection: Moving targets covered by strong stationary clutter can be detected by weak variations in the backscattered signals caused, for example, even by small motions. Again, large fractional and absolute bandwidths are beneficial for penetration of opaque objects, the registration of movements and target separation.

1.2.2 Radiation Regulation

UWB devices radiate electromagnetic energy over a large spectral band. Hence, they will increase the level of the noise background. In order to avoid any impact on existing communications systems, the maximum radiation level and the frequency bands of a licence-free operation of UWB devices are restricted by law.

Operational modes requiring deviations from the corresponding radiation rules need consultation with the local authorities for radio communication.

By definition, a UWB device must occupy instantaneously a spectral band which has either a fractional bandwidth $b_f \geq 0.2$ or an absolute bandwidth $B \geq 500\,\text{MHz}$ in the United States and $B \geq 50\,\text{MHz}$ in Europe. Hence, UWB sensors using swept or stepped sine waves do not belong to UWB devices in the strong sense of the definition. Nevertheless, we will also consider sine wave approaches here since there is no physical reason which prohibits their use for wideband measurements.

The international rules for licence-free operation of UWB devices largely orientate at the maximum allowed perturbation level for unintentional radiation which is often simply termed as 'part 15 limits'. The 'part 15' (exactly 47 CFR §§15) is a section of US Federal Communications Commission (FCC) rules and regulations. It mainly regards unlicensed transmissions. It is a part of Title 47 of the Code of Federal Regulations (CFR). The part 15 limit refers to an effective isotropically radiated power (for definition see Section 4.8.2) of $\text{EIRP} = -41.3\,\text{dBm/MHz} \mathrel{\hat{=}} 74\,\text{nW/MHz}$ which corresponds to an effective strength of the electric field of about $E_{\text{rms}} \approx 500\,\mu\text{V}/\left(\text{m}\sqrt{\text{MHz}}\right)$ at a distance of 3 m from the radiator. Hence, the spectral power density of a licence-free UWB radiation is worldwide limited to a maximum value of $-41.3\,\text{dBm/MHz}$. It represents an average value. The peak power level may be typically 40 dB higher than the average spectral power density level. It is determined over a frequency band of 50 MHz width. Its absolute maximum is 0 dBm. Depending on the region and the intended device application, the approved frequency bands for UWB operation are different. The following figure summarizes some of these rules (Figures 1.2–1.4).

The very first obligatory rules for UWB radiation were introduced by the FCC in the United States in 2002. As can be seen from Figure 1.2, the FCC rules provide a large frequency band from 3.1 to 10.6 GHz for UWB radiation with maximum power limitations as for 'part 15 limits'. These frequencies are very well suited for communication purposes but less useful for many sensing applications due to the bad wave penetration into substances.

Other countries modified these rules according to their situation. The main promoter of UWB regulation was the communication industry since it expected a huge market for high-data-rate short-range communication. But the initial effort to develop such communication devices has been largely reduced following a typical hype cycle.

For many people working in this field, UWB technique is linked with the frequency band 3.1–10.6 GHz (FCC–UWB band) and the use of pulse signals. But these simplifications and reductions are not true. They may have historical roots:

- The large influence of the communication industry mainly moulded the public image of UWB technology.
- The first UWB regulation was performed in the United States, resulting in the well-known FFC–UWB band.
- The early and most of the current UWB systems – radio devices, high-resolution radars, ground-penetrating radars (GPRs) and so on – used pulses.

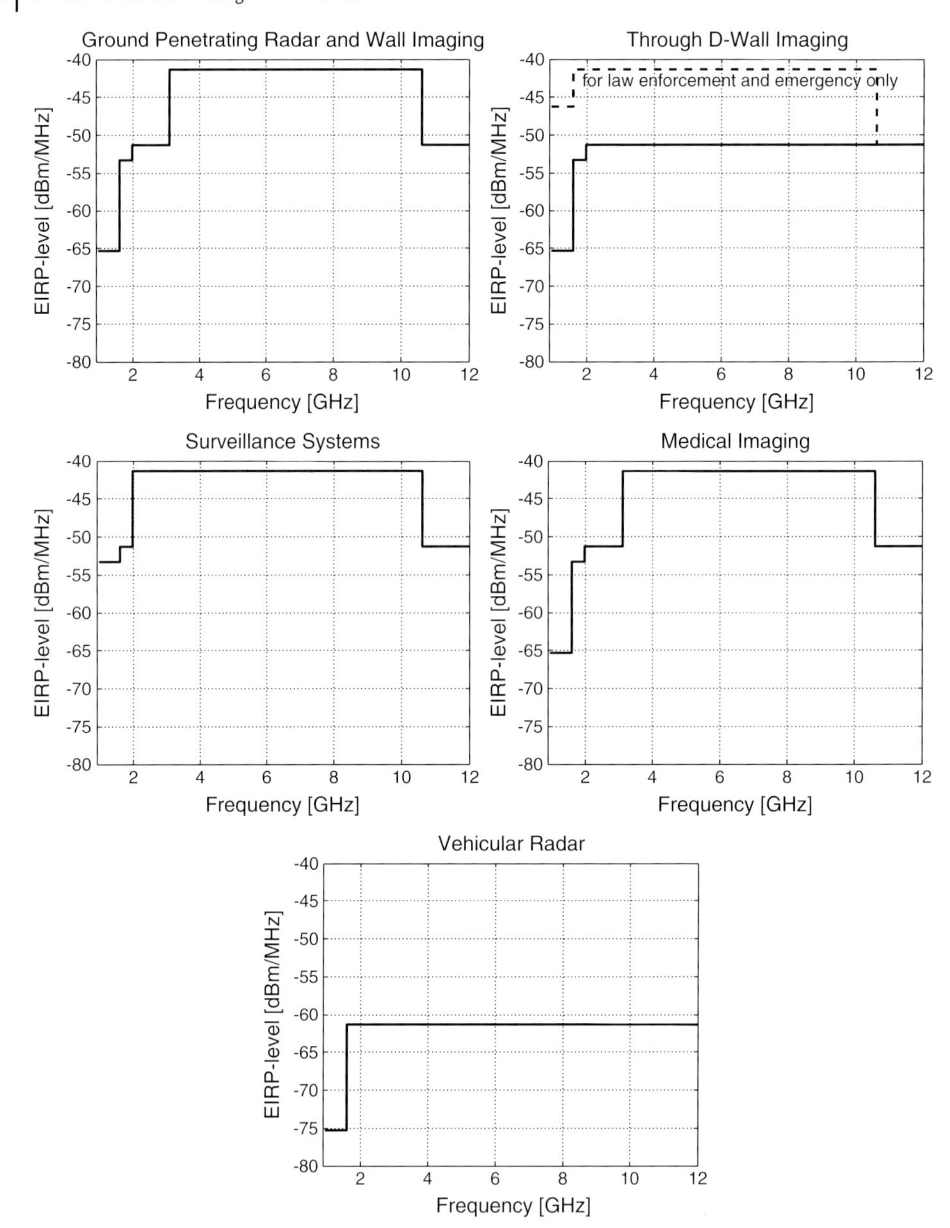

Figure 1.2 FCC-radiation limits for the United States [12]. The spectral power of the radiated signals is measured at 1 MHz resolution bandwidth.

We will not impose such restrictions ourselves here. First, not all countries allow such wide UWB band as the FCC. Second, UWB sensing needs, in certain circumstances, the whole frequency spectrum without gaps, for example, for sounding of hidden objects. Hence, we need exceptional rules for specific applications (e.g. in

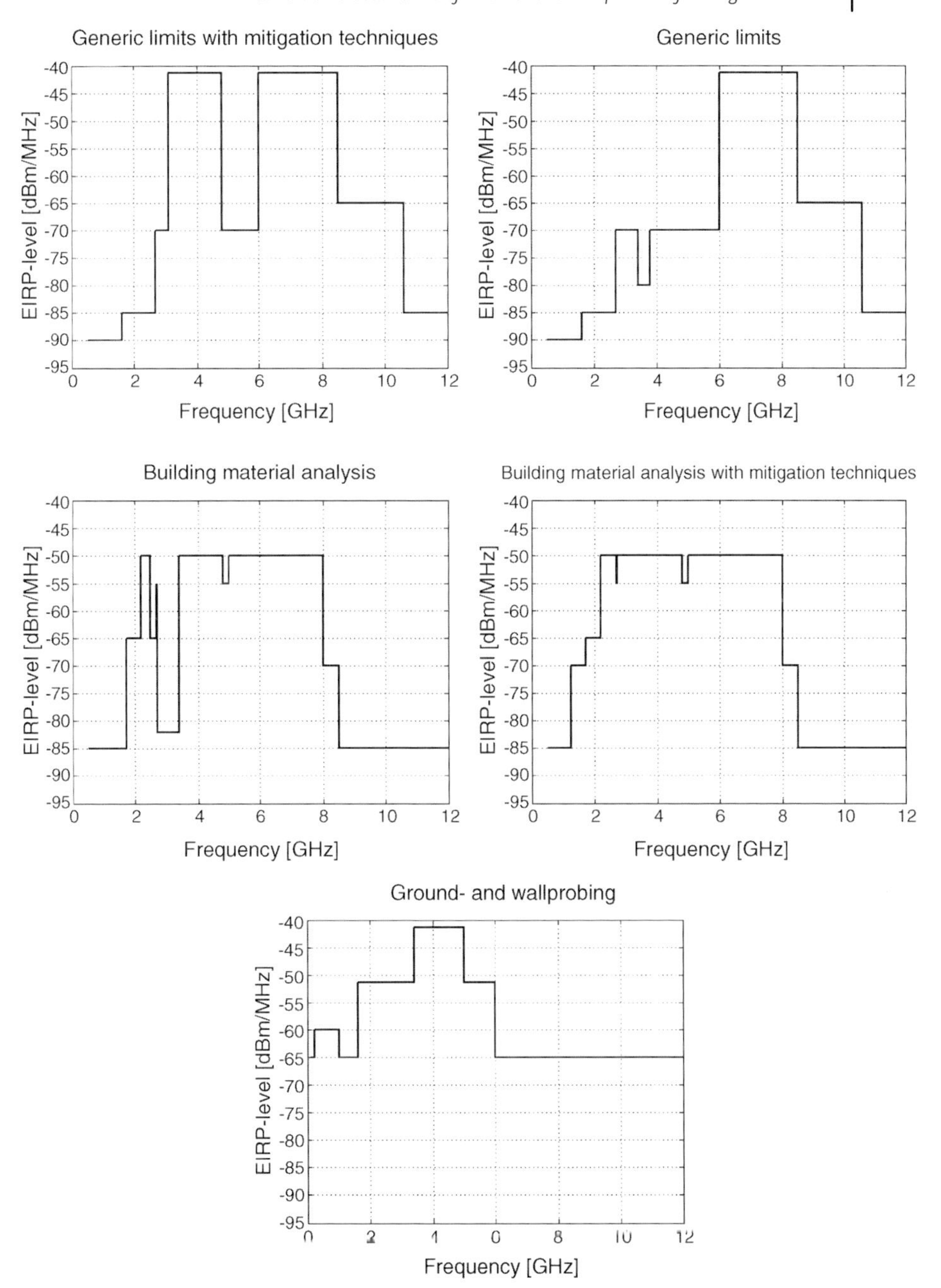

Figure 1.3 ECC-radiation limits in Europe for selected UWB devices [23, 24, 26, 29].

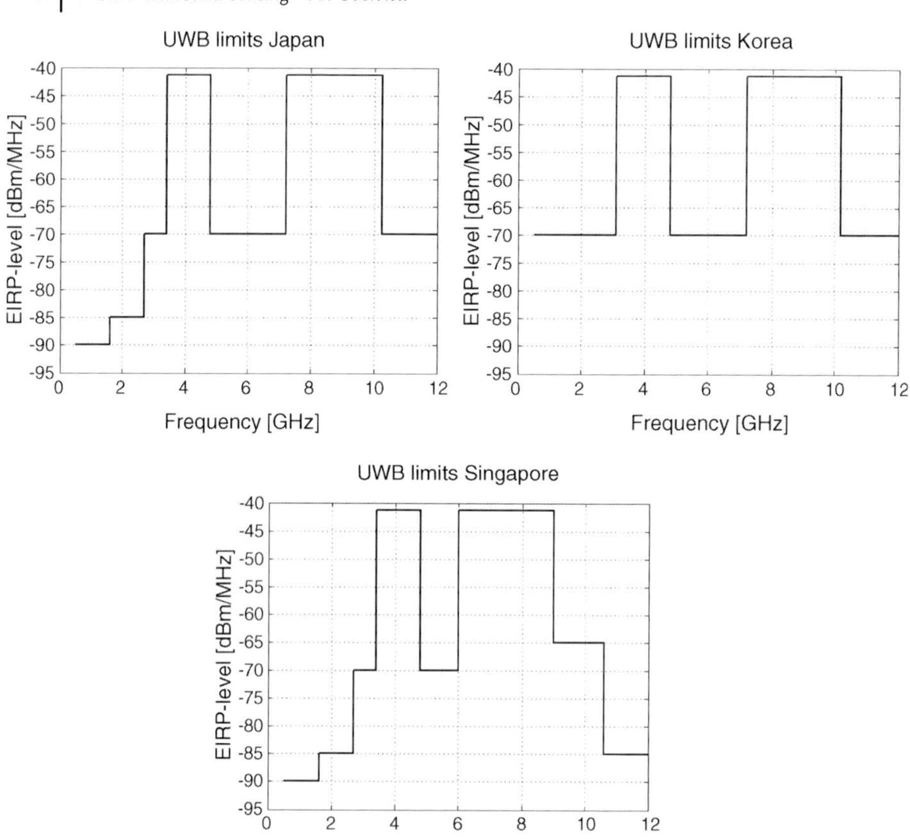

Figure 1.4 Selected radiation limits for some Asian countries: MIC-radiation limits in Japan for indoor device [30], for South Korea [31] and Singapore [32].

public interest – search and rescue) or the measurement environment has to be shielded appropriately to avoid interference with other devices (e.g. medical or industrial applications). Third, modern technology of electronic devices promotes the generation of a larger family of UWB signals than only pulses. Hence, one will have more flexibility in selecting adequate sensing principles for his application.

For detailed discussions on regulation and definition issues or UWB interference with other communication devices the interested reader may refer to Refs [11–30].

1.2.2.1 Implication of UWB Radiation on Biological Tissue

The spectrum of electromagnetic waves, which is in technical use, extends from several kHz to about 10^{20} Hz corresponding to wavelengths of kilometres to nanometres or picometres. The spectral range used by UWB sensors (about 100 MHz – 10 GHz) seems to be comparatively small, which largely reduces the impact of the radiation on biological tissues.

The physical impact of electromagnetic radiation on substances can roughly be classified into heating, ionization and cracking of molecules. Classically, the interaction of electromagnetic field and matter is described by two fundamental approaches – one deals with waves and the other with a kind of 'particles', that is the photons.

Heating can be explained by a wave model; that is, an electric wave (field) penetrates the body and moves charged particles (ions, molecules with dipole moment etc.). This can lead to relaxation phenomena (see, e.g., water relaxation, Section 5.3) which results in heating as it is exploited in microwave ovens. The specific heating power P_V caused from a homogenous electric field E due to dielectric losses is given by (ε'' is the imaginary part of permittivity, f is the frequency):

$$P_V = 2\pi \int_{f_l}^{f_u} f\varepsilon''(f) E_{rms}^2(f) df \quad [W/m^3] \tag{1.5}$$

Using exemplarily water as the main substance of tissue, we get an induced heating power of about 8 mW/m^3 and 8 μW/kg if we expose the tissue to an UWB field corresponding to the FCC limits (i.e. 7 GHz bandwidth and −41.3 dBm/MHz power spectral density). This power is extremely low and absolutely harmless. It should be noted that the radiated power of a cell phone is more than thousand times higher than that permitted for an UWB device according to the comparatively relaxed FCC rules. The power entry in case of the medically used diathermy (commonly used for muscle relaxation) covers several watts per kilogram tissue.

Ionization and molecule destruction may be caused by electric sparks or highly energetic photons. An electric spark creates a 'conductive tunnel' of high current density which leads to local overheating and consequently to material destruction. The creation of electric sparks in isolators needs electrical fields of several kV/mm. Such high fields occur inside high-power short pulse devices [1] but are far from any reality in UWB short-range applications. The exposition of various organisms and biological tissue by pulse-shaped electric fields did not indicate any harmful impact [33–37].

Photon energy becomes harmful for biological tissue if the radiation frequency exceeds that of the visible light, that is ultra-violet, X-ray and gamma rays. The photon energy W is calculated as follows ($h = 6.62 \times 10^{-34}$ J s $= 4.13 \times 10^{-15}$ eV s – Planck constant):

$$W = hf \tag{1.6}$$

Hence, the photon energy of the still harmless blue light (380 nm) is about $W \approx 3$ eV (eV – electron volt) while that of a 10 GHz microwave is $W \approx 40\,\mu$eV, that is 100 000 times less. Consequently, short-range UWB sensing is an absolutely harmless sensing technique which can be applied preferentially for medical purposes and also in private homes.

1.3
A Brief History of UWB Technique

Some people date the beginning of ultra-wideband techniques to the famous experiments of Heinrich Hertz to prove the existence of electromagnetic waves [38]. But this is quite vague and was not proposed by him. Even if he used a transmitter principle – the spark gap – as it can be found also in modern high-power pulse systems, the actual and intentional UWB research started in the 1950s and the 1960s. This activity was essentially driven by three communities:

- **Radar:** Ground and foliage penetration as well as high-resolution imaging and target classification.
- **Communications:** High data rate, multi-path immunity, licence-free operation.
- **High power:** EMP simulations and non-lethal weapons.

At that time the theoretical basis and the technical conditions to operate a very wideband system started to develop. First, wideband devices (sampling oscilloscopes, network analysers, time domain reflectometers) were launched on the market. The key points of this development were the understanding and the development of antennas for non-sinusoidal signals, pulse sources and wideband receivers. Later, the handling and processing of digitized data gained more and more importance, permitting to manage the large data and to extract the wanted information from the measurements more efficiently.

In the early phase of development, terms such as baseband, carrier-free, impulse, time domain, non-sinusoidal and so on were commonly used to describe wideband techniques. The currently used notation 'ultra-wideband' was introduced by the DARPA (US Defence Advanced Research Projects Agency) around 1990; not everybody is satisfied with this definition (see, e.g., [16]). According to DARPA rules, devices or signals are called ultra-wideband if they cover a fractional bandwidth $b = 2(f_u - f_l)/(f_u + f_l) \geq b_0$ larger than a given threshold b_0. f_u, f_l are the upper and lower cut-off frequencies of the spectral band occupied by the sounding signal. Initially, the threshold was fixed to $b_0 = 0.25$ at -20 dB cut-off level. Later in 2002, it was reduced to $b_0 = 0.2$ at -10 dB cut-off level by the FCC and other regulation authorities (e.g. ECC for Europe, MIC for Japan) made further modifications.

The early UWB radar activities were directed, on one hand, to develop high-resolution military radars with improved target recognition capabilities. On the other hand, the good penetration of UWB signals into opaque objects was exploited by establishing ground-penetrating radar (around 1974). Finally, in the last decade of twentieth century, the first low-power, low-cost and lightweight UWB devices were constructed, opening the field of numerous civil applications. We will neither delve deeper into UWB history here nor discuss the pros and cons of the various aspects of the whole UWB technique. We will mainly consider issues on low-power sensing application. The interested reader can find more information on the mentioned topics in Refs [1, 4, 39–52].

The books published so far on UWB technique are aimed at radars and localization [8, 6, 30, 53–56], ground penetration [9] and through wall radar [57], medical sensing [58] and communications [59–74]. Some pioneered works can be found in Refs [3, 75, 76]. A summary on UWB antennas is given in Refs [77–79], and Refs [80–83] report about integrated UWB devices.

The growing interest in UWB techniques and adjacent topics is also expressed in a number of international conferences and workshops such as the annual conference ICUWB (IEEE International Conference on Ultra-Wideband, former UWBST) and the bi-annual conferences GPR (International Conference on Ground Penetrating Radar) and AMEREM/EUROEM (American Electromagnetics/European Electromagnetics) which publishes the book series 'Ultra-Wideband Short-Pulse Electromagnetics' [84–92].

1.4 Information Gathering by UWB Sensors

The sensing technique should never be considered uncoupled from its final destination as part of a control loop depicted in Figure 1.5. Such loops can be components of an engine control or the regulation of industrial processes. But sophisticated sensors may also control or monitor social and medical procedures (e.g. vital data capturing for law enforcement, patient supervision, assistance of handicapped people etc.). The task of the sensor is to provide reasonable information which allows to initiate purposeful actions. Sophisticated concepts of control loops consist of several layers providing sufficient flexibility to react to different situations. Such flexibility mainly results from two aspects:

- The sensor which observes the object/process of interest is not restricted to a simple transformation of a physical value to another one as it is done by classical sensor devices. Rather, the sensor may be divided in two parts. One part is the sensor head which initially only provides a neutral observation signal. The

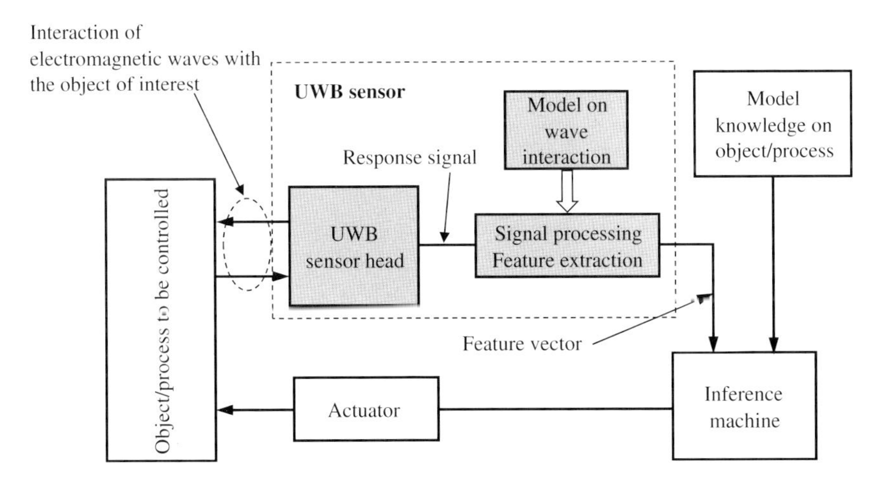

Figure 1.5 Generic control loop using UWB sensors for the observation of an object/process of interest.

second part concerns the data processing which extracts the wanted information from the observations depending on the object/process of interest and depending on the target of the intended actions. In case of UWB sensors, one is often able to extract several characteristic values from the captured data, not only one quantity as in case of many classical sensors. Such information is often supplied as the so-called features vector, that is a set of numbers summarizing some characteristic properties of the object/process of interest.

- The simple determination of the difference between observation and reference as used in primal control loops is replaced by a more complex inference system which includes knowledge about the nature of the object/process of interest and the objective of the actions to be controlled. The inference system may also work on the observations from a number of spatially distributed sensors of the same type (sensor array) and/or it may fuse the information gained from sensors of different kind collecting various physical quantities of the object/process of interest. Additionally, one may insert a 'man in the loop' which either supervises the whole system or contributes to feature extraction and inference.

There is no rigid and clear separation between sensor and inference system. This mainly concerns the level of the models and a priori knowledge, which are required to extract reasonable features and deduce the correct actions. Nevertheless, here in this book, we want to restrict to the working principle of the sensor elements and to the extraction of appropriate features. But we should always have in mind that the selection of appropriate sensors and data processing strategies depends on the structure and objective of the whole control loop.

UWB sensors exploit electromagnetic waves of extremely low power for sensing the scenario under test as illustrated in Figure 1.6. It is well known that the wave

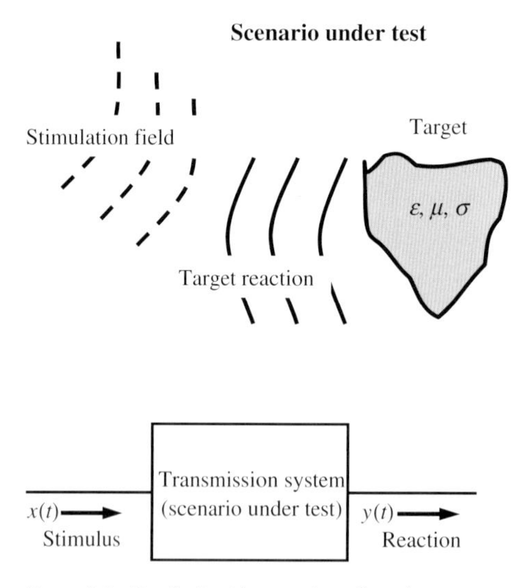

Figure 1.6 Symbolized interaction of an electromagnetic wave with an arbitrary object and formalization of the interaction by a transmission system.

propagation is influenced by objects located within the propagation path. The wave interaction is described by Maxwell's laws and the corresponding boundary conditions. Consequently, the deformation of the stimulation field depends on the geometry of the object as well as the substances from which it is built (expressed by the material parameters permittivity ε, permeability μ and conductivity σ). Therefore, there is some hope to infer the cause (e.g. the geometric structure of the scenario) of such deformations from the captured field. The way of deduction is directed opposite to the chain of reaction; thus, a so-called inverse problem has to be solved in order to gain information about the scenario under test.

Since we will restrict ourselves to the macroscopic world and test objects with linear behaviour against electromagnetic fields, the chain of reaction implies that a certain action on an object or scenario leads to a specific reaction. Hence we can predict the dynamic reaction of a system if we know both its structure and the stimulation signal. Such a prediction is also referred as a solution of a forward problem for which there always exists a solution and this solution will be unique.

The inversion of this problem, that is either to conclude the stimulus of the system from the knowledge of the system behaviour and the system reaction (type 1), or to determine the system behaviour from the known stimulus and the system reaction (type 2), is usually not unique. From a mathematical point of view, inverse problems are often ill-posed problems. According to Jacques Hadamard, well-posed problems should have the following properties:

- A solution exists.
- The solution must be unique.
- The solution depends continuously on the data.

Ill-posed problems violate at least one of these conditions.

We will only deal with the type 2 of inverse problem, that is to look for the system behaviour in order to get some indications about the type and state of the investigated scenario and we mainly have to combat against the last two implications of an ill-posed problem. That is, we may need some regularization techniques in order to avoid data disruption and we have to accept that our endeavour to characterize or to identify an unknown scenario/object under test may lead to the right result only with a certain probability.

The probability of a correct solution improves with increasing prior knowledge about the test scenario and the number of disjoint features gained from the sensors. The better one can confine the variability of a test scenario, the less disjoint features are required for a reliable characterization and identification of the tested system. The uniqueness of information collected from the test objects depends strongly on the diversity of the observations (measurements). Such diversity of information capturing basically reflects in the following points:

- **Capturing of quantities of different nature:** Exploitation of different interaction phenomena (electrical, sound, vibration, chemical, gas and so on; active and passive sensors) reduces cross-ambiguities since different objects show very rarely identical behaviour with respect to all interaction phenomena.

- **Diversity in space–time:** All interaction phenomena take place in space and time. Hence, stimulation of test objects and capturing of their reaction should be performed at several locations and we need dedicated stimulation procedures, that is sounding signals of an appropriate time evolution. As we will see later, we can generalize the term 'signal of appropriate time evolution' by the term 'frequency diversity', that is signal bandwidth which is a point of major concern in ultra-wideband sensing.
- **Diversity in observation time:** The properties and the behaviour of a test scenario may change over time. Often the time variations of these properties (i.e. the 'history' of the property variations) are also characteristic for a specific scenario or test object. Thus, the observation of a scenario over a long duration may bring additional information and reliability.

A discussion on diversity in sensor principles will not play any role in what follows since we will restrict ourselves to UWB sensing which is based on purely electromagnetic interactions. But diversity in space–time and observation time will be essential for many UWB sensing applications. UWB array, UWB-MiMo[9] array, UWB scanning and others are the practical counterparts of the abstract term 'space–time diversity'.

The term 'time' needs still some clarification since it is used in different contexts. Basically, we have three procedures running in parallel if we are dealing with UWB sensing.

1) The interaction of the sounding signal or wave with the test objects.
2) The capturing of measurement data created from the stimulated objects.
3) The temporal variation of the test scenario.

Usually all these procedures are overlapped and mixed together. But fortunately, in UWB sensing, we can consider them separately. We apply sounding waves which travel with the speed of light ($c \approx 30\,\mathrm{cm/ns}$) and we are restricted to short-range sensing. Hence, the interaction of sounding signals with the scenario under test extends only over the nanosecond range. The time frame within which such interactions take place is called interaction time t (sometimes also called 'fast time').

Compared to that, the observed objects and scenario change their behaviour quite smoothly so that they can be considered stationary over the interaction time. A temporal change in properties of the scenarios and objects usually includes a mass transport of any kind (movement of a body, mechanical oscillations, variation of substance composition etc.). These phenomena proceed slower, by many orders, than the propagation of electromagnetic waves so that repetition rates in the range of 1–1000 measurements per second or even less are sufficient in most cases to observe the ('historical') evolution of the test scenario. The time frame within

9) MiMo – multiple input and multiple output; the term MiMo array assigns an antenna array/arrangement which includes radiators that may act as transmitter (multiple output) or receiver (multiple input). Typically, the individual radiators may be operated independently from each other.

which the scenarios under test change (remarkably and hence observably) their behaviour is called observation time T (sometimes also called 'slow time').

As the examples show, there is still a big gap between the duration of wave interaction and the actual required measurement interval. This gap can be used to extend the duration of data recording in order to increase the statistical confidence in the measurement (e.g. by repeating the measurement several times) or/and to simplify the receiver circuits by applying stroboscopic sampling. The time we finally need to capture a whole data set is called recording time T_R. Its duration must be equal to or larger than the interaction time of the sounding signal and it must be equal to or shorter than the time interval between two succeeding measurements. Hence, its maximum value will be fixed either by the Nyquist theorem concerning the time variability (bandwidth) of the scenario to be observed or by the Doppler effect[10] which is linked to the speed of a body moving through the observed scenario.

After having the data captured from the test scenario, the measurements have to be processed appropriately. Figure 1.7 depicts a possible way to extract information about the scenario under test. For that purpose, we illustrate the interaction of the stimulus signal with the scenario by a transmission system (black-box model) which describes the dynamics of interaction (see Figure 1.6). Starting from the measurement of the stimulus and the reaction of the observed scenario, one can first determine a characteristic function which is either the impulse response function $g(t)$ or frequency response function $\underline{G}(f)$ (they are introduced in Section 2.4). This step is also called system identification. It provides a so-called non-parametric system description.

One can extract meaningful parameters from the determined functions $g(t)$ or $\underline{G}(f)$ (note that some methods of parameter extraction may also work directly on the captured data). This step supposes a model about the interactions of the test objects with the sounding field since otherwise the determined functions cannot be interpreted in a meaningful way. Model structure and the resulting parameter set are summarized by the term parametric system description. The parameter estimation leads to a drastic reduction in data amount since the number of parameters are usually orders less than the number of data samples in $g(t)$ or $\underline{G}(f)$. Parametric and non-parametric system identifications are counted among the ill-conditioned problems which require appropriate regularization.

Finally, depending on the actual application, some of these parameters or even a combination of them will shape up as the most descriptive quantities. They represent the features. These characteristic numbers will find access to an inference or classification process running at higher layers of signal processing which we will exclude from our further considerations due to its strong dependence on the actual application.

10) The limitation of the recording time due to the Doppler effect has only to be respected as long as the measurement scenario should be considered as stationary during the recording time. If one abandons this requirement, the recording time may be extended at the expense of more complex receivers or data processing (see Sections 3.3.6 and 3.4.4).

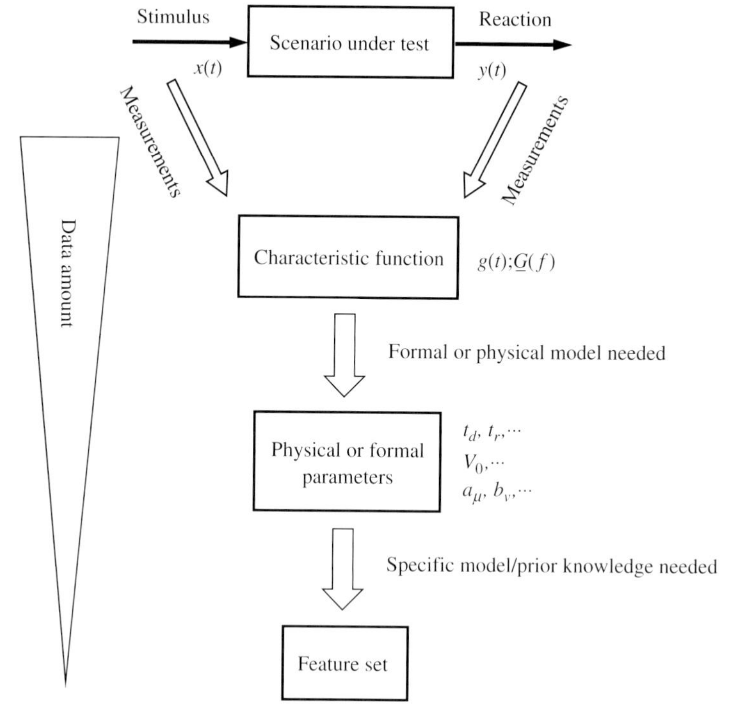

Figure 1.7 Typical approach to extract information about a scenario under test. The mentioned functions and parameters are introduced in Chapter 2.

The capturing of the test system behaviour may be based on different measurement arrangements (see Figure 1.8). One of them involves a small (ideally point shaped) measurement volume (Figure 1.8a) in which only the stray field components at the end of the probe contribute to the measurement result. This method will be applied if only the material parameters $(\varepsilon, \mu, \sigma)$ at a certain location are of interest. Scanning the probe across a surface may lead to high-resolution images due to the small interaction volume of the probing field. Such measurements are usually referred to as impedance spectroscopy or dielectric[11] spectroscopy since the measurement results are usually given in dependence of the frequency. The interaction of the sounding field and the material under test is based on near field effects. These fields rapidly decay with the distance from the probe which leads to the small interaction volumes.

A second method provides a line-shaped measurement volume (Figure 1.8b). Here, a non-shielded transmission line, for example a two- or three-wire line, is surrounded by the material of interest. The inserted sounding wave is guided through the line whereas its propagation is largely influenced by the embedding material. Therefore, the injected wave will be partially reflected, some parts are absorbed and others are transmitted so that conclusions concerning the

11) This notation supposes a relative permeability $\mu_r = 1$ which is valid for most substances.

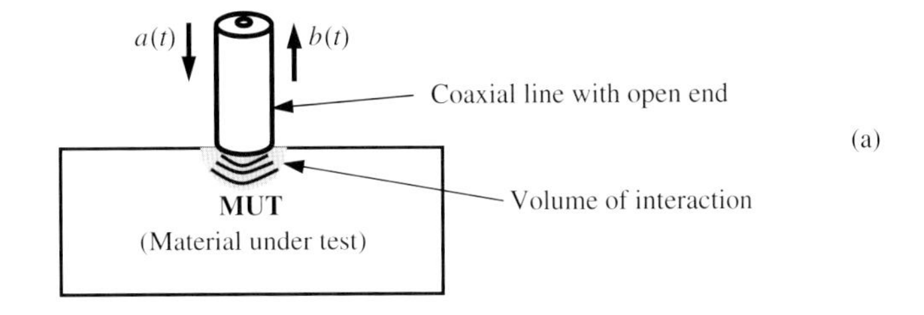

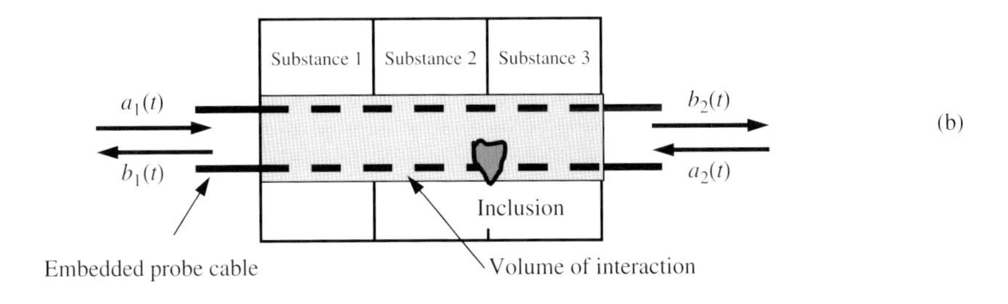

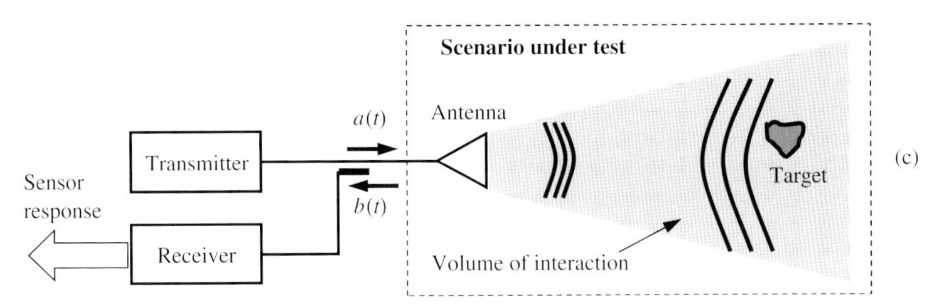

Figure 1.8 Basic measurement arrangements. *a* and *b* represent guided waves which are incident on and emanate from the scenario under test.

embedding material can be drawn by evaluating the deformations which the inserted signal has sustained by travelling through the cable. These techniques are called TDR and TDT (time domain transmission).

A further approach is depicted in Figure 1.8c. It is based on sounding with free waves which are generated and received by antennas. Depending on the antenna characteristics, the observation volume is either cone shaped or omnidirectional. This method permits a contactless sounding over a certain distance. It is usually referred as a radar technique. If the waves are penetrating soil, walls or other construction elements and so on to check hidden targets, it is also called GPR, SPR (surface-penetrating radar) or TWR (through wall radar).

In order to give a vision on the future potential of UWB sensing, let us compare this technical sensing approach with two comparable methods which were 'engineered' by the nature itself – the human eye and the sonar of a bat. All these methods apply scattering of electromagnetic or acoustic waves as source of information. Hence, the 'philosophy' behind the UWB radar sensors is very close to that of the human eyes or the bats' echolocation. So we can learn a lot from the nature for a better understanding of problems and challenges of UWB sounding.

Our eyes (in connection with our brain) which exploit electromagnetic waves as UWB sensors are our major tool to capture information about the environment. If the UWB sensing technique and its further development are seen under such a light, a great deal of new applications and sensing approaches will be expected within the next years. Its field of use will span from simplest tasks as distance measurements to sophisticated target recognition problems applied under relaxed as well as harsh conditions.

However, there are some important differences between both concepts:

- **The different wavelength of the sounding waves:** The human eyes exploit light scattering which involves electromagnetic waves having a wavelength between 380 nm (blue light) and 780 nm (red light). UWB radar sensors use microwaves having a wavelength ranging from centimetres to decimetres. Therefore, light waves can better map the (outer) geometry of items of daily life. These objects are typically thousands of times larger than the wavelength of light. In contrast to that, microwaves can better penetrate materials (except metal), so they are able to look 'behind things'.
- **Illumination of the scene:** The human eye needs an external source for the sounding waves that is the light of the sun or of a lamp. The eye is not synchronized to that source. The UWB radar sensors provide its own sounding signal on which it is synchronized.
- **Data capturing:** In the case of the eye, the data capturing is non-coherent due to the lack of synchronization with the light source. The receptors in the eye only gather light power. It is roughly divided into three spectral ranges –red, green and blue. The receptors are arranged in a dense array (the retina), which finally provides a projection of the waves that are backscattered from the objects within the considered scene. The image creation is done by a lens having a diameter of few millimetres. It is large compared to the wavelength resulting in a very fine angular resolution of about 0.02[12] degrees. The colour impression comes from frequency-dependent material parameters or subtle structured surfaces. The lack of synchronization with the sounding waves avoids direct range information. Therefore, the object distance has to be estimated by auxiliary methods (estimation of apparent object size or angle estimation by two eyes) which are less precise than range estimation by round-trip time.

In contrast, the UWB radar sensor works in a synchronous way. It captures the actual time shape of the scattered electromagnetic field. Therefore, it provides immediately precise range information via the travelling time of the waves. The modification of the time shape of the backscattered signal is caused due to the geometric structure, the surface texture and the material composition of the observed objects. Depending on the application, UWB sensors appear as single sensor or multi-sensor (sensor array) system. UWB imaging always requires a sensor array

12) This corresponds to a 1-cent coin seen from 50 m distance.

approach. The image creation is usually done by signal processing avoiding any lenses. The resolution of an UWB radar image is orders below that of an optical image due to the longer sounding signals. An UWB array which would have an angular resolution (in the far field) comparable to that of an eye would cover a field of about 900 m of diameter, which is far from any realistic arrangement. However, a UWB radar system is able to provide real 3D-images of an investigated volume, it can image optically hidden objects, and it is robust against dust, smoke and fog. Since the size of the imaged objects is often in the same order as the spatial extension of the sounding wave, diffraction and resonance phenomena complicate the UWB imaging. Consequently, one should not try to see UWB radar images with the 'human eye' (i.e. in a purely geometric way); rather, we should better understand the pretended image artefacts.

- **Information extraction:** The data capturing by the eyes or by any other sensor is meaningless if no appropriate conclusions can be extracted. Here, the human eye has advantage over technical sensors since its 'data' are processed by the brain, the most excellent tool to extract information.

In the second example mentioned above – the echolocation of a bat – the similarities to UWB sensing are to be seen in a comparable[13] wavelength of both types of sounding waves and the comparatively simple sensor arrangement – one transmitter (mouth) and two receivers (ears) – creating an interferometric radar/sonar. In the case of the bat (and other mammals, for example, whales apply the same principle), we can recognize the performance of this principle of autonomous orientation, localization and target classification. Here, the data capturing and 'image' formation is strongly dependent on the movement of the 'sensor platform' in order to gain the spatial diversity to reconstruct an 'image' of the environment [93, 94]. Certainly the brain will again play a particular role within the whole sensor system, which underlines the increasing importance and challenges of signal processing of ultra-wideband data. The bat has learned during its evolution to interpret backscattered signals correctly even if they are affected by resonance effects and diffraction phenomena. So we can hope to improve considerably the UWB sensing technique in the future by adopting their 'thinking' as our own.

The above-mentioned comparison of UWB sensing with the human eye (with some analogies also to optical imaging and image processing) should not lead to the conclusion that UWB sensing will compete with optical or infrared sensors. It is, in contrast, far from that. First of all, image creation is only one option of UWB sensors. As mentioned, in the case where images were created, they are usually not comparable with optical images and difficult to interpret. The only reason we referred to the human eye and the sonar of the bat is to allude to achievable performance of sensors which are based on similar physics – that is electromagnetic

13) The wavelengths of the sounding waves generated by a bat are about 5–20 time shorter than in the case of an ordinary UWB sensor. If an UWB sensor would operate within the frequency band from 20 to 100 GHz, it would provide sounding wave of identical wavelength than some bat species. In contrast, the wavelength ratio between light and UWB signals is about 10^5.

or acoustic wave scattering – in order to motivate further research and encourage future developments of UWB sensing. The largest development potentials will be found in the field of scenario adaptive sensor electronics and application-specific data processing and information extraction.

Summarising the previous discussion the applications of UWB sensors are mainly to be seen under the following aspects:

- **Investigation of material parameters or mixture of substances within the microwave range:** Within the considered frequency range (i.e. about 100 MHz–10 GHz), mainly dipole relaxations of less heavy molecules occur. Here, water is a very important candidate for microwave measurements. It is found in food staff, building material and so on and it is one of the most important substances of biological tissue. Therefore, quality control, non-destructive testing in civil engineering (NDTCE) and medical engineering will be corresponding fields of application.
- **High-resolution distance measurements:** This is the classical field of radar applications. But in contrast to the conventional radar and due to the high bandwidth as well as the low-power emission, these sensors are small and lightweight and therefore well suited for industrial, automotive or other civil sensing tasks including home and office applications. UWB sensors are able to separate objects within the centimetre range and they can register target movements down to the micrometre range.
- **Detection and imaging of obscured objects or structures:** Most materials (except metal) provide a reasonable penetration of dm- and cm-waves. Clothes and some plastic foils or layers may be penetrated even by mm-waves. Applications are to be seen in archaeology, geology, non-destructive testing in civil engineering, homeland security, rescue operations, medical engineering, foreign body detection and so forth.
- ***Target recognition and tracking*:** The large bandwidth of the sounding signal leads to a wave interaction which is specific for every target. Therefore, a target leaves a 'fingerprint' in the scattered signal which can be used for recognition if corresponding mapping rules are known. By distributing several UWB sensors over an observation area and combining their data appropriately, the movement of targets (and possibly their behaviour) may be tracked. Ambient assisted living (AAL, i.e. assistance of handicapped people), homeland security, labour protection and localization are some catchwords of possible operational areas.
- ***Autonomous orientation*:** A relatively simple[14)] sensor arrangement is able to provide excellent performance in orientation and classification based on sophisticated sensing strategies and exploiting sounding waves whose wavelengths are spread over nearly one decade. Approaching such performances by technical radar sensors would lead to a much wider variety of applications areas such as remote sensing, robotics, transportation and stock-keeping, sensor networks, homeland security and counter-terrorism.

14) The word 'simple' should be seen here in the sense of the number of involved sensor components – one transmitter and two receivers – but it does not refer to adaptivity and flexibility of the sensing strategy.

References

1 Prather, W.D., Baum, C.E., Torres, R.J. *et al.* (2004) Survey of worldwide high-power wideband capabilities. *IEEE Trans. Electromagn. C*, **46** (3), 335–344.

2 Immoreev, I.J. and Taylor, J.D. (2002) Future of radars. 2002 IEEE Conference on Ultra Wideband Systems and Technologies, 2002. Digest of Papers, pp. 197–199.

3 Harmuth, H.F. (1981) *Nonsinusoidal waves for Radar and Radio Communication*, Academic Press, New York.

4 Immoreev, I. (2009) Ultrawideband radars: features and capabilities. *J. Commun. Technol. Electron.*, **54** (1), 1–26.

5 Politano, C., Hirt, W., Rinaldi, N. *et al.* (2006) Regulation and standardization, in *UWB Communication Systems: A Comprehensive Overview* (eds M.G. Di Benedetto, T. Kaiser, A.F. Molisch *et al.*), Hindawi Publishing Corporation, pp. 447–492.

6 Taylor, J.D. (2001) *Ultra-Wideband Radar Technology*, CRC Press, Boca Raton, FL.

7 Baum, C.E., Rothwell, E.J., Chen, K.M. *et al.* (1991) The singularity expansion method and its application to target identification. *Proc. IEEE*, **79** (10), 1481–1492.

8 Astanin, L.Y. and Kostylev, A.A. (1997) *Ultrawideband Radar Measurements Analysis and Processing*, The Institution of Electrical Engineers, London, UK.

9 Daniels, D.J. (2004) *Ground Penetrating Radar*, 2nd edn, Institution of Electrical Engineers, London.

10 Chang, P.C., Burkholder, R.J. and Volakis, J.L. (2010) Adaptive CLEAN with target refocusing for through-wall image improvement. *IEEE Trans. Antenn. Propag.*, **58** (1), 155–162.

11 FCC (2002) Revision of Part 15 of the Commission's Rules Regarding Ultra-Wideband Transmission Systems.

12 FCC (GPO) (2005) Title 47, Section 15 of the Code of Federal Regulations SubPart F: Ultra-wideband, October 2005; available under: http://www.gpo.gov/fdsys/pkg/CFR-2005-title47-vol1/content-detail.html.

13 Brunson, L.K., Camacho, J.P., Doolan, W.M. *et al.* (2001) *Assessment of Compatibility Between Ultrawideband Devices and Selected Federal Systems.* U.S. Department of Commerce, National Telecommunications and Information Administration (NTIA).

14 Hirt, W. (2007) The European UWB radio regulatory and standards framework: overview and implications. IEEE International Conference on Ultra-Wideband, 2007. ICUWB 2007, pp. 733–738.

15 Politano, C., Hirt, W. and Rinaldi, N. (2006) Regulation and standardization, in *UWB Communication Systems: A Comprehensive Overview* (eds M.G. Di Benedetto, T. Kaiser, A.F. Molisch *et al.*), Hindawi Publishing Corporation, pp. 447–492.

16 Sabath, F., Mokole, E.L. and Samaddar, S.N. (2005) Definition and classification of ultra-wideband signals and devices. *Radio Sci. Bull.*, **313**, 12–26.

17 Shively, D. (2003) Ultra-wideband radio – The New Part 15. *Microwave J.*, **46** (2), 132–146.

18 Singapore, I. (2007) Technical specification for ultra wideband (UWB) devices, http://www.ida.gov.sg/doc/Policies%20and%20Regulation/Policies_and_Regulation_Level2/IDATSUWB.pdf.

19 Siwiak, K. (2005) UWB emission mask characteristics compared with natural radiating phenomena, http://timederivative.com/2005-04-032rX-UWB&black-body-radiation.pdf.

20 Wannisky, K.E. and United States General Accounting Office (2002) *Federal Communications Commission Ultra-Wideband Transmission Systems*, U.S. General Accounting Office.

21 ECC (2005) The protection requirements of radiocommunications systems below 10.6 GHz from generic UWB applications. ECC Report 64 and Annex.

22 ECC (2008) Final report on UWB in response to a request from the European Commission.

23 ECC (2007) ECC/DEC/(07)01: ECC Decision of 30 March 2007 on building material analysis (BMA) devices using UWB technology.

24 ECC (2006) ECC Decision of 1 December 2006 on the conditions for use of the radio spectrum by ground- and wall-probing radar (GPR/WPR) imaging systems.
25 ECC (2008) The impact of object discrimination and characterization (ODC) applications using ultra-wideband (UWB) technology on radio services.
26 ECC (2006) ECC Decision of 1 December 2006 on the harmonised conditions for devices using ultra-wideband (UWB) technology with low duty cycle (LDC) in the frequency band 3.4–4.8 GHz.
27 (2005) FCC 05-08, Petition for Waiver of the Part 15 UWB Regulations Filed by the Multi-band OFDM Alliance Special Interest Group.
28 Rahim, M.A. (2010) *Interference Mitigation Techniques to Support Coexistence of Ultra-Wideband Systems*. Jörg. Vogt. Verlag.
29 Luediger, H. and Kallenborn, R. (2009) Generic UWB regulation in Europe. *Frequenz*, **63** (09–10), 172–174.
30 Sahinoglu, Z., Gezici, S. and Güvenc, I. (2008) *Ultra-wideband Positioning Systems: Theoretical Limits, Ranging Algorithms, and Protocols*, Cambridge University Press.
31 Radio Spectrum Policy and Planning Group (2008) Spectrum allocations for ultra wide band communication devices: a discussion paper, Ministry of Economic Development, New Zealand.
32 I. Singapore (2007) IDA's Decision and Explanatory Memorandum on the regulatory framework for devices using ultra-wideband technology.
33 Jauchem, J.R., Ryan, K.L., Frei, M.R. *et al.* (2001) Repeated exposure of C3H/HeJ mice to ultra-wideband electromagnetic pulses: lack of effects on mammary tumors. *Radiat. Res.*, **155**, 369–377.
34 Jauchem, J.R., Seaman, R.L., Lehnert, H.M. *et al.* (1998) Ultra-wideband electromagnetic pulses: lack of effects on heart rate and blood pressure during two-minute exposures of rats. *Bioelectromagnetics*, **19**, 330–333.
35 Seaman, R. (2007) Effects of exposure of animals to ultra-wideband pulses. *Health Phys.*, **92** (6), 629–634.
36 Simicevic, N. (2008) Dosimetric implication of exposure of human eye to ultra-wideband electromagnetic pulses. Asia-Pacific Symposium on Electromagnetic Compatibility and 19th International Zurich Symposium on Electromagnetic Compatibility, APEMC 2008, Singapore, pp. 208–211.
37 Simicevic, N. (2007) Exposure of biological material to ultra-wideband electromagnetic pulses: dosimetric implications. *Health Phys.*, **92** (6), 574–583.
38 Cichon, D.J. and Wiesbeck, W. (1995) The Heinrich Hertz wireless experiments at Karlsruhe in the view of modern communication. International Conference on 100 Years of Radio, 1995, pp. 1–6.
39 Barrett, T.B. (2000) History of ultrawideband (UWB) radar communications: pioneers and innovators. Progress in Electromagnetics PIERS 2000, Cambridge, MA.
40 Barrett, T.B. (2001) History of ultra wideband communications and radar: Part II, UWB radars and sensors. *Microwave J.*, **44** (2), 22–53.
41 Barrett, T.B. (2001) History of ultra wideband communications and radar: Part I, UWB Communications. *Microwave J.*, **44** (1), 22–56.
42 Kahrs, M. (2003) 50 years of RF and microwave sampling. *IEEE Trans. Microwave Theory*, **51** (6), 1787–1805.
43 Rytting, D. (2008) ARFTG 50 year network analyzer history. Microwave Symposium Digest, 2008 IEEE MTT-S International, pp. 11–18.
44 Mokole, E.L. (2007) Survey of Ultra-wideband Radar, in *Ultra-Wideband, Short-Pulse Electromagnetics 7* (eds F. Sabath, E.L. Mokole, U. Schenk *et al.*), Springer, pp. 571–585.
45 Shirman, Y.D., Almazov, V.B., Golikov, V.N. *et al.* (1991) Concerning first Soviet research on ultrawideband radars (in Russian). *Radiotechnika I*, 96–100.
46 Vickers, R.S. (1991) Ultra-wideband radar-potential and limitations. in Microwave Symposium Digest, 1991, *IEEE MTT-S International*, **1**, 371–374.
47 Yarovoy, A.G. and Ligthart, L.P. (2004) Ultra-wideband technology today. 15th International Conference on

Microwaves, Radar and Wireless Communications, vol. 2, 2004. MIKON-2004, pp. 456–460.

48 Leonard E. Miller (2003) *Wireless Communication Technologies Group National Institute of Standards and Technology Gaithersburg*, Maryland, available at: http://www.antd.nist.gov/wctg/manet/NIST_UWB_Report_April03.pdf

49 Astanin, L.Y. and Kostylev, A.A. (1993) The current status of research into ultrawideband radars in Russia, IEE Colloquium on Antenna and Propagation Problems of Ultrawideband Radar, pp. 3/1–3/6.

50 Fontana, R.J. (2004) Recent developments in short pulse electromagnetics or UWB the Old-Fashioned Way? Presented to Microwave Theory and Techniques Society Northern VA Communications Society, March 9, 2004.

51 Fontana, R.J., Foster, L.A., Fair, B. *et al.* (2007) Recent advances in ultra wideband radar and ranging systems. IEEE International Conference on Ultra-Wideband, 2007. ICUWB 2007, pp. 19–25.

52 Fang, G. (2007) The research activities of ultrawide-band (UWB) Radar in China. IEEE International Conference on Ultra-Wideband, 2007. ICUWB 2007, pp. 43–45.

53 Taylor, J.D. (1995) *Introduction to Ultra-Wideband Radar Systems*, CRC Press, Boca Raton.

54 Hawa, Y. (2009) *Study of Fast Backprojection Algorithm for UWB SAR*, VDM Verlag.

55 Kyamakya, K. (2005) Joint 2nd Workshop on Positioning, Navigation and Communication 2005 (WPNC '05) & 1st Ultra-Wideband Expert Talk 2005 (UET '05), Shaker Verlag GmbH.

56 Noel, B. (1991) *Ultra-Wideband Radar: Proceedings of the First Los Alamos Symposium*, CRC Press, Boca Raton, FL.

57 Amin, M.G. (2011) Through-The-Wall Radar Imaging CRC Press.

58 Merkl, B. (2010) *The Future of the Operating Room: Surgical Preplanning and Navigation using High Accuracy Ultra-Wideband Positioning and Advanced Bone Measurement*, LAP Lambert Academic Publishing.

59 Oppermann, I., Hämäläinen, M. and Iinatti, J. (2004) *UWB Theory and Applications*, John Wiley & Sons, Ltd, Chichester, UK.

60 Siriwongpairat, W.P. and Liu, K.J.R. (2008) *Ultra-Wideband Communications Systems: Multiband OFDM Approach*, IEEE Press, Hoboken, NJ.

61 Siwiak, K. and McKeown, D. (2004) *Ultra-Wideband Radio Technology*, John Wiley & Sons, Ltd, Chichester, UK.

62 Aiello, R. and Batra, A. (2006) *Ultra Wideband Systems. Technologies and Applications*, Newnes.

63 Ghavami, M., Michael, L. and Kohno, R. (2007) *Ultra Wideband Signals and Systems in Communication Engineering*, John Wiley & Sons, Inc., New York.

64 Heidari, G. (2008) *WiMedia UWB: Technology of Choice for Wireless USB and Bluetooth*, John Wiley & Sons, Ltd, Chichester, UK.

65 Jogi, S. and Choudhary, M. (2009) *Ultra Wideband Demystified Technologies, Applications, and System Design Considerations*, River Publishers.

66 Kaiser, T., Molisch, A.F., Oppermann, I. *et al.* (2006) *Uwb Communication Systems: A Comprehensive Overview*, Hindawi Publishing Corporation.

67 Kaiser, T. and Zheng, F. (2010) *Ultra Wideband Systems with MIMO*, John Wiley & Sons, Ltd, Chichester, UK.

68 Nekoogar, F. (2004) *Ultra-Wideband Communications: Fundamentals and Applications*, Prentice Hall International.

69 Nikookar, H. and Prasad, R. (2008) *Introduction to Ultra Wideband for Wireless Communications*, Springer, Netherlands.

70 Reed, J.H. (2005) *An Introduction to Ultra Wideband Communication Systems*, Prentice Hall International.

71 Shen, X., Guizani, M. and Qiu, R.C. (2006) *Ultra-Wideband Wireless Communications and Networks*, John Wiley & Sons, Ltd, Chichester, UK.

72 Vereecken, W. and Steyaert, M. (2009) *UWB Pulse-based Radio: Reliable Communication over a Wideband Channel*, Springer, Netherlands.

73 Verhelst, M. and Dehaene, W. (2009) *Energy Scalable Radio Design: for Pulsed UWB Communication and Ranging*, Springer-Verlag, GmbH.

74 Wang, J. (2008) *High-Speed Wireless Communications: Ultra-wideband, 3G Long Term Evolution, and 4G Mobile Systems*, Cambridge University Press.

75 Harmuth, H.F. (1984) *Antennas and Waveguides for Nonsinusoidal Waves*, Academic Press, Inc., Orlando.

76 Miller, E.K. (1986) *Time Domain Measurements in Electromagnetics*, Van Nostrand Reinhold Comapny Inc., New York.

77 Schantz, H.G. (2005) *The Art and Science of Ultrawideband Antennas*, Artech House, Inc., Norwood.

78 Becker, J., Filipovic, D. and Schantz, H. (2008) *Ultra-Wideband Antennas*, Hindawi Publishing Corp.

79 Allen, B. (2007) *Ultra-wideband: Antennas and Propagation for Communications, Radar and Imaging*, John Wiley & Sons, Ltd, Chichester, UK.

80 Paulino, N., Goes, J. and Garcao, A.S. (2008) *Low Power UWB CMOS Radar Sensors*, Springer.

81 Dederer, J. (2009) *Si/SiGe HBT ICs for Impulse Ultra-Wideband (I-UWB) Communications and Sensing*, Cuvillier.

82 Gharpurey, R. and Kinget, P. (2008) *Ultra Wideband: Circuits, Transceivers and Systems*, Springer, Berlin.

83 Safarian, A. and Heydari, P. (2008) *Silicon-Based RF Front-Ends for Ultra Wideband Radios*, Springer, Berlin.

84 Bertoni, H.L. (1993) *Ultra-Wideband, Short-Pulse Electromagnetics*, Springer-Verlag, GmbH.

85 Felsen, L.B. and Carin, L. (1995) *Ultra-Wideband, Short-Pulse Electromagnetics 2*, Springer, Berlin.

86 Baum, C.E., Stone, A.P. and Carin, L. (1997) *Ultra-Wideband, Short-Pulse Electromagnetics 3*, Springer, Berlin.

87 Heyman, E., Mandelbaum, B. and Shiloh, J. (1999) *Ultra-Wideband Short-Pulse Electromagnetics 4*, Kluwer Academic/Plenum Publishers, New York.

88 Smith, P.D. and Cloude, S. (2002) Ultra-wideband, Short-Pulse Electromagnetics 5, Kluwer Academic/Plenum Publishers, pp. xi, 751.

89 Mokole, E.L., Gerlach, K.R. and Kragalott, M. (2004) *Ultra-Wideband, Short-Pulse Electromagnetics 6*, Springer, Berlin.

90 Sabath, F., Mokole, E.L., Schenk, U. *et al.* (2007) *Ultra-Wideband Short-Pulse Electromagnetics 7*, Springer, Berlin.

91 Stone, A.P., Baum, C.E. and Tyo, J.S. (2007) *Ultra-Wideband, Short-Pulse Electromagnetics 8*, Springer, New Mexico.

92 Sabath, F., Giri, D., Rachidi, F. *et al.* (2010) *Ultra-Wideband, Short Pulse Electromagnetics 9*, Springer-Verlag, GmbH.

93 Sachs, J., Zetik, R., Peyerl, P. *et al.* (2005) Autonomous orientation by ultra wideband sounding. International Conference on Electromagnetics in Advanced Applications (ICEAA), Torino, Italy.

94 Vespe, M., Jones, G. and Baker, C.J. (2009) Lessons for radar. *IEEE Signal Proc. Mag.*, **26** (1), 65–75.

2
Basic Concepts on Signal and System Theory

2.1
Introduction

The aim of this chapter is to give the reader a short review of the most important and basic aspects of signal and system theory. As even mentioned, the goal of ultra-wideband sensing is to pick up information about passive objects or scenarios. Passive objects means that they do not emit any energy. But any collection of data requires an energy exchange (even if it is very small) between target and measurement device. Therefore, in the case where passive objects are investigated, they have to be activated first in order to evoke a measurable reaction. Figure 2.1 symbolizes graphically this approach. Here, we summarize the objects and scenarios under test by the generic term 'system'. Such a system generates an observable signal $y(t)$ if it is stimulated by a signal $x(t)$ which carries a certain amount of energy. In the case of UWB sensors, the physical nature of the two signals is that of a voltage, an electrical current or an electromagnetic field. We will always suppose that the injected energy is so small that the test object is not influenced in its behaviour. Hence, we cannot observe saturation effects and the thermodynamic equilibrium of the object is not influenced. That is, the test object behaves linear with respect to the stimulus signal (see also Section 2.4.1).

Except for sine wave signals,[1] the evolution of the signal $y(t)$ deviates mostly from that of the stimulus $x(t)$. This modification is driven by the (dynamic) behaviour of the system, which represents a characteristic property of the system under test. That is, it leaves his 'fingerprint' in the output signal which can be used to recognize, to quantify or to qualify the tested scenario. The questions arising are: What kind of information ('fingerprint') we can extract? How to do this in the best way?

In order to answer these questions, we will start with the introduction of wideband signals and the methods of their description and characterization. Then the behaviour of systems is discussed, first on the basis of the very generic assumption of linearity and afterwards by respecting also the energetic interaction between the different components of the measurement chain. By the way, the behaviour of

1) In linear systems, they are only subjected a phase shift and amplitude variation.

Handbook of Ultra-Wideband Short-Range Sensing: Theory, Sensors, Applications, First Edition. Jürgen Sachs.

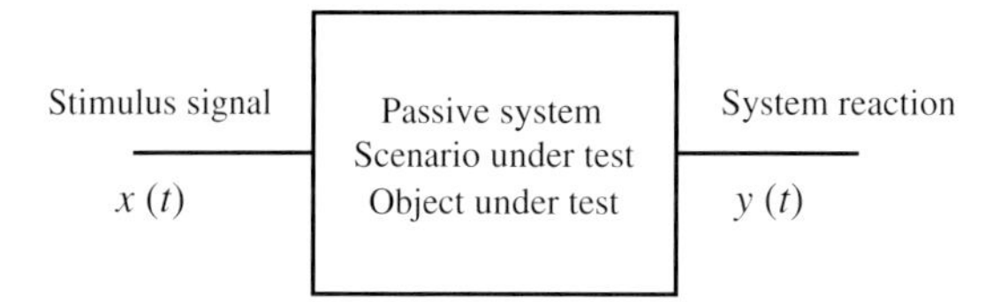

Figure 2.1 Illustration of stimulation and reaction of a passive object or scenario.

the measurement device/sensor can be modelled by the same method as the measurement objects itself. Perturbations from the idealized models and their impact onto the measurement results will be considered next and finally – since modern sensor device applying digital principles – the basic ideas of data sampling are summarized.

2.2
UWB Signals, Their Descriptions and Parameters

2.2.1
Classification of Signals

2.2.1.1 Types of Stimulus Signals

A typical feature of UWB signals is the occurrence of abrupt chances of the signal amplitudes which lead to their large bandwidth. In contrast to that narrowband signals have a sine wave-like appearance with comparative slow variations of the signal envelope. Figure 2.2 classifies the UWB signals with respect to their temporal and amplitude distribution. They can be periodic or non-periodic and they may have distributed their energy over a large time or it is concentrated within a short duration. For measurement purposes, periodic signals are preferably used to stimulate the test systems,

- since they are easy to generate,
- since they result in stable measurements within a short measurement time and
- since they lead to relaxed technical demands on the measurement devices as we will see later.

Nevertheless, under certain conditions, also random signals are useful particularly if low probability of detection and interception are of importance for the sensor operation. Sensing principles based on random signals may work in an active or passive mode. In the active mode, the scene is illuminated by the sensor device itself as it is done by all the sensors using periodic signals. In the passive mode, the sensors exploiting sounding signals which originates from foreign sources (e.g. TV- and radio stations) or which are radiated from the test object itself by thermal stimulated microwaves. Passive ultra-wideband devices require sophisticated equipment (e.g. real-time data capturing) and they may cause longer measurement time as for periodic signals.

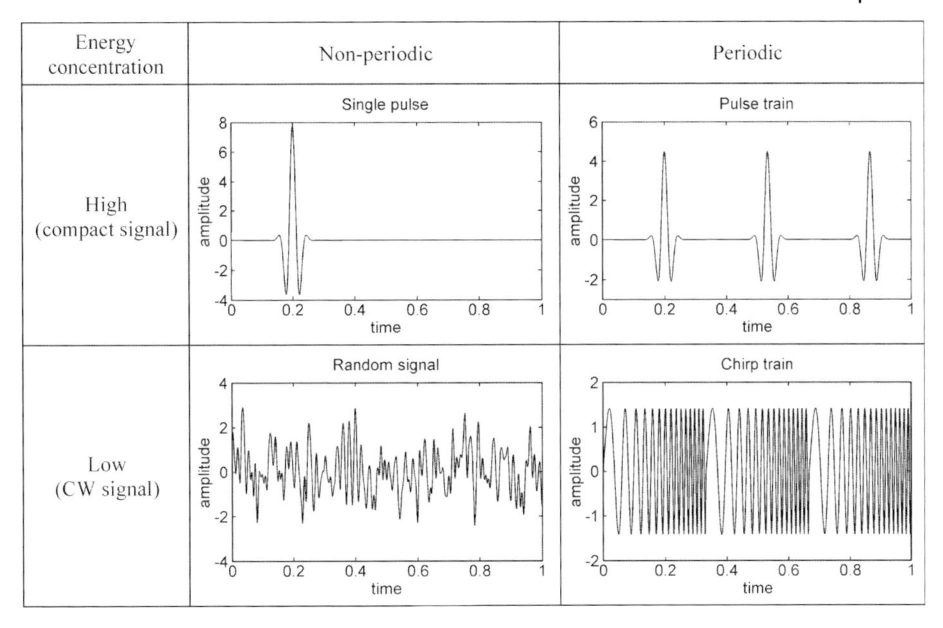

Figure 2.2 Classification of ultra-wideband signals. The amplitude of the signals is normalized to unit power or energy.

A signal of finite duration, we will call 'energy-signal' due to its final energy. Examples are the single pulse, a limited number of periods from a chirp signal or a pseudo-noise sequence, a single burst of random signal and so on. All the other signal classes of Figure 2.2 are of infinity duration and therefore of infinity energy. We will call them power signals since they may be distinguished by their mean power.

We will, however, underline that every process – that is also a measurement – is of finite duration. Therefore, the signal energy is finally the crucial quantity and the model of a periodic signal in a strong sense is only theoretical model. Periodicity in our sense means that at least during the measurement process the stimulus signal is repeated regularly in time.

2.2.1.2 **Random Process**

All signals occurring in practical measurements are affected by random perturbations which we will summarize by the term noise. The reasons for the noise may be manifolds, for example thermal noise of electronic devices and antennas, flicker noise, radio jamming, jitter and so forth. Since one can treat random noise only by statistical methods, it is a usual approach to consider a whole ensemble of the same measurement consisting, for example from N realizations. That is, instate of only one measurement signal $x(t)$, one deals with an ensemble of signals which we express by the notation $\underset{\sim}{x}(t) = [x_1(t)\ x_2(t) \cdots x_N(t)]$. This ensemble is assigned as random process. In our case, the individual realizations $x_i(t)$ of the random process typically consist of a deterministic part $x(t)$ which is the same in all

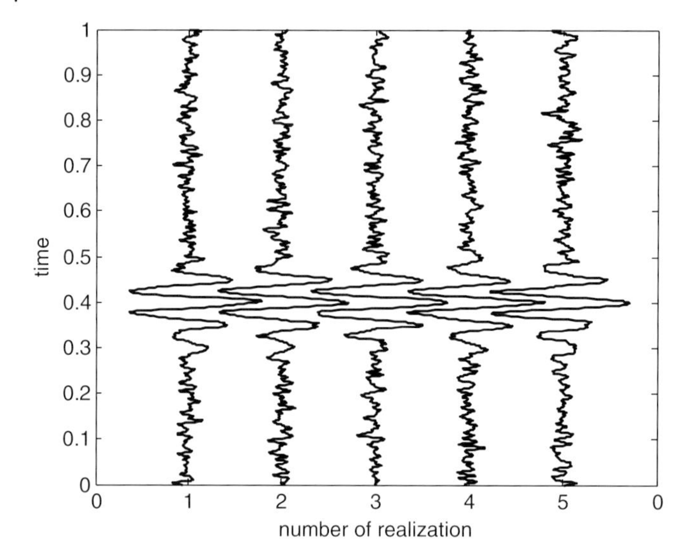

Figure 2.3 Illustration of a random process including deterministic and stochastic signal parts.

realizations and a purely random quantity $n_i(t)$ which is different from realization to realization:

$$\underset{\sim}{x}(t) = x(t) + \underset{\sim}{n}(t) \tag{2.1}$$

In our measurements, we are typically interested in $x(t)$ while $\underset{\sim}{n}(t)$ models the unwanted perturbations which are supposed to have zero mean. Figure 2.3 illustrates a random process as written in (2.1). The actual capturing of a random process is usually done by repeating the measurements N-times and by considering every measurement as one realization. This approach supposes that the measurement scenario is stationary over all N measurements and that the random contributions are uncorrelated between the individual measurements.

2.2.1.3 **Analogue and Digital Signals**

Since we only consider macroscopic effects, the involved physical qualities can be considered as analogue signal, that is they are continuous in time and magnitude. Practically, one can only deal with digitized signals since the data amount of analogue signals tends to infinity. Digital signals are discrete in time and magnitude and hence of limited data amount. They can be stored and flexibly processed by computers. Processing of analogue data is mostly restricted to simple operations (filtering, multiplication, integration, differentiation etc.) which are less flexible to adapt in their parameters. Analogue storage of UWB signals and signal delays over more than a few nanoseconds is practically impossible and very expensive. But many analogue operations work at higher frequencies and larger bandwidth than digital one, they often need simple electronic components and they are often more efficient with respect to power consumption. Analogue integral operations are replaced in the digital domain by matrix multiplications.

Hence, the device conception of an UWB sensor has to trade carefully analogue processing against digital processing. This decision cannot be made by general rules rather it must be done in connection with the actual application of the UWB sensor and the requirements to its performance.

We will express analogue signals by a symbol with the time argument in round brackets: $x(t)$. Sampled signals use square brackets for the time argument: $x[n\Delta t_s] \hat{=} x[n]$ (Δt_s is the sampling interval). They are often expressed in vector notation: $\mathbf{x} = [x[1] \quad x[2] \quad \cdots \quad x[n] \quad \cdots]^T$.

We will not make a big difference between the notations of analogue and sampled data. Mostly we will deal with the analogue conventions even if many signal processing operations in UWB sensors are done by numerical procedures.

2.2.2 Signal Description and Parameters of Compact Signals in the Time domain

2.2.2.1 Basic Shape Parameters

The natural and most convenient illustration of UWB signals is their representation in the time domain, that is to depict the signal amplitude in dependence of the time as measured by an oscilloscope and symbolized in Figure 2.2. There are many different types and shapes of UWB signals (see, e.g. Section 2.3) which results in a large number of parameters quantifying such time functions. At this point, we will first concentrate on a few of the most important parameters of pulse-shaped functions because they are of general interest in the UWB technique. For more pulse parameters the reader is referred to Ref. [1].

Figures 2.4 and 2.5 illustrate some shape parameters of simple, little jagged single pulses and of two pulse trains. As specified, the parameters are typically

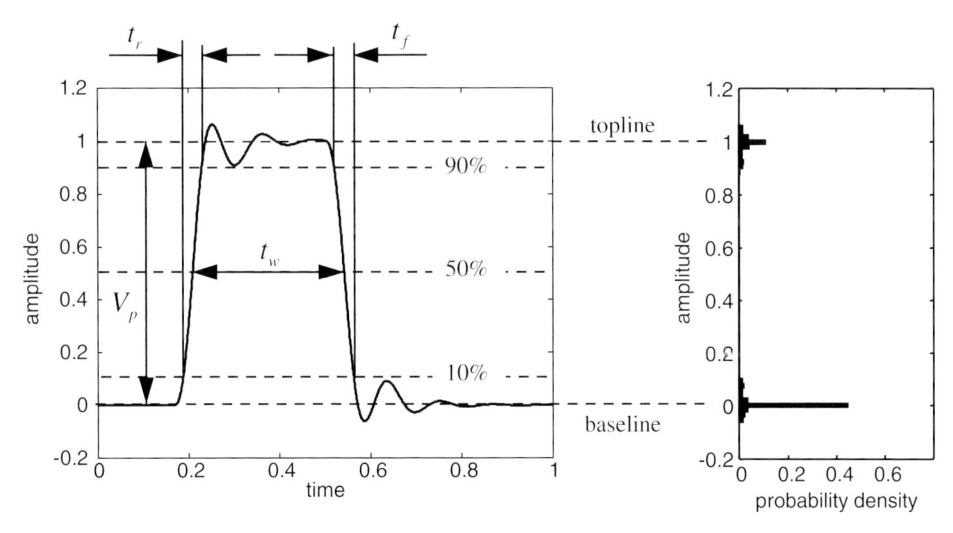

Figure 2.4 Some shape parameters of a pulse-shaped time function: t_r is the rise time, t_f is the fall time, t_w is the pulse width or pulse duration and V_p is the pulse amplitude.

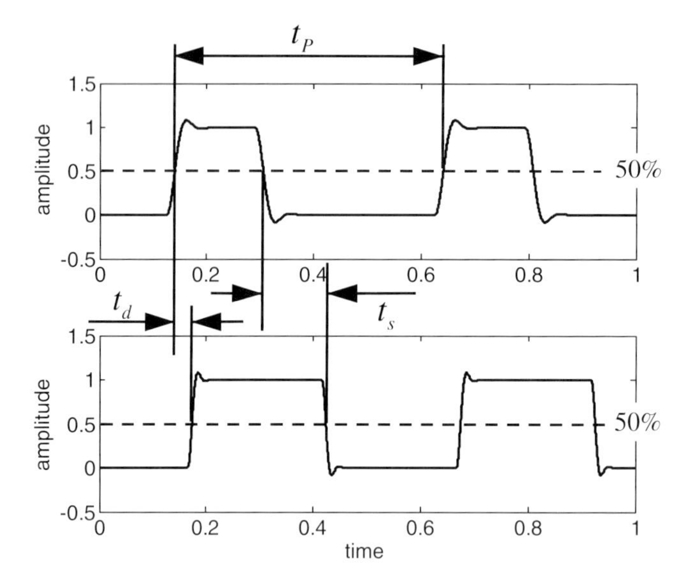

Figure 2.5 Periodic signal and parameters describing the relation between two time functions: t_P is the signal period, t_d is the delay time and t_s is the storage time.

referred to threshold values of 0%, 10%, 50%, 90% or 100%. In some cases, the threshold values 10% and 90% are replaced by 20% and 80% which, however, has to be displayed explicitly. The automatic determination of the threshold values is based on the histogram of the pulse form (for histogram determination, see Section 2.2.3). The two amplitude values of the highest probability are defined as the baseline (0%) and the topline (100%) of the pulse function.

If, however, any impulse is transmitted by an antenna or a high-pass system, its DC value will be zero. Therefore, the waveform will consist of a few oscillations and the above-mentioned definition and determination of pulse parameters fails. In order to evade these difficulties, one introduces the signal envelope which can be quantified in the same way as simple pulses. A simple way to identify the (amplitude) envelope $x_{\text{env}}(t)$ of an oscillating pulse $x(t)$ applies the absolute value of the analytic signal $\underline{x}_{\text{a}}(t)$.

The analytic signal is a complex-valued helical function that imaginary part is built from the Hilbert transform $\text{HT}\{x(t)\}$ of the original signal:

$$\underline{x}_{\text{a}}(t) = x(t) + jx_{\text{H}}(t) \quad \text{with } x_{\text{H}}(t) = \text{HT}\{x(t)\} \tag{2.2}$$

The Hilbert transform is defined as follows:

$$\text{HT}\{x(t)\} = \frac{1}{\pi}\text{PV}\int_{-\infty}^{\infty}\frac{x(t)}{t-\xi}\text{d}\xi = x(t) * \frac{1}{\pi t} \tag{2.3}$$

Herein, PV $\int \cdots \mathrm{d}\xi$ refers to the Cauchy principal[2)] value of the integral in order to solve the singularity at $t - \xi = 0$. The symbol $*$ means convolution. It will be introduced in Section 2.4.2. (2.3) is also known under Kramers–Kronig relation. As we will see in Section 2.2.5, a real-valued 'physical' signal $x(t)$ is described by a spectrum of positive and negative frequencies. In contrast to that, the spectrum of the analytic signal $\underline{x}_a(t)$ is restricted to only positive frequencies. This property allows the calculation of (2.2) and (2.3) via

$$\underline{x}_a(t) = 2\,\mathrm{IFT}\{u_{-1}(f)\mathrm{FT}\{x(t)\}\} \tag{2.4}$$

FT{ } and IFT{ } represent the Fourier transformation and the inverse Fourier transformation. They will be introduced in Section 2.2.5. $u_{-1}(f)$ is the unit step or Heaviside function (see Annexes A.2 and B.1, Table B.1). From the analytic signal we can determine various signal features:

$$\text{Envelope: } x_{\mathrm{env}}(t) = |\underline{x}_a(t)| = \sqrt{\underline{x}_a(t)\underline{x}_a^*(t)} = \sqrt{x^2(t) + x_{\mathrm{H}}^2(t)} \tag{2.5}$$

Instantaneous phase:

$$\varphi_i(t) = \arg \underline{x}_a(t) = \arctan \frac{x_{\mathrm{H}}(t)}{x(t)} \tag{2.6}$$

Instantaneous frequency:

$$f_i(t) = \frac{1}{2\pi}\frac{\mathrm{d}\varphi_i(t)}{\mathrm{d}t} \tag{2.7}$$

Note that the instantaneous phase $\varphi_i(t)$ has to be unwrapped before it can be derivated. For a deeper discussion of complex representation of real signals and the Hilbert transform, the reader can refer to Ref. [2].

Here, we are mainly interested in the envelope. Figure 2.6 demonstrates an example in which the pulse width t_w represents the width of the envelope. This value is also assigned as the FWHM-value (full width at half maximum) or FDHW-value (full duration at half maximum).

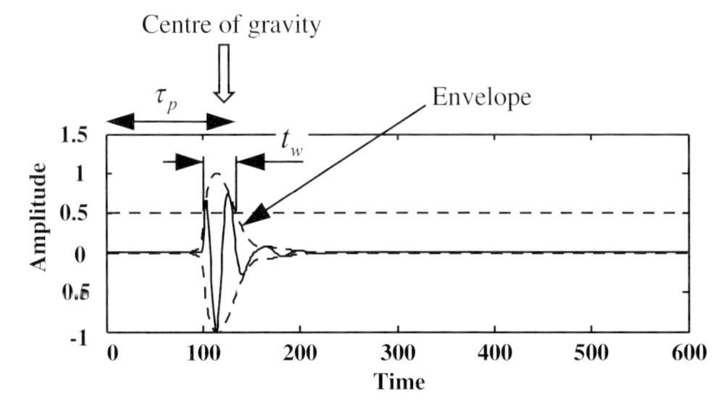

Figure 2.6 Time position and envelope of an oscillating pulse.

2) In the particular case it is: $\mathrm{PV}\int_{-\infty}^{\infty} \frac{x(t)}{t-\xi}\mathrm{d}\xi = \lim_{\varepsilon \to 0}\left(\int_{-\infty}^{t-\varepsilon} \frac{x(t)}{t-\xi}\mathrm{d}\xi + \int_{t+\varepsilon}^{\infty} \frac{x(t)}{t-\xi}\mathrm{d}\xi\right)$.

Besides the numerous shape characteristics, several integral parameters are of practical importance. These parameters do not refer to individual aspects of the waveform rather they consider the pulse function as a whole. Some of these features are defined and their background is mentioned.

2.2.2.2 L_p-norm

In the case of a vector, the norm is a measure of its length or dimension. As we will see now, there are several ways to define them. Corresponding we can do with a sampled signal as it can be considered as a vector too. In connection with signals, the norm can be seen as an expression of the signal 'force' or 'strength' and its impact or effect to stimulate a system.

Extending the vector-norm to continuous signals, the L_p-norm of a function $x(t)$ for real numbers of p is defined by

$$\|x(t)\|_p = \sqrt[p]{\int_{-\infty}^{\infty} |x(t)|^p \mathrm{d}t} \quad |p| \geq 1 \tag{2.8}$$

which is convenient for energy (time limited) signals. Applied to power signals, (2.8) always gives infinity values so it is proposed to modify it to

$$\|x(t)\|_p = \lim_{t_x \to \infty} \sqrt[p]{\frac{1}{2t_x} \int_{-t_x}^{t_x} |x(t)|^p \mathrm{d}t} \quad |p| \geq 1 \tag{2.9}$$

for signals of infinite duration. Figure 2.7 illustrates the p-norms for two energy signals (single wideband pulse, single narrowband pulse) and four different power signals (pulse train, random noise, chirp and M-sequence) as function of the number p. All signals used in the example have either unit energy or unit power.

Only a few of the norms have actually a physical interpretation. One of the most important L_p-norms is the L_2-norm (Euclidian norm). For a power signal, it refers to its rms value and its square is proportional to the average active power P effectuated by the signal. If the signal $x(t)$ is a time varying voltage across a resistor R_0, the realized mean power P is

$$\begin{aligned} P &= \frac{x_{rms}^2}{R_0} \quad \text{with } x_{\mathrm{rms}} = \|x(t)\|_2 = \lim_{t_x \to \infty} \sqrt{\frac{1}{2t_x} \int_{-t_x}^{t_x} |x(t)|^2 \, \mathrm{d}t} \\ &= \lim_{t_x \to \infty} \sqrt{\frac{1}{2t_x} \int_{-t_x}^{t_x} x^2(t) \, \mathrm{d}t} \quad \text{if } x(t) - \text{real} \end{aligned} \tag{2.10}$$

Corresponding holds for signals of limited duration $t \leq t_{\mathrm{w,max}}$. Their energy is

$$\mathfrak{E} = \frac{1}{R_0} \int_{-t_{\mathrm{w,max}}}^{t_{\mathrm{w,max}}} x^2(t) \, \mathrm{d}t = \frac{1}{R_0} \int_{-\infty}^{\infty} |x(t)|^2 \, \mathrm{d}t = \frac{\|x(t)\|_2^2}{R_0} \quad \text{if } x(t) = 0 \text{ for } |t| > t_{\mathrm{w,max}} \tag{2.11}$$

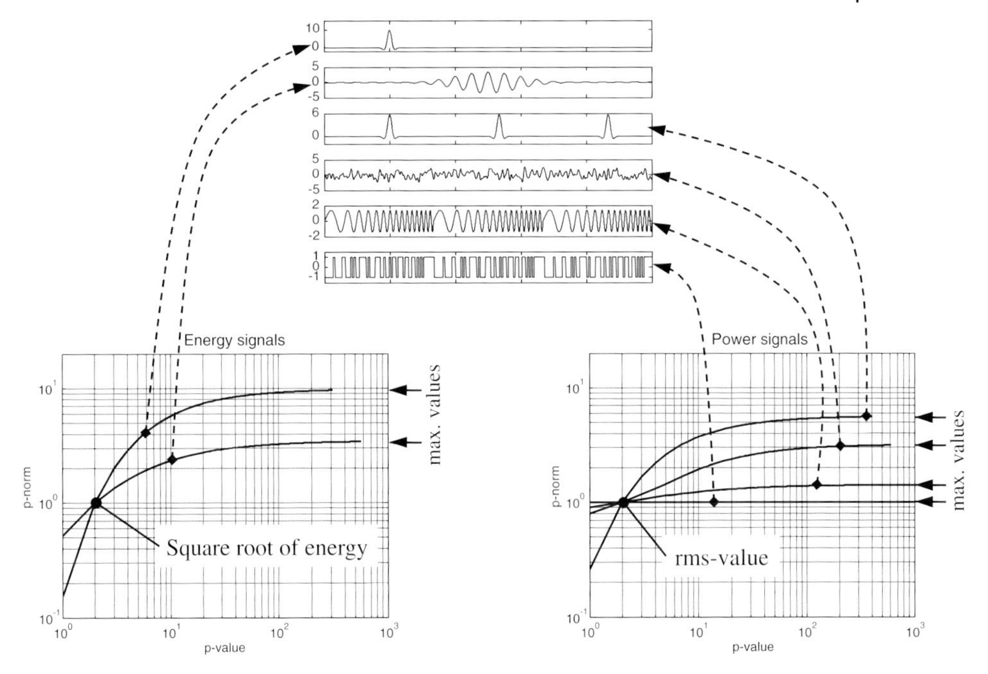

Figure 2.7 Evolution of the L_p-norm as function of the exponent p for several types of signals (see upper graph). Take heed of the ordinate scaling in the upper plots for the different signals.

The L_1-norm represents the area beneath the rectified waveform and the L_∞- or $L_{-\infty}$-norm are simply the maximum and minimum values of that 'rectified' signal. As Figure 2.7 demonstrates, the L_p-norm tends to the maximum value of the signal by increasing the number p. Except the mention of some useful inequalities, we will not extend our considerations on L_p-norms. Comprehensive discussions on this topic can be found in Refs [3–5].

Minkowski inequality

$$\|x(t) + y(t)\|_p \leq \|x(t)\|_p + \|y(t)\|_p \tag{2.12}$$

Hölder's inequality

$$\|x(t)y(t)\|_1 \leq \|x(t)\|_p \, \|y(t)\|_q; \quad \frac{1}{p} + \frac{1}{q} = 1 \tag{2.13}$$

Cauchy–Schwarz inequality

$$\|x(t)y(t)\|_2 \leq \|x(t)\|_2 \, \|y(t)\|_2 \tag{2.14}$$

Young's inequality

$$\|x(t) * y(t)\|_r \leq \|x(t)\|_p \, \|y(t)\|_q; \quad \frac{1}{p} + \frac{1}{q} = 1 + \frac{1}{r} \tag{2.15}$$

2.2.2.3 **Shape Factors**

Figure 2.7 indicates that the L_p-norms of a signal are subjected to more or less strong variations in dependence from p. We can use this observation to introduce a set of ratios quantifying the global shape of signals. Herein, the L_2-norm will act as reference quantity due to its linkage with power and energy. In a first step, this leads us to generalized shape factor as defined by

$$\mathrm{SF}_p = \frac{\|x(t)\|_p}{\|x(t)\|_2} \tag{2.16}$$

where only the generalized shape factors for $p = 1$ and $p \to \infty$ have a technical meaning. That is, SF_1 corresponds to the inverse of the well known form-factor and SF_∞ is usually assigned as the crest-factor (CF) or the peak-to-average power ratio (PAPR):

$$\mathrm{SF}_\infty = \mathrm{CF} = \sqrt{\mathrm{PAPR}} = \frac{\|x(t)\|_\infty}{\|x(t)\|_2} = \frac{x_{\mathrm{peak}}}{x_{\mathrm{rms}}} \quad \text{or} \quad \mathrm{CF[dB]} = 20\log\frac{\|x(t)\|_\infty}{\|x(t)\|_2} \tag{2.17}$$

The crest-factor definition provided by (2.17) refers to power signals for which it leads to a dimensionless quantity. In order to have a corresponding value for energy signals, we slightly modify (2.17) by insertion of the recording time T_R, that is the duration over which the signal is actually captured:

$$\mathrm{CF} = \frac{\|x(t)\|_\infty \sqrt{T_\mathrm{R}}}{\|x(t)\|_2} \tag{2.18}$$

This modification relates the maximum possible energy which a signal of a given peak voltage can carry $\mathfrak{E}_{\max} \propto x_{\mathrm{peak}}^2 T_\mathrm{R}$ to the actual provided energy $\mathfrak{E} \propto \int_{T_\mathrm{R}} x^2(t)\mathrm{d}t = \|x(t)\|_2^2$

The crest-factor is an indication of the signal 'compactness'. If it tends to large values, the signal has concentrated all its power/energy within a short duration (i.e. the signal is pulse like, it has a compact form). On the contrary, a signal with the crest-factor of 1 (0 dB) has distributed all its energy equally over its duration. Hence, such a signal would have the lowest possible amplitude for a given power. Low crest-factor signals prevent test objects and sensor electronics from saturation and damage. Such signals are of major interest for sensing purposes while high crest-factor signals are of advantage for certain stress tests.

For completeness, we will mention still some other shape-related parameters which are needed in later chapters. So far, the introduced ratios do only relate signal magnitudes regardless of the speed by which the signal changes its amplitudes. Now, we will include the last mentioned aspect too. The simplest parameter relating to the speed of amplitude variation is the slew-rate. It indicates the steepest signal slope. Using the above-introduced L_p-norm notation, it can be defined as

$$\mathrm{SR} = \left\| \frac{\partial x(t)}{\partial t} \right\|_\infty = \|\dot{x}(t)\|_\infty = \max\left\{\frac{\partial x(t)}{\partial t}\right\} \tag{2.19}$$

Relating to pulse amplitude and rise time, we can approximately write ($V_{\mathrm{p}}^{\pm}$ is the magnitude of the positive and negative peak of a bipolar pulse)

$$\mathrm{SR} \approx \begin{cases} \dfrac{V_{\mathrm{p}}}{t_{\mathrm{r}}} & \text{unipolar pulse} \\ \dfrac{V_{\mathrm{p}}^{+} - V_{\mathrm{p}}^{-}}{t_{\mathrm{r}}} & \text{bipolar pulse} \end{cases} \tag{2.20}$$

In connections with the signal robustness against jitter, a further global signal parameters and characteristic function may be useful. It includes signal derivations since any temporal signal variation is sensitive to jitter. We define the slope–amplitude coherence function as

$$\mathrm{SACF}(\tau) = \frac{\left\| \dot{x}(t)x(t+\tau) \right\|_2^2}{\left\| \dot{x}(t) \right\|_\infty^2 \left\| x(t) \right\|_2^2} \tag{2.21}$$

For a given signal, the SACF-function indicates if jitter induced noise is spread over the whole signal or if the jitter remains concentrated around the signal edges after performing impulse compression. The aim of this function will be clearer in Section 4.7.3 where we will estimate the influence of random noise and jitter onto the precision of a UWB distance measurement.

2.2.2.4 **Time Position**

The time position of a pulse-shaped signal may be defined by several ways. One possibility is to refer to the time point at which, for example, the rising edge crosses a threshold. Another option uses a reference signal at canonical time position. This signal is shifted in time as long as it best matches the actual signal. The resulting time shift assigns the position of the measured signal (see Sections 2.2.4 and 4.7). Finally, one can determine the centre of gravity of the pulse. In Figure 2.6, this is illustrated by the time τ_{p}. Its calculation is based on a first-order moment. Taking the ideas behind the L_p-norm as a basis, one can define a whole set of centres of gravity τ_{p}

$$\tau_{\mathrm{p}} = \frac{\int_{-\infty}^{\infty} t|x(t)|^p \mathrm{d}t}{\int_{-\infty}^{\infty} |x(t)|^p \mathrm{d}t} \qquad p \geq 1 \tag{2.22}$$

Correspondingly, τ_1 refers to the time location of the gravity centre of the pulse area, τ_2 represents the energetic centre of the pulse and $\tau_\infty = \tau_{\mathrm{peak}}$ simply gives the position of the maximum. The energetic centre of a single pulse can also be written as

$$\tau_2 = \frac{\int_{-\infty}^{\infty} t x^2(t) \mathrm{d}t}{\int_{-\infty}^{\infty} x^2(t) \mathrm{d}t} = \frac{\int_{-\infty}^{\infty} t p(t)\, \mathrm{d}t}{\int_{-\infty}^{\infty} p(t)\, \mathrm{d}t}$$

in which $p(t) \sim x^2(t)$ represents the instantaneous power of the signal. It should be noted that the energetic centre tends to biased estimations in case of short and noise corrupted signals (see Figures 4.60–4.62).

2.2.2.5 Integral Values of Pulse Duration

Effective Pulse Width The effective width t_{eff} of a pulse describes the duration of a time interval in which most of the pulse energy is concentrated. Its definition is based on a central moment of second order:

$$t_{\text{eff}}^2 = \frac{\int_{-\infty}^{\infty} (t-\tau_2)^2 x^2(t)\mathrm{d}t}{\int_{-\infty}^{\infty} x^2(t)\mathrm{d}t} = \frac{\int_{-\infty}^{\infty} (t-\tau_2)^2 p(t)\mathrm{d}t}{\int_{-\infty}^{\infty} p(t)\mathrm{d}t} \tag{2.23}$$

Again, according to the L_p-norms, (2.23) could be generalized to

$$t_{\text{p}}^2 = \frac{\int_{-\infty}^{\infty} (t-\tau_2)^2 |x(t)|^p \mathrm{d}t}{\int_{-\infty}^{\infty} |x(t)|^p \mathrm{d}t} \qquad p \geq 1$$

however without any deeper physical meaning.

Rectangular Width For simple estimations, it could be useful to replace the actual waveform by a rectangular pulse of the same amplitude V_{p} as the original one. The width $t_{1,\text{rect}}$ of the rectangular pulse is chosen in such a way that is has the same area as the actual impulse $x(t)$. This definition is only useful for simple impulses as indicated in Figure 2.4 or for the envelope of an oscillating pulse. For a single pulse, the rectangular width $t_{1,\text{rect}}$ is given by

$$V_{\text{p}}\, t_{1,\text{rect}} = \int_{-\infty}^{\infty} x(t)\mathrm{d}t \tag{2.24}$$

The definition (2.24) supposes that the pulse function approaches (more or less) monotonically the time axis. If the monotony is largely violated, one can consider an equivalent rectangular pulse of the same energy having the equivalent rectangular width $t_{2,\text{rect}}$:

$$V_{\text{p}}^2\, t_{2,\text{rect}} = \int_{-\infty}^{\infty} x^2(t)\mathrm{d}t = \|x(t)\|_2^2 \Rightarrow t_{2,\text{rect}} = \frac{\|x(t)\|_2^2}{\|x(t)\|_\infty^2} \tag{2.25}$$

Since we supposed here a single pulse (energy signal), we can relate (2.25) to (2.18) which joins the crest-factor CF and the rectangular pulse width $t_{2,\text{rect}}$:

$$\mathrm{CF}^2 = \frac{\|x(t)\|_\infty^2}{\|x(t)\|_2^2} T_{\text{R}} = \frac{T_{\text{R}}}{t_{2,\text{rect}}} \tag{2.26}$$

More parameters of pulse-shaped functions are to be found in Section 2.4.2. Since integral parameters (see e.g. (2.8), (2.9), (2.22)–(2.24)) refer to a number of data points they should be generally less sensitive against random perturbations due to the averaging effect than pulse parameters defined by threshold values as shown in Figures 2.4 and 2.5. For densely sampled data, this observation is certainly correct. However, integral parameters lose incrementally their advantages by reducing the sampling rate to the physical minimum (Nyquist sampling).

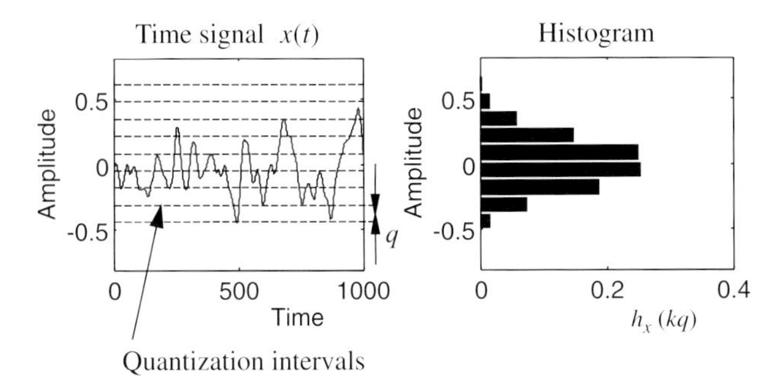

Figure 2.8 Formation of the histogram of a time signal.

In connection with the resolution of UWB radar sensors, we will discuss some of these issues in more detail (see Section 4.7).

2.2.3
Statistical Signal Descriptions

2.2.3.1 Probability Density Function and Its Moments

In many cases it is not necessary or not possible to know the time evolution of a signal. Rather it is sufficient and advisable to deal with probability quantities. The probability density function (PDF) represents the basic tool for such types of considerations.

In order to illustrate the meaning of the PDF, we will consider the operation of an analogue-to-digital converter (ADC) (see Figure 2.8). The device captures continuously N-samples of an arbitrary signal. Meanwhile we are counting the numbers of samples that have been captured for every quantization interval k. The numbers n_k of data samples captured for every quantization interval are plotted in a bar graph as depicted in Figure 2.8.

Here, we always suppose equally spaced quantization intervals of the width q and we also normalize the number n_k to the total number N of captured samples:

$$h_x(k) = \begin{cases} 0 & k < k_{\min} \\ \dfrac{n_k}{N} & k_{\min} \le k \le k_{\max} \\ 0 & k > k_{\max} \end{cases} \quad \text{with } N = \sum_{k_{\min}}^{k_{\max}} n_k \tag{2.27}$$

$h_x(k)$ is called the histogram[3] of the signal $x(t)$. It represents the probability that the amplitude of an arbitrarily captured value of $x(t)$ falls into the kth interval $x_k = q(k-0.5)\cdots q(k+0.5)$. $k_{\min} = \text{round}\{\min\{x(t)\}/q\}$ and $k_{\max} = \text{round}\{\max\{x(t)\}/q\}$ indicate the numbers of the lowest and highest quantization interval. It should be noted that for gathering a waveform histogram, it is not necessary to respect the sampling theorem. The data may be sampled with an arbitrary rate

3) We consider for histogram the numbers of events normalized to the total number. Note that usually the normalization to the total number is not done.

but one has to avoid any synchronism between the signal and data capturing. Furthermore, in order to gain a stable estimate of the histogram, the number N of gathered values must be sufficiently high.

By approaching the quantization interval q to zero and by increasing the number of gathered values to infinity, one ends up in the PDF of the signal $x(t)$:

$$p_x(x) = \lim_{\substack{q \to 0 \\ N \to \infty}} h_x(k) \tag{2.28}$$

From (2.27) and (2.28), it follows immediately

$$\sum_{k_{\min}}^{k_{\max}} h_x(k) = \frac{\sum_{k_{\min}}^{k_{\max}} n_k}{N} = \frac{N}{N} = 1 \Rightarrow \int_{-\infty}^{\infty} p_x(x)\mathrm{d}x = 1 \tag{2.29}$$

The probability that an arbitrary value of $x(t)$ is smaller than a threshold ξ is given by

$$H_x(\xi = k_\xi q) = \sum_{k_{\min}}^{k_\xi} h_x(k) \tag{2.30}$$

or

$$P_x(\xi) = \int_{-\infty}^{\xi} p_x(x)\mathrm{d}x \tag{2.31}$$

Moments and Central Moments Loosely spoken, the moments represent feature values representing position and shape of the PDF. We distinguish between moments α_p and central moments β_p of pth order:

$$\alpha_p = \int_{-\infty}^{\infty} x^p\, p_x(x)\mathrm{d}x \quad p = 1, 2, 3, \cdots \tag{2.32}$$

$$\beta_p = \int_{-\infty}^{\infty} (x - \alpha_1)^p\, p_x(x)\mathrm{d}x \quad p = 2, 3, 4 \cdots \tag{2.33}$$

Mainly the first four moments are of interest in our case whereas the first two are the most important. They may be interpreted in a slightly different way depending on how one looks at the signals. For that purpose, we will distinguish two modes of consideration. First, we will only regard a single signal and second, we will refer to a random process.

2.2.3.2 Individual Signal

The PDF of an individual signal can be determined by the way described above (compare Figure 2.8). Note the PDF was gained by merging data samples gathered at different time points. In this case, one often talks also about the waveform PDF or the waveform histogram. Here, the moments of the first two orders have the following physical meaning:

Mean value; DC value:

$$\alpha_1 = \bar{x} = \lim_{\tau\to\infty} \frac{1}{2\tau} \int_{-\tau}^{\tau} x(t)\mathrm{d}t = \int_{-\infty}^{\infty} x p_x(x)\mathrm{d}x \tag{2.34}$$

Squared rms value (it represents the total average power of the signal):

$$\alpha_2 = x_{\mathrm{rms}}^2 = \overline{x^2} = \|x(t)\|_2^2 = \lim_{\tau\to\infty} \frac{1}{2\tau} \int_{-\tau}^{\tau} x^2(t)\,\mathrm{d}t = \int_{-\infty}^{\infty} x^2 p_x(x)\mathrm{d}x \tag{2.35}$$

AC power:

$$\begin{aligned}\beta_2 &= \overline{x_{\mathrm{ac}}^2} = \overline{x^2} - \bar{x}^2 = \|x(t) - \bar{x}\|_2^2 = \alpha_2 - \alpha_1^2 \\ &= \lim_{\tau\to\infty} \frac{1}{2\tau} \int_{-\tau}^{\tau} (x(t) - \bar{x})^2\,\mathrm{d}t = \int_{-\infty}^{\infty} (x - \bar{x})^2 p_x(x)\,\mathrm{d}x\end{aligned} \tag{2.36}$$

Amplitude Probability Besides the waveform PDF, other representations may be appropriate for signals. For estimating, for example, the mutual interference between electronic devices, the choice of an optimal operational dynamic range or an appropriate number format for signal processing, it is mainly of interest to know the frequency of the large signal magnitudes. For that purpose, we introduce the amplitude probability $P_{\mathrm{A}.x}(\xi)$ which gives the probability that an absolute value of the signal $x(t)$ exceeds the threshold ξ, that is $|x(t)| \geq \xi$:

$$P_{\mathrm{A}.x}(\xi) = 1 - \int_{0}^{\xi} p_{|x|}(x)\mathrm{d}x \tag{2.37}$$

As depicted in Figure 2.9, the amplitude probability corresponds to the indicated area below the probability density function $p_{|x|}(x)$ of the rectified signal. The resulting curves are usually plotted into a Rayleigh plot having an axis scaling which leads to a straight line for normal distributed waveforms (abscissa scaling: $-0.5\log(-\ln P_{\mathrm{A}.x}(x))$; ordinate scaling: $20\log(\xi/x_{\mathrm{ref}})$, with x_{ref} – reference value). For illustration, the figure shows a Rayleigh plot for three idealized signals – a short rectangular pulse, a waveform of Gaussian distribution and a bipolar pulse sequence having a crest-factor of 1 (e.g. M-sequence).

2.2.3.3 Random Process

As mentioned in Section 2.2.1, we will restrict ourselves to the case for which the random process $\underset{\sim}{x}(t)$ consists of a deterministic signal perturbed by a random effects. We have modelled this by an ensemble of measurements (Figure 2.3). The randomness within the ensemble of measured curves for a fixed time point t_0 can be described by an appropriate PDF $p_{\underset{\sim}{x}(t_0)}(x)$, that is we have now one PDF for every time point. For a given point in time, the PDF indicates the scattering of random

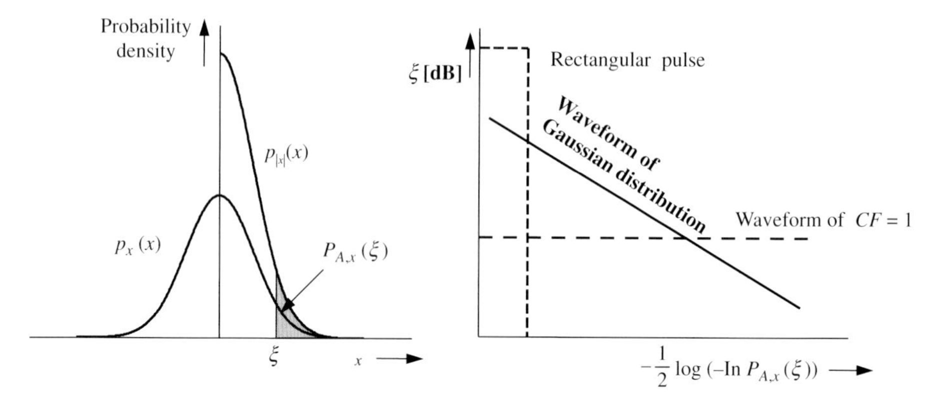

Figure 2.9 Illustration of the amplitude probability and its representation in a Rayleigh plot for idealized waveforms.

values around the wanted value of the deterministic signal part. Figure 2.10 depicts the rising edge of a pulse as a random process and it shows its probabilistic description[4] by a set of PDFs. The example shows a typical measurement situation in which one can observe increased noise at the signal edges (for the reason, see Section 2.6.3). Hence, the PDFs are flat and wide there, while the PDFs within horizontal signal parts are elevated and narrow.

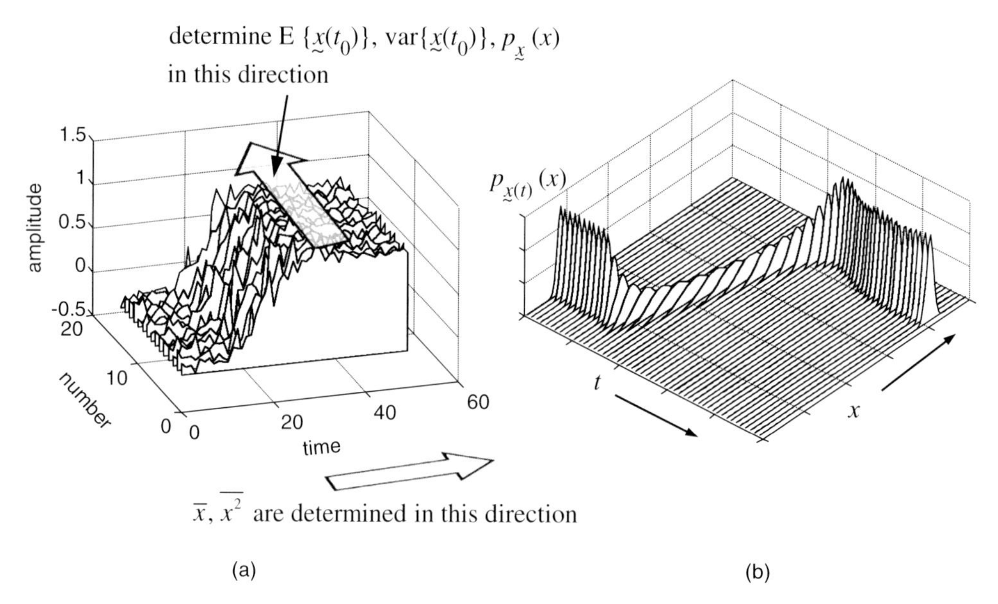

Figure 2.10 Illustration of a random process by an ensemble of measurements (a) and by its time-dependent PDFs (b). The example shows a rising pulse edge. (compare also Figures 2.11 and 2.12).

4) This description is also assigned as weak description.

For us, two PDF moments are of major interest:

Expected value $\mu(t_0) = \mathrm{E}\{\underset{\sim}{x}(t_0)\}$:

$$\alpha_1(t_0) = \mu(t_0) = \mathrm{E}\{\underset{\sim}{x}(t_0)\} = \lim_{N\to\infty} \frac{1}{N}\sum_{n=1}^{N} x_n(t_0) = \int_{-\infty}^{\infty} x\, p_{\underset{\sim}{x}}(t_0)(x)\mathrm{d}x \tag{2.38}$$

As the name imply, this quantity relates to the most probable value. It usually corresponds to the wanted value of the deterministic part of the random process, that is $\mathrm{E}\{\underset{\sim}{x}(t)\} = x(t)$ for $\underset{\sim}{n} \sim N(0, \sigma^2)$ (see (2.1) and (2.40)).

Variance $\sigma^2(t_0) = \mathrm{var}\{\underset{\sim}{x}(t_0)\}$:

$$\begin{aligned}\beta_2(t_0) = \sigma^2(t_0) = \mathrm{var}\{\underset{\sim}{x}(t_0)\} &= \lim_{N\to\infty} \frac{1}{N}\sum_{n=1}^{N} (x_n(t_0) - \mu(t_0))^2 \\ &= \int_{-\infty}^{\infty} (x - \mu(t_0))^2 p_{\underset{\sim}{x}}(t_0)(x)\mathrm{d}x \\ &= \mathrm{E}\{\underset{\sim}{x}^2(t_0)\} - \mathrm{E}^2\{\underset{\sim}{x}(t_0)\}\end{aligned} \tag{2.39}$$

The variance and its square root (standard deviation σ) represent a measure of the uncertainty or the reliability of a measurement.

We will mainly deal with electronic noise as source of random errors. Such noise results from the superposition of many individual random processes, for example the thermal movement of charge carriers. Following the central limit theorem, this will lead to the well-known Gaussian-PDF (also called normal distribution) [6]:

$$p_{\underset{\sim}{x}}(x) = \frac{1}{\sqrt{2\pi}\sigma} \mathrm{e}^{-\frac{1}{2}\left(\frac{x-\mu}{\sigma}\right)^2} \tag{2.40}$$

It has its maximum value at μ and the variance is σ^2. A short form notation for a random process with normal distribution is $\underset{\sim}{x} \sim N(\mu, \sigma^2)$. Under this condition about 67% of the 'noise events' are scattered around μ within an interval of the width $\pm\sigma$. In an interval $\pm 2\sigma$, we will find about 95% of events and 99% all noise incidents are localized within the interval $\mu \pm 3\,\sigma$. It is a usual practice to deal with a normalized version of the variance instead of the absolute value. This leads us to the signal-to-noise ratio:

$$\mathrm{SNR}(t_0) = \frac{\mu^2(t_0)}{\sigma^2(t_0)} = \frac{\mathrm{E}^2\left\{\underset{\sim}{x}(t_0)\right\}}{\mathrm{var}\left\{\underset{\sim}{x}(t_0)\right\}} \triangleq \frac{x^2(t_0)}{\mathrm{var}\left\{\underset{\sim}{x}(t_0)\right\}} \tag{2.41}$$

Note that after (2.41), the SNR-value depends on the time position corresponding to the variation of the signal amplitude and variance. One omits often this dependency and takes the maximum value to quantify the quality of a measurement, that is

$$\mathrm{SNR} = \max\{\mathrm{SNR}(t)\} = \frac{\left\|\mathrm{E}\left\{\underset{\sim}{x}(t_0)\right\}\right\|_\infty^2}{\mathrm{var}\left\{\underset{\sim}{x}(t_0)\right\}} \triangleq \frac{\|x(t)\|_\infty^2}{\sigma_\mathrm{n}^2} \tag{2.42}$$

But this only holds if the noise contribution does not depends on time, that is $\sigma^2(t) = \sigma_n^2$ which also means that the quadratic mean values in time direction $\overline{\underset{\sim}{n}_i^2}$ and ensemble direction σ_n^2 must be the same. A random signal or process, respectively, having this property is said to be ergodic. Applying (2.35) to an arbitrary realization of a zero mean random process $\underset{\sim}{n}(t)$, we can write under such condition

$$\overline{n_i^2} = \lim_{\tau\to\infty} \frac{1}{2\tau} \int_{-\tau}^{\tau} n_i^2(t)\mathrm{d}t = \sigma_n^2 = n_{\mathrm{rms}}^2 \tag{2.43}$$

Hence, the noise variance σ_n^2 in (2.42) is equal to the average noise power $\overline{n^2}$ and $\|x(t)\|_\infty^2$ is the signal peak power.

Ergodicity requires that all the signal statistics can be estimated from an arbitrary single waveform of the random process ensemble via time averaging. Ergodic signals will greatly simplify the mathematical analysis. The moments of an ergodicity process must be stationary up to fourth order which involves that the statistics must be independent of the selected time origin [7]. Note that neither the random process as depicted in Figure 2.10 nor its purely random parts are ergodic since the noise contribution at the signal edge is elevated compared to the flat impulse regions.

Finally, two examples shall demonstrate actual measurements of a random process. We will take the same measurement situation as depicted in Figure 2.10, that is the recording of a rising pulse edge. In practice, pulses are affected by additive noise and jitter (see Section 2.6.3). This leads to time-dependent noise so that the simplifications of (2.43) do not hold in such cases.

Voltage and Time Histogram In the case of periodic pulses, the stability of their edges is an important quality feature. In order to determine this behaviour, the repeatability of the intersection point of the edges with a threshold is measured and represented in corresponding histograms. Figure 2.11 illustrates the capturing of the PDF at time position t_0 which we placed in the middle of the edge. Due to the

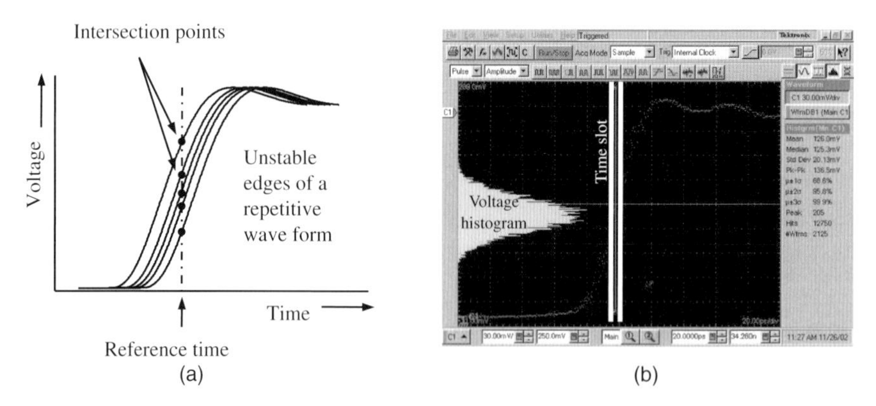

Figure 2.11 Voltage histogram: voltage distribution of the intersection points. (a) Principle and (b) measurement example of a jittered waveform.

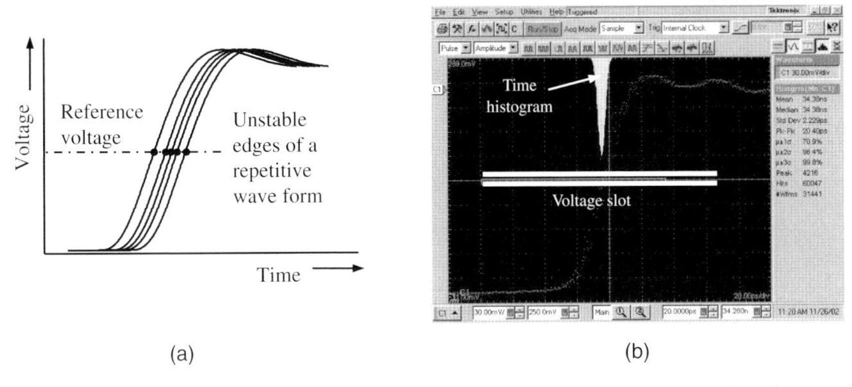

Figure 2.12 Time histogram: time distribution of the intersection points. (a) Principle and (b) measurement example of a jittered waveform.

limited resolution of the oscilloscope, we have to replace actually the reference 'time line' by a thin time slot and the histogram procedure counts the voltage of every data point which falls inside this slot. By stepping the slot through the whole screen, we would gain the time-dependent PDFs as depicted in Figure 2.10.

Basically, it is also possible to rotate the reference slot by 90°. Now, the histogram procedure determines the temporal distribution of the threshold crossing (Figure 2.12).

Finally we will allude the reader to Ref. [8] if he is interested in a more comprehensive but yet compact introduction of statistical signal theory.

2.2.4 Signal Description of Continuous Wave (CW) UWB Signals

The technique of ultra-wideband sensing has frequently to deal with pulse-shaped time functions, for example stimulus or measurement signals. The design challenges for the UWB sensors and sometimes also a risk for the measurement objects is the handling of short pulses. In order to carry sufficient energy, they need high peak power which may lead to saturation effects or even destruction. Therefore, it would be nice to distribute the signal energy over a large time reducing system overloading. UWB sounding signals which have spread their energy over a large duration often looks quite random and unstructured. Hence, superficially considered, they appear less useful for measurement purposes since they are difficult to interpret. Wideband chirps and crest factor optimized multi-sine signals (see Section 2.3) reveal indeed some noticeable structure. But also with them, it will be hardly possible to detect signal variations caused by the interaction with a measurement object. Hence, we need a tool which transforms us such types of signals into understandable shapes.

The correlation function is such a tool. It transforms any wideband signal in a short pulse-like function so that we can keep our signal descriptions from Section 2.2.2. One distinguishes two types of correlation functions – the auto-correlation function (ACF) and the cross-correlation function (CCF). The correlation functions are powerful instruments of the signal theory to describe generic properties of signals. The definition of the correlation function is not unique in the textbooks. This may concern the handling of the mean value and some types of normalization. This ambiguity will, however, not affect the basic conclusions of the correlation approach. We will refer to definitions which are usually applied in signal and system theory.

2.2.4.1 **Auto-Correlation Function**

The auto-correlation function is a mean to describe the affinity/relationship of an individual waveform $x(t)$ and a random process $\underset{\sim}{\mathbf{x}}(t)$ with its time-shifted version. The auto-correlation function for a waveform, we define as follows:

Energy signal (signal of finite duration):

$$C_{xx}(\tau) = \overline{x(t)x(t+\tau)} = \int_{-\infty}^{\infty} x(t)x(t+\tau)\mathrm{d}t \tag{2.44}$$

Power signal (signal of infinite duration):

$$C_{xx}(\tau) = \overline{x(t)x(t+\tau)} = \lim_{t_x \to \infty} \frac{1}{2t_x} \int_{-t_x}^{t_x} x(t)x(t+\tau)\mathrm{d}t \tag{2.45}$$

Figure 2.13 illustrates the effect of the correlation function if it is applied at different types of signals. Obviously, the ACF of a widespread UWB signal becomes a short, pulse shaped appearance comparable to that of an actual pulse. Thus, we can handle a widespread UWB signal like a pulse if we relate to its ACF. The more abrupt the signal changes its amplitude the narrower the correlation peak will be. The width of the correlation peak at half maximum is called coherence time τ_{coh}. That is, a time shift smaller than τ_{coh} can be accepted by the signal without major loss on mutual coherence. If one goes, however, beyond this value, there is no further relation between the signal $x(t)$ and its shifted version $x(t+\tau)$.

The maximum value of the ACF appears at $\tau = 0$ and it equals the total power and energy of the signal (see (2.10) and (2.11)):

$$\|C_{xx}(\tau)\|_{\infty} = C_{xx}(\tau = 0) = \|x(t)\|_2^2 \tag{2.46}$$

In the examples of Figure 2.13, all three signals have the same total energy but the time-stretched signals have quite lower amplitudes. Therefore, these signals do not stress the systems as the impulse. The auto-correlation function is symmetric $C_{xx}(\tau) = C_{xx}(-\tau)$ and we can apply the already introduced pulse parameters on it due to its pulse-like appearance.

For periodic signals, the ACF is cyclic with the same period as the signal. The ACF is subjected to an offset of $\bar{x}^2$ if the signal has a DC component. If the DC value is removed, one also speaks from the covariance function:

$$\mathrm{cov}_{xx}(\tau) = C_{xx}(\tau) - \bar{x}^2$$

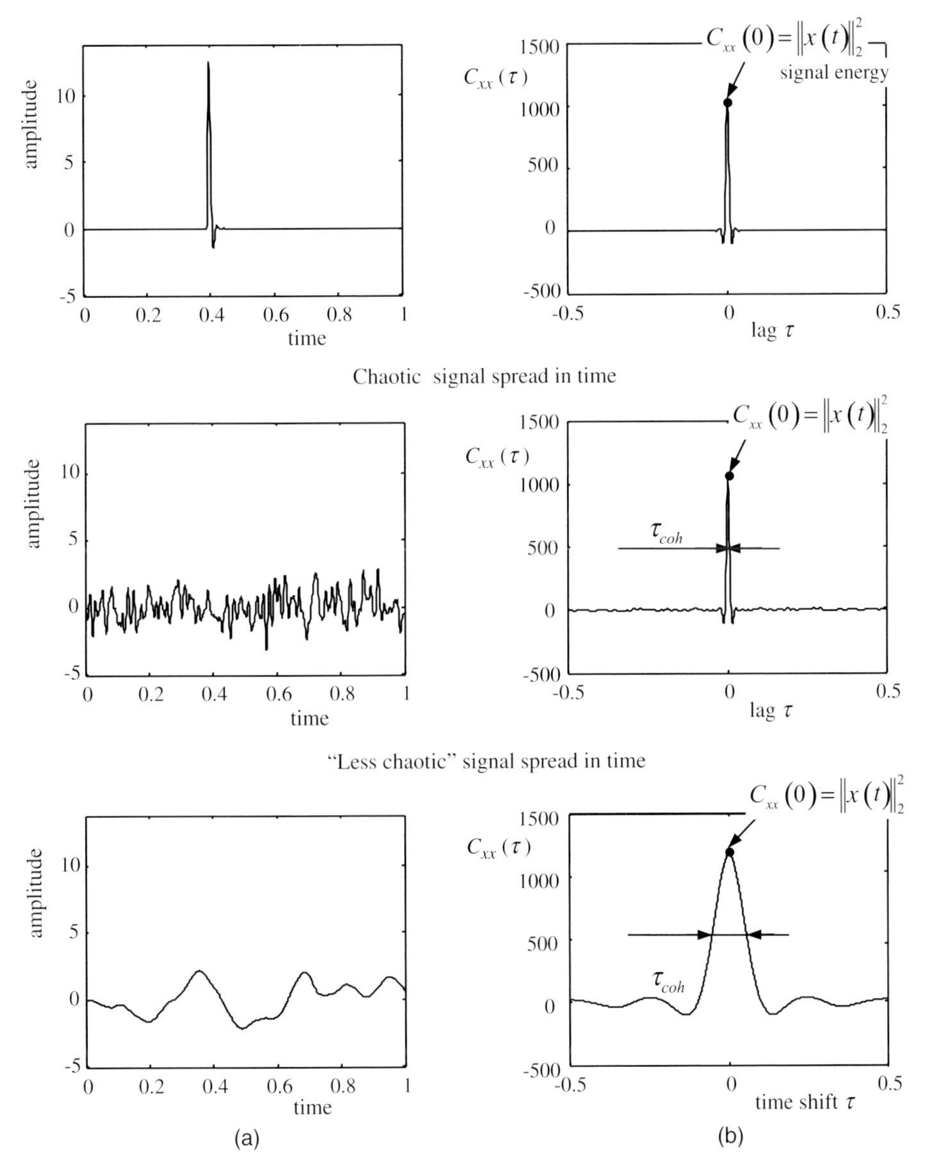

Figure 2.13 Illustration of the auto-correlation functions for different types of signals with limited extension in time (energy signal). Note that all signals carry the same energy even if their maximum amplitudes are different. (a) Time signals and (b) corresponding auto-correlation function.

In the most applications of the UWB sensor technique, the DC value of the signal is of less importance. Often, it is perturbed by offset and drift events of the electronics. Therefore, it is meaningless in the most cases and will be removed from the data. Hence, we will not distinguish here between correlation- and covariance function. We always consider the correlation function of pure AC signals.

In summary, a long chaotic signal is compressed by subjecting it to an autocorrelation procedure. Hence, we now have both, a signal with distributed energy (leading to a gentle stimulation of the systems under test) and the pulse-shaped ACF as we need it for time domain techniques of UWB data processing. The ACF is gained by correlating (i.e. comparing the similarity) of two time-shifted versions of the same signal. If the ACF main peak is very pointed, the considered signal is sensitive to small time shifts. That is, the signal is well suited for high-resolution delay time measurements.

In the case of a stochastic process, the ACF is given by

$$\begin{aligned} C_{\underset{\sim}{x}\underset{\sim}{x}}(t_1, t_2) &= \lim_{N\to\infty} \frac{1}{N} \sum_{i=1}^{N} x_i(t_1) x_i(t_2) = \mathrm{E}\left\{ \underset{\sim}{x}(t_1) \underset{\sim}{x}(t_2) \right\} \\ &= \int_{-\infty}^{\infty} \int_{-\infty}^{\infty} x_1 x_2 p_{\underset{\sim}{x}(t_1)\underset{\sim}{x}(t_2)}(x_1, x_2) \mathrm{d}x_1 \, \mathrm{d}x_2 \end{aligned} \tag{2.47}$$

Herein, $p_{\underset{\sim}{x}(t_1)\underset{\sim}{x}(t_2)}(x_1, x_2)$ represents the joint probability density function.[5] Its discrete counterpart is a two-dimensional histogram function $h_{\underset{\sim}{x}(t_1)\underset{\sim}{x}(t_2)}(k_1, k_2)$ indicating the probability that two data samples captured at time t_1 and t_2 fall into the quantization intervals k_1 and k_2 (i.e. $\{x_i(t_1) \in [q(k_1 - 0.5) \cdots q(k_1 + 0.5)]\}$ $\cap\{x_i(t_2) \in [q(k_2 - 0.5) \cdots q(k_2 + 0.5)]\}$, q is the width of the quantization interval).

If the joint PDFs are independent of time and only depend on the time difference $t_1 - t_2$, that is $p_{\underset{\sim}{x}(t_1+\Delta t)\underset{\sim}{x}(t_2+\Delta t)}(x_1, x_2) = p_{\underset{\sim}{x}(t_1)\underset{\sim}{x}(t_2)}(x_1, x_2) = p_{\underset{\sim}{x}(t_1)\underset{\sim}{x}(t_1+\tau)}(x_1, x_2)$, one speaks from a stationary process resulting in

$$\begin{aligned} &\mu(t) = \mu \\ &\sigma^2(t) = \sigma^2 \\ &C_{\underset{\sim}{x}\underset{\sim}{x}}(t_1, t_2) = C_{\underset{\sim}{x}\underset{\sim}{x}}(\tau); \quad \tau = t_1 - t_2 \end{aligned} \tag{2.48}$$

Note that a stationary process is ergodic. Pure random noise $\underset{\sim}{n}(t)$ usually behaves in such a way.

2.2.4.2 **Cross-Correlation Function**

The cross-correlation function of two waveforms is defined accordingly to the autocorrelation function by referring on two signals instead of one. It provides a

5) For the sake of completeness it should be mentioned that the complete statistic description of a random process needs still further joint PDFs of higher order, that is $p_{\underset{\sim}{x}(t_1)\underset{\sim}{x}(t_2)\underset{\sim}{x}(t_3)\cdots}(x_1, x_2, x_3, \ldots)$ and so on.

measure of the dependencies between both functions. In extension to (2.44) and (2.45), the CCF is determined from

Energy signal (signal of finite duration):

$$C_{xy}(\tau) = C_{yx}(-\tau) = \overline{x(t)y(t+\tau)} = \int_{-\infty}^{\infty} x(t)y(t+\tau)\mathrm{d}t \tag{2.49}$$

Power signal (signal of infinite duration):

$$C_{xy}(\tau) = C_{yx}(-\tau) = \lim_{t_x \to \infty} \frac{1}{2t_x} \int_{-t_x}^{t_x} x(t)y(t+\tau)\mathrm{d}t \tag{2.50}$$

For two stochastic processes, we get

$$\begin{aligned} C_{\underset{\sim}{x}\underset{\sim}{y}}(t_1, t_2) &= \lim_{N \to \infty} \frac{1}{N} \sum_{i=1}^{N} x_i(t_1)y_i(t_2) = \mathrm{E}\left\{ \underset{\sim}{x}(t_1)\underset{\sim}{y}(t_2) \right\} \\ &= \int_{-\infty}^{\infty} \int_{-\infty}^{\infty} xy p_{\underset{\sim}{x}(t_1)\underset{\sim}{y}(t_2)}(x, y)\mathrm{d}x\,\mathrm{d}y \end{aligned} \tag{2.51}$$

The CCF is not symmetric. Permuting $x(t)$ and $y(t)$, it is possible to build two different waveform CCFs which have the same but reversed shape with respect to the lag τ.

The correlation functions are an extremely useful tools for different measurement purposes:

- The CCF allows a precise determination of small time shifts between two (similar or identical) signals even if these signals are spread over a large duration.
- From a mixture of different signals, the correlation function is able to emphasize the presence of a signal of known shape. Here, the idea is to correlate the signal mixture with a reference waveform which has the shape of the wanted signal. If the signal is part of the mixture, the correlation function will give a peak. In the other case, the correlation function will provide only small values since typically the reference waveform does not match the remaining waveforms. Such an approach is called matched filtering.
- The use of the correlation technique permits the stimulation of test objects with arbitrary signals in order to characterize them. This increases the flexibility of available measurement and sensing procedures (see Sections 3.3 and 3.6 for details).
- They give a measure of the similarity of two signals which can be expressed, for example, by the fidelity value.

Fidelity A signal $x(t)$ is usually subjected to some distortions if it passes through a transmission system. We call $y(t)$ the output signal. In order to quantify the shape distortions between both signals by a single parameter, we refer to the overall difference of the normalized signals. Since we are interested only in the shape disparity between both signals, we have also to remove a possibly existing time lag, which leads us to

$$\begin{aligned} \Delta_{xy} &= \min_{\tau} \int (x_n(t+\tau) - y_n(t))^2 \mathrm{d}t \\ x_n(t) &= \frac{x(t)}{\sqrt{\int x^2(t)\mathrm{d}t}} = \frac{x(t)}{\sqrt{C_{xx}(0)}} \quad y_n(t) = \frac{y(t)}{\sqrt{\int y^2(t)\mathrm{d}t}} = \frac{y(t)}{\sqrt{C_{yy}(0)}} \end{aligned} \tag{2.52}$$

Factoring out (2.52) and disregarding a change in sign as a distortion, we get

$$\Delta_{xy} = 2 \min_{\tau} \left(1 - \left| \int x_n(t+\tau) y_n(t) \mathrm{d}t \right| \right) \tag{2.53}$$

Hence, Δ_{xy} minimizes if the integral takes its maximum. We will call this value the fidelity FI_{yx} of $y(t)$ with respect to $x(t)$[9]:

$$\mathrm{FI}_{yx} = \max_{\tau} \left| \int x_n(t+\tau) y_n(t) \mathrm{d}t \right| = \frac{\| C_{yx}(\tau) \|_{\infty}}{\sqrt{C_{xx}(0) C_{yy}(0)}} = \frac{\| C_{yx}(\tau) \|_{\infty}}{\| x(t) \|_2 \| y(t) \|_2} \tag{2.54}$$

The Cauchy–Schwarz inequality $\left(\int x(t) y(t) \mathrm{d}t \right)^2 \leq \int x^2(t) \mathrm{d}t \int y^2(t) \mathrm{d}t$ or Young's inequality (see (2.15), $p = q = 2$ and $r \to \infty$), respectively, implies that the fidelity value is between 0 and 1. If it is one, both signals have identical shape, but not necessarily the same amplitude and time position. Note that (2.54) simply represents the maximum of the normalized CCF. The term fidelity will be used in connection with time domain descriptions of antennas (see Section 4.6).

2.2.5 Frequency Domain Description

Physical processes are running in space and time. Hence, the primordial description of signals and systems is given in the time domain. However, the decomposition of arbitrary-shaped time functions into a set of elementary signals may often make life easier. So in many cases, the solution of differential equations describing the behaviour of the systems of interest will be drastically simplified or the understanding of some physical phenomenon is more intuitive by referring to convenient than to arbitrary-shaped signals. Furthermore, by exploiting the redundancies in natural signals, one can represent them in a concise and data economic form if they are referred to the proper basis [10]. And finally, some data processing routines are running faster and more efficient if they are based on the decomposed signals. There exist many types of such 'convenient' functions. Examples for analytical functions are sine waves, damped sine waves, Bessel functions, Legendre polynomials, Walsh functions, wavelets and so on to mention only a few.

The basic idea is to decompose the arbitrary-shaped function $f(\xi)$ in a set of 'convenient' functions $g_n(\xi)$, that is:

$$f(\xi) = a(0) g_0(\xi) + a(1) g_1(\xi) + a(2) g_2(\xi) + \cdots \tag{2.55}$$

The related coefficients $a(n)$ are characteristic for that function so it will be possible to reconstruct $f(\xi)$ from the knowledge of $a(n)$. The direct use of $f(\xi)$ is often related to the notation 'original domain' and dealing with $a(n)$ is related to the term 'image domain'.

In order to have a unique relation between the original function $f(\xi)$ and its related image function $a(n)$, one mostly (but not always) require that the function basis $g_n(\xi)$ are forming an orthogonal set of equations:

$$\int g_n(\xi)g_m(\xi)d\xi = \begin{cases} 0 & \text{if} \quad n \neq m \\ \|g_n(\xi)\|_2^2 & \text{if} \quad n = m \end{cases} \tag{2.56}$$

This immediately leads to the conditional equation which determines the image function $a(n)$. Due to the orthogonality of $g_m(\xi)$ and applying (2.55), we get

$$\int f(\xi)\, g_n(\xi)\, \mathrm{d}\xi = \sum_m a(m) \int g_n(\xi)g_m(\xi)\mathrm{d}\xi = a(n)\, \|g_n(\xi)\|_2^2 \tag{2.57}$$

2.2.5.1 The Fourier Series and Fourier Transformation

For brevity, we will only consider the 'convenient' sine and cosine functions. Their big advantage is that they will not change their time shape by passing a linear system[6] (see Section 2.4.3) which reflects in largely simplified calculations.[7] Following the general rule (2.55), we can decompose a time signal $x(t)$ in a (theoretically infinite) number of sine waves and cosine waves by

$$\begin{gathered} x(t) = \sum_{n=0}^{\infty} a(n)\cos 2\pi n\,\Delta f t + b(n)\sin 2\pi n\,\Delta f t = \sum_{n=0}^{\infty} X(n)\sin(2\pi n\,\Delta f t + \varphi(n)) \\ \text{with} \quad X(n) = \sqrt{a^2(n) + b^2(n)}; \quad \varphi(n) = \arctan\frac{b(n)}{a(n)} \end{gathered} \tag{2.58}$$

The functions $\sin(n\xi), \sin(m\xi)$ and $\cos(n\xi), \cos(m\xi)$ are forming an orthogonal basis, so that the representation of $x(t)$ by the image functions $a(n)$ and $b(n)$, and $X(n)$ and $\varphi(n)$ is unique. If we finally use the Euler notation $\mathrm{e}^{jx} = \cos x + j\sin x$, we end up in the commonly used decomposition of time domain functions into complex-valued phasors e^{jx}. The image domain is usually called frequency domain since the functions $g_n(\xi) = \left[\sin(n\xi), \cos(n\xi), \mathrm{e}^{jn\xi}\right]$ are only distinguished by their frequency. Note that the frequencies $n\,\Delta f$ have positive and negative values if one applies the commonly used phasor-based notation. This is expressed by the term two-sided spectrum. The negative part of the spectrum is redundant in the case of physically real signals (see (2.60) or Annex B.1, Table B.2). Hence, it is often sufficient to deals only with the positive part of the spectrum (one-sided spectrum), for example done in (2.58). If ambiguities in the notation of spectral quantities concerning one- or two-sided consideration may arise, we will label two-sided spectral quantities by a double dot, for example $\underset{..}{B}$ and one-sided quantities by a single dot, for example $\underset{.}{B}$.

6) Exactly spoken this is only true if the system is in the steady state.

7) As we will see in Section 2.4.3, the impact of a sounding signal onto a system can be modelled by multiplications instead of a convolution as required for arbitrary time signals.

The decomposition (2.58) of time functions into sine waves and phasors is called Fourier transform. We will discuss some particularities depending on the type of the signal.

Periodic power signal: If the signal under consideration is an infinite periodic signal $x(t) = x(t + nt_P)$, it can be represented by a sum of sine waves of those periods that are an integer fraction of the original signal period $t_P = 1/f_0$. Herein, f_0 is the repetition rate of the original waveform.

$$x(t) = \sum_{n=-\infty}^{\infty} \underline{X}(nf_0)e^{j2\pi nf_0t} \quad \text{with } \underline{X} = Xe^{j\varphi} \tag{2.59}$$

(2.59) is the Fourier series of $x(t)$. Following from (2.57), the complex phasor amplitudes $\underline{X}(nf_0)$ result from

$$\underline{X}(nf_0) = \frac{1}{t_P} \int_{-t_P/2}^{t_P/2} x(t) \cdot e^{-j2\pi nf_0t}\, dt \tag{2.60}$$

where we have $\underline{X}(nf_0) = \underline{X}^*(-nf_0)$ for real-valued time signals.
As expected from (2.59), the physical dimension of the sine wave (phasor) amplitudes is identical to that of the original. (2.60) represents the complex spectrum of $x(t)$, that is it depicts amplitude and phase as function of the frequency of the sinusoids (see Figures 2.14 and 2.15 for illustration). Since only sine waves with the multiple of the repetition rate f_0 appear, one speaks from a discrete spectrum. The average power carried by every spectral line is given by

$$\underset{..}{\Phi}(nf_0) = \frac{\underset{..}{\Psi}(nf_0)}{R_0} = \frac{1}{R_0}\underset{..}{X}(nf_0)\underset{..}{X}^*(nf_0) \tag{2.61}$$

$\Phi(nf_0)$ [W] is called the power spectrum of $x(t)$ and $\Psi(nf_0)$ [V^2] we will assign as auto-spectrum. If the considered signal is a voltage, then R_0 represents an ohmic resistance supplied by that voltage. Often R_0 is simply omitted and $\Psi(nf_0)$ is often assigned as power spectrum too. The power spectrum preserves the amplitude information but it loses the phase. Summing over all spectral lines gives finally the total signal power (see also (2.10)):

$$\begin{aligned} P &= \sum_{n=-\infty}^{\infty} \Phi(nf_0) = \frac{1}{R_0} \sum_{n=-\infty}^{\infty} \underline{X}(nf_0)\underline{X}^*(nf_0) \\ &= \frac{1}{R_0}\left(X^2(0) + 2\sum_{n=2}^{\infty} \underline{X}(nf_0)\underline{X}^*(nf_0)\right) = \frac{1}{R_0}\left(\bar{x}^2 + 2\sum_{n=2}^{\infty} \underline{X}(nf_0)\underline{X}^*(nf_0)\right) \\ &= \frac{1}{R_0 t_P} \int_{-t_P/2}^{t_P/2} x^2(t)\, dt = \frac{x_{\text{rms}}^2}{R_0} \end{aligned} \tag{2.62}$$

This equation is also known under the term Parseval's theorem (for periodic signals). The sum $2\sum_{n=2}^{\infty} \underline{X}(nf_0)\underline{X}^*(nf_0) = x_{\text{ac}}^2$ covers the whole AC power carried by the signal (see also (2.36)).

Instead of (2.61), we also often find the relation $\underset{.}{\Psi} = \underset{.}{X}\,\underset{.}{X}^*/2$. This formula relates to single-sided spectra which we indicate by a single dot below the

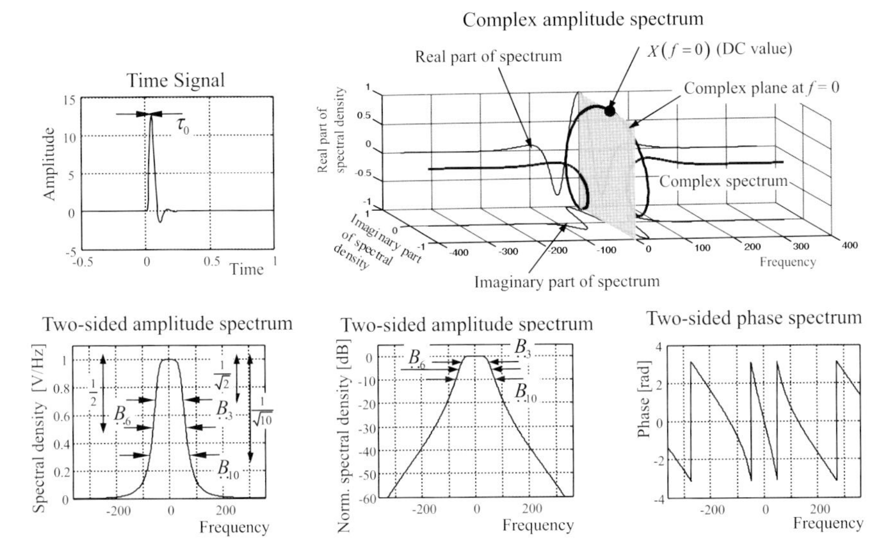

Figure 2.14 Spectrum of a single low-pass pulse represented in various graphs. The original signal is shown left above. Aside, the complex spectrum in 3D-cordinates is illustrated. The projections onto the $\Re\{\underset{\cdot}{X}\} - f$-plane and $\Im\{\underset{\cdot}{X}\} - f$-plane are also shown. The lower graphs depict the amplitude (linear and logarithmic scaled) and phase spectrum.

symbol, while (2.61) refers to two-sided spectra marked by a double point. The relations between single- and double-sided spectrums are for amplitude quantities $\underset{\cdot}{X}(f)|_{f\geq 0} = (1/2)\underset{\cdot\cdot}{X}(f)|_{f\geq 0} = (1/2)\underset{\cdot\cdot}{X}^*(f)|_{f\leq 0}$ and for power quantities $\underset{\cdot}{\Psi}(f) = \underset{\cdot\cdot}{\Psi}(f) + \underset{\cdot\cdot}{\Psi}(-f) = 2\,\underset{\cdot\cdot}{\Psi}(f)$, which can be easily find from Fourier

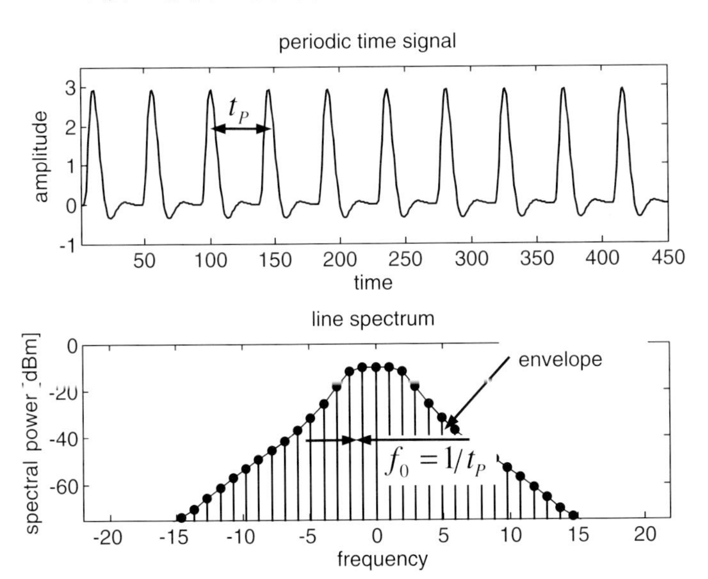

Figure 2.15 Spectrum of a periodic signal. Only the amplitude spectrum is shown.

transform. We are mainly dealing with double-sided spectral quantities wherefore the symbol labelling by single or double dot may be omitted. However, we will come back to this kind of labelling in connection with the bandwidth term since in this regard some danger of confusion exists.

Energy signal: In the case of a time-limited signal, the sum in (2.59) has to be replaced by an integral leading to the inverse Fourier transformation:

$$x(t) = \int_{-\infty}^{\infty} \underline{X}(f) e^{j2\pi f t} df \tag{2.63}$$

and (2.60) converts to the Fourier transformation:

$$\underline{X}(f) = \int_{-\infty}^{\infty} x(t) e^{-j2\pi f t} dt \tag{2.64}$$

Now, the frequencies of the sinusoids decomposing the original waveform $x(t)$ are infinitely dense (i.e. $\Delta f \to 0;\ n\Delta f \to f$). Therefore, the complex spectrum $\underline{X}(f)$ has a continuous shape and is of dimension [V/Hz] if $x(t)$ is a voltage signal, that is the spectrum represents a spectral amplitude density. The power relations (2.61) and (2.62) are modified to

$$\Phi(f) = \frac{\Psi(f)}{R_0} = \frac{1}{R_0} \underline{X}(f) \underline{X}^*(f) \tag{2.65}$$

$\Phi(f)$ is the energy density spectrum (dimension [W s/Hz] = [J/Hz]) and Parseval's theorem is written as

$$\mathfrak{E} = \int_{-\infty}^{\infty} \Phi(f) df = \frac{\|\underline{X}(f)\|_2^2}{R_0} = \frac{1}{R_0} \int_{-\infty}^{\infty} x^2(t) dt = \frac{\|x(t)\|_2^2}{R_0} \tag{2.66}$$

Non-periodic power signal: A non-periodic signal of infinite length is typically of random nature which disallows the creation of a complex spectrum. But its autocorrelation function $C_{xx}(\tau)$ is deterministic which can be used to describe such signals in the frequency domain, that is we can write

$$\Psi_{xx}(f) = \int_{-\infty}^{\infty} C_{xx}(\tau) e^{-j2\pi f \tau}\, d\tau \tag{2.67}$$

$\Psi_{xx}(f)$ is called the auto-spectrum (dimension [V^2/Hz]) of the random signal $x(t)$. The auto-spectrum is a real-valued function since $C_{xx}(\tau)$ is symmetric and it is proportional to the power density spectrum $\Phi(f) = \Psi_{xx}(f)/R_0$ (dimension [W/Hz]) if $x(t)$ represents a voltage across the resistor R_0. Consequently, the average signal power is calculated by

$$P = \frac{C_{xx}(\tau = 0)}{R_0} = \int_{-\infty}^{\infty} \Phi(f) df \tag{2.68}$$

Extending (2.67) to a cross-correlation function leads to the so-called cross-spectrum:

$$\underline{\Psi}_{yx}(f) = \int_{-\infty}^{\infty} C_{yx}(\tau) e^{-j2\pi f \tau} d\tau \tag{2.69}$$

Note that the cross-spectrum is a complex function in contrast to the auto-spectrum. The meaning and significance of the cross-spectrum will be obvious in connection with the description and measurement of linear systems (see Sections 2.4.3 and 2.6.1).

For deterministic signals, it can be shown that

$$\begin{aligned} \Psi_{xx}(f) &= \underline{X}(f)\,\underline{X}^*(f) \\ \underline{\Psi}_{xy}(f) &= \underline{X}(f)\,\underline{Y}^*(f) \end{aligned} \tag{2.70}$$

The similarity, the (linear) relation or affinity, between two spectra can be expressed by the (spectral) coherence function. It is defined by

$$\mathrm{coh}(f) = \sqrt{\frac{\underline{\Psi}_{xy}(f)\underline{\Psi}^*_{xy}(f)}{\Psi_{xx}(f)\Psi_{yy}(f)}} \tag{2.71}$$

2.2.5.2 **Some Properties and Parameters of a Spectrum**

Figure 2.14 illustrates the spectrum of a signal. For demonstration a single pulse was used. Since its spectral band starts at low frequencies, we call it low-pass or baseband pulse. Figure 2.14 represents the spectrum by different graphs where commonly only the dB-scaled spectrum is in use. It gives an impression of the power of sine waves which contributes to the original waveform.

The (average) steepness of the phase spectrum is a measure of the temporal shift of the time function with respect to a reference time. By increasing τ_0, the phase curves will become steeper. In 3D coordinates, the complex spectrum looks like a 'corkscrew'. The 'length of the corkscrew' depends on the bandwidth and its 'thread pitch' depends on the time shift τ_0.

The spectra of periodic signals are represented in the same way with the exception that they are built from discrete lines (see Figure 2.15). The envelope of the lines is proportional to the spectrum of the single waveform.

An oscillating, DC-free impulse has a spectrum which is separated in two parts of identical amplitude (but inverse phase) characteristics which are placed around the centre (or carrier) frequencies $\pm f_c$ (Figure 2.16). Since this spectrum corresponds to a typical frequency response function (FRF) of a band-pass filter, we will call such an impulse as band-pass pulse. In a simple case, such pulse can be generated by multiplying a baseband pulse with a sine wave of frequency f_c. Hence, the baseband pulse represents the envelope of the band-pass pulse.[8)]

For deeper consideration on the Fourier transform, the reader is addressed to Annex B.1. It summarizes some important rules to handle the Fourier transform and it has a table for the conversation between time and frequency domain for a few elementary signals.

The most important parameter of a spectrum is its bandwidth B. It specifies the range of sine wave frequencies which most contribute to 'the construction' of the original time signal. There are different definitions of bandwidth. The most

8) Note the envelope of the time signal will be complex valued in the general case. See Section 3.3.6, Figure 3.56 for further explications.

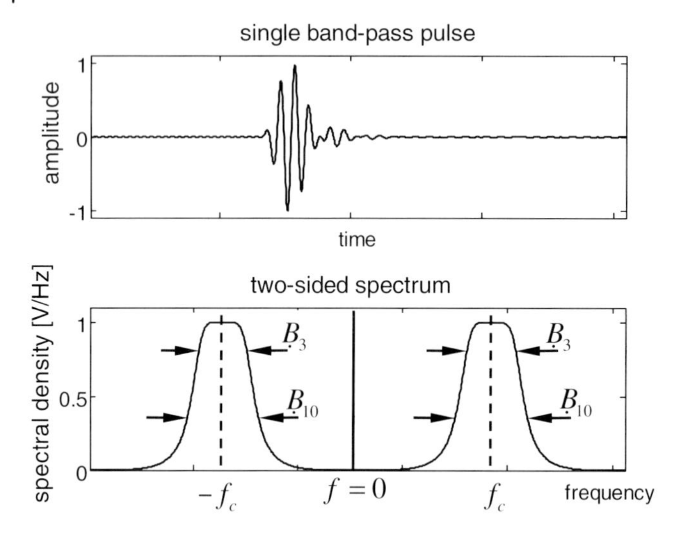

Figure 2.16 Typical spectrum of an oscillating impulse.

convenient one considers the width of the spectral function referred to a threshold. The threshold level is mostly chosen at the half of the maximum spectral power or the half of the maximum amplitude. For the first one, this corresponds to a level of $-3\,\text{dB}$ (or $1/\sqrt{2}$ for a linear scaled amplitude spectrum) below the maximum (see for example Figures 2.14 and 2.16). The second one leads to a $-6\,\text{dB}$ level (or 1/2 for a linear scaled amplitude spectrum, not shown in the figures). In the case of the ultra-wideband technique, the reference level is fixed to about one-third (exactly it is $1/\sqrt{10}$) of the maximum amplitude level (or a tens of maximum power), that is to $-10\,\text{dB}$. Thus B_3, B_6 or B_{10} are the half, fourth or tens power bandwidth.

Considering Figures 2.14 and 2.16, one will observe differences in the definition of the bandwidth, that is they refer to a two-sided bandwidth in case of a low-pass pulse and to a one-sided bandwidth for a band-pass pulse. In order to avoid ambiguities, we will write a single-sided bandwidth by one dot $\underset{\cdot}{B}$ beneath the symbol and a double-sided bandwidthby two dots $\underset{\cdot\cdot}{B}$.

Symmetric pulses whose peak is located at $t = 0$ have a purely real spectrum (see Annex B.1, Table B.2). Further we suppose that they approach (more or less) monotonically the time axis. In such cases, it may be convenient to replace the actual amplitude spectrum by an equivalent rectangular spectrum having the same maximum value $X_{\max}$ and covering the same area, that is we can write for baseband pulses:

$$X_{\max}\,\underset{\cdot\cdot}{B}_{1,\text{rect}} = \int_{-\infty}^{\infty} X(f)\,\mathrm{d}f \tag{2.72}$$

The resulting width of the rectangular spectrum is called equivalent rectangular bandwidth $\underset{\cdot\cdot}{B}_{1,\text{rect}}$. A corresponding approach we already used to simplify the time shape of pulses (see (2.24)) resulting in an equivalent rectangular pulse width t_{rect}. For a band-pass pulse, we can restrict ourselves to the positive frequencies in the integral which would lead to the single-sided rectangular bandwidth. Note that this approximation makes only sense if the spectrum does not contain too many and too strong side lobes.

Referring to energetic aspects, we can define an effective bandwidth B_{eff} by applying second-order moment as we already did for the effective pulse width t_{eff} (see (2.23)):

$$B_{\text{eff}}^2 = \frac{\int_{-\infty}^{\infty} f^2 \Psi_{xx}(f) \mathrm{d}f}{\int_{-\infty}^{\infty} \Psi_{xx}(f) \mathrm{d}f} = -\frac{1}{4\pi^2 C_{xx}(0)} \left. \frac{\partial^2 C_{xx}(\tau)}{\partial \tau^2} \right|_{\tau=0} \tag{2.73}$$

Respecting Parseval's theorem and the differentiation rule of Fourier transform (Annex B.1, Table B.2), the effective bandwidth can be related to the second derivative of ACF. Note that (2.73) is only valid for baseband pulses. For band-pass pulses, one has to apply the central moments (i.e. $\int_0^\infty (f - f_c)^2 \Psi_{xx} \, \mathrm{d}f$) and the envelope of the ACF.

A second approach is to follow the idea behind (2.72) for power values instead of amplitude quantities, so that we get

$$X_{\max}^2 \, \underset{..}{B}_{2,\text{rect}} = \int_{-\infty}^{\infty} |\underline{X}(f)|^2 \mathrm{d}f = \|x(t)\|_2^2 \tag{2.74}$$

2.2.5.3 Time-Bandwidth Products

Time-bandwidth products are simple rules of thumb to switch between time and frequency domain. They are useful for practical purposes since they give fast and simple relations between both methods of signal description. We will introduce several of these relations which are each for a specific class of problems. In Section 2.3 some idealized UWB signals are discussed and their actual time-bandwidth products are evaluated (see also Ref. [7] for further discussions).

Pulse-Like Waveforms Under the condition mentioned in connection with (2.72) and combining this equation with (2.24), one can show via the Fourier transform that

$$\underset{..}{B}_{1,\text{rect}} t_{1,\text{rect}} = 1 \tag{2.75}$$

If the pulse/spectrum consists of (a low number of) oscillations or the integrals tend to zero, one cannot approximate them by a rectangular function. Under these constraints, we can apply the power-related equivalent width for which also approximately holds

$$\underset{..}{B}_{2,\text{rect}} \, t_{2,\text{rect}} \approx 1 \tag{2.76}$$

if the pulse represents a compact waveform. These are very simple rules of thumb indicating that the shorter an impulse is the wider is its spectrum. It can be used to estimate roughly the bandwidth demands to transmit a low-pass pulse of a given duration with only minor violations of its shape. The relation (2.75) is a general feature of pulse waveforms. The quantity of equivalent rectangular widths

t_{rect}, $\underset{..}{B}_{rect}$ are usually close to the pulse width t_w and to the -3dB bandwidth $\underset{..}{B}_3$ which are often easier to determine in practice.

With some minor changes, the same approach can also be used for band-pass pulses (fractional bandwidth $b < 1$). But now, we have to take the duration of the pulse envelope $t_w \mathrel{\hat{=}}$ FDHM and the single-sided bandwidth $\underset{.}{B}_3$ of the spectrum as depicted in Figure 2.16. According to (2.75), we get

$$t_w \underset{.}{B}_3 \approx 1 \tag{2.77}$$

The determination of the pulse envelope was introduced in (2.5).

UWB Radio Pulse We will understand under a radio signal a signal which has passed a radio transmission channel. For that purpose, the electric signal has to be transformed in an electromagnetic wave and to reconverted back into the electric domain. Following to Rayleigh's postulate, the force lines of any electromagnetic field in free space should be closed [11]. That is, a radiator of finite dimension cannot emit a steady-state electromagnetic field. Hence, we have to require a DC-free receive signal (see also radiation from an antenna: Sections 4.4. and 5.5):

$$\bar{x} = \int_{-\infty}^{\infty} x(t)\mathrm{d}t = X(0) = 0 \tag{2.78}$$

Practically this means that at least some small spectral parts around $f = 0$ are missed. As long as these parts do not cause too much energy loss, we use the full two-sided bandwidth like in (2.75) and neglect the missing spectral power. Tables 2.2 and 2.6 show two examples for such signals.

However, with increasing size of the spectral gap at low frequencies, the TB products tend successively from (2.75) (or (2.76)) to the relation (2.77). The width of the actual physically occupied spectrum of a low-pass pulse is only the half of its two-sided bandwidth. Hence, a baseband pulse has roughly half the duration of a band-pass pulse if both occupy the same physical bandwidth. Consequently, the width of an UWB radio pulse has a characteristic duration in between these two cases depending on the gap size at low frequencies. A demonstration of such behaviour is found in Section 2.3.6.

Time Spread Wideband Signals Wideband signals of long duration T have a short pulse-like auto-correlation function as depicted in Figure 2.13. Its characteristic width is the coherence time τ_{coh}. The transformation of the auto-correlation function to the frequency domain leads to a spectrum of the characteristic width B_3. Correspondingly to (2.75) or (2.77), we can write in accordance to pulse-like waveforms

$$\text{Base-band signal:} \quad \tau_{coh} \underset{..}{B}_3 \approx 1 \tag{2.79}$$

$$\text{Band-pass signal:} \quad \tau_{env,coh} \underset{.}{B}_3 \approx 1 \tag{2.80}$$

As for the pulse-like waveforms above, we can also distinguish here between baseband and band-pass signals. Wideband band-pass signals which are spread over a large time have an oscillating ACF like the pulse in Figure 2.16 except its asymmetry (the ACF is always symmetric). So, it can be handled like the oscillating pulse.

Table 2.1 Characteristic parameters of a unipolar rectangular pulse.

Parameter	Value
Duty cycle	$d_c = \dfrac{t_w}{t_P}$
Rectified value	$\lVert x(t) \rVert_1 = V \dfrac{t_w}{t_P} = V d_c$
rms value	$\lVert x(t) \rVert_2 = V \sqrt{\dfrac{t_w}{t_P}} = V \sqrt{d_c}$
Crest factor	$\mathrm{CF} = \sqrt{\dfrac{t_P}{t_w}} = \dfrac{1}{\sqrt{d_c}}$
Slew rate	$\mathrm{SR} = \infty$
Effective pulse width	$t_{\mathrm{eff}} = \dfrac{t_w}{\sqrt{12}}$
Rectangular width	$t_{1,\mathrm{rect}} = t_w$
	$t_{2,\mathrm{rect}} = t_w$
3 dB bandwidth	$t_w \underset{\cdot\cdot}{B}_3 \approx 0.886$
6 dB bandwidth	$t_w \underset{\cdot\cdot}{B}_6 \approx 1.21$
10 dB bandwidth	$t_w \underset{\cdot\cdot}{B}_{10} \approx 1.48$
Rectangular bandwidth	$\underset{\cdot\cdot}{B}_{1,\mathrm{rect}}$ – not applicable
	$t_w \underset{\cdot\cdot}{B}_{2,\mathrm{rect}} = 1$
Effective bandwidth	B_{eff} – not applicable
Time-bandwidth product	$\mathrm{TB} = t_{2,\mathrm{rect}} \underset{\cdot\cdot}{B}_{2,\mathrm{rect}} = 1$

Table 2.2 Characteristic parameters of a bipolar rectangular pulse.

Parameter	Value
Duty cycle	$d_c = \dfrac{2t_w}{t_P}$
Rectified value	$\lVert x(t) \rVert_1 = 2V \dfrac{t_w}{t_P} = V d_c$
rms value	$\lVert x(t) \rVert_2 = V \sqrt{\dfrac{2t_w}{t_P}} = V \sqrt{d_c}$
Crest factor	$\mathrm{CF} = \sqrt{\dfrac{t_P}{2t_w}} = \dfrac{1}{\sqrt{d_c}}$
Slew rate	$\mathrm{SR} = \infty$
Effective pulse width	$t_{\mathrm{eff}} = \dfrac{t_w}{\sqrt{3}}$
Rectangular width	$t_{1,\mathrm{rect}}$ – not applicable
	$t_{2,\mathrm{rect}} = 2t_w$
3 dB bandwidth	$t_w \underset{\cdot\cdot}{B}_3 \approx 1.16$
6 dB bandwidth	$t_w \underset{\cdot\cdot}{B}_6 \approx 1.33$
10 dB bandwidth	$t_w \underset{\cdot\cdot}{B}_{10} \approx 1.48$
Rectangular bandwidth	$\underset{\cdot\cdot}{B}_{1,\mathrm{rect}}$ – not applicable
	$t_w \underset{\cdot\cdot}{B}_{2,\mathrm{rect}} \approx 0.952$
Effective bandwidth	B_{eff} – not applicable
Time-bandwidth product	$\mathrm{TB} = t_{2,\mathrm{rect}} \underset{\cdot\cdot}{B}_{2,\mathrm{rect}} \approx 1.9$

Now, we can conclude that a signal with a short auto-correlation function must have a large bandwidth. As even mentioned, the shortness of the auto-correlation depends on the abruptness and randomness of the signal time shape. That is, a signal which contains steep transitions between adjacent amplitude values will have a wide spectrum.

Time spread wideband signals permit still to define a second time-bandwidth product. Namely, the product of their bandwidth $\underset{..}{B}_3$ with its actual duration t_D which is much larger than 1 for these types of signals:

$$t_D \underset{..}{B}_3 = TB \gg 1 \tag{2.81}$$

If one speaks about the time-bandwidth product TB, very often only the relation (2.81) is in mind. It relates to an improvement of the signal-to-noise ratio due to compression of the time spread wideband signal via correlation (usually also referred as correlation gain). Annex B.5 provides some deeper details. Hence, a key figure of a time spread wideband signal is its time-bandwidth product corresponding to (2.81).

Surprisingly, a wideband signal which is spread over a large duration is not called 'time spread wideband signal' but rather it is known under the term 'spread spectrum signal' even if a short pulse also has a spread spectrum. Therefore, we will avoid the term 'spread spectrum signal' here.

Uncertainty Relation We have seen from the raw estimates (2.75) and (2.77) that the more concentrated the pulse is the more spread out its spectrum. It will not be possible to shrink both pulse width and bandwidth. The theoretical limit of the concurrent compaction of a time signal and its spectrum is expressed by the uncertainty principle [12]:

$$B_{\text{eff}}\, t_{\text{eff}} \geq \frac{1}{4\pi} \tag{2.82}$$

The equal sign only holds for time signals shaped like a Gaussian function (see Table 2.5). For all other signals we get larger values than $(4\pi)^{-1}$. Therefore, the Gaussian pulse is also called 'economic' pulse since it needs the smallest bandwidth for a given duration.

We can find a corresponding uncertainty principle in quantum mechanics (Heisenberg uncertainty principle) where the momentum and position waveform of particles are also Fourier pairs.

Voltage Step Finally, we will still mention an often-used approximation for the bandwidth $\underset{.}{B}_3$ which is linked to the rising edge of a step function. If we approximate the actual pulse by the step response of a first-order low pass, we get for the bandwidth rise time relation:

$$t_r \underset{.}{B}_3 = \frac{\ln 9}{2\pi} \approx 0.35 \tag{2.83}$$

Herein, we used t_r as the 10–90% rise time, $h(t) = 1 - e^{-t/\tau}$ as the step response of a first-order low pass and $\underline{G}(f) = 1/(1 + j2\pi f \tau)$ as its frequency response function. From the last one, we get $2\pi \underset{.}{B}_3 \tau = 1$ concerning the bandwidth.

2.2.6 Doppler Scaling and Ambiguity Function

The use of correlation functions (as introduced in (2.44) and below) is a basic approach to pre-process measurement signals. Up to now, we tacitly supposed that the measurement object does not vary in time during capturing the data. But this is not always true. If we consider, for example, a radar measurement with a moving target, then a permanent change of the backscattered signals due to the movement has to be respected. We will shortly discuss these variations and introduce a modification of the correlation function (see also Ref. [13]). We will consider Figure 2.17 for this purpose. Suppose for simplicity an idealized moving reflector. It completely reflects the incident wave which will be returned back to the antenna after the round trip time τ. As long as the reflector does not move, the time shape of the reflected signal should be identical to that of the incident wave (except a change in sign). The question we try to answer is how the moving reflector will affect the time evolution of the reflected signal. The effects of propagation loss and spatial spreading of the waveform are of minor interest here. Their impact can be summarized in the amplitude A of the received signal if necessary. We will omit this here. Dispersion effects are excluded from the consideration, too. They would cause additional waveform distortions due to the propagation.

The round-trip time – antenna-target-antenna – is given by $\tau = 2r/c$ (c is the speed of light). Since the target distance r vary due to the movement, the round-trip time τ depends on time t

$$\tau(t) = \frac{2r(t - \tau(t)/2)}{c} \tag{2.84}$$

Note that the target distance which counts here is the distance when the wave is actually reflected. This is half the round-trip time before the reception (see Figure 2.17). (2.84) holds for any radial movement of the target including time

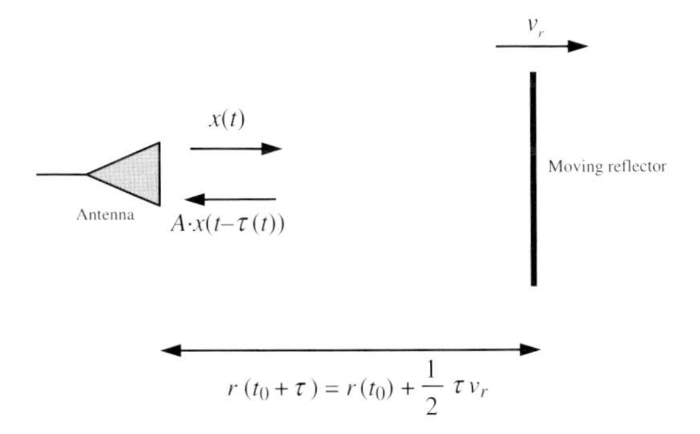

Figure 2.17 Simple model of a moving reflector. $r(t_0)$ refers to the target distance when the sounding pulse is launched and $r(t_0 + \tau)$ is the distance resulting from round trip time measurement at the receiver position at the time $t_0 + \tau$. But note that this value actually corresponds to the distance at time $t_0 + \tau/2$.

variable accelerations. Developing it in a Taylor series and choosing time zero so that $\tau(t = t_0 = 0) = \tau_0$ gives [13, 14]

$$\tau(t) = \tau_0 + \frac{d\tau}{dt}(t - \tau_0) + \frac{1}{2}\frac{d^2\tau}{dt^2}(t - \tau_0)^2 + \cdots$$

By restricting to a constant radial speed v_r (i.e. omitting all terms of higher order), this leads to

$$\frac{d\tau}{dt} = \frac{2}{c}\frac{dr(\xi)}{d\xi}\frac{d\xi}{dt} = \frac{2v_r}{c}\left(1 - \frac{1}{2}\frac{d\tau}{dt}\right) \quad \text{with } \xi = t - \frac{1}{2}\tau; \quad v_r = \frac{dr(\xi)}{d\xi}$$

which for the receive signal finally results in

$$Ax(t - \tau(t)) = Ax\left(\frac{c - v_r}{c + v_r}(t - \tau_0)\right) = \sqrt{s}x(s(t - \tau_0)) \tag{2.85}$$

Here, $s = (c - v_r)/(c + v_r) \approx 1 - 2(v_r/c)$ (for $v_r \ll c$) is a scaling factor describing a time dilation ($0 < s < 1$) or compression ($s > 1$) of the scattered signal due to the movement of the target. If the target approaches the antenna, its velocity v_r is negative hence s will be larger than 1. For the inverse case, s is smaller than 1. A motionless target does not affect the backscattered signal since $s = 1$. The time dilation or compression of the scattered wave will distribute the signal energy over a larger or smaller time. In order to meet energy conservation, the amplitude of the scattered wave is scaled with $\sqrt{s}$ in (2.85).

The time scaling of a signal due to the motion of a target is usually referred to as Doppler effect. The Doppler effect is often considered as a pure frequency shift (Doppler shift, Doppler frequency). But this interpretation is only valid for narrowband signal and does not hold for wideband signals.

Figure 2.18 illustrates the Doppler effect for a narrowband (pure sine wave) and a wideband signal of long duration. Actually, in both cases the signals are stretched (or compressed) like a rubber band. In the narrowband case, this simply affects a change in frequency of the carrier (Figure 2.18a) while in the wideband case transmit and receive signal de-correlate with increasing speed and signal duration. This is illustrated by the two points emphasized each in the two upper graphs of the wideband signal (Figure 2.18b). These points indicate two states of the signal. Due to the target movement the original time distance between both states within the sounding wave will vary ($\Delta t_1 \neq \Delta t_2$) and therefore transmitter and receiver signal will de-correlate with growing speed and signal duration.

Insertion of the Doppler-affected signal into the auto-correlation function (2.44) leads to the (auto) wideband ambiguity function [13]:

$$\chi_{xx}(\tau, s) = \sqrt{s}\int_{-\infty}^{\infty} x(t)x(s(t - \tau))dt \tag{2.86}$$

For $s = 1$, it is identical to the auto-correlation function. The ambiguity function is a two-dimensional function which characterizes the sensitivity of a signal against time shift (i.e. distance of a target) and Doppler scaling (i.e. speed of a

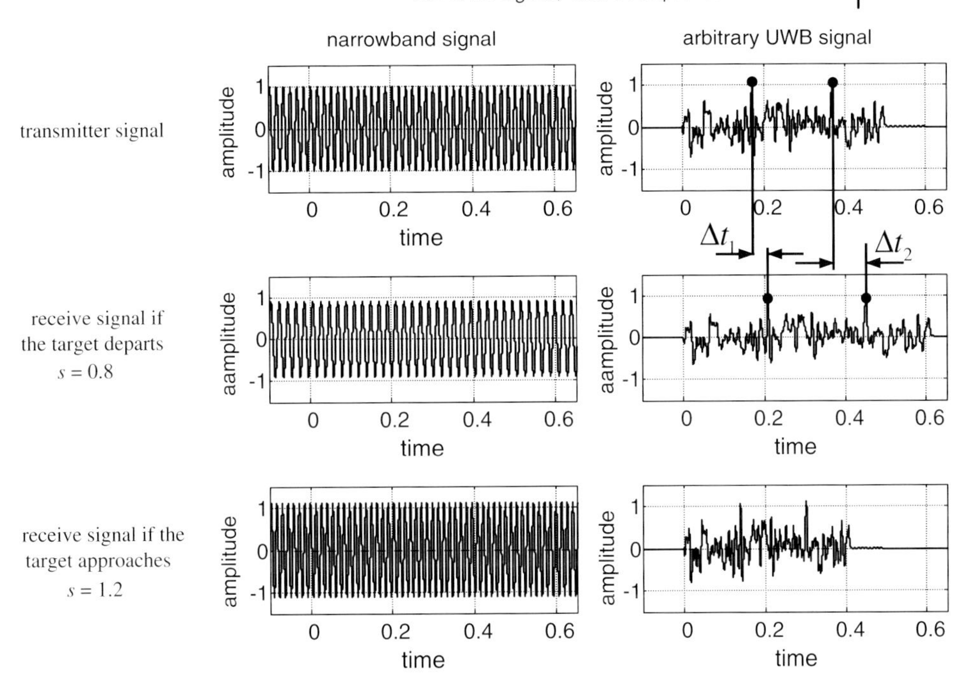

Figure 2.18 The effect of a moving target onto the scattered signal (Doppler spread). Note that the signal variations due to Doppler are drawn exaggerated in order to demonstrate the effect.

target). Figure 2.19 depicts some basic types of ambiguity functions by simplified drawings.

The ideal (but never realizable) ambiguity function is the delta function $\chi_{xx}(\tau, s) = \delta(\tau, s)$. A signal of such property would enable a perfect determination of range and velocity by a single measurement (but this would contradict the uncertainty relation (2.82)).

Noise-like and pseudo-noise waveforms have a thumbtack ambiguity function. Its width in τ or s direction can be controlled by the bandwidth and duration of the sounding signals (see also example below). Narrowband signals tend to a bad range but a good Doppler resolution. A limiting case is a pure sine wave having a delta-plane $\chi_{xx}(\tau, s) = \delta(s)$ as ambiguity function. As depicted in Figure 2.19c, there is no range resolution but perfect Doppler resolution. An extreme short sounding pulse has the ambiguity function according to Figure 2.19d, that is $\chi_{xx}(\tau, s) = \delta(\tau)$. It provides perfect range resolution but no Doppler sensitivity. Finally, Figure 2.19e shows a knife-edge ambiguity function leading to a mutual dependence between range and target speed, that is $\chi_{xx}(\tau, s) = \delta(\tau, s)$ with $\tau = ks$ where k is a constant depending on the sounding signal. Time spread wideband signals are showing such behaviour.

Figure 2.20 gives a further example for an ambiguity function. It relates to a repetitive waveform. Here, the scaling factor s is replaced by an arbitrary scaled target speed v_r. The drawing represents only a simple situation. Often, the

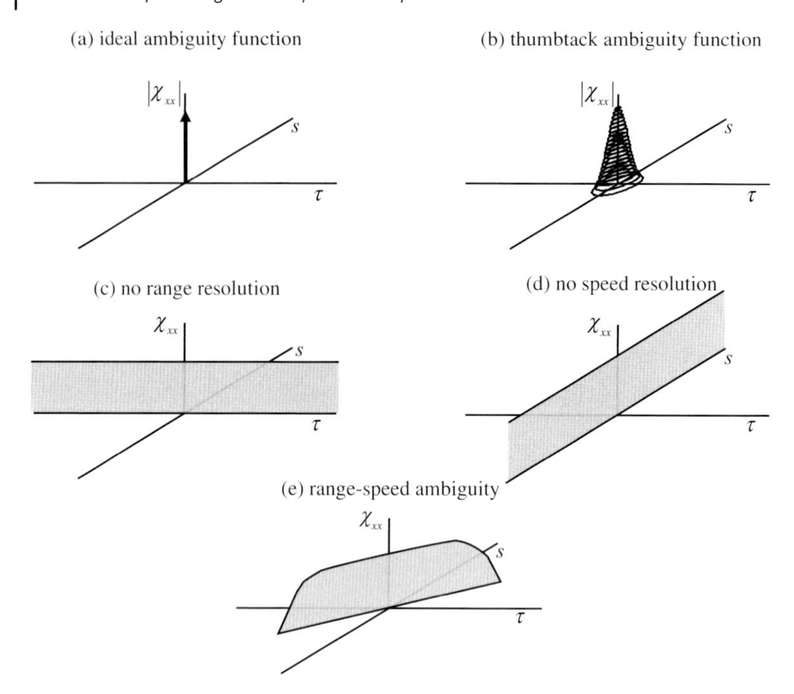

Figure 2.19 Some basic types of ambiguity functions (simplified drawings) [14].

individual peaks of the ambiguity function are additionally surrounded by side lobes which are mainly caused by the band limitation of the signals.

In the considered case, the time on target (the duration over which the signal interacts with the object of interest) consists of several periods $t_{\text{ot}} = Nt_{\text{P}}$ of a time

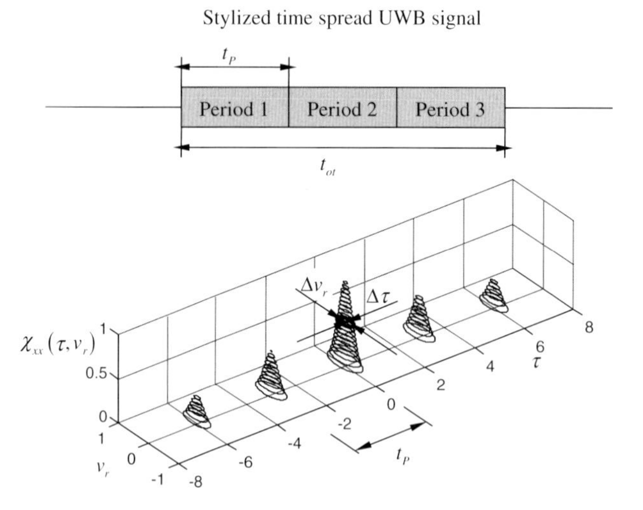

Figure 2.20 Typical shape of an ambiguity function for a baseband signal of large bandwidth consisting of three identical signal sections.

spread wideband signal. Pseudo-random codes are typical representatives of such signals. It is obvious from the figure that there is no unique mapping between τ and r if the round-trip time exceeds the period of the sounding signal ($\tau \geq t_P$). The maximum allowed distance for a unique measurement is called the unambiguous range. It is given by

$$r_{ua} = \frac{t_P c}{2} \tag{2.87}$$

Due to the periodicity of $x(t)$, the ambiguity function is periodic in τ-direction with the signal period, hence the round-trip time restriction to $\tau = 0 \cdots t_P$ for an unambiguous measurement. Non-periodic signals (e.g. random noise) do not limit the unambiguous range. Generally spoken, the time distance between two adjacent peaks in the ambiguity function should be larger than the settling time of the considered scenario.

The ambiguity function at $s = 0$ ($v_r = 0$) represents the auto-correlation function of the signal $\chi_{xx}(\tau, v_r = 0) = C_{xx}(\tau)$. The width $\Delta\tau$ of this function is quantified by the coherence time $\Delta\tau = \tau_{coh}$. Referring to (2.79) or (2.80), it is approximately inverse proportional to the bandwidth B of the signal $x(t)$. $\Delta\tau$ is a measure to distinguish closely located targets, usually assigned as range resolution (see Section 4.7.2 for details):

$$\delta_r \approx \frac{1}{2}\Delta\tau c = \frac{1}{2}\tau_{coh} c \tag{2.88}$$

Furthermore, the accuracy of a distance measurement as well as the sensitivity to detect weak target movements depends on this figure.

Applying baseband signals of a large bandwidth, the ambiguity function will be concentrated around the s- or v_r-axis and we will not have further peaks in s- (v_r-) direction (that is, Doppler ambiguity will not appear). As obvious from Figure 2.18, transmit and receive signal progressively de-correlate with increasing mismatch of scaling factor and target speed Δv_r. Such de-correlation is negligible if the target displacement is less than the range resolution during time on target, that is we get

$$\delta_r \geq \Delta v_r t_{ot} \tag{2.89}$$

Using (2.89) and (2.81), we result in the usual expression [13, 14]

$$\Delta v_r \leq \frac{1}{2} c \frac{\tau_{coh}}{t_{ot}} \approx \frac{1}{2} c \frac{1}{t_{ot} B_{\ddot{3}}} = \frac{c}{2\mathrm{TB}} \tag{2.90}$$

Here, we used the time-bandwidth product TB of the sounding signal as an expression for the ratio of temporal resolution τ_{coh} and the interaction time t_{ot} with target. If one selects for the time on target $t_{ot} \leq \delta_r / v_{r,max}$, which is also termed as narrowband condition (or narrowband processing), one must not respect the de-correlation between receive and transmit signal. In the opposite case, one speaks about wideband processing for which the de-correlation has to be respected. Wideband processing requires a two-dimensional search in τ- and s-direction leading to increasing complexity of the sensor electronics. An example for an UWB

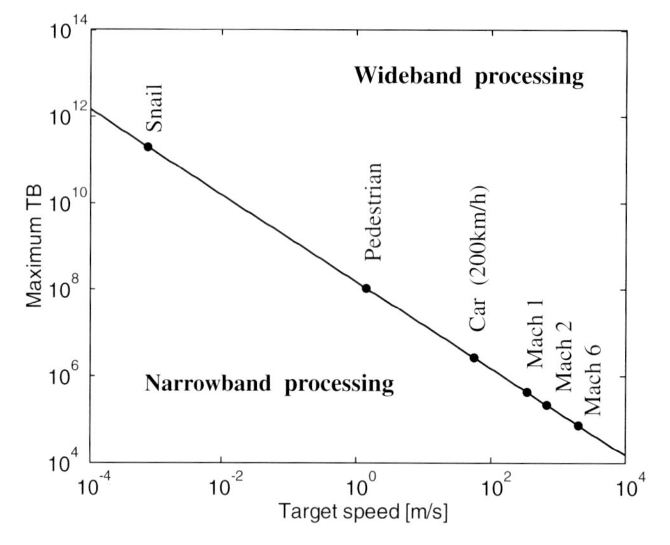

Figure 2.21 Maximum time-bandwidth product as function of radial target speed.

sensor with direct Doppler evaluation is introduced in Section 3.3.6. But mostly one tries to avoid Doppler influence. This is done either by signals which are not sensitive to Doppler (e.g. short pulses) or by signals which do not de-correlate (e.g. linear frequency modulated sine waves – FMCW, Section 3.4.4) and one illuminates the target only over a short time t_{ot}. By that way, the two-dimensional search can be replaced by two separate searches in τ and s. More on narrowband and wideband ambiguity functions can be found in Ref. [15].

Figure 2.21 illustrates (2.90) for a speed interval of technical interest. It depicts the border between narrowband and wideband processing for the product of bandwidth and time on target $\text{TB} = t_{\text{ot}}\ddot{B}_3$.

Let us finally come back to the range-Doppler coupling as already figuratively drawn in Figure 2.19e. Figure 2.22 shows a more realistic situation. It illustrates

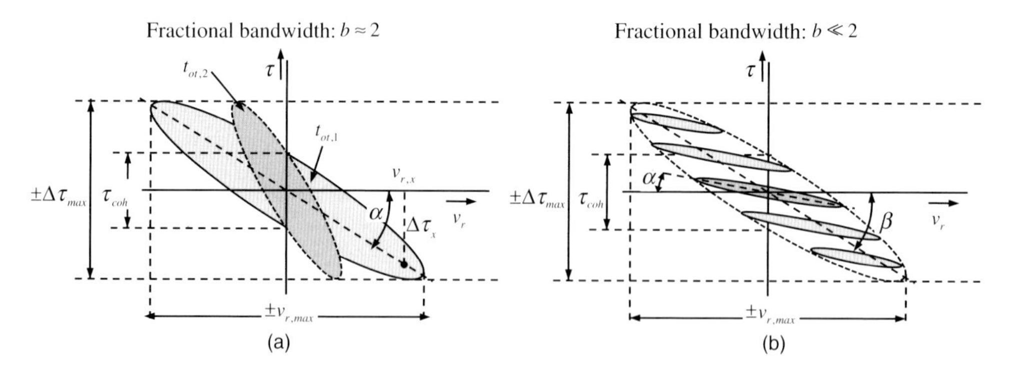

Figure 2.22 Horizontal slice at half power level of typical wideband ambiguity functions. The relation between scaling factor and range velocity is $v_r = c(1 - s)/2$.

the typical contours of the wideband ambiguity function kernel at $-3\,\text{dB}$ level under different conditions.

Wideband case: The fractional bandwidth of the sounding signal is close to its maximum $b \approx 2$. The Kernel of the ambiguity function consists typically of one compact global peak. The two slices in Figure 2.22a refer to sounding signals of the same type and bandwidth but for different observation length $t_{\text{ot},1} < t_{\text{ot},2}$. Obviously, the longer the time on target is the lower the target speed must be to prevent de-correlation between transmit and receive signal.

Reduced fractional bandwidth: The fractional bandwidth is considerably below its maximum $b \ll 2$. If the fractional bandwidth b decreases the auto-correlation function shows more and more oscillations, hence the compact maximum of the ambiguity function will be destroyed. That is, the global peak splits into multiple spikes as demonstrated in Figure 2.22b.

As to be seen from the figure, the $-3\,\text{dB}$ contour area is slanted by an angle α or β. This leads to additional range errors for moving targets. The left drawing gives an example. If the target moves at a speed of $v_{\text{r},x}$ the measurement will result in an additional range error of $\Delta r_x = \Delta\tau_x c/2$. The slant angle mainly depends on the duration t_{ot} of the interaction between signal and target as well as the type of the applied signal, thus we can arrange

$$\begin{aligned} \frac{\tan(\alpha,\beta)}{[\text{s}^2/\text{m}]} &= \frac{\Delta\tau_x}{v_{\text{r},x}} = \frac{2\Delta r_x}{v_{\text{r},x}c} = k_{\alpha,\beta}t_{\text{ot}} \\ \Delta r_x &= \frac{1}{2}k_{\alpha,\beta}c v_{\text{r},x}t_{\text{ot}} \quad \text{for } \left|v_{\text{r},x}\right| \le v_{\text{r,max}} \end{aligned} \tag{2.91}$$

where k_α and k_β are coefficients depending on the type of signal and its bandwidth. Simulations for two time-stretched wideband signals lead to following values for range-Doppler coupling (see also Sections 2.3.5 and 2.3.6):

	$k_\alpha c/2$	$k_\beta c/2$
Chirp $b = 2$	0.7	%
Chirp $b = 1$	0.6	2.1
M-sequence $b = 2$	0.5	%

A comprehensive treatise of signals and their ambiguity function can be found in Ref. [2]. The discussions there refer indeed to the narrowband ambiguity function. Nevertheless, it gives a first impression of the wideband behaviour of the corresponding signal type.

2.3 Some Idealized UWB Signals

This chapter gives a short tabulate overview of idealized wideband signals. We will consider periodic and aperiodic signals. In the case of periodic signals, we should have in mind that a measurement is always of final duration thus only a limited number of signals periods can be used for data capturing. The work with idealized signals can be useful to find simple approximations and signal parameters. In practice, we have to deal with more or less serious deviations from the idealization which mainly express by infinite signal bandwidth (signals with rectangular edges and sharp bends) or violation of the causality (Gauss pulse, sinc pulse). Some rules to calculate the following relations for the idealized signals are to be found in Annexes 7.1 and 8.1. The consideration of shape parameters related to the jitter sensitivity of the signals (refer to (2.21)), we will shift to a later point (Section 4.7.3). The shown spectral functions refer to two-sided spectra even if only the part for positive frequencies is drawn. Some examples will consider periodic signals having a line spectrum. Some others will deal with energy signals or random signals which have a continuous spectrum.

2.3.1 Rectangular Unipolar and Bipolar Pulse Trains

Rectangular pulse trains are very simply theoretical signals. Some functions are summarized in Figures 2.23 and 2.24 as well as in Tables 2.1–2.2. The maxima of the side lobes of the amplitude spectrum decay only with f^{-1}. Therefore, the side lobes in the power spectrum decay with f^{-2}. Due to their abrupt change of the amplitude values, such signals would have an infinitely wide spectrum which can never be realized in practice. Actually, one has always to count with a limited rise time of the pulses and some overshot or ringing as well. Hence, approximations of real signals by rectangular functions should be made with some care. The rectangular pulse train is written as (see also Annex B.1, Table B.1 for definition of rect x; the rectangular function is also assigned as normalized boxcar function):

$$x(t) = \operatorname{rect}\left(\frac{t - nt_{\mathrm{P}}}{t_{\mathrm{w}}}\right) \tag{2.92}$$

In comparison to the unipolar pulse, the bipolar pulse doubles its duration by approximately keeping the bandwidth hence its TB product is twice of the unipolar pulse.

2.3.2 Single Triangular Pulse

The triangular pulse results from the convolution[9] of a rectangular pulse with itself:

$$x(t) = V \operatorname{tri}\left(\frac{t}{t_0}\right) = V \operatorname{rect}\left(\frac{t}{t_0}\right) * \operatorname{rect}\left(\frac{t}{t_0}\right) \tag{2.93}$$

9) The convolution is explained in Section 2.4.2.

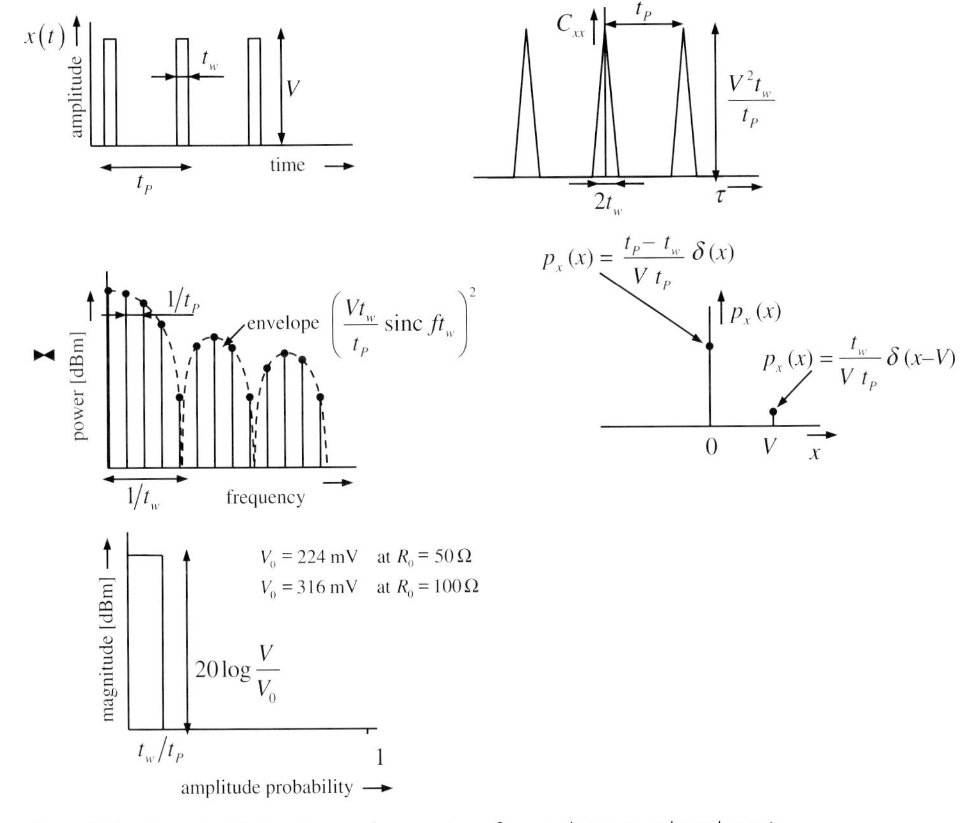

Figure 2.23 Characteristic functions and parameters of a unipolar rectangular pulse train.

Its amplitude spectrum is given by

$$\underline{X}(f) = t_0\, V \sin c^2(t_0 f) \tag{2.94}$$

It resembles the spectrum of the rectangular spectrum. However its side lobes decay stronger, that is with f^{-2} in the amplitude spectrum and f^{-4} in the power spectrum. Figure 2.25 depicts the shape of the time and correlation function as well as the power spectrum of a single pulse. In the case of a periodic repetition, the spectrum is composed from spectral line corresponding to Figures 2.23 or 2.24. The correlation function is built from a set of parabolic arcs. The probability functions given in Figure 2.25 and the characteristic values of Table 2.3 relate to a limited recording time T_R.

2.3.3 Sinc Pulse

The sinc pulse is written as

$$x(t) = V \operatorname{sinc}\left(\frac{t}{t_0}\right) = V\frac{\sin \pi t/t_0}{\pi t/t_0} \tag{2.95}$$

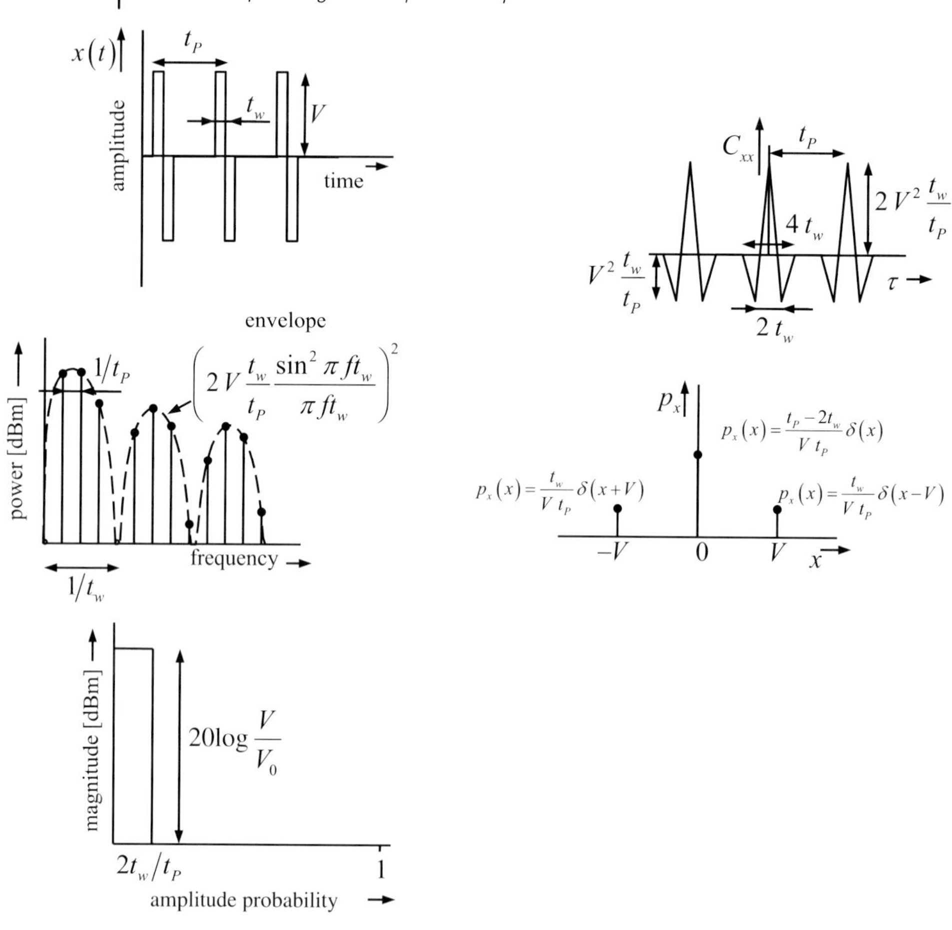

Figure 2.24 Characteristic functions of a bipolar rectangular pulse train.

Its first two derivations are

$$\dot{x}(t) = \frac{V}{t_0}\left(\frac{\cos \pi y}{y} - \frac{\sin \pi y}{\pi y^2}\right); \quad y = \frac{t}{t_0} \tag{2.96}$$

$$\ddot{x}(t) = \frac{V}{t_0^2}\left(\frac{2\sin \pi y}{\pi y^3} - \frac{2\cos \pi y}{y^2} - \frac{\pi \sin \pi y}{y}\right); \quad y = \frac{t}{t_0} \tag{2.97}$$

The pulse has an ideal rectangular amplitude and auto-spectrum:

$$\underline{X}(f) = V t_0 \operatorname{rect}(t_0 f) \tag{2.98}$$

Therefore, the shape of its ACF is identical to the original signal, that is

$$C_{xx}(\tau) = V^2 t_0 \operatorname{sinc}\frac{\tau}{t_0} \tag{2.99}$$

The sinc pulse uses the spectral band of given width $\underset{..}{B} = t_0^{-1}$ to full capacity (Table 2.4). However, it has to pay for it by very weak decay in time domain. The

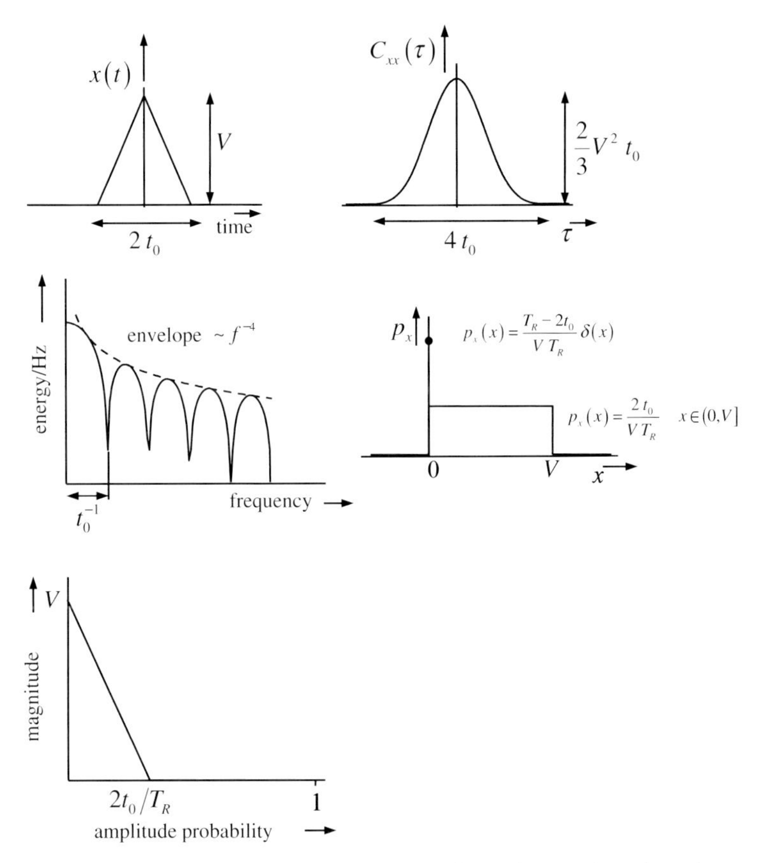

Figure 2.25 Some of the characteristic functions of a single triangular pulse. (The axes of the amplitude probability function are linearly scaled).

identical shape of signal and ACF follows immediately from the rectangular spectrum and (2.67), (2.70) (Figure 2.26).[10)]

2.3.4 Gaussian Pulses

The Gaussian pulse is likely used in theory and it is often tried to approximate it by real waveforms because it is the most 'economic' pulse. That is, it has the narrowest bandwidth for a given pulse width (see also uncertainty relation (2.82)). We will consider here only single Gaussian pulses with different degrees of differentiation. The comprehension of periodic waveform is straightforward and similar to that described above. However, one should respect the truncation errors which are caused from the infinitely long decay of the individual waveform.

10) Note that relation (2.67) is not restricted to random signals.

Table 2.3 Characteristic parameters of a triangular pulse.

Parameter	Value
L_1-norm	$\Vert x(t)\Vert_1 = V\, t_0$
Energy	$\Vert x(t)\Vert_2^2 = \frac{2}{3} V^2 t_0$
Crest factor (see (2.18))	$\mathrm{CF} = \sqrt{\frac{3T_{\mathrm{R}}}{2t_0}}$
Slew rate	$\mathrm{SR} = \frac{V}{t_0}$
Pulse width	$t_{\mathrm{w}} = t_0$
Effective pulse width	$t_{\mathrm{eff}} = \sqrt{\frac{2}{5}} t_0$
Rectangular width	$t_{1,\mathrm{rect}} = t_0$
	$t_{2,\mathrm{rect}} = \frac{2}{3} t_0$
3 dB bandwidth	$t_{\mathrm{w}}\, \underset{\cdot\cdot}{B}_3 \approx 0.638$
6 dB bandwidth	$t_{\mathrm{w}}\, \underset{\cdot\cdot}{B}_6 \approx 0.886$
10 dB bandwidth	$t_{\mathrm{w}}\, \underset{\cdot\cdot}{B}_{10} \approx 1.11$
Rectangular bandwidth	$t_0\, \underset{\cdot\cdot}{B}_{1,\mathrm{rect}} = 1$
	$t_0\, \underset{\cdot\cdot}{B}_{2,\mathrm{rect}} \approx \frac{2}{3}$
Effective bandwidth	$t_0\, B_{\mathrm{eff}} \approx 0.276$
Uncertainty relation	$t_{\mathrm{eff}}\, B_{\mathrm{eff}} \approx 0.174$
Time-bandwidth product	$\mathrm{TB} = t_{1,\mathrm{rect}}\, \underset{\cdot\cdot}{B}_{1,\mathrm{rect}} = 1$

The basic relations of the Gaussian pulse are as follows:

$$\text{Time shape: } x(t) = V\, \mathrm{e}^{-\pi\left(\frac{t}{t_0}\right)^2} \tag{2.100}$$

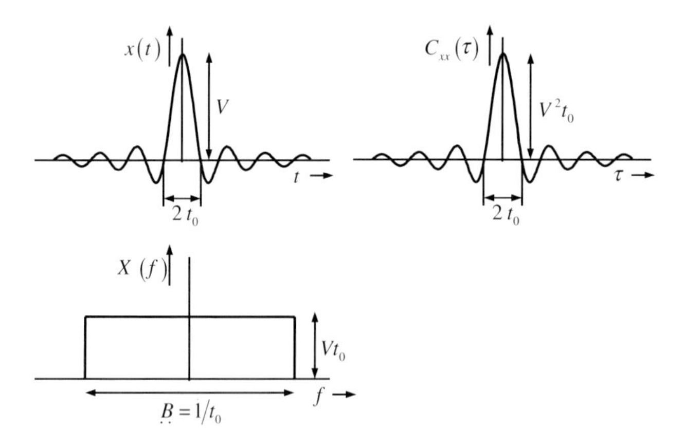

Figure 2.26 Some characteristic functions of the sinc pulse.

Table 2.4 Characteristic parameters of the sinc pulse.

Parameter	Value
L_1-norm	$\lVert x(t)\rVert_1 \to \infty$
Energy	$\lVert x(t)\rVert_2^2 = V^2 t_0$
Crest factor	Not applicable
Slew rate	$\mathrm{SR} \approx 1.37 \dfrac{V}{t_0}$
Slew rate of first derivative	$\ddot{x}(0) = \dfrac{\pi^2}{3}\dfrac{V}{t_0^2} \approx 3.29 \dfrac{V}{t_0^2}$
Pulse width	$t_\mathrm{w} = 1.2 t_0$
Effective pulse width	Not applicable
Rectangular width	$t_{1,\mathrm{rect}} = t_0$
	$t_{2,\mathrm{rect}} = t_0$
Bandwidth	$t_0 \ddot{B} = 1$
Time-bandwidth product	$\mathrm{TB} = t_0 \ddot{B} = 1$
Effective bandwidth	$\ddot{B} = 2\sqrt{3} B_\mathrm{eff}$

$$\text{Amplitude spectrum: } \underline{X}(f) = V t_0 \,\mathrm{e}^{-\pi (t_0 f)^2} \tag{2.101}$$

$$\text{Auto-correlation: } C_{xx}(\tau) = \frac{V^2 t_0}{\sqrt{2}} \mathrm{e}^{-\frac{\pi}{2}\left(\frac{t}{t_0}\right)^2} \tag{2.102}$$

Figure 2.27 and Table 2.5 give an overview of characteristic functions and parameters of a single Gaussian pulse. The indicated probability density function refers to

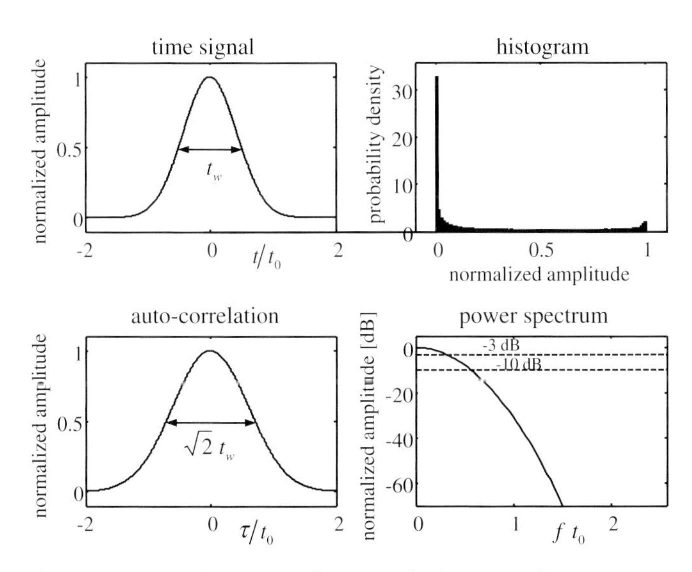

Figure 2.27 Some characteristic functions of a Gaussian pulse.

Table 2.5 Characteristic parameters of a Gaussian pulse.

Parameter	Value
L_1-norm	$\Vert x(t)\Vert_1 = V t_0$
Energy	$\Vert x(t)\Vert_2^2 = \frac{1}{2} V^2 t_0$
Crest factor	$\mathrm{CF} \approx \sqrt{\frac{2T_\mathrm{R}}{t_0}}$
Slew rate $\left(\frac{\partial x}{\partial t} @0.61\ V\right)$	$\mathrm{SR} = \frac{V}{t_0}\sqrt{2\pi}\,\mathrm{e}^{-0.5} \approx 1.52\frac{V}{t_0}$
Slew rate of first derivative	$\ddot{x}(0) = \frac{2\pi V}{t_0^2}$
Pulse width	$t_\mathrm{w} = t_0 2\sqrt{\frac{\ln 2}{\pi}} \approx 0.94\, t_0$
Effective pulse width	$t_\mathrm{eff} = \frac{t_0}{2\sqrt{\pi}}$
Rectangular width	$t_{1,\mathrm{rect}} = t_0$ $t_{2,\mathrm{rect}} = \frac{1}{2} t_0$
Rise time (10–90%)	$\frac{t_\mathrm{r}}{t_0} = \frac{\sqrt{\ln 10} - \sqrt{\ln 10 - \ln 9}}{\sqrt{\pi}}$ $t_\mathrm{r} \underset{\cdot\cdot}{B}_3 \approx (0.447) \approx 0.5$
L dB cut-off bandwidth	$t_0 \underset{\cdot\cdot}{B}_\mathrm{L} = 2\sqrt{\frac{L \ln 10}{20\pi}}$
3 dB bandwidth	$t_0 \underset{\cdot\cdot}{B}_3 = 2\sqrt{\frac{\ln 2}{2\pi}} \approx 0.664$
6 dB bandwidth	$t_0 \underset{\cdot\cdot}{B}_6 = 2\sqrt{\frac{\ln 4}{2\pi}} \approx 0.94$
10 dB bandwidth	$t_0 \underset{\cdot\cdot}{B}_{10} = 2\sqrt{\frac{\ln 10}{2\pi}} \approx 1.21$
60 dB bandwidth	$t_0 \underset{\cdot\cdot}{B}_{60} \approx 2.97$
80 dB bandwidth	$t_0 \underset{\cdot\cdot}{B}_{80} \approx 3.42$
100 dB bandwidth	$t_0 \underset{\cdot\cdot}{B}_{100} \approx 3.83$
Rectangular bandwidth	$t_0 \underset{\cdot\cdot}{B}_{1,\mathrm{rect}} = 1$ $t_0 \underset{\cdot\cdot}{B}_{2,\mathrm{rect}} = \frac{1}{\sqrt{2}}$
Effective bandwidth	$t_0 B_\mathrm{eff} = \frac{1}{2\sqrt{\pi}}$
Uncertainty relation	$t_\mathrm{eff} B_\mathrm{eff} = \frac{1}{4\pi}$
Time-bandwidth product	$\mathrm{TB} = t_{1,\mathrm{rect}} B_{1,\mathrm{rect}} = 1$

the time segment as shown in the figure (upper left). It corresponds to a recording time of $T_\mathrm{R} = 4t_0$.

The integral of the Gaussian pulse provides a step function with a smooth rising edge. Such signal may be useful for the investigation of step-response behaviour as applied in

TDR (time domain reflectometry) devices. The Gaussian step function is given by

$$x_{\text{step}}(t) = \frac{V}{t_0}\int \mathrm{e}^{-\pi\left(\frac{t}{t_0}\right)^2}\,\mathrm{d}t = \frac{1}{2}\left(1 + \operatorname{erf}\left(\sqrt{\pi}\frac{t}{t_0}\right)\right) \tag{2.103}$$

where $\operatorname{erf}(x) = \frac{2}{\sqrt{\pi}}\int\limits_0^x \mathrm{e}^{-t^2}\,\mathrm{d}t$ represents the error function.

The first few time derivations of the Gaussian pulse and their maxima are

$$\text{First derivation: } \dot{x}(t) = -\frac{2\pi V}{t_0}\frac{t}{t_0}\mathrm{e}^{-\pi\left(\frac{t}{t_0}\right)^2}$$

$$\frac{\|\dot{x}(t)\|_\infty}{V/t_0} = \sqrt{\frac{2\pi}{e}} \tag{2.104}$$

$$\text{Second derivation: } \ddot{x}(t) = -\frac{2\pi V}{t_0^2}\left(1 - 2\pi\left(\frac{t}{t_0}\right)^2\right)\mathrm{e}^{-\pi\left(\frac{t}{t_0}\right)^2} \tag{2.105}$$

$$\|\ddot{x}(t)\|_\infty = \frac{2\pi V}{t_0^2}$$

$$\text{Third derivation: } \dddot{x}(t) = \frac{4\pi V}{t_0^3}\left(3\left(\frac{t}{t_0}\right) - 2\pi\left(\frac{t}{t_0}\right)^3\right)\mathrm{e}^{-\pi\left(\frac{t}{t_0}\right)^2} \tag{2.106}$$

$$\|\dddot{x}(t)\|_\infty = \frac{4V\sqrt{3\pi^3\left(3-\sqrt{6}\right)}}{t_0^3}\mathrm{e}^{-\left(\frac{3-\sqrt{6}}{2}\right)}$$

$$\text{Fourth derivation} \quad \ddddot{x}(t) = \frac{(2\pi)^2 V}{t_0^4}\left(3 - 12\pi\left(\frac{t}{t_0}\right)^2 + (2\pi)^2\left(\frac{t}{t_0}\right)^4\right)\mathrm{e}^{-\pi\left(\frac{t}{t_0}\right)^2}$$

$$\|\ddddot{x}(t)\|_\infty = \frac{3(2\pi)^2 V}{t_0^4} \tag{2.107}$$

The spectrum of these waveforms can be easily determined from (2.101) applying the derivation rule of the Fourier transform (see Annex B.1, Table B.2). For illustration, the time evolution, spectrum, ACF und PDF are plotted in Figures 2.28–2.31. The Rayleigh plot of the amplitude probability function of the Gaussian pulse and its derivations is shown in Figure 2.32 and finally the characteristic parameters of the first-order derivation is summarized in Table 2.6.[11)]

2.3.5
Binary Pseudo-Noise Codes

There are different types of binary pseudo-noise codes, for example Barker code, M-sequence, Golay-Sequence, Gold-code, Kasami-code and so on [16].

11) The double-sided bandwidth $\underset{\cdot\cdot}{B}$ neglects the spectral irruption at $f \approx 0$. The single-sided bandwidth $\underset{\cdot}{B}$ respects the missing spectral parts, hence $\underset{\cdot\cdot}{B} > 2\underset{\cdot}{B}$.

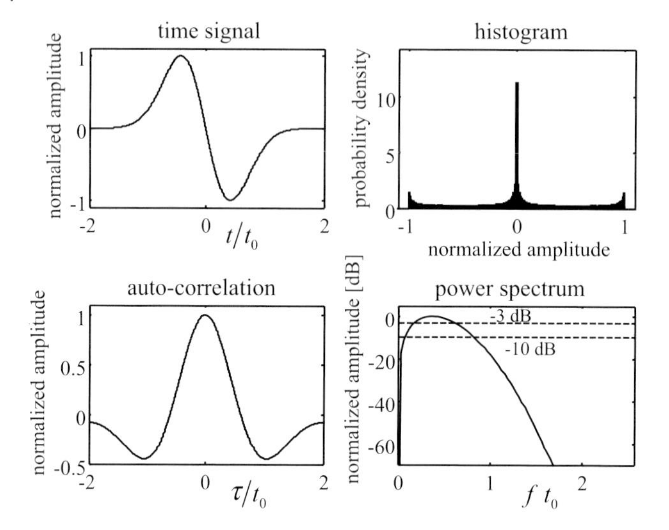

Figure 2.28 Characteristic functions of the first derivative of a Gaussian pulse.

They are applied in many fields, for example in the communication and (large range) radar technique or for global positioning systems and others. However, up to now, their use in the UWB sensing technique is still quite limited even if these signals dispose some interesting properties. Here, we will concentrate on some basics features of the maximum length binary sequence or shortly termed M-sequence. In Section 3.3.4, a new ultra-wideband approach based on such signals will be discussed in more details. For a given signal power, an M-sequence is the signal which has the lowest amplitude. Furthermore, it has

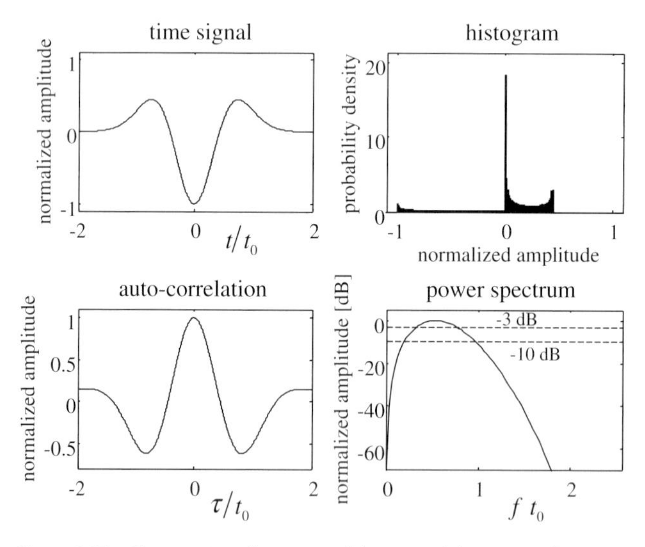

Figure 2.29 Characteristic functions of the second derivative of a Gaussian pulse.

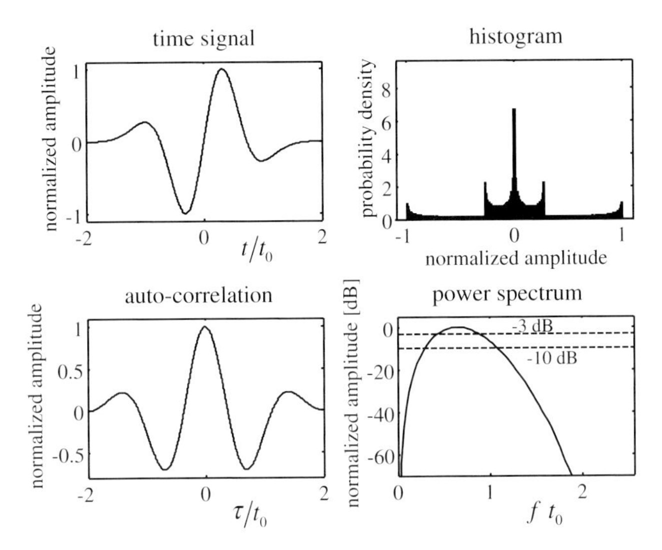

Figure 2.30 Characteristic functions of the third derivative of a Gaussian pulse.

the shortest auto-correlation function of all pseudo-random codes. Such codes are typically generated by digital shift registers which are pushed by a clock generator of the repetition rate f_c.

Figure 2.33 summarizes some basic properties of M-sequences. For illustration purposes, only a very short M-sequence was selected. An M-sequence is composed from N elementary pulses (termed as chips). This number depends on the order n of the shift register:

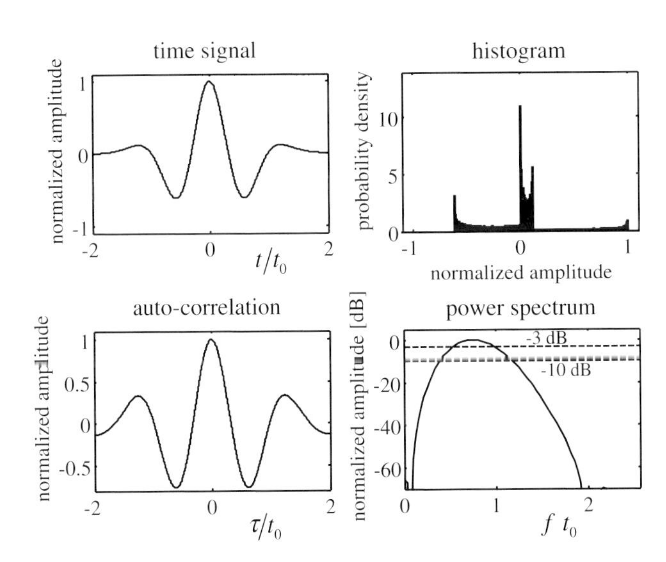

Figure 2.31 Characteristic functions of the fourth derivative of a Gaussian pulse.

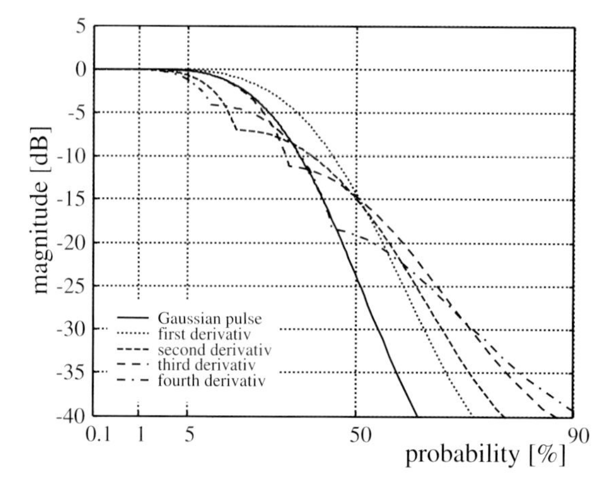

Figure 2.32 Amplitude probability of the Gaussian pulse and its derivations for a recording time of $T_R = 4t_w$.

$$N = 2^n - 1 \tag{2.108}$$

Hence, the duration of one period is

$$t_p = \frac{2^n - 1}{f_c} = (2^n - 1)t_c \tag{2.109}$$

where t_c is the chip duration. Apparently, the polarity of the individual chips varies randomly within the period of the M-sequence, which leads to N_T transitions between the positive and negative voltage level:

$$N_T = 2^{n-1} \tag{2.110}$$

Except of minor differences, the envelope of the M-sequence spectrum is shaped like that of rectangular pulses:

$$\Psi(f) = \begin{cases} \dfrac{V^2}{N} & f = 0 \\ V^2 \dfrac{N+1}{N^2} \operatorname{sinc}^2 f t_c & f \neq 0 \end{cases} \tag{2.111}$$

This involves that nearly 77.4% of the total AC signal power is concentrated within the spectral band $[-f_c/2, f_c/2]$ (Table 2.7).[12), 13)]

The M-sequence is a signal which spreads its energy over the whole period. Additionally, one can p-times repeat the measurements resulting in an overall time

12) The rectangular energetic width $t_{2,\text{rect}}$ is omitted here, since $t_{1,\text{rect}} = t_{\text{ACF,rect}}$ of the ACF already relates to the signal power.

13) $B_{1,\text{rect}}$ of the ACF already relates to a power quantity. Hence, we omitted $B_{2,\text{rect}}$ since it is meaningless.

Table 2.6 Characteristic parameters of the first derivative of a Gaussian pulse.

Parameter	Value
L_1-norm	$\lVert\dot{x}(t)\rVert_1 = 2\,V$
Energy	$\lVert\dot{x}(t)\rVert_2^2 = \dfrac{\pi V^2}{\sqrt{2}t_0}$
Peak value	$\lVert\dot{x}(t)\rVert_\infty = \dfrac{\sqrt{2\pi}V}{\sqrt{e}t_0}$
Crest factor	$\mathrm{CF} = \sqrt{\dfrac{2\sqrt{2}T_\mathrm{R}}{\mathrm{e}t_0}} \approx 1.02\sqrt{\dfrac{T_\mathrm{R}}{t_0}}$
Slew rate	$\mathrm{SR} = \dfrac{2\pi V}{t_0^2}$
Pulse width	$t_\mathrm{w} \approx 1.53 t_0$
Effective pulse width	$t_\mathrm{eff} = \dfrac{t_0}{2\sqrt{\pi}}$
Rectangular width	$t_{1,\mathrm{rect}}$ – not applicable
	$t_{2,\mathrm{rect}} = \dfrac{\mathrm{e}t_0}{2\sqrt{2}} \approx 0.961 t_0$
3 dB bandwidth	$t_0 \underset{\cdot\cdot}{B}_3 \approx 1.31$
	$t_0 \underset{\cdot}{B}_3 \approx 0.46$
6 dB bandwidth	$t_0 \underset{\cdot\cdot}{B}_6 \approx 1.53$
	$t_0 \underset{\cdot}{B}_6 \approx 0.64$
10 dB bandwidth	$t_0 \underset{\cdot\cdot}{B}_{10} \approx 1.76$
	$t_0 \underset{\cdot}{B}_{10} \approx .80$
Fractional bandwidth	$b \approx 1.68$
Rectangular bandwidth	$\underset{\cdot\cdot}{B}_{1,\mathrm{rect}}$ – not applicable
	$t_0 \underset{\cdot\cdot}{B}_{2,\mathrm{rect}} = \dfrac{e}{2\sqrt{2}} \approx 0.961$
Effective bandwidth	$t_0 B_\mathrm{eff} = \dfrac{1}{2}\sqrt{\dfrac{3}{\pi}}$
Uncertainty relation	$t_\mathrm{eff} B_\mathrm{eff} = \dfrac{\sqrt{3}}{4\pi}$
Time-bandwidth product	$\mathrm{TB} = t_{2,\mathrm{rect}} B_{2,\mathrm{rect}} = \dfrac{\mathrm{e}^2}{8} \approx 0.924$

on target of

$$t_\mathrm{ot} = pNt_\mathrm{c} = \frac{p(2^n - 1)}{f_\mathrm{c}} \tag{2.112}$$

For radar measurements of moving targets, the duration of this time segment determines whether Doppler scaling affects the M-sequence or not. The ambiguity function depicted in Figure 2.34 gives an example for a seventh-order M-sequence pushed by a 9 GHz clock. The recording time extends over $p = 128$ synchronous repetitions (synchronous averaging). Obviously, a target movement will not affect the sounding signal as long as the speed remains below 20 km/s.

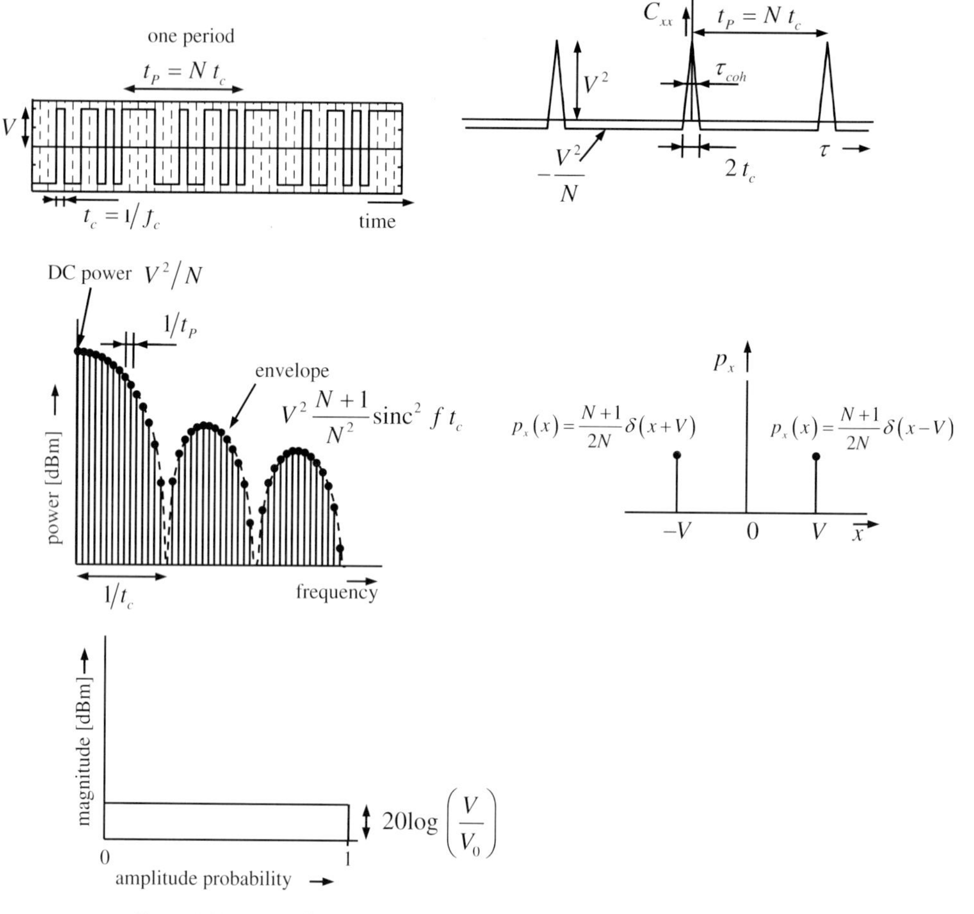

Figure 2.33 Train of M-sequence signals. The example shows an M-sequence of order 4.

There exist many other binary sequences which could be basically applied for UWB purposes. The reason why we restrict ourselves here mainly to M-sequences is their proper auto-correlation function which does not provide side lobes (at least for the theoretical case of an unlimited bandwidth). This makes them useful for radar applications avoiding the suppression of week targets by the side lobes of strong reflecting objects. The weak DC component V^2/N is often accounted as side lobe by other authors. But actually it will only affect the DC component of a measured impulse response function (IRF) and will not degrade the detectability of weak targets (compare also Annex B.9).

Alternatively, the complementary Golay-sequences are binary codes with perfect triangular auto-correlation function without DC component. Thus, their application will lead to correct measurements results also in case of DC-coupled test objects. Annex B.9 gives a short introduction. We will not

Table 2.7 Characteristic parameters of the M-sequence of order n.

Parameter	Value
Rectified value	$\|x(t)\|_1 = V$
rms value	$\|x(t)\|_2 = V$
Crest factor	$\mathrm{CF} = 1$
Slew rate	$\mathrm{SR} = \infty$
Width of ACF	$\tau_{coh} = t_c$
Effective ACF width	$t_{eff} = \sqrt{\frac{2}{5}} t_c$
Rectangular ACF width	$t_{1,rect} = t_{ACF,rect} = t_c$
3 dB bandwidth	$t_c \ddot{B}_3 \approx 0.886$
6 dB bandwidth	$t_c \ddot{B}_6 \approx 1.21$
10 dB bandwidth	$t_c \ddot{B}_{10} \approx 1.48$
Rectangular ACF bandwidth	$t_c \ddot{B}_{1,rect} = t_c \ddot{B}_{ACF,rect} = 1$
Effective ACF bandwidth	$t_c B_{eff} \approx 0.276$
Signal time-bandwidth product	$\mathrm{TB} = t_p B_{1,rect} = N = 2^n - 1$

penetrate deeply into their behaviour and application here since they provide comparable results as the M-sequence for most of the applications considered in this book. However, the technical implementation of Golay-sensors is a bit more demanding[14] than that for M-sequences even if nearly identical concepts may be used.

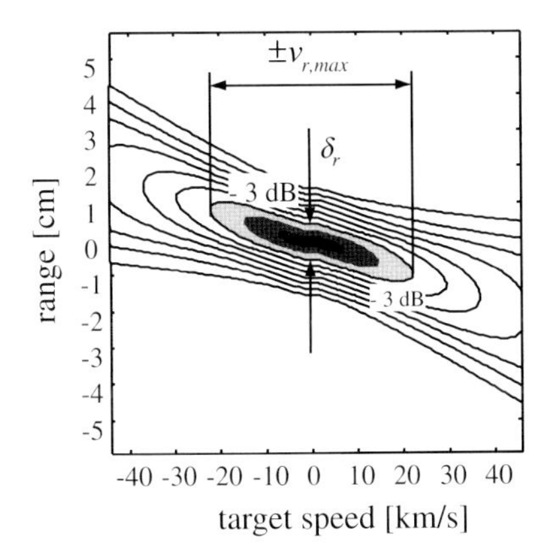

Figure 2.34 Central part of the ambiguity function of an M-sequence of order $n = 7$ and the clock rate $f_c = 9$ GHz for a time on target and recording time $t_{ot} = T_R = 1.8$ µs. δ_r refers to the range resolution and $v_{r,max}$ gives the maximum target speed which not yet affects the M-sequence by Doppler spreading.

14) This remark mainly relates to the question of the generation of the sequence and the sampling clock in case of very wideband sensors.

2.3.6 Chirp

The chirp is a sine wave with increasing/decreasing frequency. Typically, the frequency variation is chosen to be linear in most of the practical cases

$$f(t) = f_0 + at \quad \text{with } a = \frac{f_1 - f_0}{t_D} \tag{2.113}$$

where f_0 is the staring frequency, f_1 is the stop frequency, t_D is the chirp duration and a is the chirp rate [Hz/s]. A single chirp can be written as a complex exponential (note that the actual physical signal is represented either by $x(t) = Re\{\underline{x}(t)\}$ or by $x(t) = \mathrm{Im}\{\underline{x}(t)\}$):

$$\underline{x}(t) = V\,\mathrm{rect}\left(\frac{t - t_D/2}{t_D}\right)\mathrm{e}^{j\varphi(t)} \tag{2.114}$$

Since phase and frequency are related by

$$f(t) = \frac{1}{2\pi}\frac{\mathrm{d}\varphi(t)}{\mathrm{d}t} \Leftrightarrow \varphi(t) = 2\pi\int_0^t f(\xi)\mathrm{d}\xi = 2\pi\left(f_0 t + \frac{at^2}{2}\right) \tag{2.115}$$

(2.114) results in

$$\underline{x}(t) = V \cdot \mathrm{rect}\left(\frac{t - t_D/2}{t_D}\right)\mathrm{e}^{j2\pi\left(f_0 t + \frac{at^2}{2}\right)} \tag{2.116}$$

The average power of the chirp is $P \approx V^2/2$. With increasing bandwidth and low cut-off frequency f_0 it approaches the exact value of a sine wave. The amplitude spectrum is given by the convolution of a rectangular pulse spectrum with the spectrum of the complex exponential (see also Annex B.3)

$$\underline{X}(f) = V\underline{X}_{\mathrm{pulse}}(f) * \underline{X}_{\mathrm{expn}}(f) \tag{2.117}$$

with

$$\underline{X}_{\mathrm{pulse}}(f) = t_D\,\mathrm{sinc}(f t_D)\mathrm{e}^{j\pi f t_D}$$

and

$$\underline{X}_{\mathrm{expn}}(f) = \frac{1}{\sqrt{a}}\mathrm{e}^{-j\pi\left(\frac{(f-f_0)^2}{a} - \frac{1}{4}\right)}$$

Detailed calculations of the spectrum and the ACF are quite cumbersome. We will not repeat this here. The interested reader can find corresponding calculations in Ref. [17]. We will rather illustrate some typical chirp properties by referring to two examples (see Figure 2.35) relating two chirps of TB = 2000 each and a fractional bandwidth of $b = 2$ and $b = 1$. The upper example relates to the case $b = 2$, that is the spectrum starts at very low frequencies. In order to give an illustrative view of the time shape, only the first 20% of the actual time signal are drawn. The spectrum is nearly rectangular shaped which consequently leads to a sinc(x)-shaped ACF. The probability density functions of a sine wave and a chirp look quite similar (but they are not identical). Hence, the crest-factor is about CF $\approx$ 1.4 like for the sine.

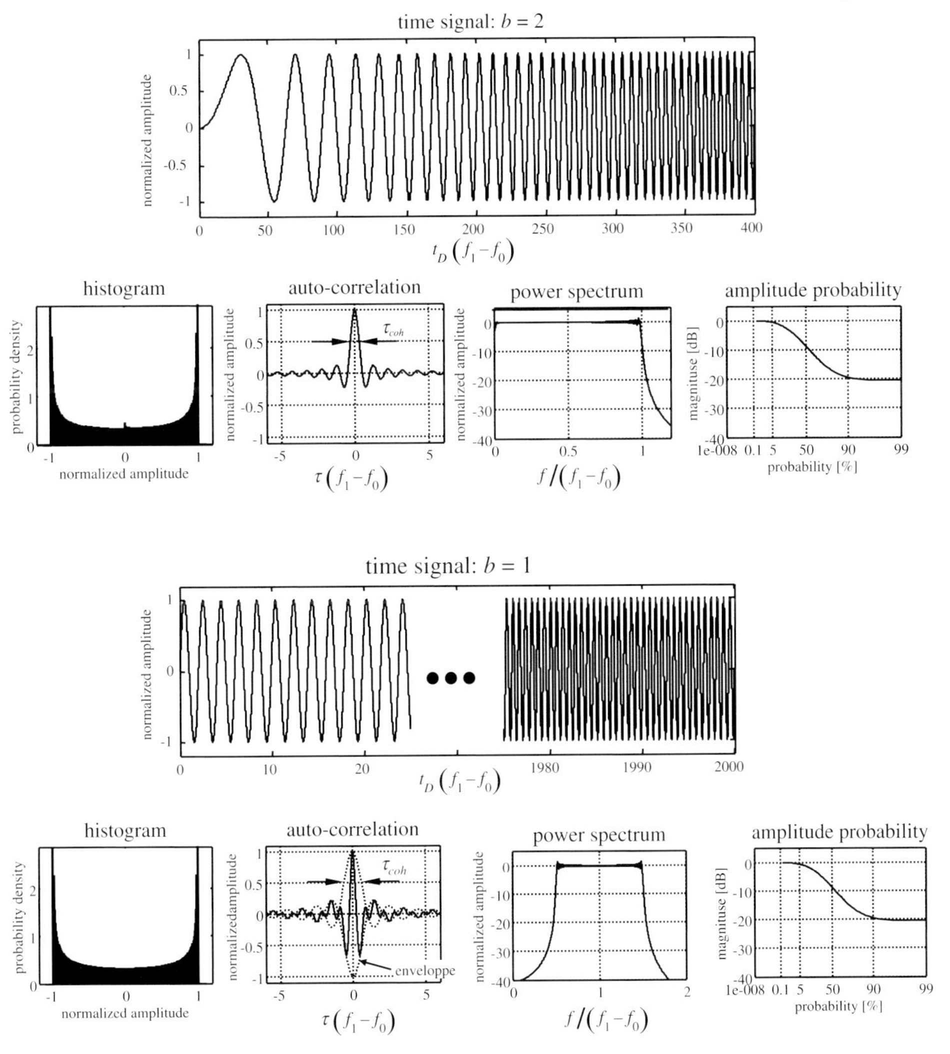

Figure 2.35 Two chirp signals and their characteristic functions. The signals are of identical time bandwidth product (TB = 2000) but different fractional bandwidth.

For the second example ($b = 1$), the beginning and the end of the chirp signal are depicted. As shown in the drawings, the relative frequency variations are smaller now than before even if the absolute frequency variation is the same in both cases. The ACF tends to more oscillations like a band-pass pulse, so that the pulse width t_{coh} indicating the width of the envelope is about twice the coherence width of the first chirp (compare Section 2.2.5.3). However, the main lobe of the ACF is shorter than the main lobe for the case $b = 2$ since in average higher frequencies are involved. As long as the ACF pulse does not contain too many oscillations this behaviour may be exploited to improve, for example, the radar resolution in connection with super-resolution techniques.

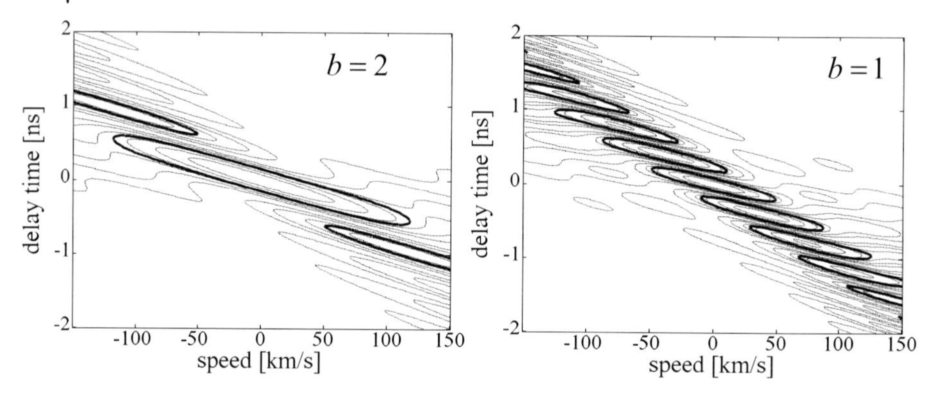

Figure 2.36 The Kernel of the ambiguity function of chirp signals: TB = 2000 and frequency sweep: 2 GHz.

Sensor devices using chirp signals usually have a quite high start frequency f_0 so that their behaviour approaches to that considered in the second example. Hence, the time bandwidth product should relate to the single-sided bandwidth (see also the remarks in Section 2.2.5.3):

$$\mathrm{TB} = t_\mathrm{D} \underset{\cdot}{B} = t_\mathrm{D}(f_1 - f_0) \tag{2.118}$$

Finally, Figure 2.36 depicts the kernel of the ambiguity function for our two example chirps. As expected, the ambiguity function kernel of the signal with high fractional bandwidth is much more concentrated than the kernel for the signal of lower fractional bandwidth. Neglecting range-Doppler coupling, we have unique conditions in the case $b = 2$ as long as a target moves slower than 50 km/s.[15]

Obviously, the global structure of the ambiguity function resembles the knife-edge form as depicted in Figure 2.19e. Hence, no de-correlation between sounding and measurement signal will appear even in the case of a high-target speed. This will lead to simple sensor electronics. But we will observe large range errors in case of moving targets. In order to compensate this error, one uses up- and down chirps which allows separating distance and speeding measurement (see Section 3.4.4).

2.3.7 Multi-Sine

The multi-sine is the most flexible concept to create a measurement signal. At least in theory, it can be flexibly adapted to the measurement problem, that is the spectrum and the amplitude probability can be matched to certain needs [17, 18]. The idea is to create the spectrum of the signal in the computer according to the actual

15) Note that the indicated speed values for short-range sensor applications are typically not of interest. These high values only appear since we used for illustration purposes a very short sweep time $t_\mathrm{D} = 1\,\mu\mathrm{s}$ for the chirp. Sensors applying chirp signals (FMCW-sensor) typically have sweep times in the millisecond range, that is more than 1000 times larger than applied here which would bring down the speed values to more realistic values.

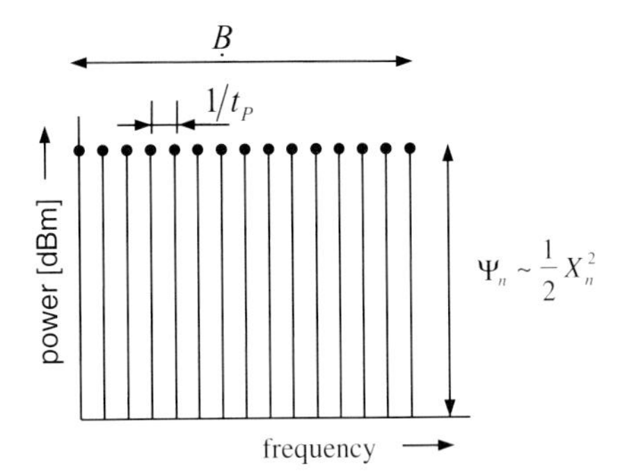

Figure 2.37 Spectrum of the intended multi-sine signal. Basically, the required spectrum could have any shape.

requirements, to calculate the time shape of the signal via inverse Fourier transform and to transfer these data to an arbitrary waveform generator. It will periodically emit the calculated waveform. It is also possible to apply the designed signal only once and to repeat the measurement with a different time signal which has the same spectrum (e.g. by varying only the phase before the transformation in the time domain). This will randomize the measurement procedure. But it will also lead to an extended measurement time since averaging is required over several ensembles of captured data (see Section 2.3.8).

The multi-sine approach is limited by the performance of the available high-speed arbitrary waveform generators particularly with respect to their limited sampling rate and DAC resolution. The maximum sampling rate restricts the bandwidth (currently a few GHz with best devices on market) and the DAC resolution limits the dynamic range of the sounding signal (generation of unwanted spectral components).

Figures 2.38 and 2.39 shall illustrate the multi-sine method for a simple example. We require a stimulus signal of a constant low-pass spectrum over the (physical) width $\underset{\cdot}{B}$ and a signal period of t_P. The resulting time-bandwidth product of the signal will be $2\underset{\cdot}{B}t_P$ since $\underset{\cdot}{B}$ relates to a one-sided bandwidth. Figure 2.37 exhibits the spectrum which we did require. The average power of the signal will be

$$P \approx X_0^2 + \frac{1}{2}\sum_{n=1}^{N} X_n^2$$

in which $N = \underset{\cdot}{B}t_P$ is the number of spectral lines (positive frequencies only) and X_n is the magnitude of the individual sine waves composing the waveform.

Before the time domain signal can be calculated, the phase of all spectral lines has to be fixed. This can be done quite freely. Figure 2.38 depicts three time signals which have identical power spectrum but different phase angles. In the last example, an optimization algorithm was used to select the phase value of the different spectral lines in such a way that the resulting time signal will have the lowest possible crest-factor [19]. As expected from the identical power spectrum of all signals, the auto-correlations functions are identical. Hence, a radar sensor applying any of these

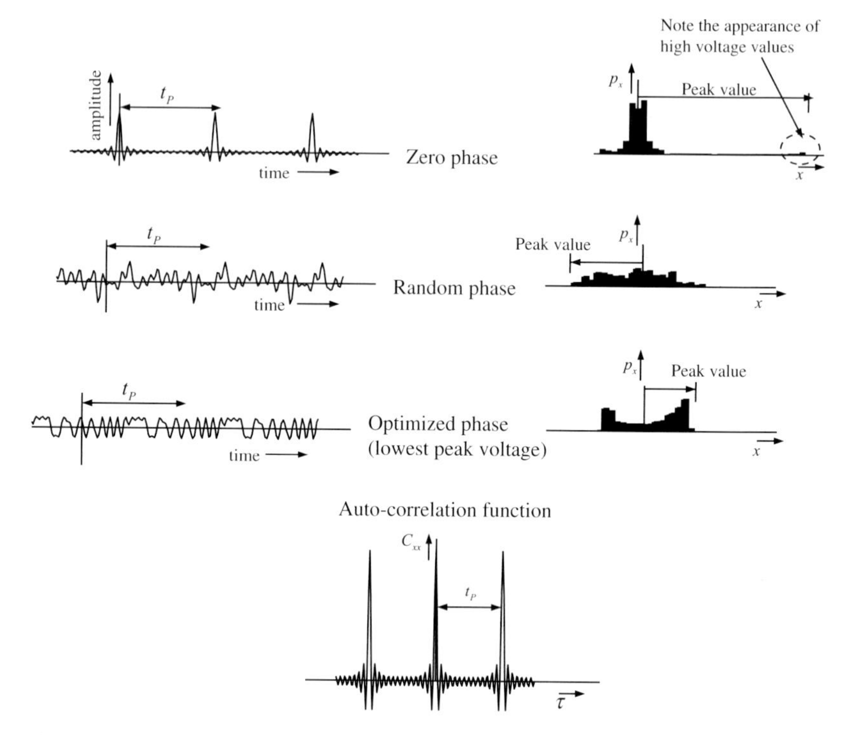

Figure 2.38 Effect of phase selection onto the time shape and amplitude distribution of a multi-sine signal. The auto-correlation function will be not affected by the phase.

signals would have the same range resolution. The shape of the ACF is given by the power spectrum of the signal. Since the power spectrum can freely be designed, one can create signals of a wanted ACF shape. For completeness, Figure 2.39 depicts the amplitude probability of all three signals illustrating their different amplitude behaviour. Obviously the crest-factor optimized signal has the lowest amplitude values

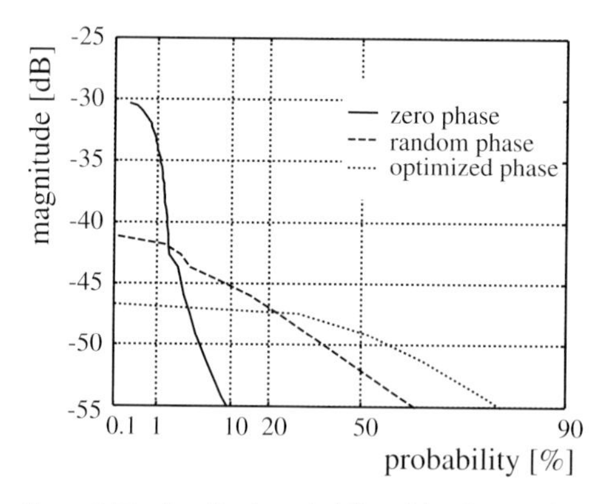

Figure 2.39 Amplitude probability of the three multi-sine signals from Figure 2.38.

causing to the lowest interference with other electronic devices and it prevents overloading or saturation effects in the measurement receiver and the object under test.

2.3.8 Random Noise

Due to its random nature, the emission of random noise is difficult to detect. This may be of advantage if a camouflaged operation of the sensor is in the foreground. Sensor-independent radiation sources can also be used to illuminate a test scenario as long as one is able to obtain an appropriate reference signal. The signals emitted by radio stations or natural noise sources can be considered as random in our case. Passive radar is a prominent example of such approach by which the radiation of radio and TV stations is used for sounding purposes. Currently, the bandwidth of such systems is, however, far from UWB requirements so that we will not go deeper into this topic here.

The ambiguity function of random noise has the thumbtack form (see Figure 2.19). That is, noise provides unique measurement results for range and speed whereas range and speed resolution are determined by the bandwidth $\underset{..}{B}_N$ and the exploited signal duration t_D (see Figure 2.40). Some additional discussions on the noise ambiguity function can be found in Ref. [20].

A further point which counts for noise signals is the minimal mutual interception between sensors of the same type due to the vanishing cross-correlation of two independent noise sources. But noisy stimulus signals have also some serious disadvantages because their randomness requires a comparatively long measurement time in order to gain stable measurements. We will point to that inconvenience.

A random signal is characterized mainly by two features – its spectrum and its amplitude distribution. Here, we will consider exemplarily band-limited white noise of zero mean with a Gaussian (i.e. normal distributed) $x(t) \sim N(0, \sigma^2)$ or uniform amplitude distribution $x(t) \sim U(x_{\min}, x_{\max})$. The term band-limited white noise means that the spectral power is frequency independent within a limited frequency band of the (two sided) width $\underset{..}{B}_N$. Figure 2.40 depicts two examples of the time signal and the amplitude distribution. Both signals have the same power but their maximum voltage is differently high. In theory, the Gaussian noise should have a crest-factor of infinity but for practical purposes one can count with CF ≈ 3. In that case, one neglects that about 1‰ of signal values are bigger than $3\,x_{\mathrm{rms}}$.

As to be seen in Figure 2.40, a time segment of sufficient duration t_D was taken to observe the signal. This is a usual approach since the measurement time is always limited and for many applications it should be actually as short as possible.

However, a too short recording time t_D leads to some problems. Since the signal within the considered time interval is completely random, all signal parameters or functions are also at least partially of random nature. Hence, one is only able to determine the estimations of these parameters which are quantified by the most probable value (i.e. the expected value) and a measure of its confidence (i.e. the variance). Figure 2.41 gives a simplified example for the auto-correlation function of stationary random noise.

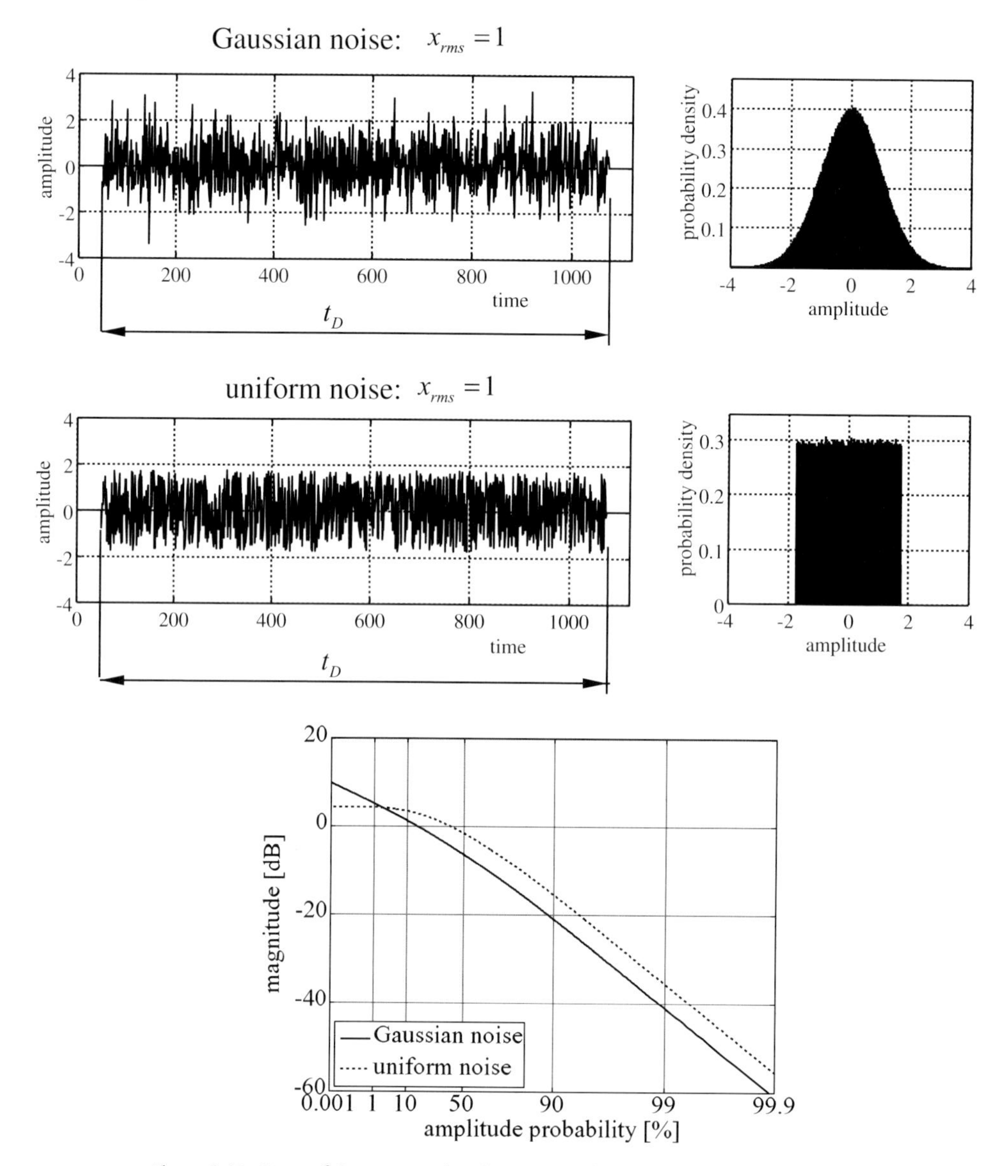

Figure 2.40 Burst of Gaussian and uniform noise of the same power. Left: Typical images in the time domain. Right: Amplitude probability density. Bottom: Amplitude probability.

The expected auto-correlation function of band-limited white noise represents a short pulse centred at $\tau = 0$. The width of the pulse depends on the noise bandwidth $\underset{..}{B}_N$. For time lags $\tau \gg \underset{..}{B}_N^{-1}$, the auto-correlation should approach zero since not any coherency exist between the signal and its shifted version. But actually, we will find a more or less random ACF base which we can see if the measurements were repeated several times.

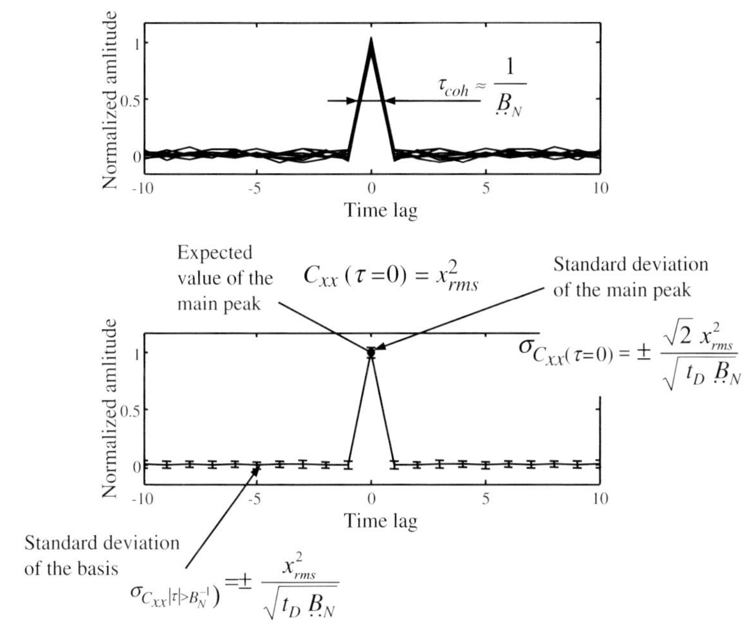

Figure 2.41 Ensemble of auto-correlation functions (only the section around the main peak is shown) determined from different data sets of random noise. The data were taken from different time-limited sections of band-limited white Gaussian noise.

A simple analysis for normal distributed noise (see Annex B.4) gives a result as symbolized in the lower part of Figure 2.41. The expected shape of the ACF and a corresponding confidence interval of width $2\,\sigma_{C_{xx}}$ (standard deviation) are faced. It is interesting to note that the variance of the main peak is twice the variance of the ACF base. The relative variance of the main peak is

$$\frac{\sigma^2_{C_{xx}(\tau=0)}}{C^2_{xx}(\tau=0)} = \frac{1}{\mathrm{SNR}(\tau=0)} = \frac{2}{\ddot{B}_{\mathrm{N}} t_{\mathrm{D}}} \tag{2.119}$$

and that of the ACF base is

$$\frac{\sigma^2_{C_{xx}\left(|\tau|>\ddot{B}_{\mathrm{N}}^{-1}\right)}}{C^2_{xx}(\tau=0)} = \frac{1}{\mathrm{SNR}\left(|\tau|>\ddot{B}_{\mathrm{N}}^{-1}\right)} = \frac{1}{\ddot{B}_{\mathrm{N}}\, t_{\mathrm{D}}} \tag{2.120}$$

Both equations underline the importance of a large time-bandwidth product $\mathrm{TB} = t_{\mathrm{D}} \ddot{B}_{\mathrm{N}}$ to gain a stable measurement.

We supposed in our consideration a signal segment of final duration t_{D}. Since the determination of the correlation functions requires a time shift by τ, it remains only a signal segment of duration $t_{\mathrm{D}} - |\tau|$ to perform the product and the averaging. This time interval will decrease with rising time lag τ reflecting in growing performance degradation of the ACF variance. Hence, it is recommended that $|\tau|_{\max} \ll t_{\mathrm{D}}$.

As we have seen from the previous discussion, noise signals require a long recording time by nature in order to gain sufficiently stable measurement results. Hence, the achievable measurement speed will be comparatively small and the amount of data to be handled will be quite large. Further impacts of random test signals onto the sensor performance are discussed in Section 3.6.

In summary one can state that periodic test signals as M-sequence, chirp, multi-sine or others are preferable if measurement speed, simple data handling and simple sensor electronics are in the foreground of interest. Such signals do not cause randomness so that measurement time and data amount can be adapted to the requirements of the test scenario and must not be constrained by the test signal. On the other hand, the use of random stimulation signals is of advantage if low probability of intercept and camouflaged operations are of major interest or the ambiguity function must have the thumbtack shape.

2.4
Formal Description of Dynamic Systems

2.4.1
Introduction

The final goal of an ultra-wideband sensor is to gain information from a device, material or scenario under test by stimulating them with electromagnetic power which is distributed over a large bandwidth. We will summarize all these measurement objects under the term system here. Its behaviour has to be quantified and qualified by the measurement. As even depicted in Figure 2.1, this is done by capturing the system reaction caused by a specific stimulation. Figure 2.42 illustrates some test scenarios which we consider by the same approach independently from the actual measurement task.

The questions which we will follow here are as follows:

- What are the characteristic values or functions which are quantifying or qualifying a system?
- How to stimulate the systems so that they are 'willing to disclose' their behaviour as comprehensive as possible?
- How to extract the identified characteristic values or functions of the system under test from the actual measurement data?

These questions are not only of interest for the system under test but also for the measurement device/sensor itself since it is necessary to evaluate its utility for a certain measurement task. It should also be mentioned that the systems under test as considered in Figure 2.42 may cover the actual objects of interest as well some auxiliary components of the measurement arrangement. Such components may be antennas, probes or any applicators which enables the electromagnetic interaction with the wanted objects. These parts of the system under test which do not relate to the wanted object are often summarized under the term measurement bed.

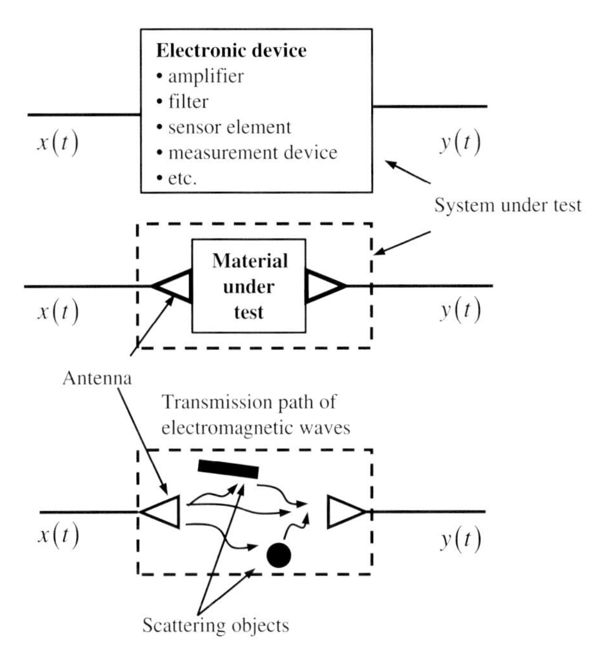

Figure 2.42 Illustration of a system under test referring to different measurement tasks.

At first, we will suppose that the measurement bed does not affect the behaviour of target so its influence can be neglected in what follows. Later, we will come back to the problem of the measurement bed and we will try to remove its actual effect onto the objects of interest.

In order to be able to define characteristic functions and values of the systems of interest, we will start from three very general assumptions:

1) The system under test may be stimulated by injecting a certain amount of energy (carried by the signal $x(t)$) and it is possible to observe its (dynamic) reaction $y(t)$. Often, both signals are also called input or output signal.
2) We suppose that the system under test behaves linearly. What this means is explained hereafter. For the measurement objects we have in mind here, this assumption is usually fulfilled as long as the stimulation signals do not lead to saturate the objects. That is, the peak and average power of the stimulation signal should not exceed a certain limit (depending on the type of measurement object). Due to the strong power limitations by the radio regulation laws, ultra-wideband sensors are usually far from any saturation effects of the test objects. But linearity must also hold for the sensor electronics itself. Therefore, the system designer has to pay attention to operate the electronics in the linear range (see Section 2.6.5 for characterizing non-linear deviations).
3) In order to simplify the considerations, we will further require that the system under test does not (or only little) change its behaviour during the measurement of a single response function of the test system, that is the system behaviour

must be time invariant (at least during the measurement). Since the considered objects typically change their properties with time, we are in limited in the duration for measurement data recording. If we deal with a scenario which includes continuously moving objects, the recording time T_R is limited by the target speed (compare (2.90), Section 2.2.6). For oscillatory variations of the scenario, the repetition rate of the measurements must meet the Nyquist criteria, that is the measurement rate must be at least twice the bandwidth of the scenario variations.

2.4.2 Time Domain Description

2.4.2.1 Linearity

On macroscopic level, the interaction of the electromagnetic field with matter is modelled by the set of Maxwell's equations. These are differential equations describing the dynamic that is the temporal behaviour of the interaction. If the field strength is not too high, the electrical parameters of matter are independent of the field strength in most cases that we are interested here, that is $\varepsilon \neq \varepsilon(\mathbf{E}, \mathbf{H})$, $\mu \neq \mu(\mathbf{E}, \mathbf{H})$ and $\sigma \neq \sigma(\mathbf{E}, \mathbf{H})$. Therefore, an object or scenario consisting of lumped or distributed items can be described by a set of linear differential equations. As even mentioned, the restriction to linearity can be generally supposed for sensor applications since one not intends to influence the behaviour of system under test by the sounding signal. This is also valid for the measurement device and the sensor elements themselves.

Supposing for simplicity that our system of interest is stimulated by one signal $x(t)$ and we only observe one reaction $y(t)$. We symbolize this by a black-box model having one input and one output as depicted in Figures 2.1, 2.42 and 2.43.

Herein, $f\{\cdots\}$ may be any linear function $L\{\cdots\}$ which describes the behaviour of the system. A function is called linear, if it meets the law of superposition:

$$L\{x_1(t) + x_2(t)\} = L\{x_1(t)\} + L\{x_2(t)\} \tag{2.121}$$

As Figure 2.44 and (2.121) show, there is no difference for a linear system if a composite signal, for example $x(t) = x_1(t) + x_2(t)$ is transferred as a whole or if the individual parts $x_1(t)$ and $x_2(t)$ are transferred individually and superimposed at the end. These two signals can also sequentially stimulate the system if it behaves time invariant.

The practical importance of this relation is that the system reaction $y(t)$ to an arbitrary input signal $x(t)$ can be determined from the knowledge of the system

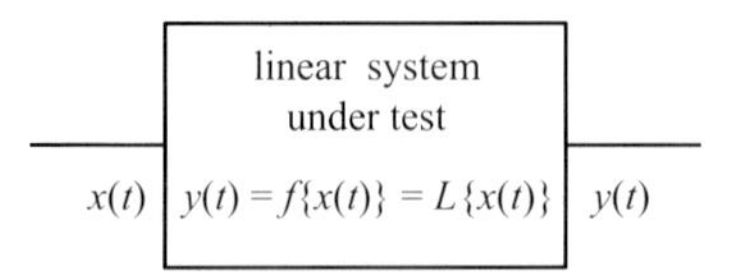

Figure 2.43 Linear scenario under test symbolized by a black box model with single input and single output.

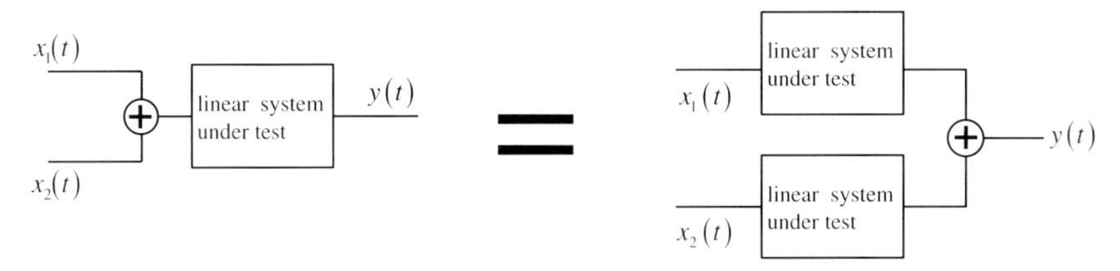

Figure 2.44 Demonstration of the law of superposition.

behaviour with respect to a 'standardized input signal'. For that purpose, the arbitrary input signal $x(t)$ is decomposed in a sum of standardized signals. Since the system reaction to the standard signals is supposed to be known, the output signal $y(t)$ can be determined by recomposing the system reaction by the same rules as the input signal was decomposed. From the law of superposition, we make use in what follow to find characteristic functions which describes the behaviour of a linear system. For a deeper discussion on the linearity of systems, the reader can refer to Ref. [21].

2.4.2.2 **The Impulse Response Function or the Time Domain Green's Function**

Let us consider a system that dynamic behaviour can be described by a linear equation which is subjected to certain boundary conditions:

$$L\{y(t)\} = x(t) \tag{2.122}$$

$L\{\cdots\}$ is a linear differential operator, $x(t)$ is a known stimulation signal and $y(t)$ is the unknown system reaction. Supposing furthermore that $y = g(t,\tau)$ is a solution of (2.122) if the system was stimulated by a Dirac function $x = \delta(t-\tau)$ (i.e. a short pulse launched at time τ), that is

$$L\{g(t,\tau)\} = \delta(t-\tau) \tag{2.123}$$

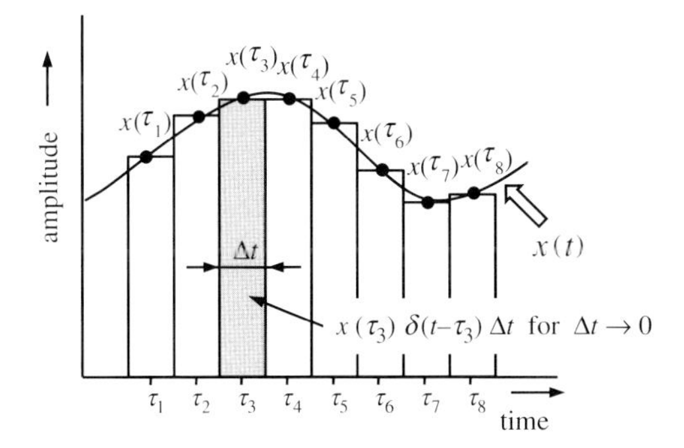

Figure 2.45 Decomposition of the signal $x(t)$ into a sum of short impulses which tends to Dirac functions for $\Delta t \to 0$.

in which the Dirac function (also Delta function, Dirac distribution) is defined, for example, as

$$\delta(t) = \begin{cases} 0 & t \neq 0 \\ \infty & t = 0 \end{cases} \quad \text{and} \quad \int_{-\infty}^{\infty} \delta(t) dt = 1 \tag{2.124}$$

Some more discussions of the Delta function are given in Annex A.2.

As illustrated in Figure 2.45, the arbitrary signal $x(t)$ can be decomposed in a set of time-shifted and -weighted Dirac functions by

$$x(t) = \int \delta(t-\tau)\, x(\tau) d\tau \tag{2.125}$$

That is, the input signals $x(t)$ is represented by an infinite number of time-shifted Dirac functions $\delta(t-\tau_i)$ now. Hence, (2.122) can be modified by (2.125) and (2.123) to

$$L\{y(t)\} = \int \delta(t-\tau) x(\tau) d\tau = \int L\{g(t,\tau)\} x(\tau) d\tau \tag{2.126}$$

Due to the supposed linearity of the system, the effect of every individual Dirac function onto the system can be considered separately. Therefore, as graphical illustrated in Figure 2.44, integral and linear operator on the right-hand side of (2.126) can be changed

$$L\{y(t)\} = \int L\{g(t,\tau)\} x(\tau) d\tau \Rightarrow L\{y(t)\} = L\left\{ \int g(t,\tau) x(\tau) d\tau \right\} \tag{2.127}$$

from which finally follows:

$$y(t) = \int g(t,\tau) x(\tau) d\tau \tag{2.128}$$

(2.128) represents a basic relation of system theory. It is the so-called convolution (for time variable systems). The relation will be simplified by considering only time invariant systems (as suppose above for simplicity):

$$y(t) = \int g(t-\tau) x(\tau) d\tau = \int g(\tau) x(t-\tau) d\tau = g(t) * x(t) \tag{2.129}$$

In the case of time invariance, the system behaviour is independent of the time point τ of stimulation. Therefore, the response function $g(t,\tau)$ will always have the same shape and hence it can be replaced by

$$g(t,\tau) = g(t-\tau)$$

as we did in (2.129). This equation represents a fundamental law of the theory of linear systems. For convenience, the integral is usually shortened by the star sign $*$ and it is also referred as convolution product.

Following to (2.123), we can gain $g(t)$ of a system under test if we hit it with a Dirac function $\delta(t)$. Technically, one tries to approach a Dirac function by a short pulse. Hence, $g(t)$ is also called impulse response function (IRF). For illustration of the response function $g(t)$, we will consider two scenarios in

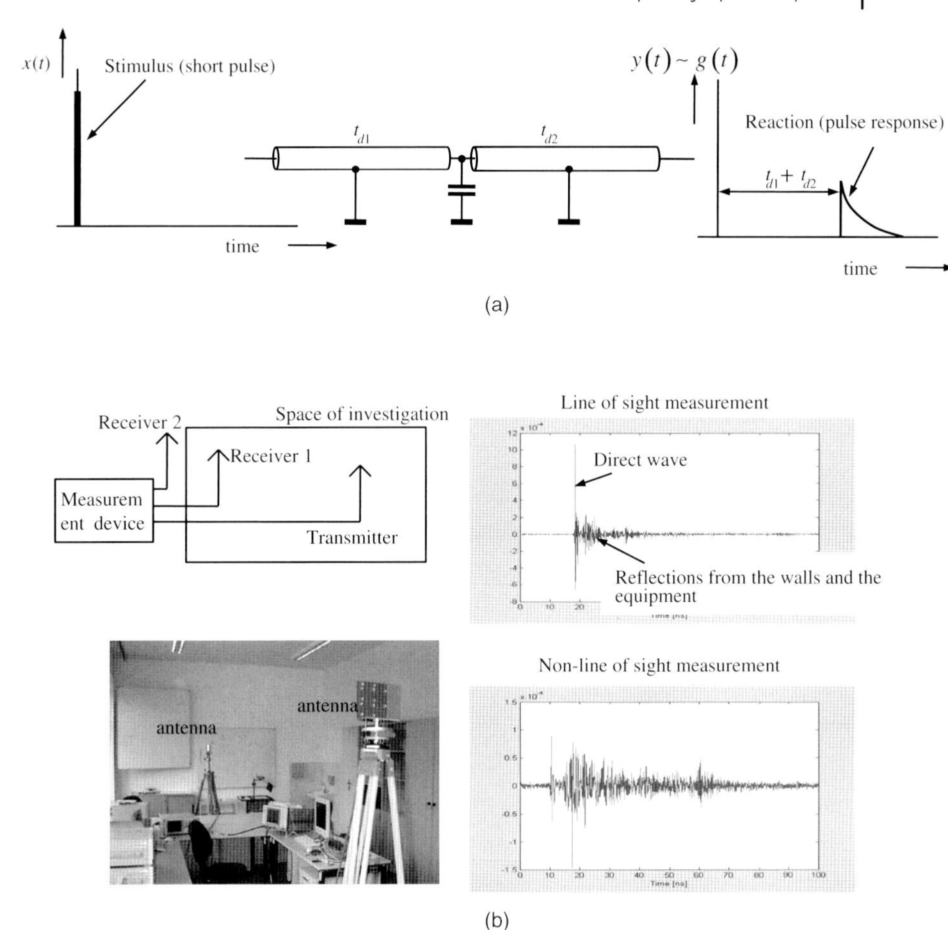

Figure 2.46 Illustration of the measurement of an impulse response function. (a) Cable measurement and (b) measurement of indoor wave propagation under line of side and non-line-of-sight conditions.

Figure 2.46. In the first case, we have a very simple shape of the pulse response. Here, we deal with a damaged cable. Supposing, the defect provokes an additional capacity to ground. If we inject a very short pulse $x(t) \sim \delta(t)$, the output signal $y(t)$ will be an exponentially decaying voltage delayed by the propagation time through the cable. This signal represents the pulse response $y(t) \sim g(t)$ of the considered test object. We can conclude from this measurement that the cable was possibly equeezed leading to the additional capacity, that is from a simple measurement we got the information which we are interested in.

The second example shows an apparently chaotic (but nevertheless deterministic) impulse response. Its shape is caused by multiple reflections of the sounding waves within the room under investigation. This function cannot be interpreted as easily as in the previous case. In order to gain more information, the measurements have to be repeated at several positions. We will discuss related problems in Section 4.10.

As we have seen, the IRF $g(t)$ may be more or less Gordian depending on the measurement object. But it always starts at a certain time and settle down after a while. This leads us to two generic properties of $g(t)$ for linear passive[16] systems:

It is stable: That is, the L_1-norm of $g(t)$ is finite and the response function tends to zero with increasing time.

$$\begin{aligned} \|g(t)\|_1 &= \int_0^\infty |g(t)| \mathrm{d}t < \infty \\ \lim_{t\to\infty} g(t) &= 0 \end{aligned} \tag{2.130}$$

It is causal: That means the system cannot react before the stimulation has been taken place. Supposing the stimulation at the time point $\tau = 0$, it holds

$$g(t) = \begin{cases} g(t) & \text{for } t > 0 \\ \equiv 0 & \text{for } t \leq 0 \end{cases} \tag{2.131}$$

This condition will also have some impacts onto the frequency response function of the system (see (2.149)).

Due to the significance of the convolution for the description of the system behaviour, we will summarize some important properties and aspects here. Annex B.2 gives some simple examples and further rules for the manipulation of convolution products.

- If two systems are connected in chain, the impulse response of the chain is given by the convolution of the individual impulse responses:

$$g_{\mathrm{tot}}(t) = g_1(t) * g_2(t) \tag{2.132}$$

The commutative law holds for convolution

$$g_1(t) * g_2(t) = g_2(t) * g_1(t) \tag{2.133}$$

hence, the order of interconnection of linear transmission systems does not matter.[17]

- The convolution of a function with a Dirac pulse has no effect on the shape of the function (see (2.125) and Figure 2.45). The practical consequence is $g_1(t) * g_2(t) \approx g_1(t)$ if $g_2(t)$ is pulse shaped and short compared to any variations or oscillations within $g_1(t)$.

16) The term passive means that the considered system has no internal energy sources. That is, if it emits any signal (energy) it can only be caused by an external stimulation (i.e. injection of a finit amount of energy).

17) This conclusion has to be considered with care. It is only valid if not any retroaction between the involved systems exists. See also Section 2.5.

- The convolution of a function $x(t)$ with the time reversed version of $y(-t)$ represents the cross-correlation function[18] between $x(t)$ and $y(t)$:

$$C_{xy}(t) = \int x(\xi)\, y(\xi + t)\mathrm{d}\xi = x(t) * y(-t) \tag{2.134}$$

and

$$C_{yx}(t) = \int x(\xi + t)\, y(\xi)\mathrm{d}\xi = x(-t) * y(t) \tag{2.135}$$

If $y(t) = x(-t)$, the convolution represents an auto-correlation. As we saw above (Figure 2.13 or (2.79)), the auto-correlation $C_{xx}(\tau)$ leads to a short pulse-like function if the signal $x(t)$ is of long duration and covers a wide spectral band. Hence, if it is possible to build a system whose impulse response is equivalent to the time shape of a specific signal $g(t) \sim x_0(-t)$ then the output signal will be a short pulse if it is stimulated by the signal $x_0(t)$. Otherwise, if the input signal does not match the shape of the impulse response, $x(t)$ and $g(t)$ de-correlate and we can observe only small output signals. A system having such an impulse response is called matched filter.

- We can modify the convolution integral (2.129) by an additional convolution with the time inverted stimulus signal. Applying (2.134), one can introduce a further convolution referring to correlation functions instead of the pure signals:

$$y(t) * x(-t) = g(t) * x(t) * x(-t) \quad \Rightarrow \quad C_{yx}(t) = g(t) * C_{xx}(t) \tag{2.136}$$

and

$$y(t) * y(-t) = g(t) * x(t) * y(-t) \quad \Rightarrow \quad C_{yy}(t) = g(t) * C_{xy}(t) \tag{2.137}$$

  Mainly (2.136) is gladly exploited for measurement purposes (see Sections 2.6.1.7, 3.3 and 3.6)

- The output signal $y(t)$ of a system can also be considered as a weighted or masked version of the input signal $x(t)$, whereas $g(t)$ represents the weighting function. Therefore, it should be possible to enhance or suppress particular signal components in $x(t)$. For that purpose the signal $x(t)$ is subjected to a transmission system of a specific impulse response function $g(t)$. A system intentionally designed for such a purpose is usually called a filter. A narrowband filter is, for example, designed to enhance sine wave components of a certain frequency in $x(t)$ and to suppress all the rest. Matched filters as described above are doing the same but for signals having a more complicated time shape than a simple sine.

So far the considered equations describe the behaviour of a system with one input and one output, the so-called SiSo systems. But it is easy to extend the description to systems with numerous in- and outputs, the so-called MiMo systems (Figure 2.47). Due to the linearity of the considered scenarios, the transmission

18) We will omit the distinction between energy and power signals for brevity. But note that the correlation functions for power signals need a normalization by the integration window, that is $\lim_{T\to\infty} \frac{1}{T}\int_T \cdots dt$ since otherwise these integrals tend to infinity.

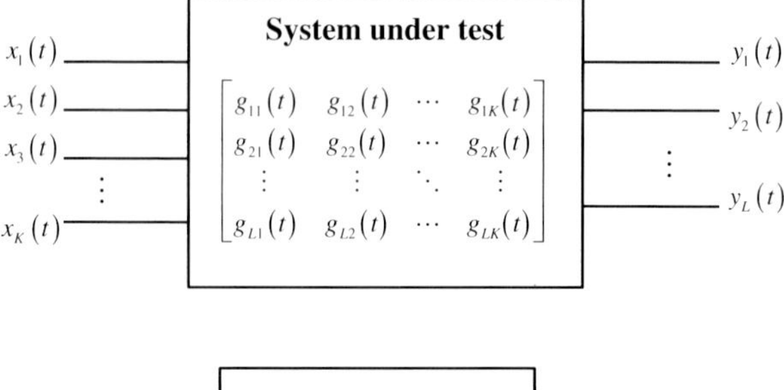

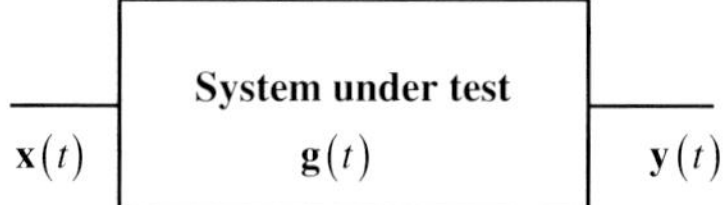

Figure 2.47 System with multiple inputs and multiple outputs: $[L, K]$-MiMo System.

behaviour of every input–output pair can be quantified separately by an impulse response $g_{lk}(t)$. The effect of all inputs $k = 1 \cdots K$ onto a specific output l is finally given by the superposition of all corresponding transmission pairs $g_{lk}(t)$ $(k = 1 \cdots K)$ as executed by (2.138)

$$\begin{aligned}
y_1(t) &= \int g_{11}(t-\tau)x_1(t)\mathrm{d}\tau + \int g_{12}(t-\tau)x_2(t)\mathrm{d}\tau + \cdots + \int g_{1K}(t-\tau)x_K(t)\mathrm{d}\tau \\
y_2(t) &= \int g_{21}(t-\tau)x_1(t)\mathrm{d}\tau + \int g_{22}(t-\tau)x_2(t)\mathrm{d}\tau + \cdots + \int g_{2K}(t-\tau)x_K(t)\mathrm{d}\tau \\
&\vdots \\
y_L(t) &= \int g_{L1}(t-\tau)x_1(t)\mathrm{d}\tau + \int g_{L2}(t-\tau)x_2(t)\mathrm{d}\tau + \cdots + \int g_{LK}(t-\tau)x_K(t)\mathrm{d}\tau
\end{aligned} \tag{2.138}$$

Using matrix notation, this results in an expression close to (2.129)

$$\mathbf{y}(t) = \mathbf{g}(t) * \mathbf{x}(t) \tag{2.139}$$

with

$$\begin{aligned}
\mathbf{x}(t) &= \begin{bmatrix} x_1(t) & x_2(t) & \cdots & x_K(t) \end{bmatrix}^T \\
\mathbf{y}(t) &= \begin{bmatrix} y_1(t) & y_2(t) & \cdots & y_L(t) \end{bmatrix}^T \\
\mathbf{g}(t) &= \begin{bmatrix} g_{11}(t) & g_{12}(t) & \cdots & g_{1K}(t) \\ g_{21}(t) & g_{22}(t) & \cdots & g_{2K}(t) \\ \vdots & \vdots & \ddots & \vdots \\ g_{L1}(t) & g_{L2}(t) & \cdots & g_{LK}(t) \end{bmatrix}
\end{aligned}$$

Hence, a system with K input and L output channels is characterized by a $[L, K]$-matrix of impulse response functions.

What are the conclusions of the above considerations concerning our goal to characterize the behaviour of a system or to recognize an object?

As (2.128) and (2.129) imply, the functions $g(t, \tau)$ or $g(t)$ are characteristic functions of a linear and a linear time invariant (LTI) system since based on their knowledge the response signal $y(t)$ to any arbitrary stimulation $x(t)$ can be determined.

The functions $g(t, \tau)$ or $g(t)$ are the Green's function of the system under test. They are named in honour to George Green a British mathematician and physicist. These functions contain all information (in more or less encrypted form) about the system under test which it is 'willing' to reveal by electromagnetic sounding.

As requested from (2.123), the practical determination of $g(t)$ supposes the stimulation of an unknown system by a signal having a shape as close as possible to a Dirac function.[19] Since a $\delta(t)$-like signal is called an impulse, the function $g(t)$ is typically named as IRF and the denotation Green's function is usually not applied in this field.

At this point two questions arise:

1) In order to gain the IRF $g(t)$, the system has to be stimulated with a Dirac pulse what is obviously not possible in practice due to its infinity amplitude and zero time duration. Hence, what are the consequences and limitations of practical measurements? These aspects will be discussed in Chapter 3.
2) The IRF will be a meaningless function as long as it will not be interpreted by a useful method in order to extract the wanted information about the system under test. Consequently, we have to ask how we can do this, which approaches we can follow? We will review the general approach.

2.4.2.3 **Extraction of Information from the Impulse Response Function**

The IRF shape is a consequence of dynamic processes going on within the considered system. These processes quantify and qualify the state of the system or objects interacting with the sounding signal. They may even give some hints of what kind the system or object is. Hence, appropriate parameters quantifying the time shape of the impulse response function are a possible source of information about the measurement object.

Since the impulse response of typical objects considered here has a pulse-like shape, we can apply the parameters of pulse waveforms as even introduced in Section 2.2.2. Additionally, some further ones are introduced here below. Vice versa, they can also be applied to characterize pulse-shaped signals.

Figure 2.48a depicts an IRF which apparently shows only a small undershoot. By switching to a logarithmic scale, it will be obvious that an oscillating process is continued. That is, the system needs some time to settle down. The settling is caused by an internal energy flux between different storage elements (capacities, inductivities) and by internal reflection and propagation events. Hence, the manner of settling is a characteristic property of a system.

19) It should be mentioned that the stimulation by a Dirac function is not the only way to gain a characteristic function of a system. An alternative way, usually applied in time domain reflectometry, is the use of step function (Heaviside function). The resulting system function is called step response $h(t)$. The approach to introduce step function is nearly identical as demonstrated above for Dirac pulses. Therefore, it will not be discussed here. The step signal is given by the integral of the Dirac pulse $u(t) = \int \delta(t)dt$. Consequently, the step response $h(t)$ of the system can be calculated from the impulse response by $h(t) = \int g(t)dt$. Since both functions relate together, we are not able to gain more information about the system by measuring both. Hence, they represent only a way of different presentation of measurement results.

(a)

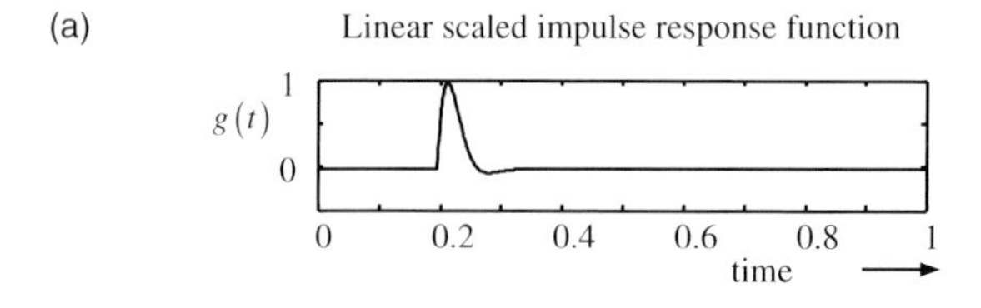

(b)

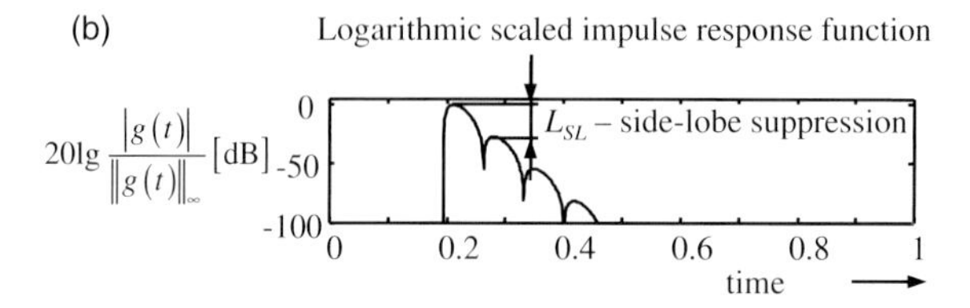

(c)

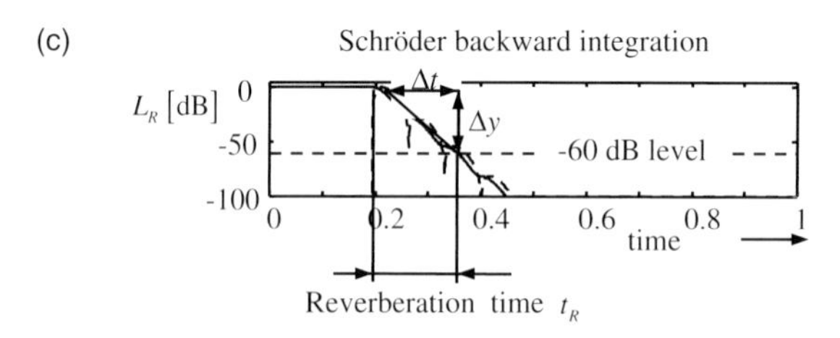

Figure 2.48 Settling of an impulse response function. (a) Linear scaled IRF, (b) logarithmic scaled IRF and (c) average IRF decay.

The side-lobe level or the side-lobe suppression is one possibility to label it (see Figure 2.48b for definition). A further simple way to quantify the settling is to determine an average value referring to the rate of amplitude diminishment. This can be given either by the decay rate or by the reverberation time.[20] For that purpose, Schröder backward integration [22, 23] is applied to gain a smooth descending function which can be interpreted as an average decay rate of energy. Herein backward integration means that the integration is executed inversely in time, which can also be expressed by

$$L_{\mathrm{R}}(t) = 10\log\left(\frac{\lim\limits_{T_{\mathrm{R}}\to\infty}\int\limits_0^{T_{\mathrm{R}}} g^2(\tau)\mathrm{d}\tau - \int\limits_0^{t} g^2(\tau)\mathrm{d}\tau}{\|g(\tau)\|_\infty^2}\right) \qquad t \le T_{\mathrm{R}} \tag{2.140}$$

The reverberation time t_{R} results from the $-60\,\mathrm{dB}$ threshold crossing of $L_{\mathrm{R}}(t)$ (see Figure 2.48c) and the decay rate DR (typically expressed in dB/ns) is given by the average descent:

$$\mathrm{DR} = \frac{\Delta y}{\Delta t} \quad [\mathrm{dB/ns}] \tag{2.141}$$

20) The reverberation time originates from acoustics to characterize the echo behaviour of rooms.

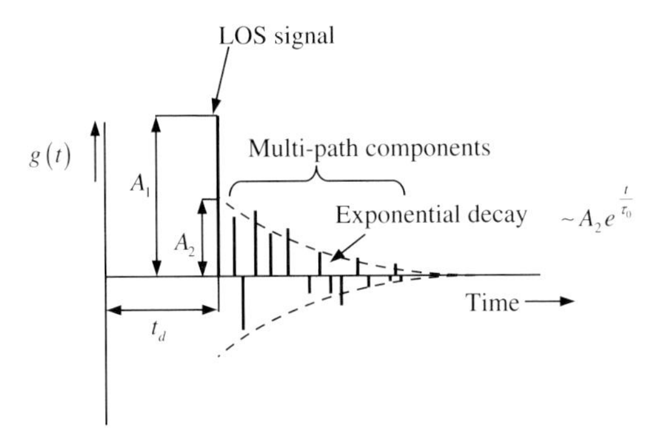

Figure 2.49 Schematic IRF of a multi-path radar and transmission channel. Compare Figure 2.46 for an actual measurement.

The practical implementation of the Schröder backward integration needs some alertness since measurement noise can destroy the whole approach. The left integral in (2.140) tends to infinity in the presence of noise and by extending the recording time T_R. Therefore, the integrations should be applied only over signal parts which exceed the noise level.

The settling of the pulse response as depicted in Figure 2.48 is not only a characteristic feature of the scenario under test. It has also to be respected if IRF measurements are performed with periodic signals. In order to avoid time aliasing, the IRF should be settled down to noise level before a new period of the sounding signal starts.

Figure 2.49 demonstrates a second example which refers to the typical behaviour of a radar or transmission channel in an environment rich of multi-path propagation (e.g. indoor conditions). The time interval t_d relates to the propagation time between transmit and receive antenna. The exponential decay has the same meaning as above in Figure 2.48. If there are no dominant scatterer, it can be determined by the same way, for example via Schröder backward integration. Decay rate DR and time constant τ_0 of the exponential decay relate by

$$\mathrm{DR}\,[\mathrm{dB/ns}] = \frac{20 \log \mathrm{e}}{\tau_0\,[\mathrm{ns}]} \tag{2.142}$$

Furthermore, we can introduce a ratio assigning the strength of the line-of-sight (LOS) path in relation to the multi-path components:

$$L = 20 \log \frac{A_1}{A_2} \tag{2.143}$$

The parameters of a waveform and an IRF considered up to here should only be understood as examples to illustrate the methods. Depending on the specific task, different and new parameters can be introduced. All the parameters which were defined above are related to more or less formally motivated shape properties of the waveforms from which one tries to gain information about the actual behaviour of the system under test. In Section 2.4.4, a more sophisticated method of parameter extraction will be introduced. There, the resulting parameters will be better linked to the internal processes going on within the systems of interest.

So far, our comprehension of the impulse response for system characterization was simply based on a black box which was only to be supposed of linear and time invariant behaviour. No further knowledge was required, hence the name black box. But this is in contradiction with the extraction of information from the measured response function $g(t)$. In order to interpret the IRF shape correctly, some prior knowledge about the measurement scenario and understanding of physical processes is inevitably needed. If these assumptions are fulfilled, appropriate shape parameters of $g(t)$ can be linked with the specific behaviour of the actual object. While the definition and extraction of shape parameters is a generic task, the interpretation of their meaning and their linkage to the object behaviour needs some restrictions since a certain event or distortion in the impulse responses may be caused by several effects. Fortunately quite often (but not always), such a priori knowledge is immediately given by the experimental arrangement.

In summary, the system behaviour cannot be correctly assigned without a minimum of knowledge of the experimental conditions. Referring to these aspects, we will classify our measurement tasks into three groups:

Recording selected state parameters of an object or scenario: Examples are the measurement of a distance or a thickness, the determination of moisture content, recording a movement trajectory and so on, that is to capture continuous quantities. In such a case, one or several shape parameters of the impulse response function (or related expressions as we will see later) are mapped by an appropriate model to the wanted quantity. Many times, these models are based on a physical understanding of the processes taking place within the measurement scenario. The selected shape parameters should be as reliable as possible in order to have a precise measurement of the required quantity.

Detection of an object or a specific state of a scenario: If the sudden emergence of an object or the achievement of a specified state of the test scenario are reflected in specific shape parameters of $g(t)$, then the object is detected and the state achieved if the corresponding shape parameters cross a threshold. The challenge of the task is to identify reasonable parameters (for object detection via radar measurement, it is, for example, the amplitude of the return pulse) and to select a threshold which leads to a high detection rate at low false decisions.

Recognition of an object or a specific state: The interaction of wideband sounding signals with an object or scenario may be quite multifaceted. Therefore, the object/scenario may influence several shape parameters of $g(t)$ in a very specific manner. The shape parameters which are the most sensible within a certain scenario, we will call feature values. If every object or state of the scenario influences the selected features in a different but approximately know way, it is possible to assign a measurement result to a specific object (or state of the system under test). Figure 2.50 illustrates the approach by a simple example.

Supposing two features A and B (e.g. amplitude and effective pulse width) of an impulse response function were extracted from the measurements and we have to distinguish between three kinds of objects (object 1, 2 and 3) which must be detected and classified by a radar measurement. That is, one has to detect whether

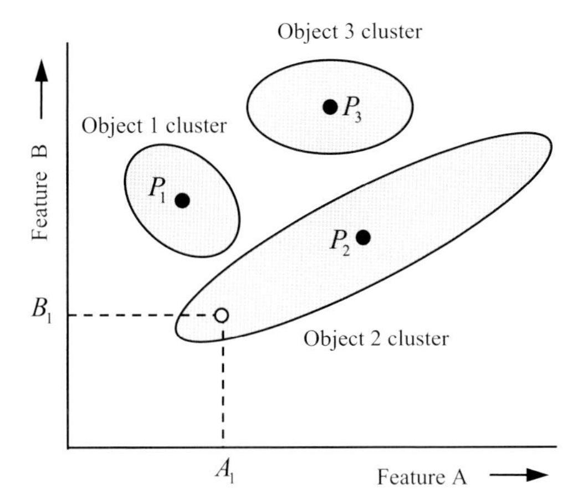

Figure 2.50 Illustration of object recognition by classification.

there is a target and if so of which type it is – 1, 2 or 3. For that purpose it is required to know a priori the areas of the feature space around which the features of the three objects are clustered. If the measurement leads to the feature values A_1 and B_1, the detected object is of type 2 according to Figure 2.50. The assignment of a measurement to a certain object may be based on the shortest Mahalanobis distance. It represents a weighted distance measure between the point $[A_1, B_1]$ and the centres of gravity P_1, P_2, P_3 of the different clusters. The weighting is performed in dependence from the variance of the related cluster spaces [24].

The boundaries of the feature clusters for the different objects must either be determined from physical models or they has to be gained from a learning procedure based on known objects by applying neural networks. The features selected from the IRF should provide reasonable discrimination performance. That is, the cluster spaces of the various objects should not overlap. Since we can assign much more features to an UWB signal than to a narrowband signal, UWB sounding will have better recognition capabilities than narrowband sensors. More on the topics of classification and neural networks can be found, for example, in Refs [25–30].

2.4.3 The Frequency Response Function or the Frequency Domain Greens Function

The description of the dynamic behaviour of linear systems in the frequency domain is quite more common for electrical engineers than the description in the time domain as it was introduced in Section 2.4.2. Certainly, this has historical roots but it is also caused by the fact that the calculations in the frequency domain are rather simple compared to that in the time domain. Nevertheless, both methods to describe the system behaviour are useful and every approach has its own advantages and disadvantages. Some phenomena can be better understood in the time domain and others are easier to interpret in the frequency domain.

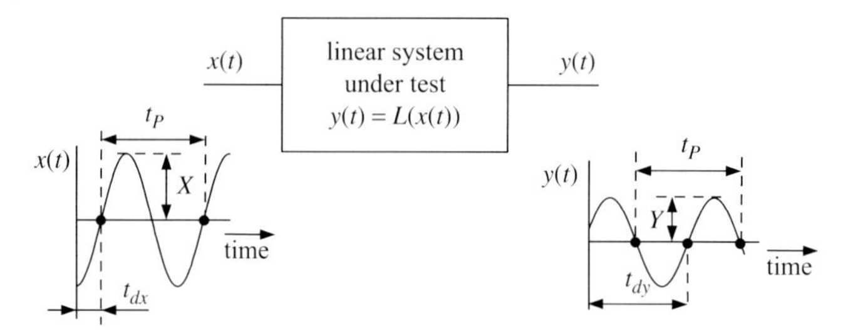

Figure 2.51 Basic approach to determine the frequency response function.

Compared to the term impulse response function (implying system stimulation by impulses), the term frequency response function (FRF) is somewhat misleading since it means the response of a linear system stimulated by pure sine waves. However, the measurement method requires the stimulation with sine waves of largely different frequencies, therefore its name frequency response measurement.

The principle is depicted in Figure 2.51 for a sine wave of frequency $f_0 = 1/t_P$. Obviously, the measurement device has only to capture the amplitude ratio $G(f_0)$ between the output and input signal

$$G(f_0) = \frac{Y}{X} \quad f_l \leq f_0 \leq f_u \tag{2.144}$$

and the mutual phase difference $\varphi(f_0)$

$$\varphi(f_0) = 2\pi \frac{t_{dy} - t_{dx}}{t_P} = 2\pi f_0 \, \Delta t_d \tag{2.145}$$

for every frequency f_0 within a given interval $[f_l, f_u]$. It is not necessary to record the whole time evolution of the output signal. In contrast to an arbitrary-shaped time signal, the shape of a sine wave will not be affected by a linear system as long as we consider the steady state. However, the measurements have to be repeated for all frequencies of interest.

The data set of the two quantities $G(f)$ and $\varphi(f)$ is sufficient to characterize the system response. $G(f)$ is the amplitude (frequency) response. It is often given in logarithmic scale by $G(f)\,[\mathrm{dB}] = 20 \log G(f)$. $\varphi(f)$ is the phase (frequency) response. Finally, both functions are joined to the complex-valued frequency response function:

$$\underline{G}(f) = G(f)\, e^{j\varphi(f)} = \Re\{\underline{G}(f)\} + j\Im\{\underline{G}(f)\} = I(f) + jQ(f) \tag{2.146}$$

in which $\Re\{\underline{G}(f)\} = I(f)$ is the real part or in-phase component of the FRF and $\Im\{\underline{G}(f)\} = Q(f)$ is the imaginary part or the quadrature component. Similar to the impulse response function, the frequency response function represents the Greens function of the system under test for the frequency domain.

Frequency response $\underline{G}(f)$ and impulse response $g(t)$ are mutual related via the Fourier transform. Performing the Fourier transform of both sides of the convolution (2.129) leads to

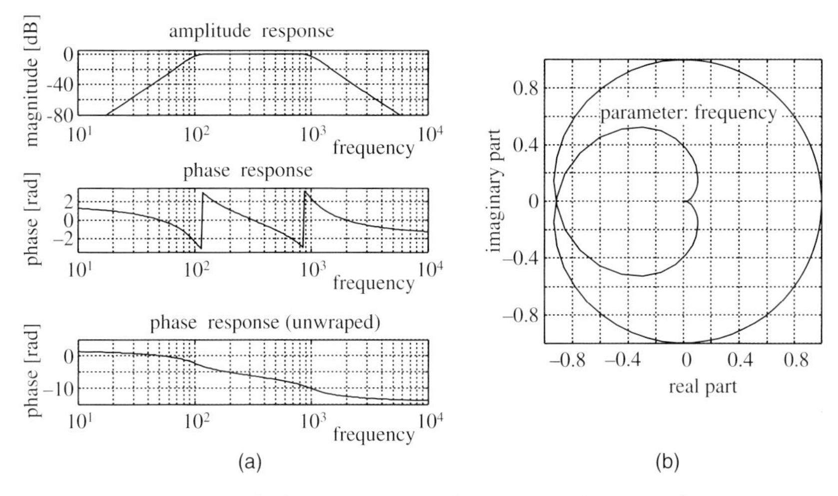

Figure 2.52 Typical plots of a frequency response function. Only the positive frequencies are shown: (a) Bode plot and (b) Nyquist plot.

$$\begin{aligned}\int y(t)\,\mathrm{e}^{-j2\pi f t}\,\mathrm{d}t &= \int\left(\int g(\tau)\,x(t-\tau)\mathrm{d}\tau\right)\mathrm{e}^{-j2\pi f t}\,\mathrm{d}t\\ &= \int g(\tau)\left(\int x(t-\tau)\,\mathrm{e}^{-j2\pi f t}\,\mathrm{d}t\right)\mathrm{d}\tau\\ \text{with}\quad \xi &= t-\tau\\ &= \int g(\tau)\mathrm{e}^{-j2\pi f \tau}\,\mathrm{d}\tau \int x(\xi)\mathrm{e}^{-j2\pi f \xi}\,\mathrm{d}\xi\end{aligned}$$

which finally gives (see Annex B.1, Table B.2)

$$\underline{Y}(f) = \underline{G}(f)\,\underline{X}(f) \tag{2.147}$$

with

$$\underline{G}(f) = \int g(t)\mathrm{e}^{-j2\pi f t}\mathrm{d}t \tag{2.148}$$

Figure 2.52 shows the two most common graphic FRF representations. The Bode plot depicts the magnitude and phase as function of the frequency. Magnitude and frequency are logarithmically scaled. The phase is typically bounded to $\pm\pi$ which results in a ambivalent non-continuous function as illustrated in the middle. Unwrapping the phase function overcomes this limit and leads to a continuous function that is often easier to handle (see, e.g. (2.155)). The Nyquist plot depicts the FRF in the complex plan. Further methods are to plot the real and the imaginary part of $\underline{G}(f)$ versus the frequency and the Nichols plot. It plots the magnitude in dB-scale against the phase in rad or deg.

2.4.3.1 **Properties of the Frequency Response Function and the Utility of the Frequency Domain**

- Impulse response function $g(t)$ and frequency response function $\underline{G}(f)$ can be mutually converted by the Fourier transform. Therefore, both functions carry the same information about the system under test. There exists no preference on the favoured use of one of them. Only the method on how the information about the system under test is extracted and which kind of information is wanted will privilege either $g(t)$ or $\underline{G}(f)$.

- As seen from above (see (2.147)), the convolution in the time domain is replaced by a simple product of complex-valued quantities in the frequency domain. In order to accelerate the numerical calculation of a convolution, it is advisable to
  1) transform the time functions into the frequency domain,
  2) perform the multiplication and
  3) transform the resulting complex spectrum back into the time domain.

  Due to the numerical efficiency of the fast Fourier transform (FFT), the numerical overall effort is much less than the straight way to solve the convolution integral.
- Since the impulse response of a physical system is a real-valued function, the frequency response function is conjugate complex: $\underline{G}(f) = \underline{G}^*(-f)$.
- Since the impulse response of a physical system is causal, the real and imaginary parts of the frequency response function are not independent of each other. A causal IRF may be decomposed in an even and odd part:

$$g(t) = g_{\text{even}}(t) + g_{\text{odd}}(t)$$

where $g_{\text{even}}(t) = \operatorname{sgn}(t) g_{\text{odd}}(t)$.[21] Consequently, the FRF is composed from two parts corresponding to

$$\underline{G}(f) = G_{\text{even}}(f) + jG_{\text{odd}}(f)$$

in which $G_{\text{even}}(f) = \Re\{\underline{G}(f)\}$ is a symmetric real-valued spectral function and $G_{\text{odd}}(f) = \Im\{\underline{G}(f)\}$ is a pure imaginary anti-symmetric spectral function (see Annex B.1, Table B.2). Both are mutually dependent, since

$$\begin{aligned} g_{\text{even}}(t) = \operatorname{sgn}(t)\, g_{\text{odd}}(t) \quad \xrightarrow{\text{FT}} \quad G_{\text{even}}(f) &= \frac{1}{j\pi f} * G_{\text{odd}}(f) \\ &= \frac{1}{j\pi} \text{PV} \int_{-\infty}^{\infty} \frac{G_{\text{odd}}(\xi)}{f - \xi} \, d\xi \end{aligned} \tag{2.149}$$

The integral on the right-hand side is called the Hilbert transform $\text{HT}\{G_{\text{odd}}(f)\}$. Finally, we can summarize for the FRF of all physically real systems

$$\begin{aligned} \Re\{\underline{G}(f)\} &= \frac{1}{\pi} \text{PV} \int_{-\infty}^{\infty} \frac{\Im\{\underline{G}(\xi)\}}{f - \xi} \, d\xi \\ \Im\{\underline{G}(f)\} &= -\frac{1}{\pi} \text{PV} \int_{-\infty}^{\infty} \frac{\Re\{\underline{G}(\xi)\}}{f - \xi} \, d\xi \end{aligned} \tag{2.150}$$

(2.150) is also known under Kramers–Kronig relation. For minimum phase systems, this equation can also be modified for amplitude $|\underline{G}(f)|$ and phase

21) This condition supposes that $g(t)$ is not composed from a Dirac pulse at the origin which holds for physically realisable systems.

values $\varphi = \arg\underline{G}(f)$. That is, the knowledge of either of these four quantities ($\Re\{\underline{G}(f)\}$, $\Im\{\underline{G}(f)\}$, $|\underline{G}(f)|$ or $\arg\underline{G}(f)$) would be sufficient to determine the missing part of the FRF and hence to calculate the complete impulse response. If the FRF is completely determined, Hilbert transform may be used for measurement error corrections ensuring causality. Finally, one should note that the integral does not converge at $f \to \xi$. Therefore, the Hilbert transform is defined via the Cauchy principal value expressed by the notation PV $\int \cdots d\xi$ (refer also to footnote 2). Some more discussion on Hilbert transform and its use for waveform reconstruction is to be found in Refs [31–34].

- From Parseval theorem (see Annex B.1, Table B.2) it follows

$$\int_{-\infty}^{\infty} g^2(t)dt = \int_{-\infty}^{\infty} \underline{G}(f)\,\underline{G}^*(f)df \qquad (2.151)$$

Expanding (2.147) with the complex conjugate of the signal spectra $\underline{X}^*(f)$ or $\underline{Y}^*(f)$ result in

$$\underline{Y}(f)\,\underline{X}^*(f) = \underline{G}(f)\,\underline{X}(f)\,\underline{X}^*(f) \rightarrow \underline{\Psi}_{yx}(f) = \underline{G}(f)\,\Psi_{xx}(f) \qquad (2.152)$$

and

$$\underline{Y}(f)\,\underline{Y}^*(f) = \underline{G}(f)\,\underline{X}(f)\,\underline{Y}^*(f) \rightarrow \Psi_{yy}(f) = \underline{G}(f)\,\underline{\Psi}_{xy}(f) \qquad (2.153)$$

Herein, $\underline{\Psi}_{yx}(f) = \underline{\Psi}^*_{xy}(f)$, $\Psi_{yy}(f)$ and $\Psi_{xx}(f)$ represent the cross- and auto-spectrum of the measurement signals $x(t)$ and $y(t)$ (see Section 2.2.5 (2.70)). The technical implementation of (2.152) (and (2.153)) will be discussed in Section 3.4 for sine wave excitation and Section 3.6 for random noise signals. The time domain equivalence of these equations are (2.136) and (2.137).

- Systems with several input- and output-channels can be considered by a comparable approach as introduced for the MiMo–IRF ((2.138) and (2.139), Figure 2.47):

$$\begin{bmatrix} \underline{Y}_1(f) \\ \underline{Y}_2(f) \\ \vdots \\ \underline{Y}_L(f) \end{bmatrix} = \begin{bmatrix} \underline{G}_{11}(f) & \underline{G}_{12}(f) & \cdots & \underline{G}_{1K}(f) \\ \underline{G}_{21}(f) & \underline{G}_{22}(f) & \cdots & \underline{G}_{2K}(f) \\ \vdots & \vdots & \ddots & \vdots \\ \underline{G}_{L1}(f) & \underline{G}_{L2}(f) & \cdots & \underline{G}_{LK}(f) \end{bmatrix} \begin{bmatrix} \underline{X}_1(f) \\ \underline{X}_2(f) \\ \vdots \\ \underline{X}_K(f) \end{bmatrix} \qquad (2.154)$$

$$\mathbf{Y}(f) = \mathbf{G}(f)\,\mathbf{X}(f)$$

2.4.3.2 Parameters of the Frequency Response Function

Like the specifics of the IRF of a system, the form of its FRF is a consequence of internal dynamic processes which can be used to quantify and qualify the system under test. That is, in order to be able to assign specific information to the investigated systems, appropriate characteristic values have to be estimated from the measurements. The philosophy to apply them to the FRF is the same as for quantifying the spectral shape of signals (see Section 2.2.5). Typical terms in that connection are bandwidth, ripple, side lobe, gain cutoff, pass-band and stop-band attenuation and so on. Depending on the actual measurement problem, these definitions can be arbitrarily extended. Here, we will add exemplarily only one quantity – the group delay – since the basic approach of FRF shape parameter definition is straightforward and does not need further discussions.

The group delay refers to the phase response of the system under test. It is defined by

$$\tau_g(f) = -\frac{1}{2\pi}\frac{\mathrm{d}\varphi(f)}{\mathrm{d}f} \quad \text{with } \varphi(f) = \arg\{\underline{G}(f)\} \tag{2.155}$$

It is a measure of the speed of energy transport through the system under test. The physical background is to register the time delay which a modulated narrow-band pulse (burst of long duration at carrier frequency f) is subjected when it passes the test object. Supposing a system of low or constant attenuation, a frequency-independent group delay $\tau_g(f) = \text{const.}$ will not affect the signal shape of a wideband pulse. If, however, the group delay depends on the frequency, the shape of the passing waveform will be destroyed.

2.4.4 Parametric System Descriptions

As mentioned several times above, the parameters gained from the characteristic functions IRF or FRF of a system refer to more or less geometrical motivated quantities which describe their shape. Now, we will deal with a different approach giving a better view inside the dynamic behaviour of the system of interest. This approach is often summarized by the term parametric system description. Herewith it is intended to express that the considered parameters are directly linked to the system dynamic.

2.4.4.1 Differential Equation

As before, we suppose linear time invariant systems but we further assume that they consist of lumped elements. It is well known that the dynamics of such systems can be modelled by ordinary differential equations with constant coefficients. Figure 2.53 gives a simple example of a linear electrical circuit with lumped elements.

For an arbitrary system of this type, we can generalize the differential equation to the form

$$\sum_{n=1}^{N} a_{N-n}\frac{\mathrm{d}^n y(t)}{\mathrm{d}t^n} = \sum_{m=0}^{M} b_{M-m}\frac{\mathrm{d}^m x(t)}{\mathrm{d}t^m}; \quad a_0 = 1 \tag{2.156}$$

This equation is often normalized so that $a_0 = 1$. The coefficient a_n and b_m represent the parameters of the equation which are finally responsible for the specific dynamic behaviour of the considered system. M and N determine to the order of the differential equation. They relate to the number of elements which are able to

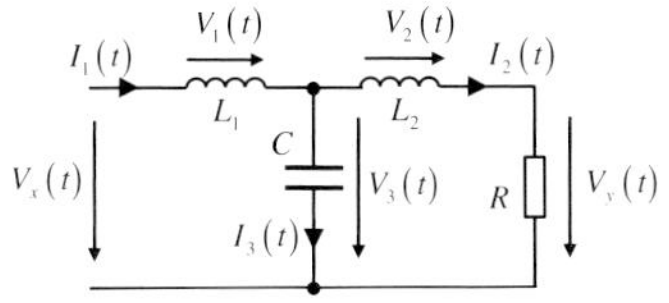

$$\left.\begin{array}{l} V_x = V_1 + V_3 \\ V_3 = V_2 + V_y \\ I_1 = I_2 + I_3 \\ V_1 = L_1 \dfrac{dI_1}{dt} \\ V_2 = L_2 \dfrac{dI_2}{dt} \\ I_3 = C \dfrac{dV_3}{dt} \\ V_y = R\, I_2 \end{array}\right\} \Rightarrow \frac{d^3V_y}{dt^3} + \frac{R}{L_2}\frac{d^2V_y}{dt^2} + \left(\frac{1}{L_1C} + \frac{1}{L_2C}\right)\frac{dV_y}{dt} + \frac{R}{L_2L_1C}V_y = \frac{R}{L_2L_1C}V_x$$

Figure 2.53 Example of simple circuit with lumped elements and the corresponding differential equation describing the dynamic behaviour of the voltage transfer from V_x to V_y.

store energy (L and C). In other words, every system is specified by a set of parameters (coefficients) a_n and b_m. In order to gain information about an unknown object, the extraction of these parameters is hence an alternative approach compared to the shape parameters of $g(t)$ or $\underline{G}(f)$ as discussed in the previous Section 2.4.2 or 2.4.3.

The solution of (2.156) is usually split in two parts:

$$y(t) = y_h(t) + y_{in}(t) \tag{2.157}$$

Here, $y_h(t)$ is the solution of the homogeneous equation $\sum_{n=0}^{N} a_{N-n}((\mathrm{d}^n y(t))/\mathrm{d}t^n) = 0$ and $y_{\mathrm{in}}(t)$ is a particular solution of the whole (inhomogeneous) equation. The homogeneous solution $y_{\mathrm{h}}(t)$ leads to the eigenmodes of the system, that is it represents the system behaviour under free (externally unforced) conditions. They are characteristic for an individual system. Hence, the determination of $y_{\mathrm{h}}(t)$ is of special interest in system recognition. Interesting to note that the eigenmodes only depend on the coefficients a_n. The particular solution $y_{\mathrm{in}}(t)$ depends on both, the system specifics (expressed by a_n, b_m) and the stimulus $x(t)$ which excites/forces the system under test. Correspondingly, both components of $y(t)$ are termed forced $y_{\mathrm{in}}(t)$ and unforced solution $y_{\mathrm{h}}(t)$ as well.

The conception leading to the description as demonstrated in (2.156) is denoted as a grey box model since these parameters are better linked to the internal behaviour of a system as this is the case for a black box model. The term grey box shall express that such a model permits a better (but not a really clear) view 'inside' the unknown system. Nevertheless, it will not reveal the actual internal structure of the system under test. Actually, it is even a major problem of practical applications to fix the right order $[M, N]$ of the model. A model which reveals the internal structure is termed as white box model. A white model cannot be extracted from measurements if there is no comprehensive prior knowledge.

(2.156) may be graphically illustrated by a Mason- or signal flow graph as depicted in Figure 2.54. Annex B.7 gives a short introduction to the handling of Mason graphs. The figure shows two canonical forms. One is based on differentiators. The other one uses only integrators. The implementation of a system simulation by differentiator

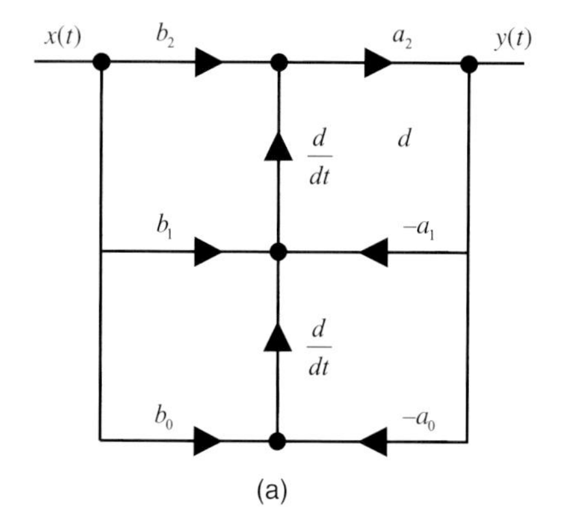

(a)

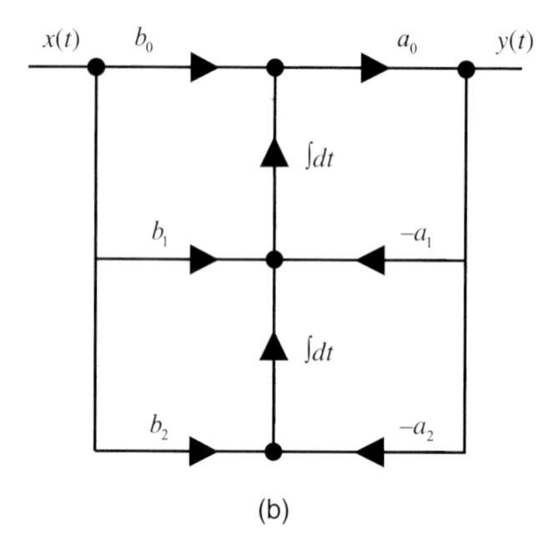

(b)

Figure 2.54 Flow graph representation of (2.156). The graphs refer to an equation of second order. (a) Using differentiators and (b) using integrators.

flow graph (Figure 2.54a) will suffer from noise which always corrupts the signals in practice. 'When such a signal is differentiated, the derivative of the usually rapidly varying noise will 'drown out' the derivative of the signal' ([21], page 35). Therefore, the flow graphs of linear systems are mostly based on integrators. The peculiar numbering of the coefficients in (2.156) was already selected to get a common index numbering for the integrator type flow graph.

2.4.4.2 The Laplace Transform

The handling of derivations or integrations can be avoided by subjecting (2.156) a Fourier transform or a Laplace transform. The Laplace transform may be considered as an extension of the Fourier transform which is preferably used for the transformation of one-sided functions, for example the impulse response function. While the Fourier transform, the function $x(t)$ decomposes in a sum of sine waves, the Laplace transform decomposes it in a sum of exponentially increasing or decreasing (damped) sine waves. Mathematically, this is expressed by exponential functions of complex frequency

$$s = \sigma + j2\pi f \tag{2.158}$$

The Laplace transform and its inverse are defined by

$$\underline{X}(s) = \mathrm{LT}\{x(t)\} = \int_0^\infty x(t)\mathrm{e}^{-st}\,\mathrm{d}t \tag{2.159}$$

$$x(t) = \mathrm{ILT}\{\underline{x}(s)\} = \frac{1}{j2\pi} \int_{\sigma-j\infty}^{\sigma+j\infty} \underline{X}(s)\mathrm{e}^{st}\,\mathrm{d}s \quad \text{for } \sigma < \sigma_0 \tag{2.160}$$

σ_0 is a real number describing the limit of the region of convergence. In the case where $x(t)$ has properties as a stable impulse response function (see (2.130)), all singularities of $\underline{X}(s)$ are in the left half-plane of s. Then σ can be set to zero and (2.160) becomes identical to the inverse Fourier transform. Some elementary properties of the Laplace transform are summarized in Annex B.1, Table B.3.

2.4.4.3 Transfer Function

Applying the rules of Laplace transform to the differential equation (2.156), we can simplify it to an algebraic equation:

$$\underline{Y}(s) \sum_{n=0}^{N} a_{N-n}\, s^n = \underline{X}(s) \sum_{m=0}^{M} b_{M-m}\, s^m \tag{2.161}$$

which can be modified to

$$\underline{G}(s) = \frac{\underline{Y}(s)}{\underline{X}(s)} = \frac{\sum_{m=0}^{M} b_{M-m} s^m}{\sum_{n=0}^{N} a_{N-n} s^n} \tag{2.162}$$

$\underline{G}(s)$ is a rational function in s representing the transfer function of the considered system. Figure 2.55 depicts an example of a transfer function pictured over the complex plane. Obviously, the emphasized curve at $\sigma = 0$ represents the FRF $\underline{G}(f)$. The zeros of the numerator and denominator polynomials provide singular points in the complex s-plane being decisive for the overall shape of $\underline{G}(s)$. The transfer function is a nice tool to design lumped systems (e.g. filters) of a desired behaviour.

However, the determination of the coefficients a_n, b_m of $\underline{G}(s)$ from measurements is not straightforward. Methods providing such coefficients from measurement data are assigned as parameter estimation. They require some assumptions or prior knowledge about the object and they have to respect the imperfectness of measurement data due to noise. A usual approach is to match the parameters of the system model to the measured function via curve fitting. This may be done either in the frequency domain (using a system model based, for example, on (2.162)) or in the time domain (based on system description (2.156)).

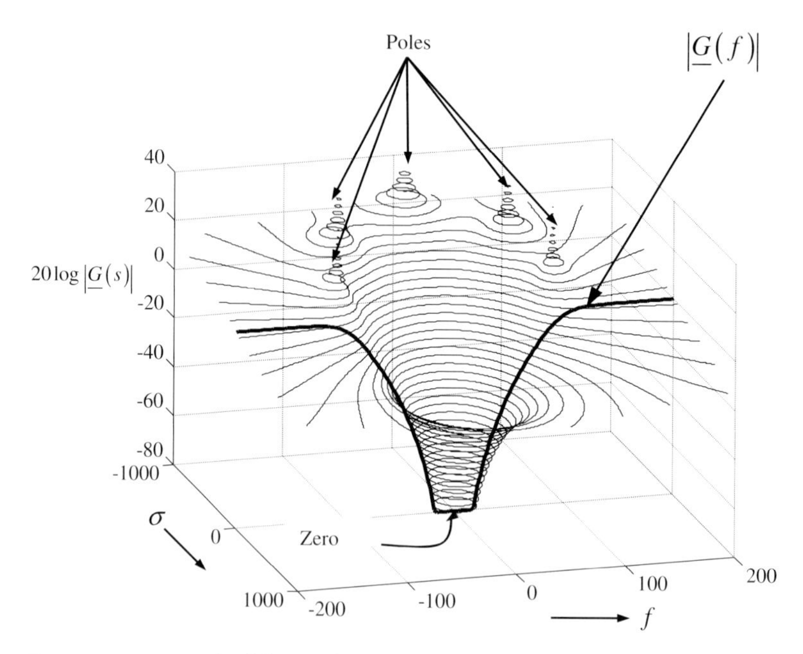

Figure 2.55 Magnitude of the transfer function $|\underline{G}(s)|$ plotted over the complex s-plane. The emphasized line represents the magnitude of the frequency response function $|\underline{G}(f)|$. The 'zero' indicates the location where the numerator polynomial is zero and the 'poles' refer to the locations where the denominator polynomial has its zeros.

An example shall demonstrate the method. Supposing we have measured the FRF $\underline{G}(f)$ of an object and we want to know the coefficients a_n, b_m of a model system which would provide the same FRF $\hat{\underline{G}}(s, a_n, b_m)|_{\sigma=0} = \hat{\underline{G}}(f, a_n, b_m)$. We determine the wanted coefficients by minimizing a cost function $J(\boldsymbol{\theta})$ ($\boldsymbol{\theta} = [a_n, b_m]$ is the parameter vector) in dependence from a_n, b_m.

$$\boldsymbol{\theta} = [a_n, b_m] = \arg\min_{\boldsymbol{\theta}} J(\boldsymbol{\theta}) \Rightarrow \frac{\mathrm{d}J(\boldsymbol{\theta})}{\mathrm{d}\boldsymbol{\theta}} = 0 \tag{2.163}$$

Herein, the cost function is a measure of the mutual deviations between $\underline{G}(f)$ and $\hat{\underline{G}}(f, \boldsymbol{\theta})$. We can consider the following expression as such a deviation:

$$J(\boldsymbol{\theta}) = \int_{-\infty}^{\infty} \left|\underline{G}(f) - \hat{\underline{G}}(f, \boldsymbol{\theta})\right|^2 \mathrm{d}f; \quad \boldsymbol{\theta} = [a_n, b_m] \quad n \leq N, \; m \leq M \tag{2.164}$$

If we have found the minimum of $J(\boldsymbol{\theta})$, we end up with a set of parameters $\boldsymbol{\theta} = [a_n, b_m]$ characterizing the dynamic behaviour of the measurement object. These parameters can be used as a feature set for object classification (compare Figure 2.50).

More on these topics can be found, for example, in Ref. [18] for frequency domain parameter estimation and in Refs [35, 36], for time domain approaches.

Furthermore, MATLAB® and SIMULINK provide a rich offer on processing tools for corresponding problems.

Polynomial and rational functions may be represented by various notations. Hence, (2.162) can be written in different forms:

Zero-pole-gain form:

$$\underline{G}(s) = g\frac{\prod\limits_{m=1}^{M}(s - s_{0m})}{\prod\limits_{n=1}^{N}(s - s_{\text{pn}})} \tag{2.165}$$

Residue form:

$$\underline{G}(s) = \sum_{n=1}^{N}\frac{r_n}{s - s_{\text{pn}}} + \sum_{\nu=1}^{M-N} k_\nu s^\nu \tag{2.166}$$

s_{0m} is the zeros of the numerator polynomial, s_{pn} is the zeros of the denominator polynomial and therefore the poles of the rational function, g is the gain factor, r_n is the residue and k_ν is the coefficient of the remainder polynomial (only if $M > N$). Further possibilities are given by decomposing $\underline{G}(s)$ in second-order sections or by applying a lattice structure. Since numerator and denominator polynomials have only real coefficients, their roots are either real and/or they appear in pairs of complex conjugates.

The zero-pole-gain notation is very illustrative since it gives immediately an impression on the shape of $\underline{G}(s)$ by plotting the poles and zeros into the complex s-plane (compare Figure 2.55). All poles of a stable system must be placed in the left half of the complex plain, that is $\sigma_{\text{pn}} < 0$.

The residue form (2.166) results from a partial fraction decomposition of the rational function (2.162) (note that in the case of repeated poles additional terms of the form $r_{m,k}/(s - s_{\text{pn}})^k$ appear, which we did not respect in (2.166). $k = 1 \cdots K$, K is the multiplicity of the nth pole). The residue form is of major interest for understanding the dynamic behaviour of a system and it may be the base to recognize specific objects. It can be shown,[22] that the poles are either real valued or they appear in pairs of complex conjugate. The inverse Laplace transform applied to the individual terms within the first sum in (2.166) leads to an exponential function with a complex argument (see Annex B.1, Table B.3):

$$\frac{r_n}{s - s_{\text{pn}}} \underset{ILT\{\}}{\overset{LT\{\}}{\rightleftarrows}} r_n\,\mathrm{e}^{s_{\text{pn}}t} = u_{-1}(t) r_n\,\mathrm{e}^{\sigma_{\text{pn}}t}\,\mathrm{e}^{j2\pi f_{\text{pn}}t}$$

$$\frac{r_{m,k}}{(s - s_{\text{pn}})^k} \underset{ILT\{\}}{\overset{LT\{\}}{\rightleftarrows}} u_{-1}(t)\frac{r_{m,k}}{(k-1)!}t^{k-1}\,e^{s_{pn}t}$$

By joining the terms of a conjugate complex pair, we get a real-valued function representing a decaying sine wave of frequency f_{pn} and the decay rate σ_{pn}.

22) This behaviour finally results from the zeros of polynomials with real coefficients which are either real valued or appear in pairs of complex conjugate.

Such an oscillation of a system is referred as a natural mode, as an eigenmode or simply as mode. Following (2.166) the overall dynamic behaviour of a system can be decomposed into individual modes. An oscillation mode is a very specific property of a system or object. Hence, the knowledge of the complete set of all modes or even the knowledge of a sub-set of these modes can lead to a classification or recognition of certain objects. As to be seen from (2.165) and (2.166) the modes are quantified by the zeros of the denominator polynomial of (2.162) which is only linked to the coefficients a_n of the differential equation (2.156). This underlines the earlier mentioned fact that the homogeneous solution $y_h(t)$ is of specific interested.

2.4.4.4 State Space Model

The most powerful description of the dynamic behaviour is given by the state space representation. It may be extended also to non-linear and time variant system behaviour what is however out of scope here. The term state space model shall imply that this system model refers to the energetic states of internal storage elements, that is capacitors or inductors in the case of an electrical circuit with lumped elements. The energy stored in an element is $\mathfrak{E} = \int V(t)\, I(t)\mathrm{d}t$ that is $\mathfrak{E}_C = C \int V(t)\dot{V}(t)\mathrm{d}t = C\, V^2(t)/2$ for a capacitor and $\mathfrak{E}_L = L \int I(t)\dot{I}(t)\mathrm{d}t = LI^2(t)/2$ for an inductor. Thus, the 'natural' variables to describe the energetic states are the current running through an inductor and the voltage across a capacitor. For our circuit cited in Figure 2.53, we will get following matrix expressions:

$$\frac{\mathrm{d}}{\mathrm{d}t}\begin{bmatrix} I_1 \\ I_2 \\ V_3 \end{bmatrix} = \begin{bmatrix} 0 & 0 & -1/L_1 \\ 0 & -R/L_2 & 1/L_2 \\ 1/C & -1/C & 0 \end{bmatrix}\begin{bmatrix} I_1 \\ I_2 \\ V_3 \end{bmatrix} + \begin{bmatrix} 1/L_1 \\ 0 \\ 0 \end{bmatrix} V_x$$
$$V_y = \begin{bmatrix} 0 & R & 0 \end{bmatrix}\begin{bmatrix} I_1 \\ I_2 \\ V_3 \end{bmatrix} \tag{2.167}$$

Herein, the vector $[I_1 \quad I_2 \quad V_3]$ is called the state vector. The Mason graph of the circuit gained from the state space description (2.167) is shown in Figure 2.56.

The reference to the energetic state of an actual circuit element requires the knowledge of the internal structure of the system. Since it is commonly not known

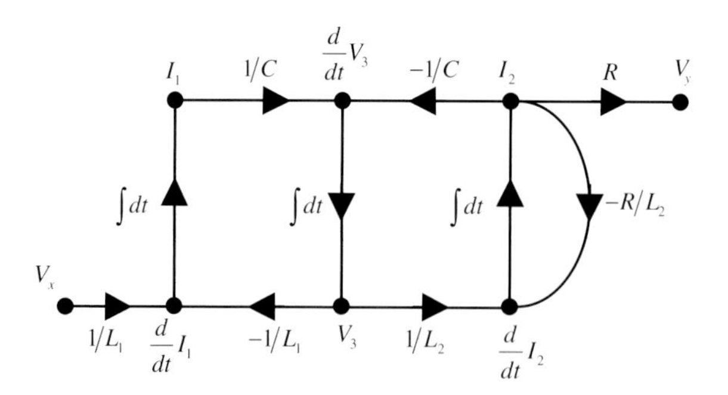

Figure 2.56 Signal flow graph related to (2.167) and the circuit shown in Figure 2.53.

in measurement tasks, the applied states refer usually to an arbitrary linear combination of the physically real internal states. Consequently, every system may be represented by a large number of state space models which, however, all show the same dynamic behaviour between in- and output. One of the possible models structures will actually reflect the internal structure (which one is however difficult to predict). Other (more or less canonical) structures will be of particular interest for system identification or classification.

A state space description can also be formally gained by transforming the Nth-order linear differential equation (2.156) into a set of N-linear differential equations of first order. The following example demonstrates this. For brevity, we suppose $b_M \neq 0, b_{M-1} \cdots b_0 = 0$. First, we substitute the differential quotients by the variables q_n. They are called the states:

$$
\begin{aligned}
y &= q_0 \\
\frac{dy}{dt} &= q_1 = \frac{dq_0}{dt} \\
\frac{d^2y}{dt^2} &= q_2 = \frac{dq_1}{dt} \\
&\vdots \\
\frac{d^{N-1}y}{dt^{N-1}} &= q_{N-1} = \frac{dq_{N-2}}{dt}
\end{aligned}
$$

In connection with (2.156), this leads to the following system of N differential equations of first order

$$
\begin{aligned}
\frac{dq_0}{dt} &= q_1 \\
\frac{dq_1}{dt} &= q_2 \\
&\vdots \\
\frac{dq_{N-1}}{dt} &= -a_N q_0 - a_{N-1} q_1 \cdots - a_1 q_{N-1} + b_M x
\end{aligned}
$$

which we convert into matrix notation in order to gain the state space description:

$$
\frac{d}{dt}\begin{bmatrix} q_0 \\ q_1 \\ \vdots \\ q_{N-1} \end{bmatrix} = \begin{bmatrix} 0 & 1 & \cdots & 0 \\ 0 & 0 & \cdots & 0 \\ \vdots & \vdots & \ddots & 1 \\ -a_N & -a_{N-1} & \cdots & -a_1 \end{bmatrix} \cdot \begin{bmatrix} q_0 \\ q_1 \\ \vdots \\ q_{N-1} \end{bmatrix} + b_M x \quad \text{and} \quad y = q_0
$$

If this approach is generalized for systems with K inputs and L outputs and no restrictions concerning the coefficients b_m are applied, we end up in the usual form of the state space equations:

$$
\text{Time domain:} \quad \begin{aligned} \frac{d}{dt}\mathbf{q}(t) &= \mathbf{A}\,\mathbf{q}(t) + \mathbf{B}\,\mathbf{x}(t) \\ \mathbf{y}(t) &= \mathbf{C}\,\mathbf{q}(t) + \mathbf{D}\,\mathbf{x}(t) \end{aligned} \tag{2.168}
$$

$$\text{Laplace domain:} \quad \begin{aligned} s\underline{\mathbf{Q}}(s) &= \mathbf{A}\,\underline{\mathbf{Q}}(s) + \mathbf{B}\,\underline{\mathbf{X}}(s) \\ \underline{\mathbf{Y}}(s) &= \mathbf{C}\,\underline{\mathbf{Q}}(s) + \mathbf{D}\,\underline{\mathbf{X}}(s) \end{aligned} \tag{2.169}$$

Herein, $\mathbf{x}$ is the input or control vector $[1, K]$, $\mathbf{y}$ is the output or observation vector $[1, L]$, $\mathbf{q}$ is the state vector $[1, N]$, $\mathbf{A}$ is the state or system matrix $[N, N]$, $\mathbf{B}$ is the input or control matrix $[N, K]$; $\mathbf{C}$ is the output or observation matrix $[L, N]$ and $\mathbf{D}$ is the feed through matrix $[L, K]$.

Removing the state vector from (2.169), one can relate MiMo-transfer functions and state space parameters by

$$\underline{\mathbf{G}}(s) = \left(\underline{\mathbf{Y}}(s)\,\underline{\mathbf{X}}^T(s)\right)\left(\underline{\mathbf{X}}(s)\,\underline{\mathbf{X}}^T(s)\right)^{-1} = \mathbf{C}(s\mathbf{I} - \mathbf{A})^{-1}\mathbf{B} + \mathbf{D} \tag{2.170}$$

If we finally apply inverse Laplace transform onto (2.170), we get the MiMo-pulse response functions expressed by state space parameters.[23)]

$$\begin{aligned} \mathbf{g}(t) &= \mathbf{C}\,\mathrm{e}^{\mathbf{A}t}\,\mathbf{B} + \mathbf{D}\,\delta(t) = \mathbf{C}\,\boldsymbol{\Phi}(t)\,\mathbf{B} + \mathbf{D}\,\delta(t) \\ \boldsymbol{\Phi}(t) &= \mathrm{e}^{\mathbf{A}t} = \sum \frac{(\mathbf{A}t)^n}{n!} \end{aligned} \tag{2.171}$$

The matrix exponential $\boldsymbol{\Phi}(t)$ is also called the fundamental matrix. The system matrix $\mathbf{A}$ depends exclusively on the parameters a_n of the differential equation. Therefore, it is linked to the homogeneous solution of the differential equations and describes the unforced (natural) behaviour of the considered system.

According to the flow graph of the differential equation in Figure 2.54, one can find a comparable graphical illustration for the state space model. Figure 2.57 shows a generalized block schematics and a detailed structure of a second-order SiSo system.

As mentioned above, the considered states $\mathbf{q}(t)$ or $\mathbf{Q}(s)$ are arbitrarily selected and they usually reflect a linear combination of the actual physical states of the system. By a linear transformation, they can be converted in a new set of states $\mathbf{q}_T(t)$ or $\mathbf{Q}_T(s)$:

$$\mathbf{q}(t) = \mathbf{T}\,\mathbf{q}_T(t) \quad \text{or} \quad \mathbf{Q}(s) = \mathbf{T}\,\mathbf{Q}_T(s) \tag{2.172}$$

where $\mathbf{T}$ is an arbitrary transformation matrix. For the state space equation (2.169) ((2.168)), this leads to

$$\begin{aligned} s\underline{\mathbf{Q}}_T(s) &= \mathbf{T}^{-1}\mathbf{A}\mathbf{T}\,\underline{\mathbf{Q}}_T(s) + \mathbf{T}^{-1}\mathbf{B}\,\underline{\mathbf{X}}(s) = \mathbf{A}_T\,\underline{\mathbf{Q}}_T(s) + \mathbf{B}_T\,\underline{\mathbf{X}}(s) \\ \underline{\mathbf{Y}}(s) &= \mathbf{C}\mathbf{T}\,\underline{\mathbf{Q}}_T(s) + \mathbf{D}\,\underline{\mathbf{X}}(s) = \mathbf{C}_T\,\underline{\mathbf{Q}}_T(s) + \mathbf{D}\,\underline{\mathbf{X}}(s) \end{aligned} \tag{2.173}$$

23) The conversion is based on the identities $(1-x)^{-1} = \sum x^n;\ |x| < 1$ and $s^{-n} \overset{ILT}{\underset{LT}{\rightleftarrows}} u_{-1}(t)(t^{n-1}/(n-1))$ (see Annex B.1, Table B.3) which has to be extended to matrix notation.

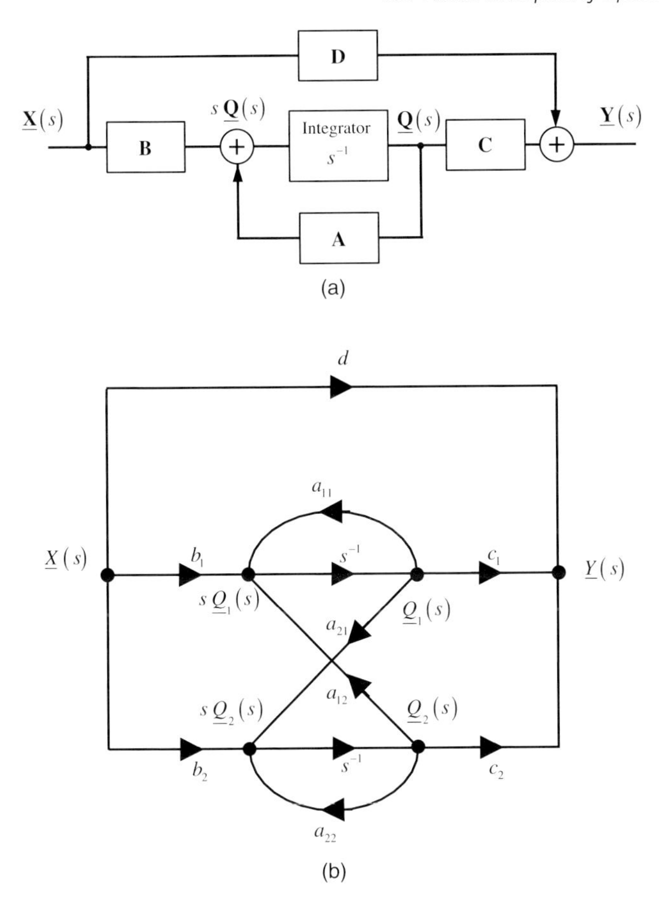

Figure 2.57 Generalized flow graph (a) of a state space model in the s-domain and detailed structure for a SiSo system (b).

with

$$\mathbf{A}_T = \mathbf{T}^{-1}\mathbf{A}\mathbf{T}$$
$$\mathbf{B}_T = \mathbf{T}^{-1}\mathbf{B}$$
$$\mathbf{C}_T = \mathbf{C}\mathbf{T}$$

It is of special interest to find a transformation matrix $\mathbf{T}$ which results in a diagonal system matrix $\mathbf{A}_T$. The mathematical procedure to gain this matrix is based on the eigenvalue decomposition of the initial matrix $\mathbf{A}$. The result of such decomposition is depicted in Figure 2.58 for the example of the second-order system above. Obviously, the different states of the system are decoupled now and we have the same representation as described by the residue form (see (2.166)). That is, besides the feed through branch, the individual braches symbolize the different eigenmodes (natural modes) of the system.

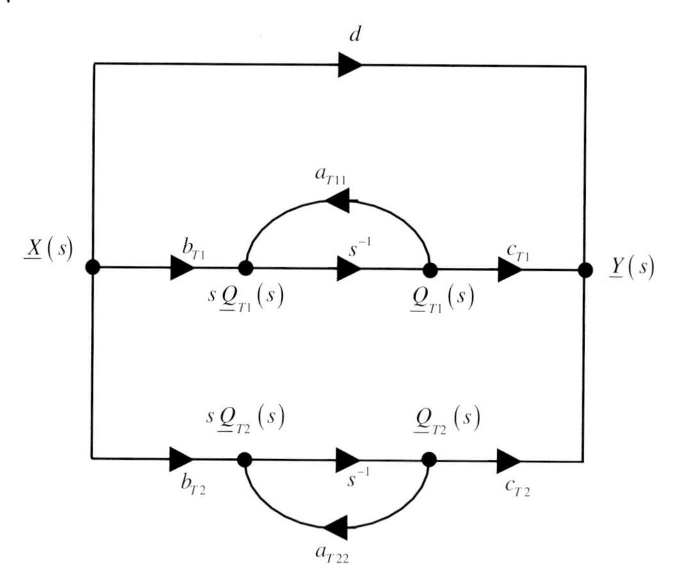

Figure 2.58 Modal representation of a second-order SiSo system.

If we extend the modal representation to MiMo systems (e.g. three inputs and two outputs), we will come for time domain notation to

$$\frac{\mathrm{d}}{\mathrm{d}t}\begin{bmatrix} q_{T1}(t) \\ q_{T2}(t) \\ q_{T3}(t) \\ q_{T4}(t) \end{bmatrix} = \begin{bmatrix} a_{T11} & 0 & 0 & 0 \\ 0 & a_{T22} & 0 & 0 \\ 0 & 0 & a_{T33} & 0 \\ 0 & 0 & 0 & a_{T44} \end{bmatrix} \begin{bmatrix} q_{T1}(t) \\ q_{T2}(t) \\ q_{T3}(t) \\ q_{T4}(t) \end{bmatrix} + \begin{bmatrix} b_{T11} & b_{T12} & b_{T13} \\ b_{T21} & b_{T22} & b_{T23} \\ b_{T31} & b_{T32} & b_{T33} \\ b_{T41} & b_{T42} & b_{T43} \end{bmatrix} \begin{bmatrix} x_1(t) \\ x_2(t) \\ x_3(t) \end{bmatrix}$$

$$\begin{bmatrix} y_1(t) \\ y_2(t) \end{bmatrix} = \begin{bmatrix} c_{T11} & c_{T12} & c_{T13} & c_{T14} \\ c_{T21} & c_{T22} & c_{T23} & c_{T24} \end{bmatrix} \begin{bmatrix} q_{T1}(t) \\ q_{T2}(t) \\ q_{T3}(t) \\ q_{T4}(t) \end{bmatrix} + \begin{bmatrix} d_{11} & d_{12} & d_{13} \\ d_{21} & d_{22} & d_{23} \end{bmatrix} \begin{bmatrix} x_1(t) \\ x_2(t) \\ x_3(t) \end{bmatrix}$$

An equivalent description of the system is given by a set of IRFs as discussed in Section 2.4.2:

$$\begin{bmatrix} y_1(t) \\ y_2(t) \end{bmatrix} = \begin{bmatrix} g_{11}(t) & g_{12}(t) & g_{13}(t) \\ g_{21}(t) & g_{22}(t) & g_{23}(t) \end{bmatrix} * \begin{bmatrix} x_1(t) \\ x_2(t) \\ x_3(t) \end{bmatrix}$$

Both notations describe the dynamic behaviour of the system of interest. How this result can be interpreted from the viewpoint of a measurement which typically has the goal to quantify, to classify or recognize a specific object?

- The set of IRFs results more or less directly from the measurement while the state space parameters need a procedure of parameter estimation in order to gain them from the measured data.
- It may be hardly visible from the direct IRF measurements that the individual IRFs of an MiMo object are mutually dependent. But the state space representation immediately shows such mutual linkage. As obvious from (2.171), the system matrix **A** is part of every pulse response, that is the individual impulse responses are only weighted sums of the different eigenmodes of the system.
- In order to classify objects or systems, we need a set of parameters which is unique for an individual object. In the above-mentioned example, this is the set of parameters a_{nn} describing the eigenmodes of that system. The values of a_{nn} are identical to the poles s_{pn} of the transfer function (2.165) or (2.166). Obviously, only the indicated eigenmodes can be exited independently from which input the system is stimulated or from which output it is observed. Referring to the example above, the different modes are found by decomposing the measured impulse response functions $g_{lk}(t)$ into two[24] damped sine waves where every of the six impulse response functions constitutes from the same modes. However, every mode contributes differently to the individual impulse response function.
- The eigenmodes are represented by the dynamic behaviour of the four states $q_1(t)\ldots q_4(t)$. As to be seen from the above state space representation, every of the three input signals $x_1(t)\ldots x_3(t)$ is linked with all four states via the parameters b_{nk}. These parameters express the capability of input $x_k(t)$ to stimulate the state (mode) $q_n(t)$. If the value of b_{nk} is zero, the kth-input is not able to stimulate the nth-mode.
- Corresponding holds for the output. Every output is more or less linked to the different modal states via the parameters c_{ln}. If, however, the value of c_{ln} is zero, the nth-mode is not part of the IRF measured at the lth-output, that is it cannot be observed there.
- In the case, we measure a MiMo system and we can find groups of input and output channels which are not mutually linked via identical modes, one can expect that the MiMo system is composted from separated parts which do not interact with each other.

In recapitulation, the dynamic behaviour of a linear system or object built from lumped elements is determined by a finit set of eigenmodes which can be used to identify the system or object under test. If the system under test disposes of several in- and output channels, the same set of eigenmodes holds for every input–output combination. However, the available modes may be more or less developed in the different impulse response functions. It may even happen that an individual mode cannot (or only weakly) be excited by a certain input and it cannot be observed by a particular output.

24) Note that the coefficients a_{Tnn} are forming pairs of complex conjugate (see previous paragraph 'Transfer function'). Hence, four coefficients may relate to two damped sine waves in maximum.

2.4.5
Time Discrete Signal and Systems

Most of ultra-wideband sensors involve digital signal processing. The use of numerical approaches for signal manipulations and information extraction requires the reduction of the continuous signals to time discrete signals. This operation is called sampling. Figure 2.59 depicts the approach. It can be modelled by the multiplication of the time continuous signal $x(t)$ with a Dirac comb $\Delta_{\Delta t_s}(t) = \sum_{n=-\infty}^{\infty} \delta(t - n\,\Delta t_s)$:

$$\mathbf{x}[n] = x(t)\,\Delta_{\Delta t_s}(t) = x(t) \sum_{n=-\infty}^{\infty} \delta(t - n\,\Delta t_s) = \sum_{n=-\infty}^{\infty} x(n\,\Delta t_s) \tag{2.174}$$

The Dirac comb has a comb spectrum with line spacing $f_s = 1/\Delta t_s$. The frequency f_s is called the sampling rate. Since a multiplication in the time domain results in a convolution in the frequency domain, the resulting spectrum of the sampled data is a periodic replica of the original spectrum with the period f_s. Obviously, the sub-spectra do not overlap as long as the two-sided bandwidth of the analogue signal is smaller than the sampling rate:

$$\underset{..}{B} = 2\underset{.}{B} \le f_s \tag{2.175}$$

Under this assumption, the sampled data $\mathbf{x}[n]$ can be re-converted in an analogue signal by low-pass filtering (reconstruction filter) without any losses of information. The relation (2.175) is also known under the term sampling theorem, Nyquist

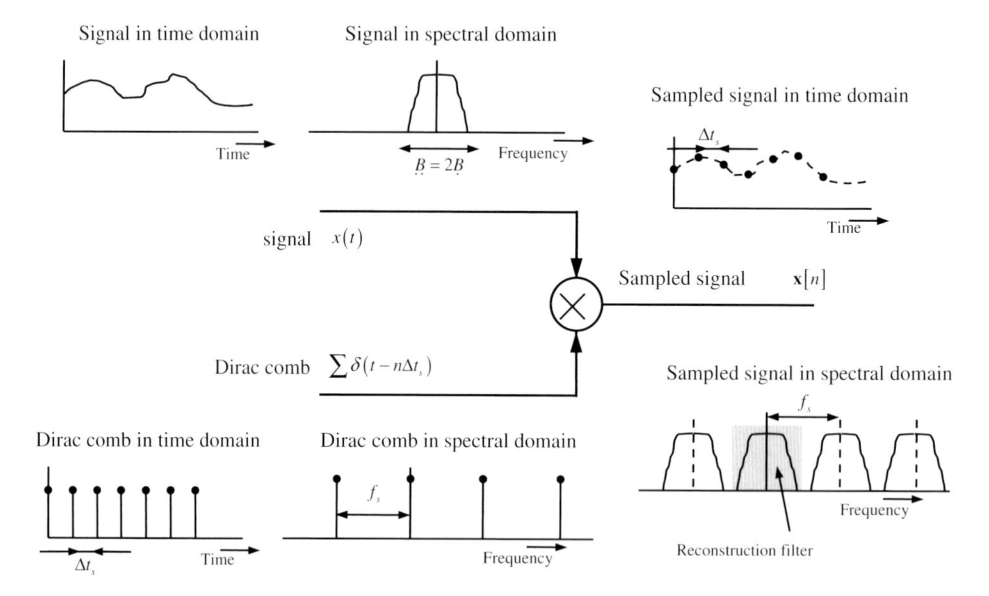

Figure 2.59 Sampling of a continuous waveform by multiplication with a Dirac comb. All signals are drawn in time and spectral domain.

theorem, Shannon theorem and so on. The lowest possible sampling rate f_s for a given signal we will call Nyquist rate.[25)]

If the sampling theorem is not correctly met, the partial-spectra will overlap causing frequency aliasing since the partial-spectra cannot be separated correctly anymore. Since in measurement systems the bandwidth of the captured signals is often not known a priory, one has to select either a sufficient high sampling rate or one has to force down spectral components violating (2.175) by anti-aliasing filters.

In the case of a measurement, the number N of sampled data is finit and the signal is usually represented by vector notation:

$$\mathbf{x}[n] = \begin{bmatrix} x[1] & x[2] & x[3] & \cdots & x[N] \end{bmatrix}^T \tag{2.176}$$

If we suppose a periodic signal (period t_P), from which one period is sampled (hence $N = t_P/\Delta t_s$), the integral signal transformations (as Fourier transformation, convolution, correlation etc.) can be expressed by matrix calculations. Attention has to be paid, if the periodic signal acts as stimulus of a test object. In that case, the period length has to exceed the settling time of the object in order to avoid overlapping of adjacent periods of the system reaction. If this is not respected time aliasing errors appear.

Some specific topics with respect to sampled data are shortly summarized.

2.4.5.1 **Discrete Fourier Transform**

Replacing $\mathrm{d}t \Rightarrow \Delta t_s$ and $t_P = 1/f_0 = N \cdot \Delta t_s = N/f_s$, we can transfer the integral form (2.60) to a discrete version of the Fourier transform:

$$\begin{aligned} \underline{X}[m] &= \frac{1}{N}\sum_{n=0}^{N-1} x[n] e^{-j\frac{2\pi}{N}mn}, \quad m = 0 \cdots N-1 \\ &= \frac{1}{N}\sum_{n=0}^{N-1} x[n]\underline{F}^{mn}, \quad \underline{F} = \mathrm{e}^{-j\frac{2\pi}{N}} \end{aligned} \tag{2.177}$$

This is the usual notation for the discrete Fourier transform. The index m relates to the frequency of the corresponding spectral line $f = mf_0$, where for $m \geq N/2$ the negative part of the spectrum is assigned. The correct mapping between m and the positive and negative frequencies is given by

$$f = f_0\left(m - N\,\mathrm{floor}\left(\frac{2m}{N}\right)\right); \quad m \in [0, N-1] \tag{2.178}$$

where floor(x) rounds x to the nearest integer towards $-\infty$.

25) Do not confuse Nyquist rate with Nyquist frequency $f_s/2$.

Representing the sum in (2.177) by a matrix product, the discrete Fourier transformation is written as

$$\begin{bmatrix} \underline{X}[0] \\ \underline{X}[1] \\ \underline{X}[2] \\ \vdots \\ \underline{X}[N-1] \end{bmatrix} = \frac{1}{N} \begin{bmatrix} \underline{F}^{0\cdot 0} & \underline{F}^{0\cdot 1} & \underline{F}^{0\cdot 2} & \cdots & \underline{F}^{0\cdot (N-1))} \\ \underline{F}^{1\cdot 0} & \underline{F}^{1\cdot 1} & \underline{F}^{1\cdot 2} & \cdots & \underline{F}^{1\cdot (N-1))} \\ \underline{F}^{2\cdot 0} & \underline{F}^{2\cdot 1} & \underline{F}^{2\cdot 2} & \cdots & \underline{F}^{2\cdot (N-1))} \\ \vdots & \vdots & \vdots & \ddots & \vdots \\ \underline{F}^{(N-1)\cdot 0} & \underline{F}^{(N-1)\cdot 1} & \underline{F}^{(N-1)\cdot 2} & \cdots & \underline{F}^{(N-1)\cdot (N-1)} \end{bmatrix} \cdot \begin{bmatrix} x[0] \\ x[1] \\ x[2] \\ \vdots \\ x[N-1] \end{bmatrix}$$

$$\underline{\mathbf{X}} = \frac{1}{N} \underline{\mathbf{F}}\, \mathbf{x} \tag{2.179}$$

and the inverse transformation is

$$\mathbf{x} = \underline{\mathbf{F}}^{-1} \underline{\mathbf{X}} \tag{2.180}$$

Concerning the normalization to N, the notation of the discrete Fourier transformation is not unique in the literature. In MATLAB, the normalization is, for example, done for the inverse transformation. Sometimes it is even completely omitted: $\underline{\mathbf{X}} = \underline{\mathbf{F}}\,\mathbf{x}$ and $\mathbf{x} = \underline{\mathbf{F}}^{-1} \underline{\mathbf{X}}$. See Annex A.6 for some properties of the matrix $\mathbf{F}$.

The numerical effort to calculate the matrix product in (2.179) and (2.180) can be roughly reduced by the factor $N/\text{lb}\, N$ if the algorithm of the FFT is applied. For large signals, the FFT/IFFT cuts down the computation time by several orders of magnitude. The FFT is one of the most applied and truly great developed algorithms [37, 38].

2.4.5.2 Circular Correlation and Convolution

Corresponding to the approach used for the discrete Fourier transform, the integral form of the correlation (2.49) converts for sampled data to

$$C_{xy}[m] = \Delta t_s \sum_{n=0}^{N-1} x[m] y[\text{mod}(n+m, N)]; \quad m = 0 \cdots N-1$$

$$\begin{bmatrix} C_{xy}[0] \\ C_{xy}[1] \\ C_{xy}[2] \\ \vdots \\ C_{xy}[N-1] \end{bmatrix} = \Delta t_s \begin{bmatrix} y[0] & y[1] & y[2] & \cdots & y[N-1] \\ y[1] & y[2] & y[3] & \cdots & y0 \\ y[2] & y[3] & y[4] & \cdots & y[1] \\ \vdots & \vdots & \vdots & \ddots & \vdots \\ y[N-1] & y[0] & y[1] & \cdots & y[N-2] \end{bmatrix} \cdot \begin{bmatrix} x[0] \\ x[1] \\ x[2] \\ \vdots \\ x[N-1] \end{bmatrix}$$

$$\mathbf{C}_{xy} = \Delta t_s \mathbf{Y}_{\text{L}} \mathbf{x} \tag{2.181}$$

In order to provide the time shift and due to the assumed periodicity of the signals, the data sample index in $\mathbf{Y}_{\text{L}}$ rotates cyclically by shifting left from raw to raw. This is expressed by the modulus index operation $\text{mod}(x, N)$. The matrix $\mathbf{Y}_{\text{L}}$ is a circulant matrix of Hankel matrix type (see Annex A.6).

Based on the convolution of continuous functions (2.129), the version for sampled data follows correspondingly to previous cases:

$$\begin{aligned}
z[m] &= \Delta t_s \sum_{n=0}^{N-1} y[\mathrm{mod}(m-n, N)]x[n]; \quad m = 0 \cdots N-1 \\
\begin{bmatrix} z[0] \\ z[1] \\ z[2] \\ \vdots \\ z[N-1] \end{bmatrix} &= \Delta t_s \begin{bmatrix} y[0] & y[N-1] & y[N-2] & \cdots & y[1] \\ y[1] & y[0] & y[N-1] & \cdots & y[2] \\ y[2] & y[1] & y[0] & \cdots & y[3] \\ \cdots & \cdots & \cdots & \ddots & \cdots \\ y[N-1] & y[N-2] & y[N-3] & \cdots & y[0] \end{bmatrix} \cdot \begin{bmatrix} x[0] \\ x[1] \\ x[2] \\ \vdots \\ x[N-1] \end{bmatrix} \\
\mathbf{z} &= \Delta t_s \mathbf{Y}_\mathrm{D} \mathbf{x}
\end{aligned} \tag{2.182}$$

The matrix $\mathbf{Y}_\mathrm{D}$ is a circulant matrix too. It is a special Toeplitz matrix. $\mathbf{Y}_\mathrm{D}$ and $\mathbf{Y}_\mathrm{L}$ are related by

$$\mathbf{Y}_\mathrm{L}\mathbf{P} = \mathbf{Y}_\mathrm{D}\mathbf{J} \tag{2.183}$$

where $\mathbf{J}$ represents the reflection matrix and $\mathbf{P}$ is the shift matrix (see Annex A.6). The matrix operations in (2.181) and (2.182) are very time consuming. Since the FFT routine works very efficient, it is quite faster to calculate the correlation as well as the convolution via the FFT and IFFT using appropriate calculation rules from Annex B.1, Table B.2 which are also expressed in the eigenvalue decomposition of the circulant matrices as shown in Annex A.6. Neglecting the normalization by N or Δt_s we finally end up in

$$\text{Convolution:} \quad \begin{aligned} \mathbf{z}[N] &= \mathrm{IFFT}\{\mathrm{FFT}\{\mathbf{y}[N]\} \circ \mathrm{FFT}\{\mathbf{x}[N]\}\} \\ &= \mathrm{IFFT}\{\underline{\mathbf{Y}}[N] \circ \underline{\mathbf{X}}[N]\} \end{aligned} \tag{2.184}$$

$$\text{Correlation:} \quad \begin{aligned} \mathbf{C}_{xy}[N] &= \mathrm{IFFT}\{\mathrm{FFT}\{\mathbf{x}[N]\} \circ \mathrm{FFT}\{\mathbf{y}[N]\}^*\} \\ &= \mathrm{IFFT}\{\underline{\mathbf{X}}[N] \circ \underline{\mathbf{Y}}^*[N]\} \end{aligned} \tag{2.185}$$

The symbol $\circ$ stands for the so-called Schur or Hadamard product, that is the entry wise product of vector or matrix elements and $\underline{Y}^*$ denotes conjugate complex.

2.4.5.3 Data Record Length and Sampling Interval

The number N_s of sampled data is a crucial point in many sensor applications since the data amount to be handled and the required processing speed largely determine the costs of the digital part of a sensor module as well as its power consumption. The number of data points N_s to be captured by one 'shot' is given by the sampling rate f_s and the recording time T_R that is $N_s = f_s T_\mathrm{R}$. Under a 'measurement shot' we will understand the collection of a full data set providing a single response function as IRF or FRF. First, we will consider the impact of the recording time before we switch to the role of the sampling rate.

At this point, we refer exclusively to real-time sampling and we will also exclude averaging. Therefore, recording time T_R and time window length t_win of the observable IRF are identical.[26)] They should be larger than the settling time t_{sett} of the test scenario in order to avoid time aliasing $T_R = t_{win} \geq t_{set}$. The issue which we will follow now concerns the impact of recording time limitation onto the measurement result. Here we have to distinguish three cases:

System stimulation by a single pulse: As already mentioned, the data acquisition has to cover duration of at least $T_R \geq t_{sett}$ to avoid truncation of the IRF.

System stimulation by periodic wideband signals: This is the best one can do with respect to error avoidance. Note, before the measurement can be started, one has to await the DUT settling after the signal source was switched on in order to achieve a periodic DUT reaction. The signal period should be longer than the IRF settling $t_P \geq t_{set}$ and the signal has to be captured over exactly one period $T_R = t_{win} = t_P$ (or an integer number of periods $T_R = pt_P$ in case of averaging). Figure 2.60 illustrates by a simple example the implications if the observation length does not exactly coincide with the period duration.

We consider a mixture of three sinusoids from which we took data over either 2 or 2.08 periods. As the FFT spectra show, there is not any impact observable if data collection coincides with the signal period. But if one deviates from that rule, spurious spectral components are generated due to signal truncation and it can be hard to detect the original spectral lines. We shortly consider why it is.

Let us call our original periodic signal $x(t) = x(t + nt_P)$, then actually captured signal $x_w(t)$ may be expressed by

$$x_w(t) = x(t)w(t) \tag{2.186}$$

where $w(t)$ represents a window function which is in the simplest case a rectangular function $w(t) = \operatorname{rect} t/T_P$. Performing the Fourier transform, we will result in the convolution of the original spectrum with the spectrum of the window function

$$\underline{X}_w(f) = \underline{X}(f) * \underline{W}(f) \tag{2.187}$$

Assuming a rectangular window, we get a sinc function having its zeros at T_R^{-1}. If these zeros coincide with the line spacing of the original signal, we will not have any mutual interaction between different spectral signal components. This is illustrated in Figure 2.61a representing the spectrum of a single sine wave. In the case of mismatch $T_R \neq nt_P$, every spectral component of the signal will generate spurious lines as depicted in Figure 2.61b which all superimpose leading to the scrambled spectrum from Figure 2.60 (below on the right)

System stimulation by random (non-periodic) wideband signals: Truncation effects cannot be avoided due to the inevitable limitation of the recording length. Figure 2.62 illustrates the consequences. The capability to separate adjacent spectral lines degrades due to the bandwidth B_{win} of the window function and weak signals are buried by the side lobes of the window as we could already observe in Figure 2.60. The bandwidth of the window function is roughly

26) In the case of a p-fold averaging and a m-fold sub-sampling recording time and IRF duration are related by $T_R = mpt_{win}$.

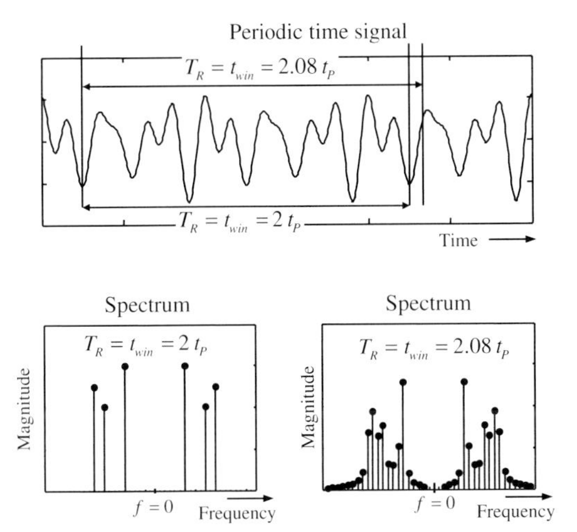

Figure 2.60 Periodic signal and its FFT spectrum for different recording lengths.

inversely to the window length $\underset{..}{B}_{\mathrm{win}} t_{\mathrm{win}} \approx 1$ (see Section 2.2.5). Therefore, the duration of the captured time segment determines the frequency resolution of the measurement. The side-lobe level decreases faster for window functions with smooth edges for the expense of a reduced spectral resolution. There are several window functions available (see MATLAB) allowing balancing between frequency resolution and side-lobe level for specific applications.

As mentioned above, the sampling interval and sampling rate $\Delta t_{\mathrm{s}} = f_{\mathrm{s}}^{-1}$ are the second issue influencing the data amount to be handled. The classical theory (which we will follow here) says that the lowest sample number N_{s} without violation of frequency aliasing (and hence without information loss) results from a sampling frequency close to the Nyquist rate of the signal to be captured, that is $f_{\mathrm{s}} \cong \underset{..}{B} = 2\underset{.}{B}$ in case of a baseband signal. This sampling rate assumes signals of full degree of freedom. But many natural processes confine the degree freedom – one is speaking from a sparse scene – which can be used to optimize the sampling

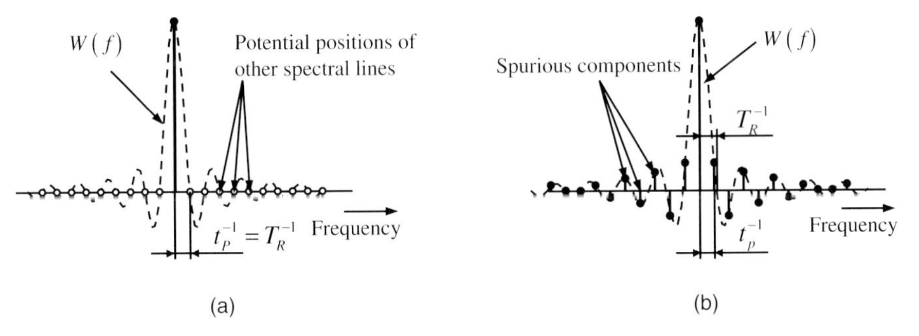

Figure 2.61 Generation of spurious spectral components due to mismatch between signal period and recording length. (a) Correct recording length and (b) generation of spurious components due to truncation.

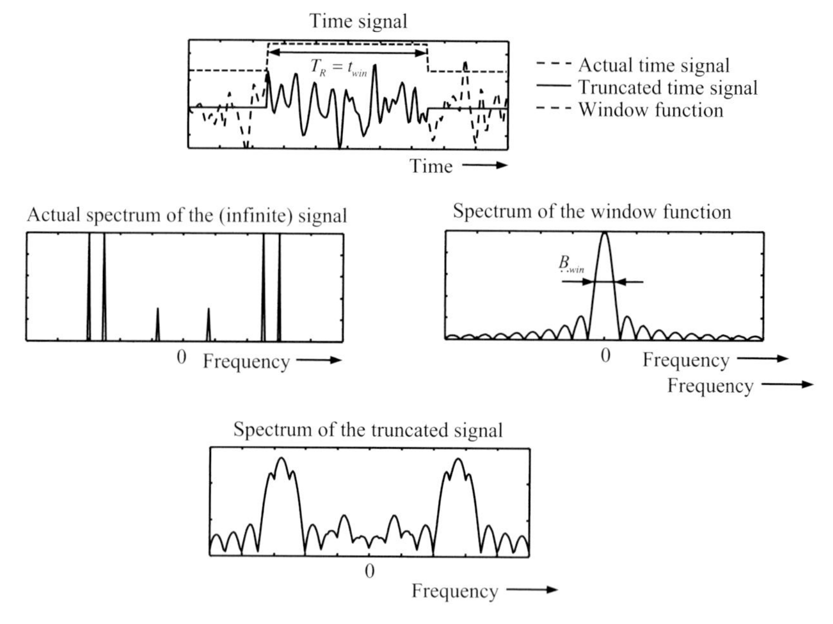

Figure 2.62 Effect of limitation of the record length onto the signal spectrum. For demonstration purposes a rectangular window was selected in order to illustrate the creation of side lobes.

rate and thus the data amount without loss on information. The corresponding technique and processing routines are called compressive sampling or compressed sensing [10, 39].

The human eye is however not satisfied with a simple plot of suchlike gathered signals since it is not able to interpolate the samples appropriately. The left part of Figure 2.63 illustrates that fact. It refers to as signal closely sampled to the Nyquist rate. The graph on the right looks much friendlier since

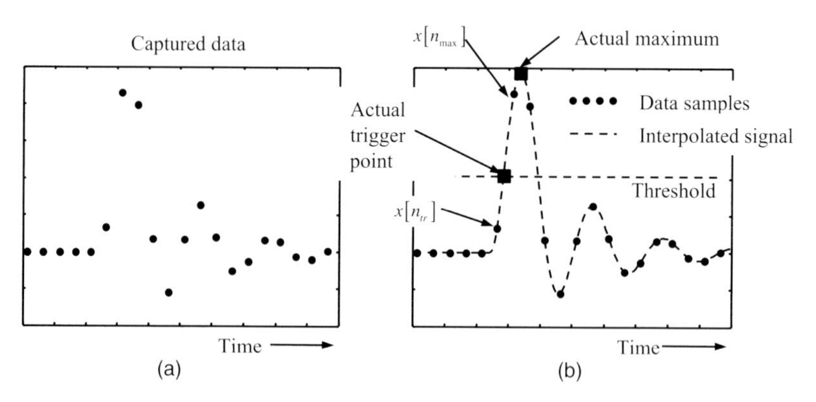

Figure 2.63 The effect of sampling onto data visualization and the extraction of shape parameters. (a) Data points which were sampled close to its Nyquist rate and (b) time function after interpolation.

additional points are added. They are gained from interpolation. The interpolated signal is identical to the original signal as long as the Nyquist theorem is met. For that purpose, the sampled data are first transformed via FFT into the frequency domain. Then zeros are padded above the Nyquist frequency ($f_s/2$), that is the 'Nyquist frequency' of the new data set was artificially increased. Since by assumption the signal has no signal energy above $f_s/2$, the padded zeros only increase virtually the sampling rate but do not affect the actual spectrum. The transformation back to the time domain finally gives a signal with a higher density of data points. The signal looks indeed better but it contains exactly the same information as before.

The determination of characteristic pulse shape parameters may require densely sampled data too. This mainly concerns parameters which are based on threshold crossing or maximum/minimum values.

Figure 2.63 illustrates the deviations between the wanted values and the values resulting from Nyquist sampled data ($x[n_{tr}]$ or $x[n_{max}]$). Also here, interpolation of the Nyquist sampled data will reduce measurement errors. However, in order to avoid unneeded memory space and processor load, over-sampled data should only be used where necessary.

A final example, we will mention here concerns the continuous time shifting of a sampled signal. Precise time shifting may be required for high-resolution range measurements and curve fitting (see Section 4.7). A simple time shift by multiple of the sampling interval Δt_s will not be sufficient in such cases. There are two simple ways to perform a sub-sample time shift. The first one deals again with FFT interpolation. This would result in over-sampled data with finer sampling interval. Now, the signal can be shifted by the step size of the over-sampled data. The second version – applying the time shift rule of the Fourier transform (see Annex B.1, Table B.2) – will even allow continuous (cyclic) time shift $\Delta t = n_d \, \Delta t_s$, $n_d \in \mathbb{R}$. For that purpose, the sampled signal is subjected a FFT, its spectrum is multiplied by the phasor $e^{-j2\pi \, m \, n_d/N}$ and finally it is transformed back into the time domain:

$$\mathbf{x}_{shift}[N] = \mathrm{IFFT}\{\mathrm{FFT}\{\mathbf{x}[N]\} \circ \mathbf{Ph}\} \quad n \in \mathbb{N};\ n_d \in \mathbb{R} \tag{2.188}$$

The vector $\mathbf{Ph}$ contains the phasor terms providing the phase shift of the individual spectral lines. The construction of the phase vector $\mathbf{Ph}$ needs some care since the particularities concerning the negative part of the spectrum has to be respected (see Section 2.4.5.1). Using the relation (2.178), the phase vector $\mathbf{Ph}$ may be written as

$$\begin{aligned}\mathbf{Ph} &= \begin{bmatrix} 1 & e^{-j2\pi\frac{n_d}{N}} & e^{-j2\pi\frac{2n_d}{N}} & e^{-j2\pi\frac{3n_d}{N}} & \cdots & e^{j2\pi\frac{3n_d}{N}} & e^{j2\pi\frac{2n_d}{N}} & e^{j2\pi\frac{n_d}{N}} \end{bmatrix} \\ &= \begin{bmatrix} e^{-j2\pi\frac{n_d m_{ph}}{N}} \end{bmatrix} \quad \text{with} \quad m_{ph} = m - N \,\mathrm{floor}\left(\frac{2m}{N}\right); \quad m = 0 \cdots N-1\end{aligned} \tag{2.189}$$

2.5
Physical System

2.5.1
Energetic Interaction and Waves

We only supposed linear and time invariant behaviour for all systems which we considered up to now. This was sufficient to develop formally some methods of modelling purely based on a mathematical formalism for linear differential equations without referencing to deeper physical conditions. Such considerations do not respect retroactions between real physical objects due to any energetic interaction. But every interaction between real physical objects and therefore also between sensor and test object, cascaded amplifiers and so on is always linked to an exchange of energy.

We will extend our considerations also to some of such aspects. At this juncture, we will follow two different approaches. One deals with current and voltage signals. It is preferentially applied if the coherence length of the sounding signals is large compared to a typical dimension d of the test object (i.e. $\tau_{coh}c \approx c/\underset{..}{B} \gg d$). The second one applies normalized waves typically used for the inverse case, that is $\tau_{coh}c \approx c/\underset{..}{B} \approx d$ or $\leq d$.

In general, the energy and power transfer is linked to the product of a physical force and flow quantity. In the domain of electromagnetism, the force quantity is either the voltage V or the electrical field E and the flow quantity refers to the current I or the magnetic field H. Let us restrict it to V and I which is sufficient if we do not account for free space propagation at this point.

Figure 2.64 depicts an electrical line which is able to transport electrical power. First we will refer to the total current and voltage which can be measured, for example at position $x = 0$ and then we consider their individual components.

For the total current and voltage, the instantaneous power flux is given by

$$p(t) = V(t)\, I(t) \tag{2.190}$$

For $p(t) > 0$ the power flux goes from left to right, for $p(t) < 0$ it points in the inverse direction. The total average power flux[27] P is expressed by following relations:

$$\text{Time domain} \quad P = \overline{V(t)I(t)} = \lim_{t_x \to \infty} \frac{1}{t_x} \int_{-t_x/2}^{t_x/2} V(t)I(t)\mathrm{d}t \tag{2.191}$$

27) We have to deal for time-limited signals with the flux of total energy which is given by $E = \int_{T_R} V(t)I(t)\mathrm{d}t$.

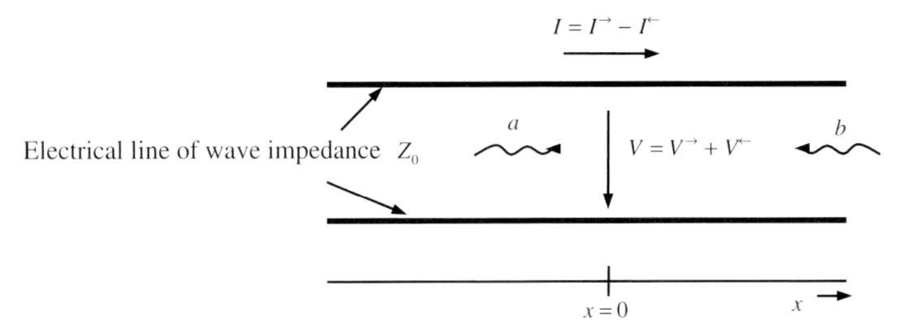

Figure 2.64 Schematized electrical line performing energy transport via guided waves.

Frequency domain[28)]

$$\underline{\Phi}_{VI}(f) = \underline{V}(f)\underline{I}^*(f) = \underline{\Phi}^*_{VI}(-f) \tag{2.192}$$

whereas $P = \int_{-\infty}^{\infty} \underline{\Phi}_{VI}(f)\mathrm{d}f$.

$\underline{\Phi}_{VI}(f)$ is the voltage–current cross-spectrum (see also Section 2.2.5 (2.70)). Its real part refers to the active power at frequency f and the imaginary part represents the reactive power which does not cause a wanted effect.

Now let us consider the individual components of current and voltage. It is well known that the electrical and magnetic field and their integral quantities V and I propagate with $c_0 = 30\,\mathrm{cm/ns}$ in vacuum and about $c \approx 20\,\mathrm{cm/ns}$ in a cable. Concerning a cable, there are two possibilities[29)] to propagate – from left to right, that is $I^{\rightarrow}$, $U^{\rightarrow}$ and from right to left $I^{\leftarrow}$, $U^{\leftarrow}$. This we can expressed in the time and frequency domain by

$$\begin{aligned} I^{\rightarrow}(t,x) &= I^{\rightarrow}\left(t - \frac{x}{c}\right) = I_0^{\rightarrow}(t, x=0) * \delta\left(t - \frac{x}{c}\right) \\ \underline{I}^{\rightarrow}(f,x) &= \underline{I}_0^{\rightarrow}(f, x=0)\mathrm{e}^{j2\pi f\left(t-\frac{x}{c}\right)} = \underline{I}_0^{\rightarrow}(\omega, x=0)\mathrm{e}^{j(\omega t - kx)} \end{aligned} \tag{2.193}$$

$$\begin{aligned} I^{\leftarrow}(t,x) &= I^{\leftarrow}\left(t + \frac{x}{c}\right) = I_0^{\leftarrow}(t, x=0) * \delta\left(t + \frac{x}{c}\right) \\ \underline{I}^{\leftarrow}(f,x) &= \underline{I}_0^{\leftarrow}(f, x=0)\mathrm{e}^{j2\pi f\left(t+\frac{x}{c}\right)} = \underline{I}_0^{\leftarrow}(\omega, x=0)\mathrm{e}^{j(\omega t + kx)} \end{aligned} \tag{2.194}$$

The total current is hence composed from two current waves (for details on wave propagation, see Chapter 5)

$$I(t,x) = I^{\rightarrow}(t,x) - I^{\leftarrow}(t,x) \tag{2.195}$$

and correspondingly the total voltage results from the superposition of two voltage waves:

$$V(t,x) = V^{\rightarrow}(t,x) + V^{\leftarrow}(t,x) \tag{2.196}$$

28) See also remarks concerning to (2.61).
29) We will restrict ourselves to the fundamental cable/waveguide mode here.

where

$$\begin{aligned} V^{\rightarrow}(t,x) &= V^{\rightarrow}\left(t-\frac{x}{c}\right) = V_0^{\rightarrow}(t,x=0) * \delta\left(t-\frac{x}{c}\right) \\ \underline{V}^{\rightarrow}(f,x) &= \underline{V}_0^{\rightarrow}(f,x=0)\mathrm{e}^{j2\pi f\left(t-\frac{x}{c}\right)} = \underline{V}_0^{\rightarrow}(\omega,x=0)\mathrm{e}^{j(\omega t-kx)} \end{aligned} \tag{2.197}$$

$$\begin{aligned} V^{\leftarrow}(t,x) &= V^{\leftarrow}\left(t+\frac{x}{c}\right) = V_0^{\leftarrow}(t,x=0) * \delta\left(t+\frac{x}{c}\right) \\ \underline{V}^{\leftarrow}(f,x) &= \underline{V}_0^{\leftarrow}(f,x=0)\mathrm{e}^{j2\pi f\left(t+\frac{x}{c}\right)} = \underline{V}_0^{\leftarrow}(\omega,x=0)\mathrm{e}^{j(\omega t+kx)} \end{aligned} \tag{2.198}$$

The voltage and the current waves, which propagate in the same direction, are not independent of each other. They are related by the wave impedance of the cable, that is

$$V^{\rightarrow} = Z_0\, I^{\rightarrow} \quad \text{and} \quad V^{\leftarrow} = Z_0\, I^{\leftarrow} \tag{2.199}$$

Therefore it is meaningless to distinguish between current and voltage wave and one usually joins them in the normalized waves:

$$a = \sqrt{Z_0}\, I^{\rightarrow} = \frac{1}{\sqrt{Z_0}}\, V^{\rightarrow} \tag{2.200}$$

and

$$b = \sqrt{Z_0}\, I^{\leftarrow} = \frac{1}{\sqrt{Z_0}}\, V^{\leftarrow} \tag{2.201}$$

The dimension of the normalized waves is $\left[\sqrt{\mathrm{W}}\right]$ and usually a denotes the wave travelling towards an object and b assigns the wave which leaves an object. The total voltage and current at the feeding line become now

$$I = \frac{a-b}{\sqrt{Z_0}} \quad \text{and} \quad V = \sqrt{Z_0}(a+b) \tag{2.202}$$

The wave impedance Z_0 depends on the geometrical configuration of the cable and the dielectric properties of the material which separates both conductors. A standard value for laboratory equipment is $50\,\Omega$. We will suppose for simplicity that the cables are lossless and free of dispersion (Z_0 will be real valued and frequency independent) and that only one wave mode can propagate on the cable. In the case of, for example a coaxial cable, the last point is fulfilled if the upper frequency of the wave is below a cut-off value. This cut-off value depends on the cable diameter, that is the smaller the diameter the higher the frequency cut-off.

From (2.191) and (2.202), the power flux expresses by

$$P = \lim_{t_x \to \infty} \frac{1}{t_x} \int_{-t_x/2}^{t_x/2} \left(a^2(t) - b^2(t)\right)\mathrm{d}t \tag{2.203}$$

and in the spectral domain we get[30]

$$\Phi(f) = \underline{a}(f)\underline{a}^*(f) - \underline{b}(f)\underline{b}^*(f) \tag{2.204}$$

2.5.2 *N*-Port Description by IV-Parameters

Following the consideration from the previous paragraph, a physical system and test object which interacts with its environment (i.e. measurement devices) through n ports deals with $2n$ signals, two at every port, that is either one flux and one force quantity or two waves – incident and emanating. Figure 2.65 illustrates a two-port device and the externally accessible quantities which are two current and two voltage signals.

Since we suppose a linear device, a linear relation must exist between these four signals. Hence, we can formally express their mutual dependence by any system of linear equations. Taking into account also arbitrary linear combinations of the signals, it exist an infinite number of possibility to write down these equations. The three most common notations are given here below (above – frequency domain description and left – time domain description):

$$\begin{bmatrix} \underline{V}_1(f) \\ \underline{V}_2(f) \end{bmatrix} = \begin{bmatrix} \underline{Z}_{11}(f) & \underline{Z}_{12}(f) \\ \underline{Z}_{21}(f) & \underline{Z}_{22}(f) \end{bmatrix} \begin{bmatrix} \underline{I}_1(f) \\ \underline{I}_2(f) \end{bmatrix}$$

or (2.205)

$$\begin{bmatrix} V_1(t) \\ V_2(t) \end{bmatrix} = \begin{bmatrix} Z_{11}(t) & Z_{12}(t) \\ Z_{21}(t) & Z_{22}(t) \end{bmatrix} * \begin{bmatrix} I_1(t) \\ I_2(t) \end{bmatrix}$$

$$\begin{bmatrix} \underline{I}_1(f) \\ \underline{I}_2(f) \end{bmatrix} = \begin{bmatrix} \underline{Y}_{11}(f) & \underline{Y}_{12}(f) \\ \underline{Y}_{21}(f) & \underline{Y}_{22}(f) \end{bmatrix} \begin{bmatrix} \underline{V}_1(f) \\ \underline{V}_2(f) \end{bmatrix}$$

or (2.206)

$$\begin{bmatrix} I_1(t) \\ I_2(t) \end{bmatrix} = \begin{bmatrix} Y_{11}(t) & Y_{12}(t) \\ Y_{21}(t) & Y_{22}(t) \end{bmatrix} * \begin{bmatrix} V_1(t) \\ V_2(t) \end{bmatrix}$$

$$\begin{bmatrix} \underline{V}_2(f) \\ \underline{I}_2(f) \end{bmatrix} = \begin{bmatrix} \underline{A}_{11}(f) & \underline{A}_{12}(f) \\ \underline{A}_{21}(f) & \underline{A}_{22}(f) \end{bmatrix} \begin{bmatrix} \underline{V}_1(f) \\ \underline{I}_1(f) \end{bmatrix}$$

or (2.207)

$$\begin{bmatrix} V_2(t) \\ I_2(t) \end{bmatrix} = \begin{bmatrix} A_{11}(t) & A_{12}(t) \\ A_{21}(t) & A_{22}(t) \end{bmatrix} * \begin{bmatrix} V_1(t) \\ I_1(t) \end{bmatrix}$$

30) See also remarks concerning to (2.61).

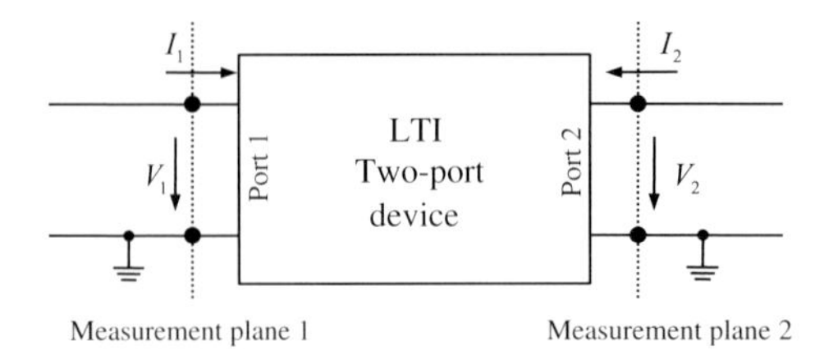

Figure 2.65 Two-port system with measurement signals. The term measurement plane shall indicate the geometric location of voltage and current capturing.

(2.205) describes the two-port behaviour by the so-called impedance parameters, (2.206) relates to the admittance parameters and (2.207) defines the chain or ABCD-parameter representation. Hence, a two-port is completely characterized by a set of four frequency or time domain functions and an n-port device consequently by n^2 functions. Examples of the conversion from one parameter set to another are given in Annex B.6. Note that the convolution in (2.205) to (2.207) has to be handled corresponding to (2.138).

The term 'parameters' for the matrix entries in this conjunction is somewhat ambiguous since the **Z**-, **Y**- or **A**-parameters are actually functions in t or f. That is, every entry of the matrices in (2.205) to (2.207) represents either a FRF or an IRF as we have discussed them in Section 2.4. Referring to the notation of the previous chapter every one-port device represents a SiSo system and every n-port device belongs to the MiMo systems.

It should be noted that the n-port parameters as defined above must not necessarily relate to a physical meaningful or illustrative interpretation since they are only formally deduced from the linearity of the test object. The different n-port parameters can be represented by a state space model with n inputs and n outputs if the test object consists of lumped elements (see Section 2.4.4.4). Hence, all n-port IRFs or FRFs are composed of the same set of modes which, however, contribute with different intensity (including missed modes) to the individual parameter functions.

Quite a few of modern electronic RF-components and sensor elements which are part of ultra-wideband sensors are also based on differential circuit conception. Therefore, we will shortly introduce the basics of their system description. For more details, the reader can refer to Ref. [40]. The differential technique simplifies the construction and feeding of very wideband antennas, it reduces problems concerning electromagnetic interference, power supply noise, non-ideal circuit ground, non-linear distortions and others. But it also implicates some problems with the device characterization.

A port of a differential device must have two feeding lines which are related to the ground (see Figure 2.66). Therefore, we can assign two voltage and two current signals to every modal-port. A modal two-port device either represents a four-port device if all port variables are considered independently from each other, that is in the frequency domain we have

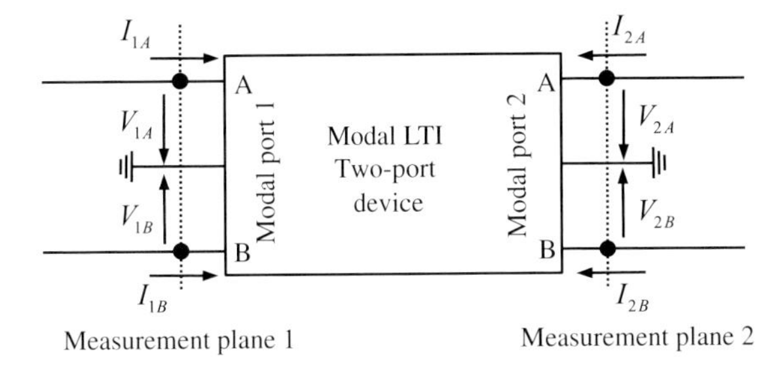

Figure 2.66 Modal two-port device.

$$\begin{bmatrix} \underline{V}_{1A}(f) \\ \underline{V}_{1B}(f) \\ \underline{V}_{2A}(f) \\ \underline{V}_{2B}(f) \end{bmatrix} = \begin{bmatrix} \underline{Z}_{1A1A}(f) & \underline{Z}_{1A1B}(f) & \underline{Z}_{1A2A}(f) & \underline{Z}_{1A2B}(f) \\ \underline{Z}_{1B1A}(f) & \underline{Z}_{1B1B}(f) & \underline{Z}_{1B2A}(f) & \underline{Z}_{1B2B}(f) \\ \underline{Z}_{2A1A}(f) & \underline{Z}_{2A1B}(f) & \underline{Z}_{2A2A}(f) & \underline{Z}_{2A2B}(f) \\ \underline{Z}_{2B1A}(f) & \underline{Z}_{2B1B}(f) & \underline{Z}_{2B2A}(f) & \underline{Z}_{2B2B}\nu \end{bmatrix} \begin{bmatrix} \underline{I}_{1A}(f) \\ \underline{I}_{1B}(f) \\ \underline{I}_{2A}(f) \\ \underline{I}_{2B}(f) \end{bmatrix}$$

or it can be considered as a device of two-modal ports where every modal port can be driven by two independent modes – balanced and unbalanced referred to the ground. The conversion between both approaches is given by

$$\text{balanced} \qquad V_\text{b} = V_A - V_B \quad \text{and} \quad 2I_\text{b} = I_A - I_B \tag{2.208}$$

$$\text{unbalanced} \quad 2V_\text{u} = V_A + V_B \quad \text{and} \quad I_\text{u} = I_A + I_B \tag{2.209}$$

The balanced (differential or odd) mode is usually the wanted signal mode for such systems whereas the unbalanced (common or even) mode is unintended and should be suppressed as much as possible. The robustness of differential systems against electromagnetic interference, power supply noise, even order distortions and floating ground potential mainly comes from their ability to suppress unbalanced signals.

For the balanced mode, the port voltages V_A and V_B have the same magnitude but inverse polarity. Hence, the ground potential is in-between. The same holds for the port currents I_A and I_B so there is no current flow over the ground. Thus, the quality of the ground electrode has not any longer a dominant influence onto the circuit behaviour.

The modal two-port parameters are given now by, for example,

$$\begin{bmatrix} \underline{V}_{b1}(f) \\ \underline{V}_{b2}(f) \\ \underline{V}_{u1}(f) \\ \underline{V}_{u2}(f) \end{bmatrix} = \begin{bmatrix} \underline{Z}_{bb11}(f) & \underline{Z}_{bb12}(f) & \underline{Z}_{bu11}(f) & \underline{Z}_{bu12}(f) \\ \underline{Z}_{bb21}(f) & \underline{Z}_{bb22}(f) & \underline{Z}_{bu21}(f) & \underline{Z}_{bu22}(f) \\ \underline{Z}_{ub11}(f) & \underline{Z}_{ub12}(f) & \underline{Z}_{uu11}(f) & \underline{Z}_{uu12}(f) \\ \underline{Z}_{ub21}(f) & \underline{Z}_{ub22}(f) & \underline{Z}_{uu21}(f) & \underline{Z}_{uu22}(f) \end{bmatrix} \begin{bmatrix} \underline{I}_{b1}(f) \\ \underline{I}_{b2}(f) \\ \underline{I}_{u1}(f) \\ \underline{I}_{u2}(f) \end{bmatrix} \tag{2.210}$$

See Annex B.6 for the conversion from four-port to modal two-port parameters. Since we deal with linear systems here, we can use these conversation rules to determine the modal parameters (or mixed-mode parameters) from classical four-port measurements with grounded (single-ended) measurement devices.

The parameters of the modal two-port are arranged into sub-matrices describing the system behaviour for purely balanced modes at the in- and output ports (sub-matrix upper left), for the unbalanced modes (sub-matrix lower right) as well as for mode coupling from balanced to unbalanced (sub-matrix lower left) and unbalanced to balanced (sub-matrix upper right). For an ideal differential device, all sub-matrices should have zero entries besides the one in the upper left.

2.5.3 *N*-Port Description by Wave Parameters

Using (2.200)–(2.202), the total voltage and current at every port can be decomposed into normalized waves as depicted Figure 2.67. In the case of a linear device, the waves at all ports are linearly dependent from each other so that we can establish an arbitrary linear system of equations as already exercized above for *IV*-parameters. These equations describe the behaviour of the device with respect to incident a_i and emanate b_j waves. Two kinds of equations are common, which are given here again in frequency and time domain notation, respectively

$$\begin{bmatrix} \underline{b}_1(f) \\ \underline{b}_2(f) \end{bmatrix} = \begin{bmatrix} \underline{S}_{11}(f) & \underline{S}_{12}(f) \\ \underline{S}_{21}(f) & \underline{S}_{22}(f) \end{bmatrix} \begin{bmatrix} \underline{a}_1(f) \\ \underline{a}_2(f) \end{bmatrix}$$
$$\text{or} \tag{2.211}$$
$$\begin{bmatrix} b_1(t) \\ b_2(t) \end{bmatrix} = \begin{bmatrix} S_{11}(t) & S_{12}(t) \\ S_{21}(t) & S_{22}(t) \end{bmatrix} * \begin{bmatrix} a_1(t) \\ a_2(t) \end{bmatrix}$$

In short form this is written as
$$\begin{aligned} \underline{\mathbf{b}}(f) &= \underline{\mathbf{S}}(f)\underline{\mathbf{a}}(f) \\ \mathbf{b}(t) &= \mathbf{S}(t) * \mathbf{a}(t) \end{aligned} \tag{2.212}$$

$$\begin{bmatrix} \underline{b}_1(f) \\ \underline{a}_1(f) \end{bmatrix} = \begin{bmatrix} \underline{T}_{11}(f) & \underline{T}_{12}(f) \\ \underline{T}_{21}(f) & \underline{T}_{22}(f) \end{bmatrix} \begin{bmatrix} \underline{a}_2(f) \\ \underline{b}_2(f) \end{bmatrix}$$
$$\text{or} \tag{2.213}$$
$$\begin{bmatrix} b_1(t) \\ a_1(t) \end{bmatrix} = \begin{bmatrix} T_{11}(t) & T_{12}(t) \\ T_{21}(t) & T_{22}(t) \end{bmatrix} * \begin{bmatrix} a_2(t) \\ b_2(t) \end{bmatrix}$$

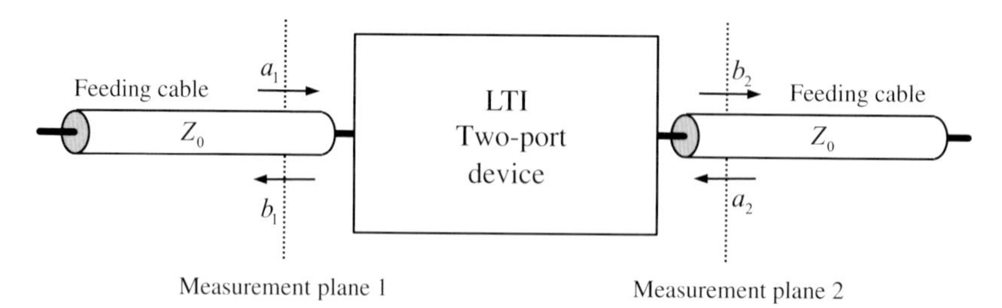

Figure 2.67 Two-port device with normalized waves.

The first equation (2.211) refers to the scattering parameters (S-parameter) and the second (2.213) to the transmission parameters (T-parameters, transfer parameters, cascade parameters). Both system descriptions are preferably used in the domain of RF- and microwave engineering. Annex B.6 summarizes the conversion between S- and T-parameters as well as their relation to some of the IV-parameters.

The T-parameters are convenient if one deals with cascaded systems. The S-parameters are the only ones which have an illustrative physical interpretation since they are linked to the actual energy flow. The S-parameter[31] $\underline{S}_{ii} = \left(\underline{b}_i/\underline{a}_i\right)\big|_{\underline{a}_j=0}$; $(i \neq j)$ represents a reflection coefficient since the emanate wave b_i can only be caused by reflection of the incident wave a_i because no waves are injected in all other ports by assumption $(a_j = 0; (j \neq i))$. In contrast, the S-parameter $\underline{S}_{ij} = \underline{b}_i/\underline{a}_j\big|_{\underline{a}_i=0}$; $(i \neq j)$ refers to a signal transmission from port j to port i.

For all passive linear networks and measurement objects (exception magnetized ferrites or plasma), it holds reciprocity, that is the transmission from port 1 to 2 is equal to the transmission in the opposite direction. For scattering parameters of an n-port, this is expressed by

$$\begin{aligned} \underline{\mathbf{S}}(f) &= \underline{\mathbf{S}}^T(f) \\ \mathbf{S}(t) &= \mathbf{S}^T(t) \end{aligned} \tag{2.214}$$

For the T-parameters of a two-port device holds in case of reciprocity:

$$\begin{gathered} \mathbf{det}(\underline{\mathbf{T}}(f)) = 1 \\ T_{11}(t) * T_{22}(t) - T_{12}(t) * T_{12}(t) = \delta(t) \end{gathered} \tag{2.215}$$

Lossless n-port devices meet following S-parameters conditions:

$$\begin{gathered} \underline{\mathbf{S}}(f)\underline{\mathbf{S}}^{\mathrm{H}}(f) = \mathbf{I} \\ \mathbf{S}(t) * \mathbf{S}^T(-t) = \delta(t)\mathbf{I} \end{gathered} \tag{2.216}$$

Often, the S-parameter representation of n-port devices is symbolized by a Mason- or signal flow graph as depicted in Figure 2.68. It is an illustrative method

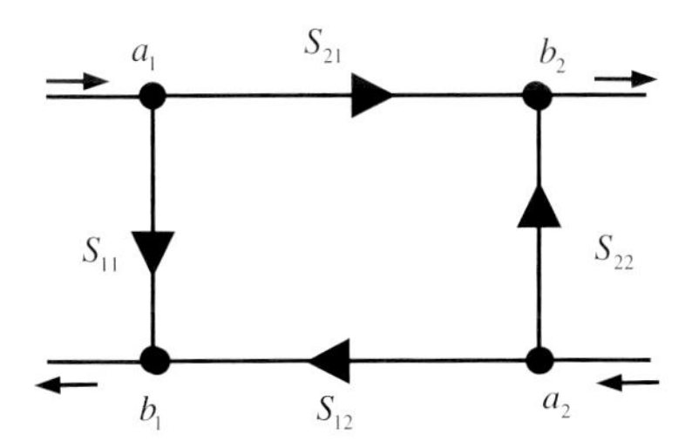

Figure 2.68 Mason graph of a two-port device. The path weight S_{ij} symbolize the strength of a signal path while the arrow indicates the signal flow direction.

31) Note the quotient $\underline{S}_{ij} = \underline{b}_i/\underline{a}_j$ is only allowed for frequency domain description. In the time domain, S_{ij} has to be de-convolved out of b_i.

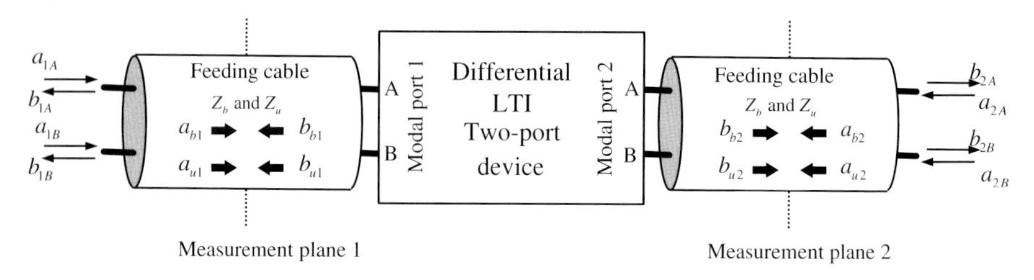

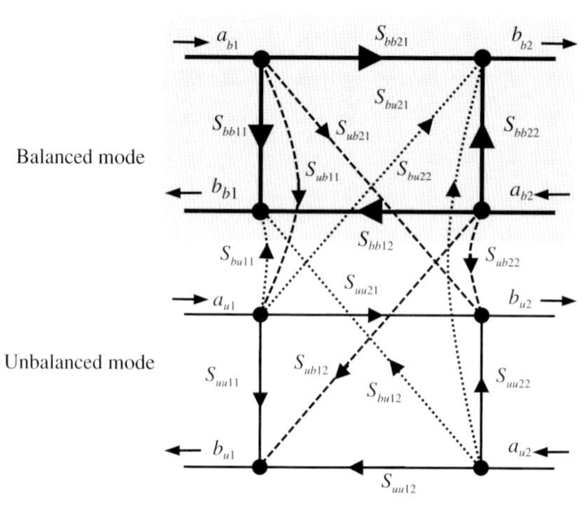

Figure 2.69 Modal two-port device fed by normalized waves and corresponding Mason graph. The usually wanted part of device parameters for the purely balanced (differential) modes is emphasized. The remaining transmission paths should be ideally zero.

to analyse the signal flow in complicated systems. Annex B.7 delves more deeply into the method.

Finally, we will still examine the S-parameters of differential system as depicted in Figure 2.69 for a modal two-port device. As before in the case of IV-parameters, we can consider either every feeding line of the device separately or we do it pairwise.

In the first case, we end up in a four-port description:

$$\begin{bmatrix} \underline{b}_{1A}(f) \\ \underline{b}_{1B}(f) \\ \underline{b}_{2A}(f) \\ \underline{b}_{2B}(f) \end{bmatrix} = \begin{bmatrix} \underline{S}_{1A1A}(f) & \underline{S}_{1A1B}(f) & \underline{S}_{1A2A}(f) & \underline{S}_{1A2B}(f) \\ \underline{S}_{1B1A}(f) & \underline{S}_{1B1B}(f) & \underline{S}_{1B2A}(f) & \underline{S}_{1B2B}(f) \\ \underline{S}_{2A1A}(f) & \underline{S}_{2A1B}(f) & \underline{S}_{2A2A}(f) & \underline{S}_{2A2B}(f) \\ \underline{S}_{2B1A}(f) & \underline{S}_{2B1B}(f) & \underline{S}_{2B2A}(f) & \underline{S}_{2B2B}(f) \end{bmatrix} \begin{bmatrix} a_{1A}(f) \\ \underline{a}_{1B}(f) \\ \underline{a}_{2A}(f) \\ \underline{a}_{2B}(f) \end{bmatrix}$$

for which the normalized waves are referred to the wave impedance Z_0.

In the case of a pairwise feeding, we deal with two types of incident waves a_b and a_u as well as with two types of emanate waves b_b and b_u at every modal-port. The

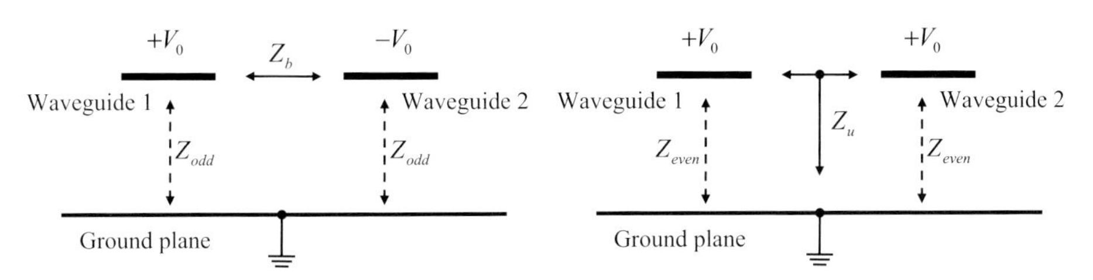

Figure 2.70 Coupled microstrip line under balanced and unbalanced conditions and corresponding wave impedances. Usually, the ground electrode and the lines are manufactured on a board of low loss dielectric. Hence, the permittivity of the space between the electrodes is different from the permittivity of the upper half space.

waves with subscript b are balanced with respect to ground; the waves with subscript u belong to the unbalanced mode. In that case, the S-matrix can be written as

$$\begin{bmatrix} \underline{b}_{b1}(f) \\ \underline{b}_{b2}(f) \\ \underline{b}_{u1}(f) \\ \underline{b}_{u2}(f) \end{bmatrix} = \left[\begin{array}{cc|cc} \underline{S}_{bb11}(f) & \underline{S}_{bb12}(f) & \underline{S}_{bu11}(f) & \underline{S}_{bu12}(f) \\ \underline{S}_{bb21}(f) & \underline{S}_{bb22}(f) & \underline{S}_{bu21}(f) & \underline{S}_{bu22}(f) \\ \hline \underline{S}_{ub11}(f) & \underline{S}_{ub12}(f) & \underline{S}_{uu11}(f) & \underline{S}_{uu12}(f) \\ \underline{S}_{ub21}(f) & \underline{S}_{ub22}(f) & \underline{S}_{uu21}(f) & \underline{S}_{uu22}(f) \end{array}\right] \begin{bmatrix} \underline{a}_{b1}(f) \\ \underline{a}_{b2}(f) \\ \underline{a}_{u1}(f) \\ \underline{a}_{u2}(f) \end{bmatrix} \tag{2.217}$$

Such S-parameters are also referred as mixed-mode S-parameters [40]. Now however, attention has to be paid for the wave impedance of the different modes and also for their propagation speed [41, 42]. The propagation speed of both modes can be different since the spatial distribution of the electrical field is different. In the case of PCB waveguides (see Figure 2.70), for one mode the electrical field may be bounded more to the dielectric board than for the other. Therefore, both modes may travel with different speed.

The balanced modes a_b and b_b is connected with the wave impedance Z_b and the unbalanced modes a_u and b_u with Z_u. Unfortunately, the notation of wave impedance is not unique in the literature, since also the odd and even mode impedances Z_odd and Z_even are in use. Figure 2.70 opposes the impedance definitions for the example of a coupled microstrip line.

If the coupled line is purely fed by a balanced mode, the wave 'sees' the impedance Z_b between both lines. This is the same as twice the impedance Z_odd between an individual line and ground:

$$Z_\mathrm{b} = 2Z_\mathrm{odd} \tag{2.218}$$

For the unbalanced mode, the total impedance Z_u which the wave is 'seeing' is that between both lines and ground. Since Z_even is defined between one line and ground, it follows:

$$Z_\mathrm{u} = \frac{1}{2} Z_\mathrm{even} \tag{2.219}$$

The strength of the coupling between the two lines is characterized by the coupling factor

$$k = \frac{Z_{\text{even}} - Z_{\text{odd}}}{Z_{\text{even}} + Z_{\text{odd}}} \tag{2.220}$$

The direct measurement of mixed-mode parameters is often a crucial task due to the lack of corresponding measurement devices. But as long as the differential device behaves linearly, their measurement can be performed via the determination of the four-port parameters with single-ended measurement devices and a subsequent transformation by modal decomposition (see Annex B.6). In the case shown there, it is recommended to design the feeding lines of the differential device according to

$$Z_0^2 = Z_b Z_u = Z_{\text{odd}} Z_{\text{even}} \tag{2.221}$$

in which Z_0 is the characteristic impedance of the measurement device. Many often, one simply uses two uncoupled lines of identical length and the impedance Z_0, hence

$$Z_b = 2Z_0 \quad \text{and} \quad Z_u = \frac{1}{2} Z_0$$

By that choice, one is also sure to have the same propagation speed for the balanced and unbalanced mode since the structure of the electrical field within the feeding lines is identical for both modes.

2.5.4 Determination of *N*-Port Parameters

We have seen from previous discussions that we can lead back any *N*-port description to a formal MiMo system of *N* input and *N* output signals. We have modelled its input–output behaviour in time or frequency domain by the expressions (2.139) and (2.154):

$$\mathbf{y}(t) = \mathbf{g}(t) * \mathbf{x}(t) \tag{2.222}$$

$$\underline{\mathbf{Y}}(f) = \underline{\mathbf{G}}(f)\,\underline{\mathbf{X}}(f) \tag{2.223}$$

Herein, $\mathbf{x}(t), \underline{\mathbf{X}}(f)$ represents a set of N stimulus signals which may be either a current, a voltage or a wave. They are given either in time or frequency domain. $\mathbf{y}(t), \underline{\mathbf{Y}}(f)$ is a corresponding set of system responses and the wanted system functions are $\mathbf{g}(t), \underline{\mathbf{G}}(f)$ which may be *Z*-, *Y*-, *A*-, *S*-parameters or others.

The objective of the measurement is to determine the whole set of characteristic IRFs $\mathbf{g}(t)$ or FRFs $\mathbf{G}(f)$ which covers N^2 independent functions ($N(N+1)/2$ functions in case of passive networks due to their reciprocity). We will go now into the matter how this can be done. Basically, we have several options. In order to be the most general, we will assume that the wanted functions are mutually independent. Note that this must not always be the case as we have seen from the discussions about state space model (Section 2.4.4 – modal decomposition) and the new developments concerning compressed sensing (Section 2.4.5.3). Therefore, the expense to measure the DUT behaviour may be reduced under specific conditions compared to the approaches we will discuss below.

Stimulation with a set of uncorrelated wideband signals: All ports of the DUT are stimulated in parallel and its reaction is also measured at all ports. Certainly, the system reactions represent a superposition of the stimulation from all ports so that the question arises how to separate them. For that purpose, we convert (2.222) in a correlation relation:

$$\mathbf{y}(t) * \mathbf{x}^T(-t) = \mathbf{g}(t) * \mathbf{x}(t) * \mathbf{x}^T(-t) \tag{2.224}$$

where the convolution product of two signal vectors represent a matrix of correlation functions. If the stimulation signals at the individual ports are selected in such a way that they are mutually uncorrelated and of large bandwidth, we yield

$$\mathbf{x}(t) * \mathbf{x}^T(-t) = \begin{bmatrix} x_1(t) \\ x_2(t) \\ \vdots \\ x_N(t) \end{bmatrix} * \begin{bmatrix} x_1(-t) & x_2(-t) & \cdots & x_N(-t) \end{bmatrix}$$
$$\approx \delta(t)\,\mathbf{I} \quad \text{for} \quad C_{nm}(t) = x_n(t) * x_m(-t) = \begin{cases} \delta(t); & n = m \\ 0; & n \neq m \end{cases}$$

so that (2.224) can be simplified to

$$\mathbf{y}(t) * \mathbf{x}^T(-t) \propto \mathbf{g}(t) \Rightarrow C_{\mathrm{y}_m \mathrm{x}_n}(t) \propto g_{mn}(t) \tag{2.225}$$

The challenge of the measurement method is to provide the set of N uncorrelated UWB signals. Binary PN-signals, for example Gold-codes approximately fulfil this condition (see also Section 3.3.6). Larger flexibility to create uncorrelated signals is given by the multi-sine concept (see Sections 2.3.7 and 3.5) which, however, requires elaborate arbitrary waveform generators.

N -fold repetition of the measurement: Again all ports are stimulated and measured in parallel as before. But now, we repeat the measurement N-times with a different set of stimulation signals in each case. The different signals are handled by extending the signal vectors $\mathbf{x}(t)$ and $\mathbf{y}(t)$ to matrices in which every column represents an individual measurement (assigned by the second index), that is

$$\mathbf{X}_m(t) = \begin{bmatrix} \mathbf{x}_1(t) & \mathbf{x}_2(t) & \cdots & \mathbf{x}_N(t) \end{bmatrix} = \begin{bmatrix} x_{11}(t) & x_{12}(t) & \cdots & x_{1N}(t) \\ x_{21}(t) & x_{22}(t) & \cdots & x_{2N}(t) \\ \vdots & \vdots & \ddots & \vdots \\ x_{N1}(t) & x_{N2}(t) & \cdots & x_{NN}(t) \end{bmatrix}$$

$$\mathbf{Y}_m(t) = \begin{bmatrix} \mathbf{y}_1(t) & \mathbf{y}_2(t) & \cdots & \mathbf{y}_N(t) \end{bmatrix} = \begin{bmatrix} y_{11}(t) & y_{12}(t) & \cdots & y_{1N}(t) \\ y_{21}(t) & y_{22}(t) & \cdots & y_{2N}(t) \\ \vdots & \vdots & \ddots & \vdots \\ y_{N1}(t) & y_{N2}(t) & \cdots & y_{NN}(t) \end{bmatrix}$$

We insert these matrices in (2.222) resulting in a set of N^2 linear equations.

$$\mathbf{Y}_m(t) = \mathbf{g}(t) * \mathbf{X}_m(t) \tag{2.226}$$

To solve these equations, we apply again (2.224) and require from our set of test signals that they meet a corresponding condition as above:[32)]

$$\mathbf{y}_m(t) * \mathbf{x}_m^T(-t) = \mathbf{g}(t) * \mathbf{x}_m(t) * \mathbf{x}_m^T(-t) \propto \mathbf{g}(t) \quad \text{for } \mathbf{C}_{\mathbf{x}_m\mathbf{x}_m}(t) \propto \mathbf{I}\delta(t) \tag{2.227}$$

Note that $\mathbf{C}_{\mathbf{x}_m\mathbf{x}_m}(t)$ represents a matrix of correlation functions. Its diagonal is formed from a sum of auto-correlation functions and the outer diagonal elements are sums of cross-correlations between the individual stimulus signals where they should mutually cancel out.

In order to simplify the measurement arrangement, we will assume that the individual stimulation signals all have identical shape. They are only modified by their amplitude. Thus, $\mathbf{x}_m(t)$ can be expressed by

$$\mathbf{x}_m(t) = x_0(t)\begin{bmatrix} A_{11} & A_{12} & \cdots & A_{1N} \\ A_{21} & A_{22} & \cdots & A_{2N} \\ \vdots & \vdots & \ddots & \vdots \\ A_{N1} & A_{N2} & \cdots & A_{NN} \end{bmatrix} = x_0(t)\mathbf{A} \tag{2.228}$$

so that the side condition $\mathbf{C}_{\mathbf{x}_m\mathbf{x}_m}(t) \propto \mathbf{I}\delta(t)$ in (2.227) is fulfilled if the amplitudes of the stimulation signals are selected to meet the condition

$$\mathbf{A}\,\mathbf{A}^T = \mathbf{I} \tag{2.229}$$

Section 3.3.6 demonstrates a practical implementation of such principle.

For DUT stimulation with pure sinusoids, we modify (2.226) and (2.227) to

$$\underline{\mathbf{Y}}_m(f)\underline{\mathbf{X}}_m^{\mathrm{H}}(f) = \underline{\mathbf{G}}(f)\underline{\mathbf{X}}_m(f)\underline{\mathbf{X}}_m^{\mathrm{H}}(f) \tag{2.230}$$

which we can solve by

$$\underline{\mathbf{G}}(f) = \underline{\mathbf{Y}}_m(f)\underline{\mathbf{X}}_m^{\mathrm{H}}(f)\left(\underline{\mathbf{X}}_m(f)\underline{\mathbf{X}}_m^{\mathrm{H}}(f)\right)^{-1} \tag{2.231}$$

Herein, $\underline{\mathbf{X}}_m$ represents a matrix of complex amplitudes related to the sine waves stimulating the DUT. Ideally, we would require in accordance to (2.229) or the side condition in (2.227) that

$$\underline{\mathbf{X}}_m(f)\underline{\mathbf{X}}_m^{\mathrm{H}}(f) \propto \mathbf{I} \tag{2.232}$$

which would indeed provide best conditions for matrix inversion in (2.231). However, we can relax this condition as long as $\underline{\mathbf{X}}_m(f)\underline{\mathbf{X}}_m^{\mathrm{H}}(f)$ behaves not as bad if it is inverted. The quality of the matrix inversion in (2.231) can be assessed by the condition number $\kappa\left(\underline{\mathbf{X}}_m\right)$ of the stimulation matrix $\underline{\mathbf{X}}_m$. The condition number κ of a matrix is defined by the ratio between largest $\sigma_{\max}$

32) By the way, (2.225) and (2.227) are closely related since we can cascade in time the N-signals applied to any port and consider them as a single one which would lead to the first approach.

and smallest singular value $\sigma_{\min}$ [38, 43] (see Annex A.6 for singular value):

$$\kappa(\underline{\mathbf{X}}_m) = \frac{\sigma_{\max}(\underline{\mathbf{X}}_m)}{\sigma_{\min}(\underline{\mathbf{X}}_m)} \tag{2.233}$$

It is a measure indicating how the random errors in $\underline{\mathbf{X}}_m$ transfer to the random errors in $\underline{\mathbf{G}}$. $\kappa(\underline{\mathbf{X}}_m)$ tends to unity, if $\underline{\mathbf{X}}_m\underline{\mathbf{X}}_m^{\mathrm{H}}$ tends to an indentity matrix. If the condition number is too far from unity, a better set of stimulation signals should be applied, and noise dominates the measurement.
We have seen, condition (2.232) must not perfectly met to solve (2.231). Related to our time domain consideration above, we can conclude that the stimulation signals must not be perfectly de-correlated. But if so, we have to apply de-convolution (which is actually done by (2.231) in frequency domain) and cannot further deal with the simplifications in (2.227).
Sequential stimulation of the ports: The simplest solution to guarantee de-correlated sounding signals is to relinquish simultaneous stimulation of all ports rather they are stimulated successively. Hence, all matrices of stimulus signals introduced above are purely diagonal:

$$\begin{aligned} \mathbf{x}_m(t) &= x_0(t)\,\mathbf{I} \\ \underline{\mathbf{X}}_m(f) &= \underline{X}_0(f)\,\mathbf{I} \end{aligned}$$

The measurements of course are performed at all ports in parallel during this procedure as before.
Over-determined measurement: Under specific conditions it can be useful to repeat the measurements more often than required by always taking a different set of test signals. The reason could be, for example, noise suppression by keeping low probability of interference or detection. This leads to a set of measurement matrices consisting of more than N raws:

$$\left.\begin{aligned} \mathbf{x}_m(t) &= [\mathbf{x}_1(t) \quad \mathbf{x}_2(t) \quad \cdots \quad \mathbf{x}_N(t) \quad \cdots \quad \mathbf{x}_M(t)] \\ \mathbf{y}_m(t) &= [\mathbf{y}_1(t) \quad \mathbf{y}_2(t) \quad \cdots \quad \mathbf{y}_N(t) \quad \cdots \quad \mathbf{y}_M(t)] \end{aligned}\right\}; \quad M > N$$

and

$$\left.\begin{aligned} \underline{\mathbf{X}}_m(t) &= [\underline{\mathbf{X}}_1(t) \quad \underline{\mathbf{X}}_2(t) \quad \cdots \quad \underline{\mathbf{X}}_N(t) \quad \cdots \quad \underline{\mathbf{X}}_M(t)] \\ \underline{\mathbf{Y}}_m(t) &= [\underline{\mathbf{Y}}_1(t) \quad \underline{\mathbf{Y}}_2(t) \quad \cdots \quad \underline{\mathbf{Y}}_N(t) \quad \cdots \quad \underline{\mathbf{Y}}_M(t)] \end{aligned}\right\}; \quad M > N$$

so that the resulting input–output relations are over-determined. We can solve such over-determined equation in the Gaussian sense applying (2.227) or (2.231) as long as we make sure that the condition number of the stimulation matrix keeps values close to unity. The matrix $\underline{\mathbf{X}}_m^{\mathrm{H}}\left(\underline{\mathbf{X}}_m\underline{\mathbf{X}}_m^{\mathrm{H}}\right)^{-1}$ represents a Moore–Penrose pseudo-inverse [38, 43] (see also Annex A.6). For further remarks on the determination of the DUT behaviour based on noisy data, see the last part of the following chapter.

2.6
Measurement Perturbations

Let us recall the purpose of our considerations. It is to quantify and qualify the behaviour of test objects or scenarios. We will do this by stimulating the objects of interest and to measure their reaction. Hence, as to be seen from the previous considerations, we need to gather signals in the time or frequency domain. That is, we have to capture at predefined time instants or frequencies the amplitude of one or more signals.[33] But both the collection of the amplitude value as well as the appointment of the measurement moment and the measurement frequency are corrupted by errors. In other words, the abscissa and ordinate values of our recorded data set are partially erroneous.

These errors may be of random or systematic (i.e. deterministic) nature. We will discuss their impact onto the measurement and how to deal with them. We will distinguish between random errors (additive random noise; quantization noise; jitter and phase noise) as well as deterministic errors of linear or non-linear nature.

2.6.1
Additive Random Noise and Signal-to-Noise Ratio

Up to a certain degree, every data recording is subjected to some randomness. At this point, we will only respect additive and ergodic perturbations $n(t) \sim N(0, \sigma^2)$ superimposed to the wanted (deterministic) signal $x_0(t)$:

$$x_{\mathrm{cap}}(t) = x_0(t) + n(t) \tag{2.234}$$

Here, $x_{\mathrm{cap}}(t)$ represents the actually captured signal. Later, we will extend our consideration also to random errors of the 'time axis' (refer to Sections 2.6.3 and 3.2.3).

The random signal $n(t)$ may have several physical reasons, for example the thermal movement of the charge carrier, scattering of electrons at crystal impurities or carrier generation and recombination (flicker noise), shot noise or others as well as manmade perturbations caused by radio emitters and so on.

To exemplify, we will shortly consider as a source of noise the thermal agitation of free charges which is usually referred as thermal or Johnson–Nyquist noise. Since the signal source and the noise source are independent, we can suppose that the wanted signal $x_0(t)$ and the noise $n(t)$ are not correlated. Since thermal noise can be considered as an ergodic signal, we can state

$$C_{xn}(\tau) = \int x_0(t)n(t+\tau)\mathrm{d}t = \mathrm{E}\left\{x_0(t)\underset{\sim}{\mathbf{n}}(t+\tau)\right\} = 0 \tag{2.235}$$

33) Note that frequency domain measurements are usually linked to the capturing of two signals namely the real and imaginary part or the absolute value and the phase even if we only deal with one measurement channel.

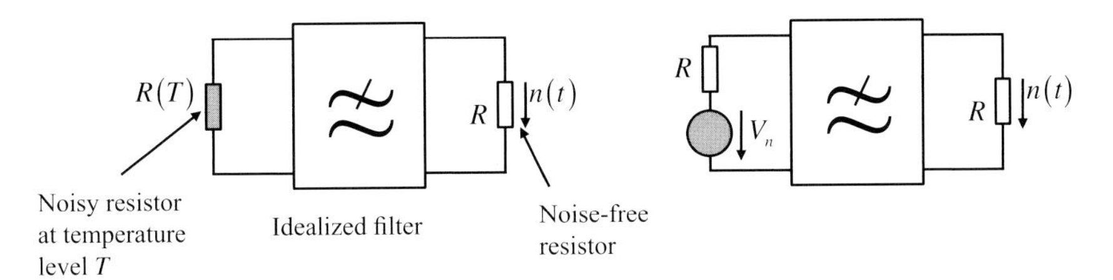

Figure 2.71 Basic circuit model with noisy resistor. Noise sources are grey filled. In the right part, the noisy resistor is modelled by a noisy voltage source and an idealized noise-free resistor.

Furthermore, it is known that thermal noise is Gaussian distributed with zero mean $n(t) \sim N(0, \sigma^2)$ so that its average power P_{n} can be expressed by its variance σ^2 (see (2.43))

$$P_{\text{n}} \sim n_{\text{rms}}^2 = \|n(t)\|_2^2 = \overline{n^2(t)} = \sigma^2 \tag{2.236}$$

The spectral power density of thermal noise is frequency independent which is usually expressed by the term white noise (at least if we restrict ourselves to a frequency band below some tens of THz). So finally, $n(t)$ is assigned as 'additive white Gaussian noise (AWGN)'.

As source of thermal noise, we usually consider a resistor R at temperature level T (see Figure 2.71). The available noise power spectral density of this resistor is given by

$$\underset{\cdot}{\Phi}_{\text{n}} = 2\underset{\cdot\cdot}{\Phi}_{\text{n}} = kT \tag{2.237}$$

Herein, $k = 1.38 \times 10^{-23}$ W s/K is the Boltzmann's constant and T is the absolute temperature in K. For the standard temperature $T_0 = 290$ K ($\approx 17\,^\circ$C), the spectral density of the available thermal noise power is

$$\underset{\cdot}{\Phi}_{\text{n0}}(T_0 = 290\ \text{K}) = 4 \times 10^{-21}\ \text{W/Hz} = -174\,\text{dBm/Hz}$$

The available power of a signal source refers to the maximum power which can be transferred to a load. This situation is given if the internal resistor of the source is equal to the load resistor as it is demonstrated in Figure 2.71. For simplicity, the load resistor is considered to be noise free here. It represents, for example, the input impedance of a measurement receiver. In dependence from the bandwidth $\underset{\cdot}{B}_n$ of the input filter of the receiver, the power dissipation in the load resistor is

$$P_{\text{n}} = \frac{n_{\text{rms}}^2}{R} = \frac{\sigma^2}{R} = \underset{\cdot}{\Phi}_{\text{n}}(T)\underset{\cdot}{B}_{\text{n}} = kT\underset{\cdot}{B}_{\text{n}} \tag{2.238}$$

and hence the open-circuit voltage of the noise source results in

$$V_{\text{n,rms}} = 2n_{\text{rms}} = \sqrt{4kT\underset{\cdot}{B}_{\text{n}}R} \tag{2.239}$$

Obviously, the measurement values will be increasingly affected by noise if the bandwidth of the data capturing system increases. This is an important observation namely for ultra-wideband sensors since there the bandwidth must be large by principle.

We will introduce some concepts how to deal with additive random perturbations. In order to stay simple, we will suppose initially a frequency-independent perturbation as in the case of thermal noise. But we will allow an arbitrary strength of the perturbations caused by any effects and not only by thermal ones.

2.6.1.1 **Signal-to-Noise Ratio (SNR)**

The SNR-value is an important quality measure of a measurement or rather a measurement device. It indicates the corruption of the captured data by random errors and noise, respectively. There are several ways to define a SNR-value. We will use the ratio between the peak power of the non-disturbed signal and the average power of the noise:

$$\mathrm{SNR}_0 = \frac{P_{\mathrm{peak}}}{P_{\mathrm{n}}} = \frac{\|x_0(t)\|_\infty^2}{\|n(t)\|_2^2} = \frac{x_{\mathrm{peak}}^2}{\sigma^2} \tag{2.240}$$

(2.240) demonstrates various notations of the same issue. Usually, the SNR-value is given in dB, that is $\mathrm{SNR}\,[\mathrm{dB}] = 10 \log \mathrm{SNR}$. The subscript 0 shall indicate that the SNR-value refers to the original receive signal which is not yet subjected to any data processing.

Figure 2.72 illustrates the destruction of the waveform for different degrees of perturbation. The example for $\mathrm{SNR} = 0\,\mathrm{dB}$ refers to the case in which the signal peak power and noise average power are the same. Obviously, it will be hardly possible to detect the rectangular pulses buried by the noise. Note that for practical purposes, the noise voltage is roughly distributed between the limits of $\pm 3n_{\mathrm{rms}} = \pm 3\sigma$. Therefore, the amplitude variations of the noise exceed the swing of the wanted signal.

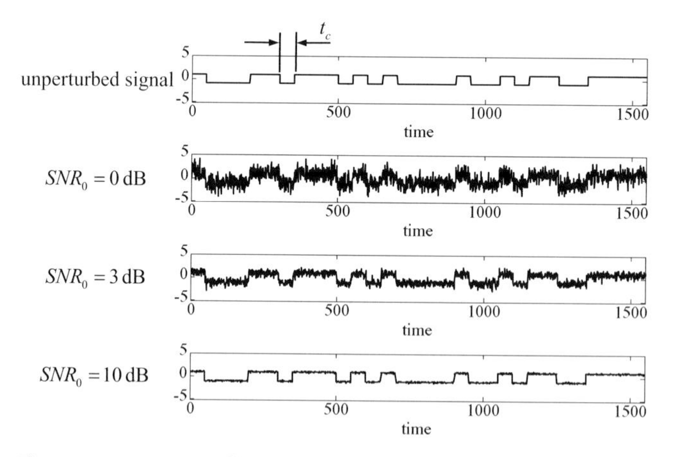

Figure 2.72 Sequence of bipolar rectangular pulses perturbed by white Gaussian noise for various SNR-values.

The random errors are mostly determined by the measurement conditions which cannot be influenced in many cases. That's why the noise has to be accepted as it is (we suppose here of course that the measurement conditions are carefully selected in order to avoid unnecessary perturbations). Here, the question arises, what can we do in order to improve the situation? We will refer to three methods.

2.6.1.2 **Sliding Average**

If the temporal fluctuations in the noise signal behave more abrupt than the variations in the signal of interest, the noise can be reduced by limiting the bandwidth. Abrupt signal variations cause a wide spectrum, that is the noise has a large bandwidth and the signal occupies only a small part of the spectrum. Hence, we can reduce the noise power by cutting out the useless spectral parts. Figure 2.73a demonstrates the principle of signal smoothing applying averaging within a sliding window (low-pass filtering). On the right-hand side (part b), the noise reduction for the $\mathrm{SNR}_0 = 0\,\mathrm{dB}$ case (Figure 2.72) is shown for several integration window lengths t_I

For the rough signal, the rms value of the noise results from $n_{\mathrm{rms}}^2 = \sigma^2 = \underset{..}{\Psi}_\mathrm{n}\,\underset{..}{B}_\mathrm{n}$ in which $\underset{..}{\Psi}_\mathrm{n}$ is the spectral noise power density and $\underset{..}{B}_\mathrm{n}$ is the actual bandwidth of the noise. This bandwidth is given, for example, by the bandwidth of the

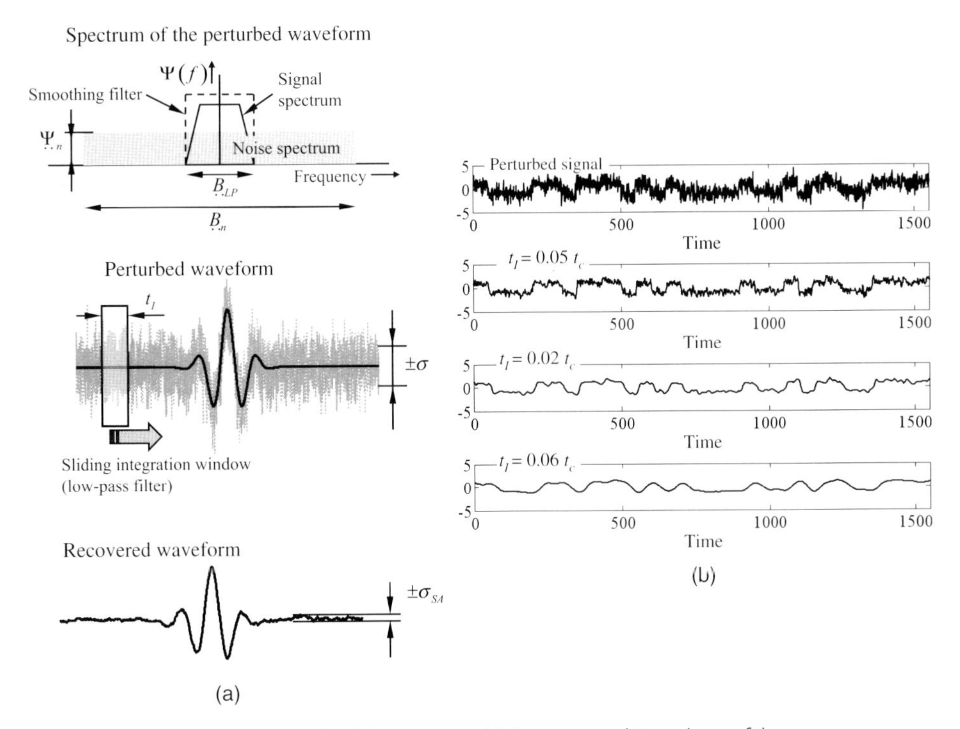

Figure 2.73 Signal improvement by sliding average. (a) Spectrum and time shape of the perturbed signal and after application of sliding average and (b) examples from Figure 2.72 for different window lengths t_I.

measurement receiver front-end. If we reduce the bandwidth by a smoothing filter to $\underset{..}{B}_{LP}$, the remaining variance of the signal is

$$\sigma_{SA}^2 = \underset{..}{\Psi}_n \underset{..}{B}_{LP} = \sigma^2 \frac{\underset{..}{B}_{LP}}{\underset{..}{B}_n} \tag{2.241}$$

The smoothing filter may be a low-pass filter of any type having the equivalent noise bandwidth of $\underset{..}{B}_N = \underset{..}{B}_{LP}$. Given the filter FRF or IRF, the equivalent noise bandwidth calculates from (compare also with (2.74))

$$\underset{..}{B}_N G^2(f=0) = \int_{-\infty}^{\infty} |\underline{G}(f)|^2 \, df = \int_{-\infty}^{\infty} g^2(t) \, dt = \|g(t)\|_2^2 \tag{2.242}$$

A smoothing filter with rectangular weighting function $g(t) = \mathrm{rect}(t/t_I)/t_I$ is also called short-time integrator or boxcar integrator since its output signal is given by integration over the (finite and short) time t_I:

$$\bar{x}_{t_I}(t) = \frac{1}{t_I} \mathrm{rect}\left(\frac{t}{t_I}\right) * x(t) = \frac{1}{t_I} \int_{t-t_I/2}^{t+t_I/2} x(\xi) d\xi$$

It may be implemented by FIR filter with identical coefficients. The equivalent noise bandwidth is $\underset{..}{B}_N = t_I^{-1}$ due to the rectangular IRF for which holds $\underset{..}{B}_{2,\mathrm{rect}} t_I = \underset{..}{B}_N t_I \approx 1$ (see Table 2.1). Hence, we get

$$\sigma_{SA}^2 = \frac{\sigma^2}{\underset{..}{B}_n t_I} = \frac{\sigma^2}{\mathrm{TB}} \tag{2.243}$$

By referring the bandwidth of the low-pass filter to an integration time we were able to apply a TB product to indicate the noise reduction. The right part of Figure 2.73 demonstrates the effect of smoothing by a sliding rectangular window of different length. It is clearly visible that an increasing window length reduces the noise but it will also affect the signal shape if its length becomes too large. This will limit the maximum TB product. Of course, one can apply for noise suppression other low-pass filters as a sliding averager but this will not basically change the situation. The shape distortions of the signal $x_0(t)$ will only be different but they will increase proportionally to the strength of the noise suppression as well.

One may achieve a better noise suppression by keeping the shape of $x_0(t)$ if one tries to decompose the perturbed signal in a set of functions which are better suited to describe the signal $x_0(t)$ than it would be possible by sine waves where the latter one is presumed if 'usual' filters are applied. In the case $x_0(t)$ represents an IRF, wavelet decomposition can provide better results. Here the unknown signal $x_{cap}(t)$ is decomposed by an appropriately selected reference function, the so-called mother wavelet, that is it is represented by a sum of mother wavelets which are scaled and shifted in time and amplitude (see also (2.55) for the basic idea of signal decomposition). The noisy parts of the signal will be mainly reflected by the wavelet components of large time scaling and they will have small amplitudes. By excluding wavelet components of too large time scaling and too small amplitudes, one can gain a signal exempt from noise after applying the inverse wavelet transform. The approach is called wavelet de-noising. For more details, the reader is referred to Refs [44, 45] or to MATLAB which provides a rich collection of wavelet tools and mother wavelet.

Finally, it should be underlined that noise reduction by any of the above-mentioned methods can be used for reduction of data amount by down sampling. This may be of major concern in many UWB applications due to the quantity of data to be handled. The filter approaches will not provide reasonable noise improvement in case of data sampling close to the Nyquist rate.

2.6.1.3 Synchronous Averaging

Synchronous averaging (also referred as stacking) supposes a periodic repetition of the captured signal in order to permit stacking of a number of measurements from which the average is finally taken. We model the captured signal as $x_p(t) = x_0(t - mt_p) + n(t)$. Herein, $x_0(t)$ is the wanted periodic signal of period t_p and $n(t)$ is random noise. In order gain a noise reduced version of the signal, we stack a number of periods as illustrated in Figure 2.74. Usually, one can suppose that random errors from period to period are uncorrelated. Considering the captured signal as random process $\underset{\sim}{x}_p(t)$, expected value, variance and signal to noise ration of the averaged (stacked) signal are (refer to Annex A.3):

$$\begin{aligned}
\bar{x}_p(t') &= \frac{1}{p}\sum_{m=0}^{p-1} x_p(t - mt_p); \quad t' = \operatorname{mod}(t, t_p) \\
E\left\{\bar{\underset{\sim}{x}}_p(t')\right\} &= E\left\{\frac{1}{p}\sum_{m=0}^{p-1}\left(x_0(t - mt_p) + \underset{\sim}{n}(t)\right)\right\} = x_0(t') \\
\operatorname{var}\left\{\bar{\underset{\sim}{x}}_p(t')\right\} &= \operatorname{var}\left\{\frac{1}{p}\sum_{m=0}^{p-1}\left(x_0(t - mt_p) + \underset{\sim}{n}(t)\right)\right\} = \frac{1}{p^2}\sum_{m=0}^{p-1}\operatorname{var}\left\{\underset{\sim}{n}(t)\right\} = \frac{\sigma^2}{p} = \sigma_p^2
\end{aligned} \tag{2.244}$$

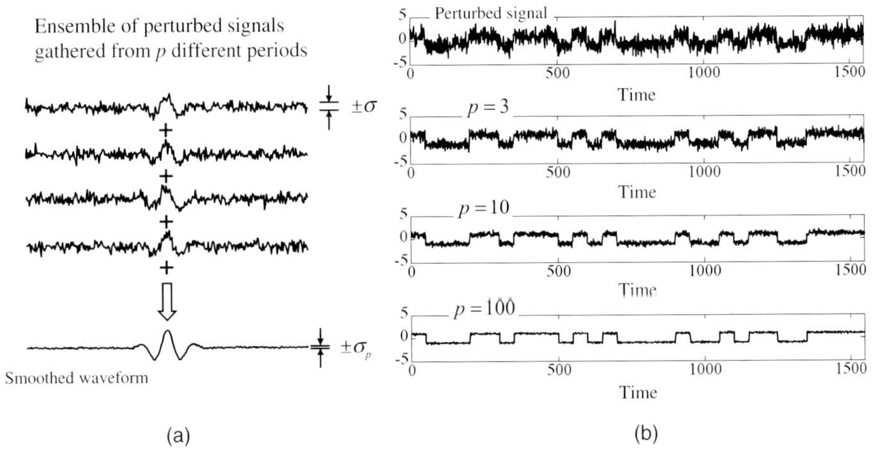

Figure 2.74 Noise reduction by synchronous averaging. (a) Principle and (b) examples from Figure 2.72 for different averaging numbers.

$$\begin{aligned} \mathrm{SNR_{av}} &= p\,\mathrm{SNR_0} \\ \mathrm{SNR_{av}[dB]} &= g_{\mathrm{av}} + \mathrm{SNR_0[dB]} \end{aligned} \tag{2.245}$$

The number p represents the number of repeated measurements. It is often also expressed by the term averaging gain: $g_{\mathrm{av}}[\mathrm{dB}] = 10 \log p$.

Figure 2.74b refers also to the $\mathrm{SNR_0} = 0\,\mathrm{dB}$ case from Figure 2.72 to illustrate the noise improved with increasing averaging number p. In contrast to the previous example (Figure 2.73), the level of noise reduction does not affect the signal shape. Therefore, any SNR-value can be gained in principle by synchronous averaging. However, averaging increases the measurement time by a factor of p; thus, now the limitation in performance improvement is given by the maximum allowed recording time T_R or time on target t_{ot} (see also (2.89)).

A further restriction is caused from time unstable measurement devices. If the period of the signals cannot be exactly guaranteed over a sufficient long time, the efficiency of averaging will degrade since the mutual match of signal periods will be disturbed by time drift. In the case of ultra-wideband sensors, time stability may be an important issue.

Assuming the data amount of the captured data is H_0 [bit], synchronous averaging reduces the data amount to

$$H_{\mathrm{av}} = H_0 \frac{\mathrm{lb}\, p}{p} \tag{2.246}$$

2.6.1.4 **Matched Filter/Correlator**

With the third method, we like to detect a signal with known shape in a noisy environment. For that purpose, we compare a reference signal of the wanted shape with the disturbed measurement signal. Since the actual time position of the receive signal is not known a priori, one has to search for an optimal match between both signals. Hence, one of the signals is shifted in time as long as the best match is achieved – therefore the name matched filtering. If both signals match together in shape and time position, we get a high peak at the filter output which overtops the noise as illustrated in Figure 2.75. The degree of peak elevation over the noise level gives a performance figure on how reliable the detection of the searched signal is.

We presume a signal mixture consisting of the wanted signal $x_0(t)$ and other components as noise or interference signals. Technically, the search for the best match may be done by two approaches – first by a correlator or second by a filter with a very specific impulse response, that is $g(t) = x_0(-t)$. Figure 2.76 illustrates both methods. The correlator searches for the best match by delaying the reference waveform by τ. The second method uses a filter through which the receive signal is propagating. If the impulse response of the filter coincides with the searched signal, a high signal peak will appear at the output.

In the case of digital operation, the matched filter simply represents a FIR filter of length M. Its coefficients $b[m]$ are identical to the time samples of the wanted waveform but inversely arranged, that is $b[m] = x_0[M - m];\ m \in [0, M - 1]$. The use of these methods for ultra-wideband purposes may be very beneficial but their

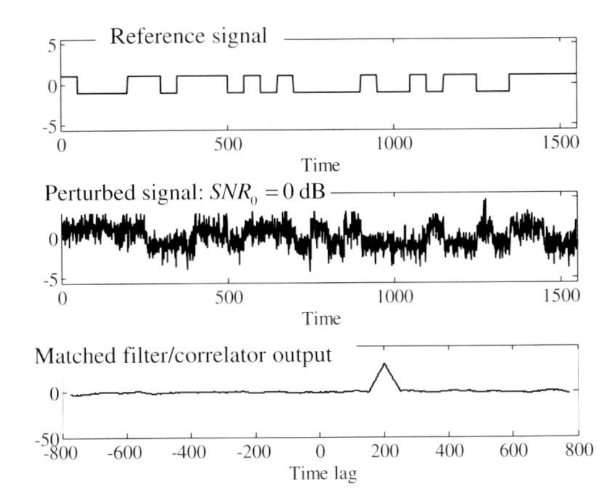

Figure 2.75 Example which shows the matching of a reference waveform with a perturbed measurement signal. For a perfect match, the time shapes of wanted and reference signal must be identical and their time positions must coincide. In the shown case the reference signal is shifted about 200 time samples against the reference waveform.

practical implementation is also somewhat challenging. Some of these issues will be discussed in Chapter 3.

Following Annex B.4, the signal-to-noise ratio of the output signal $y(t)$ can be expressed by the relation

$$\mathrm{SNR}_y = \frac{\|y(t)\|_\infty^2}{\sigma_y^2} = \frac{\|x_0(t)\|_2^2}{\sigma^2} t_1 \underset{\cdot\cdot}{B}_{\mathrm{n}} \tag{2.247}$$

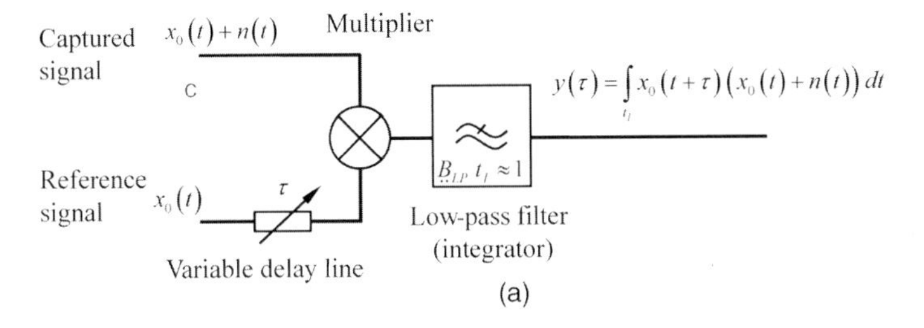

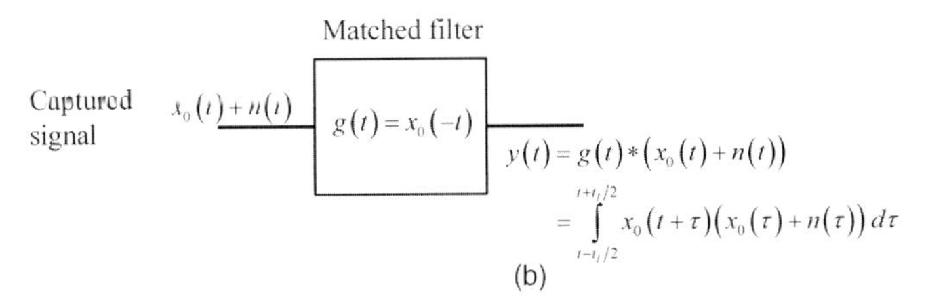

Figure 2.76 Two realizations of the matched filter approach by correlation (a) and a filter of IRF $g(t) = x_0(-t)$ (b).

$\underset{\cdot\cdot}{B}_{\mathrm{n}}$ is the bandwidth of the signals and t_{I} is either the integration time of the correlator or the duration of the matched filter IRF. Integration time and bandwidth of the low-pass filter in the correlator are approximately related by $t_{\mathrm{I}} \underset{\cdot\cdot}{B}_{\mathrm{LP}} \approx 1$.

If we further combine synchronous averaging with correlation/matched filtering, we achieve from (2.247) and (2.244) for the total SNR-value:

$$\mathrm{SNR}_{\mathrm{tot}} = p\frac{\|y(t)\|_{\infty}^{2}}{\sigma_{y}^{2}} = \frac{\|x_0(t)\|_2^2}{\sigma^2} p t_{\mathrm{I}} \underset{\cdot\cdot}{B}_{\mathrm{n}} = \frac{\mathfrak{E}}{\underset{\cdot\cdot}{\Phi}_{\mathrm{n}}} \tag{2.248}$$

$\underset{\cdot\cdot}{\Phi}_{\mathrm{n}} = \underset{\cdot\cdot}{\Psi}_{\mathrm{n}}/R_0 = \sigma^2/\underset{\cdot\cdot}{B}_{\mathrm{n}} R_0$ is the spectral density of the noise and $\mathfrak{E} = \|x_0(t)\|_2^2 t_{\mathrm{D}}/R_0$ is the totally captured energy of the considered signal segment of duration $t_{\mathrm{D}} = p t_{\mathrm{I}}$ which equals the recording time $T_{\mathrm{R}} = t_{\mathrm{D}}$ in a properly designed device. The energy stored by the integrator or the filter is $\mathfrak{E}_{\mathrm{I}} = \|x_0(t)\|_2^2 t_{\mathrm{I}}/R_0$ which is p-times accumulated by synchronous averaging. Basically it does not matter whether correlation/matched filtering or synchronous averaging is performed first so that for the technical realization one should start with the simpler procedure.

The simple relation (2.248) is of major importance for a measurement. It tells us that we have to capture as much as possible energy from the process under observation in order to get a reliable measurement result. Introducing (2.240) into (2.248), we get an improvement ratio of the noise performance referred to the captured signal having an SNR-value of SNR_0:

$$\mathrm{SNR}_{\mathrm{tot}} = \frac{\|x_0(t)\|_2^2}{\sigma^2} T_{\mathrm{R}} \underset{\cdot\cdot}{B}_{\mathrm{n}} = \frac{\mathrm{TB}}{\mathrm{CF}^2} \mathrm{SNR}_0 \tag{2.249}$$

Here, we replaced $T_{\mathrm{R}} \underset{\cdot\cdot}{B}_{\mathrm{n}}$ by the general TB product as introduced in (2.81). It is also termed as processing gain

$$10 \log \mathrm{TB} = g_{\mathrm{proc}}[\mathrm{dB}] = g_{\mathrm{av}} + g_{\mathrm{comp}} = 10 \log p + 10 \log t_{\mathrm{I}} \underset{\cdot\cdot}{B}_{\mathrm{n}} \tag{2.250}$$

which is composed from the averaging gain and compression gain. If we further respect that every real technical receiver will not be able to capture lossless the energy supplied by the signal, we end up in

$$\mathrm{SNR}_{\mathrm{tot}} = \eta \frac{\mathfrak{E}}{\underset{\cdot\cdot}{\Phi}_{\mathrm{n}}} = \frac{\eta\,\mathrm{TB}}{\mathrm{CF}^2} \mathrm{SNR}_0 \tag{2.251}$$

where η refers to the efficiency of the receiver which describes its capability to actually accumulate the provided energy $\mathfrak{E}_{\mathrm{acc}} = \eta \mathfrak{E}$ ($\mathfrak{E}_{\mathrm{acc}}$ – by the receiver accumulated energy, $\mathfrak{E} = \|x_0(t)\|_2^2 T_{\mathrm{R}}/R_0$ – from the signal provided energy).

What are the conclusions from the previous consideration?

- The measurement receiver should have a reasonable efficiency η. As we will see later, this is a critical point for most ultra-wideband receivers since most of the currently used receiver conceptions suffer from data loss by sub-sampling or sequential frequency stepping.

- The recording time T_R of the signal should be as long as possible to achieve reliable measurement results. In this connection, it does not matter if the recording time is purely spent for a single (but long) correlation $T_R = t_I$ or the signal is stacked (synchronously averaged) p times before performing the correlation over a shorter signal of duration t_I which also leads to $T_R = pt_I$. Synchronous averaging prerequisites periodic signals which should have a period length of $t_P = t_I$. Since synchronous averaging is much easier to handle as correlation, periodic signals are to be preferred for many measurement tasks. In case of digital processing, averaging should be performed before the correlation since it reduces the length of the data vectors. The specification of T_R and t_I for a specific measurement scenario is subjected to some constraints:
  - T_R should be short enough to ensure approximately time invariance during data capturing (see (2.89)) if the scenario contains moving objects.
  - In scenarios with oscillating variations, the Nyquist theorem with respect to the bandwidth of these oscillations has to be respected. That is, the measurement rate $r_m = T_R^{-1}$ must be at least two times higher than the rate of the scenario oscillations.
  - $t_I = t_P$ determines the unambiguity range for radar sensing (see (2.87)), hence it cannot be arbitrarily reduced.
  - $t_P = f_0^{-1}$ determines the line spacing of the signal spectrum. In order to fully exploit the radiation power permitted by the UWB radiation rules the line spacing should be smaller than, for example, $f_0 \leq 1\,\text{MHz}$ according to FCC rules.
- Obviously, the magnitude of the average power P of the signal is of interest and not its peak power value. That is, high peak amplitude $\|x(t)\|_\infty$ of the signal does not bring reasonable SNR improvements as long as signal peaks only appear rarely in the signal (as e.g. in pulse trains with small duty cycles). In such a case, the crest factor CF tends to a large number which partially (or even completely) compensates the effect of noise reduction by correlation.
- The random perturbations result from external and internal sources: $\underset{..}{\Psi}_n = \underset{..}{\Psi}_{n,ext} + \underset{..}{\Psi}_{n,int}$. The internal noise refers to perturbations generated by the measurement device itself. This property is quantified by the noise factor F or the system noise temperature T_S which will be introduced below. The external interference is linked to the actual measurement scenario. It may have different reasons, for example antenna noise, electromagnetic interference by other devices, random fluctuations of the measurement scenario and so on

(2.251) indicates the SNR-value for the best case, that is if the receive signal is exactly known. If, however, an ultra-wideband measurement is done, the receive signal is composed from the (known) stimulus $x_0(t)$ convolved with the (unknown) impulse response $g(t)$ of the test object and we like to know how strong the wanted $g(t)$ is corrupted by noise.

Let us consider for that purpose the signal flow in Figure 2.77. We introduce a time-dependent SNR-value

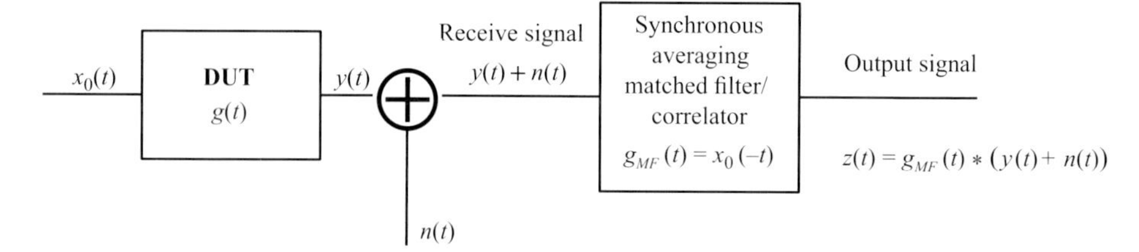

Figure 2.77 Schematics of a perturbed correlation measurement of an unknown device under test (DUT).

$$\mathrm{SNR}_z(t) = \frac{z^2(t)}{\sigma_z^2} \tag{2.252}$$

where $z^2(t)$ is the instantaneous power of the wanted signal (including synchronous averaging) and σ_z^2 refers to the noise level at the output of the correlator or matched filter (note that in this case the matched filter will not be anymore matched to its actual input signal). Restricting ourselves to periodic signals and assuming that the bandwidth of the stimulus B_x is larger than the bandwidth of the DUT B_{DUT}, we can approximately write applying the law of commutativity of the convolution:

$$\begin{aligned} z(t) &= x_0(t) * g(t) * g_{\mathrm{MF}}(t) = x_0(t) * x_0(-t) * g(t) = C_{xx}(t) * g(t) \\ &\approx C_{xx}(0)\tau_{\mathrm{coh}} g(t) \approx \frac{C_{xx}(0)}{\underset{..}{B}_x} g(t) \\ \text{for} \quad C_{xx}(t) &\approx \begin{cases} C_{xx}(0) = p\|x_0(t)\|_2^2 & \text{for } |t| \leq \tau_{\mathrm{coh}}/2 \\ 0 & \text{for } |t| > \tau_{\mathrm{coh}}/2 \end{cases} \end{aligned} \tag{2.253}$$

Insertion of (2.248) in (2.253) provides

$$\mathrm{SNR}_z(t) \approx \tau_{\mathrm{coh}}^2 g^2(t) \mathrm{SNR}_{\mathrm{tot}} \approx g^2(t) \frac{\mathrm{SNR}_{\mathrm{tot}}}{B_x^2} \tag{2.254}$$

As expected, we will have a large SNR-value if $g(t)$ has a large magnitude and the SNR-value will degrease if $g(t)$ settles down.

Applying (2.242), the average SNR-value is

$$\overline{\mathrm{SNR}_z} = \frac{1}{t_{\mathrm{I}}} \int_{t_{\mathrm{I}}} \mathrm{SNR}_z(t)\mathrm{d}t = \frac{\mathrm{SNR}_{\mathrm{tot}}}{\underset{..}{B}_x^2} \frac{1}{t_{\mathrm{I}}} \int g^2(t)\mathrm{d}t = \mathrm{SNR}_{\mathrm{tot}} G^2(0) \frac{\underset{..}{B}_{ng}}{\underset{..}{B}_x^2 t_{\mathrm{I}}} \tag{2.255}$$

Herein, $G(f = 0)$ is the DC gain of the DUT, $\underset{..}{B}_{\mathrm{ng}}$ is the equivalent noise bandwidth of the DUT and $t_{\mathrm{I}} = t_{\mathrm{P}}$ is the integration time of the correlator. This equations shows that we will have in average the best SNR-value if the bandwidth of the DUT and the sounding signal are nearly equal $\underset{..}{B}_{\mathrm{ng}} \approx \underset{..}{B}_x$ and if the signal period of the sounding signal is nearly equal to the settling time of the measurement object $t_{\mathrm{sett}} \approx t_{\mathrm{P}} = t_{\mathrm{I}}$.

As considered so far, the matched filter philosophy was based on matching a filter response to the shape of a wanted signal in order to emphasize it under a mixture of arbitrary signals. Certainly, due to commutativity of convolution, we can also invert the approach by asking for a stimulus signal $x(t)$ which best matches a given transmission channel of IRF $g_0(t)$. Again, the best match is achieved if the output signal is a short pulse, that is if we select the stimulus that holds $x(-t) = g_0(t)$. Such procedures are usually referred as time-reversal techniques [46]. They can be applied to focus the stimulation energy on a selected transmission path or to highlight wanted transmission behaviour within a test scenario. In the case of a radar channel, this could be used to indicate the presence or absence of a specific target. The technical challenge is to provide very wideband signals with pre-defined but arbitrary shape (see also Section 3.5).

2.6.1.5 Device Internal Noise

If a signal is passing through any electronic device, internal noise will be added. The degree of noise accumulation is expressed by the noise factor F or the noise figure NF:

$$F = \frac{\mathrm{SNR_{in}}}{\mathrm{SNR_{out}}} \quad \text{and} \quad \mathrm{NF} = 10 \log F \tag{2.256}$$

Here, the SNR-values refer to the quotient of available signal and noise power at the input and output of the device where $\mathrm{SNR_{in}}$ is identical to the SNR-value of the signal source. One assumes that the input noise is provided purely by thermal agitation in the source resistor. Its temperature is defined to $T_0 = 290$ K. Hence for a device of bandwidth $\underset{\cdot}{B}_\mathrm{n}$, the available noise power is $P_{\mathrm{n}0} = \Phi_{\mathrm{n}0}\,\underset{\cdot}{B}_\mathrm{n} = kT_0\underset{\cdot}{B}_\mathrm{n}$ (see (2.238)). By Thévenin's theorem, we can summarize all internal noise sources to a single one which we can place at the output or input of the considered device. Supposing an available power gain g [47] of the device, the overall noise power at the output may be expressed either by $P_{\mathrm{n,out}} = gP_{\mathrm{n}0} + P_{\mathrm{n,int}}^{(\mathrm{out})}$ or by $P_{\mathrm{n,out}} = g\left(P_{\mathrm{n}0} + P_{\mathrm{n,int}}^{(\mathrm{in})}\right)$. In the first case, the internal noise source is placed at the output port and in the second case the additional noise is modelled to come from the input. Insertion in (2.256) leads to

$$F = \frac{x_{\mathrm{rms}}^2/P_{\mathrm{n}0}}{gx_{\mathrm{rms}}^2/P_{\mathrm{n,out}}} = \frac{P_{\mathrm{n,out}}}{gP_{\mathrm{n}0}} = \frac{gP_{\mathrm{n}0} + P_{\mathrm{n,int}}^{(\mathrm{out})}}{gP_{\mathrm{n}0}} = \frac{P_{\mathrm{n}0} + P_{\mathrm{n,int}}^{(\mathrm{in})}}{P_{\mathrm{n}0}} = \frac{P_{\mathrm{n,in}}}{P_{\mathrm{n}0}} \tag{2.257}$$

Consequently, the total output noise power level of a device having a noise factor F is given by

$$\begin{aligned} P_{\mathrm{n,out}} &= FgkT_0\underset{\cdot}{B}_\mathrm{n} \\ L_{\mathrm{n,out}} &= -174\ \mathrm{dBm/Hz} + \mathrm{NF} + g[\mathrm{dB}] + 10\log\frac{\underset{\cdot}{B}_n}{1\ \mathrm{Hz}} \end{aligned} \tag{2.258}$$

If we relate the internal noise to the device input, we get an overall input noise power level of

$$\begin{aligned} P_{\mathrm{n,in}} &= FkT_0\underset{\cdot}{B}_\mathrm{n} \\ L_{\mathrm{n,in}} &= -174\ \mathrm{dBm/Hz} + \mathrm{NF} + 10\log\frac{\underset{\cdot}{B}_n}{1\ \mathrm{Hz}} \end{aligned} \tag{2.259}$$

Thus, the sensitivity of the device will be reduced by its noise factor F compared to the physically inevitable perturbations due to thermal agitation within the signal source of temperature T_0.

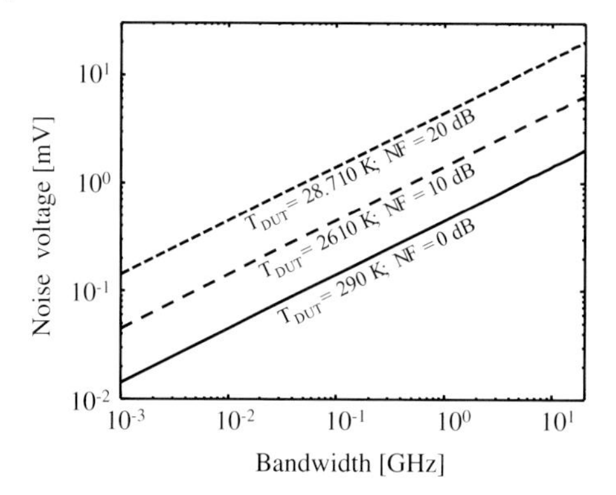

Figure 2.78 Thermal noise as function of the physical bandwidth $\underset{\cdot}{B}$ and noise figure and noise temperature in a 50 Ω system.

A further concept to describe the noise behaviour uses the system noise temperature T_{DUT}, that is one presumes that the additional noise is provided by a resistor of temperature T_{DUT}. This approach is more concise than the noise factor since the noise performance of the considered device is independent of the actual source temperature T_{s}. As we have seen from (2.257) the total input noise decomposes in two parts – one is caused by the source and the second by the device itself:

$$P_{\text{n,in}}(T_{\text{s}}, T_{\text{DUT}}) = P_{\text{nS}} + P_{\text{n,int}}^{(\text{in})} = k\underset{\cdot}{B}_{\text{n}}(T_{\text{s}} + T_{\text{DUT}}) = k\underset{\cdot}{B}_{\text{n}} T_{\text{sys}} \qquad (2.260)$$

These two power values can be expressed by the noise provided from two resistors having the temperature T_{s} and T_{DUT}. T_{sys} is the total equivalent noise temperature. Supposing a source temperature of $T_{\text{s}} = T_0$, we can bridge the gap between noise factor and noise temperature:

$$P_{\text{n,in}}(T_0, T_{\text{DUT}}) = k\underset{\cdot}{B}_{\text{n}}(T_0 + T_{\text{DUT}}) = FkT_0\underset{\cdot}{B}_{\text{n}} = kT_0\underset{\cdot}{B}_{\text{n}} + (F-1)kT_0\underset{\cdot}{B}_{\text{n}}$$

and hence we get

$$T_{\text{DUT}} = (F-1)T_0 \qquad (2.261)$$

Figure 2.78 gives an impression of the magnitude which has to be expected for the noise voltage in very wideband devices.

2.6.1.6 **Quantization Noise**

An ADC (Figure 2.79) maps a continuous signal to a discrete one where the conversation speed is controlled by the sampling clock having the repetition rate $f_{\text{s}} = \Delta t_{\text{s}}^{-1}$.

The transfer characteristic of an ideal bi-polar ADC is shown in Figure 2.80. If N is its number of bits, it is able to distinguish the following number of quantization intervals:

$$\text{Mid-tread characteristic:} \quad n_{\text{q}} = 2^N - 1$$

$$\text{Mid-riser characteristic:} \quad n_{\text{q}} = 2^N$$

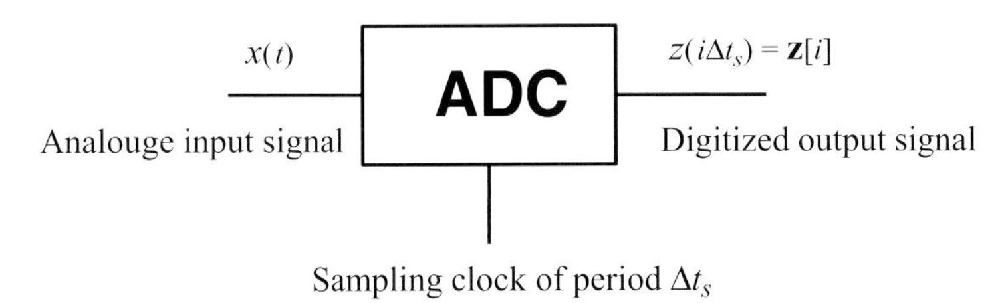

Figure 2.79 Analogue-to-digital converter.

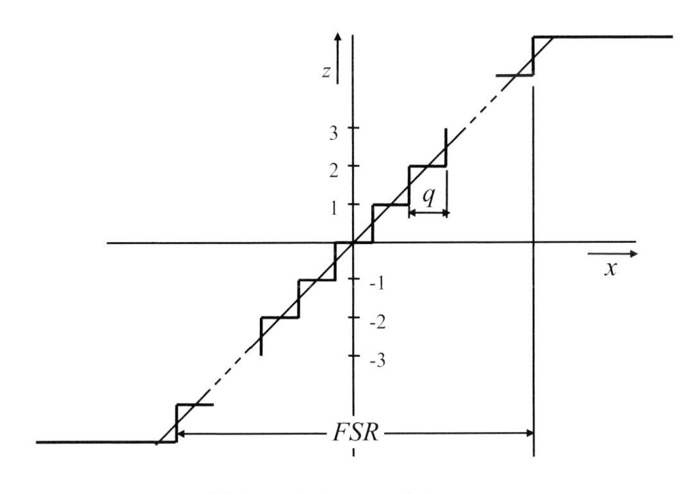

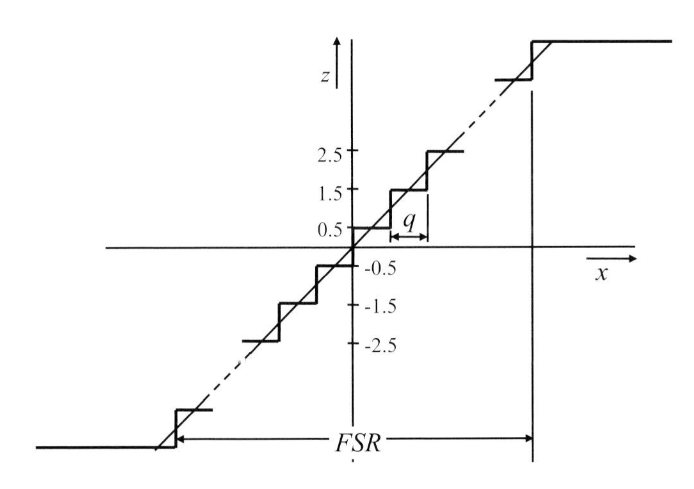

Figure 2.80 Ideal quantizer transfer functions.

For a given full-scale range (FSR) of the ADC, the width of its quantization steps q are

$$q = \frac{\text{FSR}}{n_\text{q}} \approx \frac{\text{FSR}}{2^N} \tag{2.262}$$

The deviations between the digitized signal $\mathbf{z}[i]$ and the actual input signal $x(t)$ represent the quantization error

$$e_\text{q}(t) = x(t) - z(i\,\Delta t_\text{s}) \tag{2.263}$$

Obviously from Figure 2.80, its maximum value is $\pm 0.5q$ in the case of an ideal ADC. Figure 2.81 depicts the quantization error as function of the (continuous) value of the input quantity for the ideal ADC as well as the typical appearance of the error in the case of a real ADC.

Assuming the input signal $x(t)$ does not have any synchronism with the sampling clock f_s of the ADC, we can expect randomness for the quantization error. Therefore, the rms value of the quantization error e_q can be calculated via its second-order moment:

$$e_{\text{q,rms}}^2 = \sigma_\text{q}^2 = \int_{-\infty}^{\infty} e_\text{q}^2 p_\text{q}(e_\text{q})\,\text{d}e_\text{q} \tag{2.264}$$

This value can be seen as the variance σ_q^2 of a random process generated by the quantization effect. Therefore, we will deal with it like we did with random perturbations as discussed above. This leads us to an equivalent circuit of a digitizing unit

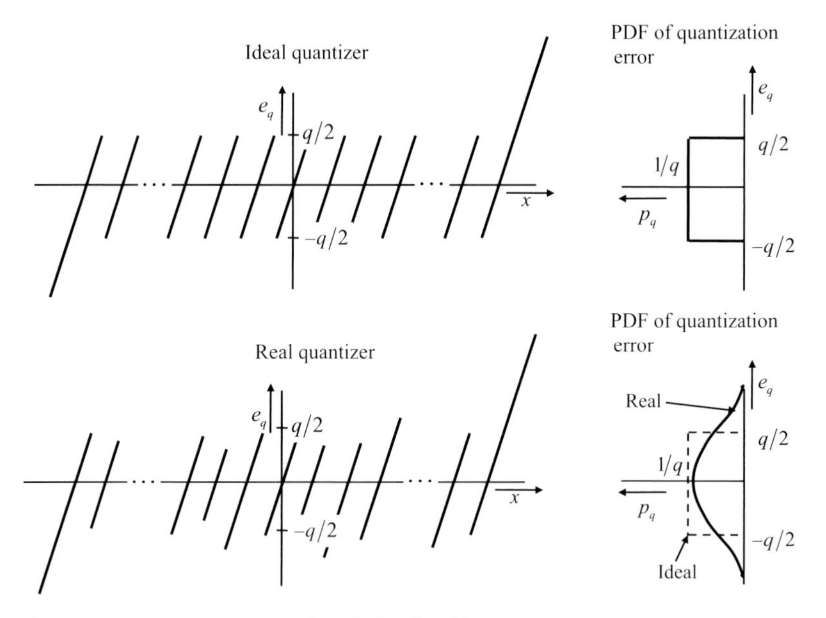

Figure 2.81 Quantization error of an ideal and real ADC.

Statistical model of a quantizer

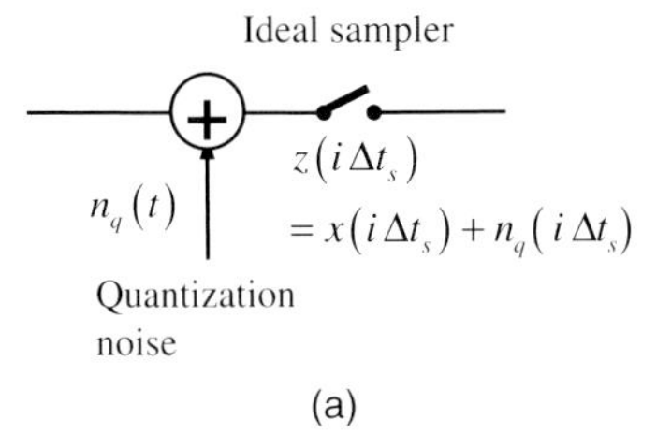

Quantizer fed by a perturbed signal

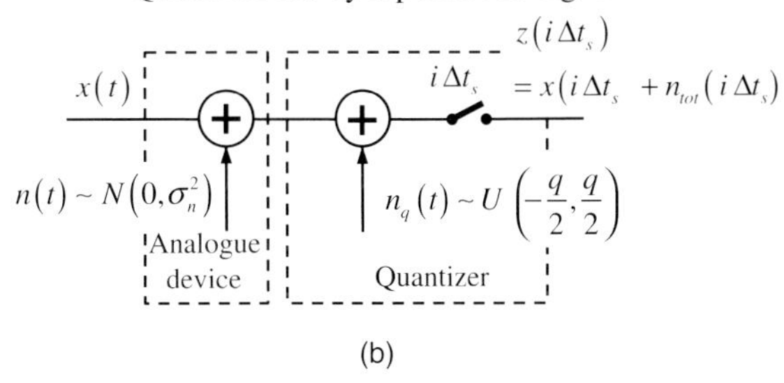

Figure 2.82 Quantization noise model.

as depicted in Figure 2.82 in which also a combination of 'analogue' and 'digital' noise is indicated. Combining both noise sources leads to an overall noise of the variance:

$$\sigma_{\text{tot}}^2 = \sigma_{\text{n}}^2 + \sigma_{\text{q}}^2 \tag{2.265}$$

Further details can be found in Ref. [8] which given a comprehensive overview on quantization effects.

From (2.264) and the quantization error distribution as shown in Figure 2.81, it follows for an ideal ADC.

$$\sigma_{\text{q.ideal}}^2 = \frac{q^2}{12} \leq \sigma_{\text{q}}^2 \tag{2.266}$$

This value represents a lower bound for the effective quantization errors of a real quantizer. The impact of the actual effective quantization error σ_{q} onto the ADC resolution is expressed by the effective number of bits ENOB:

$$\text{ENOB} = N - \text{lb}\frac{\sigma_{\text{q}}}{\sigma_{\text{q.ideal}}} \tag{2.267}$$

It indicates the actually reliable bit number of an ADC.

Usually the ADC is fed by pre-amplifiers or other electronic devices which always add noise (see Figure 2.82b). We will expand the ENOB-concept to the whole receiver by joining all noise sources. Like the SNR-value (see (2.240)), the

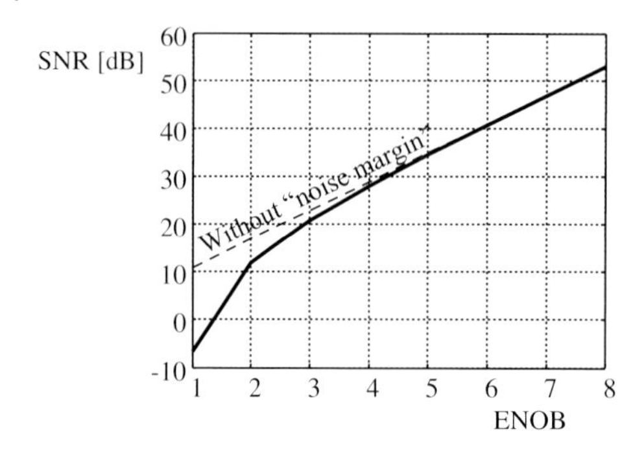

Figure 2.83 Effective number of bits versus signal-to-noise ratio. Herein, $\sigma_n/\sigma_{tot} = 1$, that is $\sigma_q \ll \sigma_n$, was supposed in order to indicate the maximum deviations between complete and approximated relation (2.269).

ENOB-value is a quality figure of a measurement relating maximum signal strength to random errors. Therefore, both feature values are mutually related. This shall be shown in what follows. For that purpose, we will refer to the total noise of rms value σ_{tot} in (2.265). Let us further suppose that the analogue input noise is Gaussian distributed (hence, a negligible quantity of only 3‰ of all noise events have an amplitude larger than $3\sigma_n$) and that the ADC does not yet provide a hard clipping of the overall signal if the peak value x_{peak} appears at its input. Hence, we can state succinctly for the full-scale range of the ADC:

$$\mathrm{FSR} \geq 2\left(x_{peak} + 3\sigma_n\right) \tag{2.268}$$

The term $3\sigma_n$ represents a 'noise margin' which avoids signal clipping even if large noise peaks add to the signal. Inserting (2.268) and (2.262) into (2.267) yields

$$\mathrm{ENOB} = -\mathrm{lb}\frac{\sqrt{12}\sigma_{tot}}{\mathrm{FSR}} = -\mathrm{lb}\frac{\sqrt{12}\sigma_{tot}}{2\left(x_{peak} + 3\sigma_n\right)}$$

and referring to the SNR definition (2.240), we finally end up in

$$\begin{aligned}\mathrm{SNR}\,[\mathrm{dB}] &= 20\log\left(\sqrt{3}2^{\mathrm{ENOB}} - 3\frac{\sigma_n}{\sigma_{tot}}\right) \\ &\approx 10\log 3 + \mathrm{ENOB}\,20\log 2 \approx 4.8 + 6\mathrm{ENOB} \quad \text{for ENOB} > 3\end{aligned} \tag{2.269}$$

Figure 2.83 depicts both the complete relation as well as its approximation which holds above about 3 bits. In the literature, the relation between ENOB and SNR is usually given by SNR [dB] = 1.76 + 6.02 ENOB. This relationship bases, however, on a different definition of the SNR-value[34] and it supposes sine wave signals. Furthermore, it neglects the 'noise margin'.

Finally, we should point to remember the assumption which was made to establish the treatment above: It was the randomness of the quantization error. This

34) $\mathrm{SNR} = \dfrac{\text{average signal power}}{\text{average noise power}}$.

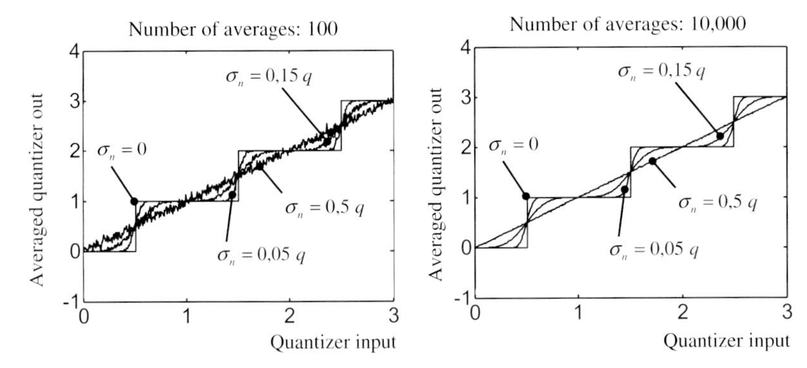

Figure 2.84 Simulated ADC-transfer characteristic for different degrees of signal corruption by Gaussian noise.

assumption does not hold, if there is a synchronism between the sampling clock and the signal to be gathered. But just that is often the case in ultra-wideband sensors where the generator of the sounding signal and the receiver are strongly synchronized. This implies, for example, that synchronous averaging will not work (with respect to quantization errors) since by repeating the measurement all data samples corresponding to the same time position will always result in the same quantization interval. Hence, associated data samples are mutually correlated and averaging does not lead to any improvement of the noise performance.

The situation changes if the captured signal is adequately contaminated by additional random noise $\sigma_n \approx \sigma_{tot} > \sigma_q$ (see Figure 2.82b). In that case, the errors of the periodically repeated measurements are not any more correlated and therefore synchronous averaging will work again. If the averaging number is large enough, the resolution of the measurement processes can finally even be better than the width of the quantization interval of the ADC would allow it. Hence for ultra-wideband sensor applications, the resolution of an ADC should be at least in the order of the noise background.

Figure 2.84 gives an impression about the performance improvement of quantization by averaging a noisy signal. We can observe that a pure deterministic signal ($\sigma_n = 0$) will not profit from averaging since the resulting transfer characteristics is identical to the ADC quantization. The staircase function tends to flatten by increasing the noise power and we already arrive at a straight line for $\sigma_n \geq q/2$. In the case of Gaussian noise about 99.7% of all noise samples are scattered over an interval of $\pm 3\sigma_n$. Hence, we can conclude that at least three steps of the ADC transfer characteristic must be involved in the averaging process.

The reduction of quantization errors by averaging of noise waveforms (which is also called dithering) may attempt oneself to reduce the ADC resolution in favour of the averaging number. This leads in the limit to a one-bit ADC, that is a simple comparator [48]. Such devices may be operated at a high clock rates so that the achievable averaging number can be quite high even if short recording times are required. But unfortunately, as we will show in what follows, this approach

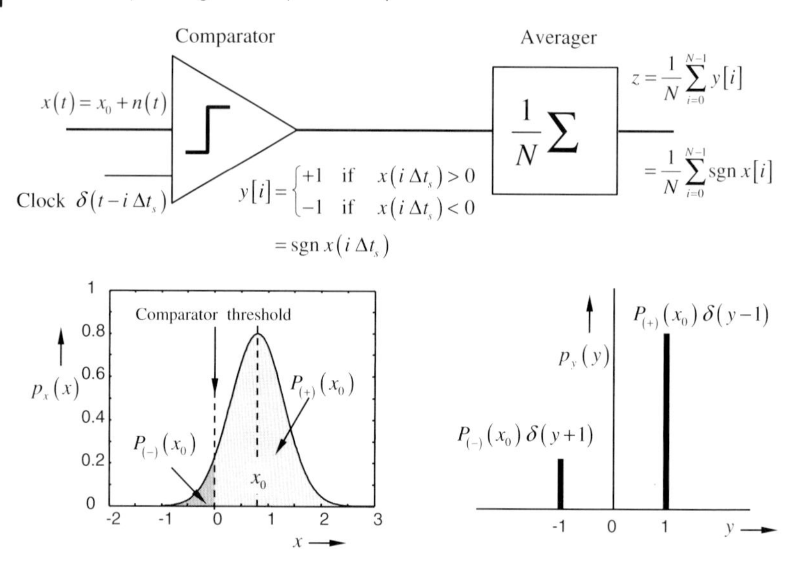

Figure 2.85 Basic schematics of a polarity receiver (1-bit receiver) (top) and PDF examples for the input signal $x(t)$ (bottom, left) and the polarity signal $y[i] = \text{sign}(x(i\,\Delta t_s))$ (bottom, right).

generates non-linear distortions of the averaged receive signal which will limit the dynamic range of the device. Let us consider Figure 2.85 for that purpose.

We suppose a DC value x_0 which is superimposed by Gaussian noise $n(t) \sim N(0, \sigma_n^2)$. The goal is to determine the value of x_0 as precise as possible simply by referring to the sign of the sampled signal $x(i\,\Delta t_s) = x_0 + n(i\,\Delta t_s)$. The sign of the test signal is determined by a comparator which is clocked with the sampling rate $f_s = \Delta t_s^{-1}$. If the input voltage is negative of any value, the comparator provides $y[i] = -1$. In the case of any positive voltage, it provides $y[i] = +1$. The probability that the comparator provides $+1$ or -1 at its output can be deduced from the PDF of the input signal $x(t)$ (see Figure 2.85). The area under the PDF left from the threshold represents the probability that $y[i] = -1$, that is

$$P_{(-)}(x_0) = P(x < 0) = \int_{-\infty}^{0} p_x(x)\mathrm{d}x = \frac{1}{\sqrt{2\pi}\sigma_n} \int_{-\infty}^{0} \mathrm{e}^{-\frac{1}{2}\left(\frac{x-x_0}{\sigma_n}\right)^2} \mathrm{d}x \tag{2.270}$$

Correspondingly, the probability to get $y[i] = +1$ is

$$P_{(+)}(x_0) = P(x > 0) = \int_{0}^{\infty} p_x(x)\mathrm{d}x = \frac{1}{\sqrt{2\pi}\sigma_n} \int_{0}^{\infty} \mathrm{e}^{-\frac{1}{2}\left(\frac{x-x_0}{\sigma_n}\right)^2} \mathrm{d}x \tag{2.271}$$

These observations leads to the PDF $p_y(y)$ of the polarity signal (comparator output) as depicted in Figure 2.85. We consider $\underset{\sim}{y}[i]$ as a random process. Its PDF can be used to determine expected value and variance of the output quantity z. The expected value of z which shall lead us to the wanted measurement value x_0 is estimated from (refer also to Annex A.3)

$$\hat{z} = \overline{\text{sgn}(x)} = \text{E}\{\underset{\sim}{y}\} = \int_{-\infty}^{\infty} y p_y(y)\text{d}y = P_{(+)} - P_{-} \qquad (2.272)$$

Insertion of (2.270) and (2.271) in (2.272) and exploiting symmetry properties of the Gaussian PDF, we finally result in

$$\hat{z} = \overline{\text{sgn}(x)} = \text{E}\{\underset{\sim}{y}\} = \frac{2}{\sqrt{2\pi}\sigma_n}\int_{0}^{x_0} \text{e}^{-\frac{1}{2}\left(\frac{x}{\sigma_n}\right)^2}\text{d}x = \text{erf}\left(\frac{x_0}{\sqrt{2}\sigma_\text{n}}\right) \qquad (2.273)$$

The function $\text{erf}(x)$ represents the error function which we develop in a Taylor series:

$$\begin{aligned} \text{erf}(x) &= \frac{2}{\sqrt{\pi}}\sum_{n=0}^{\infty}\frac{(-1)^n}{(2n+1)n!}x^{2n+1} \\ \Rightarrow \hat{z} &= \text{erf}\left(\frac{x_0}{\sqrt{2}\sigma_\text{n}}\right) \approx \sqrt{\frac{2}{\pi}}\frac{x_0}{\sigma_\text{n}}\left(1 - \frac{1}{6}\left(\frac{x_0}{\sigma_\text{n}}\right)^2 + \cdots\right); \quad \frac{x_0}{\sqrt{2}\sigma_\text{n}} \ll 1 \end{aligned} \qquad (2.274)$$

from which follows the wanted measurement value:

$$x_0 \approx \sqrt{\frac{\pi}{2}}\sigma_\text{n}\frac{\hat{z}}{1-e} \approx \sqrt{\frac{\pi}{2}}\sigma_\text{n}\hat{z}(1+e_\text{nl}) \quad \text{with } e_\text{nl} = \frac{1}{6}\left(\frac{x_0}{\sigma_\text{n}}\right)^2 \ll 1 \qquad (2.275)$$

Herein, e_nl represents a deviation from the linear relation we wished for. Equation (2.275) conducts us to two conclusions:

- The determination of the absolute value of x_0 requires the exact knowledge of the (external and device internal) noise power. Under practical constraints this will be barely possible, so that only a normalized measurement value x_0/σ_n can be determined. In order to be comparable over a series of measurements, one has to assure constant noise conditions during the whole measurement campaign.
- The noise power should be much higher than the signal power in order to minimize non-linear distortion. If the captured signal is additionally subjected a correlation afterwards, the non-linear distortions should be reduced as best as possible in order to avoid spurious peaks (leading, e.g. to false targets) in the correlation function (see Section 3.3.5 for discussion of related problems in the case of M-sequence stimulation).

The variance of z is given by

$$\text{var}\{z\} = \frac{1}{N^2}\text{var}\left\{\sum \underset{\sim}{y}[i]\right\} = \frac{1}{N}\left(\text{E}\{\underset{\sim}{y}^2[i]\} - \text{E}^2\{\underset{\sim}{y}[i]\}\right) = \frac{1-\hat{z}^2}{N} \qquad (2.276)$$

since from (2.273) $\text{E}\{\underset{\sim}{y}[i]\} = \hat{z}$ and $\text{E}\{\underset{\sim}{y}^2[i]\} = \int_{-\infty}^{\infty} y^2 p_y(y)\text{d}y = P_{(-)} + P_{(+)} = 1$.

In order to keep linearity of the measurement process, we have to meet $\hat{z} \ll 1$. Hence, the variance of the measured quantity is about $\text{var}(z) \approx N^{-1}$.

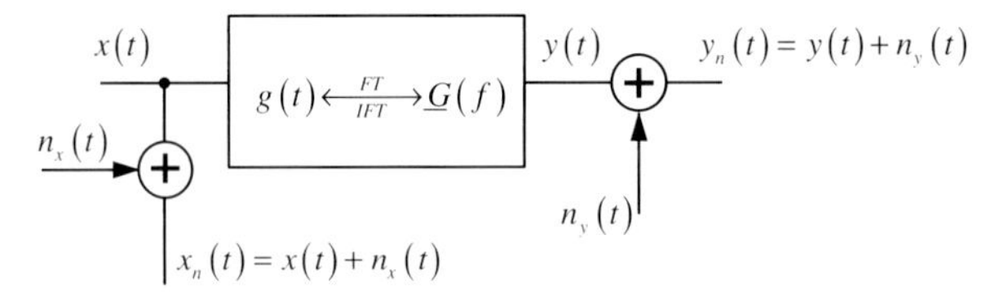

Figure 2.86 Error in variable model.

2.6.1.7 IRF and FRF Estimation from Noisy Data

So far, we have only investigated the impact of noise onto a single signal. We will close the discussion on noise effects by considering still some basic aspects of IRF and FRF determination under noisy condition. Figure 2.86 depicts the model of our measurement arrangement. For shortness, we only consider a simple SiSo system. Extensions to MiMo systems are straightforward by swapping to matrix representations.

The objective of the measurement is to determine either $g(t)$ or $\underline{G}(f)$ which are defined by the input–output relations

$$y(t) = g(t) * x(t) \quad \overset{\text{FT}}{\underset{\text{IFT}}{\longleftrightarrow}} \quad \underline{Y}(f) = \underline{G}(f)\underline{X}(f) \tag{2.277}$$

Unfortunately, the unperturbed signals are not accessible by the measurements so that we have to deal with the noise corrupted signals $x_{\text{n}}(t)$ and $y_{\text{n}}(t)$ or their spectra. Consequently, we only get an estimation of the wanted DUT behaviour:

$$y_{\text{n}}(t) = \hat{g}(t) * x_{\text{n}}(t) \quad \overset{\text{FT}}{\underset{\text{IFT}}{\longleftrightarrow}} \quad \underline{Y}_{\text{n}}(f) = \hat{\underline{G}}(f)\underline{X}_{\text{n}}(f) \tag{2.278}$$

The question which we will follow now is how to reduce as much as possible the influence of the measurements noise onto the estimations $\hat{g}(t)$ or $\hat{\underline{G}}(f)$.

The most general approach with respect to the applicable types of wideband signals is given by the relations (2.136), (2.137) as well as (2.152), (2.153). They do not pose any restrictions concerning the test signals which may be periodic or non-periodic. In most of practical cases, we can assume that the stimulus signal and both random noise perturbations are mutually uncorrelated, that is $C_{n_x x}(t) = C_{n_y x}(t) = C_{n_x n_y}(t) = 0$. Further, we presume band-limited white noise of bandwidth $\ddot{B}_{\text{N}}$ and variance $\|n_x(t)\|_2^2 = \sigma_x^2$ and $\|n_y(t)\|_2^2 = \sigma_y^2$. Thus, if we adopt these conditions to above-mentioned relations, we yield for the expected values

Version 1 in time and frequency domain: [35]

$$\begin{aligned} \mathrm{E}\{\underset{\sim}{y}_n(t) * \underset{\sim}{x}_n(-t)\} &= \mathrm{E}\{\hat{g}_1(t) * \underset{\sim}{x}_n(t) * \underset{\sim}{x}_n(-t)\} \\ C_{yx}(t) &= \hat{g}_1(t) * (C_{xx}(t) + C_{n_x n_x}(t)) \\ C_{yx}(t) &\approx \hat{g}_1(t) * \left(C_{xx}(t) + \sigma_x^2 \delta(t)\right) \approx \left(x_{\text{rms}}^2 + \sigma_x^2\right)\hat{g}_1(t) \end{aligned} \tag{2.279}$$

35) The correct shape of the ACF from band-limited white noise is $C_{n_x n_x}(t) = \sigma_x^2 \operatorname{sinc} \ddot{B}t$ which we have approximated by $\sigma_x^2\, \delta(t)$ since we deal with large bandwidth. Corresponding holds for the stimulus signal which is of the same bandwidth as the noise.

$$\begin{aligned}
\mathrm{E}\{\underset{\approx}{Y}_{\mathrm{n}}(f)\underset{\approx}{X}^{*}_{\mathrm{n}}(f)\} &= \mathrm{E}\{\underline{\hat{G}}_1(f)\underset{\approx}{X}_{\mathrm{n}}(f)\underset{\approx}{X}^{*}_{\mathrm{n}}(f)\} \\
\underline{\Psi}_{yx}(f) &= \underline{\hat{G}}_1(f)(\Psi_{xx}(f) + \Psi_{n_x n_x}(f)) \\
\underline{\Psi}_{yx}(f) &\approx \underline{\hat{G}}_1(f)\left(\Psi_{xx}(f) + \frac{\sigma_x^2}{\ddot{B}}\right)
\end{aligned} \tag{2.280}$$

Version 2 in frequency and time domain:

$$C_{yy}(t) + \sigma_y^2\delta(t) \approx \hat{g}_2(t) * C_{xy}(t) \tag{2.281}$$

$$\Psi_{yy}(f) + \frac{\sigma_y^2}{\ddot{B}} \approx \underline{\hat{G}}_2(f)\underline{\Psi}_{xy}(f) \tag{2.282}$$

We can observe that for both versions at least one of the noise terms will be suppressed. Since typically the output noise σ_y^2 is more serious, the relation (2.279) or (2.280) is mostly implemented by the measurement receivers or the data pre-processing. Corresponding examples are given by Figures 3.70, 371 and 3.90. The second noise term cannot be suppressed by the proposed approaches, so that we will result in a biased estimation of the IRF or FRF depending on the strength of the remaining noise term.
In order to establish a bias-free-IRF and -FRF estimation, we apply periodic signals. By this 3rd version, we perform averaging before any other operation and determine afterwards the wanted functions, that is
Version 3; first step synchronous averaging:

$$\begin{aligned}
\bar{x}(t') &= \sum_{p=0}^{M-1} x_{\mathrm{n}}(t - pt_{\mathrm{P}}); \quad t' = \mathrm{mod}(t, t_{\mathrm{P}}) \\
\bar{y}(t') &= \sum_{p=0}^{M-1} y_{\mathrm{n}}(t - pt_{\mathrm{P}});
\end{aligned} \tag{2.283}$$

where $\bar{x}(t')$ and $\bar{y}(t')$ are bias-free estimations of the involved signals. If the stimulus signal is a train of short pulses, we can simply state

$$\hat{g}_3(t') \propto \bar{y}(t'); \quad \bar{x}(t') \approx \delta(t') \tag{2.284}$$

In case of time spread signals and for the frequency domain, we will end up in

$$\hat{g}_3(t') \propto C_{\bar{y}\bar{x}}(t'); \quad C_{\bar{x}\bar{x}}(t') \approx \delta(t') \tag{2.285}$$

$$\underline{\hat{G}}_3 = \frac{\underline{\bar{Y}}(f)\,\underline{\bar{X}}^*(f)}{\underline{\bar{X}}(f)\,\underline{\bar{X}}^*(f)}; \quad \underline{\bar{X}}(f) = \mathrm{FT}\{\bar{x}(t')\};\ \underline{\bar{Y}}(f) = \mathrm{FT}\{\bar{y}(t')\} \tag{2.286}$$

These estimations tend asymptotically to the correct IRF and FRF. UWB measurement methods applying short pulse excitation (Section 3.2) or pseudo-noise test signals (Sections 3.3 and 3.5) exploit this procedure. Version 3 will be our favoured method of noise suppression since it works bias free and occupies only modest processing capacity due to the simple averaging algorithm.
Basically, there exist still two further possibilities which we will still mention for completeness. They are restricted to frequency domain implementations [49, 50].

Version 4 determines the arithmetic mean of the FRF:

$$\underline{\hat{G}}_4(f) = \frac{1}{M}\sum_{p=0}^{M-1}\frac{\underline{Y}_{n,p}(f)}{\underline{X}_{n,p}(f)} = \frac{1}{M}\sum_{p=0}^{M-1}\underline{G}_{n,p}(f)$$

$$\underline{X}_{n,p}(f) = \mathrm{FT}\{x_n(t)\};\ \underline{Y}_{n,p}(f) = \mathrm{FT}\{y_n(t)\};\quad t \in pt_p + [0, t_p) \qquad (2.287)$$

It provides a bias-free estimation but it does not converge in case of low SNR-values of the input signal.

Version **5 applies the geometric mean** which is nothing but the arithmetic mean of the dB-scaled magnitude and the phase of the FRF. It may lead to less biased estimations as (2.280) or (2.282):

$$\begin{aligned}
\underline{\hat{G}}_5(f) &= \hat{G}_5(f)e^{j\hat{\varphi}_5(f)} = \sqrt[M]{\prod_{p=0}^{M-1}\underline{G}_{n,p}(f)} \\
\Rightarrow \hat{G}_5(f)[\mathrm{dB}] &= \frac{1}{M}\sum_{p=0}^{M-1}G_{n,p}(f)[\mathrm{dB}] \\
\Rightarrow \hat{\varphi}_5(f) &= \frac{1}{M}\sum_{p=0}^{M-1}\varphi_{n,p}(f)
\end{aligned} \qquad (2.288)$$

As obvious from Figure B.4 in Annex B.4.4, the geometric mean actually provides an unbiased estimation of the output signal of an IQ-demodulator if the perturbations are mainly caused by phase noise. The variance of all methods is roughly the same. As it is well known, it can be reduced by increasing the number of averages.

2.6.2 Narrowband Interference

As their name implies, UWB sensors possess receivers which operate over a huge bandwidth. In many cases, such sensors have to share the operational frequency band with other usually narrowband transmission system. Therefore, one has to count with interferences. Figure 2.87 illustrates the situation in the spectral domain and Figure 2.88 gives an example of a real radio transmission collected from an ultra-wideband receiver during the measurement. We can see narrowband transmissions with constant carrier frequency and transmissions based on frequency hopping.

The actual impact of these narrowband interferers onto the sensor depends on the working principle of its electronics. However, a rough estimate of the influence can be given. For that purpose, we will suppose that the data capturing rate of the ultra-wideband sensor as well as the transmit signal are not correlated with the narrowband interferer. In that case, the interferer is randomized over the whole measurement bandwidth $\underset{\cdot\cdot}{B}$ by the data capturing process. Hence, the jammer behaves like random noise which final affects the

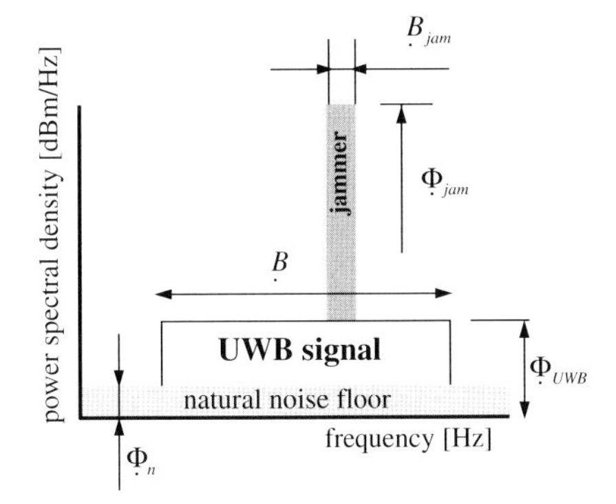

Figure 2.87 Narrowband interference of an ultra-wideband sensor.

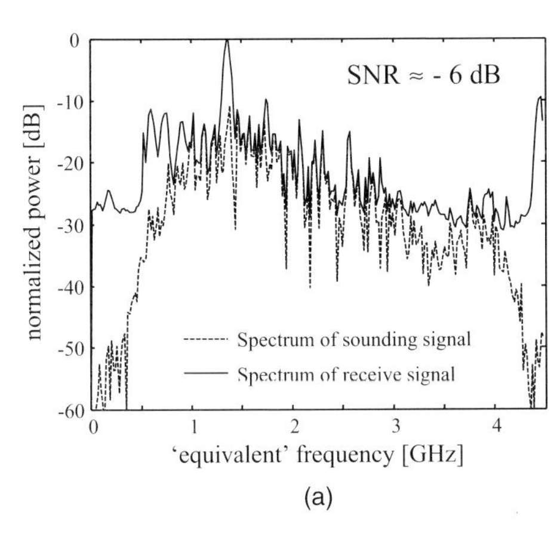

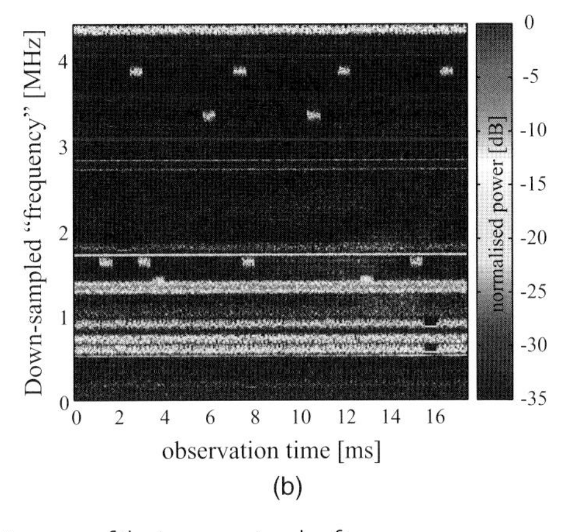

Figure 2.88 Example of narrowband interference captured from a sub-sampling receiver working with a digitizing rate of 17 MHz. (a) (Average) spectrum represented in dependence from the (equivalent) RF-frequency (note that the frequency mapping of the jammers is not correct due to sub-sampling). (b) Spectrogram of the jammer signals after removing the ultra-wideband sounding signal. The signals are represented as function of the actual frequency. But also here, the frequency mapping is wrong due to the sub-sampling approach.

SNR-value of the measurement data by its average PSD $\overline{\Phi}_{\text{jam}}$:[36]

$$\text{SNR} = \frac{\mathfrak{E}}{\Phi_{\text{n}} + \overline{\Phi}_{\text{jam}}} = \frac{\eta \Phi_{UWB} B T_{\text{R}}}{\Phi_{\text{n}} + \eta \dfrac{B_{\text{jam}}}{B} \Phi_{\text{jam}}} \tag{2.289}$$

36) The erroneous frequency mapping of the jammers due to the sub-sampling receiver is out of interest here since we are not interested in these signals aside from their suppression. However a low receiver efficency due to sub-sampling will degrade the caputerd power of the jammer by the same ratio as for the wanted signal.

As to be seen from (2.289), one can counteract the degradation of the SNR-value by

- increasing the spectral power Φ_{UWB} of the stimulus,
- enlarging the bandwidth and
- extending the recording time, for example by synchronous averaging or correlation.

In practice however, a complete randomization of the narrowband perturbation is usually not achievable. Thus, Moiré effects are partially seen in the measurement data.

Furthermore, an at least theoretical possibility to combat the narrowband perturbations is to filter them out by tracking or prediction filter. If the interferers are not stable in time (compare the frequency hopping in the spectrogram of Figure 2.88), a digital filter implementation is usually preferable in order to be able to follow the variations. This method would improve the SNR-value since the perturbations are cancelled from the data.

2.6.3
Jitter and Phase Noise

Jitter and phase noise are random errors of the abscissa values of our measured time or frequency domain functions. Both terms describe the same phenomenon from different viewpoints. Jitter can be intuitively seen as a variation in the zero-crossing times of a signal (or general threshold crossing). Several definitions of jitter are in use. One distinguishes between deterministic and random jitter. The first one refers to reproducible irregularities of the chip length in coded pulse sequences – also referred as data jitter. An example is given in Figure 3.46. But here, this type of jitter is out of interest. With respect to random jitter, we will consider trigger jitter and cycle jitter assigning random variations of a time point or a time interval. Further definitions can be found in Ref. [51].

2.6.3.1 Trigger Jitter

Ultra-wideband sensors require a sophisticated and well-coordinated timing of stimulus generation and data capturing. Hence, random timing errors will limit the performance of the sensor as additive noise it does. Here, we will consider two aspects: How jitter is generated and what does it provoke?

A very popular approach to control the timing of electronic circuits is based on threshold crossing which releases a trigger event as depicted in Figure 2.89. This trigger event can be used to start the generation of a stimulus signal or to capture a data sample. Since both the trigger waveform and the threshold are corrupted by noise, the switching event is subject to trigger jitter. From Figure 2.89, the variance φ_{j}^2 of these temporal variations results in

$$\varphi_{\mathrm{j}}^2 = \mathrm{var}\{\Delta\underset{\sim}{\tau}_{\mathrm{j}}\} = \frac{\sigma_0^2 + \sigma_x^2}{(\mathrm{d}x/\mathrm{d}t)^2} \geq \frac{\sigma_0^2 + \sigma_x^2}{\mathrm{SR}^2} \qquad (2.290)$$

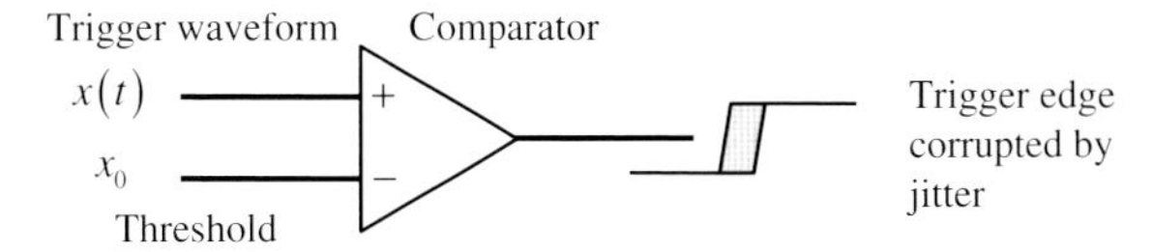

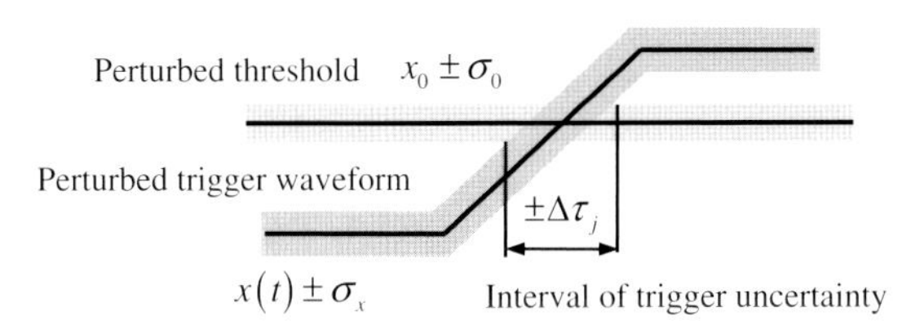

Figure 2.89 The principle of edge triggering and jitter generation by noisy reference signals. The area of uncertainty due to the noise is illustrated by the grey interval around the unperturbed signals.

Obviously, the susceptibility for random timing errors depends on the slope of the trigger waveform. Therefore, the threshold should be placed in a steep region of the trigger edge whose slew rate $\mathrm{SR} = \|\mathrm{d}x/\mathrm{d}t\|_{\infty}$ should be as high as possible. If we suppose an ideal noise-free threshold, the trigger jitter induced by a rising (falling) edge of amplitude X_0 and a 10–90% rise time of t_r is

$$\varphi_\mathrm{j} = \frac{5t_\mathrm{r}}{4\sqrt{\mathrm{SNR}}} \quad \text{with } \mathrm{SNR} = \frac{X_0^2}{\sigma_x^2} \tag{2.291}$$

For a sine wave of the same amplitude and frequency f_0, the random jitter results in

$$\varphi_\mathrm{j} = \frac{1}{2\pi f_0\sqrt{\mathrm{SNR}}} \tag{2.292}$$

(2.291) and (2.292) give an estimate of the robustness of the trigger point against additive noise. The repeated start of a waveform by an instable trigger cause a fluctuating signal which appears on the screen of an oscilloscope as depicted in Figure 2.90a.

In the case of data gathering Figure 2.90b, an instable control of the sampling point (triggering of sampling gate) translates into amplitude noise since the captured value will be assigned to the intended sampling point and not to the actual one.

Consequently, the entire noise of a digitized waveform estimates from the total additive noise and the jitter induced noise by

$$\sigma_\mathrm{entire}^2 = \sigma_\mathrm{tot}^2 + \left(\frac{\mathrm{d}x}{\mathrm{d}t}\right)^2 \varphi_\mathrm{j}^2 \tag{2.293}$$

That is, actually, the overall noise will always depends on the waveform. Interestingly to note that trigger jitter only affects signal sections with strong amplitude

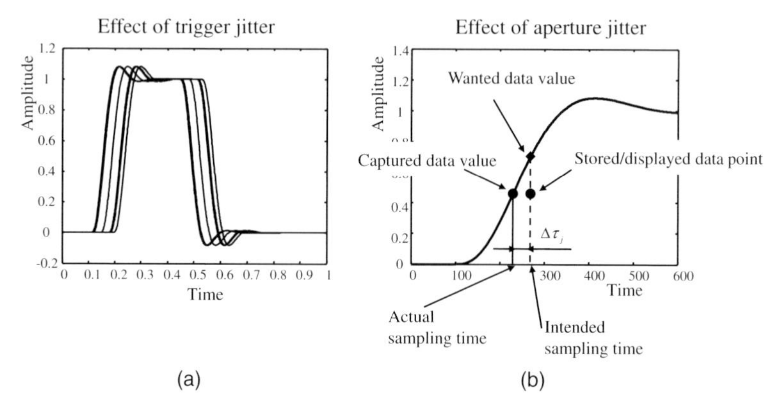

Figure 2.90 Effect of time jitter on triggering a signal generator and on data capturing (aperture jitter) exemplified for a single data point.

variations (i.e. edges) but smooth or flat signal parts are not subjected to that kind of perturbations. Figure 2.91a illustrates this behaviour for a step like waveform. As long as the trigger jitter is caused from a random process, synchronous averaging of the gathered data will reduce the corruption by noise. However, the shape of the averaged waveform $\overline{x}(t)$ will be masked by the jitter-PDF $p_{\text{jitter}}(t)$ via convolution:

$$\overline{x}(t) = x(t) * p_{\text{jitter}}(t) \tag{2.294}$$

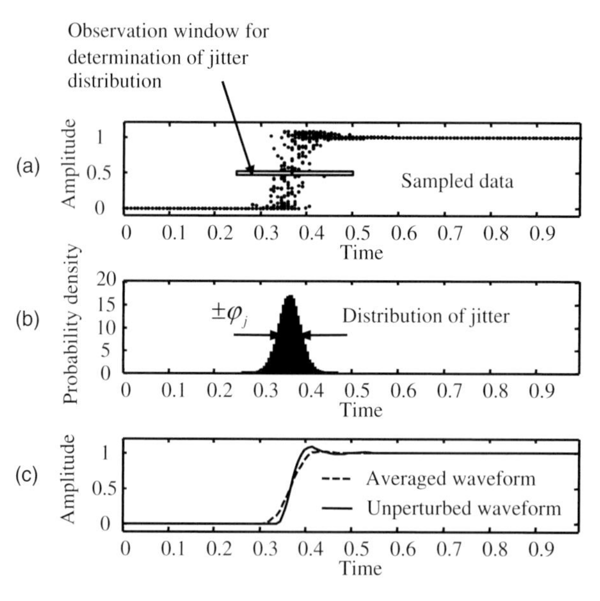

Figure 2.91 Effect of aperture jitter on an impulsive waveform. The upper graph represents an ensemble of measurements. Compare also Figure 2.12.

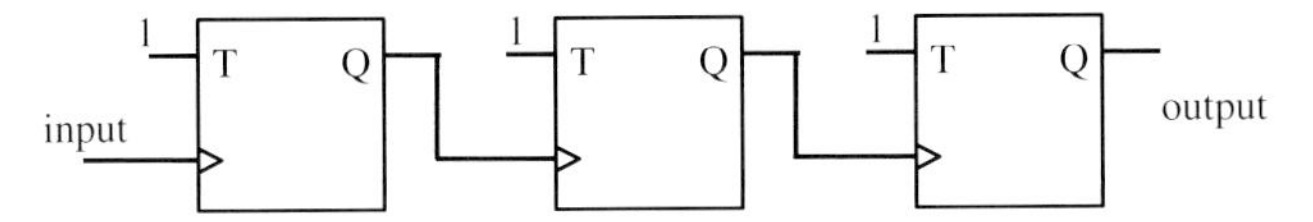

Figure 2.92 Simple binary divider built from cascaded T-flip-flops as an example for jitter accumulation due to cascading of threshold switching devices.

Figure 2.91 demonstrates such signal smoothing by averaging. From (2.294) follows that the maximum usable bandwidth of a jittered waveform is $\underset{\cdot\cdot}{B}_{\max} \approx \varphi_{\mathrm{j}}^{-1}$. In contrast to that behaviour, we have seen for additive noise that synchronous averaging provides the original waveform after a sufficient number of averages. A masking effect as with jitter could not be observed for pure additive noise. Up to a certain extent, the effect of random jitter can be removed by de-convolving the jitter-PDF out of the averaged data [52].

Cascading a number of switching circuits (e.g. comparator, Schmitt trigger, flip-flop and so on as illustrated in Figure 2.92) will lead to a jitter accumulation:

$$\varphi_{\mathrm{j,tot}}^2 = \varphi_{\mathrm{j},1}^2 + \varphi_{\mathrm{j},2}^2 + \varphi_{\mathrm{j},3}^2 + \cdots \tag{2.295}$$

2.6.3.2 Phase Noise

Timing instabilities not only arise in connection with threshold triggering but also in oscillators. The timing of an UWB sensor device is often controlled by such oscillators. Hence, its spectral purity will determine largely the overall sensor performance. If we assume a sine wave reference generator of the wanted frequency f_{c0}, the temporal variations of its output signal $x(t)$ can be modelled as random phase modulation. For brevity, we omit other deficiencies as additive noise and random amplitude modulations:

$$x(t) = X_0 \sin(\phi(t)) = X_0 \sin\left(2\pi f_{\mathrm{c0}} t \pm \Delta\phi(t)\right) \tag{2.296}$$

Instantaneous phase and frequency are linked by

$$\phi(t) = 2\pi \int_0^t f(\xi)\mathrm{d}\xi = 2\pi \left(f_{\mathrm{c0}} t + \int_0^t \Delta f(\xi)\mathrm{d}\xi \right) = 2\pi f_{\mathrm{c0}} t + \Delta\phi(t) \tag{2.297}$$

where $\Delta f(t)$ and $\Delta\phi(t)$ assign random frequency and phase fluctuations of the carrier signal. These fluctuations are called phase noise and typically they are quite small $|\Delta\phi(t)| \ll \pi$ so that (2.296) can be approximated as:

$$x(t) \approx X_0 \sin 2\pi f_{\mathrm{c0}} t + X_0\, \Delta\phi(t) \cos 2\pi f_{\mathrm{c0}} t \tag{2.298}$$

Expressing this signal in the spectral domain, we obtain

$$\begin{aligned} \underset{\cdot\cdot}{\Psi}_{xx}(f) &= \Psi_0 \delta(f \pm f_{\mathrm{c0}}) + \underset{\cdot\cdot}{m}_{\phi}(f) * \Psi_0 \delta(f \pm f_{\mathrm{c0}}) \\ &= \Psi_0 \left(\delta(f \pm f_{\mathrm{c0}}) + \underset{\cdot\cdot}{m}_{\phi}(f \pm f_{\mathrm{c0}}) \right) \end{aligned} \tag{2.299}$$

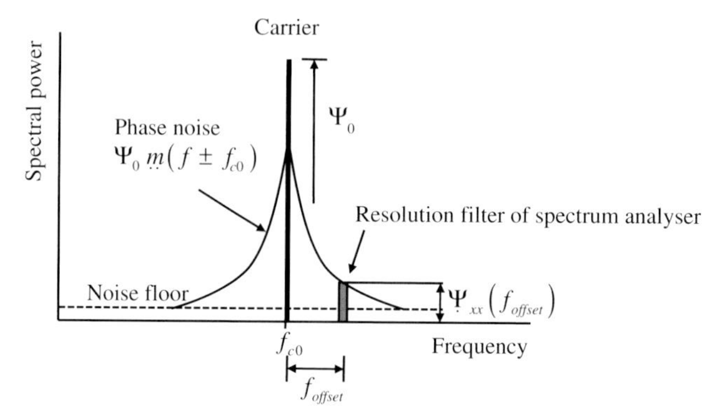

Figure 2.93 Spectrum of a sine wave signal affected by phase noise. Only the positive frequencies are shown.

Figure 2.93 depicts a typical spectrum of a sine wave showing the mentioned instabilities. Ideally, the spectrum of an oscillator signal is represented by a Dirac delta $\delta(f - f_{c0})$ located at f_{c0}. But obviously, the phase noise spectrum $\underset{\cdot\cdot}{m}_\phi(f)$ provokes widening of the frequency line.

The power spectrum $\underset{\cdot\cdot}{\Psi}_{xx}(f)$ results from Fourier transform of the auto-correlation function $C_{xx}(\tau)$ of the oscillator signal $x(t)$. $\Psi_0 = X_0^2/4$ corresponds to the power of the carrier and

$$\underset{\cdot\cdot}{m}_\phi(f) = \int_{-\infty}^{\infty} C_{\Delta\phi\,\Delta\phi}(\tau)\mathrm{e}^{-j2\pi f\tau}\,\mathrm{d}\tau \tag{2.300}$$

represents the phase noise spectrum of dimension $\left[\mathrm{rad}^2/\mathrm{Hz}\right]$ following from the auto-correlation function of the phase noise $C_{\Delta\phi\,\Delta\phi}(\tau) = \lim_{T\to\infty}\frac{1}{T}\int_T \Delta\phi(t)\Delta\phi\,(t+\tau)\mathrm{d}t$. A widely used quality figure to characterize the spectral purity of an oscillator is the ratio $\underset{\cdot\cdot}{\Psi}_{xx}\left(f_{\text{offset}}\right)/\Psi_0$ which is typically expressed by the phase noise level in dBc/Hz:

$$L_\phi\left(f_{\text{offset}}\right)\,[\mathrm{dBc/Hz}] = -10\log\frac{\underset{\cdot\cdot}{\Psi}_{xx}\left(f_{\text{offset}}\right)}{\Psi_0} = -10\log \underset{\cdot\cdot}{m}_\phi\left(f_{\text{offset}}\right) \tag{2.301}$$

The spectral shape of the phase noise spectrum $\underset{\cdot\cdot}{m}_\phi(f)$ is usually modelled by five independent noise processes which show different frequency behaviour [53]:

- Random walk frequency modulation $\underset{\cdot\cdot}{m}_\phi(f) \propto |f|^{-4}$
- Flicker frequency modulation $\underset{\cdot\cdot}{m}_\phi(f) \propto |f|^{-3}$
- White frequency modulation $\underset{\cdot\cdot}{m}_\phi(f) \propto |f|^{-2}$
- Flicker phase modulation $\underset{\cdot\cdot}{m}_\phi(f) \propto |f|^{-1}$
- White phase modulation $\underset{\cdot\cdot}{m}_\phi(f) \propto |f|^{0}$

For convenience, we will only refer to an oscillator spectrum shaped according to the Cauchy–Lorentz distribution [51]:

$$\underset{\cdot\cdot}{\Psi}_{xx}(f) = \frac{\Psi_0}{\pi}\frac{a}{a^2 + \left(f \pm f_{c0}\right)^2} \tag{2.302}$$

a is a factor relating to the strength of phase noise. The normalization corresponding to (2.301) gives us the phase noise spectrum:

$$\underset{\cdot\cdot}{m}_{\phi}(f_{\text{offset}}) = \frac{\dot{\Psi}_{xx}(f - f_{c0})}{\Psi_0} = \frac{1}{\pi}\frac{a}{a^2 + f_{\text{offset}}^2} \quad \text{whereas} \quad \int_{-\infty}^{\infty} \underset{\cdot\cdot}{m}_{\phi}(f)\mathrm{d}f = 1 \tag{2.303}$$

It avoids any singularity by approaching the carrier frequency (as above-mentioned models would evoke) while maintaining the asymptotic frequency behaviour of white frequency modulation [51]. Further it keeps the total signal power independent of phase noise. The strength of the phase noise is given by the constant a [Hz]. Under the applied assumptions, it can be determined from $a = \pi f_{\text{offset},0}^2 \underset{\cdot\cdot}{m}_{\phi}(f_{\text{offset},0})$ using a single metering of the spectrum (see Figure 2.93) at an offset frequency which has to be far enough to the carrier, that is $2\pi f_{\text{offset},0} \underset{\cdot\cdot}{m}_{\varphi}(f_{\text{offset},0}) \ll 1$. The phase noise auto-correlation function is obtained from inverse Fourier transform of (2.303):

$$C_{\Delta\phi\,\Delta\phi}(t) = \frac{u_{-1}(t)\mathrm{e}^{-at} + u_{-1}(-t)\mathrm{e}^{at}}{2} \tag{2.304}$$

2.6.3.3 Cycle Jitter

We are still interested in the variations $\Delta\tau_{\mathrm{P}}(t)$ of the sine wave period caused by the phase noise. We can establish following relation between period and frequency of the perturbed sinusoid:

$$t_{\mathrm{P}0} + \Delta\tau_{\mathrm{P}}(t) = \frac{1}{f_{c0} + \Delta f(t)} \approx \frac{1}{f_{c0}}\left(1 - \frac{\Delta f(t)}{f_{c0}}\right) \tag{2.305}$$

$t_{\mathrm{P}0}$ and f_{c0} are the period and the frequency of the unperturbed carrier. Based on this equation, we can mutually relate the variance of frequency noise $\sigma_{\Delta f}^2 = \operatorname{var}\{\Delta f(t)\}$ and cycle jitter $\varphi_{\mathrm{c}}^2 = \operatorname{var}\{\Delta\tau_{\mathrm{P}}(t)\}$ (it is also denoted as sample variance [53]) assuming ergodic behaviour of the perturbations

$$\varphi_{\mathrm{c}}^2 = \frac{\sigma_{\Delta f}^2}{f_{c0}^4} \tag{2.306}$$

We will find the frequency noise variance $\sigma_{\Delta f}^2$ from (2.297), (2.303) and the rules of Fourier transform ($\underset{\cdot\cdot}{m}_{\mathrm{f}}(f)$ [$\mathrm{Hz}^2/\mathrm{Hz}$] is the frequency noise spectrum):

$$\underset{\cdot\cdot}{m}_{\phi}(f) = \frac{\underset{\cdot\cdot}{m}_{f}(f)}{f^2} \tag{2.307}$$

To perform frequency counting, we observe the sine wave over the duration t_{w}. We express the measured waveform as windowed version of the original signal

$$x_{\mathrm{m}}(t) = x(t) * \frac{1}{t_{\mathrm{w}}}\operatorname{rect}\frac{t}{t_{\mathrm{w}}} \tag{2.308}$$

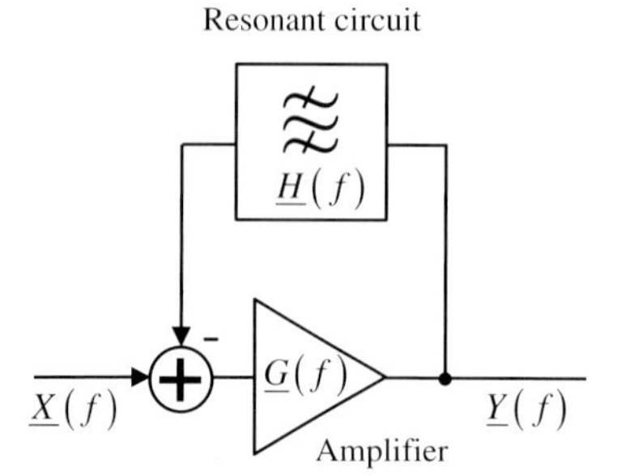

Figure 2.94 Block schematics of an amplifier with a feedback network.

and calculate the frequency noise power (variance) $\sigma_{\Delta f}^2$ of the signal segment. For that purpose, we have to weighing the frequency noise spectrum with the frequency response function of the window:

$$\begin{aligned}\sigma_{\Delta f}^2 &= \int_{-\infty}^{\infty} \underset{\cdot\cdot}{m}_{\mathrm{f}}(f)\left|\underline{G}_{\mathrm{w}}(f)\right|^2 \mathrm{d}f = \int_{-\infty}^{\infty} \underset{\cdot\cdot}{m}_{\mathrm{f}}(f)\operatorname{sinc}^2 f t_{\mathrm{w}}\,\mathrm{d}f \\ &= \frac{1}{(\pi t_{\mathrm{w}})^2}\int_{-\infty}^{\infty} \underset{\cdot\cdot}{m}_{\phi}(f)\sin^2 \pi f t_{\mathrm{w}}\,\mathrm{d}f\end{aligned} \tag{2.309}$$

Applying to the Cauchy–Lorentz spectrum (2.303), we finally yield

$$\sigma_{\Delta f}^2 = \frac{a}{\pi}\int_{-\infty}^{\infty} \frac{\sin^2 \pi f t_{\mathrm{w}}}{(\pi t_{\mathrm{w}} a)^2 + (\pi t_{\mathrm{w}} f)^2}\mathrm{d}f \approx \frac{a}{\pi^2 t_{\mathrm{w}}}\int_{-\infty}^{\infty}\frac{\sin^2 x}{x^2}\mathrm{d}x = \frac{a}{\pi^2 t_{\mathrm{w}}}; \quad a t_{\mathrm{w}} \le 0.01 \tag{2.310}$$

We will mainly use sine wave generators as time reference (i.e. clock generators) in UWB devices. Hence, we are interested in periodic events separated by one or multiple periods. Therefore, the time window length may also be written as $t_{\mathrm{w}} = p t_{\mathrm{P}} = p/f_{\mathrm{c}0};\ p \in \mathbb{N}$. From this the side condition in (2.310) is also $\underset{\cdot\cdot}{m}_{\phi}(f_{\mathrm{offset},0}) \ll 0.01 f_{\mathrm{c}0}/p\pi f_{\mathrm{offset},0}^2$ which is already fulfilled from pure quality oscillators for not too long time window. Using (2.306), we can finally find the cycle jitter of a sine wave generator:

$$\varphi_{\mathrm{c}}^2 = \frac{f_{\mathrm{offset},0}^2}{\pi p f_{\mathrm{c}0}^3}\underset{\cdot\cdot}{m}_{\phi}\left(f_{\mathrm{offset},0}\right) \tag{2.311}$$

We can observe from this relation that the best jitter performance is achieved with generators of low phase noise if they operate at high frequency. Since time base implementation is a conceptual issue for UWB sensing devices, above-mentioned aspects have to be respected appropriately. The M-sequence principle as discussed in Section 3.3.4 makes explicitly use from the conclusions of (2.311).

For additional discussions on phase noise or jitter and their impact, the reader is pointed to Ref. [54] and Sections 3.4.4, 4.7.3 and Annex B.4.4.

2.6.3.4 **Oscillator Stability**

It is an interesting question to ask for the design criteria of an oscillator which should provide an oscillation of high purity. The reader is referred to the paper [55] for a more deeper look at the theory. Here, we will only discuss some basic aspects which may impact the device concept of UWB sensors. Let's consider for that purpose an amplifier with a feedback network as depicted in Figure 2.94. The overall transfer characteristic of this network is given by

$$\frac{\underline{Y}(f)}{\underline{X}(f)} = \frac{\underline{G}(f)}{1 + \underline{G}(f)\underline{H}(f)}$$

From this we can see that the network tends to oscillate if the open loop gain will become

$$\underline{G}(f)\underline{H}(f) = -1$$

or separated in amplitude and phase conditions

$$\begin{aligned} G(f)H(f) &= 1 \\ \phi_G(f) + \phi_H(f) &= (2n+1)\pi \end{aligned}$$

An oscillator is built in such a way that it automatically runs into this condition after switching the power on.

Since the oscillator circuit automatically locks to above conditions, any variation of $\underline{G}(f)$ or $\underline{H}(f)$ leads to a change in the oscillation frequency. In this connection, the phase ϕ_G and ϕ_H of both building blocks show the most sensitive reaction. Let us suppose, for example, that the transfer characteristics depend beside the frequency f still on other quantities such as temperature T, ageing a, vibrations v and so on (that is $\phi = \phi(f, T, a, v, \cdots)$). Therefore, any phase variation affects the phase condition by

$$\left(\frac{\partial \phi_G}{\partial f} + \frac{\partial \phi_H}{\partial f}\right)\Delta f + \left(\frac{\partial \phi_G}{\partial T} + \frac{\partial \phi_H}{\partial T}\right)\Delta T + \left(\frac{\partial \phi_G}{\partial a} + \frac{\partial \phi_H}{\partial a}\right)\Delta a + \left(\frac{\partial \phi_G}{\partial v} + \frac{\partial \phi_H}{\partial v}\right)\Delta v + \cdots = 0$$

which results in an unwanted frequency variation

$$\Delta f = -\frac{(\partial\phi_G/\partial T + \partial\phi_H/\partial T)\Delta T + (\partial\phi_G/\partial a + \partial\phi_H/\partial a)\Delta a + (\partial\phi_G/\partial v + \partial\phi_H/\partial v)\Delta v + \cdots}{(\partial\phi_G/\partial f + \partial\phi_H/\partial f)} \tag{2.312}$$

What do we learn from (2.312) with the aim to stabilize an oscillator?

- External influences (e.g. temperature, vibrations, etc.) should be minimized by corresponding counter measures.
- The circuits should be designed in a way that they are robust against external influences, that is $\partial\phi/\partial T$; $\partial\phi/\partial a$; $\partial\phi/\partial v \cdots \to 0$. This also includes minimum internal noise.

- The sensitivity of phase variations against frequency should be as large as possible (i.e. $\partial\phi_G/\partial f + \partial\phi_H/\partial f \to \infty$). First and foremost, this condition can be met by a low-loss feedback network since the quality factor[37] Q of a passive network and its phase variation and group delay are mutually related: $\partial\phi_H/\partial f|_{f_{c0}} = -2\pi\tau_g(f_{c0}) \approx -2Q/f_{c0}$.

Concerning stable sensor timing, two points should be highlighted. First, an oscillator of best jitter performance must be a single tone device since only then it is possible to provide a high Q feedback network. Second, oscillators of low operational frequencies must provide a reasonable better phase noise level $L_\phi(f_{\text{offset}})$ than a high frequency version in order to gain the same jitter performance (see (2.311)). This has to be respected appropriately for the layout of the timing system of high performance UWB sensor devices. As a general rule of thumb, we can state that two oscillators of the operational frequencies f_1 and f_2 provoke nearly the same timing jitter φ_j if their phase noise levels mutually relate as $L_{\phi 1}(f_{\text{offset}}) \approx L_{\phi 2}(f_{\text{offset}}) + A \lg f_2/f_1$ where the constant A depends on the dominant phase noise characteristics. The constant equals $A = 30$ for our example of Cauchy–Lorentz-spectrum (2.311). Further discussions on phase noise and jitter can be found, for example, in Refs [56–60].

2.6.4
Linear Systematic Errors and their Correction

Unfortunately, it is not possible to construct wideband RF-devices with nearly ideal behaviour. Signal leakage, device internal reflections, port mismatch and frequency-dependent components and so on result in erroneous frequency behaviour of the measurement device and in a device impulse response having unwanted side lobes and spurious components. We will summary all this imperfections by the term 'internal' or 'device clutter'.

Since internal clutter represents systematic deviations from a wanted behaviour, one can try to remove them from the measured quantities because the evoked errors remains the same in all measurements. Due to the supposed linearity and time invariance, the usual philosophy to eliminate the internal clutter consists of two steps. First, the device imperfections must be identified and second they must be de-convolved from the measured data in order to gain the correct measurement of the unknown object under test.

The device imperfections are very multi-faceted, therefore it is usually not possible and practical to describe individually the properties of all errors by physical models. Rather the common strategy is to decompose the real measurement device into an ideal one and a virtual transmission system which is assigned to cause all errors whatever their reasons are. Figure 2.95 depicts the approach for an example

37) The quality factor of a network is defined as $Q = 2\pi f \frac{\text{energy stored}}{\text{power loss}} = 2\pi \frac{\text{energy stored}}{\text{energy loss per cycle}}$.

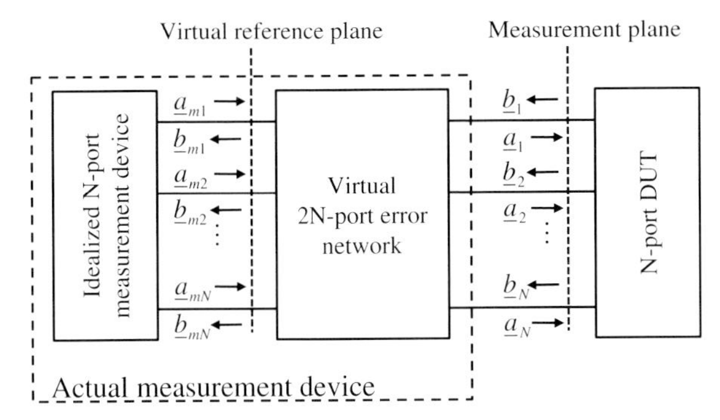

Figure 2.95 Formal error model for an *N*-port measurement. The indicated example refers to the determination of scattering parameters by capturing the normalized waves a_{mj} and b_{mj}.

of *S*-parameter measurement. The method is not restricted to this type of measurements and it does not require prior knowledge about the physical reasons of the device errors. As even mentioned, the only assumptions to be made are linearity and time invariance of the device imperfections.

All of the errors which are provoked by the inadequacy of the measurement device and the sensor are merged by the virtual error system. Therefore, the formal error network must represent a $2N$-port system if the DUT has N-measurement ports. A measurement or sensor system which interacts over N-ports with the device under test has apparently to deal with $4N^2$ error terms. We are interested in *S*-parameters which represent normalized quantities and not absolute values. Thus, one of the error terms can be selected freely (typically it is set to unity) and we finally get in maximum $4N^2 - 1$ independent error terms. Every of these error terms represent either a FRF or an IRF depending on the considered domain – frequency or time. That is, a complete correction of a one-port measurement requires 3 error functions, a two-port measurement needs 15 error functions, a full correction of a three-port measurement even expects 35 error terms and so on. The handling of so many error functions causes some problems. Furthermore, the efficiency of error correction will decrease with increasing number of error functions because every term is also subjected to random errors. Hence for practical purposes, only the most important errors will be respected.

Additionally it has to be mentioned that an error correction of high quality does only work if the complete *S*-Matrix of the DUT is determined (see below). The practical implication of this fact leads to the requirement to measure at every port of the DUT the incoming and outgoing waves (or the voltage and the current if one is interested in impedance or admittance parameters). Nowadays, this is however usually not yet done by UWB sensors due to the additional hardware effort. But error correction is a widely used approach under laboratory conditions in connection with network analysers. Hence, the method will also find his way into the UWB sensor technique for the future time.

We will first shortly summarize the general idea of internal clutter reduction by the demonstrated formal and virtual error network and then the approach shall be illustrated more deeply for one- and two-port devices. For a comprehensive view on issues of error correction, the reader can refer to Refs [61–64]. We will also underline that the philosophy of error correction as discussed below is a general approach which is not only restricted to correct guided wave measurements. An example to extend the method to waves propagating in free space is given in Refs [65, 66].

Due to the simpler calculations, we prefer frequency domain notation here and we will use scattering parameters without loss of generality. Furthermore, we merge all port signals at the measurement and virtual reference plane to column vectors:

$$\text{Wanted signals:}\quad \begin{array}{l} \underline{\mathbf{A}} = \left[\underline{a}_1 \quad \underline{a}_2 \quad \underline{a}_3 \quad \cdots \quad \underline{a}_N\right]^T \\ \underline{\mathbf{B}} = \left[\underline{b}_1 \quad \underline{b}_2 \quad \underline{b}_3 \quad \cdots \quad \underline{b}_N\right]^T \end{array}$$

$$\text{Measured signals:}\quad \begin{array}{l} \underline{\mathbf{A}}_m = \left[\underline{a}_{m1} \quad \underline{a}_{m2} \quad \underline{a}_{m3} \quad \cdots \quad \underline{a}_{mN}\right]^T \\ \underline{\mathbf{B}}_m = \left[\underline{b}_{m1} \quad \underline{b}_{m2} \quad \underline{b}_{m3} \quad \cdots \quad \underline{b}_{mN}\right]^T \end{array}$$

The error network behaves linearly and time invariant. Therefore, we can establish an arbitrary system of linear equations for the port signals which we will express by the matrix relation

$$\begin{bmatrix} \underline{\mathbf{A}} \\ \underline{\mathbf{B}} \end{bmatrix} = \begin{bmatrix} \underline{\mathbf{W}} & \underline{\mathbf{X}} \\ \underline{\mathbf{Y}} & \underline{\mathbf{Z}} \end{bmatrix} \begin{bmatrix} \underline{\mathbf{A}}_m \\ \underline{\mathbf{B}}_m \end{bmatrix} \tag{2.313}$$

Herein, the sub-matrices $\underline{\mathbf{W}}, \underline{\mathbf{X}}, \underline{\mathbf{Y}}$ and $\underline{\mathbf{Z}}$ are composed from the error terms and they are of dimension $[N \times N]$. For the chosen example, the WXYZ-matrix represents an inverse T-matrix (see (2.213)). But basically we are not restricted to use this type of matrix representation. In the case of an ideal measurement device, the WXYZ-matrix would be an identity matrix.

The wanted S-parameters are defined by $\underline{\mathbf{B}} = \underline{\mathbf{S}}\underline{\mathbf{A}}$ and for the measured (and erroneous) ones hold $\underline{\mathbf{B}}_m = \underline{\mathbf{S}}_m \underline{\mathbf{A}}_m$. Insertion in (2.313) gives

$$\underline{\mathbf{Y}} + \underline{\mathbf{Z}}\,\underline{\mathbf{S}}_m = \underline{\mathbf{S}}\,\underline{\mathbf{W}} + \underline{\mathbf{S}}\,\underline{\mathbf{X}}\,\underline{\mathbf{S}}_m \tag{2.314}$$

This equation provides the general basis for the error correction. It is a matrix equation. That is, an individual entry of the wanted $\underline{\mathbf{S}}$-matrix depends usually on all entries of the other matrices. Hence, as mentioned above, the whole $\underline{\mathbf{S}}_m$-matrix has to be measured even though only a single entry of the $\underline{\mathbf{S}}$-matrix is of interest.

(2.314) has a twofold meaning. On the one hand it is needed for the determination of the error terms and on the other hand it provides the adjusted measurement results. It should also be mentioned that (2.314) is only one of a number of forms to mutually relate $\underline{\mathbf{S}}$ and $\underline{\mathbf{S}}_m$.

We will clarify the approach by two examples. In the first one, we consider only one-port devices. But we do not restrict our discussion to S-parameters rather we will include other parameter types too, for example Z-parameters in order to show the generality of error correction. The second example deals with a two-port device.

Here, we will make use of a modified version of (2.314) since it will lead to simpler calculations for the considered case.

One-port error correction – scattering parameters: In the case of a one-port device, the matrix equation (2.314) simplifies to a scalar one, in which one of the four error terms (e.g. $\underline{W}$) can be factored out. With the remaining three error terms we finally get

$$\begin{aligned} \frac{\underline{Y}}{\underline{W}} + \frac{\underline{Z}}{\underline{W}}\underline{S}_m &= \underline{S} + \frac{\underline{X}}{\underline{W}}\underline{S}\,\underline{S}_m \\ \underline{E}_1 + \underline{E}_2\underline{S}_m &= \underline{S} + \underline{E}_3\underline{S}\,\underline{S}_m \end{aligned} \tag{2.315}$$

If the error terms $\underline{E}_1, \underline{E}_2, \underline{E}_3$ of the measurement and sensor device are known, the correct reflection coefficient $\underline{S}$ of the DUT can be calculated from the erroneous measurement $\underline{S}_m$ via (2.315):

$$\underline{S} = \frac{\underline{E}_1 + \underline{E}_2\underline{S}_m}{1 + \underline{E}_3\underline{S}_m}. \tag{2.316}$$

But beforehand, the error terms must be determined based on reference measurements with devices of exactly known behaviour. This action is usually referred as calibration and the reference objects are the calibration standards. Since we are looking for three error terms, we need at least three (or even more) measurements $\underline{S}_{mi}$ with a corresponding number of known objects $\underline{S}_i$. Using (2.315), all the measurements can be summarized in the following matrix equation:

$$\begin{aligned} \begin{bmatrix} 1 & \underline{S}_{m1} & -\underline{S}_1\underline{S}_{m1} \\ 1 & \underline{S}_{m2} & -\underline{S}_2\underline{S}_{m2} \\ \vdots & \vdots & \vdots \end{bmatrix} \begin{bmatrix} \underline{E}_1 \\ \underline{E}_2 \\ \underline{E}_3 \end{bmatrix} &= \begin{bmatrix} \underline{S}_1 \\ \underline{S}_2 \\ \vdots \end{bmatrix} \\ \underline{\mathbf{M}}_{\mathrm{cal}}\,\underline{\mathbf{E}} &= \underline{\mathbf{S}}_{\mathrm{cal}} \end{aligned} \tag{2.317}$$

Herein, the complex-valued column vector $\underline{\mathbf{E}}$ contains all error elements which we are looking for. The vector $\underline{\mathbf{S}}_{\mathrm{cal}}$ merges the a priori known reflection coefficients of the calibration standards and the entries of the matrix $\underline{\mathbf{M}}_{\mathrm{cal}}$ originate from the calibration measurements and the calibration standards. Hence in the usual case with three calibration measurements, the error vector $\underline{\mathbf{E}}$ calculates to

$$\underline{\mathbf{E}} = \underline{\mathbf{M}}_{\mathrm{cal}}^{-1}\,\underline{\mathbf{S}}_{\mathrm{cal}} \tag{2.318}$$

As long as the measurement port consists of a coaxial RF-connector, the most convenient calibration standards are the short (ideally $\underline{S} = -1$), the open (ideally $\underline{S} = 1$) and the match (ideally $\underline{S} = 0$). Actual calibration standards deviate, however, from the ideal values. In the case of commercial standards, this is typically respected by correction functions provided by the manufacturer (e.g. [67]).

If the measurement plane is not defined by an industry standard (i.e. standard RF-connectors), one has to deal with self-made calibration devices adapted to the actual measurement environment. In principle any one-port system can be used for calibration as long as its behaviour is well known and the applied standards show sufficient diversity in their characteristics. If this is only inadequately possible, the use of a larger number of calibration measurements may be helpful. But then (2.318) has to be modified according to (2.319) since otherwise matrix inversion cannot be performed (see also [38] or Annex A.6 for pseudo-inverse):

$$\underline{\mathbf{E}} = \left(\underline{\mathbf{M}}_{\mathrm{cal}}^{\mathrm{H}} \underline{\mathbf{M}}_{\mathrm{cal}}\right)^{-1} \underline{\mathbf{M}}_{\mathrm{cal}}^{\mathrm{H}} \underline{\mathbf{S}}_{\mathrm{cal}} \tag{2.319}$$

We can assess the quality of the matrix inversion in (2.318) or (2.319) by the condition number $\kappa(\underline{\mathbf{M}}_{\mathrm{cal}})$ of the calibration matrix $\underline{\mathbf{M}}_{\mathrm{cal}}$. If it is too far from unity, a better set of calibration standards should be applied, and the sounding signal contains too little energy.

After the errors of the measurement device are determined by the above-described method, the measurement values of an unknown device can be removed from the internal clutter by applying (2.316). However attention should be paid, since (2.316) represents a de-convolution which often leads to ill-conditioned problems. Such a problem arises if the spectral power of the stimulus signal breaks down within any spectral interval. Figure 2.96 illustrates the phenomenon in comparison to a network analyser measurement. Since most UWB sensors apply wideband signals, we have always to count with such problem because the spectral signal power must vanish in the upper frequency range in order to respect the Nyquist theorem. Consequently, above a certain frequency, the measurement values are useless and they should be eliminated by a window function $W(f)$ as depicted in (2.320) and Figure 2.96. Network analyser measurements are not affected by such effects since they stimulate the measurement objects up to the band limit with approximately the same power. The

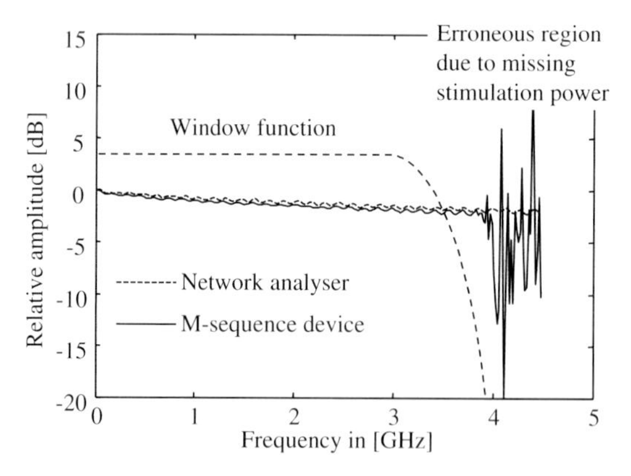

Figure 2.96 Generation of random errors due to ill-conditioned error correction within the upper frequency band of an UWB sensor (M-sequence device).

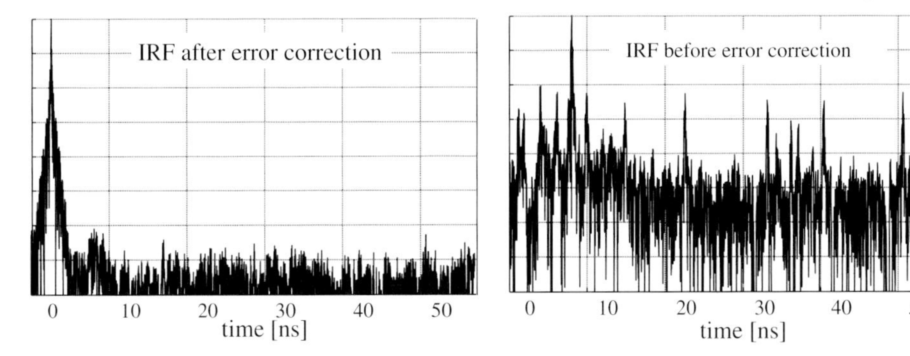

Figure 2.97 Demonstration of IRF error correction for a M-sequence radar of 12 GHz bandwidth.

corrected measurement values are given by[38)]

$$S = \frac{\underline{E}_1 + \underline{E}_2 \underline{S}_{\mathrm{m}}}{1 + \underline{E}_3 \underline{S}_{\mathrm{m}}} W \tag{2.320}$$

If one is interested in the corrected FRF of the DUT, it may be sufficient to apply simply a rectangular window to cut out the dirty part of the FRF.

The determination of an error corrected IRF needs a careful selection of the window function since it is not only required to suppress the ill-conditioned signal components but also to reduce artificial side lobes of the corrected IRF. Therefore, also network analyser data has to be subjected windowing before they are transformed into the time domain.

Figure 2.97 shows an example of such an IRF correction for a 12 GHz bandwidth M-sequence radar [68, 69]. The decay of the error corrected main lobe – which is actually composed from a number of narrow side lobes – is mainly determined by the choice of the window function.

A typical window function W has values of unity for frequency intervals of low perturbations and their values should go down to zero with decreasing spectral power of the stimulus and hence decreasing signal-to-noise ratio. As mentioned, the choice of an appropriate window function may be critical since it influences the shape of the impulse response function gained from the inverse Fourier transform of the measurement S-parameters. As a general rule of thumb one can state that a window function having wide top and short/steep edges provides a narrow pulse with strong side lobes in the time domain whereas a smooth window with a short top and long/soft edges results in an impulse of a slightly widened main peak but strongly decaying side lobes. In many cases, the time domain side lobes provoked by the window function are of non-causal nature. MATAB® provides a number of different standard window functions. It is also possible to suppress ill-conditioned parts in dependence of the SNR-value by applying a Wiener-Filter approach or Tikhonov regularization [70] (see also Section 4.8 for an example).

38) Do not confuse the window function $W(f)$ with the error term $\underline{W}(f)$ in equation (2.315).

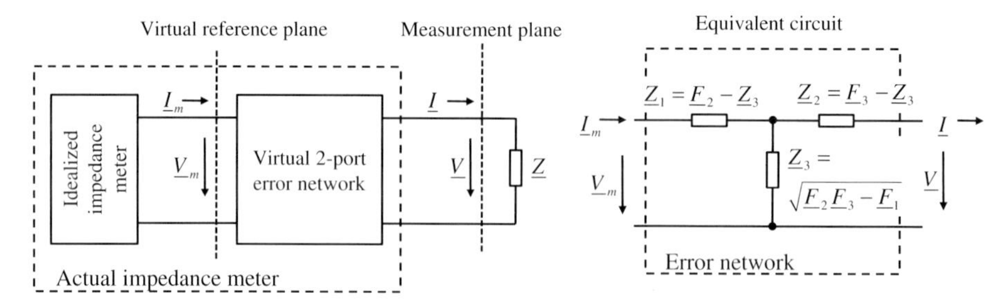

Figure 2.98 Error model for an impedance measurement and a possible equivalent circuit for the error network.

One-port error correction – impedance parameters: Nowadays, the above demonstrated error correction is usually applied in network analysers which are designed to measure scattering parameters. The method, however, is of generic nature and not only restricted to *S*-parameter measurements. Let us consider an impedance measurement for that purpose. As Figure 2.98 demonstrates, the real impedance metre is decomposed in the same way as shown in Figure 2.95. The actual measured impedance value is $\underline{Z}_{\mathrm{m}} = \underline{V}_{\mathrm{m}}/\underline{I}_m$ and the wanted one is $\underline{Z} = \underline{V}/\underline{I}$. Using a chain matrix representation to describe the error network

$$\begin{bmatrix} \underline{V}_{\mathrm{m}} \\ \underline{I}_{\mathrm{m}} \end{bmatrix} = \begin{bmatrix} \underline{A}_{11} & \underline{A}_{12} \\ \underline{A}_{21} & \underline{A}_{22} \end{bmatrix} \begin{bmatrix} \underline{V} \\ \underline{I} \end{bmatrix}$$

we get

$$\underline{A}_{12} + \underline{A}_{11}\,\underline{Z} = \underline{A}_{21}\,\underline{Z}\underline{Z}_{\mathrm{m}} + \underline{A}_{22}\,\underline{Z}_{\mathrm{m}}$$

and finally we will end up after having factoring out, for example $\underline{A}_{21}$:

$$\underline{F}_1 + \underline{F}_2\underline{Z} = \underline{Z}\underline{Z}_{\mathrm{m}} + \underline{F}_3\underline{Z}_{\mathrm{m}}$$

This equation is comparable to (2.315). Therefore, we can deal with it in the same way as even show in order to find the values of the error terms and to provide the corrected measurement result.

In the two examples above, we simply used formal two-port parameters of an arbitrary type in order to describe the error networks. But we also can model it by physically reliable structures as a Mason graph representation for *S*-parameter measurements or an equivalent *T*- or π-network for the impedance measurement. The last case is demonstrated in Figure 2.98 which also includes the transformation from the deduced error terms to the equivalent elements.

Two-port error correction: Figure 2.99 shows a slightly extended version of Figure 2.95 adapted to two-port *S*-parameter measurements. The waves travelling towards the DUT $\underline{a}_i$ or being emanated from it $\underline{b}_j$ are separated by any approach introduced in Chapter 3. Thus, the actually captured voltage signals are proportional to the wanted normalized waves. The errors provoked by the

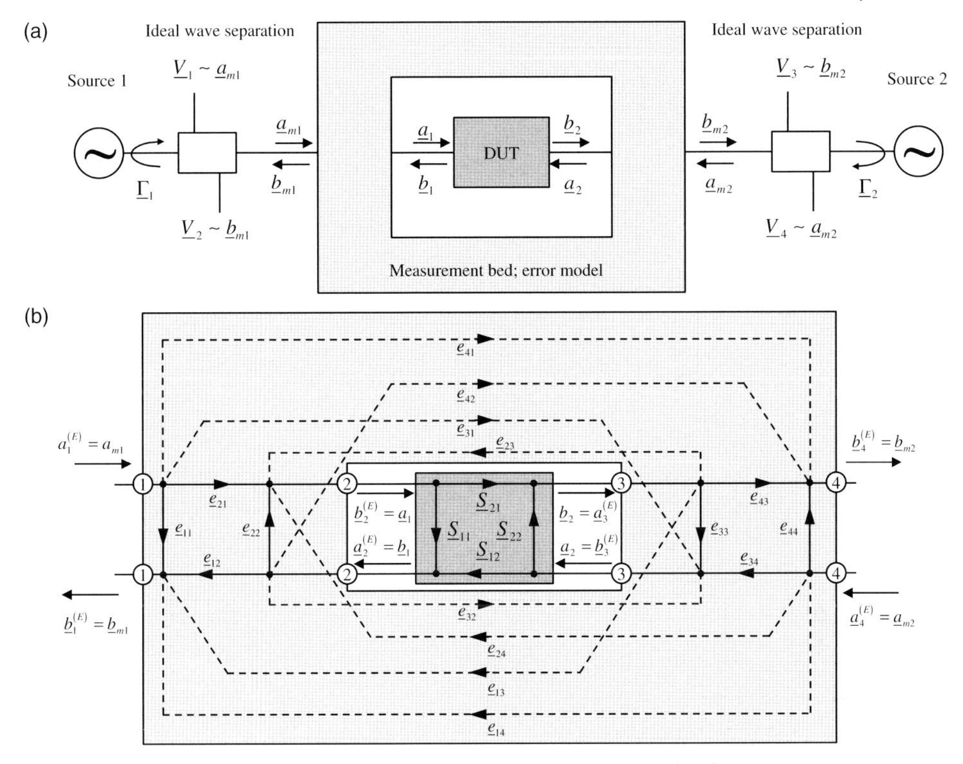

Figure 2.99 Formal error model for a two-port measurement. (a) Measurement schematics with measurement bed and (b) Mason graph of the measurement arrangement for *S*-parameters.

measurement arrangement are summarized by the virtual measurement bed surrounding the DUT. It represents a four-port system which is composed from 16 scattering parameters $\underline{e}_{ij}$. The related Mason graph is depicted in Figure 2.99. It also contains the port numbering of the error model.

In order to be able to determine all *S*-parameters, the DUT must be stimulated from both sides (refer also to Section 2.5.4). Therefore, we introduce two signal generators. In a first step source 1 is active and source 2 is off. It is called forward measurement and provides the normalized waves $\underline{a}_{m11}, \underline{a}_{m21}, \underline{b}_{m11}, \underline{b}_{m21}$. During the reverse measurement, source 2 is on and source 1 is off. The captured waves we call $\underline{a}_{\mathrm{m}12}, \underline{a}_{\mathrm{m}22}, \underline{b}_{\mathrm{m}12}, \underline{b}_{\mathrm{m}22}$.

For a reason, we will see later, we are interested in the knowledge of the *T*-parameters of the DUT which can be converted afterwards into the more convenient *S*-parameters (see Annex B.6), that is we can write for the wanted quantities

$$\begin{bmatrix} \underline{b}_1 \\ \underline{a}_1 \end{bmatrix} = \begin{bmatrix} \underline{T}_{11} & \underline{T}_{12} \\ \underline{T}_{21} & \underline{T}_{22} \end{bmatrix} \begin{bmatrix} \underline{a}_2 \\ \underline{b}_2 \end{bmatrix} \Rightarrow \underline{\mathbf{L}} = \underline{\mathbf{T}}\,\underline{\mathbf{R}} \tag{2.321}$$

Since our access is restricted to the ports ① and ④, we only succeed to measure the *T*-parameters $\underline{\mathbf{T}}_m$ which can be calculated from the captured waves as follows:

$$\begin{bmatrix} \underline{b}_{\mathrm{m}11} & \underline{b}_{\mathrm{m}12} \\ \underline{a}_{\mathrm{m}11} & \underline{a}_{\mathrm{m}12} \end{bmatrix} = \underline{\mathbf{L}}_{\mathrm{m}} = \underline{\mathbf{T}}_{\mathrm{m}} \begin{bmatrix} \underline{a}_{\mathrm{m}21} & \underline{a}_{\mathrm{m}22} \\ \underline{b}_{\mathrm{m}21} & \underline{b}_{\mathrm{m}22} \end{bmatrix} = \underline{\mathbf{T}}_{\mathrm{m}} \underline{\mathbf{R}}_{\mathrm{m}} \tag{2.322}$$

Note that this equation is able to deal with two incidents waves in every measurement step so that the generator mismatch $\underline{\Gamma}_1$ and $\underline{\Gamma}_2$ can be respected and it will not be part of the error terms. It also means that the generator mismatch is allowed to vary during the measurement cycle. Such variations may appear by switching from forward to reverse or reverse to forward measurement.

The *S*-parameters $\underline{\mathbf{E}}_S$ of the error network as depicted by the Mason graph in Figure 2.99 are defined by

$$\begin{bmatrix} \underline{b}_1^{(E)} \\ \underline{b}_2^{(E)} \\ \underline{b}_3^{(E)} \\ \underline{b}_4^{(E)} \end{bmatrix} = \begin{bmatrix} \underline{e}_{11} & \underline{e}_{12} & \underline{e}_{13} & \underline{e}_{14} \\ \underline{e}_{21} & \underline{e}_{22} & \underline{e}_{23} & \underline{e}_{24} \\ \underline{e}_{31} & \underline{e}_{32} & \underline{e}_{33} & \underline{e}_{34} \\ \underline{e}_{41} & \underline{e}_{42} & \underline{e}_{43} & \underline{e}_{44} \end{bmatrix} \begin{bmatrix} \underline{a}_1^{(E)} \\ \underline{a}_2^{(E)} \\ \underline{a}_3^{(E)} \\ \underline{a}_4^{(E)} \end{bmatrix} \Rightarrow \underline{\mathbf{B}}^{(E)} = \underline{\mathbf{E}}_S \underline{\mathbf{A}}^{(E)} \tag{2.323}$$

This notation is not convenient for our *T*-parameter descriptions as selected in (2.321) and (2.322). Hence, we define a new error matrix $\underline{\mathbf{E}}_T$ as follows:

$$\begin{bmatrix} \underline{a}_2^{(E)} \\ \underline{b}_2^{(E)} \\ \underline{b}_3^{(E)} \\ \underline{a}_3^{(E)} \end{bmatrix} = \underline{\mathbf{E}}_T \begin{bmatrix} \underline{b}_1^{(E)} \\ \underline{a}_1^{(E)} \\ \underline{a}_4^{(E)} \\ \underline{b}_4^{(E)} \end{bmatrix} \quad \Rightarrow \quad \begin{bmatrix} \underline{\mathbf{L}} \\ \underline{\mathbf{R}} \end{bmatrix} = \underline{\mathbf{E}}_T \begin{bmatrix} \underline{\mathbf{L}}_{\mathrm{m}} \\ \underline{\mathbf{R}}_{\mathrm{m}} \end{bmatrix} \tag{2.324}$$

where we used the identities which follows from the Mason graph in Figure 2.99:

$$\left[\underline{a}_2^{(E)} \; \underline{b}_2^{(E)} \; \underline{b}_3^{(E)} \; \underline{a}_3^{(E)}\right]^T = \left[\underline{b}_1 \; \underline{a}_1 \; \underline{a}_2 \; \underline{b}_2\right]^T = \left[\underline{\mathbf{L}}^T \; \underline{\mathbf{R}}^T\right]^T$$
$$\left[\underline{b}_1^{(E)} \; \underline{a}_1^{(E)} \; \underline{a}_4^{(E)} \; \underline{b}_4^{(E)}\right]^T = \left[\underline{b}_{\mathrm{m}1} \; \underline{a}_{\mathrm{m}1} \; \underline{a}_{\mathrm{m}2} \; \underline{b}_{\mathrm{m}2}\right]^T = \left[\underline{\mathbf{L}}_{\mathrm{m}}^T \; \underline{\mathbf{R}}_{\mathrm{m}}^T\right]^T$$

The conversion between both notations may be done via matrix calculus in which $\mathbf{Q}_j$ represent transformation matrices:

$$\begin{bmatrix} \underline{a}_2^{(E)} \\ \underline{b}_2^{(E)} \\ \underline{b}_3^{(E)} \\ \underline{a}_3^{(E)} \end{bmatrix} = \begin{bmatrix} 0 & 0 & 0 & 0 \\ 0 & 1 & 0 & 0 \\ 0 & 0 & 1 & 0 \\ 0 & 0 & 0 & 0 \end{bmatrix} \begin{bmatrix} \underline{b}_1^{(E)} \\ \underline{b}_2^{(E)} \\ \underline{b}_3^{(E)} \\ \underline{b}_4^{(E)} \end{bmatrix} + \begin{bmatrix} 0 & 1 & 0 & 0 \\ 0 & 0 & 0 & 0 \\ 0 & 0 & 0 & 0 \\ 0 & 0 & 1 & 0 \end{bmatrix} \begin{bmatrix} \underline{a}_1^{(E)} \\ \underline{a}_2^{(E)} \\ \underline{a}_3^{(E)} \\ \underline{a}_4^{(E)} \end{bmatrix} \Rightarrow \begin{bmatrix} \underline{\mathbf{L}} \\ \underline{\mathbf{R}} \end{bmatrix} = \mathbf{Q}_1 \underline{\mathbf{B}}^{(E)} + \mathbf{Q}_2 \underline{\mathbf{A}}^{(E)} = (\mathbf{Q}_1 \underline{\mathbf{E}}_S + \mathbf{Q}_2) \underline{\mathbf{A}}^{(E)}$$

$$\begin{bmatrix} \underline{b}_1^{(E)} \\ \underline{a}_1^{(E)} \\ \underline{a}_4^{(E)} \\ \underline{b}_4^{(E)} \end{bmatrix} = \begin{bmatrix} 1 & 0 & 0 & 0 \\ 0 & 0 & 0 & 0 \\ 0 & 0 & 0 & 0 \\ 0 & 0 & 0 & 1 \end{bmatrix} \begin{bmatrix} \underline{b}_1^{(E)} \\ \underline{b}_2^{(E)} \\ \underline{b}_3^{(E)} \\ \underline{b}_4^{(E)} \end{bmatrix} + \begin{bmatrix} 0 & 0 & 0 & 0 \\ 1 & 0 & 0 & 0 \\ 0 & 0 & 0 & 1 \\ 0 & 0 & 0 & 0 \end{bmatrix} \begin{bmatrix} \underline{a}_1^{(E)} \\ \underline{a}_2^{(E)} \\ \underline{a}_3^{(E)} \\ \underline{a}_4^{(E)} \end{bmatrix} \Rightarrow \begin{bmatrix} \underline{\mathbf{L}}_{\mathrm{m}} \\ \underline{\mathbf{R}}_{\mathrm{m}} \end{bmatrix} = \mathbf{Q}_3 \underline{\mathbf{B}}^{(E)} + \mathbf{Q}_4 \underline{\mathbf{A}}^{(E)} = (\mathbf{Q}_3 \underline{\mathbf{E}}_S + \mathbf{Q}_4) \underline{\mathbf{A}}^{(E)}$$

This finally leads to

$$\underline{\mathbf{E}}_T = (\mathbf{Q}_1\underline{\mathbf{E}}_S + \mathbf{Q}_2)(\mathbf{Q}_3\underline{\mathbf{E}}_S + \mathbf{Q}_4)^{-1}. \tag{2.325}$$

Now, we decompose the new error matrix into sub-matrices and join (2.324) with (2.321) and (2.322)

$$\begin{bmatrix} \underline{\mathbf{L}} \\ \underline{\mathbf{R}} \end{bmatrix} = \begin{bmatrix} \underline{\mathbf{E}}_T^A & \underline{\mathbf{E}}_T^B \\ \underline{\mathbf{E}}_T^C & \underline{\mathbf{E}}_T^D \end{bmatrix} \begin{bmatrix} \underline{\mathbf{L}}_\mathrm{m} \\ \underline{\mathbf{R}}_\mathrm{m} \end{bmatrix} \quad \Rightarrow \quad \underline{\mathbf{E}}_T^A \underline{\mathbf{T}}_\mathrm{m} + \underline{\mathbf{E}}_T^B = \underline{\mathbf{T}}\,\underline{\mathbf{E}}_T^C \underline{\mathbf{T}}_\mathrm{m} + \underline{\mathbf{T}}\,\underline{\mathbf{E}}_T^D \tag{2.326}$$

This will give us an equation of same nature and identical conclusions as already introduced by (2.314).
We will try to solve this equation with respect to the error terms [71]. For that purpose, we will assume a measurement device for which the internal crosstalk can be neglected. By proper shielding the measurement receivers and generators, these errors can often be reduced to an insignificant level. That is, we can omit all transmission paths in the Mason graph which are drawn by dashed lines. Now only eight error terms remain and the error matrices simplifies to

$$\underline{\mathbf{E}}_S = \begin{bmatrix} \underline{e}_{11} & \underline{e}_{12} & 0 & 0 \\ \underline{e}_{21} & \underline{e}_{22} & 0 & 0 \\ 0 & 0 & \underline{e}_{33} & \underline{e}_{34} \\ 0 & 0 & \underline{e}_{43} & \underline{e}_{44} \end{bmatrix}$$

$$\underline{\mathbf{E}}_T^A = \frac{1}{\underline{e}_{12}} \begin{bmatrix} 1 & -\underline{e}_{11} \\ \underline{e}_{22} & -\Delta_A \end{bmatrix} = \frac{1}{\underline{e}_{12}} \underline{\mathbf{E}}_{T0}^A \quad \underline{\mathbf{E}}_T^B = \begin{bmatrix} 0 & 0 \\ 0 & 0 \end{bmatrix}$$

$$\underline{\mathbf{E}}_T^C = \begin{bmatrix} 0 & 0 \\ 0 & 0 \end{bmatrix} \qquad \underline{\mathbf{E}}_T^D = \frac{1}{\underline{e}_{43}} \begin{bmatrix} -\Delta_D & \underline{e}_{33} \\ -\underline{e}_{44} & 1 \end{bmatrix} = \frac{1}{\underline{e}_{43}} \underline{\mathbf{E}}_{T0}^D \tag{2.327}$$

$$\begin{aligned} \Delta_A &= \underline{e}_{11}\underline{e}_{22} - \underline{e}_{21}\underline{e}_{12} \\ \Delta_D &= \underline{e}_{33}\underline{e}_{44} - \underline{e}_{34}\underline{e}_{43} \end{aligned}$$

Here, we used (2.325) to convert $\underline{\mathbf{E}}_S$ to $\underline{\mathbf{E}}_T$. The corresponding error model is called eight-term error model even if only seven errors are actually independent of each other as we will see below.
Inspecting the remaining Mason graph, it will be obvious that the network can be separated into three cascaded parts. Cascaded networks are preferentially handled by T-parameters. This was the reason why we switched to them. We need a sufficient number of calibration standards to determine the unknown error terms. There is a choice of various types of standards [61], but we will consider only one of them for illustration.
If we initially start with three calibration standards of the form

$$\underline{\mathbf{S}} = \underline{\mathbf{S}}_{\mathrm{cal},j} = \begin{bmatrix} \underline{\Gamma}_j & 0 \\ 0 & \underline{\Gamma}_j \end{bmatrix}; \quad j = 1, 2, 3$$

The left and the right part of the error network are disconnected. Thus, they can be considered separately and it remains two one-port error models which cover three error terms each. As we have seen above, one-port error models can be

calibrated by using open, short and match. If one connects only such types of devices at port ② and ③ , one actually provokes a two-port S-matrix of the above shown structure since any transmission between port ② and ③ is avoided.
Referring to the Mason graph in Figure 2.99 and the calculation rules from Annex B.7, we find

$$\underline{S}_{11}^{(E)} = \underline{S}_{\mathrm{m}11} = \frac{\underline{e}_{11} - \Delta_A \underline{S}_{11}}{1 - \underline{e}_{22}\underline{S}_{11}}; \quad \underline{S}_{11} = \underline{\Gamma}_j, \quad \underline{S}_{21} = \underline{S}_{12} = 0 \tag{2.328}$$

$$\underline{S}_{44}^{(E)} = \underline{S}_{\mathrm{m}22} = \frac{\underline{e}_{44} - \Delta_D \underline{S}_{22}}{1 - \underline{e}_{33}\underline{S}_{22}}; \quad \underline{S}_{22} = \underline{\Gamma}_j, \quad \underline{S}_{21} = \underline{S}_{12} = 0 \tag{2.329}$$

These equations have the same meaning as (2.315) or (2.316) and the error terms $\underline{e}_{11}, \underline{e}_{22}, \Delta_A$ or $\underline{e}_{33}, \underline{e}_{44}, \Delta_D$ has to be determined by the same approach as discussed in connection with these equations.
Looking back on (2.327), we can observe that we have just nearly solved the two-port calibration by two simple one-port calibrations. Only $\underline{e}_{12}$ and $\underline{e}_{43}$ need still some attention. For this purpose, we connect port ② and ③ via a simple transmission network (typically a short line or an attenuator) having the transmission matrix $\mathbf{T}_{\mathrm{cal},4}$. Using (2.326) and (2.327), we come to relation (2.330) where $\underline{\mathbf{T}}_{\mathrm{m},4}$ is the measured transmission matrix and $\underline{\mathbf{E}}_{T0}^{A}, \underline{\mathbf{E}}_{T0}^{D}$ are just known from the three-term calibration:

$$\frac{\underline{e}_{12}}{\underline{e}_{43}} \underline{\mathbf{T}}_{\mathrm{cal},4} = \underline{\alpha}\, \underline{\mathbf{T}}_{\mathrm{cal},4} = \underline{\mathbf{E}}_{T0}^{A} \underline{\mathbf{T}}_{\mathrm{m},4} \left(\underline{\mathbf{E}}_{T0}^{D}\right)^{-1} \tag{2.330}$$

This equation indicates that the terms $\underline{e}_{12}$ and $\underline{e}_{43}$ dos not appear separately so that it would be sufficiently to know only their complex ratio $\underline{\alpha}$ for a full calibration. This gives in conclusion the just mentioned number of seven unknown error terms. We know from Section 2.5.3, Eq. (2.215) that the determinate of the **T**-matrix from any passive network is one, hence (2.330) simplifies and has to be solved for $\underline{\alpha}$ in order to gain the still missing error term:

$$\begin{aligned} \underline{\alpha}^2 &= \left(\frac{\underline{e}_{12}}{\underline{e}_{43}}\right)^2 = \det\left(\underline{\mathbf{E}}_{T0}^{A} \underline{\mathbf{T}}_{\mathrm{m},4} \left(\underline{\mathbf{E}}_{T0}^{D}\right)^{-1}\right) \\ \underline{\alpha} &= \pm\sqrt{\frac{\det\left(\underline{\mathbf{E}}_{T0}^{A}\right)\det\left(\underline{\mathbf{T}}_{\mathrm{m},4}\right)}{\det\left(\underline{\mathbf{E}}_{T0}^{D}\right)}} \end{aligned} \tag{2.331}$$

The sign ambiguity provokes a $\pm\pi$-phase uncertainty of $\underline{\alpha}$ which can, however, be resolved from an approximate knowledge of the delay time of the transmission standard.
The eight-term error correction as shown above is not restricted to two-port measurements. It may be extended to any port number as long as the crosstalk can be neglected. That is, one has to subjected every port separately to a three-term calibration and finally all combinations of transmissions between the ports have to be measured applying a short cable or attenuator.

2.6.5 Non-Linear Distortions

The reasons and the implications of non-linear distortions are very complex. There is no theory in cohesive form as for linear networks which can be applied for all types of non-linearity. We even were in contact with non-linear distortions without appointing explicitly to that fact. Quantization errors caused by the transfer characteristic of an ADC represent actually non-linear deviations. There we could show that this type of non-linearity can be modelled by noise if no synchronism between the captured signal and the ADC clock exists.

Now, we will summarize a further approach which deals with transfer characteristics showing only smooth and soft deviations from a straight line. Figure 2.100 depicts a typical shape of such a transfer function as it can be observed by saturation effects in amplifiers or other devices. The method which we will consider is very popular to handle non-linear distortions. But one should become aware of its simplifications and restrictions because it is assumed that the regarded devices are frequency independent and the stimulation signals are sine waves. Both prerequisites are usually not fulfilled under the conditions of UWB sensors. However, as we will see later, the philosophy behind this method can also be applied to qualify UWB devices by a simple and practical technique.

The use of Volterra series would avoid the indicated limitations. The method is quite extensive and will hence not be considered here further. The interested reader can refer to Refs [72, 73] for more details.

A smooth transfer characteristic as shown in Figure 2.100 can be modelled by a truncated Taylor series within the operational range of interest:

$$V_2 = c_0 + c_1 V_1 + c_2 V_1^2 + c_3 V_1^3 + \cdots \quad \text{for } |V_1| \leq V_{1\max} \tag{2.332}$$

The coefficient c_0 refers to an offset value. It is mostly out of interest since it is suppressed by AC-coupled electronics or it is affected by DC drift of amplifiers. c_1 represents the (wanted linear) voltage gain of the stage and all other coefficients $c_2, c_3, \ldots$ are responsible for the strength of the non-linearity. If this device is stimulated by a sine wave $V_1(t) = V_0 \sin 2\pi f_0 t$, the output signal $V_2(t)$ is composed of a mixture of harmonics. Table 2.8 and Figure 2.101 summarize the behaviour of the first three ones. In Table 2.8, the strength of the different signal components is

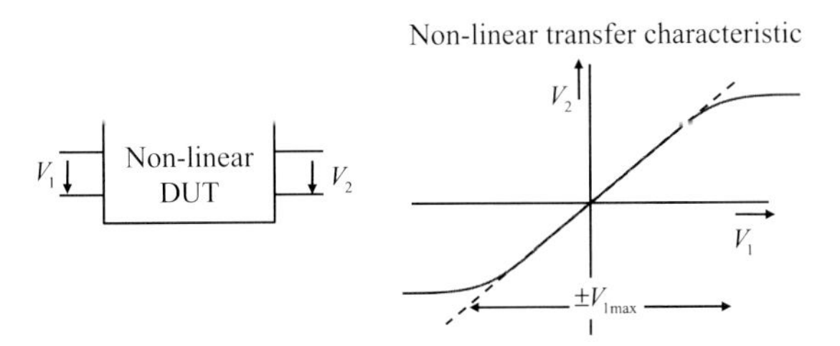

Figure 2.100 Non-linear device and its transfer function showing saturation effects for large input signals.

Table 2.8 Composition of the output signal due to non-linear distortions for stimulation with single tone signal.

		Frequency	Amplitude		Level [dBm] (without compression terms)
			Without compression terms	Compression terms	
Input signal		f_0	V_0		$L_1 = 10\log\frac{V_0^2}{2R_0P_0}$
Output signal	First harmonic	f_0	c_1V_0	$\frac{3}{4}c_3V_0^3 + \frac{5}{8}c_5V_0^5 + \cdots$	$L_{21} = L_1 + 20\log\lvert c_1\rvert$
	Second harmonic	$2f_0$	$\frac{c_2}{2}V_0^2$	$\frac{1}{2}c_4V_0^4 + \frac{15}{32}c_6V_0^6 + \cdots$	$L_{22} = 2L_1 + 10\log\frac{c_2^2R_0P_0}{2}$
	Third harmonic	$3f_0$	$\frac{c_3}{4}V_0^3$	$\frac{5}{16}c_5V_0^5 + \frac{21}{64}c_7V_0^7 + \cdots$	$L_{23} = 3L_1 + 20\log\frac{\lvert c_3\rvert R_0P_0}{2}$

The generation of DC components is not respected here.

given in absolute values and as power level L_i[dBm]:

$$L_i[\text{dBm}] = 10\log\frac{P_i}{P_0} = 10\log\frac{V_i^2}{2R_0P_0} \tag{2.333}$$

P_0 is the reference power of 1 mW and R_0 refers to the characteristic impedance of the measurement environment. It is typically 50 or 100 Ω depending on unbalanced or balanced measurement.

As Table 2.8 and Figure 2.101 show, the level of the non-linear distortions is growing two or three times faster than the linear output component if we increase the power of the input signal. The slope of the distortion power as function from

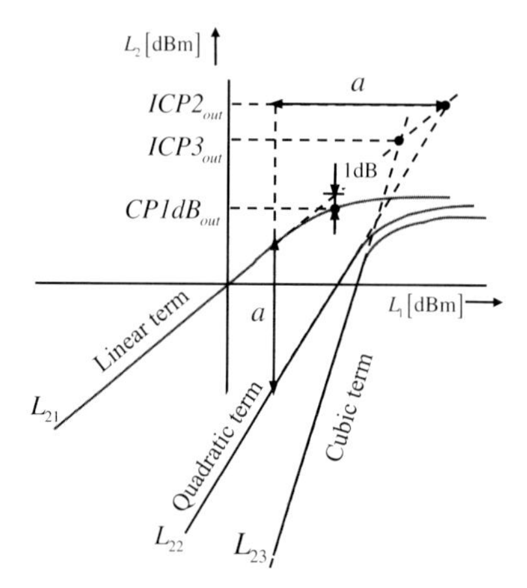

Figure 2.101 Level diagram for a non-linear device.

the input voltage depends on the order of non-linearity. This behaviour is of general nature and not only restricted to sine wave stimulation. Hence, we can use a comparable method for the characterization of ultra-wideband sensors.

Referring to (2.332), the characteristic values describing the susceptibility of a DUT to provoke non-linear distortions are given by the values of the coefficients $c_2, c_3, \ldots$. But in praxis, they are never been found. Rather, one uses the intercept points ICP2 and ICP3 as depicted in Figure 2.101 since from them it is very simple to predict the strength of the non-linear distortions. If the signal level is, for example, a[dB] below the intercept point, then the second-order distortion is a[dB] (see Figure 2.101) and the third-order distortion is $2a$[dB] below the primary signal level. The relations between the intercept points and the coefficients of the Taylor series are

$$\mathrm{ICP2}_{\mathrm{out}}[\mathrm{dBm}] = 10\log\frac{2c_1^4}{c_2^2 R_0 P_0} \tag{2.334}$$

$$\mathrm{ICP3}_{\mathrm{out}}[\mathrm{dBm}] = 10\log\frac{2|c_1^3|}{|c_3| R_0 P_0} \tag{2.335}$$

A further parameter is the 1 dB compression point (CP1 dB). It indicates the maximum signal power which can just be handled by the device. As a rule of thumb, the CP1 dB-value can be ranged about 14 dB below ICP3.

In connection with ultra-wideband systems, these parameters should be used with care. As (2.332) implies, non-linear distortions depend on the amplitude of a signal and not from its average power. But all parameters ICP2, ICP3 and CP1 dB are defined on the basis of power values for sine waves. By following relation, they can be very roughly adapted to other types of signals:

$$A[\mathrm{dBm}] = A_{\sin}[\mathrm{dBm}] + 20\log\frac{\mathrm{CF}_{\sin}}{\mathrm{CF}} = A_{\sin}[\mathrm{dBm}] + 3\,\mathrm{dB} - 20\log\mathrm{CF} \tag{2.336}$$

Herein, the symbol A refers to one of the three values ICP2, ICP3 or CP1 dB and CF is the crest factor.

2.6.6 Dynamic Ranges

The primary goal of an ultra-wideband sensor is to determine the impulse response function of a scenario under test. The above-discussed device inadequacies permit us to define several sensitivity figures for such a measurement.

We refer the definitions of dynamic range to the impulse response function since it is of major interest in UWB sensing. In order to make comparable these definitions for any UWB principle, we only consider pulse functions at the in- and output of the DUT (Figure 2.102). This requires to virtually modifying UWB systems which use time spread signals. Applying the law of commutativity of convolution, we can exchange correlator and DUT, so we have comparable signals in any case at the ports of the DUT.

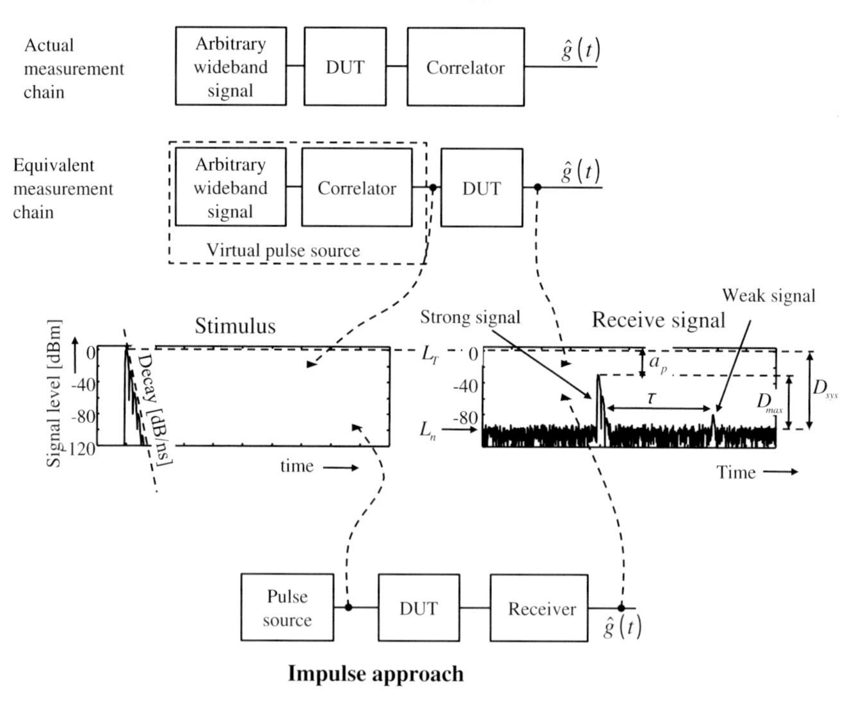

Figure 2.102 Testing the sensor dynamic range by a reference device consisting of signal paths with strong and weak attenuation. The DUT IRF affected by the imperfections of signal source and receiver is called $\hat{g}(t)$.

The sensitivity of a device refers to its ability to capture small signals and to distinguish them from perturbations. Therefore, the strongest attenuation which can just be observed from a sensor results from the difference between the stimulation level L_{T} and the perturbation level L_{p}. As we have seen above, these perturbations may be of different cause:

- Random noise (including jitter).
- Linear distortions (device internal clutter).
- Non-linear distortions.

Random noise can usually be considered to be independent of the strength of the receive signal. But the remaining two perturbation sources depend on the strength of the captured signal. Consequently, the question of system dynamic must be answered for the weakest signal in presence of a large one. Figure 2.102 illustrates the situation. Obviously, we have to deal with two major dynamic figures: D_{sys} refers to the whole sensor device and D_{max} concerns the receiver only.

For completeness a third point has to be mentioned, namely the decay rate of the stimulus signal. This property mainly determines the performance of the device to separate two closely located peaks (i.e. small τ) of a strong and a weak signal

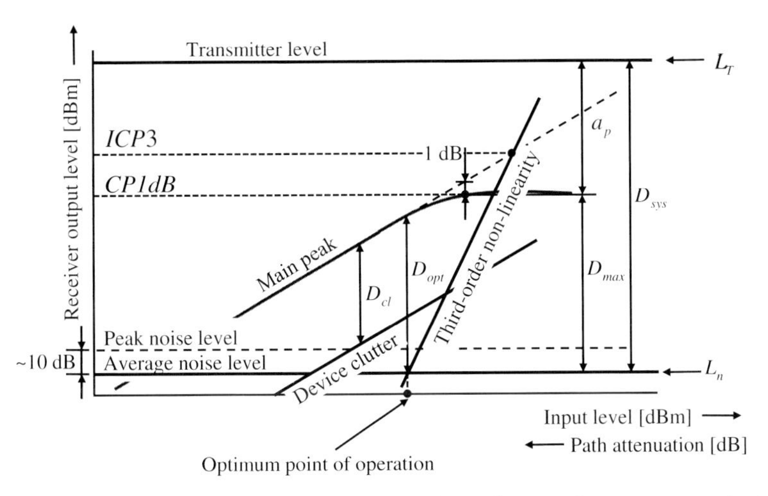

Figure 2.103 Level diagram of wanted signals and perturbations of an UWB sensor device.

without overwhelming. This aspect is mainly a question of system bandwidth which was already discussed in Section 2.4.2.

Due to the dependence of the perturbation level from the total strength of the input signal, the detectability of a weak signal depends on several conditions. Figure 2.103 summarizes all influencing quantities and it depicts characteristic dynamic range parameters. The figure represents the run of the peak level for signals of different nature (non-linear signals, clutter, noise) as function of the input level and as function of the main path attenuation for a given stimulation level L_T.

As expected, the main peak level increases linearly with decreasing attenuation of the main signal path. Above the 1 dB compression point CP1 dB, the receiver tends to saturate and a further reduction of the signal attenuation will not increase the level of the captured data.

Clutter Free Dynamic Range D_{cl}: We have considered the device internal clutter as a linear effect.[39] Hence, it follows in parallel the main signal – at least within the range of linear operation. The clutter-free dynamic range D_{cl} is given by the difference between the main signal peak and the maximum peak caused by clutter. D_{cl} can be improved by device calibration (see Figure 2.97). The degree of amelioration is mainly limited by the quality of the calibration standards as well as the noise and the non-linear distortions in the calibration data.

Spurious-free dynamic range D_{opt}: We will call a signal as spurious signal if it is caused by a non-linear effect. The third-order non-linearity is typically the most critical one. It leads to signal peaks which rise two times faster than the main peak – even if the signals are non-sinusoidal. Non-linear effects appear at large signal levels. The optimum dynamic range D_{opt} is

39) Sometimes internal clutter is referred to all systematic device internal perturbations. In this case, it would also include non-linear distortions.

given in the case where noise and spurious are identical. For calibration purposes, the sensor should operate in the optimum point of operation since it represents the largest range of linearity.

Maximum signal-to-noise ratio D_{max}: It is given by the relation between maximum signal level (fixed at the 1 dB compression point) and the noise level L_n. Note that the noise rms value is typically used as reference. For comparison, the noise peak level is also plotted in Figure 2.103. It is roughly 10 dB above the rms level (disregarding about 3 promille of noise events). In radar applications, D_{max} mainly limits the sensitivity to detect weak moving targets in the presence of strong static scatterer. Note that in this case the device internal clutter and the spurious signals may be completely removed by high-pass filtering since they are usually static.

In accordance with previous discussions, we can estimate D_{max} by adapting (2.251) as follows:

$$D_{max} = \frac{\eta_r \mathrm{TB}}{\mathrm{CF}^2} D_R \tag{2.337}$$

Herein, D_R assigns the actual dynamic range of the receiver circuitry. It relates the maximum acceptable input voltage to the noise floor. We can express this by two ways:

$$D_R = \frac{V_{max}^2}{\sigma^2} = \frac{V_{1\,dB}^2}{\sigma_n^2} = \frac{V_{1\,dB}^2}{R_0 F k T_0 \underset{\cdot}{B}_N} = \frac{(\mathrm{FSR}/2)^2}{\sigma_{tot}^2} \approx 3\,2^{2\,\mathrm{ENOB}} \tag{2.338}$$

In the first case, the receiver is considered as an analogue device characterized by the quantities: input 1 dB compression point CP1 dB[dBm] = 10 log $V_{1\,dB}^2 / R_0$ 1 mW, noise factor F, equivalent noise bandwidth $\underset{\cdot}{B}_N$ and input impedance R_0. The second description considers the receiver as a virtual analogue-to-digital converter (see Section 2.6.1.6) quantified by an effective number of bits which includes all noise sources. The quantity of D_R is largely determined by the performance of the circuit elements and the implemented circuit structure.

Insertion of (2.338) in (2.337) leads to an equation which we have separated in several terms to underline the different factors determining the overall dynamic of the sensor:

$$D_{max} = \frac{2}{kT_0} \underbrace{\frac{1}{\mathrm{CF}^2}}_{\text{type of test signal}} \underbrace{T_R}_{\text{recording time}} \underbrace{\eta_r}_{\text{sub-sampling}} \underbrace{\frac{V_{1\,dB}^2 / R_0}{F}}_{\text{receiver circuit}} \tag{2.339}$$

Here, we supposed that the bandwidth of the sounding signal and the bandwidth of the receiver are matched so that the TB product in (2.338) can be approximated by TB = $T_R \underset{\cdot\cdot}{B} \approx 2 T_R \underset{\cdot}{B}_N$. The efficiency of data capturing is assigned by η_r. It is mainly determined by the sub-sampling method mostly applied in UWB sensing. Equation (2.339) gives us a rough guideline to assess the different sensor principles. It is interesting to note that the maximum achievable dynamic range of a UWB sensor does not depends on the operational bandwidth if the sensor system is optimally configured.

System performance D_{sys}: It indicates the weakest signal path which just can be observed by the sensor system. It results from D_{max} and the attenuation a_p of the strongest transmission path.

2.7 Summary

The chapter has reviewed the major concepts of signal and system description in time and frequency domain with the reason to understand and assess the various principles of UWB sensing which will be discussed in the next chapter. Moreover, it should provide a basic tool set to process and interpret UWB sensor data. In this connection, we restrict ourselves to the elementary approaches of signal and system characterization.

Nowadays, signal and system description is mostly performed in the frequency domain which often simplifies calculations since the considered sine wave signals are not subjected a shape variation due to the test objects. For narrowband systems, this is a usual and sufficient method. Basically, the corresponding approaches are valid and applicable also for ultra-wideband applications but they will often avoid an intuitive comprehension of the actions and events running within the test objects or sensor devices. Therefore, considerations in the time domain based on pulse excitation play an important role too since they are more intuitive for the human being. However, the mathematics behind is more complicated because non-sinusoidal signals lose their original shape if they interact with a dynamic system. This we have expressed by the convolution.

Interaction phenomena of arbitrary wideband signals (noise, PN-codes, multi-sine etc.) with dynamic systems are not anymore intuitively comprehensible. Here, the correlation did help us since it could transform complicated wideband signals into simple pulse-shaped functions so that we can deal with these functions as with pulses.

References

1 Miller, E.K. (1986) *Time Domain Measurements in Electromagnetics*, Van Nostrand Reinhold Company Inc., New York.

2 Rihaczek, A.W. (1996) *Principles of High-Resolution Radar*, Artech House, Inc., Boston, London.

3 Baum, C. (1985) *Norms of Time-Domain Functions and Convolution Operators*, Air Force Weapon Laboratory.

4 Baum, C. (1988) *Energy Norms and 2-Norms*, Air Force Weapons Laboratory.

5 Baum, C. (2001) *Relationships between Time- and Frequency-Domain Norms of Scalar Functions*, Air force Research Laboratory.

6 Johnson, O. (2004) *Information Theory and the Central Limit Theorem*, Imperial College Press, London.

7 Marple, S.L. (1987) *Digital Spectral Analysis with Applications*, Prentice-Hall, Inc., Englewood Cliffs, NJ.

8 Widrow, B. and Kollár, I. (2008) *Quantization Noise: Roundoff Error in Digital Computation, Signal Processing,*

Control, and Communications, Cambridge University Press, Cambridge, UK.

9 Lamensdorf, D. and Susman, L. (1994) Baseband-pulse-antenna techniques. *IEEE Antenn. Propag.*, **36** (1), 20–30.

10 Candes, E.J. and Wakin, M.B. (2008) An introduction to compressive sampling. *IEEE Signal Proc. Mag.*, **25** (2), 21–30.

11 Astanin, L.Y. and Kostylev, A.A. (1997) *Ultrawideband Radar Measurements Analysis and Processing*, The Institution of Electrical Engineers, London, UK.

12 Hazewinkel, M. (ed.) (2002) Encyclopaedia of mathematics, in *Encyclopaedia of Mathematics*, Springer-Verlag, Berlin.

13 Weiss, L.G. (1994) Wavelets and wideband correlation processing. *IEEE Signal Proc. Mag.*, **11** (1), 13–32.

14 Skolnik, M.I. (1989) *Radar Handbook*, McGraw-Hill Professional.

15 Ruggiano, M. and van Genderen, P. (2007) Wideband ambiguity function and optimized coded radar signals. European Radar Conference, 2007. EuRAD 2007, pp. 142–145.

16 Zepernick, H.J. and Finger, A. (2005) *Pseudo Random Signal Processing – Theory and Application*, John Wiley & Sons, Inc., New York.

17 Hein, A. (2004) *Processing of SAR Data: Fundamentals, Signal Processing, Interferometry*, Springer, Berlin.

18 Pintelon, R. and Schoukens, J. (2001) *System Identification: A Frequency Domain Approach*, IEEE Press, Piscataway, NJ.

19 Ning, Z. and Pierre, J.W. (2008) Time-limited perturbation waveform generation by an extended time-frequency domain swapping algorithm. 51st Midwest Symposium on Circuits and Systems, 2008. MWSCAS 2008, pp. 149–152.

20 Axelsson, S.R.J. (2006) Generalized ambiguity functions for ultra wide band random waveforms. International Radar Symposium, 2006. IRS 2006, pp. 1–4.

21 Kailath, T. (1979) *Linear Systems*, Prentice Hall.

22 Schroeder, M.R. (1965) New method for measuring reverberation time. *J. Acoust. Soc. Am.*, **37**, 409–412.

23 Faiget, L., Ruiz, R. and Legros, C. (1996) The true duration of the impulse response used to estimate reverberation time. Acoustics, Speech, and IEEE International Conference on Signal Processing, vol. 2, 1996. ICASSP-96. Conference Proceedings, pp. 913–916.

24 McLachlan, G.J. (1992) *Discriminant Analysis and Statistical Pattern Recognition*, Wiley Interscience, New York.

25 Bishop, C.M. (2006) *Pattern Recognition and Machine Learning*, Springer, Berlin.

26 Duda, R.O., Hart, P.E. and Stork, D.G. (2001) *Pattern Classification*, John Wiley & Sons, Inc., New York.

27 Fukunaga, K. (1991) *Statistical Pattern Recognition*, Academic Press, New York.

28 Pavel, M. (1993) *Fundamentals of Pattern Recognition*, Dekker, New York.

29 Schuermann, J. (1996) *Pattern Classification – A Unified View of Statistical and Neural Approaches*, John Wiley & Sons, Inc., New York.

30 van der Heijden, F., Duin, R.P.W., de Ridder, D. *et al.* (2005) *Classification, Parameter Estimation and State Estimation: An Engineering Approach using MATLAB*, John Wiley & Sons, Inc., New York.

31 Tesche, F.M. (1992) On the use of the Hilbert transform for processing measured CW data. *IEEE Trans. Electromagn. Compat.*, **34** (3), 259–266.

32 Tesche, F.M. (1993) Corrections to ‘On the use of the Hilbert transform for processing measured CW data’ (Aug. 1992, 259–266). *IEEE Trans. Electromagn. Compat.*, **35** (1), 115.

33 Pyati, V.P. (1993) Comment of on the use of the Hilbert transform for processing measured CW data. *IEEE Trans. Electromagn. Compat.*, **35** (4), 485.

34 James, J.R. and Andresic, G. (1993) Comments on ‘On the use of the Hilbert transform for processing measured CW data’. *IEEE Trans. Electromagn. Compat.*, **35** (3), 408.

35 Ljung, L. (1999) *System Identification: Theory for the User*, Prentice Hall PTR, New Jersey.

36 Söderström, T. and Stoica, P. (1994) *System Identification*, New edition, Prentice-Hall.

37 Van Loan, C. (1992) *Computational Frameworks for the Fast Fourier Transform (Frontiers in Applied Mathematics)*, Society for Industrial Mathematics, Philadelphia.

38 Press, W.H., Teukolsky, S.A., Vetterling, W.T. *et al.* (2007) *Numerical Recipes 3rd Edition: The Art of Scientific Computing*, Cambridge University Press, New York.

39 Ender, J. (2010) Do we still need Nyquist and Kotelnikov? – Compressive sensing applied to radar. 8th European Conference on Synthetic Aperture Radar EUSAR 2010, Aachen, Germany.

40 Eisenstadt, W.R., Stengel, B. and Thomson, B.M. (2006) *Microwave Differential Circuit Design using Mixed-Mode S-Parameters*, Artech House, Boston, London.

41 Smolyansky, D. and Corey, S. (2000) Characterisation of differential interconnects from time domain reflectometry measurements. *Microwave J.*, **43** (3), 68–80.

42 Smolyansky, D.A. and Corey, S.D. (2002) Computing self and mutual capacitance and inductance using even and odd TDR measurements. IEEE Conference on Electrical Performance of Electronic Packaging, pp. 117–122.

43 Golub, G.H. and Van Loan, C.F. (1996) *Matrix Computations*, 3rd edn, Johns Hopkins University Press, Baltimore, MD.

44 Nason, G.P. (2008) *Wavelet Methods in Statistics with R*, Springer.

45 Donoho, D.L. (1995) De-noising by soft-thresholding. *IEEE Trans. Inf. Theory*, **41** (3), 613–627.

46 Fink, M. and Prada, C. (2001) Acoustic time-reversal mirrors. *Inverse Probl.*, **17**, R1–R38.

47 (2010) *Fundamentals of RF and Microwave Noise Figure Measurements*, Agilent Technologies; Application Note 57-1 http://cp.literature.agilent.com/litweb/pdf/5952-8255E.pdf.

48 Reeves, B.A. (2010) Noise augmented Radar system, US Patent US 7, 341 B2.

49 Pintelon, R., Schoukens, J. and Renneboog, J. (1988) The geometric mean of power (amplitude) spectra has a much smaller bias than the classical arithmetic (RMS) averaging. *IEEE Trans. Instrum. Meas.*, **37** (2), 213–218.

50 Schoukens, J. and Pintelon, R. (1990) Measurement of frequency response functions in noisy environments. *IEEE Trans. Instrum. Meas.*, **39** (6), 905–909.

51 Rick, P. (2001) *Overview on Phase Noise and Jitter*, http://cp.literature.agilent.com/litweb/pdf/5990-3108EN.pdf.

52 Andrews, J.R. (2009) *Removing Jitter From Picosecond Pulse Measurements*, PICOSECOND PULSE LABS Application Note AN-23.

53 Allan, D., Hellwig, H., Kartaschoff, P. *et al.* (1988) Standard terminology for fundamental frequency and time metrology. Proceedings of the 42nd Annual Frequency Control Symposium, 1988, pp. 419–425.

54 Demir, A., Mehrotra, A. and Roychowdhury, J. (2000) Phase noise in oscillators: a unifying theory and numerical methods for characterization. *IEEE Trans. Circuits-I.*, **47** (5), 655–674.

55 Lee, T.H. and Hajimiri, A. (2000) Oscillator phase noise: a tutorial. *IEEE J. Solid-St. Circ.*, **35** (3), 326–336.

56 (1976) *Understanding and Measuring Phase Noise in the Frequency Domain*, Hewlett-Packard Application Note AN 207, http://www.home.agilent.com/agilent/facet.jspx?to=80030.k.1&c=178004.i.2&lc=eng&cc=US.

57 (2004) *Clock (CLK) Jitter and Phase Noise Conversion*, Maxim Application Note 3359, http://www.maxim-ic.com/app-notes/index.mvp/id/3359.

58 (2009) *Time Domain Oscillator Stability Measurement Allan Variance*, Rohde & Schwarz, Application Note, http://www2.rohde-schwarz.com/file_11752/1EF69_E1.pdf.

59 (2006) *Using Clock Jitter Analysis to Reduce BER in Serial Data Applications*, Agilent Technologies, Application Note, http://www.home.agilent.com/agilent/facet.jspx?to=80030.k.1&c=178004.i.2&lc=eng&cc=US.

60 Wehner, D.R. (1995) *High Resolution Radar*, 2nd edn, Artech House, Norwood, MA.

61 Rytting, D.K., *Network Analyzer Error Models and Calibration Methods*, in RF & Microwave Measurements for Wireless Applications (ARFTG/NIST Short Course Notes), 1996. can be down loaded from http://cpd.ogi.edu/IEEE-MTT-ED/Network%20Analyzer%20error%20Models%20and%20Calibration%20Methods.pdf.

62 Butler, J.V., Rytting, D.K., Iskander, M.F. *et al.* (1991) 16-term error model and calibration procedure for on-wafer network analysis measurements. *IEEE Trans. Microwave Theory*, **39** (12), 2211–2217.

63 Rytting, D.K. (2001) Network analyzer accuracy overview. 58th ARFTG Conference Digest-Fall, pp. 1–13.

64 Ferrero, A., Sampietro, F. and Pisani, U. (1994) Multiport vector network analyzer calibration: a general formulation. *IEEE Trans. Microwave Theory*, **42** (12), 2455–2461.

65 Wiesbeck, W. and Kahny, D. (1991) Single reference, three target calibration and error correction for monostatic, polarimetric free space measurements. *Proc. IEEE*, **79** (10), 1551–1558.

66 Pancera, E., Bhattacharya, A., Veshi, E. *et al.* (2009) Ultra wideband impulse radar calibration. IET International Radar Conference, 2009, pp. 1–4.

67 (2001) *Specifying Calibration Standards for the Agilent 8510 Network Analyzer*, Product Note 8510-5B, http://cp.literature.agilent.com/litweb/pdf/5956-4352.pdf.

68 Herrmann, R., Sachs, J. and Peyerl, P. (2006) System evaluation of an M-sequence ultra wideband radar for crack detection in salt rock. International Conference on Ground Penetrating Radars (GPR), Ohio (Columbus).

69 Herrmann, R., Sachs, J., Schilling, K. *et al.* (2008) 12-GHz Bandwidth M-sequence radar for crack detection and high resolution imaging. International Conference on Ground Penetrating Radar (GPR), Birmingham, UK.

70 Daniels, D.J. (2004) *Ground Penetrating Radar*, 2nd edn, Institution of Electrical Engineers, London.

71 Ferrero, A. and Pisani, U. (1992) Two-port network analyzer calibration using an unknown 'thru'. *IEEE Microwave Guided Wave Lett.*, **2** (12), 505–507.

72 Schetzen, M. (2006) *The Volterra And Wiener Theories of Nonlinear Systems*, Krieger Publishing Company.

73 Bendat, J.S. (1990) *Nonlinear System Analysis and Identification from Random Data*, New York.

3
Principle of Ultra-Wideband Sensor Electronics

3.1
Introduction

The aim of the ultra-wideband sensor electronics is to stimulate a device or scenario under test with electromagnetic energy and to register the reaction. The kernel and the particularity of ultra-wideband sensors and at the same time the challenge of their technical implementation is the large fractional bandwidth of operation. The larger the bandwidth of operation the more information about the target can be extracted from the measured data. Depending on the application, the bandwidth of interest may extend from a few hundred MHz to several GHz or even to tens of GHz. Further aspects relate to the sensitivity of the sensor. It is typically expressed by the achievable signal-to-noise ratio (SNR) which characterises the amplitude uncertainty of the captured voltage samples. But also the time point of data capturing is subject to random errors (it is called jitter) affecting the overall sensor performance. As we will see, the SNR-value largely depends on the efficiency of the UWB receiver while the jitter is a question of internal sensor timing. The efficiency of most of UWB principles is often very weak so that considerable improvement potentials for future developments exist.

In this chapter, we will discuss some approaches how measurements can be done over such a large bandwidth. Since short-range sensor applications are in the focus of our interest, we will exclude all high-power short pulse principles.

There are several UWB approaches suitable for sensor applications. All of them must be able to provide a stimulus to excite the test objects and to capture the object reaction which is usually converted into voltage signals by appropriate sensor elements (applicators). The different UWB principles vary in the applied stimulus signal, the method of voltage capturing and the method of internal timing control. The selection of the best sensor principle for the use under industrial or daily conditions has to be made on the basis of various requirements and conditions. The following listing summarizes some of these issues:

- **Operational frequency band $[f_l, f_u]$, bandwidth $\underset{\cdot}{B} = f_u - f_l$:** It determines the resolution of propagation time measurements which is linked to the precision of distance and movement measurements as well as the spatial resolution of radar

Handbook of Ultra-Wideband Short-Range Sensing: Theory, Sensors, Applications, First Edition. Jürgen Sachs.

imaging. The capability to recognize targets depends on it as well as the wave penetration into or through different material. Furthermore, the radio regulation issues do also concern this point.

- **Period duration of the stimulus $t_P = f_0^{-1}$:** It limits the maximum observalble settling time of the scenario under test. In the case of radar measurements, it corresponds to the round-trip time and the range of unambiguous target localization. Further, it influences the resolution of frequency domain measurements (i.e. the spacing between adjacent spectral lines). In connection with the bandwidth it determines the amount of data to be handled per IRF or FRF.
- **Duration of data recording, measurement rate $T_R = m_R^{-1}$:** It influences the maximum speed of a target which can be tracked by the sensor without the need to respect Doppler spread during data capturing. In the case of a scenario with arbitrary time variability, m_R has to respect the Nyquist theorem. That is, it must be at least twice as large as the bandwidth of the scenario variation is.[1] Measurement rate, bandwidth and period duration will essentially determine the data throughput to be handled by the sensor system.
- **Noise, jitter and drift:** These random errors limit the sensitivity to register small and weak targets as well as their movement. In particular, the detection of shallow buried targets will be affected by jitter since it converts to noise in case of wave reflections at a boundary. Time and amplitude instabilities will further reduce the performance of the systematic errors correction. Moreover, sensors with pure time drift will restrict the maximum number of synchronous averaging and hence the noise suppression.
- **Power consumption:** Power consumption is a very important issue for the case of autonomous sensor operation due to limited battery capacity.
- **Technical implementation:** Here various aspects must be respected. This concerns the number of measurement channel (maximum dimensions of a sensor array), the mechanical and electrical robustness, the size and weight of the sensor system, its handling, the purchase price and the operating expense, the commercial availability of components, the requirements concerning telecommunication and computer infrastructure and many things more.
- **Interference, probability of intercept and robustness against jamming:** Ultra-wideband sensors emit electromagnetic waves. Even if their radiated power is extremely small, they may interfere with other devices under critical circumstances. Inversely they are strongly affected by other electromagnetic sources like mobile phones or any radio transmitter. Sometimes the sensor operation should be even as concealed as possible.

Clearly a single sensor system cannot meet all these conditions simultaneously. But usually not all requirements have the same priority. Their weighting strongly depends on the intended application. Since every sensor approach has its strong

1) For illustration, we refer to vital data capturing. In the case of heartbeat measurements, we have to respect a maximum beat rate of about 200 bpm (beats per minute). If we want to know, the heart movement up to the 10th harmonics, the measurement rate must be $m_R \approx 67$ Hz.

sides as well as its weaknesses, the application scenario and the appliers rating of the requirements and operating conditions will finally favour the use of a certain sensor principle.

The function principles of ultra-wideband sensors are mostly named after the signal type which is used to stimulate the scenario under test. Here, we will distinguish between following methods:

- Short pulse sensors (Section 3.2).
- Sensors based on pseudo-noise (PN) excitation (Sections 3.3 and 3.5).
- The different sine wave approaches (Section 3.4).
- Random noise sensors (Section 3.6).

The measurement receivers in all this sensor conceptions capture a voltage signal which is nearly almost converted into the numerical domain for further processing. The various methods to gauge a voltage differ in the timing of data acquisition and the nature of the captured quantity – integral or instantaneous value. In Sections 3.2.3 and 3.2.4, we will introduce the most common principles applied in UWB sensors. A general overview of basic sampling techniques for wideband signals can be found in Ref. [1], page 45 ff.

For the purpose to introduce and demonstrate the various working principles, we will assume always the same measurement task in what follows. In order to keep simple, it is aimed to perform a transmission measurement. The goal of such a measurement is to determine the system reaction in the time or frequency domain (i.e. to measure the IRF or FRF). Later, in Section 3.7, we will extend the simple transmission measurements to other sensing tasks (e.g. impedance measurement, determination of scattering parameters etc.).

3.2 Determination of the System Behaviour by Pulse Excitation

3.2.1 Basic Principle

The impulse principle is the direct method to determine the IRF of a DUT because it represents the practical implementation of Green's approach as introduced in Section 2.4.2. Many people even confine the UWB technique to this measurement method which is however a too strong simplification and disregards the diversity of UWB methods.

Figure 3.1 depicts the basic structure of a pulse system, which is the most applied principle in UWB sensors up to now.

A short pulse $x(t)$ stimulates the scenario under test. Its reaction $y(t)$ is captured by a sampling receiver and mostly transformed into the numerical domain resulting in the sampled data $\mathbf{y}[n]$. As long as the stimulating pulse is short enough, the shape of the captured signal $y(t)$ is proportional to the wanted impulse response function $g(t)$. In several cases (e.g. for time domain reflectometry (TDR)), a step

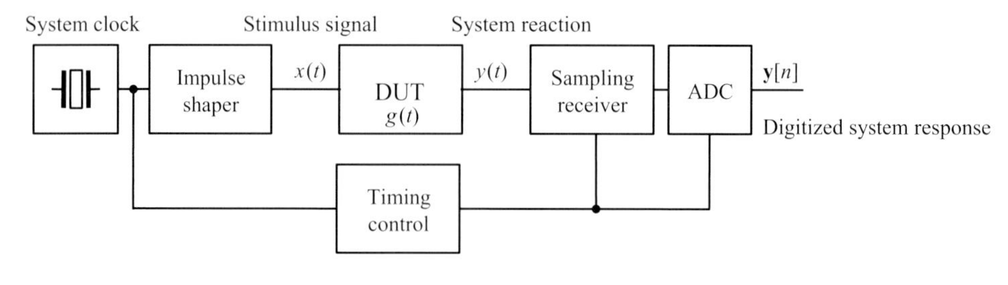

$$y(t) = g(t) * x(t)$$
$$y(t) \sim g(t) \text{ for } x(t) \rightarrow \delta(t) = u_0(t)$$
$$y(t) \sim h(t) \text{ for } x(t) \rightarrow u(t) = u_{-1}(t) = \int \delta(t) dt$$

Figure 3.1 Basic sensor structure using impulse excitation. $\delta(t)$ is the Dirac delta and $u_{-1}(t)$ is the Heaviside function (see also Annex 7.2).

function (Heaviside function) $u(t) = u_{-1}(t) = \int \delta(t)\, dt$ is preferred as stimulus. Therefore, the receive signal is proportional to the step response of the test object $y(t) \sim h(t)$. Impulse and step response are mutually related by

$$h(t) = \int g(t) dt = u_{-1}(t) * g(t) \quad \Leftrightarrow \quad g(t) = \frac{d}{dt} h(t) = u_1(t) * h(t) \tag{3.1}$$

As we figured out in Section 2.6.1 (2.248), (2.251), the susceptibility of a measurement to noise depends on the injected stimulation energy. Since an impulse carries energy only during a very short time interval, its amplitude must appropriately high. This is the most critical issue of the principle. The generation of high voltage pulses, the high field exposure of the test objects and the risk to over modulate the receiver may provide serious challenges for the system layout. One of the strong points of the pulse method is the opportunity of simple technical implementations as long as the required parameters are not too demanding. Furthermore, the measured data represents directly the wanted response function of the DUT without the need of any further processing. This allows an intuitive evaluation of the measurements.

Besides the digital part for data processing, the key components of pulse systems are the pulse shaper [2], the sampling receiver (or sampling gate) and the timing circuit which controls the data capturing. These parts largely determine the performance of the UWB sensor. The pulse generators typically provide nanosecond and sub-nanosecond pulses. This would require data gathering rates beyond tens of GHz in order to meet the Nyquist theorem. Such measurement rates are not feasible in sensor devices which are subject to strong conditions as low costs, low-power consumption, small size and so on. Therefore, the signals are always recorded by an approach which is denoted by terms as sub-sampling, stroboscopic sampling or equivalent time sampling.

3.2.2 Pulse Sources

3.2.2.1 Monolithically Integrated Pulse Sources

Monolithic integrated solutions of pulse sources are of profit for mass-product sensors or if high bandwidth and/or small sensor size are required. A seeming drawback of monolithic integration is the limited output voltage swing and peak output power particularly if the circuits are implemented in a low-cost semiconductor technology like SiGe [3]. However, the power limitation due to the UWB regulation issues usually suppress the desire for high-power pulses so that just a voltage swing in the order of 1 V or even less may be sufficient.

The standard switching stage of monolithically integrated circuits is an ECL-cell as depicted in Figure 3.2. It represents a transistor bridge. In the switching mode, the total current I provided by the current source toggles between both branches of the bridge, that is $I_1 = I$ or $I_2 = I$. Since the accurate ECL-stage biasing prevents saturation of the transistors, the unit permits short switching time in the order of 10–20 ps (measured on chip) depending on the applied semiconductor technology.

The output voltage swing is

$$\Delta V_2 = V_2^+ - V_2^- \approx 2\,I\,R \tag{3.2}$$

In order to match a certain frequency mask, the pulse shape has additionally to be adapted appropriately. This is often done by an n-fold derivation of square wave pulses having a very short rise time. Here, the derivation products of the falling edge are typically clipped out in order to avoid double pulses of inverted shape [4].

If the signal power is not yet sufficient, an additional power amplifier has to boost the pulses. Referring to Refs [5, 6], the maximum output power of state-of-the-art UWB power amplifiers is less than 20 dBm (sine waves; 50 Ω environment) which corresponds to a peak-to-peak voltage of about 6 V. However, the chip

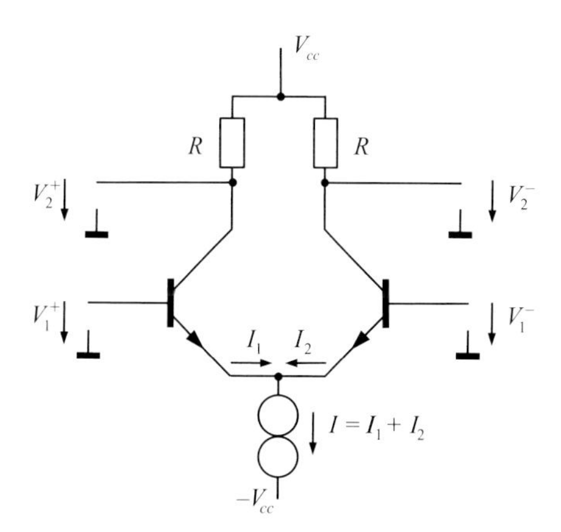

Figure 3.2 Basic schematics of an ECL cell.

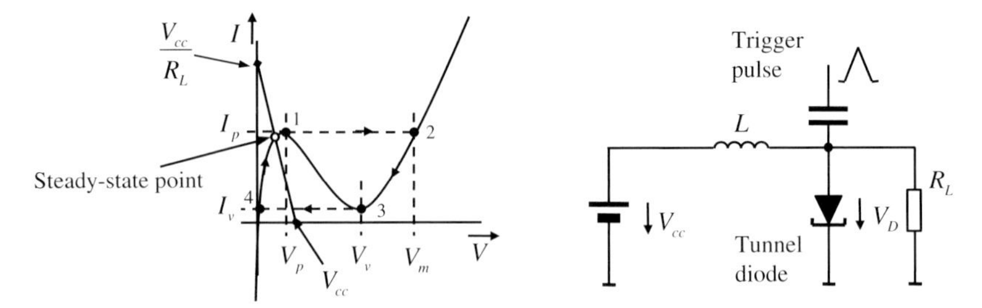

Figure 3.3 *IV*-characteristic of a tunnel diode and basic circuit of a pulse generator.

area of such amplifier is often larger than the area occupied by the rest of the UWB sensor (see Section 6.2 for an integtrated RF-UWB-head).

Circuit integration is a longsome and costly procedure so that circuit implementations by lumped elements may also be of interest. An example of a sub-nanosecond pulse source based on PCB implementation is given in Ref. [7].

3.2.2.2 Tunnel Diode

The tunnel diode is a strongly doped semiconductor device having an *IV*-characteristic of negative slope. This can be exploited to build very fast switching devices. Figure 3.3 shows the *IV*-characteristic of the diode and the basic circuit schematic for pulse generation.

The operational point of the diode in the steady state is placed somewhat below the peak of the hump. The diode voltage is boosted beyond the hump (point 1) by a trigger pulse (which may be quite smooth). This brings the diode in an unstable negative (differential) resistance region which forces the current down. But the inductor loaded to the current I_p avoids any fast variation of the current. Hence, the diode voltage V_D jumps to point 2 of the *IV*-characteristic. This transition is very fast resulting in rising edges down to 20 ps rise time and a voltage step of about 250 mV. The maximum achievable slew rate is about 15 V/ns with tunnel diode switches [1]. During the discharge of the inductor, the operational point goes slowly down to point 3 and from there the voltage jumps back to point 4. This transition is also quite fast but usually not of interest for most applications. Finally, the inductor is re-charged and the operational point moves into the steady-state position where it is awaiting a new trigger event.

The tunnel diode approach is often used to provide step waveforms as they are applied in time domain reflectometry. The duration t_w of the pulse is limited by the discharge time of the inductor (from point 2 to point 3 in the *IV*-characteristic).

3.2.2.3 Avalanche Transistor

Pulse generators for higher output voltages exploit the avalanche effect. Typical voltages of these pulse shapers are in the range of 20 V to several hundred volts. But it also possible to provide a few kilovolts by cascading some avalanche stages.

The root idea is depicted on the left-hand side of Figure 3.4. A cable of characteristic impedance $Z_0 = R_L$ is loaded via a high-resistance network to the voltage V_{cc}.

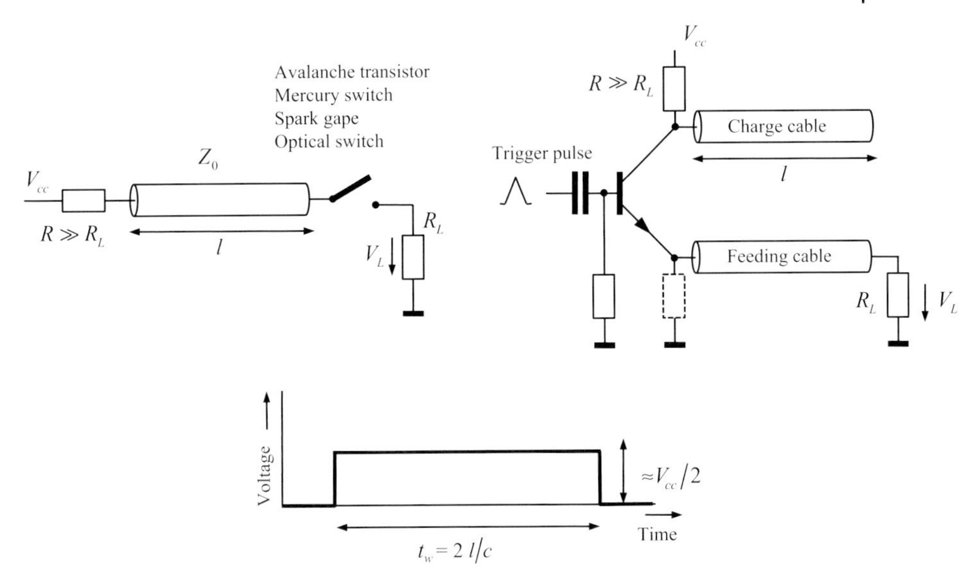

Figure 3.4 Principle and basic circuit of an avalanche pulse generator.

During loading the switch is open. If the switch is closed, the cable discharges over R_L which results in an impulse of the amplitude $V_{cc}/2$ and a duration which corresponds twice the delay time of the cable. The cable can also be replaced by a capacity. Then the discharge pulse is of exponential shape.

As denoted in Figure 3.4, there are several possibilities to implement the switch. For sensor applications, only the avalanche transistor seems to be feasible (see Figure 3.4, right) since the other methods are too expensive or too error-prone. In the steady state, the collector-base junction is reverse-biased by a voltage close to the breakthrough value. That is, a very intense electrical field exists across the pn-junction. If in this state the collector current rises above a certain limit (e.g. by a weak base current pulse), the electrons are very quickly accelerated causing local overheating or ionization. This produces new electrons which force overheating so we result in a thermal feedback mechanism which leads to an abrupt rise of the collector current while the collector voltage drops down (second breakdown avalanche mode). That is, the transistor switches on with a very high slew rate (about 200 V/ns [1]).

During this time, the transistor is subject to a huge power exceeding its average power dissipation by orders of magnitude. Therefore, the repetition rate of avalanche pulsers has to be restricted to low values to avoid overheating. Typically the maximum repetition rate is between 100 kHz and some MHz. Unfortunately, the breakthrough process is not exactly reproducible. Therefore, avalanche generators tend to produce random jitter in the 10 ps range. A detailed description of the second breakdown avalanche mode and its exploitation for pulse generator design is given in Ref. [8]. Further information can be found in Refs [9–11].

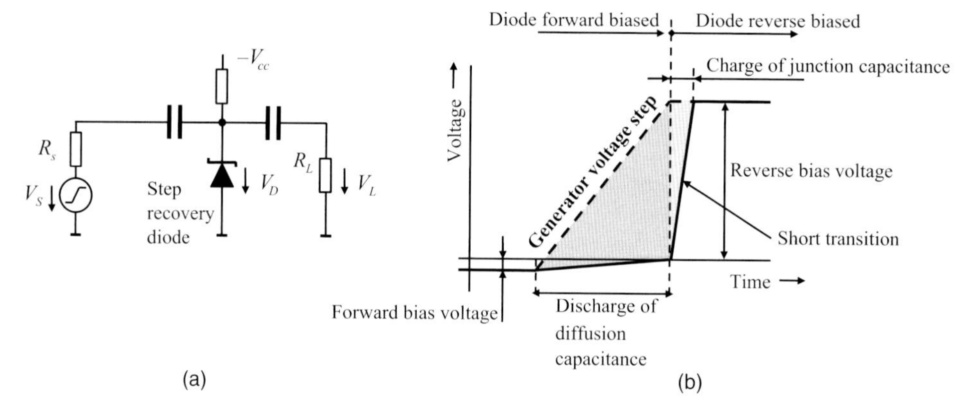

Figure 3.5 Impulse sharpening by step recovery diode. (a) Circuit diagram and (b) schematized voltages.

3.2.2.4 **Step Recovery Diode (Snap-Off Diode)**

The particularity of a step recovery diode is the large ratio between its diffusion and junction capacitance. The (large) diffusion capacitance dominates the behaviour of the forward biased diode while under reverse bias conditions only the (small) junction capacitance takes effect. This property is used to sharpen the transition of a rising edge. Figure 3.5 demonstrates the principle.

In the steady state, the diode is forward biased by V_{CC}. Thus, its diffusion capacitance is well charged. A pulse whose rising edge should be shortened tries to force the diode in reverse direction. Before it can do that, it has to discharge the large diffusion capacitance. In a well-designed circuit, the pulse has to spend all its charge accumulated in the rising edge to complete the discharge of the diffusion capacitor. During this time, the voltage V_D across the diode tends slowly to zero while the pulse amplitude reaches its maximum. After the voltage V_D has passed the zero volt level, the diode becomes reversed biased. Now only the teeny junction capacitance is active. It can be loaded very fast up to the generator voltage V_S. Figure 3.5 symbolizes the two phases by areas in dark and light grey. The charge transport related to the dark grey area must be absorbed by the diffusion capacitance while only the few charge behind the small light grey area has to be applied by the feeding voltage V_S to load the junction capacitor. This needs only a short time leading to the steep rising edge of the pulse.

The maximum voltage step is limited by the breakdown voltage of the diodes. It ranges from 10 to about 100 V. The achievable slew rate is about 400 V/ns [1, 12].

3.2.2.5 **Non-Linear Transmission Line**

From the classical telegrapher's equations, it is known that the speed of a wave propagating along a transmission line is given by $c = 1/\sqrt{LC}$. The idea concerning the non-linear transmission line (see Figure 3.6) is to apply a voltage-dependent capacitance (and/or a current-dependent inductance) in order to manipulate the propagation speed in dependence from the signal magnitude. If we replace the capacitances

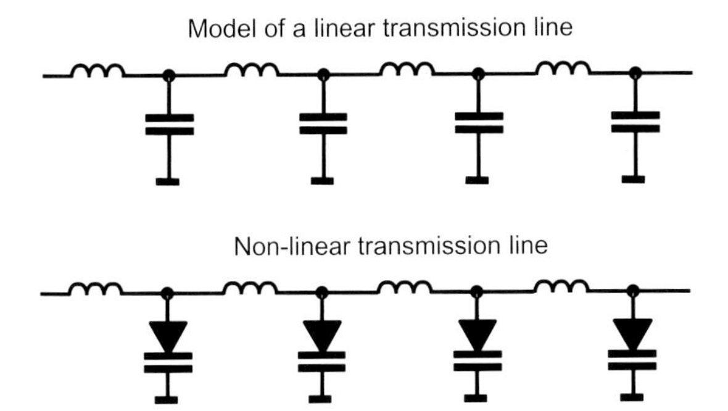

Figure 3.6 Equivalent circuit of a linear transmission line and circuit of a non-linear transmission line.

of a linear transmission line by the pn-junction capacitance of reverse biased varactor diodes, we have a decreasing line capacitance for increasing voltage. That is, the propagation speed will rise at higher voltages. Hence, a proper design supposed, the leading edges of a pulse will be compressed while propagating through the line [13, 14]. The behaviour of a non-linear transmission line can be compared with an ocean wave getting steeper and steeper while approaching the beach.

With this technique, it is possible to provide impulses with about 4 ps rise time at amplitudes of more than 5 V [15] resulting in a slew rate of about 1.25 kV/ns. Furthermore, the principle of non-linear transmission line permits to control (over a comparatively small range) the average delay time of a pulse by biasing the diodes with a reverse DC voltage.

3.2.3 Voltage Capturing by Sub-Sampling (Stroboscopic Sampling)

3.2.3.1 Preliminary Remarks

Real-time digitizing of very wideband signals as applied in UBW sensors provokes a huge amount of data. Their analogue to digital conversion, storage and processing requires an enormous effort of technical equipment causing considerable power dissipation. Modern real-time oscilloscopes are indeed able to generate and handle 8-bit data at sampling rates of tens of GHz. But at full sampling speed, they are only able to operate over a few milliseconds in maximum to fill their memory.

For sensor applications we have in mind here, such devices are out of scope. We need continuously working systems of low power and acceptable costs. In order to reduce the technical effort of the sensors, we have to reduce the data throughput: At least we have to find out a reasonable compromise between technical effort and data quantity to be handled in dependency on the actual application. Fortunately, we are usually able to repeat continuously the stimulation of our test objects so that we are not forced to gather the system reaction by a single shot. Therefore, we can stretch the data gathering to reduce the effective data throughput. Basically this can be done by two ways. First, we reduce the actual sampling rate below the Nyquist rate which is allowed under specific constraints (see Section 3.2.3.3) and second one can serialize the voltage quantization by bit-wise data conversion (see Section 3.2.4). Furthermore, the word length of high-speed data should be optimized to the required signal-to-noise ratio and digital signals of large crest factor should be avoided in order to place as much signal 'energy' as possible in short digital numbers.

Data capturing of very wideband signals has to deal with two key aspects. One is to take an instantaneous voltage sample over a sufficiently short time and the second concerns the precise timing to take this voltage sample. We will first consider some methods on how to capture RF-voltages, we will summarize the principles of timing control and finally some aspects of serializing the voltage quantization will be discussed.

3.2.3.2 Principles of Voltage Sampling

There are many circuits to perform gathering of voltage samples. Here, we will discuss a few principles. They distinguish in the complexity of the circuit layout and their performance.

Additive Sampling This method represents the simplest approach. The basic idea is depicted in Figure 3.7.

A short pulse of sufficient large amplitude is added by an appropriate timing (see Section 3.2.3.3) to the RF-signal. After cutting out only the upper part of the sum-signal by a threshold device, a pulse train remains which is modulated by the RF-voltage. Smoothing this pulse train by a low-pass filter (LPF) or integrator leads finally to the wanted low-frequency waveform (equivalent time waveform) which is of the same shape as the RF-signal but stretched in time. This time is often termed as 'equivalent time'. The down-converted signal can be processed further by analogue low-frequency circuits or it may be converted by standard ADCs.

The working principle of additive sampling circuits can be discussed by two ways. Figure 3.8 illustrates one of them. Here, the original RF-signal and its down-converted version are separated by a threshold device. Obviously, the threshold voltage must be larger than the RF-peak-voltage $V_{TH} > \|V_{RF}(t)\|_{\infty}$ in order to

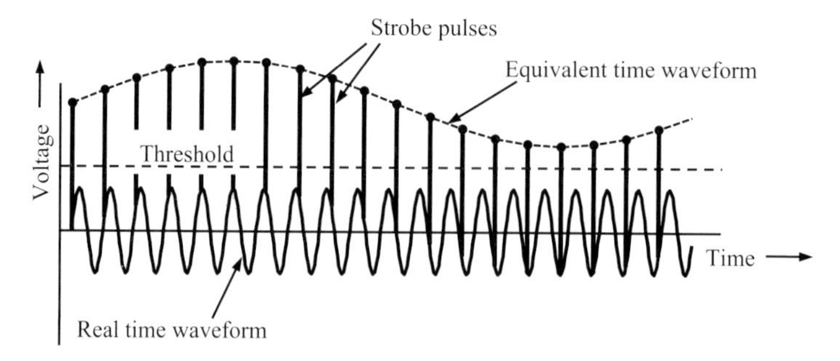

Figure 3.7 Generation of a low-frequency signal by adding short pulses to a high-frequency waveform.

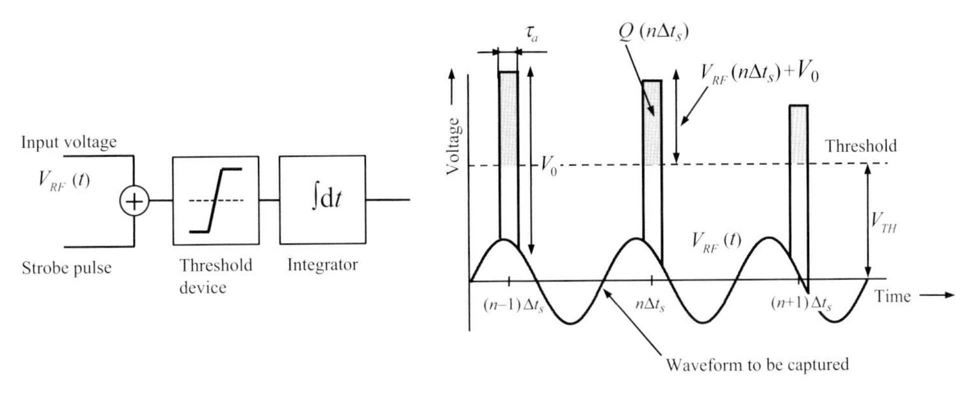

Figure 3.8 Additive sampling: separation of RF- and down-converted signal by a threshold.

guarantee a complete separation. Consequently, the amplitude of the strobe pulse has to respect the condition $V_0 \geq V_{TH} + \|V_{RF}(t)\|_\infty \geq 2\|V_{RF}(t)\|_\infty$ which permits also negative signal components to cross the threshold.

The integrator smoothes the remaining pulses so we get back finally an analogue waveform. For that purpose, it distributes temporally the charge $Q(n\Delta t_s)$ of the sum-signal which is able to overcome the threshold during the aperture time τ_a:

$$Q[n] = Q(n\,\Delta t_s) \propto \tau_a(V_0 - V_{TH} + V_{RF}(n\,\Delta t_s)) \tag{3.3}$$

That is, the output voltage of the circuit constitute from a constant part which depends on the strobe pulse and the threshold voltage as well as a variable part proportional the wanted RF-signal. After removing the DC component, one gets the wanted down-converted signal whose amplitude is as bigger as larger the duration τ_a of the strobe pulse is. But we are interested in the instantaneous values of the RF-voltage. Therefore, we should avoid any integration effect during the aperture time. Depending on the bandwidth of the RF-signal, this limits the aperture time to (compare Section 2.3.1, Table 2.1)

$$\tau_a B \leq 0.886 \quad \text{or} \quad \tau_a B \leq 0.443 \tag{3.4}$$

So far, we supposed ideal rectangular strobe pulses which are actually not available. A real pulse will have finite rise- and fall-times. In order to keep simple, we respect this by triangular strobe pulse as shown in Figure 3.9. In this case, (3.3) has to be modified by a signal-dependent aperture time τ_x:

$$Q[n] = Q(n\,\Delta t_s) \propto \frac{1}{2}\tau_x(V_0 - V_{TH} + V_{RF}(n\Delta t_s)) \tag{3.5}$$

leading to

$$Q[n] = Q(n\,\Delta t_s) \propto \frac{1}{2}\tau_a\frac{(V_0 - V_{TH})^2}{V_0} + \frac{V_0 - V_{TH}}{V_0}\tau_a\,V_{RF}(n\,\Delta t_s) + \frac{\tau_a}{2\,V_0}V_{RF}^2(n\,\Delta t_s) \tag{3.6}$$

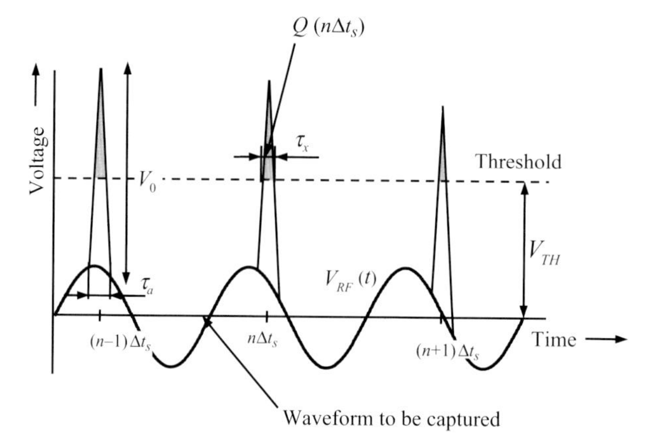

Figure 3.9 Additive sampling with triangular strobe pulses.

That is, we get a quadratic term perturbing the measurement result. Its suppression finally requires strobe pulse amplitude V_0 much larger than the RF-voltage $V_0 \gg \|V_{\mathrm{RF}}\|_\infty$ in the practical case.

As promised, we will consider the working principle still by a second way. For that purpose, we refer to Figure 3.10. It will lead to the same results as summarized by (3.3)–(3.6). We will suppose an idealized voltage-controlled switch and rectangular strobe pulses. The IV-characteristic of the switch is given in the upper diagram of Figure 3.10. The switch will be closed if the voltage V becomes larger than a threshold value and it will be open if the voltage is below that level.

The idea behind the switch is to deal with two different time constants of RC-networks, a short time constant $\tau_{\mathrm{S}} = R_{\mathrm{S}} C_{\mathrm{H}}$ and a time constant $\tau_{\mathrm{H}} = R_{\mathrm{L}} C_{\mathrm{H}}$ which is orders larger than τ_{S}, hence $R_{\mathrm{S}} \ll R_{\mathrm{L}}$. Additionally, one requires that the aperture time τ_{a} is much less than the time constants

$$\tau_{\mathrm{a}} \ll \tau_{\mathrm{S}} \ll \tau_{\mathrm{H}} \tag{3.7}$$

so that the RF-source can load the capacitor only to a low percentage.

Two basic circuits are known – the serial and parallel connection. If the strobe pulse is applied, the switch will close and the hold capacitor C_{H} collects the charge $Q(n\,\Delta t_{\mathrm{s}})$ of the nth sample, which is in both cases

$$\begin{aligned} Q(n\,\Delta t_{\mathrm{s}}) &\approx Q_0 + \frac{V_{\mathrm{RF}}(n\,\Delta t_{\mathrm{s}})}{R_{\mathrm{S}}}\tau_{\mathrm{a}}; \quad \tau_{\mathrm{a}} \ll \tau_{\mathrm{S}} \\ Q_0 &\approx \begin{cases} (I_0 - V_{TH}/R_{\mathrm{S}})\tau_a\text{-serial} \\ -V_{TH}\,\tau_a/R_{\mathrm{S}}\text{-parallel} \end{cases} \end{aligned} \tag{3.8}$$

This charge provides a voltage across the hold capacitor of

$$V_{\mathrm{C}}(n\,\Delta t_{\mathrm{S}}) = \frac{Q_0}{C_{\mathrm{H}}} + \frac{\tau_{\mathrm{a}}\,V_{\mathrm{RF}}(n\,\Delta t_{\mathrm{S}})}{R_{\mathrm{S}} C_{\mathrm{H}}} = V_0 + \frac{\tau_{\mathrm{a}}}{\tau_{\mathrm{S}}} V_{\mathrm{RF}}(n\,\Delta t_{\mathrm{S}}) \tag{3.9}$$

If the strobe pulse is switched of, this voltage decays slowly with the time constant τ_{H} towards zero. Typically, we have $\tau_{\mathrm{S}} = 10 \cdots 100\tau_{\mathrm{a}}$ so that the wanted AC part in $V_{\mathrm{C}}(n\,\Delta t_{\mathrm{s}})$ is about 10–100 times smaller than the RF-voltage. Hence, a low noise low-frequency amplifier is mandatory for further processing of the signal.

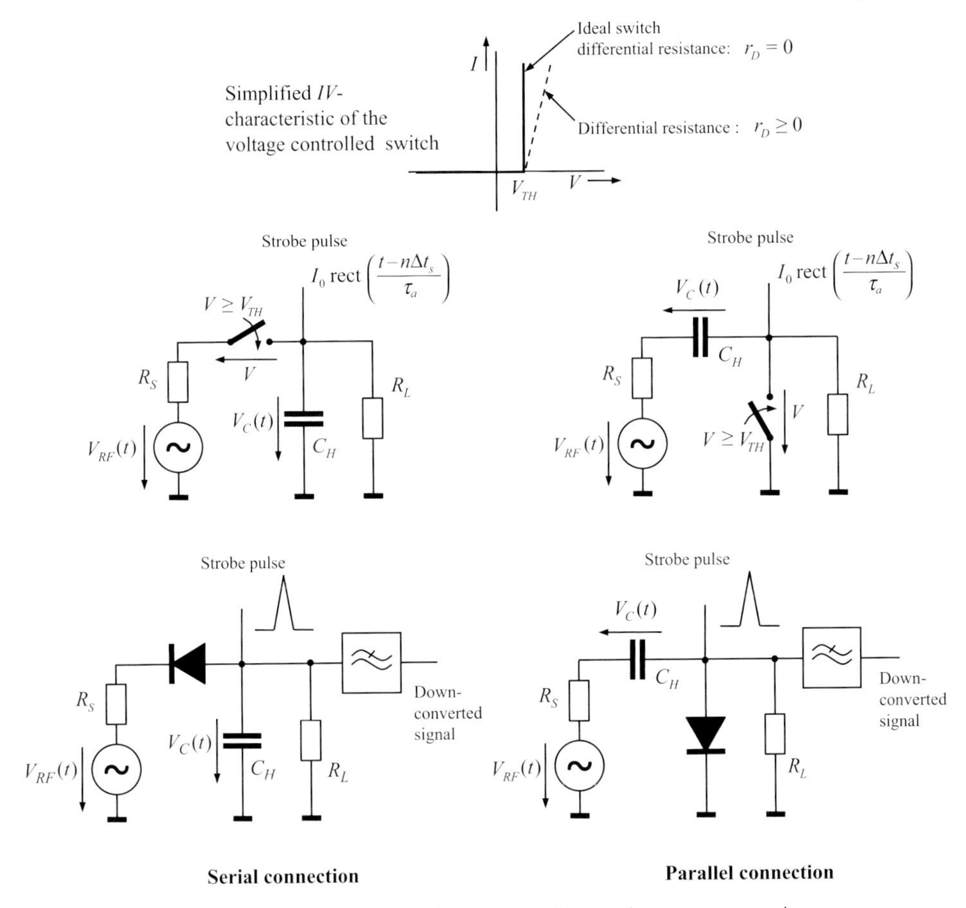

Figure 3.10 Additive sampling by voltage-controlled switches. The two basic circuits are shown – serial and parallel connection – by idealized voltage-controlled switches (above) and by diodes (below).

In reality, the voltage-controlled switches are Schottky junction diodes (also called hot carrier diodes). These diodes perform very fast switching since they do not have recovery time as usual pn-junction diodes. The strobe pulse controlling the diodes will not be rectangular (as indicated in the figure) under real conditions, so that the aperture time will depend on the input signal as already mentioned above.

Two examples of practically implemented circuits are shown in Figure 3.11. Both circuits apply parallel connection of the switching element. The schematic on the left-hand side includes some parts of the pulse-forming circuit which provides the strobe pulse. The strobe signal is a simple square wave whose leading edge is shortened by the step-recovery diode (see Section 2.2.4). The small inductor L_S derivates the square wave leading to a short strobe pulse which opens the diode. In the second case (right), it is also a step-recovery diode (not shown) which provides a waveform of a short leading edge. But here, the strobe pulse is created by differentiation

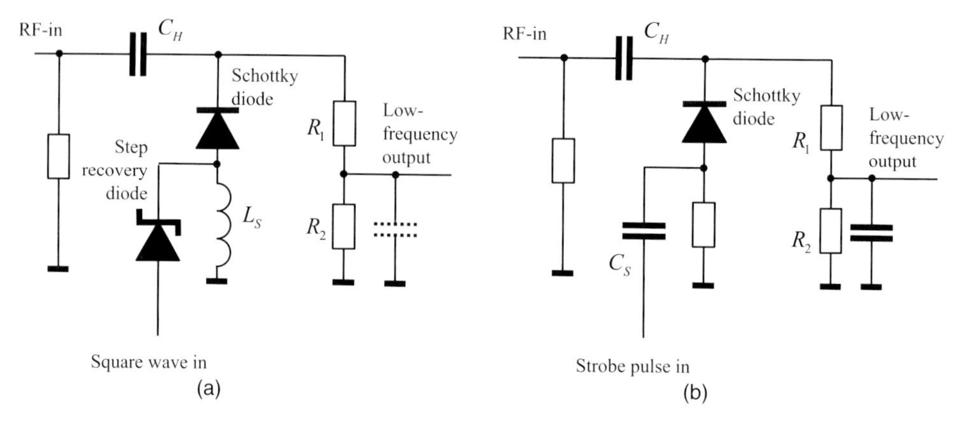

Figure 3.11 Simplified schematics of additive sampling circuits according to Wichmann (a, [16]) and McEwan (b, [17]).

via a small capacitance. In both cases, the captured signal is decoupled by a voltage divider R_1/R_2 preventing the injection of too strong strobe pulse at the input of the low-frequency amplifier.

Corresponding to (3.9), the strobe pulse only causes an additional DC value V_0 superimposed to the wanted quantity $\tau_a V_{RF}(n\,\Delta t_S)/\tau_S$ in the ideal case. Theoretically, it can be simply removed by subtraction of a constant value. However, any variations in shape or amplitude of the strobe pulse will be transferred immediately to the measurement signal due to their additive superposition. Since the strobe pulse must be of much larger amplitude than the RF-signal, its absolute noise level or drift may be dominant. Hence, the stability of the strobe pulse will limit the overall performance of the circuit. One tries to remove this inconvenience by applying multiplicative sampling.

Multiplicative Sampling There are several approaches and operational modes of multiplicative sampling. The basic ideas are depicted in Figure 3.12. Herein, switching is considered as a multiplication by 0 or 1, that is the switch is like a gate which permits or prohibits the RF-voltage the access to the hold capacitor. The circuit is close to that of the serial connection from Figure 3.10 except the behaviour of the switch. Here, it does not create a voltage drop by changing its state. It is also possible to implement the circuit by a parallel connection in a corresponding way as shown above. Since this is a straightforward procedure, we will omit this for brevity here.

As in the example above, the main point is to load C_H to a voltage which is proportional to the RF-voltage at a well-defined time point and to give an analogue or digital detection circuit enough time to capture the voltage across this hold capacitor. Hence, the time constant of discharge $\tau_H = C_H R_L$ must be sufficiently large as already required above for additive sampling. Depending on the switch mode of the gate, one distinguishes two types of operation.

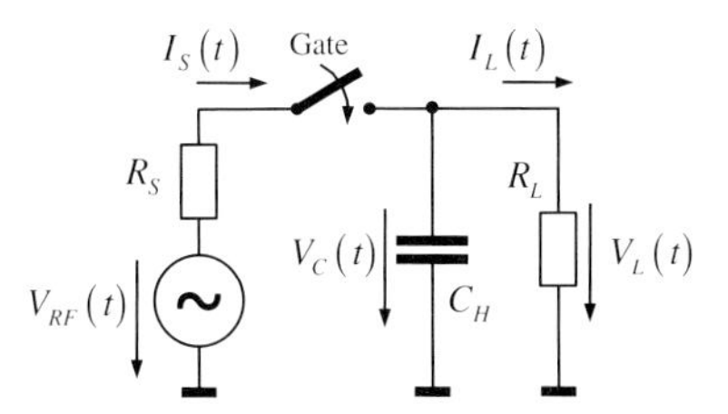

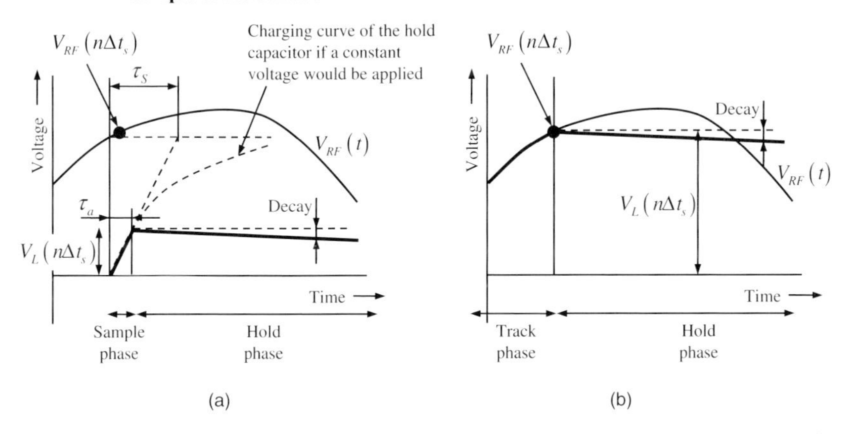

Figure 3.12 The principle of multiplicative sampling (serial connection) and the two modes of operation. (a) Sample & Hold (S&H) and (b) Track & Hold (T&H).

Sample & Hold (S&H) Mode The gate switch is open most of the time. It is closed only for a short time τ_a and the hold capacitor starts to store some charge Q resulting in the voltage $V_C(t)$ across the capacitor which is equal to the voltage $V_L(t)$ across the load resistor. Since the time constant of the charging circuit is much larger than the aperture time $\tau_S = R_S C_H \gg \tau_a$, the current $I_s(t)$ charging the capacitor can be supposed as constant (see Figure 3.12). If we further suppose that the RF-voltage does not vary during the aperture time (i.e. $\tau_a \underset{..}{B} \ll 1$; $\underset{..}{B}$– bandwidth of the RF-signal) we can write for the voltage sample captured at time point $t = n\,\Delta t_s$

$$V_C(n\,\Delta t_s) = V_L(n\,\Delta t_s) = \frac{Q(n\,\Delta t_s)}{C_H} \approx \frac{1}{C_H} \int\limits_{n\,\Delta t_s - \frac{\tau_a}{2}}^{n\,\Delta t_s + \frac{\tau_a}{2}} I_S(t)\mathrm{d}t \quad \text{for } R_S \ll R_L$$

$$= \frac{1}{C_H R_S} \int\limits_{n\,\Delta t_s - \frac{\tau_a}{2}}^{n\,\Delta t_s + \frac{\tau_a}{2}} V_{RF}(t)\mathrm{d}t \approx \frac{\tau_a}{\tau_S} V_{RF}(n\,\Delta t_s) = \sqrt{\eta_{S\,\&\,H}}\, V_{RF}(n\,\Delta t_s) \quad \text{for } \tau_a \underset{..}{B} \ll 1 \tag{3.10}$$

Hence, we get the same result as in (3.9) except the DC term V_0 which was a major source of perturbations as mentioned above. Additionally, we introduced the conversion gain $g_{S\&H}$ of the sampling circuit which can also be interpreted as a conversion efficiency $\eta_{S\&H}$. We will define them by relating the instantaneous power of the captured signal to the instantaneous power provided by the RF-signal:

$$\begin{aligned} \eta_{S\&H} &= \frac{V_L^2(n\,\Delta t_s)}{V_{RF}^2(n\,\Delta t_s)} = \left(\frac{\tau_a}{\tau_S}\right)^2 \\ g_{S\&H} &= 10\lg \eta_{S\&H} \end{aligned} \quad (3.11)$$

Since the time constant of the charging circuit is much larger than the aperture time $\tau_S \gg \tau_a$, the efficiency of S&H-circuits are usually comparatively small.

The practical implementation is based on additive sampling circuits which are operated in differential or bridge modes so that the strobe pulses mutually cancel out. Also here, one has to respect the finite rise time of the strobe pulse which leads to non-linear effects during charging the hold capacitor as already discussed above. But now, these effects compensate mutually as long as they are of even order. In the idealized case of a triangular strobe pulse, the non-linearity would be even completely annihilated. This will be not the case for real strobe pulse, so that the requirement for large pulse amplitudes remains valid but it will be less critical as for additive sampling circuits.

Figures 13–15 present some examples of S&H-circuits. Figure 3.13 deals with a two diode sampling bridge. The strobe pulses are typically created from square waves originating from a single source. After steepening their leading edge by step-recovery diodes, they are differentiated by a small capacitance, a small inductance or a short line in order to get the short strobe pulses. Then they are split by a BALUN in two symmetrical pulses of identical shape. Due to the symmetry of the bridge, they will cancel out at the signal line. Since both pulses originate from one source, noise and other perturbations will be suppressed due to the differential operation. The magnitude of the admissible RF-voltage can be increased beyond the threshold voltage of the diodes (diode voltage forward drop) by biasing them, so that they cannot be opened solely by the input signal.

Figure 3.13 also shows two simplified equivalent circuits for the two switching states of the diode bridge. During the sample phase, the diodes are on and the hold capacitor can be charged by a time constant $\tau_S = (R_S + (R + r_D)/2)C_H$. In the hold phase, the diodes are off so that the hold capacitor cannot be charged. It can only be discharged over the input resistor of the following amplifier. However, the gate-diodes will also provoke cross-coupling via their junction capacitance C_j. It may be compensated by a blow-by path as e.g. used in Tektronix sampling bridges for sequential sampling. (see Refs [21, 22] for details).

The bandwidth of the S&H-circuits is mainly determined by parasitic elements of switching elements and the pulse width of the strobe pulse. An example to

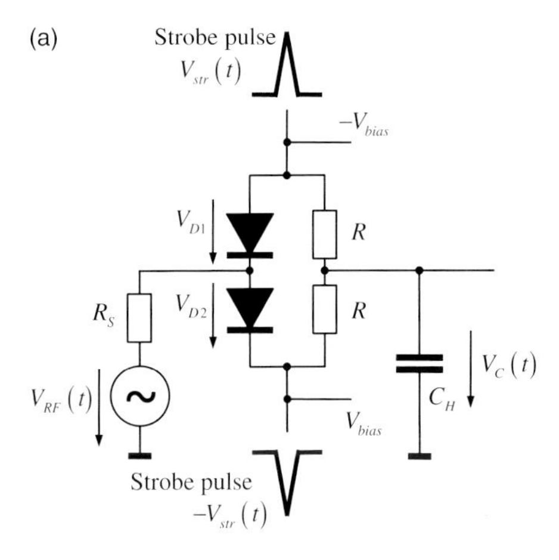

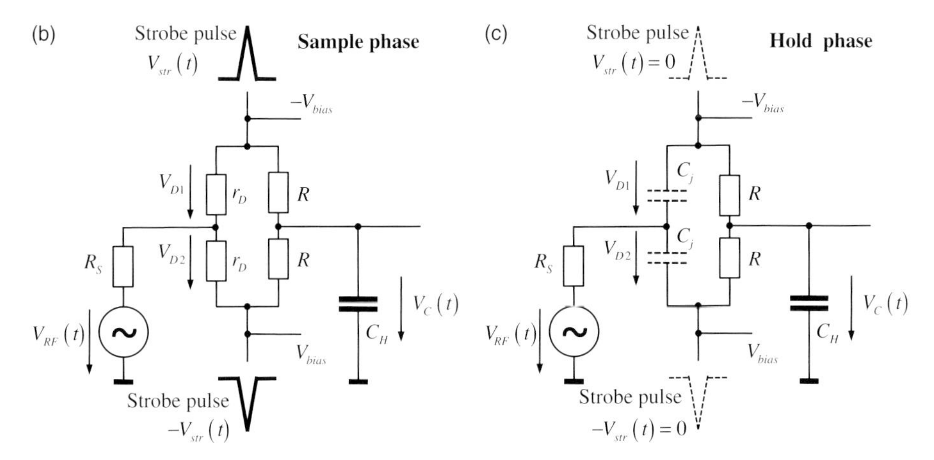

Figure 3.13 Working principle and examples of the two diode sampling bridge. Basic circuit schematics (a) and simplified equivalent circuits for the sample (b) and the hold phase (c). See also [18] for a practical circuit of one of the first implementations. Photograph and schematics of the Picosecond Pulse Labs (PSPL) 100-GHz-Through-line sampler (d) [19, 20].

optimize the sampling circuit under such conditions is given in Ref. [23]. It refers to one of the first microwave samplers. New and commercially available sampling devices provide a bandwidth up to 100 GHz. They use non-linear transmission lines (shock lines) for strobe pulse generation [19, 24]. Figure 3.13 (below right) gives an example of such a circuit. It is a two diode bridge where the second half of the bridge is built from the input ports of a differential amplifier. The short stub lines (propagation time $\tau_L \approx 2$ ps) at the outer sides perform strobe pulse shaping by differentiation of the incoming pulse edges which are steepened by the already mentioned non-linear transmission line.

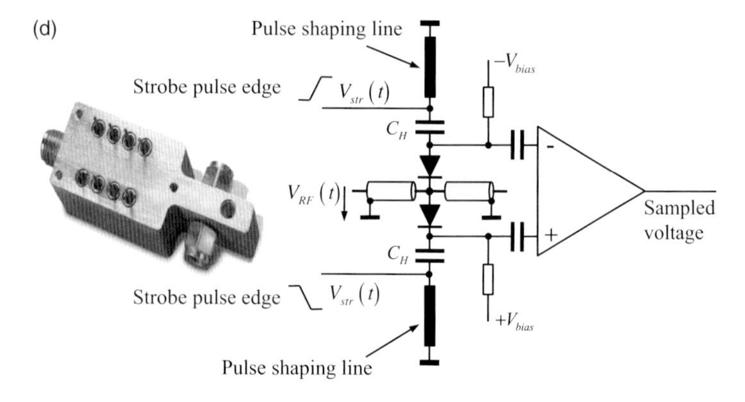

Figure 3.13 *(Continued)*

Figure 3.14 shows further diode bridges with more than two diodes. The four-diode bridge improves the cross-coupling and reduces the serial resistance for charging the hold capacity. The six-diode circuit further reduces cross-coupling during the hold phase by shunting the RF-voltage across the bridge diagonal. Both circuits are limited to a lower bandwidth compared to the two diode bridges due to additional parasitic elements.

So far, the strobe pulse was always required to be a symmetric signal. But as shown in Figure 3.15, it is also possible to invert the approach. Now, the strobe pulse is single ended and the RF-signal is differential. For UWB sensor electronics, these circuits are of large interest since the circuits are very simple and the RF-signals are often symmetric so that wideband antennas may be directly connected without intersection of BALUNs.

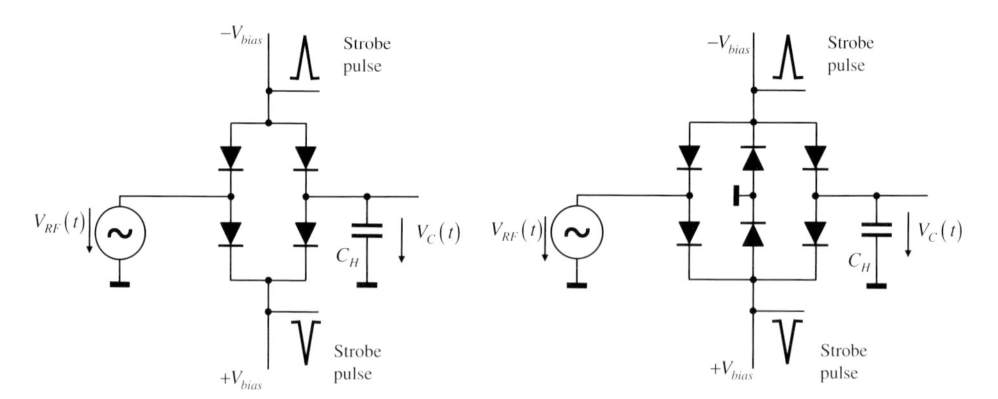

Figure 3.14 Basic schematics of four and six diode gate circuits.

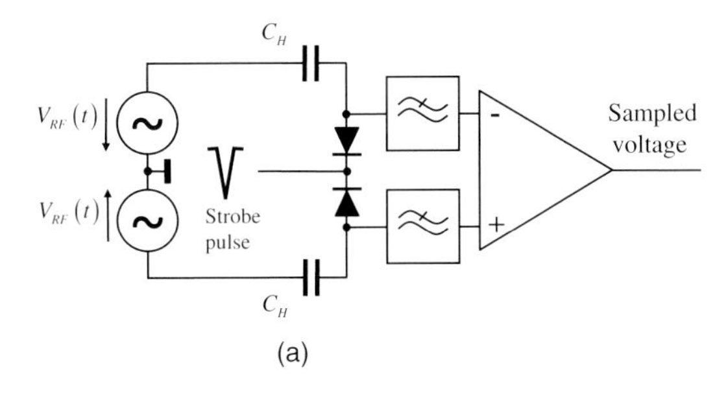

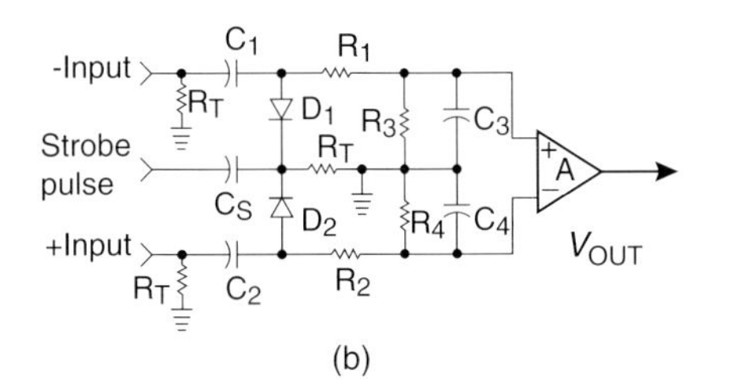

Figure 3.15 Sampling bridges with symmetric signal line and single ended strobe pulse line. (a) Schematic diagram and (b) circuit example [17] (compare Figure 3.11, see also Ref. [16] for the symmetric Wichmann sampler).

Further details on analysis and characterization of sampling units can be found in Refs [25–27].

Track & Hold (T&H) Mode For the S&H-mode, we have seen above at (3.10) and in Figure 3.12 that the voltage $V_C(t)$ across the hold capacitor represents only a fraction of the input voltage $V_{RF}(t)$ due to the short aperture time τ_a. The T&H-mode follows a different approach of switch control. Here, the switch is closed over a long time (during the acquisition and track phase) and one supposes that the voltage across the hold capacitor can follow the input RF-voltage $V_C(t)|_{track} = V_{RF}(t)$ (see Figure 3.12). The switch opens (hold state) at the sampling instance $n\,\Delta t_s$ and the voltage V_C remains unchanged $V_C(t)|_{hold} = V_{RF}(n\,\Delta t_s)$ except a weak decay. During the hold phase, a slowly working voltage or charge detector is able to capture the measurement value as before in the S&H-mode.

One should however respect that the actual hold voltage deviates from the ideal case in Figure 3.12. These deviations are illustrated in Figure 3.16. Since the hold voltage needs some time to settle down, the AD-conversion should start appropriately delayed to the T&H-clock. As obvious from the figure, the RF-input-voltage may appear at the output port of the T&H-circuit during the track phase. This limits the acquisition time of the ADC to half the sampling interval Δt_s. Cascading two T&H stages and to operate them with inverted clock

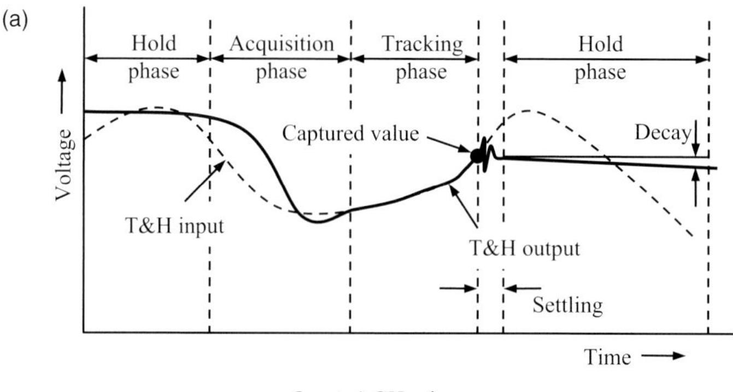

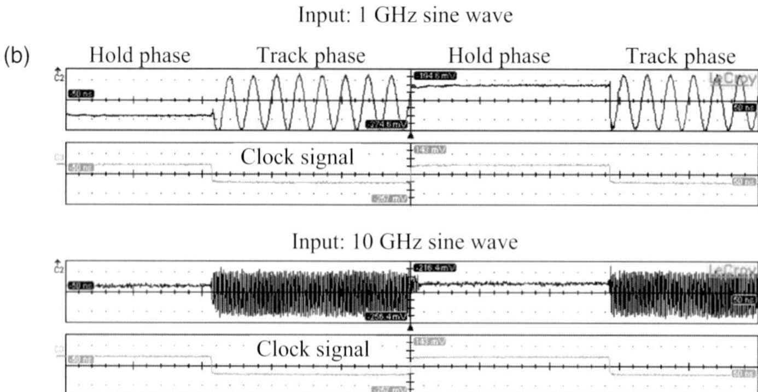

Figure 3.16 Typical behaviour of the voltage across the hold capacitor for operation in the T&H-mode. (a) Schematic drawing and (b) measured output voltage and clock signal.

signals avoids penetration of the RF-signal and doubles the available time for data capturing.

The conversion efficiency and conversion gain of a T&H-circuit is

$$\eta_{\mathrm{T\&H}} \approx 1 \Rightarrow g_{\mathrm{T\&H}} \approx 0\,\mathrm{dB} \tag{3.12}$$

and its bandwidth can be roughly estimated by

$$\underset{.}{B}_3 < \frac{1}{2\pi(R_s + r_D)C_H} \tag{3.13}$$

Here, we referred to the simplified circuits in Figure 3.12 or 3.17. r_D represents the parasitic resistance of the switch if it is on. A S&H-circuit designed for the same analogue bandwidth would have an aperture time of roughly $\tau_a \approx 1/2\underset{.}{B}_3$. Since the hold capacitance C_H cannot be made arbitrarily small, the source impedance R_S usually limits the analogue bandwidth. Therefore, T&H-circuits are always equipped with input amplifiers maintaining a low output impedance to permit high-speed charge transfer to the hold capacitor.

(a)

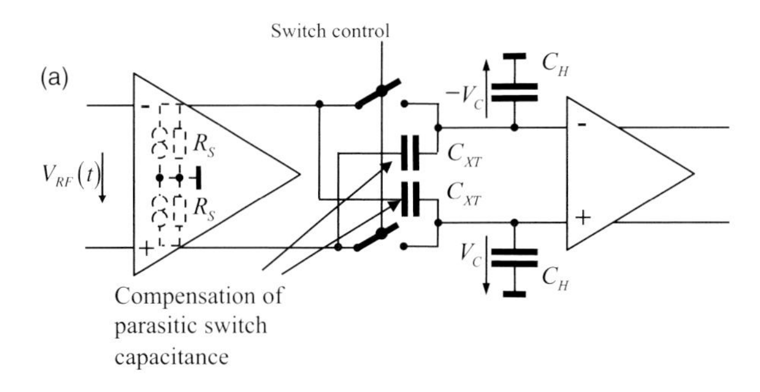

(b)

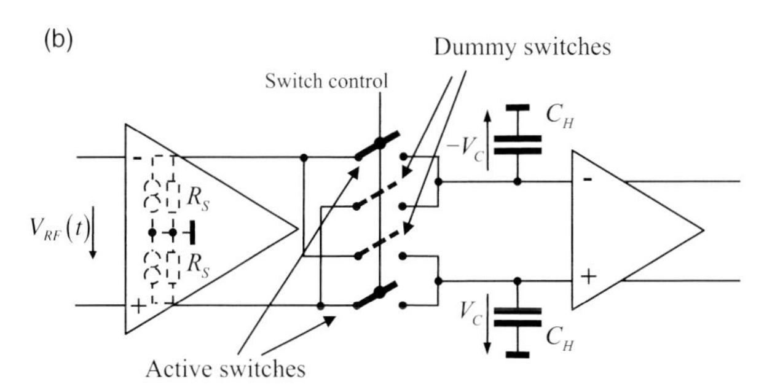

Figure 3.17 Schematic of differential T&H-device with switch leakage compensation. (a) Emulation of switch leakage by a capacitance and (b) application of two identical switch pairs.

The switching circuits applied here are based on either diode bridges (see Figure 3.14) or transistor switches. Two diode bridges are less applicable due to the resistors which lead to large time constant in the track mode. Monolithically integrated T&H-circuits for microwave applications are typically designed as symmetric circuits in order to master cross-coupling between the different stages. As depicted in Figure 3.17, differential signalling provides also some options to compensate the leakage of the open switch. Two versions are shown. One of them uses compensation capacitances C_{XT} emulating the switch behaviour in the off-mode [28]. The other one uses two identical switch pairs where one is active and the other is always in the off state (see Section 6.2).

Feedback Sampling As we have seen in Figure 3.9, the actual aperture time of the sampling gate switches depends on the value of the captured voltage sample. This causes an unwanted non-linear behaviour which restricts the RF-voltage to values typically below $\pm 100\,\mathrm{mV}$ in case of diode bridges. If, however, the voltage difference between two consecutive samples is not too large, we can modify the sampling scheme in such a way that not the absolute voltage is captured rather only the amplitude difference to the previous sampling point. For that purpose the captured voltage is fed back to the input as demonstrated in Figure 3.18.

Actually, the feedback doubles the overall voltage $\Delta V(t)$ across the gate switch. But at the crucial moment when the switch is actuated, $\Delta V(n\,\Delta t_s)$ takes small

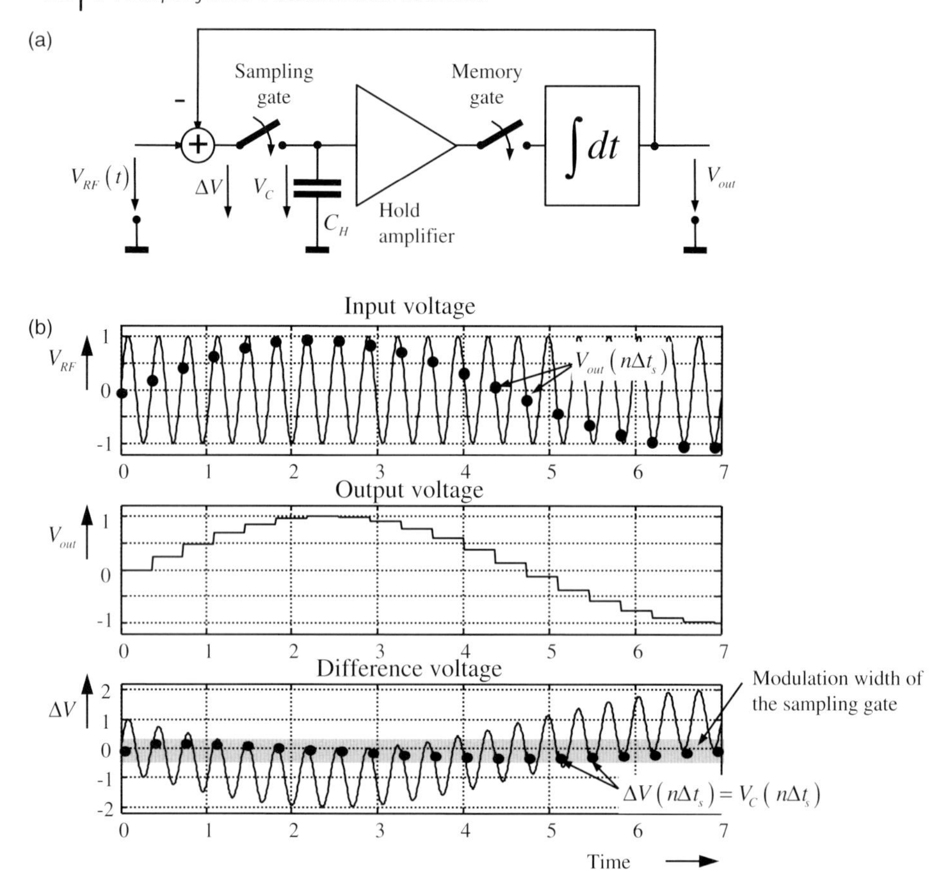

Figure 3.18 Sampling circuit with analogue feedback (a) and typical signals (b).

values so that the sampling gate switch is only loaded with the small voltage difference to the previous sample. This value is as smaller the denser the data are sampled. That is, the equivalent sampling rate[2] must largely exceed the requirements of the Nyquist theorem, that is

$$f_{\mathrm{eq}} \gg \underset{\cdot\cdot}{B} = 2\underset{\cdot}{B} \tag{3.14}$$

We will call this sampling mode 'equivalent time oversampling'.

In the case of diode bridges, the feedback voltage is usually superimposed the bias voltage which reverses the sampling diodes. By that, the diode voltage V_{D} is nearly independent of the input voltage level in the sampling moment. Thus, the gate impulse opens the diodes always over the same duration so that the sampling receiver is now able to handle RF-signals of typically $\pm 1\,\mathrm{V}$ magnitude. In order to get the amplitude of the input signal $V_{\mathrm{RF}}(n\Delta t_{\mathrm{s}})$, the signal subtractions has to be

2) See term 'equivalent sampling rate' will be introduced in Section 3.2.3.3.

revoked. This is done by the integrator whose output signal provides finally the wanted voltage. Occasionally, a memory switch is inserted to reset the hold stage before the next sample.

Basically, the width of the modulation area can be reduced to quite small values so that even a comparator will be sufficient for data conversion. A corresponding approach is illustrated in Section 3.2.4, Figure 3.28.

Product Detector Fast analogue multipliers are a further method to down-convert RF-voltages to a lower spectral band. They may be used like a sampling converter as discussed above. But they are note restricted to unipolar gate pulses to perform the down-conversion. An example of such an approach can be found in Refs [29, 30]. Furthermore, product detection is to be found in demodulators of sine wave devices (see Section 3.4.2) or correlators (see Section 3.3.3).

The classical approach to multiply two RF-voltages uses diode ring mixers. They may be built for large bandwidth but they suffer from conversion loss and harmonic generation. Recent monolithic circuit technologies promote the design of very fast analogue transistor multiplier which is usually referred as Gilbert cell. It represents a bridge of two differential amplifiers whose current source is controlled by a second voltage (Figure 3.19a). Such circuits permit the creation of various fundamental analogue signal processing structures. In connection with an integrator, it will form the so-called product detector (Figure 3.19b) which we exploit to convert a wideband RF-signal into a signal of narrow bandwidth. The figure gives a simple example. The output voltage of the product detector yields

$$V_4(t) = \frac{1}{t_I} \int_{t-t_I/2}^{t+t_I/2} V_3(\xi)\mathrm{d}\xi = \frac{g_M}{t_I} \int_{t-t_I/2}^{t+t_I/2} V_1(\xi)\, V_2(\xi)\mathrm{d}\xi \tag{3.15}$$

Herein, t_I refers to the integration time of the low-pass. In the case of two periodic signals $V_1(t + nt_P)$ and $V_2(t + nt_P + \tau)$ which may be mutually time shifted by τ, we end up in a DC voltage for a proper choice of the integration time t_I:

$$V_4(\tau) = \frac{g_M}{t_I} \int_{t_I} V_1(t + nt_P)\, V_2(t + nt_P + \tau)\mathrm{d}t; \quad t_I = m\, t_P;\ m \in \mathbb{N} \text{ or } t_I \gg t_P \tag{3.16}$$

Equation (3.16) represents a correlation function as introduced in Section 2.2.4, thus the output voltage of the product detector is a gauge of the concurrence of its two input signals.

The voltage $V_2(t)$ is typically assigned as reference waveform. Basically, it can be a time signal of arbitrary shape. If we select a short rectangular or Gaussian pulse, the product detector works as an S&H-device. But it is also possible to use other reference signals $V_2(t)$ with the goal to emphasize or suppress specific shape elements of the incoming waveform $V_1(t)$. The conversion of the resulting output voltage $V_4(t)$ or $V_4(\tau)$ into the digital domain is less challenging since its bandwidth is restricted to the inverse of the integration time t_I.

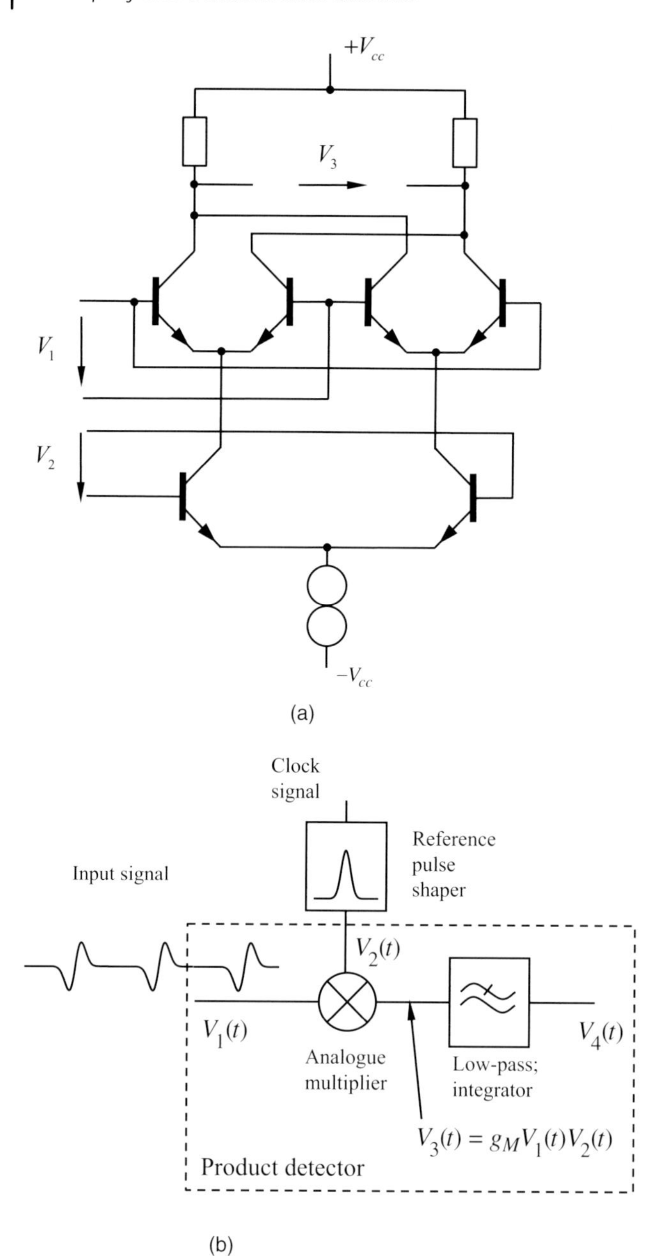

Figure 3.19 RF-voltage capturing by product detector. (a) Schematic of the Gilbert cell and (b) product detector with reference generator ($g_M\left[V^{-1}\right]$ – multiplier gain).

Finally, we will point out an ambiguous use of the terms product detector and correlation. Sometimes, the product detection is put on the same level with correlation. In fact, (3.16) represents a correlation function. However, one usually anticipates a certain correlation gain by referring to a correlation procedure. But a correlation gain can only appear if two wideband signals which are spread in time have nearly identical shape. In the case of short pulse-like signals for $V_1(t)$ and $V_2(t)$, the 'correlation gain' provided by a product detector will tend to 1 (0 dB) [29].

3.2.3.3 Timing of Data Capturing by Sub-Sampling

As we have seen above, it is possible to capture a voltage sample during a very short time interval τ_a which is an inevitable prerequisite to measurement very wideband signals (see (3.4)). But from Section 2.45 (2.175), we also know that a wideband signal requires an appropriately high sampling rate $f_s \geq \underset{\cdot\cdot}{B} = 2\underset{\cdot}{B}$ in order to avoid frequency aliasing. Since the bandwidth of typical ultra-wideband sensors is beyond several GHz, the required data capturing rate exceeds technical feasibility. Either there do not exist appropriate ADC circuits or they are too power hungry or costly. Additionally, the accrued data stream poses a real challenge to be processed continuously.

Fortunately, ultra-wideband sensors use almost always recurrent stimulus signals (noise sensors are exceptions). This opens up the opportunity to distribute the data gathering over several periods of the signal which largely relaxes the technical requirements onto the receiver which has to pick-off the voltage, to store and to process the data. However, attention has to be paid for an appropriate timing of the data capturing in order to meet the sampling theorem even with the lowered sampling rate. Several approaches are in use. We will consider the most convenient ones.

Sequential Sampling by Dual-Ramp Approach The intuitively simplest method of timing control is sequential sampling as depicted in Figure 3.20. It supposes repetitive signals which must not be necessarily periodic. The idea is to take only one voltage sample from every period (or every N periods) of the RF-signal. In order to capture the full data set, the sampling event is shifted slightly from period to period by a short time slot Δt_{eq}. Its value determines the density of data samples of the captured waveform. Hence, it has to meet the Nyquist theorem while the actual sampling distance Δt_s is less important in this context. In our case, the condition for correct sampling can be modified to

$$f_{eq} = \frac{1}{\Delta t_{eq}} = \frac{1}{\Delta t_s - N t_P} \geq 2\,\underset{\cdot}{B} = \underset{\cdot\cdot}{B} \tag{3.17}$$

Herein, B is the bandwidth of the RF-signal and f_{eq} represents a virtual sampling rate which we call equivalent sampling frequency. N is the number of periods skipped between two consecutive samples. The example shown in Figure 3.20 refers to $N = 1$. For real-time sampling, N equals zero.

Following Figure 3.20, the control of the time shift Δt_{eq} is done via two voltage ramps. A trigger circuit (comparator 1) establishes synchronism to the signal of

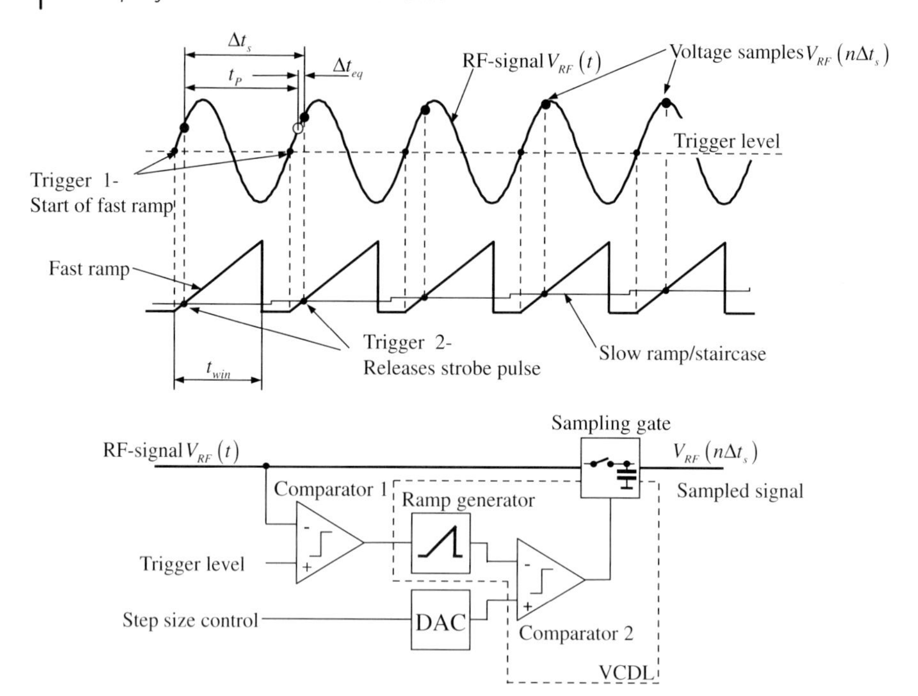

Figure 3.20 Timing control of sequential sampling by a voltage-controlled delay line (VCDL). The signal to be delayed starts a linear ramp. The comparator releases a delayed edge if the ramp has achieved the control voltage (slow ramp).

interest. It starts a ramp generator (fast ramp) synchronously to the signal. The time duration of that ramp fixes the length of the observation window t_{win}. A second ramp (or staircase voltage) moves slowly up so that the points of intersection with the fast ramp are slightly delayed from period to period. The intersection is detected by the second comparator which finally releases the sampling event. The step size of the staircase voltage and the slew rate of the fast ramp determine the value of equivalent sampling frequency $f_{\text{eq}} = 1/\Delta t_{\text{eq}}$. If the staircase keeps constant its value over M periods, the same data point will be collected M times and lastly averaged. Sometimes that approach is also called boxcar integration since it behaves like short-time integration with rectangular weighting function (see Section 2.6.1). But it will not affect the bandwidth of the recovered waveform because it acts only onto data samples of identical time position like synchronous averaging.

The slower the staircase growth the denser the sampling points will be. It is possible to bring down the virtual time resolution Δt_{eq} to the order of picoseconds which corresponds to an equivalent sampling frequency close to 1 THz permitting a signal bandwidth of hundreds of GHz. However, this is not feasible with common sampling units and moreover not necessary in usual applications of ultra-wideband sensors. There, the bandwidth is typically below 10 GHz.

The actual choice of the equivalent sampling frequency f_{eq} influences several practical aspects of the sensor implementation. Naturally, it has to be larger than the Nyquist rate of the captured signal $f_{eq} \geq 2\underset{\cdot}{B} = \underset{\cdot\cdot}{B}$. Often, namely in experimental setups with sampling oscilloscopes, it is however selected much larger by the operator as required. Thus, we get a densely sampled version of the signal (equivalent time oversampling). This will be good for visualization but it is bad for the data volume. Indeed, the human eyes need oversampling since they are not well matched to recognize curves which are digitized close to the Nyquist rate. Sampling units using the feedback principle (Figure 3.18) or delta converters (Figure 3.28) need oversampled data too in order to keep the voltage steps small. But oversampling does not provide more information about the test objects. The information content of a signal is given by its actual bandwidth and not by the density of the samples. Therefore, oversampled data can be low-pass filtered without affecting the signal shape. If carefully done, it will reduce the noise level and we can throw useless data samples away by down-sampling[3] (compare Section 2.6.1, Figure 2.73). In the best case, we will end up in a data set closely sampled to the Nyquist rate.

The dual-ramp method is widely used in pulse sensor systems. Nevertheless it suffers from some inadequacies which mainly result from noise on the voltage ramps as well as time variable offset and non-linearity. Any deviation of the ramps from a linear slope causes non-equidistant sampling which leads to distortions of the signal shape during recording since an equidistant sampling grid was assumed. A slowly varying voltage offset shifts the whole ensemble of the trigger events which controls the sampling gate. The observation of such a signal on a screen causes the impression as it moves slowly along the time axis. Hence, the precision of time measurements will be affected and the number of synchronous averaging is limited due to smearing effects.

Furthermore, we have seen in Section 2.6.3 that signal corruption by time jitter depends on the strength of additive noise and the slope of the trigger edge. If we assume an ideal trigger comparator but noisy ramp voltages of identical noise level, the sampling jitter φ_j can be estimated from (2.290) to

$$\varphi_j = t_{win}\sqrt{\frac{2}{\mathrm{SNR}_{ramp}}} \tag{3.18}$$

Herein, t_{win} represents the length of the time window and SNR_{ramp} is the signal-to-noise ratio of both voltage ramps which was supposed to be identical. A purposeful number N_s of sampling points per scan and hence an optimum equivalent sampling frequency $f_{eq,opt}$ results from

$$N_s = \frac{t_{win}}{\varphi_j} = t_{win} f_{eq,opt} = \sqrt{\frac{\mathrm{SNR}_{ramp}}{2}} \leq \underset{\cdot\cdot}{B} t_{win} \quad \text{for } \Delta t_{eq} = \varphi_j \tag{3.19}$$

Note this relation implies that the average uncertainty of a sampling point is equal to the expected time distance Δt_{eq} between two adjacent samples. This also implies that the maximum deviations from the wanted point in time even are in

3) Simply spoken down sampling is nothing but cutting out the upper part of the FFT spectrum (which does not contain signal energy) and back transformation of the reduced spectrum. Refer also Section 2.4.5 for more discussions of sampling interval.

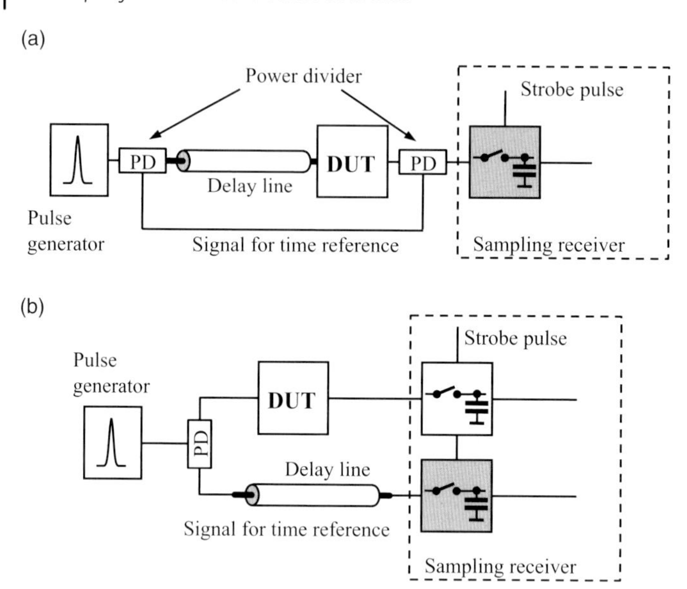

Figure 3.21 Correction of time drift by referring to a signal of known propagation time. The reference sampling unit is marked by a white box and the measurement head by a grey box.

the order of $\pm 3\,\varphi_{\mathrm{j}}$ (Gaussian jitter supposed). Equation (3.19) also indicates the maximum useful time-bandwidth product $TB_{\max} \leq \underset{\cdot\cdot}{B} t_{\mathrm{win}}$ which can be handled by the sampling receiver.

Using reference measurements, the timing of data sampling can be improved with respect to drift, jitter and non-linearity as well. The achievable performance improvement is largely determined by the SNR-value and the linearity of the sampling gates as well as the fidelity of the reference signals. Figures 3.21 and 3.22 summarize two approaches. Some methods assume multi-channel sampling receivers in which all sampling gates are controlled by the same strobe pulse so that the inadequacy of its timing is identical in all receive channels.

The general idea to corrected data corrupted by timing errors is

1) to capture in one or two reference channels well-known reference signals,
2) to determine the deviations between the measured and the reference signal and
3) to use the recovered deviations to correct the unknown signal of interest.

In the first two examples as depicted in Figure 3.21, only the time drift is corrected. For that purpose, a signal of known and stable delay is used as time reference. That is, time zero of the measurement is not determined as usual by the time axis of the receiver (e.g. oscilloscope) rather it is referred to a trigger point provided by the reference waveform. Typically one takes the threshold crossing of the steepest edge of the reference signal. The first example does not require extra receiving channels for correction. But it will lower the available time window length for the actual measurement signal. The task of the delay line is to separate reference pulse and measurement signal in time. Its delay time should be selected appropriately.

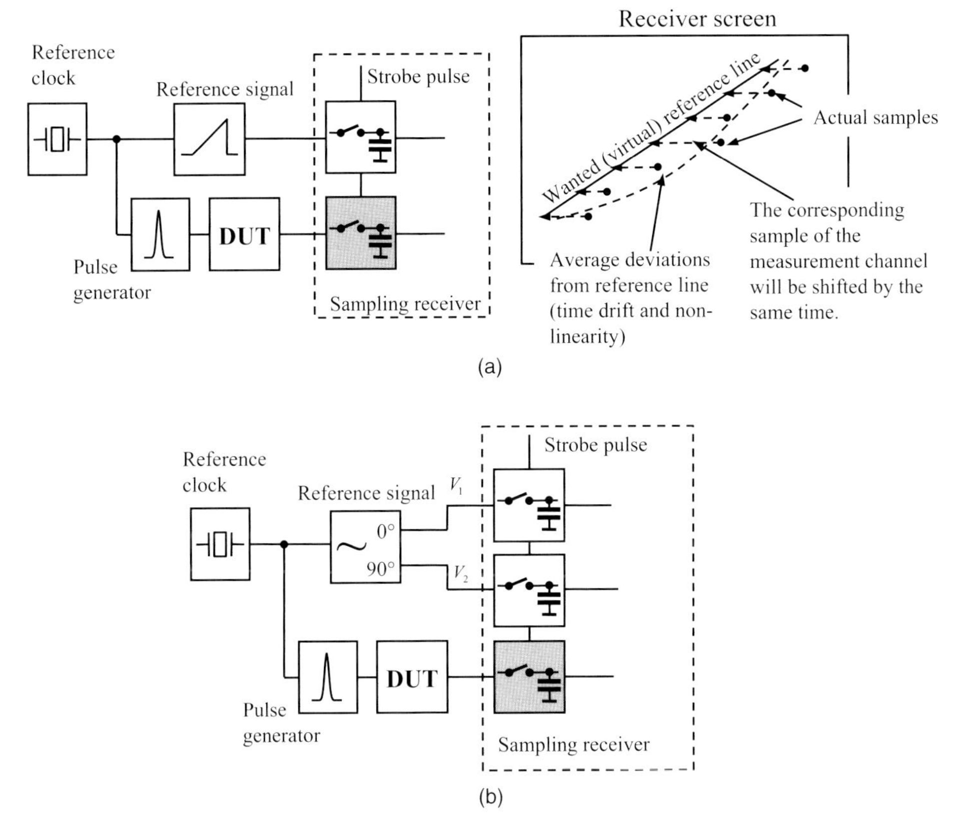

Figure 3.22 Correction of drift, jitter and non-linear time axis by well-known reference signals. (a) Basic idea using an ideal sawtooth and (b) feasible implementation with sine waves as reference.

A second channel to capture the reference signal is required if narrowing the time window is not acceptable (Figure 3.21b). Here, the delay line is used to compensate for DUT internal delays since one needs both signals within the same time window.

The third method intended to correct all timing errors uses an ideal sawtooth voltage which is captured by the reference channel (Figure 3.22). Supposing this receiver has low noise level and it is working in its linear range, then the captured signal should ideally look like a straight diagonal line on the device monitor. But actually we will get deviations as depicted in the figure. Random outliers result from jitter, an offset position is caused by drift and the deformations from a straight line are due to non-linear ramps. All these deviations can be determined point by point by comparing the wanted time position (given by the virtual reference line) with the actual ones of the captured samples. After having estimated the differences, the samples of the measurement channel can be time shifted into the correct position. Interesting to note that jitter may be largely removed by that approach since all channels suffer from the same timing errors due to the common strobe pulse.

But unfortunately one is not able to provide sawtooth signals of the required duration and fidelity. Only sine waves can be generated with a sufficient purity. In order to gain for simple data correction a continuously rising signal (as the sawtooth it did), one has to uses a trick (Figure 3.22b). Two sine waves mutually shifted by 90° are captured and one calculates *arctan* V_2/V_1. This would give a straight line of constant slope which is used for the calibration purposes as described above. Jitter caused by the stimulus generator cannot be compensated because it is independent of the reference signal. Details on the actual calibration procedure can be found in Ref. [31] which is also respecting harmonic distortions of the reference signals.

A further approach to crusade against offset-drift and non-equidistant time samples due to non-linear ramp is to insert the voltage-controlled delay line (VCDL) into a delay locked loop (DLL). Figure 3.23 depicts the basic principle of the DLL.

The control criterion of the DLL is to minimize the averaged output voltage of the phase detector. The loop is locked if $V_{\mathrm{LP}} = 0$ from which follows that

$$\bar{\tau}_0 = \tau_{\mathrm{D}} - \tau_x \tag{3.20}$$

Hence, the edges of the reference clock and the delayed clock coincide. In the case the reference clock does not provide a stable delay τ_0, the DLL will lock to the average value $\bar{\tau}_0$ due to the lag smearing of the control loop. This smearing effect can be used to digitally control the delay of a sampling circuit. The principle is illustrated by a simplified schematic in Figure 3.24 [32].

The master clock pushes the pulse shaper which provides stimulus pulses of the repetition rate f_{c}. The DUT response shall be captured within a time window of duration $t_{\mathrm{win}} = 1/2f_{\mathrm{c}}$ by stroboscopic sampling. The VCDL delays successively for that purpose the original master clock. In order to control the delay, the switch at the reference input of the phase detector toggles with a specific pattern between both switch settings. If it stays, for example permanently in the upper position, the delay will be zero $\tau_{\mathrm{D}} = 0$; if it stay in the lower position, we get $\tau_{\mathrm{D}} = t_{\mathrm{P}}/2 = t_{\mathrm{win}}$

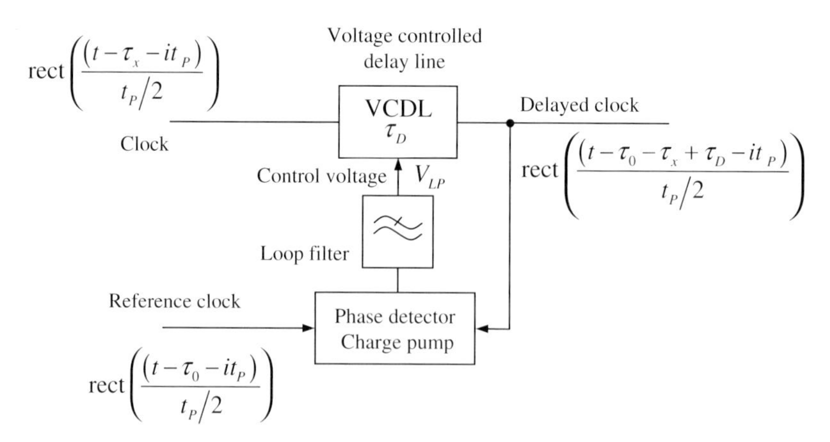

Figure 3.23 Structure of delay locked loop (DLL).

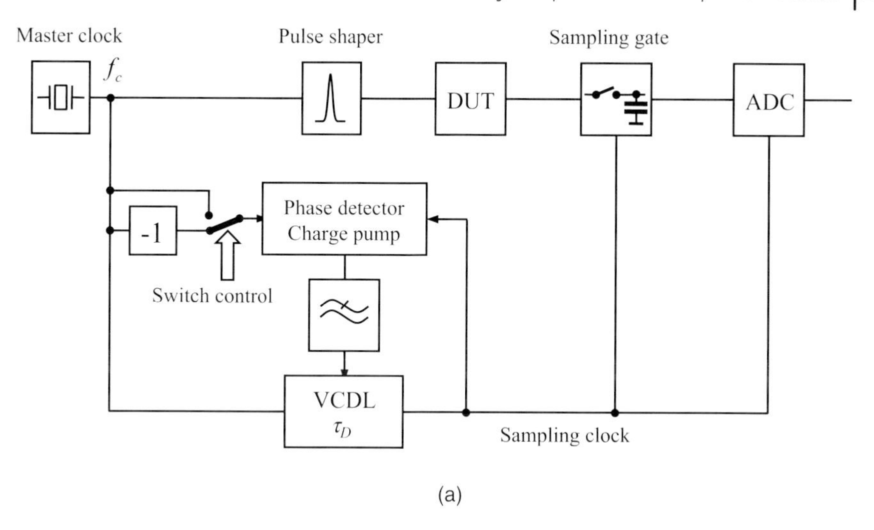

(a)

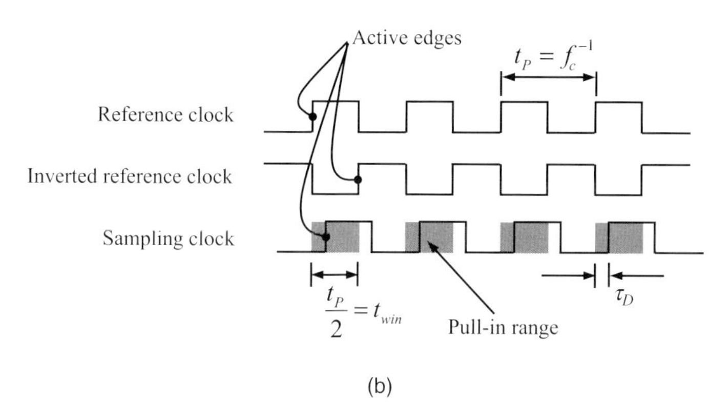

(b)

Figure 3.24 Pulse-based UWB sensor with DLL-controlled sequential sampling. (a) Basic schematic and (b) control signals. We presume that the rising edge triggers the pulse shaper, the sampling gate as well as the phase detector.

and if it toggles every period, the delay will be $\tau_D = t_P/4 = t_{win}/2$. The sampling clock delay as exemplified in Figure 3.24 is achieved by staying three periods in the upper position and one in the lower.

The bandwidth of the low-pass filter depends on the required number N_s of time steps (and hence from the time bandwidth product of the sensor). The larger this number is, the lower the cut-off frequency must be and the more time is needed to settle the DLL. Further, one can expect that the switching cycle is not completely suppressed by the loop filter so that spurious may appear which cause additional jitter of the sampling clock.

Harmonic Mixing and Interleaved Sampling Another timing approach for sampling heads results from considering sub-sampling in the frequency domain.

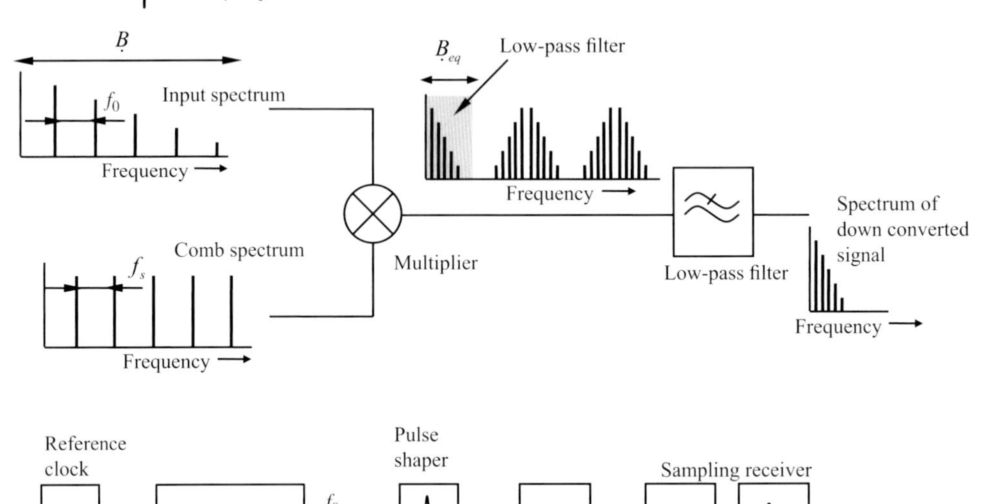

Figure 3.25 Sub-sampling described in the spectral domain. Signal model and corresponding block schematic. The multiplier symbolizes one of the above-mentioned sampling circuits (see Section 3.2.3.2).

Sub-sampling can be seen as the multiplication of a periodic signal with a Dirac comb (strobe pulses) which equals the convolution of the spectra of these two signals. As depicted in Figure 3.25, both signals have a line spectrum spread over a large bandwidth. The Dirac comb provides even a comb spectrum of infinite bandwidth. But practically it is limited by the aperture time of the sampling gate (see (3.4)), that is the width of the strobe pulses. The spacing between the spectral lines corresponds to their repetition rates f_0 and f_s. These spectra are mixed where every harmonic of the input signal mixes with every harmonic of the comb spectrum – therefore the term harmonic mixing.

If the frequency f_s of the comb signal is chosen appropriately, the spectral lines of the signal mixture are arranged in separated groups as shown in Figure 3.25. The appearance of every of such group is identical to the spectral shape of the input signal except the bandwidth over which they extend. The input signal has the single-sided bandwidth $\underset{\cdot}{B}$ while for the individual spectral groups we got $\underset{\cdot}{B}_{eq}$. If we cut-out the group of spectral lines emphasized by the grey box by a low-pass filter, we obtain a time signal of identical shape as the input signal but stretched in time. That is, we have a low-frequency signal now which is easy to digitize.

The simplest way to identify an appropriate sampling frequency f_s is given for sequential sampling. If we refer to Figure 3.20, the sampling frequency f_s must be slightly smaller than the repetition rate f_0 of the input signal. From (3.17) we simply gain

$$f_{eq} = \frac{1}{\Delta t_s - t_P} = \frac{f_0 f_s}{f_0 - f_s} \approx \frac{f_0^2}{\Delta f} \geq 2B = \underset{..}{B} \quad \text{with} f_s = f_0 - \Delta f \tag{3.21}$$

That is, the frequency slip Δf between the repetition rates of the stimulus and the strobe pulse determines the maximum observable signal bandwidth. The smaller it is the larger will be the equivalent sampling rate. Nowadays, the two frequency f_0 and f_s are provided either by DDS-circuits (DDS – direct digital synthesis) [33] or via phase locked loop (PLL)-frequency synthesis [34]. These techniques permit very fine trimming of both frequencies. Due to the reference to a single stable clock, there is no drift between both control signals and non-equidistant sampling interval may be avoided.

The sine waves triggers two impulse generators – one stimulates the device under test, the second provides the strobe pulse for the sampling gate. Typical values of f_0 and f_s are in the range of a few MHz or even below. These rates are limited by either the maximum toggle rate of the pulse generators, the conversion speed of the ADC or the settling time of the DUT.

The length of the observation time window is $t_{win} = 1/f_0$. It covers N_s data samples, which results from

$$N_s = \frac{t_{win}}{\Delta t_{eq}} = \frac{t_P}{\Delta t_s - t_P} = \frac{f_s}{f_0 - f_s} \approx \frac{f_0}{\Delta f} \tag{3.22}$$

As mentioned, the method requires triggering of two pulse generators by two sine waves of nearly the same frequency. If we assume two identical pulsers and that their jitter is only caused by the uncertainties of the trigger events due to additive noise of the trigger signals, we get from (2.292) for the random jitter φ_j of the sampling receiver

$$\varphi_j = \sqrt{\varphi_{j,1}^2 + \varphi_{j,2}^2} = \frac{1}{\pi f_0 \sqrt{2\,\mathrm{SNR}_{tr}}} = \frac{t_{win}}{\pi \sqrt{2\,\mathrm{SNR}_{tr}}} \tag{3.23}$$

Herein, $\varphi_{j,1}$, $\varphi_{j,2}$ are the individual trigger jitter of the pulse generators and SNR_{tr} is the signal-to-noise ratio of both trigger sine waves. This value is roughly six times better than the corresponding value of the dual-ramp approach (see (3.18)). Hence, devices with a roughly six times larger time-bandwidth product may be built under the same noise conditions.

The time concept according to Figure 3.25 is not restricted solely to sequential sampling. Since the sampling rate f_s may be selected quite freely, also other timing concepts are feasible. So it is not stringently required to capture one sample per period. It is also possible to ignore several periods of the test signal (compare (3.17) for $N > 1$) or to capture more than one sample per period. The only thing that has to be respected is the compliance of the Nyquist theorem.

Here, we will shortly discuss the second case, that is to capture more than one sample per period. (The first case is meaningless for sensor applications since it is useless to stimulate the objects faster than one is able to measure.) It is called interleaved sampling. In order to deduce an appropriate sampling rate f_s, we suppose that the equivalent observation time window t_{win} contains N_s of samples, so that we yield

$$N_s = \frac{t_{win}}{\Delta t_{eq}} = \frac{t_P}{\Delta t_{eq}} = \frac{f_{eq}}{f_0} = t_P f_{eq} \geq 2\, t_P \underset{\cdot}{B} = t_P \underset{\cdot\cdot}{B}; \quad N_s \in \mathbb{Z} \tag{3.24}$$

Hence, there is a lower bound of the sample number N_s which results from the Nyquist theorem and the duration of the time window.

Assuming that a period of our measurement signal is represented by N_s voltage samples. Hence, a complete cycle of data gathering has to care for the collection of the whole set of samples $n = 0 \cdots N_s - 1$. In which order they are collected does not matter as long as we know it. An example shall illustrate the procedure. Saying that one period of a signal covers 7 data points and we collect only every third or fourth one, for example

.........① 2 3 ④ 5 6 ⑦ 1 2 ③ 4 5 ⑥ 7 1 ② 3 4 ⑤ 6 7.........,

.........① 2 3 4 ⑤ 6 7 1 ② 3 4 5 ⑥ 7 1 2 ③ 4 5 6 ⑦ 1 2 3 ④ 5 6 7.........

then we need three or four periods to have them finally all. Indeed they are in the wrong order, but we can re-order them.

In general: If we capture every mth sample, we need m signal periods to get all N_s samples as long as m is not a divisor of N_s. In order to keep the Nyquist theorem, the time interval between the virtual sampling points must be Δt_{eq}. Hence, the actual interval between two consecutive samples is $\Delta t_s = m\,\Delta t_{eq}$ which leads us together with (3.24) to the sampling rate:

$$f_s = \frac{N_s}{m} f_0 \quad \text{for } m < N_s + 1;\ N_s, m \in \mathbb{Z};\ \frac{N_s}{m} \notin \mathbb{Z} \tag{3.25}$$

For $m = N_s + 1$, we have sequential sampling. The case $m = N_s - 1$ leads to time reversed sequential sampling. Reformulation of (3.25) leads to $f_c = m f_s = N_s f_0 = f_{eq}$. Respecting further that N_s and m are integer numbers, the sampling scheme described by (3.25) may be implemented as shown in Figure 3.26. Here, we use a stable RF-clock generator. Its frequency equals the equivalent sampling rate. The actual pulse repetition and sampling rates f_0, f_s are provided by digital frequency dividers (pre-scaler). Hence, we get two trigger signals comparable to those applied in Figure 3.25. But now, we have rectangular

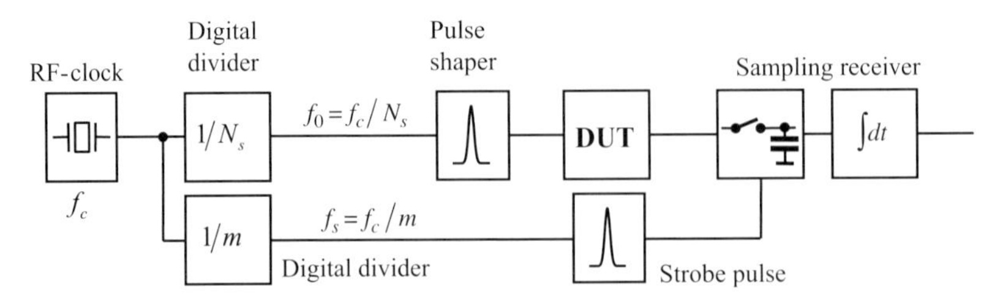

Figure 3.26 Timing control via digital frequency divider.

signals of steep edges instead of sine waves which largely reduce trigger jitter due to additive noise. Section 3.3.4 deals with an UWB sensing device which applies this approach.

Interleaved sampling allows higher measurement speed (refer also to compound sampling [35]). However, consecutive samples may have a large difference in their amplitude since they are not collected in the right order. Therefore, classical feedback samplers as introduced above (Figure 3.18) are not applicable with this timing approach.

Random Sampling Random sampling is the third of the basic timing methods for data capturing. It is not common in ultra-wideband sensing. Therefore, we will only discuss the basic idea behind the principle. It is depicted in Figure 3.27. The input signal $V_{RF}(t)$ is captured by strobe pulses of fixed and known sampling rate f_s which is, however, independent of period and bandwidth of the signal under test. The captured data samples must be stored in the data memory for later disposition. The sampling process proceeds randomly (hence the name of the principle) since there are no timing relations between the input signal and the strobe pulses. So far, one only knows the time difference between two consecutive samples which is $\Delta t_s = f_s^{-1}$. Thus, the stored data samples are still useless since they cannot be correctly assigned to the signal $V_{RF}(t)$. Therefore, we still have to establish a timing relation between the test signal and the strobe pulses.

In order to gain such reference, one determines the time lag Δt_i between a trigger event caused by the signal to be observed and the next strobe pulse[4)] (see Figure 3.27b). This is often done by charging a capacitor with a constant current. As depicted in the figure, the voltage across the capacitor is a measure of the elapsed charging time Δt_i. This value is stored in the time base memory and we can start to organize the correct time alignment of the data samples. Let us do this by an example (Figure 3.27c). We like to observe the signal within the time span t_{win} whereas the observation window should start some moments t_d before the trigger point. From the knowledge of the time intervals $t_{win}, t_d, \Delta t_s, \Delta t_i$, the device can calculate the time position of all captured samples. Related to the first trigger event in our example, the sample points 3, 4 and 5 are placed inside the observation window. The points 1 and 2 are excluded. In the cases of the second and third trigger event, the points 7, 8, 9, and 12 and 13 are located within the time interval of interest while the points 6, 10, 11, 14 are rejected, and all the rest of it.

Since the sampling process is random, one needs to wait some time in order to have sufficiently dense sample points. We add in average $\bar{n} = t_{win}/\Delta t_s$ new samples to the just existing ones for every trigger event. It should be noted that the time window t_{win} may be much shorter than the sampling interval Δt_s without violation the principle. Only the recording time will rise up since the probability to place a sample within the time window will go down. In the very unlikely case that the strobe rate f_s and signal periodicity $t_p = f_0^{-1}$ are sub-harmonic of each other (i.e.

4) In practice, one often takes the second strobe pulse after the trigger event in order to avoid dead time problems of the switches.

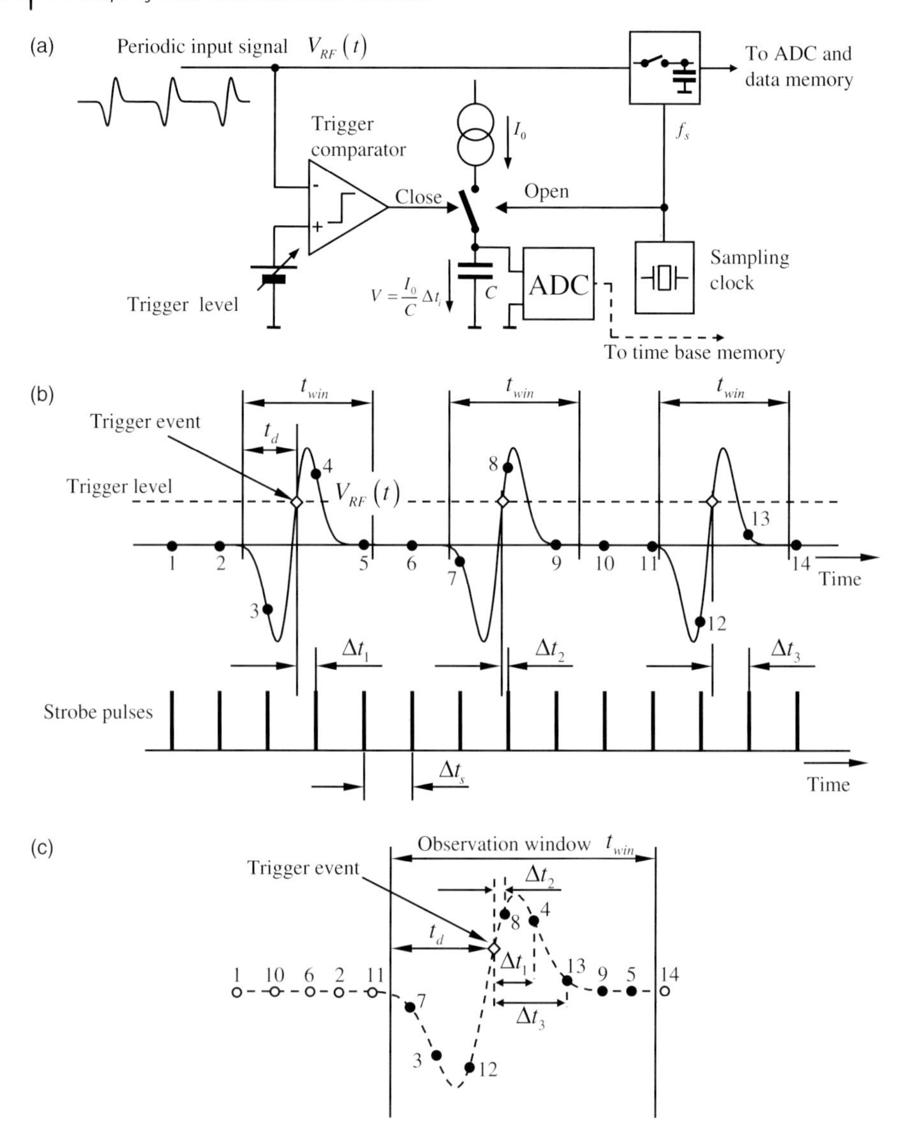

Figure 3.27 Random sampling. (a) Schematic of timing circuit for data capturing; (b) typical signals and (c) screenshot of reconstructed samples.

$f_s/f_0 \in \mathbb{Z}$ or $f_0/f_s \in \mathbb{Z}$) the approach will fail because the points are sampled always at the same positions and the displayed curve remains only sparsely occupied.

Indeed, we know the sampling interval now but it will be non-equidistant. This increases the data amount to be handled since additionally to the voltage value also the time position has to store for every sample. Furthermore, most of the data processing routines assume equidistant sampling so that extra difficulties or effort are pre-assigned for subsequent steps. Consequently, that timing approach is less suited for sensor applications which require economic solutions.

Sampling Efficiency The efficiency of a sampling receiver is a crucial point. It largely affects the sensitivity of the sensor devices which is usually quantified by the signal-to-noise ratio. Following to Section 2.6.1 namely (2.251), it is given by the ratio from signal energy accumulated by the receiver $\mathfrak{E}_{acc}$ to the noise power spectral density $\underset{..}{\Phi}_n$: $\mathrm{SNR} = \mathfrak{E}_{acc}/\underset{..}{\Phi}_n = \eta_{sr}\mathfrak{E}/\underset{..}{\Phi}_n$. Typically, the captured energy $\mathfrak{E}_{acc}$ is only a fraction η_{sr} of the actually provided energy $\mathfrak{E}$ of the signal. η_{sr} is a measure quantifying the efficiency of the sampling receiver. It is defined by the ratio

$$\eta_{sr} = \frac{\mathfrak{E}_{acc}}{\mathfrak{E}} \tag{3.26}$$

In order to estimate the efficiency value, we assume that a signal of the average power P is observed over the recording time T_R. Hence, the energy provided by the signal is $\mathfrak{E} = P\,T_R$. The average energy captured by one sample is $\mathfrak{E}_S = \eta_S P\tau_a \approx \eta_S P/\underset{..}{B}$ (η_S is the conversion efficiency of the sampling gate (see (3.11), (3.12)); τ_a is the aperture time and $\underset{..}{B} = 2\underset{.}{B}$ is the bandwidth of the sampling gate). The receiver takes $N_s = T_R/\Delta t_s$ samples during the recording time. Thus, it accumulates the energy $\mathfrak{E}_{acc} = N_s\,\mathfrak{E}_S$ resulting in an overall efficiency value of

$$\eta_{sr} = \eta_S \frac{\tau_a}{\Delta t_s} = \eta_S \tau_a f_s = \frac{\eta_S f_s}{\underset{..}{B}} = \frac{\eta_S f_s}{2\underset{.}{B}} \tag{3.27}$$

In the ideal case of real-time sampling, the analogue bandwidth of the receiver $\underset{..}{B}$ would be equal to the bandwidth $\underset{..}{B}_{RF} = \underset{..}{B}$ of the RF-signal and the sampling rate should be $f_s = \underset{..}{B}_{RF}$. As expected, this results in a maximum efficiency value $\eta_{sr} = \eta_S$ which only depends on the conversion losses of the sampling gate.

For sequential sampling, only one data point is taken from every period that is $f_s \approx t_P^{-1}$. The equivalent sampling rate has to meet $f_{eq} \geq \underset{..}{B}_{RF}$. Hence, we get from (3.27)

$$\eta_{sr} = \eta_S \frac{f_s}{f_{eq}} \tag{3.28}$$

if the bandwidth of signal and sampling gate are identical $\underset{..}{B}_{RF} = \underset{..}{B} = \tau_a^{-1}$. Since the actual sampling rate is typically orders less than the equivalent sampling rate, sampling receivers have an extremely bad efficiency as the following example shows.

A typical time domain reflectometer for laboratory purposes works with tunnel diode generator to stimulate the test objects. It has a repetition rate of about $f_0 \approx 200\,\mathrm{kHz}$ which equals the sampling rate $f_s \approx f_0$ due to the sequential sampling receiver. The overall bandwidth of the system is about $\underset{.}{B} \approx 20\,\mathrm{GHz}$, thus $\eta_{sr} \approx 5 \times 10^{-6}\eta_S$. Such devices use typically S&H-circuits for voltage capturing. If we further assume that $\tau_S \approx 10\tau_a$ (see (3.11)), that is the hold capacitor is charged to about 10% of the steady-state value, we end up in an overall efficiency of $\eta_{sr} =$

5×10^{-8} representing an appalling small value. As we can see from Section 2.6.6 (2.337), the receiver efficiency has immediate influence on the achievable dynamic range of a measurement and therefore onto the quality of the captured data.

This is all the more deplorable as the power emission of ultra-wideband sensors is strictly limited by law so it will not be possible to compensate the losses by larger stimulation signals. The improvement of this situation will be a key activity for future device developments. Nevertheless, if the bandwidth of the sensor device is in the foreground of interest, one has to accept the technical weaknesses. Notwithstanding that situation, even current ultra-wideband sensors show an amazing good sensitivity as some examples at the end of this book will show.

3.2.4
Voltage Capturing by 1 bit Conversion

As already mentioned, we can further reduce the requirements at the sensor electronics by reducing the word length of the voltage converter down to a single bit in the most extreme case. That is, the ADC degenerates to a simple comparator. In order to capture the required information from the test signal, one has to compensate the extreme low data quantity per sample by a sufficient large number of samples which however will strongly increase the recording time. This may be out of concern for application scenarios with very slow time variation. In the case of fast varying scenarios, the recording time is limited. The only chance to reduce it is to be satisfied with a lower SNR-value or to increase drastically the sampling rate f_S. The last point is technically feasible since comparators may work quite fast. The same is with algorithms designed for a short length of digital words if they are implemented in FPGAs or ASICs. We will discuss two approaches of that kind here.

In the case of sequentially and quite densely sampled data, the feedback approach (see Figure 3.18) can be simplified and implemented by a modified delta modulator. The principle which is also assigned as binary sampling is depicted in Figure 3.28 [36–38]. The goal of the circuit is that the integrator output $V_{out}(t)$ follows the RF-signal $V_{RF}(t)$ but at a lower rate. For that purpose, a fast comparator refers the RF-input $V_{RF}(t)$ to the integrator output $V_{out}(t)$. If $V_{out}(t) > V_{RF}(t)$, the comparator generates a negative voltage $-V_1$ and V_{out} goes linearly down due to the integration. In the opposite case $(V_{out}(t) < V_{RF}(t))$, V_{out} rises linearly. Hence, $V_{out}(t)$ oscillates around $V_{RF}(t)$ as long as the slew rate of the RF-signal is smaller than the integrator slope. This is what a conventional delta modulator is doing.

The trick here is to activate the comparator only ones a period t_P of the RF-signal $V_{RF}(t)$ but successively shifted by the small time interval Δt_{eq} as already discussed above in connection with sequential sampling. By that way, one gains a low-frequency replica of the RF-voltage. Thus after some additional analogue filtering, the voltage $V_{out}(t)$ may be captured by a low-speed ADC. But actually it is not needed since the bit stream $V_1[n]$ provided by the comparator already carries all

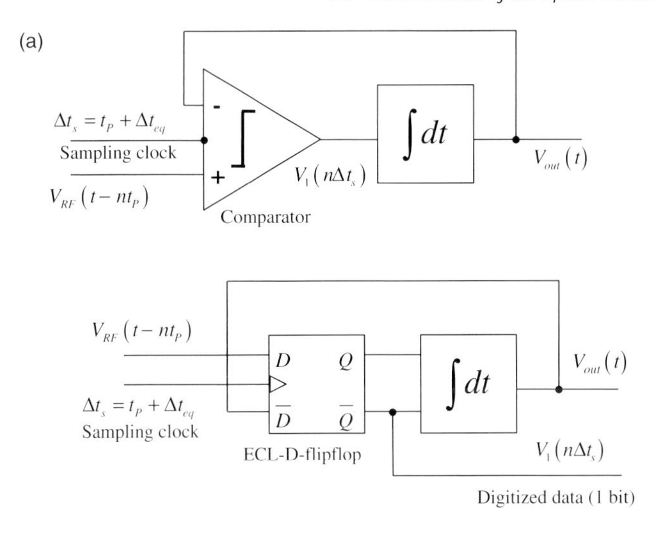

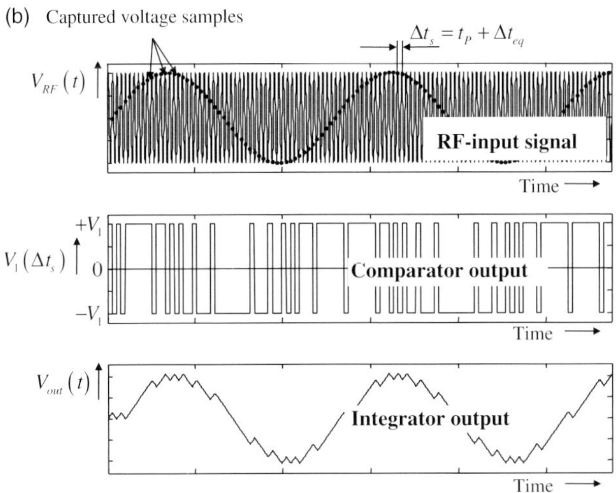

Figure 3.28 Modified delta-modulator applying equivalent time oversampling. (a) Basic schematic and practical implementation by D-FF and (b) typical signals.

information about the captured signal. Thus, the down-converted signal can be numerically reconstructed by low-pass filtering and down-sampling as it is usual in $\Delta\Sigma$ converters [39]. The strength of the remaining ripples decrease by narrowing adjacent time samples (reducing Δt_{eq}). A practically implemented circuit [37] uses a differential D-flip-flop as trigger able comparator. The timing control of the circuit is based on the principle depicted in Figure 3.25.

The second approach deals with a controllable threshold voltage which is swept over the amplitude range of the RF-signal. [40, 41] discuss two methods, from which we will only consider one since they are quite similar. Figure 3.29 depicts the basic idea. A comparator compares the RF-signal $V_{RF}(t)$ with the threshold

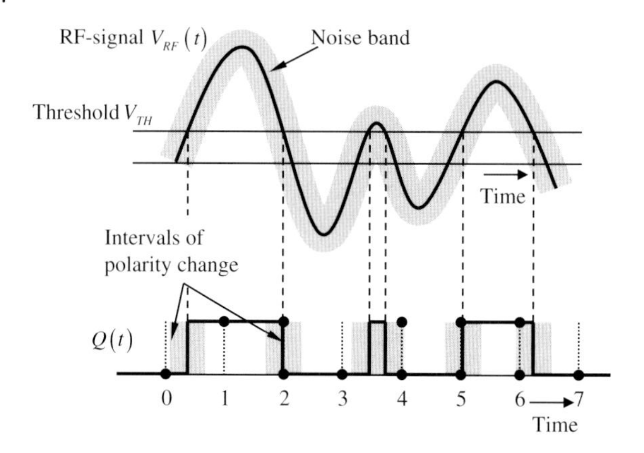

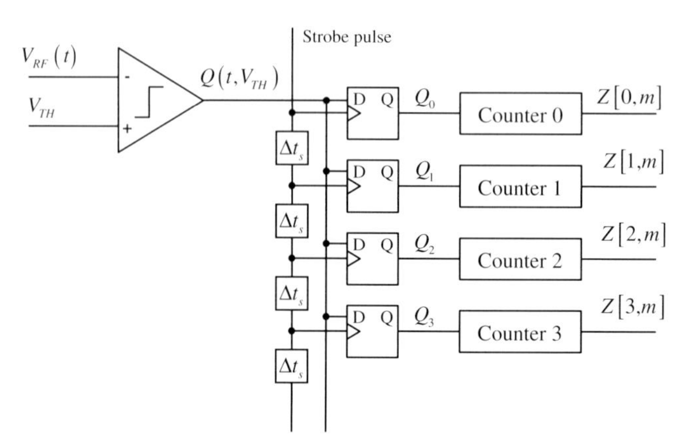

Figure 3.29 Serial voltage conversion.

voltage V_{TH}. Its output signal provides a binary sequence for which holds

$$Q(t,\ V_{\mathrm{TH}}) = \begin{cases} 1 & V_{\mathrm{RF}}(t) > V_{\mathrm{TH}} \\ 0 & V_{\mathrm{RF}}(t) < V_{\mathrm{TH}} \end{cases} = \frac{1}{2}(\mathrm{sgn}(V_{\mathrm{RF}}(t) - V_{\mathrm{TH}}) + 1) \tag{3.29}$$

An example of such signal is given in the figure. Since the RF-signal is periodically repeated, the signal $Q(t)$ is eventually sampled by any appropriate sub-sampling procedure. In the present case, one applies a number of sampling units which capture in parallel[5] the $Q(t)$ at different time positions. This is done by D-flip-flops which take one sample at a period t_{P} of the receive signal. All flip-flops are activated by one shot but temporally staggered.

While keeping constant the threshold voltage V_{TH}, the measurement is repeated N times. Since the RF-voltage is usually corrupted by noise, the comparator

5) The parallel capturing is not obligatory. It only reduces measurement time compared to a serial query.

switches randomly. The time interval within this may be happen is marked by the grey regions around the signal edges of $Q(t)$. For every measurement, a counter i registers if its D-flip-flop was $Q_i = 0 \quad (V_{RF}(i\,\Delta t_s) < V_{TH})$ or $Q_i = 1 \quad (V_{RF}(i\,\Delta t_s) > V_{TH})$. If the counter i indicates $Z[i] = 0$ events after closing the N measurements, the RF-voltage was always lower than the threshold. In our examples, this applies for the samples 0, 3 and 7. In the case, it counts $Z[i] = N$ events, the RF-voltage was always beyond the threshold (sample 1) and if the counter provides any number in between $Z[i] \in (0, N)$ the RF-voltage is in the range of the threshold (samples 2, 4, 5, 6). These samples are of major interest since we can determine the corresponding RF-voltage from the counted number as shown in what follows.

We model the periodic RF-signal by

$$V_{RF}(t) = V_{RF,0}(t + pt_P) + n(t); \quad n(t) \sim N\left(0, \sigma_n^2\right);\ p \in [0, N-1] \tag{3.30}$$

where $n(t)$ assigns the additive noise and p counts the measurement repetitions. We will repeat the measurements N times for each threshold level. The threshold voltage can be changed by steps of size q

$$V_{TH} = mq; \quad m \in [-M, M] \tag{3.31}$$

leading to the full-scale range $\mathrm{FSR} = 2Mq$ of the converter and an equivalent bit number of $H[\mathrm{bit}] = \mathrm{lb}(2M+1) \approx 1 + \mathrm{lb}M$ if $2M+1$ is the total number of threshold levels.

Supposing the counter at sample position i has counted $Z[i, m]$ events if the threshold voltage $V_{TH,m} = mq$ was applied. This number can be expressed as follows:

$$\begin{aligned} Z[i,m] &= \sum_{p=0}^{N-1} Q(i\,\Delta t_s + pt_P, mq) \\ &= \sum_{p=0}^{N-1} Q_i[p,m] = N\overline{Q_i[m]} \\ &= \frac{N}{2}\left(\overline{\mathrm{sgn}\left(V_{RF}(i\,\Delta t_s) - V_{TH,m}\right)} + 1\right) \end{aligned} \tag{3.32}$$

Assigning the expected value of the mean of the Signum function by $E\left\{\overline{\mathrm{sgn}\left(V_{RF}(i\,\Delta t_s) - V_{TH,m}\right)}\right\} = \hat{z}[i,m]$ and using (2.273) Section 2.6.1, we yields

$$\hat{z}[i,m] = \mathrm{E}\left\{\overline{\mathrm{sgn}\left(V_{RF}(i\,\Delta t_s) - V_{TH,m}\right)}\right\} = 2\frac{Z[i,m]}{N} - 1 = \mathrm{erf}\left(\frac{V_{RF,0}(i\,\Delta t_s) - V_{TH,m}}{\sqrt{2}\sigma_n}\right) \tag{3.33}$$

and for its variance, we get (see (2.255))

$$\mathrm{var}\left\{\overline{\mathrm{sgn}\left(V_{RF}(i\,\Delta t_s) - V_{TH,m}\right)}\right\} = \frac{1 - \hat{z}^2[i,m]}{N} \tag{3.34}$$

If we have available N_s counters, a time signal of N_s samples can be gathered at one shot. The whole procedure requires $N(2M+1)$ repetition of the measurements leading to the recording time

$$T_R = N(2M+1)t_p \tag{3.35}$$

and a data matrix $Z[i,m]$ of dimension $[N_s, 2M+1]$ from which we still have to extract the wanted data vector $V_{RF}[i]$. For that purpose, we select in each case only the smallest entry per raw of the data matrix $\hat{z}[i,m] = 2Z[i,m]/N - 1$, that is

$$m_i = \arg\min_m |\hat{z}[i,m]| = \arg\min_m \left| Z[i,m] - \frac{N}{2} \right| \tag{3.36}$$

This implies that the threshold voltage $V_{TH} = m_i q$ approaches best the RF-voltage $V_{RF}[i\,\Delta t_s]$ so that the noise toggles the D-FF with nearly the same probability between both states. We can now approximate the error-function in (3.33) by a linear relation (see Section 2.6.1.6, (2.275)) since all quantities $\hat{z}[i,m_i]$ are close to zero. So that we finally yield

$$V_{RF}[i] = \sqrt{\frac{\pi}{2}}\sigma_n\,\hat{z}[i,m_i] + m_i q = \sqrt{2\pi}\sigma_n\left(\frac{Z[i,m_i]}{N} - 1\right) + m_i q \tag{3.37}$$

As we have shown in Figure 2.84, we should at least respect $2\sigma_n \geq q$ in order to get a linear transfer characteristic of the analogue-to-digital conversion. From (2.275) and (3.34), we can estimate the variance of the captured voltage

$$\sigma^2_{V_{RF}} = \frac{\pi}{2N}\sigma_n^2 \tag{3.38}$$

and also the maximum achievable dynamic range (see Section 2.6.6) supposing a perfect circuit implementation.

$$D_{max} = \frac{V^2_{RF,max}}{\sigma^2_{V_{RF}}} = \frac{\left(\frac{1}{2}FSR\right)^2}{\sigma^2_{V_{RF}}} = \frac{2Nq^2M^2}{\pi\sigma_n^2} = \frac{2}{\pi}M^2N = \frac{4MT_R}{4\pi t_P} \quad \text{if } q = 2\sigma_n \tag{3.39}$$

As expected, the number M of quantization intervals and the recording time T_R will determine the device dynamic. If the noise level is larger than assumed in (3.39), the achievable dynamic will be reduced if the other conditions are remain fixed.

3.2.5
Peculiarities of Sensors with Pulse Excitation

Most of currently applied ultra-wideband sensors stimulate their test objects with pulses. This measurement approach is widely used and well developed compared to other ones. The principle – namely the reaction of the test objects – is intuitively comprehensible due to the simple stimulation signal. The advantage of this stimulus signal is that it allows following the propagation and the temporal development of energy distribution within the test scenario which makes it often easier to understand the functioning of things.

Nevertheless it is affected by some weaknesses as the need of stimulus signals of comparatively high peak power, the susceptibility to drift and jitter and the usually low sampling rate so far sequential timing principles are applied. Typical peak voltages of stimulus pulses range from less than 1 V for to more than 100 V (e.g. in several GPR devices) and their repetition rate may be usually found between 100 kHz and 10 MHz.

A deeper discussion of the performance of various wideband radar sensor concepts mainly with respect to specific aspects of range resolution will be given in Sections 4.7.2 and 4.7.3. Compared to the ultra-wideband principles which we will introduce in what follow, the peak amplitudes of the sounding signal for the pulse principle may be quite high. Therefore, one should consider the maximum field exposition of the test objects if small sensor electrode geometries are applied as, for example for impedance spectroscopy or extreme near field measurements.

But pulse-based UWB sensor electronics has also some inevitable advantages in favour of the other methods. They may be implemented by simple circuits and components based on PCB wiring technique if the technical requirements are not too demanding. Hence, the manufacturing will be cost-effective even for low number of pieces. Device costs for large-scale applications are, however, better manageable by monolithic circuit integration. It is possible too, even in low-cost technologies as SiGe or CMOS. However, one has to abstain from large amplitudes of the stimulus pulse due to the limited breakthrough voltage. Usually this does not provide serious problems for short-range applications. A complete design example of a monolithic integrated pulse-based low-power sensor is exercised in Ref. [32].

The spectral band occupied by the pulse sensor depends on the shape of the stimulus pulse. Quite often, one trims the pulse generator design to approximate a Gaussian shape in order to join short pulse duration with minimum bandwidth (refer to Section 2.2.5, (2.82)). A band-pass spectrum, as required by the UWB radiation rules, is achieved by an n-fold differentiation of a Gauss-pulse. A second option to place the stimulus energy into any frequency band is the classical principle of pulse-Doppler radar. Here, a carrier signal of the wanted frequency is modulated by the short UWB pulse and the receive signals is typically shifted back into the baseband by IQ-down-conversion. The principle is depicted in Figures 3.54 and 3.58 for an M-sequence sensor which can simply be transferred in a pulse modulated system by replacing the shift register with a pulse shaper and the binary divider with an appropriate sampling time control (e.g. Figure 3.20, 3.25 or 3.26). Ref. [42] exemplifies an experimental UWB system of the mentioned structure.

Finally, we will still point to two options which are not feasible with CW-UWB principles. These are time gating and time variable gain.

Time gating: Many ultra-wideband sensors are used for (short range) radar applications. These devices generally contain at least two antennas,[6] one for

6) Basically, one can also operate mono-static radar which uses the same antenna for transmit and receive. But then, the antenna feed point return tends to overload the receiver frontend. Since the return loss of a wideband antenna is usually smaller than the cross-coupling between two antennas, one often separates receive and transmit channel in UWB radars.

transmitting and one for receiving. Since these antennas are often mounted close together, there is a considerable crosstalk signal which tends to saturate the receiver. Such signal is usually out of interest because it does not carry any information about the target. Therefore, it will be gated out so it does not stress the receiver. For that purpose, the strobe pulses do not open the sampling gate while the crosstalk signal affects the receiver antenna, that is the observation time widow is started after the crosstalk is settled down. The selection of the gating window length must be done carefully since it leads to a blind range. In the case, the target is close to the antenna, gating cannot be applied.

Time variable gain: Like gating, the time variable gain control of the receiver is useful in radar applications too. The strength of a target echo reduces with increasing distance to the antennas and hence with increasing round trip time (see Section 4.5.2, (4.60) for details). In order to keep the amplitude of the target signal roughly independent of the target distance, the receiver gain is increased with advancing sampling time. Figure 3.30 shows the circuit schematics and

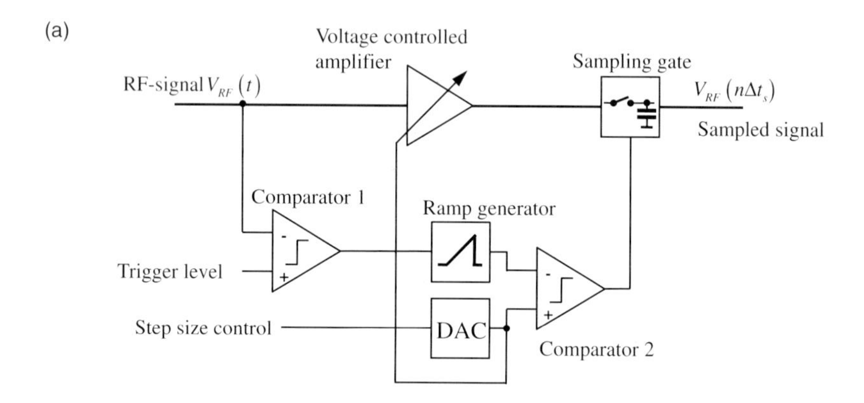

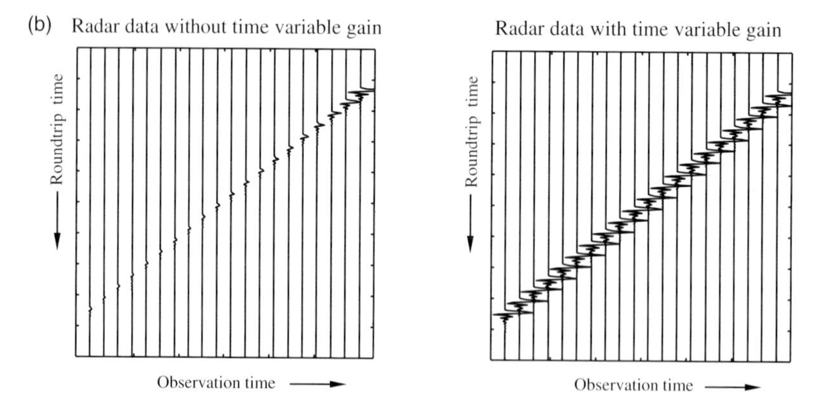

Figure 3.30 Modification of sequential-sampling unit (see Figure 3.20) for time variable gain (a) and its effect onto the captured data (b).

demonstrates the effect by a simple example. A target moves with constant speed towards the antennas. A number of measurements were released and recorded side by side as depicted in the bottom of Figure 3.30. On the left-hand side, the measurements were done with constant receiver sensitivity. The round-trip time will be shorter and shorter from shot to shot while the amplitude of the return signal increases. On the right-hand side, signal parts with large round-trip time are subject a strong amplification while signals of a short delay are not amplified so that the target leaves finally a trace of constant amplitude if the gain control is properly adjusted.

It should be noted that time variable gain may be implemented also after analogue-to-digital conversion. But this will not have any effect onto the dynamic range of the measurement.

3.3 Determination of the System Behaviour by Excitation with Pseudo-Noise Codes

We have seen from the previous chapter that the need for (comparatively) large peak power signals represents some inconvenience of the impulse measurement technique. Therefore, it is quite obvious to go away from excitations by strong single or infrequent 'shocks' and to pass instead to a stimulation of the DUT by many subtle 'pinpricks'. Binary pseudo-noise codes represent pulse signals which are more careful with DUT. These signals have spread their energy homogenously over the whole signal length. Hence, their peak voltage remains quite small by keeping sufficient stimulation power. This will spare the measurement objects and the requirements onto the receiver dynamic are more relaxed. First applications of such signals are reported in Ref. [43] for acoustic measurements of impulse response functions.

The price to pay for that voltage reduction is the chaotic structure of the receive signal which cannot be further interpreted immediately by a human being as it was possible for the system response after impulse excitation. However, the insertion of a correlation into the signal processing chain will retrieve the interpretability of the system response. The technical challenge is how to perform the correlation of a very wideband signal.

We will first summarize some aspects of UWB-PN-code generation. Then, an analogue implementation of the wideband correlation will be introduced. Finally, a very flexible approach of a digital ultra-wideband correlation will be discussed which represents a unique solution for the wideband impulse compression up to now. Several extensions of a basic principle will be introduced and the potential of further improvements will be considered.

3.3.1 Generation of Very Wideband PN-Codes

PN-codes are periodic signals which constitute of a large number of individual pulses ostensible randomly distributed within the period. These pulse sequences join the properties of two signal classes – the periodic-deterministic signals and random signals. This makes them interesting for ultra-wideband applications. There is a great deal of different pseudo-noise codes: Barker Code, M-sequence, Gold-Code, Kasami-code; Golay-code etc. [44, 45]). The amplitude of PN-signals may be bounded to two levels (0 and 1 or −1 and 1) – summarized by the term binary PN-code – or they may have three (−1, 0, 1) and more signal levels. Without loss in generality concerning the electronic principles, we will concentrate here on M-sequences since they have the best properties to measure the impulse response function of an object and they are simple to generate. Their basic properties are summarized in Section 2.3.5. Their application for UWB measurements was first discussed in Ref. [46].

Comparable properties are achieved by complementary Golay-sequences – see the comparison in Annex B.9 – even if the electronic implementation will be a bit more elaborate compared to M-sequence devices if a large bandwidth is required. As in case of M-sequences, they were first applied for acoustic measurements [47]. An early UWB implementation is shown in Ref. [48].[7] Here, we will not further deal with Golay-sequences. Based on the M-sequence concepts discussed below, it will be straightforward to implement corresponding sensor electronics for Golay-sequences.

The most simple and stable way to generate M-sequences is given by digital shift registers which includes appropriate feedbacks. Figure 3.31 depicts the basic structure of such shift registers. The unit delays are provided by flip-flops which are synchronized to a clock generator working at the clock rate $f_c = \Delta t_c^{-1}$. The weighting factors a_n are either 0 or 1. They have to be selected carefully in order to gain actually an M-sequence. The theory of PN-code generation is closely connected with Galois-fields and primitive polynomials. Here, we will only refer to the outcome of corresponding investigations. The reader who is more interested in that topic can refer to Ref. [44]. Annex B.9 gives an overview of the required feedback structure to generate M-sequences of different order. With growing shift register length, the number of allowed feedback connections and hence the number of different M-sequences of the same length increases. The time shapes of these sequences are different but they are not uncorrelated [49]. Hence, a set of M-sequences is not straight suited for multi-channel stimulations due to the mutual interference. Such interference can be minimized by switching to other codes (e.g. Gold-codes) or to introduce some simple modifications (see Section 3.3.6.2 for more details). The use of other codes will, however, affect the quality of the measured impulse response function since their auto-correlation functions are not as perfect as for M-sequences or complementary Golay-sequences.

The number N of flip-flops in the shift register will determine the length of the M-sequence. Within one period, the shift register runs through all $2^N - 1$ possible combinations of the flip-flop except the case in which all flip-flops are set to zero $q_i[n] = 0; i \in [1, N]$. This would cause a signal interruption and must be avoided

7) The comparison between M- and Golay-sequences in this reference should be considered with carefulness.

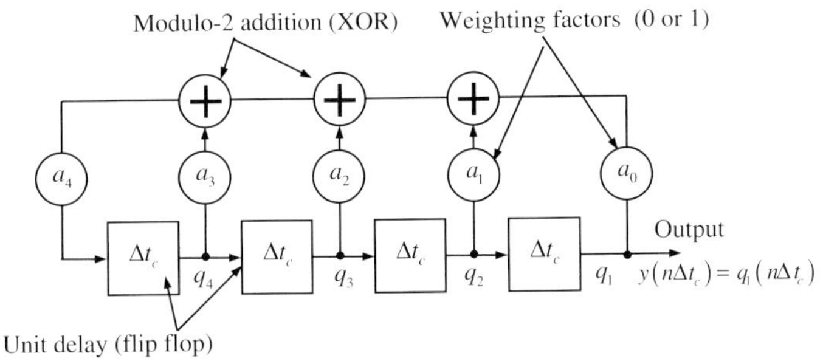

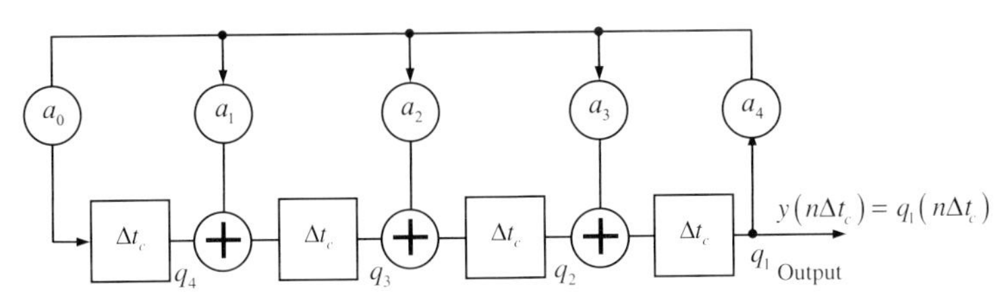

Galois feedback generator

Figure 3.31 Digital shift register with the two basic feedback structures. Note the inverse order of the weighting factor numbering for both canonical forms. The indicated signals $q_i(n\,\Delta t_c)$ represent the states of the corresponding flip-flops.

during operation but can be used to disable the shift register. By pre-selecting the flip-flop states $q_i[n = 0]$ to a certain pattern before the shift register is enabled, one can arbitrarily shift the starting point of the M-sequence over its whole period. We will use this below to control the time lag of an UWB correlator.

The maximum toggle rate of the shift register and hence the maximum bandwidth of the M-sequence is limited by the internal delay of the flip-flops and the XOR-gates. Therefore, it is recommended to implement the shift register by monolithic integration since it will reduce the physical length of the propagation path. Except for the cases which have only one XOR-gate in the loop, the Galois structure is generally to favour. The Fibonacci structure deals with cascaded XOR-gates which increase the dead time before a new change of the flip-flop states is allowed. Some further tricks to increase the bandwidth of M-sequence circuits are found in Ref. [44].

Figure 3.32 show an example of an M-sequence generator. Its output voltage is about 250 mV which is sufficient for many applications due to the large time-bandwidth product of the waveform, i.e. the compression gain of the waveform. The shift register length is 9. The feedback structure is of Fibonacci type. Here, this does not cause speed losses compared to the Galois structure due to the single XOR-gate. The additional XOR-gate connected with the output line is not required for basic sensor applications. But it opens up some interesting features for extended system conceptions (see Section 3.3.6). The maximum toggle rate of such

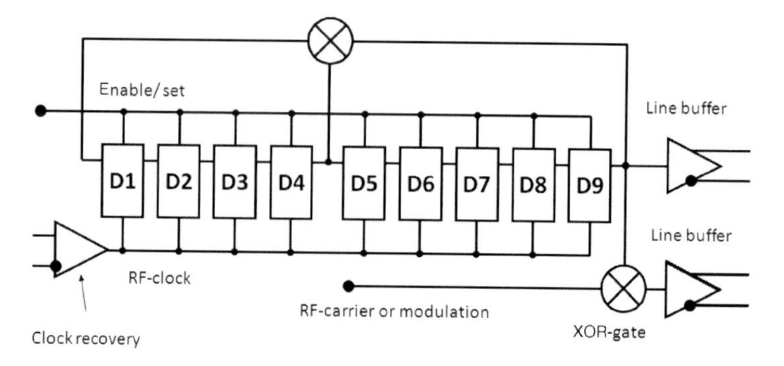

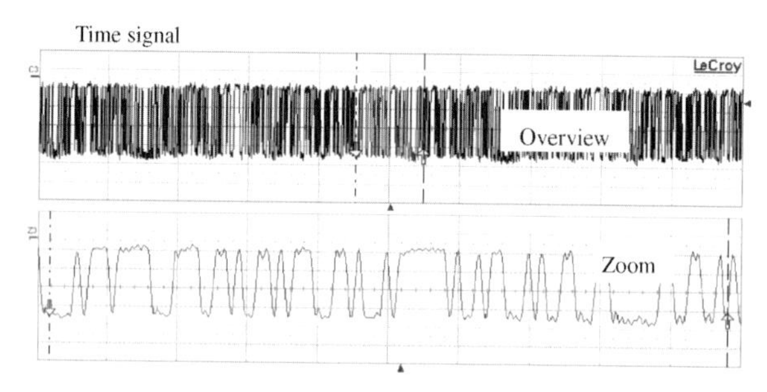

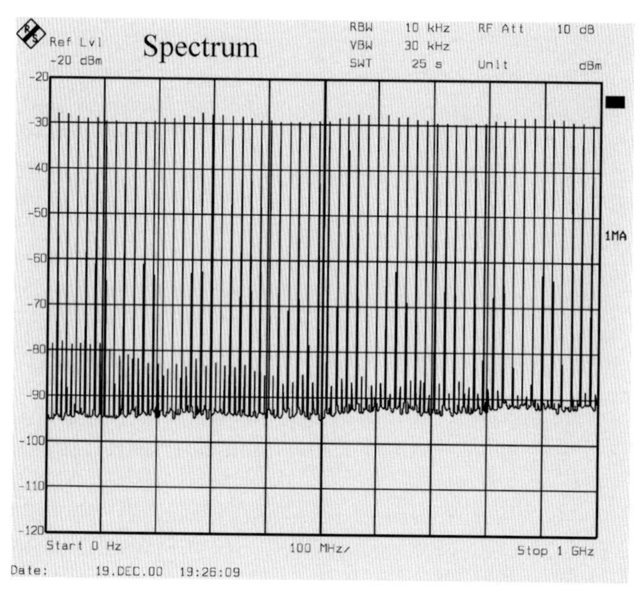

Figure 3.32 Schematic and measured signal (above: time signal (overview and zoomed version), below: lower part of the spectrum) of an ultra-wideband shift register of ninth order. The shift register is disabled by setting all flip-flops to 0. It is enabled by setting the initial state of all flip-flops to 1. Additionally, the shift register is equipped with an XOR-gate for modulation purposes.

a shift register exceeds 20 GHz, which permits the construction of ultra-wideband sensors of up to 10 GHz bandwidth.

3.3.2
IRF Measurement by Wideband Correlation

The measurement principles which we will introduce below are usually not bounded to the exclusive use of M-sequences. Basically, one can also select other pseudo-random waveforms if the application does it require. But for shortness, we will restrain on simple M-sequences for illustration of the working principles.

Recalling us the primal purpose of our measurement – it was to determine the impulse response function of a test object. As we know, this is not directly possible by a time extended signal as the M-sequence it is. Rather the captured measurement values have to be subject an impulse compression in order to gain the wanted impulse response. In the case of ultra-wideband sensing, the impulse compression has to be performed by a wideband correlation as depicted in Figure 3.33.

The measurement approach follows from the system description by convolution and the definition of correlation. Convolving (2.129) with the time inverted stimulus $x(-t)$ we get

$$y(t) * x(-t) = g(t) * x(t) * x(-t)$$

Using (2.134), it can be rewritten by correlation functions as

$$C_{yx}(t) = g(t) * C_{xx}(t) \tag{3.40}$$

and hence we can simplify (for a detailed consideration of (3.40) and (3.41) see Annex B.9)

$$C_{yx}(t) \propto g(t) \quad \text{for } C_{xx}(t) \approx \delta(t) \tag{3.41}$$

That is, the wanted IRF $g(t)$ is proportional to the cross-correlation function $C_{yx}(t)$ between DUT response and stimulus as long as the auto-correlation function of the stimulus is short enough. As we know from Section 2.2.5 the 'peakedness' of $C_{xx}(t)$ does not depends on the actual time shape of the waveform $x(t)$ but rather from the width of its power spectrum. That is, the only what we need is a stimulus signal of sufficient bandwidth independently on how this signal looks like. By that approach we gain a high flexibility in the selection of stimulus signals. But compared with the impulse approach, we have to pay for that flexibility with higher

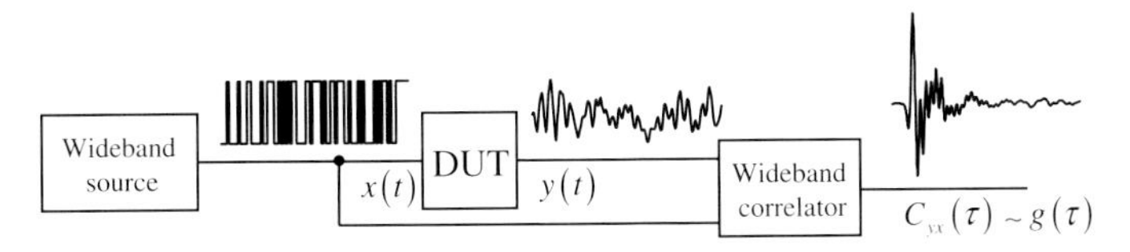

Figure 3.33 Impulse response measurement by correlation.

complexity of the sensor electronics since the cross-correlation function has to be determined in order to have access to the wanted IRF.

The central questions we have to answer in what follows are

- How to implement technically a wideband correlation?
- What its technical expense?
- What are the technical limitations?

For that purpose, we will considerer several solutions which are matched to specific signals. Here, we will start with an analogue version of the wideband correlation for pseudo-random codes.

3.3.3 The Sliding Correlator

Assuming we apply M-sequences for system stimulation, an electronic correlator has to execute following equation (see (2.49)):

$$C_{ym}(\tau) = \frac{1}{t_1} \int_{t_1} m(t+\tau) \cdot y(t) \mathrm{d}t$$

Figure 3.34 depicts a simple arrangement for that purpose. It is called sliding correlator and uses two shift registers of identical behaviour. One provides the actual stimulus signal and the second serves as reference to perform the correlation. Signal multiplication and integration is performed by a product detector. Annex B.4 shortly summarizes its basic properties. The integration may be done by an RC-low-pass as indicated in the figure. In the initial approach (Figure 3.34 (a)), the mutual time shift between the measurement $y(t)$ and the reference signal $m(t)$ is provoked by a slightly different clock rate for both shift registers. The frequency difference Δf must be small enough so that de-correlation between both signals will be avoided. The maximum admissible value of Δf depends on the TB product of the test signal. This timing approach has some similarities to the sampling control from Figure 3.25.

In a modified version of the correlator, the starting time of the reference waveform is successively swept over the whole signal period by acting on the initial states of one of the two shift registers (Figure 3.34b). In this case, the correlator does not really 'slide' through the required interval of the time lag τ rather it jumps in steps of $\Delta t_c = f_c^{-1}$ so we will get a sampled version of the correlation function.

As we know from Section 2.3.5, nearly 80% of the energy of an M-sequence is bounded to the spectral band from DC to half the clock rate $f_c/2$ of the shift register. Hence, it is recommended to select the clock rate f_c for the shift register a bit larger than $2B_{DUT}$. This selection has two consequences. First, the whole spectral band in which the DUT is operating is covered by nearly constant spectral power of the stimulus. This limits the influence of measurement noise since spectral parts in which the noise dominates the stimulus are excluded. Second, the restriction to the frequency band $[0, f_c/2)$ leads to a minimum sampling rate of f_c in order to

(a)

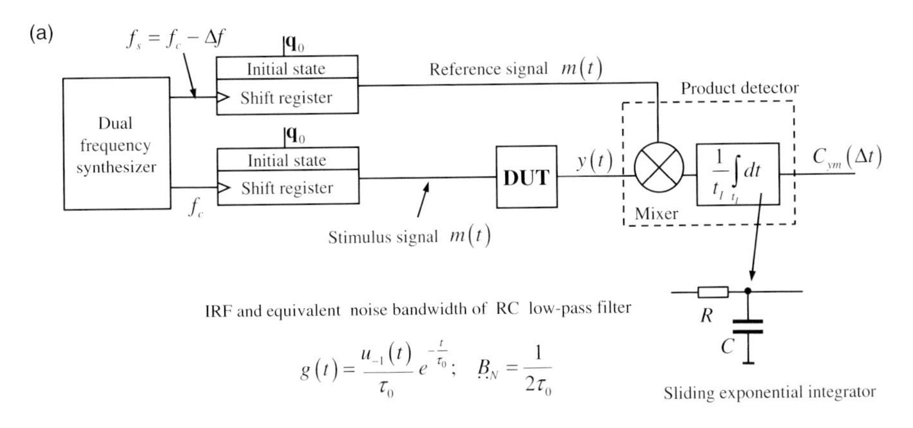

(b)

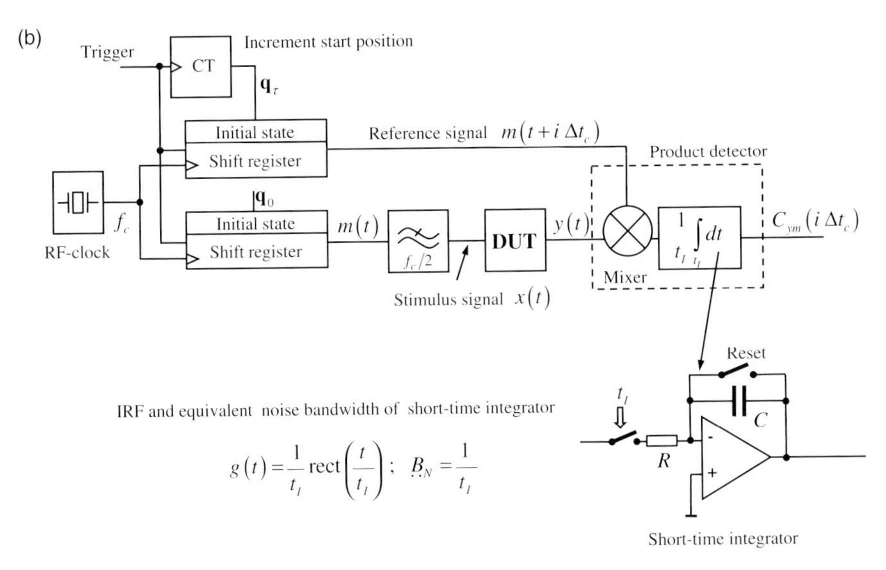

Figure 3.34 Sliding correlation using slightly different clock rate (a) and variation of the initial shift register states (b). The output voltage of the product detector is typically digitized and further processed.

meet the Nyquist theorem. That is, it is sufficient to calculate the correlation function with a sampling interval $\Delta t_s = \Delta t_c = 1/f_c$ which equals the chip duration of the M-sequence.

By that assumption, the correlation can be implemented in a likewise simple way as mentioned above. We use again two identical shift registers of the order n that are pushed now by the same RF-clock f_c. One shift register provides the stimulus signal. The initial states $\mathbf{q}_0$ of this register are kept the same over the whole measurement. Its output signal should be band limited to frequencies below $f_c/2$ in order to ensure the Nyquist theorem. The second register generates the reference signal of identical shape. But via the control of the initial states $\mathbf{q}_\tau$, the starting point of the M-sequence is successively incremented leading to the wanted variation of the time lag $\tau = i\,\Delta t_c$ between reference and measurement signal. Since the shift register has $2^n - 1$ applicable states (note that the state vector $\mathbf{q} = [0 \quad 0 \quad 0 \quad \cdots]^T$ is not allowed), we can collect by that way $N_s = N = 2^n - 1$ samples of the correlation function $C_{ym}(i\,\Delta t_c); i \in [0, N_s - 1]$, that is one sample per chip of the M-sequence (N – number of chips per period; N_s – number of captured samples). The time needed to capture a single data sample depends on the integration time t_I of the product detector. Both shift registers have to run in parallel with constant time lag $\tau = i\,\Delta t_c$ during this time. The integrator of the product detector is crucial for the noise suppression of the correlator. It largely determines the overall sensitivity of the sensor device. The measurement procedure is running stepwise so that we can apply here a short-time integrator (see Figure 3.34b) which has the best performance for a given integration time. Its integration time t_I should be an integer multiple p of the M-sequence period $t_I = p t_P = (2^n - 1) p\,\Delta t_c = N p\,\Delta t_c$ to avoid truncation errors. If the integration unit is a simple RC-network (sliding exponential averaging), its time constant $\tau_0 = RC$ must be $\tau_0 = t_I/2$ in order to provide the same performance of noise suppression as a short-time integrator. This leads to a settling time of the RC-low-pass of about $t_{sett} \approx 5\tau_0 = 2.5 t_I$ which one has to wait before the measurement value is valid, that is the recording time will be extended by a factor 2.5 without profit of additional noise reduction.

Identical noise suppression supposed, the recording time of a correlator based on short-time integration and RC-low-pass filtering can be estimated as

$$T_R = \begin{cases} N_s t_I & \text{for short-time integrator} \\ 2.5 N_s t_I & \text{for RC-low-pass filter} \end{cases} \tag{3.42}$$

The theoretical minimum of the recording time is $T_R = t_I$. It is provided from correlators which are built from N_s parallel working branches dealing each with a different time lag. Such a structure would exploit perfectly the provided signal energy to determine the correlation function. Its efficiency would only be degraded by the efficiency η_m of the mixing circuits. Since for practical reasons usually only one product detector is involved, the measurement has to be serialized which brings down the efficiency η_{cr} of the correlation receiver to

$$\begin{aligned} \eta_{cr} &= \frac{\mathfrak{E}_{acc}}{\mathfrak{E}} = \frac{\eta_m}{\gamma N_s} \\ \gamma &= 1 - \text{short-time integrator} \\ \gamma &= 2.5 - \text{RC-low-pass} \end{aligned} \tag{3.43}$$

This efficiency value is comparable with the efficiency of sequential-sampling if the equivalent sampling rate is selected close to the Nyquist rate $f_{eq} \approx \underset{\cdot\cdot}{B}_{RF}$ (see (3.27)). Nevertheless, opposed to pulse signals with CF $\gg 1$, we will gain a better dynamic range now since the crest factor here is close to unity CF ≈ 1 (refer to (2.337) or (2.339) in Section 2.6.6). The improvement factor compared to pulse excitation will be as better as larger the time-bandwidth product of the captured signal is.

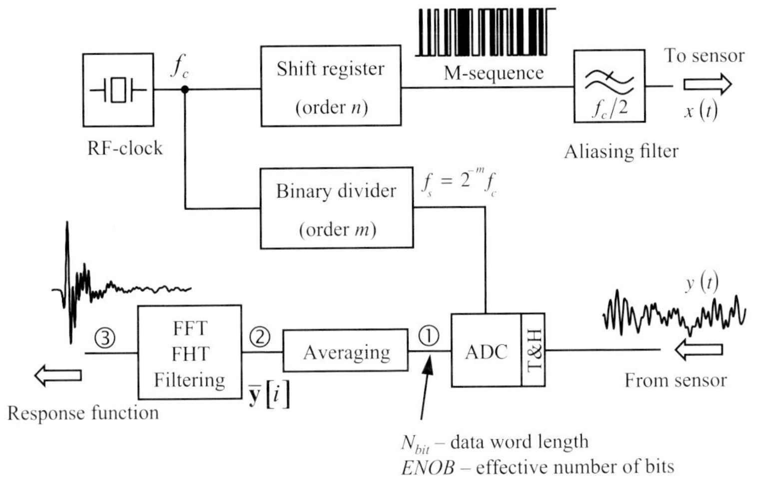

Figure 3.35 Basic circuit diagram of the digital ultra-wideband correlator.[8] (The placement of the anti-aliasing filter depends on the measurement environment. Typically, it is placed at the in- and output of the sensor device).

3.3.4 Basic Concept of Digital Ultra-Wideband PN-Correlation

Digital ultra-wideband correlation joins sliding correlator and stroboscopic sampling. The principle works for many different PN-codes. Without restriction in generality, we exclusive refer to M-sequences. Figure 3.35 depicts the principle, which was introduced first time in Ref. [46].

As in the case of the sliding correlator the device under test is stimulated by the M-sequence which is provided by the shift register. But now, the receive data are immediately sampled and converted into the digital domain. This needs a huge sampling rate in order to meet the Nyquist theorem. Due to the periodic nature of the M-sequence, sub-sampling can be applied fortunately. The data gathering is distributed over several periods by capturing the signal with the lower sampling rate f_S. This simplifies the ADCs as well as the data handling and it will reduce the power consumption of the circuits.

The question is how to organize a precise and at the same time simple sub-sampling control. The solution is already given by the principle depicted in (3.25) and Figure 3.26 whereas the particularity of the length of the M-sequence promotes extremely simple technical solutions.

8) Note that the schematic is not restricted to the solely use of M-sequences. In case of arbitrary sequences, the shift register must be replaced by a memory from which the code is cyclically readout. The sampling control remains the same if the code length is $N = 2^n \pm 1$. If the code length is $N = 2^n$, the divider which controls interleaved sampling must provide a dividing factor of $2^m \pm 1$.

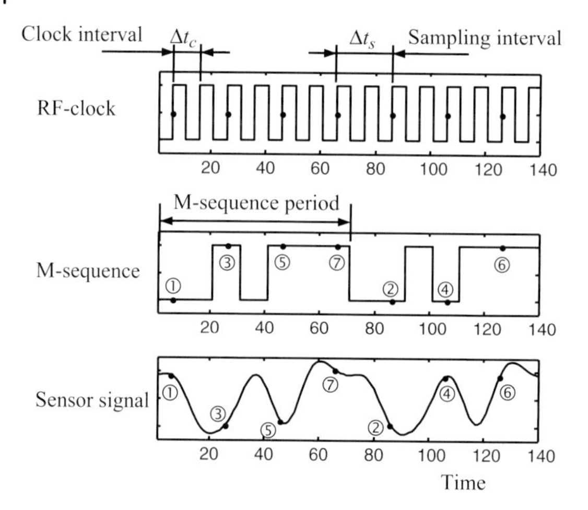

Figure 3.36 Sub-sampling control by binary divider.

As we have seen in the previous Section 3.3.3, one data sample for every chip is sufficient to capture. Since the number of chips of an M-sequence is always one less than a power of two ($N_s = N = 2^n - 1$), sub-sampling can be simply controlled via a binary divider. Figure 3.36 demonstrates the principle. It is an interleaved sampling approach since more than one sample per period is captured. For introduction of the principle, we will suppose an M-sequence of order 3. It consists of seven chips, that is a complete measurement will cover seven data samples. The T&H-circuit and the ADC shall be controlled via a one stage binary divider.

With every beat of the clock generator one new chip of the M-sequence is generated but only with every second beat a voltage value is captured. Hence, only the data samples relating to an odd chip number are captured at the beginning. But then, after passing the first period, the data samples with even chip numbers are collected. Thus, after two periods all data are recorded to meet the Nyquist theorem. The order of the data samples is scrambled. But this can be simply removed. If a binary divider of two stages is used one needs four periods for a whole measurement and so on. This principle works for all orders of M-sequences and all orders of binary dividers. The equivalent sampling rate remains always equal to the RF-clock: $f_{\mathrm{eq}} = f_{\mathrm{c}}$.

What are the advantages of such method of sampling control?

- The whole system is controlled by a single monotone RF-clock generator. Typical frequencies are in the range of 2–20 GHz depending on the intended application. These generators can be designed for very pure and stable signals (phase noise). The RF-clock generator provides the 'time axis' of the sensor device.
- Shift register and binary divider are triggered synchronously by the RF-clock. Thus, the trigger events are activated by steep signal edges leading to robust jitter behaviour. Signal generation and sampling control are working exactly in parallel.
- The output of the binary divider provides pulses with steep edges (typically 20–30 ps). Therefore, the time point of data capturing is controlled very precisely and it is robust against jitter and drift.

- The binary divider runs through all its states before it activates a new data sampling. That results in a perfect equidistant sampling since internal imperfections of the binary divider affects always in the same way.
- Supposing a stable clock generator, the sensor 'time axis' is absolute linear also for a large number of samples per scan. Hence, there are no distortions caused by transforming the data into the frequency domain and the approach is useful for high-resolution long-range applications since there is no upper limit for the time-bandwidth product (compare with (3.19) and (3.23) dual ramp and dual sine sampling control).
- The actual sampling rate f_s can simply be matched to the application requirements:
  - **Low sampling rate:** Simple ADC, simple digital hardware, low-power consumption; but low measurement rate resulting in low SNR-values and measurement scenario of low time variance.
  - **High sampling rate:** Improved SNR value and applicable in scenarios of high time variability; but elevated demands on the digital electronics and power consumption.
- RF-clock rate and PN-code (programmable shift register assumed) can be changed during operation leading to further randomization of the test signal. This results in lower probability of intercept and a lower probability of detection.
- The method of interleaved sampling does not limit the data capturing rate by the measurement principle as such (as it is done, e.g. in the case of sequential sampling). Rather more the development state of circuit technology will restrict the performance figures, for example the receiver efficiency. That is, there exists a large potential for further improvements. Nowadays, the bottleneck of the measurement speed is the AD-conversion and the handling of the digital data stream.
- The equivalent sampling rate f_{eq} is close to the theoretical minimum. Therefore, the data throughput is reduced to a minimum.

The digital M-sequence receiver is based on stroboscopic sampling, hence its efficiency η_{mr} is given by

$$\eta_{mr} = 2^{-m}\eta_S \tag{3.44}$$

where m represents the number of stages in the binary divider and η_S is the efficiency of the applied sampling gate. As expected, the efficiency is as better as shorter the binary divider is. But a short divider leads to a high data capturing rate f_s which causes high computational burden for the subsequent digital pre-processing.

The captured data of a digital M-sequence device are usually meaningless without such a processing because they appear completely random for a human being. The wanted results are typically the IRF or FRF of the DUT. They have to be calculated from the captured data before the measurements can be meaningful interpreted. The question is how this can be done in the most efficient way permitting both a high sampling rate and a continuous processing of the incoming data stream. As long as the processing covers a number of cascaded linear algorithms (e.g. filtering, cross-correlation, convolution, averaging, Fourier transform etc.), their order may be changed due to commutative low. Hence, the simplest algorithm which additionally leads to a reduction of the data throughput should be the first behind the ADC. Synchronous averaging is such an algorithm. It can be implemented in FPGAs for a very high operational speed.

In order to estimate the data throughput, we suppose an ADC of N_{bit}-resolution from which the lowest bits are typically affected by noise. The actual useful information is represented by the ENOB-number of the receiver. It includes both the quantization noise and the thermal noise of the receiver (compare Section 2.6.1.6). In order that averaging works, the actual bit number N_{bit} should be selected sufficiently larger than the effective number of bits: $\text{ENOB} + 2 - 3 \leq N_{\text{bit}}$.

Bit rate C_{B1} and data quantity H_1 behind the ADC (point ① in Figure 3.35) are given by

$$\begin{aligned} C_{\text{B1}}[\text{bits/s}] &= f_{\text{s}} N_{\text{bit}} \\ H_1[\text{bits/IRF}] &= C_{\text{B1}} T_{\text{R}} = p N_{\text{s}} N_{\text{bit}} \end{aligned}$$

Herein, we referred H_1 to the amount of data which are captured to determine one IRF or FRF. Whereas one measurement cycle covers a data set of p periods of the measurement signal[9] recorded during the time T_{R} (p - number of synchronous averaging). After synchronous summing of the data samples (point ② Figure 3.35)), bit rate and data quantity are reduced to

$$\begin{aligned} C_{\text{B2}}[\text{bits/s}] &= \frac{f_{\text{s}}}{p}(N_{\text{bit}} + \text{lb}\, p) \\ H_2[\text{bit}] &= N_{\text{s}}(N_{\text{bit}} + \text{lb}\, p) \end{aligned}$$

That is, the total number of data samples reduces by the factor p and the word length of the data samples increases by lb p bits if any data cutting is avoided. Synchronous averaging performs noise suppression by the factor $\sqrt{p}$. Hence, the digital data at point ② in Figure 3.35 have an effective resolution of

$$\text{ENOB}_2 = \text{ENOB} + \frac{1}{2}\text{lb}\, p$$

so that finally the word length $N_{2,\text{bit}} \approx \text{ENOB}_2 + 2 - 3$ should be sufficient for the data stream at the averager output. The extra number of 2–3 bits is to assure dithering effects within subsequent processing steps.

For a large factor p, synchronous averaging leads to a considerable data reduction and noise suppression. Unfortunately, the averaging factor cannot be increased arbitrarily since the recording time T_{R} would be extended proportionally. But it is limited due to time variability of the scenario under test. In the case of radar measurements, one tries to avoid signal deformations by Doppler. Thus, we have to meet the condition (supposing $\underset{\approx}{B}_{\text{DUT}} = f_{\text{eq}} = f_{\text{c}}$)

$$\frac{c}{\Delta v_{\text{r}}} \geq 2TB = 2T_{\text{R}}\underset{\approx}{B}_{\text{DUT}} = 2\,T_{\text{R}} f_{\text{c}} = 2(2^n - 1)2^m p \approx p2^{m+n+1} \qquad (3.45)$$

9) Note that the number of stimulus periods elapsing during T_{R} is $p\,2^m$.

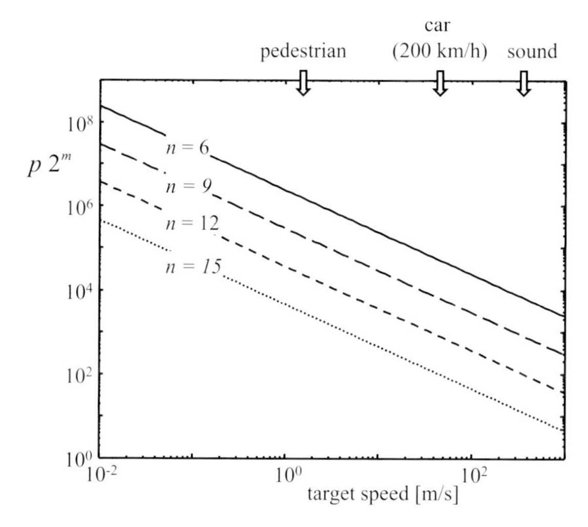

Figure 3.37 Maximum sub-sampling factor 2^m and number p of synchronous averaging in dependence from target speed.

which follows from (2.90). Figure 3.37 illustrates relation (3.45) for M-sequences of different orders.

After data reduction by synchronous averaging, the data processing has to be specified for the intended application. At this, the first processing step typically concerns one of the following procedures:

- Impulse compression by digital correlation with the ideal M-sequence.
- Transformation into the frequency domain.
- Removal of stationary data.

3.3.4.1 Digital Impulse Compression

In the case of radar sensing, we need the IRF $g(t)$ of the DUT and scenario under test. Hence, we have to calculate the cross-correlation between receive signal $y(t)$ and actual stimulus signal $x(t)$ since $g(t) \propto C_{yx}(t)$. If we accept initially some deviations from the actual DUT IRF, we can also apply the correlation between $y(t)$ and the ideal M-sequence $m(t)$. This would require only one receiver channel and the digital correlation may be implemented by a fast algorithm. This processing strategy is to be preferred if a high update rate of the measurements is required and the exact shape of the IRF is less important (e.g. if only the determination of the round-trip time is of interest)

In order to determine the cross-correlation $C_{ym}(t)$ between $y(t)$ and $m(t)$, we take the averaged data samples $\bar{\mathbf{y}}[i]; i \in [0, N_s - 1]$ and proceed as described by (2.181). The data samples $\mathbf{m}[i]; i \in [0, N_s - 1]$ of the M-sequence are a priori known. They are arranged in the circulant matrix $\mathbf{M}_{\text{circ}}$, so that the samples $\mathbf{C}_{ym}[i]$ of the cross-correlation function are determined by

$$\mathbf{C}_{ym} = \mathbf{M}_{\text{circ}} \bar{\mathbf{y}} \tag{3.46}$$

The immediate calculation of the matrix product (3.46) is a numerical expensive procedure. But it can be drastically speeded up by converting $\mathbf{M}_{\text{circ}}$ into a Hadamard matrix $\mathbf{H}$ [50, 51] (see also Annex A.6 for properties of $\mathbf{H}$).

$$\mathbf{C}_{ym} = \mathbf{PHQ}\bar{\mathbf{y}} \quad \text{with} \quad \mathbf{M}_{\text{circ}} = \mathbf{PHQ} \tag{3.47}$$

The involved matrices and vectors have the following dimensions: $C_{ym}[N_s, 1]$, $\mathbf{y}[N_s, 1]$, $\mathbf{M}_{\text{circ}}[N_s, N_s]$, $\mathbf{H}[N_s + 1, N_s + 1]$, $\mathbf{P}[N_s, N_s + 1]$ and $\mathbf{Q}[N_s + 1, N_s]$. The transformation matrices $\mathbf{P}$ and $\mathbf{Q}$ are sparsely occupied, i.e. the only one entry per raw is 1, the rest are zeros. Their construction is straightforward [50] and their product with a vector is nothing than re-sorting the data samples. This is a numerically inexpensive procedure.

The Hadamard matrix can be recursively constructed:

$$\mathbf{H}_1 = 1; \quad \mathbf{H}_2 = \begin{bmatrix} 1 & 1 \\ 1 & -1 \end{bmatrix}; \quad \mathbf{H}_{2n} = \begin{bmatrix} \mathbf{H}_n & \mathbf{H}_n \\ \mathbf{H}_n & -\mathbf{H}_n \end{bmatrix}$$

Its product with a vector may be graphically illustrated by a butterfly operation as demonstrated in Figure 3.38. The procedure is called the fast Hadamard transform (FHT) or the fast Walsh–Hadamard transform. The butterfly operations are well known from the FFT. But in contrast to the FFT algorithm, the FHT-butterfly only includes sum or difference operations which can be executed at high processing speed by simple arithmetic logic units for integer operations. The calculation of the matrix via FHT involves only $n2^n$ integer operations permitting the determination of several ten thousands of IRFs per second with modern FPGAs. The adder needs a word length of about $N_{2,\text{bit}} + n \approx \text{ENOB}_2 + 2 \cdots 3 + n$ bits (n– order of M-sequence). Due to the signal compression the amplitude of data samples will grow requiring the n additional bits in order to avoid clipping effects. In the case of an M-sequence (i.e. CF = 1), the correlation leads to a noise suppression by the factor $\sqrt{N_s} = \sqrt{2^n - 1}$. Hence, we can expect at position ③ in Figure 3.35 an ENOB-number of

$$\text{ENOB}_{3,\text{FHT}} \approx \text{ENOB}_2 + \frac{1}{2} n \approx \text{ENOB} + \frac{1}{2}(n + \text{lb}\, p) \tag{3.48}$$

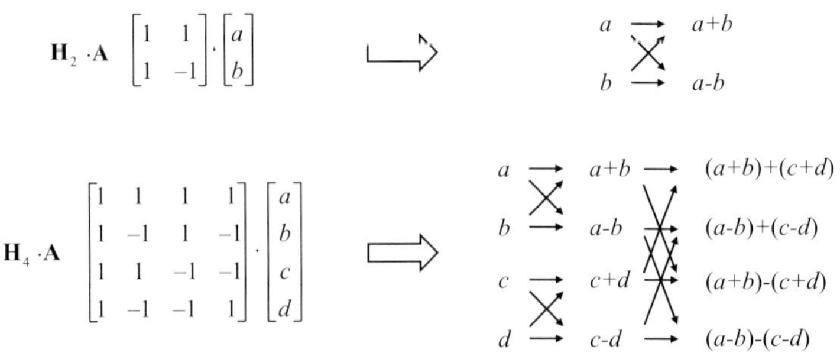

$$\mathbf{H}_2 \cdot \mathbf{A} \quad \begin{bmatrix} 1 & 1 \\ 1 & -1 \end{bmatrix} \cdot \begin{bmatrix} a \\ b \end{bmatrix} \Rightarrow \quad a \to a+b, \quad b \to a-b$$

$$\mathbf{H}_4 \cdot \mathbf{A} \quad \begin{bmatrix} 1 & 1 & 1 & 1 \\ 1 & -1 & 1 & -1 \\ 1 & 1 & -1 & -1 \\ 1 & -1 & -1 & 1 \end{bmatrix} \cdot \begin{bmatrix} a \\ b \\ c \\ d \end{bmatrix} \Rightarrow \quad \begin{matrix} a \to a+b \to (a+b)+(c+d) \\ b \to a-b \to (a-b)+(c-d) \\ c \to c+d \to (a+b)-(c+d) \\ d \to c-d \to (a-b)-(c-d) \end{matrix}$$

Figure 3.38 The Hadamard butterfly.

which is equivalent to the maximum signal-to-noise ratio (see Section 2.6.6) of digital UWB M-sequence correlator:

$$\begin{aligned} D_{3,\max} &= \mathrm{SNR}_{\max} = 3\,2^{2\mathrm{ENOB}_3} = 3\,2^{2\mathrm{ENOB}}\,2^n p \\ D_{3,\max}[\mathrm{dB}] &= 4,8 + 6\mathrm{ENOB} + 3n + 10\lg p \end{aligned} \tag{3.49}$$

Insertion of (3.45) in (3.49) gives an estimation of the maximum achievable dynamic range of an UWB radar sensor under the constraints of a time variable scenario:

$$D_{3,\max} = 3\,2^{2\mathrm{ENOB}} T_{\mathrm{R}} f_{\mathrm{s}} \leq 3\,2^{2\mathrm{ENOB}-(m+1)} \frac{c}{\Delta v_{\mathrm{r}}} \tag{3.50}$$

whereas the recording time is expressed as $T_{\mathrm{R}} = (2^n - 1)p/f_{\mathrm{s}} \approx N_{\mathrm{s}} p/f_{\mathrm{s}}$.

3.3.4.2 Transformation into the Frequency Domain

Under certain conditions, data representation in the frequency domain is to be preferred compared to the time domain. Such situations may appear, if

- **The FRF of the DUT is required but not the IRF:** Impedance spectroscopy is a typical sensor application dealing with spectral data instead of time domain functions.
- **The subsequent data processing requires calibrated data:** If the exact shape of the DUT IRF is required, systematic errors have to be removed from the measurements. This is done by calibration which de-convolves parasitic device internal propagation paths from the measured signal. De-convolution is preferentially performed in the frequency domain. For details on device calibration the reader can refer to Sections 2.6.4 and 3.3.5.
- **The FHT algorithm is not available for impulse compression:** The determination of the correlation function C_{ym} for arbitrary periodic sounding signals is based on (3.46) which is computational inefficient. In order to accelerate the correlation processing, the widespread standard procedure of FFT-algorithms may be applied too as shown in (2.178).

The FFT acts on the averaged data $\bar{\mathbf{y}}[i]$ as before the FHT. It is a very efficient algorithm which can run at the sensor internal FPGA or DSP for high-speed applications or it is executed at a host computer, for example under MATLAB® in case of experimental sensor investigations.

The FFT algorithm was initially designed for a number of data samples which equals a power of two. Under this condition, it has its highest speed. Unfortunately, our data set always covers one sample less. New FFT algorithms are able to work on arbitrary sample numbers. They use prime factorization and apply specific butterfly operations adapted to the occurring primes. If the involved prime numbers are small, the computational time will be short otherwise it will rise. Figure 3.39 compares the time t_1 to execute an FFT of $N_1 = 2^n - 1$ samples (i.e. to calculate the spectrum of an M-sequence) with the time t_2 needed by a 2^n-FFT. We can observe that the loss in processing speed is not dramatic with respect to the maximum

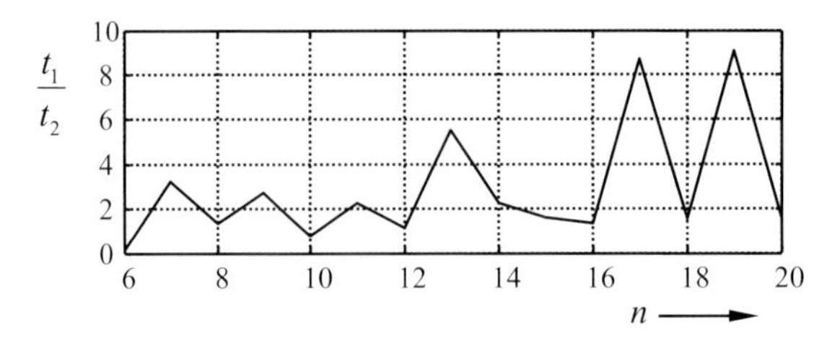

Figure 3.39 Normalized FFT execution time extension for M-sequences of different orders (MATLAB R2009b; Version 7.9.0.529).

processing speed. The prime factorization for M-sequences up to order 20 is given in Annex B.9. Referring to Figure 3.39, it confirms the fact that large prime numbers will require larger processing time.

3.3.4.3 **Removal of Stationary Data**

The measurement data of an UWB sensor are often composed from a multitude of components. In the case of a radar measurement, we have, for example the wanted reflection from the target but we also have lots of unwanted reflections from the environment. In many of the applications, the not intended reflections will actually dominate the receive signal. But fortunately they are often stationary so that they can be easily removed.

Figure 3.40 demonstrates a simplified situation. It depicts a two-dimensional data set of impulse response functions $g(t,T) \propto C_{ym}(t,T)$ represented as Wiggel plot. We will call it a radargram. It constitutes from IRFs which are measured at different time points. We suppose that any variations within the test scenario are negligible during the recording time of a single IRF but they will be visible from the repetitions of the measurement. The 'short time' during which the

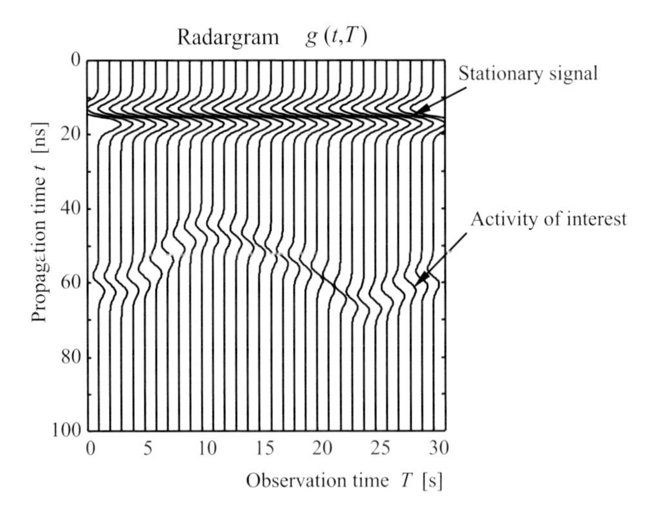

Figure 3.40 Radargram of impulse response functions for a time variable scenario. See Figure 3.49 and (3.57) for details on reconstruction of a radargram from the captured data.

electromagnetic waves are interacting with the test objects is called interaction time or propagation time t. The 'long time' over which the variations of the test object are observed we call observation time T.

The radargram in Figure 3.40 shows two traces. The time variable one concerns to the wanted target and the second one represents reflections from a stationary object which is out of interest in our case. The stationary signal may be known from prior measurements (i.e. $g_0(t)$) or it may be gained from horizontal averaging over the observation time T_w: $\bar{g}_{T_\mathrm{w}}(t) = \frac{1}{T_\mathrm{w}} \int_{-T_\mathrm{w}/2}^{-T_\mathrm{w}/2} g(t,\tau)\mathrm{d}\tau$. After subtracting it from the whole radargram, the wanted IRF $\tilde{g}(t,T)$ of the time variable target is emphasized:

$$\tilde{g}(t,T) = g(t,T) - g_{\mathrm{bg}}(t) \tag{3.51}$$

The approach is also referred to as background or clutter removal where $g_{\mathrm{bg}}(t)$ represents the background signal which may be selected by either $g_{\mathrm{bg}}(t) = g_0(t)$ or $g_{\mathrm{bg}}(t) = \bar{g}_{T_\mathrm{w}}(t)$.

Background removal reduces the energy of the remaining signals which we can exploit to minimize the bit stream and data amount to be handled. The degree of data reduction largely depends on the actual scenario. The order of feasible data diminution shall be demonstrated by a measurement example. The task is to detect the breathing of a person (movement of the chest) hidden by a wall (more on this topic can be found in Sections 4.8, 6.5 and 6.7). Scattering at the wall represents the stationary signal dominating the whole radargram while the chest movement provides weak variations in the back scatter signal.

Figure 3.41 opposes different radargrams from the captured data. The radargram Figure 3.41 (a) deals with the data (i.e. IRFs) as they are provided by an arbitrary UWB radar sensor. Obviously, the wall reflections hide completely the target. After removing them (Figure 3.41 (b)), the periodic movement of the chest becomes visible. The receive data of an M-sequence sensor after removing the wall reflections are shown in Figure 3.41 (c). They indicate the time variance due to breathing but

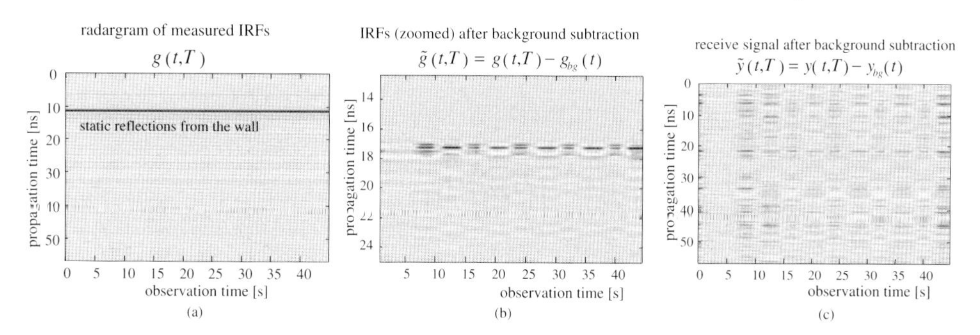

Figure 3.41 Radargram (false colour representation) of a breathing person behind a wall measured with an M-sequence UWB sensor. (a) Original data after impulse compression; (b) IRFs exempted from wall reflections and (c) captured signal exempted from wall reflections before impulse compression.

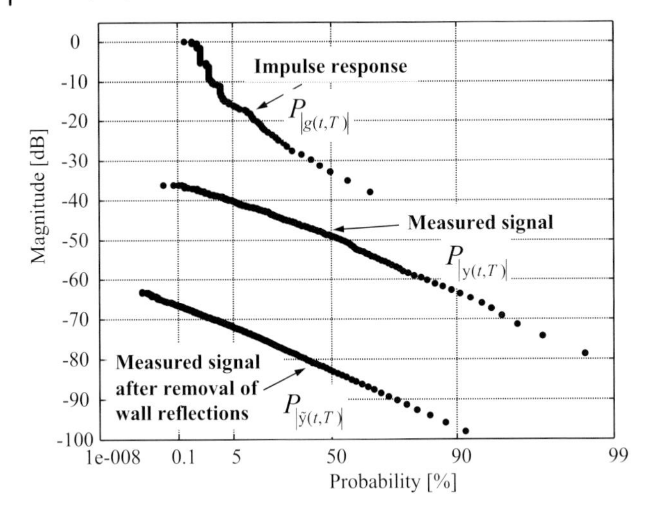

Figure 3.42 Amplitude probability of various signals from the person detection scenario. The measured signal refers to the actually captured voltage of an M-sequence sensor. The impulse response is gained from correlation with the ideal M-sequence and removal of wall reflections is simply done by subtracting the horizontal average of the captured data.

they do not show the round-trip time to the target since correlation was not yet performed. All three radargrams contain the same information about the target but they comprise differently large data volumes.

Figure 3.42 depicts the amplitude probability of the radargrams. The larger the magnitude of a signal is the more bits are required to represent the data samples correctly. If we take the pulse response $g(t, T)$ as reference, we find that the maximum amplitude of the actual receive signal $y(t, T)$ of an M-Sequence device is about 36 dB below and the magnitude of $\tilde{y}(t, T)$ is yet more than 60 dB smaller. Hence, for the shown example and in the case of an optimal system layout, we need about 10 bits per sample less to store or to process if we deal with $\tilde{y}(t, T)$ compared to $g(t, T)$.

Removal of stationary data is a simple arithmetic procedure. Hence in the case of high-speed and bit rate critical applications, it is recommended that the sensor internal FPGA performs such data reduction prior to further processing or data transfer. The procedure will not reduce the actual information content.

In summary, the digital ultra-wideband correlator represents a flexible device conception. It is possible to adjust the sensor features to several conditions by acting on four device parameters (see Figure 3.43) and the digital pre-processing:

- **Clock rate f_c of the shift register:** It determines the bandwidth and the time resolution of the sensor. In the case of radar applications it is responsible for the range resolution.
- **Shift register length n:** In connection with the clock rate, it determines the duration t_D over which the reaction of the object under test can be observed. It corresponds to the period t_P of the stimulus and should at least cover the settling time

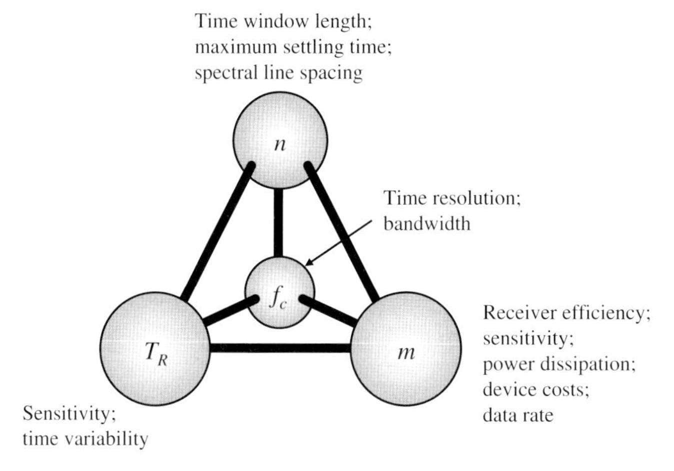

Figure 3.43 Design tetrahedron of a digital ultra-wideband correlator.

of the scenario under test. For the case of radar applications, it fixes the unambiguous range and it defines the spacing of spectral lines for frequency domain measurements.

- **Binary divider length** m: It relaxes the speed requirements onto the receiver. A large dividing factor reduces the data throughput which simplifies the digital part, reduces power dissipation and system costs due to simpler electronics. But it also reduces sensitivity since the receiver efficiency is largely decreased. Thus, the user has to decide what is in the foreground of his interest.
- **Recording time** T_{R}: The length of the recording time can be controlled by the number p of averaging. It is limited by the time variability of the scenario under test. Target speed and acceleration, and the rate of mechanical object oscillations are the dominant parameters restricting the recording time in radar measurements. A large recording time is desirable for suppression of random errors.
- **Pre-processing:** An M-sequence sensor is typically equipped with an FPGA and/or DSP for data pre-processing as
  - data reduction by averaging and background removal
  - digital impulse compression or correlation
  - data conversion into frequency domain and
  - others like error correction – see Sections 2.6.4 and 3.3.5; detection; round-trip time estimation and so on.

Reasonable sensor internal pre-processing can help to reduce the numerical load of the main processor which becomes important for applications with large sensor array.

Thanks to the sensor structure, all mentioned parameters can be digitally controlled which would allow changing the device features even during the operation (an appropriate system layout assumed).

3.3.5 Some Particularities of PN-Sequence Devices

The use of PN-sequence correlation (analogue or digital) in UWB sensing devices often causes some signal phenomenon which may lead to misinterpretations if one looks on it with the eyes of the classical pulse technique. The pulse base of an IRF (CCF) determined by PN-sequence correlation often shows irregular corruptions which are mostly interpreted as random noise as illustrated in Figure 2.97. But actually, the deviations are of deterministic nature which can be removed from the data by appropriate calibration as also depicted in that figure. The major causes of these distortions are imperfections of M-sequence generation. We will refer to three effects resulting in such deviations.

Limitation of spectral bandwidth: The first reason is the spectral limitation of the M-sequence to the frequency band from DC to $f_c/2$. The ideal M-sequence has an unbounded spectrum which we have to limit in order to meet the Nyquist theorem. Figure 3.44 illustrates the impact of spectral bounding by simple

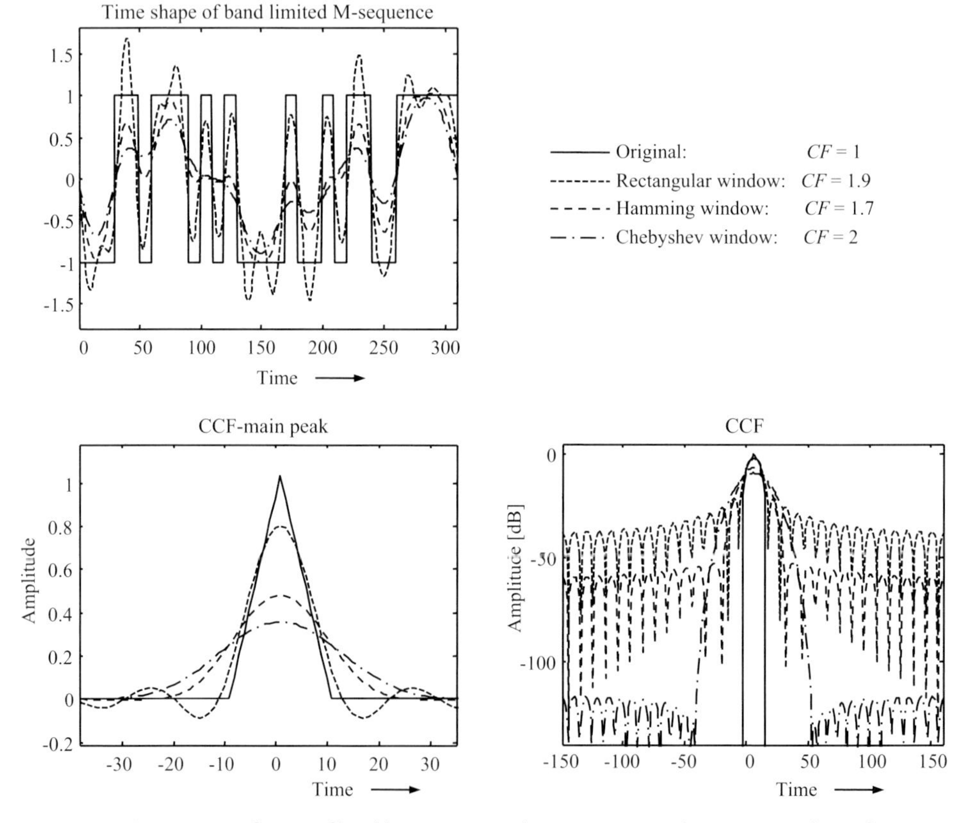

Figure 3.44 Influence of band limitation onto the M-sequence and its cross-correlation function. For illustration purposes, only a short M-sequence of order 5 was taken. The crest factors resulting from the different types of band limitation are also given.

simulations. For that purpose, the Fourier spectrum of an ideal M-sequence was subject by several window functions (filters) and the resulting spectrum was transformed back into the time domain. The figure shows how the band limitation affects the time shape of the M-sequence as well as the cross-correlation of the band-limited M-sequence with the ideal M-sequence reference.

Obviously, a hard limit (rectangular window) will cause strong side lobes of the compressed M-sequence. By smoothing the frequency cutting (examples Hamming and Chebyshev window), the side lobe level can be reduced but only at the expense of the main lobe width. In the case of a digitized UWB sensor, one can freely select the window function during the digital post-processing. In order to gain the largest flexibility, the analogue anti-aliasing filter should cut the spectrum as close as possible at the Nyquist frequency by steep filter edges. Indeed, this will lead to an intrinsic device CCF with strong side lobes. But corresponding to the actual need, they may be reduced by appropriate digital post-processing. It should also be mentioned that filtering and windowing increases the crest factor (see also Figure 3.44) so that the maximum achievable dynamic range $D_{\max}$ (see (3.49), (3.50)) will be slightly reduced.

The two other reasons which affect the purity of the CCF result from non-linear distortion of the M-sequence. We will consider two phenomena – amplitude distortions by saturation effects and deterministic jitter.

Amplitude distortion: Figure 3.45 depicts a simulation indicating the influence of non-linear distortions of second and third order. The upper plot acts as reference for an unaffected CCF. It represents the correlation function of the unperturbed but band-limited M-sequence (by Hamming window) $m_H(t)$ with the ideal M-sequence $m(t)$. The involved side lobes are caused from the band limitation.

The lower parts of the figure show the cross-correlations function $C_{mm_H^{(2)}}(\tau)$ for quadratic and cubic distortions $C_{mm_H^{(3)}}(\tau)$. We can observe an increasing average level of the ripples and the appearance of some dominant spurious spikes. This behaviour coincides generally with the picture of actual measurement signals as in the already mentioned Figure 2.97.

Basically, the reaction of a signal $x(t)$ to non-linear distortions depends on its higher order correlation functions:

$$\begin{aligned}
&\text{bi-correlation}: \quad C_{xxx}(\tau_1,\tau_2) = \overline{x(t)x(t+\tau_1)x(t+\tau_2)}\\
&\text{tri-correlation}: \quad C_{xxxx}(\tau_1,\tau_2,\tau_3) = \overline{x(t)\cdot x(t+\tau_1)\cdot x(t+\tau_2)\cdot x(t+\tau_3)}, \cdots
\end{aligned}$$

It was shown by Ref. [51] that the higher order correlation functions have either the value 1 or N^{-1} as we already know it from the first-order correlation function of the ideal M-sequence (see Figure 2.33). But in contrast to the first-order correlation, the higher order functions have several maxima, for example the bi-correlation in the $\tau_1 - \tau_2$-plane looks like a sparsely occupied nail bed. This is the reason why an M-sequence localize the non-linear distortions into some dominant peaks which are distributed by a certain pattern over the whole period of the CCF. The pattern depends on the order of non-linearity and the actual code of the PN-sequence.

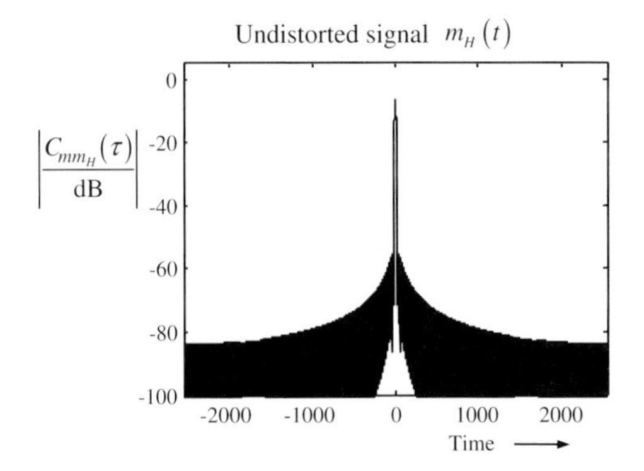

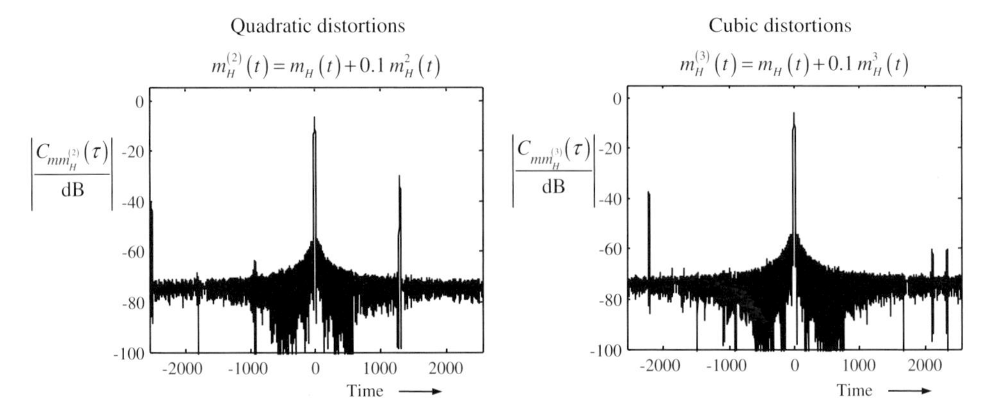

Figure 3.45 The effect of amplitude distortions onto the cross-correlation function of a ninth order M-sequence.

Deterministic jitter: The pulse duration of the individual chips of an actual M-sequence is irregular but reproducible (therefore, the name deterministic jitter). This causes non-linear effects as well. The impacts on the cross-correlation function are demonstrated by the simulations in Figure 3.46. The upper part of the figure compares the eye pattern of the band-limited M-sequence. In the left, we see an ideal (but band-limited) sequence and at the right-hand side the signal is additionally affected by deterministic jitter. In the example, the mean value of the jitter was about 10% of the chip duration of the ideal M-sequence ($\Delta t_{c,rms} \approx 0.1\,\Delta t_c$). The impact of the chip deformation onto the spectrum and the correlation function are depicted in the both plots below. We can observe again an increased spurious level which pretends to be random noise. But actually it is not random noise which can simply be shown by synchronous averaging. It will not lead to a reduction of the spurious level due to the deterministic nature of the signals.

All three deterministic effects discussed above, we will formally summarize by the bloc schematics as depicted in Figure 3.47. We divide both the stimulus

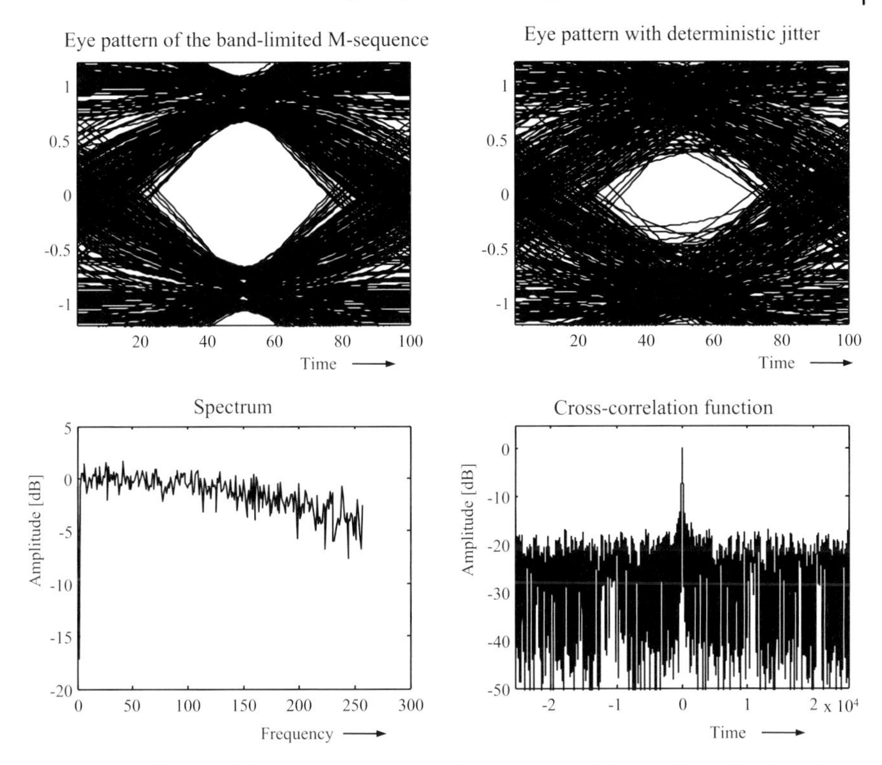

Figure 3.46 Influence of deterministic jitter onto the spectrum and the correlation function of a ninth order M-sequence. The average variations of the chip duration.

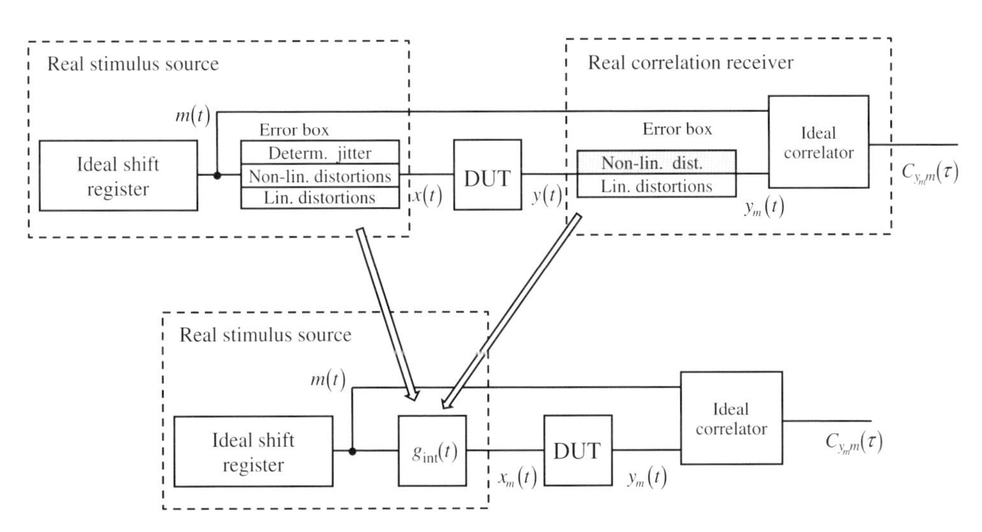

Figure 3.47 Error model of an M-sequence correlator. Except the non-linear distortions in the receiver, all errors can be joined by the single function $g_{int}(t)$.

generator and the correlation receiver in two parts – an ideal device and an error box. On the transmitter side, we have to deal with three different sources which affects the ideal M-sequence $m(t)$. Fortunately, the actual stimulus $x(t)$ of the DUT keeps the same as long as any retroaction from the DUT to the shift register is avoided. Therefore, we can formally establish a linear relation between $m(t)$ and $x(t)$ even if this does not reflects the physical reality.

On the receiver side, we have to deal with linear and non-linear distortions. Since the signal $y(t)$ is unknown and strongly dependent from the DUT, the handling of non-linear distortions will be quite complicated. Therefore, we will suppose that the receive power is sufficiently below the ICP3 value of the correlator which allows to disregard non-linear distortions. Now, only the linear distortions of the receiver remain, which we can join with the error box of the transmitter due to the commutative law of convolution. Hence, we can describe the behaviour of the correlator by an intrinsic impulse response functions $g_{int}(t)$ which covers all systematic deviations of the device. The wanted IRF of the DUT $g_{\mathrm{DUT}}(t)$ and the measured correlation function $C_{y_{\mathrm{m}}m}(t)$ are related by

$$C_{y_{\mathrm{m}}m}(t) = g_{int}(t) * g_{\mathrm{DUT}}(t) * C_{mm}(t) \tag{3.52}$$

$C_{mm}(t)$ is the a priori known auto-correlation of the ideal M-sequence which we typically approximate by a Delta function $\delta(t)$. That is, we can extract the cleared impulse response function $g_{\mathrm{DUT}}(t)$ from the measurements $C_{y_{\mathrm{m}}m}(t)$ by de-convolving the intrinsic impulse response function $g_{int}(t)$. Equation (3.52) shows that device imperfections can be removed from the measured data by linear methods of error-correction. However, one should remember that (3.52) only relates to a simplified error model which does not respect energetic interactions between measurement device and DUT. A full correction of all device internal errors has to follow the approach discussed in Section 2.6.4.

3.3.6 System Extensions of Digital PN-Correlator

The digital PN-correlator represents a RF-measurement principle joining analogue RF-technique and high-speed digital circuitry. It will be shown that the basic principles introduced in Section 3.3.4 can be flexibly extended in order to meet the requirements of different sensor applications. We will introduce some basic examples of modified sensor architectures.

3.3.6.1 Improving the Sampling Efficiency

The improvement of the sampling efficiency will be one of the major requirements for future UWB sensor development. This concerns all wideband sensing principles independent of their actual functioning. We have seen above that low efficiency degrades sensitivity and extends recording time. Basically, this drawback could be compensated by enlarging the stimulation power. But for UWB sensing this is not an option due to the strong limitation of radiated power. Hence, the only way is to improve the efficiency.

What are the reasons which limit the sampling rate and how the situation can be improved? The sensor electronics is built from sub-components whose performance is limited by physical laws and technical constraints. The figure of Merit FoM is a characteristic quantity which roughly summarizes the boundaries within a component may be operated. The figure of Merit shall express the expense in terms of energy and costs which is required to achieve a certain effect. We will restrict ourselves exclusively to energetic aspects here so that the FoM value can be generalized by following ratio:

$$\mathrm{FoM} = \frac{\text{energy expenditure}}{\text{main parameters of intended effect}}$$

This quality figure can be applied to hardware components as well as to algorithms. Some examples shall illustrate the philosophy. The definitions are not unique. They are usually matched or modified to the specific needs of the respective consideration.

Flash analogue-to-digital converter:

$$\mathrm{FoM}_{\mathrm{ADC}} = \frac{P}{2^{\mathrm{ENOB}} f_s} \,[\mathrm{J/conversion}] \tag{3.53}$$

P is the power dissipation of the ADC, $ENOB$ is its effective word length and f_s is the sampling rate).

2R-R-digital-to-analogue converter:

$$\mathrm{FoM}_{\mathrm{DAC}} = \frac{P}{\mathrm{ENOB}\, f_s} \,[\mathrm{J/conversion}] \tag{3.54}$$

The flash ADC needs one circuit unit (comparator, MUX etc.) per quantization interval. The 2R-R DAC needs one circuit unit per bit. This makes different definitions of the FoM-number reasonable.

Amplifier:

$$\mathrm{FoM}_{\mathrm{ampl}} = \frac{P}{G B\, \mathrm{CP}\, 1\, \mathrm{dB}} \left[\frac{\text{power-efficency}}{\mathrm{Hz}}\right] \tag{3.55}$$

P is the power dissipation of the amplifier, G is the power gain in linear units, B is the bandwidth and CP 1 dB is the 1 dB compression point in linear units.

Arbitrary integer operation:

$$\mathrm{FoM}_{\mathrm{op}} = \frac{P}{r(N+M)} \,[\mathrm{J/operation}] \tag{3.56}$$

P is the power dissipation of the arithmetic unit, r is the processing rate and N, M are the word length of the operands.

The design goal is to implement circuits and algorithms which provide FoM-values as small as possible. The quantity of the FoM-value depends on the circuit technology and the cleverness of the designer who implemented the wanted function.

Polarity receiver; 1-bit receiver: Let restrict our further consideration to the ADC since it seems to be the bottleneck of efficiency improvement. If we suppose that the FoM-value and the maximum power dissipation are fixed by the semiconductor technology and the heat transport, we can only exchange the sampling rate f_s for the ENOB-value as it is obvious from (3.53). That is, if one designs an ADC which provides one or two bit less than a reference type it may be operated at double or fourfold sampling rate by keeping the power consumption. In the extreme case, the ADC may be of 1 bit, that is the receiver is simply a comparator, which operates in real-time sampling mode $f_s = f_c$ [52]. The required dynamic range of the sensor must be achieved by the processing gain which can take high values now due to the large number of captured samples. However, this simple straightforward strategy has some essential drawbacks.

1) The averaged signal tends to non-linear distortions (see (2.273) and Figure 3.48), if the receiving signal is not sufficiently randomized that is $y_0 \approx \sigma_n$. Thus, the correct representation of the data samples would require the numerical expensive inversion of the error function.
2) In a mixture from strong and weak signals, the strong ones are favoured while the weak ones are chocked due to the reduced differential sensitivity $\tan\alpha = d\bar{y}/dy$ (Figure 3.48).

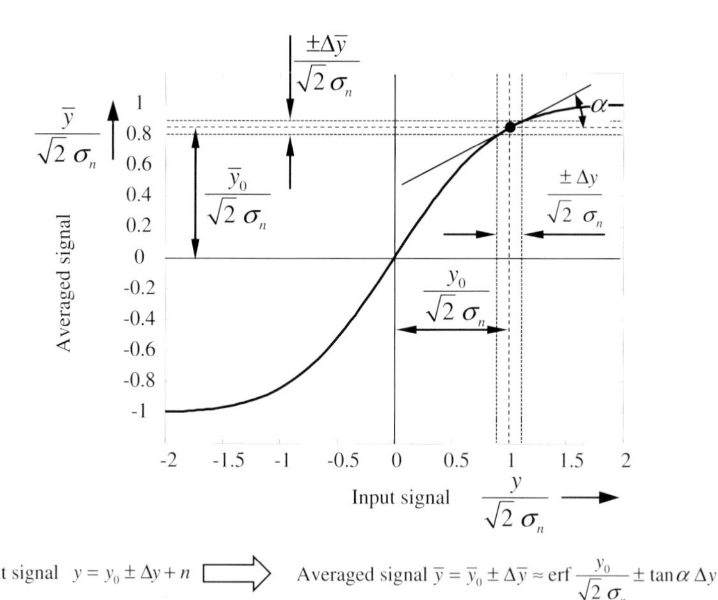

Figure 3.48 Transfer characteristic of a 1-bit receiver (polarity receiver) for randomized input voltage. The signal to be captured is supposed to consist from a dominant value y_0, a weak component Δy and the noise n having the variance σ_n^2. Note that weak signal parts Δy are as much suppressed as dominant the largest signal component y_0 is.

3) A strongly randomized input signal reduces the non-linearity but it requires a larger amount of averaging in order to gain stable estimations of the data samples. This counteracts the sampling rate enhancement.
4) The ultimate device feature is the dynamic range $D_{3,\max}$. If we keep it constant for a given recording time T_R, every reduction of the ADC resolution by one bit has to be compensated by the fourfold sampling rate (refer to (3.50)). On the other hand, the FoM-value ((3.53)) permits only a boost by a factor of two for every bit which is reduced. Thus, we will end up in a negative balance.

Feedback receiver: The previous example has shown us that an increase of sampling rate at the cost of ADC resolution is unproductive. Thus, the question arises if there are still other opportunities to improve the sampling efficiency without affecting the receiver resolution. A possible loophole is based on two assumptions:

1) DA-converters are typically much less power hungry than AD-converters of comparable performance.
2) The time variability of the test scenarios is negligible compared to the short interaction time of the sounding waves with the targets.

Before we enter into details, we will separate the captured voltage samples $V_{in}(i\,\Delta t_s)$ in two independent processes. This is allowed as long as data recording is fast enough (see (3.45)). Figure 3.49a illustrates the time history of the

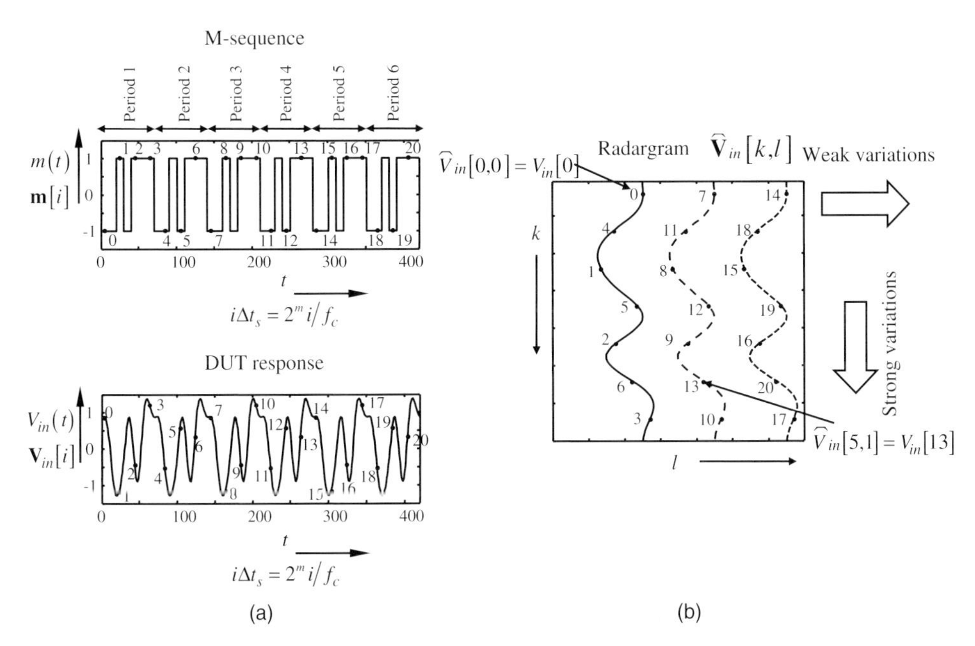

Figure 3.49 Illustration of the measurement procedure to capture three periods of the DUT response. (a) Time history of data capturing by sub-sampling with $f_s = f_c/2$ and (b) reordering of the captured data in a radargram (impulse compression was not yet performed).

measurement procedure of a device applying an M-sequence of third order ($n = 3$) and sub-sampling by the factor 2 ($m = 1$). In the reordered data $\overline{\mathbf{V}}_{\text{in}}[k, l]$ (like a radargram) in Figure 3.49 (b), we have rearranged the sequentially sampled data vector $\mathbf{V}_{\text{in}}[i]$ by the correct order:

$$\begin{aligned} \overline{\mathbf{V}}_{\text{in}}[k, l] &= \mathbf{V}_{\text{in}}[i] \\ k = \operatorname{mod}(2^m, 2^n - 1); \quad l &= \operatorname{floor}\left(\frac{i}{2^n - 1}\right) \end{aligned} \tag{3.57}$$

Herein, the index i stands for the actual sampling time:

$$\text{sampling time vector :} \qquad \mathbf{t}_{\text{s}}[i] = \Delta t_{\text{s}} i = \frac{2^m}{f_{\text{c}}} i \tag{3.58}$$

the index k relates to the interaction time (short time, equivalent time):

$$\text{short time vector :} \qquad \mathbf{t}[k] = \Delta t_{\text{c}} k = \frac{k}{f_{\text{c}}} \tag{3.59}$$

and the index l refers to the observation time:

$$\text{observation time vector :} \qquad \begin{aligned} \mathbf{T}[l] &= \Delta T l \\ \text{with} \quad \Delta T &= (2^n - 1)\Delta t_{\text{s}} = \frac{2^m(2^n - 1)}{f_{\text{c}}} \end{aligned} \tag{3.60}$$

The interaction time t for short-range sensing is usually in the range of tens of nanoseconds while detectable variations within the scenario under test need at least milliseconds or seconds under typical conditions. Hence, we have orders of magnitude in between so that unexpected fluctuations from measurement to measurement are negligible. Even the update time between two measurements $T_{\text{R}} = (2^n - 1)\Delta t_{\text{s}}$ is very short compared to the time variations in the test scenario. An implemented M-sequence system $f_{\text{c}} = 9\,\text{GHz}, n = 9; m = 9$ has a recording time of $T_{\text{R}} = 30\,\mu\text{s}$. Therefore, the data samples arranged on a horizontal line $k = k_0$ in the radargram matrix show only weak amplitude variations, that is

$$\overline{\mathbf{V}}_{\text{in}}[k_0, l + 1] \approx \overline{\mathbf{V}}_{\text{in}}[k_0, l] \pm \varepsilon \pm 3\sigma_n \tag{3.61}$$

Herein, $3\sigma_n$ represents (approximately) the maximum noise amplitude and ε refers to small signal variations due to the weak time variance of the measurement object. Figure 3.41 shows a practical example of this behaviour and Figure 3.42 quantifies the strength of the expected effect. It indicates that the variations ($\varepsilon \pm 3\sigma_n$) are about 30 dB below the magnitude of the actual receive signal ($\overline{\mathbf{V}}_{\text{in}}[k, l]$). Thus, new information about the DUT within a consecutive measurement shot is limited to the relative small amplitude interval $\pm(\varepsilon + 3\sigma_n)$. Consequently, we could reduce the full-scale range of the ADC roughly to this interval without loss on data. We already discussed such an approach for densely sampled data (see Figure 3.18 – sequential feedback sampling). There, two consecutive voltage samples did not differ so much so that the preceding sample could be used as reference for the feedback loop.

In the case of interleaved and Nyquist sampling, the difference between two successive samples may cover the whole FSR of the ADC but two consecutive measurement shots are quite similar. A measurement shot comprises a full set of data samples to calculate one IRF or FRF. Hence, the feedback has to be established between subsequent shots instead of subsequent samples. This requires a modification of the feedback principle as depicted in Figure 3.50.

The idea is to store and to integrate the captured difference signal in a 'rotating' bank of $2^n - 1$ prediction filters which provide for every measurement cycle the correct feedback quantity. For that purpose, we arrange all involved data in matrices (radargrams) as depicted in Figure 3.49 so that only horizontally arranged samples of $\overline{\mathbf{Y}}[k,l]; k = \text{const.}$ and $\overline{\mathbf{X}}[k,l]$ will be mutually related. We can write for an arbitrary raw k_0 of the radargrams

$$\begin{aligned} \overline{\Delta V}[k_0,l] &= V_{\text{in}}(i\,\Delta t_s) - \overline{Y}[k_0,l]q_2 \\ V_{\text{in}}(t) &= V(t) + n(t) \end{aligned} \tag{3.62}$$

whereas the input voltage is composed from the wanted and periodic signal $V(t)$ and noise $n(t)$ and the feedback voltage provided by the DAC is $\overline{Y}[k_0,l]q_2$. q_2 assigns the quantization interval of the DAC. The conversion between the indices i and k,l is the same as already introduced in (3.57)–(3.60). The operation of the (idealized) receiver ADC can be modelled as

$$\overline{X}[k_0,l] = \operatorname{sgn}\left(\overline{\Delta V}[k_0,l]\right)\min\left\{\operatorname{round}\frac{\left|\overline{\Delta V}[k_0,l]\right|}{q_1}, \frac{\text{FSR}_{\text{ADC}}}{2}\right\} \tag{3.63}$$

The width of its quantization interval is given by q_1 and FSR_{ADC} represents the full-scale range. A input signal exceeding $\pm\text{FSR}_{\text{ADC}}/2$ will be clipped from the

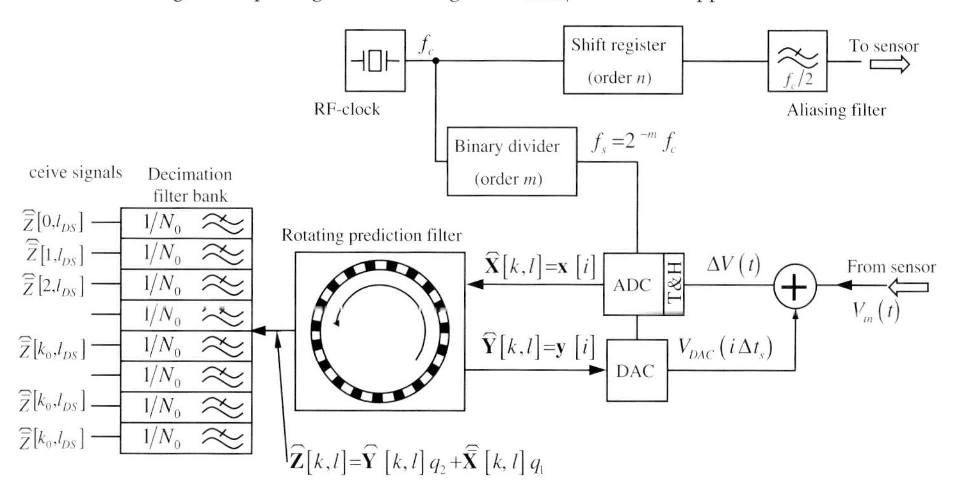

Figure 3.50 Circuit schematic of interleaved feedback sampling for digital ultra-wideband correlator. Note that the method is not mandatorily restricted to PN-code devices.

ADC (see Figure 3.52). The prediction filter takes over the role of the integrator from Figure 3.18. It provides the feedback quantity

$$\overline{Y}[k_0, l+1] = \text{round}\left(\overline{\overline{X}}[k_0, l] + \overline{Y}[k_0, l]\right) \tag{3.64}$$

in which the ADC output is usually low-pass filtered by an exponential filter (α – forgetting factor):

$$\overline{\overline{X}}[k_0, l] = (1-\alpha)\,\overline{X}[k_0, l] + \alpha\overline{\overline{X}}[k_0, l-1]; \quad \alpha \in [0, 1). \tag{3.65}$$

The digitized receive signal finally yields

$$\overline{Z}[k_0, l] = \overline{Y}[k_0, l]q_2 + \overline{\overline{X}}[k_0, l]q_1 \tag{3.66}$$

which is appropriately low-pass filtered and down-sampled leading to the radargram $\overline{\overline{\mathbf{Z}}}[k, l_{DS}]; l_{DS} = \text{floor}(l/N_0)$ (N_0 – down-sampling ratio, decimation factor). The filter cut-off and the down-sampling rate are dependent from the time variability of the DUT. In order to gain the radargram of impulse response functions, $\overline{\overline{\mathbf{Z}}}[k, l_{DS}]$ has to be subject a column wise impulse compression.

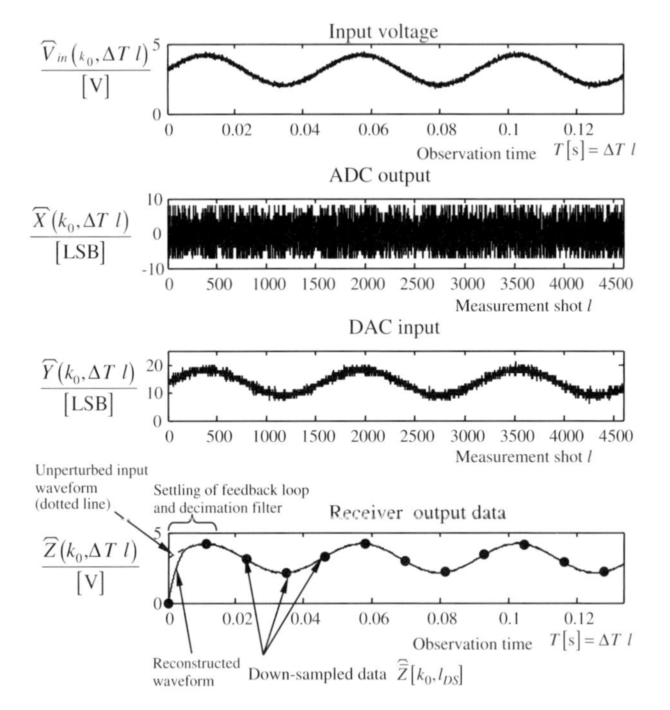

Figure 3.51 Simulated signals in dependence from the observation time of an interleaved feedback sampling loop for a fixed interaction time $t_0 = k_0\,\Delta t_c$. (ADC: $q = 0,05$ V, $\text{FSR}_{\text{ADC}} = \pm 0,4$ V (corresponds to 4 bit); DAC: $q = 0,248$ V; input noise: $\sigma_n = 0,07$ V; forgetting factor: $\alpha = 0,6$; decimation filter: sliding averager $T_I = 11,6$ ms, decimation factor $N_0 = 400$).

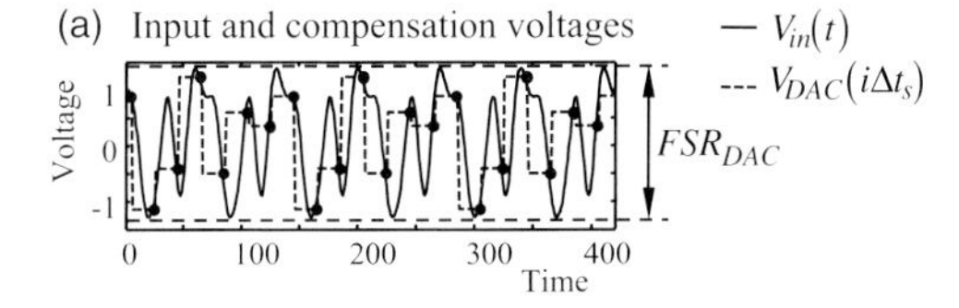

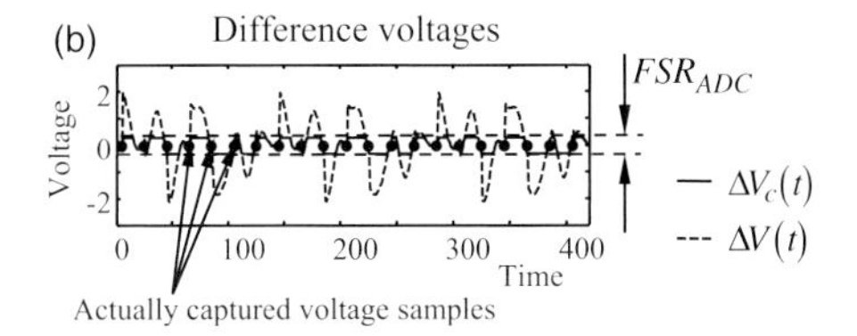

Figure 3.52 Typical voltage signals at the summation point of the interleaved feedback sampler.

Figure 3.51 illustrates some signals for an arbitrary raw k_0 of the radargrams. A sinusoidal time variation of the test scenario was supposed. It was modelled by

$$V_{\text{in}}(t) = V_0 + V_1 \sin 2\pi f_{\text{DUT}} t + n(t); \quad n(t) \sim N\left(0, \sigma_n^2\right)$$

As it can be seen from the bottommost diagram, the noise is reduced and the down-sampled data can follow nicely the input waveform after the control loop and the decimation filter are engaged. The timing in the example is based on the parameters of a standard M-sequence device ($f_{\text{c}} = 9\,\text{GHz}, n = 9, m = 9$) providing an update rate of the measurements of $m_{\text{R}} = 34.4\,\text{kHz}$ ($\Delta T = 29\,\mu\text{s}$).

Figure 3.52 illustrates the data recording from a different perspective. Now, we are considering the consecutive capturing of the voltage samples. The upper graph represents the input voltage $V_{\text{in}}(t)$ and the samples which shall be captured. Due to the feedback loop the DAC 'knows' the voltage of the subsequent sample in advance. Therefore, it jumps right after a sampling event is finished to the next expected voltage. In the case, the feedback loop is working well, the DAC voltage and the input voltage have nearly the same value ($V_{\text{in}}(i\,\Delta t_{\text{s}}) \approx V_{\text{DAC}}(i\,\Delta t_{\text{s}})$ at the time point when the next sampling procedure is actuated. Hence, the difference voltage $\Delta V(i\,\Delta t_{\text{s}}) = V_{\text{in}}(i\,\Delta t_{\text{s}}) - V_{\text{DAC}}(i\,\Delta t_{\text{s}}) \approx 0$ which will be captured by the ADC has a quite low value (see Figure 3.52b). Though note that the maximum value of the difference voltage $\|\Delta V(t)\|_\infty = 2\|V_{\text{in}}(t)\|_\infty$ is even two times larger than the input voltage because the feedback loop only works for the sampling point and not continuously in time. But since we are only interested in the signal parts around zero voltage, $\Delta V(t)$ may be clipped without affecting linearity or sensitivity of the device. The voltage interval of interest for data capturing is assigned by FSR_{ADC} in Figure 3.52. The clipping is done either by a separate clipping amplifier or by the ADC itself. Storage effects of the clipping device and settling of the DAC voltage within the sampling interval Δt_{s} has carefully to be considered in the actual system design. The full-scale range of the DAC must be able to handle the bipolar input voltage, hence

$$\mathrm{FSR_{DAC}} \geq 2\|V_{\mathrm{in}}(t)\|_{\infty} \tag{3.67}$$

What is the profit of the feedback principle compared to the forward sampling? First, we can augment the dynamic of the receiver by increasing the maximum input voltage. If we suppose that the full-scale range of the ADC $\mathrm{FSR_{ADC}}$ covers the quantization interval q_2 of the DAC, the total resolution in bit of the feedback receiver equals the sum of the ADC and DAC resolution in the ideal case. Practically, $\mathrm{FSR_{ADC}}$ should overlap a few quantization intervals of the DAC so that we approximately yield

$$\mathrm{ENOB_{feedback}} \approx \mathrm{ENOB_{ADC}} + \mathrm{ENOB_{DAC}} - \underbrace{(1 \cdots 2)}_{\text{due to overlap}} \tag{3.68}$$

Second, the ADC may again degenerate to a comparator as discussed above in paragraph 'Polarity Receiver' which could be applied to maximize the sampling rate due to the fast operation of comparators and DACs. However by the feedback principles, the comparator is now always driven by a small noisy voltage without offset (i.e. $y_0 \to 0$, see Figure 3.48) avoiding the difficulties of the simple polarity receiver. The overall resolution of the device is determined by $\mathrm{ENOB_{DAC}}$ and further sensitivity improvements must be provided by the processing gain which will be as higher as larger the sampling rate will be.

Third, one can try to find the optimum device configuration for specific conditions. We will illustrate such a consideration by a strongly simplified example. For that purpose and to keep simple, we assume that ADC and DAC are built from switching circuits which behave nearly comparably. The figure of merit of a simple switch can be defined by

$$\mathrm{FoM_0} = \frac{P_0}{f_s}$$

where P_0 is the power dissipation of the switch and f_s is the switch rate. A DAC of N_{DAC} bit resolution needs at least N_{DAC}-switch stages. Thus, its power consumption can approximately be estimated by[10] (compare (3.54))

$$P_{\mathrm{DAC}} = N_{\mathrm{DAC}} f_s \,\mathrm{FoM_0}$$

A flush ADC of N_{ADC} bit resolution requires in minimum $2^{N_{\mathrm{ADC}}} - 1$ comparators which are counted here as switches. The gates of the internal ADC–MUX are not respected for simplicity. Hence, we get for the power demand (compare (3.53))

$$P_{\mathrm{ADC}} = (2^{N_{\mathrm{ADC}}} - 1) f_s \,\mathrm{FoM_0}$$

10) Here, we deal with the bit number of the devices which supposes ideal behaviour. The effective bit number should be used if device imperfections have to be included into the consideration.

A convenient design criterion could be to operate ADC and DAC at the same power so that the resolution between both devices should be roughly distributed as

$$P_{\text{ADC}} = P_{\text{DAC}} \quad \Rightarrow \quad 2^{N_{\text{ADC}}} - 1 \approx N_{\text{DAC}}$$

Supposing the availability of a 4-bit high-speed ADC, the compatible DAC should have about 15 bits. If we select the technically feasible and not too demanding quantities $N_{\text{ADC}} = 4$ and $N_{\text{DAC}} = 8$, we yield in the best case an overall resolution of $N_{\text{feedback}} \approx 10$ bit by respecting an ADC–DAC overlap of 2 bits (see (3.68)).

Let us take a simple forward sampling device as reference. It has to apply a 10-bit ADC in order to get the same resolution performance. Using a reference ADC of the same technology, we can expect roughly the same figure of merit. If both ADCs are operated at the same power, the sampling rates of both systems relate as

$$\frac{f_{\text{s}}}{f_{\text{s, ref}}} = \frac{2^{N_{\text{feedback}}} - 1}{2^{N_{\text{ADC}}} - 1} \approx 2^{N_{\text{DAC}} - (1 \cdots 2)}$$

In the case of our example, the sampling rate and hence the measurement rate too could be improved by a factor of 64 or we could extend the averaging to reduce the noise by a factor of 8 (equal to 3 bit more overall resolution) if we keep the measurement rate as before.

3.3.6.2 MiMo-Measurement System

Multiple input and multiple output (MiMo) sensor devices are required for distributed sensor arrangements (e.g. an antenna array) or if the n-port parameters of a DUT have to be determined. The capability of error correction (see Section 2.6.4) also supposes a sufficient number of measurement channels.

Due to its simple structure and the robust timing, the digital PN-correlator can simply be extended to a device of an arbitrary number of input and output channels. Figure 3.53 depicts the bloc schematics.

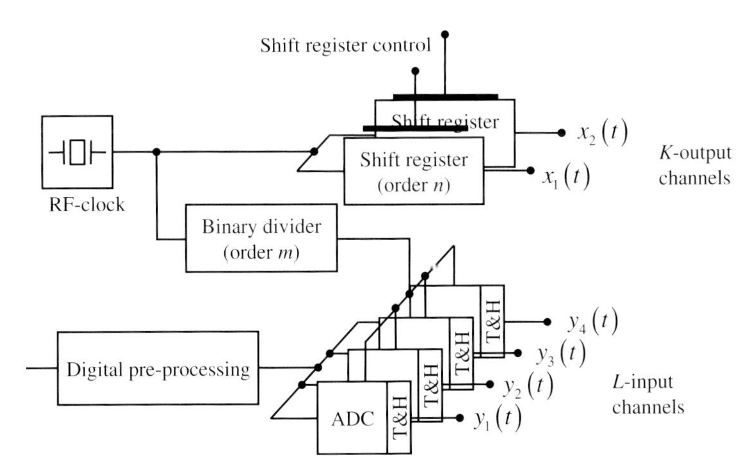

Figure 3.53 Basic structure of a multi-channel sensor device.

A MiMo-measurement device of L input and K output channels must be able to characterize the scenario under test by $L\,K$ impulse response functions as it was introduced in Figure 2.47, Section 2.4.2. In order to gain these IRFs from the measurements, we have to apply correlation since we deal with wideband signals of long duration. In accordance to (2.136) and (2.139), we can write

$$\mathbf{C}_{yx}(t) = \mathbf{y}(t) * \mathbf{x}^T(-t) = \mathbf{g}(t) * \mathbf{x}(t) * \mathbf{x}^T(-t) = \mathbf{g}(t) * \mathbf{C}_{xx}(t) \tag{3.69}$$

with

$$\mathbf{x}(t) = \begin{bmatrix} x_1(t) & x_2(t) & \cdots & x_K(t) \end{bmatrix}^T$$

$$\mathbf{y}(t) = \begin{bmatrix} y_1(t) & y_2(t) & \cdots & y_L(t) \end{bmatrix}^T$$

$$\mathbf{C}_{yx}(t) = \begin{bmatrix} y_1(t) * x_1(-t) & y_1(t) * x_2(-t) & \cdots & y_1(t) * x_K(-t) \\ y_2(t) * x_1(-t) & y_2(t) * x_2(-t) & \cdots & y_2(t) * x_K(-t) \\ \vdots & \vdots & \ddots & \vdots \\ y_L(t) * x_1(-t) & y_L(t) * x_2(-t) & \cdots & y_L(t) * x_K(-t) \end{bmatrix}$$

$$\mathbf{C}_{xx}(t) = \begin{bmatrix} x_1(t) * x_1(-t) & x_1(t) * x_2(-t) & \cdots & x_1(t) * x_K(-t) \\ x_2(t) * x_1(-t) & x_2(t) * x_2(-t) & \cdots & x_2(t) * x_K(-t) \\ \vdots & \vdots & \ddots & \vdots \\ x_K(t) * x_1(-t) & x_K(t) * x_2(-t) & \cdots & x_K(t) * x_K(-t) \end{bmatrix}$$

If the correlation matrix of the stimulus signals is a diagonal matrix of δ-functions, (3.69) simplifies and the correlation matrix $\mathbf{C}_{yx}(t)$ immediately represents the wanted set of impulse response functions $\mathbf{g}(t)$:

$$\mathbf{C}_{yx}(t) \propto \mathbf{g}(t) \quad \text{if} \quad \mathbf{C}_{xx}(t) \approx \begin{bmatrix} \delta(t) & 0 & \cdots & 0 \\ 0 & \delta(t) & \cdots & 0 \\ \vdots & \vdots & \ddots & \vdots \\ 0 & 0 & \cdots & \delta(t) \end{bmatrix} \tag{3.70}$$

Equation (3.70) requires that all stimulus signals $x_1(t) \cdots x_K(t)$ must have a large bandwidth and additionally they must be mutually uncorrelated. A set of Gold-codes meets approximately these conditions [44]. However, the different codes are not perfectly uncorrelated and their auto-correlation functions are affected by side lobes so that they are less perfect as for M-sequences. But by a clever choice of specific Gold-codes, one is able to relieve a certain time segment of the response functions from perturbing side lobes [53, 54].

We will not consider this approach in details rather we will stay at M-sequences due to their better auto-correlation and simpler handling. But in this case, we cannot apply (3.70) because different M-sequences of the same order are not completely uncorrelated rather we have to de-convolve $\mathbf{C}_{xx}(t)$ from $\mathbf{C}_{yx}(t)$. This may lead to numerical problems if the non-diagonal elements of $\mathbf{C}_{xx}(t)$ becomes too large. Hence, one should select amongst all possible M-sequences of a given length a subset whose mutual correlation is as low as possible. Corresponding considerations can be found in Ref. [55]. A MiMo system using this approach measures at maximum speed. But additional numerical expense to perform the de-convolution has to be respected and it needs either a set of different shift registers or programmable registers.

Another option is, in order to keep the measurement system simple, to restrict on K identical shift registers which provide all the same type of M-sequence namely $m(t)$. Nonetheless, what can we do to gain all IRFs if we do not have available a set of uncorrelated stimulus signals? The solution is to repeat the measurement at least K times and to perform simple modifications of the test signals each time. Without loss in generality, we will exemplify the approach for $K = 2$ using arbitrary test signals first. Later, we will refer to the M-sequence.

Let us assume that the signals $x_{11}(t)$ and $x_{21}(t)$ stimulate the scenario within a first measurement step. Here, the first index indicates the number of the stimulation port and the second index stands for the number of the measurement. Thus, in a second measurement run we deal with $x_{12}(t)$ and $x_{22}(t)$ and so forth. These signals can be arranged in the matrix

$$\mathbf{x}(t) = \begin{bmatrix} x_{11}(t) & x_{12}(t) \\ x_{21}(t) & x_{22}(t) \end{bmatrix}$$

and the same can be done for the measured signals

$$\mathbf{y}(t) = \begin{bmatrix} y_{11}(t) & y_{12}(t) \\ y_{21}(t) & y_{22}(t) \\ \vdots & \vdots \\ y_{L1}(t) & y_{L2}(t) \end{bmatrix}$$

Since (3.69) also holds under this condition, we get for $\mathbf{C}_{xx}(t)$

$$\mathbf{C}_{xx}(t) = \begin{bmatrix} x_{11}(t) * x_{11}(-t) + x_{12}(t) * x_{12}(-t) & x_{11}(t) * x_{21}(-t) + x_{12}(t) * x_{22}(-t) \\ x_{21}(t) * x_{11}(-t) + x_{22}(t) * x_{12}(-t) & x_{21}(t) * x_{21}(-t) + x_{22}(t) * x_{22}(-t) \end{bmatrix}$$

Corresponding to (3.70), $\mathbf{C}_{xx}(t)$ should be a diagonal matrix of δ-functions. Since $\mathbf{C}_{xx}(t)$ constitutes now from four different time signals, it should not be a serious problem to meet this condition by a proper choice of these signals. That is, the four stimulus signals must be either uncorrelated[11)] or their cross-correlation must mutually compensate.

We will try to achieve this by using identical shift registers at every stimulation port. For that purpose, the shift register as depicted in Figure 3.32 is used. It opens two possibilities to control the output signal. The first one is to switch it on/off by

11) It should be noted that (3.69) can also be solved for $\mathbf{g}(t)$ if $\mathbf{C}_{xx}(t)$ is not ideally diagonal. This would require de-convolution which is best performed in the frequency domain, that is $\mathbf{C}_{xx}(t) \xrightarrow{FT} \underline{\mathbf{\Psi}}_{xx}(f)$. We have already discussed corresponding problems in Sections 2.5.4 and 2.6.4. There, we have seen that the condition number $\kappa(\underline{\mathbf{\Psi}}_{xx}(f))$ may be a mean to assess the quality of the inversion. Note also that only the diagonal elements of $\underline{\mathbf{\Psi}}_{xx}(f)$ represent actually auto spectra.

acting on the enable/set input. The second option is to invert the polarity of the M-sequence $m(t)$ by feeding the XOR-gate with either 0 (non-inversion) or 1 (inversion). We can express this by

$$\text{Version 1:} \quad \mathbf{x}(t) = m(t)\begin{bmatrix} 1 & 0 \\ 0 & 1 \end{bmatrix} \quad \Rightarrow \quad \mathbf{C}_{xx}(t) = C_{mm}(t)\begin{bmatrix} 1 & 0 \\ 0 & 1 \end{bmatrix} \tag{3.71}$$

$$\text{Version 2:} \quad \mathbf{x}(t) = m(t)\begin{bmatrix} 1 & 1 \\ 1 & -1 \end{bmatrix} \quad \Rightarrow \quad \mathbf{C}_{xx}(t) = 2\,C_{mm}(t)\begin{bmatrix} 1 & 0 \\ 0 & 1 \end{bmatrix} \tag{3.72}$$

Generalizing these approaches to an arbitrary number K of input channels, version 1 implies that all shift registers are sequentially activated within the measurement procedure. This general philosophy (independently from the type of the used stimulus signal) is the usual method to measure n-port devices in our days (e.g. by network analysers).

For the second version, the matrix in (3.72) indicating the polarity of the test signals must be a Hadamard matrix since for it holds $\mathbf{H}\,\mathbf{H}^T = K\,\mathbf{I}$ ($\mathbf{H}$ – Hadamard matrix of order$K$; $\mathbf{I}$ – identity matrix). This method only works if a Hadamard matrix of the order K exists. Version 2 increases the overall stimulation power since all transmitters are always active. Consequently it improves the dynamic range of the measurements compared to version 1 provided that the receivers are not saturated by the higher power. If the power of the individual transmitters has to be reduced by any reason (e.g. if any constructive interference of signals at the receiver inputs leads to saturation effects), the advantages of version 2 in favour of version one will vanish.

3.3.6.3 **Up–Down-Conversion**

The device conceptions considered so far are intended to operate in the baseband that is from (nearly) DC to $f_c/2$. But many applications and the radio regulation as well require a wideband operation within an arbitrary frequency band. In this chapter, we will introduce some conceptions – based on homodyne and heterodyne principles – which will give a high flexibility in selecting the operational frequency band. The intended approaches require mixing devices which are typically restricted to low peak power operations. Hence, the use of M-sequences is of advantages compared to short pulse modulations due to the lower burden by voltage peaks.

Figure 3.54 depicts a first example which is based on a homodyne concept. In imitation to the term FMCW (*F*requency *M*odulated *C*ontinuous *W*ave)-sensor we will call it the MSCW (*M-S*equence modulated *C*ontinuous *W*ave)-principle. The idea is to shift the M-sequence spectrum $\underline{M}(f)$ to a higher frequency band by multiplying it with a carrier signal of the frequency f_0 which leads to an operational band $|f_0 \pm f_c/2|$ for the stimulus signal $x(t)$ (see Annexes B.1 and B.2, Table B.2). Hence, we have increased the physical bandwidth of the sounding signal which equals the two-sided bandwidth of the M-sequence $\ddot{B}_x = \ddot{B}_m$. Nevertheless, the coherence time of both signals will be untouched by up-conversion (compare

(a)

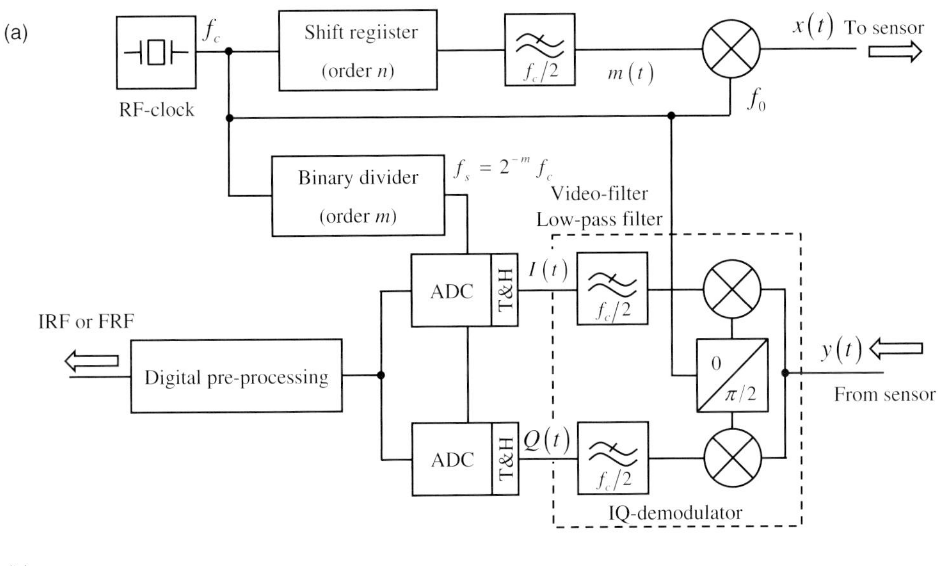

(b)

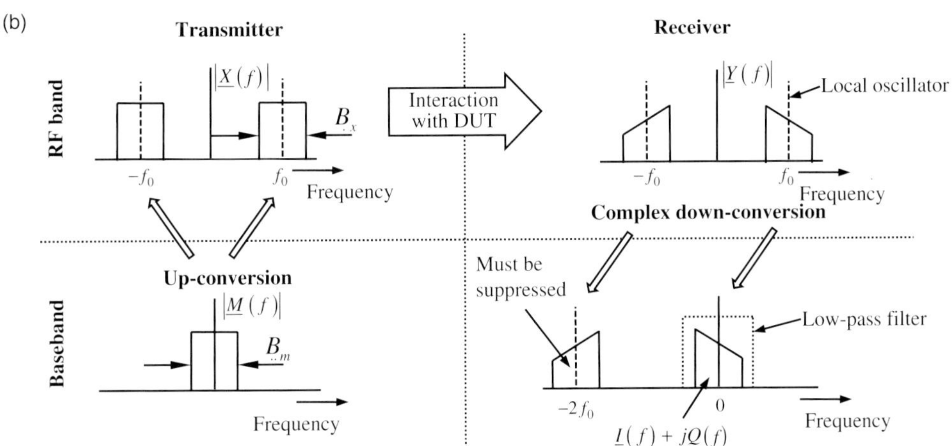

Figure 3.54 MSCW-principle using a homodyne IQ-concept for direct down-conversion. (a) Circuit schematics and (b) simplified spectra (see also Section 3.4.2.1).

Section 2.2.5; (2.80)). The product of carrier signal and M-sequence is either performed by analogue multipliers as ring mixers or Gilbert cells or it is simply done by a fast XOR-gate as already implemented in the shift register sample from Figure 3.32. In the last case, the low-pass filter in the transmitter chain must be removed.

Very simple devices can be built if the master clock f_c feeds the shift register and the mixer/XOR-gate (compare Figure 3.54). If one selects, for example $f_0 = f_c \approx 7\,\text{GHz}$, the operational frequency band fortunately coincide with the FCC-band (3.1–10.6 GHz). Figure 3.55 depicts the spectrum of an implemented

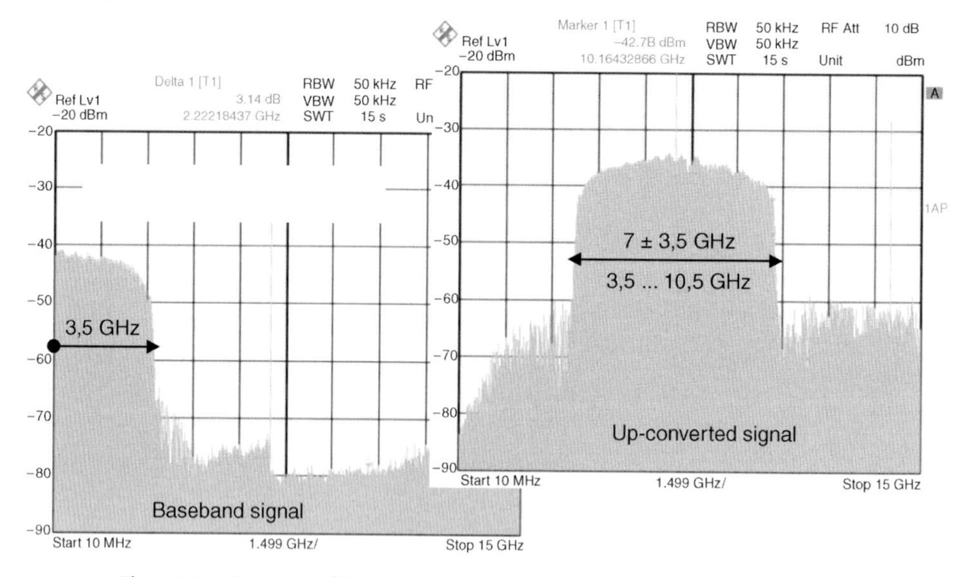

Figure 3.55 Spectrum of the baseband and the up-converted signal for $f_c = f_0 = 7$ GHz.

system for FCC conform operation. Corresponding can be done to meet the ECC-band (6–8.5 GHz). Here, the RF-clock has to be selected to $f_c \approx 2.41$ GHz while the carrier frequency results from the tripled clock rate $f_0 = 3f_c$.

In order to apply the same data capturing approach as described before, the receive signal $y(t)$ must be shifted back into the baseband (Figure 3.54). For that purpose, it is multiplied with a local oscillator (LO) signal which is identical to the carrier for the up-conversion in order to assure coherence between transmitter and receiver.

As the rules of the Fourier transform imply (see Annex B.1, Table B.2), the frequency shift can be done mathematically by multiplying the inputs signal $y(t)$ with the complex exponential $e^{j2\pi f_0 t}$. As Figure 3.54 depicts, this procedure leads to two sub-spectra which are centred around $f = 0$ Hz and $f = -2f_0$. The first one represents the spectrum of the wanted down-converted signal and the second one is out of interest. It will be suppressed by low-pass filters.

However in the real world, time signals can never be complex. Hence, we split the complex multiplication in two real operations replacing the exponential function by a sine and a cosine that is we have to provide two LO-signals mutually delayed by quarter period. Finally, we get two measurable real-time signals $I(t)$ and $Q(t)$ which we formally combine to the complex time signal $\underline{z}(t)$ representing the complex-valued envelope of the captured RF-signal.

In the case, the Video-filter is a first order low-pass (e.g. RC-low-pass), we have:

$$\underline{z}(t) = I(t) + jQ(t) = \frac{1}{\tau_0} u_{-1}(t) e^{-t/\tau_0} * \left(\cos 2\pi f_0 t + j\, y(t) \sin 2\pi f_0 t\right) \tag{3.73}$$

if the Video-filter is a short-time integrator of integration time t_{I}, we get:

$$\underline{z}(t) = I(t) + jQ(t) = \mathrm{rect}\left(\frac{t}{t_{\mathrm{I}}}\right) * \left(y(t)\cos 2\pi f_0 t + j\, y(t)\sin 2\pi f_0 t\right) \tag{3.74}$$

The circuit which performs (3.73) and (3.74) are called IQ-demodulator. The signal $I(t)$ represents the component of $y(t)$ which is in-phase to the carrier signal and $Q(t)$ is called the quadrature component that is it is that part of $y(t)$ which is 90° to the carrier.

If the complex signal $\underline{z}(t)$ is correlated with the original M-sequence, we finally get the wanted complex impulse response $\underline{g}_0(t)$ of the DUT:

$$\underline{g}_0(t) \sim \underline{z}(t) * m(-t) = I(t) * m(-t) + j\, Q(t) * m(-t) \tag{3.75}$$

Note that the IRF $\underline{g}_0(t)$ only represents the impulse response of the DUT if it has pass band behaviour and its bandwidth is not larger than the bandwidth of the stimulus.

How the complex IRF $\underline{g}_0(t)$ can be interpreted in relation to the actual and physically correct (real valued) impulse response $g(t)$ of the DUT? From the consideration above, the relation between the real and complex-valued IRF is given by

$$\begin{aligned} &g(t) = \Re\{\underline{g}_0(t) \cdot \mathrm{e}^{j2\pi f_0 t}\} \\ &\text{with} \quad \underline{g}_0(t) = g_0(t)\mathrm{e}^{j\varphi_{0\mathrm{i}}(t)} \end{aligned} \tag{3.76}$$

where $g_0(t) = g_{\mathrm{env}}(t)$ represents the envelope of the physically real IRF $g(t)$ as depicted in the upper graph of Figure 3.56 and $\varphi_{0\mathrm{i}}(t)$ is the instantaneous phase. Hence, the complex IRF $\underline{g}_0(t)$ can be considered as the complex envelope of the physically real IRF $g(t)$. The term envelope was introduced in Section 2.2.2 based on the analytic signal. If we relate (2.2), (2.5) and (2.6) to our current problem, we can write analytic IRF:

$$\underline{g}_{\mathrm{a}}(t) = g(t) + jg_{\mathrm{H}}(t) = g(t) + jHT\{g(t)\} = g_{\mathrm{env}}(t)\mathrm{e}^{j\varphi_{\mathrm{i}}(t)} = \underline{g}_{\mathrm{env}}(t)\mathrm{e}^{j2\pi f_0 t} \tag{3.77}$$

from which we can find

$$\begin{aligned} \underline{g}_{\mathrm{env}}(t) &= g_{\mathrm{env}}(t)\mathrm{e}^{j\varphi_{0\mathrm{i}}(t)} \\ g_{\mathrm{env}}(t) &= g_0(t) = |\underline{g}_{\mathrm{a}}(t)| \\ \varphi_{\mathrm{i}}(t) &= \arg(\underline{g}_{\mathrm{a}}(t)) = \varphi_{0\mathrm{i}}(t) + 2\pi f_0 t \\ f_{\mathrm{i}}(t) &= \frac{1}{2\pi}\frac{\mathrm{d}}{\mathrm{d}t}\varphi_{\mathrm{i}}(t) \end{aligned} \tag{3.78}$$

This confirms the consistency of the terms complex IRF and complex envelope. In (3.78), $\varphi_{\mathrm{i}}(t)$ refers to the instantaneous phase of the analytic IRF while $\varphi_{0\mathrm{i}}(t)$ is the instantaneous phase of the complex envelope/IRF. $\varphi_{0\mathrm{i}}(t)$ quantifies the deviation of the instantaneous frequency $f_{\mathrm{i}}(t)$ of the actual DUT compared to the carrier frequency f_0 of the measurement device. If holds $f_0 = \overline{f_{\mathrm{i}}(t)}$, the phase fluctuations within the complex IRF would be minimal. The lower graphs in Figure 3.56 illustrate the complex IRF from different view angles. For clarification, the 3D-plot also includes the projections of the complex function $\underline{g}_0(t)$ onto to the complex plane, the in-phase-time plane and the quadrature-phase-time plane. As obvious from Figure 3.56, high-frequency

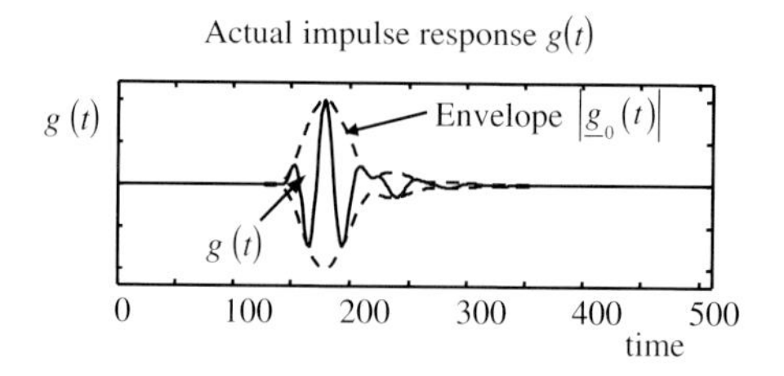

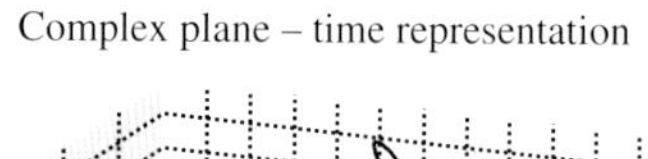

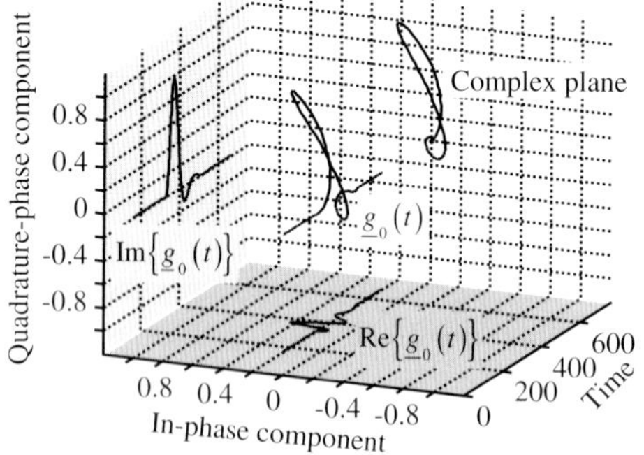

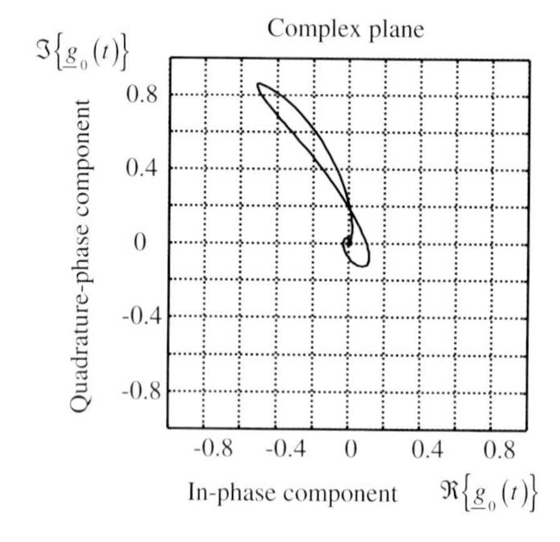

Figure 3.56 Different representations of the complex impulse response of a band-pass DUT.

variations of the real-valued IRF are removed by the use of the complex-valued base-band function $\underline{g}_0(t)$. It is smoother and consequently it is satisfied with a lower density of sampling points which reduces the data volume to be handled. However, the achievable data reduction for large fractional bandwidth DUTs will be less significant as for narrowband DUTs. Some measurements which show the impact of dispersive transmission channels onto the complex IRF are exemplified in [56]. Another example is depicted by the movie "Fig 3.56_complex_IRF" (to be downloaded from Wiley-homepage). It shows the radar return of a target which departs from the antenna.

The technical challenge to implement a device structure as depicted in Figure 3.54 results from the requirement of identical transmission behaviour for the

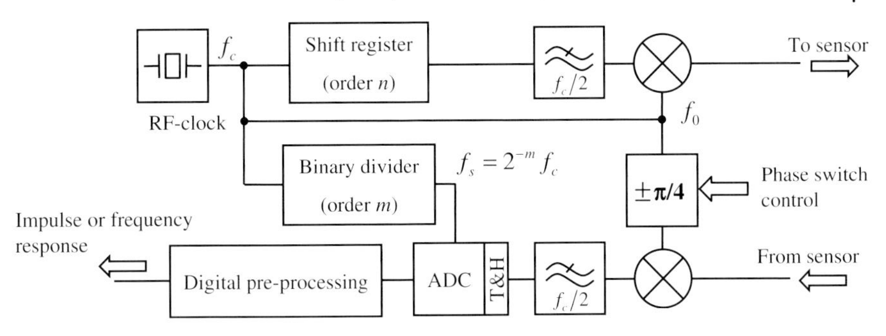

Figure 3.57 Wideband MSCW-principle using sequential IQ-down-conversion.

I- and Q-path. Except from monolithic integration, this is hardly to be guaranteed for a signal bandwidth exceeding several GHz. In order to reduce the I/Q imbalance to a minimum, it is recommended to perform the down-conversion sequentially as depicted in Figure 3.57. For that purpose, the measurement is repeated two times – first the phase shifter is set to $\varphi_0 - \pi/4$ and then to $\varphi_0 + \pi/4$ (φ_0 is an unknown but constant internal phase delay of the LO-circuit). During these two measurement steps, the DUT should not change its behaviour, that is any movement should be negligible. This approach leads to identical (linear) signal deformations for the I- and Q-path which can be removed by usual system calibration (see Section 2.6.4).

Finally, a second up–down-converter stage as depicted in Figure 3.58 further increases the flexibility to operate in arbitrary frequency bands. This approach is typically used to insert the main signal amplification between the two converter stages with the goal to reduce the influence of $1/f$-noise in the IQ-down-converter.

3.3.6.4 **Equivalent Time Oversampling**

So far, the equivalent sampling rate was selected to be equal to the clock rate of the shift register $f_{\text{eq}} = f_c$. This is an optimum choice since it provides the lowest data amount and the captured data are restricted to a spectral band which is well occupied by signal power so that needles noise accumulation is avoided. But this rigid

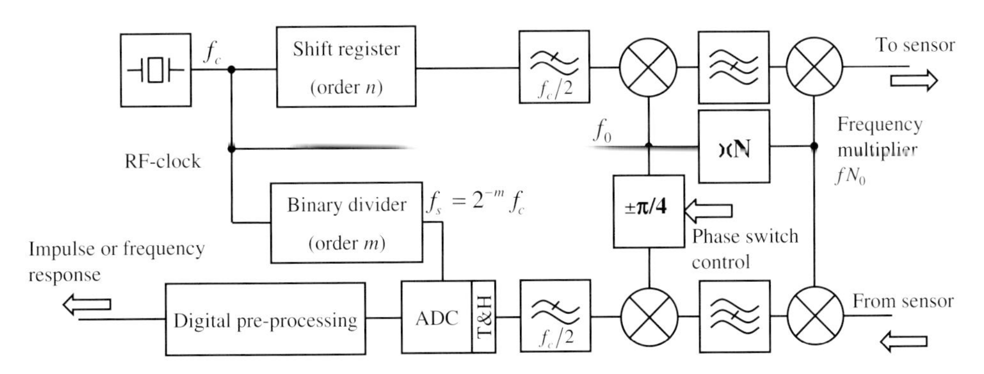

Figure 3.58 Circuit schematic of an MSCW-device using two conversation stages [57].

system layout limits the flexibility of the device conception. As we have seen from Figure 3.35, we need an analogue anti-aliasing filter in order to limit the signal bandwidth to $f_c/2$. Hence, we cannot change the clock rate and bandwidth of the M-sequence without replacing the anti-aliasing filter. Furthermore, the quality of this analogue filter will largely determine the suppression of aliasing components and it will cause pulse shape distortions due to steep transitions from pass- to stop band.

Therefore, we will try to replace analogue by digital filtering which would result in a simpler implementation (since no analogue filter is required) and the device bandwidth as well as the filter characteristic could be adapted to the actual application without variations in the hardware. We will also see that this concept even permits to increase the usable bandwidth beyond the maximum value given by the maximum toggle rate of the shift register.

If the anti-aliasing filter is rejected, the data gathering system must be able to handle the full bandwidth of the receive signal. That is, as long as spectral parts of the stimulus signal exceed the noise level they have to be captured correctly without violation of the Nyquist theorem. In this regard, the term noise level has to be considered with care since it does not mean the noise level of the actual receive signal rather it refers to the remaining noise after averaging has been applied. Figure 3.59 illustrates what is meant for an example of an idealized M-sequence spectrum.

The spectral part of the signal emphasized by the grey area is typically used by the basic M-sequence approach. As mentioned above, the signal contains about 80% of its energy within this frequency band $\underset{\cdot}{B}_0$. By limiting the receiver bandwidth to that value, the equivalent sampling rate can be selected to $f_{eq} = f_c = 2\underset{\cdot}{B}_0$ what we have done so far. If we renounce that band limitation, the observation bandwidth has to cover all spectral parts which still contain deterministic signal components. Hence, the equivalent sampling rate must be increased. For the

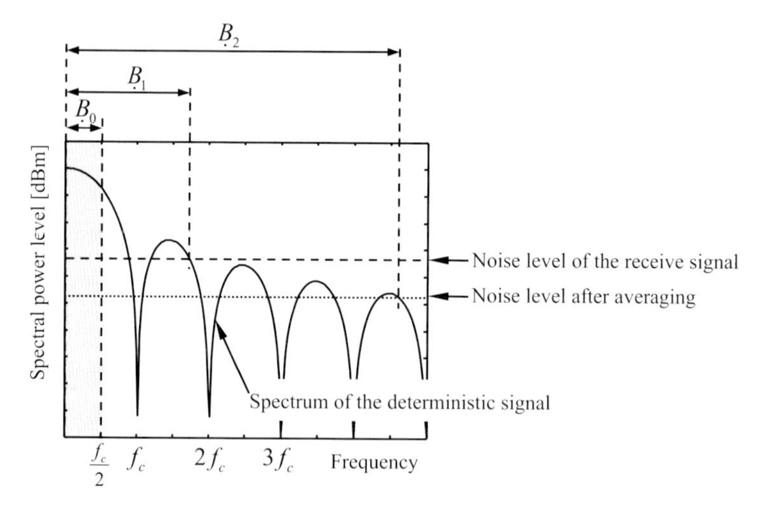

Figure 3.59 Illustration of bandwidth requirements for data capturing.

indicated examples this means $f_{eq} \geq 2B_2 \gg f_c$ and $f_{eq} \geq 2\,B_1 > f_c$ whether averaging was applied or not during data capturing.

At this point the question arises how to increase the equivalent sampling rate. By the hitherto principle, one voltage sample was captured from every chip of the M-sequence leading to the equivalent sampling rate $f_{eq} = f_c$. Now we need several samples per chip and we will do it by the same way as before except a simple modification as depicted in Figure 3.60.

Supposing the equivalent sampling rate must be three times larger than the clock rate $f_{eq} = 3f_c$. In that case, we divide the measurement process into three steps. Step one works as usual. The receiver captures (with or without averaging) a set of $2^n - 1$ data samples by sub-sampling. As before, the sampling points are located at $i\,\Delta t_{s0} = i2^m/f_c$ and the corresponding equivalent time spacing of the captured samples is $\Delta t_{eq} = 1/f_{eq} = 1/f_c$. By the second step, the sampling

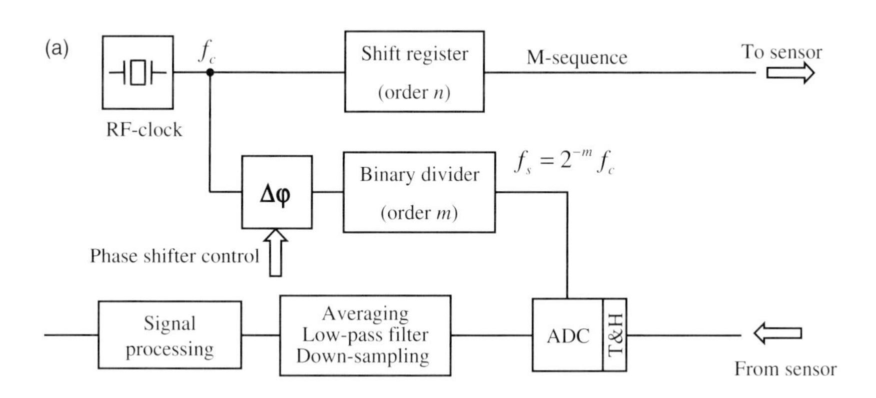

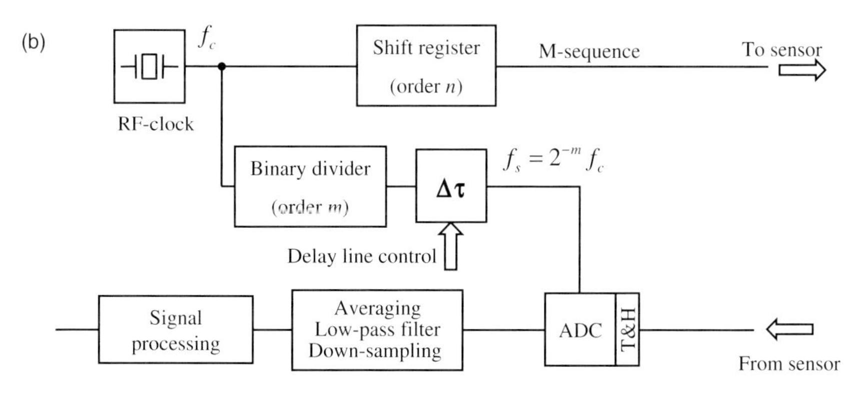

Figure 3.60 Two modifications of the basic M-sequence circuit to increase the equivalent sampling rate.

time points are shifted by a third of the clock period, that is $i\,\Delta t_{s1} = (i\,2^m + 1/3)/f_c$ and the measurements procedure is repeated as before. This leads to a second data set of $2^n - 1$ voltage samples having the same equivalent sampling rate as in the first run. But this set of samples is taken $\Delta t = 1/3f_c$ apart from the first one. The small Δt-shift is managed either by a controllable phase shifter (Figure 3.60(a)) or by delay line (Figure 3.60(b)). Finally, in a third step, the sampling time points are shifted again so that sampling takes place at $i\,\Delta t_{s2} = (i2^m + 2/3)/f_c$. After interleaving all three data sets, we will end up in a digitized signal of the threefold equivalent sampling rate $f_{eq} = 3f_c$. By that way, the equivalent sampling rate can be increased to nearly arbitrary large values.

We achieved the increased flexibility of sampling control by introducing an analogue component into the sampling control circuit. May this degrade the jitter performance of the device due to the additional devices? Generally this is not the case. Concerning the phase shifter, the signals before and after the device are sine waves of identical frequency. Hence, it will marginally affect the trigger jitter of the binary divider (compare (2.292)) as long as the phase shifter does not provide additional noise. The controllable delay line may work by the slow ramp – fast ramp principle as depicted in Figure 3.20. This will affect the slope of the trigger edge for the T&H- or S&H-circuit resulting in lower jitter robustness (compare (2.291), (3.18)). Since, however, the maximum sweep range of the delay line is quite short ($\Delta t \approx 1/f_c$), it will also not be a serious source of additional jitter.

Actually implemented M-sequence devices which are working on the basic principle as shown in Figure 3.35 have an additional low-pass filter in the receiver chain in order to suppress outer band noise. Since we have renounced such filter here, the noise floor will increase correspondingly to the receiver bandwidth which is now determined by the analogue bandwidth of the T&H- or S&H-circuits. But, we are able to reject the additional noise by catching up the band limitation with digital low-pass filtering whereas now bandwidth and impulse response of such filters may be flexibly adapted to the actual need. The price we have to pay for the larger flexibility is a somewhat lower averaging factor (supposing the same measurement time) since the measurements have to be repeated several times. Due to the intrinsic band limitation of the T&H- or S&H-circuits, oversampling factors of 4 till 8 are mostly sufficient. Hence, the actual loss in noise performance is practically restricted to a factor of 2–3.

Figure 3.61 depicts a further version to increase the equivalent sampling rate. Here, several ADC channels are working in parallel so that we will not have averaging losses as in the preceding case. Simple technical solutions of the delay lines are short micro-strip lines if the sampling unit is implemented on printed circuit boards or one exploits the propagation delay of logic gates in the case of a monolithic integration. In order to keep an equidistant sampling, the delay lines have to be trimmed precisely corresponding the given clock rate f_c.

For the introduced sampling method, the usable bandwidth of the receiver is only limited by the analogue bandwidth of the T&H- or S&H-circuits. Therefore, if their bandwidth is large enough, the device concept can be further extended by allocation of additional spectral power beyond the $f_c/2$ point as shown in

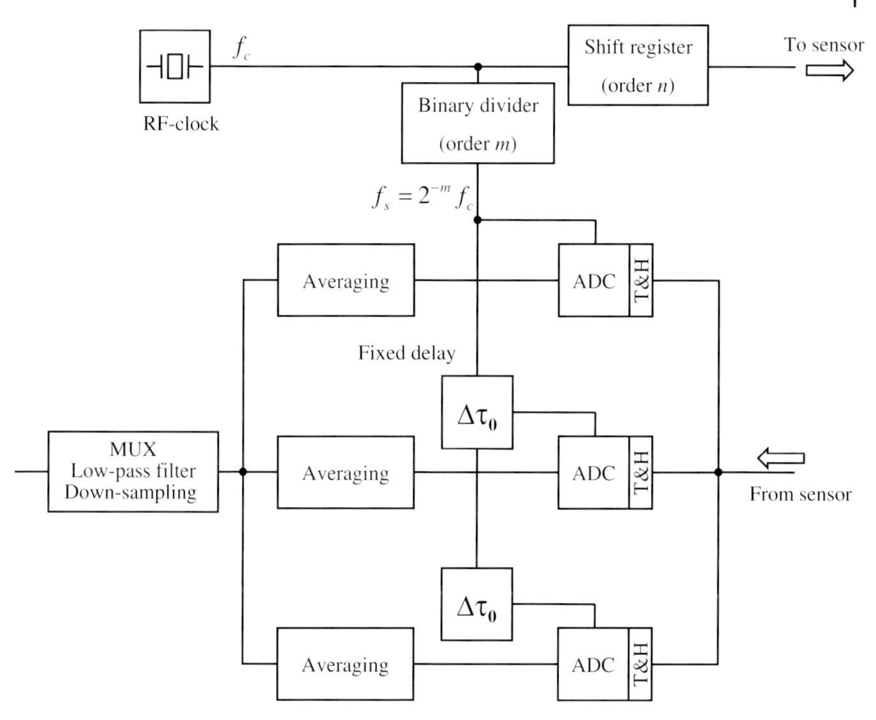

Figure 3.61 Increasing the equivalent sampling rate by factor 3 with time cascaded data capturing ($f_c\,\Delta\tau_0 = 1/3$).

Figure 3.62. In the shown example the M-sequence is mixed with its own clock rate (see also Figures 3.54 and 3.55) resulting in a spectral band $f_c/2 \cdots 3f_c/2$ and additionally it is superimposed with the original M-sequence (spectral band $0 \cdots f_c/2$). The resulting signal occupies a spectrum from $0 \cdots 3f_c/2$ but it is not any more an M-sequence. Therefore, in place of the Hadamard transform a correlation with the actual transmit signal (labelled as 'Reference' in Figure 3.62) has to be applied for impulse compression. The mixing and superposition of the different signals do not provide an absolute flat spectrum wherefore the device intrinsic impulse response will show some deviations from the wanted shape. Since they are of deterministic nature, they can be removed by calibration as discussed in Section 2.6.4. Figure 3.62 depicts the intrinsic impulse response of an implemented device after error removal [58, 59].

If the power divider is not inserted in the transmission path, the operational frequency band is $f_c/2 \cdots 3f_c/2$. Above (Figure 3.54), we used IQ-down-conversion for capturing of such signal. As we have seen here, one can also apply equivalent time oversampling for the same purpose whereas IQ-down-conversion has to be implemented by digital means now if required for data reduction.

3.3.6.5 **Beam Steering and Doppler Bank**

Two further device modifications have mainly in mind specific radar applications. The first refers to steer the beam of an antenna arrays without to turn it. For simplicity, we

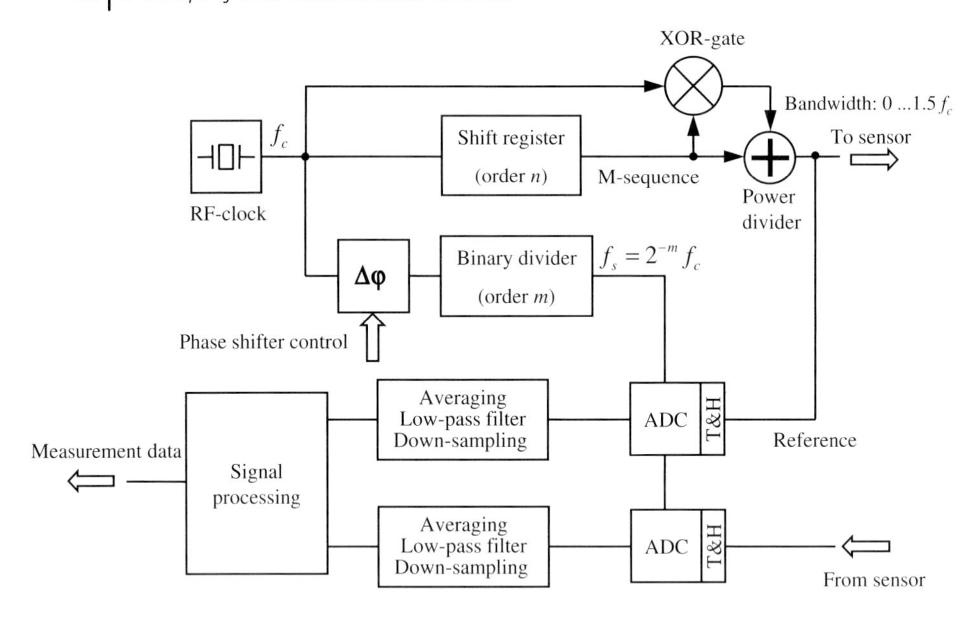

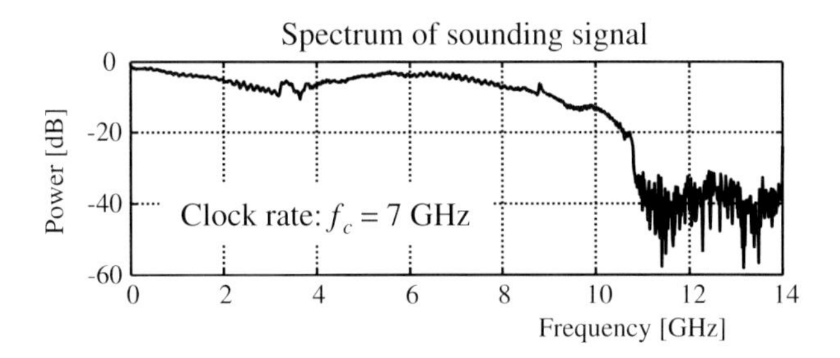

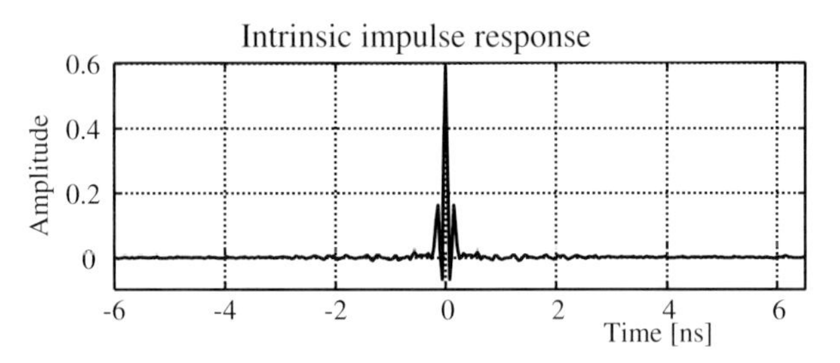

Figure 3.62 M-sequence device with expanded bandwidth and equivalent time oversampling [60].

will suppose that the individual radiators have an isotropic radiation characteristic. In the case of an UWB receiving array, the steering is done by appropriately processing the receive signals (see Section 4.10) which is typically executed in the digital domain. If, however, the transmit power should be concentrated within a small volume in space, one has to make sure that the waves emanating the different radiators actually

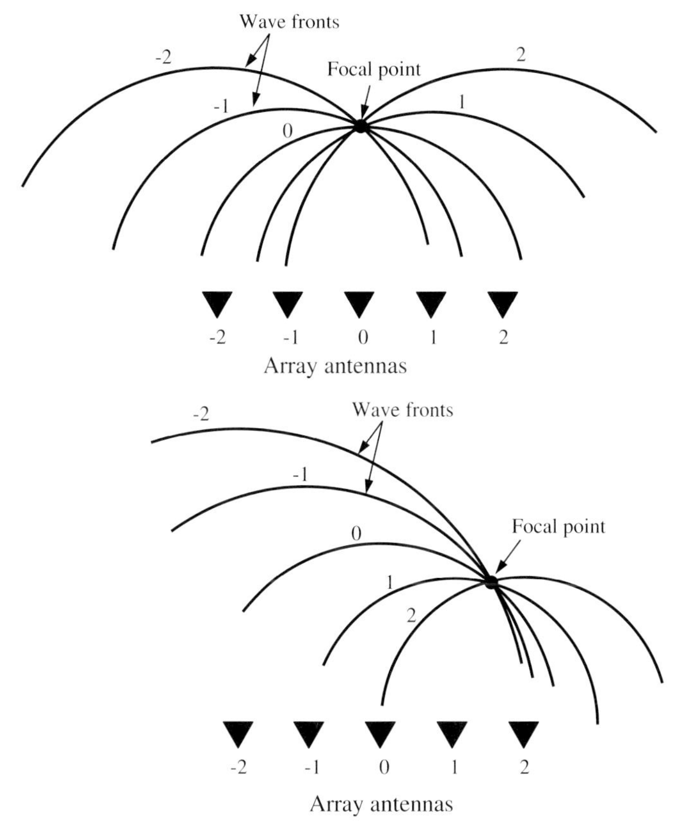

Figure 3.63 Illustration of beam steering by an array of UWB radiators. The indicated wavefronts relate to pulse signals.[12] They are numbered corresponding the radiator from which they are launched. The larger the radius of a wavefront is, as earlier the wave has been started.

interfere at the right position. We can control this by launching the waves from the individual antennas at different time points so that after an individual propagation time they all 'meet' at the right position. Figure 3.63 demonstrates the steering in the near field of a linear array of five radiators by a simple example. Further fundamental reflections on time domain beam steering can be found in Refs [61–63]. As obvious from the example, the performance of beam steering will largely depend on the precise launching of waves. That means that the signals which stimulate the different radiators have to be shifted carefully in time.

Since shift register chips made from SiGe are potentially low-cost components, one can equip every antenna element with its own signal generator which is directly integrated into the antenna feed. Since all shift registers are triggered by

12) We used here pulse signals for demonstration of the interference principle even if we deal here with PN-sequences. But the interference patterns of PN-sequences are not interpretable by a human being. If we would subject the PN-pattern a correlation procedure as explained in Section 4.3, Figures 4.6 and 4.7, we would result in an identical interference pattern as shown in the figure.

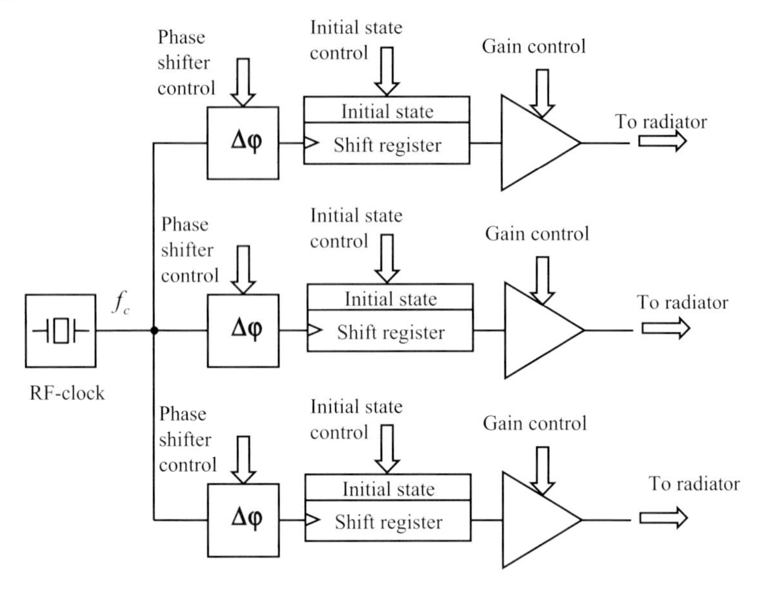

Figure 3.64 Steering of the radiation beam for an UWB antenna array.

the same RF-clock, the synchronism of the whole array is guaranteed. The steering is performed by selecting adequate trigger points for the different shift registers.

Corresponding to Figure 3.64, we have two opportunities to control the mutual timing between the radiations. A coarse pre-selection can be done by controlling the initial states of the shift register. This method allows discrete steps $\Delta t = 1/f_c$ of time shifting. The fine-tuning within these intervals, is finally be made by phase shifters. In order to control both the direction and the shape of the beam, the amplitude of the individual radiations may be controlled too. For that purpose, attenuators or variable gain amplifiers have to be additionally inserted. These components must provide signals of identical time shape independently from their output amplitude.

The second modification relates to measurements on high-speed targets. Here, two aspects are of major interest. First, one likes to know radial distance *and* speed of the target. Second, we have seen form (3.45) that a moving target will limit the recording time T_R and consequently the maximum achievable dynamic range of the sensor as shown in (2.337). In order to improve the sensor sensitivity and to get a precise speed measurement, the time on target has to be extended. Basically this can be done by two ways. By the first one, the radar return is periodically captured with a short recording time so that Doppler scaling does not affects the impulse compression. After rearranging the data in a radargram (Figure 3.30 gives an example for constant radial speed), one can follow the target trace by applying tracking filters (moving target indication – MTI – see Section 4.8).

Here, we will discuss a second approach resulting from a simple variation of the basic M-sequence concept. Since we like to increase the recording time above the value given by (3.45), we have to make sure that de-correlation between transmit and receive signal is avoided. This is done as follows: We launch at time t_0 a train of M-sequences extended over the duration t_D. We take from the sounding signal N_s equally distributed samples as reference for the later correlation. Our target should have a constant radial velocity v_r and its distance is $r_0 = r(t_0)$ at time t_0 (see Figure 2.17 for definition of the scenario). The first return signal entering the receive antenna arrives at $t_1 = t_0 + 2\,r_0/(c - v_r)$ while the very last part of the M-sequence train bounces the antenna at $t_2 = t_0 + t_D + 2(r_0 + v_r\,t_D)/(c - v_r)$. Thus, the duration of the receive signal is

$$t_{DR} = t_2 - t_1 = t_D \frac{c + v_r}{c - v_r} \approx t_D\left(1 + 2\frac{v_r}{c}\right) \tag{3.79}$$

In order to avoid any de-correlation with the transmit signal, we have to capture from the receive signal the identical number N_s of equally spaced samples. This leads to two distinct sampling rates for the transmitter f_{sT} and receiver f_{sR} which are related by

$$f_{sT} t_D = f_{sR} t_{DR} = N_s \tag{3.80}$$

If we take the transmitter sampling rate $f_{sT} = f_s = 2^{-m} f_c$ as reference, we yield the required offset of the receiver sampling rate:

$$f_{sR} = f_s + \Delta f_s \quad \Rightarrow \quad \Delta f_s = -2 f_s \frac{v_r}{c} \tag{3.81}$$

Equation (3.81) should not be confused with the frequency shift by narrowband Doppler. Actually it refers to the growing variations of the line spacing $n\,\Delta f_s$ of the comb spectrum (n– number of the harmonics) in a harmonic mixer (see Figure 3.25 for illustration of harmonic mixing).

The sampling rate adjustment by (3.81) will be able to compensate the Doppler scaling exactly. But we can also observe from (3.45) that the receiver is able to tolerate a target movement as long as the speed is limited to the interval $\pm\Delta v_r$. Hence, it would be sufficient to perform the search of the unknown target speed in a number of intervals $v_r = (2n_D \pm 1)\Delta v_r$; $n_D \in [-N_D, N_D]$ which are also referred as Doppler cells. The sampling rate offset for the individual Doppler cells results from

$$\Delta f_{sn} = -4\,n_D f_s \frac{\Delta v_r}{c} = -\frac{4 n_D f_s}{2 T_R f_c} = -\frac{n_D}{2^{m-1} T_R} \tag{3.82}$$

where $T_R \approx t_D$ refers to the required time to capture the data for one impulse response function (including averaging) and 2^m represents the factor of sub-sampling. n_D is a positive number for targets going away from the antenna and it is negative if they approach. The overall number of Doppler cells is $2N_D + 1$. It depends on the maximum targets speed $v_{r,max}$ and the required speed resolution Δv_r:

$$N_D = \frac{|v_{r,max}|}{2\,\Delta v_r} - \frac{1}{2} = \frac{T_R f_c\,|v_{r,max}|}{c} - \frac{1}{2} \tag{3.83}$$

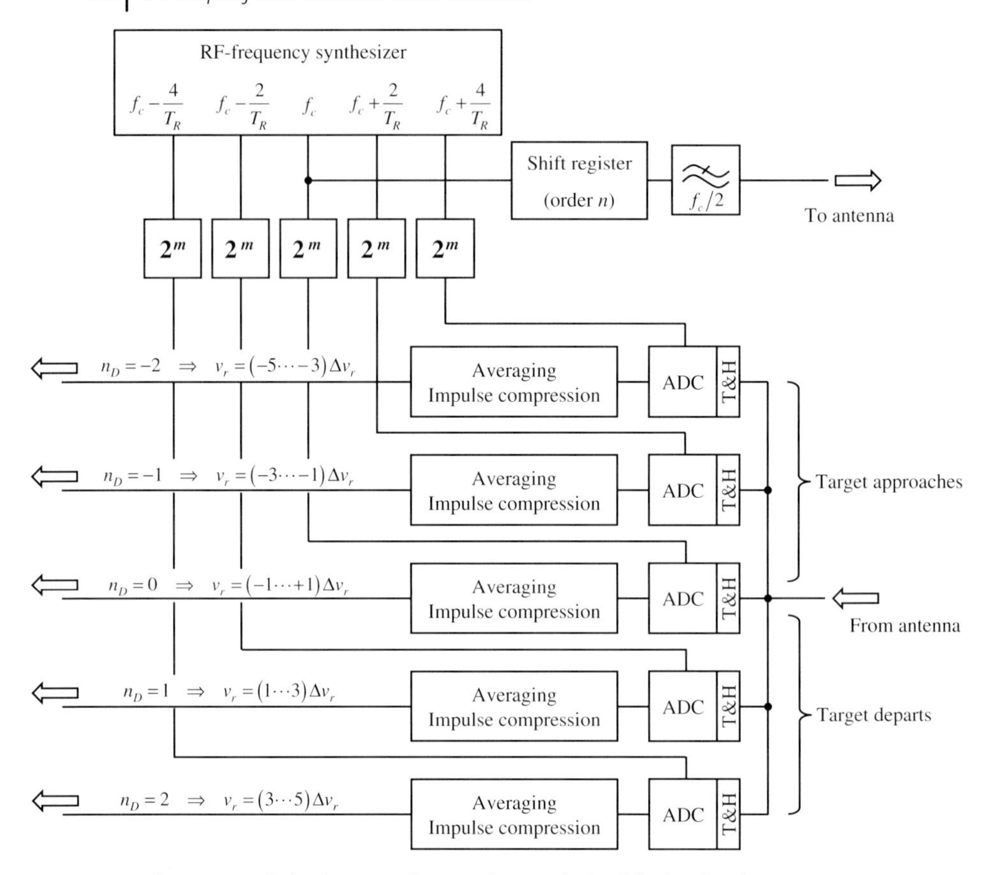

Figure 3.65 Block schematic of a PN-radar Doppler bank for baseband operations.

Figure 3.65 depicts the practical implementation of (3.82) for baseband M-sequence radar. It is built from a number of Doppler channels. The synthesis of their sampling rates is best done at the RF-clock level since it provides the best jitter performance. That channel which best meets the target speed provides the largest correlation peak so that by the time position of the peak and the number n_D of Doppler channel distance and speed can be determined in parallel. Range δ_r and speed resolution δ_v are given by (compare (3.45))

$$\delta_\mathrm{r} \approx \frac{\tau_\mathrm{coh} c}{2} \approx \frac{c}{f_\mathrm{c}} \tag{3.84}$$

$$\delta_\mathrm{v} = \Delta v_\mathrm{r} \approx \frac{c}{2T_\mathrm{R} f_\mathrm{c}} \approx \frac{\delta_\mathrm{r}}{2T_\mathrm{R}} \tag{3.85}$$

τ_coh is the coherence time of the sounding signal.

The example in Table 3.1 illustrates some values which can be expected if an M-sequence device operates at a clock rate of about 10 GHz.

Table 3.1 M-sequence Doppler bank ($f_c = 10\,\text{GHz} \Rightarrow \delta_r = 3\,\text{cm}$, maximum presumed target speed $v_{r,\max} = 300\,\text{m/s}$).

Recording time $T_R \approx t_D$	Measurement rate $m_r = 1/T_R$	Velocity resolution $\delta_v = c/2\,T_R f_c$	Clock rate offset $\Delta f_{cn} = 1/T_R$	Number of Doppler channels $2N_D+1$
100 ms	10 IRF/s	±0.15 m/s	20 Hz	2001
10 ms	100 IRF/s	±1.5 m/s	200 Hz	201
1 ms	1000 IRF/s	±15 m/s	2 kHz	21
100 μs	10 000 IRF/s	No Doppler scaling	20 kHz	1

3.3.6.6 Transmitter–Receiver Separation

The correct operation of any UWB sensor requires a rigid synchronization between signal source and receiver. As long as transmitter and receiver are located closely together, synchronization is usually not a problem. However, this will not be always feasible. Some methods to decouple the transmitter from the receiver will be shortly mentioned.

Separate RF-clock: Transmitter and receiver have their own time base (Figure 3.66). The synchronization signals will be subject a frequency lag those quantity and time stability will depend on the quality of the two RF-sources. Due to data gathering by sub-sampling, one needs a comparatively long time to capture the complete data set for at least one period. Thus, even in case of high-quality RF-clock generators, one runs the risk of de-correlation by direct relating transmitter and receiver. On the contrary, the correlation will not be affected if it is performed between two receive channels which are synchronized by the same

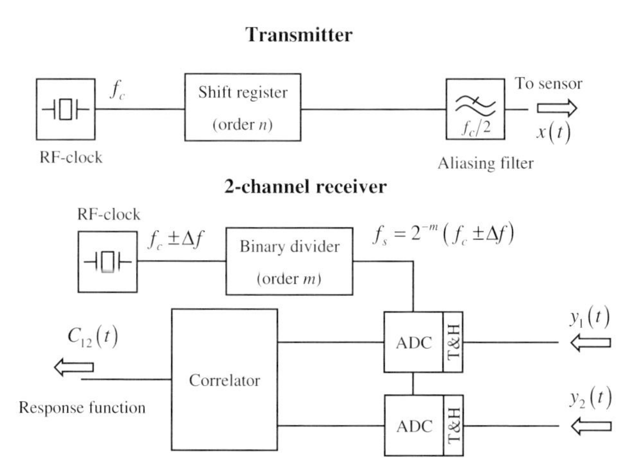

Figure 3.66 Transmitter and receiver operate with separate time base (RF-clock).

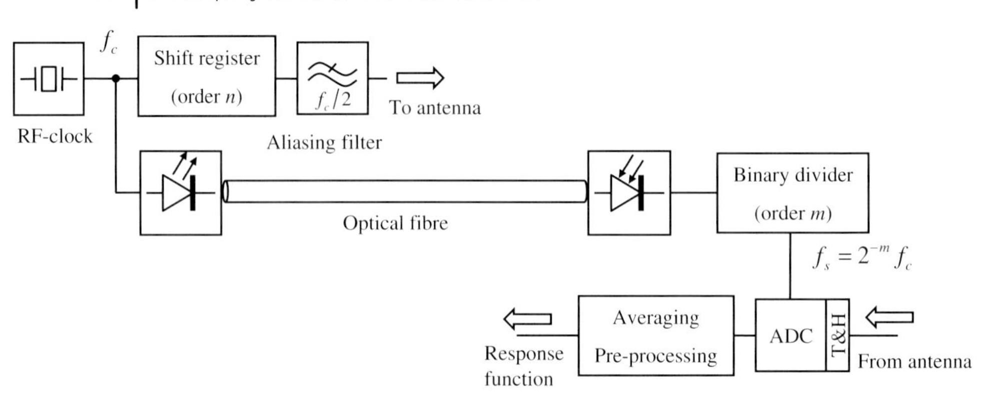

Figure 3.67 Transmitter and receiver clock are synchronized via an optical link.

timing source. Indeed, one looses the absolute time reference to the transmitter but the time difference of arrival (TDOA) between $y_1(t)$ and $y_2(t)$ can be exactly measured. Synchronous averaging of the receiving data cannot be performed before the correlation due to the lack of an exact knowledge of the signal period.

Synchronization via optical fibre (Figure 3.67): Optical fibres provide lower attenuation than coaxial cables and they are free from any leakage of electromagnetic fields. Hence, this type of synchronization is useful if a certain distance between transmitter and receiver has to be bridged or if in case of shielding investigations the leakage of coax-cables is perturbing the measurement. Synchronous averaging of the captured data is permitted since receiver and transmitter are rigidly coupled. Absolute delay time measurements are not possible with the configuration in Figure 3.67 since only the RF-clock is synchronized.

Synchronization via wireless link (Figure 3.68): Synchronization via wireless link provides the largest degrees of freedom for a separate placement of receiver and transmitter. Main sources of trouble may come from fading and external radio sources. Both effects can be reduced by a microwave link with narrow beamwidth and narrow bandwidth. Furthermore, one has to keep stable line of

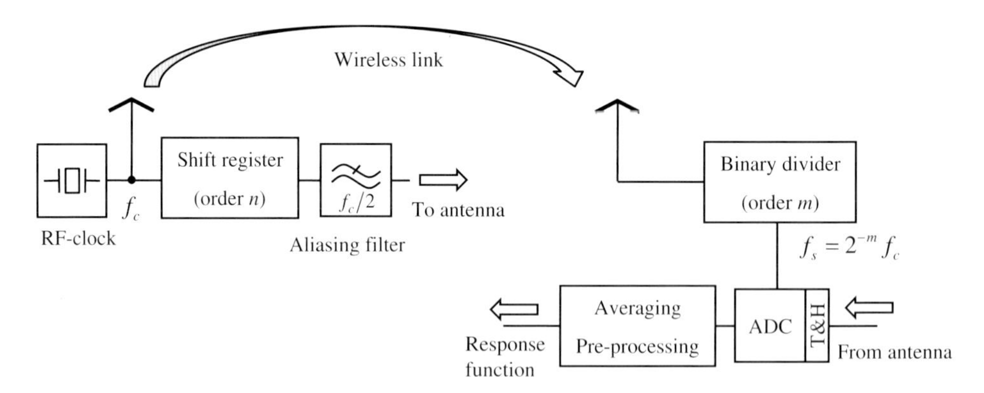

Figure 3.68 Transmitter and receiver are synchronized via wireless link.

sight condition for the synchronization signal even if transmitter or receiver are moved. The Doppler effect will not affect the capturing of the direct wave signal since it acts contemporaneously and in the same sense onto the synchronization as well as line of sight signal. However, reflections from the surrounding will be subject the Doppler effect as before.

Synchronization by early–late locked loop (Figure 3.69): The objective of this type of synchronization is to lock the receiver exactly onto the sounding signal. We presume that transmitter and receiver are based on shift registers providing the same PN-code and the clock rates of both systems are quite close but not identical. The switch in Figure 3.69 is placed at the 'search' position. Hence, we have the situation as depicted in Figure 3.34a in which the output voltage of the low-pass filter slowly slides over the whole correlation function. In our case, correlator ② takes over the task to measure this voltage. It is symbolized by a small

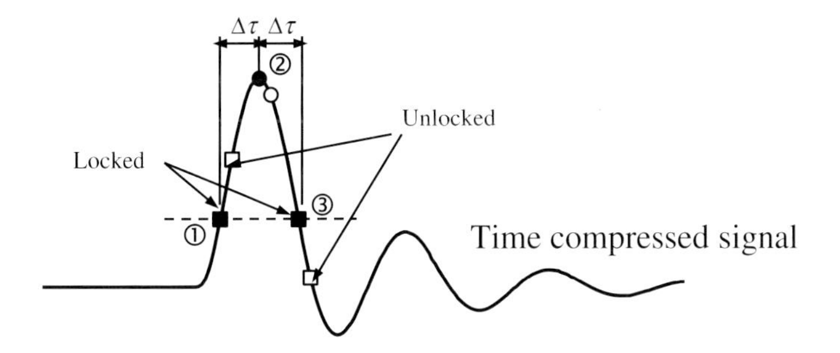

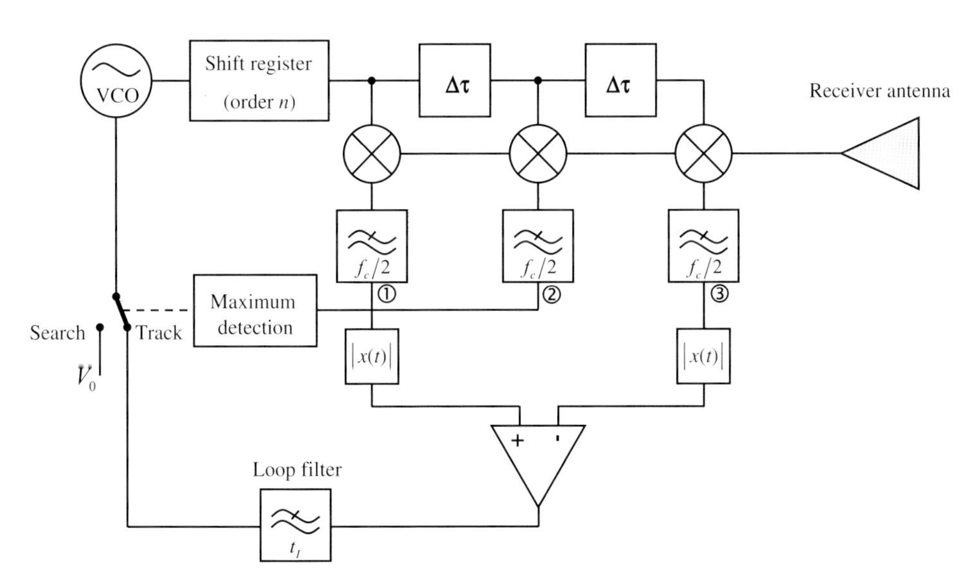

Figure 3.69 Receiver synchronization by early–late locked loop.

circle in Figure 3.69. If this voltage achieves at its maximum level, we have the best match between receive signal and internal reverence provided by the shift register. In order to avoid further sliding and to keep the synchronism, the switch will be toggled in the 'track' position. Now, the voltage-controlled oscillator (VCO) is controlled by the difference voltage provided by two further correlation circuits ① and ③. One of them provides an advanced sample and the other one a delayed sample referred to the correlator ②. They are assigned by small squares. The loop is locked if the difference voltage is zeros, that is if delayed and advanced sample have the same position at the rising and falling edge of the correlation function. This is symbolized by the black squares.

Under multi-path conditions, the correlation function is composed of a multitude of peaks and the early–late locked loop would be locked coincidentally to any of these peaks. In order to catch the biggest one, a second loop is usually applied pursuing the search while the first one is just locked [64].

Basically, the early–late locked loop can also be implemented in the basic M-sequence device concept (Figure 3.35) without extension of the RF-circuitry. This requires an appropriate FPGA emulating the analogue circuit as depicted in Figure 3.69 and the FPGA finally controls the clock generator via an additional DAC. If one can expect only a small lag between the clock rates in the unlocked modus, the control loop may be even operate on the averaged data which largely relaxes the requirements on the FPGA.

3.4 Determination of the System Behaviour by Excitation with Sine Waves

3.4.1 Introduction

The use of sine waves to characterize the transmission behaviour of a device or a sensor concerns to the oldest measurement principles. Amongst all the other measurement approaches, it permits the device characterization over the largest bandwidth and it is one of the most sensitive measurement techniques. Vector network analysers (VNA) are often applied under laboratory conditions for ultra-wideband sensing. Some field deployable systems are described for example in Refs [65–68].

However, by definition, all measurement methods using sine wave excitation are not counted amongst the ultra-wideband approaches since they require signals of a wide instantaneous bandwidth. In the classical sine wave techniques, the test objects are sequentially stimulated at a multitude of frequencies. The individual measurement at a given spectral line is indeed a narrowband measurement. However, the overall result will be a wideband measurement leading to comparable conclusions about the test objects as the other ultra-wideband principles it does. Hence for short-range sensing, the sine wave approaches are of the same interest as the other ones. The only thing what has to be supposed is linearity and time invariance of the objects of interest.

Several sine wave-based measurement approaches are in use. Here, we will restrict to some fundamental aspects of these techniques which concern the basic modes of object stimulation and the procedures to capture the measurement signals. The object stimulation is done either by the variation of the sine wave frequencies in discrete (and typically but not necessarily equidistant) steps or by a continuous sweep over the spectral band of interest. The first method is usually denominated as SFCW (stepped frequency continuous wave)-method while the second is called the FMCW-principle. The data recording is either based on heterodyne or homodyne receiver conceptions. They convert the RF-signals to lower frequencies where they can easily be captured.

As introduced in Section 2.4.3, the frequency domain measurement simply involves the capturing of an amplitude $G(f_0)$ and phase value $\varphi(f_0)$ (or an in-phase/real part $I(f_0)$ and a quadrature/imaginary part $Q(f_0)$) at a given frequency f_0:

$$\underline{G}(f_0) = G(f_0) \cdot e^{j\varphi(f_0)} = I(f_0) + jQ(f_0)$$

Since the device under test as well as the measurement device has been settled down at every frequency step before the data can be taken, the achievable measurement speed is in many cases slower as for wideband measurement approaches.

We will summarize the basic receiver structures first which permits the capturing of RF-voltages by low-speed ADCs. A more comprehensive view on receiver structures can be found in Ref. [69]. Then the two basic modes of object stimulation – SFCW and FMCW – will be considered and compared. Finally, the multi-sine approach will be discussed. It will speed up the measurement by reducing the settling time and it actually concerns to the UWB approaches since it applies test signals of large instantaneous bandwidth.

3.4.2
Measurement of the Frequency Response Functions

3.4.2.1 Homodyne Receiver

A signal source provides a sine wave of known frequency f_0 which stimulates the DUT. The task of the receiver is to determine amplitude and phase variations caused by the DUT (compare Figure 2.51). Here, the idea is to convert the RF-voltage down to the DC range where it can easily be captured. Figure 3.70 depicts the basic concept in which the stimulation signal also acts as LO-signal in the down-converter. It makes use of following identities:

$$\frac{1}{t_I}\int_{t_I} \cos 2\pi f_0 t \cos(2\pi f_0 t + \varphi)\,dt = \mathrm{rect}\left(\frac{t}{t_I}\right) * \left(\cos 2\pi f_0 t \cos(2\pi f_0 t + \varphi)\right) = \frac{1}{2}\cos\varphi$$

$$\frac{1}{t_I}\int_{t_I} \sin 2\pi f_0 t \cos(2\pi f_0 t + \varphi)\,dt = \mathrm{rect}\left(\frac{t}{t_I}\right) * \left(\sin 2\pi f_0 t \cos(2\pi f_0 t + \varphi)\right) = -\frac{1}{2}\sin\varphi$$

The integration time, we select either as an integer multiple of the half sine wave period or we make it much larger than a half period, that is $2t_I f_0 = n$, $n \in \mathbb{N}$ or $2t_I f_0 \gg 1$.

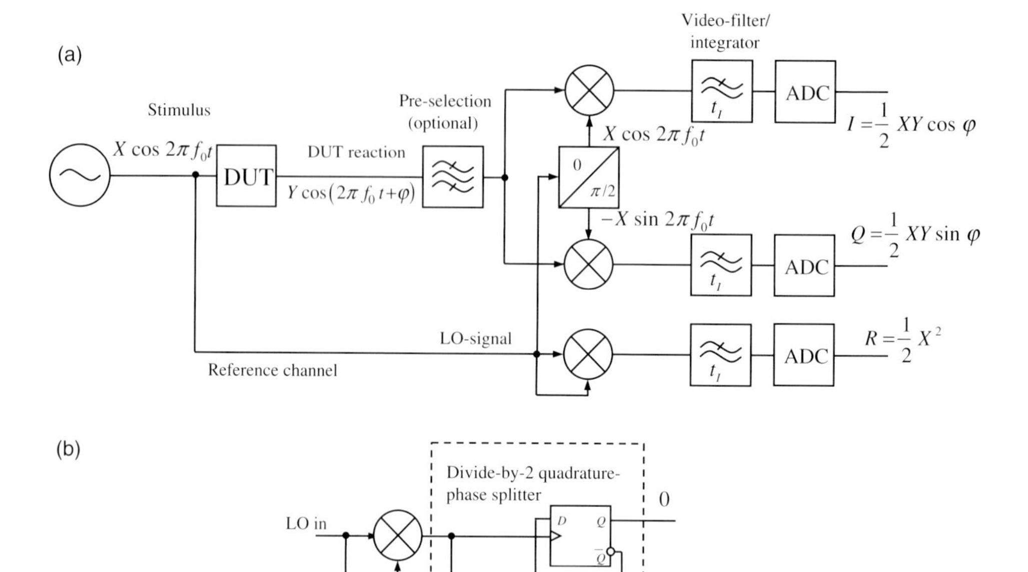

Figure 3.70 Basic concept for sine wave measurements using homodyne frequency down-conversion (a) and an example for a wideband quadrature-phase splitter (b).

Technically they are implemented by multipliers and integrators/video-filters. The resulting voltages are of low frequency. They can be captured by high-resolution ADCs.

If we express the involved signals by complex notation, the functioning of the ideal homodyne receiver may be summarized by following expressions:

$$\begin{aligned} I + jQ &= \frac{1}{2}\underline{y}(t)\underline{x}^*(t) = \frac{1}{2}XY(\cos\varphi + j\sin\varphi) \\ R &= \frac{1}{2}\underline{x}(t)\underline{x}^*(t) = \frac{1}{2}X^2 \\ \text{with} \quad \underline{x}(t) &= X\,e^{j2\pi f_0 t}; \quad \underline{y}(t) = Y\,e^{j(2\pi f_0 t + \varphi)} \end{aligned} \tag{3.86}$$

Thus, the wanted FRF of the test object at the frequency f_0 results in

$$\underline{G}(f_0) = \frac{I(f_0) + jQ(f_0)}{R(f_0)} \tag{3.87}$$

which is nothing then (2.152) $\underline{G}(f_0) = \frac{\underline{Y}(f_0)\underline{X}^*(f_0)}{\underline{X}(f_0)\underline{X}^*(f_0)}$.

Due to its simple structure, it seems that the concept can easily be implemented even by monolithic integration [70, 71]. Actually, for narrowband applications, one makes largely use of this simplicity. But in the case of UWB applications, the implementation of the $\pi/2$ phase shifter needs some care. Usual approaches apply wideband quadrature hybrids, polyphase filters (Ref. [72] gives a short introduction to analogue polyphase filters) or divide-by-2 quadrature-phase splitters as exemplified in Figure 3.70. Often, one also renounces the measurement of the complex-valued transfer function and is satisfied with its real part only. This is the usual approach applied in FMCW-devices if they operate over a larger bandwidth.

Modern DDS-circuits operate up to several GHz meanwhile [73, 74]. They offer new perspectives to solve the IQ-down-conversion by feeding the I- and Q-chain with two DDS circuits providing two identical signals of appropriate phase shift. Both DDS-circuits have to be driven by the same reference clock for that purpose to assure exact synchronism. However, by the best knowledge of the author, a technical implementation of two out of phase DDS circuits in ultra-wideband sensor devices for homodyne down-conversion was not yet reported.

Despite its simplicity, the homodyne receiver concept evokes some deficiencies affecting the sensitivity and stability of the measured data [69]:

- LO-leakage creates in-band interference since stimulus and LO-frequencies are identical.
- LO-self-mixing generates a DC offset which immediately affects the down-converted I- and Q-signals which are located around zero frequency.
- Backscattering of the LO-leakage from the device under test leads to a DC offset which depends on the properties of that device.
- The principle assumes identical transmission behaviour of the I- and Q-path and an exact $\pi/2$ phase shift between the two LO-signals. Over a large bandwidth, this cannot be guaranteed. Hence, I/Q imbalance will additionally contaminate the measurement values. More discussions on this topic and a proposal of quadrature error correction by singular value decomposition are to be found in Refs [75, 76].
- Flicker noise ($1/f$-noise) may drastically affect the SNR of the I- and Q-signals.
- The mixers are sensitive to even order distortions.
- The receiver does not reject harmonic distortions or other spurious components of the sounding signal.

3.4.2.2 **Heterodyne Receiver**

In contrast to the homodyne receiver, the conception of the heterodyne receiver is more comprehensive which leads to a better performance but also to increased device costs, larger power consumption and dimensions. It is the sensor conception providing the highest sensitivity. The generic receiver structure of network analysers – the classical wideband measurement device – is based on such approach. Figure 3.71 illustrates the basic device structure.

The wanted measurement result is the same as expressed by (3.86) and (3.87) for the homodyne down-conversion. But now, the IQ-demodulator works at the fixed intermediate frequency (IF) f_{if}. It is placed in a spectral band convenient for circuit implementations with favourable technical performance. That is, phase shifter, I/Q imbalance and main signal amplification can be best handled in the IF-band. Flicker noise does not appear in the IF-band since it is restricted to low frequencies

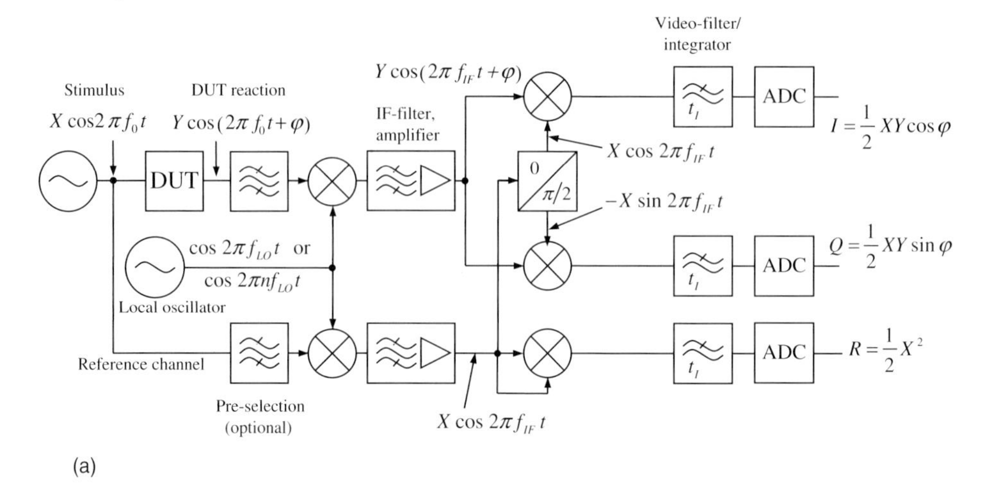

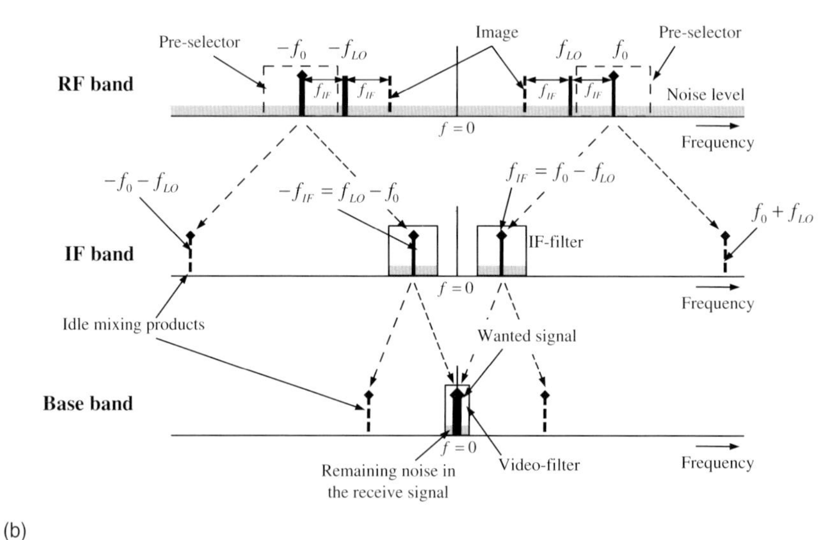

Figure 3.71 RF-measurement device based on heterodyne down-conversion. (a) Basic block schematic and (b) typical signal spectra. The LO-signal is either $\cos 2\pi f_{LO} t$ for fundamental mixing or it is a spectral comb $\cos 2\pi n f_{LO} t$ in the case of harmonic mixing (see Figure 3.72)

only. Furthermore, the imbalance of the I- and Q-channels is independent of the frequency of the sounding signal simplifying the error correction [77]. For low IF-receivers, it is also possible to implement the IQ-down-conversion completely in software. This would further reduce the problems with the I/Q imbalance and filtering would be more flexible [69]. For that purpose, the signals have to be sampled immediately behind the IF-filters.

In order to keep the intermediate frequency fixed $f_{if} = |\pm f_0 \pm f_{LO}| = \text{const}$, we need a mixing stage in front of the receiver whose LO-frequency f_{LO} is tracked in parallel to the frequency f_0 of the stimulus signal. The down-converted IF-signals pass the IF-filters which are mainly responsible for the potentially high receiver sensitivity. They perform several tasks:

- They reject the unwanted mixing products (up-converted frequencies).
- They avoid the penetration of LO-leakage into the IQ-down-converter.
- They prevent the IQ-stage form DC voltage caused from LO-leakage and self-mixing.
- They suppress harmonic distortions caused by the mixers.
- They (mostly) reject harmonic distortions and spurious signals of the stimulus.

The degree of noise suppression is determined either by the IF- or video (low-pass)-filter whichever has the narrowest bandwidth. Usually the reference channel is much less affected by noise than the measurement channel. Therefore, the IQ-demodulator does not raise the spectral power density of the noise (except by its intrinsic noise generation). Hence, it does not matter which of the two filters will finally determine the remaining noise band. In the example illustrated in Figure 3.71, the video-filter is responsible for the sensitivity of the device. Under these constraints, the system designer has some flexibility to select the filters. If, however, the reference channel is affected by noise too or one has to anticipate lots of spurious signals in the IF-stage, the IF-filter should be as narrow as possible to increase the device sensitivity.

The first mixer can be operated in two modes. In the first one, the LO-signal is a simple sine wave (as depicted in Figure 3.71; fundamental mixing) and in the second one, the LO-signal is a train of Dirac pulses (comb spectrum) of appropriate repetition rate. This type of down-conversion is referred to as harmonic mixing which is often used in wideband network analysers. It permits very wideband operation but it exacerbates the noise performance. Figure 3.72 illustrates the principle. The LO-frequency is controlled in such a way that the mixing product of one of its

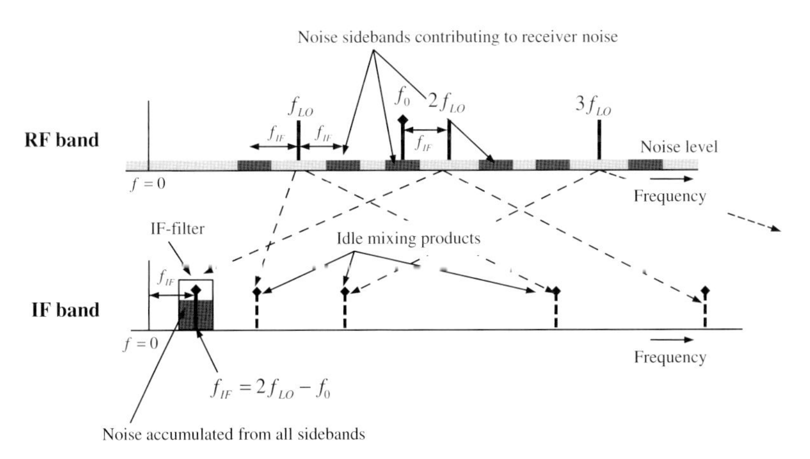

Figure 3.72 Harmonic mixing.

harmonics with the input signal falls into the IF-band, that is $f_{\mathrm{IF}} = nf_{\mathrm{LO}} \pm f_0$. It is the second harmonic in the considered example. All other mixing products are out of interest and are suppressed by the IF-filter.

The input filters (pre-selectors) are not required by the principle but they are of profit if the device under test is subject to external interferences which provide perturbing signals of various frequencies, for example from radio or TV-stations. The pre-selectors suppress possible image signals (refer to Figure 3.71), avoid overmodulation of the first mixers due to heavy interference with out-of-band disturb signals and reject LO-leakage. In order to meet these tasks, the filters should not be too wideband. But then, they must be tuneable over the whole frequency band of interest which is a challenging task for ultra-wideband devices. Therefore, they are typically omitted or replaced by fixed filters of an appropriate pass band.

The price to be paid for the improved sensitivity of the heterodyne principle is the increased complexity of the frequency allocation for the LO-signals as well as the larger number of mixers and filters.

3.4.3
Sine Wave Sources of Variable Frequency

The choice of the sine wave source strongly influences the performance and the costs of the ultra-wideband sensor. The technical key parameters to be accounted for are as follows:

- **The tuning range:** According to the ultra-wideband applications discussed here, the sine wave generators have to cover a tuning range of large fractional bandwidth.
- **The frequency accuracy** refers to the repeatability and stability of the frequency source. The repeatability means the ability of the generator to hit always the same and the required frequency in repetitive measurements. The stability or drift relates to slow time frequency variations due to temperature, ageing and so on. The frequency accuracy is important in applications which involve system calibration, super resolution techniques and precise range measurements.
- **Phase noise** means the short time fluctuations of the frequency. They mainly influence the sensor sensitivity to detect small target movements and it degrades the usability of the measured data in super resolution techniques.
- **The purity** of the sine wave is typically degraded by harmonics or other spurious signals. They cause unwanted signal components in the receiver pretending, for example false targets and causing clutter in radar applications.
- **Tuning speed and settling time** are important issues if the recording time is subject to strong restrictions. Herein, the tuning speed refers to the maximum chirp rate in the case of a continuous frequency variation. The term settling time is connected with a discontinuous frequency variation. The settling time has to be awaited before the next measurement can be started since the generator needs some time to engage the final frequency of every step.

Two basic approaches to generate a sine wave over a large frequency band are in use. The first one applies the DDS and the second is based on VCO.

Direct digital synthesis: A high-speed DAC converts a waveform stored in a RAM into an analogue RF-voltage. Basically, such device is structured like an arbitrary waveform generator as depicted in Figure 3.82. The circuit is controlled by a stable single tone clock which is the reason for the high timing/frequency accuracy. The principle permits very fine steps of frequency variation (depending on the RAM-size) and it does not suffer from settling time in case of frequency variation. The main disadvantages results from the restricted DAC resolution which leads to spurious signals. Meanwhile, the upper frequency of commercially available DDS circuits goes beyond some GHz and the maximum chirp rate reaches very high values too [74]. Thus, they are attractive for ultra-wideband applications. However, these circuits are often quite power hungry.

Voltage- or current-controlled oscillators: Oscillators with tuneable resonator circuits are the classical devices to provide sine waves of controllable frequency. The frequency variation is either done via varactor diodes or small YIG spheres. In the first case, the reverse bias voltage controls the capacity of a resonant circuit while in the second case the strength of a magnetic field influences the inductivity. YIG oscillators permit a large tuning range, their resonance circuit is of high Q-factor (and hence low phase noise) and they show good linearity between control current and frequency. But they are quite expensive, which reduces their desirability for sensor applications.

The technical performance of varactor-tuned oscillators is usually worse as the YIG devices, but they are of lower cost. Their tuning range is typically below 2 $f_u/f_l = (C_{min}/C_{max})^2 \leq 2$ so that a full frequency sweep will not exceed a fractional bandwidth of $b = 2(f_u - f_l)/(f_u + f_l) \leq 2/3$). This is not always sufficient for ultra-wideband sensors. Furthermore, their tuning characteristic is non-linear and they are prone to phase noise.

The performance of a varactor as well as of a YIG tuned oscillator can largely be improved, if it is inserted in a control loop – the so-called PLL – which locks it to an accurate reference frequency. Figure 3.73 depicts the principle (for details on PLL, see Ref. [78]).

The VCO is driven by a voltage which the phase detector (it is just a multiplier in simple cases) creates from the phase and frequency difference f_1 and f_2 of its two input signals. One of them is provided from an accurate single tone reference oscillator and the second one comes from the VCO to be controlled. If the control loop is locked, the frequencies of both input signals of the phase detector are identical, that is

$$f_1 = f_2 \Rightarrow f_{VCO} = \frac{N}{M} f_0 \quad (3.88)$$

That is, the VCO-frequency f_{VCO} is now rigidly linked to the stable reference frequency f_0. By acting on the factors M and N of the digital dividers, the VCO-frequency may be precisely adjusted in steps of $\Delta f_{VCO} = f_1 = f_0/M$.

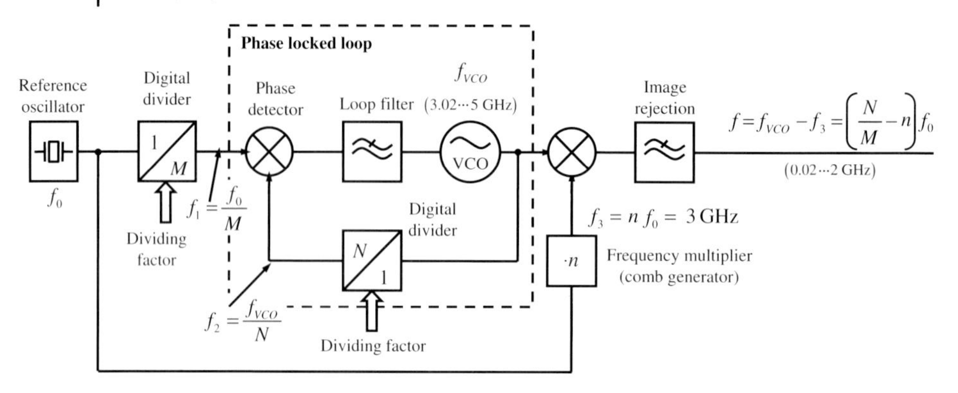

Figure 3.73 Generic schematic for frequency synthesizes with an integer N-PLL.

In order to achieve a large tuning range for the output frequency, the VCO operates in a higher frequency band than required. Figure 3.73 gives an example. The desired tuning range should be 20–2000 MHz, that is $f_u/f_l = 100$ and $b = 1.96$. A VCO cannot immediately provide such a range. Therefore, it is designed for a higher band but with keeping the required sweep interval (e.g. 3–5 GHz). Now, we have a frequency ratio of $f_u/f_l \approx 1.7$ which is technically feasible. This band is finally down-converted by the auxiliary frequency f_3 into the band of interest. The auxiliary frequency must as stable as the reference generator since any of its variations are copied into the band of interest. Hence, it is usually derived from the reference generator. This is done by a frequency multiplier (comb generator) (e.g. based on step recovery diodes) in our example which generates a LO-signal of $f_3 = 3\,\text{GHz}$.

As we can see from (3.88) and Figure 3.73, the loop operates at the frequency $f_1 = f_0/M$. In an integer N-PLL, it fixes the minimum step size $\Delta f_{\text{VCO}} = f_1$ of the output frequency. Its value is determined by the specific application namely the duration[13] $t_{\text{win}} = \Delta f_{\text{VCO}}^{-1}$ of the impulse response to be observed (refer to Section 2.4.5.3). Hence, f_1 must be often quite low and the dividing factor N has to take large values. Thus, we have to deal with a loop filter of low cut-off frequency resulting in long settling time and in consequence slow sweep speed. Furthermore, noise from the phase detector and reference generator appear N-fold enforced in the VCO-signal.

So there is a trade-off between frequency resolution, sweep time and phase noise. Several approaches were developed to work at a high loop frequency without losing frequency resolution. One possibility is to deal with fractional N-PLLs [78, 79]. Here, the divider in the feedback loop switches between different dividing factors during the operation. If, for example after every fifth signal transition of the divider output the dividing factor is jumps from N to $N+1$ (i.e. $N\ N\ N\ N\ N+1\ N\ N\ N\ N\ N+1\ N\ \cdots$), the resulting

13) It depends on the settling time of the test scenario and from the radar unambiguity range.

(average) dividing factor is $N + 0.2$. By controlling the occurrence of dividing factor alterations, one is able to achieve a narrow frequency grid of the VCO-signal. Attention has to be paid for the generation of spurious components caused by the divider alterations.

Another option is to apply the integer N-PLL for coarse trimming (if necessary) of the VCO-frequency and DDS for inertial-less fine trimming. Thanks to the loop filter, DDS spurious are largely suppressed in case of proper design (Figure 3.74).

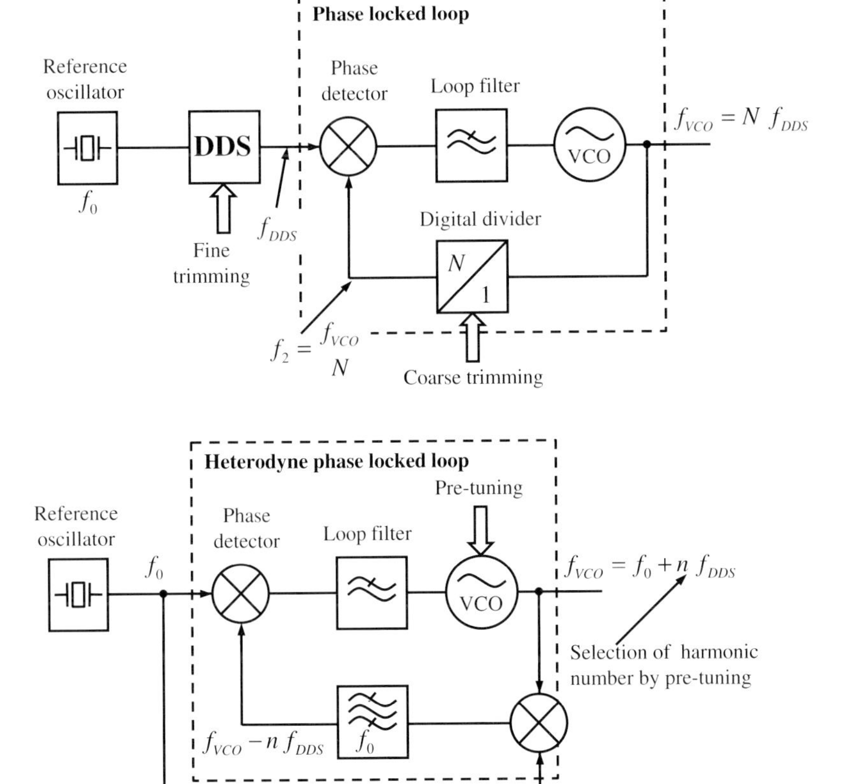

Figure 3.74 Two options to reduce settling time of frequency synthesizers by fine trimming via direct digital synthesis (DDS).

3.4.4 Operational Modes

The sine wave sources may be operated in two basic modes – a discrete or continuous frequency variation. It sweeps between a lower f_l and upper frequency f_u with a comparatively low speed allowing the test object and measurement device to achieve the steady state.

3.4.4.1 Stepped Frequency Continuous Wave (SFCW)

The signal source generates a train of adjacent narrowband pulses by successively varying the carrier frequency $f_c \in [f_l, f_u]$ as depicted in Figure 3.75. Such signal is assigned as stepped frequency continuous wave and the related modulation schema is called frequency shift keying (FSK). Network analysers typically apply this mode of frequency control.

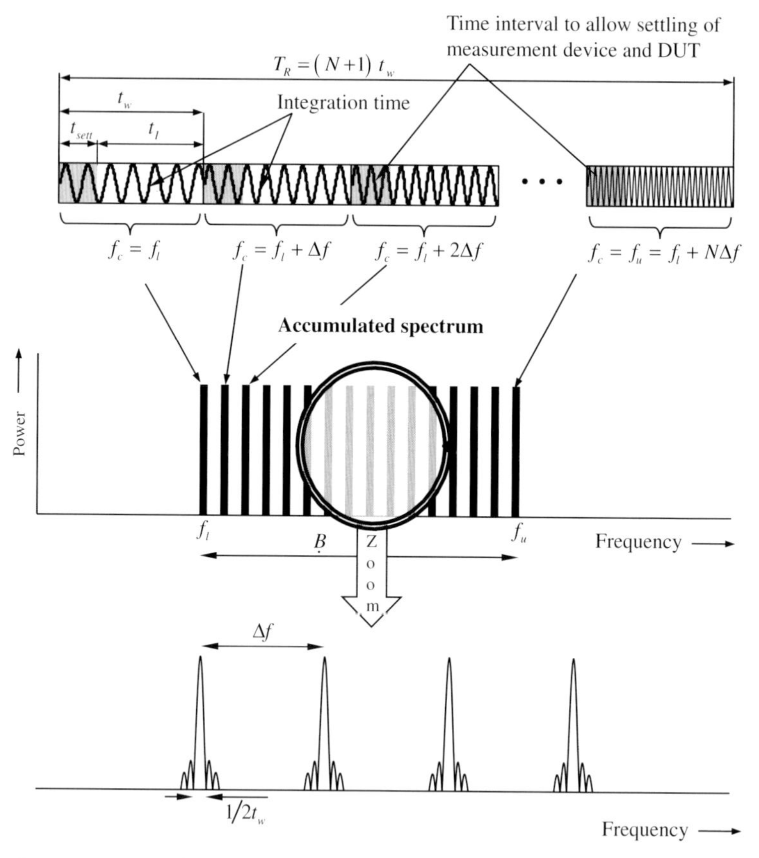

Figure 3.75 Train of adjacent narrowband pulses with stepwise increasing carrier frequency and related spectrum accumulated over T_R. Note that the sequence of frequency variation can be basically selected arbitrarily as long as the test scenario behaves time invariant [80].

The signal generated in this way stimulates the scenario under test. We give the measurement device as well the measurement object some time[14] (t_{sett}) to settle down at each pulse before we start the signal recording which is either based on homodyne or heterodyne down-conversion. It will result in the quantities $I(f_c), Q(f_c), R(f_c)$ for every carrier frequency. The time available for capturing of these data is assigned by t_I. It is best exploited for noise suppression if we use a short-time integrator (see Sections 2.6.1 and 3.3.3).

We can model the stimulus signal actively used (i.e. neglecting the part of settling) for data capturing by following expression:

$$\begin{aligned} x(t) &= \sum_{n=0}^{N} \text{rect}\left(\frac{t - n\,t_w}{t_I}\right) \cos\left(2\pi\left(f_l + n\,\Delta f\right)t\right) \\ &= \Re\left\{\sum_{n=0}^{N} \text{rect}\left(\frac{t - n\,t_w}{t_I}\right) e^{j2\pi\left(f_l + n\,\Delta f\right)t}\right\} \end{aligned} \tag{3.89}$$

where $N+1$ assigns the number of spectral lines included in the stimulus signal and $\Delta f = (f_u - f_l)/N$ represents the line spacing. Absolute and fractional bandwidths of the excitation signal are given by $\underset{\cdot}{B} = f_u - f_l$ and $b = \underset{\cdot}{B}/f_m$ with the centre frequency $f_m = (f_u + f_l)/2$. It should be mentioned that the signal (3.89) is usually not account ultra-wideband even if $b \geq 0.2$ since its instantaneous bandwidth is narrow. We simplify the spectrum of a single pulse to a line spectrum even if every pulse actually provides a narrow sinc-shaped spectrum as emphasized in the zoomed part of Figure 3.75.

Supposing ideal behaviour of the homodyne and heterodyne receiver, we can estimate the intrinsic impulse response of the measurement device by applying the rules of Fourier transform onto the stimulus signal. If we allow complex-valued time functions (see analytic signal Sections 2.2.2 and 3.3.6.3), we get[15]

$$\underline{g}_a(t) \propto \text{sinc}\left(\underset{\cdot}{B}\,t\right) e^{j2\pi f_m t} = \text{sinc}\left(\frac{t}{\tau_0}\right) e^{j2\pi f_m t} = \underline{g}_0(t) e^{j2\pi f_m t}; \quad \underset{\cdot}{B}\tau_0 = 1 \tag{3.90}$$

The envelope $\underline{g}_0(t)$ of the analytic IRF is a real-valued sinc function of the characteristic width $\tau_0 = \underset{\cdot}{B}^{-1}$. In the case of non-ideal receivers, it will degenerate to a complex-valued envelope showing more or less deviations from the sinc function. The deviations can be removed by appropriate system calibration covering either a full calibration as discussed in Section 2.6.4 or respecting only the I/Q-imbalance [75, 76].

14) In case of a radar scenario, the settling time corresponds to the round-trip time of the most far target $t_{sett} = 2r_{ua}/c$ (r_{ua} is the unambiguity range; c is the speed of light). In the case of short-range sensing, single and multiple reflections of the environment often provide large signal contributions. Here, we have to wait until all these reflections are died out before we can start the measurement.

15) Equation (3.90) is based on the assumption of a continuous spectrum allocation within the spectral band $[f_l, f_u]$. Actually, we have however a spectrum of $N_s = N+1$ lines. Hence, the correct expression for the impulse response would be

$$\begin{aligned} \underline{g}_a(t) &= \frac{\sin \pi \underset{\cdot}{B} t}{\sin\left(\pi \underset{\cdot}{B} t / N_s\right)} e^{j2\pi f_m t} \\ &\approx N_s \frac{\sin \pi \underset{\cdot}{B} t}{\pi \underset{\cdot}{B} t} e^{j2\pi f_m t}; \quad \frac{\pi \underset{\cdot}{B} t}{N_s} \ll 1. \end{aligned}$$

The physical real-valued intrinsic IRF of the ideal device is given by

$$g(t) = \Re\{\underline{g}_a(t)\} = \text{sinc}\left(\underset{\cdot}{B}t\right)\cos 2\pi f_m t \tag{3.91}$$

Following the discrete Fourier transform, the frequency spacing Δf of the sounding signal determines the time window length t_{win} within the IRF of a DUT may be observed, that is

$$t_{win}\,\Delta f = 1 \tag{3.92}$$

It should be selected larger than the settling time of the scenario $t_{win} \geq t_{set}$ and the maximum expected round trip time of a radar measurement in order to avoid time aliasing or range ambiguities. Figure 3.76 illustrates a typical intrinsic IRF of a SFCW-device. Obviously its envelope suffers from strong ringing which decays quite slowly. It is due to the abrupt truncation of the stimulus spectrum. We already discussed relating effects in previous chapters (Sections 2.4.5.3 and 3.3.5), where we have seen that ringing may be suppressed by adequate shaping of the signal spectrum before applying inverse Fourier transform. Figure 3.76 gives an example for the Hann-window. We can observe a stronger decay of the side lobes, however, at the expense of a wider pulse width. Finally, the user has to decide what type of window is appropriate in his actual application – does he need a sharp main lobe or a strong decay.

The digitized version of the IRF covers $N_s = 1 + \underset{\cdot}{B}/\Delta f = N + 1$ complex-valued samples if we restrict ourselves to the complex-valued envelope $\underline{g}_0(t)$ whereas the real-valued IRF $g(t)$ from (3.91) requires $N_s \geq 2 f_u t_{win}$ real-valued data samples in minimum in order to meet the Nyquist theorem.

As in the previous device conceptions, the efficiency of data capturing depends on the energy losses evoked by the SFCW-receiver. Referring to Figure 3.75 and supposing identical power P of all narrowband pulses, the overall energy provided

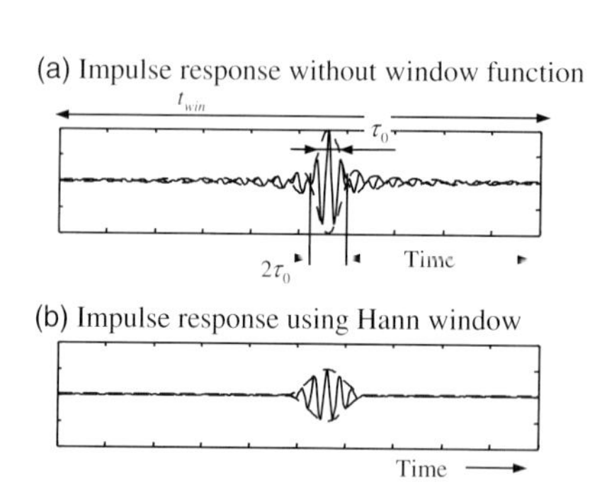

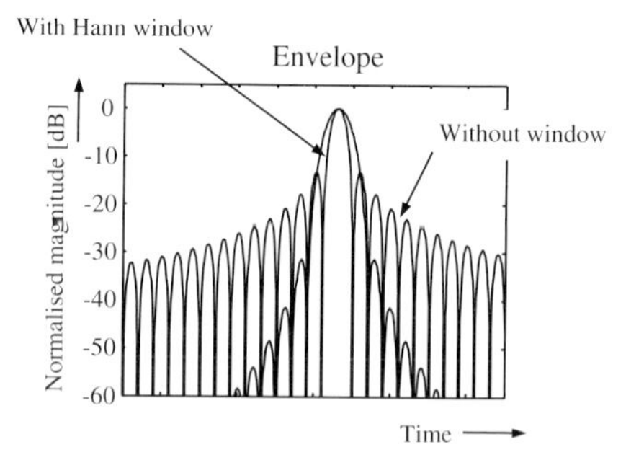

Figure 3.76 Intrinsic impulse response function of a SFCW-sensor. Two versions are shown – (a) the original IRF without any manipulation of the signal spectrum before conversion into the time domain and (b) shaping the signal spectrum with the Hann-window before applying inverse Fourier transform. The actual IRFs are shown left-hand side and their envelope in logarithmic scale is depicted on the right.

by the signal source during the recording time T_R yields

$$\mathfrak{E} = PT_R$$

while the actually captured energy is

$$\mathfrak{E}_{acc} = \eta_m (N+1) t_I P$$

if a short-time integrator is used for energy accumulation. Other types of video filtering will lower the amount of accumulated energy (see (3.42) and (3.43)). η_m is the efficiency of the first mixers. This gives us for the efficiency of the SFCW-device:

$$\eta_{SFCW} = \frac{\mathfrak{E}_{acc}}{\mathfrak{E}} = \eta_m \frac{(N+1) t_I}{T_R} = \eta_m \frac{t_I}{t_{sett} + t_I} \tag{3.93}$$

Since we can select mixers of high efficiency, the overall efficiency mainly depends on the relation between settling and integration time. The settling time can roughly be estimated from

- the settling time of signal source (it mainly depends on the PLL-loop (τ_{PLL}), see Figures 3.73 and 3.74; the settling time of the VCO can usually be omitted),
- the settling time of the DUT ($\tau_{DUT} \le t_{win}$), which should be shorter than the observation length of the IRF,
- the settling time of the pre-selector (τ_{pre}) and
- the settling time of the IF-filter (τ_{IF}).

$$t_{sett}^2 \approx \begin{cases} \tau_{PLL}^2 + \tau_{DUT}^2 + \tau_{pre}^2 + \tau_{IF}^2 \le \tau_{PLL}^2 + t_{win}^2 + \tau_{pre}^2 + \tau_{IF}^2 & \text{heterodyne} \\ \tau_{PLL}^2 + \tau_{DUT}^2 + \tau_{pre}^2 \le \tau_{PLL}^2 + t_{win}^2 + \tau_{pre}^2 & \text{homodyne} \end{cases} \tag{3.94}$$

The largest of the indicated time constants will determine the overall settling time. In the case of proper device design, the pre-selector has larger bandwidth than the IF-filter and the PLL-loop works at high frequencies (DDS fine trimming or fractional N-PLL required). Hence, their time constants can be dropped in (3.94). If we further select the IF-bandwidth larger than the line spacing $\dot{B}_{IF} \ge \Delta f = t_{win}^{-1}$, that is $\tau_{IF} < t_{win}$, (3.93) and (3.94) reduce to

$$\eta_{SFCW} \approx \eta_m \frac{t_I}{t_{win} + t_I} \quad \text{for } t_{sett} \approx t_{win} \tag{3.95}$$

Hence, the efficiency tends to unity for $t_I \gg t_{win}$, leading to a recording time of $T_R \approx N_0 t_I \approx t_{win} \dot{B} t_I$.

As expected, the ambiguity function of the frequency-shifted narrow-band pulse train $x(t)$ (see (3.89)) is sensitive to range-Doppler coupling. Referring to the definition (2.91), Figure 2.22 and the considerations in Refs [81] (Chapter 5) and [82], we can approximately write

$$\frac{\tan\alpha}{s^2/m} = \frac{\Delta\tau}{v_r} \approx \frac{2f_m}{c}\frac{t_w}{\Delta f} \approx \frac{2f_m}{c}\frac{T_R}{\dot{B}} = \frac{2}{c}\frac{T_R}{b} \tag{3.96}$$

v_r is the range velocity; c is the speed of light. In the case of a moving target, the I- and Q-signals of a homodyne or heterodyne receiver oscillates with the Doppler frequency

$$f_d = -2(f_1 + n\,\Delta f)\frac{v_r}{c} \tag{3.97}$$

Therefore, the duration t_I of the integration should meet the condition $f_d t_I \approx f_d t_w \ll 1$ in order to avoid signal cancellation by the video-filter.

That is, if high efficient receivers are required in radar measurements with moving targets, one has to accept the maximum target speed $v_r \ll c/2\,t_I f_u \approx c/2\,t_w f_u$ and a range error of $\Delta r \approx \pm c\,\Delta\tau/2 = v_r T_R/b$ provoked by range-Doppler coupling (3.96). The range error may be reduced by shortening the recording time T_R by either lowering the integration time t_I (reducing the SNR-value) or by exciting a set of sine waves of different frequencies in parallel.

An example for the last approach is given in Refs [82, 83] in which eight sine waves are generated in parallel requiring however large technical effort. Apparently, the method will bring some benefit concerning the device performance since the measurement speed can be octuplicated by keeping the integration time t_I. But this conclusion is not inevitably correct. By generating eight sine waves in parallel, one may risk constructive interference of all signals in the amplifiers of the transmitter and receiver. To avoid amplifier saturation, the magnitude of every sine wave has to be downsized by a factor of eight if one takes not care of an appropriate phase of the individual sine wave components (refer to Figure 2.38). Hence, we lose 64-times of SNR-value and of signal power per spectral line, while the octuplicate reduction of t_I in a single sine wave device leads to the same recording time but only to eight-times lowered SNR-value.

Under certain conditions – assuming point targets and limited number of targets – the range-Doppler ambiguity can be resolved by a clever choice of frequency coding. Some examples can be found in Refs [84–86] which however only refer to automotive radar applications of comparatively small fractional bandwidth. A major problem of these methods is the handling of the numerous ghost targets. We will shortly introduce the basic ideas of these approaches in the next section.

Before closing the discussions on the SFCW-principle, we will still mention the influence of additive random noise and of phase noise onto the device performance. A measurement shot of the SFCW-sensor comprises N_s measurements of noise affected IQ-data pairs which have usually to be converted into time domain by Fourier transform. The question is how the noise will impinge upon the time domain waveform. We will answer the question separately for additive and phase noise by a simplified considerations.

Additive random noise: The receiver data for the nth frequency step are written as random process in complex form $\underset{\approx}{Z}[n] = (I_0[n] + \Delta\underset{\sim}{I}[n]) + j(Q_0[n] + \Delta\underset{\sim}{Q}[n])$, whereas the variance of the perturbations is given by (see Annex B.4.4 (B.26))

$$\sigma_{\mathrm{I}}^2[n] = \sigma_{\mathrm{Q}}^2[n] = \operatorname{var}\left\{\Delta \underset{\sim}{I}[n]\right\} = \operatorname{var}\left\{\Delta \underset{\sim}{Q}[n]\right\} = \frac{1}{8} X_0^2[n] \frac{\sigma_{\mathrm{n}}^2}{\underset{\sim}{B}_{\mathrm{n}} t_{\mathrm{I}}} \tag{3.98}$$

$X_{0,n}$ is the magnitude of the nth reference sine wave, σ_{n}^2 represents the noise power and $\underset{\sim}{B}_{\mathrm{n}}$ is the equivalent noise bandwidth. It corresponds either to the bandwidth of the pre-selector in homodyne receivers (see Figure 3.70) or to the IF-bandwidth of a heterodyne receiver (see Figures 3.71 and 3.72).

In order to get the time domain response $\underset{\sim}{z}(t)$ (being proportional to the wanted IRF of the DUT), we calculate the inverse Fourier transform after having performed N_{s} measurements at different frequencies $f[n] = f_1 + (n-1)\Delta f;\, n \in [1, N_{\mathrm{s}}]$.

$$\begin{aligned} &\underset{\sim}{z}(t) = \sum_{n=-N_{\mathrm{s}}}^{N_{\mathrm{s}}} \underset{\sim}{Z}[n] e^{j2\pi f[n]\, t} \\ &\underset{\sim}{Z}[0] = 0; \quad \underset{\sim}{Z}[n] = \underset{\sim}{Z}^*[-n]; \quad f[n] = \operatorname{sgn}(n)\left(f_1 + (|n| - 1)\Delta f\right) \end{aligned} \tag{3.99}$$

The time signal will be a real-valued function so that symmetry properties of the Fourier transform may be applied to gain the $\underset{\sim}{Z}[n]$ values for negative frequencies, that is $\underset{\sim}{Z}[n] = \underset{\sim}{Z}^*[-n]$. Since the noise contributions $\Delta \underset{\sim}{I}[n], \Delta \underset{\sim}{Q}[n]$ are of zero mean, the expected value of the time domain function is as follows (refer to Annex A.3 for calculation rules and Annex B.4.4 for IQ-demodulator):

$$\begin{aligned} \mathrm{E}\left\{\underset{\sim}{z}(t)\right\} &= z_0(t) = \sum_{n=-N_{\mathrm{s}}}^{N_{\mathrm{s}}} \left(I_0[n] + jQ_0[n]\right) e^{j2\pi f[n]t} = \sum_{n=-N_{\mathrm{s}}}^{N_{\mathrm{s}}} \frac{X_0[n] Y_0[n]}{2} e^{j(2\pi f[n]t + \varphi[n])} \\ &= \sum_{n=-N_{\mathrm{s}}}^{N_{\mathrm{s}}} \underline{Z}_0[n] e^{j2\pi f[n]t} \quad \text{with } \underline{Z}_0 = I_0 + jQ_0 = \frac{1}{2} X_0 Y_0\, e^{j\varphi} \end{aligned} \tag{3.100}$$

Further, we can guess identical variance of all data samples and that the noise terms $\Delta \underset{\sim}{I}[n], \Delta \underset{\sim}{Q}[n]$ are independent except the coincidence between negative and positive frequencies as depicted in (3.99). Therefore, following relations hold (refer to Annex A.3 and (3.98)):

$$\begin{aligned} \mathrm{E}\left\{\Delta \underset{\sim}{I}[n] \Delta \underset{\sim}{I}[m]\right\} &= \mathrm{E}\left\{\Delta \underset{\sim}{Q}[n] \Delta \underset{\sim}{Q}[m]\right\} = \mathrm{E}\left\{\Delta \underset{\sim}{I}[n] \Delta \underset{\sim}{Q}[m]\right\} \\ &= \mathrm{E}\left\{\Delta \underset{\sim}{I}[n] \Delta \underset{\sim}{Q}[n]\right\} = 0; \quad |n| \neq |m| \end{aligned} \tag{3.101}$$

$$\mathrm{E}\left\{\Delta \underset{\sim}{I}[n] \Delta \underset{\sim}{I}[m]\right\} = \mathrm{E}\left\{\Delta \underset{\sim}{Q}[n] \Delta \underset{\sim}{Q}[m]\right\} = \frac{1}{8} X_0^2[n] \frac{\sigma_{\mathrm{n}}^2}{\underset{\sim}{B}_{\mathrm{n}} t_{\mathrm{I}}}; \quad |n| = |m| \tag{3.102}$$

$$\begin{aligned} \mathrm{E}\left\{\Delta \underset{\sim}{I}^2[n] + \Delta \underset{\sim}{I}^2[-n]\right\} &= \operatorname{var}\left\{\Delta \underset{\sim}{I}[n] + \Delta \underset{\sim}{I}[-n]\right\} = 4\sigma_{\mathrm{I}}^2[n] \\ \mathrm{E}\left\{\Delta \underset{\sim}{Q}^2[n] + \Delta \underset{\sim}{Q}^2[-n]\right\} &= \operatorname{var}\left\{\Delta \underset{\sim}{Q}[n] + \Delta \underset{\sim}{Q}[-n]\right\} = 4\sigma_{\mathrm{Q}}^2[n] \end{aligned} \tag{3.103}$$

and by representing the product of two polynomials as convolution of the original coefficient sequence (also assigned as Cauchy product) we yield for the variance of the time function:

$$
\begin{aligned}
\sigma_z^2 &= \operatorname{var}\left\{\underset{\sim}{z}(t)\right\} = \mathrm{E}\left\{\left(\underset{\sim}{z}(t)-z_0(t)\right)\left(\underset{\sim}{z}(t)-z_0(t)\right)^*\right\}\\
&= \mathrm{E}\left\{\left(\sum_{n=-N_s}^{N_s}\left(\Delta \underset{\sim}{I}[n]+j\,\Delta \underset{\sim}{Q}[n]\right)\mathrm{e}^{j2\pi f[n]t}\right)\left(\sum_{n=-N_s}^{N_s}\left(\Delta \underset{\sim}{I}[n]-j\,\Delta \underset{\sim}{Q}[n]\right)\mathrm{e}^{-j2\pi f[n]t}\right)\right\}\\
&= \mathrm{E}\left\{\left(\sum_{n=-N_s}^{N_s}\sum_{k}\left(\Delta \underset{\sim}{I}[n]+j\,\Delta \underset{\sim}{Q}[n]\right)\left(\Delta \underset{\sim}{I}[n-k]-j\,\Delta \underset{\sim}{Q}[n-k]\right)\right)\right\}\\
&= \mathrm{E}\left\{\sum_{n=-N_s}^{N_s}\Delta \underset{\sim}{I}^2[n]+\Delta \underset{\sim}{Q}^2[n]\right\}\\
&= \sum_{n=1}^{N_s}\mathrm{E}\left\{\Delta \underset{\sim}{I}^2[n]+\Delta \underset{\sim}{I}^2[-n]\right\}+\mathrm{E}\left\{\Delta \underset{\sim}{Q}^2[n]+\Delta \underset{\sim}{Q}^2[-n]\right\}\\
&= 4\sum_{n=1}^{N_s}\left(\sigma_{\mathrm{I}}^2[n]+\sigma_{\mathrm{Q}}^2[n]\right)\\
&= \frac{\sigma_{\mathrm{n}}^2}{\ddot{B}_{\mathrm{n}}t_{\mathrm{I}}}\sum_{n=1}^{N_s}X_0^2[n]
\end{aligned}
\qquad (3.104)
$$

Obviously, we get a time-independent perturbation level of power σ_z^2 across the time function $z_0(t)$. According to (2.42), the SNR-value of the time domain waveform is

$$
\mathrm{SNR}_n = \frac{\left\|\mathrm{E}\left\{\underset{\sim}{z}(t)\right\}\right\|_\infty^2}{\operatorname{var}\left\{\underset{\sim}{z}(t)\right\}} \qquad (3.105)
$$

Here, we introduced the index n for the SNR-value in order to identify the source of perturbation as additive noise. The best case estimation of (3.105) presumes frequency-independent magnitude of the stimulus signal $X_0[n] = X_0$ and a DUT which causes a delay at constant attenuation, that is $Y_0[n] = Y_0$. Under these conditions, we finally find

$$
\mathrm{SNR}_n = \frac{\left(\sum\limits_{n=-N_s}^{N_s}((X_0[n]\,Y_0[n])/(2))\right)^2}{\left(\sigma_{\mathrm{n}}^2/\ddot{B}_{\mathrm{n}}t_{\mathrm{I}}\right)\sum\limits_{n=1}^{N_s}X_0^2[n]} = \frac{\ddot{B}_{\mathrm{n}}t_{\mathrm{I}}N_s Y_0^2}{\sigma_{\mathrm{n}}^2} \qquad (3.106)
$$

Phase noise: Phase noise in SFCW sensors comes from random phase fluctuations $\Delta\varphi(t)$ of the frequency synthesizer (refer to Section 2.6.3) as well as from frequency-dependent phase variations of the sensor electronics. The appearance in the time domain data of both effects is comparable. But while the first effect is actually random, the second one is of systematic nature and can hence be removed by calibration. We will deal only with the random errors in what follows.

For that purpose, we consider the output signal of the IQ-demodulator in polar form. Thus, we obtain for the nth frequency step $\underline{Z}[n] = Z[n]\mathrm{e}^{j\varphi[n]}$. If phase noise is involved, we will model the output signal of the IQ-demodulator as a random process. Following to Annex B.4.4, it can be written as

$$\underset{\sim}{\underline{Z}}[n] = Z_0[n]\mathrm{e}^{j\left(2\pi\tau f[n] - \Delta\underset{\sim}{\phi}_c[n,\tau]\right)} = \underline{Z}_0[n]\mathrm{e}^{-j\,\Delta\underset{\sim}{\phi}_c[n,\tau]} \tag{3.107}$$

Here, we supposed a delay of τ between the sine waves $x(t) = X_0 \sin 2\pi f[n]t$ and $y(t) = Y_0 \sin 2\pi f[n](t+\tau)$. $\Delta\underset{\sim}{\phi}_c[n,\tau]$ is the cumulative phase noise [87] of the nth frequency step:

$$\Delta\underset{\sim}{\phi}_c[n,\tau] = \Delta\underset{\sim}{\phi}(n, t-\tau) - \Delta\underset{\sim}{\phi}(n,t) \tag{3.108}$$

from which we assume ergodicity[16] ($\sigma^2_{\Delta\phi_c}[n] = \sigma^2_{\Delta\phi_c}$) and normal distribution of zero mean $\Delta\underset{\sim}{\phi}_c \sim N\left(0, \sigma^2_{\Delta\phi_c}\right)$. The variance of cumulative phase noise $\sigma^2_{\Delta\phi_c}$ can be determined from the phase noise spectrum of the sounding signal. As shown in Annex B.4.4 ((B.235)–(B.237)), it depends on the delay time τ. Referring to (B.238), expected value and variance of the IQ-demodulator output $\underset{\sim}{\underline{Z}}[n]$ at a single frequency step are given by

$$\mathrm{E}\left\{\underset{\sim}{\underline{Z}}[n]\right\} = \underline{Z}_0[n]\mathrm{e}^{-\frac{1}{2}\sigma^2_{\Delta\phi_c}[n]} \tag{3.109}$$

$$\mathrm{var}\left\{\underset{\sim}{\underline{Z}}[n]\right\} = \underline{Z}_0^2[n]\left(1 - \mathrm{e}^{-\sigma^2_{\Delta\phi_c}[n]}\right) \tag{3.110}$$

According to (3.99), the expected value of the time signal results from the Fourier series to

$$\begin{aligned}
\mathrm{E}\left\{\underset{\sim}{z}(t)\right\} &= \mathrm{E}\left\{\sum_{n=-N_s}^{N_s} \underline{Z}_0[n]\mathrm{e}^{j2\pi f[n]\,t}\,\mathrm{e}^{-j\,\Delta\phi_c[n,\tau]}\right\} \\
&= \sum_{n=-N_s}^{N_s}\left(\underline{Z}_0[n]\mathrm{e}^{j2\pi f[n]\,t}\,\mathrm{E}\left\{\mathrm{e}^{-j\,\Delta\phi_c[n,\tau]}\right\}\right) \\
&= \sum_{n=-N_s}^{N_s}\left(\underline{Z}_0[n]\mathrm{e}^{j2\pi f[n]\,t}\,\mathrm{e}^{-\frac{1}{2}\sigma^2_{\Delta\phi_c}[n]}\right) \\
&= z_0(t)\,\mathrm{e}^{-\frac{1}{2}\sigma^2_{\Delta\phi_c}}
\end{aligned} \tag{3.111}$$

This result indicates that phase noise will additionally attenuate the receive signal and that far targets are stronger affected than close ones in case of radar measurements since $\sigma^2_{\Delta\phi_c}$ increases with round-trip time.

The estimation of the variance is based on the same game as demonstrated in (3.104). From the physical nature of phase noise, we can again assume

16) It should be respected that ergodicity may possibly only approximately fulfilled in the case of very wideband scanning. Phase noise is a relative measure to quantify the time stability of a sine wave. It relates the fluctuations to the signal period. That means a sinusoid of a large frequency must provoke less temporal fluctuations than a low-frequency signal if both have the same phase noise. Thus, the signal source must be able to meet this requirement even if the frequency varies over several octaves.

independence of the perturbations between the individual frequency steps. Therefore, following identities may be established:

$$\begin{aligned}
&\mathrm{E}\left\{\left(\underset{\approx}{Z}[n]-\mathrm{E}\left\{\underset{\approx}{Z}[n]\right\}\right)\left(\underset{\approx}{Z}[m]-\mathrm{E}\left\{\underset{\approx}{Z}[m]\right\}\right)^{*}\right\}\\
&=\underline{Z}_0[n]\underline{Z}_0^{*}[m]\mathrm{E}\left\{\left(\mathrm{e}^{-j\,\Delta\underset{\sim}{\phi}_{\mathrm{c}}[n,\tau]}-\mathrm{e}^{-\frac{1}{2}\sigma^2_{\Delta\phi_{\mathrm{c}}}[n]}\right)\left(\mathrm{e}^{+j\,\Delta\underset{\sim}{\phi}_{\mathrm{c}}[m,\tau]}-\mathrm{e}^{-\frac{1}{2}\sigma^2_{\Delta\phi_{\mathrm{c}}}[m]}\right)\right\}\\
&=\underline{Z}_0[n]\underline{Z}_0^{*}[m]\mathrm{E}\left\{\mathrm{e}^{-j\left(\Delta\underset{\sim}{\phi}_{\mathrm{c}}[n,\tau]-\Delta\underset{\sim}{\phi}_{\mathrm{c}}[m,\tau]\right)}-\left(\mathrm{e}^{j\,\Delta\underset{\sim}{\phi}_{\mathrm{c}}[m,\tau]}+\mathrm{e}^{-j\,\Delta\underset{\sim}{\phi}_{\mathrm{c}}[n,\tau]}\right)\mathrm{e}^{-\frac{1}{2}\sigma^2_{\Delta\phi_{\mathrm{c}}}}+\mathrm{e}^{-\sigma^2_{\Delta\phi_{\mathrm{c}}}}\right\}\\
&=\begin{cases}Z_0^2[n]\left(1-\mathrm{e}^{-\sigma^2_{\Delta\phi_{\mathrm{c}}}}\right); & n=m\\ Z_0^2[n]\,\mathrm{e}^{j4\pi\tau f[n]}\left(\mathrm{e}^{-2\sigma^2_{\Delta\phi_{\mathrm{c}}}}-\mathrm{e}^{-\sigma^2_{\Delta\phi_{\mathrm{c}}}}\right); & n=-m\\ 0; & |n|\neq|m|\end{cases}
\end{aligned} \tag{3.112}$$

Equation (3.112) implies ergodicity ($\sigma^2_{\Delta\phi_{\mathrm{c}}}[n]=\sigma^2_{\Delta\phi_{\mathrm{c}}}[m]=\sigma^2_{\Delta\phi_{\mathrm{c}}}$); symmetry properties of Fourier transform ($\Delta\phi_{\mathrm{c}}[n]=-\Delta\phi_{\mathrm{c}}[-n]$; $f[n]=-f[-n]$) and it made use of (A.31) and (B.238). Thus, exploiting again Cauchy product of two sums, we can find for the variance of the time domain signal

$$\begin{aligned}
\sigma_z^2&=\mathrm{var}\left\{\underset{\sim}{z}(t)\right\}=\mathrm{E}\left\{\left(\underset{\sim}{z}(t)-\mathrm{E}\left\{\underset{\sim}{z}(t)\right\}\right)\left(\underset{\sim}{z}(t)-\mathrm{E}\left\{\underset{\sim}{z}(t)\right\}\right)^{*}\right\}\\
&=\mathrm{E}\left\{\left(\sum_{n=-N_s}^{N_s}\underline{Z}_0[n]\mathrm{e}^{j2\pi f[n]\,t}\left(\mathrm{e}^{-j\,\Delta\underset{\sim}{\phi}_{\mathrm{c}}[n,\tau]}-\mathrm{e}^{-\frac{1}{2}\sigma^2_{\Delta\phi_{\mathrm{c}}}[n]}\right)\right)\cdots\right.\\
&\qquad\left.\cdots\left(\sum_{n=-N_s}^{N_s}\underline{Z}_0^{*}[n]\mathrm{e}^{-j2\pi f[n]\,t}\left(\mathrm{e}^{j\,\Delta\underset{\sim}{\phi}_{\mathrm{c}}[n,\tau]}-\mathrm{e}^{-\frac{1}{2}\sigma^2_{\Delta\phi_{\mathrm{c}}}[n]}\right)\right)\right\}\\
&=\mathrm{E}\left\{\sum_{n=-N_s}^{N_s}\sum_{k}\underline{Z}_0[n]\,\underline{Z}_0^{*}[n-k]\mathrm{e}^{j2\pi(f[n]-f[n-k])\,t}\cdots\right.\\
&\qquad\left.\cdots\left(\mathrm{e}^{-j\left(\Delta\underset{\sim}{\phi}_{\mathrm{c}}[n,\tau]-\Delta\underset{\sim}{\phi}_{\mathrm{c}}[n-k,\tau]\right)}-\mathrm{e}^{-\frac{1}{2}\sigma^2_{\Delta\phi_{\mathrm{c}}}}\left(\mathrm{e}^{-j\,\Delta\underset{\sim}{\phi}_{\mathrm{c}}[n,\tau]}+\mathrm{e}^{j\,\Delta\underset{\sim}{\phi}_{\mathrm{c}}[n-k,\tau]}\right)+\mathrm{e}^{-\sigma^2_{\Delta\phi_{\mathrm{c}}}}\right)\right\}
\end{aligned} \tag{3.113}$$

Following to (3.112), the inner sum only contributes for two specific cases of k, namely if holds $n=n-k$ and $-n=n-k$. All other terms give zero. Considering a DUT which only provokes a delay but not any other signal distortions, we yield for these two particular cases

$$\begin{aligned}
\left.\sigma_z^2\right|_{k=0}&=\left(1-\mathrm{e}^{-\sigma^2_{\Delta\phi_{\mathrm{c}}}}\right)\sum_{n=-N_s}^{N_s}Z_0^2[n]=2N_sZ_0^2\left(1-\mathrm{e}^{-\sigma^2_{\Delta\phi_{\mathrm{c}}}}\right)\\
\left.\sigma_z^2\right|_{k=2n}&=2\left(\mathrm{e}^{-2\sigma^2_{\Delta\phi_{\mathrm{c}}}}-\mathrm{e}^{-\sigma^2_{\Delta\phi_{\mathrm{c}}}}\right)\sum_{n=1}^{N_s}Z_0^2[n]\cos(4\pi\tau f[n])\\
&=2Z_0^2\left(\mathrm{e}^{-2\sigma^2_{\Delta\phi_{\mathrm{c}}}}-\mathrm{e}^{-\sigma^2_{\Delta\phi_{\mathrm{c}}}}\right)\sum_{n=1}^{N_s}\cos(4\pi\tau f[n])
\end{aligned}$$

so that we finally get

$$\sigma_z^2 = \sigma_z^2\big|_{k=0} + \sigma_z^2\big|_{k=2n} = 2N_s Z_0^2\left(1 - e^{-\sigma_{\Delta\varphi_c}^2}\right)\left(1 - e^{-\sigma_{\Delta\varphi_c}^2}\frac{1}{N_S}\sum_{n=1}^{N_s}\cos(4\pi\tau f[n])\right) \tag{3.114}$$

We can observe from this equation that the phase noise-induced perturbations are independent on time t, that is they are scattered over the whole time signal. The strength of the perturbation does however depend on the delay time τ between reference and measurement signal. First, this results from the behaviour of the cumulative phase noise variance (see Annex B.4.4 (B.237)) $\sigma_{\Delta\phi_c}^2 \approx 4\pi a\tau$ and second, it is caused from the sum term in (3.114) whose impact we will shortly investigate. Since the frequency grid of the measurement is usually quite dense, we can approximate the sum by an integral

$$\begin{aligned}\frac{1}{N_S}\sum_{n=1}^{N_s}\cos(4\pi\tau f[n]) &= \frac{1}{N_S}\sum_{n=1}^{N_s}\cos\left(4\pi\tau\left(f_1 + (n-1)\Delta f\right)\right) \approx \frac{1}{N_2\,\Delta f}\int_{f_l}^{f_u}\cos 4\pi f\tau\, df \\ &= \cos 4\pi\tau f_0 \frac{\sin 2\pi\tau B}{2\pi\tau B} = \cos 4\pi\tau f_0 \frac{\sin 2\pi b\tau f_0}{2\pi b\tau f_0}\end{aligned}$$

Herein, $f_l, f_u, f_0 = (f_u + f_l)/2$ are the lower, upper and centre frequency, $B = f_u - f_l$ is the absolute bandwidth of the measurement and $b = B/f_0$ is the fractional bandwidth. For extreme near field applications, the sum can be simplified to[17)]

$$\frac{1}{N_S}\sum_{n=1}^{N_s}\cos(4\pi\tau f[n]) \approx 1 - 2(\pi\tau f_0)^2\left(\frac{b^2}{3} + 4\right) \quad \text{for} \quad \tau f_0 \le 0.1 \tag{3.115}$$

and its actual evolution in dependency on $f_0\tau$ is depicted in Figure 3.77 indicating that it tends rapidly to zero for large fractional bandwidth. Thus, the sum term may be neglected in many practical applications.
We can further approximate $1 - e^{-\sigma_{\Delta\varphi_c}^2} \approx \sigma_{\Delta\phi_c}^2$ for short-range sensing due to the comparatively short delay τ, so that we can observe that the phase noise provoked variance will rapidly growth with increasing delay time:

$$\sigma_z^2 \approx 16\, N_s Z_0^2 a(\pi\tau)^3 f_0^2\left(\frac{b^2}{3} + 4\right) \quad \text{for } \tau f_0 \le 0.1 \tag{3.116}$$

17) Note that the delay time τ has to respect the overall delay in the x- and y-channels before IQ-demodulation, that is $\tau = \tau_{DUT} + \tau_y - \tau_x$. $\tau_{DUT} + \tau_y$ represents the overall delay between signal source and y-input of the IQ-demodulator and τ_x relates to the overall delay between signal source and x-input of the IQ-demodulator. Hence, the condition (3.115) can only be met if the two input channels of the receiver are well balanced with respect to the internal delay.

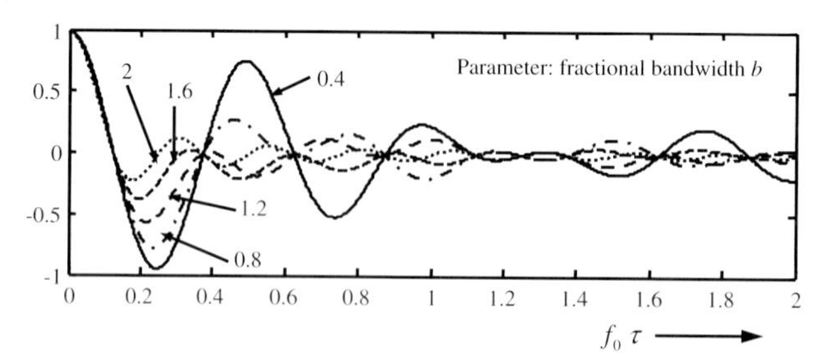

Figure 3.77 $(1/N_s)\sum_{n=1}^{N_s}\cos(4\pi\tau f[n])$ as function of the delay time τ and operational frequency band (given by the centre frequency f_0 and the fractional bandwidth b).

But this growth will be restricted only to the very near range. Finally it will stabilize to a linear increase as function of the delay:

$$\sigma_z^2 \approx 2N_s Z_0^2 \sigma_{\Delta\phi_c}^2 = 8N_s Z_0^2 a\pi\tau \tag{3.117}$$

In (3.116) and (3.117), we used the identity $\sigma_{\Delta\phi_c}^2 = 4a\pi\tau$ deduced in Annex B.4.4 (B.237) where the constant $a = \pi f_{\text{offset}}^2 \underset{\cdot\cdot}{m}_\phi(f_{\text{offset}})$ has to be determined from the phase noise spectrum $\underset{\cdot\cdot}{m}_\phi(f)$ of the sounding signal at an appropriate offset frequency f_{offset}. According to (3.117) and (3.111),[18] the phase noise affected SNR-value is

$$\text{SNR}_{\Delta\phi} = \frac{\left\| \text{E}\left\{ \underset{\sim}{z}(t) \right\} \right\|_\infty^2}{\text{var}\left\{ \underset{\sim}{z}(t) \right\}} \approx \frac{N_s}{2a\pi\tau}; \quad \text{since } e^{-\frac{1}{2}\sigma_{\Delta\phi_c}^2} \approx 1 \tag{3.118}$$

What do we learn from (3.118)? The perturbation level caused from phase noise affected sounding signals does not depend on the signal power while additive noise perturbations depend on it. Hence, there is a level of receive signal for which both perturbations are equal. From (3.106) and (3.118), we can conclude

$$\text{SNR}_\text{n} = \text{SNR}_{\Delta\phi} \Rightarrow Y_{0,\text{opt}}^2 = \frac{\sigma_\text{n}^2}{2a\pi\tau \underset{\cdot\cdot}{B}_\text{n} t_\text{I}} \tag{3.119}$$

Consequently, for a given test scenario, a further enhancement of stimulation power $X_0^2/2$ will not improve the SNR-value if the optimum receiver level $Y_{0,\text{opt}}$ is already reached.

UWB short-range devices are very often applied as radar sensors under strong multi-path conditions. That involves, for example the search of a weak target in the presence of strong scatters. In case of phase noise dominance, these scatters will determine the perturbation level and hence the device sensitivity to detect weak targets.

18) If the DUT does not provoke linear phase distortions as we did presume, the peak value of $z_0(t)$ corresponds to the coherent superposition of all spectral components, that is $\|z_0(t)\|_\infty = 2N_s Z_0$.

The discussion of (3.119) with respect to specific test scenarios needs further analysis in order to find optimum operational conditions for the sensor device. Since the signal strength caused from an unwanted scatterer depends on the distance and hence on the round-trip time τ, the conclusions concerning optimum transmitter level are scenario dependent. The reader can refer to Section 4.7.3 for further discussions concerning the impact of phase noise and jitter onto the measurement performance of an UWB device.

3.4.4.2 Continuous Frequency Variation

Now, we will replace the discrete frequency variations of a SFCW-device by continuous frequency variation [79, 88–90]. This operation mode is well known under the term FMCW-principle. FMCW-sensors are basically able to measure the FRF of a DUT as we did with the SFCW approach. However, these sensors are mostly found in radar applications performing range measurements. Hence, we will take a range measurement as example to illustrate the FMCW-principle.

To keep simple, we assume a target at distance r to the sensor which involves a round-trip time of $\tau = 2\,r/c$. Shape distortions of the backscattered signal are out of scope here. Thus, we can model our test scenario by a simple IRF and FRF:

$$g(t) = \delta(t - \tau) \underset{\text{IFT}}{\overset{\text{FT}}{\longleftrightarrow}} \underline{G}(f) = \mathrm{e}^{-j2\pi f \tau}$$

Generally, a FMCW-sensor is composed from a homodyne or heterodyne receiver and a sine wave source which linearly varies the frequency f of the sounding signal. The frequency variation shall occur quite slowly so that the test scenarios stay nearly in their steady state. We are interested in the resulting output signal of the receiver and how to use it for our goal. For that purpose, we will discuss three cases – a stationary target at distance r_0 and the sounding of moving targets with narrowband as well as with wideband sweep.

The sounding signal is a linear chirp that shifts its frequency from f_l to f_u during the time interval t_w, that is

$$\begin{aligned} f(t) &= f_\mathrm{l} + at \\ \text{with} \quad a &= \frac{f_\mathrm{u} - f_\mathrm{l}}{t_\mathrm{w}} = \frac{\dot{B}}{t_\mathrm{w}} \end{aligned} \tag{3.120}$$

The chirp rate a quantifies the speed of frequency variation. Applying complex notation, the sounding signal may be written as (see also Section 2.3.6)

$$\underline{x}(t) = X\,\mathrm{e}^{j2\pi\left(f_\mathrm{l} t + \frac{1}{2} a t^2\right)} \quad t \in [0, t_\mathrm{w}] \tag{3.121}$$

Case 1: stationary target at distance r_0: Following our assumption, the receive signal $y(t)$ is subject only a delay of $\tau_0 = 2\,r/c$. Hence, we yield

$$\underline{y}(t) = x(t - \tau_0) = X\,\mathrm{e}^{j2\pi\left(f_\mathrm{l}(t-\tau_0) + \frac{1}{2} a (t-\tau_0)^2\right)} \tag{3.122}$$

Using (3.86), the homodyne or heterodyne receiver provides the two signals $I(t)$ and $Q(t)$ according to

$$I(t) + jQ(t) = \frac{1}{2}\underline{y}(t) \cdot \underline{x}^*(t) = \frac{1}{2}X^2\, \mathrm{e}^{j\pi a\tau_0^2}\, \mathrm{e}^{-j2\pi(f_1+at)\tau_0} \tag{3.123}$$

Insertion of (3.120) yields

$$2\frac{I(t) + jQ(t)}{X^2} = \mathrm{e}^{j\pi a\tau_0^2}\, \mathrm{e}^{-j2\pi f\tau_0} = \mathrm{e}^{j\pi a\tau_0^2}\, \underline{G}(f) \tag{3.124}$$

which implies that with exception of the phase term $\mathrm{e}^{-j\pi a\tau_0^2}$ the temporal shape of the signals $I(t)$ and $Q(t)$ correspond to the spectral shape of the FRF of the DUT.[19] The undesired phase term in (3.124) is negligible if a sufficiently slow chirp rate a is selected. (3.123) provides the corresponding condition

$$a\tau_{0,\max} \ll 2f_1 \tag{3.125}$$

The phase term is of minor interest for radar measurements since they only intend to determine the round-trip time τ_0. For the purpose of range estimation, we rewrite (3.123) in the form

$$\begin{aligned} I(t) + jQ(t) &= \frac{1}{2}X^2\, \mathrm{e}^{j(-2\pi a\tau_0 t + \varphi_0)} = \frac{1}{2}X^2\, \mathrm{e}^{j(2\pi f_\mathrm{B}\, t + \varphi_0)} \\ &= \frac{1}{2}X^2\left(\cos\left(2\pi f_\mathrm{B} t + \varphi_0\right) + j\sin\left(2\pi f_\mathrm{B} t + \varphi_0\right)\right) \\ \text{with} \quad \varphi_0 &= \pi a\tau_0^2 - 2\pi f_1\tau_0 \\ f_\mathrm{B} &= -a\tau_0 \end{aligned} \tag{3.126}$$

We can conclude from this equation that

- the signals $I(t)$ and $Q(t)$ are pure sine waves for a single target of the beat frequency

$$f_{\mathrm{B}0} = |a\tau_0| = \left|a\frac{2r_0}{c}\right| \tag{3.127}$$

  Hence, the target distance can be calculated from the knowledge of the frequency $f_{\mathrm{B}0}$. The frequency estimation is either done by subjecting one of both signals the Fourier transform or any other parametric or non-parametric methods of spectral analysis [91].
- The duration of the beat signals equal roughly t_w, so that we can approximately write

$$\begin{aligned} I(t) &\propto \mathrm{rect}\left(\frac{t - t_\mathrm{w}/2}{t_\mathrm{w}}\right)\cos 2\pi f_{\mathrm{B}0} t \\ Q(t) &\propto \mathrm{rect}\left(\frac{t - t_\mathrm{w}/2}{t_\mathrm{w}}\right)\sin 2\pi f_{\mathrm{B}0} t \end{aligned} \tag{3.128}$$

19) Equation (3.124) is valid for all types of linear DUT and not only restricted to a delay system as supposed in our example.

If we perform Fourier transform on the complex time signal $I(t) + jQ(t)$ to estimate the beat frequency f_{B0}, we will result in a one-sided spectrum of the form $\mathrm{sinc}\left((f - f_{B0})t_w\right)$. The location of its maximum will give us the wanted beat frequency.

In the case of two targets, we get two overlapping sinc functions. They can only be distinguished if both beat frequencies are separated at least by

$$\begin{aligned} \delta_f &= |f_{B1} - f_{B2}| \geq \frac{1}{t_w} \\ \text{with} \quad \delta_f &= \frac{2a\delta_r}{c} \Rightarrow \delta_r = \frac{c}{2t_w a} = \frac{c}{2B} \end{aligned} \tag{3.129}$$

which leads to the well-known relation of radar range resolution δ_r (see also Section 4.7.2).

- Since for range measurement only the estimation of beat frequency is required, IQ-demodulation is not mandatorily needed. Thus, we can reduce the sensor electronics to the components as exhibited in Figure 3.78. If we also apply Fourier transform for beat frequency estimation, we have to deal with a two-sided spectrum now since we have available only a real-valued time signal. The shape of the spectrum composes from two sinc functions $\propto \mathrm{sinc}\left((f - f_{B0})t_{win}\right) + \mathrm{sinc}\left((f + f_{B0})t_{win}\right)$. These two sinc functions may overlap if the beat frequency becomes too small. If this is happen, the side lobes of one sinc function affect the maximum location of the other one. Thus, the beat frequency will be erroneous determined which leads to slightly wrong distance measures. If two sinc functions are separated by, for example ± 3 side lobes (i.e. $f_{B0} \geq 3/t_w$) the range error is $\Delta R \leq 0.1\,\delta_r$, that is the range accuracy will drop by coming too close to the sensor.
- The cut-off frequency of the video-filter must excide the largest beat frequency. Hence, the integration time has to meet the condition

$$f_{B,\max}\, t_I = \frac{2ar_{\max}t_I}{c} < 1 \tag{3.130}$$

As Figure 3.78 depicts, the structure of a FMCW-sensor may be very simple, since only the transmitter and receiver signal has to be mixed. The

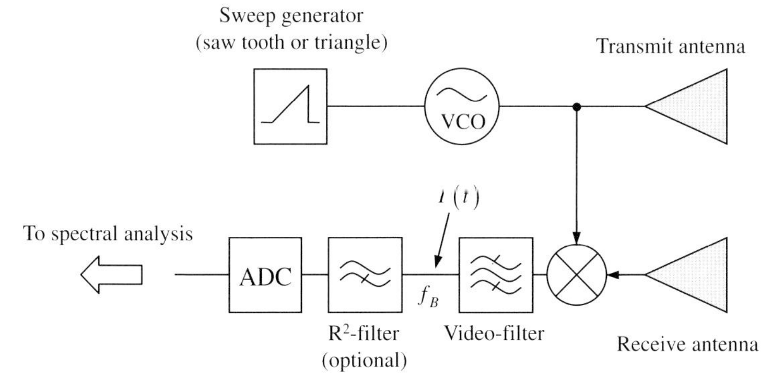

Figure 3.78 Simplified circuit schematic of a FMCW-radar.

resulting in-phase signal $I(t)$ is a sine wave of frequency f_B which is proportional to the target range as demonstrated above. The video-filter suppresses unwanted mixing products (images). It can be a simple low-pass filter those upper cut-off frequency is determined from the maximum target range as can seen from (3.130). But typically that filter also has a lower cut-off in order to suppress phase noise of the VCO, leakage of the sweep signal and direct transmitter–receiver coupling. Directly coupled signals have a short delay leading to low beat frequencies which can be suppressed by appropriate high-pass filtering. The lower cut-off frequency of the filter will determine the blind range of the radar.

The second filter is optional. It is often termed as R^2-filter. It serves to compensate the spreading loss of the radar signal (omitted in our demonstrating example). We will see in the next chapter that the strength of the receive signal decreases by r^2 (r is the target distance). Since the beat frequency f_B increases with r, the spreading loss can be compensated by a second-order high-pass filter which is performing a twofold differentiation.

The FMCW-approach is prone to any deviations of the VCO-signal from the expected chirp (3.121). Discrepancy in linear frequency sweep provoke unstable beat frequency leading to a blurred spectrum of the IQ-signal and hence degraded range resolution. Since the VCO-characteristic is usually non-linear, it has to be compensated anyhow. Several approaches are in use. The simplest one applies a sweep voltage of inverse non-linearity. Further, the VCO may be operated in a high-resolution PLL as exemplified in Ref. [79] or one corrects the VCO-non-linearity by referring to a known reference delay (see Ref. [92] for an example).

In addition, any device internal frequency response variations (amplifier FRF, antenna FRF, device internal reflections etc.) lead to modulation of the sounding signal

$$\underline{x}(t) = X(1 + m(t))e^{j2\pi\left(f_1 t + \frac{1}{2}at^2\right)}$$

which causes side lobes and background clutter in the Fourier transformed beat signal.

Case 2: moving target and frequency sweep of low fractional bandwidth: A moving target of radial speed v_r provokes additional shift of the beat frequency. As long as the frequency sweep is much smaller than the centre frequency f_m, the beat frequency can modelled approximately by

$$f_B = f_d - a\tau_0 = -\frac{2}{c}\left(ar_0 + f_m v_r\right) \tag{3.131}$$

This equation shows range-velocity ambiguities which cannot be solved from a single measurement of beat frequency. We trace (3.131) in range-velocity coordinates as depicted in Figure 3.79a. Thus, the actual values of r_0 and v_r are placed

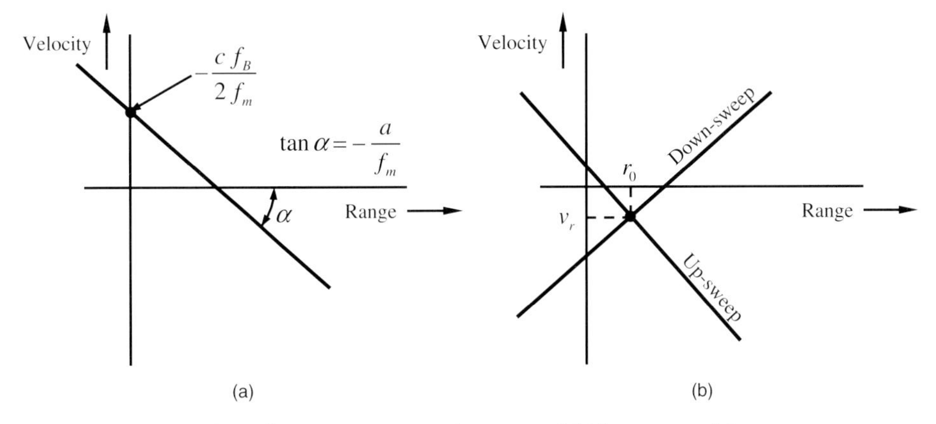

Figure 3.79 Range-velocity diagram. (a) For a single sweep and (b) for an up- and down-sweep.

anywhere at the line. In order to solve the ambiguity, we need a second measurement with a different chirp rate which provides a second line. The intersection finally gives us the wanted quantities of range and speed (Figure 3.79b). Preferentially, the two lines should be orthogonal in order to minimize uncertainties of the intersection point. This involves two sweeps in different directions, that is up- and down-sweep.

Figure 3.80 depicts the corresponding sweep procedure by referring to the instantaneous frequency at several points of the device. It illustrates the procedure considered in Figure 3.79b. The transmitter frequency is modulated by a triangular time function. In the case of a motionless target, the transmitted waveform will be returned back at the receiver after the round-trip time τ_0. Hence, the complete modulation scheme is shifted by τ_0 in time direction leading to a constant frequency difference $f_{B0} = -a \cdot \tau_0$ within the time interval $t_{w,eff}$. Thus, the actual time for spectrum analysis is a bit shorter than assumed in (3.129) and corresponding holds for the effective available bandwidth. Since however $\tau_0 \ll t_w$, (3.129) gives reasonable results. A moving target will add the Doppler frequency onto f_{B0}. That is, it will shift the whole modulation schema by f_d in frequency direction, that is $f_{B\pm} = \pm f_{B0} + f_d$.

If the measurement scenario has involved two moving targets, the range-velocity diagram (Figure 3.79) is composed from four straight lines and four intersection points. Two of these points represent the real targets and the other two concerns to ghost targets. In order to resolve these new ambiguities, one extends the modulation schema by further chirp rates. Figure 3.81 gives an example of a multi-target scenario. A real target is identified by a common intersection of three lines. Single intersections are omitted. Nevertheless, as indicated in the figure, there remains still some risk of ghost target detection which can be reduced by adding still more chirp rates or a plausibility check involving the history of the radar scene.

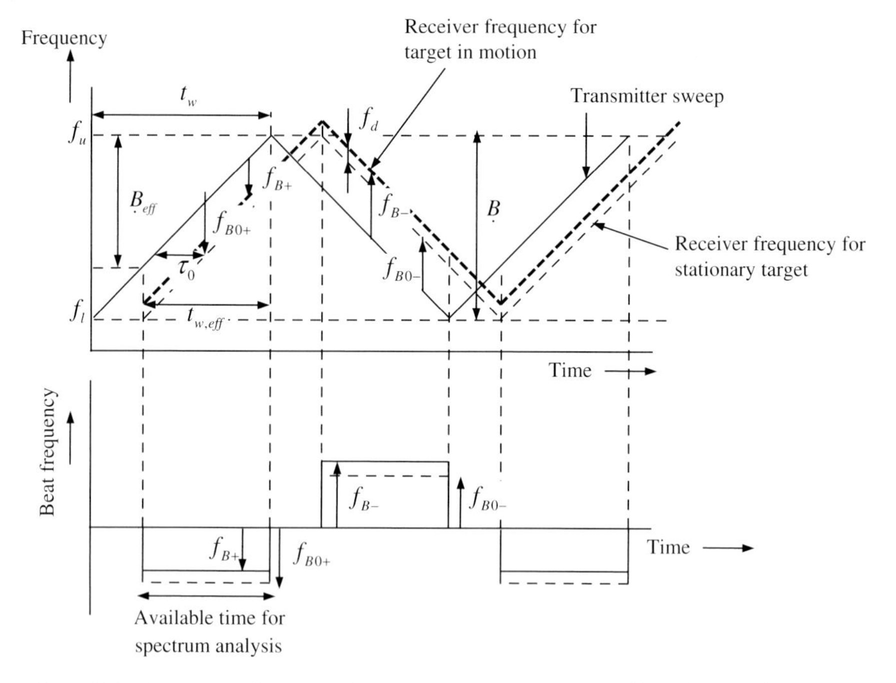

Figure 3.80 Sweep procedure to resolve range-velocity ambiguity for single target. The index + stands for the up-sweep and index – for the down-sweep. Note that the device depicted in Figure 3.78 is not able to distinguish between positive and negative beat frequencies. It only measures the absolute value.

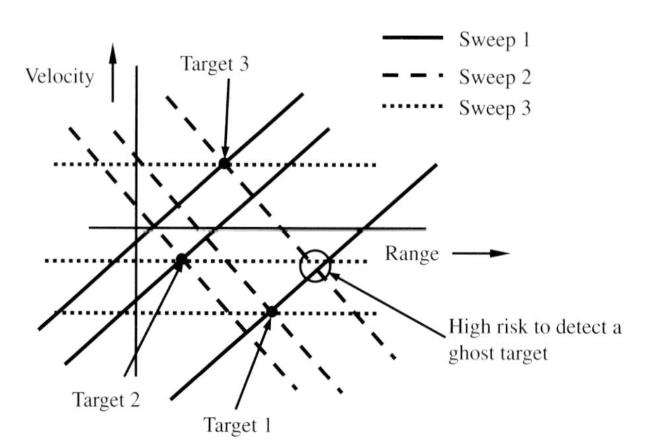

Figure 3.81 Range-velocity diagram for multi-target scenario and three different chirp rates. Note, sweep 3 has the chirp rate zero (i.e. it is a pure sine wave indicating only the Doppler shift).

Case 3: moving target and frequency sweep of large fractional bandwidth:
The model (3.122) for the receive signal has to be modified for large fractional bandwidth. From (2.86), we find

$$\underline{y}(t) = \sqrt{s}\, x(s(t-\tau_0)) = \sqrt{s}\, X\, e^{j2\pi\left(f_1 s(t-\tau_0)+\frac{1}{2}as^2(t-\tau_0)^2\right)} \tag{3.132}$$

in which $s = (c - v_r)/(c + v_r)$ is the scaling factor and τ_0 is the round trip time. This leads to the receiver output signal

$$I(t) + jQ(t) = \frac{1}{2}\underline{y}(t) \cdot \underline{x}^*(t) = \frac{1}{2}\sqrt{s}X^2\, e^{j(2\pi f_B\, t+\varphi_0)} \tag{3.133}$$

The initial phase φ_0 is out of interest here and the beat frequency f_B is given by

$$f_B(t) = f_1(s-1) + \frac{1}{2}at(s^2-1) - as^2\tau_0 = f_{Bl} + a_B t \tag{3.134}$$

with

$$f_{Bl} = \left(4\frac{v_r}{c} - 1\right)a\tau_0 - 2f_1\frac{v_r}{c}; \quad a_B = -2a\frac{v_r}{c}; \quad s \approx 1 - 2\frac{v_r}{c}; \quad s^2 \approx 1 - 4\frac{v_r}{c}$$

Obviously, the beat signal represents a chirp too. If we estimate f_{Bl} and a_B from the measurement by an appropriate procedure, we can determine target range and velocity from a single sweep now. The range-velocity separation will be as better as larger the fractional bandwidth of the sounding signal will be.
In spite of its drawbacks, the FMCW-principle is a powerful measurement approach as long as the fractional bandwidth must not be too large and the resolution requirements are not too demanding. It is very robust against Doppler with respect to target detection. There is indeed a large range-Doppler coupling but the signal amplitude – crucial for target detection – will not be affected by Doppler effect. The most critical RF-components (VCO and mixer) can be implemented by standard technology (small fractional bandwidth assumed) and the receiver efficiency is very high compared to sub-sampling receiver. We can observe from Figure 3.80 that the available time for energy accumulation is $t_{w,eff} \approx t_w$ per sweep. The complete recording time will be $T_R \approx N_a t_w$ if the measurement has to be repeated N_a-times with different chirp rates to resolve range-Doppler ambiguities. So that with η_m (mixer efficiency), the overall efficiency can be written as

$$\eta_{FMCW} = \eta_m \frac{t_{w,eff}}{T_R} \approx \frac{\eta_m}{N_a} \tag{3.135}$$

3.5
The Multi-Sine Technique

The multi-sine approach is a measurement technique optimized to FFT processing. As already mentioned in Section 2.3.7, the idea behind the multi-sine technique is to provide a periodic stimulus signal (period t_p) which can be flexibly synthesized

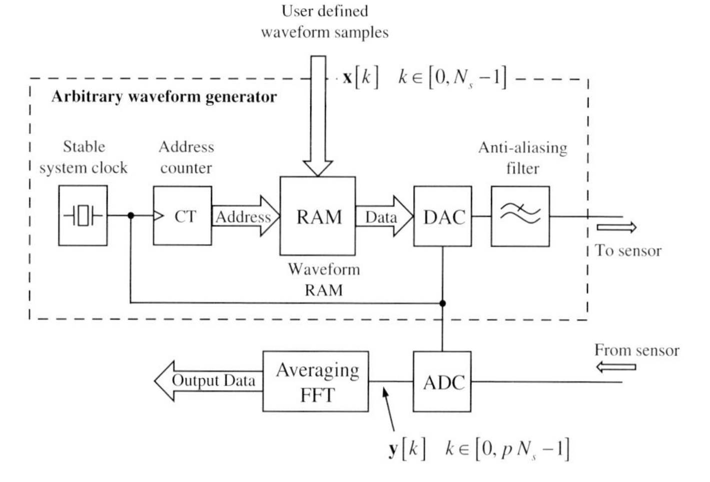

Figure 3.82 Generic conception of a multi-sine device using Nyquist sampling. The RAM is organized as ring buffer which is cyclically addressed by the address counter. Thus, the signal will be periodically repeated and the receiver may capture a number p of periods from the system reaction to perform averaging.

by the user with respect to spectrum and amplitude distribution in order to meet specific aspects of the measurement scenario [93, 94]. The multi-sine technique belongs to the pseudo-noise approaches since it allows for generation of arbitrary periodic signals. Figure 3.82 depicts a generic device structure for baseband measurements. They key components are the wideband arbitrary waveform generator, the wideband data acquisition system and the data pre-processing. Except the technical solutions for the generation of the wideband multi-sine, the device conception and philosophy is close to that of M-sequence devices. That is, sub-sampling can be correspondingly used in order to relax the requirements onto data capturing and processing speed. The easiest way to do this is again to insert a binary divider in the ADC clock line and to apply $N_s = 2^m \pm 1$ voltage samples per signal period (see (3.25)). This is always possible since the user is responsible to create the test signal beforehand and to transfer it to the waveform RAM.

The data pre-processing typically covers FFT and synchronous averaging. The operational spectral band can be shifted to any frequency applying the same methods as discussed for the M-sequence (Figures 3.54, 3.57 and 3.58, [95]).

While the data capturing may work at lower clock rates (due to sub-sampling), the stimulus generation has to deal with the actual spectral band of interest in order to stimulate the test object in the right way. In the case of wideband and ultra-wideband applications, this puts strong requirements to the clock rate of the arbitrary waveform generator. The larger flexibility of signal construction has to be paid by higher power consumption, higher non-linear distortions due to the low bit number of the high-speed DAC and higher system cost.

Non-linear distortions caused by the transfer characteristic of the DAC provoke signal components pretending to be noise. But they will not disappear by

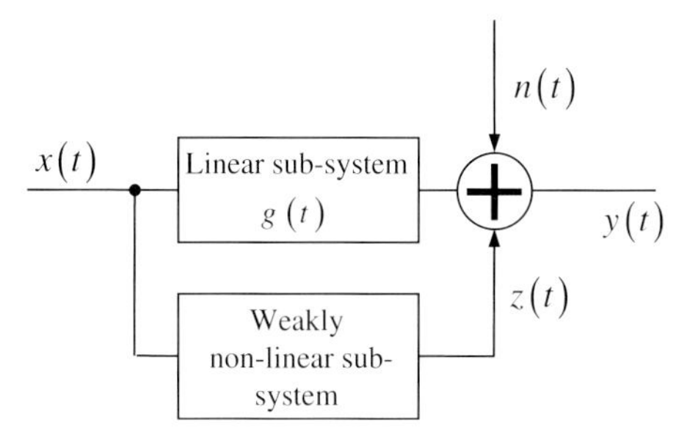

Figure 3.83 Decomposition of the measurement device and the system under test in a linear sub-system and a weakly non-linear system.

synchronous averaging since they are of deterministic nature. We already got to know a corresponding phenomenon in connection with the M-sequence where we introduced an intrinsic impulse response $g_{int}(t)$ (which has to be determined from a calibration measurement) to correct the measurement data. As long as the signal shape is kept always the same, such a calibration technique may be applied also here since we can suppose that the generated signals are stable in time. A further method is to capture the actual stimulus signal by a second receiver channel which serves for reference.

The flexibility to generate a large number of different signals with identical power spectrum opens up a new and interesting method to combat on the one hand the impact of these non-linear distortions or on the other hand to investigate weakly non-linear test objects (see Ref. [93] for details).

We will restrict our discussion to the suppression of DAC quantization errors. For that purpose, we decompose the transmission channel in a linear path and a non-linear path[20] (see Figure 3.83). That is, non-linear effects of the DAC (or even the DUT) contribute to the non-linear sub-system which provokes the signal $z(t)$. If we also respect random noise $n(t)$, the captured signal $y(t)$ may be written as

$$y(t) = g(t) * x(t) + z(t) + n(t) \tag{3.136}$$

Herein, $x(t)$ represents the ideal test signal which is free of quantization errors and random noise. Our goal is to separate the different signal components by an appropriate measurement procedure. First, we calculate the cross- and auto-correlation of the receive signal presuming $C_{xx}(t) \approx x_{\mathrm{rms}}^2\,\delta(t)$:

$$Cyx(t) = y(t) * x(-t) \approx x_{\mathrm{rms}}^2\, g(t) + C_{zx}(t) + C_{nx}(t) \tag{3.137}$$

$$\begin{aligned} C_{yy}(t) = y(t) * y(-t) &\approx x_{\mathrm{rms}}^2\, g(t) * g(-t) + C_{zz}(t) + C_{nn}(t) + \cdots \\ &+ g(-t) * C_{zx}(t) + g(t) * C_{xz}(t) + g(-t) * C_{nx}(t) \\ &+ g(t) * C_{xn}(t) + C_{nz}(t) + C_{zn}(t) \end{aligned} \tag{3.138}$$

20) Exactly spoken, we split the signal path into correlated and uncorrelated signals. The correlated signals are determined by linear transmission effects (and negligible non-linear effects in case of weak non-linearity, i.e. the compression terms do not yet determine the scenario - compare Table 2.8). The uncorrelated signals are provoked by non-linear effects.

$C_{nn}(t)$ and $C_{zz}(t)$ are the auto-correlation functions of the random noise and of the non-linear distortions. They will give an impression on how strong the measurement is affected by these perturbations. The remaining correlation functions refer to cross-correlations.

The determination of the wanted IRF $g(t)$ is based on (3.137) assuming $C_{xx}(t) \to \delta(t)$ and the strength of the perturbations (noise and non-linearity) can be estimated from (3.138). But these two equations are still affected by a number of cross-correlation functions which are not known so far. We can neglect all correlations with random noise $C_{nx}(t) = C_{nz}(t) = 0$ due to independence from noise and test signal. Thus, it only remains the perturbing term $C_{zx}(t)$ which requires additional considerations. This term represents the correlation between $z(t)$ and $x(t)$. Since in our case $z(t)$ is caused from DAC quantization effects, it cannot be assumed that $x(t)$ and $z(t)$ are completely uncorrelated (i.e. $C_{zx}(t) \neq 0$)

At this point, we will shortly come back to quantization errors. We have seen in Section 2.6.1 that quantization effects are usually modelled as random errors with the consequence that they can be arbitrarily reduced by averaging. But this is only applicable if no synchronism is in the captured data. Consequently, this assumption does not hold for measurements with periodic signals as we do it consider here. However as we could observe in Figure 2.81, additional random noise in the measurement data could change the situation, at least in the case of ADC quantization effects.

In order to do the same with the DAC, we have to interrupt the synchronism by randomizing the (digital) input signal. Hence, we could repeat the measurements M-times by applying a set of test signals $x_m(t) = x_0(t) + n_m(t)$ where we simply add random numbers to a given signal $x_0(t)$. Since we know these random numbers, we will not lower the SNR-value. But we will affect the ACF of the test signal which complicates the determination of the IRF $g(t)$. It can be avoided by only randomizing the phase and keeping constant the spectral power within the whole set of test signals. Hence, we can write for the samples of the stimulus

$$\begin{gathered} \mathbf{x}_m[k] = \sum_{i=-N}^{N} X[i] e^{j\left(2\pi i f_0 k\,\Delta t_s + \varphi[i,m]\right)}; \quad k \in [0, N_s - 1]; \quad N_s = \frac{1}{f_0\,\Delta t_s} \geq 2N+1 \\ X[i] = X[-i] \\ \varphi[0,m] = 0; \quad \varphi[i,m] = -\varphi[-i,m] \quad i \in [1,N]; \quad \varphi[i,m] \sim U(-\pi,\pi) \end{gathered} \tag{3.139}$$

The randomized test signal will cause a randomized non-linear signal which can be suppressed by averaging over the ensemble of test signals, so that we finally yield[21)]

$$\bar{C}_{yx}(t) = \frac{1}{M}\sum_{m=0}^{M-1} C_{y_m x_m}(t) \approx g(t) \quad \text{since} \quad \frac{1}{M}\sum_{m=0}^{M-1} C_{z_m x_m}(t) \to 0 \tag{3.140}$$

$$\bar{C}_{yy}(t) = \frac{1}{M}\sum_{m=0}^{M-1} C_{y_m y_m}(t) \approx g(t) * g(-t) + C_{zz}(t) + C_{nn}(t) \tag{3.141}$$

21) Smooth non-linearity as amplifier saturation can be modelled by Taylor series. In that case $\bar{C}_{zx}(t)$ does not necessarily tend to zero, so that the IRF estimation will be biased.

Figure 3.84 shows the flow graph of the procedure. The measurement is repeated M-times always with a different test signal. One can capture the data over one or more periods after the measurement scenario has been settled down (usually this need one period). Averaging can be performed by two different loops. On the one hand, it can be done within a p-fold repetition of an identical signal (inner averaging loop, not shown in the figure). This would lead only to noise suppression in the receiver. On the other hand, averaging over the whole ensample of test signals (outer averaging loop) provides additionally the wanted suppression of DAC quantization effects. These two loops can be used to separate the strength of noise and non-linear effects if wanted. This might be of interest for investigations of slightly non-linear test objects. If one is only interested in a cleared IRF or FRF and the random noise level (including quantization noise of the ADC) is not larger than the effective quantization level of the DAC, the separation in two loops only extends the measurement time without improvement of perturbation suppression. Furthermore, the calculation of correlation functions runs faster via frequency domain, therefore we applied spectral functions for signal processing in Figure 3.84.

For illustration, Figure 3.84 shows a simulated example based on a 6-bit DAC of ENOB $= 4,4$ bit. We omitted random noise so that the perturbation level of the IRFs is solely caused from the non-linear distortions of the DAC improved by the processing gain of the sounding signal. As we can observe the averaging over the ensemble of measurement signals leads to the expected reduction DAC-induced errors.

If the procedure is operated with both loops, the overall noise suppression (including ADC quantization noise) is proportional to $\sqrt{pM}$ while the reduction of DAC quantization noise is about $\sqrt{M}$. The efficiency of the measurement procedure is given as in the other cases of measurement principles by the energetic losses during data capturing. If we suppose p-repetitions of the inner loop and M-repetitions of the outer loop, the available signal energy is $\mathfrak{E} = T_{\mathrm{R}}P = (p+1)Mt_{\mathrm{p}}P$. The actually accumulated energy is $\mathfrak{E}_{\mathrm{acc}} = \eta_{\mathrm{S}}pMt_{\mathrm{p}}P$ since we always have to wait for one signal period before we can start the inner loop. η_{S} is the efficiency of the sampling gate. It results for the efficiency of real time data capturing:

$$\eta_{\mathrm{ms}} = \eta_3 \frac{p}{1+p} \tag{3.142}$$

Hence, an extensive use of the inner loop provides the largest efficiency. The outer loop will not influence the efficiency. It will only extend the measurement time. Hence, if we can renounce the outer loop, we get the best efficiency for the shortest recording time of all measurement methods, that is we need either a high-resolution DAC or a second receive channel for capturing the actual stimulus signal.

Basically the multi-sine approach is a pseudo-random technique because arbitrary-shaped (periodic) signals can be generated. We used the term 'multi-sine' since the signal behaviour was mainly designed in the spectral domain where any periodic signal is represented by a set of spectral lines, that is sinusoids. Needless

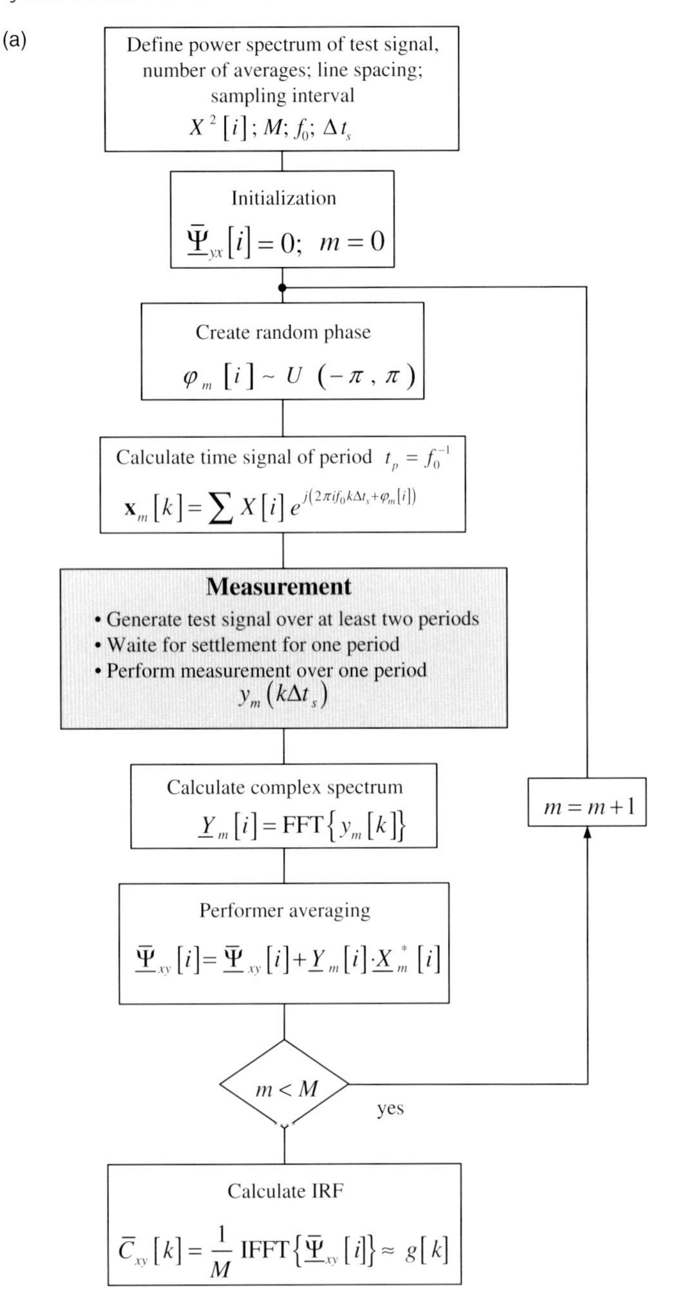

Figure 3.84 Flow graph to reduce the influence of the DAC quantization errors (a). The part (b) depicts a simulated example for a 6-bit DAC of ENOB = 4.4 bit. The time-bandwidth product of the test signal is TB = 256.

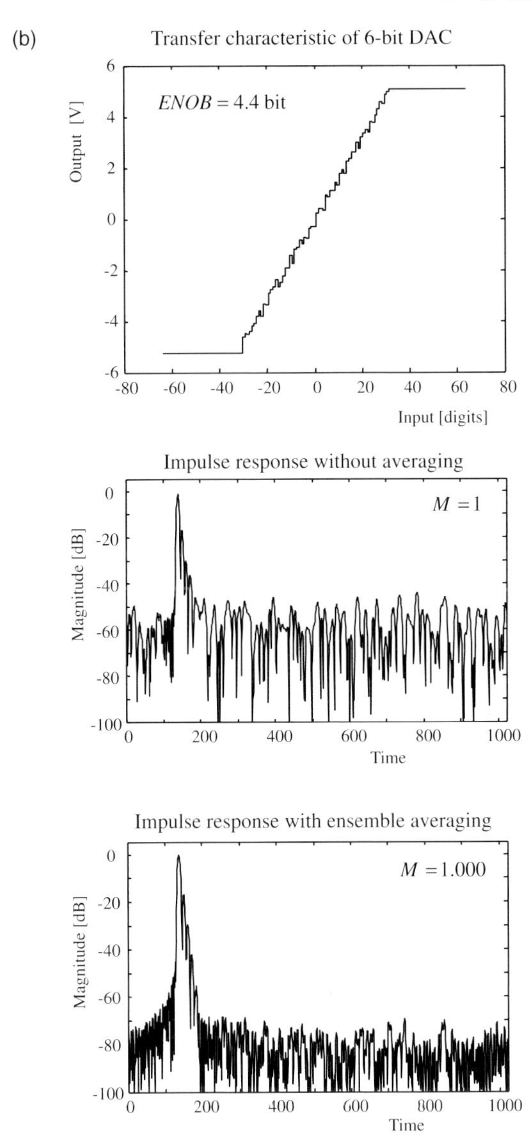

Figure 3.84 *(Continued)*

to say that the design philosophy of pseudo-random signals is not solely restricted to optimize spectral properties.

Another optimization criterion could be for example to emphasize a certain type of radar targets by matching the sounding signal to their specific scattering behaviour. A related approach is to concentrate sounding energy in a multi-path rich environment at predefined positions in space. As we have already illustrated in Section 2.6.1.4, the idea here is to consider the radar and propagation path as a matched filter of IRF $g(t)$ which is supposed to be known. If we feed the scenario under test with a signal (wave) of time reversed IRF shape $x(t) \propto g(-t)$, a simple receiver will provide us a high peak voltage if the specific target is present in the scenario or if the receiver is located at the predefined place. The corresponding method is usually assigned as time-reversal technique. It was first introduced in acoustic sounding [96]. There the technical challenges of signal generation are quite more relaxed as for UWB sensing [97, 98].

3.6 Determination of the System Behaviour with Random Noise Excitation

The stimulation of test objects by random signals has some charm but it also suffers from some weaknesses and technical problems. For a given average signal power, random noise provides the lowest probability from all types of signals to interfere with other communication or sensor system. It also means that the probability to detect a random noise sensor has the lowest possible value since, for example a (single unit) radar warning receiver has no opportunity to lock on any code. Hence, there exist some interests to apply such sensors for concealed operations. Nevertheless also random noise radars leave its marks since they appear as hot spots in the background noise. Finally, random noise signals still have another smart property. Their ambiguity function has a thumbtack character (see Figure 2.19) that is there are no range and velocity ambiguities.

Further it is interesting to note that random noise sensors must not necessarily have their own signal source. Theoretically they work with any radiation provided for example from radio stations, thermal sources or others. A radar system exploiting external sources of sounding signals is called passive radar. It cannot be detected by electronic means but it is able to detect random noise sources as well as passive objects. Furthermore, it does not load an already 'electromagnetic active' environment with additional radiation. A historical abridgment of this sensing technique is given in Ref. [99]. Currently, this sensing principle has not an actual importance for UWB sensing since except thermal microwave radiation no other 'public' UWB sources are available.

The principle to determine the behaviour of linear time invariant system (i.e. $g(t)$ or $\underline{G}(f)$) by random noise excitation is based on correlation either performed in time or frequency domain:

$$C_{yx}(t) = g(t) * C_{xx}(t) \tag{3.143}$$

or

$$\underline{\Psi}_{yx}(f) = \underline{G}(f) \cdot \Psi_{xx}(f) \tag{3.144}$$

As known from previous discussions of the correlation approach, one assumes that the auto-correlation function $C_{xx}(t)$ of the noise is much shorter than the impulse response $g(t)$ which is comparable with the requirement that the width of the noise power spectrum $\Psi_{xx}(f)$ must be larger than the bandwidth of the test object.

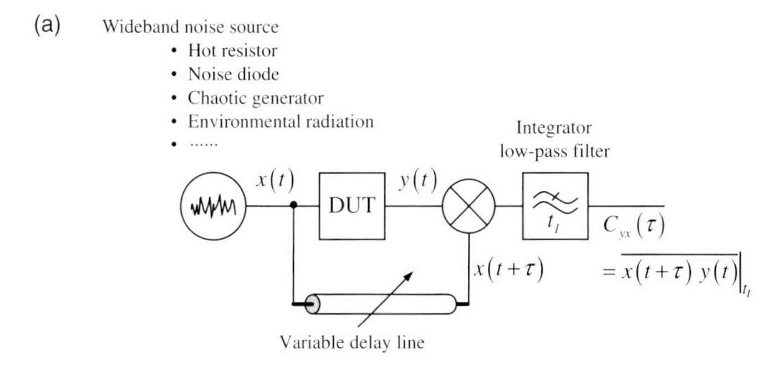

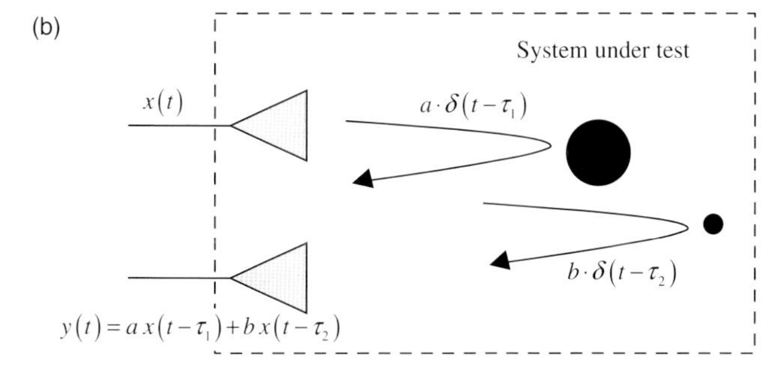

Figure 3.85 Generic correlation receiver (a) and simplified multi-path scenario (b).

In order to discuss the general constraints of the sensing approach, we will refer to the generic structure of a correlation receiver which directly follows from the definition (2.49) of the correlation function $C_{yx}(\tau) = \int y(t) \cdot x(t+\tau)\mathrm{d}t$. It is depicted in Figure 3.85 including a simplified test scenario for illustration of the expected effects.

Obviously, a correlator simply consists of a product detector, a controllable delay line and a low-pass filter performing the integration. The device functioning is as follows: One adjusts the delay line to the retardation τ_0 and integrates over the time t_I the signal mixture from input and retarded reference signal. After the integration is closed, the output voltage of the integrator corresponds to $C_{yx}(\tau_0)$. As next, we can select a new time lag by acting on the delay line. The variation of the delay time can be done stepwise or in a slow continuous fashion to give sufficient time for low pass filtering.

Even if we have to deal with a very simple structure it provokes some technical challenges which are issues of the practical device implementation and the handling of test signal randomness. The most serious of them is the variable delay line. It is very hardly to implement such a device for analogue wideband signals and a large tuning range. Cascading RF-cables of different length is an often used approach which is however a quite expensive and bulky solution. Additionally, it suffers from growing signal deformations with increasing time shift τ due to cable

Idealized cross-correlation function
(infinite integration time)

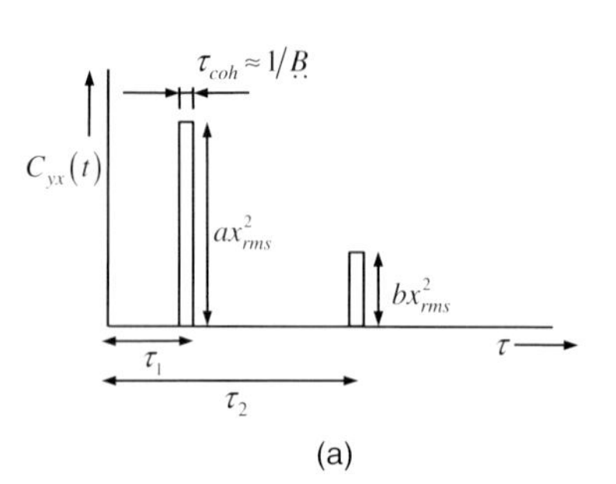

Uncertainty of the cross-correlation function
(due to finite integration time)

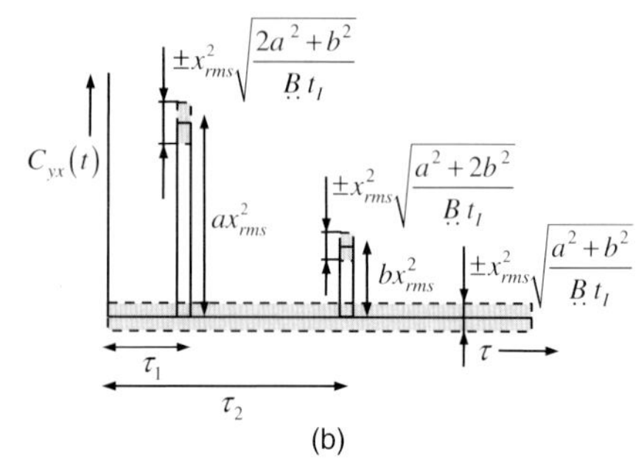

Figure 3.86 Schematized system response of the scenario depicted in Figure 3.85. (a) Infinite integration time and (b) finite integration time. The solid lines represent the expected value and the grey boxes symbolize the 68% confidence interval of the correlation function. For details see Annex B.4.1.

loss and dispersion. We will refer below to some modified approaches which mitigate the handicap of such analogue delay lines.

But first we will address the impact of randomness onto the device behaviour. For illustration of the problem, we will consider the simple example in Figure 3.85. It concerns an elementary radar scenario with two targets of different size. To be simple, we assume that both targets perfectly reflect the incoming wave without affecting their time shape.[22] The only differences should be that the reflections are different in strength and round trip time. The idealized impulse response function of the scenario under test is given by two Dirac pulse of different weight and located at τ_1 and τ_2, respectively.

$$g(t) = a\,\delta(t - \tau_1) + b\,\delta(t - \tau_2) \quad \text{with } a \gg b;\, \tau_1 \neq \tau_2$$

If the bandwidth of the sounding signal is restricted to a limited value $\underset{..}{B}$, than the resulting pulses in the correlation function will spread roughly over the duration τ_{coh}. Figure 3.86a exhibits schematically the cross-correlation function $C_{yx}(t)$ as it can be expected from the considered scenario if the integration time would tend to infinity $t_{\text{I}} \to \infty$. Since we always have to limit the integration time t_{I}, some randomness will be left in our measurement even in the perfect case of not any additional perturbations. Figure 3.86b depicts the impact of integration time limitation on the resulting cross-correlation function. The given relations suppose a normal distributed sounding signal $x(t) \sim N\left(0, x_{\text{rms}}^2\right)$ of the bandwidth $\underset{..}{B}$.

What are the practical consequences of this randomness?

- Obviously, the integration time t_{I} plays an important role to clean the measured curve from random fluctuations. In contrast to the measurement approaches discussed in the previous chapters, these fluctuations are not caused by

22) As the next chapter it will show, this is not a correct assumption. But it will simplify to concentrate on the key point which is intended to be considered here.

perturbations of the measurement process but rather from the test signal itself. That is, in the case of the pulse, pseudo-random or sine wave techniques, device internal noise or external radio interference is suppressed proportionally to the observation time. Of course, this is also valid for the random noise technique but the interferer suppression will not be visible for the user as long as the 'stimulating noise' is stronger than the 'perturbation noise'.

- We have introduced an efficiency value η in previous chapters in order to have a parameter which quantifies the performance of the measurement principle to suppress uncorrelated perturbations. We will use the same parameter here, but with a different meaning. Now it refers the capability of the measurement device to suppress the randomness induced by its own stimulus signal.
- As Figure 3.86 demonstrates, a strong peak (i.e. a dominant signal path within the scenario under test) leads automatically to strong random fluctuations (also assigned as clutter or side lobes) which are spread over the whole correlation function. That is, we have to deal with an increasing danger to lose weak targets in the presence of a strong signal path independently from what their mutual distance is. The antenna coupling will, therefore, often limit the device sensitivity in radar applications because it is typically the largest signal. The situation may be somewhat relaxed by using random signal of lower degree of freedom which must be artificially generated by very wideband arbitrary generators.[23] A further approach is to describe the IRF/FRF of the strongest signal paths by an appropriate system model (e.g. parametric model, assumption of point scatters etc.), to estimate the signal components which they provoke and to subtract them from the receive signal [100]. Since the strong random signals are removed now, the overall variance of the remaining signal components will be lowered and small targets will have a better chance to be detected. The improvement potential depends on the quality of the system model which of course always underlies some restriction.
- Tracking of moving targets has to be considered with some care. Random noise has a thumbtack ambiguity function and typically we need a long observation time to reduce the variance in the data. That is, we only achieve a maximum correlation gain if the time delay τ and the scaling factor s of the reference waveform coincides with the received waveform. A typical correlator as depicted in Figure 3.85 is not able to scale the signal. Hence, it only works for low-speed targets. The extension of the device conception for large target speeds must involve a two-dimensional search in τ and s. The practical implementation could be based on real-time data capturing and a digital correlator bank which handles in parallel the correlation between the receive signal and the noisy stimulus

23) If arbitrary waveform generators come into the game, a multi-sine concept which is based on synchronism of transmit and receive signal is to be preferred in favour of a pure random noise approach since it avoids truncation effects and large variance of the data. If the randomness of the transmit signal is a strong design criteria, the multi-sine technique offers comparable performance as a true random noise sensor since the time shape of the sounding signal may be quickly changed if one respects some elementary conditions (see Figure 3.84).

Every branch of the correlator bank has to apply a reference signal of different scaling factor s. For a given delay bin τ_0 and Doppler bin s_0, an alternative could be to sweep slightly the delay time $\tau = \tau_0 + s_0 t \quad t \in [0, t_I]$ of the correlator in Figure 3.85 while the low-pass filter is performing integration. This procedure has to be repeated for all delay and Doppler bins of interest. Both methods are technically very ambitious for noise signals of large fractional bandwidth and long integration interval. Hence, existing noise sensor concepts are typically restricted to a modest bandwidth which allows observing the target motion by Doppler frequency (compare Figure 3.87). Though, the target must not leave the range bin $\Delta r \approx c/2\underset{\cdot}{B}$ (assuming band-pass noise) during the Doppler measurement.

We will give a short overview of possible device structures which perform cross-correlation or cross-spectrum measurements. Since most of the basic principles and device components were already discussed in connection with other ultra-wideband measurements approaches, we will restrict ourselves only on specific aspects here which concern particularities of the treatment of noise signals. The reader may find more discussion on noise sensors and noise radar in following papers [97, 98, 101–112].

3.6.1 Time Domain Approaches

The basic time domain approach was already depicted in Figure 3.85. By that approach, the cross-correlation function is captured successively be stepping through the length of the delay line. For every delay step, we have to wait at least the integration time t_I before the next step can be done. Supposing N_s delay steps of the width Δt_s, we can observe the cross-correlation function (IRF) over the time interval t_{win}:

$$t_{win} = N_s \, \Delta t_s \tag{3.145}$$

The related receiver bandwidth is given by $\underset{\cdot\cdot}{B} = 2\underset{\cdot}{B} = \Delta t_s^{-1}$. The time T_R needed to record all data is

$$T_R = N_s \gamma t_I \tag{3.146}$$

in which t_I represents the integration time and the quantity γ is an efficiency parameter of the integration circuits (see Section 3.3.3). Due to the affinity with the sliding correlator, we can use the same considerations concerning the efficiency of the measurement principle and we will end up in a result as already known from (3.43)

$$\eta_C = \frac{\eta_m \, \Delta t_s}{\gamma T_R} = \frac{\eta_m}{\gamma N_s} \tag{3.147}$$

The correlation circuit from Figure 3.85 mainly suffers from two problems: the signal multiplication in the baseband suffering from flicker noise (see also Section

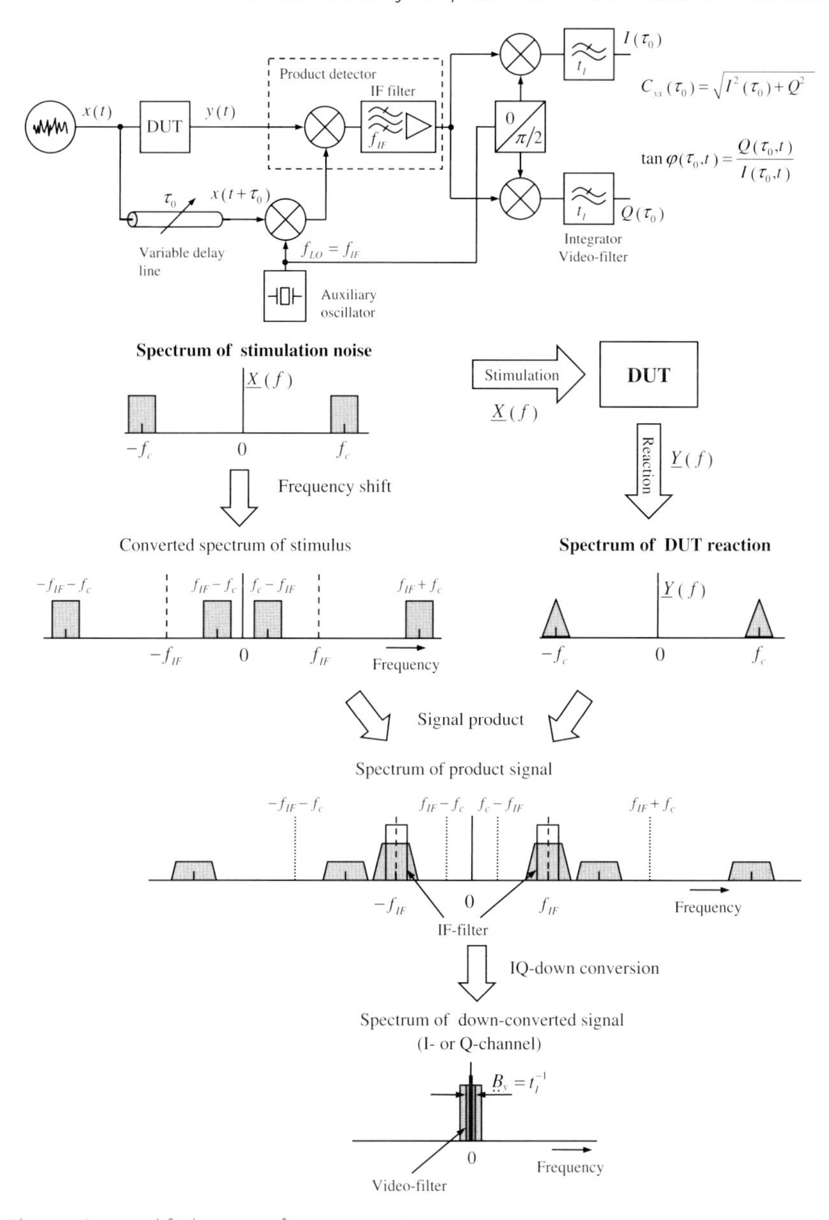

Figure 3.87 Modified version of a cross-correlator avoiding direct mixing within the baseband. The device is extended by an IQ-demodulator for Doppler measurements. The lower part illustrates the mixing schematic in the spectral domain for random noise band-pass stimulus.

3.4.2.1) and the technical implementation of the variable delay line which must have a large dispersion-free bandwidth and low losses.

The first point may be addressed by performing the main amplification and the multiplication in an intermediate frequency band f_{IF} as depicted in Figure 3.87 [109, 110]. For that purpose, the spectrum of the retarded reference signal $x(t-\tau_0)$ is shifted into the IF-band where it is multiplied with the device output $y(t)$. The IF-filter serves for image rejection and noise reduction. The IQ-demodulator finally provides the wanted value of the correlation function for the time lag τ_0:

$$C_{yx}(\tau_0) = \sqrt{I^2(\tau_0) + Q^2(\tau_0)} \tag{3.148}$$

In the case the duration of dwell at the time lag τ_0 is larger than the integration time t_I of the video-filters, the in-phase and quadrature-phase signals become time dependent if a target at the round-trip time τ_0 is moving. The resulting IQ-vector rotates in the complex plane with the Doppler frequency f_D as long as the target does not drop out of the τ_0 related range bin:

$$\begin{aligned} \tan\varphi(\tau_0,t) &= \frac{Q(\tau_0,t)}{I(\tau_0,t)} \\ f_D(\tau_0) &= \frac{1}{2\pi}\frac{d\varphi_{unw}(\tau_0,t)}{dt} \end{aligned} \tag{3.149}$$

The phase must be unwrapped before it can be differentiated, hence the symbol φ_{unw}. A performance analyses of Doppler measurements with random noise sensors is given in Ref. [113]. The Doppler principle applied in Figure 3.87 assumes narrowband conditions, that is a remarkable timescaling due to too large bandwidth or target speed would lead to degradation of the measurement results.

The second point concerns the replacement of the delay line by a simpler version. Figure 3.88 presents a generic device structure which assumes that both signals $x(t+\tau)$ and $y(t)$ are sampled and afterwards multiplied. In that case, only the

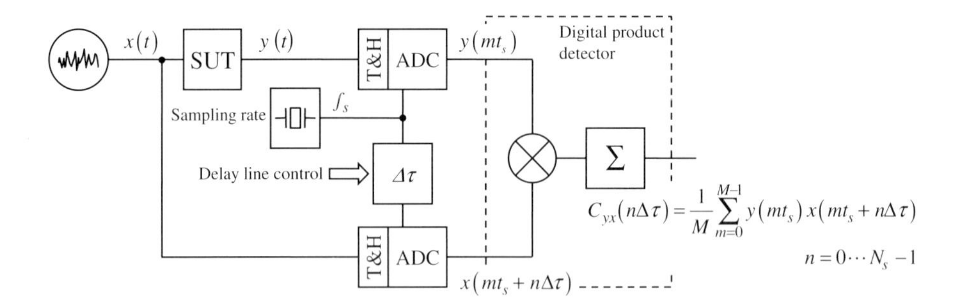

Figure 3.88 Digital correlation using sub-sampling. The time interval between two samples is $t_s = 1/f_s$ whereas the correlation function is sampled by the density $\Delta\tau$.

strobe pulses which control the sampling gates must be delayed. This task is much simpler to solve and any method mentioned in Section 3.2.3.3 is feasible. It is interesting to note that the actual sampling rate f_s can be selected arbitrarily since the bandwidth of the measurement device depends on the analogue bandwidth of the sampling gates and the step size $\Delta\tau$ of the delay line but not from f_s. The approach can be considered as sub-sampling for which the equivalent sampling rate is $f_{eq} = 1/\Delta\tau$. Needless to say that the efficiency of the sampling receiver corresponding to (3.147) will additionally lowered by the sub-sampling factor $\Delta\tau f_s$.

In case of less demanding applications (with respect to dynamic range or measurement speed), further simplifications of the correlation circuit may be achieved by restricting the data conversion to 1-bit resolution. We already discussed comparable approaches in connection with other UWB principles (Sections 3.2.4 and 3.3.6.1). Here, we have two possibilities to replace analogue devices by binary ones. Figure 3.89 illustrates them.

Relay correlator: Here, the correlation is performed between the DUT reaction $y(t)$ and the clipped reference signal which may be mathematically expressed by the signum function $\operatorname{sgn} x(t)$. Basically, the delay may be performed before or after clipping. Preferentially it will be done after clipping since one can apply here fast binary shift registers. The multiplier works in switching mode which is

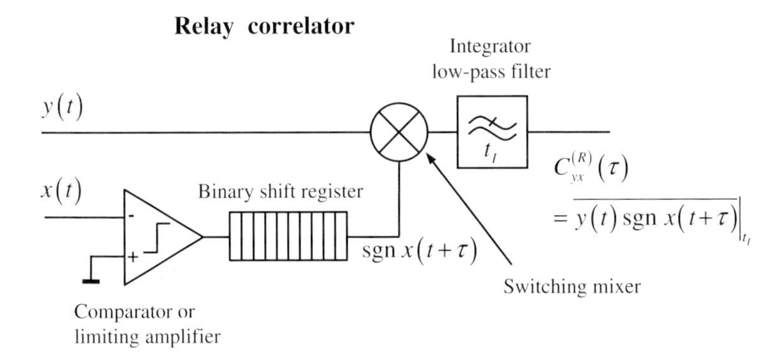

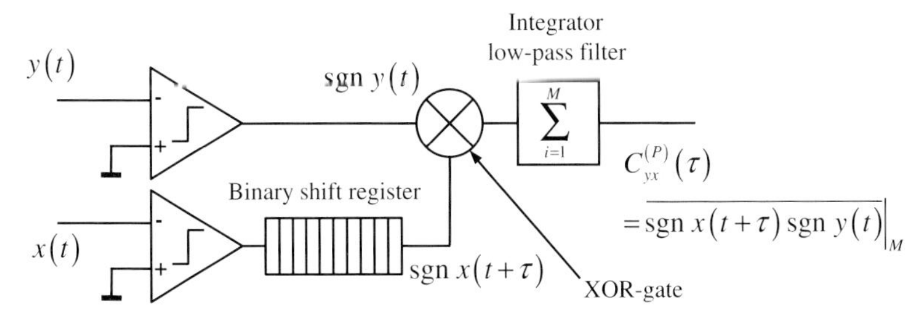

Figure 3.89 Binary correlators.

the usual mode of operation for double balanced mixers. The actual captured correlation function $C_{yx}^{(R)}(\tau)$ relates to the actually wanted by following relation:

$$C_{xy}(\tau) = \sqrt{\frac{\pi}{2}} x_{rms} \, C_{yx}^{(R)}(\tau) \tag{3.150}$$

Polarity-coincidence correlator: Now, one completely set the analogue character of the signals aside and only refers to their polarity so that all processing, that is delay, multiplication, integration, can be done by fast digital electronic. Assuming a stimulus $x(t)$ of Gaussian PDF, the wanted cross-correlation function results from

$$C_{yx}(\tau) = x_{rms} y_{rms} \sin\left(\frac{\pi}{2} C_{yx}^{(P)}(\tau)\right) \tag{3.151}$$

The polarity correlator may also be converted in an analogue correlator if the input signals are superimposed by auxiliary voltages before clipping. A detailed analysis of quantization effects in correlation devices including relay- and polarity-coincidence correlator is given in Ref. [114].

3.6.2
Frequency Domain Approaches

The principles are based on relation (3.144). They provide the complex-valued FRF which can be converted into the time domain IRF via inverse Fourier transform. We will mention two basic conceptions – a digital one and an analogue method.

Processing of digitized noise data: We assume digitized signals $x[i]$ and $\mathbf{y}[i]$ captured by high-speed digitizer, for example real-time oscilloscopes. In order to gain a reliable system response $\mathbf{g}[i]$, both signals have to be captured with sufficiently high sampling rate $f_s = \Delta t_s^{-1}$ over a sufficiently long time T_R. In the case of UWB applications, the sampling rate may be beyond tens of GHz. Sub-sampling does fail here. The recording time T_R must largely exceed the intended observation length t_{win} of the IRF in order to reduce the variance sufficiently. Hence, the data recorder must have available a large high-speed data memory.
After having captured the data, the IRF must be calculated which is based on correlation processing as depicted in (3.143). Since we can usually select test signals of sufficiently large bandwidth, it is allowed to replace the auto-correlation function in (3.143) by a Dirac function $\delta(t)$ so that convolution can be omitted. But the calculation of the cross-correlation function still remains. It will be numerically quite expensive if it is immediately performed in the time domain. Applying FFT-processing leads to much faster calculations namely if the IRF must be observed within a large time window t_{win} involving many samples. The basic processing steps to elude time domain processing may be summarized by

$$\mathbf{g}[i] \propto \mathbf{C}_{yx}[i] = \text{IFFT}\{\text{FFT}\{\mathbf{y}[i]\} \circ (\text{FFT}\{\mathbf{x}[i]\})^*\} \tag{3.152}$$

However this equation is to be taken with care due to two reasons:

- Let us assume that the impulse response function $g(t)$ has to be determined within the time window t_{win}. For radar applications this means that we will observe a target with round trip time up to t_{win}. That is, the first coherent signal components originating from far-off targets will arrive the receiver at about t_{win}. Hence, the duration of data segment to be processed must be much longer as the actually needed duration t_{win}. In case one considers only a short time segment in the order of t_{win} a far target is not able to place backscattered energy into the receive signal so it would be discriminated against others with a shorter round trip time. Furthermore, a time segment of too short length leads to biased estimations of the frequency response function (see Figure 2.60).
- The spectrum of a time-limited part of a random signal is random and therefore the correlation function too. In order to stabilize the estimates of the cross-spectrum and the cross-correlation, one has to perform averaging. Welch's method is a very popular method to do this [115–117]. For that purpose, the signals are captured over a long recording time T_R. Afterwards, they are portioned into M segments $\mathbf{x}_m[i]$ and $\mathbf{y}_m[i]$ of equal duration whereas adjoining segments may also overlap. Usually, the cut segments are additionally weighted by a window function $\mathbf{w}[i]$ in order to reduce truncation effects. From that the impulse response finally estimates to

$$\mathbf{g}[i] \propto \mathbf{C}_{yx}[i] = \text{IFFT}\left\{\sum_{m=0}^{M-1} \text{FFT}\{\mathbf{y}_m[i] \circ \mathbf{w}[i]\} \circ (\text{FFT}\{\mathbf{x}_m[i] \circ \mathbf{w}[i]\})^*\right\}$$

$$i = [1 \cdots N_{win}] + m(N_{win} - N_{overlap}); \quad m \in [0, M]; \quad M = \text{floor}\frac{N_s - N_{overlap}}{N_{win} - N_{overlap}} \tag{3.153}$$

$N_s = T_R f_s$ is the number of sampled data; $N_{win} = t_{win} f_s$ is the length of IRF and $N_{overlap}$ is the sample number of overlapping of adjacent signal segments.

Theoretically, the approach permits continuous data capturing without loss on energy of the received signals. Hence, the receiver efficiency will be close to one. This is absolutely feasible but only for signals of comparatively low bandwidth. In the case of ultra-wideband signals, the situation looks quite more demanding. Recently, we have commercially available digitizers exceeding 10 GHz of bandwidth and hundreds of Mbits of memory space. Hence, UWB applications of the discussed approach are technically feasible. These devices are however very expensive, power hungry and of large size, so that their use is banned to the laboratory or to special applications. Furthermore, it is hardly possible to process the captured data online with the speed of their capturing. Therefore, large gaps have to be introduced in the measurement process which will drastically force down the efficiency.

Random noise network analyser: By extending the concept of a usual spectrum analyser (heterodyne receiver) corresponding to Figure 3.90, the device could be used to measure the complex frequency response function of a device under test

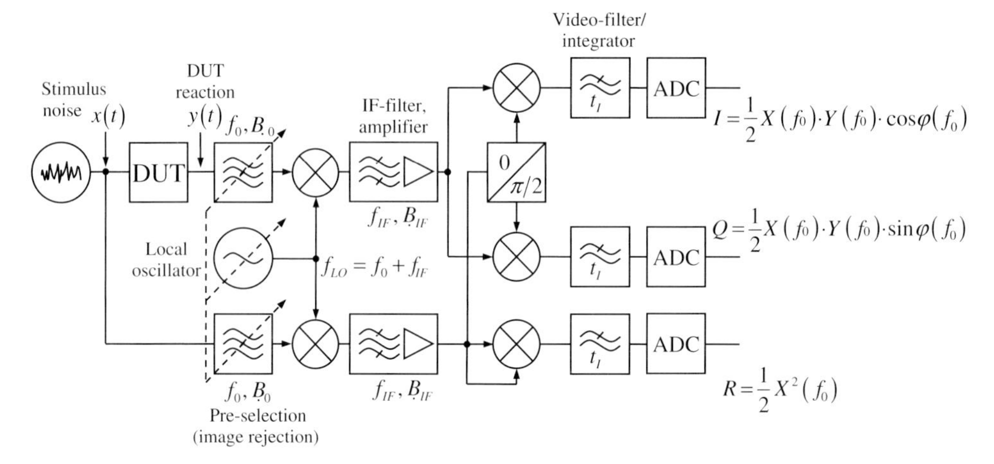

Figure 3.90 Generic structure of random noise network analyser (vector heterodyne receiver).

as it does a network analyser. The idea is to cut out a small portion of the noise spectra by a heterodyne concept and to determine the cross- and the auto-power of the signals passing the IF-filters. The technical challenge of such concept is to achieve acceptable imbalance between the two signal paths over a large bandwidth. In order to gain the frequency response function, the local oscillator is stepped over the desired frequency band. Its step size $\Delta f_{\text{LO}} = \Delta f$ determines the spectral resolution of the measurement and hence the duration t_{win} of the resulting IRF after FFT processing ($t_{\text{win}} = 1/\Delta f$). The bandwidth B_{IF} of the IF-filter should be smaller than Δf in order to avoid spectral overlapping between adjacent lines. The video-filters suppress again the randomness of the measurement signal.
The frequency response function of the DUT results from

$$\underline{G}(f_0) = \frac{I(f_0) + jQ(f_0)}{R(f_0)} \tag{3.154}$$

where the quantities $I(f_0)$; $Q(f_0)$ and $R(f_0)$ are captured form high-resolution ADCs. All transmission channels within the measurement device should behave identically in order to avoid amplitude and phase mismatch. Remaining imbalances can be removed by a calibration routine as described in Section 2.6.4 as long as the device is stable over time.
The estimation of the receiver efficiency is based on the relations (3.93) and (3.94) which we already introduced for the SFCW-sensor. However, there are some differences concerning to the resulting value. In case of the SFCW-sensor, the main task of the IF-filter is image rejection and the video-filter has to suppress first of all the double IF-frequency appearing at the output of the IQ-mixers. Both tasks do not require narrowband filters of long settling time. Extension of the integration time of the video-filter indeed better suppresses additive noise but a long integration time is not required by principle.

The situation is completely different for system stimulation by noise. Now the IF-filter must cut-out a narrow frequency band of the noise that has to be analysed and the video-filter has to suppress first and foremost the randomness in the stimulation signal. Both measures require narrow filters of long settling time degrading the receiver efficiency and the measurement speed.

3.7 Measuring Arrangements

All measurement approaches discussed in the previous Sections 3.2–3.6 are aimed to determine the transmission behaviour of a device under test either for the time or the frequency domain. We did not make a big difference between both domains since in the case of measurements with large fractional bandwidth one can swap between both via Fourier transform. We also did not consider the measurement environment; we only were concentrated on how to capture and to process the data in order to gain either $g(t)$ or $\underline{G}(f)$ of an arbitrary wideband transmission system. We will discuss some aspects on how to insert the sensor electronics into measurement circuits in what follows. Here, we will not distinguish between the different measurement approaches since they are all doing the same with respect of their outcome.

3.7.1 Capturing of Voltage and Current

A basic point in this connection is the fact that all receiver principles and methods of data recording considered above are built to capture a voltage between two fixed points within an electric circuit. Figure 3.91 depicts some fundamental variants on how to measure a voltage or a current. Three points are important to be discussed:

- **Single ended or differential measurement:** Single ended voltage capturing should be applied if the sensor element has to be connected via cables with the electronics. Differential measurements are preferable if the sensor element operates differentially and the wiring can be made by two identical and rigid lines manufactured together on a PCB. A differential voltage capturing is suggested in connection with wideband antennas which are mostly differentially fed. Differential feeding via individual cables is crucial due to possible tolerance of their electric length. Cables which differ only by 100 μm in their geometric length will for example cause an imbalance of about −40 dB for a sensor device of 10 GHz bandwidth (cable celerity of 5 ns/m assumed).
- **Feed trough measurement or line termination:** Feed trough measurements serve to observe the voltage on a line connecting a signal source and a sensor element or a load. The measured voltage will strongly depend on the location where the feed trough head is placed. In the case of a simple T-connection as depicted in Figure 3.91, the receiver input impedance must be as high as

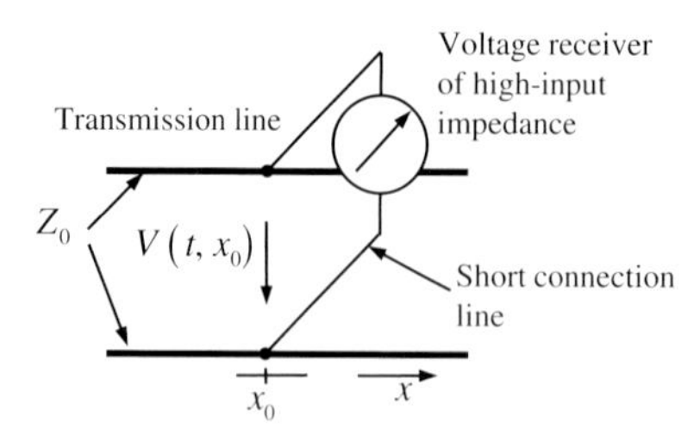

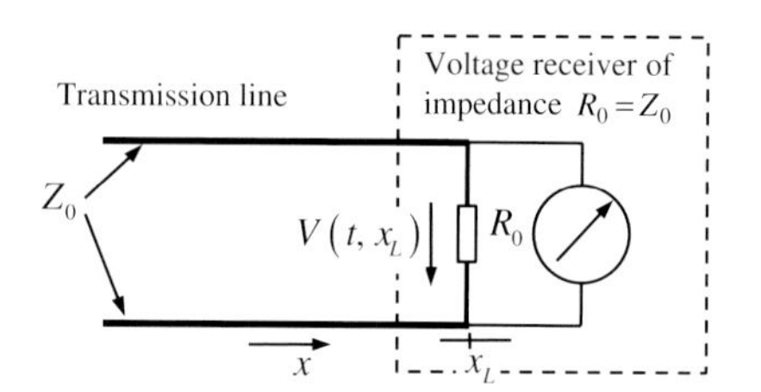

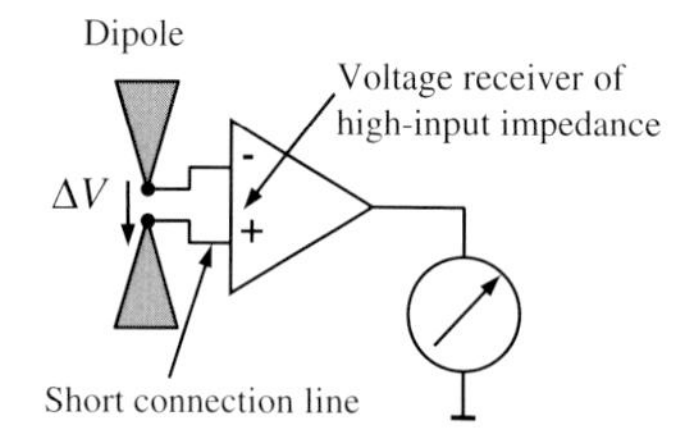

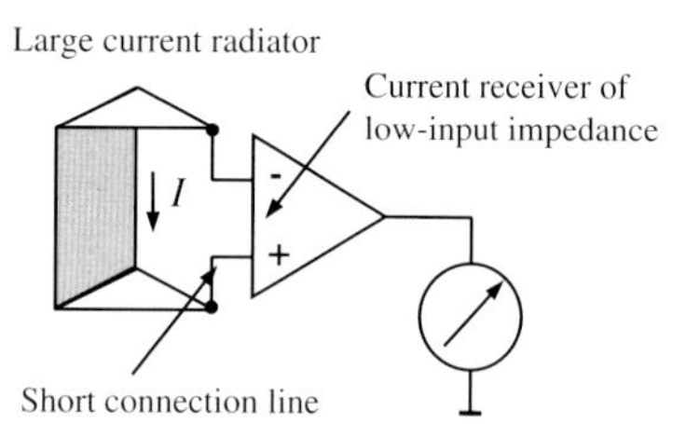

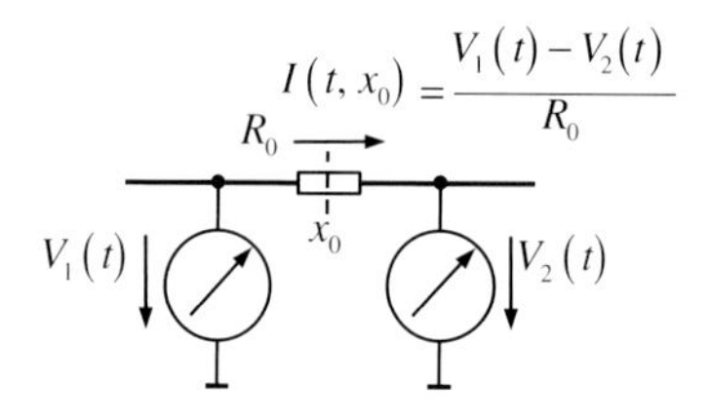

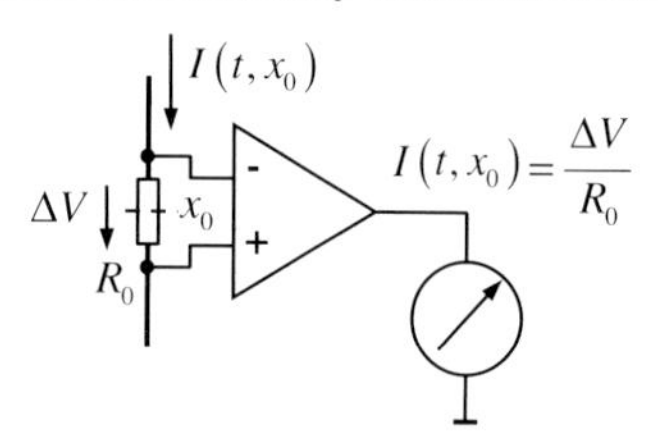

Figure 3.91 Basic arrangements to measure voltage and current.

possible since the signals travelling along the connection line should not be affected by the measurement device. The second option is to measure and to absorb the energy travelling towards a preferentially matched load, that is the measurement device has to terminate the line.

- **Input impedance of the receiver:** The sensor electronics must match the cable impedance (typically 50 Ω) if the spot of voltage capturing is only accessible via cable of considerable length (that is $L > t_r c/10$; t_r is the rise time of the sounding signal; c is the propagation speed at the cable). Otherwise one risks multiple reflections perturbing the measurements (see consideration below). High or low impedance devices (for voltage or current measurement) are feasible if the sensor electronics is directly placed at the measurement spot avoiding any cables. Such solutions largely profits from monolithically integrated sensor devices.

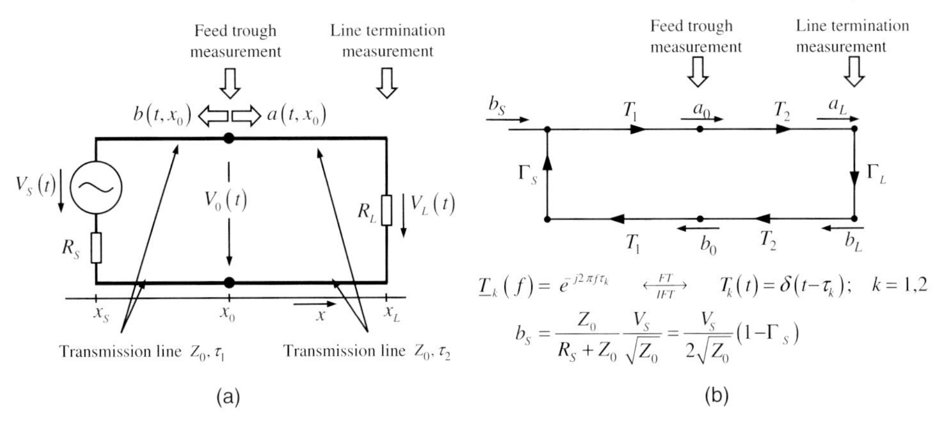

Figure 3.92 Elementary measurement arrangement. (a) Electrical circuit and (b) Mason graph. V_s and Γ_L or R_L carry the wanted sensor information and one of both voltages $V_0(t) = V(t, x_0)$ or $V_L(t) = V(t, x_L)$ represents the actual captured signal.

3.7.2 Basic Measurement Circuit

The elementary measurement circuit consist of signal source, load and measurement device (see Figure 3.92). Depending on the sensor principle, either the behavior of the signal source or of the load includes the wanted information about the test object. Due to its importance, we will shortly analyse the basic behaviour of this circuit. In order to keep it simple, we will restrict ourselves to purely restive load and source impedances. The example considers both a feed trough measurement and a line termination measurement as well. It is supposed that the voltage pick-up in the feed trough head does not affect the characteristic impedance of the transmission line and the line termination should also not be influenced by the voltage measurement device. The connecting cables have the characteristic impedance Z_0 and their one-way propagation delay is τ_1 and τ_2.

To shorten the notation, we assign all signals related to position x_0 by index 0 and the index L applies to all quantities at line termination. Following (2.196) in Section 2.5.1, the captured voltages can be written as

Feed trough:

$$V_0(t) = V(t, x_0) = \sqrt{Z_0}(a(t, x_0) + b(t, x_0)) = \sqrt{Z_0}(a_0(t) + b_0(t)) \tag{3.155}$$

Line termination:

$$V_L(t) = V(t, x_L) = \sqrt{Z_0}(a(t, x_L) + b(t, x_L)) = \sqrt{Z_0}(a_L(t) + b_L(t)) \tag{3.156}$$

Resulting from the Mason graph in Figure 3.92, we yield for the normalized waves in the frequency domain (see Annexes B.7 and B.8)

$$\begin{aligned} \underline{a}_0 &= \frac{\underline{b}_S \underline{T}_1}{1 - \underline{T}_1^2 \underline{T}_2^2 \Gamma_S \Gamma_L}; & \underline{b}_0 &= \frac{\underline{b}_S \Gamma_L \underline{T}_1 \underline{T}_2^2}{1 - \underline{T}_1^2 \underline{T}_2^2 \Gamma_S \Gamma_L} \\ \underline{a}_L &= \frac{\underline{b}_S \underline{T}_1 \underline{T}_2}{1 - \underline{T}_1^2 \underline{T}_2^2 \Gamma_S \Gamma_L}; & b_L &= \frac{\underline{b}_S \underline{\Gamma}_L \underline{T}_1 \underline{T}_2}{1 - \underline{T}_1^2 \underline{T}_2^2 \Gamma_S \Gamma_L} \end{aligned} \tag{3.157}$$

We can modify the loop using the equality $\frac{1}{1-x} = \sum_{n=0}^{\infty} x^n |x| < 1$:

$$\underline{\text{loop}}(f) = \frac{1}{1 - \underline{T}_1^2 \underline{T}_2^2 \Gamma_S \Gamma_L} = \sum_{n=0}^{\infty} \left(\underline{T}_1^2 \underline{T}_2^2 \Gamma_S \Gamma_L \right)^n$$

which allows us to express (3.157) also in the time domain:

$$\begin{aligned} &a_0(t) = b_S(t - \tau_1) * \text{loop}(t); \qquad b_0(t) = \Gamma_L b_S(t - \tau_1 - 2\tau_2) * \text{loop}(t) \\ &a_L(t) = b_S(t - \tau_1 - \tau_2) * \text{loop}(t); \qquad b_L(t) = \Gamma_L b_S(t - \tau_1 - \tau_2) * \text{loop}(t) \\ &\text{with} \quad \text{loop}(t) = \sum_{n=0}^{\infty} (\Gamma_S \Gamma_L)^n \cdot \delta(t - 2\,n(\tau_1 + \tau_2)) \end{aligned} \tag{3.158}$$

Figure 3.93 illustrates (3.156) and (3.158). It shows the voltage across the line termination under several conditions. The given values of the source and load reflection refer to a line impedance of $Z_0 = 50\,\Omega$. We used a step voltage for

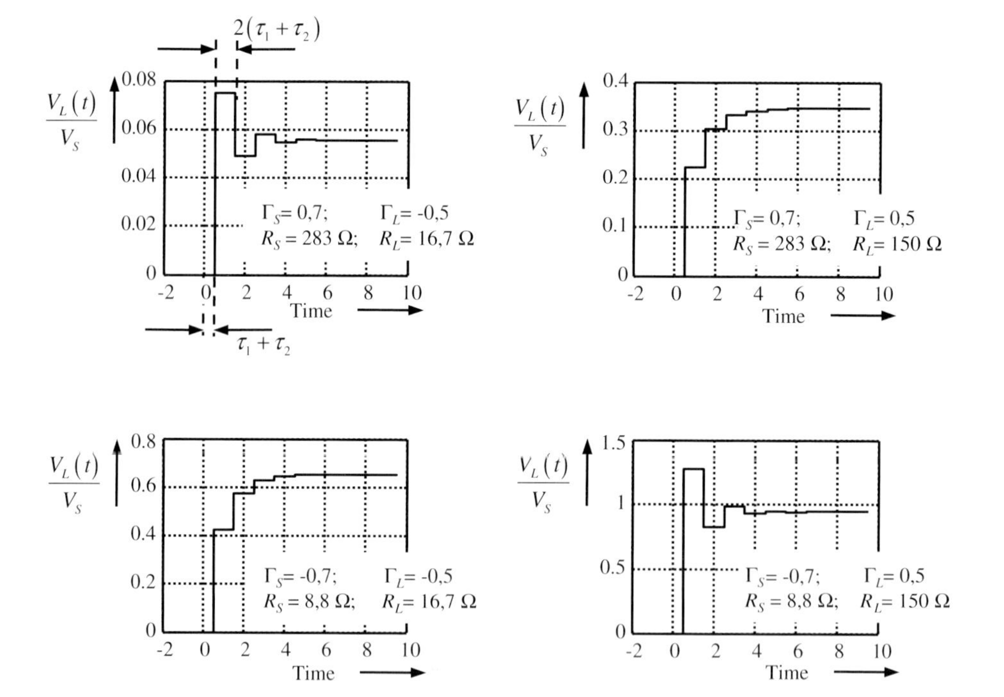

Figure 3.93 The voltage over the line termination R_L if the voltage source provides a voltage step of amplitude V_S.

stimulation in order to demonstrate illustratively how the voltage approaches its final value. Obviously, multiple reflections between load and source retard the voltage equilibrium. The decay rate of the reflections depends on the mismatches at both terminations as well as the length of the connecting cable.

Exploiting the equality $\sum_{n=0}^{N} x^n = \dfrac{1 - x^{N+1}}{1 - x}$ $|x| < 1$, the magnitude of the voltage $V_L(t)$ after N circulations in the loop is given by following expression:

$$V_{L,N} = \sqrt{Z_0} b_S (1 + \Gamma_L) \sum_{n=0}^{N} (\Gamma_S \Gamma_L)^n = \frac{1}{2} V_S (1 - \Gamma_S)(1 + \Gamma_L) \frac{1 - (\Gamma_S \Gamma_L)^{N+1}}{1 - \Gamma_S \Gamma_L} \tag{3.159}$$

Consequently, the first voltage step arriving at the receiver has the magnitude

$$V_{L,0} = \frac{1}{2} V_S (1 - \Gamma_S)(1 + \Gamma_L) = \frac{2 R_L Z_0}{(R_S + Z_0)(R_L + Z_0)} V_S$$

Note, under certain conditions (i.e. $(1 - \Gamma_S)(1 + \Gamma_L) > 2$), the measured voltage $V_{L,0}$ may be even higher than the open-circuit voltage V_s of the signal source – compare Figure 3.93. The ensuing voltage steps can be expressed by

$$V_{L,N} = V_{L,0} \frac{1 - (\Gamma_L \Gamma_S)^{N+1}}{1 - \Gamma_L \Gamma_S}$$

and finally the voltage aspires the value

$$V_{L,\infty} = \frac{V_{L,0}}{1 - \Gamma_L \Gamma_S} = \frac{(1 - \Gamma_S)(1 + \Gamma_L)}{2(1 - \Gamma_L \Gamma_S)} V_S = \frac{R_L}{R_L + R_S} V_S$$

since $lim_{N \to \infty} (\Gamma_L \Gamma_S)^N = 0 \quad \text{for } |\Gamma_L| < 1; |\Gamma_S| < 1$

This is the value, which we would expect from simple steady-state consideration of the electrical circuit. The voltage approaches monotonically to its final value if $\Gamma_S \Gamma_L > 0$ otherwise it approaches alternately.

In concordance to (2.141) and (2.142), we can describe the voltage settling by a decay rate DR and a time constant τ_D:

$$\begin{aligned} \mathrm{DR} &= \frac{20 \lg(|(V(t_1) - V_\infty)/(V(t_2) - V_\infty)|)}{t_2 - t_1} \\ &= \frac{20 \lg((|V_{L,N_1} - V_{L,\infty}|)/(|V_{L,N_2} - V_{L,\infty}|))}{2(N_1 - N_2)(\tau_1 + \tau_2)} = \frac{20 \lg|\Gamma_s \Gamma_L|}{2(\tau_1 + \tau_2)} \end{aligned} \tag{3.160}$$

The decay rate quantifies the speed (usually given in dB/ns) to approach the final state. It relates the voltage variation during $\Delta N = N_1 - N_2$ loop cycles. Since the envelope of the voltage steps converges exponentially to the final value, it is also feasible to apply a time constant to quantify the settling process, that is $V(t) - V_\infty \propto e^{-t/\tau_D}$:

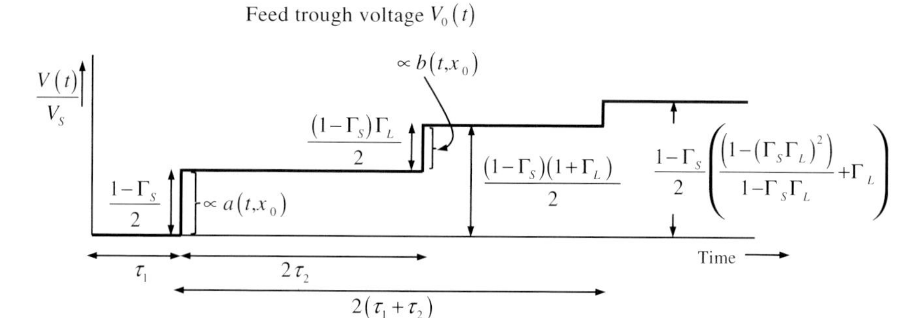

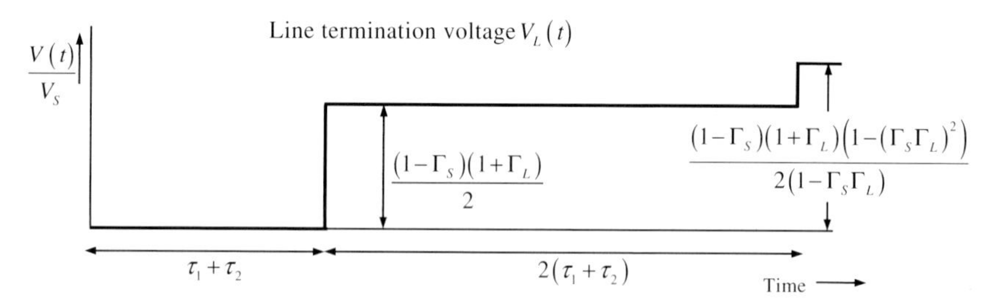

Figure 3.94 The first reflection cycle of the feed trough $V_0(t)$ and the line termination voltage $V_L(t)$ in comparison.

$$\begin{aligned}
\tau_D &= -\frac{20\lg e}{\text{DR}} = -\frac{40(\tau_1+\tau_2)\lg e}{20\lg|\Gamma_S\Gamma_L|} \approx 17.4\frac{\tau_1+\tau_2}{\text{RL}_S+\text{RL}_L} \\
\text{RL}_S &= -20\lg|\Gamma_S| \quad \text{(source return loss)} \\
\text{RL}_L &= -20\lg|\Gamma_L| \quad \text{(load return loss)}
\end{aligned} \tag{3.161}$$

The voltage $V_0(t)$ captured by a feed trough measurement head may be deduced similarly to $V_L(t)$. There is only one exception, namely the voltage $V_0^-(t)$ connected to the backward travelling wave $b(t, x_0)$ is delayed by $2\tau_2$ referred to the voltage $V_0^+(t)$ of the forward travelling wave $a(t, x_0)$. Figure 3.94 depicts the difference between line termination and feed trough measurement for the first reflection cycle.

What did we learn from the considerations above?

1) If $V_S(t)$ is the wanted sensor signal, than the received signal $V_L(t)$ (line termination measurement) will be a copy of $V_S(t)$ if at least one of the both reflection coefficients will be zero. In the case of feed trough measurement, the load must be matched to the cable ($\Gamma_L = 0$) in order to get $V_0(t) \propto V_S(t)$.
2) If the line is neither matched by the load nor by the source, the measurement circuit behaves like a first-order low-pass system. Its time constant τ_D is given by (3.161) and the cut-off frequency f_u of the measurement arrangement is approximately (τ – one-way delay time of the cable; $\tau = \tau_1 + \tau_2$ for our example in Figure 3.92)

$$f_{\mathrm{u}} = \frac{1}{2\pi\tau_{\mathrm{D}}} = -\frac{20\,log|\Gamma_{\mathrm{S}}\Gamma_{\mathrm{L}}|}{80\pi\tau\,log\,e} \approx \frac{\mathrm{RL_S} + \mathrm{RL_L}}{109\tau} \tag{3.162}$$

From (3.162) follows that the reflection coefficients must be small and the connecting cable has to be short in order to avoid band limitation due to the connecting circuit.

3) If we perform a corresponding signal analysis as above for step signals of finite rise time t_{r}, we will discover that mismatch has no influence on voltage settling if the cable propagation is short against the rise time $\tau \ll t_{\mathrm{r}}$. That is, high or low ohmic sensor elements may also be operated at high frequency and large bandwidth as long as connecting cables are short enough. That means the cable length should be limited to a few millimetres if we deal with baseband signals of 10 GHz bandwidth.
4) Even under unmatched conditions, $V_{\mathrm{L}}(t)$ will be copy of $V_{\mathrm{s}}(t)$ if multiple reflections may excluded by gating. In order to avoid overlapping between consecutive reflections, the settling time t_{sett} of $V_{\mathrm{s}}(t)$ must be shorter than twice the cable delay $t_{\mathrm{eff}} < 2\,\tau$.
5) The feed trough voltage measurement enables to distinguish between forward and backward travelling wave $a(t)$ and $b(t)$ (see Figure 3.94). This represents the simplest approach to measure the reflection coefficient Γ_{L}. The method is called time domain reflectometry and time domain network analyser (TDNA).

3.7.3 Methods of Wave Separation

We have seen above that the measurement devices are only able to capture voltage signals at a given point within a transmission line. In exceptional cases, it could also be possible to have available current receiver if the device is designed for low-input impedance. But the usual approach to characterize the test objects in the microwave range is based on scattering parameters as introduced in Section 2.5.3. This requires the knowledge of the normalized waves propagating on the transmission lines which fed the test object (see Figure 2.67). Therefore, we still have to answer the question how to determine the required set of normalized waves by pure voltage (current) measurements. Since the captured voltage (current) always results from the superposition of forward and backward travelling waves (refer (2.196)), the waves must be separated appropriately before they can be captured by a voltage (current) receiver.

3.7.3.1 Wave Separation by Time Isolation

The simplest approach to separate the waves was already introduced above under point 4 of the numeration list. Figure 3.95 shows an example of a two-port time domain network analyser. Here, we used feed trough voltage capturing for illustration. Other options are to insert pick-off Tees or power divider into the feeding cables and to capture the voltage via line termination measurement. Note that the conversation factors K between the waves and the voltage are different for every

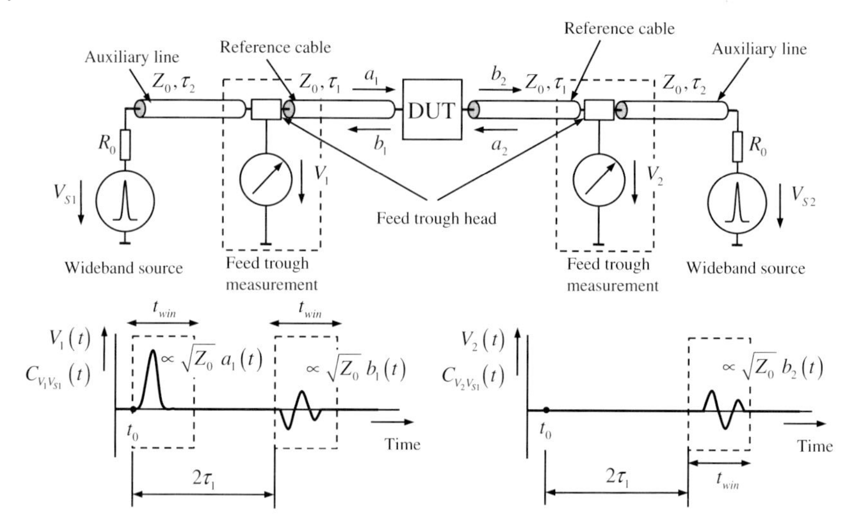

Figure 3.95 Wave separation by time isolation via transmission lines – time domain network analyser.

principle of voltage pick-up. But it is easy to identify the minor differences between the three circuits by referring the scattering matrix of the pick-off Tee and the power divider as given in Annex B.8. Table 3.2 summarizes the conversion factors for the three versions of voltage picking.

The reference line is used as impedance reference and to isolate in time the forward and backward travelling waves. Therefore, one measurement head is sufficient to capture both waves if the reference lines are sufficiently long. The auxiliary lines are used to move reflections from voltage pick-up and source out of the observation window. They can be omitted if both are well matched.

The reflection $S_{11}(t)$ and transmission behaviour $S_{21}(t)$ of the DUT is analysed by gating out the different signal components from the captured voltage as depicted in Figure 3.95 where we suppose that the second signal source is switched off $V_{S2} = 0$. Inversely, the determination of $S_{12}(t)$ and $S_{22}(t)$ requires $V_{S1} = 0$, that is

Table 3.2 Voltage-wave conversion for different voltage pick-up principles.

	Feed trough head	Pick-off Tee	Power divider
$x_{1a}(t)$	$\sqrt{Z_0} b_S(t)$	$\sqrt{Z_0}\dfrac{2}{2a+3} b_S(t)$	$\dfrac{1}{2}\sqrt{Z_0} b_S(t)$
$y_{1b}(t)$	$\sqrt{Z_0} s_{11}(t) * b_S(t)$	$\sqrt{Z_0}\dfrac{2(2a+1)}{(2a+3)^2} s_{11}(t) * b_S(t)$	$\dfrac{1}{4}\sqrt{Z_0} s_{11}(t) * b_S(t)$
$y_{2b}(t)$	$\sqrt{Z_0} s_{21}(t) * b_S(t)$	$\sqrt{Z_0}\dfrac{2(2a+1)}{(2a+3)^2} s_{21}(t) * b_S(t)$	$\dfrac{1}{4}\sqrt{Z_0} s_{21}(t) * b_S(t)$
K	$1 \triangleq 0\,\text{dB}$	$\dfrac{2a+1}{2a+3} = 0.9 \triangleq -0.91\,\text{dB}$ $a = 17/2 \Rightarrow R = 425\,\Omega$	$\dfrac{1}{2} \triangleq -6\,\text{dB}$
$b_{S1,2} = \dfrac{Z_0}{R_0 + Z_0}\dfrac{V_{S1,2}}{\sqrt{Z_0}} = \dfrac{V_{S1,2}}{2\sqrt{Z_0}}(1 - \Gamma_S)$			

one swaps the active source from left to right. The extension to multi-port test objects is straightforward.

Gating assumes signal components of finite length namely shorter than twice the cable delay, that is $2\tau_1$. Here, we have to distinguish two cases:

- **The stimulus signal is a short pulse:** Under that constraint, we can immediately take the captured voltage to gate out the appropriate parts as depicted in Figure 3.95 and to determine the related time domain scattering parameters:

  Gating:

$$\begin{aligned} x_{1a}(t) &= V_1(t); \quad t \in [t_0, t_0 + t_{win}] \\ y_{1b}(t) &= V_1(t); \quad t \in [t_0 + 2\tau_1, t_0 + 2\tau_1 + t_{win}] \\ y_{2b}(t) &= V_2(t); \quad t \in [t_0 + 2\tau_1, t_0 + 2\tau_1 + t_{win}] \end{aligned}$$

  S-parameter estimation:

$$\left.\begin{aligned} y_{1b}(t) = K\, S_{11}(t) * x_{1a}(t) \quad &\Rightarrow \quad y_{1b}(t) \propto K\, S_{11}(t) \\ y_{2b}(t) = K\, S_{21}(t) * x_{1a}(t) \quad &\Rightarrow \quad y_{2b}(t) \propto K\, S_{21}(t) \end{aligned}\right\} \text{ if } x_{1a}(t) \approx \delta(t) \tag{3.163}$$

  Herein, K is a conversion factor depending on the method of voltage picking. Table 3.2 gives a summary.

  In the case the side condition $x_{1a}(t) \neq \delta(t)$ is only inadequately met, the simplification in (3.163) are not allowed so that the scattering parameters must be determined via de-convolution. It is preferentially executed in frequency domain in compliance with ill-conditioned problems (compare Figure 2.96 and Section 4.8.3). Equation (3.164) exemplarily summarizes the procedure for the input reflection of the DUT:

$$\begin{aligned} y_{1b}(t) = K S_{11}(t) * x_{1a}(t) &\xrightarrow{\text{FT}} \underline{Y}_{1b}(f) = K \underline{S}_{11}(f) \underline{X}_{1a}(f) \\ \Rightarrow \hat{\underline{S}}_{11}(f) &= \frac{\underline{Y}_{1b}(f)}{K \underline{X}_{1a}(f)} W(f) \xrightarrow{\text{IFT}} \hat{S}_{11}(t) \end{aligned} \tag{3.164}$$

  Herein, $W(f)$ represents a window function that is aimed to suppress ill-conditioned frequency bands and $\hat{S}_{11}$ means the estimated scattering parameter of the DUT.

- **The stimulus is an arbitrary wideband signal:** Since the signal may be spread over time, gating is not feasible. Therefore, time compression, for example by correlation has to be performed before gating can be applied. Hence, the gating procedure has to be modified by

$$\begin{aligned} x_{1a}(t) &= C_{V_1 V_{S1}}(t); \quad t \in [t_0, t_0 + t_{win}] \\ y_{1b}(t) &= C_{V_1 V_{S1}}(t); \quad t \in [t_0 + 2\tau_1, t_0 + 2\tau_1 + t_{win}] \\ y_{2b}(t) &= C_{V_2 V_{S1}}(t); \quad t \in [t_0 + 2\tau_1, t_0 + 2\tau_1 + t_{win}] \end{aligned}$$

  The rest of the procedure, that is (3.163), is the same as for pulse-shaped signals.

Traditionally, the TDR or the TDVA are working with step voltages or short pulses. But they are not restricted on these signals. Any continuous wideband

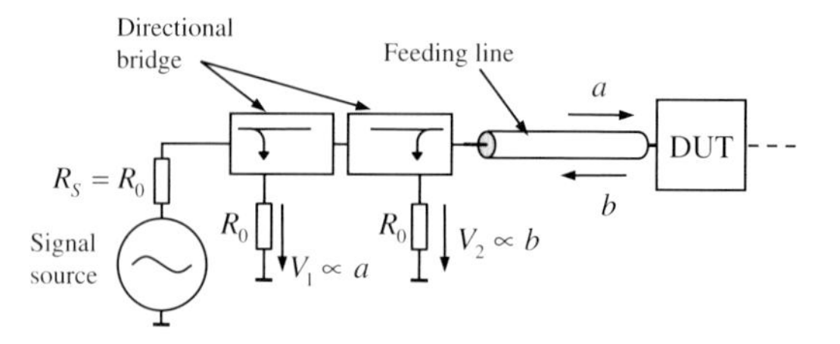

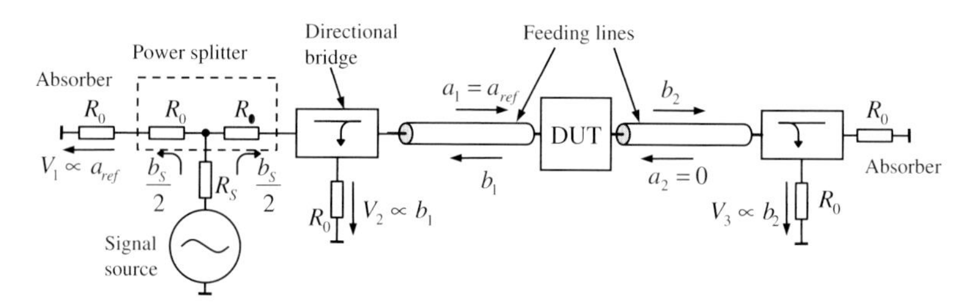

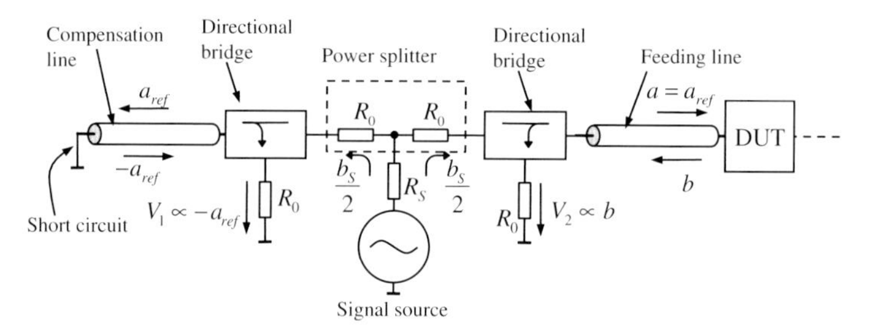

Figure 3.96 Basic circuits for wave separation by directional coupler.

signal inclusively swept or stepped sine waves can be used for stimulation as long as their bandwidth is sufficiently large. They only have to be compressed, for example by correlation or FFT processing in order to result in short time domain functions on which the gating procedure according to Figure 3.95 can be applied. The length of the reference cables should be selected sufficiently long in order to avoid truncation effects due to gating.

If step signals are used to stimulate the objects, some attention has to be paid on the signal processing via Fourier transformation since signal truncation inevitably leads to corruptions in the frequency space. They can be avoided by simple tricks introduced by A.M. Nicolson, W.L. Gans and N.S. Nahman [118–120].

3.7.3.2 Wave Separation by Directional Couplers

Directional couplers exploit wave interference to separate forward and backward travelling waves. Annex B.8 gives two examples – a line and resistive coupler – and summarizes their idealized behaviour. Due to the internal losses, resistive couplers have a lower efficiency than line couplers. On the other hand, line couplers require sophisticated design and large dimensions for high fractional bandwidth. Resistive couplers are not band limited by principle (neglecting imperfections of technical implementation), but they require differential voltage capturing. Commercially available restive bridges uses wideband transformers for the voltage pick-up. These are the components which typically limit the bandwidth of resistive bridges. Monolithically integrated resistive couplers with differential amplifiers are less restrictive concerning the bandwidth. They operate from DC upto tens of GHz [121].

Figure 3.96 depicts some basic arrangements for wave separation by directional couplers. The first method uses two couplers – one for every propagation direction – inserted into the feeding path of the device under test. The captured voltages are proportional to the corresponding waves. In case of a multi-port device, the exhibited circuit has to be placed at any port so that one is able to capture all injected and emanating waves by stepwise activating the signal sources. Note that the sources act also as absorbers if they are switched off. A corresponding procedure is described in more details in Sections 2.5.4 and 3.3.6.2; refer also to Figure 2.99.

In the second and third approach, the stimulation wave is split up in two equal parts. One of them propagates over the measurement path and stimulates the DUT. The second one is injected into the reference path. The waves b_1 and b_2 returning from and passing the DUT are measured via the directional couplers as before. They provide us the voltages V_2 and V_3. Note that the wave b_1 returning back to the source will be partially reflected from the power splitter so that it acts as an additional incident wave now. Since however the power splitter transfers the same amount of energy to the reference path as it reflects (i.e. $S_{22} = S_{33} = S_{32} = S_{23} = 0.25$; see Annex B.8), the reference wave a_{ref} will correctly indicate the incident wave of the DUT. A power divider should not be applied to split the source power in two paths since it only behaves correctly if the source and reference port are ideally matched.

The measurement of the reference wave a_{ref} may be done by two options. In the simplest one, the voltage $V_1(t)$ across the absorber of the reference path is measured (version 1). The second method provokes a total reflection of the reference wave and one captures the reflected wave by an additional coupler (version 2). The benefits of this method against version 1 are the mutual compensation of the coupler FRFs and feeding cable effects if identical directional bridges and RF cables are used. A drawback of both versions is the need of different circuits for reflection and transmission measurements.

3.7.3.3 Wave Separation by Voltage Superposition

Voltage measurements across an impedance can also be used for wave separation. Figure 3.97 depicts two versions. One uses an ohmic resistor and the second method determines the voltage at two different locations. The voltage pick-up can

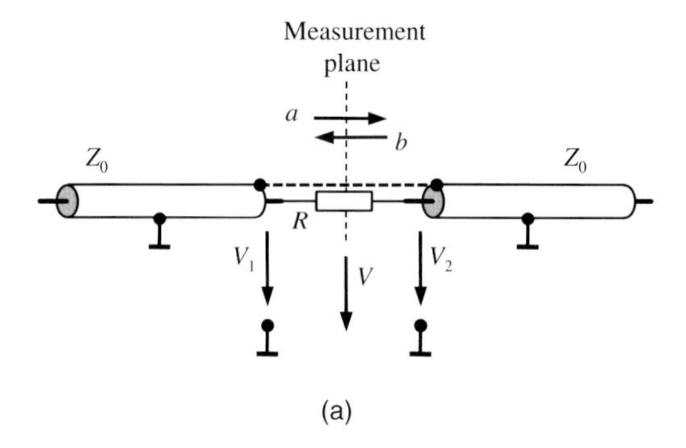

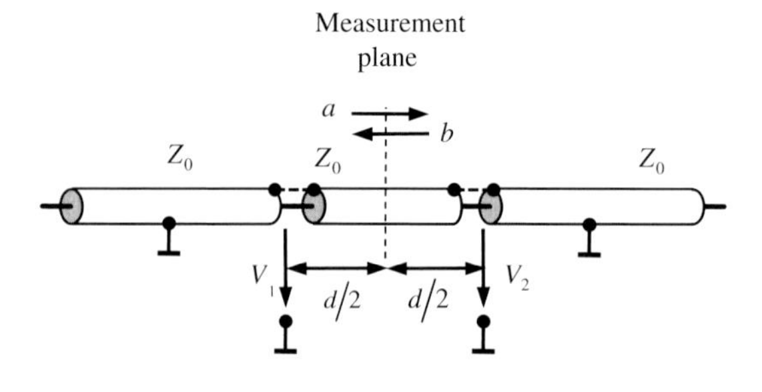

Figure 3.97 Wave separation by voltage measurements via series elements. (a) Series resistor and (b) line segment of length d.

be implemented by feed trough heads, pick-off Tees or power dividers whereas the last one would provoke additional signal attenuation. The equations below suppose feed trough heads but it is straightforward to extend them for the two other methods of voltage capturing.

Series resistor: Decomposition of the total current I and the total voltage V at the reference plane in its forward and backward travelling components yield

$$\begin{aligned} I &= \frac{V_1 - V_2}{R} = I^{\rightarrow} - I^{\leftarrow} = \frac{a-b}{\sqrt{Z_0}} \\ V &= \frac{V_1 + V_2}{2} = V^{\rightarrow} + V^{\leftarrow} = \sqrt{Z_0}(a+b) \end{aligned} \tag{3.165}$$

so that the normalized waves result in

$$\begin{aligned} a &= \frac{\sqrt{Z_0}}{2R}\left[\left(1+\frac{R}{2Z_0}\right)\cdot V_1 - \left(1-\frac{R}{2Z_0}\right)\cdot V_2\right] \\ b &= \frac{\sqrt{Z_0}}{2R}\left[\left(1+\frac{R}{2Z_0}\right)\cdot V_2 - \left(1-\frac{R}{2Z_0}\right)\cdot V_1\right] \end{aligned} \tag{3.166}$$

These equations hold for both time and frequency domain.

Line segment of length *d*: The short segment delays the waves propagating through the lines. This affects the total voltages differently in dependence from the direction of propagation. In the frequency domain, this can be summarized by

$$\begin{bmatrix} \underline{V}_1 \\ \underline{V}_2 \end{bmatrix} = \sqrt{Z_0} \begin{bmatrix} e^{j\pi f d/c} & e^{-j\pi f d/c} \\ e^{-j\pi f d/c} & e^{j\pi f d/c} \end{bmatrix} \cdot \begin{bmatrix} \underline{a} \\ \underline{b} \end{bmatrix} \tag{3.167}$$

Inverting this equation leads to the wanted normalized waves:

$$\begin{aligned} \underline{a} &= \frac{\underline{V}_1 e^{j\pi f d/c} - \underline{V}_2 e^{-j\pi f d/c}}{j2\sqrt{Z_0}\sin(\pi f d/c)} \\ \underline{b} &= \frac{\underline{V}_2 e^{j\pi f d/c} - \underline{V}_1 e^{-j\pi f d/c}}{j2\sqrt{Z_0}\sin(\pi f d/c)} \end{aligned} \tag{3.168}$$

Note that this measurement arrangement is limited to an operational frequency range in which the dominator does not approach too close to zero. The wave separation by voltage superposition works with narrowband as well as wideband signals. In order to get a corresponding relation for the time domain, we write down (3.167) in the related form

$$\begin{aligned} V_1(t) &= \sqrt{Z_0}(a(t+\tau) + b(t-\tau)) \\ V_2(t) &= \sqrt{Z_0}(a(t-\tau) + b(t+\tau)) \end{aligned} \tag{3.169}$$

with $\tau = d/2c$. If the rise time t_r of the waveform is longer than to the propagation delay τ, the waves may be decomposed in Taylor series which we cut after the first term, that is $a(t \pm \tau) \approx a(t) \pm \dot{a}(t)\tau$ and $b(t \pm \tau) \approx b(t) \pm \dot{b}(t)\tau$. After some straightforward manipulations, we finally yield

$$\begin{aligned} a(t) &= \frac{1}{2\sqrt{Z_0}}\left(V_m(t) + \frac{c}{d}\int \Delta V(t)dt\right) \\ b(t) &= \frac{1}{2\sqrt{Z_0}}\left(V_m(t) - \frac{c}{d}\int \Delta V(t)dt\right) \\ V_m &= \frac{V_1 + V_2}{2}; \quad \Delta V = V_1 - V_2 \end{aligned} \tag{3.170}$$

which results also from inverse Fourier transform of (3.168) under the condition $fd/c \ll 1$.

3.7.3.4 Capturing of E- and H-Field

We will finalize our consideration of wave separation with a last method. We capture the voltage $V(t, x_0)$ and the current $I(t, x_0)$ at the same position x_0 via the strength of the electric and magnetic field provided by the waveguide. Figure 3.98 illustrates the principle.

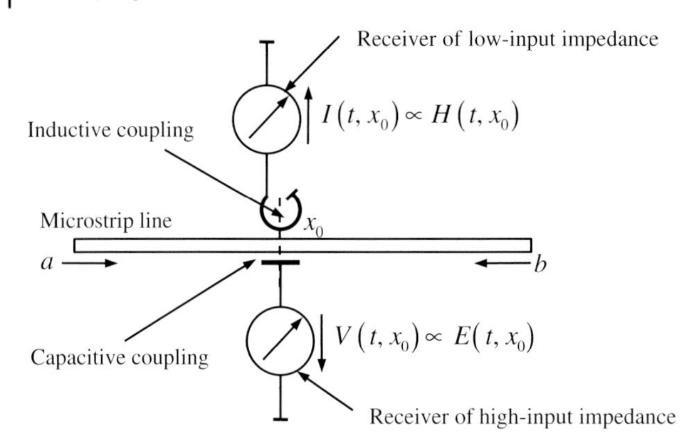

Figure 3.98 Contactless capturing of voltage and current via field coupling.

Using (2.202), we can establish the relations

$$\begin{aligned}
\underline{K}_{\mathrm{V}}(f)\,\underline{V}(f, x_0) &= \sqrt{Z_0}\left(\underline{a}(f, x_0) + \underline{b}(f, x_0)\right) \\
\underline{K}_{\mathrm{I}}(f)\,\underline{I}(f, x_0) &= \frac{1}{\sqrt{Z_0}}\left(\underline{a}(f, x_0) - \underline{b}(f, x_0)\right) \\
&FT \updownarrow IFT \\
K_{\mathrm{V}}(t) * V(t, x_0) &= \sqrt{Z_0}(a(t, x_0) + b(t, x_0)) \\
K_{\mathrm{I}}(t) * I(t, x_0) &= \frac{1}{\sqrt{Z_0}}(a(t, x_0) - b(t, x_0))
\end{aligned} \tag{3.171}$$

which immediately leads to the wanted normalized waves. K_{I} and K_{V} are frequency-dependent coefficients describing the strength of E- and H-field coupling, respectively.

Finally it should be mentioned that all methods of voltage/current capturing and wave separation discussed above are subject to multitude of error influences caused by mismatches, imbalances, unwanted frequency dependencies and so on. We did not consider them in detail since it is cumbersome and less promising to analyse individually all these errors with the goal to improve the measurement accuracy by a corresponding error model. Rather a formal error model is applied which leads to largely improved measurement results after calibration of the measurement system without knowledge of the exact reason of the (linear) errors. The reader is reminded of Section 2.6.4 for this generalized error correction approach.

3.8
Summary

The chapter was aimed for summarizing various exciting principles of UWB sensor electronics and to allude to some future developments. The sensor principles are typically denoted by the type of sounding signal which stimulates the objects under test. Here, we considered sub-nano second pulses, binary and analogue pseudo-random signals, for example M-sequence or multi-sine, frequency modulated sine waves and random noise. By definition, sine waves with a wide frequency sweep but slow modulation rate do not belong to UWB signals since their instantaneous bandwidth is narrowband. Nevertheless, we considered corresponding measurement approach since this it does not make any difference for LTI test scenarios which type of sounding signal is applied.

Except for the random noise approaches, we mostly dealt with periodic sounding signals which explicitly include pseudo-random signals. The advantage of periodic signals are that no truncation effects occur at data capturing (as long as always an integer number of signal periods are captured) and the measurement time will be short by principle since long integration times are not required due to the deterministic nature of the signals. For suppression of random perturbations, one can apply synchronous averaging which leads to bias-free estimations. Drawbacks may be the ambiguity caused from too short signal periods (leading to time aliasing) and the probability of detection and probability of interference.

Random noise signals are of benefit with respect to these aspects but they suffer from comparatively slow measurement rates since one has to wait for stabilization of measurements due to the signal randomness. If both measurement speed and interference and concealed operation are of major interest, pseudo-random approaches often provide a reasonable compromise. Ambiguities due to signal periodicity are less critical for short-range applications since one can always select signals of adequate period duration.

In the case of industrial and private sensor applications, technical parameters (range, sensitivity, reliability etc.) as well as economic aspects and handling (device costs, power consumption, physical dimensions, robustness etc.) are typical in the foreground of interest. Depending on the weighting of specific application aspects, the user can and has to select amongst several UWB principles which all have particular advantages and disadvantages.

Pulse-based systems may be implemented by simple and hence cost effective and power efficient circuits. Currently, it is the most applied method of UWB sensing. Their main drawback is the need to handle spiky signals which either have low energy (leading to insensitive sensors) or they require the handling of high peak voltages complicating the circuitry.

Binary pseudo-random code sensors provide powerful signals at low voltage level. Hence, they are potentially sensitive. On the one hand, the low signal levels enable monolithic circuit integration but on the other hand, the monolithic integration is indispensable since otherwise the short gate delays for the PN-code generation are not feasible. Monolithic integration is a handicap for small lot sizes but of advantage for large number of items.

Sine wave concepts are very sensitive namely if they are based on the heterodyne concept since random noise perturbations and non-linear distortions may be widely suppressed by narrowband filters. Homodyne receiver structures will forfeit sensitivity but their implementation is much simpler and integration friendly. A synthesizer-based sine wave generation is very precise but elaborate and expensive whereas a free running continuous frequency sweeper (VCO) is less demanding but also less precise.

The generation of arbitrary-shaped signals (we summarized corresponding approaches by the term multi-sine technique) is certainly the most flexible technique to adapt the sensor device to the actual measurement scenario. Here, the technical challenge is the generation of high-quality UWB signals. This requires fast DAC circuits and memories. Such components are typically quite power hungry and they are restricted by principle to comparatively low number of quantization intervals.

Finally we can state that all measurement approaches applying test signals of large instantaneous bandwidth exploit sub-sampling methods for data capturing in order to master the data flood with acceptable technical means. Clearly, extensive sub-sampling will bring down the receiver efficiency erasing the achievable SNR-value from the one which could be physically possible. Narrowband approaches (well designed for short internal settling time) provide high receiver efficiency and comparatively low data throughput as long as the measurement rate will not be too demanding.

In UWB array applications for strongly time variable scenario, data handling and processing may become a serious problem for commercially available processing units and standard software environment. In order to reduce such problems on host unit level, the UWB sensors should not generate unessential data. This may be achieved by adapting the equivalent sampling rate to the physical minimum (Nyquist sampling) and sensor internal pre-processing and data reduction based on FPGAs. Future trends of sensor development will presumably lead to conjointly integration of sensor and processing electronic on chip and package level.

References

1 Miller, E.K. (1986) *Time Domain Measuremenst in Electromagnetics*, Van Nostrand Reinhold Comapny Inc., New York.

2 Andrews, J.R. (2008) Picosecond pulse generation: techniques and pulsers capabilities; Application Note AN-19, http://www.picosecond.com/objects/AN-19.pdf].

3 Maiti, C.K. and Armstrong, G.A. (2001) *Applications of Silicon-Germanium Heterostructure Devices*, Institute of Physics Publishing, Bristol.

4 Dederer, J., Trasser, A. and Schumacher, H. (2006) SiGe impulse generator for single-band ultra-wideband applications. Third International SiGe Technology and Device Meeting, 2006. ISTDM 2006, pp. 1–2.

5 Sewiolo, B., Fischer, G. and Weigel, R. (2009) A 12-GHz high-efficiency tapered traveling-wave power amplifier with novel power matched cascode gain cells using SiGe HBT transistors. *IEEE Trans. Microwave Theory*, **57** (10), 2329–2336.

6 Sewiolo, B., Fischer, G. and Weigel, R. (2009) A 15GHz bandwidth high efficiency power distributed amplifier for ultra-wideband-applications using a low-cost SiGe BiCMOS technology. IEEE Topical Meeting on Silicon Monolithic Integrated Circuits in RF Systems, 2009. SiRF '09, pp. 1–4.

7 Gerding, M., Musch, T. and Schiek, B. (2004) Generation of short electrical pulses based on bipolar transistors. *Adv. Radio Sci.*, **2**, 7–12.

8 Baker, R.J. (1991) High voltage pulse generation using current mode second breakdown in a bipolar junction transistor. *Rev. Sci. Instrum.*, **62**, 1031–1036.

9 Mitchell, W.P. (1968) Avalanche transistors give fast pulses. *Electron. Des.*, **6**, 202–209.

10 Thomas, S.W., Griffith, R.L. and Teruya, A.T. (1991) Avalanche transistor selection for long term stability in streak camera sweep and pulser applications. *Proc. SPIE*, **1358**, 578.

11 Rein, H.M. and Zahn, M. (1975) Subnanosecond-pulse generator with variable pulsewidth using avalanche transistors. *Electron. Lett.*, **11** (1), 21–23.

12 Pulse and waveform generation with step recovery diodes, Application note AN 918, Hewlett-Packard, http://www.hp.woodshot.com/hprfhelp/5_downld/lit/diodelit/an918.pdf (1984).

13 Afshari, E. and Hajimiri, A. (2005) Nonlinear transmission lines for pulse shaping in silicon. *IEEE J. Solid-St. Circ.*, **40** (3), 744–752.

14 Case, M.G. (1993) Nonlinear transmission lines for picosecond pulse, impulse and millimeter-wave harmonic generation, in *Electrical and Computer Engineering*, University of California Santa Barbara.

15 Picosecond Pulse Labs (2006) A new breed of comb generators featuring low phase noise and low input power. *Microwave J.*, **49** (5), 278.

16 Wichmann, G. (2006) Non-intrusive inspection impulse radar antenna, USA. US patent 7,042,385 B1

17 McEwan, T.E. (1994) *Ultra-Wideband Receiver*, University of California, California.

18 Barrett, T.B. (2001) History of ultra wideband communications and radar: Part I, UWB communications. *Microwave J.*, **44** (1), 22–56.

19 Agoston, A., Pepper, S. Norton, R. *et al.* (2003) 100 GHz through-line sampler system with sampling rates in excess of 10 G samples/second. IEEE MTT-S International Microwave Symposium Digest, vol. 3, 2003, pp. 1519–1521.

20 Agoston, A., Pepper, S. Norton, R. *et al.* (2003) High-speed sampler modules for making 40 Gb/s eye diagram measurements. 62nd ARFTG Microwave Measurements Conference, 2003. Fall 2003, pp. 45–52.

21 Thomann, W., Knorr, S.G. and Chang, M.-P.I. (1991) A 40 GHz/9 ps sampling head for wideband applications. 21st European Microwave Conference, 1991, pp. 830–835.

22 Andrews, J.R. and DeWitte, G.J. (1984) Construction of a broadband universal sampling head. *IEEE Trans. Nucl. Sci.*, **31** (1), 461–464.

23 Grove, W.M. (1966) Sampling for oscilloscopes and other RF systems: DC through X-band. G-MTT International Symposium Digest, 1966, pp. 191–196.

24 Whiteley, W.C., Kunz, W.E. and Anklam, W.J. (1991) 50GHz sampler hybrid utilizing a small shockline and an internal SRD. IEEE MTT-S International Microwave Symposium Digest, vol. 2, 1991, pp. 895–898.

25 Williams, D.F. and Remley, K.A. (2001) Analytic sampling-circuit model. *IEEE Trans. Microwave Theory*, **49** (6), 1013–1019.

26 Williams, D.F., Remley, K.A. and DeGroot, D.C. (1999) Nose-to-nose response of a 20-GHz sampling circuit. 54th ARFTG Conference Digest-Spring, pp. 1–7.

27 Williams, D.F., Clement, T.S. Remley, K. A. *et al.* (2007) Systematic error of the nose-to-nose sampling-oscilloscope calibration. *IEEE Trans. Microwave Theory*, **55** (9), 1951–1957.

28 Borokhovych, Y., Gustat, H. Tillack, B. *et al.* (2005) A low-power, 10 GS/s track-and-hold amplifier in SiGe BiCMOS technology. Proceedings of the 31st European Solid-State Circuits Conference, 2005. ESSCIRC 2005, pp. 263–266.

29 Dederer, J., Schleicher, B. Trasser, A. *et al.* (2008) A fully monolithic 3.1–10.6 GHz UWB Si/SiGe HBT impulse-UWB correlation receiver. IEEE International Conference on Ultra-Wideband, 2008. ICUWB 2008, pp. 33–36.

30 Leib, M., Menzel, W. Schleicher, B. *et al.* (2010) Vital signs monitoring with a UWB radar based on a correlation receiver. Proceedings of the Fourth European Conference on Antennas and Propagation (EuCAP), 2010, pp. 1–5.

31 Hale, P.D., Wang, C.M. Williams, D.F. *et al.* (2006) Compensation of random and systematic timing errors in sampling oscilloscopes. *IEEE Trans. Instrum. Meas.*, **55** (6), 2146–2154.

32 Paulino, N., Goes, J. and Garcao, A.S. (2008) *Low Power UWB CMOS Radar Sensors*, Springer.

33 Leib, M., Schmitt, E. Gronau, A. *et al.* (2009) A compact ultra-wideband radar for medical applications. *Frequenz*, **63** (1–2), 1–8.

34 Gerding, M., Musch, T. and Schiek, B. (2006) A novel approach for a high-precision multitarget-level measurement system based on time-domain reflectometry. *IEEE Trans. Microwave Theory*, **54** (6), 2768–2773.

35 Guo, S., Sun, S., Zhang, Z. A novel equivalent sampling method using in the digital storage oscilloscopes Instrumentation and Measurement Technology Conference, 1994.

36 Becker, R. (2004) Spatial time domain reflectometry for monitoring transient soil moisture profiles, Institut für Wasser und Gewässerentwicklung Universität Karlsruhe (TH).

37 Huebner, C., Schlaeger, S. Becker, R. *et al.* (2005) Advanced measurement methods in time domain reflectometry for soil moisture determination, in *Electromagnetic Aquametry* (ed. K. Kupfer), Springer, Berlin, Heidelberg.

38 Kupfer, K. (2005) *Electromagnetic Aquametry: Electromagnetic Wave Interaction with Water and Moist Substances*, Springer, Berlin, New York.

39 Schreier, R. and Temes, G.C. (2005) *Understanding Delta-Sigma Data Converters*, John Wiley & Sons, Inc., Hoboken.

40 Hjortland, H.A., Wisland, D.T. Lande, T. S. *et al.* (2006) CMOS impulse radar. 24th Norchip Conference, 2006, pp. 75–79.

41 Hjortland, H.A., Wisland, D.T. Lande, T. S. *et al.* (2007) Thresholded samplers for UWB impulse radar. IEEE International Symposium on Circuits and Systems, 2007. ISCAS 2007, pp. 1210–1213.

42 Yunqiang, Y. and Fathy, A.E. (2009) Development and implementation of a real-time see-through-wall radar system based on FPGA. *IEEE Trans. Geosci. Remote Sensing*, **47** (5), 1270–1280.

43 Schroeder, M.R. (1979) Integrated-impulse method measuring sound decay without using impulses. *J. Acoust. Soc. Am.*, **66** (2), 497–500.

44 Zepernick, H.J. and Finger, A. (2005) *Pseudo Random Signal Processing – Theory and Application*, John Wiley & Sons, Inc.

45 Andres, T.H. and Staton, R.G. (1977) Golay sequences, in *Lecture Notes in Mathematics; Combinatorial Mathematics V* (ed. C.H.C. Little), Springer, Berlin, Heidelberg, New York.

46 Sachs, J., Peyerl, P. and Roßberg, M. (1999) A new principle for sensor-array-application. IEEE Instrumentation and Measurement Technology (IMTC), Venice, Italy, pp. 1390–1395.

47 Foster, S. (1986) Impulse response measurement using Golay codes. IEEE International Conference on Acoustics, Speech, and Signal Processing, ICASSP, pp. 929–932.

48 Vazquez Alejos, A., Muhammad, D. and Ur Rahman Mohammed, H. (2007) Ground penetration radar using Golay sequences. IEEE Region 5 Technical Conference, 2007, pp. 318–321.

49 Sarwate, D.V. and Pursley, M.B. (1980) Crosscorrelation properties of pseudorandom and related sequences. *Proc. IEEE*, **68** (5), 593–619.

50 Xiang, N. (1991) A mobile universal measuring system for the binaural room-acoustic modelling-technique, Bundesanstalt für Arbeitsschutz.

51 Alrutz, H. (1983) *Über die Anwendung von Pseudorauschfolgen zur Messung an linearen Übertragungssystemen*. Thesis. Mathematisch-Naturwissenschaftliche Fachbereiche, Georg-August-Universität zu Göttingen, Göttingen.

52 Reeves, B.A. (2010) Noise augmented radar system, US Patent 7,341 B2.

53 Sakamoto, T. and Sato, T. (2007) Code-division multiple transmission for high-speed UWB radar imaging with array antennas. IEEE Antennas and Propagation Society International Symposium, 2007, pp. 429–432.

54 Sakamoto, T. and Sato, T. (2009) Code-division multiple transmission for high-speed UWB radar imaging with an antenna array. *IEEE Trans. Geosci. Remote Sensing*, **47** (4), 1179–1186.

55 Tirkel, A.Z. (1996) Cross correlation of m-sequences-some unusual coincidences. IEEE 4th International Symposium on Spread Spectrum Techniques and Applications Proceedings, vol. 3, 1996, pp. 969–973.

56 J. Sachs, S. Wöckel, R. Herrmann et al., (2007) "Liquid and moisture sensing by ultra wideband pseudo noise sequences," Measurement Science and Technology, vol. 18, pp. 1074–1087.

57 Kmec, M., Müller, J. Rauschenbach, P. *et al.* (2008) Integrated cm- and mm-Wave UWB transceiver for M-sequence based sensors. European Electromagnetics (EUROEM), Lausanne, Switzerland.

58 Herrmann, R., Sachs, J. and Peyerl, P. (2006) System evaluation of an M-sequence ultra wideband radar for crack detection in salt rock. International Conference on Ground Penetrating Radars (GPR), Ohio, Columbus.

59 Herrmann, R., Sachs, J. Schilling, K. *et al.* (2008) 12-GHz bandwidth M-sequence radar for crack detection and high resolution imaging. International Conference on Ground Penetrating Radar (GPR), Birmingham, UK.

60 Herrmann, R. (2011) M-sequence based ultra-wideband radar and its application to crack detection in salt mines, Faculty of Electrical Engineering and Information Technology, Ilmenau University of Technology (Germany), Ilmenau.

61 Harmuth, H.F. (1981) *Nonsinusoidal waves for Radar and Radio Communication*, Academic Press, New York, London, Toronto, Sydney, San Francisco.

62 Harmuth, H.F. (1984) *Antennas and Waveguides for Nonsinusoidal Waves*, Academic Press, Inc., Orlando.

63 T. Kaiser, and F. Zheng, Ultra Wideband Systems with MIMO. John Wiley & Sons, 2010.

64 Zetik, R., Sachs, J. and Thomä, R. (2003) From UWB Radar to UWB Radio. International Workshop on Ultra Wideband Systems (IWUWBS), Oulu, Finland.

65 Noon, D.A. (1996) *Stepped-Frequency Radar Design and Signal Processing Enhances Ground Penetrating Radar Performance*, Department of Electrical and Computer Engineering, University of Queensland, Queensland.

66 Eide, E.S. and Hjelmstad, J.F. (1999) A multi antenna ultra wideband ground penetrating radar system using arbitrary waveforms. IEEE 1999 International Geoscience and Remote Sensing Symposium, vol. 3, 1999. IGARSS '99 Proceedings, pp. 1746–1748.

67 Eide, E.S. (2000) *Radar Imaging of Small Objects Closely Below the Earth Surface*, Department of Telecommunication, Norwegian University, of Science and Technology, Trondheim, Norway.

68 Parrini, F., Pieraccini, M. Spinetti, A. *et al.* (2009) ORFEUS project: the surface GPR system. European Radar Conference, 2009. EuRAD 2009, pp. 93–96.

69 Mirabbasi, S. and Martin, K. (2000) Classical and modern receiver architectures. *IEEE Commun. Mag.*, **38** (11), 132–139.

70 Analog Devices (2008) 700 MHz to 2.7 GHz Quadrature Demodulator ADL 5382 (data sheet and Application Note).

71 Bardin, J.C. and Weinreb, S. (2008) A 0.5–20 GHz quadrature downconverter. IEEE Bipolar/BiCMOS Circuits and Technology Meeting, 2008. BCTM 2008, pp. 186–189.

72 Behbahani, F., Kishigami, Y. Leete, J. *et al.* (2001) CMOS mixers and polyphase filters for large image rejection. *IEEE J. Solid-St. Circ.*, **36** (6), 873–887.

73 *DSM300 Series – 3 GHz DDS-Based Programmable Linear or Non-Linear Frequency Chirping Generator*, Product note, Euvis Inc.

74 *2 to > 8 Gsps Arbitrary Waveform Generation Modules*, Product note, Euvis Inc.

75 Noon, D.A., Longstaff, D. and Stickley, G. F. (1994) Correction of I/Q errors in

homodyne step frequency radar refocuses range profiles. IEEE International Conference on Acoustics, Speech, and Signal Processing, vol. 2, 1994. ICASSP-94, pp. II/369–II/372.

76 Noon, D.A., Longstaff, I.D. and Stickley, G.F. (1999) Wideband quadrature error correction (using SVD) for stepped-frequency radar receivers. *IEEE Trans. Aero. Elec. Sys.*, **35** (4), 1444–1449.

77 Churchill, F.E., Ogar, G.W. and Thompson, B.J. (1981) The correction of I and Q errors in a coherent processor. *IEEE Trans. Aero. Elec. Sys.*, **AES-17** (1), 131–137.

78 Banerjee, D. (2006) *PLL Performance, Simulation and Design Handbook*, 4th edn, National Semiconductor.

79 Musch, T. (2005) Advanced radar signal synthesis based on fractional-n phase locked loop techniques. *Frequenz*, **60** (1–2), 6–10.

80 Lukin, K., Mogyla, A. Palamarchuck, V. *et al.* (2010) Stepped-frequency noise radar with short switching time and high dynamic range. 11th International Radar Symposium (IRS), 2010, pp. 1–4.

81 Rihaczek, A.W. (1996) *Principles of High-Resolution Radar*, Artech House, Inc., Boston, London.

82 van Genderen, P., Hakkaart, P. van Heijenoort, J. *et al.* (2001) A multi frequency radar for detecting landmines: design aspects and electrical performance. 31st European Microwave Conference, 2001, pp. 1–4.

83 Genderen, P.V. (2003) Multi-waveform SFCW radar. 33rd European Microwave Conference, 2003, pp. 849–852.

84 Mende, R. and Rohling, H. (1997) A high performance AICC radar sensor-concept and results with an experimental vehicle. Radar 97 (Conf. Publ. No. 449), pp. 21–25.

85 Rohling, H. and Meinecke, M.M. (2001) Waveform design principles for automotive radar systems. 2001 CIE International Conference on Radar Proceedings, pp. 1–4.

86 Rohling, H. and Moller, C. (2008) Radar waveform for automotive radar systems and applications. IEEE Radar Conference, 2008. RADAR '08, pp. 1–4.

87 Wehner, D.R. (1995) *High Resolution Radar*, 2nd edn, Artech House, Norwood, MA.

88 Stove, A.G. (1992) Linear FMCW radar techniques. *IEE Proc. F Radar Signal Process.*, **139** (5), 343–350.

89 Vossiek, M., Kerssenbrock, T.V. and Heide, P. (1998) Novel nonlinear FMCW radar for precise distance and velocity measurements. IEEE MTT-S International Microwave Symposium Digest, vol. 2, 1998, pp. 511–514.

90 Waldmann, B., Weigel, R. Gulden, P. *et al.* (2008) Pulsed frequency modulation techniques for high-precision ultra wideband ranging and positioning. IEEE International Conference on Ultra-Wideband, 2008. ICUWB 2008, pp. 133–136.

91 *MATLAB; Signal Processing Toolbox*, User's Guide, The Mathworks Inc. available at http://www.mathworks.com/help/pdf_doc/signal/signal_tb.pdf

92 Nalezinski, M., Vossiek, M. and Heide, P. (1997) Novel 24 GHz FMCW front-end with 2.45 GHz SAW reference path for high-precision distance measurements. IEEE MTT-S International Microwave Symposium Digest, vol. 1, 1997, pp. 185–188.

93 Thoma, R.S., Groppe, H. Trautwein, U. *et al.* (1996) Statistics of input signals for frequency domain identification of weakly nonlinear systems in communications. Instrumentation and Measurement Technology Conference, vol. 1, 1996. IMTC-96, pp. 2–7.

94 Nacke, T., Land, R. Barthel, A. *et al.* (2008) Process instrumentation for impedance spectroscopy – a modular concept. 11th International Electronics Conference on Biennial Balticries, 2008. BEC 2008, pp. 235–238.

95 MEDAV GmbH (2007) *RUSK Multidimensional Channel Sounder.*

96 Fink, M. and Prada, C. (2001) Acoustic time-reversal mirrors. *Inverse Probl.*, **17**, 1–38.

97 Walton, E.K. (2007) Radar system using RF noise. US Patent 7,196,657 B2 to The Ohio State University, USA.

98 Walton, E.K. (2010) The use of high speed FIFO chips for implementation of a noise radar. 11th International Radar Symposium (IRS), 2010, pp. 1–4.

99 Kuschel, H. and O'Hagan, D. (2010) Passive radar from history to future. 11th International Radar Symposium (IRS), 2010, pp. 1–4.
100 Axelsson, S.R.J. (2006) Suppression of noise floor and dominant reflectors in random noise radar. International Radar Symposium, 2006. IRS 2006, pp. 1–4.
101 Axelsson, S.R.J. (2007) Random noise radar/sodar with ultrawideband waveforms. *IEEE Trans. Geosci. Remote Sensing*, **45** (5), 1099–1114.
102 Axelsson, S.R.J. (2006) Generalized ambiguity functions for ultra wide band random waveforms. International Radar Symposium, 2006. IRS 2006, pp. 1–4.
103 Axelsson, S.R.J. (2004) Noise radar using random phase and frequency modulation. *IEEE Trans. Geosci. Remote Sensing*, **42** (11), 2370–2384.
104 Lukin, K.A. and Narayanan, R.M. (2010) Fifty years of noise radar. 11th International Radar Symposium (IRS), 2010, pp. 1–2.
105 Lukin, K.A. (2006) Radar design using noise/random waveforms. International Radar Symposium, 2006. IRS 2006, pp. 1–4.
106 Tarchi, D., Lukin, K. Fortuny-Guasch, J. *et al.* (2010) SAR imaging with noise radar. *IEEE Trans. Aero. Elec. Sys.*, **46** (3), 1214–1225.
107 Dawood, M. and Narayanan, R.M. (2001) Receiver operating characteristics for the coherent UWB random noise radar. *IEEE Trans. Aero. Elec. Sys.*, **37** (2), 586–594.
108 Dawood, M. and Narayanan, R.M. (1999) Doppler measurements using a coherent ultrawideband random noise radar. IEEE Antennas and Propagation Society International Symposium, vol. 4, 1999, pp. 2226–2229.
109 Narayanan, R.M. and Kumru, C. (2005) Implementation of fully polarimetric random noise radar. *IEEE Antennas Wireless Prop. Lett.*, **4**, 125–128.
110 Stephan, R. and Loele, H. (2000) Theoretical and practical characterization of a broadband random noise radar. IEEE MTT-S International Microwave Symposium Digest, vol. 3, 2000, pp. 1555–1558.
111 Theron, I.P., Walton, E.K. Gunawan, S. *et al.* (1999) Ultrawide-band noise radar in the VHF/UHF band. *IEEE Trans. Antennas Propag.*, **47** (6), 1080–1084.
112 Narayanan, R.M., Wei, Z. Wagner, K.H. *et al.* (2004) Acoustooptic correlation processing in random noise radar. *IEEE Geosci. Remote Sensing Lett.*, **1** (3), 166–170.
113 Zhixi, L. and Narayanan, R.M. (2006) Doppler visibility of coherent ultrawideband random noise radar systems. *IEEE Trans. Aero. Elec. Sys.*, **42** (3), 904–916.
114 Watts, D.G. (1962) A general theory of amplitude quantization with applications to correlation determination. *Proc. IEE – Part C: Monogr.*, **109** (15), 209–218.
115 Welch, P. (1967) The use of fast Fourier transform for the estimation of power spectra: a method based on time averaging over short, modified periodograms. *IEEE Trans. Audio Electroacoust.*, **15** (2), 70–73.
116 Kay, S.M. (1988) *Modern Spectral Estimation*, Prentice Hall, Englewood Cliffs, NJ.
117 Pratt, W.K. (1991) *Digital Image Processing*, John Wiley & Sons, Inc., New York.
118 Nicolson, A.M. (1973) Forming the fast Fourier transform of a step response in time-domain metrology. *Electron. Lett.*, **9** (14), 317–318.
119 Paulter, N.G. Jr and Stafford, R.B. (1993) Reducing the effects of record truncation discontinuities in waveform reconstructions. *IEEE Trans. Instrum. Meas.*, **42** (3), 695–700.
120 Ferrari, P., Duvillaret, L. and Angenieux, G. (1996) Enhanced Nicolson's method for Fourier transform of step-like waveforms. *Electron. Lett.*, **32** (22), 2048–2049.
121 F. van Raay, and G. Kompa, "A New Active Balun Reflectometer Concept for DC to Microwave VNA Applications," in Microwave Conference, 1998. 28th European, 1998, pp. 108–113.

4
Ultra-Wideband Radar

4.1
Introduction

This and the following chapters are aimed to extend our discussion to measurements at distributed systems and remote sensing over short distances. We will start with some intuitive considerations requiring only elementary knowledge of wave propagation in free space. Further, we will again presume linear behavior of the objects under test with respect to wave interaction; that is object properties are independent on the strength of incident fields.

First, we will concentrate on basic measurement scenarios with the goal to introduce the most important figures of merit of such sensors. To begin with, we will omit the vector character of the electromagnetic field and complicated wave propagation phenomena. Some of such considerations will be discussed in the next chapter.

The discussions below will extend the basic descriptions of linear systems from Chapter 2 to simple scenarios of wave propagation including antennas and canonical scattering objects such as point scatters or boundary surfaces. We will start with a short introduction to the measurement problem related to distributed systems. We will consider wave propagation and simple scattering mechanisms in time and frequency domain. Here, we need to hark back to vector and matrix calculus. Some basic rules are summarized in Annex A for the less experienced reader. Then, some methods to characterize wideband antennas for pulse radiation are discussed where we also include the scattering behavior of the antennas, since it may be important for short-range sensing due to mutual interactions between target and antenna. We continue with the key performance figures of UWB radars and we will review some issues of UWB target detection and UWB short-range imaging.

4.2
Distributed System – the Measurement Problem

In Chapter 2, we discussed the dynamic behavior of linear systems from a very general viewpoint. We used a black box model which did not allow us to have a view inside. So we were not able to recognize any internal structure. The

Handbook of Ultra-Wideband Short-Range Sensing: Theory, Sensors, Applications, First Edition. Jürgen Sachs.

observation of any interaction, that is the energy exchange, of the 'black box system' with its environment was only accessible via a limited number of pathways. We used these pathways to stimulate the system as well as to observe its reaction to our stimulation. As a result, we got a set of impulse or frequency response functions which indeed provide some formal parameters (e.g. bandwidth, rise time, decay rate etc.) of the object under test but these functions are difficult to interpret with respect to deeper details if nothing (except linearity and time invariance) is known about the object.

The situation changes, if at least some information about the internal structure of the system under test is known a priori. This we will consider now by referring to systems under test from which we have the following prior knowledge (see also Figure 4.1):

- We will deal with a small number of rigid objects of arbitrary but unknown shape. These objects may be arbitrarily localized in space where their position and orientation are not known. For the beginning, we will restrict ourselves to two types of objects: rigid bodies of small size and infinite planes.
- For stimulation of the scenario under test, we excite an electromagnetic wave at one or more points in space at known positions.
- The propagation medium behaves isotropic; that is the propagation speed of the sounding waves is independent of any direction.
- The waves are scattered by the objects involved in the test scenario. We suppose linear[1)] objects. That is, their scattering behavior is independent of the amplitude of the sounding wave.
- The involved waves are captured at one or more known positions in space.
- The distance between the different objects and antennas should be large enough so that we can omit near-field effects.

Figure 4.1 illustrates the intended situation. In the shown example, the external energy exchange with the measurement device takes place over three pathways (measurement ports) from which only one is used to stimulate the system but all ports are able to capture waves appearing at the corresponding antenna positions. The internal propagation paths are unknowns. The aim of our endeavour is to determine these. Note that the measurement antennas are part of the measurement scenario. They will affect the propagation conditions of the undisturbed scenario under test. Hence, they should be as 'invisible' as possible for the waves travelling within the test environment.

The Mason graph indicates all return and transmission paths which may theoretically be measured between the points ①, ② and ③. Every of these paths may be composed of numerous individual ray paths and propagation effects including line-of-sight connections, single and multiple reflections, diffraction and others. The scenario depicted in Figure 4.1 permits the measurement of nine

1) Here, the term linear refers to the electric properties of the material from which the objects are composed but not to any geometric properties. That is, the electric parameters σ, ε, μ of any involved substances should not depend on the strength of the sounding fields but they may depend on frequency and space.

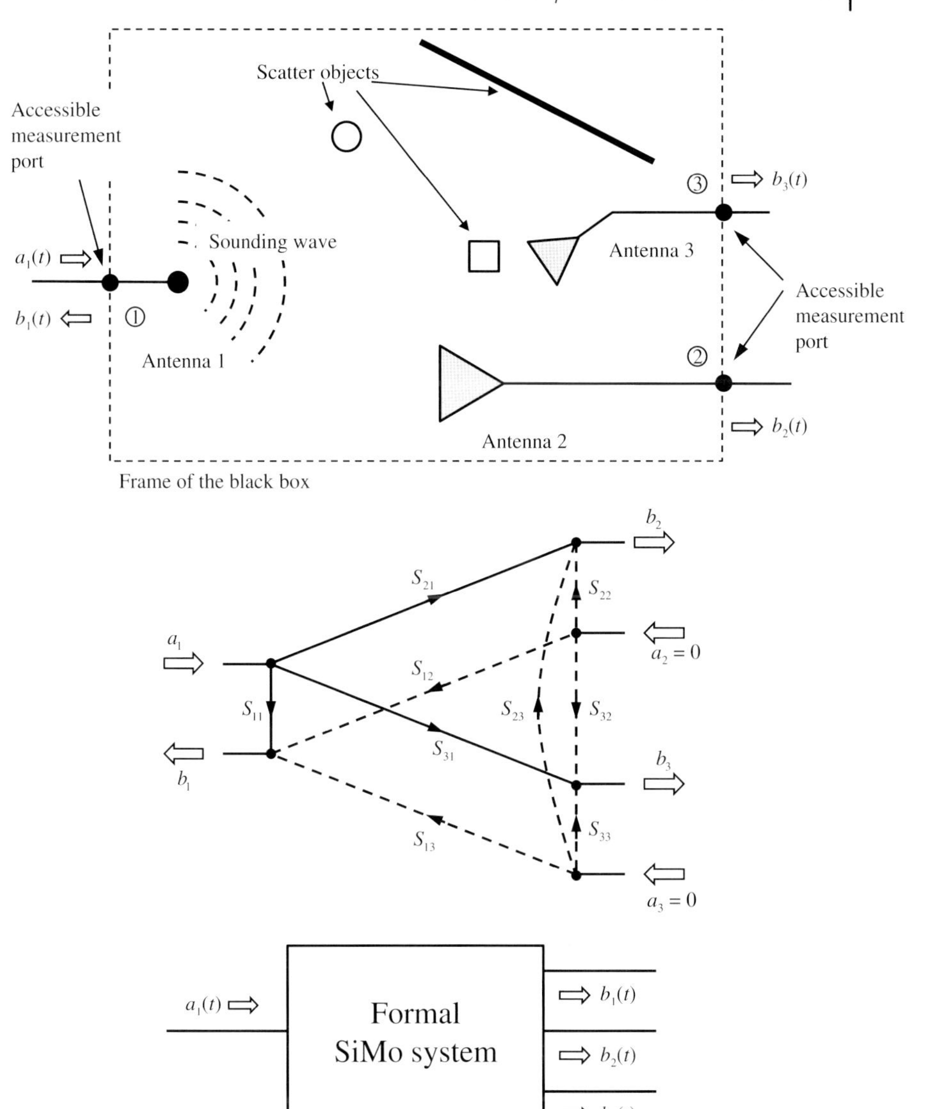

Figure 4.1 Symbolized test scenario and its input–output description by Mason graph or a formal SiMo-transmission system.

different IRFs from which only six are linearly independent as we already know from (2.214), that is $S_{ij} = S_{ji}$. But actually, our example scenario is restricted to the measurement of only three IRFs as indicated by the bold lines of the Mason graph, since two of the antennas are not equipped with a signal generator. Thus from a formal standpoint, we can symbolize the scenario by a transmission system (black box) with one input and three output channels. The interior of the black box is encircled by a dashed line in the upper drawing of Figure 4.1.

The measurable signals should be normalized waves, in our example, travelling along the feeding lines. We can determine from them the three IRFs $S_{11}(t), S_{21}(t), S_{31}(t)$ representing time domain scattering parameters, that is

$$\begin{bmatrix} b_1(t) \\ b_2(t) \\ b_3(t) \end{bmatrix} = \begin{bmatrix} S_{11}(t) \\ S_{21}(t) \\ S_{31}(t) \end{bmatrix} * a_1(t)$$

The goal of the measurement will be to extract some information about the scenario under test from the measurable IRFs $S_{ij}(t)$, for example to recognize and characterize some of the individual propagation paths comprising the scattering objects. Hence, one needs some ideas about the processes running internally in our scenario. Obviously, the system response is based on energy exchange via electromagnetic waves between the involved bodies. These bodies are characterized by their geometric shape and material composition as well as by their positions and orientations in space.

We will categorize the scattering objects into five classes as depicted in Figure 4.2:

- **Point objects** are arbitrary-shaped bodies whose characteristic dimensions are small compared to the wavelength λ_c of a narrowband carrier signal or the spatial extension of a sub-nanosecond pulse. Typically such bodies are modelled by spheres of diameter $d \leq \lambda_c = c/f_c$ or $d \leq ct_x$ (c is the propagation speed of the wave and t_x represents either the rise time or pulse width depending on the considered scattering phenomena). As shown in the upper left of Figure 4.2, a simple position vector $\mathbf{r}$ is sufficient to characterize the spatial condition of a point object. Some issues of position vector notations and vector handling are summarized in Annexes A.4. and A.5.
- **Line objects:** These objects have a large (theoretically infinite) extension in one direction but a small diameter ($d \leq \lambda_c = c/f_c$ or $d \leq ct_x$). Long straight wires, edges or long thin cavities in an otherwise homogeneous media are examples of such objects. A line object needs two specifications to be uniquely characterized with respect to its position. These are the position of an arbitrary point located at the line – given by the position vector $\mathbf{r}$– and its direction. The last one is given by either two angles with respect to a global coordinate system or a direction vector $\mathbf{e}$ of unit length (for details see Annex 7.4).
- **Planar objects:** A planar object has a small extension in only one direction. In the remaining two dimensions, it is (theoretically) infinitely extended. Its spatial position may be uniquely characterized by two vectors. One assigns the position $\mathbf{r}$ of an arbitrary point within the plane, and the second is the direction vector $\mathbf{n}$ of unit length pointing orthogonal to the plane. If $\mathbf{a}_1$ and $\mathbf{a}_2$ are two vectors of different directions located in the plane (see Figure 4.2), the normal vector $\mathbf{n}$ of the plane is given by

$$\mathbf{n} = \frac{\mathbf{a}_1 \times \mathbf{a}_2}{|\mathbf{a}_1 \times \mathbf{a}_2|} \tag{4.1}$$

Herein, the symbol $\times$ stands for the vector product (see Annex A.5).

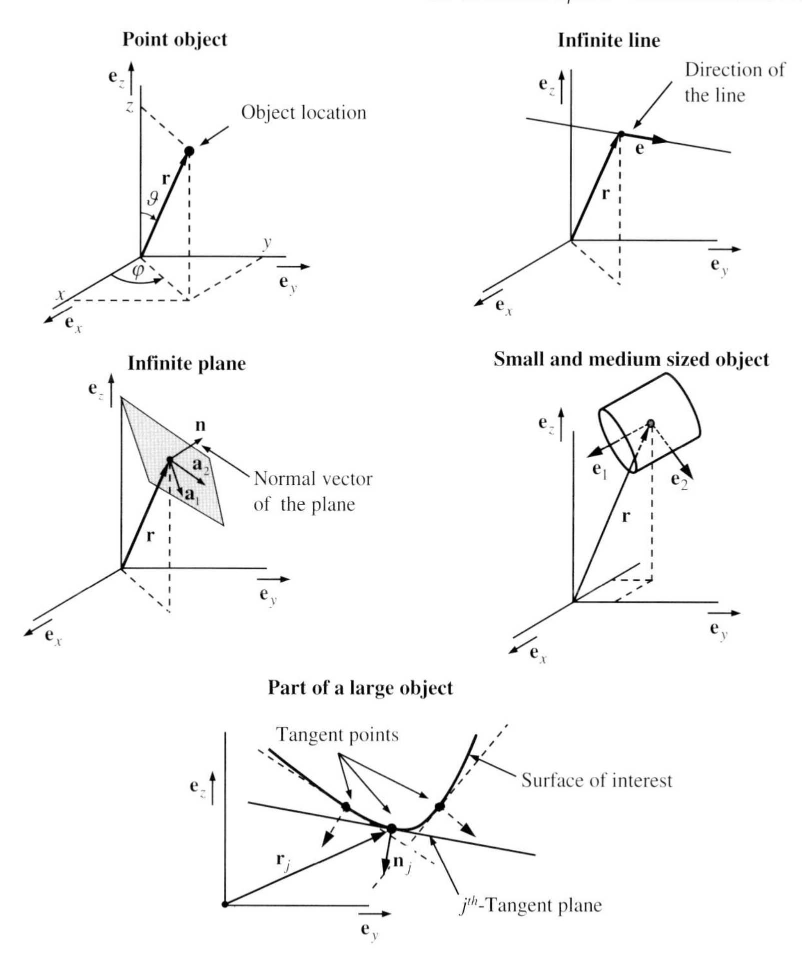

Figure 4.2 The five object classes and their position and orientation in space.

- **Small- and medium-sized objects:** Under a small object we will understand a rigid body of arbitrary shape whose characteristic dimensions d are in the order of characteristic spatial extensions of the sounding signal, that is $d \approx 0.1 \cdots 10\lambda_c$ or $d \approx 0.1 \cdots 10_{ctx}$. Due to the complex mechanisms of wave interaction, it is often hardly possible to establish a simple relation between scattering signals and the body geometry. Hence, it is a common practice to replace the actual scattering body (of finite dimensions) by a virtual point object (of infinitesimal dimensions). This virtual point object is placed at position $\mathbf{r}$ assigning a reference point of the considered body. This ersatz object should have identical scattering behavior as the real one which we formally described by an impulse response function (introduced in Section 4.5.2). The impulse response function does not depend on the position of the body as long as the wavefront curvature of the incident wave remains always the same. This simplification supposes a planar wavefront across the object, that is sufficient distance to the source of the sounding wave (see

Section 4.5 for details) and it requires to quantify the spatial orientation of the 'virtual scattering point' which is not needed (and not possible) in the case of an actual point scatterer. For that purpose, one has to fix two reference directions $\mathbf{e}_1$ and $\mathbf{e}_2$ (or three angles) to assign uniquely the spatial orientation of the body (compare Figure 4.2). These directions and the virtual scattering point may be arbitrarily selected whereas the reference directions should be orthogonal, that is $\mathbf{e}_1 \cdot \mathbf{e}_2 = 0$ (the dot $\cdot$ represents the scalar product of two vectors). Typically, the selection of these reference quantities is however geared to points and axes of symmetry or other particular characteristics (compare also Section 4.4).

- **Large objects:** Under a large object, we will understand a body encircled by a smooth surface which can be approximated by a finite number of patches approaching tangent planes. Every patch can be considered as a planar object. Hence, we can characterize a large object by a set of position and normal vectors $\mathbf{r}_j$, $\mathbf{n}_j$ assigning the location and orientation of the individual patches.

4.3 Plane Wave and Isotropic Waves/Normalized Wave

The interaction between the different bodies of the scenario results from energy exchange via waves. At that point, we restrict ourselves to the simplest types of waves – plane waves and isotropic or spherical waves, respectively. We will also omit the vector nature of the electromagnetic field as well as propagation phenomena as attenuation due to energy absorption, dispersion, diffraction and so on. These points will be discussed in deeper detail in the next chapter.

Generally speaking, a wave $W(t, \mathbf{r}')$ represents a specific state of a physical quantity (electric or magnetic field in our case) in dependence on time t and space where the considered point in space (observation point) is characterized by its position vector $\mathbf{r}'$. Since the position vector may be decomposed into three coordinates (see Annex A.4), a wave is finally described by a four-dimensional wave function in which, however, the arguments t and $\mathbf{r}'$ are not independent of each other. Rather, we can write

$$W(t, \mathbf{r}') \propto W\left(t - \frac{\mathbf{e}_0 \cdot (\mathbf{r}' - \mathbf{r}_0)}{c}\right) \propto W(t, \mathbf{r}_0) * \delta\left(t - \frac{\mathbf{e}_0 \cdot (\mathbf{r}' - \mathbf{r}_0)}{c}\right) \tag{4.2}$$

which holds for an arbitrary wave if it propagates in a homogeneous and isotropic medium free of any dispersion (see also next chapter for propagation with dispersion). Herein, c represents the propagation speed of the wave and $\mathbf{e}_0$ is a unit vector indicating the propagation direction. Note that generally $\mathbf{e}_0$ depends on $\mathbf{r}'$; that is, the wave may propagate in different directions at different positions in space.

In Section 2.5.3, we used normalized waves to describe the signal propagation along electrical lines or waveguides. Here, we will apply the same concept for free wave propagation (compare also first chapter in Ref. [1]). Under many circumstances, we can approximate $E(t, \mathbf{r}) = Z_s H(t, \mathbf{r})$ where the proportionality factor Z_s does not depends on the arbitrary position $\mathbf{r}$ and on time t. $Z_s = \sqrt{\mu/\varepsilon}$ is called the characteristic impedance or intrinsic impedance of the propagation medium. Rewriting this equation, we get a relation which resembles (2.200) of guided waves.

$$W(t,\mathbf{r}) = \frac{E(t,\mathbf{r})}{\sqrt{Z_s}} = \sqrt{Z_s}\, H(t,\mathbf{r}) \tag{4.3}$$

We call $W(t,\mathbf{r})$ the normalized wave. Its unity is $\left[\sqrt{\mathrm{W/m^2}}\right]$. Note that the relation only holds for linear, lossless, isotropic, homogeneous propagation medium (which we mostly assume in what follows) and it requires far-field condition. Furthermore, it does not respect the vector nature of the field quantities E and H (see Chapter 5 for details).

From Poynting's theorem (see also Chapter 5 for its introduction) and as consequence from (4.3), the power transfer through unit area (i.e. the magnitude of the Poynting vector) at an observation position $\mathbf{r}'$ results in:

Instantaneous power flux density:

$$\mathfrak{P}(t,\mathbf{r}') = |\mathbf{E}(t,\mathbf{r}') \times \mathbf{H}(t,\mathbf{r}')| = W^2(t,\mathbf{r}') \tag{4.4}$$

Average power flux density:

$$\overline{\mathfrak{P}}(\mathbf{r}') = \lim_{T\to\infty} \frac{1}{2T} \int_{-T}^{T} W^2(t,\mathbf{r}')\, dt \tag{4.5}$$

Power flux per unit area and Hertz:

$$\overline{\mathfrak{P}}(f,\mathbf{r}') = \underline{W}(f,\mathbf{r}')\underline{W}^*(f,\mathbf{r}') \tag{4.6}$$

Equation (4.2) implies that the time evolution $W(t) = W(t,\mathbf{r}_0)$ of a signal captured by a field probe at position $\mathbf{r}_0$ will be the same at any other observation positions $\mathbf{r}'$ except a delay (or advance) Δt:

$$\begin{aligned} \Delta t &= \frac{\mathbf{e}_0 \cdot (\mathbf{r}' - \mathbf{r}_0)}{c} = \frac{\Delta r}{c} \\ \Delta t &= \begin{cases} > 0 & \text{delayed} \\ < 0 & \text{advanced} \end{cases} \end{aligned} \tag{4.7}$$

The scalar product $\Delta r = \mathbf{e}_0 \cdot (\mathbf{r}' - \mathbf{r}_0)$ corresponds to the distance which the wave front was propagating. It equals the projection of the distance between the two positions $\mathbf{r}'$ and $\mathbf{r}_0$ onto the propagation direction $\mathbf{e}_0$, and consequently c represents the displacement speed of a certain state of the wave.

Figure 4.3 illustrates (4.2) for a plane wave. It is characterized by unique propagation direction $\mathbf{e}_0$ and constant amplitude within the whole space. Hence, the proportionality in (4.2) may be replaced by an equal sign:

Arbitrary time shape:

$$W_P(t,\mathbf{r}') = W_P\left(t - \frac{\mathbf{e}_0 \cdot (\mathbf{r}' - \mathbf{r}_0)}{c}\right) = W_P(t,\mathbf{r}_0) * \delta\left(t - \frac{\mathbf{e}_0 \cdot (\mathbf{r}' - \mathbf{r}_0)}{c}\right) \tag{4.8}$$

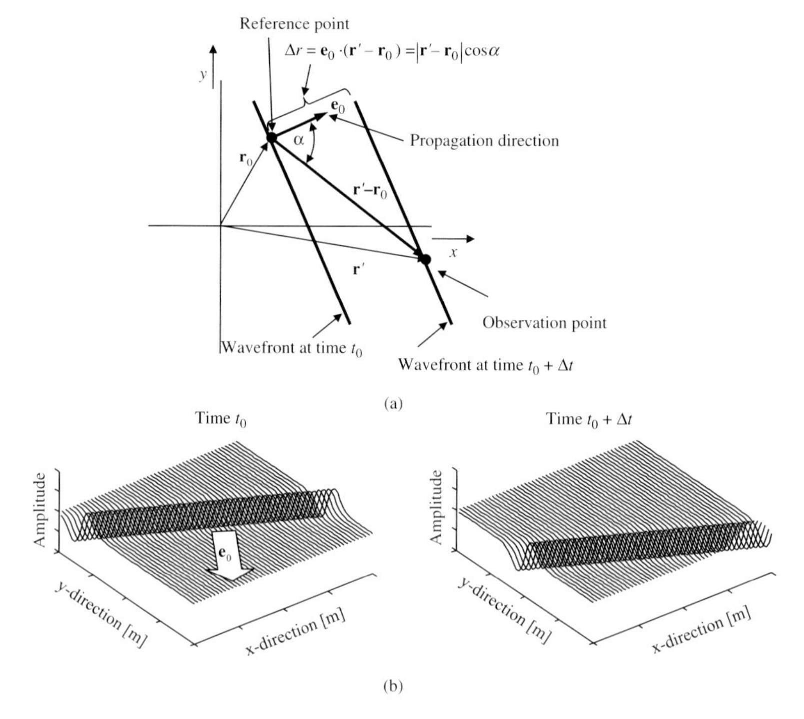

Figure 4.3 2D-illustration of plane wave propagation for a pulsed wave. (a) Definition of quantities and symbolized birds view of the wavefront for two consecutive time points. (b) 'Side view' of the wave propagation. The wave shape corresponds to the first derivative of a Gaussian waveform.

Time harmonic waves:

$$\begin{aligned}\underline{W}_{\mathrm{P}}(f,\mathbf{r}') &= \underline{W}_{\mathrm{P}}(t,\mathbf{r}_0)\mathrm{e}^{j2\pi(f\,t-p\,\mathbf{e}_0\cdot(\mathbf{r}'-\mathbf{r}_0))} = \underline{W}_{\mathrm{P}}(t,\mathbf{r}_0)\mathrm{e}^{j2\pi\left(\frac{t}{t_{\mathrm{P}}}-\frac{\mathbf{e}_0\cdot(\mathbf{r}'-\mathbf{r}_0)}{\lambda}\right)} \\ &= \underline{W}_{\mathrm{P}}(t,\mathbf{r}_0)\mathrm{e}^{j2\pi f\left(t-\frac{\mathbf{e}_0\cdot(\mathbf{r}'-\mathbf{r}_0)}{c}\right)} = \underline{W}_{\mathrm{P}}(t,\mathbf{r}_0)\mathrm{e}^{j(2\pi f\,t-\mathbf{k}\cdot(\mathbf{r}'-\mathbf{r}_0))}\end{aligned} \tag{4.9}$$

Note that for time harmonic waves, the term $\mathrm{e}^{j2\pi f t}$ is often omitted so it often reads $W_{\mathrm{P}}(t,\mathbf{r}_0)\mathrm{e}^{-j\mathbf{k}\cdot(\mathbf{r}'-\mathbf{r}_0)}$ or $\underline{W}_{\mathrm{P}}(t,\mathbf{r}_0)\mathrm{e}^{-j\,2\,\pi\,\mathbf{p}\cdot(\mathbf{r}'-\mathbf{r}_0)}$. The vectors $\mathbf{k}$ and $\mathbf{p}$

$$\mathbf{k} = \begin{bmatrix} k_x \\ k_y \\ k_z \end{bmatrix} = \mathbf{e}_0\frac{2\pi f}{c} = \mathbf{e}_0\frac{2\pi}{\lambda} = 2\pi\mathbf{p} \tag{4.10}$$

are called the wave vector and the vector of spatial frequency. They indicate the propagation direction and the spatial periodicity of a sine wave. The magnitude $k = |\mathbf{k}|$ is called the wave number. λ is the wavelength.

Due to their simplicity, plane waves are frequently used to describe wave phenomena. However, one should have in mind that these waves are physically not feasible in strong sense since they have an infinite extension which would require

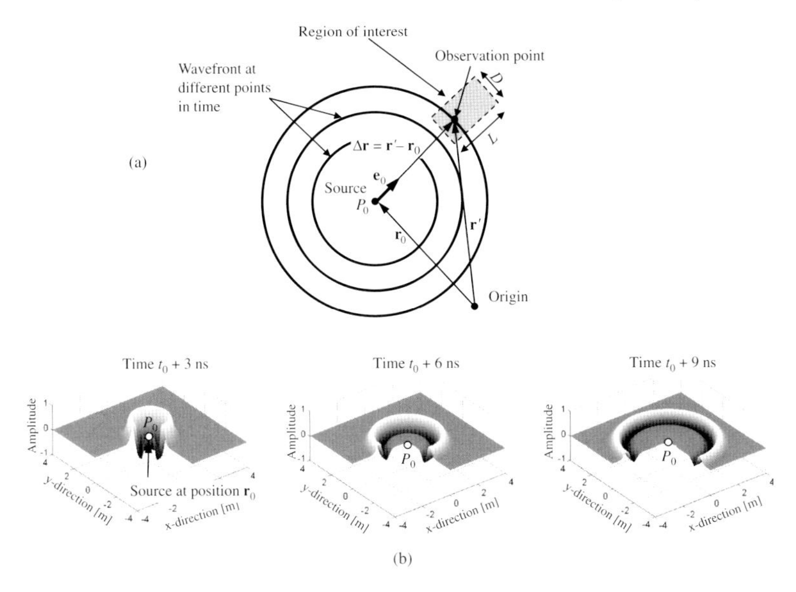

Figure 4.4 Isotropic wave (spherical wave) of a pulsed time shape. (a) Definitions of quantities. (b) Illustration of wave expansion for three different points in time after launching. A quarter of the wavefront is cut out for better visualization of the spatial distribution of wave amplitude.

an infinitely extended source and they transport an infinite quantity of energy. Furthermore, they violate the causality principle (see below (4.18)).

In a second example, we will consider the scalar isotropic wave[2] (see Figure 4.4). Although it is still idealized, it is closer at the physical reality as the plane wave is, particularly for the case of short-range sensing. This wave is thought to spring from an infinitesimal small volume enclosing the point P_0 which is pointed by the position vector $\mathbf{r}_0$. We suppose an isotropic propagation medium. Therefore, the energy emanating from point P_0 is equally spread by the wave over the whole space where the encompassed region expands in time with speed c. As obvious from Figure 4.4, the propagation direction $\mathbf{e}_0$ and the distance vector $\Delta\mathbf{r} = \mathbf{r}' - \mathbf{r}_0$ are always in parallel independently of what the observation point is. Hence, (4.2) can be modified for a spherical wave to

$$W_{\mathrm{SP}}(t, \mathbf{r}') = W_{\mathrm{SP}}\left(t - \frac{|\mathbf{r}' - \mathbf{r}_0|}{c}\right) = K W_0(t, \mathbf{r}_0) * \delta\left(t - \frac{\Delta r}{c}\right) \quad \text{with} \qquad (4.11)$$
$$\Delta r = |\mathbf{r}' - \mathbf{r}_0|$$

where the wave magnitude must depend on the distance Δr between observation and source point. It is expressed by the still unknown factor K. Its dependence on Δr can be determined from the energy conservation. Supposing the signal which generates the wave at point P_0 has the instantaneous power $p(t)$, the law of energy conservation requires that all energy collected over an

2) The correct vector field notation of a spherical is given in Section 5.5.

arbitrary closed surface S enclosing the source must be equal to the injected energy retarded by the propagation time to reach the observation points:

$$\mathfrak{E} = \underbrace{\int p(t)\mathrm{d}t}_{\text{injected energy}} = \underbrace{\int \oint \mathfrak{P}(t,\mathbf{r}_s)\mathrm{d}S\,\mathrm{d}t}_{\text{total energy of radiated wave}} = \int \oint W_{SP}^2(t,\mathbf{r}_s)\mathrm{d}S\,\mathrm{d}t \tag{4.12}$$

Herein, $\mathbf{r}_s$ represents the position vector of a point located at the surface S. To estimate the surface integral, we can select any closed surface and thus also a sphere of radius $\Delta r_s = |\mathbf{r}_S - \mathbf{r}_0|$ centred at the point source. Since the wave spreads its energy equally into space, the magnitude $W_{SP}(t,\mathbf{r}_s)$ will be constant over the sphere. Hence, it can be taken out of the integrals and we find from (4.11) and (4.12)

$$\int p(t)\mathrm{d}t = K^2 \int W_0^2(t,\mathbf{r}_0)\mathrm{d}t \oint \mathrm{d}S = K^2 4\pi\,\Delta r_s^2 \int W_0^2(t,\mathbf{r}_0)\mathrm{d}t \tag{4.13}$$

and (dispersion-free propagation supposed)

$$p(t) = K^2 W_0^2\left(t - \frac{\Delta r}{c}, \mathbf{r}_0\right) \oint \mathrm{d}S = K^2 4\pi\,\Delta r_s^2 W_0^2\left(t - \frac{\Delta r}{c}, \mathbf{r}_0\right) \tag{4.14}$$

As already mentioned, the signal power $p(t)$ can be considered as the total instantaneous source yield spread over the whole sphere. Relating $p(t)$ to the surface of a unit sphere, the quantity $W_\Omega^2(t,\mathbf{r}_0)$ $\left[\sqrt{\mathrm{W/sr}}\right]$

$$W_\Omega^2(t,\mathbf{r}_0) = \frac{p(t)}{4\pi} = K^2 W_0^2(t,\mathbf{r}_0)\Delta r_s^2 \tag{4.15}$$

can be interpreted as source yield of the point radiator per solid angle. So that finally in time and frequency domain, a spherical wave can be written as

Time domain:

$$W_{SP}(t,\mathbf{r}) = W_\Omega(t,\mathbf{r}_0) * \frac{1}{\Delta r}\delta\left(t - \frac{\Delta r}{c}\right) = \frac{W_\Omega(t - \Delta r/c, \mathbf{r}_0)}{\Delta r} \quad \text{with} \tag{4.16}$$

$$\Delta r = |\mathbf{r}' - \mathbf{r}_0|$$

Time harmonic fields:

$$\underline{W}_{SP}(f,\mathbf{r}') = \underline{W}_\Omega(f,\mathbf{r}_0)\frac{\mathrm{e}^{j2\pi f\left(t - \frac{\Delta r}{c}\right)}}{\Delta r} = \underline{W}_\Omega(f,\mathbf{r}_0)\frac{\mathrm{e}^{j(2\pi f t - k\,\Delta r)}}{\Delta r} \mathrel{\hat{=}} \underline{W}_\Omega(f,\mathbf{r}_0)\frac{\mathrm{e}^{-jk\,\Delta r}}{\Delta r} \tag{4.17}$$

Both equations can be separated into two parts. $W_\Omega(t,\mathbf{r}_0)$ and $\underline{W}_\Omega(f,\mathbf{r}_0)$ define the time evolution and the spectral composition, respectively, of the fields in the source point while the term $(1/\Delta r)\delta(t - (\Delta r/c))$ or $\mathrm{e}^{-jk\,\Delta r}/\Delta r$ expresses the propagation of the wave. With exception of a constant factor, the propagation term corresponds to the free space Green's function of a scalar field (for further details on Green's function see Section 5.7).

Figure 4.4 illustrates the propagation of a pulse wave according to (4.16). For theoretical analysis, attention has to be paid in close vicinity of the source point P_0

($\Delta r \to 0$), since there arises a singularity. Therefore, the proximate source region requires a deeper analysis of the problem, for which the reader can refer to Ref. [2].

Figure 4.4 leads us to further observation which is transcribed by the term causality. Namely it indicates that the field radiated by a source located at point $\mathbf{r}_0$ and switched on at time t_0 must be zero for all observation points whose distances exceed $\Delta r = |\mathbf{r}' - \mathbf{r}_0| > c(t - t_0)$:

$$W(t, \mathbf{r}') = \begin{cases} \equiv 0 & \text{for} \quad t \le t_0 + \dfrac{\Delta r}{c} \\ W(t, \mathbf{r}') & \text{for} \quad t > t_0 + \dfrac{\Delta r}{c} \end{cases} \tag{4.18}$$

Causality implies that electromagnetic waves propagate away from the source and that the reaction of an object to wave stimulation cannot occur before the wave arrives at this object. Plane waves are non-causal waves due to their infinite extension.

Often, one is only interested in a limited area of the wave. This may be happen if, for example the test object which interacts with the wave is of small size. If this region of interest (compare Figure 4.4) is restricted to a small area at sufficiently large distance from the source, the spherical wave can be approximated by a plane wave. Inside this zone, the curvature of the wavefront as well as the amplitude decay should be negligible. The maximum allowed approach of the region of interest to the source is given by the acceptable deviation of the wavefront from a straight line or/and the maximum permitted amplitude variation.

Referring to Figure 4.5, we will shortly estimate the deviations between plane and spherical wave within the region of interest. The wave enters the region of interest with the amplitude $\|W_{SP}(r'_{RI})\|_\infty$ (r_{RI} is the distance between source and start of the region of interest). From (4.16), we find that the magnitude decreases to $\|W_{SP}(r'_{RI} + L)\|_\infty = \|W_{SP}(r'_{RI})\|_\infty (r'_{RI}/(r'_{RI} + L))$ when the wave leaves the region. In contrast to that, the amplitude of a plane wave remains always constant. Referring the amplitude variation across the region of interest to the amplitude when the wave enters that region, we get the relative amplitude deviation Δe_L in comparison to plane wave propagation of

$$\begin{aligned} \Delta e_L &= \frac{\Delta W_{SP}}{\|W_{SP}(r'_{RI})\|_\infty} = \frac{\|W_{SP}(r'_{RI})\|_\infty - \|W_{SP}(r'_{RI} + L)\|_\infty}{\|W_{SP}(r'_{RI})\|_\infty} \\ &= \frac{L}{r'_{RI} + L} \approx \frac{L}{r'_{RI}}; \quad r'_{RI} \gg L \end{aligned} \tag{4.19}$$

The normalized curvature error relates the deviation Δx from a straight line to a characteristic spatial dimension $c\tau_P$ of the sounding wave:

$$\Delta e_C = \frac{\Delta x}{c\tau_P} \tag{4.20}$$

From Figure 4.5, it follows for Δx

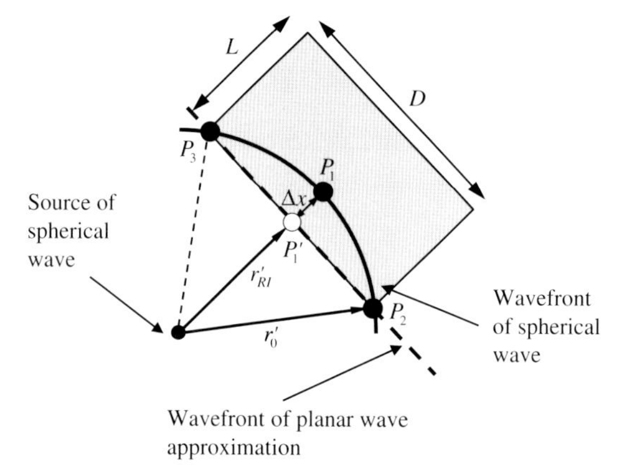

Figure 4.5 Approximation of a spherical wave by a plane wave while entering the region of interest of the width D and depth L.

$$\begin{aligned} r_0'^2 &= \left(\frac{D}{2}\right)^2 + r_{\mathrm{RI}}'^2 = \left(r_{\mathrm{RI}}' + \Delta x\right)^2 \\ \frac{D^2}{4} &= \Delta x\left(2r_{\mathrm{RI}}' + \Delta x\right) \approx 2\,\Delta x\, r_{\mathrm{RI}}'; \quad r_{\mathrm{RI}}' \gg \Delta x \end{aligned} \tag{4.21}$$

That is, if we accept a maximum curvature error of $\Delta e_{\mathrm{C,max}}$, the distance of the region of interest from the source must be larger than

$$r_{\mathrm{RI}}' \geq \frac{D^2}{8\,\Delta e_{\mathrm{C,max}} c\tau_{\mathrm{P}}} \tag{4.22}$$

The notion 'characteristic spatial dimension' needs still some clarification since it depends on the actual circumstances of the measurement scenario and the required precision of the measurement. Typical quantity applied for that purpose is the pulse duration ($\tau_{\mathrm{P}} \triangleq t_{\mathrm{w}}$) or the rise time ($\tau_{\mathrm{P}} \triangleq t_{\mathrm{r}}$) for pulsed waves. For time spread wideband signals, the corresponding quantities of the ACF should be taken, and for time harmonic signals, one takes the period $\tau_{\mathrm{P}} \triangleq t_{\mathrm{P}} = f^{-1}$.

Since we will approximate the spherical wave by a planar one, we expect that the wave has identical states in all three points P_1', P_2, P_3 as depicted in Figure 4.5. But actually the state in P_1' is slightly in advance compared to the other ones. This leads to an amplitude deviation between P_2, P_3 and P_1' whose maximum we will estimate. We define the maximum of the relative amplitude errors Δe_{D} as

$$\Delta e_{\mathrm{D}} = \frac{\left\| W_{\mathrm{SP}}\left(r_0'\right) - W_{\mathrm{SP}}\left(r_{\mathrm{RI}}'\right)\right\|_\infty}{\left\| W_{\mathrm{SP}}\left(r_{\mathrm{RI}}'\right)\right\|_\infty} \tag{4.23}$$

For small Δx, we can write $W_{\mathrm{SP}}(r_0') = W_{\mathrm{SP}}(r_{\mathrm{RI}}' + \Delta x) \approx W_{\mathrm{SP}}(r_{\mathrm{RI}}') + \left.\frac{\partial W_{\mathrm{SP}}(r)}{\partial r}\right|_{r=r_{\mathrm{RI}}'} \Delta x$. Using (4.16), we get

$$\Delta e_{\mathrm{D}} \cong \frac{\left\| \left.\frac{\partial W_{\mathrm{SP}}(r)}{\partial r}\right|_{r=r_{\mathrm{RI}}'} \Delta x \right\|_\infty}{\left\| W_{\mathrm{SP}}(r_{\mathrm{RI}}') \right\|_\infty} = \frac{r_{\mathrm{RI}}' \, \Delta x \left\| \frac{\partial}{\partial r'} \left(\frac{W_\Omega(t - r'/c)}{r'} \right)_{r=r_{\mathrm{RI}}'} \right\|_\infty}{\left\| W_\Omega(t - r_{\mathrm{RI}}'/c) \right\|_\infty}$$

Substituting $t - (r'/c) = \tau$, the differential quotient results in

$$\frac{\partial}{\partial r'} \left(\frac{W_\Omega(t - r'/c)}{r'} \right) = -\frac{W_\Omega(\tau)}{r'^2} - \frac{1}{cr'} \frac{\partial W_\Omega(\tau)}{\partial \tau}$$

If we finally apply Minkowski inequality (2.12), the relative amplitude error due to wavefront curvature within the region of interest is

$$\Delta e_{\mathrm{D}} \le \frac{\Delta x}{r_{\mathrm{RI}}'} + \frac{\Delta x}{c} \frac{\left\| \partial W_\Omega(\tau)/\partial \tau \right\|_\infty}{\left\| W_\Omega(\tau) \right\|_\infty} \tag{4.24}$$

We will consider this with respect to waves of large and narrow bandwidth. For wideband short-pulse waves, we insert the slew rate (2.19) which is defined as $\mathrm{SR} = \left\| \partial W_\Omega(\tau)/\partial \tau \right\|_\infty \approx \left\| W_\Omega(\tau) \right\|_\infty / t_{\mathrm{r}}$ in the case of a pulse-shaped wave, where t_{r} represents the rise time of the wavefront. Using (4.21) and respecting that the spatial extension of the pulse wave is much smaller than the distance r_{RI}', (4.24) leads to

$$\Delta e_{\mathrm{D}} \approx \frac{D^2}{8 r_{\mathrm{RI}}'} \left(\frac{1}{r_{\mathrm{RI}}'} + \frac{1}{c t_{\mathrm{r}}} \right) \approx \frac{D^2}{8 r_{\mathrm{RI}}' c t_{\mathrm{r}}} \tag{4.25}$$

This equals the curvature error (4.22) if the characteristic time is $\tau_{\mathrm{P}} = t_{\mathrm{r}}$.

For a band-pass pulse or a pure sine wave of frequency $f_{\mathrm{c}} = c/\lambda_{\mathrm{c}}$, we modify (4.24) to

$$\Delta e_{\mathrm{D}} \approx \frac{D^2}{8 r_{\mathrm{RI}}'} \left(\frac{1}{r_{\mathrm{RI}}'} + \frac{2\pi f_{\mathrm{c}}}{c} \right) = \frac{D^2}{8 r_{\mathrm{RI}}'} \left(\frac{1}{r_{\mathrm{RI}}'} + \frac{2\pi}{\lambda_{\mathrm{c}}} \right) \approx \frac{\pi D^2}{4 \lambda_{\mathrm{c}} r_{\mathrm{RI}}'} \quad \text{with} \quad W_\Omega(\tau) = W_0 \sin 2\pi f_{\mathrm{c}} \tau \tag{4.26}$$

This can be converted into

$$r_{\mathrm{RI}}' \ge \frac{2D^2}{\lambda_{\mathrm{c}}} \tag{4.27}$$

if we accept a relative amplitude error of $\Delta e_{\mathrm{D}} \le \pi/8 \approx 40\%$. The related curvature error is $\Delta e_{\mathrm{C}} = 1/16 \triangleq \pi/8 \approx 22.5°$ using as characteristic time τ_{P} the period of the carrier signal ($\tau_{\mathrm{P}} f_{\mathrm{c}} = 1$). (4.27) is a well-known expression but usually differently interpreted. Mostly this equation is applied to indicate the start of the far-field region of an antenna with a typical dimension of D (see also Ref. [3], p. 589).

So far, as exemplified in Figures 4.3 and 4.4, we always used for illustration wideband waves of compact time structure, that is sub-nanosecond pulse waveforms. The imagination of the time evolution and propagation of such waves is comparatively simple and comprehensible for a human being.[3] This is still true if the waves interact with obstacles in the propagation path or are emitted from an antenna structure.

In contrast, the comprehension of continuous wave propagation may be much more demanding since complicated interference pattern arising from waves of different origin may hide the temporal progression of the interaction phenomena. This may happen if sine wave, random noise, PN-code or chirp signals are emitted. Figure 4.6a demonstrates a simple example in which a band-limited M-sequence of low order is emitted from a point source. Even though we only generated a simple spherical wave and no scatterer is involved, the wave pattern provides a complicated impression which is difficult to interpret. So it is very hard to see that the represented area covers three periods of the sounding signal. A completely confusing field is depicted in Figure 4.7 even if the radiation scenario is only slightly more complicated as in the first example. It shows a snapshot of the field radiated from a long dipole if it is fed by a band-limited M-sequence.

Since we are usually dealing with wideband signals here, a simple trick may help us to trace back the wave phenomena of time-stretched signals to the association of the convenient impulse propagation. For that purpose, we consider the propagation of the 'correlated wave' $\mathfrak{W}_{\mathrm{SP}}(t, \mathbf{r})$ instead of the actual one $W_{\mathrm{SP}}(t, \mathbf{r})$:

$$\mathfrak{W}_{SP}(t, \mathbf{r}) = \frac{W_{\Omega}(t, \mathbf{r}_0)}{r} * \delta\left(t - \frac{r}{c}\right) * x(-t) = \frac{\mathfrak{W}_{\Omega}(t, \mathbf{r}_0)}{r} * \delta\left(t - \frac{r}{c}\right) \tag{4.28}$$

Here, the correlation is expressed by the convolution of the wave with the time-inverted reference signal $x(t)$ Typically, one takes as reference the feeding signal of the transmit antenna or a signal captured at an arbitrary but fixed reference point within the wave field. Figures 4.6b and 4.7b show the 'correlated waves' which are

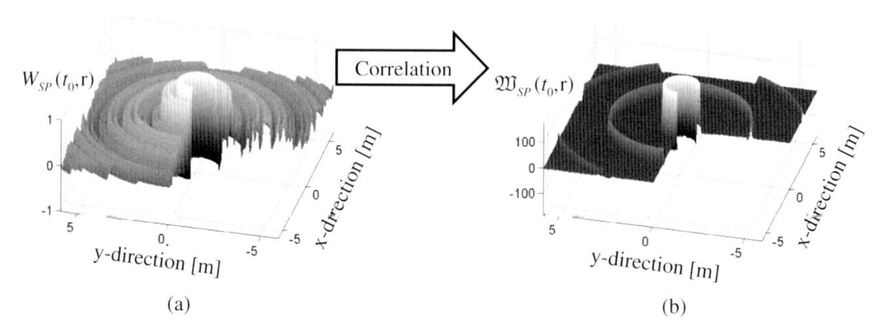

Figure 4.6 Snapshot of a spherical wave generated from a band limited M-sequence of fifth order (a) and the correlated version of this wave (b). The wave around the centre region was gated out in order to suppress the singularity in the proximity of the source. (colored figure at Wiley Homepage: Fig_4_6).

3) In contradiction, the mathematical treatment of such problems directly in time domain is rather complicated.

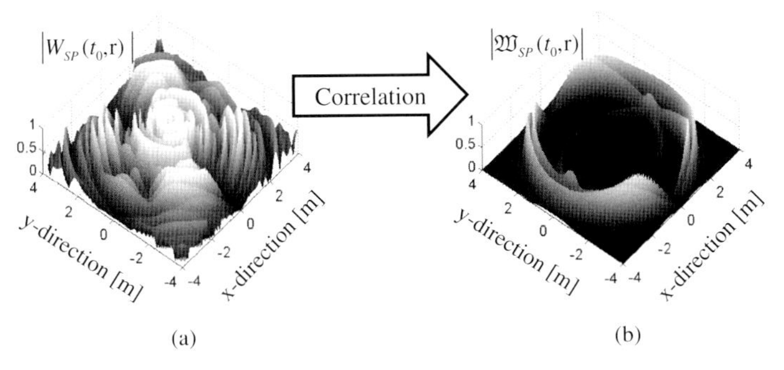

Figure 4.7 Snapshot of the field transmitted from a long dipole fed by a seventh-order M-sequence (a). The pattern on the right side concerns the same wave after correlation with the feeding signal (b). (colored figure at Wiley Homepage: Fig_4_7)

much easier to comprehend. Now, the three periods of a circular wave in Figure 4.6 can be recognized and the typical 'pulse' radiation of a long dipole is also quite visible.

The characteristic spatial width of the pulse-shaped 'correlated wave' is the coherence length λ_{coh} which is related to the coherence time τ_{coh} (refer to Figure 2.13) via

$$\lambda_{\mathrm{coh}} = \tau_{\mathrm{coh}} c \approx \frac{c}{\underset{\cdot\cdot}{B}} \tag{4.29}$$

where $\underset{\cdot\cdot}{B}$ is the bandwidth of the sounding wave and c the wave speed. Correspondingly, we can deal with the rise time of the ACF.

We will still clarify this approach of thinking by considering the generalized block schematic of the measurement arrangement (Figure 4.8a). It is applicable to all ultra-wideband measurements and field simulations of propagation scenarios as well. A sub-nanosecond pulse generator stimulates a spreading filter. This spreading filter provides a signal $x_{\mathrm{sf}}(t)$ which distributes its energy over a large duration and which has the same spectrum as the pulse $x_{\mathrm{p}}(t)$, that is

$$\begin{gathered} x_{\mathrm{sf}}(t) = g_{\mathrm{s}}(t) * x_{\mathrm{p}}(t) \\ \text{with} \quad g_{\mathrm{s}}(t) * g_{\mathrm{s}}(-t) = \delta(t) \text{ and } |\underline{G}_{\mathrm{s}}(f)| = \text{const} \end{gathered} \tag{4.30}$$

$x_{\mathrm{sf}}(t)$ may be a chirp, a M-sequence, a stepped sine, a segment of random noise and so on which is created by the (virtual) spreading filter. This signal is radiated and scattered by objects within the propagation path. As the examples in Figure 4.6 and 4.7 have shown, this leads to non-interpretable wave fields. Finally that peculiar wave is captured resulting in the signal

$$y_{\mathrm{sf}}(t) = g(t) * x_{\mathrm{sf}}(t) \tag{4.31}$$

($g(t)$ is the wanted impulse response of the test scenario) and the time spreading is again removed by a compression filter which has the same but time-inverted pulse response as the spreading filter $g_{\mathrm{c}}(t) = g_{\mathrm{s}}(-t)$, that is, we visualise

$$y_{\mathrm{p}}(t) = y_{\mathrm{sf}}(t) * g_{\mathrm{c}}(t) \tag{4.32}$$

at the screen of our measurement device.

Consequently, the scenario under test actually involves the stretched signals $x_{sf}(t)$ and $y_{sf}(t)$ while the interpretable signals for visualisation and comprehension are $x'_p(t)$ and $y_p(t)$. Writing down the transmission behavior of the complete chain, we end up in

$$\begin{aligned} y_p(t) &= g_c(t) * g(t) * g_s(t) * x_p(t) = g(t) * \{g_c(t) * g_s(t) * x_p(t)\} \\ &= g(t) * x'_p(t) \\ x'_p(t) &= x_p(t) \quad \text{since} \quad g_c(t) * g_s(t) = \delta(t) \end{aligned} \tag{4.33}$$

Thus, the final result is identical to the popular use of pulse excitation in the ultra-wideband technique. Here, we made use of the commutative law of the convolution. Following (4.33), the insertion of spreading and compression filters is theoretically without any effect and hence apparently senseless. However, there are several practical reasons which make the insertion of these (virtual) filters very useful and advantageous. Basically, all measurement approaches introduced in Chapter 3 dealing with time spread signals may be considered on this base.

Let us come back to the hardly interpretable wave propagation of stretched wideband signals. Thanks to the commutative law of convolution, we can change (at least in our minds) the order of the sub-systems of the measurement channel as depicted in Figure 4.8b. Now, we have again the easily interpretable pulse excitation

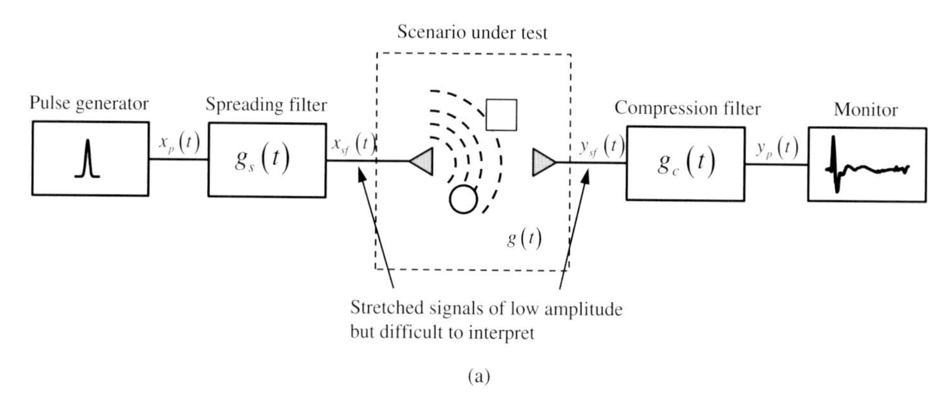

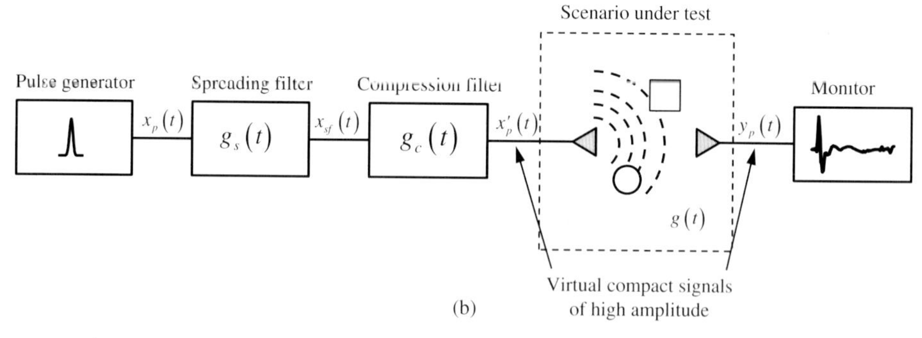

Figure 4.8 Generalized block diagram of ultra-wideband measurements (a) and the schematic with permuted order of blocks (b) leading to the same measurement result in case of linear and time invariant test scenarios.

and propagation within our test scenario (compare Figures 4.6b and 4.7b on which this approach was applied).

We will mainly deal with the propagation of pulsed waveform since this corresponds to our natural understanding of wave propagation. But as we have shown, this will not exclude any other wideband waveform because we can always convert the signal shape via correlation (filtering) as long as the phenomena of interaction behaves linearly with respect to the wave magnitude and time invariant over the duration of data capturing. Therefore, we can simply apply parameters of pulse wave propagation also for arbitrary-shaped waves if we refer to their correlation function. That is, the above-mentioned conditions to approximate a spherical wave by a planar one are generally valid provided that in (4.19), (4.22) and (4.25) comparable parameters of the correlation function are taken, that is the width of the correlation peak and its rise time.

4.4 Time Domain Characterization of Antennas and the Free Space Friis Transmission Formula

4.4.1 Introduction

The goal of this chapter is to find a formal description of the antenna which is convenient and which meets the specific challenges of ultra-wideband high-resolution short-range sensing. The antenna is part of the test scenario and will therefore affect its behavior. Particularly in the case of short-range sensing, the interaction between the targets and the antennas by multiple scattering cannot be excluded. Further, the dimensions of the antennas exceed the achievable resolution of an UWB radar. In narrowband radar, such aspects are of minor importance since the antennas are typically smaller then the radar resolution and the distances between antenna and target are large enough so that mutual interactions may be neglected. In order to meet the specifics of UWB sensing, we have to extend and refine the traditional antenna models and antenna descriptions since they were developed for narrowband problems. Here, we will do this by an intuitive way using the basic concepts of signal and system theory from Chapter 2 which we extend over aspects of free space propagation using the simple wave models introduced above. For additional discussions on the topic of time domain antenna characterization the interested reader may refer to, for example Refs [4–8]. An overview about UWB antenna principles and implemented examples are given in Refs [9–12].

In order to abstract from a specific antenna geometry, we characterize the antenna behavior by a black box model as it was discussed in Chapter 2. Black box models use impulse response functions (or frequency response function) to characterize the systems which they model. That is, they consider the reaction of a system (output signal) as a consequence of stimulation (input signal) and the systems behavior (impulse response). This is symbolized in Figure 4.9 for an antenna in

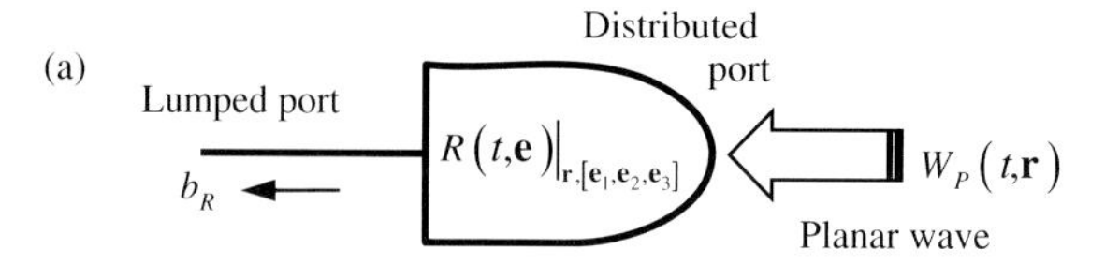

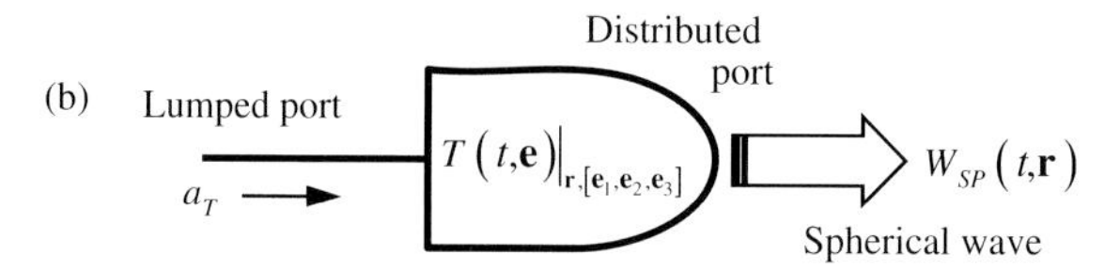

Figure 4.9 Antenna symbolized as black box model in receive mode (a) and transmit mode (b).

receive and transmit mode. There, the antenna behavior is specified by the impulse response functions $R(t, \mathbf{e})|_{\mathbf{r},[\mathbf{e}_1,\mathbf{e}_2,\mathbf{e}_3]}$ assigning the receive mode and $T(t, \mathbf{e})|_{\mathbf{r},[\mathbf{e}_1,\mathbf{e}_2,\mathbf{e}_3]}$ which refers to the transmit mode. Besides the time t, these impulse response functions depend on the propagation direction $\mathbf{e}$ of the incident and emanated wave, respectively. Furthermore, antenna position and orientation have to be quantified which we symbolize by the subscript $\mathbf{r}$, $[\mathbf{e}_1, \mathbf{e}_2, \mathbf{e}_3]$. The approach to assign the electro-dynamic behavior of an object by an impulse response can always be applied as long as the objects of interest behave linear and time invariant. Since antennas meet these conditions, there are no restrictions to apply this method formally also to them.

However, the black box models considered so far were only describing systems with lumped ports (refer to Sections 2.4 and 2.5). This bounds the signals to linear structures as waveguides and cables. These restrictions cannot be sustained anymore in scenarios with objects arbitrarily located in 3D-space. Hence, we have to extend appropriately our black box model by parameters which include also spatial aspects. Formally spoken, an antenna features two basic characteristics:

1) It is a rigid body of finite dimension. Corresponding to our classification according to Figure 4.2, we may number typical UWB antennas amongst small- and medium-sized objects.
2) It is able to convert guided waves at a lumped port to free waves emanating the distributed port and vice versa. This involves that the conversion may be dependent on the direction $\mathbf{e}$ of wave propagation.

These conditions must be appropriately and generically respected by our model. Figure 4.10 symbolizes an arbitrary-shaped body, which is able to convert bounded waves into free waves and vice versa. In order to abstract from its individual geometric shape and structure, we assign all related electric properties of that body to a reference point P_A (pointed by position vector $\mathbf{r}$). In principle, its location may be arbitrarily selected within the antenna body (or even outside the antenna body – see Figure 4.28). However, usually one tries to select

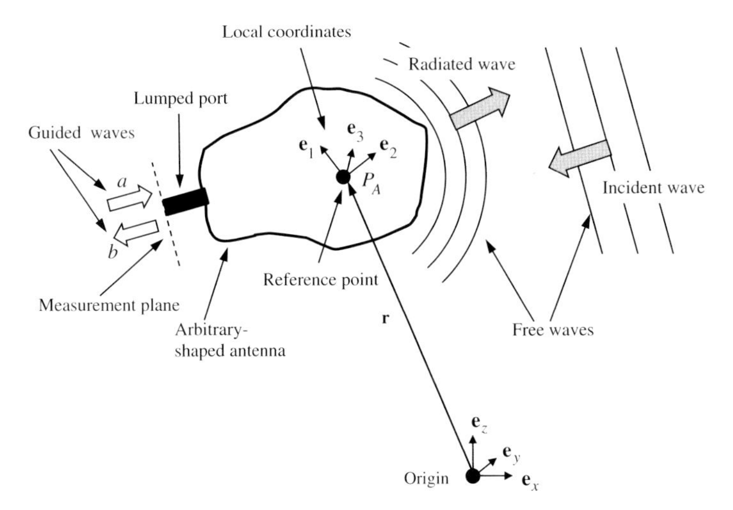

Figure 4.10 Arbitrary-shaped body which is able to convert guided electromagnetic waves to freely propagating ones and vice versa.

distinguished points such as centres of symmetry or the centre of radiation (it will be introduced below – Figure 4.32). The antenna orientation in space is defined by the local coordinate system $[\mathbf{e}_1, \mathbf{e}_2, \mathbf{e}_3]$. It can be related by Euler angles to the global one $[\mathbf{e}_x, \mathbf{e}_y, \mathbf{e}_z]$ (see Annex A.4). Basically, the local coordinates can be arbitrarily selected but usually one is geared to distinguished directions of the antenna and respects an orthogonal right-hand system, that is $\mathbf{e}_1 \cdot \mathbf{e}_2 = 0$; $\mathbf{e}_3 = \mathbf{e}_1 \times \mathbf{e}_2$.

We will introduce a simple antenna model based on IRFs (or complex-valued FRFs) which meets the requirements of ultra-wideband sensing over short distances. The modelling of the antenna by a virtual point is important inasmuch as most of ultra-wideband sensing tasks such as ranging, imaging, tracking and so on suppose point-to-point relations. However, one should be aware of the simplifications of this simple model. In fact, the model supposes sufficient distance between antenna and observation point. That means, seen from the observer, the antenna body should appear only within a small solid angle. The distances in short-range sensing are typically quite short by assumptions. Consequently, the geometric size of the antennas must be sufficiently small in order to fulfil the modelling assumptions with an acceptable error bound. Some effects of antenna size will be discussed in Section 4.6.

The antenna IRFs are sensitive to the selection of the reference point and the local coordinate system. Above, we expressed this by the subscript $\mathbf{r}$, $[\mathbf{e}_1, \mathbf{e}_2, \mathbf{e}_3]$. In order to shorten the notation, we will replace it simply by the symbol of the reference point P_A in what follows having in mind that this involves both, the definition of the antenna location and orientation, that is

$$T(t, \mathbf{e})|_{\mathbf{r},[\mathbf{e}_1,\mathbf{e}_2,\mathbf{e}_3]} \rightarrow T(t, \mathbf{e})|_{P_A};\ R(t, \mathbf{e})|_{\mathbf{r},[\mathbf{e}_1,\mathbf{e}_2,\mathbf{e}_3]} \rightarrow R(t, \mathbf{e})|_{P_A}; \quad \text{etc.}$$

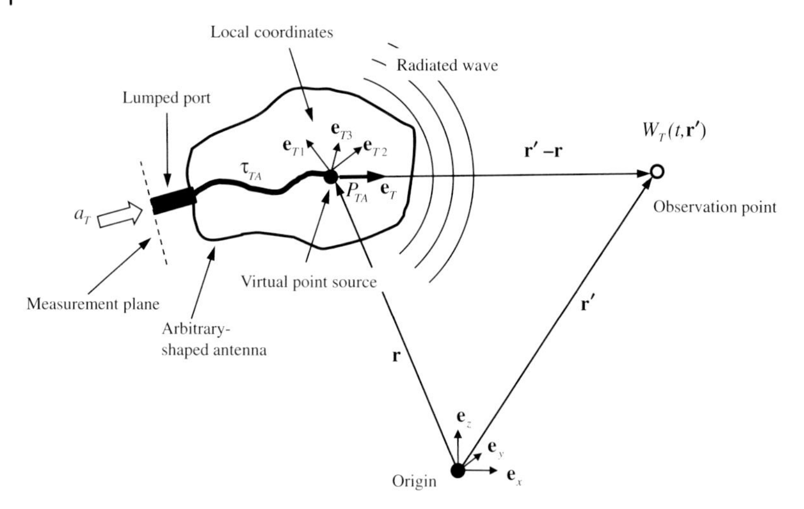

Figure 4.11 Antenna in transmit mode.

4.4.2 Antenna as Transmitter

Figure 4.11 depicts a body of arbitrary geometry acting as transmit antenna. It is enclosed by an isotropic propagation medium. The antenna is fed via the lumped port with electromagnetic energy which will be at least partially distributed in free space. How this energy is distributed, that is if there are preferential directions or frequencies, depends on the geometry and the material composition of the antenna.

As just mentioned, we will replace the actual radiating construction by a virtual point source having identical radiation properties (above a certain distance from the antenna) as the real structure. We consider this virtual point as source of the radiated waves. Its location is assigned by the position vector $\mathbf{r}$ and we give it also an orientation in space defined by the local coordinate system $[\mathbf{e}_{T1}, \mathbf{e}_{T2}, \mathbf{e}_{T3}]$.

In our model, the virtual point source is fed from a lumped port. The only possibility to capture signals by measurement devices is given at this port. Its position in space is typically not identical to the virtual point source. Hence, we have to account for a signal delay τ_{TA} between the measurement plane and the virtual radiation point P_{TA}.

Since we know that the antenna is a linear device, we can formally establish a convolution relation between the signal $a_T(t)$ at the measurement plan and the wave field $W_T(t, \mathbf{r}')$ in the observation point $\mathbf{r}'$. Basically, the input signal may be a current, a voltage or a normalized wave [13]. For demonstration, we will apply the last one here. We suppose an antenna of small dimensions, so we can approximate the radiated field by a quasi-spherical wave (assuming the observation point is far enough from the antenna – for details see Section 4.6.5). We will understand, under the term quasi-spherical wave, an electromagnetic field

- which springs from a source of finite dimensions,
- which expands with equal speed in all directions of space,
- which may have a variable strength in dependence on the propagation direction and
- which has a wavefront whose shape may slightly deviate from an expanding sphere.

To model the field $W_T(t, \mathbf{r}')$ in the observation point, we can establish a transmission channel consisting of

- the signal source providing the normalized wave $a_T(t)$ feeding the antenna,
- the antenna featuring the angle-dependent IRF $T(t, \mathbf{e}_T)$ or FRF $\underline{T}(f, \mathbf{e}_T)$ and
- an isotropic spreading of field energy.

By this way, we are able to join the point radiator which provides an ideal spherical wave (compare Figure 4.4) with the real antenna whose deviations from an isotropic power distribution and spherical wavefront are assigned to the transmit response $T(t, \mathbf{e}_T)$ or $\underline{T}(f, \mathbf{e}_T)$. Following our consideration above ((4.16) and (4.17)), we can establish a convolution relation for time and frequency domain:

$$\begin{aligned} W_T(t, \mathbf{r}') &= \underbrace{\frac{\delta(t - (|\mathbf{r}' - \mathbf{r}|/c))}{|\mathbf{r}' - \mathbf{r}|}}_{\text{propagation term}} * \underbrace{T(t, \mathbf{e}_T)|_{P_{TA}} * \delta(t - \tau_{TA})}_{\text{antenna behavior}} \underbrace{* a_T(t)}_{\text{stimulus}} \\ &= \frac{T(t, \mathbf{e}_T)|_{P_{TA}} * a_T(t - (\Delta r/c) - \tau_{TA})}{\Delta r}; \quad \Delta r = |\mathbf{r}' - \mathbf{r}| \end{aligned} \tag{4.34}$$

$$\underline{W}_T(f, \mathbf{r}') = \frac{\underline{T}(f, \mathbf{e}_T)|_{P_{TA}} \underline{a}_T(f) e^{-j(2\pi f \tau_{TA} + k\,\Delta r)}}{\Delta r} \tag{4.35}$$

where in contrast to (4.16) and (4.17) the source terms are angle dependent:

$$\begin{aligned} W_\Omega(t, \mathbf{r}, \mathbf{e}_T) &= T(t, \mathbf{e}_T)|_{P_{TA}} * \delta(t - \tau_{TA}) * a_T(t) \\ \underline{W}_\Omega(f, \mathbf{r}, \mathbf{e}_T) &= \underline{T}(f, \mathbf{e}_T)|_{P_{TA}} \underline{a}_T(f) e^{-j(2\pi f \tau_{TA})} \end{aligned}$$

Here, we are mainly interested in $T(t, \mathbf{e}_T)|_{P_{TA}}[1/s]$ and $\underline{T}(f, \mathbf{e}_T)|_{P_{TA}}$ [dimensionless] . They represent the IRF and FRF of an antenna in transmit mode related to the virtual point source P_{TA} (including position and orientation). In a homogenous environment, the field in the observation point is not dependent on the actual position in space $\mathbf{r}$; it is rather the distance $\Delta r = |\mathbf{r}' - \mathbf{r}|$ between virtual point source and observation point which is of interest. Additionally, the radiated field registered by the observer usually depends on the aspect angle under which he sees the antenna. This is expressed by the argument $\mathbf{e}_T$ in the impulse/frequency response function representing the unit direction vector $\mathbf{e}_T = (\mathbf{r}' - \mathbf{r})/|\mathbf{r}' - \mathbf{r}|$ pointing in observation direction. Figure 4.12 shows a measurement example exhibiting the angular dependence of the radiation behavior. In our initial black box model of Figure 4.9, we can express the angular dependence by replacing the distributed port by a multi-channel output in which every channel stands for a specific propagation direction.

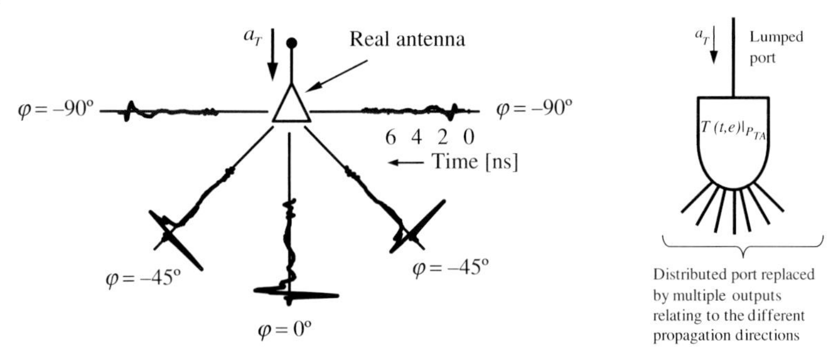

Figure 4.12 Illustration of the angle-dependent impulse response $T_{P_{\mathrm{TA}}}(t, \mathbf{e}_{\mathrm{T}})$ of an antenna in transmitting mode and its description by a SiMo-system (single input–multiple output).

4.4.3 Antenna as Receiver

Also here, we will assign the capability of an antenna to convert freely propagating waves into guided ones to a single point P_{RA} (see Figures 4.10 and 4.13). From this point, the collected wave energy propagates to the lumped antenna port where it can be captured by a measurement device. Due to the assumed linearity of the antenna, we can again formally establish a convolution relation between the exciting field $W_{\mathrm{P}}(t, \mathbf{r})$ and the antenna response $b_{\mathrm{R}}(t)$, in which the antenna behavior is modelled by the IRF $R(t, \mathbf{e})|_{P_{\mathrm{RA}}}$ and FRF $\underline{R}(f, \mathbf{e})|_{P_{\mathrm{RA}}}$:

$$b_{\mathrm{R}}(t) = R(t, \mathbf{e}_{\mathrm{R}})|_{P_{\mathrm{RA}}} * \delta(t - \tau_{\mathrm{RA}}) * W_{\mathrm{p}}(t, \mathbf{r}) \tag{4.36}$$

$$\underline{b}_{\mathrm{R}}(f) = \underline{R}(f, \mathbf{e}_{\mathrm{R}})|_{P_{\mathrm{RA}}} \underline{W}_{\mathrm{p}}(f, \mathbf{r}) \mathrm{e}^{-j2\pi f \tau_{\mathrm{RA}}} \tag{4.37}$$

As long as the incident waves can be considered (at least locally) as planar and the propagation medium is homogeneous, the antenna properties do not depend on the antenna position $\mathbf{r}$ but they may vary as function from the aspect angle under which the antenna 'sees' the incident wave. This, we have expressed again by the

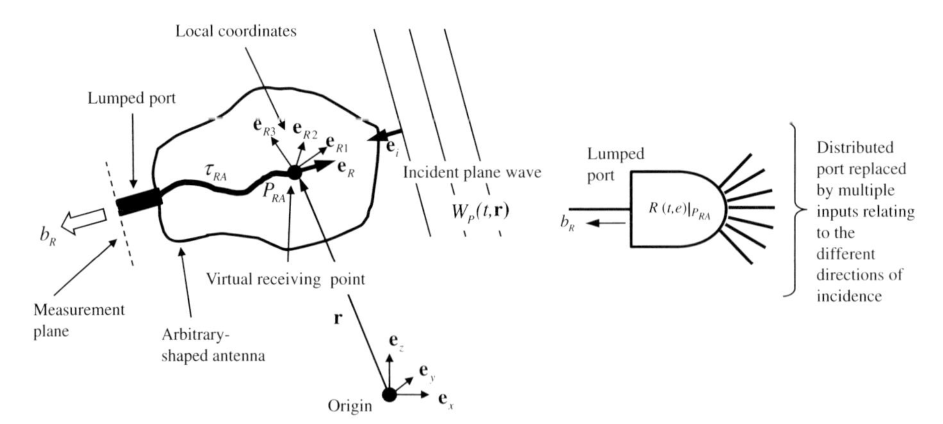

Figure 4.13 Antenna in receiving mode and its modelling by a MiSo-system.

unit vector $\mathbf{e}_R = -\mathbf{e}_i$ indicating the direction of incidence. In accordance to the transmission mode black box model, we respect the angular dependence by multiple input channels relating to the different direction of incidence (Figure 4.13). The unit of $R(t, \mathbf{e})$ is [m/s] and it is [m] for $\underline{R}(f, \mathbf{e})$ wherefore the FRF is often denoted as normalized antenna height.

4.4.4
Transmission Between Two Antennas – The Scalar Friis Transmission Formula

After having defined the characteristic functions of an antenna, we are able to establish the relation between stimulus $a_1(t)$ of the antenna 1 and receive signal $b_2(t)$ of the antenna 2 and vice versa for a scenario as depicted in Figure 4.14.

Following (4.34) (or (4.35)), the guided wave $a_1(t)$ which feeds antenna 1 creates an electromagnetic field at position $\mathbf{r}_2$ of

$$W_T(t, \mathbf{r}_2) = \frac{T_1(t, \mathbf{e}_1)|_{P_1} * a_1(t - (\Delta r/c) - \tau_1)}{\Delta r} \quad \text{with } \Delta r = |\mathbf{r}_2 - \mathbf{r}_1| \tag{4.38}$$

With the help of antenna 2, this field will be converted back into a guided wave $b_2(t)$ so it can be captured by a measurement device. If the curvature of the free wave caused from antenna 1 is negligible, (4.36) leads to

$$\begin{aligned} b_2(t) &= R_2(t, \mathbf{e}_2)|_{P_2} * W_T(t - \tau_2, \mathbf{r}_{RA}) \\ &= \frac{1}{\Delta r} T_1(t, \mathbf{e}_1)|_{P_1} * R_2(t, -\mathbf{e}_1)|_{P_2} * a_1\left(t - \frac{\Delta r}{c} - \tau_1 - \tau_2\right) \quad \text{since} \quad \mathbf{e}_2 = -\mathbf{e}_1 \end{aligned} \tag{4.39}$$

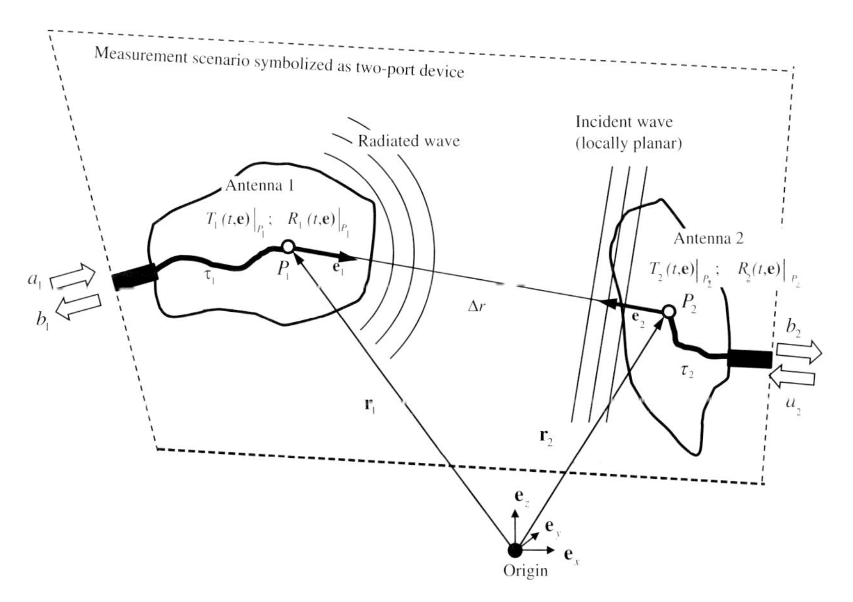

Figure 4.14 Direct transmission between two antennas symbolized by a two-port system. The example refers to the transmission from antenna 1 to antenna 2.

Corresponding holds for time harmonic fields:

$$\underline{b}_2(f) = \frac{\underline{T}_1(f, \mathbf{e}_1)|_{P_1} \underline{R}_2(f, -\mathbf{e}_1)|_{P_2}}{\Delta r} \underline{a}_1(f) \mathrm{e}^{-j2\pi f(\tau_1+\tau_2+\Delta r/c)} \tag{4.40}$$

These equations describe the transmission behavior between two arbitrary placed antennas if their mutual distance is larger than a minimum value given either by (4.22) or (4.137). They indicate that the strength of the receive signal decreases inversely with increasing antenna distance. It should be mentioned that the equation is not yet complete since we omitted the vector character of the electromagnetic field; that is, effects of antenna polarization are not respected. We will catch up with this in Section 5.5. But (4.40) is just sufficient for exact distance determination between two antennas via travelling time measurement as it is necessary, for example, for localization purposes.

For the transmission in inverse direction, we can follow the same considerations as we did in (4.39). Furthermore, we can state reciprocity[4)] for our two-port system so that the transmission in forward and backward direction must be identical, that is $S_{21}(t) = S_{12}(t)$ (see (2.214) and Annex C.1). Hence, we can write

$$\begin{aligned} T_1(t, \mathbf{e}_1)|_{P_1} * R_2(t, -\mathbf{e}_1)|_{P_2} &= T_2(t, -\mathbf{e}_1)|_{P_2} * R_1(t, \mathbf{e}_1)|_{P_1} \\ \underline{T}_1(f, \mathbf{e}_1)|_{P_1} \underline{R}_2(f, -\mathbf{e}_1)|_{P_2} &= \underline{T}_2(f, -\mathbf{e}_1)|_{P_2} \underline{R}_1(f, \mathbf{e}_1)|_{P_1} \end{aligned} \tag{4.41}$$

This equation represents a general reciprocity relation which is valid for an arbitrary antenna combination. But we may also use (4.41) to deduce the self-reciprocity of any antenna. We only need to look for a reference antenna from which we know the self-reciprocity in order to gain via (4.41) the self-reciprocity of an arbitrary antenna type [13].

The Hertzian dipole is such a radiator since it can be modelled by analytical expressions. For notations using normalized waves, the relation between transmit and receive behavior of a short dipole is given by

$$T_{\mathrm{HD}}(t, \mathbf{e}) = \frac{1}{2\pi c} \frac{\partial}{\partial t} R_{\mathrm{HD}}(t, -\mathbf{e}) = \frac{1}{2\pi c} u_1(t) * R_{\mathrm{HD}}(t, -\mathbf{e}) \tag{4.42}$$

Relation (4.42) is discussed more deeply in Section 5.5. Inserting this equation into (4.41) gives us finally the self-reciprocity relation of any passive radiator ($u_1(t)$ – represents the unit doublet – see Annex A.2)

$$T(t, \mathbf{e}) = \frac{1}{2\pi c} R(t, -\mathbf{e}) * \frac{\partial}{\partial t} \cdots = \frac{1}{2\pi c} R(t, -\mathbf{e}) * u_1(t) * \cdots \tag{4.43}$$

4) This relation is generally valid. It does not suppose empty space or far-field conditions. Exceptions are if non-linear materials, plasma or biased ferromagnetic materials are placed in the propagation path of the waves.

whose frequency domain counterpart is

$$\underline{T}(f, \mathbf{e}) = j\frac{\underline{R}(f, -\mathbf{e})}{\lambda} \tag{4.44}$$

The reciprocity relation (4.43) tells us that any antenna has, for receive and transmit mode, identical angular characteristics but the involved signals are distinguished by their time derivation. The notation of the derivation operator in (4.43) shall indicate that it can be applied to any time function involved in the transmission chain which is due to the commutative law of the convolution operation (see (4.45) for example).

Applying antenna reciprocity, (4.39) can be modified into a description containing only receive or transmit functions. Equation (4.45) gives an example in which the involved antennas are modelled solely by their receiving characteristics.

$$b_2(t) = \frac{1}{2\pi c\, \Delta r} R_1(t, \mathbf{e}_1)|_{P_1} * R_2(t, -\mathbf{e}_1)|_{P_2} * u_1(t) * a_1\left(t - \frac{R}{c} - \tau_1 - \tau_2\right) \tag{4.45}$$

It is the so-called time domain Friis transmission formula.

The above introduced antenna functions $R(t, \mathbf{e})$ and $T(t, \mathbf{e})$ are different from the commonly used forms of (narrowband) antenna description. We will shortly bridge the gap to those characteristics here. Narrowband characteristics are usually based on spectral power relations. The antenna (realized) gain $G(f, \mathbf{e})$ is the quantity which summarizes the transmitting properties by a comparable philosophy as we used to define the complex-valued antenna FRF $\underline{T}(f, \mathbf{e})$. It relates the spectral and areal power density $\overline{\mathfrak{P}}(f, \mathbf{r}')$ of the transmitted wave in the observation point $\mathbf{r}'$ to the power which is accepted by the considered antenna. Supposing a perfectly matched antenna, the accepted spectral power equals the injected one, that is $\Phi_{\text{in}}(f) = \underline{a}(f)\underline{a}^*(f)$. Using (4.6), (4.35) and placing the antenna at canonical position (i.e. $\mathbf{r} = \mathbf{0}$), we gain the relation (for details on antenna gain and related topics, see Eq. (4.103))

$$G(f, \mathbf{e}) = 4\pi r^2 \frac{\overline{\mathfrak{P}}(f, \mathbf{r})}{\underline{a}(f)\, \underline{a}^*(f)} = 4\pi\, \underline{T}(f, \mathbf{e})\, \underline{T}^*(f, \mathbf{e}) \quad \text{with} \quad \mathbf{e} = \frac{\mathbf{r}}{r} \tag{4.46}$$

In the receiving mode, we described the antenna by $\underline{R}(f, \mathbf{e})$ which has a measurement unit of [m] so that the squared version for the intended power relation has the unit of an area. Thus, one attributes an area to the antenna – called effective aperture A – over which the antenna captures the wave energy. It is interesting to note that this concept is applicable also for antennas which are not dominated by any area, for example, dipoles. Supposing the antenna is located at position $\mathbf{r}$, the available spectral power at its feeding port is

$$\Phi_{\text{out}}(f) = \underline{b}(f)\underline{b}^*(f) = A(f, \mathbf{e})\,\overline{\mathfrak{P}}(f, \mathbf{r}) \tag{4.47}$$

Compared with (4.37), this gives

$$A(f, \mathbf{e}) = \underline{R}(f, \mathbf{e})\underline{R}^*(f, \mathbf{e}) \tag{4.48}$$

Hence, the reciprocity relation for power quantities is given by

$$A(f,\mathbf{e}) = \frac{\lambda^2}{4\pi} G(f,\mathbf{e}) \tag{4.49}$$

while the original Friis transmission formula results from (4.40) to

$$\begin{aligned}\Phi_2(f) &= \frac{G_1(f,\mathbf{e})A_2(f,-\mathbf{e})}{4\pi\,\Delta r^2}\Phi_1(f) = \frac{G_1(f,\mathbf{e})G_2(f,-\mathbf{e})\lambda^2}{(4\pi\,\Delta r)^2}\Phi_1(f) \\ &= \frac{A_1(f,\mathbf{e})A_2(f,-\mathbf{e})}{(\lambda\,\Delta r)^2}\Phi_1(f)\end{aligned} \tag{4.50}$$

Note that for many purposes of ultra-wideband short-range sensing, the classical method to characterize antenna by power quantities is not sufficient and will lead to information losses in the data interpretation.

4.5
Indirect Transmission Between Two Antennas – The Scalar Time Domain Radar Equation

The indirect transmission between two antennas is of major interest for most of UWB sensor applications. It requires an object within the propagation path of the waves which scatters the waves towards one or more receiving antennas. This type of radio wave transmission is used in sensing technique to gain information about the scattering object. It is usually referred to as a radar measurement. If transmitting and receiving antennas are identical and at fixed position in space, one speaks of mono-static radar. Correspondingly bi-static radar is composed of separate antennas for transmitting and receiving, and multi-static radar, radar array or MiMo radar refers to an arrangement of a number of antennas. We will deal only with bi-static arrangements to deduce the basic relations for UWB radar. The assignment of these results to mono- or multi-static configurations is straightforward. We will consider two idealized but fundamental scenarios – wave scattering at a flat interface and scattering at small objects.

4.5.1
Wave Scattering at Planar Interfaces

The scattering scenario based on infinite plane and bi-static antenna arrangement is illustrated in Figure 4.15. Antenna 1 generates an electromagnetic field with a quasi-spherical wavefront travelling towards the plane. It acts as a mirror and reflects back the incident wave so that antenna 2 can capture it.[5] Except the mirroring, the reflected wave keeps its quasi-spherical wavefront so that we can simply

5) Since an antenna typically radiates in all directions, antenna 2 will also receive the direct wave. We will omit this wave here since it can be often suppressed by time gating due to the shorter propagation time.

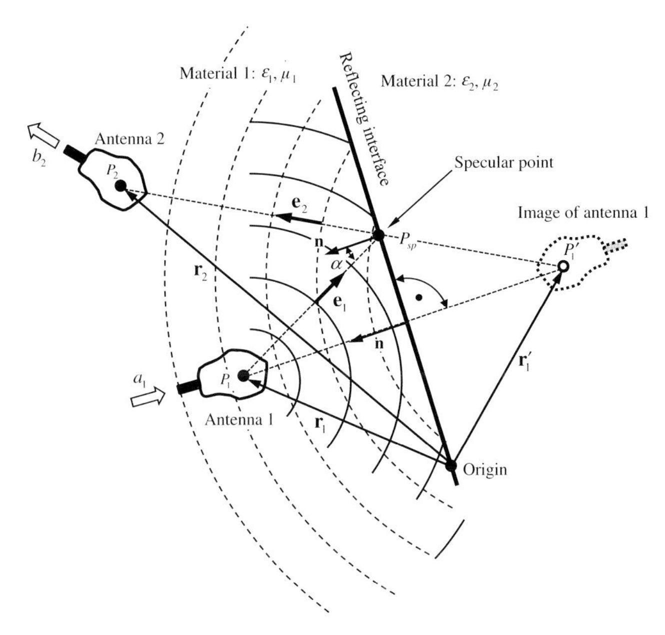

Figure 4.15 Illustration of spherical wave scattering at an infinite planar interface between two lossless materials.

model it by a wave emanating from a virtual source at the opposite side of the interface. We call this source the image antenna.

If we place the origin of the coordinate system at the reflecting plane, the image antenna is located at ($\mathbf{r}_1$ is the position of transmitting antenna; $\mathbf{n}$ is the normal vector of the scattering plane; $\mathbf{T}_{\mathrm{H}}^{(\mathbf{n})} = \mathbf{I} - 2\mathbf{n}\mathbf{n}^T$ – Householder matrix – refer to Annex A.6)[6)]

$$\mathbf{r}_1' = \mathbf{r}_1 - 2\mathbf{n}\left(\mathbf{r}_1 \cdot \mathbf{n}\right) = \mathbf{T}_H^{(\mathbf{n})}\,\mathbf{r}_1 \tag{4.51}$$

Hence, the virtual antenna distance is ($\mathbf{r}_2$ is the position of the receiving antenna)

$$\Delta r = \left|\mathbf{r}_2 - \mathbf{r}_1'\right| = |\Delta\mathbf{r}| \tag{4.52}$$

and the location $\mathbf{r}_{\mathrm{sp}}$ of the specular point P_{sp} equals the intersection of $\Delta\mathbf{r}$ and the reflection plane, that is

$$\mathbf{r}_{\mathrm{sp}} = \mathbf{r}_1' + \lambda\,\Delta\mathbf{r} \tag{4.53}$$

$\mathbf{r}_{\mathrm{sp}}$ is located within the scattering plane, so that we can write

$$\mathbf{n} \cdot \mathbf{r}_{\mathrm{sp}} = 0 \Rightarrow \lambda = \frac{\mathbf{r}_1' \cdot \mathbf{n}}{\Delta\mathbf{r} \cdot \mathbf{n}} \tag{4.54}$$

6) Expressing the dot product by matrix notation $\mathbf{n} \cdot \mathbf{r}_1 = \mathbf{n}^T\mathbf{r}_1$, we can rewrite (4.51) as $\mathbf{r}_1' = \mathbf{r}_1 - 2\mathbf{n}\mathbf{n}^T\mathbf{r}_1 = \left(\mathbf{I} - 2\mathbf{n}\mathbf{n}^T\right)\mathbf{r}_1$.

Since the curvature of the wavefront of the transmitted wave is not affected by scattering at a planar interface, we can apply the Friis transmission formula to describe the transmission from the image antenna to antenna 2. However, we have to respect the effect of the interface on the wave magnitude. We will do this by introducing the reflection coefficient $\Gamma(\mathbf{e}_1 \cdot \mathbf{n})$ which depends on the angle of incidence α and the material parameters (see Annex C.2 and Ref. [14] for details). The reflection will not affect the time evolution of the scattered wave if the material properties are frequency independent. Thus, the signal captured from antenna 2 can be modelled by

$$b_2(t) = \frac{\Gamma(\mathbf{e}_1 \cdot \mathbf{n})}{2\pi c\,\Delta r} R_1(t, \mathbf{e}_1)|_{P_1} * R_2(t, -\mathbf{e}_2)|_{P_2} * u_1(t) * a_1\left(t - \frac{\Delta r}{c} - \tau_1 - \tau_2\right) \tag{4.55}$$

Herein, the direction of incidence $\mathbf{e}_1$ and of reflection $\mathbf{e}_2$ are related by

$$\mathbf{e}_2 = \mathbf{e}_1 - 2(\mathbf{e}_1 \cdot \mathbf{n})\mathbf{n} = \mathbf{T}_{\mathrm{H}}^{(\mathbf{n})}\,\mathbf{e}_1 = \mathbf{e}_1 - 2\cos\alpha\,\mathbf{n} \tag{4.56}$$

(4.55) represents the radar equation for scattering at a plane surface and (4.56) is nothing but the vector notation of the well-known reflection law – the angle of incident equals the reflection angle. The length Δr of the propagation path is given by the distance between the points $P_1 \to P_{\mathrm{sp}}$ plus $P_{\mathrm{sp}} \to P_2$. It is identical with the distance $P_1' \to P_2$.

It should be noted that the simple transmission model (4.55) relates to point sources which will be more or less violated if large antennas and short distances are involved (see Section 4.6.5). Furthermore, a rough surface will disturb the reflection law (4.56) so that waves from different incident angles may achieve the receive antenna. Figure 4.16 illustrates the diffuse reflection at such an interface. Since the length of the propagation paths is different, the superposition of all wave components by antenna 2 will destroy the original time evolution of the transmitted signal in dependence of the directional characteristic of both antennas $(T_1(t, \mathbf{e}_1), R_2(t, -\mathbf{e}_2))$ and the surface $(\Gamma(t, \mathbf{e}_1, \mathbf{e}_2))$.

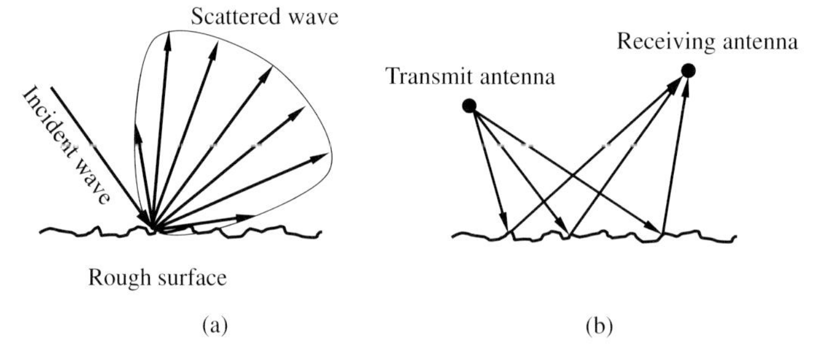

Figure 4.16 Reflection at a rough surface. (a) Directional diagram of the reflection. The length of the arrows relates to the intensity of the reflection. The reflection coefficient depends on both the direction of incidence $\mathbf{e}_1$ and the observation direction $\mathbf{e}_2$ and it will become dependent on time t due to the different length of the propagation paths. (b) Some of the transmission paths between both antennas.

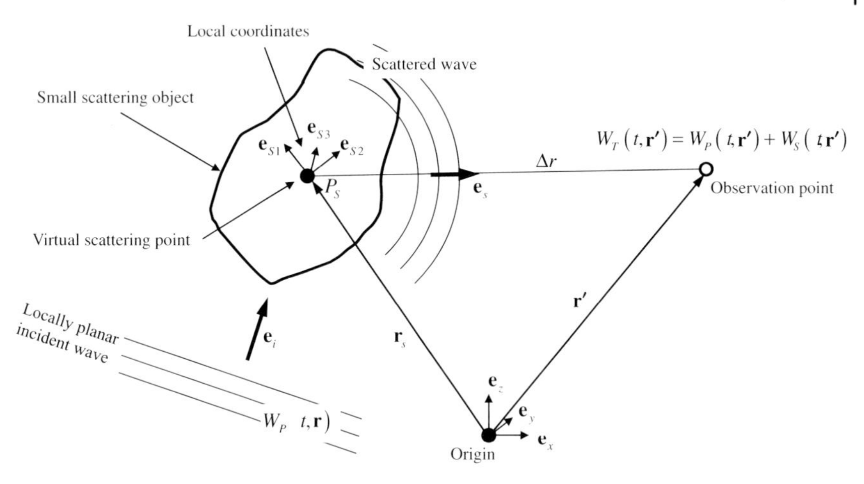

Figure 4.17 Scattering at a small object (directions are fixed corresponding the forward scattering alignment convention [15]).

4.5.2
Wave Scattering at Small Bodies

We will refer to a scenario as depicted in Figure 4.17. A transmitting antenna (not shown in the figure) generates an electromagnetic sounding field $W_P(t, \mathbf{r})$. It is placed sufficiently far away from the scattering object. Thus, we can consider it as locally planar when it bounces against the body which is localized at position $\mathbf{r}_S$. Depending on its properties, the illuminated body will partially or totally reradiate the energy collected from the incident field, and will shadow some regions of space from that wave. Since by assumption the object is small, the scattered field behaves like a quasi-spherical wave which we can model as the radiated field of a transmitting antenna. The only difference is that the scatterer is not 'fed' by a lumped port but rather by a freely propagating wave.

We are interested in the field at the observation point $\mathbf{r}'$. The total field $W_T(t, \mathbf{r}')$ in that point can be pictured as the superposition of the incident field of the planar wave and the field scattered by the object

$$W_T(t, \mathbf{r}') = W_P(t, \mathbf{r}') + W_S(t, \mathbf{r}') \tag{4.57}$$

where we are only attracted to the scattered one $W_S(t, \mathbf{r}')$. Due to the linear behavior of the scatterer with respect to the electromagnetic field, we can formally model the causal chain by convolving the incident wave with an impulse response function describing the target reaction, as we know it from antenna modelling. At this we 'shrink' again the actual body to a point which we assign the scattering properties. For that purpose, we have to fix the reference point and orientation of the considered object as we did before with the antennas. The corresponding relations are closely related to (4.34) or (4.35), so that the quasi-spherical wave at the observation point $\mathbf{r}'$ may be expressed as

$$W_S(t,\mathbf{r}') = \underbrace{\frac{\delta(t-(\Delta r/c))}{\Delta r}}_{\text{propagation term}} * \underbrace{\Lambda(t,\mathbf{e}_i,\mathbf{e}_s)|_{P_S} * W_p(t,\mathbf{r}_s)}_{\text{"source" term}} = \frac{\Lambda(t,\mathbf{e}_i,\mathbf{e}_s)|_{P_S} * W_p(t-\Delta r/c,\mathbf{r}_s)}{\Delta r}$$

$$\text{with} \quad \Delta r = |\mathbf{r}' - \mathbf{r}_s|; \quad \mathbf{e}_s = \frac{\mathbf{r}' - \mathbf{r}_s}{|\mathbf{r}' - \mathbf{r}_s|} \tag{4.58}$$

$$\underline{W}_S(f,\mathbf{r}') = \frac{\underline{\Lambda}(f,\mathbf{e}_i,\mathbf{e}_s)|_{P_S}\, \underline{W}_p(f,\mathbf{r}_s)}{\Delta r} e^{-jk\,\Delta r} \tag{4.59}$$

Here, we have split again the scattered wave into two parts: an ideal isotropic wave (propagation term) which springs up from the virtual scattering point $\mathbf{r}_s$ and an angular-dependent source term $\Lambda(t,\mathbf{e}_i,\mathbf{e}_s)|_{P_S}$ [m/s] or $\underline{\Lambda}(f,\mathbf{e}_i,\mathbf{e}_s)|_{P_S}$ [m] describing time (or spectral) evolution and angular distribution of the new wave caused by scattering of the incident field $W_P(t,\mathbf{r}_s)$. The side condition $\cdots|_{P_S}$ shall express that the scattering properties are related to a virtual reference point and orientation. The dimension of the scattering FRF tempts us to call this quantity normalized scattering length. The scattering IRF and FRF depend on shape and constitution of the scattering body as well as the directions of incidence $\mathbf{e}_i$ and observation $\mathbf{e}_S$. They contain all information about the target which is accessible by radar measurements. Therefore, its determination is of major interest in ultra-wideband and radar sensing.

After having introduced the scattering IRF and FRF for modelling the target behavior, the mathematical description of the whole transmission chain is straightforward (compare Figure 4.18). Supposing the transmit antenna having the angular-dependent IRF $T_1(t,\mathrm{e})$ or FRF $\underline{T}_1(f,\mathrm{e})$ is located at point P_1 defined by the position vector $\mathbf{r}_1$, we can calculate from (4.34) or (4.35) the field strength at target position $\mathbf{r}_s$ (point P_S). From this we are able to estimate via (4.58) or (4.59) the wave field incident on the receiving antenna at point P_2 (position vector $\mathbf{r}_2$). And finally using (4.36) or (4.37), the IRF $R_2(t,\mathbf{e})$ or FRF $\underline{R}_2(f,\mathbf{e})$ of the receiving antenna leads to wanted receiver signal $b_2(t)$ (τ_1, τ_2 are the internal antenna delay):

$$b_2(t) = \frac{1}{\Delta r_1\,\Delta r_2} T_1(t,\mathbf{e}_1)|_{P_1} * \Lambda(t,\mathbf{e}_i,\mathbf{e}_s)|_{P_S} * R_2(t,-\mathbf{e}_2)|_{P_2} * a_1(t-\tau) \tag{4.60}$$

$$\underline{b}_2(f) = \frac{1}{\Delta r_1\,\Delta r_2} \underline{T}_1(f,\mathbf{e}_1)|_{P_1} \underline{\Lambda}(f,\mathbf{e}_i,\mathbf{e}_s)|_{P_S} \underline{R}_2(f,-\mathbf{e}_2)|_{P_2} \underline{a}_1(f) e^{-j2\pi f\tau} \tag{4.61}$$

$$\text{whereat} \quad \tau = \frac{\Delta r_1 + \Delta r_2}{c} + \tau_1 + \tau_2;$$

$$\Delta r_1 = |\mathbf{r}_S - \mathbf{r}_1|; \quad \Delta r_2 = |\mathbf{r}_2 - \mathbf{r}_S|; \quad \mathbf{e}_1 = \mathbf{e}_i = \frac{\mathbf{r}_S - \mathbf{r}_1}{\Delta r_1}; \quad \mathbf{e}_s = -\mathbf{e}_2 = \frac{\mathbf{r}_2 - \mathbf{r}_S}{\Delta r_2}$$

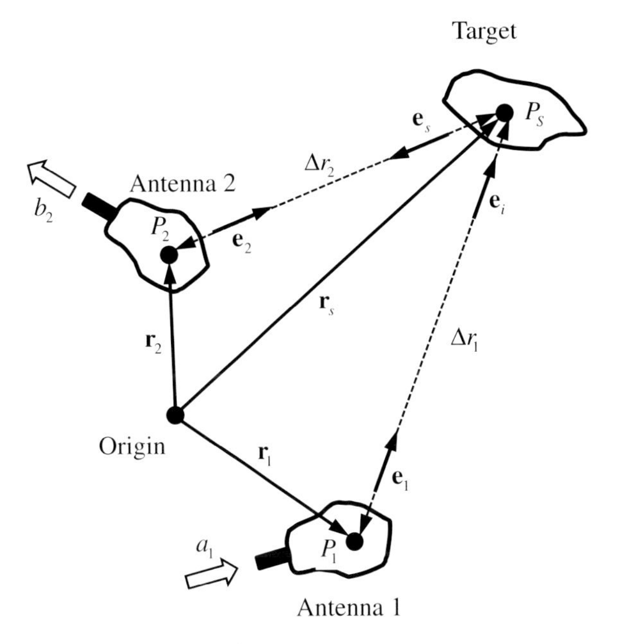

Figure 4.18 Bi-static radar measurement arrangement.

These two equations represent the scalar time and frequency domain radar equation for bi-static arrangement. In (4.60) and (4.61), we have distinguished between receiving and transmit mode of the antennas. Using the antenna reciprocity relation (4.43) or (4.44), one can unify these equations to the use of either $R(t, \mathbf{e})$ or $T(t, \mathbf{e})$.

We can also apply reciprocity to the radar equations (4.60) and (4.61), which takes the form (corresponding holds for the time domain)

$$\frac{\underline{b}_2}{\underline{a}_1} = \underline{T}_1(f, \mathbf{e}_1)|_{P_1} \underline{\Lambda}(f, \mathbf{e}_1, -\mathbf{e}_2)|_{P_S} \underline{R}_2(f, -\mathbf{e}_2)|_{P_2}$$
$$= \frac{\underline{b}_1}{\underline{a}_2} = \underline{T}_2(f, \mathbf{e}_2)|_{P_2} \underline{\Lambda}(f, \mathbf{e}_2, -\mathbf{e}_1)|_{P_S} \underline{R}_1(f, -\mathbf{e}_1)|_{P_1} \tag{4.62}$$

Taking into account (4.41), we result in the reciprocity of the scattering object

$$\underline{\Lambda}(f, \mathbf{e}_1, -\mathbf{e}_2)|_{P_S} = \underline{\Lambda}(f, \mathbf{e}_2, -\mathbf{e}_1)|_{P_S} \tag{4.63}$$

which tells us that direction of incidence and observation direction are interchangeable.

It is important to note that the amplitude of the receiving signal decreases by the product of the antenna distances. Hence, a radar measurement poses much higher requirements at the dynamic range of the sensors than a direct transmission between two antennas.

For completeness, we will also mention the power version of (4.61) which will bring us the classical narrowband radar equation. First, we have to introduce a power quantity relating to the scattering object (which we place at canonical position $\mathbf{r}_s = \mathbf{0}$ for simplicity). For that purpose, as already introduced above, we consider the scatterer as a source of radiation which provides at the observer position $\mathbf{r}'$ the power flux per unit area and Hertz of $\overline{\mathfrak{P}}_s(f, \mathbf{r}') = \underline{W}_S(f, \mathbf{r}')\underline{W}_S^*(f, \mathbf{r}')$ (refer to (4.6)). We will compare it with a virtual scatterer that isotropically distributes in space the spectral power $\Phi_i(f)$ which it has 'collected' from the incident wave. The magnitude of $\Phi_i(f)$ depends on the power flux density of the incident planar wave $\overline{\mathfrak{P}}_i(f, \mathbf{r}_s = \mathbf{0}) = \underline{W}_P(f, \mathbf{r}_s = \mathbf{0})\underline{W}_P^*(f, \mathbf{r}_s = \mathbf{0})$ and a virtual area σ over which the power is collected, that is $\Phi_i(f) = \sigma \overline{\mathfrak{P}}_i(f, \mathbf{r}_s = 0)$. According to the assumption of isotropic scattering, the captured energy will be equally redistributed leading to a (virtual) power flux density in the observation point of $\overline{\mathfrak{P}}_0(f, \mathbf{r}') = \Phi_i(f)/4\pi r'^2$. Since real and virtual scatterer should provide the same power flux density at the observation point, the virtual area σ is a characteristic gauge of reflectivity of the target

$$\overline{\mathfrak{P}}_s(f, \mathbf{r}') = \overline{\mathfrak{P}}_0(f, \mathbf{r}') \rightarrow \sigma(f, \mathbf{e}_i, \mathbf{e}_s)\left[\mathrm{m}^2\right] = 4\pi r'^2 \frac{\overline{\mathfrak{P}}_s(f, \mathbf{r}')}{\overline{\mathfrak{P}}_i(f, \mathbf{r}')} \tag{4.64}$$

Insertion of (4.59) leads to the conversion between the above introduced normalized scattering length and σ which is termed radar cross section (RCS):

$$\sigma(f, \mathbf{e}_i, \mathbf{e}_s) = 4\pi \underline{\Lambda}(f, \mathbf{e}_i, \mathbf{e}_s)\underline{\Lambda}^*(f, \mathbf{e}_i, \mathbf{e}_s) \tag{4.65}$$

Using (4.46), (4.48), (4.49), (4.61), (4.65), the radar equation can be gained in its classical form

$$\begin{aligned} \Phi_2(f) &= \frac{G_1(f, \mathbf{e}_1)\sigma(f, \mathbf{e}_i, \mathbf{e}_s)A_2(f, \mathbf{e}_2)}{(4\pi\, \Delta r_1\, \Delta r_2)^2}\Phi_1(f) \\ &= \frac{G_1(f, \mathbf{e}_1)G_2(f, \mathbf{e}_2)\sigma(f, \mathbf{e}_i, \mathbf{e}_s)\lambda^2}{(4\pi)^3(\Delta r_1\, \Delta r_2)^2}\Phi_1(f) \\ &= \frac{A_1(f, \mathbf{e}_1)A_2(f, \mathbf{e}_2)\sigma(f, \mathbf{e}_i, \mathbf{e}_s)}{4\pi\lambda^2(\Delta r_1\, \Delta r_2)^2}\Phi_1(f) \end{aligned} \tag{4.66}$$

where at $\Phi_1(f) = \underline{a}_1(f)\underline{a}_1^*(f)$; $\Phi_2(f) = \underline{b}_2(f)\underline{b}_2^*(f)$

$$\mathbf{e}_1 = \mathbf{e}_i; \quad \mathbf{e}_2 = -\mathbf{e}_s$$

These equations were developed for narrowband radar. In such cases, the spatial envelope[7)] of the sounding signal is much larger than typical dimensions of the target. Hence, the antennas as well as the scatterer will barely affect the time evolution of the signal. They will only influence the signal magnitude and consequently its power so that (4.66) expresses correctly the situation.

7) In connection with arbitrary sounding signals including random and pseudo-noise, it must be correctly called coherence length.

In ultra-wideband sensing, we usually do not have such conditions. Here, typical target and antenna dimensions are in the same order as the spatial pulse length (coherence length). Since the geometric structure of a body strongly influences the electromagnetic fields, every action – transmission/scattering/reception – leads to signal deformation as it is formally expressed by (4.60). Nevertheless, sometimes ultra-wideband scenarios are evaluated by simple power transfer considerations as in (4.66). Under certain conditions, this may be feasible but one should be aware of the large error potential and the allowed conditions of simplification.

Target ranging is the primordial radar application. For narrowband radar, range resolution is determined solely by the sounding signal, namely by its pulse or coherence width, which only depends on the signal bandwidth and the propagation speed of the wave (see Section 4.7.1). The situation is less well defined in the case of ultra-wideband radar since the signal deformations due to antennas or scatterer will affect the resolution too. This is one of the reasons why we introduced characteristic functions for antennas and scatterer which are going beyond the classical radar evaluation approach. In order to achieve range resolution smaller than the physical size of the involved bodies (antenna and scatterer), we must know their IRFs in order to have correct references for round-trip time measurements. In Section 4.7.3, we will deal in more detail with this topic. Since it is not always possible to know the IRF of all involved bodies, we will shortly discuss how their geometric size will affect the respective IRF. We will start with some basic properties of UWB scattering illustrated by means of some simple examples. The topic will be continued in Section 4.6 for UWB antennas

For introducing basic properties of scattering at small- and medium-sized objects, we first refer to scattering at a sphere. It is the only body of finite dimensions whose behavior can be analytically modelled. Annex C.3 summarizes the equations for time harmonic fields [16]. Subjecting these relations to a Fourier transform, we can obtain the related scattering IRF. Certainly, the sphere will not be the geometry of most practical relevance but in some sense its scattering behavior can be generalized allowing us to gain some insight into the scattering effects of other bodies.

Figure 4.19 illustrates the frequency dependence of the backscattering cross section $\sigma(f, \vartheta = 0, \varphi = 0)$ which we normalized to the geometric cross section of the sphere. Furthermore, the frequency is normalized via $n = ka = 2\pi(a/\lambda) = (2\pi f a)/c$ so that n equals the number of periods which can be placed around the circumference of a circle with radius a.

Inspecting Figure 4.19, we can distinguish three regions of characteristic behavior:

Rayleigh scattering: It holds if the circumference of the sphere is smaller than the wavelength and the coherence length of a wideband sounding signal. For a perfect electric conductor, the radar cross section decreases with the fourth power of the wavelength [16]:

$$\frac{\sigma}{\pi a^2} = \frac{4}{a^2} |\underline{\Lambda}|^2 \approx 9(ka)^4; \quad ka < 1 \tag{4.67}$$

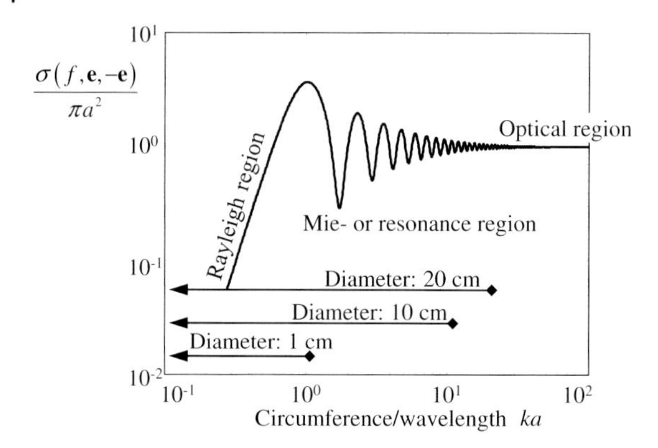

Figure 4.19 Normalized radar cross section for mono-static backscattering of a perfect conducting sphere. The arrows indicate the *ka* interval occupied by the sounding signal for the three examples illustrated in Figure 4.20.

The radar cross section of a small dielectric sphere of permittivity ε_2 embedded in a propagation medium of permittivity ε_1 takes approximately the form (slightly modified version of (10–38) in Ref. [17])

$$\frac{\sigma}{\pi a^2} \approx 4(ka)^4 \left|\frac{\varepsilon_1 - \varepsilon_2}{2\varepsilon_1 + \varepsilon_2}\right|^2; \quad ka < 1; \quad \varepsilon_1 \approx \varepsilon_2 \tag{4.68}$$

This relation supposes weak dielectric contrast between propagation medium and scatterer $\varepsilon_1 \approx \varepsilon_2$.

Referred to the volume V_{sp} of a metal sphere, we can also write

$$\sigma \approx \frac{1}{\pi}\left(\frac{9V_{\mathrm{sp}}(2\pi f)^2}{4c^2}\right)^2 \tag{4.69}$$

We interpret this equation as rough approximation of the radar cross section of a small body of any geometric shape as long as all its dimensions are smaller compared to the wavelength. If the description of the scattering body by a sphere is not precise enough (e.g. if the dimensions in *x*-, *y*- and *z*-directions are too different), approximations by discs, ellipsoids, cylinders, wires and so on may be useful. Estimations of their radar cross sections are to be found, for example, in Refs [16, 17]. A body respecting the condition of Rayleigh scattering is called point scatterer (refer also to Figure 4.2). Rewriting (4.69) in the form

$$\sigma = 4\pi|\underline{\Lambda}(f)|^2 \approx \frac{1}{\pi}\left(-\frac{9V_{\mathrm{sp}}(j2\pi f)^2}{4c^2}\right)^2 \tag{4.70}$$

we can conclude

$$\begin{gathered}\underline{\Lambda}(f) \approx -\frac{9V_{\mathrm{sp}}}{8\pi c^2}(j2\pi f)^2 \\ \downarrow \text{IFT} \\ \Lambda(t) \approx -\frac{9V_{\mathrm{sp}}}{8\pi c^2}\frac{\partial^2}{\partial t^2}\cdots = -\frac{9V_{\mathrm{sp}}}{8\pi c^2}u_2(t) * \cdots\end{gathered} \tag{4.71}$$

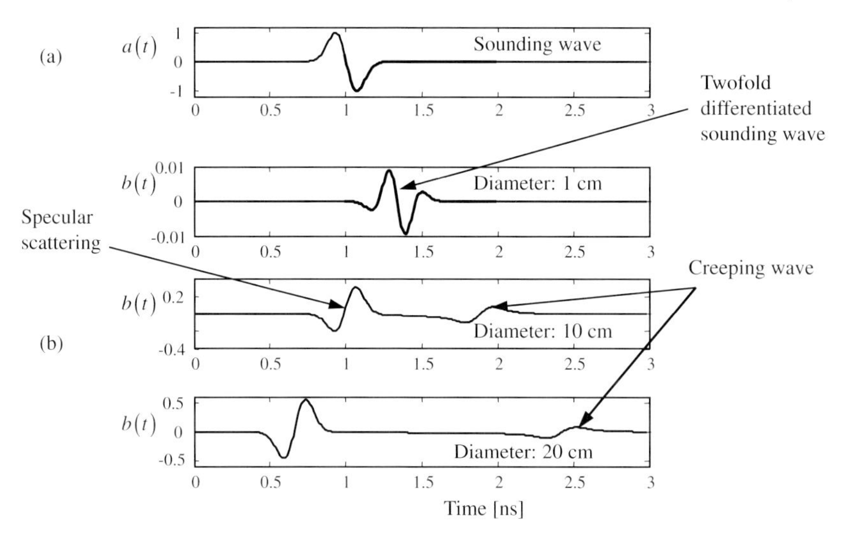

Figure 4.20 Backscattering from a sphere. (a) Time evolution of the sounding wave (differentiated Gaussian pulse). (b) Backscattered signals from spheres of different diameter.

which implies a twofold differentiation of the sounding signal that we can actually observe in Figure 4.20 for the small sphere of 1 cm diameter. The spectral band of the sounding signal occupies only the Rayleigh region in that case.

Mie- or resonant scattering: The circumference of the sphere ranges between 1 and about 20 periods of the sounding signal. In the case of a wideband signal, the coherence length is about the same factors less than the circumference. Now, creeping waves occur which travel around the body and reradiate. They may interfere destructively or constructively, with the direct reflex leading to the oscillation of the radar cross section in dependence on wavelength and sphere diameter (compare Figure 4.19). In the time domain (see Figure 4.20 for diameter 10 and 20 cm), the different wave types may be resolved due to their difference in propagation time. As we can see in the figure, the leading pulse of the reflected signal is a copy of the incident wave (except its inverse sign) like the reflection at a mirror. Therefore, it is called the specular reflection.

The decay and repetition rate of the creeping waves depend on the radius of the sphere so that basically the diameter of a spherical object may be determined via radar measurements insofar as the fractional measurement bandwidth is large enough (an example is given in Ref. [18]).

Optical scattering: The circumference of the sphere is much larger than the characteristic lengths of the sounding wave (i.e. wavelength, pulse length, coherence length). With increasing radius of the sphere, the creeping wave loses its influence within the lit region so that the oscillations of the backscattering (Figure 4.19) diminish and the specular reflex will be dominant. The curvature of the scattering objects tends to zero, which allows approximating scattering by Fresnel equations (see Section 5.4 or Annex C.2).

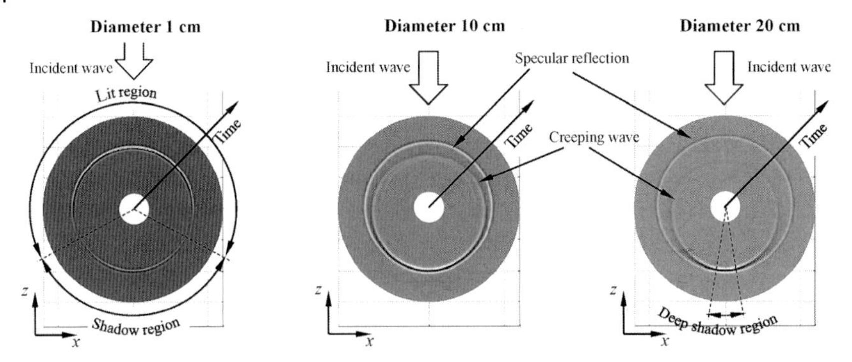

Figure 4.21 Simulation of bi-static scattering from spheres of different diameter. The plots show the time evolution of the scattered electric field in the x,z-plane $W_S(t, \mathbf{r}')$. The field is captured at a circle of 60 cm diameter around the scatterer ($r' = \sqrt{x^2 + z^2} = 30$ cm). The illustrated time interval covers 5 ns. (colored figure at Wiley Home page: Fig_4_21)

While Figures 4.19 and 4.20 only consider scattering reverse to the direction of incidence, Figure 4.21 permits a view around the sphere. For that purpose, we observe how the scattered field evolves in time. We occupied a number of observation points which are distributed along a circle in the x–z-plane. Now, we are able to observe the wave development around the sphere.

In our simplified consideration, we will roughly distinguish only between lit and shadow regions. The reader can refer to Ref. [16] for a comprehensive discussion on more regions of space connected to scattering at a sphere. In the lit region, the body will actually create a new wave travelling upwards. In the shadow region, the strength of the total field is reduced since the body blocks the propagation of the incident wave. Only the comparatively weak creeping waves are able to circle around and detach within the shadowing region. With increasing diameter, the sphere blocks nearly completely the incident wave within the deep shadow zone. Thus, the total field tends to zero there and from (4.57) follows for the deep shadow region:

$$W_P(t, \mathbf{r}') \approx -W_S(t, \mathbf{r}') \tag{4.72}$$

This implies that the forward scattered wave compensates the incident field and hence its amplitude is much larger than that of the backscattered wave which can be clearly seen in Figure 4.21 (right).

Figures 4.22–4.26 show some measurement examples to further illustrate backscattering behavior of real targets. All measurements were done with an M-sequence radar device 1–13 GHz [19] with a quasi-mono-static arrangement. The scattering objects were only rotated in the azimuth plane with the help of a turntable. Hence, we can approximately equal the direction of incidence $\mathbf{e}_\mathrm{i}$ and observation direction $\mathbf{e}_\mathrm{S}$ to describe both by the azimuth angle φ. Due to the high spatial resolution of an UWB radar, the experiments do not necessarily require an unechoing chamber since perturbing reflections can be gated out. Section 4.6.7 discusses measurement conditions for antenna measurements which can be correspondingly transferred to the measurements of the scattering behavior.

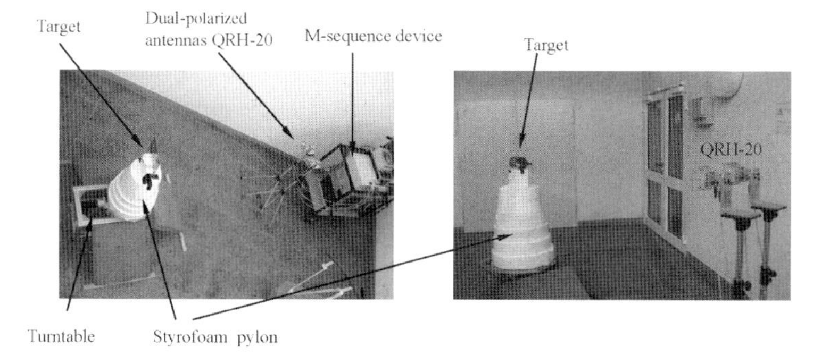

Figure 4.22 Simple arrangement of UWB backscatter measurements placed in an empty office room or vestibule. The antenna target-range is about 3 m. The targets 'see' the antennas under an angle of 6°.

Figure 4.23 shows the backscatter behavior of a first target. It is simply a small metal plate. The backscatter IRF is shown in polar coordinates where the time elapses along the radius. For the sake of better illustration, the envelope of the backscatter IRF $\Lambda(t, \varphi, \varphi)$ was applied which represents the absolute value of the analytic function $\underline{\Lambda}_a(t, \varphi, \varphi)$ (see (2.2)–(2.4)). The inner circle refers to a time point $t_1 = 10.59$ ns. The shown time interval extends over 2.81 ns. The signal magnitude is encoded in grey scale: white is zero and black represents maximum amplitude. As expected, the sheet metal acts as mirror; that is, it reflects back the wave for normal incidence. Hence, it seems that mono-static radar can only detect the plate if it is aligned perpendicular to the wave propagation. However, the edges of the plate provoke the so-called knife-edge diffraction (see, e.g. [3] or [16] for knife-edge and wedge diffraction, respectively) which scatters the waves around. This effect is much weaker than the reflection but it can be observed over nearly the whole circle. In order to quantify globally the angular dependence of scattering, we introduce a scattering directivity pattern as follows (compare also section 4.62 and 4.61):

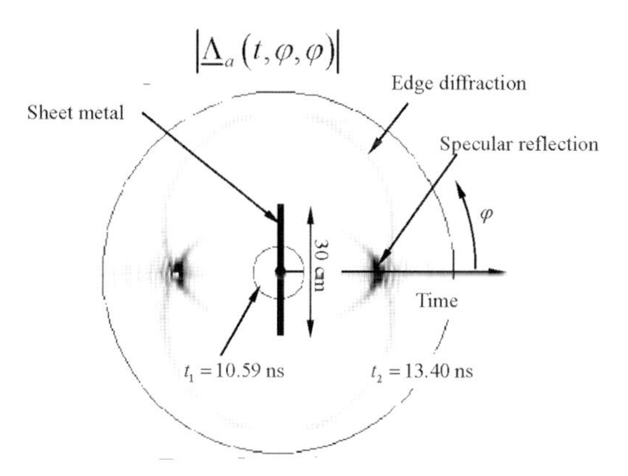

Figure 4.23 Envelope of the scattering IRF $|\underline{\Lambda}_a(t, \varphi, \varphi)|$ of sheet metal. The IRF envelope is calculated via (2.5). The sheet metal was 30 cm × 21 cm. (colored figure at Wiley homepage: Fig_4_23)

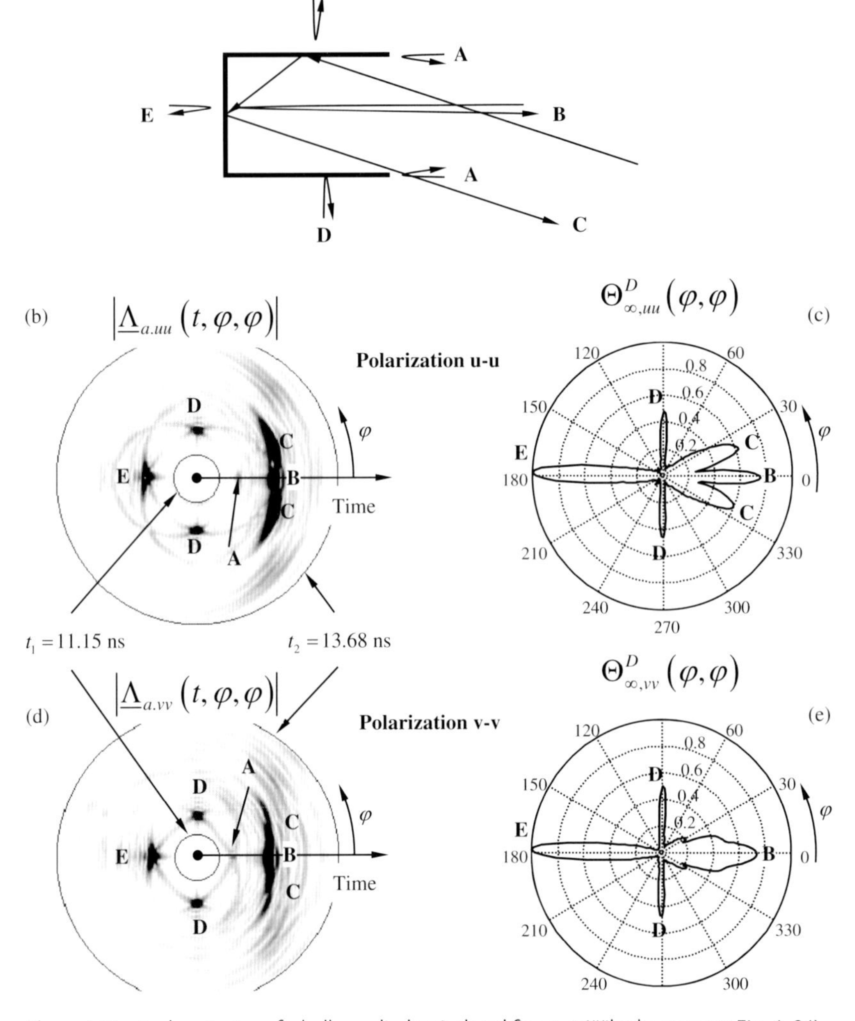

Figure 4.24 Backscattering of a hollow cylinder. (colored figure at Wiley homepage: Fig_4_24)

$$\Theta_{\mathrm{p}}^{\mathrm{D}}(\mathbf{e}_{\mathrm{i}}, \mathbf{e}_{\mathrm{s}}) = \frac{\|\Lambda(t, \mathbf{e}_{\mathrm{i}}, \mathbf{e}_{\mathrm{s}})\|_{\mathrm{p}}}{\|\Lambda(t, \mathbf{e}_{\mathrm{i}0}, \mathbf{e}_{\mathrm{s}0})\|_{\mathrm{p}}} \tag{4.73}$$

It relates any L_p-norm (with respect to t) of the angle-dependent IRF to its norm at a given direction of incidence $\mathbf{e}_{\mathrm{i}0}$ and observation $\mathbf{e}_{\mathrm{s}0}$. Typically these directions are fixed by the global maximum of backscattering amplitude.

Such curves for the infinity norm are shown in the second example as depicted in Figure 4.24. It deals with a hollow cylinder which is closed at one side. The test object has a diameter of 10 cm and a length of 11.5 cm. In Figure 4.24a, the main scattering mechanisms are illustrated. We can observe the specular reflections B, D and E, the double bounce reflection C and edge diffractions from which only one, namely A, is emphasized. Figure 4.24b, d represent the envelope of the backscatter IRF and the corresponding directional diagram are depicted in Figure 4.24c, e. Since the strongest reflection appears at the rear of the cylinder (assigned by E), we took it as reference value (denominator) for the directive pattern (4.73).

Figure 4.24 deals with two different backscatters IRFs, and moreover our last example depicted in Figure 4.25 shows even four backscatter IRFs. So

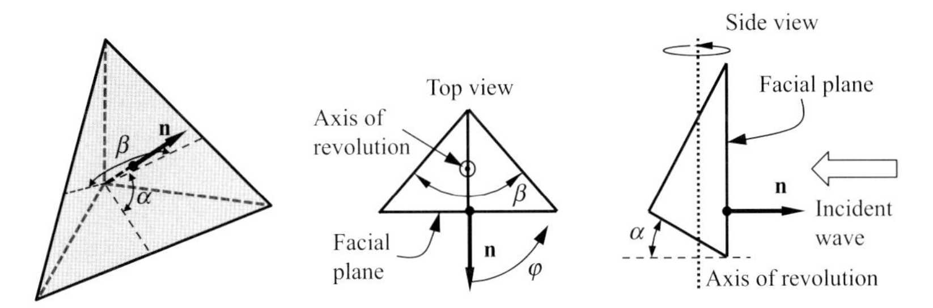

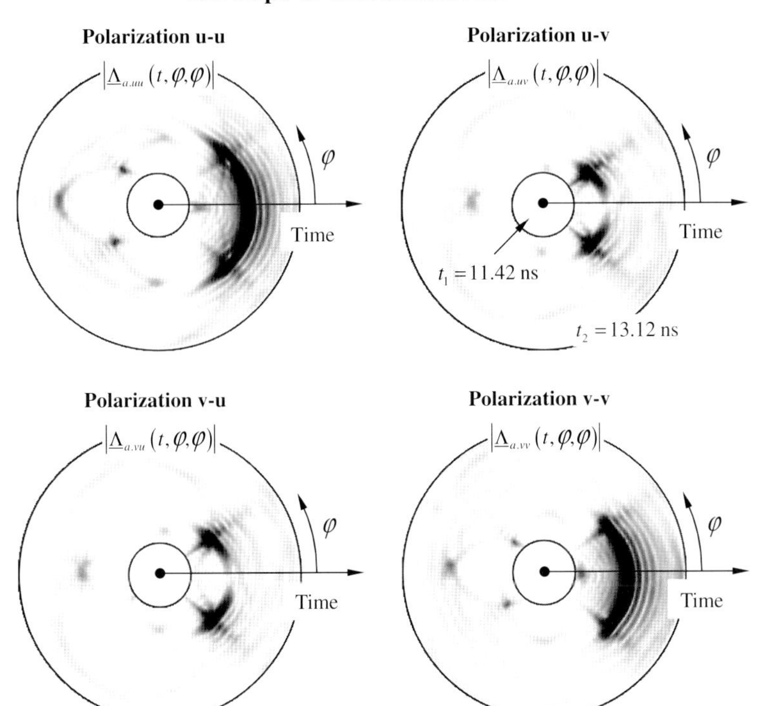

Figure 4.25 Backscattering of triangular trihedral corner reflector. The normal vector of the facial plane was located in the azimuth plane. (Angle between normal vector and side planes: $\cos\alpha = \sqrt{2/3} \triangleq 35^\circ$; angle between intersection lines of side planes and azimuth plane: $\cos\beta = 1/5 \triangleq 78^\circ$.) (colored figures at Wiley homepage: Fig_4_25)

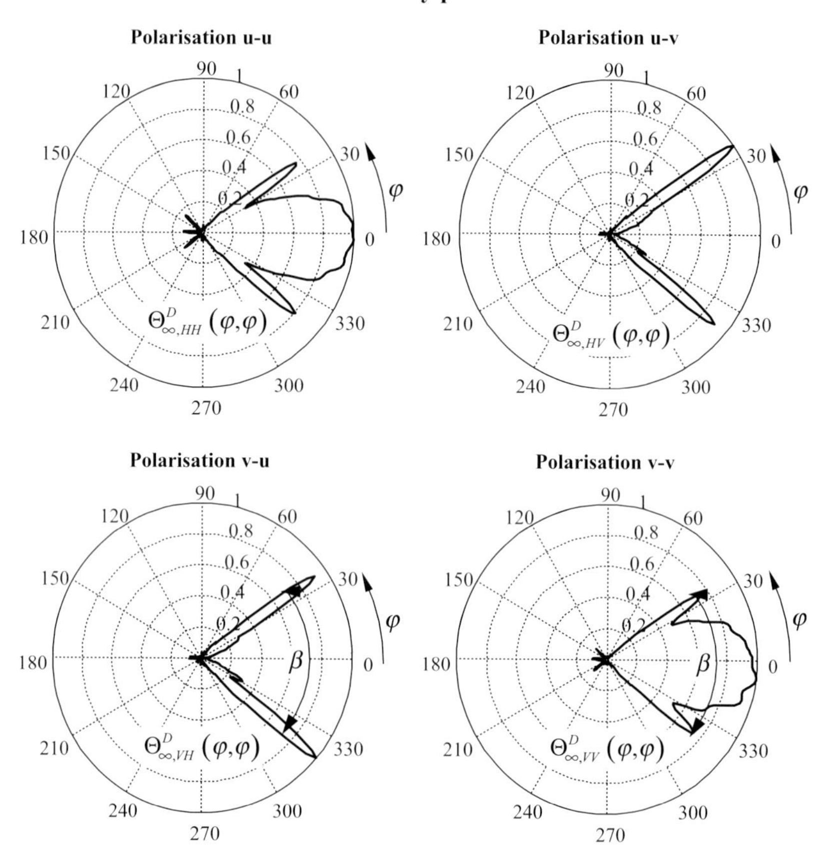

Figure 4.25 (*Continued*)

far, we did not yet introduce multiple backscatter functions since we considered electromagnetic waves simply as scalar fields, which is however not true. Actually, the electromagnetic field is of vector nature, which means that not only the propagation direction (on what we did restrict ourselves up to now) but also the direction of the field forces have to be respected. We will discuss some of these issues in Chapter 5. Corresponding phenomena are expressed by the terms field polarization or polarimetry. Here, we will not go deeper into these topics. We will take the examples only to illustrate the phenomena.

The backscatter IRF $\Lambda_{uu}(t, \varphi, \varphi)$ relates to the case in which the direction of the electric field vector of the incident wave is parallel to the azimuth plane and we only observed the field component of the backscattered electric field which is also parallel to the azimuth plane. We mark this polarization by the index u; that is, the first index of $\Lambda_{uu}(t, \varphi, \varphi)$ relates to the field orientation (i.e. polarization) of the backscattered wave while the second one gives the orientation of the incident wave. Correspondingly, $\Lambda_{vv}(t, \varphi, \varphi)$ means that the orientations of the incident and observed electric field are parallel to the plane of incidence (see Section 5.4 (5.88), (5.9) and Figure C.1). The selection of the field orientation for transmission or reception is done via the orientation of the antennas.

Figure 4.25 shows still two further versions of polarisation in which the orientation of the antennas is mixed. Index uv means that the incident wave is vertically polarized and the observed one is horizontally polarized and index vu represents the inverse case. Now, one speaks about cross-polarization while the above-mentioned examples refer to co-polarization.

The triangular trihedral corner reflector provokes between one and three reflections depending on the angle of incidence. However, we have always triple bounce reflections in our arrangement as long as the wave is incident on the facial plane (determined by the angle $\varphi \leq \pm\beta/2$). Outside this area ($\varphi > \pm\beta/2$), we can only observe wedge diffraction. Specular backscattering does not appear since at any rotation angle φ no plane will be orientated perpendicular to the incident wave. It is interesting to note that the field orientation of the backscattered wave deviates from the orientation of the incident field at the isolated angle φ; that is, we can discover cross-polar scattering components.

A final example (Figure 4.26) relates to the backscattering of a person, which is much more complicated than that of a simple geometric object. It largely depends on the body posture and it is subjected to a variance due to weak body motions of a vital person. The grey emphasized area of the backscatter signal in Figure 4.26 gives an impression of the strength of such motion if the person gives his best to keep his position. The measurement distance in the shown example was roughly the same as in the previous measurements. Consequently, the height of the person will violate the far-field condition. Hence, $\Lambda(t, \varphi, \varphi)$ will additionally depend on the range and the height of the antenna placement. The example shown in Figure 4.26 refers to scanning at the height of the person's chest.

In summary, we can state that real objects show a much more complex geometry than a sphere and hence their scattering behavior is quite more complicated. Notably, manmade objects typically exhibit symmetric construction elements and they are formed from edges, caves, planes, corners and so on which all reflect and diffract waves. Due to their symmetry, their geometric seize and the arrangement of their construction elements, eigenmodes may appear which are very specific to an individual object. They can be used to recognize or distinguish targets (see Section 4.6.4). The creeping waves of the sphere are an expression of such behavior.

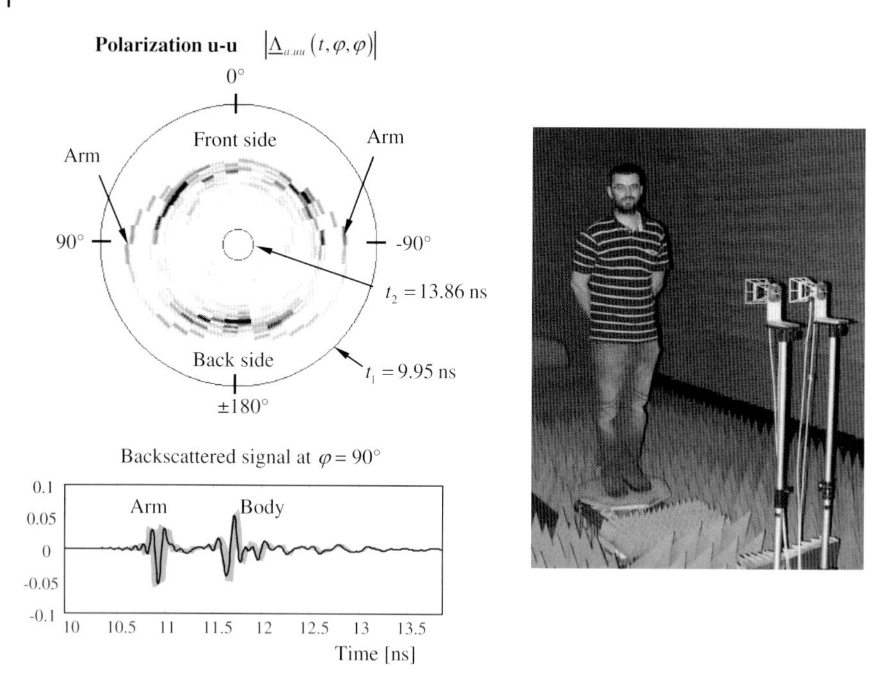

Figure 4.26 Scattering behavior of a person in rest. Note, the time axis direction is reversed compared to Figures 4.23–4.25. (colored figure at Wiley homepage: Fig_4_26)

On the other hand, resonance phenomena extend the spatial length of the backscattered wave beyond the geometric extension of the target which leads to blurring effects in UWB radar imaging. UWB imaging intends to reconstruct the geometric structure of a target or scenario under test. This supposes specular or point reflections because they may be well assigned to geometric locations. Resonant scattering which is caused from multipath propagation provokes additional pulses or ringing in the scattering IRF pretending ghost targets in microwave images since it is not possible to distinguish between single and multiple scattering by simple means.

Geometric structures of sharp curvature (point-like objects, cone points, thin wires, knife edges, slots, holes etc.) tend to scatter the incident field to more or less all directions. We can observe such diffraction effects at all objects having a surface with sharp curvatures (sharp with respect to the coherence length of the sounding wave). The diffraction theory provides the background of theoretic analysis which is usually done for time harmonic fields. Analytic solutions of the scattering problems only exist for simple geometries. They are usually quite cumbersome to find, and with only few exceptions (the sphere is one of them) the solutions represent approximations. There are quite a lot of textbooks dealing with these and related issues. A small selection of them includes [3, 16, 17, 20–24].

4.6 General Properties of Ultra-Wideband Antennas

The usual characterization of antennas by power-related quantities such as antenna gain or aperture is well adapted to narrowband applications and large transmission distances. For UWB sensing, these approaches are only partially useful. We will consider some general antenna properties mainly under the aspects of ultra-wideband short-range sensing. We already introduced the two IRFs $T(t, \mathbf{e})$ and $R(t, \mathbf{e})$ in the previous chapter which we assigned to the transfer behavior of an antenna. Now, we will complete the antenna model and discuss some basic antenna features in the time domain.

In order to come to a complete description, we consider an antenna as a formal or physical electro-dynamic system as introduced in Sections 2.4 and 2.5. Figure 4.27 illustrates the first model type. We symbolize the antenna by a formal transmission system of $N+1$ input channels and $N+1$ output channels. One of these channels relates to the feeding line of the antenna and the remaining ones symbolize the radiation into the different space regions. For that purpose, we divide the whole space surrounding the antenna into N segments of solid angle $\Delta\Omega$ so that $N\,\Delta\Omega = 4\pi$. Every of these space segment is assigned to an individual direction pointing away from the antenna. Theoretically, the number N of space segments tends to infinity but for practical purposes it will be always restricted to a finite number which leads us to the discretized model of Figure 4.27. In order to simplify the notation, we consider the antenna at canonical position; that is, its reference point is placed at the origin of the coordinate system and the reference directions are aligned according to the coordinate axes. The incident waves are plane waves which may come from different directions, and the emanated wave is a quasi-spherical wave which we observe at distance r around the antenna.

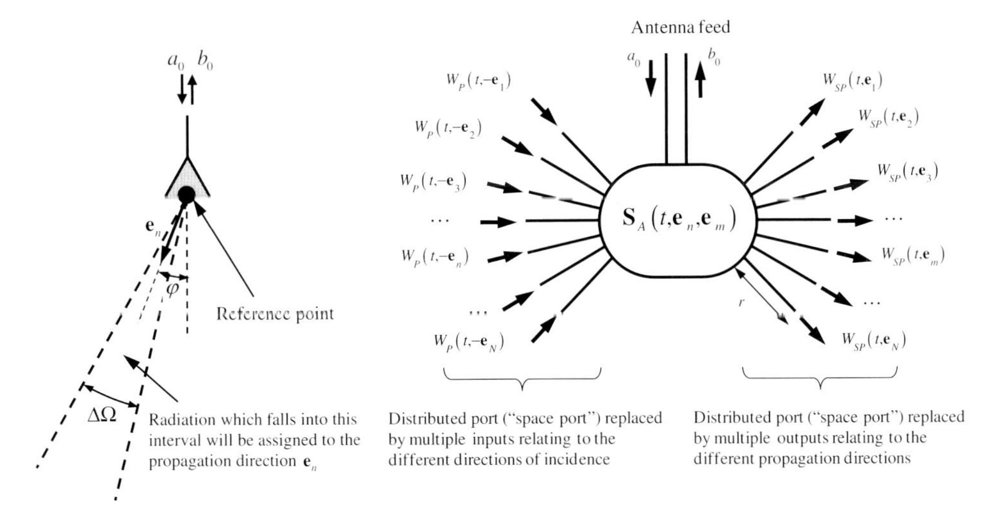

Figure 4.27 A formal MiMo system representing a discretized antenna model.

As we have seen in Section 2.5 that such MiMo system is characterized by a set of $(N+1)^2$ IRFs. Since we are dealing with normalized waves here, the IRFs represent modified time domain scattering parameters. Originally, S-parameters are defined for guided waves (Section 2.5.3). We will extend this concept here to a mixture of guided and free propagating waves. The handling of such S-parameters has to be done with some care, since the involved quantities are of different physical nature and measurement units. Based on Figure 4.27, the structure of the 'mixed mode' antenna S-matrix[8)] results in

$$\mathbf{S}_\mathrm{A}(t, \text{-}\mathbf{e}_\mathrm{i}, \mathbf{e}_\mathrm{s}) = \mathbf{S}_\mathrm{A}(t, m, n) = \begin{bmatrix} S_{00}(t) & S_{01}(t) & \cdots & S_{0n}(t) & \cdots & S_{0N}(t) \\ S_{10}(t) & S_{11}(t) & \cdots & S_{1n}(t) & \cdots & S_{1N}(t) \\ \vdots & \vdots & \ddots & \vdots & \ddots & \vdots \\ S_{m0}(t) & S_{m1}(t) & \cdots & S_{mn}(t) & \cdots & S_{mN}(t) \\ \vdots & \vdots & \ddots & \vdots & \ddots & \vdots \\ S_{N0}(t) & S_{N1}(t) & \cdots & S_{Nn}(t) & \cdots & S_{NN}(t) \end{bmatrix}$$
$$= \begin{bmatrix} \Gamma(t) & R_1(t) & R_2(t) & R_3(t) & \cdots & R_N(t) \\ T_1(t) & \Lambda_{11}(t) & \Lambda_{12}(t) & \Lambda_{13}(t) & \cdots & \Lambda_{1N}(t) \\ T_2(t) & \Lambda_{21}(t) & \Lambda_{22}(t) & \Lambda_{23}(t) & \cdots & \Lambda_{2N}(t) \\ T_3(t) & \Lambda_{31}(t) & \Lambda_{32}(t) & \Lambda_{33}(t) & \cdots & \Lambda_{3N}(t) \\ \vdots & \vdots & \vdots & \vdots & \ddots & \vdots \\ T_N(t) & \Lambda_{N1}(t) & \Lambda_{N2}(t) & \Lambda_{N3}(t) & \cdots & \Lambda_{NN}(t) \end{bmatrix} \tag{4.74}$$

Note that the indices range from zero to N where the index 0 refers to the lumped port. All space segments $\Delta\Omega$ are numbered from $1 \cdots N$ and the segments of number $n,\ m \in [1, N]$ points in direction $\mathbf{e}_n$ and $\mathbf{e}_m$ are all outward directed. At this, the index $m \mathrel{\hat{=}} \mathbf{e}_m = \mathbf{e}_\mathrm{s};\ m \in [1, N]$ symbolizes the observation direction, and the index $n \mathrel{\hat{=}} \mathbf{e}_n = -\mathbf{e}_\mathrm{i}$ represents the inverse of the direction of incidence. To have a compact notation, we join several parts of the antenna S-matrix which then take the form

$$\mathbf{S}_\mathrm{A}(t, \text{-}\mathbf{e}_\mathrm{i}, \mathbf{e}_\mathrm{s}) \mathrel{\hat{=}} \mathbf{S}_\mathrm{A}(t, m, n) = \begin{bmatrix} \Gamma(t) & \mathbf{R}(t, \mathbf{e}_n) \\ \mathbf{T}(t, \mathbf{e}_m) & \mathbf{\Lambda}(t, \mathbf{e}_n, \mathbf{e}_m) \end{bmatrix} \tag{4.75}$$

The frequency response function can be arranged correspondingly. The physical meaning of the different IRFs (FRFs) is the following:

8) Even if the antenna is a reciprocal object, the scattering matrix as defined by (4.74) and (4.75) is it not due to the differences in transmit and receiving IRF, that is $\mathbf{S}_\mathrm{A} \neq \mathbf{S}_\mathrm{A}^T$ since we have involved different wave types and reference points at the 'space' port. Reciprocity can be re-established by chaining two antennas since we have then again two well-defined locations, one for every antenna.

- The transmission from a_0 to b_0 represents the reflection $\Gamma(t)$ at the feeding port.
- The transmission from a_0 to the different space segments ('space port') $W_{SP}(t, \mathbf{e}_n)$ is the already introduced transmit IRF of the antenna which we express by a column vector in the discretized form:

$$T(t, \mathbf{e}_s) \mathrel{\hat{=}} \mathbf{T}(t, \mathbf{e}_m) = \begin{bmatrix} T_1(t) & T_2(t) & \cdots & T_m(t) & \cdots & T_N(t) \end{bmatrix}^T$$

  The index m refers to a specific radiation direction.
- Corresponding holds for the conversion from space wave (incident wave $W_P(t, \mathbf{e}_i)$) to guided wave b_0. For this case, the receiving IRF was introduced. We will represent it as row vector for the discretized notation. By convention, we defined the receiving IRF in dependence on the direction of incidence (refer to Figure 4.13). For the antenna model according to Figure 4.27, all directions point outwards. Hence, the inversion of directions in the equation below

$$R(t, -\mathbf{e}_i) \mathrel{\hat{=}} \mathbf{R}(t, \mathbf{e}_n) = \begin{bmatrix} R_1(t) & R_2(t) & \cdots & R_m(t) & \cdots & R_N(t) \end{bmatrix}$$

The transmission from the incident plane wave $W_P(t, \mathbf{e}_n)$ to the emanating wave $W_{SP}(t, \mathbf{e}_m)$ at the 'space port' may be considered like the scattering at small bodies.[9] That is, the antenna simply acts as scatterer and we describe its behavior by the scattering IRF $\Lambda(t, \mathbf{e}_i, \mathbf{e}_s)$. Hence, we get for the discretized notation

$$\Lambda(t, -\mathbf{e}_i, \mathbf{e}_s) \mathrel{\hat{=}} \mathbf{\Lambda}(t, \mathbf{e}_n, \mathbf{e}_m) = \begin{bmatrix} \Lambda_{11}(t) & \Lambda_{12}(t) & \cdots & \Lambda_{1n}(t) & \cdots & \Lambda_{1N}(t) \\ \Lambda_{21}(t) & \Lambda_{22}(t) & \cdots & \Lambda_{2n}(t) & \cdots & \Lambda_{2N}(t) \\ \vdots & \vdots & \ddots & \vdots & \ddots & \vdots \\ \Lambda_{m1}(t) & \Lambda_{m2}(t) & \cdots & \Lambda_{mn}(t) & \cdots & \Lambda_{mN}(t) \\ \vdots & \vdots & \ddots & \vdots & \ddots & \vdots \\ \Lambda_{N1}(t) & \Lambda_{N2}(t) & \cdots & \Lambda_{Nn}(t) & \cdots & \Lambda_{NN}(t) \end{bmatrix}$$

Following to (4.63), we can conclude

$$\begin{aligned} \Lambda(t, \mathbf{e}_n, \mathbf{e}_m) &= \Lambda(t, \mathbf{e}_m, \mathbf{e}_n) \\ \mathbf{\Lambda} &= \mathbf{\Lambda}^T \end{aligned}$$

Antenna backscattering is omitted in many of the discussions on antenna behavior since it is out of interest for long-distance transmissions at least if one disregards aspects of camouflage and stealth operations (see Refs [25–27]). The situation alters for short-range applications since antenna backscattering may lead to multiple reflections due to its proximity to the targets. Multiple reflections largely complicate the interpretation of measurement data. Hence, we will include antenna backscattering in our further discussion in order to understand some fundamental issues which may be helpful for designing appropriate UWB radiators for

9) The basic idea behind this consideration is the same as for the lumped port which leads to the reflection IRF $\Gamma(t)$. But an important difference should be mentioned. While for the lumped port the incident and the reflected waves are measured at the same position, the reference positions for incident and reflected (scattered) wave are different for the 'space port'. The incident wave is considered at antenna position but the scattered wave is always observed at sufficient distance from the antenna. The same is with transmit and receiving IRFs.

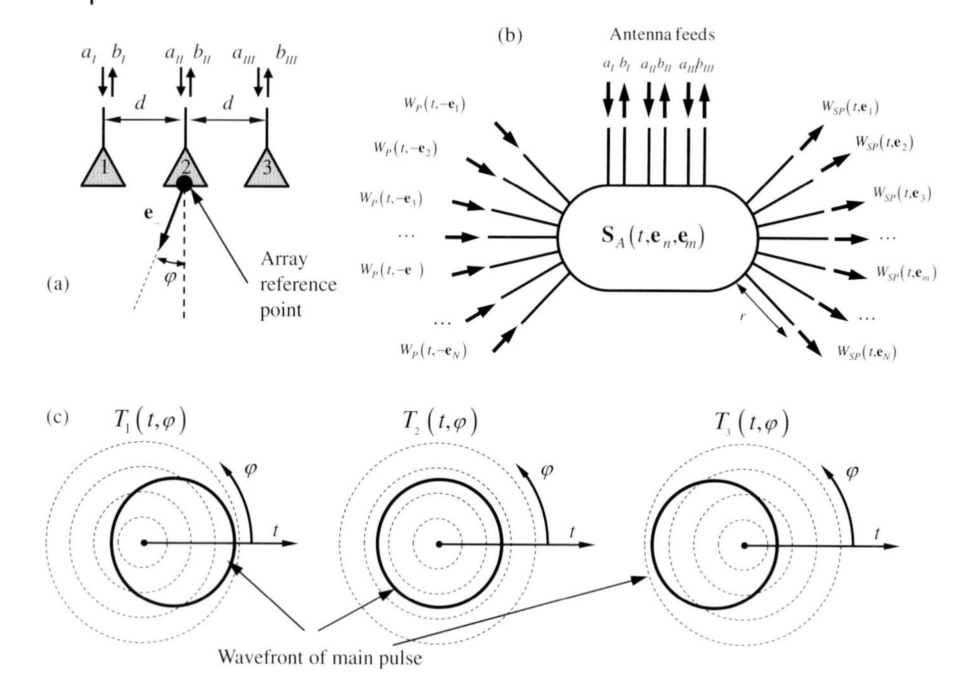

Figure 4.28 Antenna array (a) modeled as MiMo-system (b) and simplified illustration of transmission IRFs (c) for omni-directional radiators. Note, the model considers the array as a whole and hence it will refer all its properties to a single virtual point which we placed at the position of the central antenna.

short-range measurements. We will see hereafter that antenna backscattering depends (at least partially) on its radiation behavior, which may be used for antenna gain measurements under variable load conditions [28–30].

Basically, the model depicted in Figure 4.27 can be enlarged by further feeding ports to include also antenna arrays [20, 25, 29]. Figure 4.28b shows an example for $K = 3$ antennas. The basic structure of the antenna S-matrix remains the same, but the anciently scalar term $\Gamma(t)$ becomes a $[K,K]$-matrix. Its diagonal elements are the reflection IRFs of the feeding ports and the remaining terms refer to antenna cross-coupling. The transmission and receiving IRFs become: $\mathbf{T}(t, \mathbf{e}_n)$ – $[N,K]$-matrix and $\mathbf{R}(t, \mathbf{e}_n)$ – $[K,N]$-matrix. Above antenna model supposes a unique reference point for the whole array. This leads to asymmetric radiation characteristics of the individual antenna elements if the reference point and the actual radiation centre of the antenna move away. Figure 4.28c demonstrates the behavior by stylized transmit IRFs. The structure of the scattering IRFs $\mathbf{\Lambda}(t, \mathbf{e}_n, \mathbf{e}_m)$ remains the same as for the single element antenna but it counts for the 'space' port reflections as a whole; that is, it sums up all reflections of the individual radiators.

For the sake of simplicity, we address only single-port antennas and disregard multi-port antennas. First, their inclusion would be straightforward on the basis of matrix calculus, and second, it is more unlikely that antenna arrays meet the

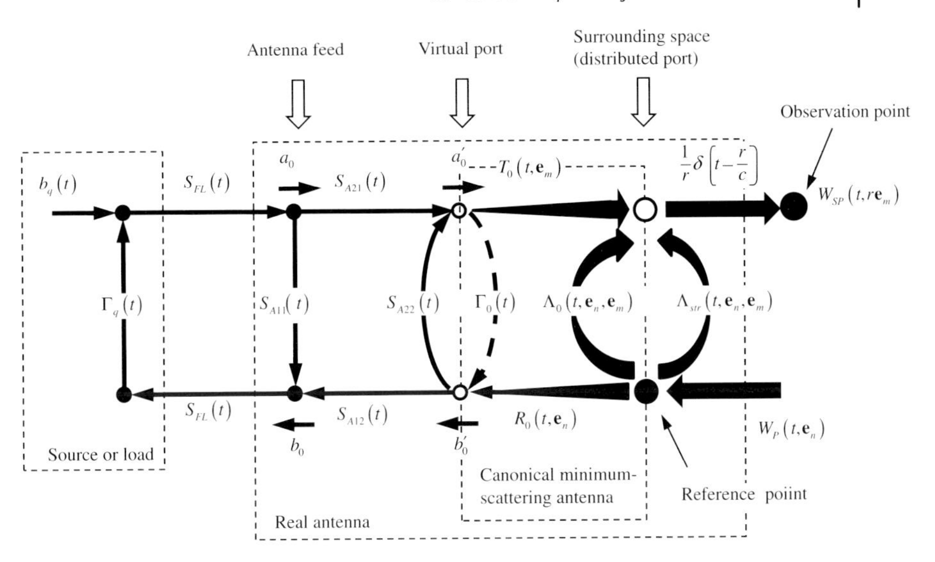

Figure 4.29 Mason graph as physical antenna model. S_{FL} is the transmission coefficient of a feeding line connecting the antenna with a generator or load. $\Lambda_{str}(t, \mathbf{e}_n, \mathbf{e}_m)$ symbolizes antenna reflection due to construction elements and $S_{Aij}(t)$ refers to antenna internal signal paths which deviate from the idealized minimum-scattering antenna.

modelling conditions for short-range applications. As already mentioned, an 'observer' must see the antenna under a small solid angle. Since the array is considered here as a single object, its overall dimensions and hence the number of antennas must be quite small.

In order to widen our view about basic antenna behavior, we divide the antenna into an ideal part – we call it the canonical minimum-scattering antenna – and some additional parts to which we assign those properties of a real antenna that deviate from the idealized one. Figure 4.29 symbolizes the signal flow by Mason graph within the decomposed antenna. Herein, the thin arrows relate to guided waves whereas the bold arrows symbolize freely propagating waves.

4.6.1
Canonical Minimum-Scattering Antenna

The canonical minimum-scattering antenna is the kernel of our antenna model. It is defined by the following points:

- It is placed at the origin of the coordinate system (hence, we will omit the reference to the antenna position in the formulas which follow).
- The antenna has no internal delay.
- It has an ideally matched feed line port: $\Gamma_0(t) = 0$.
- It is lossless.
- It has no structural reflections, $\Lambda_{str}(t, \mathbf{e}_i, \mathbf{e}_s) = 0$.

- There are no additionally internal propagation paths ($S_{A11}(t) = S_{A22}(t) = 0$; $S_{A21}(t) = S_{A12}(t) = 1$). Furthermore, we assume antenna loading Γ_q directly at its terminal (i.e. $S_{FL}(t) = 1$).

What do we understand by structural reflections? First of all an antenna is a body, typically a metallic structure, which is able to convert free into guided waves and vice versa. However, the antenna can also be considered as a body which scatters electromagnetic fields if it is situated along the propagation path of a wave. But in contrast to usual scatterers, the scattering from the antenna is composed of two parts – a 'real' scattering due to its mechanical structure and a partial re-radiation of the captured wave energy. That re-radiations appear will be obvious by considering the antenna equivalent circuit in Figures 5.19 and 5.20. There, the antenna is modelled by internal impedance whose real part $\Re\{\underline{Z}_D\}$ quantifies the power transfer[10] between guided and free wave. In the receiving mode, the captured power moves from the antenna to the receiver input load. Hence, a current has also to pass $\underline{Z}_D$. This detracts energy from the electric circuit which we can retrieve in the re-radiated field. Making the current zero (i.e. high ohmic load), the re-radiation will stop and the antenna becomes 'invisible'. The canonical minimum scattering antenna addresses only this part of scattering, which obviously depends on the antenna loading Γ_q and the power which the antenna is able to collect. For power collection, the antenna needs a spatial extension expressed by the effective antenna height $\underline{R}(f, \mathbf{e}_i)$ and the effective area $A(t, \mathbf{e}_i)$. They relate to theoretical minimal dimensions required for a given wave conversion. Practically, an antenna structure must be larger; it requires structural components which are not involved in wave conversion but which are exposed to the incident field too. Hence, we can observe additional scattering signals which exceed the theoretical minimum by re-radiation. We will summarize all scattered fields, which are not directly linked to the receiving properties of the antenna, in the term structural reflection $\Lambda_{str}(t, \mathbf{e}_i, \mathbf{e}_s)$.

For the sake of simplicity and shortness, we will confine our considerations mainly to the frequency domain. Abstaining from the discretized notation, the scattering matrix $\mathbf{S}_{CA}$ of the canonical minimum-scattering antenna is gained from the Mason graph in Figure 4.29

$$\begin{aligned} \begin{bmatrix} \underline{b}'_0(f) \\ r\, e^{j\frac{2\pi f r}{c}}\, \underline{W}_{SP}(f, r\,\mathbf{e}_s) \end{bmatrix} &= \begin{bmatrix} 0 & \underline{R}_0(f, \mathbf{e}_i) \\ \underline{T}_0(f, \mathbf{e}_s) & \underline{\Lambda}_0(f, \mathbf{e}_i, \mathbf{e}_s) \end{bmatrix} \begin{bmatrix} \underline{a}'_0(f) \\ \underline{W}_P(f, \mathbf{e}_i) \end{bmatrix} \\ \underline{\mathbf{S}}_{CA}(f, \mathbf{e}_i, \mathbf{e}_s) &= \begin{bmatrix} 0 & \underline{R}_0(f, \mathbf{e}_i) \\ \underline{T}_0(f, \mathbf{e}_s) & \underline{\Lambda}_0(f, \mathbf{e}_i, \mathbf{e}_s) \end{bmatrix} \end{aligned} \tag{4.76}$$

Since the considered antenna is lossless by assumption, the S-matrix must be unitary $\underline{\mathbf{S}}^H \underline{\mathbf{S}} = \mathbf{I}$ following (2.215). For radiation in free space, we have to adapt this condition to (assuming an infinite number of space ports):

$$\oint \underline{\mathbf{S}}_{CA}^H\, \underline{\mathbf{S}}_{CA}\, d\Omega = \mathbf{I} \tag{4.77}$$

10) If the antenna is not lossless, the real part $\Re\{\underline{Z}_D\}$ also respects the ohmic antenna losses.

This leads us to

$$\oint \underline{T}_0(f,\mathbf{e})\underline{T}_0^*(f,\mathbf{e})\mathrm{d}\Omega = 1 \tag{4.78}$$

$$\oint \underline{T}_0^*(f,\mathbf{e})\underline{\Lambda}_0(f,\mathbf{e}_\mathrm{i},\mathbf{e})\mathrm{d}\Omega = \oint \underline{T}_0(f,\mathbf{e})\underline{\Lambda}_0^*(f,\mathbf{e}_\mathrm{i},\mathbf{e})\mathrm{d}\Omega = 0 \tag{4.79}$$

$$R_0^*(f,\mathbf{e}_\mathrm{i})R_0(f,\mathbf{e}_\mathrm{i}) + \oint \underline{\Lambda}_0^*(f,\mathbf{e}_\mathrm{i},\mathbf{e})\underline{\Lambda}_0(f,\mathbf{e}_\mathrm{i},\mathbf{e})\mathrm{d}\Omega = 1 \tag{4.80}$$

Note, the integration in (4.77) and (4.80) is only with respect to the direction $\mathbf{e} = \mathbf{e}_\mathrm{s}$ of the emanating waves.

In a lossless network, the components of the scattering matrix are mutual dependent. Translated to the canonical minimum-scattering antenna, we can show that the scatter length $\underline{\Lambda}_0(f,\mathbf{e}_\mathrm{i},\mathbf{e}_\mathrm{s})$ of the antenna only depends on its radiation properties. In order to determine such dependency, we consider the case for which the antenna is 'invisible' for the electromagnetic field; that is, it does not scatter the incident wave.

The total electric field enclosing the antenna is composed of the incident field and the scattered field. That is, the incident wave must 'penetrate' the antenna without any deformation for the scattering-free case. We can formulate this condition by

$$r\,\mathrm{e}^{j\frac{2\pi f r}{c}}\underline{W}_\mathrm{SP}(f,\mathbf{e}_\mathrm{s}) = \begin{cases} \underline{W}_\mathrm{P}(f,\mathbf{e}_\mathrm{i}) & \text{for} \quad \mathbf{e}_\mathrm{s} = -\mathbf{e}_\mathrm{i} \\ 0 & \text{for} \quad \mathbf{e}_\mathrm{s} \neq -\mathbf{e}_\mathrm{i} \end{cases} \tag{4.81}$$

In the receiving mode, the wave emanating from the antenna results from the Mason graph

$$r\,\mathrm{e}^{j\frac{2\pi f r}{c}}\underline{W}_\mathrm{SP}(f,r\,\mathbf{e}_\mathrm{s}) = \left(\underline{\Lambda}_0(f,\mathbf{e}_\mathrm{i},\mathbf{e}_\mathrm{s}) + \underline{R}_0(f,\mathbf{e}_\mathrm{i})\underline{T}_0(f,\mathbf{e}_\mathrm{s})\underline{\Gamma}_\mathrm{q}(f)\right)\underline{W}_\mathrm{P}(f,\mathbf{e}_\mathrm{i}) \tag{4.82}$$

Thus, we can conclude for the scattering-free case

$$\underline{\Lambda}_0(f,\mathbf{e}_\mathrm{i},\mathbf{e}_\mathrm{s}) + \underline{R}_0(f,\mathbf{e}_\mathrm{i})\underline{T}_0(f,\mathbf{e}_\mathrm{s})\underline{\Gamma}_\mathrm{q}(f) = \delta(\mathbf{e}_\mathrm{i}+\mathbf{e}_\mathrm{s}) = \begin{cases} 1 & \text{for} \quad \mathbf{e}_\mathrm{s} = -\mathbf{e}_\mathrm{i} \\ 0 & \text{for} \quad \mathbf{e}_\mathrm{s} \neq -\mathbf{e}_\mathrm{i} \end{cases} \tag{4.83}$$

(4.83) and (4.78)–(4.80) are only met simultaneously if $\underline{\Gamma}_\mathrm{q}(f) = 1$, which coincides with our expectation from the electrical equivalent circuit above. Hence, we finally find for the antenna scattering length

$$\underline{\Lambda}_0(f,\mathbf{e}_\mathrm{i},\mathbf{e}_\mathrm{s}) = \delta(\mathbf{e}_\mathrm{i}+\mathbf{e}_\mathrm{s}) - \underline{R}_0(f,\mathbf{e}_\mathrm{i})\underline{T}_0(f,\mathbf{e}_\mathrm{s}) \tag{4.84}$$

and the whole scattering matrix of the canonical minimum-scattering antenna becomes

$$\underline{\mathbf{S}}_\mathrm{CA}(f,\mathbf{e}_\mathrm{i},\mathbf{e}_\mathrm{s}) = \begin{bmatrix} 0 & \underline{R}_0(f,\mathbf{e}_\mathrm{i}) \\ \underline{T}_0(f,\mathbf{e}_\mathrm{s}) & \delta(\mathbf{e}_\mathrm{i}+\mathbf{e}_\mathrm{s}) - \underline{R}_0(f,\mathbf{e}_\mathrm{i})\underline{T}_0(f,\mathbf{e}_\mathrm{s}) \end{bmatrix} \tag{4.85}$$

whose time domain counterpart is

$$\mathbf{S}_{\mathrm{CA}}(t, \mathbf{e}_\mathrm{i}, \mathbf{e}_\mathrm{s}) = \begin{bmatrix} 0 & R_0(t, \mathbf{e}_\mathrm{i}) \\ T_0(t, \mathbf{e}_\mathrm{s}) & \delta(t, \mathbf{e}_\mathrm{i} + \mathbf{e}_\mathrm{s}) - R_0(t, \mathbf{e}_\mathrm{i}) * T_0(t, \mathbf{e}_\mathrm{s}) \end{bmatrix} \tag{4.86}$$

If the receiver should absorb as much as possible power $P_{\mathrm{R,max}} = \int \Phi(f)\,\mathrm{d}f$ from the incident field, its reflection coefficient must be zero $\Gamma_\mathrm{q} = 0$ leading to power conversion in the receiver of

$$\Phi_{\mathrm{R,max}}(f) = \underline{b}'_0(f)\underline{b}'^*_0(f) = \underline{R}_0(f, \mathbf{e}_\mathrm{i})\underline{R}^*_0(f, \mathbf{e}_\mathrm{i})\,\overline{\mathfrak{P}}_\mathrm{P}(f, \mathbf{e}_\mathrm{i}) \tag{4.87}$$

if $\overline{\mathfrak{P}}_\mathrm{P}(f, \mathbf{e}_\mathrm{i}) = \underline{W}_\mathrm{P}(f, \mathbf{e}_\mathrm{i})\underline{W}^*_\mathrm{P}(f, \mathbf{e}_\mathrm{i})$ is power density (magnitude of pointing vector) of the incident wave. The field backscattered from the antenna under this condition is $(\mathbf{e}_\mathrm{s} \neq \mathbf{e}_\mathrm{i})$

$$\underline{W}_\mathrm{B}(f, r\,\mathbf{e}_\mathrm{s}) = \underline{W}_\mathrm{SP}(f, r\,\mathbf{e}_\mathrm{s})|_{\Gamma_\mathrm{q}=0} = -\underline{R}_0(f, \mathbf{e}_\mathrm{i})\underline{T}_0(f, \mathbf{e}_\mathrm{s})\underline{W}_\mathrm{P}(f, \mathbf{e}_\mathrm{i})\frac{\mathrm{e}^{j\frac{2\pi f r}{c}}}{r} \tag{4.88}$$

Its total power results in (respecting (4.78))

$$\begin{aligned} \Phi_\mathrm{S}(f) &= \oint \underline{W}_\mathrm{B}(f, r\,\mathbf{e}_\mathrm{s})\,\underline{W}^*_\mathrm{B}(f, r\,\mathbf{e}_\mathrm{s})\mathrm{d}S \\ &= \underline{R}_0(f, \mathbf{e}_\mathrm{i})\underline{R}^*_0(f, \mathbf{e}_\mathrm{i})\,\overline{\mathfrak{P}}_\mathrm{P}(f, \mathbf{e}_\mathrm{i}) \oint \underline{T}_0(f, \mathbf{e}_\mathrm{s})\underline{T}^*_0(f, \mathbf{e}_\mathrm{s})\mathrm{d}\Omega \\ &= \underline{R}_0(f, \mathbf{e}_\mathrm{i})\underline{R}^*_0(f, \mathbf{e}_\mathrm{i})\,\overline{\mathfrak{P}}_\mathrm{P}(f, \mathbf{e}_\mathrm{i}) = \Phi_{\mathrm{R,max}}(f) \end{aligned} \tag{4.89}$$

confirming the well-known fact that an ideal antenna under matched conditions reradiates the same energy as it provides to the load.

The conclusions for short-range applications may be twofold. If an ambitious suppression of multiple reflections is required, this can be achieved either by antennas of small height $\underline{R}_0(f, \mathbf{e}_\mathrm{i})$ (which is equivalent to small aperture A or low gain G antenna) and/or by connecting the antenna feed with high-impedance circuits. Note that the last approach implies that the receiver and transmitter electronics must be placed as close as possible to the radiation point of the antenna in order to avoid impedance transformation by the feeding cable or antenna internal propagation.

4.6.2
Spectral Domain Antenna Parameters

After having determined the scattering matrix of the canonical minimum-scattering antenna, we will define a number of useful antenna parameters. First, we stay in the spectral domain to bridge the gap between the conventionally known quantities and the introduced antenna scattering matrix. Thereafter, we will switch to the time domain where a multiplicity of antenna parameters can be defined.

Directive gain [1] and effective isotropic radiated power (EIRP): The directive gain of an antenna relates the radiated power density $\overline{\mathfrak{P}}_{\mathrm{P}}(f, r\,\mathbf{e})$ at distance r and directed towards direction $\mathbf{e}$ to the mean power density $\overline{\mathfrak{P}}_{\mathrm{P,av}}(f, r) = \Phi_{\mathrm{rad}}(f)/4\pi r^2$ radiated by the considered antenna ($\Phi_{\mathrm{rad}}(f) = \underline{a}'_0(f)\underline{a}'^*_0(f)$ – radiated power per Hz).[11)]

$$D(f, \mathbf{e}) = \frac{\overline{\mathfrak{P}}_{\mathrm{P}}(f, r\,\mathbf{e})}{\overline{\mathfrak{P}}_{\mathrm{P,av}}(f, r)} \tag{4.90}$$

With

$$\overline{\mathfrak{P}}_{\mathrm{P}}(f, \mathbf{r}) = \underline{W}_{\mathrm{SP}}(f, \mathbf{r})\underline{W}^*_{\mathrm{SP}}(f, \mathbf{r}) = \frac{\underline{T}_0(f, \mathbf{e})\underline{T}^*_0(f, \mathbf{e})}{r^2}\underline{a}'_0(f)\underline{a}'^*_0(f)$$

and

$$\overline{\mathfrak{P}}_{\mathrm{P,av}}(f, r) = \frac{1}{4\pi r^2}\oint \overline{\mathfrak{P}}_{\mathrm{P}}(f, \mathbf{r})\mathrm{d}S = \frac{\Phi_{\mathrm{rad}}(f)}{4\pi r^2}$$

we gain

$$D(f, \mathbf{e}) = 4\pi r^2 \frac{\overline{\mathfrak{P}}_{\mathrm{P}}(f, r\mathbf{e})}{\Phi_{\mathrm{rad}}(f)} = 4\pi \underline{T}_0(t, \mathbf{e})\underline{T}^*_0(t, \mathbf{e}) \tag{4.91}$$

The maximum value of the directive gain is shortly termed as directivity $D_{\max} = \|D(\mathbf{e})\|_\infty = D(\mathbf{e}_0)$. The direction $\mathbf{e}_0$ points towards the direction of maximum radiation intensity. We will call it boresight direction. Relating the directivity gain to $D_{\max}$, we get the directivity pattern or normalized gain pattern

$$\Theta(f, \mathbf{e}) = \frac{D(f, \mathbf{e})}{D_{\max}(f)} = \frac{D(f, \mathbf{e})}{D(f, \mathbf{e}_0)} \tag{4.92}$$

and finally we will still mention the effective isotropic radiated power

$$\mathrm{EIRP} = \Phi_{\mathrm{rad}}(f) D_{\max}(f) \quad \left[\frac{\mathrm{W}}{\mathrm{Hz\ sr}}\right] \tag{4.93}$$

which is a quantity to characterize the strongest field intensity per Hz and steradian provided by an antenna and a power source. Typically it is given in dBm/MHz.

Bandwidth: The antenna bandwidth relates to the width of the spectral band within the intensity of the radiated wave does not fall below a ceratin level:

$$B = \max(f) - \min(f) \quad \text{for} \quad D(f, \mathbf{e}_0) > \varrho \max(D(f, \mathbf{e}_0)) \tag{4.94}$$

Mostly, the bandwidth is only given for boresight direction and the cut-off value is typically related to a fraction ϱ of $\varrho = 1/2 \to -3$ dB, $\varrho = 1/4 \to -6$ dB or $\varrho = 1/10 \to -10$ dB where the last value is the preferred one in UWB technique.

11) Note a_0 is only a virtual signal feeding the ideal antenna. It is not the actual input signal.

Beamwidth: The beamwidth covers a solid angle $\Delta\Omega_B$ within the radiation intensity does not fall below a certain level, that is

$$\Delta\Omega_B = \int \mathbf{e}_B d\mathbf{S} \tag{4.95}$$

for all radiation directions $\mathbf{e}_B$ which meet $\Theta(\mathbf{e}_B) \geq \varrho$. $\Delta\Omega_B$ can be approximated by referring to the beamwidth in two orthogonal directions, for example $\mathbf{e}_\varphi$ and $\mathbf{e}_\vartheta$. Supposing the half-power beamwidth in both directions is $\Delta\varphi$ and $\Delta\vartheta$, we can roughly state

$$\begin{aligned} \Delta\Omega_B &\cong \Delta\varphi\,\Delta\vartheta \\ D_{max} &\cong \frac{4\pi}{\Delta\varphi\,\Delta\vartheta} \end{aligned} \tag{4.96}$$

Non-ideal radiator: So far all considerations did refer to the ideal antenna. Let us now also respect some non-ideal properties. For that purpose, we will first come back to the scattering matrix of the whole antenna as introduced in (4.75).

$$\begin{aligned} \underline{\mathbf{S}}_A(f, \mathbf{e}_i, \mathbf{e}_s) &= \begin{bmatrix} \underline{\Gamma}(f) & \underline{R}(f, \mathbf{e}_i) \\ \underline{T}(f, \mathbf{e}_s) & \underline{\Lambda}(f, \mathbf{e}_i, \mathbf{e}_s) \end{bmatrix} \\ &= \begin{bmatrix} \underline{S}_{A11}(f) & \underline{S}_{A12}(f)\,\underline{R}_0(f, \mathbf{e}_i) \\ \underline{S}_{A21}(f)\,\underline{T}_0(f, \mathbf{e}_s) & \underline{\Lambda}_0(f, \mathbf{e}_i, \mathbf{e}_s) + \underline{\Lambda}_x(f, \mathbf{e}_i, \mathbf{e}_s) \end{bmatrix} \end{aligned} \tag{4.97}$$

$\underline{\Lambda}_0(f, \mathbf{e}_i, \mathbf{e}_s)$ specifies the re-radiation of the idealised antenna (see (4.84)) and the parasitic scatter length $\underline{\Lambda}_x$ joins all unwanted scatter effects caused due to an inefficient antenna structure and internal reflections

$$\underline{\Lambda}_x(f, \mathbf{e}_i, \mathbf{e}_s) = \underbrace{\underline{\Lambda}_{str}(f, \mathbf{e}_i, \mathbf{e}_s)}_{\text{structural reflections}} + \underbrace{\underline{S}_{A22}(f)\underline{R}_0(f, \mathbf{e}_i)\underline{T}_0(f, \mathbf{e}_s)}_{\text{re-radiation due to internal mismatch}} \tag{4.98}$$

Structural reflections and internal mismatch cannot be separated by measurements. A real antenna usually suffers from ohmic or dielectric losses. Hence, the absolute values of the internal transmission coefficients $|\underline{S}_{A21}(f)|$, $|\underline{S}_{A12}(f)|$ are smaller than 1:

$$\underline{S}_{A21}(f)\underline{S}^*_{A21}(f) = \underline{S}_{A12}(f)\underline{S}^*_{A12}(f) = \eta_A \leq 1 \tag{4.99}$$

The phase of $\underline{S}_{A21}(f) = \underline{S}_{A12}(f)$ is a measure of the internal delay between antenna feed and radiation centre (see Figure 4.11 or 4.13). The feed point reflection $\underline{S}_{A11}(f) \neq 0$ is a further source of losses.

Our antenna model can also be applied for active antennas, i.e. antennas which have amplifiers integrated close to their radiation centre. But then one has to be careful with the reciprocity of the arrangement which does not longer hold for transmission paths including the amplifier. Hence, an antenna may either act as receiver or as transmitter. In that case we have

$$\begin{array}{lll} \text{Transmitting antenna}: & |S_{A21}(f)| > 1; & |S_{A12}(f)| < 1 \\ \text{Receiving antenna}: & |S_{A21}(f)| < 1; & |S_{A12}(f)| > 1 \end{array}$$

Gain and realized gain [1]: With regard to the antenna losses, we define various power spectral quantities:

$$\text{Power incident at the antenna}: \quad \Phi_{in}(f) = \underline{a}_0(f)\underline{a}_0^*(f)$$

$$\text{Power accepted by the antenna}: \quad \Phi_{ac}(f) = \left(1 - \underline{S}_{A11}(f)\underline{S}_{A11}^*(f)\right)\Phi_{in}(f)$$
$$\text{Actually radiated power}: \quad \Phi_{rad}(f) = \eta_A(f)\Phi_{ac}(f)$$

The accepted power Φ_{ac} excludes the return loss from the power budged, and Φ_{rad} disregards also internal antenna losses (expressed by the efficiency factor η_A) within the radiation balance. These spectral power quantities can be used to extend above-introduced antenna gain definition (4.91) by including several loss mechanisms of the antenna.

$$\text{Realised gain}: \quad G(f, \mathbf{e}) = 4\pi r^2 \frac{\overline{\mathfrak{P}}_p(f, \mathbf{r})}{\Phi_{in}(f)} = 4\pi \underline{T}(f, \mathbf{e})\underline{T}^*(f, \mathbf{e}) \tag{4.100}$$

$$\text{Antenna gain}: \quad G_0(f, \mathbf{e}) = 4\pi r^2 \frac{\overline{\mathfrak{P}}_p(f, \mathbf{r})}{\Phi_{ac}(f)} \tag{4.101}$$

$$\text{Directive gain}: \quad D(f, \mathbf{e}) = 4\pi r^2 \frac{\overline{\mathfrak{P}}_p(f, \mathbf{r})}{\Phi_{rad}(f)} = 4\pi \underline{T}_0(f, \mathbf{e})\underline{T}_0^*(f, \mathbf{e}) \tag{4.102}$$

Based on these definitions, the different antenna gain quantities are related by

$$G(f, \mathbf{e}) = \left(1 - \underline{S}_{A11}(f)\underline{S}_{A11}^*(f)\right)G_0(f, \mathbf{e}) = \left(1 - \underline{S}_{A11}(f)\underline{S}_{A11}^*(f)\right)\eta_A(f)D(f, \mathbf{e}) \tag{4.103}$$

Effective aperture: The effective aperture is a parameter characterizing the receive properties of the antenna. Usually, one restricts the specification of this parameter to the boresight direction even if it has the same angular dependence as the gain to which it is related via the reciprocity relation (4.44). The antenna aperture is defined as (see also (4.47))

$$\text{Effective aperture}: \quad A(f) = \frac{\underline{h}_0(f)\underline{h}_0^*(f)}{\overline{\mathfrak{P}}_p(f)} \mathrel{\widehat{=}} \frac{\lambda^2 G(f)}{4\pi} \tag{4.104}$$

It includes internal antenna losses. Exclusion of these losses leads to an efficient antenna area of

$$A_0 = \frac{\lambda^2 D}{4\pi} = \eta_A A \tag{4.105}$$

Structural efficiency: Finally, we will introduce a parameter which relates to the scattering properties of the antenna. We have seen that already an ideal antenna scatters waves in dependency on its transmission properties and the feeding port load.

A real antenna will provoke additional scattering just on the basis of its geometric structure. In many applications this is of no relevance but it is not negligible in some short-range sensing tasks due to the risk of multiple reflections between target and antenna. Hence, we need a figure of merit which allows us to assess different antenna concepts with respect to their backscatter property.

Following the Mason graph in Figure 4.29, the total scattering $\underline{\Lambda}_{tot}(f, \mathbf{e}_i, \mathbf{e}_s)$ at the 'space' port takes the form of (arguments are omitted for shortness):

$$\underline{\Lambda}_{tot} = \underbrace{\underline{\Lambda}_{str} + \underline{\Lambda}_0 + \underline{R}_0 \underline{T}_0 \underline{S}_{A22}}_{\underline{\Lambda}_1} + \underbrace{\underline{R}_0 \underline{T}_0 \underline{S}_{A21} \underline{S}_{A12} \underline{S}_{FL}^2 \underline{\Gamma}_q}_{\underline{R}\,\underline{T}\,\underline{\Gamma}'_q} = \underline{\Lambda}_1 + \underline{R}\,\underline{T}\,\underline{\Gamma}'_q \tag{4.106}$$

Herein, the apparent reflection factor $\underline{\Gamma}'_q = \underline{S}_{FL}^2 \underline{\Gamma}_q$ refers to the load reflection $\underline{\Gamma}_q$ transformed to the feed point plane.

Figure 4.30 depicts the basic behavior of the 'space' port reflections in the complex plane for different cases. The shown examples refer to a single frequency and a fixed incidence and observation direction. On the left-hand side, we find the case for an idealized minimum scattering antenna. The circle emphasized in grey colour covers the area of possible reflections which can be taken by the antenna in dependence on its apparent load reflection $\underline{\Gamma}'_q$. The circle may be measured by connecting the antenna with a shortened feeding cable of variable time delay τ. If the apparent reflection factor at the feeding port becomes one, the antenna will not re-radiate; that is, it will be invisible. Avoiding any cable, on open antenna feed 'sees' $\underline{\Gamma}'_q = 1$ for all frequencies. Hence, the antenna stays invisible at any frequency.

The other two cases refer to real antennas which will always show structural reflections. The structural reflection will be typically caused by different geometric construction elements. Hence, in addition to the re-radiated waves, we have to count for constructive or destructive interferences so that either $|\underline{\Lambda}_1| > |\underline{T}\,\underline{R}|$ or $|\underline{\Lambda}_1| < |\underline{T}\,\underline{R}|$ may be happen. It will not be possible to decide which of both currently dominates scattering from single-frequency measurements. As obvious from Figure 4.30 (right-hand side), one can always find an appropriate antenna load to make the antenna invisible as long as $|\underline{\Lambda}_1| < |\underline{T}\,\underline{R}|$ holds (see also [26, 31]). Since interference phenomena are frequency and space dependent, this condition may be maintained only within a narrow frequency band and a narrow beamwidth. A mutual compensation of scattering components is not feasible under UWB conditions.

We find from Figure 4.30 that the radius of the circle equals the distance of its midpoint to the origin in the case of the minimum scattering antenna but they are different for real antennas. Hence, we can take this disparity to distinguish a

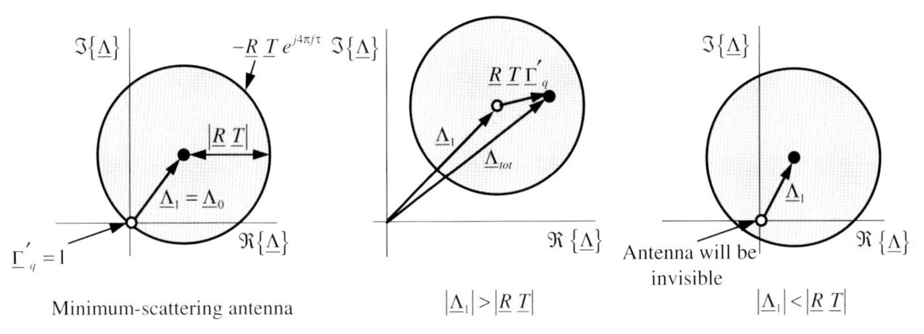

Figure 4.30 Antenna 'space' port reflections in the complex Λ-plane. The grey circle covers the area over which the antenna reflectivity may be controlled via the feeding port load. Note that the diagrams refer to particular frequency and directions $\mathbf{e}_\mathrm{i}$, $\mathbf{e}_\mathrm{S}$.

real antenna from an ideal one. However, we have to take the averaged quantities in order to equalize the effect of destructive and constructive interferences. On this basis, we define the structural efficiency as follows:

$$\Xi(\mathbf{e}_\mathrm{i}, \mathbf{e}_\mathrm{S}) = \frac{\text{quadratic mean of circle diameter}}{\text{quadratic mean of midpoint distance}} = \frac{\int_B |\underline{T}(f, \mathbf{e}_\mathrm{s})\underline{R}(f, \mathbf{e}_\mathrm{i})|^2 \mathrm{d}f}{\int_B |\underline{\Lambda}_1(f, \mathbf{e}_\mathrm{i}, \mathbf{e}_\mathrm{s})|^2 \mathrm{d}f} \tag{4.107}$$

B is the considered bandwidth.

The structural efficiency Ξ equals one for the ideal case of the minimum scattering antenna and it tends towards zero if the structural reflections dominate; that is, the average antenna radar cross section increases:

$$\overline{\sigma}_\mathrm{A} = \frac{4\pi}{B} \int_B |\underline{\Lambda}_1|^2 \mathrm{d}f \tag{4.108}$$

The measurement of the structural efficiency in frequency domain is quite cumbersome. As we will see next, time domain measurements will reach similar results much faster.

Hence a good antenna for short-range applications should provide low structural scattering. This can be achieved by reducing the metal content and high-permittivity materials for the antenna construction. Sophisticated implementations are based on active or passive field probes which are fed by optical cables for both signals and power supply [32]. Other approaches use resistively loaded antennas of low cross-sectional area with respect to the incident field (e.g. Vee-antennas [33, 34]) and antenna constructions which shift structural reflection of the antenna attachment out of the propagation time window of interest [35].

4.6.3 Time Domain Antenna Parameters

The content of this chapter is to be seen from the viewpoint of precise range determination by time delay measurements, that is $r = ct_\mathrm{d}$. Supposing the propagation speed c of the sounding wave is known, we have two fundamental problems with respect to this task. First, the range r is determined by the distance between two well-defined points in space, and second, the delay time t_d refers to two points (i.e. points in time). That is, we are referenced to points in both cases but actually we have involved geometric objects (antennas, scatter) and time signals of finite dimension or duration. In order to deal with this problem, we already defined antenna and scattering IRFs which we linked to a point in space. In Section 2.2.2, a pulse of finite duration was assigned to a single time position.

Here, we will consider how the knowledge of the antenna IRFs may be exploited for precise range measurements and how to estimate the range errors if some approximations are used. Basically, we have to consider all objects interfering with the sounding wave (receiving and transmitting antenna, scatter object) in order to get precise range information. But at first we will limit our discussion to a single antenna of finite dimensions which may operate either as transmitter or as receiver. A later extension of such consideration which respects all objects within the propagation path will be straightforward.

From the frequency domain scattering matrix $\underline{\mathbf{S}}_A(f, \mathbf{e}_i, \mathbf{e}_s)$ (see (4.97)) of a real antenna, the time domain counterpart immediately follows to

$$\begin{aligned} \mathbf{S}_A(t, \mathbf{e}_i, \mathbf{e}_s)\Big|_P &= \begin{bmatrix} S_{A11}(t) & R(t, \mathbf{e}_i) \\ T(t, \mathbf{e}_s) & \Lambda(t, \mathbf{e}_i, \mathbf{e}_s) \end{bmatrix}\Bigg|_P \\ \text{with} \quad R(t, \mathbf{e}_i) &= R_0(t, \mathbf{e}_i) * S_{A12}(t) \\ T(t, \mathbf{e}_s) &= T_0(t, \mathbf{e}_s) * S_{A21}(t) \\ \Lambda(t, \mathbf{e}_i, \mathbf{e}_s) &= \Lambda_0(t, \mathbf{e}_i, \mathbf{e}_s) + \Lambda_x(t, \mathbf{e}_i, \mathbf{e}_s) \end{aligned} \tag{4.109}$$

From now, we will not make anymore a difference between canonical and real antenna. Therefore, we re-introduce the reference point (plus orientation) P to which all IRFs have to be referred. By having measured the angel-dependent scattering parameters $\mathbf{S}_A(t, \mathbf{e}_i, \mathbf{e}_s)|_P$ related to this point, all electrical properties of the antenna are known and distance estimations could be basically done at precision in the frame of the IRF measurement error. However, this is not very practical in most of the cases as demonstrated by Figure 4.31. It implies that we need for every direction of incidence and observation the related IRF of all involved antennas. This may lead to a data volume which is difficult to handle in practice. Furthermore, the angles of incidence or of observation are not known a priori so that one has to approach iteratively to the actual angles in order to take always the correct antenna IRF for data processing. Lengthy and computational expensive procedures would be the consequences.

How we can reduce this effort for practical purposes? Let us consider a simple example. In all radar measurements with the aim to localize a target or to image a scenario, one maps round trip time to distance. Figure 4.31 illustrates a simple scenario in which the distance to two targets should be measured based on a radiator of finite dimensions. For the sake of shortness, we simplify the targets by point objects placed at position P_1 and P_2.

We will start our consideration with an antenna of unknown behavior. In order to perform a high-resolution measurement, we need at least the angle-dependent

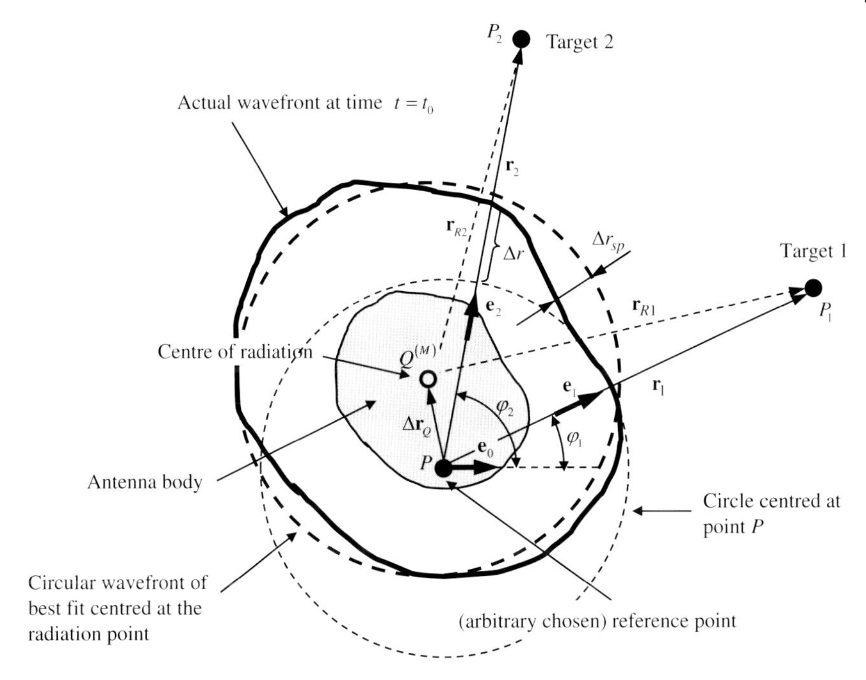

Figure 4.31 Illustration of round trip time measurement by mono-static radar. The bold line symbolizes the spatial shape of the wavefront at time t_0 after launching. The targets are supposed to be point scatterer in order to simplify the problem.

transmitting or receiving IRFs $R(t, \mathbf{e})|_P$ or $T(t, \mathbf{e})|_P$ of the antenna. For their determination, we fix a reference point P (typically located within the antenna volume) and a reference direction[12] $\mathbf{e}_0$ which we select more or less arbitrary due to missing information about the antenna.

We determine the impulse responses and discover that the wavefront of the transmitted wave is localized at a curve deviating from a circle as depicted by the bold line in Figure 4.31. Basically this would not be a problem since we know its exact shape. Hence, we can precisely calculate the distance $|\mathbf{r}_1| = \overline{PP_1}$ or $|\mathbf{r}_2| = \overline{PP_2}$ from round-trip time measurements.

In order to do this correctly, we have to respect that the wave hurries on ahead by Δr in $\mathbf{e}_2$-direction compared to $\mathbf{e}_1$-direction. $\Delta r = c\,\Delta t$ can be taken from $T(t, \mathbf{e})$- or $R(t, \mathbf{e})$-pattern if φ_1, φ_2 are known (note, we also need ϑ_1, ϑ_2 for 3D problems). Unfortunately, the angles are usually not known and they cannot be determined by the simple arrangement as depicted in Figure 4.31. Their estimation would require more than only one antenna.

We can simplify the situation if we approximate the actual wavefront by a spherical one of best approach (dotted bold circle) which we centre at point $Q^{(M)}$. We call

12) The complete assignment of the orientation in space requires two direction vectors. Since we not yet deal with polarized fields and we only consider 2D examples here, we can omit the second one for the moment.

point $Q^{(M)}$ the effective centre of radiation since it is the source of a quasi-spherical wave. Taking this point as reference, we can immediately estimate the target distances $|\mathbf{r}_{R1}|$ and $|\mathbf{r}_{R2}|$ from round-trip time measurements if we are able to tolerate the remaining deviations $\Delta r_{sp}(\mathbf{e})$ between the quasi-spherical wavefront and a real sphere.

This leads us to the questions how to determine the radiation centre from measurements and what are the parameters of the antenna IRF which determine the range error Δr_{sp}.

4.6.3.1 Effective Centre of Radiation

The determination of all time domain antenna parameters supposes the knowledge of the radiation centre. Except for electrical small antennas, it is usually not trivial to predefine the radiation centre from the antenna geometry. Hence, it has to be determined from measurements. We fix an arbitrary reference point P and direction $\mathbf{e}_0$ of the antenna for that purpose. We place the antenna at a turntable and align $\mathbf{e}_0$ with its start position and the point P with the centre of revolution. Then the radiation pattern $T(t, \mathbf{e})|_P$ or $R(t, \mathbf{e})|_P$ of the antenna under test (AUT) is measured respecting a sufficiently large observer distance r_0 (see Section 4.6.5). Basically, it does not matter whether $T(t, \mathbf{e})|_P$ or $R(t, \mathbf{e})|_P$ is measured since both are mutually related by reciprocity (4.43). Hence, we can restrict ourselves to the transmission IRF in what follows.

The result of such measurement is symbolized in Figure 4.32a in polar coordinates. Here, the radius axis has a twofold meaning. We refer it to a distance or the elapsed time. For the second case, time zero $t = 0$ is placed in the centre of the

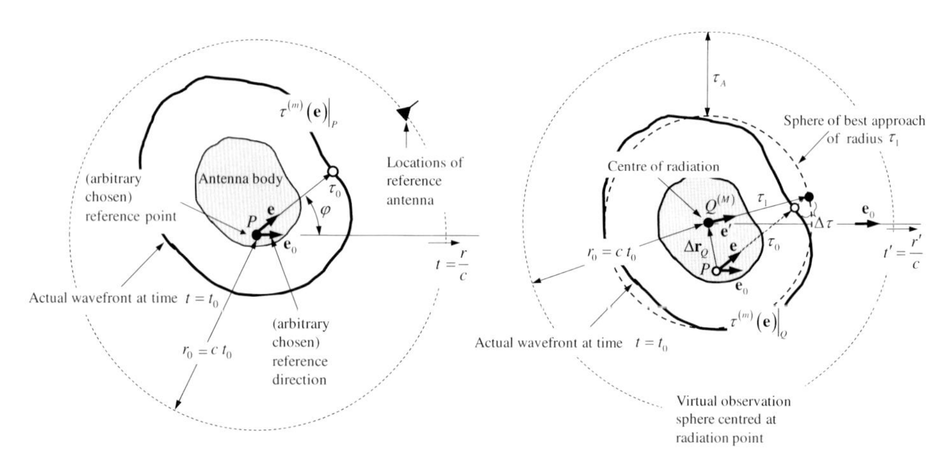

Figure 4.32 Determination of effective radiation point by centre transformation of antenna IRF. The centre transformation approximates the real wavefront by a sphere whereas it maps the empty spot at distance τ_0 from point P to the black spot at distance τ_1 from point $Q^{(M)}$.

coordinate system and time t_0 refers to a circle of radius $r_0 = ct_0$. t_0 is the time needed by an isotropic wave launched at point P and time $t = 0$ to achieve at observer positions which are localized on a circle of radius r_0 around the antenna under test. As depicted in the figure, the actual wavefront of the radiated field looks usually different.

We get this wavefront by stimulating the antenna feed with a short pulse and determining the angle-dependent propagation delay $\tau^{(m)}(\mathbf{e})\big|_P$ of the signal received by the (ideally point like) reference antenna. Time delay measurement is based on pulse position determination which can be performed by several methods so it will be hard to define a unique approach given by the nature of the problem. Therefore, we apply the superscript $^{(m)}$ to indicate the method of pulse position estimation. As introduced in Section 2.2.2, the superscript $^{(m)}$ may refer to threshold crossing ($m \triangleq 50\%$) or several centres of gravity ($m \triangleq p$) (including maximum position $m \triangleq \infty$) of pulse signals or the correlation function of time-stretched wideband signals.

As demonstrated by Figure 4.31, the eccentricity of the propagation delay $\tau^{(m)}(\mathbf{e})\big|_P$ (Figure 4.32a) complicates precise range measurements since it relates to point P. If we could refer the distance measurement to point $Q^{(M)}$ as depicted in Figure 4.32b, the delay time and distance would not depend on the observation angle given that the difference between the real propagation delay $\tau^{(m)}(\mathbf{e})\big|_Q$ and the sphere of best approach (radius τ_1) is within an acceptable error bound. The sphere of best approach models a virtual isotropic wave which best approximates the wavefront of the real wave transmitted by the antenna. Since the virtual isotropic wave is not able to reach the observer position within the time $t_0 = r_0/c$, the time lag τ_A (see Figure 4.32) must be caused by propagation delay between measurement plane and radiation centre (compare Figure 4.11). In order the gain the midpoint $Q^{(M)}$ and the internal antenna delay τ_A we centre the measured antenna IRF $T(t, \mathbf{e})|_P$ by a corresponding transformation:

$$T(t, \mathbf{e})|_{Q^{(M)}} = \frac{\tau_1}{\tau^{(m)}(e)|_P} T(t - \Delta\tau(\mathbf{e}) - \tau_A, \mathbf{e})|_P; \quad \Delta\tau(\mathbf{e}) = \frac{\Delta\mathbf{r}_Q \cdot \mathbf{e}}{c} \tag{4.110}$$

This transformation supposes the knowledge of $\Delta\mathbf{r}_Q \triangleq \Delta\mathbf{r}_Q^{(M)}$ and $\tau_A \triangleq \tau_A^{(M)}$. They can be determined by several approaches from the measured IRF wherefore we use again a superscript $^{(M)}$ to indicate the applied method.

Method1: Osculating Circle ($^{(M)} \rightarrow {}^{(oc)}$)

Typically, one considers the direction of largest wave intensity and approaches an osculating circle to the $\tau^{(m)}(\mathbf{e})\big|_P$ -curve in such a way that it touches the curve in the main beam direction. The midpoint $Q^{(oc)}$ of the osculating circle represents the virtual source of the isotropic wave. The determination of the osculating circle is based on curvature estimation which needs derivations of first and second order. They are sensitive to random errors. Furthermore, the osculating circle embraces only a small part of the wavefront $\tau^{(m)}(\mathbf{e})\big|_P$ around the main beam direction. Hence, the method is not robust against noise and the approximation may hold only within a small angle around the main beam direction.

Method 2: Error Minimization ($^{(M)} \rightarrow {}^{(\Delta\tau)}$)

Here, we refer to the deviation $\Delta\tau(\mathbf{e})$ of the wavefront $\tau^{(m)}(\mathbf{e})\big|_P$ from a circle (sphere in 3D). For sufficiently large distance between AUT and reference antenna, we can postulate $\mathbf{e} \approx \mathbf{e}'$ which leads to

$$\tau_1^2 \approx \left(\frac{\Delta r_Q}{c}\right)^2 + \left(\tau^{(m)}(\mathbf{e})\big|_P + \Delta\tau(\mathbf{e})\right)^2 - \left(\tau^{(m)}(\mathbf{e})\big|_P + \Delta\tau(\mathbf{e})\right)\frac{\Delta\mathbf{r}_Q \cdot \mathbf{e}}{c}$$
$$\Delta\tau(\mathbf{e}) \approx \tau_1 - \tau^{(m)}(\mathbf{e})\big|_P + \frac{\Delta\mathbf{r}_Q \cdot \mathbf{e}}{c}; \quad \text{since} \quad \left[\tau_1, \tau^{(m)}(\mathbf{e})\big|_P\right] \gg \left[\Delta\tau(\mathbf{e}), \frac{|\Delta\mathbf{r}_Q|}{c}\right] \tag{4.111}$$

The best coincidence between the real waveform and the isotropic wave model is achieved if we succeed to minimize the overall error in dependence on midpoint location and radius of the sphere of best approach. For that search, we can involve the whole radiation around the antenna or we respect only the strongest fields within a given beamwidth $\Delta\Omega_B$. That is, we get the radiation centre from a multi-dimensional search corresponding to

$$\text{for the whole antenna:} \quad \left[\Delta\mathbf{r}_Q^{(\Delta\tau)}, \tau_1^{(\Delta\tau)}\right] = \underset{\Delta\mathbf{r}_Q, \tau_1}{\arg\min} \int\limits_{4\pi} \Delta\tau^2(\mathbf{e})\,d\Omega \tag{4.112}$$

or

$$\text{for the main beam:} \quad \left[\Delta\mathbf{r}_Q^{(\Delta\tau)}, \tau_1^{(\Delta\tau)}\right] = \underset{\Delta\mathbf{r}_Q, \tau_1}{\arg\min} \int\limits_{\Delta\Omega_B} \Delta\tau^2(\mathbf{e})\,d\Omega \tag{4.113}$$

From this, the position vector $\mathbf{r}_Q$ of the radiation centre follows to $\mathbf{r}_Q = \mathbf{r}_P + \Delta\mathbf{r}_Q$ if $\mathbf{r}_P$ is the position vector of the arbitrary selected reference point. The internal delay time follows from $\tau_A = t_0 - \tau_1 = (r_0/c) - \tau_1$.

Method 3: Minimization of Weighted Error $^{(M)} \rightarrow {}^{(\mathrm{we})}$

If we want to rate the impact of $\Delta\tau(\mathbf{e})$ according to the angle-dependent wave intensity, we should replace the simple time difference in (4.112) or (4.113) by a moment consisting of time and L_p-norm:

$$\left[\Delta\mathbf{r}_Q^{\mathrm{we}}, \tau_1^{\mathrm{we}}\right] = \underset{\Delta\mathbf{r}_Q, \tau_1}{\arg\min} \int\limits_{\Delta\Omega_x} \left\|T(t,\mathbf{e})|_P\right\|_p \Delta\tau^2(\mathbf{e})\,d\Omega; \quad \Delta\Omega_x \,\hat{=}\, \Delta\Omega_B;\ 4\pi \tag{4.114}$$

To illustrate the concept of radiation centre, we refer to a measurement example based on rigid horn antenna (Figure 4.33). We restrict ourselves to the 2D-antenna IRF in the azimuth plane for demonstration. We select the antenna rear as the reference point which we place in the centre of the turntable axis. By this choice, we can expect that the radiation centre will not coincide with the rotation axis of the turntable; we will get an eccentric radiation pattern. In fact, Figure 4.33 affirms our expectation. After applying centre transformation, the strongest signal components are well centred in the radiation centre. We can also observe that weaker signal

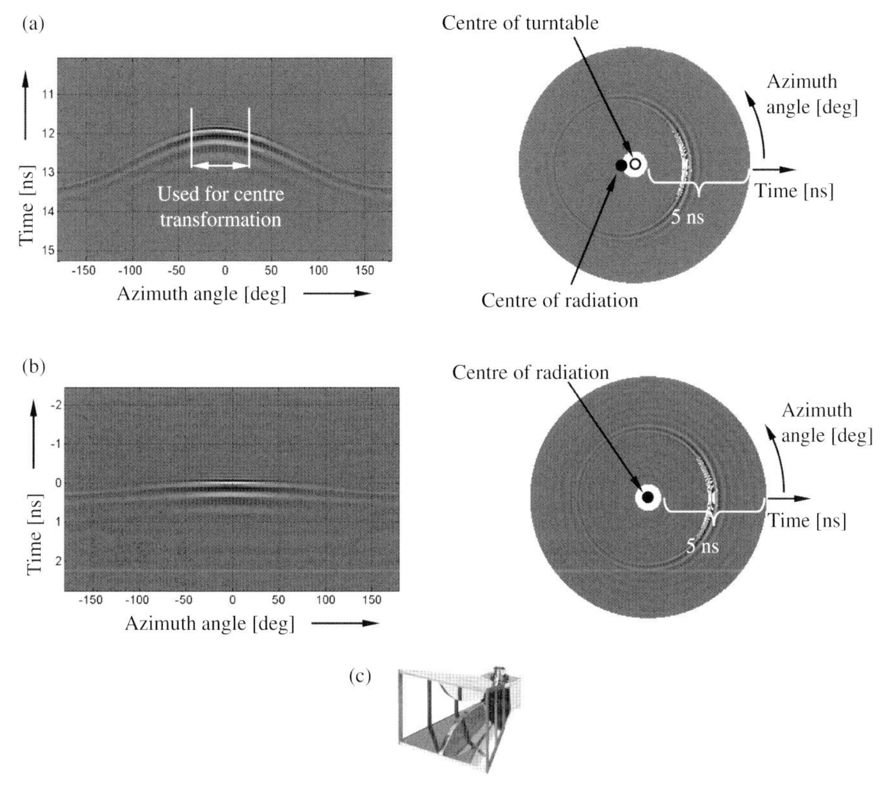

Figure 4.33 Impulse response of a DRH-20 Antenna in Cartesian (left) and polar coordinates (right). (c) Photo of the antenna showing the axis of revolution; (a) impulse response $T(t,\varphi)|_P$ referred to the centre of revolution and (b) Impulse response after centre transformation $T(t,\varphi)|_{Q^{(\Delta\tau)}}$. The radiation centre was determined via (4.113) (i.e. with respect to the main beam) and infinity norm for pulse position determination. (colored figure at Wiley homepage: Fig_4_33)

parts are not centred. From that we can conclude that the antenna has more than one radiation centre leading to angle-dependent pulse shape of the radiated wave.

In summary, we can state that a multitude of definitions and approaches exist to define a radiation centre. So the simplest way is to select a definition which best coincides with the intended data processing for the later application. In practice, the delay time τ_A is usually the most important error term for range measurements, since it includes not only antenna internal delay but also the delay caused by interconnection cables of antenna and electronics. The magnitude of $\Delta\mathbf{r}_Q$ cannot exceed the dimensions of the antenna if the reference point was not placed outside the antenna body so that it mostly remains quite small. If an absolute range error in the order of the antenna dimension is acceptable, one can renounce the elaborate determination of the radiation centre and one may be satisfied with a simple time zero determination to remove at least the internal antenna delay τ_A from the data.

We have found in the radiation centre $Q^{(M)}$ a 'dominant' point (even if it cannot be uniquely defined) to which we can assign the antenna position. We also like to replace the definition of antenna orientation by referencing to more or less 'natural' reference directions unless the antenna does not have ones. Two of such preferential directions are the orientation of the electric field in case of linear polarized antenna and the main beam direction for non-omnidirectional radiators.

Some characteristic functions will be introduced that may serve to estimate the range error if the real antenna IRF is replaced by an equivalent antenna of simplified radiation characteristic. Basically, we can define for every parameter of an antenna IRF a directivity pattern and a corresponding beamwidth so that here also the intended requirements of a later application should give the guideline for the reasonable selection of characteristic functions and parameters. Here, we will only give a selection of possible definitions (see also Refs [4, 7] for further discussion).

We define the radiation characteristic of an equivalent antenna as follows:

$$T_{eq}(t, \mathbf{e})\big|_{Q^{(M)}} = \begin{cases} T_0(t) & \text{for } \mathbf{e} \text{ within the beamwidth } \Delta\Omega_B \\ 0 & \text{for } \mathbf{e} \text{ outside the beamwidth } \Delta\Omega_B \end{cases} \tag{4.115}$$

That is, the equivalent antenna has an angle-independent IRF within a conic section of space assigned by the beamwidth $\Delta\Omega_B$ and outside of this space sector its radiation is supposed to be zero.[13] The time evolution of the IRF is taken from the real antenna after having determined its radiation centre. We will refer to two options to select an appropriate IRF $T_0(t)$ from the real antenna. First, it seems to be reasonable to take the IRF of the boresight direction $\mathbf{e}_0$

$$T_0(t) = T(t, \mathbf{e}_0)\big|_{Q^{(M)}}$$

and second, the average IRF within a limited solid angle $\Delta\Omega_0$ around the boresight direction may also be meaningful for this purpose:

$$T_0(t) = \int\limits_{\Delta\Omega_0} T(t, \mathbf{e})\big|_{Q^{(M)}} \mathrm{d}\Omega$$

4.6.3.2 **Boresight Direction and Canonical Position**

We define the boresight direction as the direction of the strongest radiation. Hence in time domain, we have several options given by the direction $\mathbf{e}_0$ which maximizes one of the L_p-norms of the antenna IRF:

$$\mathbf{e}_0 = \underset{\mathbf{e}}{argmax} \left\| T(t, \mathbf{e})\big|_{Q^{(M)}} \right\|_p \tag{4.116}$$

If the antenna is in canonical position, its radiation centre is placed at the centre of the coordinate system, the boresight direction is aligned to the azimuth axis

13) The idea behind the equivalent antenna which replaces a real antenna by a perfect conical radiator is comparable with the replacement of a pulse signal or spectrum by a rectangular function of equivalent width (see Sections 2.2.2 and 2.2.5). There, we used several options to define the equivalent widths which can be basically copied also to our consideration here.

$\varphi = 0$, and the second reference direction (e.g. polarization) coincides with the z-axis. This alignment, we will assume in what follows. Omnidirectional radiators like dipoles do not have unique direction of maximum radiation, so that at least one of the two reference directions has to be selected arbitrarily.

4.6.3.3 Time Domain Directive Gain Pattern

In imitation of (4.92), we relate the strength of the impulse response of an arbitrary direction to the strength of the main beam.

$$\Theta_{\mathrm{p}}^{\mathrm{D}}(\mathbf{e}) = \frac{\left\| T(t,\mathbf{e})\big|_{Q^{(M)}} \right\|_p}{\left\| T(t,\mathbf{e}_0)\big|_{Q^{(M)}} \right\|_p} = \frac{\left\| T(t,\mathbf{e})\big|_{Q^{(M)}} \right\|_p}{\| T_0(t) \|_p} \tag{4.117}$$

4.6.3.4 Spherical Deformation Pattern

The wavefront provided from a real antenna usually deviates from an ideal sphere even if it is placed in the radiation centre. These deviations are an important error source in high-resolution range measurements. There exist several options to parameterize such deviations. Referring to Figure 4.32b, we will only introduce one absolute and one relative quality figure which relates the deviations to the pulse width:

$$\text{Absolute}: \quad \Delta\tau_{\mathrm{p}}(\mathbf{e}) = \tau^{(m)}(\mathbf{e})\big|_{Q^{(M)}} - \tau_1 \tag{4.118}$$

$$\text{Relative}: \quad \Theta_{\mathrm{p}}^{\mathrm{SD}} = \frac{\Delta\tau_{\mathrm{p}}(\mathbf{e}) = \tau^{(m)}(\mathbf{e})\big|_{Q^{(M)}} - \tau_1}{t_{\mathrm{w}}} \approx \left(\Delta\tau_{\mathrm{p}}(\mathbf{e}) = \tau^{(m)}(\mathbf{e})\big|_{Q^{(M)}} - \tau_1 \right) \underset{\cdot\cdot}{B} \tag{4.119}$$

t_{w} is the pulse duration and $\underset{\cdot\cdot}{B}$ is the bandwidth.

4.6.3.5 Fidelity and Fidelity Pattern

The fidelity as introduced in (2.54) serves as benchmark of signal shape deformation. We use it here in a twofold manner. First, we like to have a global figure quantifying the signal deformation by passing an antenna. For that purpose, we refer to the receive signal $b(t)$ and the incident field $W_{\mathrm{P}}(t,\mathbf{e}_0)$ in boresight direction.[14)]

$$\mathrm{FI_A} = \frac{\| W_{\mathrm{p}}(-t,\mathbf{e}_0) * b(t) \|_\infty}{\| W_{\mathrm{p}}(t,\mathbf{e}_0) \|_2 \| b(t) \|_2} = \frac{\| R(t,\mathbf{e}_0) * C_{\mathrm{WW}}(t) \|_\infty}{\sqrt{C_{\mathrm{WW}}(t=0)\left[C_{\mathrm{RR}}(t,\mathbf{e}_0) * C_{\mathrm{WW}}(t) \right]_{t=0}}}$$

$$\text{using} \quad C_{\mathrm{WW}}(t) = W_{\mathrm{p}}(t,\mathbf{e}_0) * W_{\mathrm{p}}(-t,\mathbf{e}_0); \quad C_{\mathrm{RR}}(t,\mathbf{e}_0) = R(t,\mathbf{e}_0) * R(-t,\mathbf{e}_0) \tag{4.120}$$

The fidelity tends to one for all narrowband systems since the involved signals are of the same time shape, namely sinusoids. Hence, this figure is less appropriate for narrowband applications.

14) The definition can also be taken for transmit mode quantities after a minor modification which respects the derivative behavior of the transmit mode: $FI_A = \frac{\| u_{-1}(t) * W_{SP}(t,r\mathbf{e}_0) * a(-t) \|_\infty}{\| u_{-1}(t) * W_{Sp}(t,r\mathbf{e}_0) \|_2 \| a(t) \|_2}$.

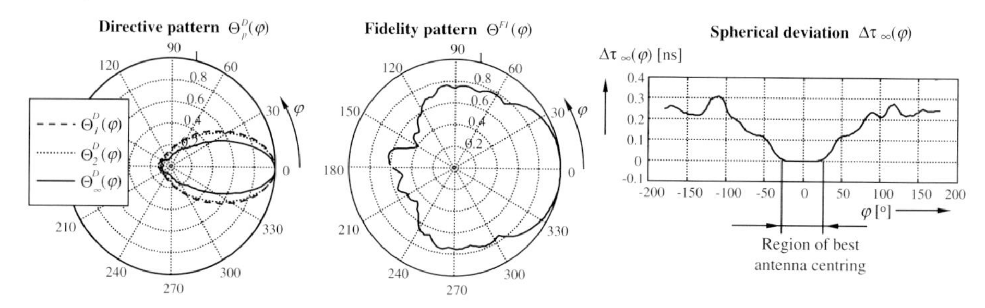

Figure 4.34 Some time domain radiation patterns in the azimuth plane for the antenna from Figure 4.33 after centre transformation. The antenna was polarized in $\mathbf{e}_z$-direction.

The second fidelity figure (fidelity pattern) is used to assess time shape as function of the radiation angle. Here, the boresight signal $b(t, \mathbf{e}_0)$ is referred to receive signals $b(t, \mathbf{e})$ of oblique incidence:

$$\Theta^{\mathrm{FI}}(\mathbf{e}) = \frac{\|b(-t, \mathbf{e}_0) * b(t, \mathbf{e})\|_\infty}{\|b(t, \mathbf{e}_0)\|_2 \|b(t, \mathbf{e})\|_2} = \frac{\|R(-t, \mathbf{e}_0) * R(t, \mathbf{e})\|_\infty}{\|R(t, \mathbf{e}_0)\|_2 \|R(t, \mathbf{e})\|_2} \tag{4.121}$$

As the fidelity value (4.120), this function too has no meaning for narrowband antennas.

Figure 4.34 depicts some of the above-introduced characteristic functions for the antenna from Figure 4.33. The directive pattern is shown for three L_p-norms in order to illustrate the differences in the estimation methods. We note a beamwidth of about $\pm 25°$ from this diagram. The fidelity pattern indicates within the same angular range only minor deviations of the signal shape. Hence, one could decide to restrict the operational space sector of the radar to that beamwidth since one can expect here the best precision. Consequently, the centre transformation (4.113) is applied. It provides the centre of an osculating circle which causes minimum deviation within the beamwidth. Actually, we can observe from the spherical deviation pattern that within the beamwidth no time difference referred to an isotropic wave exists.

4.6.3.6 Structural Efficiency Pattern

The aim of this antenna quality figure is to express the degradation of antenna backscattering due to non-ideal construction. It represents the time domain counterpart to (4.107). We define it by

$$\Xi(\mathbf{e}_\mathrm{i}, \mathbf{e}_\mathrm{s}) = \frac{\|\Lambda_0(t, \mathbf{e}_\mathrm{i}, \mathbf{e}_\mathrm{s})\|_2^2}{\|\Lambda_1(t, \mathbf{e}_\mathrm{i}, \mathbf{e}_\mathrm{s})\|_2^2} \tag{4.122}$$

It can be modified by referring to the Mason graph Figure 4.29 and (4.109):

$$\Xi(\mathbf{e}_\mathrm{i}, \mathbf{e}_\mathrm{s}) = \frac{\|R(t, \mathbf{e}_\mathrm{i}) * T(t, \mathbf{e}_\mathrm{s})\|_2^2}{\|\Lambda_1(t, \mathbf{e}_\mathrm{i}, \mathbf{e}_\mathrm{s})\|_2^2} = \frac{\left\|\Lambda_{\mathrm{tot}}^{(\mathrm{o,s})} - \Lambda_{\mathrm{tot}}^{(m)}\right\|_2^2}{\left\|\Lambda_{\mathrm{tot}}^{(m)}\right\|_2^2} = \frac{\left\|\Lambda_{\mathrm{tot}}^{(\mathrm{o})} - \Lambda_{\mathrm{tot}}^{(\mathrm{s})}\right\|_2^2}{\left\|\Lambda_{\mathrm{tot}}^{(\mathrm{o})} + \Lambda_{\mathrm{tot}}^{(\mathrm{s})}\right\|_2^2} \tag{4.123}$$

where $\Lambda_{tot}^{(o)} = \Lambda_1(t, \mathbf{e}_i, \mathbf{e}_s) + R(t, \mathbf{e}_i) * T(t, \mathbf{e}_s)$ is the antenna scattering with feeding port open: $\Gamma_q = 1$, $\Lambda_{tot}^{(m)} = \Lambda_1(t, \mathbf{e}_i, \mathbf{e}_s)$ is the antenna scattering with feeding port matched: $\Gamma_q = 0$ and $\Lambda_{tot}^{(s)} = \Lambda_1(t, \mathbf{e}_i, \mathbf{e}_s) - R(t, \mathbf{e}_i) * T(t, \mathbf{e}_s)$ is the antenna scattering with feeding port short: $\Gamma_q = -1$.

We can arbitrarily continue to define antenna pattern for other time domain parameters such as pulse width, decay and so on. But this is left for the reader who can extend the set of definitions according to the needs. The characterization of scattering responses $\Lambda(t, \mathbf{e}_i, \mathbf{e}_s)$ may follow the same philosophy. It will be a straightforward procedure which does not need further discussion. We will rather shortly address some other approaches to compact antenna and scatterer description.

4.6.4 Parametric Description of Antenna and Scatterer

So far, we tried to extract from a bulky data set of angle-dependent pulse responses $T(t, \mathbf{e}_s)$, $R(t, \mathbf{e}_i)$ or $\Lambda(t, \mathbf{e}_i, \mathbf{e}_s)$ some pulse shape–related quantities and their angular dependence in order to compact the antenna and scatterer description. Now, we will mention some other approaches to reduce the data amount or to provide features for some kind of assessments or object classification.

The basic idea behind all of these methods is the decomposition of the actual antenna or target IRF into canonical functions or the IRF of canonical objects. We like to refer to two basic methods:

1) **Eigenmode decomposition:** As depicted in Figure 4.27, we can formally grasp an antenna or scatterer as a black box model with many input and output channels with the exception that the scatter does not have the feeding channels a_0, b_0. The interaction of both types of bodies with the electromagnetic field is described by a set of linear partial differential equations. If we approximate the partial differential equations by a finite set of ordinary differential equations, we will end up in the just-mentioned black box model from Figure 4.27. As we have seen in Section 2.4, such a model may be represented in different ways. One of them describes it by parameters of an equivalent system comprising lumped elements.[15)]

   A very powerful and demonstrative parametric description is given by the state-space model (Section 2.4.4). There, we have separated the forced and the unforced behavior of the objects. Translated to our conditions here, the forced signal components (described by the matrix **D**) can be considered as direct transmissions or reflections (e.g. specular reflections) of an antenna or scatterer. The unforced behavior is characterized by the matrix **A**. It expresses the intrinsic behavior of the object and is hence of major interest in target recognition. As mentioned in Section 2.4.4, it is meaningful to transform the system matrix **A** into a diagonal form. The diagonal elements represent the natural modes (eigenmodes) of the system (poles in the complex *s*-plane). These eigenmodes are the only (damped) oscillations which the object can

15) Special attention has to be paid if signal delay dominates the behavior since it is difficult to model it by small number of lumped elements.

perform freely after it is stimulated. These modes are characteristic of the considered object, irrespective of the direction of incidence or observation.

The state-space representation has still involved two further matrices **B** and **C**. We have called them stimulation matrix (in control theory it is called control matrix) and observation matrix. They relate to the strength on how the different modes are coupled with the input and output channels. Since the channel numbering is equivalent to the directions of incidence or observation here, the entries of **B** or **C** gives an angle-dependent weight of the excitability and visibility of the eigenmodes.

Referring to (2.170) and (2.171), the eigenmode decomposition of the discretized antenna scattering matrix (4.74) or (4.75) takes the form

$$\text{Laplace domain}: \quad \underline{\mathbf{S}}_{\mathrm{A}}(s, \mathbf{e}_{\mathrm{i}}, \mathbf{e}_{\mathrm{s}}) \mathrel{\hat{=}} \underline{\mathbf{S}}_{\mathrm{A}}(s, n, m) = \mathbf{C}(s\,\mathbf{I} - \mathbf{A})^{-1}\mathbf{B} + \mathbf{D} \tag{4.124}$$

$$\text{Time domain}: \quad \mathbf{S}_{\mathrm{A}}(t, \mathbf{e}_{\mathrm{i}}, \mathbf{e}_{\mathrm{s}}) \mathrel{\hat{=}} \mathbf{S}_{\mathrm{A}}(t, n, m) = \mathbf{C}\,\mathrm{e}^{\mathbf{A}t}\,\mathbf{B} + \mathbf{D}\,\delta(t) \tag{4.125}$$

Assuming, our test object exhibits M eigenmodes and we have divided the surrounding space into N segments of solid angle $\Delta\Omega$, we have to deal with the following matrix dimensions:

Single-port antenna	Scatterer
$n, m \in [0, N]$	$n, m \in [1, N]$
$\mathbf{S}_{\mathrm{A}} - [N+1, N+1]$	$\mathbf{S}_{\mathrm{A}} - [N, N]$
$\mathbf{A} - [M, M]$	$\mathbf{A} - [M, M]$
$\mathbf{B} - [M, N+1]$	$\mathbf{B} - [M, N]$
$\mathbf{C} - [N+1, M]$	$\mathbf{C} - [N, M]$
$\mathbf{D} - [N+1, N+1]$	$\mathbf{D} - [N, N]$

The scatterer is considered here as an 'antenna' without feeding port.

Practically, the entries for the $\mathbf{A}, \mathbf{B}, \mathbf{C}, \mathbf{D}$ matrices have to be extracted from measurements. Here, one is faced with some problems. Typically the entries are found by fitting the model parameters to the behavior of the real object. The very basic idea of this approach is summarized by the two equations (2.163) and (2.164) in Section 2.4.4, which also gives some references for further studies. Singular value decomposition, matrix pencil, ARMA modelling and Prony's method are related topics to deal with algorithmic issues. MATLAB, widely spread in engineering community, supports a lot of corresponding algorithms. Nevertheless, an open issue remains. It relates to the selection of an appropriate model order which is equivalent to the expected number of eigenmodes. This must be done carefully in order to avoid over- or under-modelling. Another point relates to the measurement errors and the robustness of the fitting algorithms. A huge scientific work on system identification and parameter estimation work was done within the last decades. A discussion of corresponding topics within the frame of this book would go beyond its scope. Therefore, the interested reader is addressed to the numerously available textbooks. Specific aspects of modelling in connection with wave scattering are found under the term singularity expansion method pioneered by C. E. Baum [36, 37]. Parametric antenna modelling is summarized in Refs [38, 39] and polarimetric (see Chapter 5 for wave polarization) target decomposition can be found in Ref. [40]. Some experimental results of parametric modelling also involving human targets are presented, for example, in Ref. [41].

2) **Decomposition into spherical harmonics:** The spherical harmonics are solutions to Laplace's equations in spherical coordinates. They form an orthogonal function basis and are defined[16] as [20]

$$\underline{Y}_n^m(\vartheta,\varphi) = P_n^m(\cos\vartheta)e^{jm\varphi} \tag{4.126}$$

where $P_n^m(x)$ represents the associated Legendre function[17] of degree n and order m:

$$P_n^m(x) = (-1)^m \frac{(1-x^2)^{m/2}}{2^n n!}\frac{d^{n+m}}{dx^{n+m}}(x^2-1)^n; \quad m \in [-n,n];\ x \in [-1,1] \tag{4.127}$$

The orthogonality relation is

$$\int_0^{\pi}\int_{-\pi}^{\pi} Y_{n_1}^{m_1}(\vartheta,\varphi)Y_{n_2}^{m_2}(\vartheta,\varphi)\sin\vartheta\, d\varphi\, d\vartheta = \upsilon_0\,\delta(m_1,m_2)\delta(n_1,n_2)$$
$$\upsilon_0 = (-1)^m \frac{4\pi}{2n+1} \tag{4.128}$$

Now, we are doing the same with the spherical harmonics as we did with the Fourier transform in case of time function (see introduction to Section 2.2.5, equations (2.55) and (2.56)), that is, we decompose the radiated field (angle-dependent function) into a set of 'convenient' functions with angular arguments. Since the spherical harmonics represent a set of general solutions of spherical field problems, they are well matched to our requirements. Considering, for example, the (measured) transmit FRF of an antenna $\underline{T}(f=f_0,\mathbf{e}) = \underline{T}_0(\vartheta,\varphi)$ at the frequency $f=f_0$, the decomposition into spherical harmonics (also assigned as spherical mode expansion), takes the form

$$\underline{T}_0(\vartheta,\varphi) = \sum_{n=-\infty}^{\infty}\sum_{m=-n}^{n} a(n,m)\underline{Y}_n^m(\vartheta,\varphi) \tag{4.129}$$

16) Here, we use the angles ϑ and φ to indicate directions since it is not usual to apply the vector notion $\mathbf{e}$ in the literature: $\cos\vartheta = \mathbf{e}\cdot\mathbf{e}_z$; $e^{jm\varphi} = \left(\mathbf{e}\cdot\mathbf{e}_x + j\sqrt{1-(\mathbf{e}\cdot\mathbf{e}_x)^2}\right)^m$.

17) Since MATLAB calculates the function only for the orders $m \in [0,n]$, we give here for completeness the relation between positive and negative values of m:
$P_n^{-m}(x) = (-1)^m \frac{(n-m)!}{(n+m)!} P_n^m(x)$.

This equation is the counterpart to (2.55) for time signal decomposition. The modal coefficients $a(n,m)$ express the strength of the individual spherical modes. They are determined from the measured antenna FRF by the same approach as depicted in (2.57) whereas the orthogonality relation (4.128) has to be respected here:

$$a(n,m) = \frac{1}{v_0} \int_0^{\pi} \int_{-\pi}^{\pi} \underline{T}_0(\vartheta,\varphi) Y_n^m(\vartheta,\varphi) \sin\vartheta \, d\varphi \, d\vartheta; \quad v_0 = (-1)^m \frac{4\pi}{2n+1} \tag{4.130}$$

The calculation of the coefficients $a(n,m)$ is stopped if their values fall below a certain threshold. One gets typically a low number of spherical mode coefficients $a(n,m)$ per frequency point, which require much less memory space than the original FRF $\underline{T}_0(\vartheta,\varphi)$.

Finally, we must emphasize that the intention of (4.129) and (4.130) is to introduce the method and that they are based on the assumption of scalar waves which is not true in case of electromagnetism. Hence, in order to be more precise, the spherical harmonics and the modal coefficient extraction (4.130) have to be extended for vector waves. Such exercise is, for example, done in Refs [42, 43] for far-field antenna pattern as we supposed here.

So far, the spherical harmonic decomposition provides for every frequency point of the antenna FRF a complete set of modal coefficients. That is, the antenna properties at the frequency f_0 are not linked to its properties at a different $f_0 + \Delta f$ since the decomposition (4.130) works separately for every frequency point. However, the eigenmode decomposition above has shown that there must be some relation between different frequency components otherwise we may not find a finite number of eigenmodes. But this decomposition method was not able to give us some mutual relations concerning the angular dependencies of the radiation behavior, which is done by the spherical harmonic decomposition. Hence, it is reasonable to combine both methods as introduced by Refs [42, 43]. Thus first, the eigenvalue decomposition is performed providing a set of residuals for every mode and every direction, which are then further decomposed by spherical harmonics individually for every eigenmode so that one succeeds to describe the radiation behavior of an antenna by a relative low number of coefficients.

Further methods to decompose the behavior of an antenna or scatterer may reference to canonical object, for example dipole, quadrupole, sphere, dihedral corner, trihedral corner and so on. Here, one models the real object by an ensemble of canonical objects which should behave like the real object. For further readings, the reader can refer to Refs [18, 44].

4.6.5
Distance and Angular Dependence of Antenna Functions and Parameters

So far we assumed that radiated fields, impulse or frequency response functions and antenna parameters only depend on the aspect angle of the observer but not

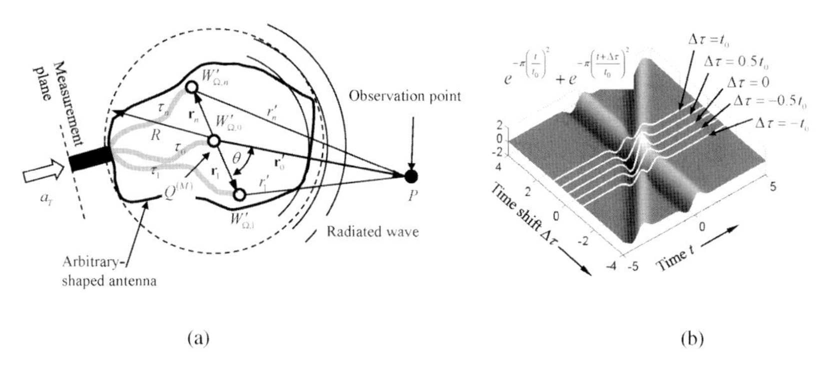

Figure 4.35 Superposition of time signals (b) and estimation of the total field in the observation point P(a).

on its distance to the antenna. Let us investigate if it is really true. We will demonstrate this by evaluating the total field of a large antenna in the observation point P at position $\mathbf{r}'_0$. The effective radiation centre $Q^{(M)}$ of the antenna is placed at the origin of the coordinate system (see Figure 4.35a). In order to estimate the total radiation of the antenna, we apply Huygens's principle. For that purpose, we fractionize the antenna in a number of infinitesimal small volume segments which all provide a spherical wave. The volume density of the source yield at position $\mathbf{r}_n \in V_A$ is denoted as $W'_{\Omega,n}(t, \mathbf{r}_n)$.

The total field in the observation point arises from the summation of the fields radiated from all point sources located within the antenna volume V_A. Assuming for the sake of simplicity that the waves can propagate freely within the antenna,[18] we get using (4.16)

$$W_{\text{SP}}(t, \mathbf{r}'_0) = \int\limits_{V_A} \frac{W'_{\Omega,n}(t, \mathbf{r}_n) * \delta(t - r'_n/c)}{r'_n} \mathrm{d}V = \int\limits_{V_A} \frac{W'_{\Omega,n}(t - r'_n/c, \mathbf{r}_n)}{r'_n} \mathrm{d}V \tag{4.131}$$

This equation imposes that we have to superimpose time functions of different mutual shift $\Delta\tau = \tau_n + r'_n/c$. Herein, τ_n is an antenna internal delay from the feeding port (measurement plane) to the individual radiation points $\mathbf{r}_n$ (symbolized by the grey lines). It is fixed by the construction of the antenna as long as we do not involve antenna arrays in our discussion. The second delay term is of major interest here since it is variable. It depends on distance and direction of the observation point.

The effect of a variable time delay $\Delta\tau$ on the superposition of time functions is depicted in Figure 4.35b. While the sum signal of sine waves is again a sinusoid (only amplitude and phase alter), the time evolution of wideband signals will diversify depending on the mutual delay as illustrated in the figure for the example of

18) This is a fairly unrealistic assumption since the metallic construction of the antenna will lead to reflection and diffraction. Our main intention is to show the effect of antenna size wherefore we will omit such effects in order to get to the point.

first derivative of a Gaussian pulse. The amplitude doubles and the time shape is preserved only for $\Delta\tau = 0$. With increasing time shift the sum signal will modify its time evolution until both signal components may again be separated for large time lags.

A major concern of UWB sensing relates to precise propagation time measurements. It requires the knowledge of the time shape of involved signals. We give a rough estimate of time shape variations depending on the observation point in order to have an indication of expected range errors. The propagation path length r_n of the individual waves can be approximated as

$$\begin{aligned} r'_n &= \sqrt{(\mathbf{r}'_0 - \mathbf{r}_n)\cdot(\mathbf{r}'_0 - \mathbf{r}_n)} = r'_0\sqrt{1 + \frac{r_n^2}{r'^2_n} - 2\frac{\mathbf{r}_n\cdot\mathbf{r}'_0}{r'^2_0}} \\ &\cong r'_0 - \frac{\mathbf{r}_n\cdot\mathbf{r}'_0}{r'_0} + \frac{1}{2}\frac{r_n^2}{r'_0}\left(1 - \left(\frac{\mathbf{r}_n\cdot\mathbf{r}'_0}{r'_0\, r_n}\right)^2\right); \quad \text{for} \quad r'_0 \gg r_n \\ &\simeq r'_0 - r_n\cos\theta + \frac{r_n^2}{2r'_0}\sin^2\theta \end{aligned} \tag{4.132}$$

Figure 4.36 shows the error $\Delta r'_n$ of the first- and second-order approximations corresponding to (4.132).

Next, we will ask for the maximum allowed difference of propagation delays which only marginally affects the time shape of the total wave in the observation point P. In order to assess the shape deformation, we deal with the relative amplitude difference Δw of the partial waves emanating from the reference point $\mathbf{r}_0 = \mathbf{0}$ and an arbitrary point $\mathbf{r}_n$ within the antenna volume:

$$\begin{aligned} \Delta w &= \frac{\left|W_{\mathrm{SP},n}(t,\mathbf{r}'_n) - W_{\mathrm{SP},0}(t,\mathbf{r}'_0)\right|}{\left\|W_{\mathrm{SP},0}(t,\mathbf{r}'_0)\right\|_\infty} \approx \frac{\left|W'_{\Omega,n}(t-\tau'_n,\mathbf{r}_n) - W'_{\Omega,0}(t-\tau'_0,\mathbf{r}_0=\mathbf{0})\right|}{\left\|W'_{\Omega,0}(t-\tau'_0,\mathbf{r}_0=\mathbf{0})\right\|_\infty} \\ &\text{for } r'_n \approx r'_0 \quad \text{with} \quad W_{\mathrm{SP},n}(t,\mathbf{r}'_n) = \frac{1}{r'_n}W'_{\Omega,n}(t-\tau'_n,\mathbf{r}_n); \quad \tau'_n = r'_n\, c \end{aligned} \tag{4.133}$$

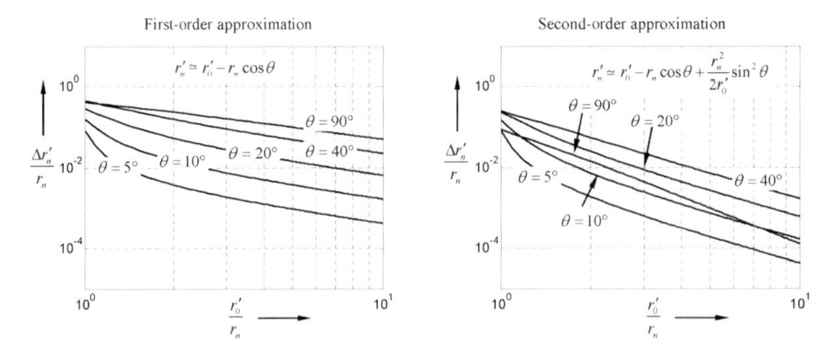

Figure 4.36 Normalized approximation error. $\Delta r'_n$ assigns the difference between exact calculation and the approximation corresponding to (4.132).

If this difference remains below a certain bound $\Delta w \le \Delta w_{\max}$, the deformations of the time evolution should be acceptable in our application. In order to get this value, we suppose spatially independent source yields $W'_{\Omega,n}(t, \mathbf{r}_n) = W'_{\Omega}(t)$. For small propagation time differences $\Delta\tau_n = \tau'_n - \tau'_0$, (4.133) can be approximated by

$$\Delta w \cong \frac{\left|\left(W'_{\Omega}(t) + \Delta\tau_n \frac{\partial W'_{\Omega}(t)}{\partial t}\right) - W'_{\Omega}(t)\right|}{\left\|W'_{\Omega}(t)\right\|_{\infty}} = \frac{\Delta\tau_n}{\left\|W'_{\Omega}(t)\right\|_{\infty}} \left|\frac{\partial W'_{\Omega}(t)}{\partial t}\right| \tag{4.134}$$

The slew rate of the waveform is $\mathrm{SR} = \left|\partial W'_{\Omega}/\partial t\right|_{\max} \approx \left\|W'_{\Omega}\right\|/t_{\mathrm{r}}$ (t_{r} is the rise time). Further, the waves originating from the border of the antenna have the largest propagation time differences:

$$\Delta\tau_{n,\max} = \left|r'_n - r'_0\right|/c = \frac{R}{c}\left(\cos\theta + \frac{R}{2r'_0}\sin^2\theta\right) \tag{4.135}$$

so that the maximum amplitude deviation may be estimated to

$$\Delta w_{\max} \cong \frac{\Delta\tau_{n,\max}}{t_{\mathrm{r}}} = \frac{R}{t_{\mathrm{r}}c}\left(\cos\theta + \frac{R}{2r'_0}\sin^2\theta\right) = \Delta w_1(\theta) + \Delta w_2\left(\theta, r'_0\right) \tag{4.136}$$

The quotient R/ct_{r} represents the ratio between antenna size and spatial extension of the waveform edges. As (4.136) shows, the field deformations depend on two terms. The first of them is only dependent on the angle θ. This term is already respected by our antenna model introduced in Section 4.4 in which we have defined the angle-dependent antenna IRFs $T(t, \mathbf{e})$ and $R(t, \mathbf{e})$. However, the second term that additionally involves the observer distance r'_0 is not respected by theses IRFs. Hence, we have to ensure that $\Delta w_2\left(\theta, r'_0\right)$ falls below an acceptable error bound $\Delta w_{2,\max}$ in order to deal with a correct wave and antenna model. This condition is met if the observer distance is larger than

$$r'_0 \ge \frac{R^2}{2t_{\mathrm{r}}c\,\Delta w_{2,\max}} \tag{4.137}$$

(4.137) describes the so-called far-field condition, which we already mentioned in (4.22). Correspondingly, we gain for sine waves if we require again $\Delta\tau_{n,\max}c \le \lambda/16$ (compare (4.27))

$$r'_0 \ge \frac{2D^2}{\lambda} = \frac{8\,R^2}{\lambda} \tag{4.138}$$

If we want to avoid the dependence of the antenna impulse response on the aspect angle, that is $\partial T(t,\theta)/\partial\theta \triangleq \partial\Delta w_1(\theta)/\partial\theta \to 0$, then the first term in (4.136) must also fall below a certain error bound:

$$R \le \Delta w_{1,\max} t_{\mathrm{r}} c \tag{4.139}$$

This condition can only be fulfilled for electrically short antennas. Their impact onto the Friis formula and radar equation will be the content of the next subchapter.

Finally we indicate that (4.134) to (4.139) do not involve antenna internal signal delay τ_n between the feeding port and the different point sources and these equations only deal with two individual field contributions originating from point sources located in the radiation centre and on the antenna periphery. Actually, the total field is built from many other point sources which are located nearer to the radiation centre. The above consideration relates to a worst-case estimation which only shows basic tendencies. Reliable estimations of angle- and distance-dependent waveform distortions have to be individualized to the corresponding antenna geometry.

Figure 4.37 shows some simple examples to illustrate the basic influence of antenna size on the radiation behavior of pulsed and sinusoidal waves. For the sake of illustration, we consider a quite idealized radiator – a large dipole of length h perfectly matched so that no reflections are created at its terminations. Since the current provides radiated fields over the entire loop, the 'back loop radiation' is separated by perfect RAM (radar absorbing material) from the 'forward loop radiation' in order to avoid mutual cancellation.

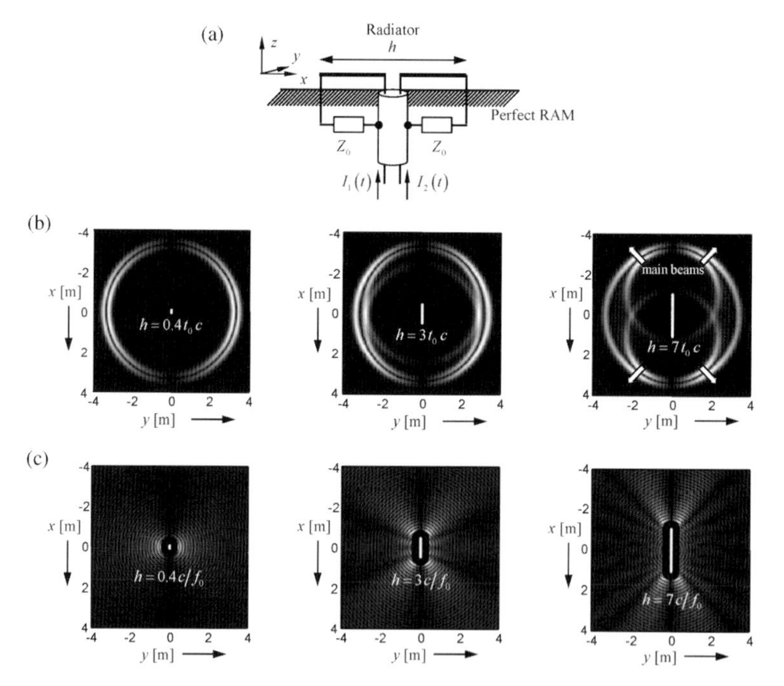

Figure 4.37 Snapshots of the radiated fields $|\mathbf{E}(t_x, x, y)|$ from dipoles of different length. (a) Schematic of the arrangement; (b) pulse stimulation $I_1(t) = -I_2(t) = I_0\, e^{\pi(t/t_0)^2}$ with $t_0 = 1$ ns and (c) sine wave stimulation $I_1(t) = -I_2(t) = I_0 \sin(2\pi f_0 t)$ with $f_0 = 1$ GHz. The near source region is cut out due to singularities of the fields for $r'_n \to 0$. (movies of wave propagation are at Wiley homepage: Fig_4_37_*)

The dipole is placed in the xy-plane and aligned along the x-axis. Since the whole loop will radiate, we separate the upper and lower parts of the radiator by an ideal shielding. The dipole is symmetrically fed, that is $I_1(t) = -I_2(t)$ and the current propagates with the speed of light along the wires. The total fields in Figure 4.37 are simulated by stringing very short dipoles ($h_0 \ll t_0 c$) together and superimposing the electric (vector) field **E** radiated from them. This approach implements (4.131) for a linear antenna. However, (4.131) represents only a simplified version based on the assumption that the individual point sources provide isotropic fields. Here, in order to avoid incorrect impression of the generated fields, we used an extended version which is based on vector waves emanating from short dipoles (see Section 5.5).

The images represent the spatial distribution of the electric field (absolute value) for the time point $t_x = 11,3\,\mathrm{ns}$ after time zero. As we observe from the short dipole, we get a nice circular wavefront of angle-independent time shape, but with decreasing amplitude by approaching the polar region. The time shape of the wave resembles the first derivative of the sounding signal (see (4.140)). With increasing length of the dipole, the time evolution of the pulsed waveform becomes angle dependent. The antenna radiates three circular waves (in the feeding point and at the terminals; see also Ref. [45]) of angle-dependent magnitude. In the case of pure sine wave stimulation, we see that the time shape of the waves is always sinusoidal independent of the radiation direction but the amplitude variations are quite pronounced. Further, the wavefront (phase front) deviates from a circular pattern.

4.6.6
The Ideal Short-Range UWB Radar Equation

It is important for many practical applications to have simple measurement conditions in order to reduce the effort for data processing and error correction which often involves elaborate sensor modelling. In order to concentrate the sounding energy within a required space segment, one needs electrically large antennas which will unavoidably disturb the angular structure of the wavefront in particular at short antenna distances. Energy concentration into a space segment is required for transmissions over large distances to combat the noise perturbations. But random noise is often not the major problem in UWB short-range imaging and high-resolution localization since clutter and the lack of exact knowledge of sounding wavefront cause much higher errors.

Electrically small radiators provide a simple and nearly perfect circular wavefront (compare Figure 4.37) so that their use may lead to better imaging results for short-range sensing due to prior knowledge of their radiation behavior. For illustration and demonstration, we refer to a simple example and experiment depicted in Figure 4.38. The radiators are dipoles whose lengths h are short compared to the pulse and coherence length of the sounding wave.

It is well known that the strength of the generated electric field is proportional to the acceleration of charge carriers which is equivalent to the temporal derivative of the current. We can approximately write for the far field of a short radiator (see

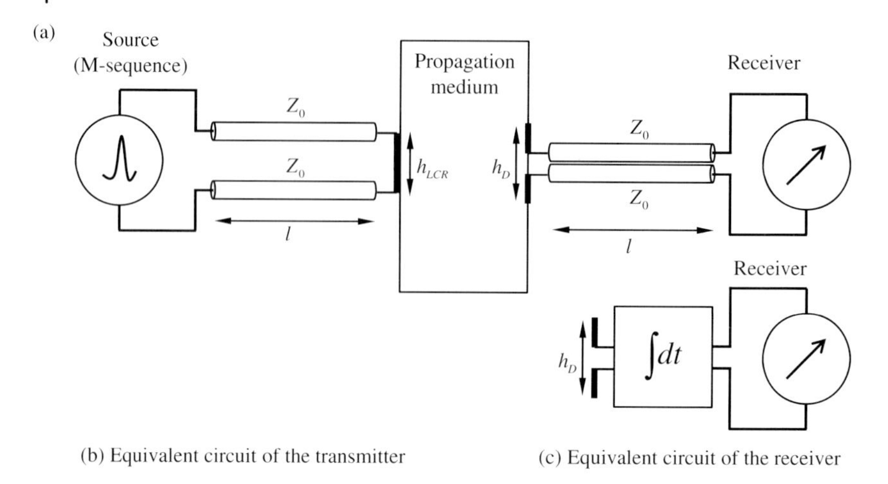

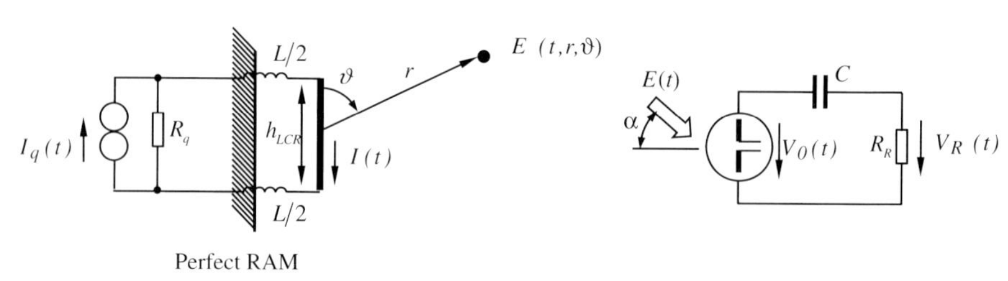

Figure 4.38 Schematic representation of transmission channel with electric short radiators (a) and equivalent representation of the radiator circuits (b) and (c).

Section 5.5 for complete equation; note that we have again omitted the vector nature of the electric field here):

$$E(t,r,\vartheta) = \frac{Z_s h_{\rm LCR}}{4\pi cr} \sin\vartheta \frac{\rm d}{{\rm d}t} I(t) = \frac{Z_s h_{\rm LCR}}{4\pi cr} \sin\vartheta u_1(t) * I(t) \tag{4.140}$$

As shown in Figure 4.38 (Figure 4.38a), we applied two different antenna types – a usual electric dipole (Figure 4.38c) as receiver and a large current radiator (LCR) (Figure 4.38b) as transmitter. The LCR relates to the opportunity to provide high antenna currents on the transmitter side. Low-cost integrated RF-circuits (e.g. SiGe-circuits) are better suited to handle high currents (10–100 mA) than voltages exceeding a few volts. Hence, the LCR potentially provides higher radiation fields than an electrically short dipole.

The large current radiator as a potentially suitable radiator for non-sinusoidal waveforms was pioneered by H. F. Harmuth [46–48] and further investigated in Refs [49–51]. The aim of an UWB radiator is to emit and receive wideband waveforms with less as possible disturbance of their time evolution.

Unfortunately, the LCR is not in practical use so far since it provides some implementation problems. The LCR should not be mixed with the short loop antenna or magnetic dipole. The idea behind the LCR is that only a short part of the wire (emphasized by the bold lines in Figure 4.38a and b) shall emit an electromagnetic wave.

Clearly, the current must flow in a loop so that actually the whole loop will radiate and not only a short part of it as wanted. The fields created by a small loop will interfere and mutually cancel so that the remaining far field becomes dependent on the second derivation of the feeding current, which we did not intend. If we, however, avoid the interference of fields by separating the 'active part' from the rest of the loop by an absorber, the radiation in one of the half spaces will follow the rule depicted by (4.140). The practical challenge is to build the antenna with the required shielding properties.

These difficulties will alleviate for radiation in a high-permittivity material. If we place the antenna immediately at the interface (for some issues on interfacial dipoles see Section 5.5), the dipole will mainly radiate into the body, which largely avoids interference with the field provided outside the test object. Therefore, we could abandon elaborated shielding.

The receiving element is as usual a short electric dipole. It provides an open source voltage of

$$V_0(t) = E(t) h_D \cos\alpha \tag{4.141}$$

where α is the angle of incidence.

We are interested in the transfer relation between the voltage $V_R(t)$ at the receiver terminal and the source current $I_q(t)$. For that purpose, we model the transmitter and receiver by simple equivalent circuits. Supposing a current source, we get for the transmitter side:

$$\begin{aligned} \underline{I}(f) &= \frac{\underline{I}_q(f)}{1 + j2\pi f \tau_{LCR}}; \quad \tau_{LCR} = \frac{L}{R_q} \\ I(t) &\approx I_q(t) \quad \text{for } \tau_{LRC} \ll t_w \end{aligned} \tag{4.142}$$

t_w is the pulse duration of stimulus and L is the inductance of the LCR.

For the receiver, we will consider two extreme cases which refer to usual input impedance (e.g. 50 or 100 Ω – differential – assigned by $^{(1)}$) and high input impedance circuits (assigned by $^{(2)}$):

$$\begin{aligned} \underline{V}_R(f) &= \frac{j2\pi f \tau_D}{1 + j2\pi f \tau_D} \underline{V}_0(f); \quad \tau_D = R_R C \\ V_R^{(1)}(t) &\approx \tau_D u_1(t) * V_0(t) \qquad \text{for} \quad \tau_D \ll t_w \\ V_R^{(2)}(t) &\approx V_0(t) \qquad \text{for} \quad \tau_D \gg t_w \end{aligned} \tag{4.143}$$

C is the static capacitance of the dipole.

Combining (4.140)–(4.143) and respecting (4.71) for point scatterer, the time domain Friis formula and radar equation take the forms:

Friis formula:

$$\begin{aligned} V_R^{(1)}(t) &= \frac{\tau_D h_D h_{LCR} Z_s}{4\pi c r} \sin\vartheta \cos\alpha \, u_2(t) * I_q\left(t - \frac{r}{c}\right) \\ V_R^{(2)}(t) &= \frac{h_D h_{LCR} Z_s}{4\pi c r} \sin\vartheta \cos\alpha \, u_1(t) * I_q\left(t - \frac{r}{c}\right) \end{aligned} \tag{4.144}$$

Radar equation:

$$\begin{aligned} V_R^{(1)}(t) &= \frac{9\tau_D h_D h_{LCR} Z_s V_s}{32\pi^2 c^3 r_1 r_2} \sin\vartheta \cos\alpha\, u_4(t) * I_q\left(t - \frac{r_1 + r_2}{c}\right) \\ V_R^{(2)}(t) &= \frac{9 h_D h_{LCR} Z_s V_s}{32\pi^2 c^3 r_1 r_2} \sin\vartheta \cos\alpha\, u_3(t) * I_q\left(t - \frac{r_1 + r_2}{c}\right) \end{aligned} \tag{4.145}$$

V_s is the volume of scatter, c is the propagation speed in the medium, Z_s is the characteristic impedance of the propagation medium and r_1, r_2 are the antenna distance.

Supposing a Gaussian-shaped stimulus pulse of width t_w and aligned antennas and scatterer ($\vartheta = 90°$; $\alpha = 0°$), we get the following transimpedance values for the transmission channels:

$$\begin{aligned} Z_{Friis}^{(1)} &= \frac{\ln 2}{2\pi} \frac{\tau_D h_D h_{LCR} Z_s}{t_w^2 c r} \approx 1.4 \frac{\tau_D h_D h_{LCR} Z_s}{t_w^2 c r} \\ Z_{Friis}^{(2)} &= \frac{\sqrt{2 \ln 2}}{2\pi\sqrt{e}} \frac{h_D h_{LCR} Z_s}{t_w c r} \approx 1.4 \frac{h_D h_{LCR} Z_s}{t_w c r} \end{aligned} \tag{4.146}$$

$$\begin{aligned} Z_{Radar}^{(1)} &= \left(\frac{3 \ln 2}{\pi}\right)^2 \frac{\tau_D h_D h_{LCR} Z_s V_s}{t_w^4 c^3 r_1 r_2} \approx 15 \frac{\tau_D h_D h_{LCR} Z_s V_s}{t_w^4 c^3 r_1 r_2} \\ Z_{Radar}^{(2)} &= \frac{9\sqrt{3(\ln 2)^3 (3 - \sqrt{6})}}{\pi^2} e^{-\frac{1}{2}(3-\sqrt{6})} \frac{h_D h_{LCR} Z_s V_s}{t_w^3 c^3 r_1 r_2} \approx 18 \frac{h_D h_{LCR} Z_s V_s}{t_w^3 c^3 r_1 r_2} \end{aligned} \tag{4.147}$$

From (4.144), we can observe that at least a single time derivative of the stimulus current occurs. It represents the minimum disturbance that a wideband signal is subjected to by transmission via antennas. This is an expected result originating from reciprocity theorem (4.43) and the fact that electric DC-fields cannot propagate. A second or even third derivative may appear due to the feeding circuit. It is caused from the high pass built from the static capacity of the dipole and the source or load resistance. Hence, if we take two dipoles – one as transmitter and one as receiver – we have to count for a third time derivative of the signals. These are two derivatives more than physically required. Replacing the transmitter dipole by a LCR, we get one derivative less since the feeding circuit only marginally influences the time signal (refer to (4.142). We can suppress a further derivative by applying receiver circuits of high input impedance and integrator circuits, respectively. Thus, we have achieved the physically best case that is delineated by $V_R^{(2)}(t)$ in (4.144) for direct transmission and in (4.145) for the radar case.

Finally, we observe that the backscattering of the antennas is minimal if the antennas are operated within high-impedance environment ($\Gamma_q \to 1$) since re-radiation is avoided for both antennas. Only structural scattering remains to contribute to possibly multiple reflections. Clearly, the condition $\Gamma_q \to 1$ cannot be maintained over a large bandwidth if the antennas are fed over cables because they provide frequency-dependent impedance transformation. Hence, antenna

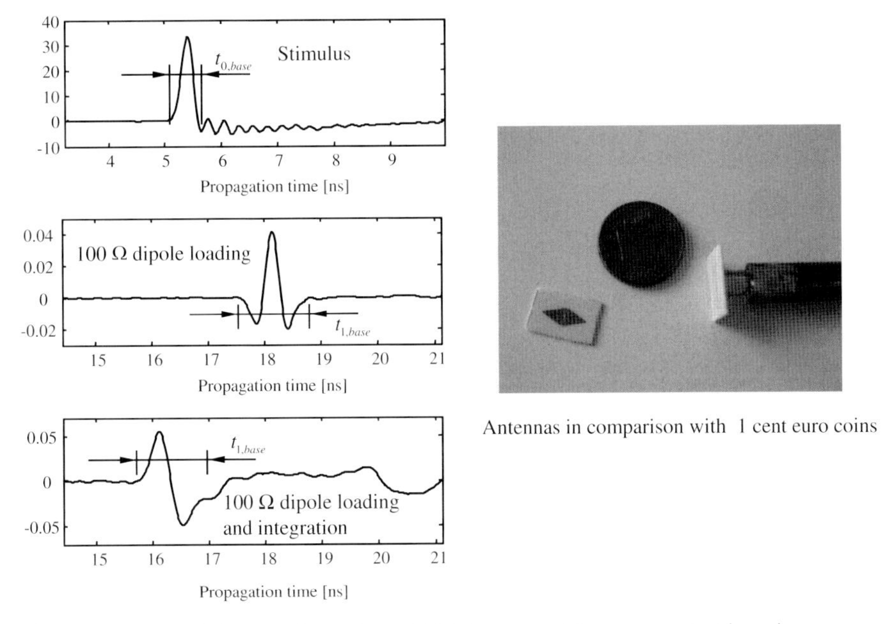

Figure 4.39 Transmission experiment through high lossy breast tissue surrogate. The radar device was an M-sequence system of 9 GHz clock rate (effective bandwidth 4 GHz) with differential RF-ports.

operation at high impedance levels is only possible with active antennas which have integrated their electronics immediately into the feeding point. An implemented example with discrete components and further discussion on this topic can be found in Refs [52, 53].

For experimental verifications, the arrangements corresponding to Figure 4.38 (top) were implemented based on an M-sequence device with balanced (differential) measurement ports and 4 GHz bandwidth. The feeding cables had a length of about $l = 40$ cm. Their role was twofold. First, they provided known source impedances of $2Z_0 = 100\,\Omega$ within the time window of about 4 ns, and second, they decoupled the fields generated by the current loop on the transmitter side without need of elaborated shielding. The integrator circuit is based on SiGe ASIC and a PCB for wiring and antenna connection. No additional attempts for shielding were performed (see Figure 4.39). The antennas were operated in direct contact with the phantom material in all experiments.

Figure 4.39 summarizes the captured signals after impulse compression with the M-sequence (refer to (3.47)). We expect a receiving signal of second derivative for operation in $100\,\Omega$ environment, which we actually achieved. The (time compressed) driving signal can be roughly considered as Gaussian pulse disregarding the trailing oscillations provoked by the anti-aliasing filter. Its pulse width is $t_w = 250$ ps, and the baseline width is about $t_{0,\text{base}} \approx 470$ ps. The second derivative of a Gaussian pulse is illustrated in Figure 2.29, which coincides well with our measured signal (Figure 4.39, middle) except the extended length of the baseline duration $t_{1,\text{base}} \approx 1.1$ ns (50% pulse width $t_{1\text{w}} = 210$ ps). We would expect 470 ps since derivation does not perform stretching. The reason for the observed prolongation is found in the dispersive propagation behavior of the test material. The material under test was breast tissue surrogate in which the water is the main source of wave dispersions (compare Figure 5.14).

The time signal (Figure 4.39, below) provided by the active antenna shows basically the expected behavior. Deviations are to be observed within the trailing part. They are caused by insufficient shielding and cable mismatch. The baseline duration will not be further extended by the time integration.

4.6.7
Short-Range Time Domain Antenna Measurements

We will show in the next chapter that UWB sounding may lead to very precise range measurements supposing the systematic errors due to antenna misalignment are negligible. This means that the centre of radiation, antenna internal propagation delay and angle-dependent IRFs must be precisely known. This attaches particular importance to the antenna calibration. Basically, antenna calibration can be done in the laboratory. But under certain conditions and high-resolution requirements, some calibration routines have to be accomplished *in situ*, that is within the actual application environment. For that purpose, the availability of different measurement approaches may be profitable. Here, we will introduce some of them. For completeness, we should state that also polarimetric properties belong to the basic behavior of antennas. So far we have not yet introduced polarimetric aspects since they are linked to the vector nature of the electric field, which we have omitted as yet. Hence, all introduced principles will not specifically respect polarimetric properties. The extension of the procedures to polarimetric measurements is however straightforward so that we can restrict ourselves here to the basic approaches. Vector fields and hence aspects of polarimetry will be introduced in Chapter 5. The reader can find a discussion on specifically polarimetric antenna measurements in Refs [29, 46, 54, 55].

4.6.7.1 Transmission Measurement Between Two Antennas

Figure 4.40 depicts the basic measurement arrangement. One antenna is the unknown AUT. The second one is a reference antenna with a desirably short impulse response. In order to get the angular dependence of the radiation behavior, the AUT is mounted on a one or two axial turntable. The axis of revolution (circular scanner) and the intersection point of the rotary axes fix the reference point P_1.

The whole arrangement is considered as a two-port device from which a transmission measurement $b_2(t) = S_{21}(t, \mathbf{e}) * a_1(t)$ is performed for any direction $\mathbf{e}$ of the AUT. Applying the Friis formula (4.45) or (4.38), we can calculate via de-convolution the desired impulse response $T(t, \mathbf{e})|_{P_1}$ or $R(t, \mathbf{e})|_{P_1}$ of the unknown antenna, if the behavior of the reference antenna including its effective centre of radiation is well known. Unfortunately, this is often not the case since no reference antennas with calibrated time domain behavior are available. One way out is to use two

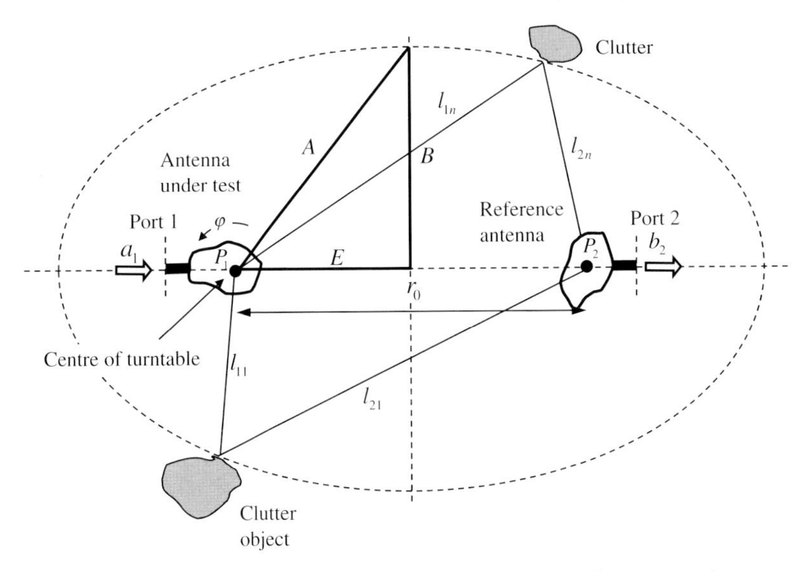

Figure 4.40 Antenna transmission measurement and evaluation of test range volume.

identical antennas from which one assumes identical behavior. The 'three-antenna approach' opens a second opportunity to solve the problem. Here, one measures the transmission behavior of all antenna combinations. One gets a set of three scattering parameters ($S_{21}^{(1-2)}(t, \mathbf{e})$, $S_{21}^{(1-3)}(t, \mathbf{e})$ and $S_{21}^{(2-3)}(t, \mathbf{e})$) from which the IRFs of the individual antennas can be extracted. Note that the bandwidth of all three antennas should be quite the same in order to reduce de-convolution artefacts.

The approach, as depicted in Figure 4.40, is the standard procedure for antenna characterization. Typically it requires an expensive unechoing chamber in order to avoid undesirable reflections from walls or other objects. Such a hall is not required for UWB antenna measurements since reflections from obstacles may be gated out of the captured signals. Nevertheless, a proper measurement needs a minimum volume of empty space. For the standard test arrangement as depicted in Figure 4.40, it depends on the desired dynamic range $L[\text{dB}]$ of the antenna measurement and the shortest measurement distance r_0 which yet fulfils far-field condition.

The time window duration t_{win} over which we observe the antenna IRF is given by

$$t_{\text{win}} \geq t_{\text{w}} + \text{DRL} \tag{4.148}$$

Herein, $t_{\text{w}} = \text{FWHM}$ is the duration of the main lobe of the AUT-IRF and $\text{DR}[\text{dB/ns}]$ is the expected decay rate of the AUT (see Section 2.4.2).

The minimum distance to keep the far-field condition can be estimated from (4.137) or determined directly from measurements; measurements seem to give more realistic values. Since there is no sharp border between near and far field, we need to define an error bound $\Delta e_{\max}$ which is acceptable for our purposes. Basically, we have two options to define such error:

Option 1: The far-field condition implies that the Friis formula (4.45) holds. It implies that the product of signal amplitude (in general: signal norm) and propagation path length is a constant $r\|b_2(t,r)\|_p = \text{const}$. Hence, we require that this law has to be respected with only minor deviations Δe_{Friis} if the antenna distance is reduced. This leads us to the requirement:

$$\Delta e_{\text{Friis}} \geq 1 - \frac{\|b(t,r_0)\|_p r_0}{\|b(t,r_{\text{far}})\|_p r_{\text{far}}} \tag{4.149}$$

r_{far} is a measurement distance which we certainly know to be in the far zone.

Option 2 makes use of the fact that the time evolution of the transmitted wave varies by approaching the antenna while it rests the same in the far field independent of the distance. Consequently, one can state to be in the far field of an antenna if the fidelity defined by (2.54) does not drop below an error bound:

$$\Delta e_{\text{FI}} \geq 1 - \frac{\|b(t,r_0) * b(-t,r_{\text{far}})\|_\infty}{\|b(t,r_0)\|_2 \|b(t,r_{\text{far}})\|_2} \tag{4.150}$$

In order to avoid clutter signals, we require that the round-trip time τ_x to any foreign object within the propagation path of the waves must be larger than

$$\tau_x \geq \frac{r_0}{c} + t_{\text{win}} \tag{4.151}$$

This condition is met if no foreign body is located within the volume of the spheroid as indicated in Figure 4.40. Its focal points are the antenna positions P_1 and P_2. Their distance is

$$\overline{P_1 P_2} = 2E = r_0 \tag{4.152}$$

The length of the semi-major axis must be $2A = l_{1n} + l_{2n} \geq \tau_x c$ (refer to Annex A.7.1) so that we get for the semi-minor axis B

$$B = \sqrt{A^2 - E^2} = \frac{1}{2}\sqrt{(l_{1n} + l_{2n})^2 + r_0^2} = \frac{1}{2}\sqrt{c^2\tau_x^2 + r_0^2} \tag{4.153}$$

Hence the total duration of the measured IRF may cover up to 10 ns if we suppose an empty laboratory space of 3 m × 3 m × 3.2 m (height, width, length) and a measurement distance of 1 m. Thus, there is an interval of about 7 ns for the actual antenna IRF.

Attention should be paid with respect to the unambiguity range of the measurement device. The consideration so far only respects clutter from the direct measurement environment. Far reflections we did not consider since they do not fall into the observation window. But this is only correct as long as time aliasing does not appear. Hence, the pulse repetition interval of the measurement device must be larger than the reverberation time of the whole laboratory room in order to be sure that no foreign reflections are convolved in our observation window.

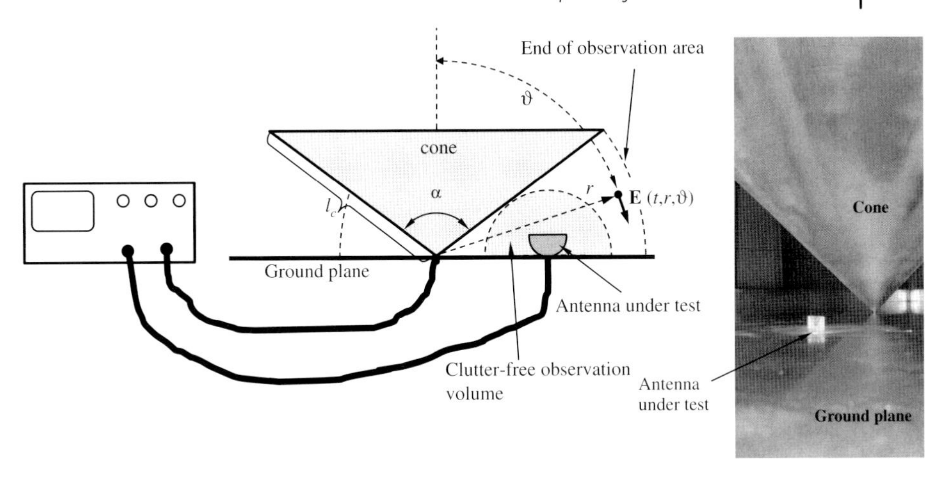

Figure 4.41 Antenna test arrangement with conical waveguide.

4.6.7.2 **Direct Measurement of Antenna Impulse Response**

The direct measurement of, for example, the receiving antenna impulse response $R(t, \mathbf{e})$ requires the exact knowledge of strength and time shape of the incident waveform $W_{SP}(t, \mathbf{r})$. The electric field propagating within a conical waveguide, as depicted in Figure 4.41, is well known, so it can be used for that purpose [56]. The antenna under test is placed in the observation area, which extends between ground plane and cone. An infinite cone assumed, the electric field inside the waveguide is given by [57]

$$W_{SP}(t, \mathbf{r}) = \frac{E_\vartheta(t, r, \vartheta)}{\sqrt{Z_s}} = \frac{V_0(t - r/c)}{r\sqrt{Z_s}\,\sin\vartheta\,\ln(cot(\alpha/4))} \quad \text{for } t_{win} = 0 \cdots \frac{l_c}{c} \tag{4.154}$$

Since infinite dimensions cannot be maintained in reality, the validation of (4.154) is restricted to the time window $[0, l_c/c]$. The characteristic impedance of the waveguide (within this time interval) is

$$Z_0 = \frac{Z_s}{2\pi}\ln\left(cot\frac{\alpha}{4}\right) \tag{4.155}$$

which results for propagation in air in $\alpha = 94^\circ, 42.8^\circ, 8.2^\circ$ for $Z_0 = 50, 100, 200\,\Omega$. The electric field is supposed to be known as long as the wave does not sustain a reflection. Therefore, the length l_c of the cone will determine the duration of the observation time window t_{win}. The AUT is placed in the middle so that we gain the time

$$t_{win} = \frac{l_c}{c} \tag{4.156}$$

before perturbing reflections of the sounding wave are back again at our measurement object.

Since the AUT causes reflections too, the effective available observation area will be limited by the distance to the cone surface; hence, we finally have

$$2d = l_c \cos\frac{\alpha}{2} \to t_{\text{win,eff}} = \frac{2d}{c} = \frac{l_c}{c}\cos\frac{\alpha}{2} \tag{4.157}$$

Since the centre of the cone (i.e. its feeding point) and the position of the AUT are precisely determined by the mechanical construction of the conical waveguide, the arrangement provides best prerequisites for precise IRF measurements. Furthermore, the ground plane promotes the exploitation of the image theorem, which allows for the measurement of the half of symmetric devices by single-ended signals.

Figure 4.42 shows a measurement example made with a mobile unit of 50 Ω feeding impedance and a measurement window of about 3 ns. Its operational bandwidth mainly depends on the feed point mismatch and the ohmic losses of cone and ground plane.

In a modified version of the method, the conical waveguide is replaced by Gigahertz transverse electromagnetic (GTEM) cells which are commercially available and simpler to construct. However, their field is less homogenous than that of the conical waveguide. Examples for antenna measurements with GTEM cell can be found, for example, in Refs [58–61].

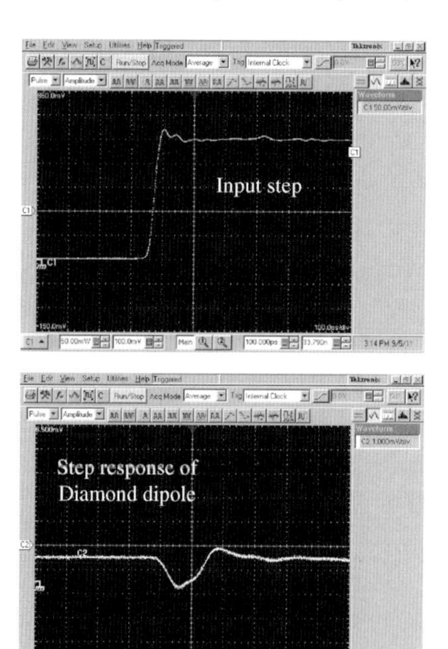

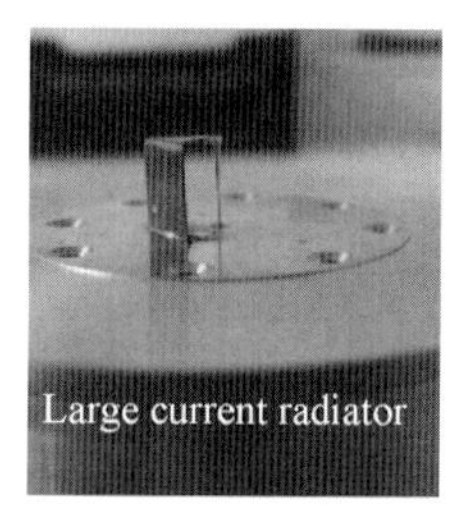

Antennas mounted at the probe fixture

Figure 4.42 Measurement example of the step response. The antennas are mounted at a probe fixture which is placed in the ground plane of the arrangement. Note the time axis scaling is different in both screen shots. The hight of the shown antennas is about 1 cm.

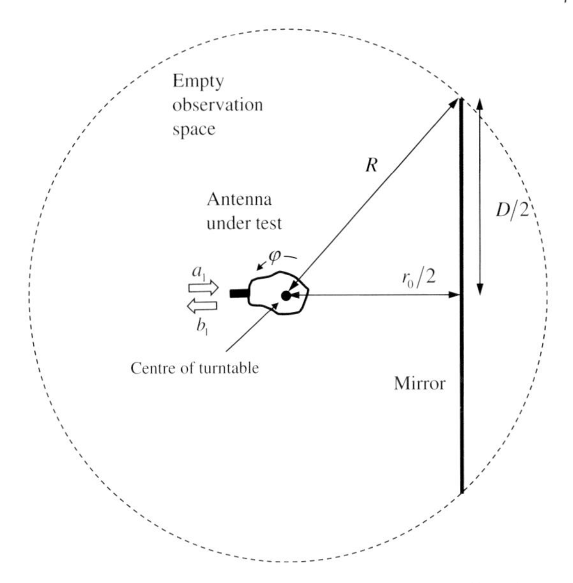

Figure 4.43 Antenna self-measurement using a perfect plane reflector.

4.6.7.3 Impulse Response Measurement by Backscattering

Applying mirroring of electromagnetic waves at a large metal plate, we can renounce the reference antenna so that the antenna under test is characterized by a self-measurement. The arrangement is shown in Figure 4.43. Considering the antenna image as reference antenna, the schematic will be comparable to Figure 4.40 except the fact the 'reference antenna' is rotated here too.

Following our notation, the normalized wave emanating the measurement port results from ($\Gamma = -1$– reflection coefficient of the sheet metal)

$$b_2(t) = \frac{-1}{2\pi c r_0} R(t-\tau, -\mathbf{e}_1)|_{P_1} * R(t-\tau, -\mathbf{e}_1)|_{P_1} * u_1(t) * a_1\left(t - \frac{r_0}{c}\right) \quad (4.158)$$

so that the wanted antenna pulse response can be determined via de-convolution from the measured signal.

The geometric conditions of the measurement arrangements are the same as above. The shortest distance antenna-reflector plane is limited by the far-field condition and the impulse duration of the sounding signal. The transmitting pulse must be settled before the leading edge of the backscattered impulse incidents into the antenna. Otherwise, a_1 and b_1 cannot be properly separated. The radius R of the observation space should allow the complete settlement of the return signal without interference from other objects. Furthermore, the sheet metal borders provoke 'knife edge' reflections (see Ref. [3] for modelling). They must be located outside the observation time window which requires a diameter of the sheet metal larger than

$$D \geq t_{\text{win}} c \sqrt{1 + \frac{2r_0}{t_{\text{win}} c}} \quad (4.159)$$

4.6.7.4 Measurement of Antenna Backscattering

The last method determines the backscattering properties of the antenna. A narrowband version of this method was discussed in Ref. [28]. We assume a slightly modified arrangement as depicted in Figure 4.40. The AUT is mounted at the turntable and its port is terminated by different loads. Typically one takes short, open, match $\Gamma_q = 0, \pm 1$ or a shortened delay line of variable length $\Gamma_q = -\delta(t-\tau)$. A reference antenna illuminates the AUT. The backscattered signals are either captured by the same antenna (mono-static arrangement) or by a second one (bi-static arrangement) placed at an offset angle with respect to the transmitter.

The AUT is regarded as pure scattering object whose total scatter length $\Lambda(t, \mathbf{e}_i, \mathbf{e}_s)$ depends on the feeding port load. As shown in Section 4.6.3 equation (4.123), the variation of backscattering as a consequence of different antenna loads can be used to determine the transmission and receiving behavior of the antenna as well as its structural scattering.

The measurement method requires the knowledge of the IRF of the reference antennas at least for boresight direction. The requirements concerning clutter-free volume of the measurement environment are the same as mentioned for the other methods of antenna characterization. Hence, usual office or laboratory spaces are acceptable for measurements as long as the observation time window does not exceed 10 ns. Larger time windows require either bigger facilities or an unechoing chamber or the reduction of environmental clutter by appropriate pre-screening (see Ref. [62]).

4.7
Basic Performance Figures of UWB Radar

4.7.1
Review on Narrowband Radar Key Figures and Basics on Target Detection

Figure 4.44 gives a very compact summary of the working principle of narrowband radar (see Refs [63–65] for a comprehensive view). The backscattered signal is captured from the antenna, amplified, possibly compressed and finally demodulated, which leads to the envelope of the radar return. Impulse compression is applied if the sounding waveform was spread before transmission in order to reduce peak power load of the transmitter. The impulse compression is performed typically in the IF domain via correlation processing which is done by either analogue (e.g. surface acoustic wave devices) or digital filters. The IRF of these filters is identical with the time-reversed sounding signals (refer to matched filter Figure 2.76). The matched filters concentrate the energy of the sounding signal within a short time interval in order to overshoot the noise and to get a clear pulse position.

The envelope of the compressed signal is subjected a detection procedure which is based on threshold crossing indicating whether there is a target or not and to trigger the round-trip time measurement. The very basic performance figures of narrowband radar are the range resolution, the range accuracy, the unambiguous

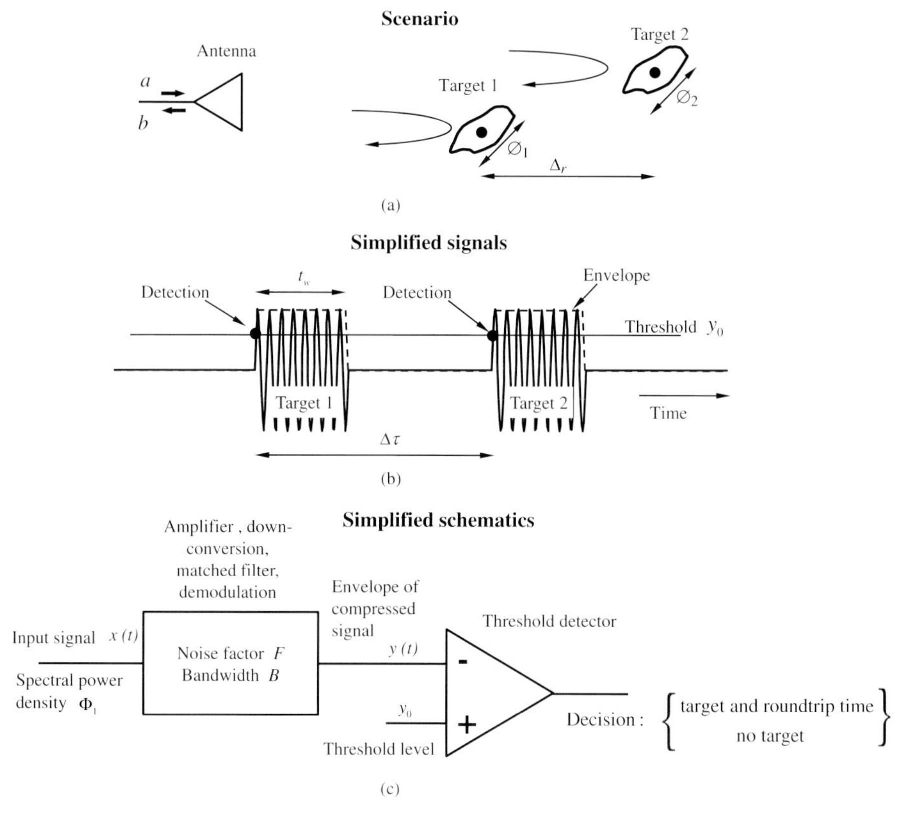

Figure 4.44 Simplified model of a mono-static narrowband radar: measurement arrangement (a); idealized return signals from two targets of identical reflectivity (b) and radar detector schematic (c).

range and the maximum detection range. We will omit here aspects of Doppler handling for the sake of shortness.

Range resolution: The range resolution δ_r describes the ability of the radar to distinguish between two closely located point targets (i.e. $\emptyset_1, \emptyset_2 \ll \delta_r$) of identical radar cross section. Referring to Figure 4.44b, the delay $\Delta\tau = 2\,\Delta r/c$ of both return signals must be larger than the pulse width (envelope width, coherence width) t_w in order to be able to separate two targets. Otherwise both signals would melt together avoiding any target separation. Thus, we must claim for the minimum distance

$$\delta_r \approx \frac{t_w c}{2} \approx \frac{c}{2B_3} \tag{4.160}$$

whereas we used relation (2.77) to convert the width of the envelope to bandwidth. The use of the bandwidth implies that any sounding signal of a given bandwidth $\underset{\cdot}{B}_3$ will provide the same range resolution independent of its time shape (anticipated pulse compression supposed).

It should be noted that typical narrowband radar pulses contain hundreds or thousands of sine wave cycles. Hence the scattered signals interfere if several densely packed scattering centres exist as illustrated in Figure 4.45. The different objects appear as single target. By changing the aspect angle, regions of constructive or destructive interference alternate, which leads to amplitude variations of the backscattered signals and hence of the radar cross section. The separation of the individual targets in Figure 4.45 is not possible, although we used a fractional bandwidth ($\cong$8%) of the sounding wave which is rather much for narrowband radar. Nevertheless, the shape of the envelope is not changed so that there is no reason to expect a multi-target scenario. In the example, we used a sine wave burst ($f_0 = 1\,\text{GHz}$) with a Gaussian envelope (bandwidth $\underset{\cdot}{B} = 80\,\text{MHz}$). The

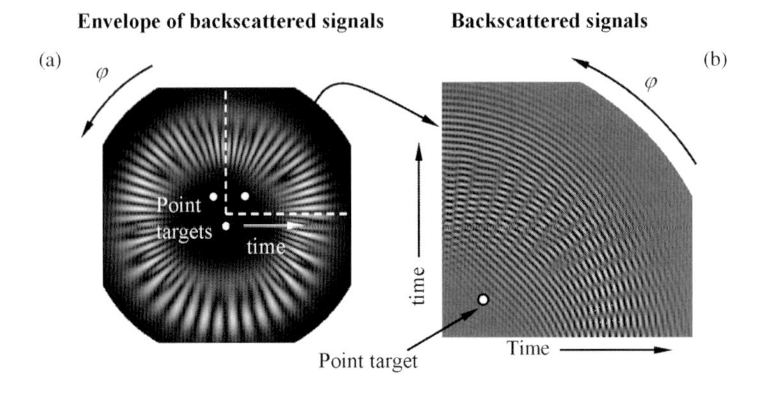

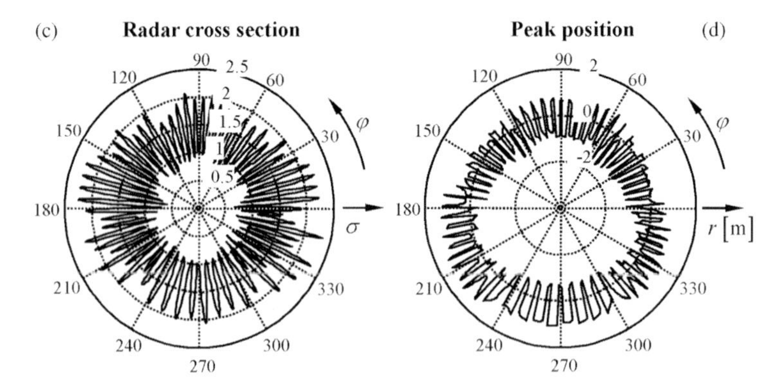

Figure 4.45 Narrowband scattering (simulation) at three point targets in dependency on azimuth angle. envelope (a) of the scattered signal (the radial axis represents time, the midpoint of the diagram is time zero); original backscattered signal (b) showing the individual sine waves (only the first quadrant is show for better visibility of the sine waves). radar cross section (c) of the three point scatter as function of azimuth angle; fluctuations of the radar range (d) gained from maximum position of the envelope. (colored figure at Wiley homepage: Fig 4_45)

targets are placed at an arbitrarily selected triangle with side length of 1.4, 1.7 and 1.8 m.

Unambiguity range: Except for random noise sounding, the transmitter signals have to be repeated periodically in order to watch continuously the scenario under test. Therefore, one has to take care that the scattered signal arrives back at the receiver before the next wave train is launched. Otherwise one would not be able to assign uniquely the measured round-trip time to the target range. Hence for a given period duration t_P of the sounding signal, the unambiguous range results to

$$r_{ua} = \frac{t_P c}{2} \tag{4.161}$$

Random noise signals have an unlimited unambiguity range since they are not repetitive.

Range accuracy: Under range accuracy (or precision), we will understand the minimum range error for a single point target that can be attained in case of noise-affected measurements.

The temporal uncertainty of threshold crossing provoked by noise was already discussed in Section 2.6.3, (2.290). It leads to the range uncertainty of

$$\delta_A = \frac{c\varphi_j}{2} = \frac{ct_r}{2\sqrt{\mathrm{SNR}}} \approx \frac{c}{4\underset{\cdot}{B}_3\sqrt{\mathrm{SNR}}} \tag{4.162}$$

Herein, φ_j refers to standard deviation of the temporal position of the envelope edges of the receive signal. The rise time of the envelope is called t_r. In case of Gaussian shape it is related to the signal bandwidth by $t_r \underset{\cdot}{B}_3 \approx 0.5$. SNR is the signal-to-noise ratio (specifically at the signal edges).

By referring to maximum position of the envelope, we get [66]

$$\delta_A = \frac{c}{2B_{\mathrm{eff}}\sqrt{\mathrm{SNR}}} \tag{4.163}$$

We will discuss this relation more deeply in Section 4.7.3.

Maximum detection range, range coverage: The detection range follows from the radar equation (4.66). In order to achieve a certain confidence in target detection (see below), a minimum SNR-value is required for the detector device, say $\mathrm{SNR}_{D,min}$. Supposing a receiving signal of roughly constant power spectral density $\Phi_2(f) \approx \Phi_{20}$ within the operational band and a receiver noise factor F (assuming that the input signal is only affected by thermal noise), the detector circuit has to deal with an SNR of

$$\mathrm{SNR}_D = \frac{\mathrm{SNR}_2}{F} = \frac{\Phi_{20}}{FkT_0} \geq \mathrm{SNR}_{D,min} \tag{4.164}$$

(T_0 is the temperature in [K] and k is the Boltzmann constant) which has to exceed the value $\mathrm{SNR}_{D,min}$ in order to meet the demands on the detection performance. We can improve the SNR value by several means. This was discussed in Section 2.6.1 and will not be further discussed here.

Insertion of (4.164) into the radar equation (4.66) leads to the wanted expression for a bi-static and mono-static radar (by suppressing the angular dependencies):

$$\begin{aligned} (r_1 r_2)^2 &\le \frac{G_1(f) G_2(f) \sigma(f) \lambda^2}{(4\pi)^3 \mathrm{SNR}_{\mathrm{D,min}} F k T_0} \Phi_{10} \\ r^4 &\le \frac{G^2(f)\, \sigma(f) \lambda^2}{(4\pi)^3 \mathrm{SNR}_{\mathrm{D,min}} F k T_0} \Phi_{10} \end{aligned} \tag{4.165}$$

Herein, $\Phi_1(f) = \Phi_{10}$ is the power spectral density of the radiated signal. We have not considered here factors such as path attenuation and jamming, which will additionally reduce the achievable ranges.

We will use this equation for a very rough range estimation of UWB sensing devices. Communication authorities of the different countries have limited the strength of electromagnetic fields radiated from UWB sensors typically to a level of $\mathrm{EIRP} \le -41.3\,\mathrm{dBm/MHz}$. We modify (4.165) by replacing the antenna gains with (4.49) and (4.93) so that we have a rule of thumb for the detection range of an UWB sensor (disregarding any processing gain; see (4.238) for an improved relation):

$$(r_1 r_2)^2 \le \frac{A \bar{\sigma}}{(4\pi)^2 \mathrm{SNR}_{\mathrm{D,min}} F k T_0} \frac{10^{\mathrm{EIRP}/10}}{1\,\mathrm{MHz}} \approx 10^5 \frac{A \bar{\sigma}}{\mathrm{SNR}_{\mathrm{D,min}}\, F} \tag{4.166}$$

Herein, A represents the aperture of the receiving antenna and $\bar{\sigma}$ is an average radar cross section of the target. Note that an increasing gain of the transmitting antenna will not improve the range of coverage since the law limits the EIRP-level and not the radiated power. The equation gives only a very rough estimation, omitting any specific aspects of UWB radar sensors. Nevertheless, it gives a picture of the limited detection range of UWB sensors if one is forced to respect the mandatory EIRP level.

Target detection: Omitting clutter for the sake of shortness, the voltage at the input of the detection circuit is pure noise in the absence of any target. If a target is present, the voltage will rise due to the reflections (see Figure 4.46). The duration of this voltage 'blob' corresponds to the coherence time of the sounding signal and its time position (typically marked by the maximum) represents the round-trip time. Following the radar equation, the 'blob' height is a measure of the target reflectivity (i.e. σ), and the time position of the maximum relates to the target distance.

The target is detected and its round-trip time is picked up if the signal level exceeds a certain marker assigned as threshold level y_0. Unfortunately, the receiving signal is ever corrupted by perturbations. Hence, target detection is always connected with the question whether the threshold crossing actually indicates a target or not. As Figure 4.46a illustrates, the threshold is crossed even if no target is present and inversely we can observe situations in which the threshold is undershoot even if a target is there. This will cause wrong decisions (depicted as the empty area under the PDFs in Figure 4.46b) as false alarms or missed targets. It is obvious that the probability of false alarms can be reduced by

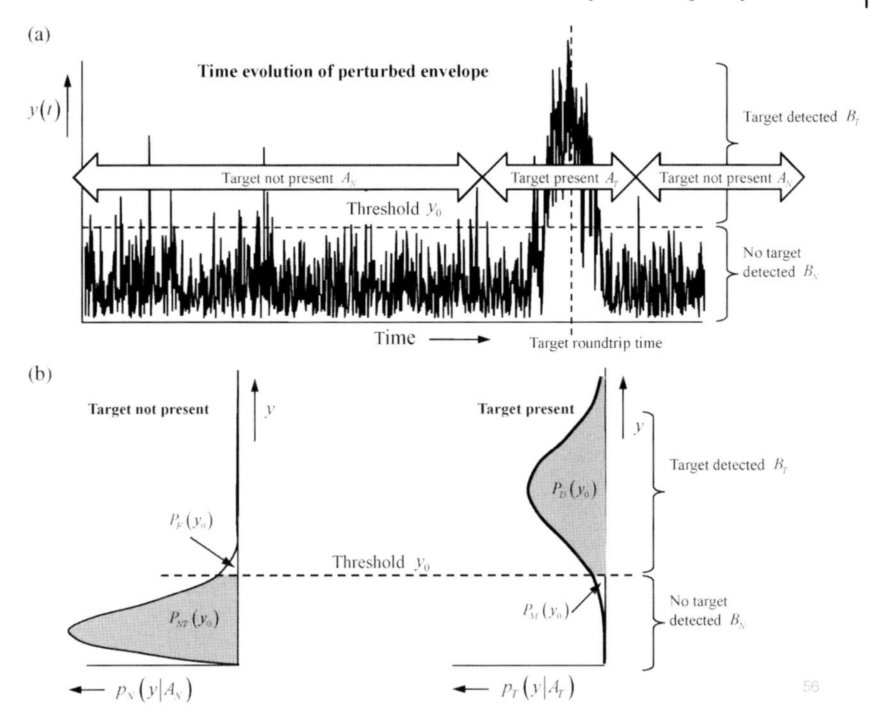

Figure 4.46 Illustration of target detection via threshold decision - detector signal (a) and probability representation (b).

elevating the threshold level. But then, the risk increases to lose a (weak) target. Thus, one has to decide by a risk analysis where the threshold has to be placed in the optimum case.

For that purpose, we will distinguish four cases which result from the condition that a target is actually present (termed as condition A_T) or not (condition A_N) and the resulting decision which may be either 'target detected' (B_T) or 'not detected' (B_N). The probability that a target is present or absent is denoted as $P(A_T)$ and $P(A_N)$ respectively; we can state that $P(A_T) + P(A_N) = 1$. We categorize our decisions by the probabilities denoted as $P(B_T|A_N)$ or $P(B_T|A_T)$ and so on. The first case means the probability to indicate the presence of a target even if there is nothing (obviously a wrong decision), and the second one refers to the probability to correctly decide the existence of a target if there is actually one.

Supposing the probability density functions of the receive signal are $p_T(y|A_T)$ and $p_N(y|A_N)$ for the target and no-target case (compare Figure 4.46b), we can find the following relations:

$$\text{True non-detection probability}: \quad P_{NT}(y_0) = \int_0^{y_0} p_N(y|A_N)\mathrm{d}y = \frac{P(B_N|A_N)}{P(A_N)} \tag{4.167}$$

$$\text{False alarm probability}: \quad P_{\mathrm{F}}(y_0) = \int_{y_0}^{\infty} p_{\mathrm{N}}(y|A_{\mathrm{N}})\mathrm{d}y = \frac{P(B_{\mathrm{T}}|A_{\mathrm{N}})}{P(A_{\mathrm{N}})} \tag{4.168}$$

$$\text{Target missed probability}: \quad P_{\mathrm{M}}(y_0) = \int_{0}^{y_0} p_{\mathrm{T}}(y|A_{\mathrm{T}})\mathrm{d}y = \frac{P(B_{\mathrm{N}}|A_{\mathrm{T}})}{P(A_{\mathrm{T}})} \tag{4.169}$$

$$\text{Probability of detection}: \quad P_{\mathrm{D}}(y_0) = \int_{y_0}^{\infty} p_{\mathrm{T}}(y|A_{\mathrm{T}})\mathrm{d}y = \frac{P(B_{\mathrm{T}}|A_{\mathrm{T}})}{P(A_{\mathrm{T}})} \tag{4.170}$$

Since $P_{\mathrm{NT}} + P_{\mathrm{F}} = 1;\ P_{\mathrm{M}} + P_{\mathrm{D}} = 1$, we can immediately find

$$P(B_{\mathrm{T}}|A_{\mathrm{N}}) + P(B_{\mathrm{N}}|A_{\mathrm{N}}) + P(B_{\mathrm{T}}|A_{\mathrm{T}}) + P(B_{\mathrm{N}}|A_{\mathrm{T}}) = 1$$

Based on the knowledge of these relations, we can ask for a criteria of an optimum threshold placement y_0. We have three options:

1) **Minimize the overall cost of the decision (Bayes criterion):** Supposing we know the costs of every decision (C_{D} is the costs of target detection, C_{M} is the costs to miss a target, C_{F} is the costs of a false alarm and C_{NT} is the costs to confirm target absence), the overall costs of the decision are

$$\begin{aligned} C_{\mathrm{tot}}(y_0) = {} & C_{\mathrm{D}} P_{\mathrm{D}}(y_0) P(A_{\mathrm{T}}) + C_{\mathrm{M}} P_{\mathrm{M}}(y_0) P(A_{\mathrm{T}}) + C_{\mathrm{F}} P_{\mathrm{F}}(y_0) P(A_{\mathrm{N}}) \\ & + C_{\mathrm{NT}} P_{\mathrm{NT}}(y_0) P(A_{\mathrm{N}}) \end{aligned} \tag{4.171}$$

so that the optimum threshold y_0 is adjusted if the costs tend to a minimum:

$$y_0 = \arg\min_{y_0} C_{\mathrm{tot}}(y_0) \tag{4.172}$$

2) **Minimize the wrong decisions (MAP – Maximum-a-posterior criterion):** That is, the threshold is fixed by solving

$$y_0 = \arg\min_{y_0} \left(C_{\mathrm{M}} P_{\mathrm{M}}(y_0) P(A_{\mathrm{T}}) + C_{\mathrm{F}} P_{\mathrm{F}}(y_0) P(A_{\mathrm{N}}) \right) \tag{4.173}$$

3) **Keep the false alarm probability below a given value (Neyman–Pearson criterion):** The threshold y_0 is to be selected so that

$$P_{\mathrm{F}}(y_0) \leq P_{\mathrm{F0}} \tag{4.174}$$

In most of the practical cases it is not possible to indicate the costs of a decision, so that usually the Neyman–Pearson criterion is a means to an end. But also here, the determination of the threshold poses some problems since the probability density functions $p_{\mathrm{T}}(y|A_{\mathrm{T}})$ and $p_{\mathrm{N}}(y|A_{\mathrm{N}})$ are mostly not or only improper known. In such cases, experimentally determined receiver operating characteristic (ROC) curves may help. They depict the false alarm versus the detection probability depending on the perturbations scenario. Figure 4.47 illustrates the typical shape of such curves. The threshold level y_0 appears as curve parameter. The lower the perturbations, the more the ROC curve will huddle

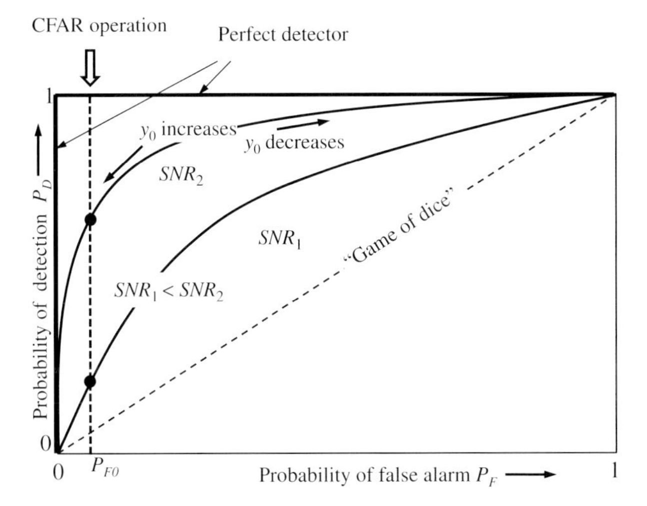

Figure 4.47 Receiver operational curve and CFAR operation.

against the upper left of the diagram. The operational mode which connects the lower left [0, 0] with upper right [1, 1] corner by a straight line is meaningless since the detector is doing a 'game of dice'.

Since the perturbations may vary over time, range and aspect angle, the threshold level is automatically adapted in the so-called constant false alarm rate (CFAR) operation mode which keeps the false alarm rate constant. The idea behind CFAR detection is illustrated in Figure 4.48. One assumes indeed variable perturbation conditions but they should not change too abruptly. Thus in CFAR detection, one analyses the signal within a short region of interest – the so-called detection cell – and compares it with signal segments (reference cells) in its neighbourhood. In this context, the decision may be made on any signal parameter or even on a bunch of parameters which seem to be appropriate for the actual application (note we are not solely restricted to amplitude values as suggested by the figures). Significant differences between detection and reference cells lead to a detection; narrow differences give no detection. By sliding the reference and detection cells over the whole signal, the detection threshold is successively updated. The various CFAR approaches can be distinguished from each other by

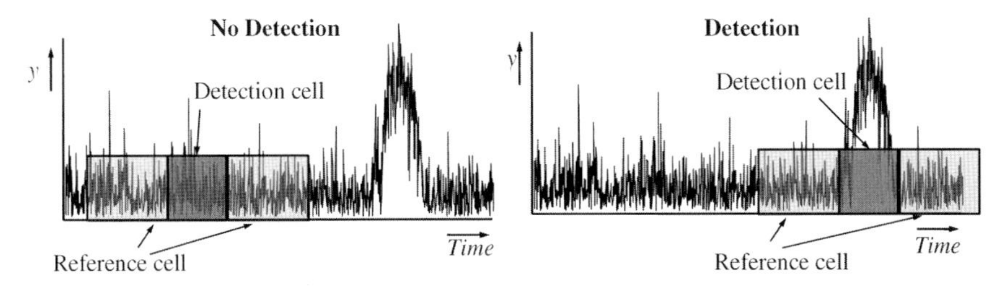

Figure 4.48 Two situations of CFAR detection.

the parameters or statistics on which they come to their decision. Clearly, the reference cells may embrace the detection cell in all directions if the data set represents multi-dimensional signals (MiMo radar, radargram, radar image etc.).

Insofar we discussed some fundamental aspects and key figures of the radar technique as it concerns narrowband measurements. Even if ultra-wideband radar has the same physical basis as the classical narrowband one, several aspects must be considered differently and more profoundly. Table 4.1 summarizes some points which separate the two radar approaches. Clearly, a keen separation cannot be established. There are always regions where both methods overlap in their performance and properties. Radar resolution will be considered in more detail specifically under the condition of ultra-wideband sensing. For more on detection issues, the reader is referred to Refs [67–71].

Table 4.1 Differences between narrow and ultra-wideband radar.

Property	Narrow band radar	Ultra-wideband radar
Signal property	$\tau_0 f_0 \gg 1$	$\tau_0 f_0 \approx 1 \cdots 5\%$
Typical target size D_0	$c \cdot \tau_0 \gg D_0 \gg c/f_0$ – single target $D_0 \leq c/f_0$ or $c\,\tau_0$ – random scatterer	
Typical antenna size D_0	$c \cdot \tau_0 \gg D_0 \gg c/f_0$	$c \cdot \tau_0 \approx c/f_0 \approx D_0$
Signal fidelity	Signal shape will not be affected by antenna and target	Signal is deformed by antenna and target
Matched filtering	Matching to transmit signal	No matching to transmit signal; signal matching requires knowledge of transmission path, that is antennas, scatterer and wave dispersion
Signal power	Up to several tens of kW of average power	Military: GW of peak power
		Commercial sensor, short-range sensing: μW to mW average power
Doppler effect	No effect on envelope; shifts carrier frequency	Compression or expansion of the whole waveform; negligible effect on short pulses
Signal separation (close targets)	Overlapped signals cannot be separated	Signal separation possible if prior knowledge is available
Detection and recognition based on	Magnitude of signal envelope and Doppler shift	Magnitude of return signal, modification of signal shape, history of signal shape variation and target position (Doppler)

c – speed of light
f_0 – centre frequency
τ_0 – pulse width or coherences width

4.7.2 Range Resolution of UWB Sensors

As already mentioned, range resolution refers to the ability of a radar device to separate densely packed objects by their distance to the sensor. Based on (4.160), we can predict that the range resolution of an ultra-wideband sensor will be orders of magnitude better than that of narrowband radar due to the short impulse duration and coherence time of the sounding signals. To illustrate the differences, Figure 4.49 depicts a simulation of the same scenario as shown in Figure 4.45 for narrowband radar. Here, we use again a Gaussian-modulated sinusoidal pulse of centre frequency $f_0 = 1$ GHz; however, the fractional bandwidth was selected more than 10 times larger. The example refers to an absolute bandwidth of $B_3 =$ 1 GHz (operational band 0.5–1.5 GHz – which is not a large bandwidth for already implemented devices).

Obviously, the signal traces separate the targets very well except for those aspect angles where two targets have the same distance to the radar antenna.

As the example has shown, we can basically use (4.160) to estimate the range resolution of ultra-wideband radar. However, we should interpret this result with some care. Remember that in the narrowband case the shape of the sounding signal is not (only marginally) changed by the target. This is different for ultra-wideband scattering. Here, the time evolution of the scattered signal will be usually destroyed except for scattering at a planar interface.

Target separation corresponding to (4.160) is based on the assumption that the spatial extension of the sounding signal exceeds the target dimensions. In the UWB case, as shown in Section 4.5.2, Eq. (4.71), the small scattering object subjects the incident field to a twofold derivation. Hence, we can state that the performance to separate targets will only marginally be changed and (4.160) stays valid since a time derivative will not affect the pulse duration or coherence time of the sounding signal.

On the other hand, if only such scatterers (or planar interfaces) are involved in the scattering scenario, we can predict the time evolution of the scattered wave since the sounding wave can be assumed to be known (second derivative for small scatterer; sign inversion for scattering at a planar interface). Consequently, we can model our receive signal with respect to size and distance of the scatterer. Thus, by

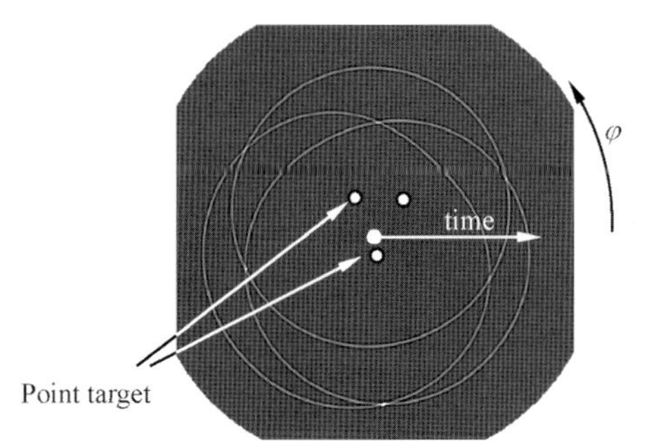

Figure 4.49 Wideband scattering at three point targets in dependency on azimuth angle.

matching the model to the real measurements, one can actually go beyond the resolution limit predicted by (4.160). Corresponding approaches are summarized under the term super-resolution technique. The following text presents a simple example for illustration.

The opposite situation appears if the target dimensions exceed the pulse width or the coherence length of the sounding signal. Then the targets themselves will be the limiting factor for the resolution. Resonance effects and multiple reflections may generate long pulse trails overlapping with the radar return of other targets. In such case, (4.160) cannot be used for performance estimation of target separation. Nevertheless, the short sounding signal is able to reveal details of the target which permit target recognition. In that sense, that is to separate individual scattering centres (bright points) of the target, (4.160) is again feasible. Furthermore, the wideband signal is able to stimulate various eigenmodes of the target (refer to Section 4.6.4) promoting target recognition.

We will come back to super-resolution approaches to illustrate the underlying philosophy. As the name implies these approaches are able to go beyond the classical resolution limit specified only by the bandwidth of sounding signal as given by (4.160). The prerequisite to improve the resolution is a valid and applicable model of the test scenario. Hence, one always needs some prior knowledge about the test scenario.

Let us consider for that purpose the separation of targets with only small difference in radar range. Figure 4.50 depicts two situations for $\Delta r \leq \delta_r$. Since the mutual target distance is smaller than the range resolution, the return signals from the targets will overlap.

The knowledge required before we can separate the two (or more) targets is the time shape of their scattering response $\Lambda_{1,2}(t, \mathbf{e}_i, \mathbf{e}_s)$ for the given directions of incidence and observation. For the sake of simplicity, we suppose K small objects of volume V_k which are all subjected to Rayleigh scattering

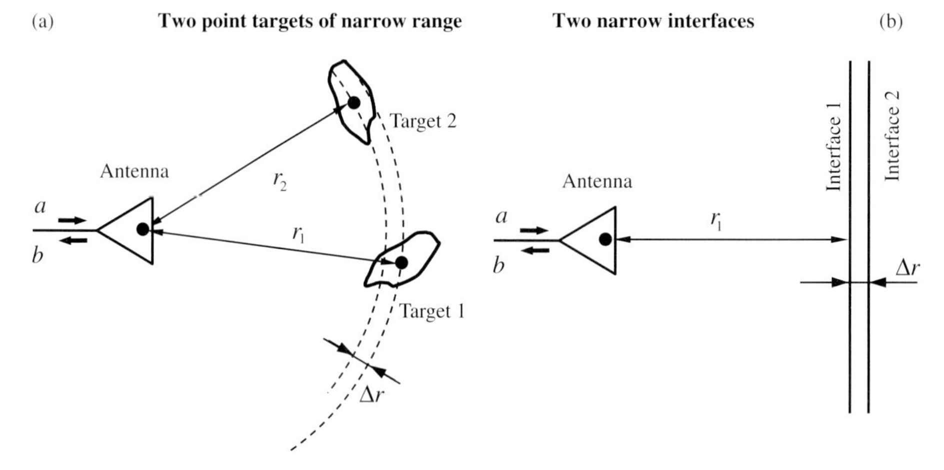

Figure 4.50 Two measurement scenarios with overlapping target returns. (a) two small targets and (b) sandwich structure.

$\Lambda_k(t, \mathbf{e}_\mathrm{i}, \mathbf{e}_\mathrm{s}) \approx -(9V_k/8\pi c^2)u_2(t)*$ (4.71) and a mono-static radar whose antenna IRF does not depend on direction $R(t, \mathbf{e}) = R(t)$. In the case of the planar sandwich structure, the scattering response simply takes the form $\Lambda_k(t, \mathbf{e}_\mathrm{i}, \mathbf{e}_\mathrm{s}) = -\Gamma_k$. To begin with we will exclude multiple reflections between the involved objects in order to reduce the degree of complexity. Since such assumptions are far from reality for layered structures, we only deal here with the small objects (Figure 4.50a). The handling of sandwich structures (Figure 4.50b) will be discussed in Section 9.1.1.

Exclusion of multiple reflections and the time domain radar equation (4.60) provide us the model of the receive signal $\hat{b}(t)$ for K arbitrarily placed bodies

$$\begin{aligned}\hat{b}(t) &= -\left(\frac{3}{4\pi}\right)^2 \frac{1}{c^3} u_3(t) * R(t) * R(t) * a_1(t) * \sum_{k=1}^{K} \frac{V_k}{r_k^2}\delta\left(t - \frac{2r_k}{c}\right) \\ &= b_0(t) * \sum_{k=1}^{K} A_k\delta\left(t - \frac{2r_k}{c}\right); \quad A_k = \frac{V_k}{r_k^2}\end{aligned} \tag{4.175}$$

Herein, we have joined all a priori known signal components in the signal $b_0(t)$. It can be gained from measurements with the arrangement corresponding to Figure 4.43 and a subsequent twofold derivation.

The wanted target volume V_k and distances r_k are to be found by matching the signal model $\hat{b}(t)$ to the actual measurement $b(t)$. This is typically done by minimizing the L_p-norm of the error between the model and the measurement. Very often, the L_2-norm is applied, which leads to following optimization rule:

$$[V_k, r_k] = \underset{V_k, r_k}{\arg\min} \int \left(b(t) - \hat{b}(t)\right)^2 \mathrm{d}t \tag{4.176}$$

The solution of (4.176) involves a multi-dimensional search which may be time-consuming and expensive. Since the different targets are independent of each other, the matching routine can be accelerated by the so-called space-alternating generalized expectation-maximization (SAGE [72]) algorithm, which serializes the search for the individual targets.

A flow chart for our example is depicted in Figure 4.51 for illustration. We assume K small scatterers and that $b_0(t)$ is a well-known pulse-like waveform. During an initialization phase (left-hand loop), we roughly determine time positions $\tau_k = 2r_k/c$ and amplitudes $A_k = V_k/r_k^2$ of the involved scatterer. For that purpose, in a first step (loop counters $m = 0$; $k = 1$), time position τ_1 and amplitude A_1 of the strongest scatterer are determined (for definition of time position refer to Section 2.2.2 and issues on time position accuracy are discussed here below). In a second step (loop counters $m = 0$; $k = 2$), the time position τ_2 and amplitude A_2 of the next powerful scatterer are determined by dealing with an updated signal $\Delta b_1(t)$ from which the first reflex is removed: $\Delta b_1(t) = b(t) - A_1 b_0(t - \tau_1)$. The third strongest maximum will be determined from $\Delta b_2(t) = \Delta b_1(t) - A_2 b_0(t - \tau_2)$ and so forth.

The parameters A_k, τ_k determined so far are still erroneous since the signals originating from the different scatters overlap. Hence, the time positions gained from $y(t) = b(t), \Delta b_1(t), \Delta b_2(t), \cdots$ must not necessarily coincide with the exact time

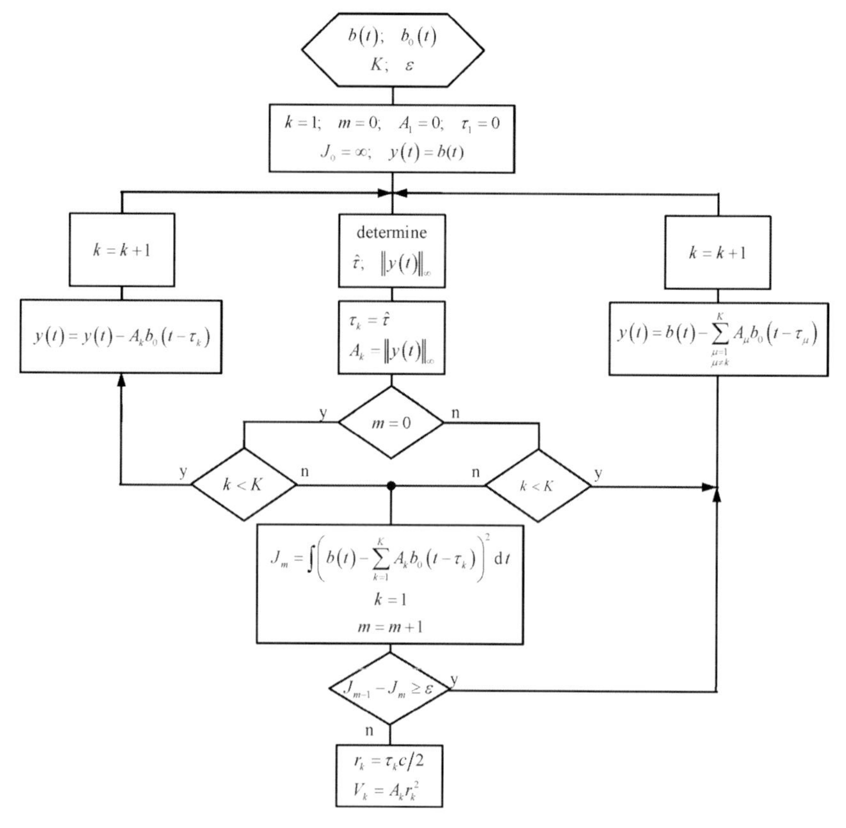

Figure 4.51 Flow chart to determine target position and volume from overlapped signals via SAGE approach. The algorithm needs the measurement data $b(t)$, the reference function $b_0(t)$, the number of targets K and the error bound ε for loop termination.

position of the backscatter signals. Therefore, we enter an iteration loop (loop counter $m = 1$; $k = 1$) to eliminate the errors of the initial guess (right-hand loop). Here again, we try to estimate successively the time position and amplitude of all scatters. Since we know approximately the time position and amplitude of all scatterers, we can now remove the signal components from all scatters except the one we just consider. By that the mutual interference between the different objects will be reduced step by step and we will end up in corrected values A_k, τ_k for all scatters from which the distance and volume can easily be calculated. The iteration loop is terminated if a new loop cycle does not provide reasonable progress in approaching the measured signal (estimated by the error bound ε).

The remaining position errors of the targets depend on deterministic and random errors. Their influence on the position accuracy of a single scatterer will be discussed in the following section.

For further illustrations on target separations, the reader may refer to Ref. [73] where a genetic optimization approach was applied instead of the SAGE method.

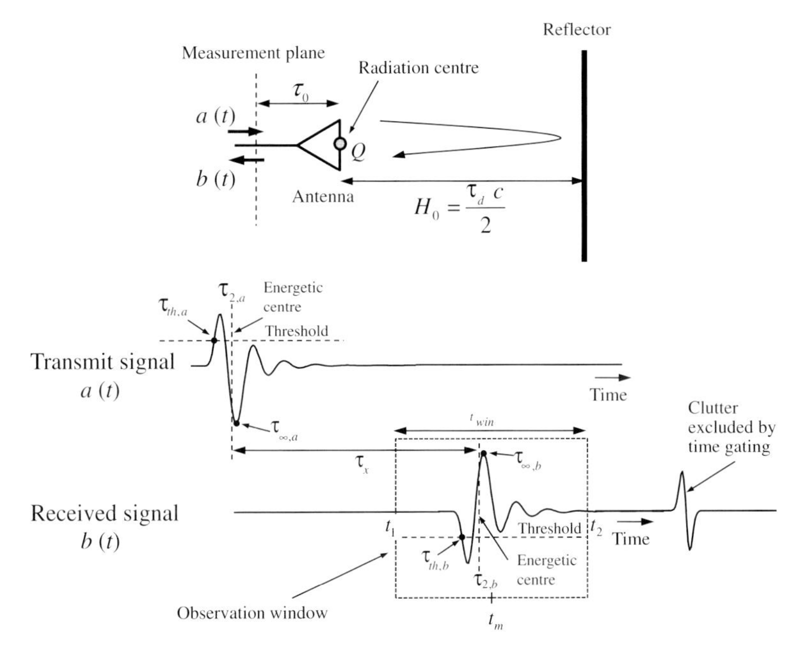

Figure 4.52 Principle of range measurement.

4.7.3 Accuracy of Range Measurement

4.7.3.1 Statement of the Problem

Ultra-wideband sensors are able to determine precisely the target range from which super-resolution techniques benefit. The performance of UWB imaging is also largely dependent on the range accuracy of a UWB measurement. Hence, we will consider this topic in more detail and try to answer the question of the physical limits of range accuracy. In order to concentrate on fundamental aspects, we will consider a very simple arrangement as depicted in Figure 4.52. The goal is to determine the distance H_0 between a perfect metallic reflector ($\Gamma = -1$) and an antenna as precise as possible via round-trip time evaluation. The accessible signals are the normalized waves $a(t)$ and $b(t)$ captured at the measurement plane.

We can model the scenario by the radar equation (4.55) for reflection at planar obstacles

$$b(t) = -\frac{1}{2\pi c H_0} R(t - \tau_0, \mathbf{e})\big|_Q * R(t - \tau_0, -\mathbf{e})\big|_Q * u_1(t) * a(t - \tau_\mathrm{d}) \quad \text{with} \quad \tau_\mathrm{d} = \frac{2H_0}{c} \tag{4.177}$$

We join all functions which can be determined by a reference measurement with a reflector at known distance: $b_0(t) = -(2\pi c)^{-1} R(t - \tau_0, \mathbf{e})\big|_Q * R(t - \tau_0, -\mathbf{e})\big|_Q * u_1(t) *$ so that we get a simplified version of (4.177)

$$b(t) = \frac{1}{H_0} b_0(t) * \delta(t - \tau_d) \tag{4.178}$$

Since our interest here is directed towards the determination of H_0, we have basically two options. One of them is to exploit the signal amplitude, that is

$$H_0 = \frac{\|b_0(t)\|_p}{\|b(t)\|_p} \tag{4.179}$$

but the better one is to deal with the delay time τ_d which has to be determined from measurable total delay time τ_x between $a(t)$ and $b(t)$. This method has a much better performance than the first one. Hence, we will only investigate the precision of delay time measurement.

For a delay time measurement, we are forced to map the two involved signals to their time positions. The difference between both positions gives us the wanted distance H_0 under the assumption of known propagation speed. The time position can only be identified for pulse-shaped signals whereas a unique definition cannot be given as already shown in Section 2.2.2. The basic options available are threshold crossing τ_{th} (typically at 50% level) and the centres of gravity (2.22) of which the energetic centre τ_2 and maximum position τ_∞ are the most important ones. Figure 4.52 illustrates the corresponding time positions and the resulting delay time τ_x:

$$\begin{aligned}
&\text{Crossing a threshold} \Rightarrow \tau_x = \tau_{th,b} - \tau_{th,a} \\
&\text{Energetic centre of gravity} \Rightarrow \tau_x = \tau_{2,b} - \tau_{2,a} \\
&\text{Maximum position} \Rightarrow \tau_x = \tau_{\infty,b} - \tau_{\infty,a}
\end{aligned}$$

The values of τ_x determined by all three methods are the same as long as both signals have identical shape but they must not have necessarily the same amplitude; that is, their fidelity (2.54) should be one, $FI_{ba} = 1$.

Applying above-introduced definitions to correlation functions, we can extend the determination of time positions also to CW signals. Their correlation functions tend to a sharp pulse for UWB signals so that we can deal with them as with real pulses. Appropriate correlation functions can be gained in three ways:

- **Cross-correlation between the two measurement signals:**

$$C_{ba}(\tau) = \frac{1}{T} \int_T a(t + \tau) b(t) \mathrm{d}t \tag{4.180}$$

The maximum position of $C_{ba}(\tau)$ immediately gives the wanted delay time τ_x. But we can also apply threshold crossing or energetic centre to localize the CCF. The last two methods will lead to a biased estimation which, however, can be appropriately respected since by assumption the time shape of the involved signals is known.

- **Cross-correlation with a device internal reference function:** In the case of an M-sequence device it is a common practice to correlate the received signals with the ideal M-sequence via Hadamard transform.

$$\begin{aligned} C_{ma}(\tau) &= \frac{1}{T}\int\limits_T m(t+\tau)a(t)\mathrm{d}t \\ C_{mb}(\tau) &= \frac{1}{T}\int\limits_T m(t+\tau)b(t)\mathrm{d}t \end{aligned} \tag{4.181}$$

These two functions may be treated like ordinary pulse signals.

- **Adjustment of two measurements signals by maximum likelihood solution:**

$$\Rightarrow \; [\tau_x, A] = \underset{\tau, A}{\arg\min}\left\{\frac{1}{T}\int\limits_T (a(t+\tau) - A\,b(t))^2\mathrm{d}t\right\} \tag{4.182}$$

Herein, the factor A represents an amplification equalizing the propagation loss of the backscattered wave compared to the reference signal. As already shown in Section 2.2.4.2, 'Fidelity', Eq. (4.182) leads to the maximization of a correlation function so it is nothing but (4.180).

In order to get the distance H_0, we need the round-trip time τ_d but actually we get τ_x from the measurements. This delay time involves additional delays caused by the measurement arrangement and a number of systematic and random deviations. We may decompose it into following parts:

$$\tau_x = \tau_d + 2\tau_0 + \Delta\tau_{ab} + \Delta\tau_A(\mathbf{e}) + \Delta\tau_Q + \Delta\tau_{cl} + \Delta\tau_j(t) + \Delta\tau_D(t) + \tau_B \tag{4.183}$$

- τ_d is the round-trip time from the antenna radiation centre to the scatter and back. This is the time which we are looking for.
- τ_0 is the internal antenna delay from the measurement port to the radiation centre.
- $\Delta\tau_{ab}$ refers to a systemic measurement error caused due to different signal shapes of $a(t)$ and $b(t)$. Exact delay time measurements must refer to identical states of both signals that are compared. This is only possible if the signals have identical shape. Hence, by directly relating $b(t)$ to $a(t)$ we have to count on deviations provoked by signal shape deformations which are evoked at least by the antennas.
- $\Delta\tau_A(\mathbf{e})$ is a systematic error arising if the antenna 'sees' the scattering plane under different aspect angles. It is provoked by a radiation pattern of the antenna which deviates from an ideal spherical wave. There are two effects involved. One respects the wavefront deviation from a sphere. This error can be estimated from the antenna's spherical deformation pattern defined by (4.118). The impact of the second one is comparable to the error $\Delta\tau_{ab}$ because the shape of the radiated wave may vary with the azimuth or inclination angle. The antenna fidelity

pattern (defined by (4.120)) is the tool to estimate the influence of the shape variations.

- $\Delta\tau_Q$ refers to tolerances of the antenna placement, that is the deviations between expected and actual radiation centre position. It is a question of mechanical precision. It may be of systematic nature – for example antennas are rigidly placed within an array – or of random nature – for example the antennas are moved for scanning.
- $\Delta\tau_{cl}$ is an error term originating from clutter. Clutter represent signal components which affect the shape of the receive signal in an unknown way. So it provokes a similar effect as the error $\Delta\tau_{ab}$.
- $\Delta\tau_j(t)$ and $\Delta\tau_D(t)$ are random errors referred to as jitter and drift. Both have comparable physical reasons. We will consider random short-time fluctuations as jitter. Smooth but nevertheless random trends of timing errors are referred to as time drift. In ultra-wideband systems, both effects are caused by trigger errors of the sampling circuits and the pulse shaper which provides the sounding signal.

  Another source of random timing errors may be the frequency instability (phase noise) of the device internal system clock. We have discussed profoundly its implication in Section 3.4.4 for sine wave approaches. We expect comparable behavior for the other measurement principles; thus, we can save ourselves for further discussions on this topic.
- τ_B represents a bias term which may arise from some algorithms of time position estimation. The determination of the centres of gravity is, for example, such a procedure which tends to bias in the presence of random noise (see Figure 4.60).

In order to avoid most of these deviations in time delay estimation, one can take several countermeasures:

- Antenna fixing at a stable frame and a constant aspect angle for all measurements; proper calibration of internal antenna delay and radiation centre.
- Reduction of device internal clutter by system calibration.
- Removing unwanted scatterer from the propagation path of the sounding wave (refer to Figure 4.43) and excluding clutter signal by time gating as exemplified in Figure 4.52.
- Performing a reference measurement with a planar reflector at known distance H_0.

Using notation (4.178), we can write for the received signal of the reference measurement:

$$b_{\mathrm{ref}}(t) = \frac{1}{H_0} b_0(t) * \delta(t - \tau_0); \quad \tau_0 = \frac{2H_0}{c} + \tau_\alpha$$

where the τ_α joins all additional and unknown delays which are caused by the test arrangement and the imperfectness of the pulse position evaluation. This signal is stored and taken for reference. If we place a target at unknown distance H, our receiving signal will be

$$b_{\mathrm{ref}}(t) = \frac{1}{H} b_0(t) * \delta(t - \tau_{\mathrm{H}}); \quad \tau_{\mathrm{H}} = \frac{2H}{c} + \tau_\alpha$$

Both signals have now identical time shape so that best prerequisites for precise time position estimation are given. Neglecting random errors, the time position estimation will lead in both cases to τ_0 and τ_{H}, whose difference is

$$\tau_{\mathrm{H}} - \tau_0 = \frac{2(H - H_0)}{c} \tag{4.184}$$

Hence, we got an exact distance value whereas the position of the reference object defines the origin of the coordinates of our arrangement. The procedure does not require the knowledge of the internal delay τ_α, but if wanted it can be determined from a second reference measurement with known distance $H_1 - H_0$.

The described measurement approach eliminates perfectly systematic errors so that only random errors $\Delta\tau_{\mathrm{j}}(t)$, drift $\Delta\tau_{\mathrm{D}}(t)$ and possibly a bias error τ_{B} of pulse position estimation remain. We will investigate their role. First, we will consider how sub-nanosecond pulses and stretched (CW) ultra-wideband signals are affected by noise and jitter. Second, we will discuss the influence of the sensor electronics over the generation of these random errors, and finally the implications of noise and jitter on the pulse position estimation will be considered.

4.7.3.2 Noise- and Jitter-Affected Ultra-Wideband Signals

We deal with the signals $x(t)$ and $y(t)$; their physical nature may be a current, a voltage, a normalized wave or anything else. $x(t)$ shall be our unperturbed reference signal, and $\underset{\sim}{y}_{\mathrm{m}}(t)$ is the perturbed measurement considered as a random process. The goal is to determine the mutual time shift between both signals.

The measurement signal is affected by random noise $\underset{\sim}{n}(t) \sim N(0, \sigma_{\mathrm{n}}^2)$ and random jitter $\Delta\underset{\sim}{\tau}_{\mathrm{j}}(t) \sim N\left(0, \varphi_{\mathrm{j}}^2\right)$. It takes the form

$$\begin{aligned}\underset{\sim}{y}_{\mathrm{m}}(t) &= y\left(t + \Delta\underset{\sim}{\tau}_{\mathrm{j}}(t)\right) + \underset{\sim}{n}(t) = y(t) + \underset{\sim}{n}(t) + \left.\frac{\partial y(\tau)}{\partial \tau}\right|_{\tau=t} \Delta\underset{\sim}{\tau}_{\mathrm{j}}(t) \\ &= y(t) + \underset{\sim}{n}(t) + \dot{y}(t)\Delta\underset{\sim}{\tau}_{\mathrm{j}}(t)\end{aligned} \tag{4.185}$$

where we replaced $\partial y/\partial t = \dot{y}$ for a compact notation. (4.185) assumes that jitter and noise are not too strong so that a Taylor-series expansion $y(t + \Delta t) = y(t) + \dot{y}(t)\Delta t + \cdots$ may be stopped after the linear term and mixed noise–jitter terms are negligible. The unperturbed receive signal shall represent a time-shifted replica of the sounding signal $y(t) = x(t - \tau_{\mathrm{d}})$. Modifications of the time shape lead to systematic deviations, which we will exclude here.

Pulse Waveforms Affected by Noise and Jitter If an UWB sensor uses short pulses for target stimulation and no additional pre-processing of the captured data, the signal as modelled in (4.185) is immediately applied for τ_{d} estimation. Such a signal is affected by noise which may depend on time and the shape of the signal. Supposing we deal with a short pulse of magnitude Y_0 and rise time t_{r} and slew rate $SR \approx Y_0/t_r$, we can divide the pulse roughly into two different regions with respect to noise interference.

- Regions of nearly constant amplitude: $\dot{y}(t) \approx 0$. The variance and the signal-to-noise ratio of the signal are determined only from additive noise:

$$\operatorname{var}\{\underset{\sim}{y}_m(t)\}|_{\partial y/\partial t \approx 0} = \sigma_n^2 \Rightarrow \mathrm{SNR_{const}} = \frac{Y_0^2}{\operatorname{var}\{\underset{\sim}{y}_m(t)\}} = \frac{Y_0^2}{\sigma_n^2} = \mathrm{SNR_0} \qquad (4.186)$$

The effects which contribute to this type of perturbations mainly originate from thermal noise, devise internal noise (noise factor F), quantization noise of the ADCs and possibly external jammers. We will summarize all of them in the $\mathrm{SNR_0}$ value.

- Regions of strong amplitude variations: $|\dot{y}(t)| \gg 0$. The overall noise is given by

$$\operatorname{var}\{\underset{\sim}{y}_m(t)\} = \sigma_n^2 + (\dot{y}(t))^2\varphi_j^2 \le \sigma_n^2 + \mathrm{SR}^2\varphi_j^2 \approx \sigma_n^2 + Y_0^2\frac{\varphi_j^2}{t_r^2} = \sigma_n^2 + \frac{Y_0^2}{\mathrm{RJR}} \qquad (4.187)$$

Herein, $\mathrm{SR} = \|\dot{y}(t)\|_\infty \approx Y_0/t_r$, the slew rate of the (unipolar) pulse and

$$\mathrm{RJR} = \frac{t_r^2}{\varphi_j^2} \approx \frac{1}{4\varphi_j^2 \underset{\sim}{B}_3^2} \qquad (4.188)$$

we call rise-time-jitter ratio. It can be considered as the 'SNR-value' of the time axis. Via $\underset{\sim}{B}_3 t_r \approx 0.5$ (for Gaussian pulse, refer to Table 2.5), the RJR value can also be related to the bandwidth of a baseband signal. The total SNR-value at signal edges is roughly given by

$$\mathrm{SNR_{edge}} = \frac{Y_0^2}{\operatorname{var}\{\underset{\sim}{y}_m(t)\}} \Rightarrow \frac{1}{\mathrm{SNR_{edge}}} = \frac{1}{\mathrm{SNR_0}} + \frac{1}{\mathrm{RJR}} \qquad (4.189)$$

Correlation Functions Affected by Noise and Jitter Correlation processing in UWB sensing plays a role for both pulse-shaped signals (Section 3.2.3.2 'Product Detector' [74]) and for time-extended signals like PN-codes (Section 3.3) or random noise (Section 3.6). We will investigate how jitter and noise affect the correlation function $C_{y_m x}(t)$ if short pulses or PN-codes are applied. We do this by estimating expected value and variance of $C_{y_m x}(t)$. Performing correlation of (4.185) with the reference $x(t)$, we can write

$$C_{y_m x}(\tau) = \underbrace{\frac{1}{t_I}\int_{t_I} y(t)x(t+\tau)\mathrm{d}t}_{\text{deterministic}} + \underbrace{\frac{1}{t_I}\int_{t_I}\left(\underset{\sim}{n}(t)x(t+\tau) + \dot{y}(t)\Delta\underset{\sim}{\tau}_j(t)x(t+\tau)\right)\mathrm{d}t}_{\text{random}} \qquad (4.190)$$

t_I is the integration time of the correlation. The different terms in (4.190) are subdivided into two groups – deterministic and random. This classification can only be made if random noise test signals are excluded. Otherwise, the left-hand-side term has to be considered as random too. Due to the always limited integration time, random test signals would lead to additional variances which we try to avoid here. Corresponding to our assumption $y(t) = x(t - \tau_\mathrm{d})$, the expected value of the correlation function (4.190) follows immediately to (see Annex B.4.2 for explanation)

$$\mathrm{E}\{C_{y_\mathrm{m}x}(\tau)\} = \frac{1}{t_\mathrm{I}}\int_{t_\mathrm{I}} x(t-\tau_\mathrm{d})x(t+\tau)\mathrm{d}t = C_{xx}(\tau - \tau_\mathrm{d}) \tag{4.191}$$

That is, as long as the reflected signal is a copy of the reference signal, the expected value of the cross-correlation function $C_{y_\mathrm{m}x}(\tau)$ equals the time shifted auto-correlation $C_{xx}(\tau - \tau_\mathrm{d})$ of the stimulus. Its maximum value can be written as

$$\left\|\mathrm{E}\{C_{y_\mathrm{m}x}(\tau)\}\right\|_\infty = \mathrm{E}\{C_{y_\mathrm{m}x}(0)\} = \|x(t)\|_2^2 \tag{4.192}$$

Referring to the considerations of Annex B.4.2, we can write for the variance of the correlation function ($\underset{\cdot\cdot}{B}_\mathrm{n}$ is the equivalent noise bandwidth)

$$\begin{aligned}\mathrm{var}\{C_{y_\mathrm{m}x}(\tau)\} &= \mathrm{var}\left\{\frac{1}{t_\mathrm{I}}\int_{t_\mathrm{I}}\left(\underset{\sim}{n}(t)x(t+\tau) + \dot{x}(t-\tau_\mathrm{d})\Delta\underset{\sim}{\tau}_\mathrm{j}(t)x(t+\tau)\right)\mathrm{d}t\right\}\\ &= \frac{\sigma_\mathrm{n}^2}{t_\mathrm{I}\underset{\cdot\cdot}{B}_\mathrm{n}}\frac{1}{t_\mathrm{I}}\int_{t_\mathrm{I}} x^2(t+\tau)\mathrm{d}t + \frac{\varphi_\mathrm{j}^2}{t_\mathrm{I}\underset{\cdot\cdot}{B}_\mathrm{n}}\frac{1}{t_\mathrm{I}}\int_{t_\mathrm{I}}\left(\dot{x}(t)x(t+\tau)\right)^2\mathrm{d}t\end{aligned} \tag{4.193}$$

$t_\mathrm{I}\underset{\cdot\cdot}{B}_\mathrm{n} = \mathrm{TB}$ represents the time-bandwidth product of the correlator. (4.193) can be further reduced if we express the integral terms by the L_2-norm:

$$\mathrm{var}\{C_{y_\mathrm{m}x}(\tau)\} = \frac{1}{\mathrm{TB}}\left(\sigma_\mathrm{n}^2\|x(t)\|_2^2 + \varphi_\mathrm{j}^2\|\dot{x}(t)x(t+\tau)\|_2^2\right) \tag{4.194}$$

We define the signal-to-noise ratio $\mathrm{SNR_C}$ of the correlation function by relating its expected value to the variance:

$$\mathrm{SNR_C}(\tau) = \frac{\left\|\mathrm{E}\{C_{y_\mathrm{m}x}(\tau)\}\right\|_\infty^2}{\mathrm{var}\{C_{y_\mathrm{m}x}(\tau)\}} = \frac{C_{xx}^2(0)}{\mathrm{var}\{C_{y_\mathrm{m}x}(\tau)\}} \tag{4.195}$$

which can also be expressed by

$$\begin{aligned}\mathrm{SNR_C}(\tau) &= \frac{\mathrm{TB}\|x(t)\|_2^4}{\sigma_\mathrm{n}^2\|x(t)\|_2^2 + \varphi_\mathrm{j}^2\|\dot{x}(t)x(t+\tau)\|_2^2} = \frac{\mathrm{TB}}{(\mathrm{CF}^2/\mathrm{SNR_0}) + ((\mathrm{SAF}^2(\tau))/(\mathrm{RJR}))}\\ &\approx \frac{\eta T_\mathrm{R}\underset{\cdot\cdot}{B}_3}{(\mathrm{CF}^2/\mathrm{SNR_0}) + ((\mathrm{SAF}^2(\tau))/(\mathrm{RJR}))}\end{aligned} \tag{4.196}$$

with

$$\mathrm{SNR}_0 = \frac{\|x(t)\|_\infty^2}{\sigma_n^2} = \mathrm{CF}^2 \frac{\|x(t)\|_2^2}{\sigma_n^2} \tag{4.197}$$

CF is the crest factor

and $$\mathrm{SAF}^2(\tau) = t_r^2 \frac{\|\dot{x}(t)x(t+\tau)\|_2^2}{\|x(t)\|_2^4} \approx \frac{\|\dot{x}(t)x(t+\tau)\|_2^2}{4\underset{..}{B}_3^2\|x(t)\|_2^4} \approx \mathrm{CF}^2\,\mathrm{SACF}(\tau) \tag{4.198}$$

where we approximated the rise time by $t_r \approx \|x(t)\|_\infty / \|\dot{x}(t)\|_\infty$ (unipolar signal expected) or $t_r\underset{..}{B}_3 \approx 0.5$ (Table 2.5). The CF- and SACF-factors are parameters which summarize global properties of the signal shape. They were introduced in Section 2.2.2 as a quantitative measure of the energy distribution within the signal and a measure of the coincidence of large amplitudes with steep edges. SACF(τ) is a function of the mutual shift between the signal and its derivative. The SAF-factor relates this quantity to the bandwidth (rise time) of the waveform.

The time-bandwidth product in (4.196) represents the theoretically maximum possible processing gain. It should be replaced by $\mathrm{TB} \approx \eta T_R \underset{..}{B}_3$ for real correlation receivers where η is the receiver efficiency, T_R is the recording time for the whole correlation signal and $\underset{..}{B}_3$ is the bandwidth of a baseband signal (refer also to (3.43) and (3.44)).

(4.196) is an extension of the usual SNR-relation (2.249) which only respects additive noise but no jitter. (4.196) expresses in compact form the importance of the signal type (expressed by CF, SAF), the receiver performance (expressed by $\mathrm{SNR}_0, \mathrm{RJR}, \eta$) and the observation conditions (respected by $\underset{..}{B}_3, T_R$) on the quality of an ultra-wideband measurement. The SNR_0- and RJR-values represent quality figures of the measurement receiver. With respect to digital implementations of the correlation receiver, we can see that the SNR-value of the correlation function is better as more data samples are involved in its determination since $N \approx \eta T_R \underset{..}{B}_3 = g_p$ (g_p is the processing gain and N is the number of data samples)).

Further, it is apparent that unfavourable signal properties as large CF- and SAF-factors will decrease the maximum possible signal-to-noise ratio. The SAF(τ)-value increases if high signal amplitudes coincide with steep signal edges. The SAF-value of compact signals depends on the time lag τ. The time positions of its maximum values are roughly in the middle of the main peak edges since there large amplitudes and high signal gradients coincide. Outside the main peak, the SAF-value tend to zero.

For a time-stretched signal (e.g. PN-codes), the SAF-function takes low values and it is nearly independent of the time lag. That is, the variance of the correlation function of a stretched signal is constant so that we cannot observe a noise accentuation at the edges of $C_{xx}(\tau)$ or $C_{yx}(\tau)$ due to jitter. Figure 4.53 shows a simulation of the SAF-function for three examples – two for compact signals (Gauss pulse and derivative Gauss pulse) and one for a stretched signal, that is a band-limited M-sequence. All signals have the same bandwidth, record length and average power. The crest factors of the signals are also given in the figure.

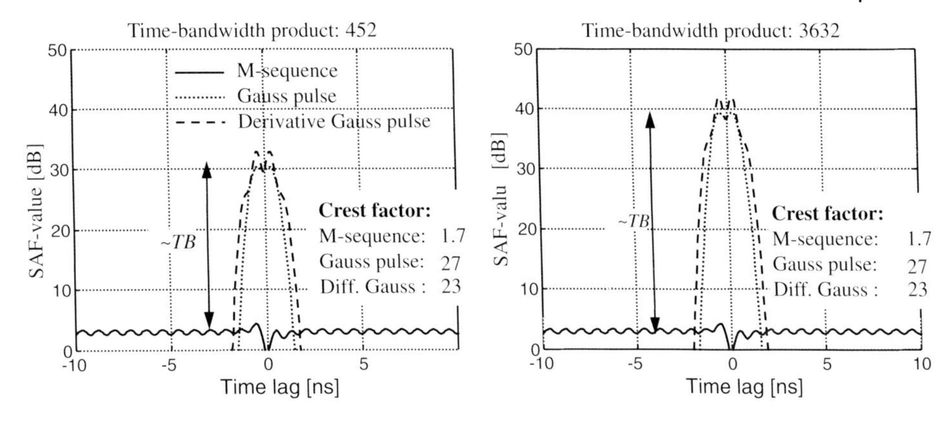

Figure 4.53 Slope–amplitude factor for different signals and time-bandwidth products. The signal bandwidth is 1 GHz. All signals have the same average power.

In Annex B.5 the peak values for the SAF-function are evaluated for simple signal approximations by analytical expressions. In spite of the strong simplifications there, the relations found in Annex B.5 show acceptable coincidence with the simulations in Figure 4.53.

In summary, we can draw the following picture about the confidence interval of a correlation function (see Figure 4.54 for illustration of the fundamental behavior).

For short pulses, correlation processing will not bring a real performance improvement (an observation which is actually not surprising). If we insert the results from Annex B.6.1 into (4.196), we obtain for pulse signals:

$$\mathrm{SNR_C}(\tau) = \begin{cases} \mathrm{SNR_0} & |\tau| \geq 2t_r \\ \dfrac{\mathrm{SNR_0 RJR}}{\mathrm{SNR_0 + RJR}} & \tau \approx 0 \end{cases} \tag{4.199}$$

which is quite the same result as for the unprocessed pulse (compare (4.186) and (4.189)). That is, a (digital or analogue) correlation processing will not provide better results since actually no energy compression can be performed.

In contrast to that, the SNR-value will be nearly independent of time if we deal with stretched wideband waveforms. This can be seen from Annex B.6.2 and (4.196), which approximately results to

$$\mathrm{SNR_C}(\tau) \approx 3\eta T_R B_3 \frac{\mathrm{SNR_0 RJR}}{\mathrm{SNR_0 + 3RJR}} \tag{4.200}$$

We can conclude with three observations:

- The 'jitter energy' is equally spread over the whole waveform for ultra-wideband CW signals. This raises the overall noise but does not elevate the noise at signal edges which are important for time position measurements.

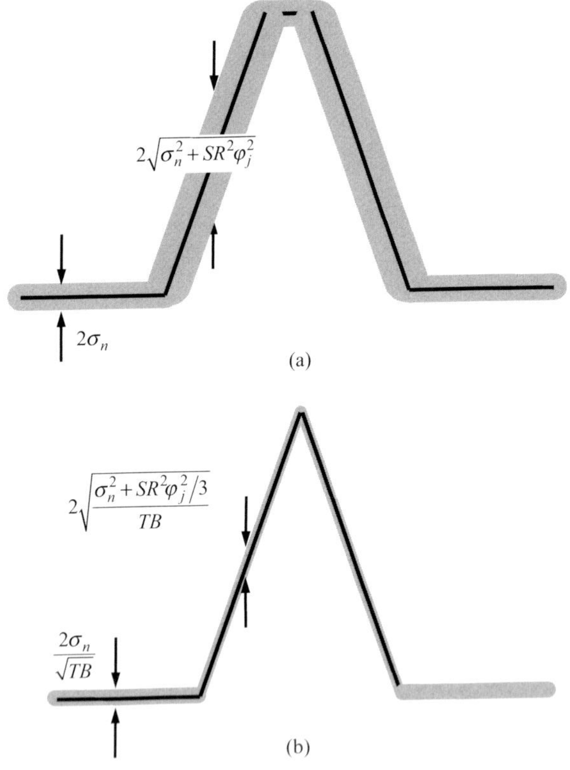

Figure 4.54 Strongly simplified picture of a correlation function and its confidence interval for short-pulse signals (a) and ultra-wideband CW signals (b) (e.g. M-sequence) for equal amplitude of correlation function.

- The overall noise is reduced by the correlation and averaging gain for UWB CW signals. Noise and jitter of pulses can only be reduced by synchronous averaging.
- From Annex B.5, we can approximately state CF $\approx$ max(SAF). Hence, jitter does not contribute to overall noise if RJR $\geq$ 10 SNR_0 for pulse and CW signals.

4.7.3.3 **Noise and Jitter Robustness of Various UWB Sensor Concepts**

After we have seen that noise and jitter affect the sensor response differently depending on the applied signal type, we will still consider how the sensor concept itself will contribute to noise and jitter robustness. Three different basic approaches are compared, distinguished by signal handling and timing.

Methods of signal handling:

- Data capturing by sequential sampling
- Analogue short-time correlation
- Digital correlation of ultra-wideband CW signal

Trigger approaches:

- Dual ramp – fast and slow ramp
- Dual sine – slightly frequency-shifted sinus for triggering of stimulus source and sampling gate
- Dual pulse – slightly frequency-shifted rectangular pulses for triggering of stimulus source and sampling gate

Basically, all three receiver operating modes may be combined with any of the timing approaches. The goal of the discussion here is to estimate the SNR_0- and RJR-value for the different device conceptions under simplified conditions. For the sake of simplicity but without restriction in generality, we introduced two noise sources in order to simulate additive noise and jitter. The jitter is always provoked by noisy threshold crossing. For details on the various sensor approaches see Chapter 3. As device under test, we take a delay line as reference object. We assume that the circuits for signal capturing and triggering are elaborated with the same care in all three examples. Their performance is given by the receiver SNR_0-value (including noise factor and quantization noise) and the trigger SNR_{tr}-value. We will exclude synchronous averaging in the consideration below. Its inclusion can be simply arranged by multiplying the achieved overall SNR-values with the number of averages. Furthermore, we do not respect efficiency losses by sub-sampling since we assume that all devices are operating with the same sub-sampling factor. Based on these assumptions, we get the following.

4.7.3.4 Short-Pulse Excitation and Dual Ramp Sampling Control

The block schematic in Figure 4.55a represents the classical structure of TDR-devices based on sequential sampling oscilloscopes. The sampling circuit is controlled by the interception of two voltage ramps. The duration of the fast ramp fixes the length of the time window t_{win}, and the slew rate of the slow voltage ramp (or staircase) determines the density of the sampling points, that is the equivalent sampling rate f_{eq}. Due to its small slew rate, the slow ramp is only drawn as horizontal threshold in Figure 4.55b.

The sampling jitter caused by the noisy ramp or threshold is

$$\varphi_j^2 = t_{win}^2 \frac{\sigma_{tr}^2}{V_R^2} = \frac{t_{win}^2}{SNR_{tr}} \tag{4.201}$$

where t_{win} is the length of the observation time window. Using (4.188), we obtain

$$\mathrm{RJR} = \frac{t_r^2}{\varphi_j^2} \approx \frac{\mathrm{SNR_{tr}}}{4t_{win}^2 B_3^2} \approx \frac{\mathrm{SNR_{tr}}}{(2\mathrm{TB})^2} \tag{4.202}$$

Herein, $\mathrm{TB} = t_{win} B_3$ represents the realized time-bandwidth product of the sampling receiver. Obviously, the jitter performance of a dual ramp sampling control degrades drastically with the time-band product of the signal to be observed. Using (4.186), (4.189) and (4.202), we get the effective SNR-value of the

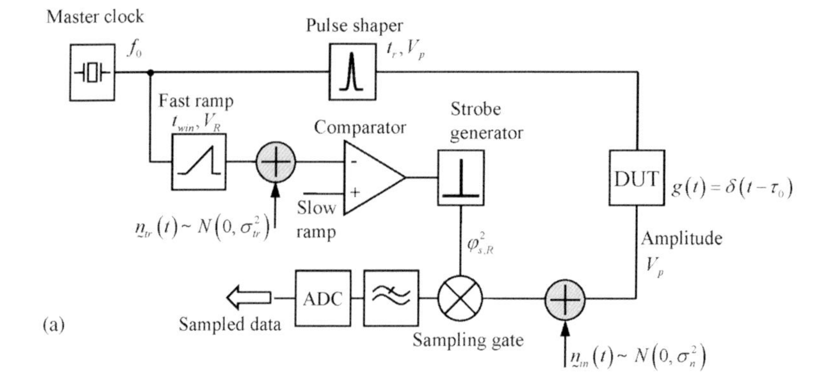

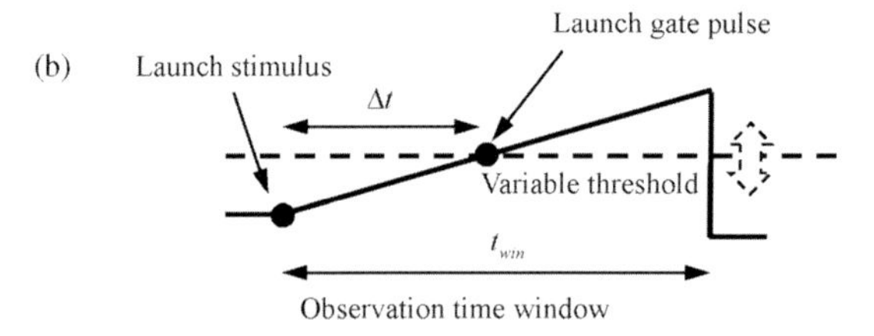

Figure 4.55 Simplified system model (a) for pulse excitation and dual ramp sampling control (b). The key parameters are indicated close to the individual blocks.

sampling receiver:

$$\mathrm{SNR} \approx \begin{cases} \mathrm{SNR}_0 & \text{at the pulse base} \\ \dfrac{\mathrm{SNR}_0 \mathrm{RJR}}{\mathrm{SNR}_0 + \mathrm{RJR}} = \dfrac{\mathrm{SNR}_0 \mathrm{SNR}_{tr}}{(2\mathrm{TB})^2 \mathrm{SNR}_0 + \mathrm{SNR}_{tr}} & \text{at the pulse edges} \end{cases} \tag{4.203}$$

Supposing comparable noise levels within the trigger and sampling circuits, we can roughly state $\mathrm{SNR}_0 \approx \mathrm{SNR}_{tr}$ which simplifies (4.203) to

$$\mathrm{SNR} \approx \mathrm{SNR}_0 \begin{cases} 1 & \text{at the pulse base} \\ (2\mathrm{TB})^{-2} & \text{at the pulse edges} \end{cases}$$

4.7.3.5 Analogue Short-Pulse Correlation and Dual Sine Timing

Figure 4.56a illustrates the basic device structure. Here, the two ramps are replaced by two sine waves of close frequency $f_0 \approx f_s$. Their period corresponds approximately to the ramp duration in Figure 4.55b. Hence, we can expect (only) slightly improved jitter behavior because of the somewhat steeper slope of the sine wave at zero crossing. The down-converter performs a multiplication with the replica of the stimulus which can be considered as a correlation of the measurement signal with the stimulus. This leads to an additional reduction of the overall bandwidth (degradation of the rise time) by a factor of $\sqrt{2}$. The bandwidth reduction also brings a noise reduction by the same factor.

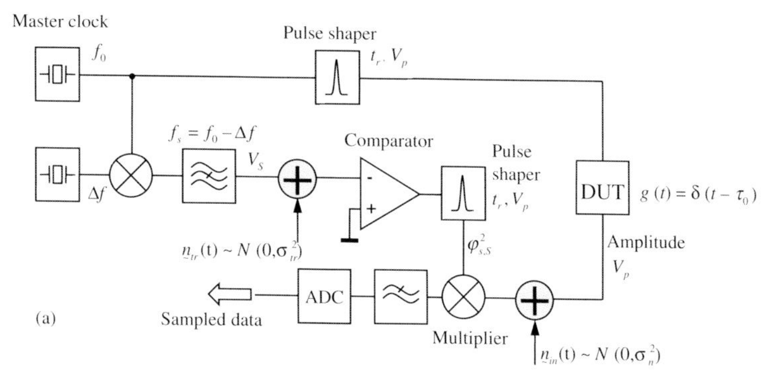

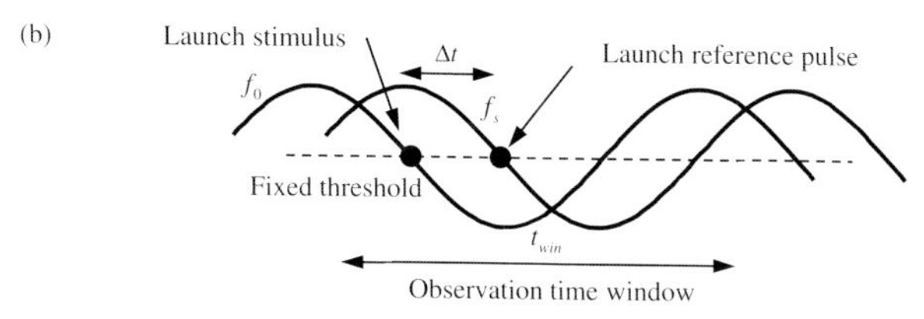

Figure 4.56 Simplified system model (a) for analogue short-pulse correlation and dual sine timing control (b).

The circuit behaves as follows. The trigger jitter caused by noisy sinusoids or noisy threshold results from (2.292) to

$$
\begin{aligned}
\varphi_j^2 &= \frac{\sigma_{tr}^2}{\left(2\pi f_s V_S\right)^2} = \frac{t_{win}^2}{(2\pi)^2 \mathrm{SNR}_{tr}} \\
\mathrm{RJR} &= \frac{t_r^2}{\varphi_j^2} \approx \frac{\pi^2 \mathrm{SNR}_{tr}}{\left(t_{win} B_3\right)^2} \approx \frac{\pi^2 \mathrm{SNR}_{tr}}{(\mathrm{TB})^2}
\end{aligned}
\tag{4.204}
$$

So that the overall signal-to-noise ratio corresponding to (4.199) is

$$
\mathrm{SNR} \approx \begin{cases} \sqrt{2}\mathrm{SNR}_0 & \text{at the pulse base} \\ \dfrac{\sqrt{2}\pi^2 \mathrm{SNR}_0 \mathrm{SNR}_{tr}}{\sqrt{2}(\mathrm{TB})^2 \mathrm{SNR}_0 + \pi^2 \mathrm{SNR}_{tr}} & \text{at the pulse edges} \end{cases}
\tag{4.205}
$$

If we again assume $SNR_0 \approx SNR_{tr}$, we achieve at:

$$
\mathrm{SNR} \approx \sqrt{2}\mathrm{SNR}_0 \begin{cases} 1 & \text{at the pulse base} \\ \dfrac{\pi^2}{\pi^2 + \sqrt{2}(\mathrm{TB})^2} & \text{at the pulse edges} \end{cases}
$$

4.7.3.6 Ultra-Wideband CW Stimulation and Dual Pulse Timing

We refer to Figure 4.57. The timing of strobe and stimulus generator is controlled via pulses of steep edges. The rise time of the edges is symbolized by the ramp in

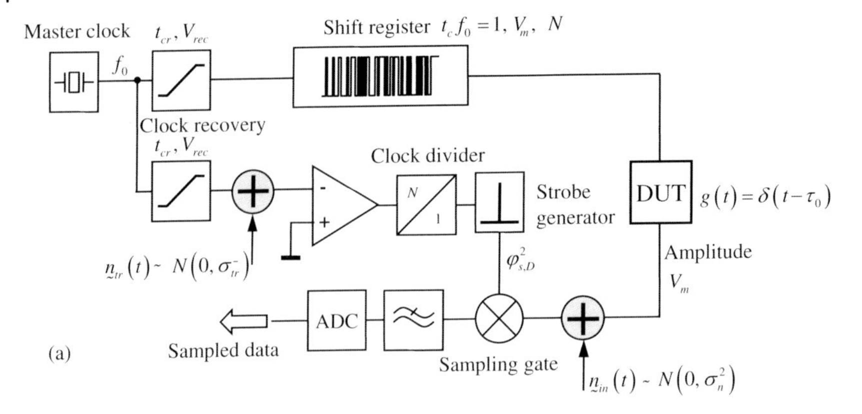

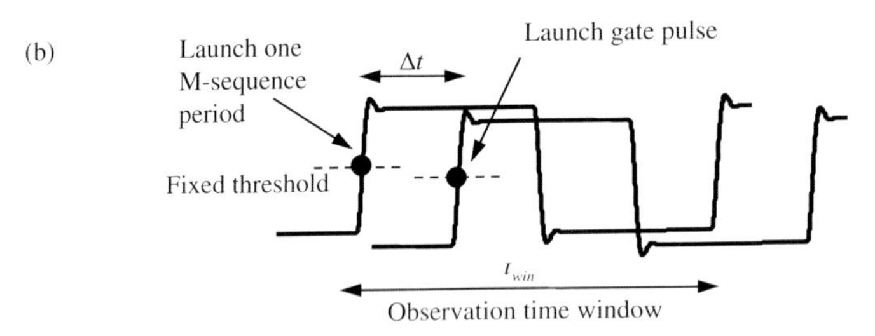

Figure 4.57 Simplified system model (a) for ultra-wideband CW stimulation (e.g. M-sequence) and dual pulse timing (b). The correlation is digitally performed. The order of the shift register is N.

the clock recovery block. The trigger signals are simplified in order to be comparable with those in Figure 4.55 and Figure 4.56. The sampling circuit is identical with that from Figure 4.55 but now it has to capture a stretched wideband signal and not a single pulse.

In case of sophisticated implementation, the rise time of all switch events within the circuit is much shorter than the chip length of the M-sequence. Hence, the jitter is mainly caused by the clock recovery stages which is fed by a sine wave of frequency f_0. Thus, from (2.292) we get

$$\varphi_{\mathrm{j}}^2 \approx \frac{1}{(2\pi f_0)^2 \mathrm{SNR}_{\mathrm{tr}}} \tag{4.206}$$

Due to anti-aliasing filtering of the sounding signal, we can roughly assume that the rise time of the individual M-sequence chips is about $t_{\mathrm{r}} \approx t_{\mathrm{c}} = f_0^{-1}$. Hence, the rise-time-jitter ratio takes the value

$$\mathrm{RJR} = \frac{t_{\mathrm{r}}^2}{\varphi_{\mathrm{j}}^2} = (2\pi)^2 \mathrm{SNR}_{\mathrm{tr}} \tag{4.207}$$

Referring to the total SNR-value (4.196) (efficiency loss by sub-sampling omitted by assumption) and applying CF- and SAF-approximations from Annex B.5.2, we approximately find

$$\mathrm{SNR} \approx \frac{2^N - 1}{(1/\mathrm{SNR}_0) + (1/3\mathrm{RJR})} \approx \mathrm{TB}\frac{3(2\pi)^2\mathrm{SNR}_0\mathrm{SNR}_{\mathrm{tr}}}{3(2\pi)^2\mathrm{SNR}_{\mathrm{tr}} + \mathrm{SNR}_0} \tag{4.208}$$

Also here, we will assume $\mathrm{SNR}_0 \approx \mathrm{SNR}_{\mathrm{tr}}$, so we get

$$\mathrm{SNR} = \mathrm{TB}\ \mathrm{SNR}_0$$

that is, jitter will have no effect on the measurements. It will neither affect the edges of the correlation function nor increase the overall noise.

In conclusion and by comparing the three UWB device concepts, we can state that wideband CW stimulation and dual pulse timing has the best robustness against random effects (noise and jitter). It outperforms the other approaches by the amazing factor of about $(\mathrm{TB})^3$ identical circuit technologies for all three conceptions supposed.

4.7.3.7 Random Uncertainty of Time Position Estimation

We have seen from (4.183) that the measurement uncertainties may have several reasons of systematic or stochastic nature. We would like to consider in a compact and illustrative way how random effects such as noise and jitter will affect the uncertainty of pulse position estimation where we will distinguish between baseband and band-pass signals. For a rigorous analysis, the reader can refer to Refs [63, 70, 75, 76].

The definition of a time position is only reasonable for compact time functions as pulses or cross-correlation functions. The options one has to fix a time position are the different centres of gravity (2.22) (including the maximum position) and crossing a threshold. We are interested in the noise behavior of the different methods. Figure 4.58 illustrated the problem. It shows a sampled baseband and band-pass signal which are affected by noise. The standard deviation of the noise is given by the error bars. The width of the error bars may depend on time, namely if jitter is not negligible. For the sake of shortness, we will not separate between jitter and random noise; rather, we consider only total noise. If required, a separate consideration would be straightforward by composing the total noise from both components as shown in the previous chapters.

The function $x(t)$ represents either a time signal or the cross-correlation function. The first one relates to pulse-shaped stimulation signals, while the second one has to be applied for wideband CW signals. Note that in the last case, the noise corruption involves the TB product of the signal and a possible jitter equalization as identified by (4.194) and (4.196), respectively.

Concerning the noise bandwidth, we will distinguish two opposite cases – first, the noise extends over the whole receiver bandwidth, and second, noise and signal bandwidth are the same. The first case appears in all sampling receivers. Even if a band-limiting filter is inserted in the input channel (e.g. anti-aliasing filter or the natural band limitation by the sampling gates), the noise of the down-converted signal will become uncorrelated since the coherence time of the input noise is much shorter than the sampling interval. The second case supposes an additional filtering of the signal after it was captured (compare Figure 2.73). This assumes,

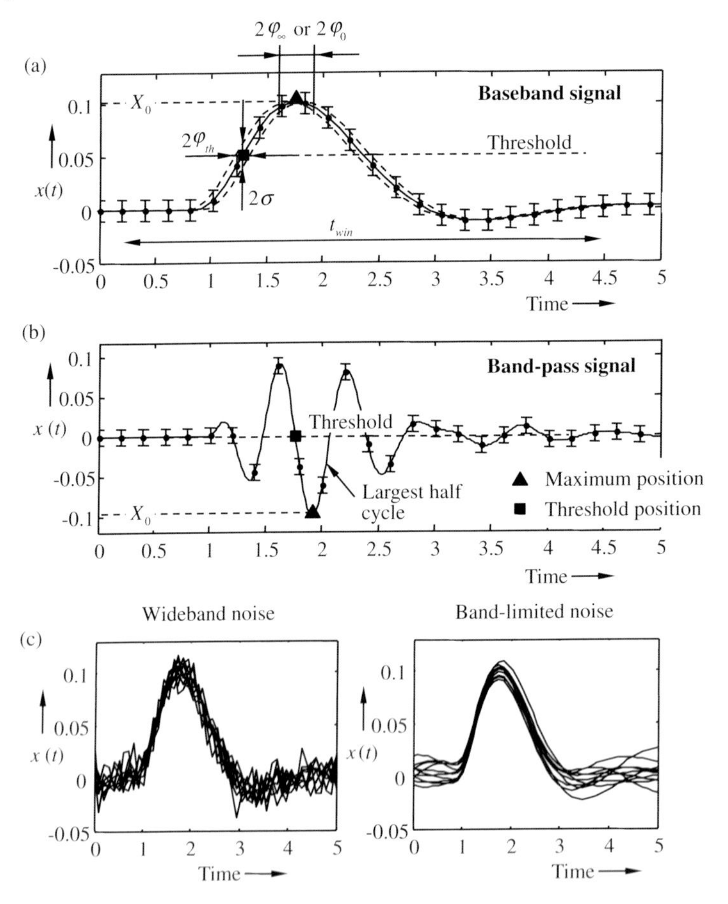

Figure 4.58 Illustration of the problem and definition of quantities. (a) Sampled baseband signal with error bars indicating the 68% confidence interval. (b) Sampled band-pass signal. (c) Signal appearance in case of wideband noise (left) and band limited noise (right) of identical variance.

however, that the bandwidth of the wanted signal is smaller than the receiver bandwidth. Note that filtering, if not carefully done, affects the time evolution of the signal which may cause additional systematic errors.

Time Position by Crossing a Threshold The time position determined from threshold crossing is assigned as τ_{th} and the related standard deviation is φ_{th}. The threshold should be placed at the steepest slope of the curve. This is typically at half magnitude of baseband signals and the first zero crossing before or after the maximum of band-pass signals (refer to square dots in Figure 4.58). Data sampling close to the Nyquist rate provokes large amplitude steps between consecutive samples so that the exact determination of threshold crossing needs adequate signal interpolation.

The standard deviation and the variance of the intersection of signal and the threshold, respectively, depend on the signal slope and the noise at that position (additive noise plus jitter-induced noise). Assuming that appropriate interpolation of sampled data was made, we can write

$$\varphi_{\text{th}}^2 = \frac{\sigma_{\text{th}}^2}{\|\partial x/\partial t\|_\infty^2} = \frac{\sigma_{\text{th}}^2}{\text{SR}^2} \approx \begin{cases} \dfrac{t_{\text{r}}^2}{X_0^2}\sigma_{\text{th}}^2 = \dfrac{t_{\text{r}}^2}{\text{SNR}} \approx \dfrac{a}{\underset{\cdot\cdot}{B}^2\,\text{SNR}} & \text{baseband signal} \\ \dfrac{\sigma_{th}^2}{\left(2\pi f_{\text{m}}\right)^2 X_0^2} = \dfrac{b_f^2}{(2\pi \underset{\cdot}{B})^2\,\text{SNR}} & \text{band-pass signal} \end{cases} \tag{4.209}$$

$\underset{\cdot}{B}$ and $\underset{\cdot\cdot}{B}$ are the one- or two-sided bandwidth of the signal, f_{m} is the centre frequency and b_{f} is the fractional bandwidth. The constant a relates the slew rate of different types of pulse signals to their bandwidth. It ranges roughly between $a = 0.53$ (for a sinc pulse – see Table 2.4) and $a = 3.8$ (for a Gaussian pulse and 60 dB bandwidth – see Table 2.5) if we place the threshold at the steepest part of the signals. The range accuracy related to (4.209) will be

$$\delta_{\text{A}} \approx \frac{\varphi_{\text{th}} c}{2} \tag{4.210}$$

As (4.209) indicates, increasing bandwidth and SNR-value will improve the performance of pulse time positioning. Additionally, we can find that small fractional bandwidth b_{f} will further reduce position uncertainties. It seems to be a paradox since narrowband signals obviously have worse resolution capabilities than wideband signals. But note that (4.209) was derived under the assumption that exactly the first zero crossing before or after the maximum was used as reference. With decreasing bandwidth the number of oscillations increases and it will be more and more critical to select correctly the largest half-cycle, so that this method will fail if we leave UWB conditions.

The registration of small target motion does not require a precise determination of threshold crossing as long as the target position must not be exactly known. This will simplify the processing of data. For motion tracking, the measurements are regularly repeated and one simply observes the evolution of individual samples placed at steep signal edges. Figure 4.59 gives an example. The radargram in the

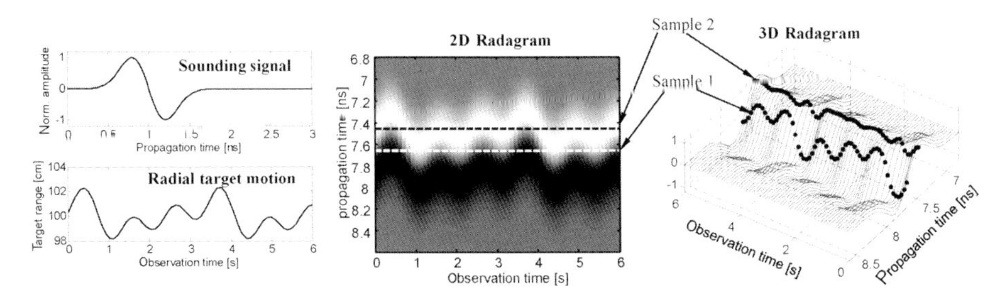

Figure 4.59 Weak motion tracking by following the time evolution of data samples placed at signal edges.

middle clearly shows the range variations of the targets. If one picks a data sample at the signal edge (sample 1 in Figure 4.59), its time evolution is proportional to the target movement as long as the amplitude variation is restricted to the linear part of the edge. This is not the case for sample 2 in our example which mostly stays within the region of the signal peak. Its voltage variation is quite weak and its time evolution represents a strongly non-linear distorted version of the target motion.

The minimum detectable target displacement δ_D is given by the sensitivity to register with some reliability a deterministic voltage variation of edge samples. This leads us to a relation corresponding to (4.210). Since the target motion affects all edges of the signal, it is recommended to insert all edge samples into the motion detection. Supposing the signal has N_0 edges with one useful sample each, we can write

$$\delta_D^2 \geq \frac{\varphi_{th}^2 c^2}{4N_0} \approx \begin{cases} \dfrac{t_r^2 c^2}{(4\cdots 8)\mathrm{SNR}} \approx \dfrac{ac^2}{(4\cdots 8)\underset{\cdot\cdot}{B}^2\mathrm{SNR}} & \text{baseband signal} \\ \dfrac{b_f^2 c^2}{8N(2\pi \underset{\cdot\cdot}{B})^2\mathrm{SNR}} \approx \dfrac{b_f^3 c^2}{8(2\pi \underset{\cdot\cdot}{B})^2\mathrm{SNR}} & \text{band-pass signal} \end{cases} \tag{4.211}$$

(4.211) is based on the assumption that a baseband signal has one or two edges which can be used for our purpose while a band-pass signal may have N oscillations with two zero crossings each where we know from (1.4) that $Nb_f = 1$. Note, however, that the variance reduction by factor N_0 only holds if the coherence time of noise is short compared to the distance between the considered data samples. N_0 tends to one, if noise and signal bandwidth are equalized by pre-filtering so that signal duration and noise coherence time are roughly the same. Nevertheless, we get an improvement since now the SNR-value increases.

We can observe from (4.211) that narrowband signals are of advantage for weak motion detection but they do not allow for precise localization of the moving target.

Maximum Position by Zero Crossing of First Derivative The determination of maximum position (assigned by the triangular spot in Figure 4.58) may be traced back to the threshold approach by referring to the first derivative of the signal. Hence, we can proceed correspondingly (refer to (4.209)) to determine the variance φ_0 of zero crossing but we have to take the noise of the differentiated signal (variance σ_d^2) and the slope of the first derivative at maximum position τ_0:

$$\varphi_0^2 = \frac{\sigma_d^2}{\left(\partial^2 x/\partial t^2\big|_{\tau_0}\right)^2} \tag{4.212}$$

We assume band-limited white noise of power spectral density Ψ_0 and bandwidth $\underset{\cdot\cdot}{B}_n$. In the case of unprocessed data, it corresponds to the equivalent sampling rate $\underset{\cdot\cdot}{B}_n = f_{eq}$ of sampling receivers. Applying differentiation rule of Fourier transformation, the power spectrum of the differentiated noise is

$$\Psi_d(f) \approx (2\pi f)^2 \Psi_n(f) = (2\pi f)^2 \begin{cases} \Psi_0 & |f| \leq \underset{\cdot\cdot}{B}_n/2 \\ 0 & |f| > \underset{\cdot\cdot}{B}_n/2 \end{cases} \tag{4.213}$$

Hence, its variance becomes for the baseband case

$$\begin{aligned}\operatorname{var}\{\dot{n}(t)\} = \sigma_{\mathrm{d}}^2 &= \int_{-\underset{..}{B}_{\mathrm{n}}/2}^{\underset{..}{B}_{\mathrm{n}}/2} \Psi_{\mathrm{d}}(f)\mathrm{d}f = (2\pi)^2\Psi_0 \int_{-\underset{..}{B}_{\mathrm{n}}/2}^{\underset{..}{B}_{\mathrm{n}}/2} f^2\,\mathrm{d}f \\ &= \frac{(2\pi)^2\Psi_0 \underset{..}{B}_{\mathrm{n}}^3}{12} = \frac{(2\pi \underset{..}{B}_{\mathrm{n}})^2\sigma_{\mathrm{n}}^2}{12}\end{aligned} \tag{4.214}$$

To get an estimate of the denominator of (4.212), we will take again the sinc and Gaussian pulse as well an arbitrary CW signal (represented by the correlation function) as example. We can find from Tables 2.4 and 2.5 and (2.73) for baseband signals:

$$\frac{1}{X_0}\frac{\partial^2 x}{\partial t^2}\bigg|_{\tau_0} = \begin{cases} \dfrac{(2\pi \underset{..}{B})^2}{12} = (2\pi B_{\mathrm{eff}})^2 & \text{sinc pulse} \\ 2(2\pi B_{\mathrm{eff}})^2 & \text{Gaussian pulse} \\ (2\pi B_{\mathrm{eff}})^2 & \text{CW signal} \end{cases} \tag{4.215}$$

Insertion of (4.214) and (4.215) in (4.212) yields

$$\varphi_0^2 = \begin{cases} \dfrac{1}{12(2\pi)^2 \mathrm{SNR}} \dfrac{\underset{..}{B}_{\mathrm{n}}^2}{B_{\mathrm{eff}}^4} & \text{sinc pulse} \\ \dfrac{1}{48(2\pi)^2 \mathrm{SNR}} \dfrac{\underset{..}{B}_{\mathrm{n}}^2}{B_{\mathrm{eff}}^4} & \text{Gaussian pulse} \\ \dfrac{1}{12(2\pi)^2 \mathrm{SNR}} \dfrac{\underset{..}{B}_{\mathrm{n}}^2}{B_{\mathrm{eff}}^4} & \text{CW signal} \end{cases} \tag{4.216}$$

If we select a sounding signal of constant power spectral density within the band limits $\pm\underset{..}{B}/2$, the related effective bandwidth follows from (2.73) as

$$B_{\mathrm{eff}}^2 = \frac{\underset{..}{B}^2}{12} \tag{4.217}$$

which also holds for the sinc pulse. Noise and signal bandwidth are identical $\underset{..}{B}_{\mathrm{n}} = \underset{..}{B}$ in the ideal case, so that (4.216) for a wideband CW signal or sinc pulse becomes

$$\varphi_0^2 = \frac{1}{(2\pi B_{\mathrm{eff}})^2 \mathrm{SNR}} \tag{4.218}$$

This relation gives the best achievable precision of pulse position estimation. In the literature (e.g. [63]), it is usually referred to as Cramér–Rao lower bound. Table 4.2 presents the achievable variances of the two methods – threshold crossing and maximum detection – for sinc and Gaussian pulses. We suppose that the signal bandwidths of the sinc and the Gaussian pulses $\underset{..}{B}$ and $\underset{..}{B}_{60}$, respectively, equal the receiver bandwidth $\underset{..}{B}_{\mathrm{R}}$. As the table shows, maximum detection based on sinc pulse waveform performs better than threshold detection by a factor 1.7 ($\approx$5 dB). In the case of Gaussian pulses, the situation is reverse. Now the threshold detection performs better by a factor of about 1.7. Since actual signals may be ranked somewhere between sinc and Gaussian behavior, both methods of pulse position estimations are more or less the same.

Table 4.2 Comparison of pulse positioning performance for different signals and methods.

	Sinc pulse	Gaussian pulse
Effective bandwidth B_{eff}	$\frac{\underset{..}{B}_{\text{R}}}{3.46}$ for $\underset{..}{B}_{\text{R}} = \underset{..}{B}$	$\frac{\underset{..}{B}_{\text{R}}}{10.3}$ for $\underset{..}{B}_{\text{R}} = \underset{..}{B}_{60}$
Maximum position by zero crossing of first derivative φ_0^2	$\frac{1}{(2\pi B_{\text{eff}})^2 \text{SNR}} \approx \frac{0.3}{\underset{..}{B}_{\text{R}}^2 \text{SNR}}$	$\frac{2.3}{(2\pi B_{\text{eff}})^2 \text{SNR}} \approx \frac{6.2}{\underset{..}{B}_{\text{R}}^2 \text{SNR}}$
Threshold crossing: φ_{th}^2	$\frac{1.75}{(2\pi B_{\text{eff}})^2 \text{SNR}} \approx \frac{0.53}{\underset{..}{B}_{\text{R}}^2 \text{SNR}}$	$\frac{1.36}{(2\pi B_{\text{eff}})^2 \text{SNR}} \approx \frac{3.6}{\underset{..}{B}_{\text{R}}^2 \text{SNR}}$
Optimum threshold level	42%	61%

For band-pass signals having a constant power spectral density Ψ_{xx0} within the frequency band[19] $|f_{\text{m}} \pm \underset{.}{B}/2|$, we find from (2.73) (note $x(t)$ is supposed to be a correlation function, b_{f} is the fractional bandwidth of the signal)

$$\frac{1}{X_0}\frac{\partial^2 x}{\partial t^2}\bigg|_{\tau_0} = (2\pi)^2 \frac{\Psi_{xx0}\int_{f_{\text{m}}-\underset{..}{B}/2}^{f_{\text{m}}+\underset{..}{B}/2} f^2\, df}{\Psi_{xx0}\int_{f_{\text{m}}-\underset{..}{B}/2}^{f_{\text{m}}+\underset{..}{B}/2} df} = (2\pi)^2\left(f_{\text{m}}^2 + \frac{\underset{.}{B}^2}{12}\right) = (2\pi f_{\text{m}})^2\left(1+\frac{b_{\text{f}}^2}{12}\right) \tag{4.219}$$

The variance of the differentiated noise takes (b_{fn} is the fractional bandwidth of noise)

$$\begin{aligned} \text{var}\{\dot{n}(t)\} = \sigma_{\text{d}}^2 &= \int_{f_{\text{m}}-\underset{..}{B}/2}^{f_{\text{m}}+\underset{..}{B}/2} \Psi_{\text{d}}(f)df = (2\pi)^2\Psi_0 \int_{f_{\text{m}}-\underset{..}{B}/2}^{f_{\text{m}}+\underset{..}{B}/2} f^2\, df \\ &= (2\pi)^2\sigma_{\text{n}}^2\left(f_{\text{m}}^2 + \frac{\underset{.}{B}_{\text{n}}^2}{12}\right) = (2\pi f_{\text{m}})^2\sigma_{\text{n}}^2\left(1+\frac{b_{\text{fn}}^2}{12}\right) \end{aligned} \tag{4.220}$$

$$\text{with} \quad \Psi_{\text{d}}(f) \approx (2\pi f)^2\Psi_{\text{n}}(f) = (2\pi f)^2\begin{cases}\Psi_0 & |f \pm f_{\text{m}}| \le \underset{.}{B}_{\text{n}}/2 \\ 0 & \text{otherwise}\end{cases}$$

so that (4.212) finally becomes for identical signal and noise bandwidth $\underset{.}{B} = \underset{.}{B}_{\text{n}}$:

$$\varphi_0^2 \approx \frac{1}{(2\pi)^2\left(f_{\text{m}}^2 + \frac{\underset{.}{B}^2}{12}\right)\text{SNR}} = \frac{1}{(2\pi f_{\text{m}})^2\left(1+(b_{\text{f}}^2/12)\right)\text{SNR}} = \frac{1}{(2\pi)^2\left(f_{\text{m}}^2 + B_{\text{eff}}^2\right)\text{SNR}}$$
$$\text{with} \quad B_{\text{eff}}^2 = \frac{\underset{.}{B}^2}{12} \tag{4.221}$$

19) We restrict ourselves to positive frequencies here.

where the effective bandwidth of a band-pass system is defined as

$$B_{\text{eff}}^2 = \frac{\int_0^\infty (f - f_\text{m})^2 \Psi(f - f_\text{m}) df}{\int_0^\infty \Psi(f) df} \tag{4.222}$$

Comparing (4.221) with the variance of the threshold detector (4.209), we find for band-pass signals of low fractional bandwidth

$$\varphi_0^2 \approx \frac{1}{(2\pi f_\text{m})^2 \text{SNR}} \approx \varphi_{\text{th}}^2; \quad b_\text{f} < 1 \tag{4.223}$$

pretending a high-range precision for large carrier frequency independently of the bandwidth of the signal. But this is not correct in general. Remember that we assumed to be able to identify uniquely the zero- or threshold crossing of a specific oscillation of the signal. Since we will increasingly face difficulties to do this by reducing the bandwidth, (4.223) is only applicable within the UWB range where the signal oscillations are restricted to a low number.

Centre of Gravity The centre of gravity τ_p defined by (2.22) is an integral measure of time position. We have extended the definition to the L_p-norm so that an infinite number of centres of gravity exist. One of these centres refers to the peak position $\tau_\infty = \tau_0$ providing a second option of its determination.

From an integral measure, one should expect higher robustness against random effects since a large number of data samples are involved, which usually reduces the variance by some processing gain. But this general observation must be taken with care. Short pulse or correlation functions which are sampled close to their Nyquist rate do not cover many data samples. Hence, we will come up with only small processing gain. For sampling rates beyond the Nyquist rate, we get more data samples. Nevertheless, the processing gain will tend to unity if the noise bandwidth is reduced to the signal bandwidth by pre-filtering. Variance reduction needs uncorrelated data samples which, however, are no longer available after low-pass filtering. Another way to get uncorrelated samples is to extend the duration of the integration time window t_{win} (compare Figure 4.58). But this also does not help, since only noise will be accumulated outside the pulse. In summary, we cannot expect a real improvement of pulse position uncertainty from the centres of gravity compared to the above-mentioned methods.

Additionally, the centres of gravity provoke considerable bias as function of the SNR-value. For demonstration, we consider the energetic centre τ_2 and we refer only to additive random noise. We can model our pulse signal (or correlation function) as

$$\underset{\sim}{x}(t) = x_0(t) + \underset{\sim}{n}(t); \quad \underset{\sim}{n}(t) \sim N(0, \sigma_n^2); \quad \text{cov}\{x_0, \underset{\sim}{n}\} = 0$$

The unperturbed signal has its centre of gravity at

$$\tau_{20} = \frac{\int_{\tau_{\text{win}} - t_{\text{win}}/2}^{\tau_{\text{win}} + t_{\text{win}}/2} t x_0^2(t) dt}{\int_{\tau_{\text{win}} - t_{\text{win}}/2}^{\tau_{\text{win}} + t_{\text{win}}/2} x_0^2(t) dt} = \frac{\int_{\tau_{\text{win}} - t_{\text{win}}/2}^{\tau_{\text{win}} + t_{\text{win}}/2} t x_0^2(t) dt}{\mathfrak{E}_x}$$

The integration was performed over a time window of duration t_{win}. The centre position of the window is $t = \tau_{\text{win}}$. $\mathfrak{E}_x$ is the total energy of the considered signal section.

The expected value of the energetic centre of gravity for the perturbed signal takes the form

$$\mathrm{E}\{\underline{\tau}_2\} = \frac{\mathrm{E}\left\{\int_{\tau_{\text{win}}-t_{\text{win}}/2}^{\tau_{\text{win}}+t_{\text{win}}/2} t(x_0(t)+\underline{n}(t))^2\,\mathrm{d}t\right\}}{\mathrm{E}\left\{\int_{\tau_{\text{win}}-t_{\text{win}}/2}^{\tau_{\text{win}}+t_{\text{win}}/2} (x_0(t)+\underline{n}(t))^2\,\mathrm{d}t\right\}} = \frac{\int_{\tau_{\text{win}}-t_{\text{win}}/2}^{\tau_{\text{win}}+t_{\text{win}}/2} t\,\mathrm{E}\left\{x_0^2(t)+2x_0(t)\underline{n}(t)+\underline{n}^2(t)\right\}\mathrm{d}t}{\mathfrak{E}_x+\mathfrak{E}_{\text{n}}}$$

$$= \frac{\int_{t_{\text{win}}} t\,x_0^2(t)\,\mathrm{d}t + \sigma_{\text{n}}^2\,t_{\text{win}}\,\tau_{\text{win}}}{\mathfrak{E}_x+\mathfrak{E}_{\text{n}}} = \frac{\tau_{20}\,\mathfrak{E}_x + \mathfrak{E}_{\text{n}}\,\tau_{\text{win}}}{\mathfrak{E}_x+\mathfrak{E}_{\text{n}}} = \frac{\tau_{20}+\tau_{\text{win}}(\mathfrak{E}_{\text{n}}/\mathfrak{E}_x)}{1+(\mathfrak{E}_{\text{n}}/\mathfrak{E}_x)} \tag{4.224}$$

Herein, $\mathfrak{E}_{\text{n}} = \sigma_{\text{n}}^2 t_{\text{win}}$ is the accumulated noise energy which continuously increases by widening the integration window whereas the signal energy $\mathfrak{E}_x$ remains constant due to the limited signal duration. Thus, the measured time position τ_2 tends towards the centre of the time window with increasing window length and noise power. Basically, the other centres of gravity τ_{p} will show corresponding behavior whereas the sensitivity to provide biased estimations reduces with increasing exponent p since large signal components are emphasized in favour of weak ones such as noise.

The examples below shall illustratively summarize the behavior of the different methods of pulse position estimation. Figure 4.60 shows how increasing noise affects the bias of the pulse position. As expected, the centres of gravity react susceptibly to increasing noise in particular for long integration window. The bias reduces by shrinking the integration window. It will actually vanish if time position τ_{p} and time window centre τ_{win} coincide. In order to achieve such conditions, an iterative search is required by which τ_{win} successively approaches τ_{p}. Threshold approach and maximum position are not affected by bias.

Figure 4.61 deals with the standard deviation of pulse position under various conditions. A sinc function was assumed as sounding signal which provides best prerequisites for stable pulse position determination. The signal was affected by random noise of different bandwidth.

We note that maximum position leads to the best results in case where signal and noise bandwidth are identical $\ddot{B}_{\text{n}} = \ddot{B}$. It is followed by threshold detection. But this order of performance may reverse if the properties of the sounding signal tend to those of a Gaussian pulse (compare Table 4.2). The stability of the centres of gravity deteriorates increasingly for long integration windows. However with rising noise bandwidth $\ddot{B}_{\text{n}} \gg \ddot{B}$, the centres of gravity outperform the other methods, while maximum detection τ_0 degrades rapidly as expected since the signal derivation emphasizes the noise. Threshold crossing does not depend on noise bandwidth.

The final example in Figure 4.62 considers a band-pass signal (sinc pulse) of fractional bandwidth $b_{\text{f}} = 0.2$. The centres of gravity are working stable over the whole SNR-range while the approaches using threshold crossing fail below a SNR $\leq$ 30 dB since the selection of the correct signal segment becomes critical in presence of strong noise. On the other hand, if this selection performs well, these two approaches provide the best result since they can work on steep signal edges.

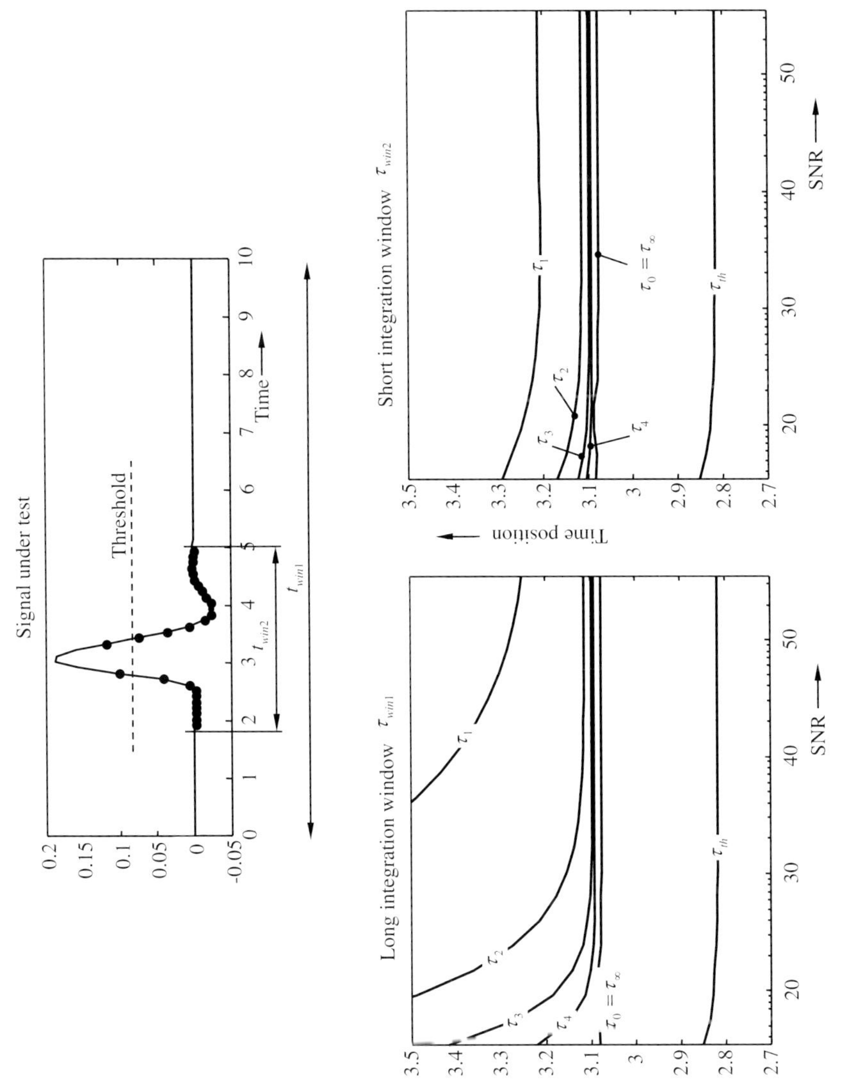

Figure 4.60 Influence of the signal-to-noise ratio onto the expected value of time position.

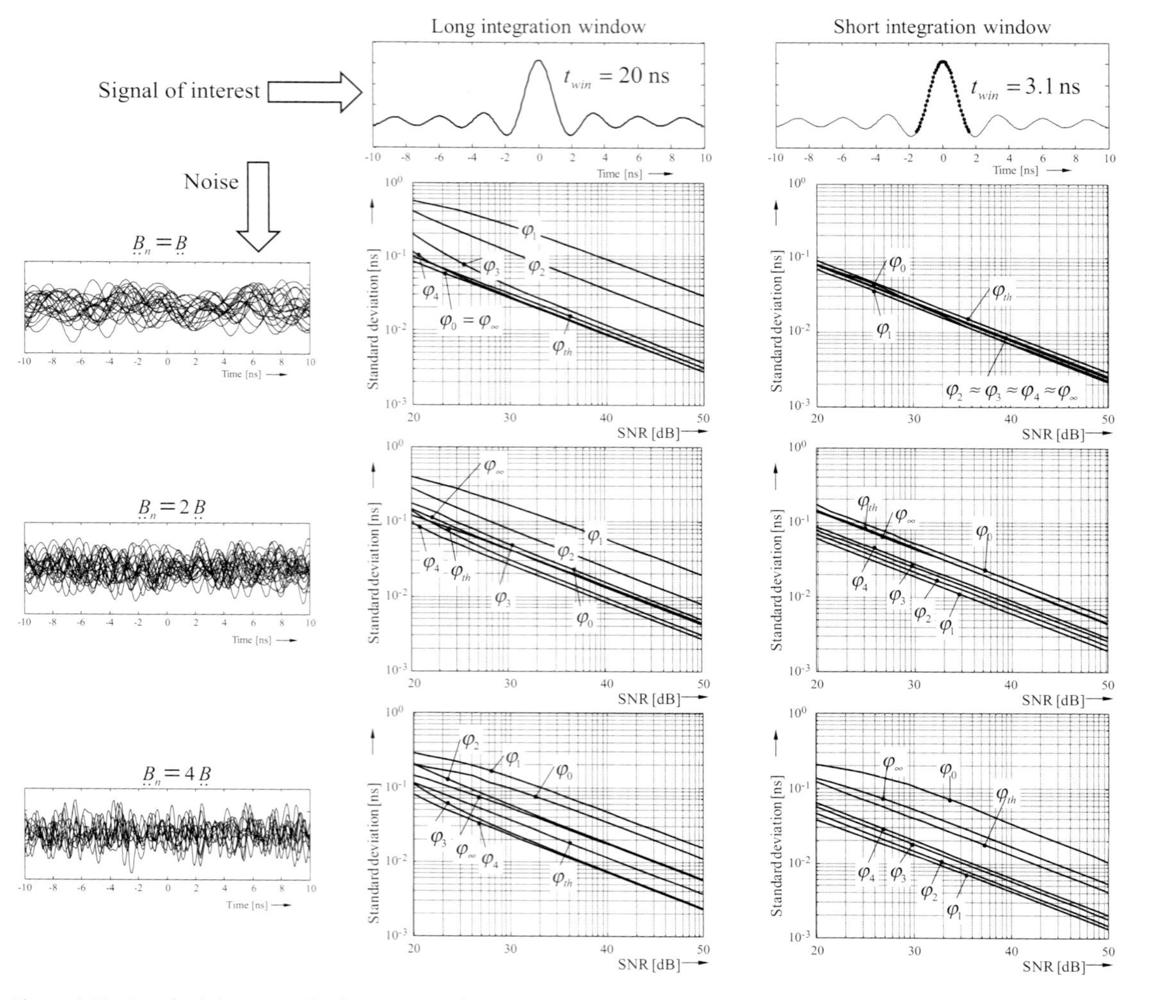

Figure 4.61 Standard deviation of pulse position of a sinc-pulse. The signal is affected by noise of different bandwidth.

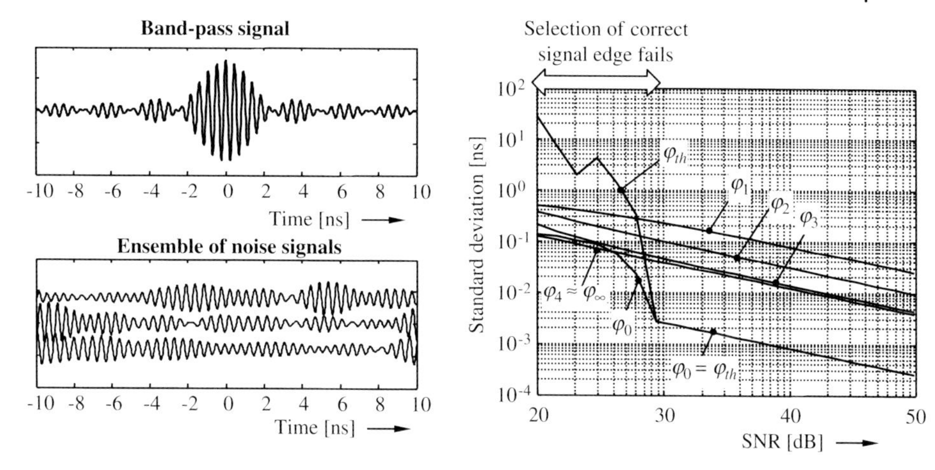

Figure 4.62 Standard deviation of pulse position of a band-pass signal having a fractional bandwidth $b_f = 0.2$. Noise and sounding signal have identical bandwidth.

4.7.3.8 Time Position Error Caused by Drift and Its Correction

We will assign slow random variations of pulse time position as drift. The main reasons for drift events are ageing of electronic components and mainly temperature variation in particular during the warm-up phase of the device.

A measurement of a signal $x(t)$ typically maps voltage as function of time to an equivalent 'device internal' function $\widehat{x}(\widehat{t})$ whereas the 'mapping rules' should be linear functions. These 'mapping rules' are realized by the working principle of the sensor device whose electronics is subjected to temporal variations and ageing which cause the drift events. They will not affect the measurement of a single IRF or FRF since the variations are quite slow. One can only note the drift by observations over a longer time, that is by repetition of the measurement. The simplest linear functions are shifting and scaling, so that we will model our drift-affected measurement by

$$\widehat{x}(\widehat{t}, T) = a(T)x(s(T)t + \Delta\tau_d(T)) + b(T) \tag{4.225}$$

where t is the interaction time (short time), T is the observation time (long time), $\widehat{t}$ is the device internal time (short time), $a(T)$ is the observation time-dependent gain, $b(T)$ is the observation time-dependent offset, $s(T)$ is the observation time-dependent timescaling and $\Delta\tau_d(T)$ is the observation time-dependent time shift.

In the case of sampling converters, $s(T)$ represents the sub-sampling factor. From a practical viewpoint, we can consider it as constant $s(T) = s_0$ as long as the sensor internal clock generator is nicely stabilized. The time shift $\Delta\tau_d(T)$ does not depend on clock generator; rather, it is usually caused by variations of trigger thresholds (i.e. amplitude variations) within the timing circuits. Hence, it behaves like gain and offset drift.

In order to remove the temporal dependencies $a(T), b(T), \Delta\tau_d(T)$ from the captured data, one performs reference measurement over stable objects and observes the temporal variations of the captured signal. Such reference measurement can be gained either from time separation of reference signal and DUT reaction (see Figures 3.21a and 3.95) or by a second receiver channel (reference channel) which is supposed to have identical drift behavior as the measurement channel (see Figure 3.21b and 3.96–3.98). The drift can be removed by the following approaches:

Method 1: Remove the DC component from reference and measurement signal and divide their complex spectra. This is typically done in all network analysers which perform relative measurements $\underline{S}_{ji}(f) = \underline{b}_j(f)/\underline{a}_i(f)$. If the sensor is configured for system calibration (see Section 2.6.4) to remove device internal clutter, such drift correction is implicitly included.

Method 2: Take the reference signal $\widehat{x}(\widehat{t}, T = 0)$ at observation time zero $T = 0$ as anchor point and determine deviations to ongoing reference measurements $\widehat{x}(\widehat{t}, T)$ by maximum likelihood estimation[20]:

$$[a(T), b(T), \Delta\tau_{\mathrm{d}}(T)] = \arg\min_{a,b,\Delta\tau_{\mathrm{d}}} \int \left(\widehat{x}(\widehat{t}, T) - \widehat{x}(\widehat{t}, T = 0)\right)^2 \mathrm{d}t \qquad (4.226)$$

Once the parameters are determined for the reference signal, they can be applied to the actual DUT response captured by the second device channel. Note that time drift correction needs sub-pixel time shift operations (see (2.188)).

Method 3: Same as method 2 but in successive steps. Since $a(T), b(T)$ and $\Delta\tau_{\mathrm{d}}(T)$ influence the measurement signal independently of each other, they can be considered separately so that no multi-dimensional search as in (4.226) is necessary. First, one determines the mutual time shift $\Delta\tau(T) = \Delta\tau_{\mathrm{d}}(T) - \Delta\tau_{\mathrm{d}}(T = 0)$ of the reference signal taken at observation time T by some method of pulse position estimation. Then, the reference signal is shifted back by $\Delta\tau(T)$ to annihilate the drift before one determines gain and offset drift. Finally, these quantities $(a(T), b(T), \Delta\tau(T))$ are applied to correct the second device channel for DUT response measurement.

Figure 4.63 shows the time drift behavior of an M-sequence device (ninth-order M-sequence, 8.95 GHz clock rate, sub-sampling factor 512) for demonstration. The pulse positions in channels A and B were measured during the warm-up phase and in thermal equilibrium. As expected, we can observe the largest variations during the warm-up phase. The absolute time drift per channel of a pulse position was roughly 1 ps during the first hour. The drift difference between both channels A and B was about 10 times less. After temperature settling, the absolute drift approaches 20–25 fs while the drift difference between both channels gets approximately 10 fs. The rms jitter is about 5 fs in the shown examples. Recent results have shown rms jitter of less than 2 fs within an observation time of about 1 hour.

20) Developing the signal in Taylor-series and terminate after the first term, the solution of (4.226) may be converted into Moore–Penrose inversion for sampled data. However, the termination of the Taylor-series after the first term may cause problems for data samples which are located close to a minimum or maximum of the signal.

(a)

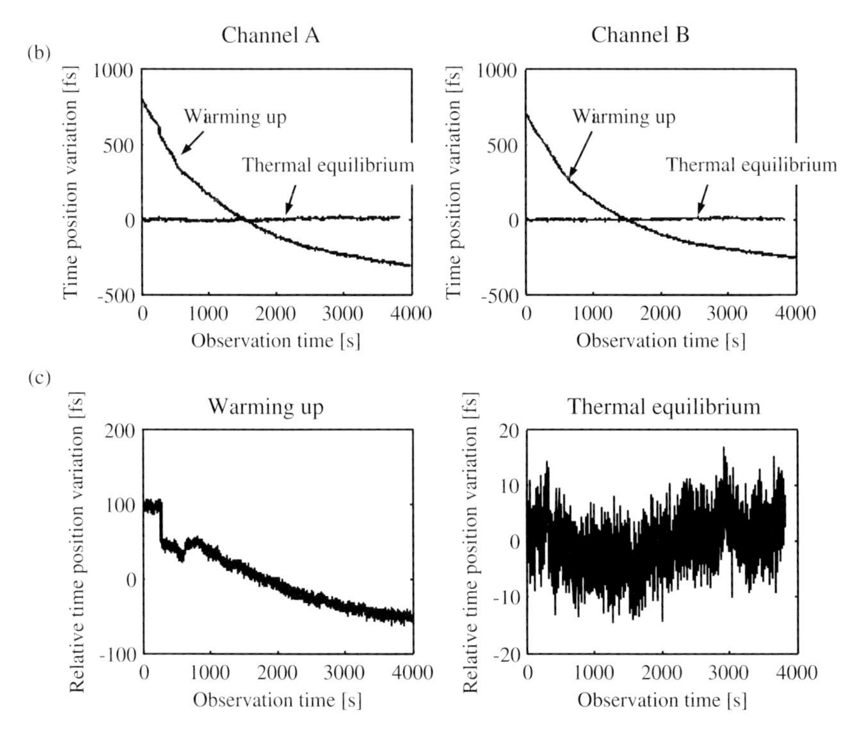

Figure 4.63 Time drift of a 1Tx/2Rx M-sequence sensor. (a) M-sequence device with power supply; (b) time drift of channel A and B and (c) time difference drift (channel A – channel B). (*Courtesy*: ILMsens.)

Summary We have considered several methods of time position estimation. It requires sharp pulse-shaped time functions which may be either a voltage pulse or a correlation function of wideband signals. The precision and hence the range accuracy of UWB radar is affected by systematic and random deviations. The systematic errors of time position are mainly due to differences in the time shape of reference and measurement signal; that is, these errors will appear if the fidelity of the involved signals is less than unity. Additionally, most of the integral method of pulse position estimation tends to provide biased estimations depending on noise.

Random errors are caused by additive noise and jitter whereas jitter mainly affects steep signal edges. But just the steep edges are the crucial signal parts which determine time position precision. Hence, a wideband correlator provides best conditions for low random time position errors since it suppresses additive noise and distributes jitter energy over the whole signal, avoiding noise accentuations at the edges. Furthermore, dual-pulse timing control of the measurement receiver largely reduces trigger jitter so that it will not affect pulse position precision.

The two most sensitive methods of pulse position estimation are based on threshold crossing. One of them determines the signal crossing at the steepest part of the leading (or falling) edge. Practically, one usually assumes that the steepest slope is at about 50% of pulse magnitude. The second of these methods determines zero crossing of the first derivative of the signal. Hence, it corresponds to the maximum position. It is the most sensitive method in case that signal and noise bandwidths are the same. If the sounding signal (or its correlation function) is a sinc function, the resulting pulse position error reaches the Cramér–Rao lower bound. Fifty percent threshold crossing does not depend on noise bandwidth; therefore, it is more robust in cases where pre-processing does not equalize signal and noise bandwidth. The performance of both methods improves with increasing centre frequency of the sounding signal. However, SNR-value and fractional bandwidth should not fall below a critical value in order to allow selection of the correct zero crossing. Adequate electronics and interpolation methods supposed the random errors of pulse position can be reduced down to 0.01–1‰ of the equivalent sampling interval in practical cases as shown in Figure 4.63.

The theoretical limits given by the Cramér–Rao lower bound are depicted in Figure 4.64 for an UWB correlation receiver dealing with sounding signals of crest factor $CF = 1$ and Nyquist sampling. The estimation supposes that the jitter-rise-time ratio does not fall below the SNR-value which is expressed here by the effective number of bits ($SNR[dB] \approx 6\,ENOB$). Sub-sampling will increase the standard deviation φ_0 of pulse position by the square root of the sub-sampling factor. Nevertheless, the resulting precision remains still quite high in case of M-sequence devices which operate typically with sub-sampling factors of 128–512.

The amazing range sensitivity of UWB sensors opens up new radar applications – for example detection of weak hidden targets, high-resolution radar imaging and detection of weak displacements. The reduction of systematic errors down to the level of random position errors will be a big challenge with respect to

- the mechanical precision and stability of the sensor arrangement,
- the knowledge of time domain antenna pattern with corresponding precision and
- temperature influence on cable length and material permittivity and so on.

For motion detection, the requirements are less critical since only relative fast temporal variations need to be detected so that temperature drift and temperature expansions are less important. Practically, such time resolution may only be exploited by

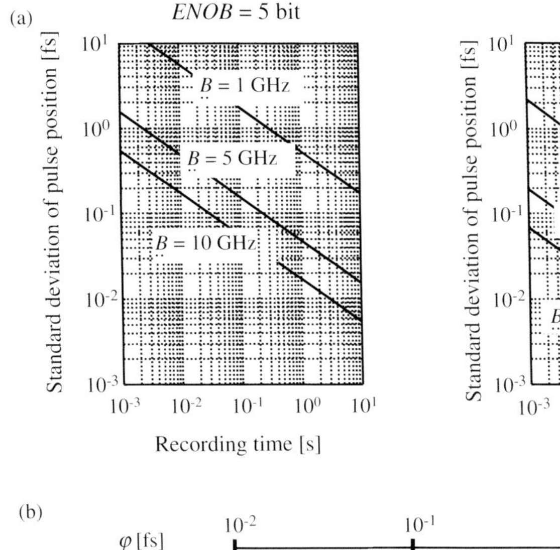

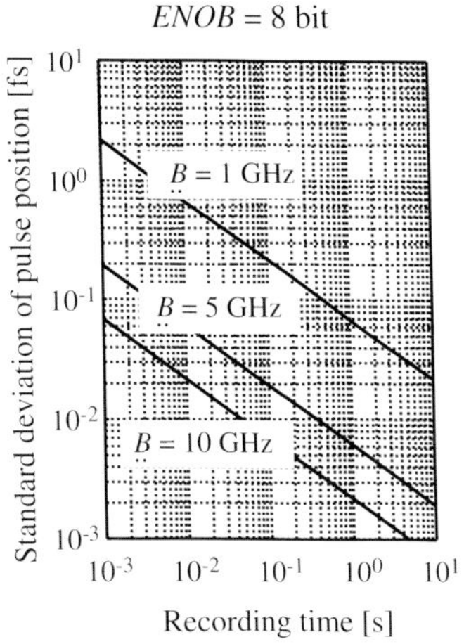

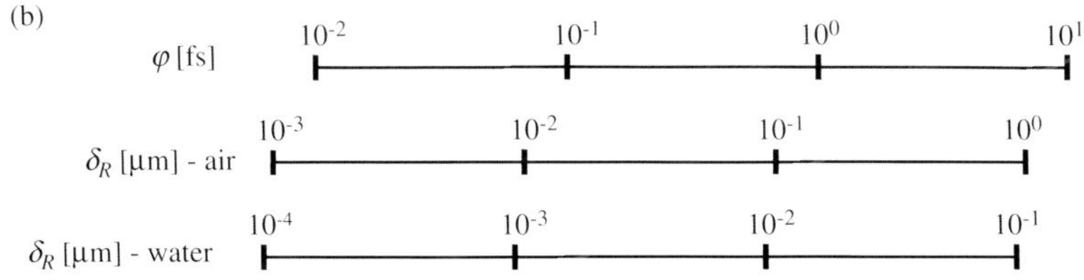

Figure 4.64 Theoretical limits (a) for random uncertainties of pulse position estimation. (b) Conversion from time to range uncertainty for two different propagation media. The conversion does not include wave dispersion of water.

fully integrated sensor devices. Let us regard RO 4003C as reasonable RF-PCB material for component wiring. The material has a thermal expansion coefficient of about 10 ppm. Therefore, a transmission line of only 2 cm length already causes a propagation time variation of 2 fs @ 1 °C temperature variation solely by length variation. Thermal permittivity variations are not yet included in this number. Hence, in order to defeat this inconvenience and to exploit the full device performance, it only remains device integration and short connections to the measurement ports.

4.8 Target Detection

4.8.1 Preliminary Remarks

The basic philosophy of target detection is to collect over a specified time (we call it time on target t_{ot}) as much as possible energy scattered from the object of interest and to concentrate this energy into a single point of time at which the decision on target is present or not has to be made. Generally one prefers methods of energy accumulation which reject noise power accumulation. However, these methods suppose specific knowledge about the target which is not always available. Hence, one can find only quasi-optimal detection conditions in many cases.

The target is detected if the captured amount of energy exceeds sufficiently the perturbation level. If this is not the case, either there is no target or it scatters too weak to be detected. Usually, the decision is made with help of an appropriately preprocessed signal which is either below a certain threshold – indicating hypothesis 1 (e.g. no target) – or above the threshold – indicating hypothesis 2 (e.g. target present). We shortly mentioned the topic in Section 4.7.1, Figure 4.46.

Here, we will deliberate on the question how to concentrate the 'target energy' in order to have a sensitive and correct decision on detection. To avoid misunderstandings and to highlight the differences to narrowband radar, we will come back to Figure 4.8 first. There it was shown that measurements by short pulses or time-stretched ultra-wideband signals can be considered in the same way if one intercalates a correlation for the second type of signals. Such correlation – effecting an impulse compression – uses the knowledge of the sounding signal which is used as reference. This procedure represents a first kind of 'energy concentration'. For short pulse-shaped signals, the procedure may be applied too but it will not bring profit with respect to noise and jitter suppression.

Such possibly inserted processing step is connected with the term of matched filtering in narrowband radar since the correlation is done via a filter whose impulse response is matched to the known time shape of the receive signal which is nearly identical to the transmit signal. Under narrowband constraints, this is an optimum procedure since it concentrates the signal energy in time as good as possible. Hence, the wanted signal is best emphasized against noise and it provides best prerequisites for secure target detection [68].

In contrast to ultra-wideband sensing, impulse compression (if necessary) alone does not provide optimum detection conditions since typically the receive signal has different shape as the transmit signal; that is, it is generally not known. So the question arises what are optimum detection conditions and how to approach them. We will consider some basic methods where we will suppose equal bandwidth of noise and baseband signal $\underset{..}{B}_{\mathrm{n}} = \underset{..}{B}$ (band-pass signals $\underset{.}{B}_{\mathrm{n}} = \underset{.}{B}$) if not otherwise declared.

Let us consider for that purpose the time domain radar equation. From (4.60) and (4.43), we yield

$$b(t) = \frac{1}{2\pi c r_1 r_2} \underbrace{R_1(t, \mathbf{e}_1) * R_2(t, -\mathbf{e}_2) * u_1(t) * a(t)}_{\text{intrinsic sensor response}} * \underbrace{\Lambda(t, \mathbf{e}_1, \mathbf{e}_2)}_{\text{target response}} \tag{4.227}$$

r_1 and r_2 are the distance between transmitter/receiver antenna and target.

Optimum detection capability would require selecting a reference signal $x_0(t)$ in such a way that fidelity FI_{bx} (see (2.54)) approaches one. From Cauchy–Schwarz inequality it can be shown that this condition is only given for

$$b(t) \sim x_0(t - \tau) \tag{4.228}$$

Or in other words, the correlation function $C_{x_0 b}(\tau) = b(\tau) * x_0(-\tau)$ has the largest peak value (all energy is perfectly concentrated) from all possible reference signals $x(t)$ and hence it is best suited for detection purposes. Therefore, the goal should be to perform correlation between receiving signal and reference signal leading to more or less compaction of the captured signal accompanied with suppression of random noise.

As we see from (4.227), the implementation of the 'compression function' $x_0(t)$ needs knowledge of the sensor device, namely the antenna IRFs $R_{1,2}(t, \mathbf{e})$, and of the sounding signal $a(t)$ but also the scatterer IRF $\Lambda(t, \mathbf{e}_1, \mathbf{e}_2)$. While the sensor-dependent part (intrinsic sensor response) may be determined from calibrations, the target-dependent part ($\Lambda(t, \mathbf{e}_1, \mathbf{e}_2)$) is the measurement goal and hence unknown in the general case. Thus, we have to discuss how it affects the detection performance. For that purpose, we decompose the reference signal $x_0(t)$ into two parts:

$$x_0(t) = \Lambda(t, \mathbf{e}_1, \mathbf{e}_2) * x_{\mathrm{ISR}}(t) \tag{4.229}$$

one relates to the mostly unknown scatterer IRF $\Lambda(t, \mathbf{e}_1, \mathbf{e}_2)$ and the second one $x_{\mathrm{ISR}}(t)$ – termed as intrinsic sensor response – joins all properties of the sensor device. In summary, we can state that we partially know the receive signal of the sensor (and hence the reference signal for the correlation) and partially not. From the known part, we can provide a concentration of energy via correlation which was spread beforehand due to long sounding signals and the ringing of the antennas.

We will mainly deal with the unknown part $\Lambda(t, \mathbf{e}_1, \mathbf{e}_2)$ and how to reduce the detection degradations as consequence of this lack of knowledge. We suppose that sensor-specific signal distortions described by the intrinsic sensor response are already removed from the measured signal. In order to avoid large data sets for $x_{\mathrm{ISR}}(t)$, the sensor response should be nearly independent of the directions $\mathbf{e}_1, \mathbf{e}_2$. This mainly concerns the antennas which may vary their pulse shape over the radiation angle. If we do neglect these pulse shape variations, the fidelity pattern $\Theta^{\mathrm{FI}}(\mathbf{e})$ of the antennas (see (4.121)) can be used to estimate the performance loss in order to judge if it is acceptable or not for a given application.

Before we enter into deeper discussions, we can already prophesy that the largest influence on the detection performance will be exerted by the sounding signal since in case of a proper selection it will provide the largest contribution to noise suppression and signal compactions if it has a large TB product. In contrast to time-stretched wideband signals, the IRF duration of antennas or scatterer are typically much shorter so that performance losses are frequently not as serious if they cannot be correctly respected due to insufficient knowledge of their behavior. Often, it is more promising to extend the time on target in order to increase the quantity of coherent (target specific) energy in favour of random noise.

4.8.2
Target Detection Under Noisy Conditions

We have seen that the concentration of backscattered energy is essential to detect a target under noisy conditions. What can be done under ultra-wideband condition to approach this goal? We will discuss several cases. For the sake of brevity, we will only presume signal perturbations by additive Gaussian noise; jitter will be omitted and assumed to be converted into additive noise in case of time-stretched wideband signals.

4.8.2.1 Detections Based on a Single Measurement

We assume a measurement of duration T_R during which all data are collected that give a single receiver signal. T_R has to meet the condition (see (2.90))

$$T_R \leq \frac{c}{2\underset{\sim}{B}v_r} \tag{4.230}$$

v_r is the range speed of a targetin order to avoid signal degradation due to Doppler effect. We model our receiving signal by random process in which the reflected signal is deterministic and the bandwidth of the sounding should be larger than the bandwidth of the scatterer IRF:

$$\underset{\sim}{b}(t) = \frac{1}{2\pi c r_1 r_2}\Lambda(t, \mathbf{e}_1, \mathbf{e}_2) * x_{\mathrm{ISR}}(t) + \underset{\sim}{n}(t); \quad \underset{\sim}{n}(t) \sim N\left(0, \sigma_n^2\right) \tag{4.231}$$

Detection Based on Known Time Shape of Receiving Function It resembles the classical narrowband case leading to optimal detection behavior since matched filtering can be thoroughly applied. Hence, the detector signal can be written as

$$\underset{\sim}{C}_{x_0 b}(t) = \underset{\sim}{b}(t) * x_0(-t) \tag{4.232}$$

The detection performance mainly depends on the achievable SNR-value which results from (2.249) to

$$\mathrm{SNR}_{x_0} = \frac{\left\| E\{\underset{\sim}{C}_{x_0 b}(t)\}\right\|_\infty^2}{\operatorname{var}\{\underset{\sim}{C}_{x_0 b}(t)\}} = \frac{1}{2\pi c r_1 r_2}\frac{\|x_0(t)\|_2^2}{\sigma_n^2} T_R \underset{\sim}{B} \tag{4.233}$$

where we can simplifying by using (2.46)

$$\begin{aligned} \|x_0(t)\|_2^2 &= \|x_{\mathrm{ISR}}(t) * \Lambda(t, \ldots) * x_{\mathrm{ISR}}(-t) * \Lambda(-t, \ldots)\|_\infty \\ &= \|C_{\mathrm{ISR}}(t) * C_\Lambda(t)\|_\infty \approx C_{\mathrm{ISR}}(0) C_\Lambda(0) \\ &= \|x_{\mathrm{ISR}}(t)\|_2^2 \, \|\Lambda(t, \ldots)\|_2^2 \end{aligned} \tag{4.234}$$

with the ACFs of scatterer response and intrinsic sensor response

$$\begin{aligned} C_\Lambda(t) &= \Lambda(t, \ldots) * \Lambda(-t, \ldots) \Rightarrow C_\Lambda(0) = \|\Lambda(t, \ldots)\|_2^2 \\ C_{\mathrm{ISR}}(t) &= x_{\mathrm{ISR}}(t) * x_{\mathrm{ISR}}(-t) \Rightarrow C_{\mathrm{ISR}}(0) = \|x_{\mathrm{ISR}}(t)\|_2^2 \end{aligned}$$

Using (4.227) and assuming sufficiently short antenna IRFs, we can further split up $C_{\mathrm{ISR}}(t)$

$$\begin{aligned} &C_{\mathrm{ISR}}(t) = C_{R_1}(t) * C_{R_2}(t) * C_{\dot{a}}(t) \\ &\Rightarrow \|C_{\mathrm{ISR}}(t)\|_\infty = C_{\mathrm{ISR}}(0) \approx C_{R_1}(0) C_{R_2}(0) C_{\dot{a}}(0) \end{aligned} \tag{4.235}$$

with the ACFs of the antenna response and the derivative of the sounding signal

$$\begin{aligned} C_{R_1}(t) &= R_1(t,\mathbf{e}_1) * R_1(-t,\mathbf{e}_1) \Rightarrow C_{R_1}(0) = \|R_1(t,\mathbf{e}_1)\|_2^2 \\ C_{R_2}(t) &= R_2(t,\mathbf{e}_2) * R_2(-t,\mathbf{e}_2) \Rightarrow C_{R_2}(0) = \|R_2(t,\mathbf{e}_2)\|_2^2 \\ C_{\dot{a}}(t) &= (u_1(t) * a(t)) * (u_1(-t) * a(-t)) = \dot{a}(t) * \dot{a}(-t) \end{aligned}$$

Supposing a baseband[21)] feeding signal $a(t)$ of rectangular power spectrum Φ_a of bandwidth $\underset{\cdot\cdot}{B}$ and applying differentiation rule of the Fourier transform, $C_{\dot{a}}(0)$ gets

$$\begin{aligned} C_{\dot{a}}(0) &= \int_{-\infty}^{\infty} \underline{\dot{a}}(f)\,\underline{\dot{a}}^*(f)\,\mathrm{d}f = (2\pi)^2\Phi_a \int_{-(\underset{\cdot\cdot}{B}/2)}^{\underset{\cdot\cdot}{B}/2} f^2\,\mathrm{d}f = \frac{\pi^2\Phi_a\underset{\cdot\cdot}{B}^3}{3} = \frac{P_a(\pi\underset{\cdot\cdot}{B})^2}{3} \\ &= \frac{\|a(t)\|_\infty^2(\pi\underset{\cdot\cdot}{B})^2}{3\mathrm{CF}^2} \end{aligned} \tag{4.236}$$

where $P_a = \Phi_a\underset{\cdot\cdot}{B}$ represents the power by which the transmit antenna is fed, $\|a(t)\|_\infty$ and CF are the peak value and the crest factor, respectively, of the normalized feeding wave. From this, (4.233) yields in the best case

$$\mathrm{SNR}_{x_0} \le \frac{\pi T_R \underset{\cdot\cdot}{B}^3}{6cr_1r_2\sigma_n^2\mathrm{CF}^2}\|R_1(t,\mathbf{e}_1)\|_2^2\|R_2(t,-\mathbf{e}_2)\|_2^2\|\Lambda(t,\mathbf{e}_1,\mathbf{e}_2)\|_2^2\|a(t)\|_\infty^2 \tag{4.237}$$

We can observe from (4.237) that short-pulse signals (i.e. large CF-value) will degrade the noise performance and hence the detectability of weak targets. In order to achieve the required confidence of the detection (defined by the detection and false alarm rate for a given target), the SNR-value must not fall below a certain value $\mathrm{SNR}_{x_0,\min}$. Thus, (4.237) can be used to give us the maximum range coverage $\bar{r}_{\max} = \sqrt{r_1 r_2}\big|_{\max}$ of an UWB sensor:

$$\bar{r}_{\max} = \sqrt{\frac{\pi T_R}{6c\mathrm{SNR}_{x_0,\min}kT_0F}\frac{\underset{\cdot\cdot}{B}}{\mathrm{CF}}}\|R_1(t,\mathbf{e}_1)\|_2\|R_2(t,-\mathbf{e}_2)\|_2\|\Lambda(t,\mathbf{e}_1,\mathbf{e}_2)\|_2\|a(t)\|_\infty \tag{4.238}$$

Herein, we assigned all noise contributions to internal receiver noise quantified by the noise factor F, that is $\sigma_n^2 = kT_0F\underset{\cdot\cdot}{B}$. Moreover, for the special cases of specular scattering $(\Lambda(t,\ldots) = \Lambda_0\,\delta(t))$ and Rayleigh scattering $\Lambda(t,\cdots) \approx -(9V_0/8\pi c^2)u_2(t)*$ (V_0 is the volume of scatter object), (4.237) leads to a coverage range of

21) Correspondingly, we get for a band-pass signal $C_{\dot{a}}(0) = 2(2\pi)^2\Phi_a \int_{f_m-(\underset{\cdot\cdot}{B}/2)}^{f_m+(\underset{\cdot\cdot}{B}/2)} f^2\,\mathrm{d}f = \frac{2}{3}\pi^2 P_a(12f_m^2 + \underset{\cdot\cdot}{B}^2)$

Specular scattering

$$\bar{r}_{\max} = \sqrt{\frac{\pi T_{\mathrm{R}}}{6c\mathrm{SNR}_{x_0,\min}kT_0F}\frac{\underset{\sim}{B}}{\mathrm{CF}}\Lambda_0\|R_1(t,\mathbf{e}_1)\|_2\|R_2(t,-\mathbf{e}_2)\|_2\|a(t)\|_\infty} \tag{4.239}$$

Rayleigh scattering

$$\bar{r}_{\max} = \sqrt{\frac{9V_0T_{\mathrm{R}}}{112c^3\mathrm{SNR}_{x_0}kT_0F}\frac{\pi^2\underset{\sim}{B}^3}{\mathrm{CF}}\|R_1(t,\mathbf{e}_1)\|_2\|R_2(t,-\mathbf{e}_2)\|_2\|a(t)\|_\infty} \tag{4.240}$$

Detection Based on Unknown Scatterer Response In this case, we only know the intrinsic sensor response; hence, the decision on target presence has to be made with the help of the function

$$\begin{aligned} \underset{\sim}{C}_{x_{\mathrm{ISR}}\mathrm{b}}(t) &= \underset{\sim}{b}(t) * x_{\mathrm{IRS}}(-t) \\ \underset{\sim}{b}(t) &= \frac{1}{2\pi c r_1 r_2}\Lambda(t,\mathbf{e}_1,\mathbf{e}_2) * x_{\mathrm{ISR}}(t) + \underset{\sim}{n}(t); \quad n(t) \sim N\left(0,\sigma_{\mathrm{n}}^2\right) \end{aligned} \tag{4.241}$$

Following the consideration in Annex B.4.2, the SNR-value of the detector signal reads

$$\mathrm{SNR}_{x_{\mathrm{ISR}}} = \frac{\left\|E\{\underset{\sim}{C}_{x_{\mathrm{ISR}}\mathrm{b}}(t)\}\right\|_\infty^2}{\mathrm{var}\{\underset{\sim}{C}_{x_{\mathrm{ISR}}\mathrm{b}}(t)\}} = \frac{1}{2\pi c r_1 r_2}\frac{\|\Lambda(t,\ldots) * C_{\mathrm{ISR}}(t)\|_\infty^2}{\sigma_{\mathrm{n}}^2 C_{\mathrm{ISR}}(0)}T_{\mathrm{R}}\underset{\sim}{B} \tag{4.242}$$

Applying Young's inequality (2.15) and respecting the observation $\|x(t)\|_1 \le \|x(t)\|_2$ (see Figure 2.7), we find

$$\begin{aligned} \|\Lambda(t,\ldots) * C_{\mathrm{ISR}}(t)\|_\infty &\le \|\Lambda(t,\ldots)\|_1 \, \|C_{\mathrm{ISR}}(t)\|_\infty \\ &= \|\Lambda(t,\ldots)\|_1 \, C_{\mathrm{ISR}}(0) \le \|\Lambda(t,\ldots)\|_2 \, \|x_{\mathrm{ISR}}(t)\|_2^2 \end{aligned} \tag{4.243}$$

so that we can establish following relation between (4.233), (4.234) and (4.243)

$$\mathrm{SNR}_{x_{\mathrm{ISR}}} \le \frac{1}{r_1 r_2}\frac{\|\Lambda(t,\ldots)\|_2^2 \, \|x_{\mathrm{ISR}}(t)\|_2^2}{\sigma_{\mathrm{n}}^2}T_{\mathrm{R}}\underset{\sim}{B} = \mathrm{SNR}_{x_0} \tag{4.244}$$

This indicates that missing knowledge of scatterer IRF may degrade the detection performance. If $\Lambda(t,\ldots)$ approaches the delta function $\delta(t)$ (e.g. in case of specular scattering), both detector function will be equal so that $\mathrm{SNR}_{x_0} = \mathrm{SNR}_{x_{\mathrm{ISR}}}$. Hence, we only have to count for minor detection degradation if the scatterer IRF is stretched in time by multiple reflections or resonances.

A Dictionary of Target Responses Exist In the case where one deals with a finite set of targets whose scatterer IRF does not strongly depend on $\mathbf{e}_1$ and $\mathbf{e}_2$, a dictionary of all possible response function $x_0^{(m)}(t)$ can be established (by reference measurement or simulations). Then, the received signal is correlated with the whole set of reference functions included in the dictionary and the correlation function with the largest maximum is taken for the decisions. Except for the search in the 'signal dictionary' for the best match, the detection performance is equal to (4.233).

Large Target with a Number of Bright Points An example of such type of scattering function is depicted in Figure 4.49 if we consider the three 'bright points' as part of a single object. In that case, the digitized target response $\Lambda(i); i \in [0, N-1]$ roughly extends over the time interval[22] $\Delta t_x = (N-1)\Delta t_s \approx D_x c/2$ (D_x is the typical target dimension, Δt_s is the sampling interval, N is the number of samples). All bright points – let us presume their number to be K in maximum – are distributed within the observation interval Δt_x. They appear as K short peaks where their time position and polarity is not prior known. In order to concentrate the energy of the pre-processed return signal (4.241), we correlate it with a synthesized set of reference samples $\Lambda_m^{(K,N)}(i); i \in [0, N-1]$ including K non-zero samples:

$$\Lambda_m^{(K,N)}(i) = \sum_{j=1}^{N} a_{jm}\delta(i-j) = \mathbf{a}_m^{(K,N)} \quad \text{with } a_{jm} \in [-1, 0, 1] \tag{4.245}$$

Herein, the column vector $\mathbf{a}_m^{(K,N)}$ represents a synthesized impulse response of length N from which K of its elements have the values -1 or 1 and the remaining entries are zero. Since the time position of the peaks is not prior known, we are forced to evaluate all variations. Thus, we have to create a dictionary of all possible reference vectors $\mathbf{a}_1^{(K,N)}, \mathbf{a}_2^{(K,N)}, \cdots \mathbf{a}_m^{(K,N)}, \cdots \mathbf{a}_M^{(K,N)}$ which have to be correlated with the pre-processed receiving signal so that we can take finally the strongest correlation function for target decision. The volume of the dictionary covers M variants of synthesized responses with

$$M = 2^K \binom{N}{K} = \frac{2^K N!}{K!(N-K)!} \tag{4.246}$$

which gives huge numbers just for small values of N and K. To some detriment of noise suppression, the dictionary volume can be reduced by correlating the absolute, squared or Hilbert transformed values of the involved signals, for example $\underset{\sim}{C}_{|x_{\mathrm{ISR}}b|}(t) = \left|\underset{\sim}{b}(t) * x_{\mathrm{IRS}}(-t) * (\pi t)^{-1}\right|$. This approach is referred as N over K detection [77].

$$M = \binom{N}{K} = \frac{N!}{K!(N-K)!}. \tag{4.247}$$

Nevertheless, even this quantity is mostly out of scope for practical applications if no prior knowledge is available. However under some constraints, a reduction of complexity can be achieved. Let us come back to the example from Figure 4.49 for illustration. We can discover there that the scattering function does not change

22) If the target shows strong multiple reflections, the observation interval Δt_x has to be extended appropriately.

abruptly over the aspect angle. That is, once one is locked on the target, the search routine must only deal with small variation in the vectors $\mathbf{a}_m^{(K,N)}$ to track the target even if the peak positions within the scatter function change due to the variation in aspect angle.

Energy Detection The technical implementation of this type of target detection is the simplest one. It requires only minor knowledge about the target, namely the approximate duration of the scatter response t_{w}. The processing steps are summarized in Figure 4.65; that is, the pulse-shaped receiving signal $x(t)$ is subjected to a square-law device and a low-pass filter (video filter). In case of time-spread wideband signals, the pulse shape has to be gained before the energy detection by preprocessing via correlation; that is, $x(t) \sim C_{x_{\mathrm{IRS}}\mathrm{b}}(t)$.

Assuming a short-time integrator as video filter $h(t) = t_{\mathrm{I}}^{-1}\,\mathrm{rect}(t/t_{\mathrm{I}})$ of integration time t_{I}, the detector signal takes the form

$$\underset{\sim}{y}(t) = \frac{1}{t_{\mathrm{I}}}\underset{\sim}{x}^2(t) * \mathrm{rect}\left(\frac{t}{t_I}\right) = \frac{1}{t_{\mathrm{I}}}\int\limits_{t-t_{\mathrm{I}}}^{t}\underset{\sim}{x}^2(\tau)\mathrm{d}\tau$$

$$\text{with} \quad \underset{\sim}{x}(t) = x_0(t) + \underset{\sim}{n}(t); \quad n(t) \sim N\left(0, \sigma_{\mathrm{n}}^2\right); \underset{\sim}{B}_{\mathrm{n}} \tag{4.248}$$

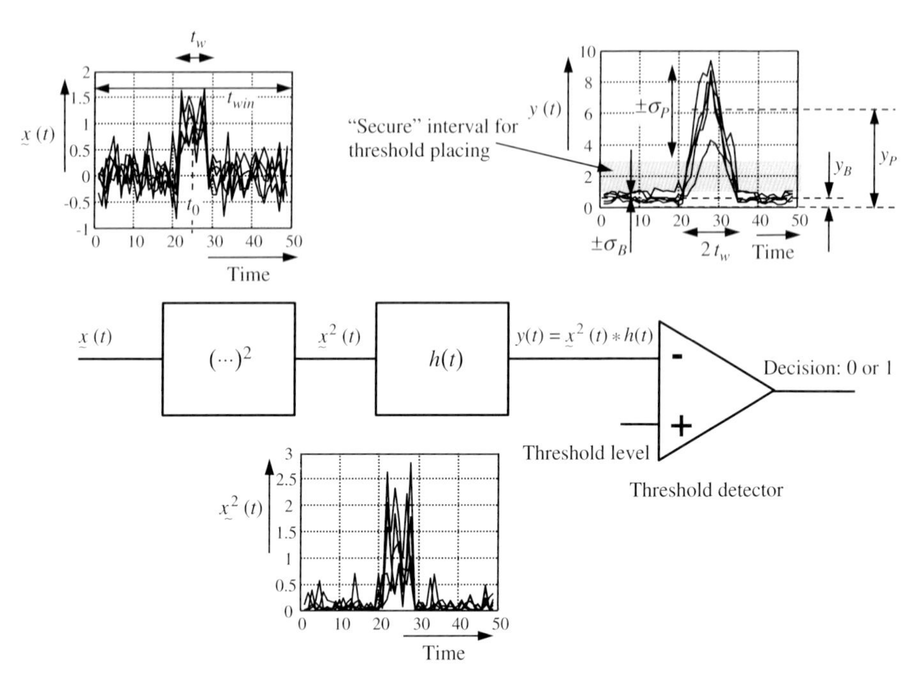

Figure 4.65 Energy detection and typical signals at selected points. The example refers to a simplified receiver signal of rectangular shape $\underset{\sim}{x}(t) \approx \mathrm{rect}((t - t_0)/t_{\mathrm{w}}) + \underset{\sim}{n}(t)$ and a noise bandwidth larger than the rectangular bandwidth of the scatterer. The video filter is a short-time integrator of integration time $t_{\mathrm{I}} = t_{\mathrm{w}}$.

where the integration time t_I must be selected much shorter than the time window length $t_{win} \gg t_I$. The energy detector is a special case of the product detector (see Annex B.4) which we also applied to determine correlation functions, however with opposite constraints concerning the integration time, that is $t_{win} \ll t_I$.

Since by assumption $x_0(t)$ is a baseband or band-pass pulse, the detector signal $\underset{\sim}{y}(t)$ will also be shaped pulse like. For the sake of brevity, we will only consider the expected value and the variance of $\underset{\sim}{y}(t)$ at the pulse base (y_B, σ_B^2) and at peak position (y_P, σ_P^2) in order to get a rough impression about the margin of detector threshold placing (see Figure 4.65).

Using calculation rules from Annex A.3, the expected value of the squared signal yields

$$\mathrm{E}\{\underset{\sim}{x}^2(t)\} = x_0^2(t) + \sigma_n^2 \tag{4.249}$$

which gives after low-pass filtering with a short time integrator

$$\mathrm{E}\{\underset{\sim}{y}(t)\} = \frac{1}{t_I}\mathrm{rect}\left(\frac{t}{t_I}\right) * \mathrm{E}\{\underset{\sim}{x}^2(t)\} = \frac{1}{t_I}\mathrm{rect}\left(\frac{t}{t_I}\right) * x_0^2(t) + \sigma_n^2 \tag{4.250}$$

If the instantaneous power of the signal is roughly equally distributed over the pulse duration t_w, $\mathrm{E}\{\underset{\sim}{y}(t)\}$ has a trapezoidal time shape except for $t_I = t_w$ where it is triangular. The peak value of the detector signal $y(t)$ may be approximated by

$$y_P = \|\mathrm{E}\{\underset{\sim}{y}(t)\}\|_\infty \approx \begin{cases} \|x_0(t)\|_2^2 + \sigma_n^2; & t_I \le t_w \\ \dfrac{t_w}{t_I}\|x_0(t)\|_2^2 + \sigma_n^2; & t_I > t_w \end{cases} \tag{4.251}$$

The pulse base results in a signal level of

$$y_B = \mathrm{E}\{\underset{\sim}{x}^2(t)\} = \sigma_n^2 \quad \text{if } x_0(t) = 0 \tag{4.252}$$

Figure 4.65 gives an impression of $\underset{\sim}{y}(t)$ for the case $t_w = t_I$, which is obviously the best selection for t_I as we can learn from (4.251).

Further, we can write for the variance of the squared signal

$$\mathrm{var}\{\underset{\sim}{x}^2(t)\} = 4x_0^2(t)\sigma_n^2 + 2\sigma_n^4 \tag{4.253}$$

After low-pass filtering with a short time integrator, the temporal shape of the variance resembles either a triangle or a trapeze depending on the length of integration time. The variance of the pulse base of $\underset{\sim}{y}(t)$ becomes

$$\sigma_B^2 = \mathrm{var}\{\underset{\sim}{y}(t)\}|_{x_0(t)=0} = \frac{2\sigma_n^4}{t_I \underset{\cdot\cdot}{B}_n} \tag{4.254}$$

and the variance at peak position τ_∞ of $\underset{\sim}{y}(t)$ is approximately

$$\sigma_{\mathrm{P}}^2 = \operatorname{var}\{\underset{\sim}{y}(\tau_\infty)\} \approx \begin{cases} \dfrac{4\,\|x_0(t)\|_2^2\sigma_{\mathrm{n}}^2 + 2\sigma_{\mathrm{n}}^4}{t_{\mathrm{I}}\underset{\cdot\cdot}{B}_{\mathrm{n}}} \approx \dfrac{4\,\|x_0(t)\|_2^2\sigma_{\mathrm{n}}^2}{t_{\mathrm{I}}\underset{\cdot\cdot}{B}_{\mathrm{n}}}; & t_{\mathrm{I}} \le t_{\mathrm{w}} \\ \dfrac{t_{\mathrm{w}}}{t_{\mathrm{I}}}\dfrac{4\,\|x_0(t)\|_2^2\sigma_{\mathrm{n}}^2}{t_{\mathrm{I}}\underset{\cdot\cdot}{B}_{\mathrm{n}}} + \dfrac{2\sigma_{\mathrm{n}}^4}{t_{\mathrm{I}}\underset{\cdot\cdot}{B}_{\mathrm{n}}} \approx \dfrac{t_{\mathrm{w}}}{t_{\mathrm{I}}}\dfrac{4\,\|x_0(t)\|_2^2\,\sigma_{\mathrm{n}}^2}{t_{\mathrm{I}}\underset{\cdot\cdot}{B}_{\mathrm{n}}}; & t_{\mathrm{I}} > t_{\mathrm{w}} \end{cases} \tag{4.255}$$

The threshold for target detection must be placed in between y_{B} and y_{P}. Both levels are corrupted by random perturbations shrinking the window width for threshold placing to get reliable detection. We quantify the detection reliability by a term closely related to a signal-to-noise ratio by

$$\mathrm{SNR}_{\mathrm{L}_2} = \frac{(y_{\mathrm{P}} - y_{\mathrm{B}})^2}{\sigma_{\mathrm{P}}^2 + \sigma_{\mathrm{B}}^2} \approx \begin{cases} \dfrac{\|x_0(t)\|_2^2\, t_{\mathrm{I}}\underset{\cdot\cdot}{B}_{\mathrm{n}}}{4\,\sigma_n^2}; & t_{\mathrm{I}} \le t_{\mathrm{w}} \\ \dfrac{\|x_0(t)\|_2^2\, t_{\mathrm{w}}\underset{\cdot\cdot}{B}_{\mathrm{n}}}{4\sigma_{\mathrm{n}}^2}; & t_{\mathrm{I}} > t_{\mathrm{w}} \end{cases} \tag{4.256}$$

We can observe from (4.256) that an extension of the integration time t_{I} above the pulse width t_{w} does not bring any profit for the reliability of target detection since the 'target energy' cannot further increase. It will only unnecessarily degrade the range resolution of the sensor since the $y(t)$-pulse will be expanded beyond $2\,t_{\mathrm{w}}$. Furthermore, this equation clarifies that an energy detector will bring only small performance gain in case of UWB sensing since the pulse width t_{w} is rather short. The noise bandwidth $\underset{\cdot\cdot}{B}_{\mathrm{n}}$ can be approximated with the equivalent sampling rate of the receiver if no additional filtering is performed before the square-law device.

L_p-Norm Detection The L_p-norm detector is composed as an energy detector (Figure 4.65) with the only difference that the square-law device is replaced by a device performing the operation $|x(t)|^p; p \in \mathbb{N}$. Hence, the energy detector is identical with the L_2-norm detector. Figure 4.66 compares the performance of the first four L_p-norm detectors. It refers to the example depicted in Figure 4.65. As quality measure of the different detectors, we used the SNR-value defined by (4.256). Figure 4.66 discloses that the L_1-detector outperforms the energy detector by a factor of about 3 and that the higher the norm, the inferior the detector performance.

4.8.2.2 Detection Based on Repeated Measurements

So far, we only considered the detection of a target on the basis of a single measurement shot. But typically, the target stays at least a certain time within the beam of the sensor antennas. Figure 4.67 illustrates the measurement scenario and shows an example of simulated data for a target moving straight away from the antenna at a speed of about 5 cm/s. The corresponding

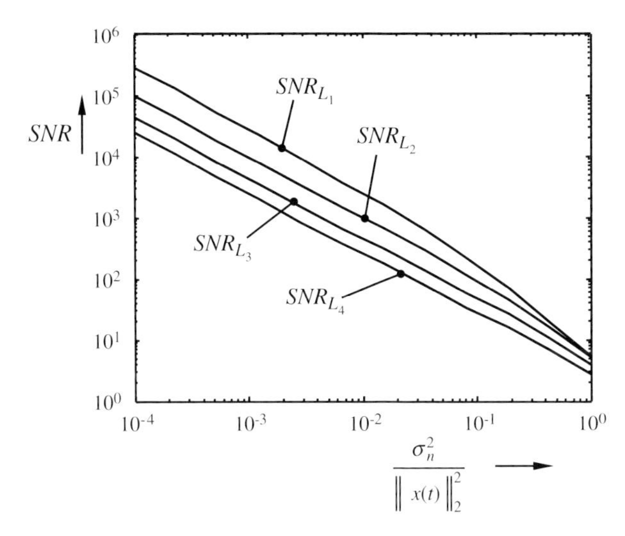

Figure 4.66 Performance of L_p-norm detectors. The examples refers to the case $t_w = t_l$ and $\underset{..}{B}_n t_l = 40$. This value is rather large under UWB conditions. It is only thought for illustration.

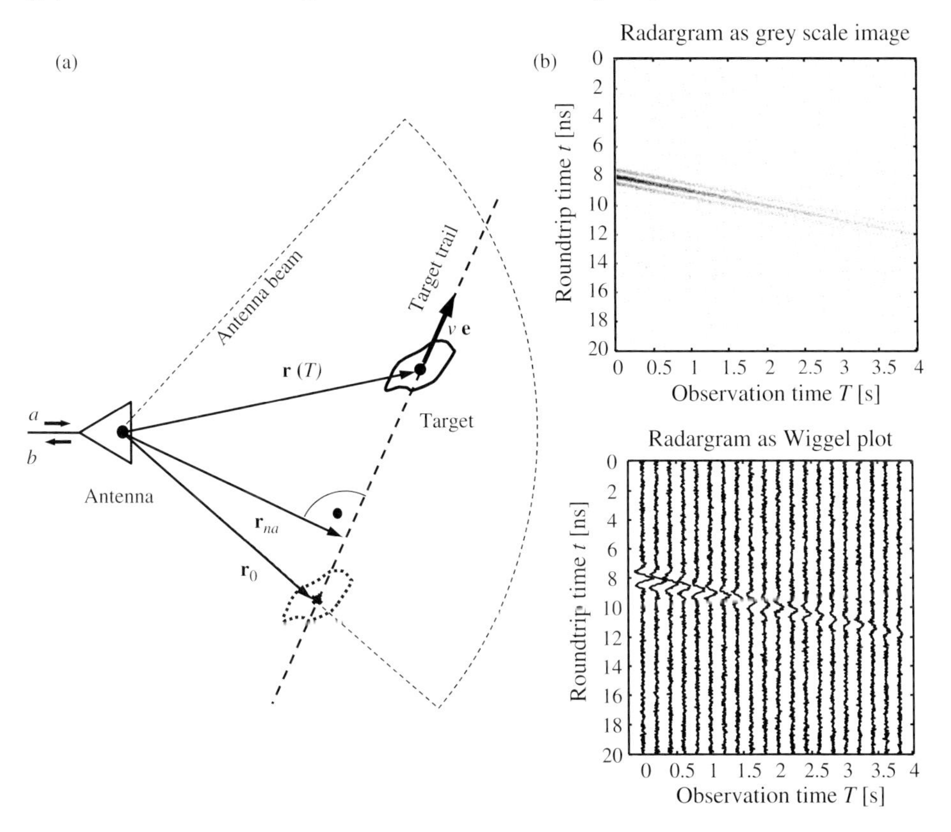

Figure 4.67 Target moving through the beam of an antenna (a) and representation of measured data by a radargram (b). The measurements are perturbed by additive noise. The noise bandwidth is much higher than the signal bandwidth in the shown case. Note the radargrams do not relate to the motion profile of the target motion shown in the example (a).

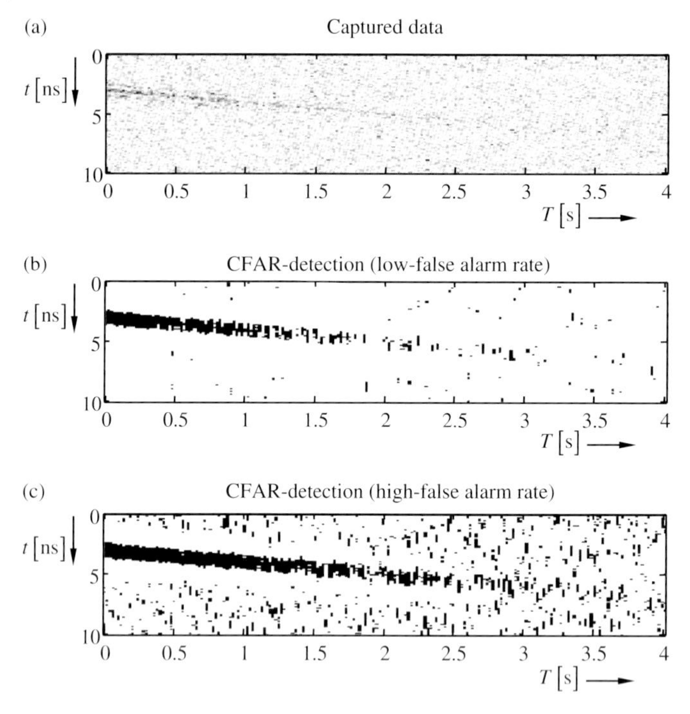

Figure 4.68 Binary image (b) and (c) of a simple CFAR detection for the example depicted in Figure 4.67 with a three times higher noise level as depicted in radargram (a).

measurements are typically represented by a radargram $x(t,T)$ and $x(t,k)$, respectively, depicting the repeated measurements by a Wiggel plot or grey scale (colour) image whereas $t = 2\,r/c$ is the round-trip time of the sounding waves and $T = k\,\Delta T = k/m_R; \quad k \in \mathbb{N}_0$ assigns the observation time (ΔT is the time interval between consecutive measurements, $m_R = \Delta T^{-1}$ is the measurement rate, that is sampling rate in observation time). The nature of the function $x(t,T)$ may be quite different. It may represent the originally measured data or some pre-processed data as herein before mentioned detector functions.

Figure 4.68 illustrates the result if the target trace is detected immediately from perturbed measurements by a simple CFAR detector (see Figure 4.48 for basics on CFAR detection). Two cases are shown. In Figure 4.68b, the detector threshold was placed at a high level which results in an acceptable number of false detections but the target trace will be already lost at an early stage. Figure 4.68c refers to low threshold level; that is, we can follow the target trace over a larger distance but the false detections are considerably high. The data on which detection is performed are shown in the upper graph to illustrate the actual noise corruption. Due to the higher noise level compared to the example in Figure 4.67, the target trace is visually lost for about $T > 2\,\mathrm{s}$.

In order to detect weak reflections, the threshold must be placed at a comparatively low level leading to numerous false detections as indicated by the salt and paper noise. Hence, the question arises how to improve the detection performance if the data trace of the target trails away. Since target range and the number of targets cannot change abruptly, one has basically two options. The first one is called 'detect before track' that was actually done in Figure 4.68. Here, detection is performed first and after some cleaning of detection data (not applied in Figure 4.68), one tries to follow the target trace in the data. Cleaning of detection data means to remove all isolated detections by appropriate methods, for example by median filtering in the simplest case. However, by narrowing the noise bandwidth, the false detections become more and more clustered, so they are difficult to remove by simple filtering.

By the second option, one tries first to follow the most probable track of the target before detection is made. Hence, the approach is called 'track before detect'. It prolongs the time on target t_{ot} and observation time T_{obs} which is used to accumulate the deterministic parts of the signal and thus to reduce the randomness of the observation. This method assumes that the time shape of the target reflection must not vary during the observation interval $t_{ot} = T_{obs}$. Furthermore, it is computationally quite expensive in case of a bulky search for unknown tracks buried in noise.

The simplest form to increase the observation time interval is horizontal low-pass filtering which may be applied if the range variation between a couple of consecutive scans is much shorter than the range resolution of the radar sensor, that is if either the target speed is small or the measurement rate is high. The maximum duration T_I of horizontal short-time integration can be estimated from (2.90). Basically, this type of filtering can be applied to any of the introduced detection signals above. Figure 4.69 visualizes the improvements by horizontal filtering if applied to different detection signals. It uses the same data as applied in Figure 4.68. The time-bandwidth products of the horizontal and vertical short-time integration are 10 and 30 for the depicted examples.

Figure 4.69 already shows us a further detection signal (bottom). The repetition of measurements allows us to introduce a new detection function which constitutes from two consecutive scans:

$$\underset{\sim}{y}_{\mathrm{ic}}(t,k) = \frac{1}{t_\mathrm{I}}\operatorname{rect}\left(\frac{t}{t_\mathrm{I}}\right) * \left(\underset{\sim}{x}(t,k-1)\underset{\sim}{x}(t,k)\right) \tag{4.257}$$

The method is called inter-period correlation processing (IPCP) [77]. Herein, $\underset{\sim}{x}(t,k) = x_0(t) + \underset{\sim}{n}(t,k)$ represents the kth measurement where one assumes that the scatterer response $x_0(t)$ does not change between two measurements. The integration time t_I should be selected corresponding to the duration t_w of the scatterer response as discussed in connection with the energy detector. Since the noise contributions of two separate measurements are usually uncorrelated, the expected value of the IPCP is bias-free (for $t_\mathrm{I} = t_\mathrm{w}$)

$$\mathrm{E}\{\underset{\sim}{y}_{\mathrm{ic}}(t,k)\} = \begin{cases} \frac{1}{t_\mathrm{I}}\operatorname{rect}\left(\frac{t}{t_\mathrm{I}}\right) * x_0^2(t) \approx \|x_0(t)\|_2^2 & \text{target} \\ 0 & \text{no target} \end{cases} \tag{4.258}$$

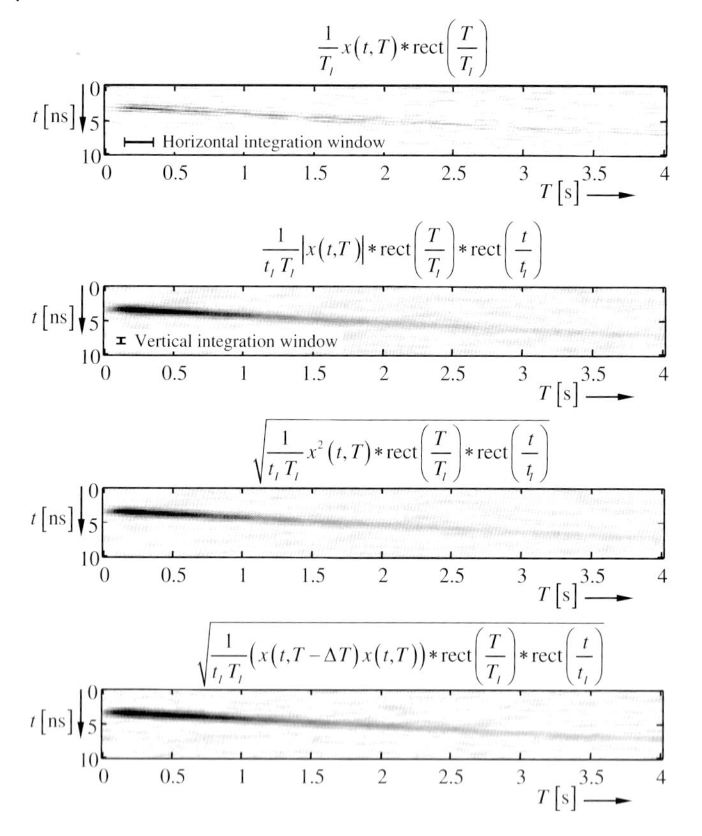

Figure 4.69 Enhancement of a slowly moving target by horizontal short-time integration using $T_l = 0.2$ s. The vertical integration window covers 0.6 ns. But in the upper most example, vertical integration over t_w is prohibited since it would cancel the signal. For illustration, the applied window lengths are depicted in the two upper graphs.

and the variance is roughly given by (for $t_l = t_w$)

$$\mathrm{var}\{\underline{y}_{ic}(t,k)\} \approx \frac{1}{t_w \underline{B}_n} \begin{cases} 2\,\|x_0(t)\|_2^2 \sigma_n^2 + \sigma_n^4 & \text{target} \\ \sigma_n^4 & \text{no target} \end{cases} \tag{4.259}$$

which is better by a factor of two referred to the energy detector (4.255). Therefore, the target trace is slightly better visible at the very bottom graph of Figure 4.69.

The potential of detection improvement by horizontal filtering is largely limited by target speed and the sensor range resolution as already mentioned above. Hence, in order to detect very weak targets buried under noise, the only chance is to increase the time on target by following the target trace over a period as long as possible. Since the target trace is usually unknown, one has to search over all possible or probable target tracks. Practically, such search can only be implemented for simple motion models. We will consider here two examples of them – linear motion of constant target speed v (refer to Figure 4.67 for the intended scenario) and periodic motion.

Example 4.1: : Linear motion of constant speed

Let us first assume a target of collision course with the antenna. Thus, the antenna-target distance r develops as follows:

$$r(T) = \frac{t_{\mathrm{rt}}c}{2} = vT + r_0 = v(T + T_0) \tag{4.260}$$

Herein, c is the speed of light, t_{rt} is the round-trip time, T is the observation time, r_0 is the target distance at $T = 0$, T_0 is the collision time and $v = \partial r/\partial T$ is the radial target speed.

The resulting radargram which we scale for target range is assigned as $x(r, T) = x(tc/2, T)$. Its appearance would be the same as for the radargrams in Figures 4.67–4.69 if the trace is not hidden by noise. If we integrate the data along the target trace, the 'target energy' will be accumulated and the random perturbations will be suppressed. The longer the integration path, the better the suppression will be. Since neither target speed v nor collision time T_0 is known, we transform the radar data into an image domain $y(v, T_0)$ where all data belonging to the correct speed and collision time collapse in a single point:

$$\begin{aligned} y(v, T_0) &= \int x(r', T')\delta(r' - v(T' + T_0))\mathrm{d}r'\,\mathrm{d}T' \\ &= \int x(v(T' + T_0), T')\mathrm{d}T' \end{aligned} \tag{4.261}$$

The Delta function in the upper integral of (4.261) assures that only those radar data are accumulated that are placed in a straight line related to speed v and collision time T_0. Figure 4.70 depicts an example of corresponding signal accumulation which was done for the radargram of Figure 4.68a for a five times elevated noise level. Detection methods as described above would completely fail under such conditions but integration along the right target track gives clear detection.

If the target flies by the antenna, the trace becomes a hyperbole. Target distance and observation time relate as follows (see also Figure 4.67)

$$\begin{aligned} \mathbf{r} &= vT\mathbf{e} + \mathbf{r}_\mathrm{T} \\ r^2 &= (vT\mathbf{e} + \mathbf{r}_\mathrm{T}) \cdot (vT\mathbf{e} + \mathbf{r}_\mathrm{T}) = v^2(T - T_0)^2 + r_0^2 \end{aligned} \tag{4.262}$$

where $\mathbf{r}_\mathrm{T}$ is the position vector of the target at $T = 0$, $r_0 = \sqrt{r_\mathrm{T}^2 + (\mathbf{e} \cdot \mathbf{r}_\mathrm{T})^2}$ is the closest approach to the antenna and $T_0 = \mathbf{e} \cdot \mathbf{r}_\mathrm{T}/v$ is the time point of

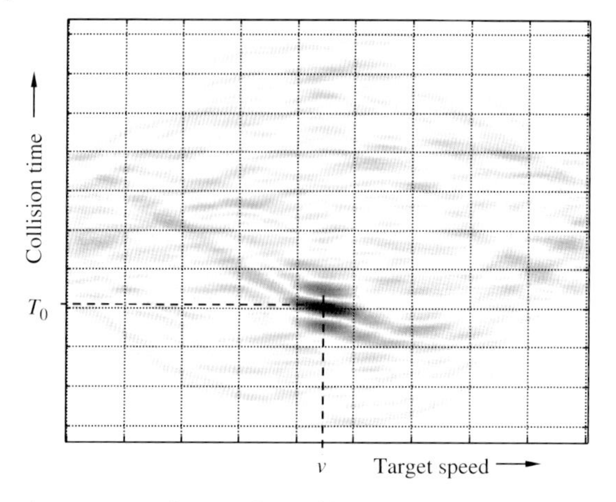

Figure 4.70 Radon transform of the radargram from Figure 4.68a. Note that the noise level supposed in this example is five times higher than shown in Figure 4.68.

closest approach. Now, we need to search for three unknown curve parameters, namely r_0, v, T_0, which is done in the same way as in (4.262).

$$\begin{aligned} y(r_0, v, T_0) &= \int x(r', T')\delta\left(r' - \sqrt{v^2(T' - T_0)^2 + r_0^2}\right) \mathrm{d}r'\, \mathrm{d}T' \\ &= \int x\left(\sqrt{v^2(T' - T_0)^2 + r_0^2}, T'\right) \mathrm{d}T' \end{aligned} \tag{4.263}$$

Basically, the method can be applied for any curved line $[r' \quad T']^T = \mathfrak{C}(a, b, c, \ldots)$ if it may be represented by a set of parameters $a, b, c, \ldots$:

$$y(a, b, c, \ldots) = \int x(r', T')\delta\left(\begin{bmatrix} r' \\ T' \end{bmatrix} - \mathfrak{C}(a, b, c, \ldots)\right) \mathrm{d}r'\, \mathrm{d}T' \tag{4.264}$$

Transformations as depicted in (4.261), (4.263) and (4.264) are usually referred to as Hough transform of a line, a hyperbole or an arbitrary curve [78]. The Hough transform of a line is also termed as Radon transform for which a computational efficient algorithm exists exploiting the projection-slice theorem [79].

Example 4.2: : Periodic motion

We refer to the scenario illustrated in Figure 4.52 where the interface or target moves periodically. Practical examples could be found at rotating machines or the organ motion of human beings or animals connected with respiration and heartbeat (see Sections 6.5 and 6.7 for more discussions on this issue). The displacement of the scatterer is modelled by M harmonics of the fundamental frequency ν_0 and deflection d_μ:

$$\mathrm{d}(T) = \sum_{\mu=1}^{M} d_\mu \sin\left(2\pi\mu\nu_0 T + \varphi_\mu\right) \tag{4.265}$$

Thus, we can write for the radargram

$$x(t,T) = b(t) * \sigma\left(t - \frac{2}{c}(r + d(T))\right) + n(t,T); \quad n(t,T) \in N\left(0, \sigma_{\mathrm{n}}^2\right) \tag{4.266}$$

Note that this equation only holds for weak targets so that multiple scattering which involves the target is negligible. A radargram example is illustrated in Figure 4.59. If the propagation environment is quit harsh, the backscattered signal $b(t)$ is expanded in time due to multipath effects and its time shape cannot be predicted. Hence, the question arises how to detect a motion-related radargram pattern with the highest sensitivity.

Basically, we can do this by minimizing the error between the actual measured data $x(t,T)$ and a parametric data model $\hat{x}(t,T,\boldsymbol{\Theta})$ whose parameter set $\boldsymbol{\Theta}$ has to be estimated:

$$\begin{aligned} &\hat{x}(t,T,\boldsymbol{\Theta}) = \hat{b}(t) * \sigma\left(t - \frac{2}{c}\left(\hat{r} + \hat{d}(T)\right)\right) \\ &\text{with} \quad \hat{d}(T) = \sum_{\mu=1}^{M} \hat{d}_\mu \sin\left(2\pi\mu\hat{\nu}_0 T + \hat{\varphi}_\mu\right) \\ &\hat{b}(t) \mathrel{\widehat{=}} \hat{b}(n\,\Delta t_{\mathrm{s}}) \mathrel{\widehat{=}} \hat{\mathbf{b}} = \begin{bmatrix} b_1 & b_2 & \cdots & b_N \end{bmatrix}^T \\ &\boldsymbol{\Theta} = \begin{bmatrix} \hat{\mathbf{b}}^T & \hat{r} & \hat{\nu}_0 & \hat{d}_1 & \cdots & \hat{d}_M & \hat{\varphi}_1 & \cdots & \hat{\varphi}_M \end{bmatrix} \end{aligned} \tag{4.267}$$

where we expressed the unknown pulse-shaped signal $\hat{b}(t)$ of duration t_{w} by a set of N data samples ($t_{\mathrm{w}} = N\,\Delta t_{\mathrm{s}}$). We observe the target over the time $T_{\mathrm{obs}} = M_{\mathrm{T}}\,\Delta T$ (M_{T} is the number of captured IRFs). Based on the measured data and our data model (4.267), we can define a cost function $J(\boldsymbol{\Theta})$ which has to be minimized with respect to the parameters $\boldsymbol{\Theta}$. Using discrete notation for the involved signals, the cost function reads

$$\begin{aligned} &J(\boldsymbol{\Theta}) = \sum_{m=1}^{M_{\mathrm{T}}} \sum_{n=1}^{N} \left(x(n,m) - \hat{x}(n,m,\boldsymbol{\Theta})\right)^2 \underset{\boldsymbol{\Theta}}{\Rightarrow} \text{minimum} \\ &\text{whereas} \quad t \to n\,\Delta t_{\mathrm{s}}; \qquad T \to m\,\Delta T \end{aligned} \tag{4.268}$$

which can also be expressed by

$$\begin{aligned} J(\boldsymbol{\Theta}) &= \mathfrak{E}_x + \mathfrak{E}_{\hat{x}} + \mathfrak{E}_{\mathrm{n}} - 2\,\Xi \\ \Xi(\boldsymbol{\Theta}) &= \sum_{m=1}^{M_{\mathrm{T}}} \sum_{n=1}^{N} x(n,m)\,\hat{x}(n,m,\boldsymbol{\Theta}) \underset{\boldsymbol{\Theta}}{\Rightarrow} \text{maximum} \\ \mathfrak{E}_x &= \sum_{m=1}^{M_{\mathrm{T}}} \sum_{n=1}^{N} x^2(n,m) \\ \mathfrak{E}_{\hat{x}} &= \sum_{m=1}^{M_{\mathrm{T}}} \sum_{n=1}^{N} \hat{x}^2(n,m,\boldsymbol{\Theta}) \\ \mathfrak{E}_{\mathrm{n}} &= M_{\mathrm{T}} N \sigma_{\mathrm{n}}^2 \end{aligned} \tag{4.269}$$

The $\mathfrak{E}$ terms represent the total energy of the backscattered signal of the modelled signal and of noise. The cost function will be minimal if the correlation function $\Xi(\boldsymbol{\Theta})$ is maximum, which involves $\mathfrak{E}_x = \mathfrak{E}_{\hat{x}}$ in the noise-free case.

The detection performance is given by the robustness of the maximum localization which depends on $\mathrm{SNR}_\Xi = \mathrm{E}^2\{\Xi_{\max}\}/\mathrm{var}\{\Xi_{\max}\}$. Corresponding estimations of the Cramér–Rao lower bound for the estimated parameters $\boldsymbol{\Theta}$ can be found in Ref. [80]. Here, we will not further follow this lane of data processing since its practical implementation will hardly be feasible with recent processing hardware.

Hence, we address a numerically efficient data processing for which we, however, have to accept some (acceptable) degradation of sensitivity [81]. The goal of the processing is to concentrate the 'target energy' included in the 2D-data record $x(t, T)$ into a single spot whereas high-dimensional signal processing should be replaced by two steps of single dimensionality. Since we have assumed periodic target deflections, the Fourier transform in observation time direction will lead to a first energy concentration:

$$\underline{X}(t, \nu) = \int\limits_{T_{\mathrm{obs}}} x(t, T)\mathrm{e}^{-j2\pi\nu T}\,\mathrm{d}T \qquad (4.270)$$

Figure 4.71 depicts a simplified example in which the target having a round-trip time of $t_0 = 6$ ns oscillates at a frequency of $\nu_0 = 3$ Hz. We can observe the appearance of harmonics in the spectrum even if the target purely oscillates sinusoidal (i.e. $M = 1$). The reason for this can be found in Figure 4.59 which explains the non-linear distortions of a 'horizontal' signal $x(t_x, T)$ depending on its location t_x at the pulse (compare time evolution of the signal at time position 1 and time position 2 in Figure 4.59).

In order to estimate the strength of the non-linear distortions, we consider a simple sinusoidal motion of the target $\mathrm{d}(T) = d_1 \sin(2\pi\nu_0 T)$. Thus, omitting noise, (4.266) yields

$$x(t, T) = b(t) * \delta\left(t - t_0 - \frac{2\,d_1 \sin 2\pi\nu_0 T}{c}\right) \qquad (4.271)$$

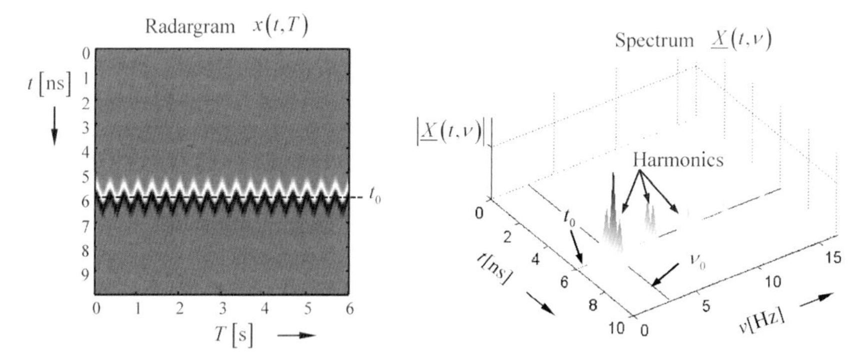

Figure 4.71 Radargram of an oscillating target and related 'observation time' spectrum. The time shape of the receiving signal $b(t)$ was assumed to be the first derivative of a Gaussian pulse. (colored figure at Wiley homepage: Fig_4_71).

Performing a first Fourier transformation in propagation time direction, we get

$$\underline{X}(f,T) = \underline{B}(f)\mathrm{e}^{-j2ft_0}\,\mathrm{e}^{-j2kd_1\,\sin\,2\pi\nu_0 T} = \underline{B}_0(f)\mathrm{e}^{-j2kd_1\,\sin\,2\pi\nu_0 T} \qquad (4.272)$$

with $k = 2\pi f/c = 2\pi/\lambda$ which is the wave number (λ is the wavelength). $\underline{B}(f)$ is the spectrum of the receiving signal if no range modulation occurs. Next, we accomplish spectral decomposition in observation time direction which leads to Fourier series whose coefficients are given by

$$\begin{aligned}\underline{X}(f,n\,\nu_0) &= \frac{1}{T_\mathrm{m}}\int\limits_{-T_\mathrm{m}/2}^{T_\mathrm{m}/2}\underline{X}(f,\xi)\mathrm{e}^{-j2\pi n\nu_0\xi}\,\mathrm{d}\xi \\ &= \underline{B}_0(f)\frac{1}{T_\mathrm{m}}\int\limits_{-T_\mathrm{m}/2}^{T_\mathrm{m}/2}\mathrm{e}^{-j(2\pi n\nu_0\xi+2kd_1\,\sin\,2\pi\nu_0\xi)}\mathrm{d}\xi\end{aligned} \qquad (4.273)$$

where $T_\mathrm{m} = \nu_0^{-1}$ is the period duration of the target motion. Using the substitution $2\pi\nu_0\xi = 2\pi\xi/T_\mathrm{m} = \tau$, we will end in (see Annex B.1, Table B.1) [82]

$$\begin{aligned}\underline{X}(f,n\nu_0) &= \underline{B}_0(f)\,\frac{1}{2\pi}\int\limits_{-\pi}^{\pi}\mathrm{e}^{-j(n\tau+2kd_1\,\sin\,\tau)}\mathrm{d}\tau \\ &= \underline{B}_0(f)\,\mathrm{J}_n(x); \quad x = -2kd_1 = -4\pi\frac{d_1}{\lambda}\end{aligned} \qquad (4.274)$$

Herein, $\mathrm{J}_n(x)$ represents the nth order Bessel function of first kind. From this, the shape of the spectral function $\underline{X}(t,n\nu_0)$ is finally gained by inverse Fourier transformation:

$$\underline{X}(t,n\nu_0) = \int\limits_{-\infty}^{\infty}\underline{B}_0(f)\,\mathrm{J}_n\left(-\frac{4\pi f d_1}{c}\right)\mathrm{e}^{j2\pi ft}\,\mathrm{d}f \qquad (4.275)$$

We can observe from (4.275) that the strength of the harmonics in the spectrum augments with increasing ratio d_1/λ.

While Figure 4.71 only refers to the idealized noise-free case, Figure 4.72 illustrates the situation where heavy noise completely impedes the detection of an oscillatory motion in the radargram and even the spectrum (i.e. the first energy accumulation) complicates a clear detection. The only thing we can do now is to coherently further concentrate deterministic signal parts which belong to the oscillating target. The first option we have is to compress the signal in propagation time direction t. Ideally, cross-correlation (matched filtering) with an appropriate reference signal would be the means of choice to do this. But unfortunately, we do not know the backscatter signal $b(t)$ under realistic conditions so that we have to settle for doing an

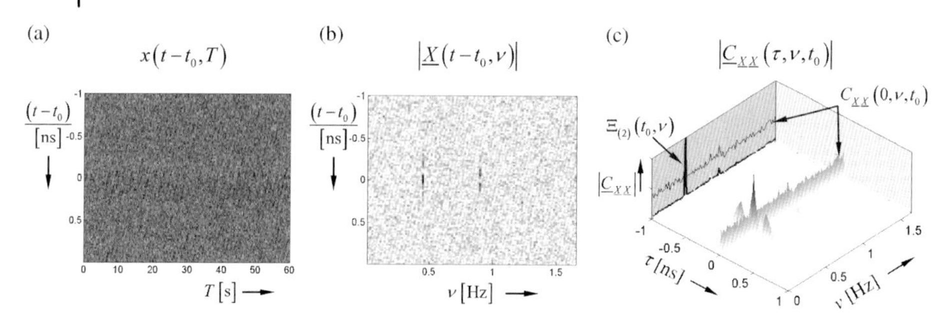

Figure 4.72 Detection of oscillating targets under heavy noise conditions. The noise was supposed to be uncorrelated (i.e. the sampled data are not subjected an additional pre-filtering). (a) Radargram; (b) 'observation time' spectrogram and (c) 'propagation time' correlation of the spectrum for time position t_0. (colored figure at Wiley homepage: Fig_4_72).

auto-correlation[23] which we perform over an integration window centred at t_c and of width t_{win}:

$$\underline{C}_{\underline{X}\underline{X}}(\tau,\nu,t_c)=\int_{t_c-(t_{win}/2)}^{t_c+(t_{win}/2)}\underline{X}^*(t,\nu)\underline{X}(t+\tau,\nu)\mathrm{d}t \tag{4.276}$$

The integration window length should be roughly equal to the (pre-estimated) duration of the scattered pulse $t_{win}\approx t_w$ in order to assure optimum accumulation of signal energy. Since the round-trip time t_0 of the target is not known, the window position t_c has to be swept over an interval within which one expects the target. At time lag $\tau=0$, the auto-correlation (4.276) has its peak value which corresponds to the total energy of the signal section falling into the time window $t_c-t_{win}/2$ to $t_c+t_{win}/2$.

Figure 4.72c illustrates the appearance of (4.276) if the integration window position coincides with the round-trip time $t_c=t_0$. The crest line $C_{\underline{X}\underline{X}}(0,\nu,t_0)=\mathfrak{E}_b(\nu,t_0)+\mathfrak{E}_n$ represents the total signal energy which is composed of noise energy $\mathfrak{E}_n$ (giving the mean level) and target energy $\mathfrak{E}_b$ (causing the peaks). For better illustration, the crest line is emphasized as projection at the rear side of the diagram. We can detect two decided peaks now relating to the first and second harmonic of the $\underline{X}(t,\nu)$ spectrum.

Inspecting $\underline{C}_{\underline{X}\underline{X}}(\tau,\nu,t_0)$ more deeply, we can discover that the auto-correlation function exhibits side lobes at harmonics of the respiration rate $n\nu_0$ originating from the deterministic signal $b(t)$. In order to further suppress randomness and in imitation of the L_p-norm definition, we

23) Note that under practical conditions, the spectral components do not form an ideal line spectrum rather they are broadened. In the case where they are broad enough, one can also take two adjacent spectral components, for example $\underline{X}(t,\nu_0)$ and $\underline{X}(t,\nu_0+\Delta\nu)$ to perform a cross-correlation. Since now the noise components in both signals are uncorrelated, the mean level of the crest line would fall down to zero and only the oscillation peaks remain.

additionally perform integration over τ getting the function

$$\Xi_{(p)}(\nu, t_0) = \int \left| \underline{C}_{\underline{X}\,\underline{X}}(\tau, \nu, t_0) \right|^p \mathrm{d}\tau; \quad p \in \mathbb{N} \tag{4.277}$$

where it seems that $p = 2$ is the most reasonable choice. $\Xi_{(p)}(\nu, t_0)$ can be subjected to a CFAR detection for automatic estimation of oscillation rate ν_0 and target range $r = rc/2$. The appearance of several peaks at a regular distance $n\nu_0$ (sufficiently large deflection amplitude d_1 assumed) can be used to improve additionally the reliability of the detection.

The two examples have illustrated the improvement of the detection performance by extending the observation time before one comes to a decision. Regardless the increased processing expense, the described methods work well only if the time shape of the receiving signal $b(t)$ does not or only marginally varies within the observation time window T_{obs}, that is the fidelity $\text{FI}(x(t, T_1), x(t, T_2)) \approx 1; \quad T_1, T_2 \in T_{\text{obs}}$ of the involved signals should be close to unity. Such variations may appear if the scatter IRF is changing – either by variation of aspect angle or by variable target geometry (e.g. for animals, human being, machines etc.) – or by multipath effects due to interference of propagation paths with nearly equal propagation time. An example of such variations is illustrated in Figure 4.73 indicating the scattering IRF of a walking person. The temporal variations are mainly caused by the motion of arms and legs as well as the alteration of the aspect angle.

4.8.3
Detection of Weak Targets Closely Behind an Interface

A major advantage of ultra-wideband sensing is its ability to detect and track hidden objects since the sounding waves can penetrate many substances (exceptions

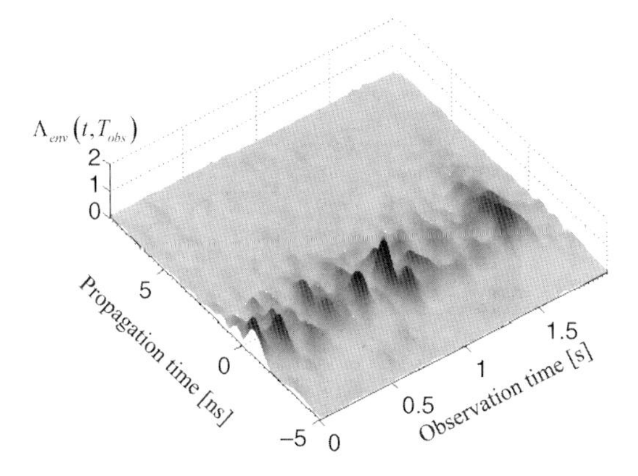

Figure 4.73 Envelope of the normalized scattering function $\Lambda(t, T_{\text{obs}})$ of a walking person.

Table 4.3 Propagation loss at 100 MHz and 1 GHz [83].

Material	Loss at 100 MHz [dB/m]	Loss at 1 GHz [dB/m]
Clay (moist)	5–300	50–3000
Loamy soil (moist)	1–60	10–600
Sand (dry)	0.01–2	0.1–20
Ice	0.1–5	1–50
Fresh water	0.1	1
Sea water	100	1000
Concrete (dry)	0.5–2.5	5–25
Brick	0.3–2.0	3–20

are, e.g., metal, saline solutions and clay or loamy soil). Table 4.3 shows the propagation loss of common materials.

We will analyse some issues determining the sensitivity to detect weakly scattering targets in the vicinity of strong scatterers. For that purpose, we will exemplarily consider two simple mono-static arrangements in which a small target is placed behind a reflecting planar interface (see Figure 4.74). As illustrated, we have to deal with two to three major signal components which we will get in case of bi-static arrangement too. The signal $b_0(t)$ stands for the feed point reflections of the antenna. It only appears as isolated signal if the antenna has sufficient distance to the first scatterer. $b_0(t)$ is a cross-talk signal if bi-static configurations are used. Since this signal does not contribute to information gathering it should be as small as possible.[24] This signal usually limits the sensor dynamic range since it leads to receiver saturation with increasing transmitter power.

The signal $b_1(t)$ represents the interface reflections. In case of the contact mode (Figure 4.74b), it is a mixture of interface and feed point reflections. Finally, we still have the target reflection $b_2(t)$. We would like to identify it also in case of strong overlap with the two other signals.

The practical applications are as follows:

- Non-destructive testing in civil engineering, that is the detection of cracks, bubbles and inclusions in non-metallic materials, disaggregation of construction material or rocks, detection of foreign bodies and so on [19, 84–86].
- Medical engineering, for example organ motion or detection and localization of affected tissue. Ref. [87] gives a wide overview of corresponding recent publications.
- Ground-penetrating radar, detection of surface close objects, for example land mines [83].
- Vermin detection [88, 89].
- Rescue of buried people [81, 90–93] and others.

24) Note that bi-static antenna configurations are more commonly used than mono-static ones, since typically the isolation between two antennas (cross-talk) is higher than the return loss.

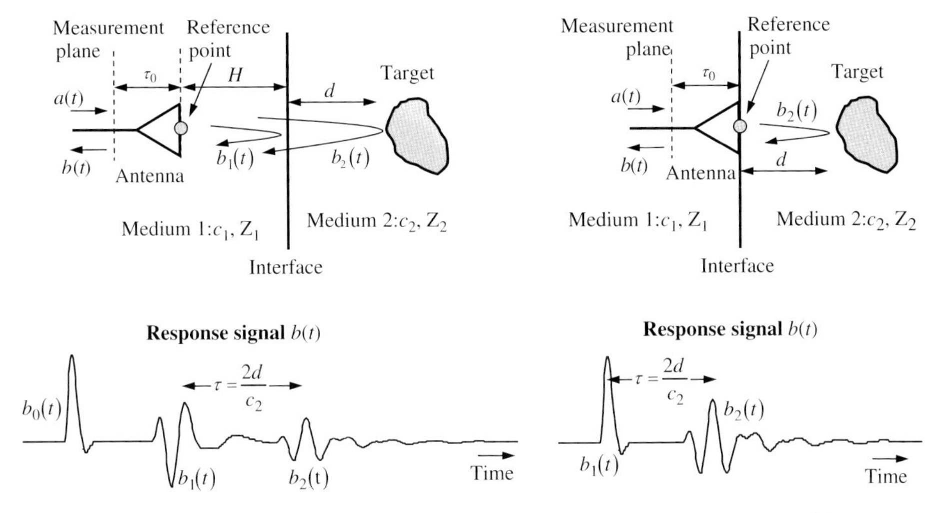

Figure 4.74 Two mono-static measurement arrangements and idealized impulse response of the test scenario. Multiple reflections (antenna–interface; target–interface; antenna–target) are neglected for the sake of simplicity.

Our goal is to indicate crucial sensor parameters affecting the sensing performance and examine some methods of target extraction from the measured data. We restrict our discussion to the simplest case where the target is placed face to face with the antenna.

4.8.3.1 Modelling of the Receiving Signal

Supposing $b_0(t)$ may be gated out by reason of sufficient time delay, the interesting parts of the receiving signal are $b(t) = b_1(t) + b_2(t)$ – the reflection from the interface $b_1(t)$ and the reflection from the target $b_2(t)$. Further components are caused by multiple reflections between interface and antenna, target and interface and target and antenna, as well as feed point reflections of the antenna. We will omit these signal parts for the sake of simplicity here (for more details on multiple reflections, see Section 9.1.1). We characterize the reflectivity of the interface by the reflection coefficient[25] $\Gamma_0 = (1-\eta)/(1+\eta)$ ($\eta = c_1/c_2 = \sqrt{\varepsilon_2/\varepsilon_1}$ is the refraction index), and the scatter response $\Lambda(t)$ relates to the target. Furthermore, we will assign $a_{\mathrm{eff}}(t)$ as the effective stimulus signal which includes the intrinsic sensor response, the antenna pulse response as well as an impulse compression if UWB-CW signals are used for stimulation. It is referred to as the antenna reference point, so that antenna internal delays must not be respected. Following Section 4.5.1, the receiving signal caused by the interface reflection can be written as

$$b_1(t) = \frac{\Gamma_0}{2H} a_{\mathrm{eff}}(t - 2\tau_{\mathrm{H}}) \quad \text{with } \tau_{\mathrm{H}} = \frac{H}{c_1} \tag{4.278}$$

25) The reflection coefficient only refers to perpendicular incidence.

The signal component which originates from the target is approximated by

$$b_2(t) \approx \frac{T_0^2 \, \mathrm{e}^{-2d\alpha_2}}{(H+d)^2} \Lambda(t) * a_{\mathrm{eff}}(t - 2\tau_{\mathrm{d}H}) \quad \text{with } \tau_{\mathrm{d}H} = \frac{H}{c_1} + \frac{d}{c_2} \tag{4.279}$$

Herein, we supposed that the propagation within medium 1 (e.g. air) is lossless. $T_0 = 2\eta/(1+\eta) = \sqrt{1-\Gamma_0^2}$ represents the transmission loss if a wave passes an interface and $\alpha_2 = -a_2[\mathrm{dB/m}]/20 \lg e$ is the loss constant of the propagation medium 2 (refer to Table 4.3 for some examples). (4.279) represents a reduced radar equation which does not respect the modification of the wavefront curvature by passing the interface (for an introduction on wave scattering, see Sections 5.4 and 5.5). In case of a missing interface ($\eta = 1$), (4.279) reduces to the ordinary time domain radar equation. Often, the second medium causes additional propagation losses. We have respected it by the exponential term $\mathrm{e}^{-2d\alpha_2}$, implying frequency-independent attenuation. This is an often made assumption which is acceptable for low losses. Mostly in case of higher losses, $\alpha_2 = \alpha_2(f)$ becomes frequency dependent and hence the propagation speed $c_2 = c_2(f)$ too as a consequence of the causality. In that case, (4.279) has to be expanded by a distance-dependent convolution term which respects attenuation and dispersion. A corresponding example is given in Section 5.4. Finally, we include device clutter$c(t)$ and random errors as additive noise $\underset{\sim}{n}(t) \sim N(0, \sigma_n^2)$ or multiplicative noise $\underset{\sim}{m}(t) \sim N\left(0, \|b_1(t)\|_2^4/\mathrm{TB}\right)$ (e. g. generated from random noise radar with limited integration time; TB is the time bandwidth product; see also Section 3.6) and jitter $\Delta\underset{\sim}{\tau}_{\mathrm{j}}(t) \sim N\left(0, \varphi_{\mathrm{j}}^2\right)$ so that we get as general signal model

$$\begin{gathered} \underset{\sim}{b}(t) = \underset{\sim}{b}_x(t) + \frac{\partial \underset{\sim}{b}_x(t)}{\partial t} \Delta\underset{\sim}{\tau}_{\mathrm{j}}(t) \\ \text{with} \\ \underset{\sim}{b}_x(t) = b_1(t) + b_2(t) + \underset{\sim}{m}(t) + \underset{\sim}{n}(t) + c(t) \end{gathered} \tag{4.280}$$

It can be simplified since typically $b_1(t)$ is the strongest component and different measurement approaches suffer differently under the mentioned random errors:

$$\underset{\sim}{b}(t) = \begin{cases} b_1(t) + b_2(t) + \underset{\sim}{n}(t) + c(t) + \dot{b}_1(t)\Delta\underset{\sim}{\tau}_{\mathrm{j}}(t) & \text{for pulse radar} \\ b_1(t) + b_2(t) + \underset{\sim}{n}(t) + c(t) & \text{for PN-sequence radar} \\ b_1(t) + b_2(t) + \underset{\sim}{n}(t) + c(t) & \text{for stepped frequency radar} \\ b_1(t) + b_2(t) + \underset{\sim}{m}(t) + c(t) & \text{for random noise radar} \end{cases} \tag{4.281}$$

4.8.3.2 Hidden Target Detection

For target detection, we need to separate $b_2(t)$ at least partially from the captured signal $\underset{\sim}{b}(t)$. In order to face the challenges, we consider a weak scatterer closely behind an interface as depicted in Figure 4.74. Figure 4.75 shows schematically the envelope of the involved signal components. We assume a pulse-shaped sounding signal $a_{\mathrm{eff}}(t)$ (UWB-CW signals are pulse compressed beforehand) which decays at

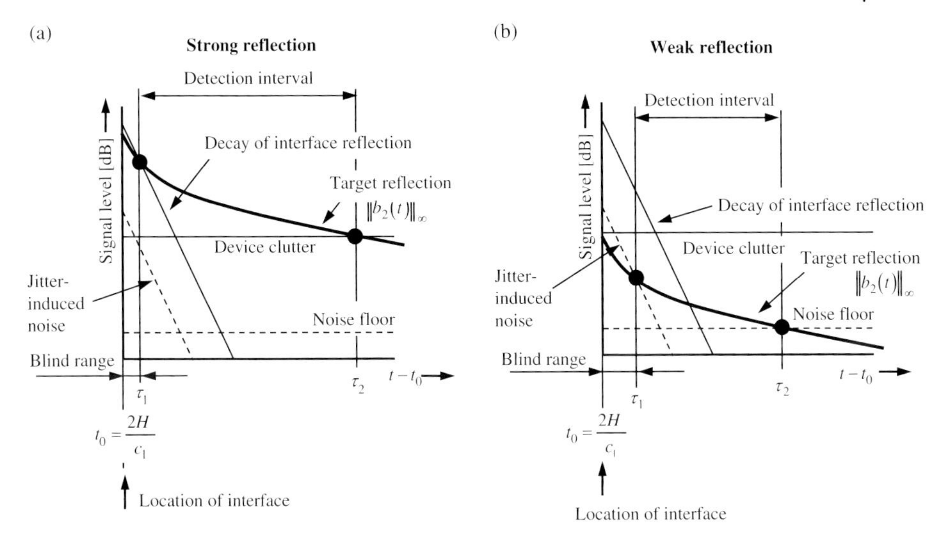

Figure 4.75 Level diagram for the different signal components in (4.280) for a strong (a) and a weak scatterer (b).

a certain rate. The reflection from a flat interface does not affect the signal shape of $b_1(t)$; hence, it decays at the same rate as $a_{\text{eff}}(t)$. Jitter-induced noise power follows the slew rate of the main signal. Thus, it decays approximately as the sounding signal. The power of device clutter and additive or multiplicative noise is considered as independent of time.

The peak value of the target return $b_2(t)$ is depicted by the bold line. We consider a strong (Figure 4.75a) and a weak target (Figure 4.75b) at different distances from the interface. As noted from (4.279), the peak value of $b_2(t)$ decays with growing distance d and thus also with propagation time $t - t_0 = 2\,d/c_2$. In the case of lossy material, the decay will be additionally strengthened.

In order to detect the target, $\|b_2(t)\|_\infty$ has to top all other signal components. For the strong target, this is the case within the two points marked in Figure 4.75a. That is, the target can be detected as long as it is placed within a depth region of $d = c_2(\tau_1 \cdots \tau_2)/2$. Immediately behind the interface, a blind range avoids detection due to the overlap with interface reflections, and at the far end, the clutter overwhelms the target signal. The weak target would not be visible at all due to these effects.

But interface reflection and device clutter represent deterministic signals. Theoretically, they can be removed by appropriate procedures down to noise level in the ideal case. If we do this, the weak target will be visible too. Blind range and maximum detection range are determined by jitter and noise now. These two limits represent the maximum achievable sensor performance. Note that well-designed UWB-CW-radar sensors (PN-sequence, stepped frequency, random noise) do not have a jitter-induced blind range since jitter is spread over the whole signal and appears as additive noise (compare (4.281) and refer Section 4.7.3). Under real-life conditions, this performance is effectively available if the interface and the antenna are stationary and the target is moving (only the motion component in range direction is of interest here). In this case, the strong interface reflex represents horizontal lines in the radargram which can be removed by horizontal high-pass filtering while the target produces a time-variable trace which is not (or only barely) influenced by the filtering as long as it moves. This approach is widely used to detect people behind a wall or buried beneath rubble and it even works for very weak motion and small targets, for example vermin. Sensitivity reductions arise if the clutter signal is non-stationary too. A separation between target and clutter is only possible if they occupy different range sectors or they have a discriminable statistics or motion profile (see Sections 6.7–6.9).

In case of static targets, one can only try to remove or suppress the interface reflection as much as possible. If its time shape is prior known or can be selected from a signal dictionary, it can be subtracted from $\underset{\sim}{b}(t)$ by the SAGE algorithm (see Figure 4.51). The achievable precision depends on the fidelity of the reference function and the precision of range measurement (see Section 4.7.3). Under perfect conditions, the strong surface reflection could be removed down to noise level.

If no prior knowledge is available, clutter and blind range must be reduced by other means. Reflections from perturbing objects within the measurement scenario are a common source of clutter. By measuring the environment beforehand without test objects, the environmental clutter can be determined from the actual data captured with the test objects. However, attention should be paid since this approach does not respect multiple reflections and shadowing effects that appear after the test object comes into the game.

4.8.3.3 **Blind Range Reduction**

Unfortunately, the pulse decay of the original generator signal is often not perfect and additionally destroyed by antenna ringing or sensor internal reflections which cannot be avoided due to technical constraints of device construction. The question arises if it is possible to reduce the blind range $\delta_{\mathrm{BR}} = \tau_1 c_2/2$ by post-processing the captured data $\underset{\sim}{b}(t)$. We will discuss this problem using some simulated examples for illustration. A practical example is given in Section 6.4.

We take the arrangement from Figure 4.74 for which we presume a flat interface with a δ-like impulse response and the target should show some oscillations. The involved signals are depicted in Figure 4.76. The impulse response of the test scenario is modelled by (multiple reflections are omitted)

$$S_{11}(t) = \Gamma_0\, \delta(t - 2\tau_{\mathrm{H}}) + g_{\mathrm{targ}}(t - 2\tau_{\mathrm{d}H}) \tag{4.282}$$

and the perturbation free receiving signal would be

$$b(t) = b_1(t) + b_2(t) = S_{11}(t) * a_{\mathrm{eff}}(t) \tag{4.283}$$

The two terms in (4.282) represent a compact notation of (4.278) and (4.279).

Furthermore, we know the actual sounding signal $a_{\mathrm{eff}}(t)$ determined, for example from a reference measurement at a perfect conducting metal plate as illustrated in Figure 4.43. The signal amplitude of $a_{\mathrm{eff}}(t)$ decays with the rate $DR_{\mathrm{a}} = \Delta L/\Delta t$

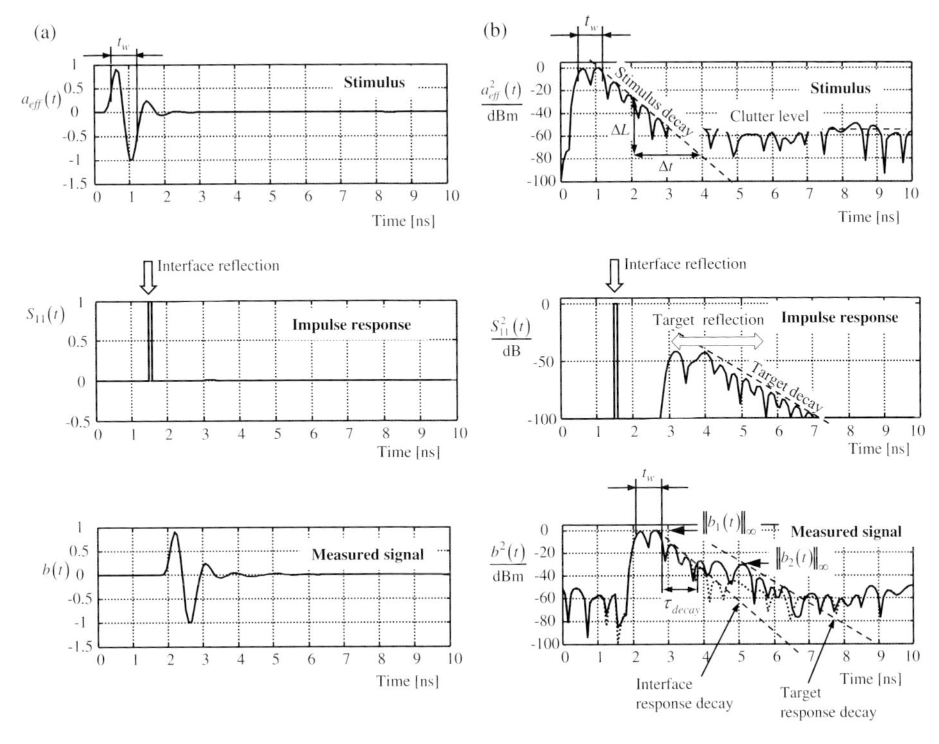

Figure 4.76 Reflections from a weak target behind an interface. (a) Linear scaled signals and (b) logarithmic scaled signals.

and passes finally in a more or less constant signal level (caused, e.g., by device internal reflections). The pulse or coherence width of $a_{\mathrm{eff}}(t)$ is t_{w}.

We took a target reflection which is 100 times below the surface reflex. In a first step, we will neglect the influence of noise. We represented every signal in a linearly and logarithmically scaled graph in Figure 4.76 in order to impart the appearance of the actual signals as well as to emphasize small signal components which are of major interest in our example.

Obviously, neither the clutter in the sounding signal $a_{\mathrm{eff}}(t)$ (Figure 4.76a, top) nor the reflection from the target (Figure 4.76a, middle) is visible in the linear scaled graphs but they are well identifiable in the dB-scaled representation. The graphs at the bottom of Figure 4.76 depict the response of the test scenario. Even in the logarithmic scale graph, it is somewhat difficult to identify the target reflection due to the masking effect of the stimulus signal. For comparison, the pure interface reflection (dashed line) is inserted too.

We can improve the sensor resolution by removing the imperfections of the effective stimulus signal from the measured response of the test scenario. This is done by de-convolving $a_{\mathrm{eff}}(t)$ out of $b(t)$ which would lead to the wanted IRF $S_{11}(t)$ of the test scenario. De-convolution may be performed in the time or frequency domain. Thanks to the FFT, the frequency domain algorithms work faster and more efficiently.

De-convolution is a procedure which works inversely to the natural cause-and-effect chain which may provoke ill-posed problems. Since these effects are quite illustrative in the frequency domain, we will restrict ourselves here to that domain. The reader can refer to Ref. [94] for an introduction to time domain de-convolution.

Let us first omit the role of measurement noise. The frequency domain notation of (4.283) is

$$\underline{b}(f) = \underline{S}_{11}(f)\underline{a}_{\mathrm{eff}}(f) \tag{4.284}$$

where the complex spectra are estimated from $\underline{b}(f) = \mathrm{FT}\{b(t)\}$; $\underline{S}_{11}(f) = \mathrm{FT}\{S_{11}(t)\}$ and $\underline{a}_{\mathrm{eff}}(f) = \mathrm{FT}\{a_{\mathrm{eff}}(t)\}$. Apparently (4.284) provides a simple solution for the de-convolution, namely

$$\hat{S}_{11}(t) = \mathrm{IFT}\left\{\frac{\underline{b}(f)}{\underline{a}_{\mathrm{eff}}(f)}\right\} = \mathrm{IFT}\left\{\frac{\mathrm{FT}\{b(t)\}}{\mathrm{FT}\{a_{\mathrm{eff}}(t)\}}\right\} \quad \text{if } \underline{a}_{\mathrm{eff}}(f) > 0 \tag{4.285}$$

Herein, $\hat{S}_{11}(t)$ is the IRF estimated from the measured data in contrast to $S_{11}(t)$ being the actual one. Goal is that $\hat{S}_{11}(t)$ deviates as little as possible from $S_{11}(t)$.

The spectral band occupied by the sounding signal or the bandwidth of the receiver, respectively, is always restricted in practical cases; that is, $|f| \leq f_{\mathrm{max}}$. Therefore, we cannot calculate the quotient between the two spectra outside of this band since only noise appears (ill-posed problem!).

What are the consequences? If we would be interested to determine the reflection from a metallic wall, we would expect $S_{11,\mathrm{W}}(t) \sim -\delta(t - 2H/c_1)$ as impulse response of the whole scenario. But actually, we would measure $b_{\mathrm{W}}(t) \sim a_{\mathrm{eff}}(t - 2H/c_1)$; that is, the receive signal is identical with the stimulus except a time delay and attenuation. Hence, we get from (4.285) for the noiseless case

$$\frac{\underline{b}_{\mathrm{W}}(f)}{\underline{a}_{\mathrm{eff}}(f)} \sim \begin{cases} -\mathrm{e}^{-j\frac{4\pi f H}{c_1}} & |f| \leq f_{\mathrm{max}} \\ 0 & |f| > f_{\mathrm{max}} \end{cases}$$

$$\Rightarrow \hat{S}_{11,\mathrm{W}}(t) \sim \mathrm{IFT}\left\{\frac{\underline{b}_{\mathrm{W}}(f)}{\underline{a}_{\mathrm{eff}}(f)}\right\} = \mathrm{IFT}\left\{\mathrm{rect}\left(\frac{f}{2f_{\mathrm{max}}}\right)\mathrm{e}^{-j\frac{4\pi f H}{c_1}}\right\} = -\mathrm{sinc}(2\pi f_{\mathrm{max}}(t - 2H/c_1)) \tag{4.286}$$

which obviously deviates from the expected behavior. Here, we supposed that the stimulus occupies a spectrum till f_{max} wherefrom it abruptly breaks down to zero.

Figure 4.77a illustrates the outcome of our gedankenexperiment. In the upper graph, the reflection from the metal wall is shown. Compared to Figure 4.76, we can state that the resulting pulse width t_{w} is improved resulting in a better separability of two nearby targets of equal reflectivity. But due to slow descending side lobes of the sinc function, the approach does not actually improve the detection performance of weak targets so far. This is demonstrated in the lower graph of Figure 4.77a. It refers to the scenario of Figure 4.76 after having applied de-convolution.

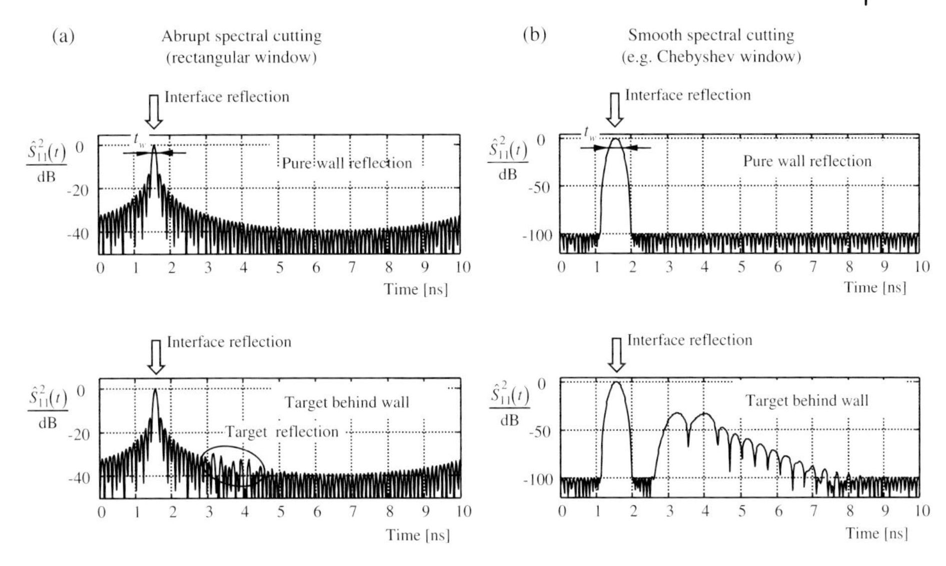

Figure 4.77 De-convolution of measurement signal and effect of the window function. The pure wall reflection refers to the example described by (4.286).

If we insert additionally a spectral weighting function $\underline{W}_{\text{sh}}(f)$ (often also assigned as window function) into (4.285) in order to smooth the edges of the spectral band, we are able to suppress the ringing caused by the abrupt spectral edges:

$$\hat{S}_{11}(t) = \text{IFT}\left\{\frac{\underline{b}(f)}{\underline{a}_{\text{eff}}(f)}\,\underline{W}_{\text{sh}}(f)\right\} \tag{4.287}$$

The effect of this additional spectral shaping is demonstrated in Figure 4.77b. In that example, we made use of a Chebyshev window with 100 dB ripple suppression to modify the spectrum for pulse shaping. Note that (4.287) and (4.285) are identical if the spectral weighting is done by rectangular function $W_{\text{sh}}(f) = \text{rect}(f/2f_{\text{max}})$, hence the notation 'rectangular' and 'Chebyshev' window in Figure 4.77.

Obviously, by selecting an appropriate window function, we were able to largely suppress the side lobes so that the estimated impulse response $\hat{S}_{11}(t)$ of our scenario clearly indicates the weak scatterer. However, we gained this improvement against the impulse width t_w which tends to increase. Hence, the sensor performance to separate two closely located targets of same reflectivity degrades. It is a general observation that for a given width of the spectral band, one has to find a reasonable compromise between the width of the main pulse and the level of the side lobes. There exist a lot of pre-defined window functions, which allows the user to apply an appropriate spectral shaping for his application [95]. However, these functions will lead to non-causal effects. If specific applications prohibit non-causal signals, the waiting function $\underline{W}_{\text{sh}}(f)$ must be implemented by a causal filter having an impulse response of the desired shape. Since de-convolution is usually applied on digitized sensor data, it is possible to emphasize specific aspects (e.g.

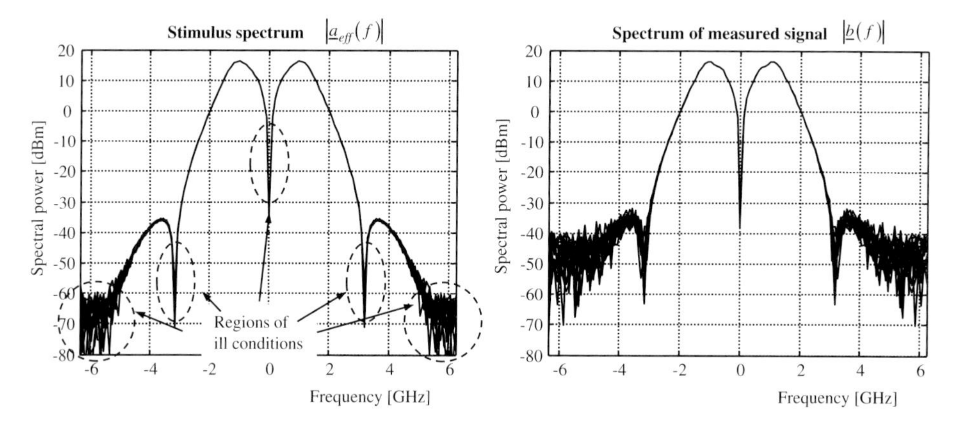

Figure 4.78 Spectra of perturbed stimulus and measurement signals. Note, the deterministic parts of both spectra differ only slightly since the target influence is very weak for the selected example.

separability of targets with equal strength of backscattering or separation of strong and weak targets) by parallel processing with different weighting functions.

The noise-affected spectra of the involved signals are illustrated in Figure 4.78. Corresponding to (4.287), we have to perform a division by $\underline{a}_{\text{eff}}(f)$ which will provoke problems if the spectral power approaches noise level. So we have to divide 'noise by noise' (a mathematical problem which resembles the indefinite expression $0/0$). The noisy regions are marked by ellipses in the spectrum of the stimulus. These regions will also appear noisy in the spectrum of the measured signal since the test object cannot react where there is no stimulus power.

The FRF resulting from such division is depicted in Figure 4.79. In order to keep the critical regions as small as possible, attention was paid to reduce the noise level as much as possible at least for the stimulus signal $a_{\text{eff}}(t)$. It is captured by a

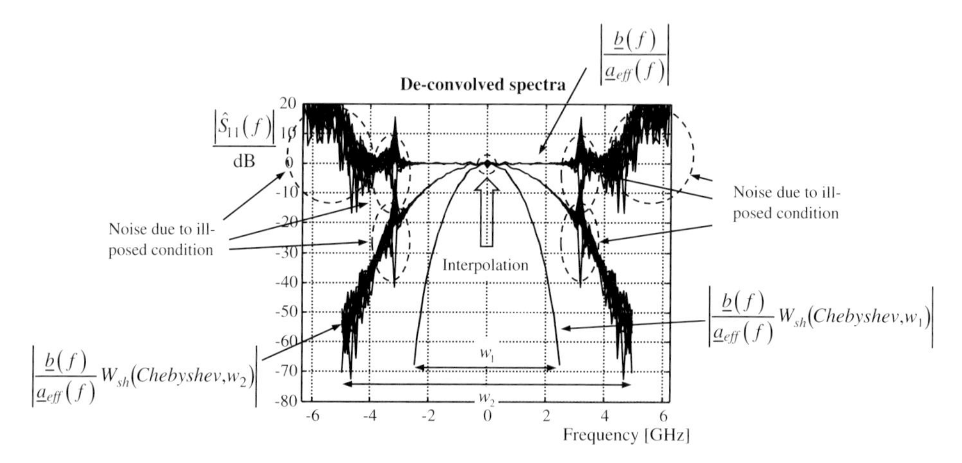

Figure 4.79 De-convolution of noise affected signals by division of their spectra.

reference measurement beforehand. Practically this involves intensive synchronous averaging requiring large recording time, which is however not critical in this case. Therefore, the noise level in the left-hand-side picture of Figure 4.78 is lower than that in the right one.

Surprisingly, the ill-conditioned region around $f = 0$ is missed in Figure 4.79, although the stimulation of the scenario at low frequencies is very weak. In the case there is only a small gap, we could try to close it via interpolation since the test scenario purports a continuous spectral trend; that is, the ill-posed problem was evaded by prior knowledge. Basically this in one approach to override 'ill-posed parts' of the response function; another one suppresses unwanted de-convolution products by regularization, which is exemplified below.

After applying the window function $W_{\mathrm{sh}}(f)$ for impulse shaping and back-transformation via IFFT, we result in the wanted estimate $\hat{S}_{11}(t)$ of the impulse response function. Figures 4.79 and 4.80 show two examples in which weighting functions of different width but of identical type were applied. In the case of the narrow window (width w_1), the ill-conditioned spectral parts were excluded before the IFFT is performed so that the noise only little affects the pulse response (Figure 4.80b). Then of course, the pulse width t_{w} will be correspondingly widened. In contrast, the wider window will lead to better time resolution but the result is strongly affected by noise (Figure 4.80a).

By keeping the wide window function and increasing noise level, the weak target will finally disappear (Figure 4.81a). But since we have not yet performed optimal de-convolution, there is chance to have the weak target back. The result of the de-

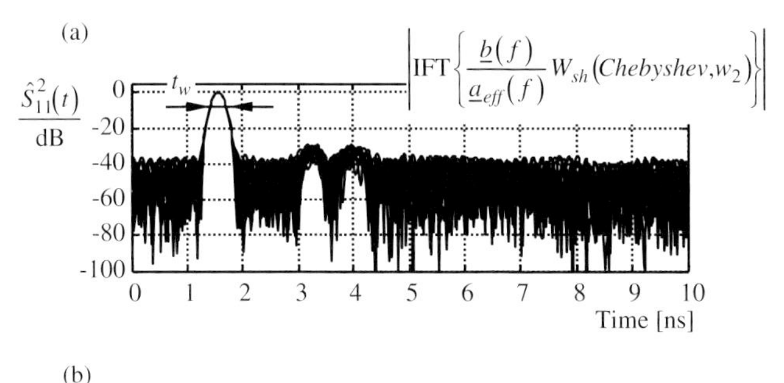

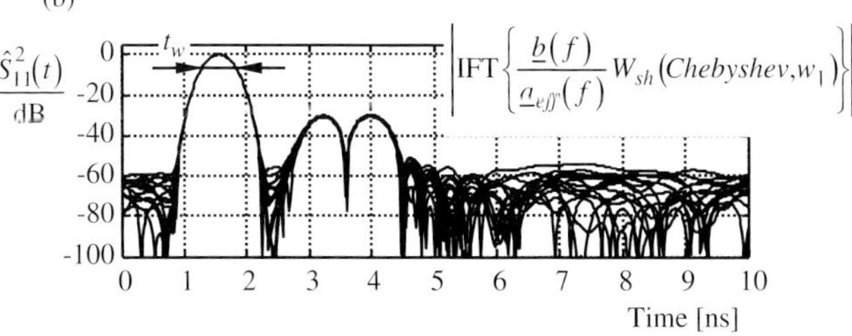

Figure 4.80 Estimation of noise affected IRF applying different window sizes. (a) Wide window function and (b) narrow window function.

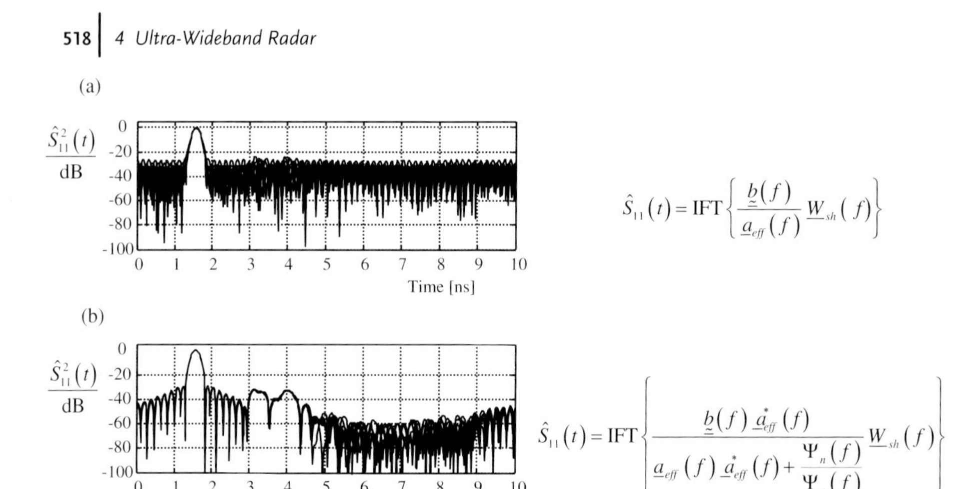

Figure 4.81 De-convolution of noise affected signals without (a) and with (b) adaptive suppression of ill-conditioned spectral division.

convolution can be improved if ill-conditioned spectral parts are suppressed depending on signal quality as it is done by Wiener de-convolution. Such procedure is also assigned as regularization [94]. Assuming that the perturbed measurement can be modelled by (Ψ_a and Ψ_n are the power spectral density of the sounding signal and noise, respectively)

$$\begin{aligned} \underset{\approx}{b}(f) &= \underline{S}_{11}(f)\,\underline{a}_{\mathrm{eff}}(f) + \underset{\approx}{n}(f) \\ \Psi_{\mathrm{a}}(f) &= \underline{a}_{\mathrm{eff}}(f)\,\underline{a}_{\mathrm{eff}}^*(f) \\ \Psi_{\mathrm{n}}(f) &= \mathrm{E}\{\underset{\approx}{n}(f)\,\underset{\approx}{n}^*(f)\} \end{aligned} \tag{4.288}$$

the improved IRF estimation of the test scenario is calculated from

$$\hat{S}_{11}(t) = \mathrm{IFT}\left\{\frac{\underset{\approx}{b}(f)\underline{a}_{\mathrm{eff}}^*(f)}{\underline{a}_{\mathrm{eff}}(f)\underline{a}_{\mathrm{eff}}^*(f) + (\Psi_{\mathrm{n}}(f)/\Psi_{\mathrm{a}}(f))}\underline{W}_{\mathrm{sh}}(f)\right\}, \tag{4.289}$$

This equation requires the determination of the frequency-dependent SNR-value $\mathrm{SNR}(f) = \Psi_{\mathrm{a}}(f)/\Psi_{\mathrm{n}}(f)$ during data capturing. If this is not possible, the SNR-value is usually replaced by an estimate. The improvements are visualized in Figure 4.81b. It indicates the reduction of randomness due to ill-conditioned division by small numbers. But also some degradation of the clutter reduction becomes visible evoked by the side lopes appearing around the main peak. These new side lobes are caused by the SNR-term which affects the weighting function $W_{\mathrm{sh}}(f)$. Hence, pulse shaping and regularization cannot be considered independently of each other.

An example of practical applications of Wiener de-convolution for UWB backscatter measurements can be found in Refs [96, 97].

4.9
Evaluation of Stratified Media by Ultra Wideband Radar

Many ultra-wideband sensing tasks can be traced back to the scattering in stratified media. Examples are ground investigations by GPR [83, 98] to recover the structuring of soil layers and wall parameter estimation for material research, non-destructive testing in civil engineering and through-wall imaging or the determination of a moisture profile. But also for other applications such as the detection of heartbeat or the movement of the chest due to breathing, stratified media may act as a simple first-order model to study the wave behavior.

4.9.1
Measurement arrangement and Modelling of Wave Propagation

The basic measurement arrangements to investigate stratified media are depicted in Figure 4.82. Both methods resemble the classical TDR measurement principle (see Section 3.7, Figure 3.95). One approach uses free waves allowing non-destructive measurements of the layer structure. We will only look at the problem of plane wave propagation with normal incidence. The reader can refer to Refs [3, 14, 17] for a detailed theoretical analysis and arrangements of larger complexity than

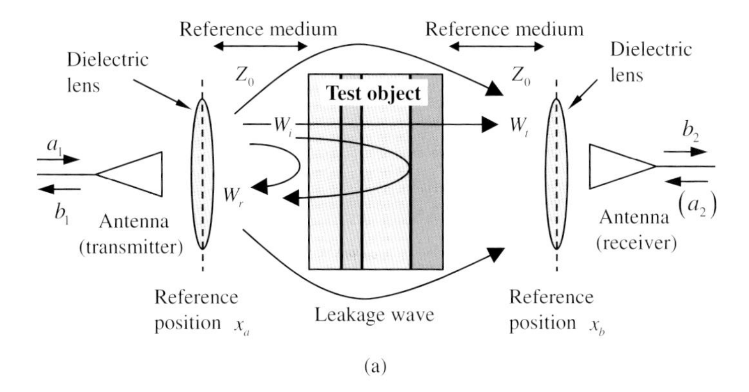

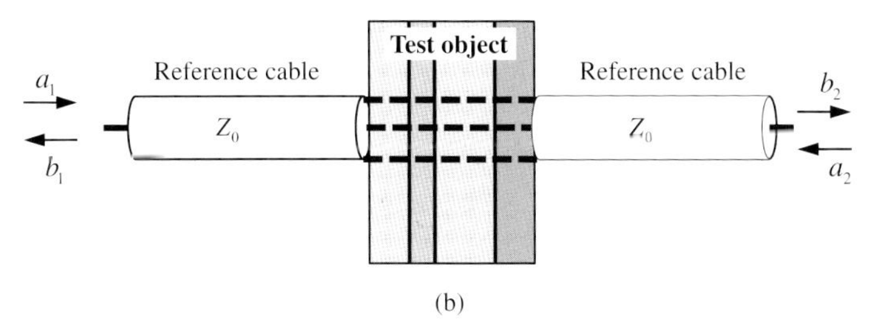

Figure 4.82 Basic measurement arrangements for characterization of stratified media applying free waves (a) or guided waves (b). If the test objects are accessible from one side only, the method reduces to pure reflection measurement.

discussed here. The second method applies guided waves travelling along a probe cable. It can be used if temporal variations of the layer material are of interest, for example in moisture profile measurements [99].

Measurements from both sides of the structure represent the optimum case since three independent measurements are provided – reflection from left and right and only one transmission (due to reciprocity) – from which the information about the layer structure can be extracted. If only one side is accessible or only one measurement channel is available, one is restricted to one reflection measurement. This will lead to purer resolutions for the deeper layers since the corresponding signals are increasingly affected by errors. Such errors can be partially reduced if one repeats the measurement by placing a strong reflector, for example sheet metal on the opposite side if the experimental conditions allow that. Ground-penetrating radar and through-wall person tracking are typical examples where one has access only from one side.

If the sounding waves are propagating in free space, some additional requirements have to be respected:

- The sounding waves have to be planar by our assumption. But antennas radiate waves with an approximately spherical wavefront. Therefore, one needs electric lenses to straighten the wavefront or the antenna distance must be large enough so that the wavefront can be approximated by a plane within the region of interest.
- The test object must be large enough in lateral dimensions in order to avoid leakage waves which bypass the test object and lateral wave modes.
- The surface of the test objects must be flat in order to avoid random scattering.

Figure 4.83a depicts a simple example consisting of four materials of different electric parameters and thicknesses. For illustration purpose here, we assume

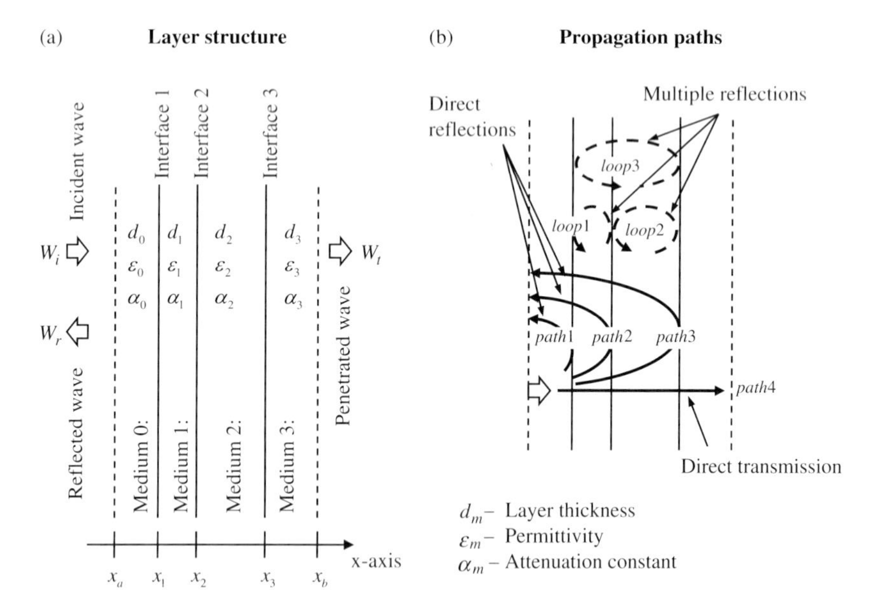

Figure 4.83 Plane wave propagation in stratified media. In the shown example, the wave enters the structure from the left side.

frequency-independent material parameters. Following our example, the sensor signals refer to position x_A and x_B. The media in which the sensors are placed (medium 0 and 3) are the reference media; hence, their parameters should be known.

The wave propagation within this structure is affected by partial reflections at every interface, and within the homogenous sections we have common propagation causing delay and attenuation. As Figure 4.83b illustrates, we have to deal with waves which take the shortest way (labelled by *path i*) and the other ones are trapped for a while in a loop oscillating between two or more interfaces (labelled as *loop j* and *loop ij*).

A detailed analysis of the wave behavior of such a layered structure can be done via scattering parameters. Graphically they are shown in Figure 4.84. Meaning and definition of the involved parameters are summarized in Table 4.4 and (4.290)–(4.292).

The wave propagation within the homogenous section m is assigned by T_{Mm} respecting attenuation by the factor $\mathrm{e}^{-\alpha_m d_m} = \mathrm{e}^{-a_m}$ and propagation delay of $\tau_m = d_m / c_m$. At every interface, the wave splits into two parts. One part travels back – this is the reflected wave (reflection coefficient $R_{0,m,m-1}$ for propagation left to right and $R_{0,m-1,m}$ for the reverse direction where $R_{0,m,m-1} = -R_{0,m-1,m}$) – and another part penetrates into the second medium – this is the transmitted wave (transmission coefficient $T_{0,m,m-1} = T_{0,m-1,m}$ which does not depend on propagation direction). The latter one is also called refracted wave which will be evident by considering oblique incidence (see Section 5.4 for details). For the simple case of normal incidence,

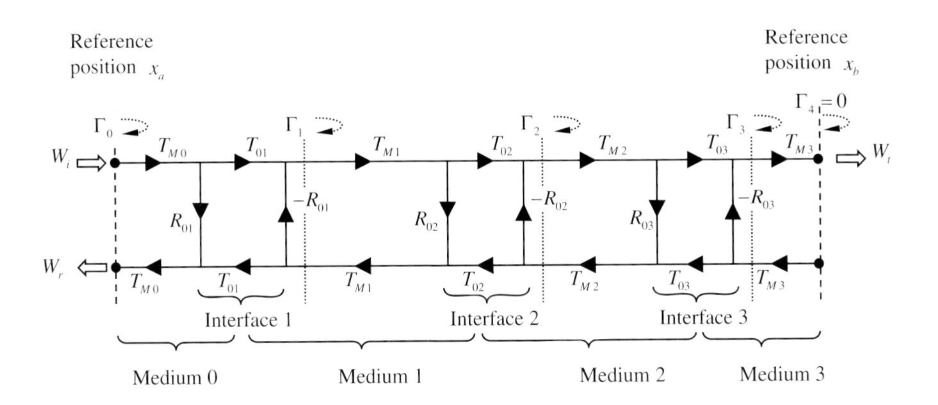

*path*1: $P_1 = T_{M0}^2 R_{01}$ *loop*1: $L_1^{(1)} = -T_{M1}^2 R_{01} R_{02}$

*path*2: $P_2 = T_{M0}^2 T_{01}^2 T_{M1}^2 R_{02}$ *loop*2: $L_2^{(1)} = -T_{M2}^2 R_{02} R_{03}$

*path*3: $P_3 = T_{M0}^2 T_{01}^2 T_{M1}^2 T_{02}^2 T_{M2}^2 R_{03}$ *loop*3: $L_3^{(1)} = -T_{M1}^2 T_{M2}^2 R_{01} R_{03} T_{02}^2$

*path*4: $P_4 = T_{M0} T_{01} T_{M1} T_{02} T_{M2} T_{03} T_{M3}$ *loop*12: $L_1^{(2)} = L_1^{(1)} L_2^{(1)} = T_{M1}^2 T_{M2}^2 R_{01} R_{02}^2 R_{03}$

Figure 4.84 Mason graph of the 4-layer media. The indicated waves refer to the incidence from left. It is supposed that the lenses do not reflect or their reflections are gated out from the considered data.

Table 4.4 Wave parameters for the structure given in Figure 4.83a and normal incidence of the sounding wave.

	Propagation left to right (medium $(m-1) \rightarrow$ medium (m)	Propagation right to left (medium $(m) \rightarrow$ medium $(m-1)$
Transmission within the layer	$\underline{T}_{Mm} = \mathrm{e}^{-\alpha_m d_m}\,\mathrm{e}^{-j2\pi f \frac{d_m}{c_m}} = \mathrm{e}^{-j(2\pi f \tau_m - ja_m)}$;	$a_m = \alpha_m d_m, \quad \tau_m = \frac{d_m}{c_m}$
Refraction index	$\eta_{m,m-1} = \eta_n = \sqrt{\frac{\varepsilon_m}{\varepsilon_{m-1}}} = \frac{\eta_{m0}}{\eta_{m-1.0}}$	$\eta_{m-1,m} = \eta_n^{-1} = \sqrt{\frac{\varepsilon_{m-1}}{\varepsilon_m}}$
Reflection at interface (n)	$R_{0,m,m-1} = R_{0n} = R_0(x_n) = \frac{1-\eta_n}{1+\eta_n}$	$R_{0,m-1,m} = -R_{0n} = -R_0(x_n) = \frac{\eta_n - 1}{\eta_n + 1}$
Transmission at interface (n)	$T_{0,m,m-1} = T_{0,m-1,m} = T_{0n} = T_0(x_n) = \frac{2\sqrt{\eta_n}}{\eta_n+1} = \sqrt{1 - R_{0n}^2}$	

The index m refers to the layers and the index n counts to the interfaces (in case the material parameters are frequency dependent, the indicated formula may be applied only for frequency domain descriptions).

the reflection R_0 and transmission coefficient T_0 are given by the same relation as for an impedance step of a transmission line (see Annex B.8). Since we suppose non-magnetic material, their values only depend on the permittivity ε_m.

In the case of infinitely extended layers, propagation speed, wave impedance, refraction index and permittivity are directly linked by the following relations (see Section 5.4):

$$\text{Refraction index}: \quad \eta_{m0} = \sqrt{\frac{\varepsilon_m}{\varepsilon_0}} \tag{4.290}$$

$$\text{Propagation speed}: \quad c_m = \frac{1}{\sqrt{\mu_0 \varepsilon_m}} = \frac{c_0}{\eta_{m0}} \tag{4.291}$$

$$\text{Intrinsic impedance}: \quad Z_m = \sqrt{\frac{\mu_0}{\varepsilon_m}} = \frac{Z_0}{\eta_m} \tag{4.292}$$

with c_0 and Z_0 propagation speed and intrinsic impedance of free space, respectively.

From the Mason graph in Figure 4.84, the waves at the reference locations x_a and x_b can be calculated via the Mason rule as mentioned in Annex B.7. Using the substitutions introduced in Figure 4.84, we yield for the reflected wave

$$\underline{W}_{\mathrm{r}}(f) = \frac{P_1 \cdot (1 - L_1^{(1)} - L_2^{(1)} - L_3^{(1)} + L_1^{(2)}) + P_2 \cdot (1 - L_2^{(1)}) + P_3}{1 - L_1^{(1)} - L_2^{(1)} - L_3^{(1)} + L_1^{(2)}} \underline{W}_{i}(f) \tag{4.293}$$

and for the transmitted wave

$$\underline{W}_{\mathrm{t}}(f) = \frac{P_4}{1 - L_1^{(1)} - L_2^{(1)} - L_3^{(1)} + L_1^{(2)}} \underline{W}_{i}(f) \tag{4.294}$$

Obviously the Mason formula becomes bulky if the number of layers increases.

For simulation purposes, we can establish a simpler recursive formula for the reflected waves. If we are looking, for example, at the overall reflection on the left-hand side, we introduce virtual reflection coefficients Γ_n for every interface as depicted in Figure 4.84. These coefficients merge all reflections occurring on the right side of interface n. Thus, by starting at the right end of the structure with $\Gamma_4 = 0$, we can estimate the reflection coefficient of the next layer towards the left hand by

$$\underline{\Gamma}_{n-1} = \underline{T}^2_{M,n-1}\left(R_{0,n} + \frac{T^2_{0,n}\underline{\Gamma}_n}{1 + R_{0,n}\underline{\Gamma}_n}\right) \quad \text{with} \quad n = N \cdots 1; \quad \Gamma_N = 0$$
$$\underline{W}_r = \underline{\Gamma}_0 \underline{W}_i \tag{4.295}$$

Of course, the same procedure but with reversed direction can be used if one looks from the opposite side into the structure.

(4.293)–(4.295) are only applicable for frequency domain notations. The related time domain notations can be deduced applying the identity $(1-x)^{-1} = \sum_{n=0} x^n$ and replacing all products of FRFs by the convolution of related IRFs (see Annex B, Table B.3 or B.6) which may however be quite laborious.

In the case of frequency-independent material parameters, the involved convolutions reduce to multiplications by factors and time shift. Under this condition, the reflected and transmitted signal in time domain may be directly deduced from the Mason graph if we simply pursue all possible tracks (including the multiple run through loops) from node A to node B. Figure 4.85 gives an ideal example for the first few components of the impulse and step response functions for reflection and transmission. In the case of cable sensors (Figure 4.82b), it is usual to apply the step response function since it approximately reflects the impedance profile of the cable sections if multiple reflections are not too dominant.

The signal components assigned by P_i in the IRF graphs refer to the direct propagation paths. They represent often the strongest components while signals which are affected by multiple reflections quickly die out. The decay rate of multiple reflections by a first-order loop is given by

$$d_k^{(1)}[\mathrm{dB/s}] = \frac{20\log\left|\underline{L}_k^{(1)}\right|}{\tau_k} \tag{4.296}$$

Herein, $20\log\left|\underline{L}_k^{(1)}\right|$ is the gain in dB of the loop k and τ_k is its roundtrip time. For higher-order loops, the corresponding sums have to be applied in (4.296). Hence, multiple reflections die out quickly out for low interface mismatch and thin layers.

Finally, the lattice diagram as depicted in Figure 4.86 provides another approach to illustrate graphically the wave propagation within a multi-layer arrangement. It allows constructing the impulse response function of the transmitted and reflected wave from the layer geometry. Every arrow represents a wave component. Its

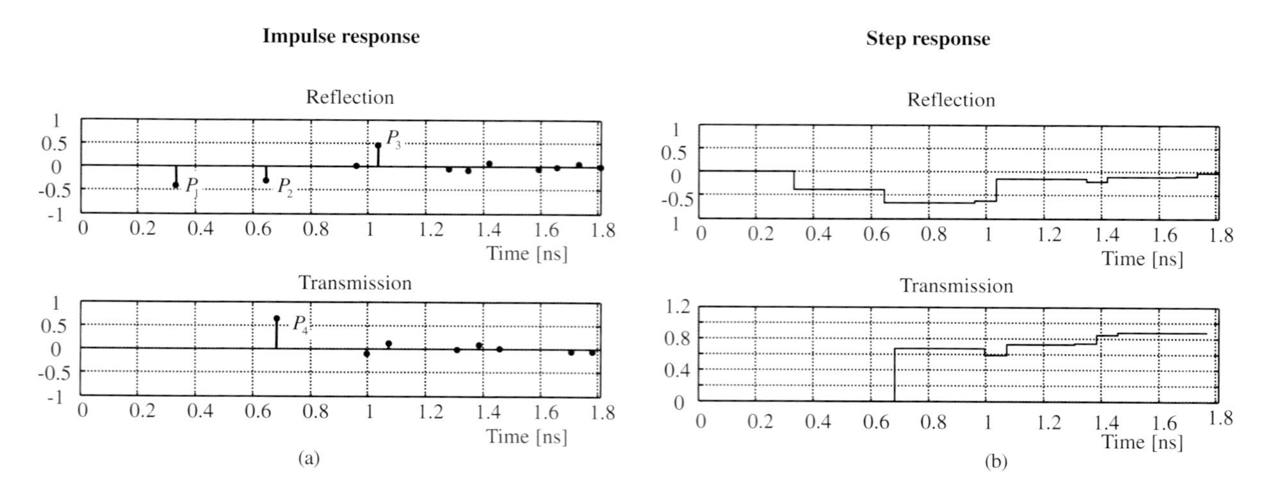

Figure 4.85 Simulated impulse response (a) and step response (b) of reflection and transmission path of a 4-layer structure: 5 cm air; 2.1 cm layer of $\varepsilon_2 = 5$; 1.3 cm layer of $\varepsilon_3 = 20$; 5 cm air; the layer material was expected to be lossless. Note the step response of the reflection signal only depicts the return signal.

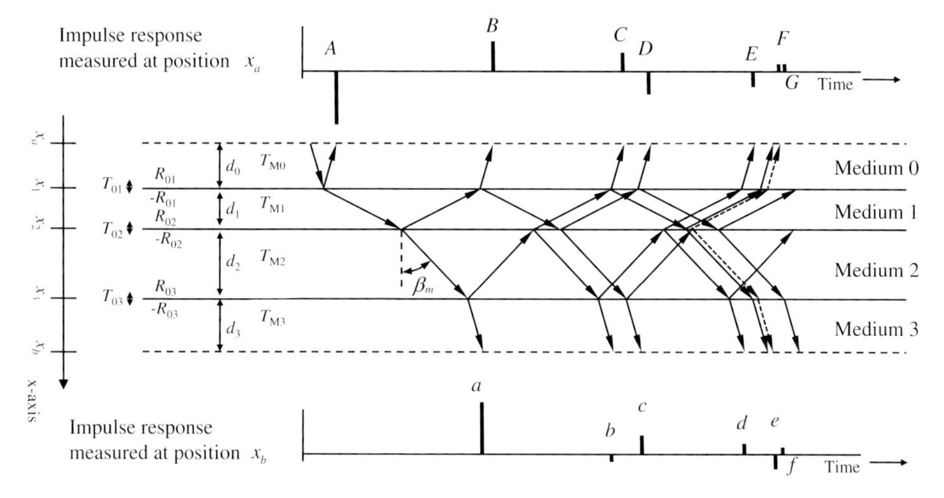

Figure 4.86 Lattice diagram for a 4-layer sandwich structure. The lattice diagram refers to normally incident wave. The slope of the different signal components is due to the propagation speed within the corresponding medium. Note, the signals F, G and e, f are plotted separately to illustrate the different propagation paths. But actually, these signals coincide in propagation time.

slope is proportional to the propagation speed within considered layer ($\tan \beta_n = c_n$). Magnitude and propagation delay of the first few signal components are summarized in Table 4.5.

Table 4.5 The first few signal components for reflection and transmission mode gained from Figure 4.86.

Wave	Magnitude	Propagation delay	Paths and loops
A	$R_{01}\, T_{M0}^2$	$2\,\tau_0$	P_1
B	$R_{02}\, T_{M0}^2\, T_{M1}^2\, T_{01}^2$	$2(\tau_0+\tau_1)$	P_2
C	$R_{03}\, T_{M0}^2\, T_{M1}^2\, T_{M2}^2\, T_{01}^2\, T_{02}^2$	$2(\tau_0+\tau_1+\tau_2)$	P_3
D	$-R_{01}\, R_{02}^2\, T_{M0}^2\, T_{M1}^4\, T_{01}^2$	$2(\tau_0+2\,\tau_1)$	$P_2\, L_1^{(1)}$
E	$-R_{02}\, R_{03}^2\, T_{M0}^2\, T_{M1}^2\, T_{M2}^4\, T_{01}^2\, T_{02}^2$	$2(\tau_0+\tau_1+2\,\tau_2)$	$P_3\, L_2^{(1)}$
$F+G$	$-2R_{01}\, R_{02}\, R_{03}\, T_{M0}^2\, T_{M1}^4\, T_{M2}^2\, T_{01}^2\, T_{02}^2$	$2(\tau_0+2\,\tau_1+\tau_2)$	$2\,P_3\, L_1^{(1)}$
	......	......	
a	$T_{M0}\, T_{M1}\, T_{M2}\, T_{M3}\, T_{01}\, T_{02}\, T_{03}$	$\tau_0+\tau_1+\tau_2+\tau_3$	P_4
b	$-R_{02}\, R_{03}\, T_{M0}\, T_{M1}\, T_{M2}^3\, T_{M3}\, T_{01}\, T_{02}\, T_{03}$	$\tau_0+\tau_1+3\,\tau_2+\tau_3$	$P_4\, L_2^{(1)}$
c	$-R_{01}\, R_{02}\, T_{M0}\, T_{M1}^3\, T_{M2}\, T_{M3}\, T_{01}\, T_{02}\, T_{03}$	$\tau_0+3\,\tau_1+\tau_2+\tau_3$	$P_4\, L_1^{(1)}$
d	$R_{02}^2\, R_{03}^2\, T_{M0}\, T_{M1}\, T_{M2}^5\, T_{M3}\, T_{01}\, T_{02}\, T_{03}$	$\tau_0+\tau_1+5\,\tau_2+\tau_3$	$P_4\, L_2^{(1)}\, L_2^{(1)}$
$e+f$	$R_{01}\, R_{03}\, T_{M0}\, T_{M1}^3\, T_{M2}^3\, T_{M3}\, T_{01}\, T_{02}$ $T_{03}\left(R_{02}^2-T_{02}^2\right)$	$\tau_0+3\,\tau_1+3\,\tau_2+\tau_3$	$P_4\left(L_1^{(2)}+L_3^{(1)}\right)$
	...	...	...

4.9.2 Reconstruction of Coplanar Layer Structure

The typical measurement tasks for the considered stratified media are to determine the number of layers or/and their thickness or/and their permittivity. Some basic approaches are summarized as follows.

Layer peeling: This approach works solely with the reflection signal from a single side. It can be applied if the propagation loss within the layers is negligible. We suppose a sounding signal (i.e. sub-nanosecond pulse, step function of short rise time or wideband signal of short coherence time) whose characteristic temporal width is shorter than the shortest propagation time of the layers. Further, we require that the parameters of medium 0 are known so that it can be used for reference.

The layer peeling algorithm works in a recursive manner penetrating step by step into the structure (see also [100, 101]). The steps are as follows (refer to Figure 4.86 and Table 4.5):

- **Step 1:** The first peak A in the pulse response provides the reflection coefficient R_{01} of the medium 1, so one can calculate its permittivity ε_1 (but only if the permittivity of medium 0 is known) and therefore one also knows its propagation speed $c_1 = c_0/\sqrt{\varepsilon_1}$ (see Table 4.4 for corresponding equations).
- **Step 2:** The round-trip time of the second pulse B results from the layer thickness d_1 and its amplitude gives us the reflection coefficient R_{02} of interface 2 after removing the transfer loss $T_{01}^2 = \left(1 - R_{01}^2\right)$ of the first interface. Now, we are able to calculate the wave speed c_2 in medium 2.
- **Step 3:** From the known reflection coefficients and layer thickness, the multiple reflections (related to the loop $L_1^{(1)}$) can be predicted. So it will be possible to identify if the next peak in the impulse response is caused by a new interface or by a wave which was trapped in an already known layer.
- **Step 4:** If the third pulse comes from multiple reflections, it will be disregarded. If not, the reflection coefficient R_{03} can be determined if one respects the transfer losses $T_{01}^2\, T_{02}^2$. We can conclude from the third reflex the thickness of layer 3 and its dielectric contrast to layer 4 and so forth.

The layer-peeling algorithm must be done with care since every removed layer will add additional flaws in the remaining data due to the inevitable measurement errors. In case of multi-layer structures, appropriate regularization techniques have to be applied in order to avoid noise accumulation. A simple method for wall parameter estimation (three-layer structure) under realistic conditions needed for through-wall tracking is described in Ref. [102]. Permittivity measurements of wall materials are found in Ref. [103].

The layer-peeling method described above assumes that the sounding signal is settled before it arrives at the next impedance step of a new layer. If this cannot be guaranteed, one has to combine layer peeling with the SAGE approach (refer to Figure 4.51) which also respects the actual shape of the sounding signal.

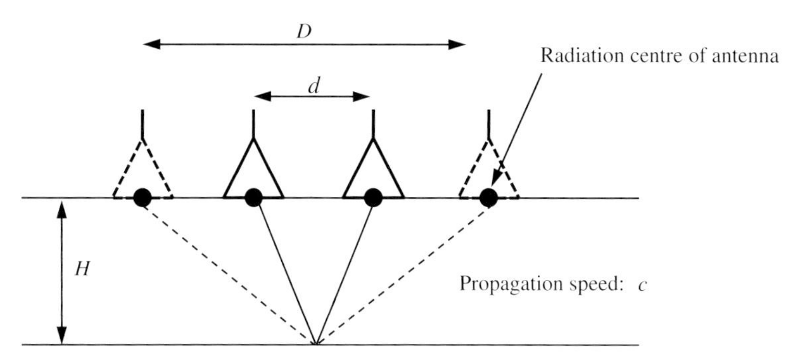

Figure 4.87 Common midpoint method for determination of layer thickness and propagation speed evaluation.

Common midpoint method: In the case the permittivity and propagation speed of the first layer are not known, the common midpoint method can be applied. It is depicted in Figure 4.87. The method supposes spherical waves. One measures the travelling time τ_d and τ_D from one antenna to the second at different antenna distances d, D. If the reflection point for both measurements is the same, layer thickness and propagation speed result from

$$H = \frac{1}{2}\sqrt{\frac{(\tau_d D)^2 - (\tau_D d)^2}{\tau_D^2 - \tau_d^2}} \tag{4.297}$$

$$c = \sqrt{\frac{D^2 - d^2}{\tau_D^2 - \tau_d^2}} \tag{4.298}$$

These equations require that the radiation centres of the antennas are placed on the first boundary. If the necessary reflection is provided by a point scatterer or an inclined boundary, the method has to be repeated at different locations in order to determine additionally either the actual position of the point scatterer or the gradient of the second boundary.

Continuous permittivity profile: We will suppose again a lossless propagation medium. But now, the test volume is not separated by sharp interfaces between two adjacent substances; rather, we have to deal with a continuous variation of the permittivity. Hence, we cannot observe individual reflection peaks but a distributed reflection.

For the sake of simplicity, we will assume that the structure under test is stimulated by a Dirac pulse and we capture the reflected signal $W_r(n)$ at a sampling rate $f_s = \Delta t_s^{-1}$. Further, we decompose the test volume in a number of slices. The thickness of the slice number m is chosen in such a way that $d_m = \tau_m c_m = \Delta t_s c_m/2$. The reflection coefficients $|R_{0,m-1,m}| = |R_{0n}| \ll 1$ of the 'virtual slice interfaces' will be very small wherefore we can approximate the transmission coefficient by $T_{0,m-1,m}^2 = T_{0n}^2 \approx 1 - R_{0n}^2/2 \approx 1$ and the loops may be neglected since they also involve second-order terms or higher. Hence, the amplitude of the captured signal sample $W_r(n)$ is directly proportional to the reflection coefficient of the 'slice interfaces' of number n. Using (c_m is the propagation speed within the mth virtual layer)

$$W_{rn} = R_{0n} = R_0(x_n) = \frac{1-\eta_n}{1+\eta_n} = \frac{c_m - c_{m-1}}{c_m + c_{m-1}} \tag{4.299}$$

we can iteratively reconstruct the permittivity profile by

$$\begin{aligned} c_m &= c_{m-1}\frac{1+W_r(m)}{1-W_r(m)}; \quad m \geq 1 \\ x_m &= x_{m-1} + \frac{c_{m-1}\Delta t_s}{2} \end{aligned} \tag{4.300}$$

Herein, it is supposed that the sounding wave is launched within a medium of known properties so that we can establish the starting values $c_{m-1} = c_0$ and $x_{m-1} = x_a$ for $m = 1$.

For oblique incidence of the sounding wave, this simple model cannot be applied since the ray path continuously changes its direction. The interested reader can find corresponding considerations in Ref. [104].

Measurement from both sides: In the general case, the assumptions of negligible propagation losses and small reflection coefficients cannot be maintained. Thus, it will be increasingly difficult to provide enough data from a single reflection measurement to unambiguously reconstruct the layer structure. This can be seen from Table 4.5 by trying to extract all reflection and absorption constants exclusively from reflection data. In our example of lossless propagation, only three signal components (A, B, C) were sufficient to reconstruct the layers. With losses, one has to resort to multiple reflections too, which will be increasingly affected by measurement errors. Under these constraints it could be helpful to measure the structure from both sides which provides two reflections and one transmission function. So finally, we can deal with up to three independent measurement signals. The extraction of the wanted information from the measured data will be simplified which can easily be proofed from Table 4.5 and statistical confidence of the results will be improved.

Matching a layer model: The most powerful but also computationally expensive method is based on matching a model of the layer structure to the measured data. A simple layer model has the form as depicted in (4.293), (4.294) or (4.295). Basically, these models may be compiled in the time or frequency domain where frequency domain models are simpler to handle and they allow straightforward extensions for frequency-dependent layer materials. Further model extensions are feasible applying more realistic sounding waves – for example plane waves may be replaced by waves from a dipole – or non-planar target structures (see [14, 16, 17]).

The layer model represents a mathematical description of the reflected or/and transmitted waves which is most often formulated in the frequency domain.

Such model may include various types of parameters such as layer thickness, permittivity values, conductivity values, relaxation times and so on of the layer materials. We join all these parameters in the vector $\mathbf{\Theta}$. If all parameters would be known, we could calculate how the test object behaves if it is stimulated by an electromagnetic wave. Hence, in the best case, we could determine three response signals:

Reflection at position x_a : $\underline{\hat{W}}_{ra}(f,\mathbf{\Theta})$

Reflection at position x_b : $\underline{\hat{W}}_{rb}(f,\mathbf{\Theta})$

Transmission : $\underline{\hat{W}}_{t}(f,\mathbf{\Theta})$

Since we will map with our model the reality, these signals should be identical with corresponding measurements $\underline{W}_{ra}(f), \underline{W}_{rb}(f), \underline{W}_{t}(f)$. In other words, if there is a mismatch between model and measurement, the parameter set $\mathbf{\Theta}$ is not yet optimally selected. In order to quantify this mismatch, we establish cost functions expressing globally the difference between measurement and model. In many cases these cost functions are the L_2-norm of the modelling error:

$$
\begin{aligned}
J_{ra}(\mathbf{\Theta}) &= \int_0^\infty \left|\underline{W}_{ra}(f) - \underline{\hat{W}}_{ra}(f,\mathbf{\Theta})\right|^2 \mathrm{d}f \\
J_{rb}(\mathbf{P}) &= \int_0^\infty \left|\underline{W}_{rb}(f) - \underline{\hat{W}}_{rb}(f,\mathbf{\Theta})\right|^2 \mathrm{d}f \\
J_{t}(\mathbf{\Theta}) &= \int_0^\infty \left|\underline{W}_{t}(f) - \underline{\hat{W}}_{t}(f,\mathbf{\Theta})\right|^2 \mathrm{d}f
\end{aligned}
\tag{4.301}
$$

Hence, the best match of model and reality is given for such parameters $\mathbf{\Theta}$ which globally minimize all cost functions

$$
\mathbf{\Theta} = \underset{\mathbf{\Theta}}{\arg\min}\left[J_{ra}(\mathbf{\Theta}), J_{rb}(\mathbf{\Theta}), J_{t}(\mathbf{\Theta})\right] \tag{4.302}
$$

leading to a multi-dimensional multi-object optimization. But often, the procedure can be simplified to the minimum search of a single cost function

$$
\mathbf{\Theta} = \underset{\mathbf{\Theta}}{\arg\min}\left(J_{ra}(\mathbf{O}) + J_{ra}(\mathbf{\Theta}) + \eta J_{t}(\mathbf{\Theta})\right) \tag{4.303}
$$

in which transmission and reflection behavior may be differently weighted, that is $\eta \neq 1$.

Basically, one can apply in (4.301) also other L_p-norms if appropriate but the L_2-norm has the advantage that it often permits to derive an exact expression for the gradient of the cost functions (an example is given in Annex 7.6

'Last Square Normal Equation'). If the minimum of the cost function cannot be determined analytically, one can access a toolbox for algorithms of minimum search (gradient methods, Monte Carlo method, genetic approaches, particle swarm optimization and others) and the user can select the best algorithms for his problem.

It is interesting to note that the search algorithms do not need necessarily complete measurement data; that is, it is often sufficient to capture only amplitude values but not the phase and/or to measure only one of the three signals $\underline{W}_{ra}(f), \underline{W}_{rb}(f), \underline{W}_{t}(f)$ and/or to perform the measurement only over a limited bandwidth. However, the higher the diversity of the captured data, the less will be the ambiguity of minimum search. Two examples of such approach can be found in Refs [105, 106] which deal with the time domain and [107] gives a frequency domain example.

4.10 Ultra-Wideband Short-Range Imaging

4.10.1 Introduction

The term 'image' or 'imaging' is mainly occupied by the visual sense or visual comprehension of human beings. Physically it is nothing but the mapping of spatially distributed properties of matter onto a graphical representation understandable by a human being. This involves an interaction of the matter to be imaged (localized at a certain position in space, say $\mathbf{r}_T$) with a sensor device localized at a different position (say $\mathbf{r}_R$). The sensor must be able to capture some sort of energy emanating from this object and assign it to the space point $\mathbf{r}_T$. Without such energy transfer, the matter could not be registered.

The simplest imaging problem (at least from the viewpoint of basic comprehension) is the determination of the spatial distribution of point objects which radiate by themselves energy (we call it point radiator) that can be captured by the sensor elements over a large distance. Such sensor elements are, for example, the rods and the cones of the retina in case of human beings, pn-junctions in case of a CCD camera or an antenna array in case of microwave imaging. Here, we are solely interested in the last type of sensors.

Apart from thermally induced radiation, the usual objects of interest behave passively; that is, one can only notice them if they are stimulated to 'radiate' by an external illumination via an auxiliary radiator. This type of radiation is typically assigned as scattering which shows a typical behavior depending on the geometry and material composition of the object. Hence with respect to imaging, we can deal with passive objects as radiating sources if they are sufficiently illuminated. The type of illumination (wavelength/coherence length, polarization, angle of incidence etc.) will force specific interaction phenomena between sounding wave and objects so that particular properties of the targets can be emphasized in the captured image. Furthermore, the active illumination of the scenario permits the construction of coherent sensor devices (commonly assigned as radar imaging), allowing exact roundtrip time measurement due to the synchronization between illuminator and receiver. Thus, distance information about the imaged objects can be gained directly and must not be determined from angular relations of stereoscopic images (resulting from non-coherent sensors as, e.g., the human eyes). Coherent imaging systems permit the construction of real 3D-images by keeping the range resolution independent of the object range.

We will first illustrate the basic idea of short-range imaging by performing a simple gedankenexperiment leading us to various imaging procedures and to some restrictions of technically implementable imaging systems. All our discussions here will be restricted to scalar fields and dispersion-free propagation knowing that this is not actually valid for electromagnetic fields. It will however simplify the considerations and keep us concentrated on basic imaging issues. Basics on vector fields and dispersion will be introduced in Chapter 5, so that the reader finally can also appreciate their influence on the imaging performance. A short summary of achievements, difficulties and future challenges of microwave imaging can be found in Ref. [108].

4.10.2 The Basic Method of Short-Range Imaging

We will start our consideration with an elementary problem. The goal is to localize (visualize) the spatial distribution of (non-interacting) point radiators within a given space region V_{obs}. For that purpose, we have one sensor available which is placed at position $\mathbf{r}_R$. The space should be isotropic and homogeneous; that is, every point source radiates an isotropic wave having an intensity which does not depend on the propagation direction.[26)]

We are looking for an idealized arrangement (we are not yet interested in a practical solution) which scans the volume of interest and allows us to collect coherently the wave energy from only one source at which our arrangement is focused. A spheroid of a reflecting surface having its focal points at the source $\mathbf{r}_T$ and the sensor position $\mathbf{r}_R$ is such an ideal arrangement which guides the waves emanating from the source at $\mathbf{r}_T$ completely towards the sensor position $\mathbf{r}_R$ where the energy must only be captured (see Figure 4.88).

For imaging purposes, shape and orientation of the spheroid have to be varied so that the focal point $\mathbf{r}_T$ moves through the observation volume V_{obs} (region of interest) while the second one remains at sensor position $\mathbf{r}_R$. If the moving focal point coincides with a radiation source, the sensor will indicate a large incident power since all waves will coherently add in point $\mathbf{r}_R$. In all other cases, the sensor will only register very weak wave energy since no accumulation happens. This is due to

26) This assumption is widely used in microwave imaging but not strictly correct. As we will see in the next chapter, point sources (e.g. electrically short dipoles) of vector fields (i.e. electric or magnetic field) provide spherical waves of angular dependent strength. The orientation of the field forces is angular dependent too and it even depends on the distance by approaching the antenna (see Chapter 5).

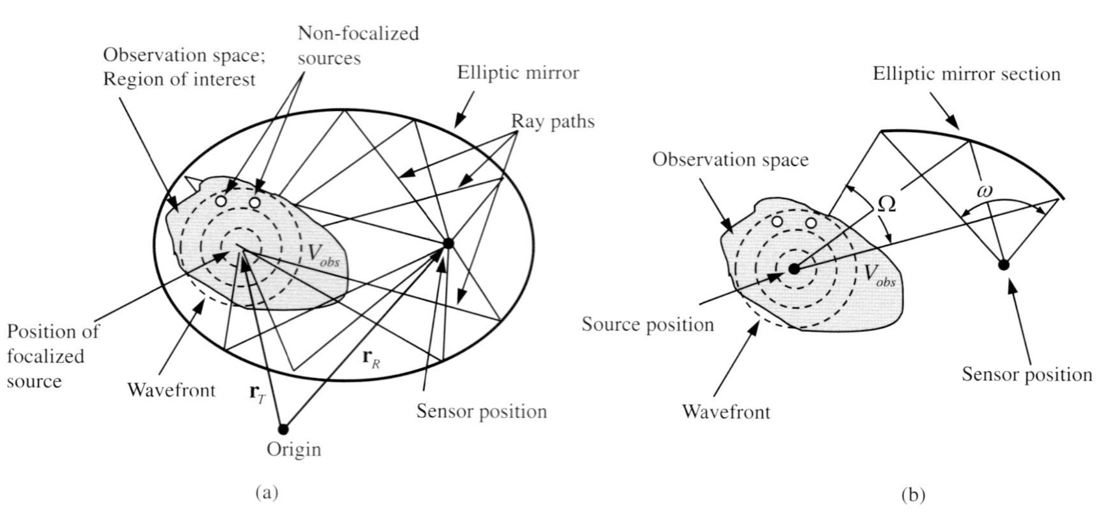

Figure 4.88 Complete (a) and partial elliptic mirror (b) to focus the wave energy of a point source at the sensor position.

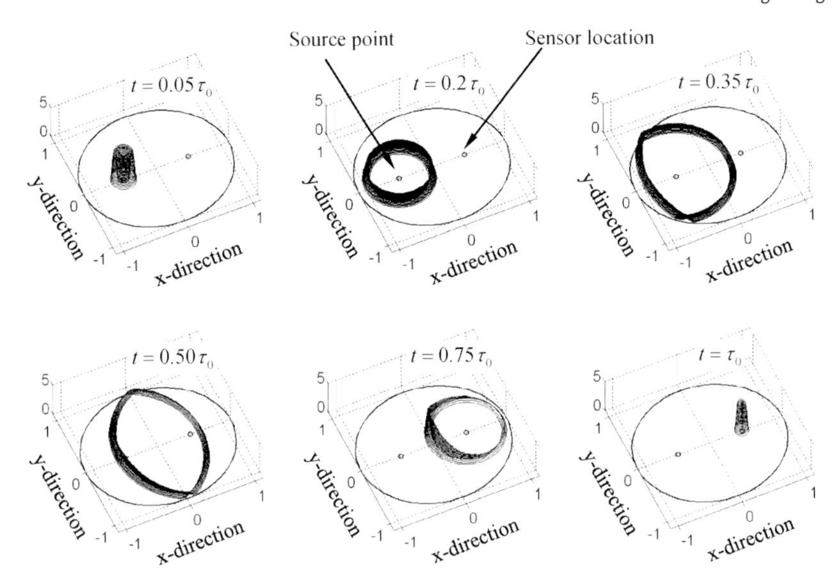

Figure 4.89 Six snapshots for pulse wave propagation within a closed elliptic mirror. Note that the simulated Gaussian-shaped wave is only for illustration. It cannot propagate in free space due to its non-zero DC component. (movie of wave propagation at Wiley homepage: Fig_4_89*)

the properties of a spheroid that rays emanating from one focal point are reflected towards the second one and the length of all rays is $2a$ where a is the length of the major axis (see Annex A.7 for details on ellipse and spheroid). Hence, the propagation time from the source to the sensor is $\tau_0 = 2a/c$. For increasing distance between source and sensor position, the elliptic mirror tends to a parabolic one.

Figures 4.89 and 4.90 illustrate the wave propagation within the spheroid for a pulsed and a sinusoidal wave. The waves collapse for both cases in the second focal point leading to a dramatically rising magnitude since the whole energy is

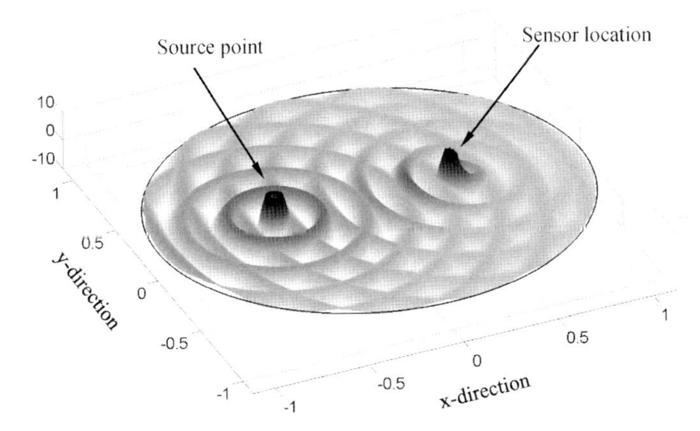

Figure 4.90 Time shot of total field within the spheroid for a sine wave source.

concentrated in this spot. If we place the sensor there, it is able to collect completely the radiated wave energy. The waves move towards the sensor with equal strength from all sides so that we have no preference of any direction. Hence, this arrangement is optimal with respect to sensitivity and spatial resolution.

As we can observe from both figures, imaging (i.e. localization of the point source) works for narrowband and wideband signals as well. But there is an important difference. For sine wave sources, we only need a simple power sensor at position $\mathbf{r}_R$ which records the permanently incoming waves. However, if the source emits a pulse wave, we need to observe the time shape of the cumulated waves in order to be able to register the maximum. On the one hand, this will complicate the sensor electronics since devices of large bandwidth are needed, but on the other hand, it opens up additional sensing capabilities since the travelling time τ_0 may also be estimated[27] providing additional information about the scenario under test. Imaging of wideband CW sources may deal with both approaches; that is, simple power sensor would be sufficient to collect the superimposing waves. But one can drastically increase the sensitivity by collecting the time shape of the collapsing wave at position $\mathbf{r}_R$ and performing a cross-correlation with the source signal (if possible). In other words, the elliptic mirror leads to a spatial concentration of the point source energy and the correlation (matched filtering) provides a temporal energy concentration. Both processes lead to optimal conditions for noise suppression and image resolution.

It is comprehensible that the typical spot size of the collapsing wave at sensor position $\mathbf{r}_R$ will relate to the spatial extension of the sounding waves which are given by the frequency f_0, the pulse width t_w or the coherence time τ_{coh} depending on the type of the radiated signal. Hence, the spatial resolution, that is the voxel size $\delta_V \approx \delta_x \delta_y \delta_z$ of the 3D-image, may roughly be expressed by

$$\delta_V \approx \delta_x \delta_y \delta_z \approx \begin{cases} \left(\dfrac{c}{f_0}\right)^3 = \lambda^3 & \text{sine wave} \\ (t_w c)^3 \approx \left(\dfrac{c}{\ddot{B}_3}\right)^3 & \text{pulse wave} \\ (\tau_{coh} c)^3 \approx \left(\dfrac{c}{\ddot{B}_3}\right)^3 & \text{UWB-CW wave} \end{cases} \tag{4.304}$$

Note that (4.304) only involves wave concentration by the elliptic mirror and does not refer to round-trip time measurements.

The complete enclosure of the volume under investigation by elliptic mirrors with controllable focal points is neither practical nor feasible. As already indicated in Figure 4.88, it will be only possible to capture the wave energy within a limited solid angle in most of the practical cases. This will lead to some degradation of the imaging performance as illustrated in Figure 4.91. It represents the spatial energy distribution of the collapsed waves for six mirrors of different size. The spatial

27) This will assume the knowledge of the time point when the wave was launched by the source. In case of independent radio transmitters, this will hardly be possible since they are typically not synchronized with the sensor. If we however illuminate the scene by ourselves as in case of radar measurements, we are able to establish the required timing reference.

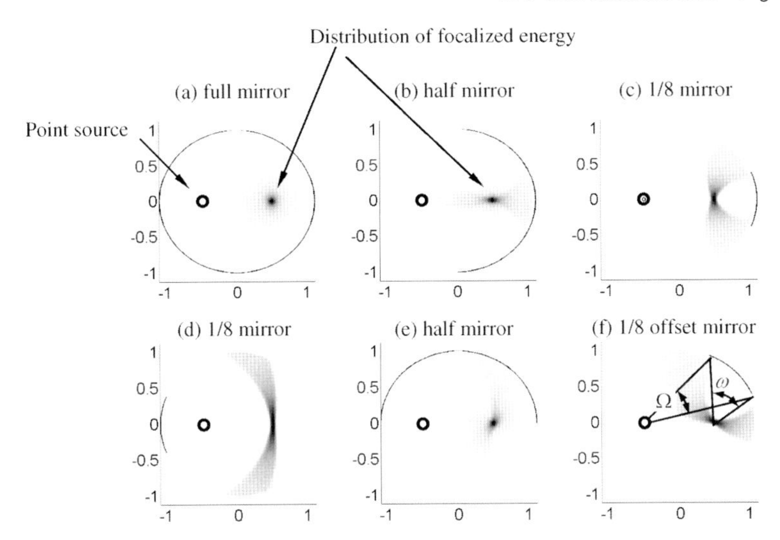

Figure 4.91 Point spread function of partial elliptic mirrors for Gaussian-shaped source signal (compare Figure 4.89). (colored figure at Wiley homepage: Fig_4_91).

distribution of the focused energy is assigned as point spread function$\Upsilon(x, y, z)$ of the related mirror (see (4.311) for definition).

The volume $\delta_v = \delta_x \, \delta_y \, \delta_z$ for which holds $\Upsilon(x_R \pm \delta_x, y_R \pm \delta_y, z_R \pm \delta_z) \geq \Upsilon(x_R, y_R, z_R)/2$ defines the 3D-resolution (i.e. the voxel size) of the imaging system where $\mathbf{r}_R = \begin{bmatrix} x_R & y_R & z_R \end{bmatrix}^T$ is the sensor focal point. The point spread function should have its maximum there.

Ideally, if the source is entirely enclosed by the mirror (Figure 4.91a), the energy will be perfectly concentrated around the sensor spot and the point spread function will only depend on the sounding signal. The related image resolution is approximately given by (4.304). If the mirror occupies only a small sector of space, the focal spot blurs leading to images of lower resolution. We can observe that blurring aggravates to the same degree as the solid angle ω or Ω under which the sensor (or the source) 'sees' the mirror becomes smaller. We will give an estimate of image resolution in the next section. Further, we discover that the signal energy available at sensor position decreases proportionally to the solid angle Ω under which the source 'sees' the mirror. Hence, the image will be increasingly affected by noise if the mirror size is reduced.

4.10.3
Array-Based Imaging

We have learned from the previous section that we need to collect coherently the waves emanating from a selected point in space. In order to construct the image, the focal point has to move successively through the volume of interest. We used a shape-adaptable elliptic mirror for that purpose. But there are two further approaches which lead to comparable results. One of them uses a lens which

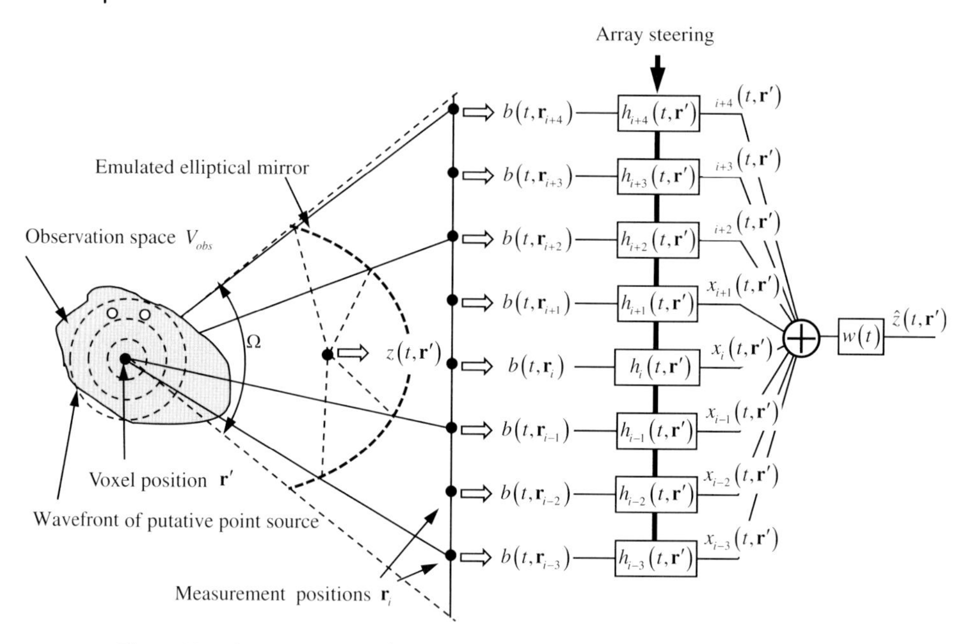

Figure 4.92 Antenna array emulating an elliptic mirror. Note that the array shape is not restricted to linear or planar arrangements. Basically the antennas may be distributed arbitrarily in space.

refracts and delays the incident waves in such a way that only partial waves coming from a source located at a specific position are accumulated at a related point behind the lens. By placing a power sensor in this point, one is able to register the radiation of that source. The observation of many radiation sources requires many sensors appropriately distributed behind the lens. This is how the human eyes work besides nearly all optical and infrared devices. Here, we will not engross this approach since lenses are impractical for ultra-wideband applications due to their large size and the related weight.

Rather, we will address a method widely used in modern radar systems as it is depicted in Figure 4.92. Here, we are aimed to emulate the behavior of the elliptic mirror by a set of receiving channels which we distribute over a certain (basically arbitrary) area. We call such arrangement antenna array. We assume that all antennas behave isotropically at least within these solid angles under which they 'watch' the region of interest. The receiving signals of all antennas are summed up to accumulate coherently the energy of the point source. However, neither the propagation time from the source to the individual antennas nor the amplitude of the incident waves corresponds to the conditions of the elliptic mirror. Hence, we have to equalize adequately the different receiving signals by time shifting and amplitude control. These corrections are individually done in every receiving channel by controllable steering filter having the IRF $h(t, \mathbf{r}', \mathbf{r}_i)$. The IRFs of these filters are individually adapted to the position $\mathbf{r}_i$ of antenna i and the focal point $\mathbf{r}'$. The image is gained by applying a whole set of IRFs which focus on all positions $\mathbf{r}'$ within the volume of interest $\mathbf{r}' \in V_{\mathrm{obs}}$. Supposing sufficiently dense arranged antennas and appropriately selected transfer functions $h(t, \mathbf{r}', \mathbf{r}_i)$, the array output signal $\hat{z}(t)$ and the signal $z(t)$ created by the virtual elliptic mirror would be the same as long as both arrangements cover identical solid angles Ω seen from the source.

We would like to focus the array at voxel position $\mathbf{r}' \in V_{\text{obs}}$. In order to gain equivalent signals $z(t, \mathbf{r}')$ and $\hat{z}(t, \mathbf{r}')$ (except an extra delay τ_0) from both arrangements, we have to provide in the summation point of the array the same conditions as we can find in the focal point of the elliptic mirror. Since for the antenna array the propagation path lengths are different for every antenna i, we have to compensate for the amplitude degradation and the delay. This is the task of the steering filters whose IRF we select as

$$h_i(t, \mathbf{r}') = h(t, \mathbf{r}', \mathbf{r}_i) = a(\mathbf{r}', \mathbf{r}_i)\delta\left(t - \tau_0 + \frac{|\mathbf{r}' - \mathbf{r}_i|}{c}\right); \quad i = 1 \cdots L \tag{4.305}$$

L is the number of array elements and τ_0 is the internal time delay of the array. It is required to keep causality. However, it is often omitted in many textbooks on array processing. $a(\mathbf{r}', \mathbf{r}_i)$ is a gain factor which compensates the spreading loss of the different propagation paths in the simplest case, that is $a(\mathbf{r}', \mathbf{r}_i) \sim |\mathbf{r}' - \mathbf{r}_i|$. If we convolve the antenna signals $b(t, \mathbf{r}_i)$ with these filter functions, array and elliptic mirror would behave equivalently.

Arranging all signals and filter functions in column vectors, we gain a compact notation of the array signal $\hat{z}(t, \mathbf{r}')$ applying dot convolution of signal and steering vector:

$$\hat{z}(t, \mathbf{r}') = w(t)\left(\mathbf{b}^T(t) * \mathbf{h}(t, \mathbf{r}')\right) = w(t)\sum_{i=1}^{L} x_i(t, \mathbf{r}') \tag{4.306}$$

$$\text{Signal vector}: \mathbf{b}(t) = \begin{bmatrix} b(t, \mathbf{r}_1) & b(t, \mathbf{r}_2) & \cdots & b(t, \mathbf{r}_i) & \cdots & b(t, \mathbf{r}_L) \end{bmatrix}^T$$

$$\text{Steering vector}: \mathbf{h}(t, \mathbf{r}') = [h_1(t, \mathbf{r}')\ h_2(t, \mathbf{r}') \cdots h_i(t, \mathbf{r}') \cdots h_L(t, \mathbf{r}')]^T$$

$w(t)$ is a window function which 'picks' of the filtered data $x_i(t, \mathbf{r}')$ at time τ_0. It is an important tool of side lobe suppression which cannot be applied in narrowband arrays due to the time-extended signals. Clearly for UWB-CW signals, the steering filters have to take over also the impulse compression in order to gain short signals which can be subjected to the window function. The effect of windowing will be obvious in the example below. The brightness of the image voxel located at position $\mathbf{r}'$ is finally given by the energy of the array signal:

$$I(\mathbf{r}') = \int \hat{z}^2(t, \mathbf{r}')\mathrm{d}t \tag{4.307}$$

The procedure described by (4.306) and (4.307) has to be repeated for all points within the observation volume V_{obs} in order to achieve the image.

Figure 4.93 illustrates the procedure for a simple scenario. It consist of two point sources placed at position $\mathbf{r}'_1$ and $\mathbf{r}'_2$ and a linear array which has equally distributed 21 antenna elements at positions $\mathbf{r}_i = [x_i \quad 0 \quad 0] \quad i = 1 \cdots 21;\ x_i \in [-1.5, 1.5]$. The signals captured by the 21 elements are depicted in the diagram Figure 4.93a.

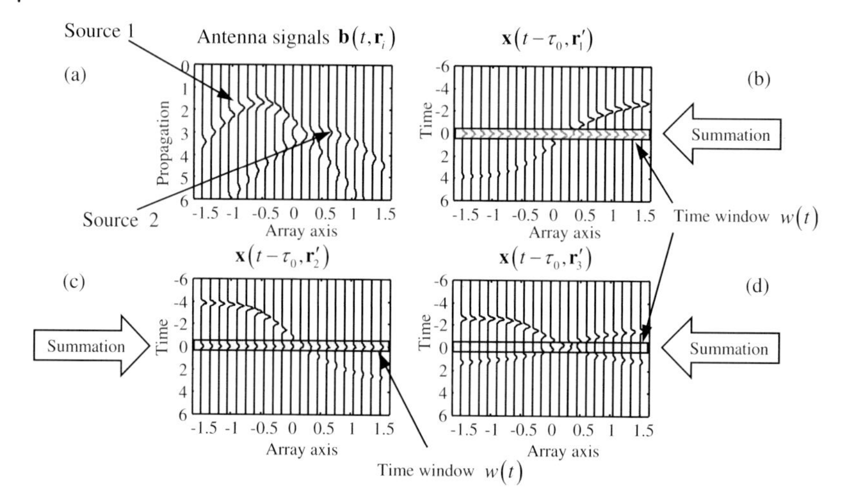

Figure 4.93 Illustration of (4.305) and (4.306) for three different focal points.

The two sources leave hyperbolic traces in the data which are caused by their different distances to the array elements.

In order to image the point $\mathbf{r}_1'$, all captured signals are 'amplified' and shifted back corresponding to (4.305) for $\mathbf{r}' = \mathbf{r}_1'$. This leads to the diagram Figure 4.93b. Supposing, additionally, that the time is known at which the source launches its signal, we can gate out the interesting part from $\mathbf{x}(t, \mathbf{r}')$ via the window function $w(t)$ which we have placed at $t = \tau_0$ in our example. Summing up all signals will emphasize only those parts which are placed within the window region. Hence, we will register a bright point at position $\mathbf{r}_1'$ indicating the source 1. The second source has only minor effect on the imaged point since most of its signal components are excluded by the window $w(t)$.

The same we have if the array is focused on $\mathbf{r}_2'$ by setting $\mathbf{r}' = \mathbf{r}_2'$ in (4.305). Now, source 2 gives a bright point in the image (Figure 4.93c). If the array spots point $\mathbf{r}_3' \neq \mathbf{r}_1', \mathbf{r}_2'$ we will not have coherent signals (Figure 4.93d) since no source is placed at this position. Hence, we will get a 'dark' image point since only minor signal parts of sources 1 and 2 achieve the summation point. Assuming very short signals, every source which is not placed at position $\mathbf{r}_3'$ may place in maximum one foreign pulse within the window. Hence in our example, the contribution of non-focused sources to the total voxel value $\hat{z}(t, \mathbf{r}')$ is less by a fraction of L, the number of antennas.

4.10.3.1 **Ultra-Wideband Radar Array**

We introduced array operation by referring to an active scenario which involves radiating point sources which we tried to localize. Our goal is, however, to image a passive scenario which constitutes from individual point scatterer. Hence, we need to illuminate additionally the scenario by one or more antennas so that the scatterers appear as 'secondary sources'. Thus, we can proceed as before.

We will assume L receiving antennas which are placed at known positions $\mathbf{r}_i$ $i \in [1, L]$ and K transmit antennas at positions $\mathbf{r}_j$ $j \in [1, K]$ where transmitting and receiving antennas may be identical (mono-static array channels; $\mathbf{r}_i = \mathbf{r}_j$) or they may be different (bi-static array channels; $\mathbf{r}_i \neq \mathbf{r}_j$). An antenna array which is built from several transmit and receive antennas is assigned as MiMo (multiple input multiple output) array.

In order to be comparable with the previous discussions on the active scenario, we exclude here any mutual interaction between the scattering objects (i.e. multipath propagation) so that we can concentrate on the direct propagation from transmitter antenna to the targets and back to the receiving antenna.

We can account for the additional propagation by modifying the steering vector:

$$h(t, \mathbf{r}', \mathbf{r}_i, \mathbf{r}_j) = h_{ij}(t, \mathbf{r}') = a(\mathbf{r}', \mathbf{r}_i, \mathbf{r}_j)\delta\left(t - \tau_0 + \frac{|\mathbf{r}_i - \mathbf{r}'| + |\mathbf{r}_j - \mathbf{r}'|}{c}\right) \quad (4.308)$$

$$i \in [1, L]; j \in [1, K]$$

$$\begin{aligned}\mathbf{h}(t, \mathbf{r}') = [\, & h_{11}(t, \mathbf{r}') \quad h_{12}(t, \mathbf{r}') \quad \cdots \quad h_{1K}(t, \mathbf{r}') \quad \cdots \\ & h_{L1}(t, \mathbf{r}') h_{L1}(t, \mathbf{r}') \cdots h_{LK}(t, \mathbf{r}')]^T\end{aligned}$$

If the antenna arrangement consists of two separate arrays – one for transmitting and one for receiving – we have in total $M = KL$ different transmission channels so that signal vector $\mathbf{b}(t)$ and the steering vector $\mathbf{h}(t, \mathbf{r}')$ are of length M. For the case that the same antennas are used for transmitting and receiving, the number of transmission channels is $M = LK = L^2$ from which, however, only $M = L(L+1)/2$ are independent due to reciprocity of the scenario. Furthermore, one often renounces mono-static measurements because of the small return loss of the antennas so that finally $M = L(L-1)/2$ remain.

The gain factor $a(\mathbf{r}', \mathbf{r}_i, \mathbf{r}_j) \sim |\mathbf{r}_i - \mathbf{r}'||\mathbf{r}_j - \mathbf{r}'|$ can be again selected to compensate the spreading loss of the signals. Below, we will generalize the concept of gain factor leading to greater flexibility of image reconstruction.

4.10.3.2 **Point Spread Function and Image Resolution**

We will ask how a point scatterer placed at position $\mathbf{r}_0 \in V_{\mathrm{obs}}$ will affect the voxel intensity $I(\mathbf{r}')$ at position $\mathbf{r}'$ for a given array geometry and sounding signal. For that purpose we assume identical antennas which provide an isotropic field[28] having a time evolution quantified by $a(t)$. Disregarding as a start the reflectivity of the point scatter and supposing isotropic and lossless propagation conditions, the signal captured by the antenna placed at point $\mathbf{r}_i$ is

$$b(t, \mathbf{r}_i, \mathbf{r}_j) = b_{ij}(t) \propto \frac{\delta(t - (|\mathbf{r}_i - \mathbf{r}_0|/c))}{|\mathbf{r}_i - \mathbf{r}_0|} * \frac{\delta(t - (|\mathbf{r}_j - \mathbf{r}_0|/c))}{|\mathbf{r}_j - \mathbf{r}_0|} * a(t) \quad (4.309)$$

$$\mathbf{b}(t) = [\, b_{11}(t) \quad b_{12}(t) \quad \cdots \quad b_{1K}(t) \quad \cdots \quad b_{L1}(t) \quad b_{L1}(t) \quad \cdots \quad b_{LK}(t)\,]^T$$

28) Isotropy is only required within the solid angle of the test volume V_{obs}.

if the sounding wave was launched from the antenna at position $\mathbf{r}_j$. Since we would like to focus on the array to the point $\mathbf{r}'$, we design our steering vector as depicted in (4.308). In accordance to (4.306), the array output yields

$$\begin{aligned}\hat{z}(t,\mathbf{r}') &= \Upsilon(t,\mathbf{r}',\mathbf{r}_0) = w(t)\left(\mathbf{b}^T(t) * \mathbf{h}(t,\mathbf{r}')\right) \\ &= w(t)a(t) * \left(\sum_{i=1}^{L}\sum_{j=1}^{K}\frac{a(\mathbf{r}',\mathbf{r}_i,\mathbf{r}_j)}{|\mathbf{r}_i-\mathbf{r}_0|\,|\mathbf{r}_j-\mathbf{r}_0|}\delta(t-\tau)\right) \\ \text{with} \quad \tau &= \tau_0 + \frac{|\mathbf{r}_i-\mathbf{r}'| + |\mathbf{r}_j-\mathbf{r}'| - |\mathbf{r}_i-\mathbf{r}_0| - |\mathbf{r}_j-\mathbf{r}_0|}{c}\end{aligned} \tag{4.310}$$

from which we can basically take the temporal mean or any L_p-norm to quantify the voxel intensity at position $\mathbf{r}'$, that is

$$\bar{\Upsilon}(\mathbf{r}',\mathbf{r}_0) = \int \Upsilon(t,\mathbf{r}',\mathbf{r}_0)\mathrm{d}t \quad \text{or} \quad \Upsilon_p(\mathbf{r}',\mathbf{r}_0) = \sqrt[p]{\int |\Upsilon(t,\mathbf{r}',\mathbf{r}_0)|^p \mathrm{d}t} \tag{4.311}$$

We call $\Upsilon(\mathbf{r}',\mathbf{r}_0)$ the point spread function[29] since it describes how the return signal of a point scatterer distributes over the image. $\Upsilon(\mathbf{r}',\mathbf{r}_0)$ should approach a Dirac function $\Upsilon(\mathbf{r}',\mathbf{r}_0) \Rightarrow \delta(\mathbf{r}'-\mathbf{r}_0)$ for high-resolution imaging array.

(4.310) and (4.311), respectively, represent usual definitions of the point spread function. But correctly spoken, we are dealing here with narrowband point spread functions since the shape variation of the reflected signal due to the point scatterer (i.e. twofold derivation) was not respected. The twofold derivation causes two additional half cycles in $b(t)$ compared to $a(t)$ but it would not extend the signal length (FWHM). The impact of such signal derivations on the point spread function is illustrated in Figure 4.94.

If $\mathfrak{O}(\mathbf{r}_0)$ assigns the object function, that is the distribution of (non-interacting) point scatterers within the observation volume, the radar image is given by

$$I(\mathbf{r}') = \int_{V_{\text{obs}}} \Upsilon(\mathbf{r}',\mathbf{r}_0)\mathfrak{O}(\mathbf{r}_0)\mathrm{d}\mathbf{r}_0 \tag{4.312}$$

Figure 4.94 illustrates the two-dimensional point spread function in the xy-plane for a linear array. Its full width at half maximum represents the achievable pixel size of the array. Figure 4.94a relates to the narrowband definition (4.310). Figure 4.94b shows the point spread function respecting UWB conditions. The same Gaussian pulse was supposed. It feeds a short dipole – provoking a first signal derivation – and then the field is reflected at a point scatterer causing two further derivations. The related point spread function shows two peaks of individually sharper pixel size. Taking both peaks together, we get about the same range resolution as shown in the left diagram. The azimuth resolution will be better.

29) Read $\Upsilon(\mathbf{r}',\mathbf{r}_0)$ as $\bar{\Upsilon}(\mathbf{r}',\mathbf{r}_0)$ or $\Upsilon_p(\mathbf{r}',\mathbf{r}_0)$.

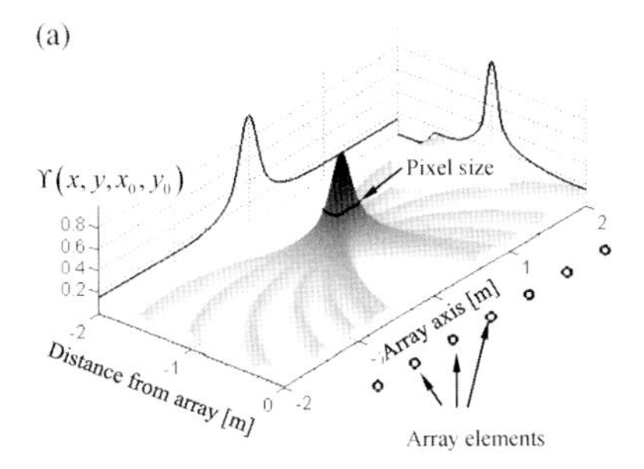

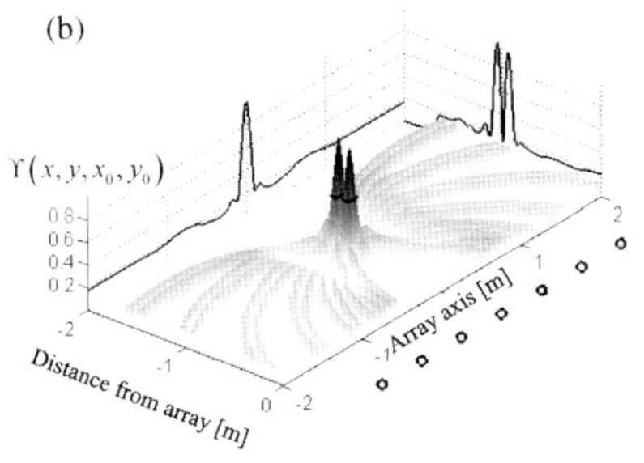

Figure 4.94 Illustration of a two-dimensional point spread function for a linear mono-static array of seven equally spaced antennas. Target location: $\mathbf{r}_0 = [0, -1]$m, sounding signal: Gaussian pulse $t_w = 0.5$ ns. (a) Without respecting time shape variations and (b) with respecting threefold derivation. (colored figure at Wiley homepage: Fig_4_94).

The sidelobes are caused from the non-coherent antenna signals (compare Figure 4.93). Since these signals cannot overlap in the case of a linear mono-static array,[30] the peak–sidelobe ratio only depends on the number of array elements:

$$\text{PSLR} = \frac{1}{L} \tag{4.313}$$

Generally, one can assume that the peak-to-sidelobe ratio increases with the number M of transmission channels, which may be much larger than the antenna number $L + K$. But unfortunately some symmetry in linear or planar MiMo arrays avoids the reduction of the sidelobe level to the ideal value $\text{PSLR} = M^{-1}$ since the number of effective array elements is reduced for some exclusive radiation directions in which these symmetries appear. This effect is also known as 'element shadowing'. The design goal of an antenna array must be to avoid such element shadowing as best as possible. The reader can find some discussions on appropriate array design in Refs [109–111]. Figure 4.95 depicts exemplarily the point spread function of two planar arrays of high symmetry.

In order to evaluate illustratively the basic conditions which determine the resolution of the radar array, we will consider the imaging approach from a different perspective. For the sake of shortness, we only deal with mono-static measurements performed from different locations in space. In the case of a single target, the receiving signals of every measurement channel i will provide a pulse whose time position depends on the roundtrip time to the target. Hence, we can estimate the target distance $r_{i\text{T}}$ but not its actual position. Since we supposed spherical sounding waves, the point scatterer may be placed everywhere on a sphere of radius $r_{i\text{T}}$ having its centre at radiator position.[31] If we repeat the measurement

30) The same antenna is used as transmitter and as receiver. Transmissions from antenna A to antenna B are not respected.

31) We have a spheroid for bi-static measurements instead of a sphere. The antenna positions equal the focal points. The antenna distance is $2e$ and the propagation time is $\tau = 2a/c$ where a represents the semi-major axis – see Annex A.7.

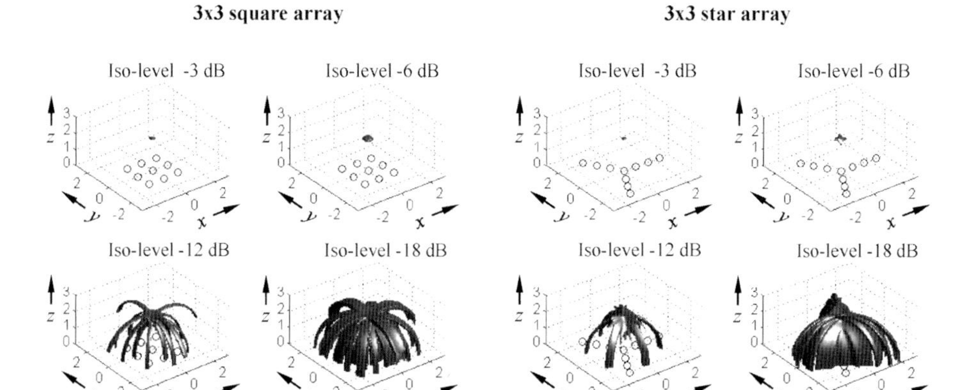

Figure 4.95 Iso-surfaces of the 3D-point spread function for two planar arrays. The array elements are placed in the *xy*-plane. They are marked by circles.

from different locations, the intersection of all related spheres will indicate the searched target position as depicted in Figure 4.96.

In contradiction to the suggestions in Figure 4.96 the radius of the spheres is only known with limited precision in reality. We will respect this by replacing the infinitely thin surface of the sphere by a 'peel' of thickness δ_r (compare Figure 4.97). If we want to estimate the localization precision of a single point scatterer, δ_r relates to the precision of pulse position estimation. Excluding systematic errors, it is given by

$$\delta_r = \pm\frac{c\varphi_0}{2} \tag{4.314}$$

where φ_0 refers to the standard deviation of pulse position estimation (compare also Figure 4.64) . The goal of imaging is to separate targets. Hence, δ_r depends on the spatial expansion of the sounding signal; that is, for baseband signals we may write

$$\delta_r = \frac{ct_w}{2} = \frac{c\tau_{coh}}{2} \approx \frac{c}{2\underset{..}{B}} \tag{4.315}$$

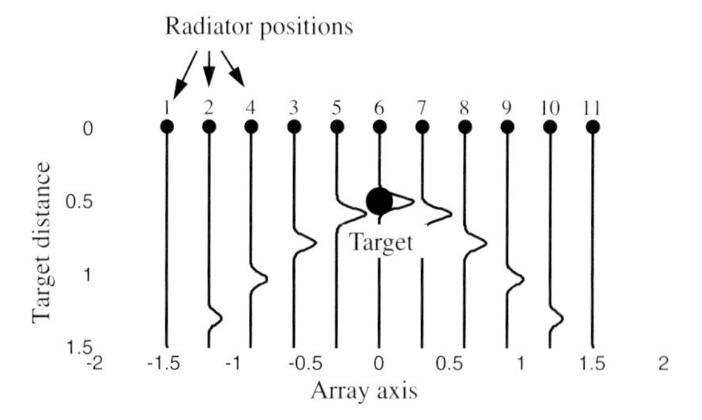

Figure 4.96 Imaging by wavefront superposition. The example shows the localization of a single target by a linear mono-static array.

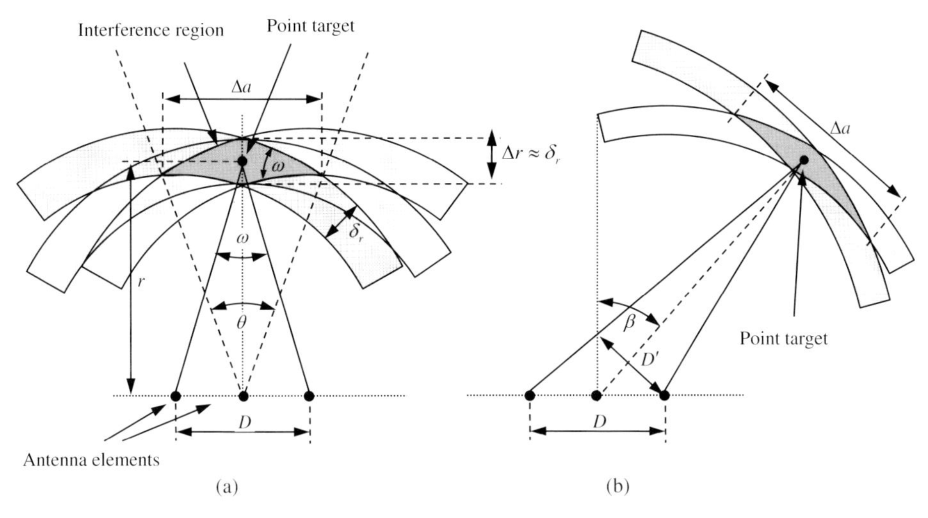

Figure 4.97 Pixel and voxel size of a linear array for sounding waves of limited pulse width. (a) Target is in boresight and (b) target is inclined to the array normal.

Seen from this viewpoint, image resolution (i.e. pixel or voxel size) may be considered as the region of overlapping of the different spheres as it is depicted in Figure 4.97 for a simple three-element array of total width D.

At some distance from the array, it follows from this graph that the width of overlapping in range direction can be approximated by

$$\Delta r \approx \delta_{\mathrm{r}} \tag{4.316}$$

That is, it is a pure question of the bandwidth of the imaging system. The width of the interference region in cross range or azimuth direction and for boresight direction is about

$$\begin{aligned} \tan\frac{\omega}{2} &\approx \frac{\delta_{\mathrm{R}}}{\Delta a} \approx \frac{D}{2r} \approx \frac{\omega}{2}; \quad r \gg D \\ \Delta a &\approx \frac{2r\delta_{\mathrm{r}}}{D} \approx \frac{2\delta_{\mathrm{r}}}{\omega} \end{aligned} \tag{4.317}$$

As already expected from Figure 4.91, the azimuth resolution will largely depend on the angle ω under which the target 'sees' the array. For target positions off the boresight direction, the effective array length seen from the target reduces to $D' \approx D\cos\beta$, leading to a degraded azimuth resolution of

$$\Delta a \approx \frac{2r\delta_{\mathrm{r}}}{D\cos\beta} \tag{4.318}$$

The angle θ may be assigned as angular resolution and beamwidth of the array. It can be roughly written as

$$\theta \approx \frac{\Delta a}{r} \approx \frac{2\delta_{\mathrm{r}}}{D\cos\beta} \quad \text{for } r \gg D \tag{4.319}$$

Relations (4.316)–(4.319) assume sufficient distance from the array. If the target approaches the array, the determination of its position will be more and more unreliable. The reader can find some estimations of resolution degradation in Annex A.7.3.

Further, we can state that in contrast to narrowband arrays, the number of antennas does not influence the image resolution. Narrowband arrays require narrow element spacing (element distance smaller than $\lambda/2 = c/2f_0$) in order to avoid ambiguities (the so-called grating lobes) due to the periodic signal structure. The same type of ambiguities appears in the ultra-wideband case too. That, is element spacing must be shorter than $ct_P/2$ where t_P is the period duration of the UWB sounding signal. Since the signal period can be always selected sufficiently long (i.e. larger than the round-trip time to the targets; the whole observation volume V_{obs} is covered by a single signal period), the elements may be very sparsely distributed over the array area without causing ambiguities by grating lobes in the point spread function.

We have however seen from Figure 4.94 that low antenna numbers will lead to large sidelobe levels which will affect the image quality if the scene contains more than only one target what is usually the case. Consequently, one actually needs a large number of array elements in order to achieve the required image contrast. Figure 4.98 illustratively compares the brightness of a radar image for a scene with four targets of identical reflectivity. The object function of the exemplified scene is

$$\mathfrak{O}(\mathbf{r}_0) = \begin{cases} 1 & \mathbf{r}_0 \in [P_1, P_2, P_3, P_4] \\ 0 & \mathbf{r}_0 \notin [P_1, P_2, P_3, P_4] \end{cases}$$

$$P_1 = (2, -2);\ P_2 = (0, -1.4);\ P_3 = (0.4, -1);\ P_4 = (-0.4, -1)$$

4.10.3.3 Steering Vector Design

As the previous discussion has shown, the focusing properties of the radar array are determined by the array geometry, the bandwidth of the involved sounding signals and the design of the steering vector $\mathbf{h}(t, \mathbf{r}')$ which 'organizes' the coherent superposition of receiving signals originating from space point $\mathbf{r}'$. Under practical conditions, the optimum choice of the steering vector is subjected to several constraints and hence there are several aspects to be respected for its implementation.

Ultra-wideband imaging is often used to investigate optically impenetrable scenarios, which means that the propagation speed of the sounding waves is only approximately known. The propagation path is often inhomogeneous which causes reflection, refraction and attenuation. This leads to clutter signals and the propagation paths to a considered space point $\mathbf{r}'$ will be difficult to predict. Furthermore, the captured data are affected by random noise and jitter due to limited performance of the sensor devices and external jamming.

Hence, the challenge is to find an optimally designed steering vector $\mathbf{h}(t, \mathbf{r}')$ whose technical implementation will not pose insuperable obstacles. A generally optimal method is hard to determine, since many things depend on the constitution of the scenario under test itself as well as the environmental conditions under

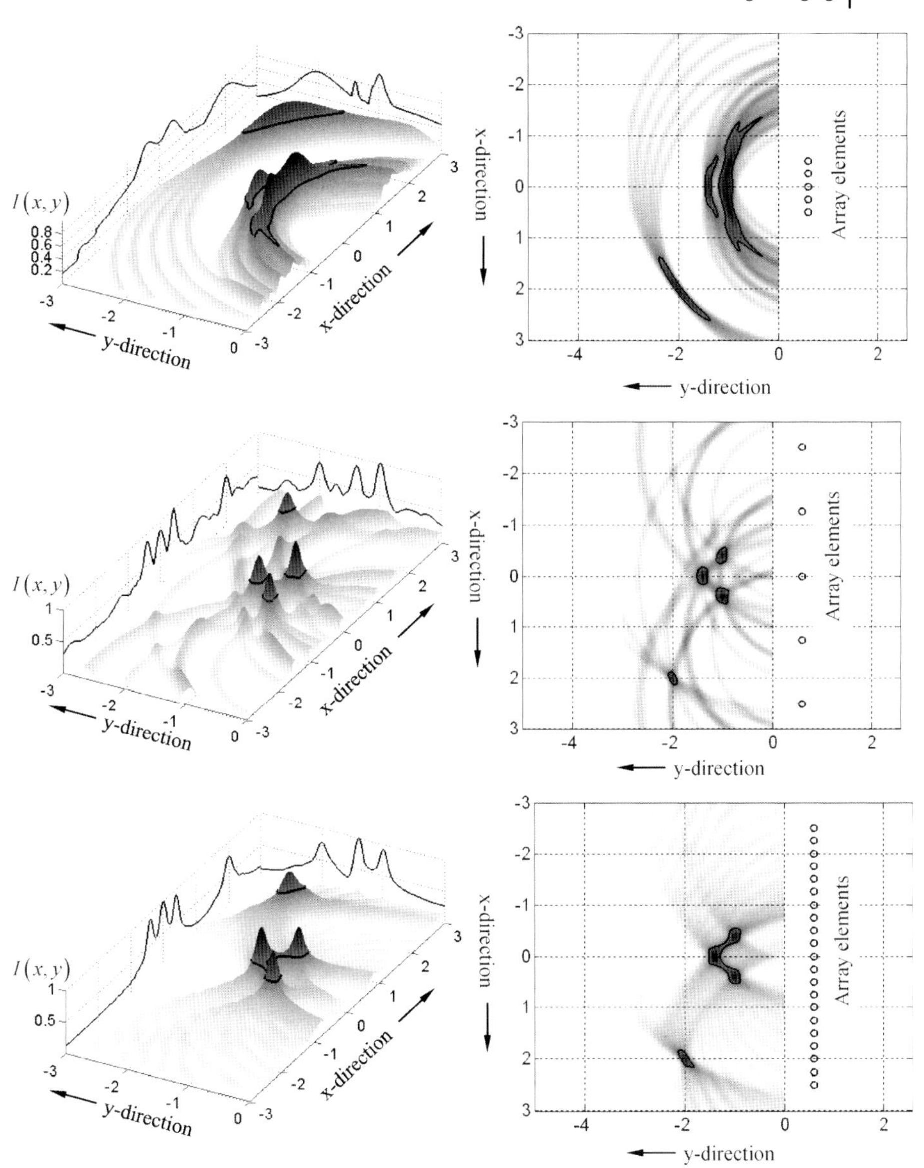

Figure 4.98 Comparison of radar images gained from differently configured linear arrays. (Signal destruction by antenna and targets are not respected compare Figure 4.94). (colored figure at Wiley homepage: Fig_4_98).

which the target is to be found (e.g. multiple reflections, target placed close to a reflecting, rough interfaces, wave attenuation and dispersion etc.). We will summarize some basic approaches for steering vector selection. For that purpose, we extend the steering filter concept and allow $h_{ij}(t,\mathbf{r}') = h(t,\mathbf{r}',\mathbf{r}_i,\mathbf{r}_j)$ being composed from several linear or non-linear subcomponents. The selection of appropriate steering functions may be guided by various intentions such as sidelobe suppression, noise suppression, clutter reduction, computational burden and so on.

Migration Stacking It is the simplest method often applied in ground-penetrating radar [83]. It is also referred to as wavefront migration or delay and sum beam forming. The method can also be assigned as Hough transform(4.264) for hyperbolic shapes in case of a linear or planar antenna array. Migration stacking can be applied for all types of array configuration and operational modes. The basic idea is to compensate the round-trip time of the individual propagation paths referred to as the considered voxel position $\mathbf{r}'$ before summing all signals. That is, the steering vector is simply given by[32)]

$$h_{ij}^{(D)}(t,\mathbf{r}') = \delta\left(t + \frac{|\mathbf{r}_i - \mathbf{r}'| + |\mathbf{r}_j - \mathbf{r}'|}{c}\right); \quad i = 1 \cdots L;\ j = 1 \cdots K \qquad (4.320)$$

in the case of a homogeneous propagation medium. If the point $\mathbf{r}'$ to be imaged is located within a second propagation medium (Figure 4.99), the correct propagation delay has to be respected so that the steering filter becomes

$$h_{ij}^{(D)}(t,\mathbf{r}') = \delta\left(t + \frac{r_1 + r_4}{c_1} + \frac{r_2 + r_3}{c_2}\right) \qquad (4.321)$$

The interface between the two media will affect the shape of the wavefront and cause reflections. Both effects have to be respected appropriately by the imaging procedure. The removal of interface reflection is a demanding task for irregular surfaces. Some corresponding procedures can be found in Refs [112–114]. A method to respect the deformation of the wavefront is given in Section 5.5 and Ref. [115].

The steering filters as introduced in (4.320) or (4.321) are of fundamental importance since they appear in all other approaches too; hence, we have marked them by

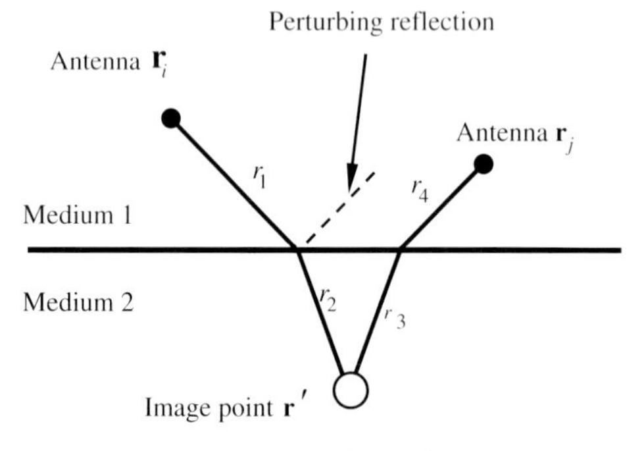

Figure 4.99 Steering vector design for propagation through an interface.

32) We omit from here on the internal delay τ_0 which is needed to keep causality of the steering functions.

the superscript $^{(D)}$. The correct choice of the propagation speed will have a large influence of the image quality.

Kirchhoff Migration Kirchhoff migration is derived from the scalar wave equation [104]. Translated to our steering vector notation, it may be expressed for monostatic operation by

$$h_{ii}(t, \mathbf{r}') = \frac{4\cos\beta}{c|\mathbf{r}_i - \mathbf{r}'|} u_1(t) * h_{ij}^{(D)}(t, \mathbf{r}'); \quad \mathbf{r}_i = \mathbf{r}_j \tag{4.322}$$

The angle β is defined in Figure 4.97 and $u_1(t)$ is the unit doublet. The Stolt- or *fg*-space migration has the same roots as the Kirchhoff migration. It provides less computational burden since it works with the fast Fourier transform [104].

Migration by De-Convolution The aim of this migration method is to approach the array response of an individual point source as close as possible to a Dirac function. That is, one has to match the steering vector to the following condition, which minimizes the quadratic error:

$$\begin{aligned} e^2 &= \left\| \mathbf{b}(t)^T * \mathbf{h}(t, \mathbf{r}') - \delta(t) \right\|_2^2 \quad \Rightarrow \quad \text{minimum} \\ \hat{\mathbf{a}}(t, \mathbf{r}') &= \underbrace{\arg\min}_{\mathbf{a}(t,\mathbf{r}')} \left(\left\| \mathbf{b}(t)^T * \mathbf{a}(t, \mathbf{r}') * \mathbf{h}^{(D)}(t, \mathbf{r}') - \delta(t) \right\|_2^2 \right) \end{aligned} \tag{4.323}$$

For that purpose, we decompose the steering vector as follows:

$$\mathbf{h}(t, \mathbf{r}') = \mathbf{a}(t, \mathbf{r}') \circledast \mathbf{h}^{(D)}(t, \mathbf{r}') \tag{4.324}$$

Herein $\circledast$ stands for the Hadamard or Schur convolution (i.e. element-wise convolution of the entries from $\mathbf{h}$ and $\mathbf{a}$), $\mathbf{h}^{(D)}(t, \mathbf{r}')$ represents time shifting to achieve coherency of waves originating from point $\mathbf{r}'$ (see (4.320) and (4.321)) and $\mathbf{a}(t, \mathbf{r}')$ is a set of IRFs aimed to de-convolve the influence of the propagation path (dispersion, propagation loss etc.) from the data. $\hat{\mathbf{a}}(t, \mathbf{r}')$ can be found by solving (4.323) which represents a least square problem. The solution will be however ill-conditioned due to the frequency band limitations of $\mathbf{b}(t)$. Hence by appropriate countermeasures, one has to force down unwanted effects. A usual method is to limit the energy $\|\mathbf{a}(t, \mathbf{r}')\|_2^2$ to a certain value so that the solution $\hat{\mathbf{a}}(t, \mathbf{r}')$ finally leads to a constrained least squares problem which may be solved via Lagrange multiplier [94, 112, 116]. Some issues of de-convolution in frequency domain have been discussed in Section 4.8.3 which we will not repeat here.

Minimum Variance Beam Forming So far, we always supposed a perfect alignment by the steering vector $\mathbf{h}^{(D)}(t, \mathbf{r}')$. This is however not always true since propagation speed, dispersion, attenuation and angular dependence of antenna impulse response are often not correctly known. Hence, the waves originating from the space point $\mathbf{r}'$ will not be perfectly aligned by the steering vector. Furthermore, clutter signals resulting from the sensor electronics or false targets and quantization errors of time shift operations add further error components to the region of interest. In order to combat the impact of these uncertainties, one can provide in a first step a coarse trimming by $\mathbf{h}(t,\mathbf{r}')$ gating out unwanted signal components and compensating time delay and attenuation based on the available knowledge of the test object. This leads to the signal vector $\mathbf{x}(t,\mathbf{r}')$ (see Figure 4.100).

In further steps, the array vector $\mathbf{x}(t,\mathbf{r}')$ is subjected to a fine trimming. In case of MiMo radar, we apply three sets of weighting functions $\mathbf{u}(t,\mathbf{r}')$, $\mathbf{v}(t,\mathbf{r}')$ and $w(t)$ which serve to suppress errors caused by coarse trimming. These functions are adaptive to the test scenario. Figure 4.100 depicts the structure of the MiMo array and the associated signal flow. For illustration, we use a receiving array of three antennas placed at $\mathbf{r}_i$ and two transmitting antennas placed at $\mathbf{r}_j$ (not shown in the figure). The measurement runs in two steps. First, transmitter 1 is active which provides the array signal vector $\mathbf{x}_1(t,\mathbf{r}')$ after coarse trimming to the image point $\mathbf{r}'$.

$$\mathbf{x}_1(t,\mathbf{r}') = \begin{bmatrix} x_{11}(t,\mathbf{r}') & x_{21}(t,\mathbf{r}') & x_{31}(t,\mathbf{r}') \end{bmatrix}^T$$

These signals are subjected to a first weighting by $\mathbf{u}(t,\mathbf{r}') = \begin{bmatrix} u_1(t,\mathbf{r}') & u_2(t,\mathbf{r}') & u_3(t,\mathbf{r}') \end{bmatrix}^T$ which may depend on the location of the voxel to be imaged. The first measurement step leads to an array output which will further be weighted by the function $v_1(t,\mathbf{r}')$. Hence, we can write

$$\begin{aligned} Y_1(t,\mathbf{r}') &= \mathbf{x}_1^T(t,\mathbf{r}')\mathbf{u}(t,\mathbf{r}') \\ z_1(t,\mathbf{r}') &= Y_1(t,\mathbf{r}')v_1(t,\mathbf{r}') \end{aligned} \tag{4.325}$$

In a second step transmitter 2 radiates and we get the array signal vector $\mathbf{x}_2(t,\mathbf{r}')$ – again after having performed a coarse trimming to the image point $\mathbf{r}'$. This vector is subjected to the same weighting by $\mathbf{u}(t,\mathbf{r}')$ but it is differently weighted by $v_2(t,\mathbf{r}')$. Arranging all array signals in the matrix $\mathbf{X}(t,\mathbf{r}') = \begin{bmatrix} \mathbf{x}_1(t,\mathbf{r}') & \mathbf{x}_2(t,\mathbf{r}') & \cdots \end{bmatrix}$ and the weighing functions $v_i(t,\mathbf{r}')$ in a column vector $\mathbf{v}(t,\mathbf{r}') = \begin{bmatrix} v_1(t,\mathbf{r}') & v_2(t,\mathbf{r}') \end{bmatrix}^T$, the brightness of the image point $\mathbf{r}'$ gets

$$\begin{aligned} I(\mathbf{r}') &= \int Z^2(t,\mathbf{r}')\mathrm{d}t \\ Z(t,\mathbf{r}') &= w(t,\mathbf{r}')\mathbf{v}(t,\mathbf{r}')^T\mathbf{Y}(t,\mathbf{r}') \\ &= w(t,\mathbf{r}')\mathbf{v}(t,\mathbf{r}')^T\mathbf{X}^T(t,\mathbf{r}')\mathbf{u}(t,\mathbf{r}') \end{aligned} \tag{4.326}$$

Equation (4.326) is called 'robust Capon beamforming' where the weighting vectors are gained from a constraint quadratic optimization which tries to maximize the energy of the back-scattered signals $s_j(t)$ and minimize a modelling error [117–121].

In case of point scatterers and isotropic propagation, the time shape of the back-scattered signal $s_j(t)$ may depend on the illuminating antenna at position $\mathbf{r}_j$ but it does not depend[33)] on the aspect angle under which the receive array antennas

33) This condition does not hold for too large targets and it will be violated by approaching the array too closely since then the angular dependency of the backscattering becomes remarkable.

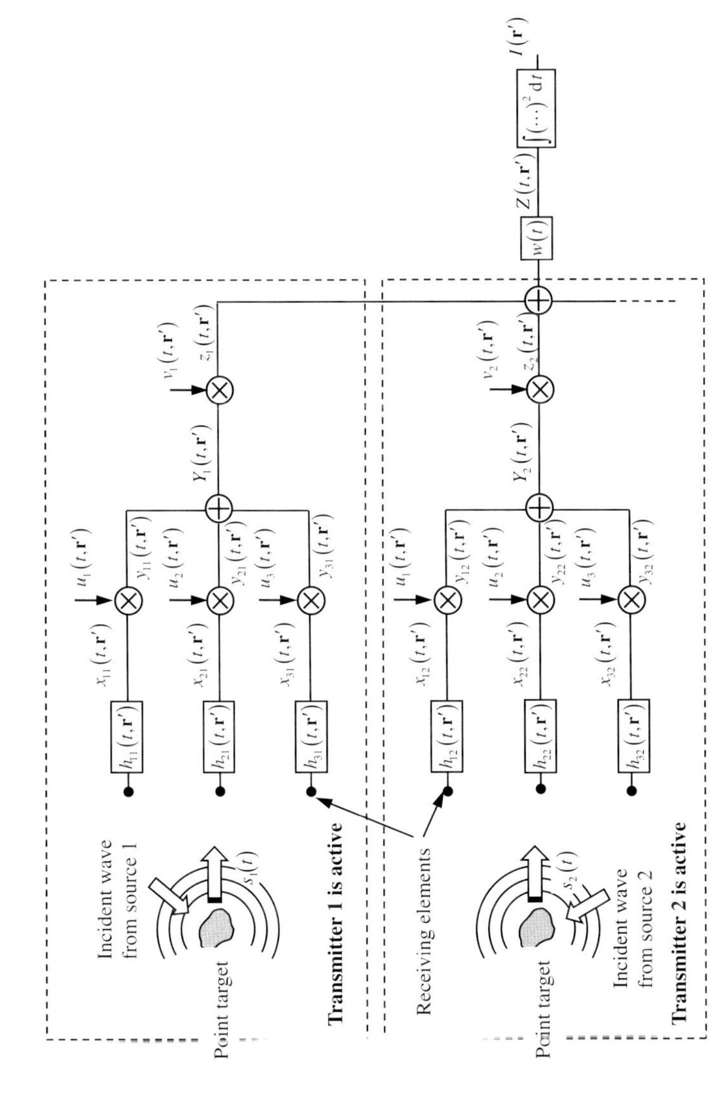

Figure 4.100 Adaptive MiMo radar imaging.

'see' the target. Hence, ideally, the array signal vector should have identical components after having applied the coarse trimming, that is

$$\mathbf{x}_j(t) = \mathbf{q}_0\, s_j(t); \quad \mathbf{q}_0 = [1 \quad 1 \quad \cdots \quad 1]^T \tag{4.327}$$

Actually, we will get some deviations which are modelled by

$$\mathbf{x}_j(t) = \mathbf{q}(t) s_j(t) + \mathbf{e}_j(t) \tag{4.328}$$

Herein, $\mathbf{e}_j(t)$ is a signal vector covering error terms as noise, clutter and modelling errors. It is subjected to both the number of the receiving as well as transmission channel. The multiplicative deviation $\mathbf{q}(t)$ is supposed to depend only on the 'scattering path', that is the paths from the scatterer via antennas till the coarse trimming $\mathbf{h}_i(t, \mathbf{r}')$. The power of the scattering signal is maximum and the error is minimum if [120]

$$\begin{aligned}
\mathbf{u}(t) &= \frac{\mathbf{R}_X^{-1}(t)\mathbf{q}_{(2)}(t)}{\mathbf{q}_{(2)}^T(t)\mathbf{R}_X^{-1}(t)\mathbf{q}_{(2)}(t)} \\
\mathbf{R}_X(t) &= \frac{1}{L}\mathbf{X}(t)\mathbf{X}^T(t); \quad \mathbf{q}_{(2)}(t) = \frac{\sqrt{L}\mathbf{q}_{(1)}(t)}{\|\mathbf{q}_{(1)}(t)\|}; \quad \mathbf{q}_{(1)}(t) = \left(\frac{1}{\lambda}\mathbf{R}_X^{-1}(t) + \mathbf{I}\right)^{-1}\mathbf{q}_0
\end{aligned} \tag{4.329}$$

Herein, $\mathbf{R}_X(t)$ represents the sample covariance matrix of coarse trimmed data.

In a second step, we optimize by the same strategy the transmitter side covering the path from the sounding signal $s(t)$ (which should be identical for all transmit antennas) till the scatterer. In accordance to (4.328), we model the differences in the transmitters by

$$\mathbf{Y}(t) = \mathbf{p}(t)s(t) + \mathbf{E}(t) \tag{4.330}$$

which equivalently leads to

$$\begin{aligned}
\mathbf{v}(t) &= \frac{\mathbf{R}_Y^{-1}(t)\mathbf{p}_{(2)}(t)}{\mathbf{p}_{(2)}^T(t)\mathbf{R}_Y^{-1}(t)\mathbf{p}_{(2)}(t)} \\
\mathbf{R}_Y(t) &= \frac{1}{K}\mathbf{Y}(t)\mathbf{Y}^T(t); \quad \mathbf{p}_{(2)}(t) = \frac{\sqrt{K}\mathbf{p}_{(1)}(t)}{\|\mathbf{p}_{(1)}(t)\|}; \quad \mathbf{p}_{(1)}(t) = \left(\frac{1}{\mu}\mathbf{R}_Y^{-1}(t) + \mathbf{I}\right)^{-1}\mathbf{p}_0 \\
&\mathbf{p}_{(0)} = [1 \quad 1 \quad 1 \quad \cdots \quad 1]^T
\end{aligned} \tag{4.331}$$

Herein, λ and μ are the regularization parameters which must be selected carefully. The MiMo array was supposed to have K transmitters and L receivers. Thus, the involved vectors and matrices have the following dimensions: $\mathbf{x}_j, \mathbf{u}, \mathbf{q}, \mathbf{e}_j \to [L, 1]$; $\mathbf{Y}, \mathbf{E}, \mathbf{p}, \mathbf{v} \to [K, 1]$; $\mathbf{X} \to [L, K]$; $\mathbf{R}_{XX} \to [L, L]$ and $\mathbf{R}_{YY} \to [K, K]$.

Non-Linear Steering Vector Under certain conditions non-linear steering vectors may improve image quality. It may emphasize specific features or it may reduce the numerical expense. There are many different options of which we will list a few.

- **Pulse position determination:** One approximates the most prominent reflection peaks of the receive signals by Delta functions before performing array processing. Equivalently one could determine the pulse position of the most prominent peaks (i.e. the round-trip time) from which one determines the target positions by calculating the intersection points of signals from different antennas. This is an approach typically used in through-wall radar (see Sections 6.8 and 6.9). The method can be extended by a matching pursuit approach by which one determines the time position of pulses of pre-defined shape.
- **Envelope processing:** Here, array processing is applied at the envelope (or simply the squared signal) instead of the original signal. This will degrade image resolution since the envelope suppresses the fine structure of the signal. However, the method is less sensitive to erroneous assumptions of the propagation speed. The actual receiving signals are bipolar; hence, they may mutually cancel out at target position if they are not correctly time aligned due to an erroneous speed estimation. The envelope is unipolar which avoids cancellation and a point target will leave a blob in the image even if time alignment could not be perfectly achieved.
- **Geometric mean array:** The voxel brightness resulting from the array processing depicted in Figure 4.92 or (4.100) can be considered as weighted arithmetic mean. Here, we propose to replace the arithmetic mean by the geometric one, that is to perform $\breve{x} = \sqrt[N]{\prod_{n=1}^{N} x_n}$ instead of $\bar{x} = (1/N)\sum_{n=1}^{N} x_n$. It was observed in Ref. [122] that this method may provide lower sidelobe levels since signal accumulation takes place only where all (!) array signals coincidentally interfere. Figure 4.101 gives an example for a simple scenario, and Figure 4.102 illustrates

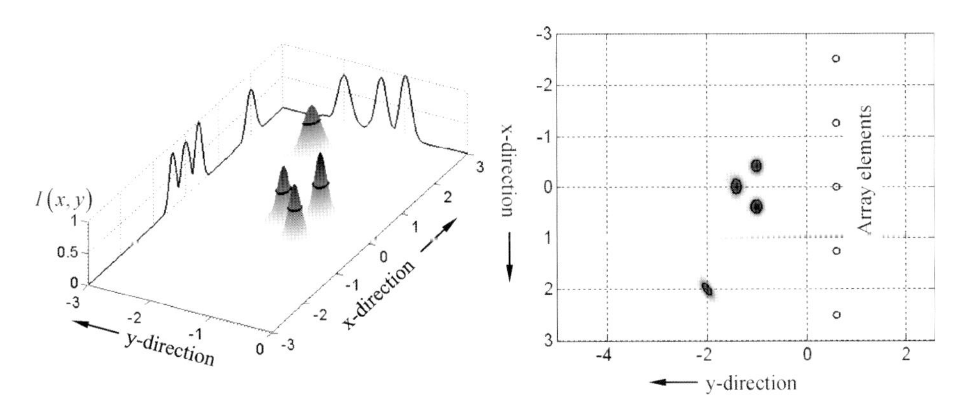

Figure 4.101 Radar image gained from geometric mean array (compare Figure 4.98 for arithmetic mean of array signals).

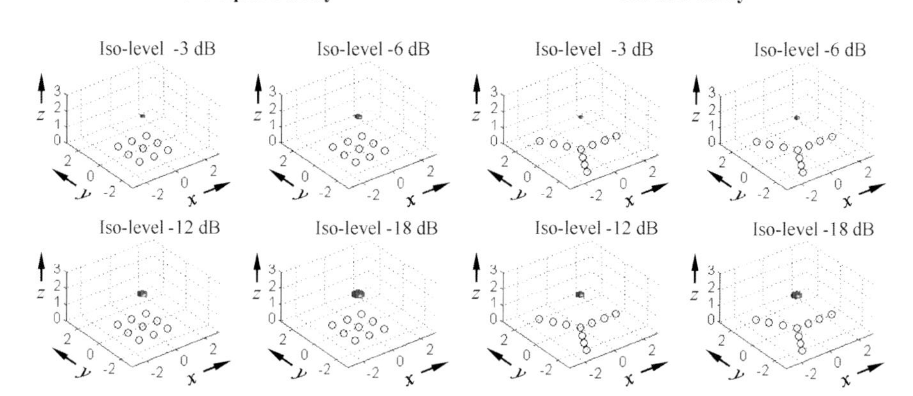

Figure 4.102 Iso-surface of 3D-point spread function for planar geometric mean arrays (compare with Figure 4.95).

the 3D-point spread function of two planar arrays for geometric mean processing. It refers to the same array geometries as applied in Figure 4.95. Obviously the array structure has less influence on the point spread function as for the 'arithmetic mean array'.

Of course there is some risk that an 'erroneous zero' in only one of the superimposed signals may destroy completely the voxel brightness. Hence, one has definitely to assure that the voxel at which the array is focusing is actually located within the beam of all involved antennas. The geometric mean array allows for extremely thinned arrays due to their large sidelobe suppression. The performance of such non-linear array processing is not yet well investigated. It needs further research.

- **Combined mean array [123,124]:** The idea is to determine the arithmetic mean of a number of antenna groups from which the geometric mean is calculated beforehand. Let us consider an example. We assume a linear array which is built from L antennas. Further, an antenna group should comprise three adjacent antennas. Thus, the output signal of the array may be written as $(x_i(t, \mathbf{r}')$, which represents the time-shifted (to focus on $\mathbf{r}'$) receive signal of the ith antenna:

$$z(t, \mathbf{r}') = \sum_{i=1}^{L-3} \sqrt[3]{\prod_{n=0}^{2} x_{i+n}(t, \mathbf{r}')} \tag{4.332}$$

Basically, many other antenna combinations for the geometric mean can be applied. In order to achieve good sidelobe suppression, the antennas involved in the geometric mean should have a large distance, however, by keeping the common illumination area.

4.10.3.4 **Sparse Scene Imaging**

We consider a MiMo radar array which is sounding a homogeneous or inhomogeneous propagation medium including a couple of dominant point

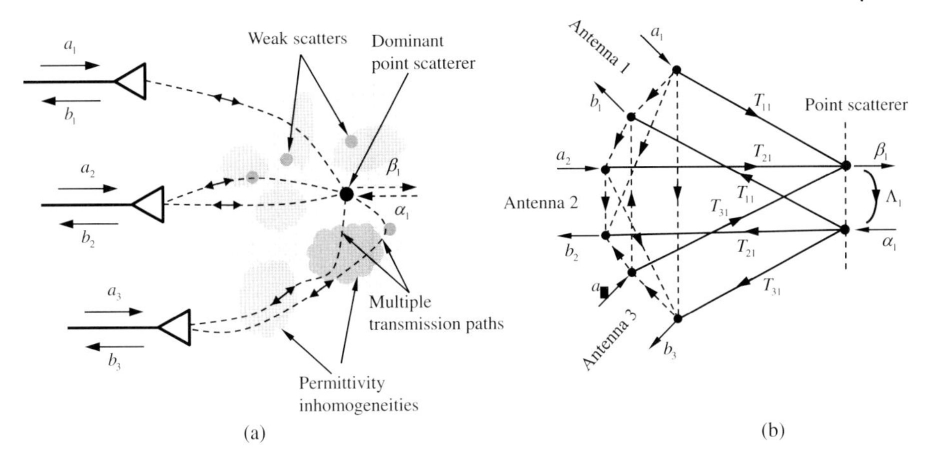

Figure 4.103 Investigation of an inhomogeneous medium including a single dominant scatterer by MiMo radar. (a) Illustration of the scenario and possible ray paths (direct antenna interaction not includes). (b) Related Mason graph (including direct antenna interaction). Note that structural reflections and re-radiation of the antennas are omitted in the Mason graph.

scatterers. We are mainly interested in these scatterers, that is their number, their strength of reflectivity and their position. Since we have involved only a low number of scatterers, we speak about a sparse scene. Figure 4.103 depicts a simple example of an inhomogeneous propagation medium. For the sake of purity, it only includes a single dominant scatterer.

We will establish an S-parameter model of the scenario under test assuming K antennas and M dominant point scatterers. We can state for a sparse scene $K > M$. Furthermore, we neglect multiple reflections between the individual point scatters (which is also assigned as the Born approximation) as well as between antennas and point scatterers. We make no assumptions about the location of the antenna elements. At this point of consideration, the actual antenna position must not yet necessarily be known.

If we perform a measurement based on the K element MiMo array, we will gain the scattering matrix $\mathbf{S}$ of the scenario. We can write for time and frequency domain notion, respectively,

$$\begin{aligned} \mathbf{b}(t) &= \mathbf{S}(t) * \mathbf{a}(t) = (\mathbf{S}_0(t) + \mathbf{S}_M(t)) * \mathbf{a}(t) \\ \underline{\mathbf{b}}(f) &= \underline{\mathbf{S}}(f)\underline{\mathbf{a}}(f) = (\underline{\mathbf{S}}_0(f) + \underline{\mathbf{S}}_M(f))\underline{\mathbf{a}}(f) \end{aligned} \tag{4.333}$$

with

$$\begin{aligned} \mathbf{a} &= \begin{bmatrix} a_1 & a_2 & \cdots & a_K \end{bmatrix}^T \\ \mathbf{b} &= \begin{bmatrix} b_1 & b_2 & \cdots & b_K \end{bmatrix}^T \end{aligned}$$

Here, we have decomposed the actually measurable scattering matrix $\mathbf{S}$ into the matrix $\mathbf{S}_0$ and $\mathbf{S}_M$. We are interested only in $\mathbf{S}_M$ which shall cover all transmission paths between antennas and the dominant scatterers. It is often referred to as multi-static data matrix or multi-static response matrix. The matrix $\mathbf{S}_0$ relates to all transmissions which do not involve the point scatterers. Hence, it mainly concerns the antenna crosstalk and antenna feed point reflections. These signal components are also assigned as background which we have to remove from the measured data before further processing.

In order to get a view inside the multi-static data matrix $\mathbf{S}_M$, we construct it from the Mason graph depicted in Figure 4.103b. There, we have introduced the transmission paths T_{mk} from antenna $k \in [1, K]$ to the position of scatterer $m \in [1, M]$ and T_{ln} from scatterer $n \in [1, M]$ to antenna $l \in [1, K]$. Such transmission path contains the transmission/receiving behavior of the antenna, the direct path between antenna and scatterer (line of sight) but also all other paths which provide energy exchange via wave propagation between the location of the antenna and the scatterer appearing in case of an inhomogeneous propagation medium. From reciprocity, we can state $T_{mk} = T_{ln}$ for $l = k; m = n$ which we already respected in Figure 4.103. We summarize the transmission paths from all antennas to the position of an individual scatterer m by the column vector

$$\mathbf{T}_m = \begin{bmatrix} T_{1m} & T_{2m} & \cdots & T_{Km} \end{bmatrix}^T$$

so that the transmissions between all antennas and the positions of all involved scatters lead to the $[K, M]$ transmission matrix (also assigned as propagation matrix)

$$\mathbf{T} = \begin{bmatrix} \mathbf{T}_1 & \mathbf{T}_2 & \cdots & \mathbf{T}_M \end{bmatrix}$$

Note that the individual entries T_{km} of that matrix represent a set of Green's functions of the test scenario involving the behavior of wave propagation as well as the radiation behavior of the antennas:

$$T_{km}(t) = G(t, \mathbf{r}_k, \mathbf{r}'_m) \tag{4.334}$$

where $\mathbf{r}_k$ is the antenna position and $\mathbf{r}'_m$ is the scatterer position.

The fields $\boldsymbol{\beta} = [\beta_1 \quad \beta_2 \quad \cdots \quad \beta_M]^T$ bouncing the individual point scatterers may be expressed by

$$\begin{aligned} \boldsymbol{\beta}(t) &= \mathbf{T}^T(t) * \mathbf{a}(t) \\ \underline{\boldsymbol{\beta}}(t) &= \underline{\mathbf{T}}^T(f)\underline{\mathbf{a}}(f) \end{aligned} \tag{4.335}$$

Correspondingly, we yield for the receiving signal of all antennas (omitting the direct transmission which we had symbolized by $\mathbf{S}_0$ – see (4.333))

$$\begin{aligned} \mathbf{b}(t) &= \mathbf{T}(t) * \boldsymbol{\alpha}(t) \\ \underline{\mathbf{b}}(f) &= \underline{\mathbf{T}}(f)\underline{\boldsymbol{\alpha}}(f) \end{aligned} \tag{4.336}$$

if $\boldsymbol{\alpha} = \begin{bmatrix} \alpha_1 & \alpha_2 & \cdots & \alpha_M \end{bmatrix}^T$ assign the waves emanating from the positions of the point scatterer. Since these waves are the reflections of the point scatterer, we can write

$$\begin{aligned}
&\boldsymbol{\alpha}(t) = \boldsymbol{\Lambda}(t) * \boldsymbol{\beta}(t) \\
&\underline{\boldsymbol{\alpha}}(f) = \underline{\boldsymbol{\Lambda}}(f)\underline{\boldsymbol{\beta}}(f) \\
&\boldsymbol{\Lambda} = \begin{bmatrix} \Lambda_1 & 0 & \cdots & 0 \\ 0 & \Lambda_2 & \cdots & 0 \\ \vdots & \vdots & \ddots & \vdots \\ 0 & 0 & \cdots & \Lambda_M \end{bmatrix}
\end{aligned} \tag{4.337}$$

where Λ_m is a measure of the reflectivity of target m which is supposed to be independent of the angle of incidence and observation. Note that this assumption only holds for small targets within a bi-static angle of about $\pm 90^\circ$ (refer to Figure 4.21). Since by assumption the scattering objects do not mutually interact, the scattering matrix $\boldsymbol{\Lambda}$ is a diagonal one. Joining (4.335)–(4.337), we achieve at the multi-static response matrix

$$\begin{aligned}
\mathbf{S}_M(t) &= \mathbf{T}(t) * \boldsymbol{\Lambda}(t) * \mathbf{T}^T(t) \\
\underline{\mathbf{S}}_M(f) &= \underline{\mathbf{T}}(f)\underline{\boldsymbol{\Lambda}}(f)\underline{\mathbf{T}}^T(f)
\end{aligned} \tag{4.338}$$

Due to reciprocity, the multi-static response matrix must be symmetric which is confirmed by (4.337) and (4.338), that is

$$\begin{aligned}
\mathbf{S}_M(t) &= \mathbf{S}_M^T(t) \\
\mathbf{S}_M(-t) &= \mathbf{S}_M^T(-t) \\
\underline{\mathbf{S}}_M(f) &= \underline{\mathbf{S}}_M^T(f) \\
\underline{\mathbf{S}}_M^*(f) &= \underline{\mathbf{S}}_M^H(f)
\end{aligned} \tag{4.339}$$

In the case of a homogenous propagation medium, (4.338) represents nothing but an ensemble of mono- and bi-static radar equations. To give an example, we consider the transmission from antenna k via scatterer m to antenna l:

$$\begin{aligned}
S_{M,lk} &= \sum_{m=1}^{M} S_{M,lk}^{(m)} = \sum_{m=1}^{M} T_{km}(t) * \Lambda_m(t) * T_{lm}(t) \\
S_{M,lk}^{(m)} &= \frac{1}{r_{mk} r_{lm}} T_{A,k}(t) * R_{A,l}(t) * \Lambda_m(t) * \delta\left(t - \frac{r_{mk} + r_{lm}}{c}\right)
\end{aligned}$$

where $T_{A,k}(t)$ is the IRF of transmitter antenna k, $R_{A,l}(t)$ is the IRF of receiver antenna l, Λ_m is the IRF of point scatterer m, r_{mk}; r_{lm} are distance between antennas and scatterer.

However, in a multi-path rich environment, the individual IRFs of transmission matrix $\mathbf{T}$ may appear quite chaotic so that a short and powerful wave emitted by one of the transmitters will be spread over large time hence forfeiting its peak power. A corresponding example is illustrated in Figure 2.46 depicting the point-to-point transmission in a laboratory space equipped with many metallic devices. Particularly the non-line-of-sight case provides a strong spreading of signal energy.

Regardless of the spreading of wave energy due to the harsh propagation environment, we will address the issue how to achieve a high peak power concentration at a specific point in space. Exemplarily, we are interested to load, for example, the point scatterer m with an electromagnetic field as strong as possible. We find from (4.335) that the total field bouncing the respective scatterer is given by

$$\begin{aligned} \beta_m(t) &= \mathbf{T}_m^T(t) * \mathbf{a}(t) \\ \underline{\beta}_m(f) &= \underline{\mathbf{T}}_m^T(f)\underline{\mathbf{a}}(f) \end{aligned} \tag{4.340}$$

where we can consider the individual transmission paths from the antennas to the scatter as spatial filters having the IRFs $T_{km}(t); k \in [1, K]$.

In order to achieve high field strength, we have to make sure that the wave fields provided by the individual antennas are time compressed and synchronously superimposed if they arrive at scatterer location. We already learned about such time compression in Section 2.6 (Figure 2.76). There, we compressed a time-extended wideband signal by an analogue or digital filter (which we called matched filter) whose IRF has the time-reversed shape of the wanted input signal. Now, we are doing the same but in the inverse manner. We select stimulation signals $a_k(t)$ whose time shapes are time-reversed versions of the related spatial filter IRFs $T_{km}(t)$, that is

$$\begin{aligned} \mathbf{a}^{(m)}(t) &= A\mathbf{T}_m(-t) \\ \underline{\mathbf{a}}^{(m)}(f) &= A\underline{\mathbf{T}}_m^*(f) \end{aligned} \tag{4.341}$$

A sensor concept which is able to generate arbitrary stimulation signals was introduced in Section 3.5, Figure 3.82. In (4.341), A assigns an amplitude factor of the stimulus signals, and the superscript (m) denotes a signal optimized to emphasize the field strength[34] at the position of scatterer m. In doing so, the fields at scatterer positions yield

$$\begin{aligned} \boldsymbol{\beta}^{(m)}(t) &= \mathbf{T}^T(t) * \mathbf{a}^{(m)}(t) = A\mathbf{T}^T(t) * \mathbf{T}_m(-t) \\ \underline{\boldsymbol{\beta}}^{(m)}(f) &= \underline{\mathbf{T}}^T(f)\underline{\mathbf{a}}^{(m)}(f) = A\underline{\mathbf{T}}^T(f)\underline{\mathbf{T}}_m^*(f) \end{aligned} \tag{4.342}$$

Generally, the set of input signal $\mathbf{a}^{(m)}(t)$ will also provide field components at the positions of the other scatterers; however, the highest magnitudes will be observed at the position of scatterer m. We speak of a well-resolved scatterer m if the field strength at its position dominates the fields at the other scatterers; that is, we would

34) Note that there are several options to optimize the array input which creates a large energy concentration at scatterer position. Here, we will only discuss a procedure related to matched filtering. Another option could be to search for stimulation signals which approach the incident field to a delta pulse $\beta_m^{(m)}(t) \to \delta(t)$. In this case, the method would resemble the Wiener de-convolution. Further examples are given in Section 4.10.3.3.

get ideally $\boldsymbol{\beta}^{(m)} = \begin{bmatrix} 0 & 0 & \cdots & \beta_m^{(m)} & \cdots & 0 \end{bmatrix}$ and $\|\beta_m^{(m)}(t)\|_\infty \gg \|\beta_n^{(m)}(t)\|_\infty$; $n \neq m$ under practical conditions. Consequently, we can state that the whole scattering scenario is well resolved if holds

$$\mathbf{T}^T(t) * \mathbf{T}(-t) \approx \mathbf{T}^T(-t) * \mathbf{T}(t) \approx \begin{bmatrix} \mathbf{T}_1^T(t) * \mathbf{T}_1(-t) & 0 & \cdots & 0 \\ 0 & \mathbf{T}_2^T(t) * \mathbf{T}_2(-t) & \cdots & 0 \\ \vdots & \vdots & \ddots & \vdots \\ 0 & 0 & \cdots & \mathbf{T}_M^T(t) * \mathbf{T}_M(-t) \end{bmatrix}$$

$$\underline{\mathbf{T}}^T(f)\underline{\mathbf{T}}^*(f) \approx \underline{\mathbf{T}}^H(f)\underline{\mathbf{T}}(f) \approx \begin{bmatrix} \underline{\mathbf{T}}_1^T(f)\underline{\mathbf{T}}_1^*(f) & 0 & \cdots & 0 \\ 0 & \underline{\mathbf{T}}_2^T(f)\underline{\mathbf{T}}_2^*(f) & \cdots & 0 \\ \vdots & \vdots & \ddots & \vdots \\ 0 & 0 & \cdots & \underline{\mathbf{T}}_M^T(f)\underline{\mathbf{T}}_M^*(f) \end{bmatrix} \tag{4.343}$$

As obvious from (4.341), one needs to know the transmission matrix $\mathbf{T}(t)$ or $\underline{\mathbf{T}}(f)$ of the scenario under test in order to construct the required set of sounding signals $\mathbf{a}^{(m)}(t)$ performing energy concentration at position of scatterer m. Below, it will be shown how this can be done using the measurable response matrix $\mathbf{S}_M(t)$ or $\underline{\mathbf{S}}_M(f)$. It is interesting to note that for the purpose of energy concentration, one is not required to know the actual position of the radiators. First investigations using this approach to concentrate energy were accomplished for ultrasound propagation. Here, one was typically aimed to concentrate the energy at the strongest scatterer. The corresponding method is often referred to as DORT (Diagonalization de l'Operateur de Retournement Temporel) [125–128]. In this book, we will not further follow this track since we are only interested in sensing tasks but not in questions of energy concentration aimed to affect the scenario under test in a singular point.

Here, we will rather deal with imaging the scenario which may be considered as a kind of 'virtual' energy or signal concentration at the location of point scatters. That is, we would like to do the same as considered above except that we want to do it only virtually (i.e. by numerical calculations). However now, we also want to know the position $\mathbf{r}'$ of best energy accumulation which was not required above. For that purpose, we assume to know the Green's function (involving also the antenna behavior) of the scenario under test. We join all the Green's functions of the (known) antenna positions $\mathbf{r}_k$; $k \in [1, K]$ and the arbitrary point $\mathbf{r}'$ of interest in the column vector

$$\mathbf{G}(\cdots, \mathbf{r}') = [G(\cdots, \mathbf{r}_1, \mathbf{r}') \quad G(\cdots, \mathbf{r}_2, \mathbf{r}') \quad G(\cdots, \mathbf{r}_K, \mathbf{r}')]^T.$$

This vector represents the IRFs of the propagation paths from the individual antennas to an arbitrary point $\mathbf{r}'$. Accordingly to (4.341), we would provide (virtually) a local energy concentration at point $\mathbf{r}'$ if we stimulate (virtually) the antenna array with a set of signals determined by

$$\begin{aligned} \mathbf{a}^{(\mathbf{r}')}(t) &= A\mathbf{G}(-t, \mathbf{r}') \\ \underline{\mathbf{a}}^{(\mathbf{r}')}(f) &= A\underline{\mathbf{G}}^*(f, \mathbf{r}') \end{aligned} \tag{4.344}$$

Further, we assume to know from measurements the multi-static response matrix $\mathbf{S}_M$ of the scenario from which we succeeded in some way to extract the transmission matrix $\hat{\mathbf{T}} = \begin{bmatrix} \hat{\mathbf{T}}_1 & \hat{\mathbf{T}}_2 & \cdots & \hat{\mathbf{T}}_m & \cdots & \hat{\mathbf{T}}_M \end{bmatrix}$ for the M dominant scatterer where the head over $\mathbf{T}$ symbolizes that it is estimated from measurements. Following above considerations, we could identify the position $\mathbf{r}'_m$ of the scatterer m by searching the virtual stimulus $\mathbf{a}^{(\mathbf{r}'_m)}(t)$ or $\underline{\mathbf{a}}^{(\mathbf{r}'_m)}(f)$ which best matches the transmission IRFs $\hat{\mathbf{T}}_m$. This, we may express by

$$\mathbf{r}'_m = \arg\max_{\mathbf{r}'} \Upsilon^m(t, \mathbf{r}') \text{ or } \mathbf{r}'_m = \arg\max_{\mathbf{r}'} |\underline{\Upsilon}^m(f, \mathbf{r}')| \tag{4.345}$$

where

$$\begin{aligned} \Upsilon^{(m)}(t, \mathbf{r}') &= \hat{\mathbf{T}}_m^T(t) * \mathbf{G}(-t, \mathbf{r}') \\ \underline{\Upsilon}^{(m)}(f, \mathbf{r}') &= \hat{\underline{\mathbf{T}}}_m^T(f) \underline{\mathbf{G}}^*(f, \mathbf{r}') \end{aligned} \tag{4.346}$$

represents the point spread function for the scatterer m. In the case of time domain processing of (4.345), one takes a windowed version of $\Upsilon^{(m)}(t, \mathbf{r}')$ accordingly to approaches depicted in Figure 4.92, 4.93 or 4.100.

The entire radar image finally results from the superposition of the point spread functions of all involved scatterers:

$$I(\mathbf{r}') = \sum_{m=1}^{M} \Upsilon^{(m)}(\cdots, \mathbf{r}') \tag{4.347}$$

In the case of well-resolved scatterers, the point spread functions do not overlap. Thus, we get for the scatters m and n

$$\Upsilon^{(m)}(\cdots, \mathbf{r}'_n) \approx \Upsilon^{(n)}(\cdots, \mathbf{r}'_m) \approx 0; \quad m \neq n \tag{4.348}$$

The need to know a priori the Green's functions of the scenario represents the main difficulty to implement the exact imaging approach[35)] based on (4.345) and (4.346). Since in case of inhomogeneous propagation media these functions are typically unidentified, one mostly settle for the free space Green's functions $\mathbf{G}_0(t, \mathbf{r}')$ which can be extended by an appropriate antenna model and the average dispersion of the propagation medium if appropriate (refer to Section 5.4 for dispersive propagation).

For further considerations, we refer to a slightly modified time reversal operator $\mathbf{C}_{\text{TRO}}$ as introduced by Ref. [125]. Since we will deal with matrix decomposition in what follows, we will restrict ourselves to frequency domain notation from this point onwards and the variable f will not be explicitly indicated. We define the time reversal operator as follows:

$$\underline{\mathbf{C}}_{\text{TRO}} = \underline{\mathbf{S}}_M \underline{\mathbf{S}}_M^* \tag{4.349}$$

35) This remark holds for all imaging approaches since steering vector design is always based on that prior knowledge.

In time domain, the $\mathbf{C}_{\text{TRO}}$ matrix represents a set of auto- and cross-correlation functions. From reciprocity of the scenario (4.339), we find the important property

$$\underline{\mathbf{C}}_{\text{TRO}} = \underline{\mathbf{S}}_M \underline{\mathbf{S}}_M^H = \underline{\mathbf{C}}_{\text{TRO}}^H \tag{4.350}$$

that is the time reversal operator is Hermitian.

Insertion of our scenario model (4.338), we achieve at

$$\underline{\mathbf{C}}_{\text{TRO}} = \underline{\mathbf{S}}_M \underline{\mathbf{S}}_M^H = (\underline{\mathbf{T}}\underline{\mathbf{\Lambda}}\underline{\mathbf{T}}^T)(\underline{\mathbf{T}}\underline{\mathbf{\Lambda}}\underline{\mathbf{T}}^T)^H = \underline{\mathbf{T}}\underline{\mathbf{\Lambda}}\underline{\mathbf{T}}^T\underline{\mathbf{T}}^*\underline{\mathbf{\Lambda}}^*\underline{\mathbf{T}}^H = \underline{\mathbf{T}}\underline{\mathbf{\Omega}}\underline{\mathbf{T}}^H \tag{4.351}$$

where $\underline{\mathbf{\Omega}}$ is a $[M, M]$ matrix

$$\underline{\mathbf{\Omega}} = \underline{\mathbf{\Lambda}}\underline{\mathbf{T}}^T\underline{\mathbf{T}}^*\underline{\mathbf{\Lambda}}^*$$

Following to (4.334), the entries of $\mathbf{T}$ represent Green's functions related to the antenna positions $\mathbf{r}_k$ and scatterer positions $\mathbf{r}'_m$ so that an equivalent notation to (4.351) is given by

$$\underline{\mathbf{C}}_{\text{TRO}} = \underline{\mathbf{G}}(\mathbf{r}, \mathbf{r}'_s)\underline{\mathbf{\Omega}}\underline{\mathbf{G}}^H(\mathbf{r}, \mathbf{r}'_s)$$

$$\text{with } \underline{\mathbf{\Omega}} = \underline{\Lambda}\underline{\mathbf{G}}^T(\mathbf{r}, \mathbf{r}'_s)\underline{\mathbf{G}}^*(\mathbf{r}, \mathbf{r}'_s)\underline{\Lambda}^*$$

$$\mathbf{T} = \underline{\mathbf{G}}(\mathbf{r}, \mathbf{r}'_s) = \begin{bmatrix} \underline{G}(\mathbf{r}_1, \mathbf{r}'_1) & \underline{G}(\mathbf{r}_1, \mathbf{r}'_2) & \cdots & \underline{G}(\mathbf{r}_1, \mathbf{r}'_M) \\ \underline{G}(\mathbf{r}_2, \mathbf{r}'_1) & \underline{G}(\mathbf{r}_2, \mathbf{r}'_2) & \cdots & \underline{G}(\mathbf{r}_2, \mathbf{r}'_M) \\ \vdots & \vdots & \ddots & \vdots \\ \underline{G}(\mathbf{r}_K, \mathbf{r}'_1) & \underline{G}(\mathbf{r}_K, \mathbf{r}'_2) & \cdots & \underline{G}(\mathbf{r}_K, \mathbf{r}'_M) \end{bmatrix} \tag{4.352}$$

In the case of well-resolved scatterers (refer to (4.343)), $\underline{\mathbf{\Omega}} = \mathbf{\Omega}^{(\text{wrs})}$ is a purely real-valued diagonal matrix:

$$\mathbf{\Omega}^{(\text{wrs})} = \begin{bmatrix} \Omega_1^{(\text{wrs})} & 0 & \cdots & 0 \\ 0 & \Omega_2^{(\text{wrs})} & \cdots & 0 \\ \vdots & \vdots & \ddots & \vdots \\ 0 & 0 & \cdots & \Omega_M^{(\text{wrs})} \end{bmatrix}$$
$$\Omega_m^{(\text{wrs})} = |\underline{\Lambda}_m|^2 \underline{\mathbf{T}}_m^T \underline{\mathbf{T}}_m^* = \sigma_m \sum_{k=1}^{K} |\underline{T}_{km}|^2 \tag{4.353}$$

where σ_m is the radar cross section of scatterer m and $|\underline{T}_{km}|$ is the path attenuation between antenna k and scatterer m

Now, we apply the time reversal operator to the multi-static response matrix gained from measurements

$$\hat{\underline{\mathbf{C}}}_{\text{TRO}} = \hat{\underline{\mathbf{S}}}_M \hat{\underline{\mathbf{S}}}_M H = \hat{\underline{\mathbf{C}}}_{\text{TRO}} H \tag{4.354}$$

and we perform eigenvalue decomposition. Since $\hat{\underline{\mathbf{C}}}_{\text{TRO}}$ is Hermitian, we result in (refer to Annex A.6 (A.130))

$$\hat{\underline{\mathbf{C}}}_{\text{TRO}} = \underline{\mathbf{Q}}\mathbf{R}\underline{\mathbf{Q}}^H; \quad \underline{\mathbf{Q}}\underline{\mathbf{Q}}^H = \underline{\mathbf{Q}}^H\underline{\mathbf{Q}} = \mathbf{I} \tag{4.355}$$

Herein, $\mathbf{R}$ represents a real-valued diagonal matrix of the eigenvalues of $\hat{\underline{\mathbf{C}}}_{\text{TRO}}$ and $\underline{\mathbf{Q}} = [\underline{\mathbf{q}}_1 \quad \underline{\mathbf{q}}_2 \quad \cdots \quad \underline{\mathbf{q}}_K]$ summarizes the related eigenvectors $\underline{\mathbf{q}}_k$ whereas we have ordered the eigenvalues by their magnitude

$$\mathbf{R} = \begin{bmatrix} R_1 & 0 & \cdots & 0 \\ 0 & R_2 & \cdots & 0 \\ \vdots & \vdots & \ddots & \vdots \\ 0 & 0 & \cdots & R_K \end{bmatrix}$$
$$R_1 \geq R_2 \geq \cdots R_k \geq \cdots R_K$$

If the scene to be observed is sparse (e.g. $M < K$ scatterers), we will ideally get only M eigenvalues, that is

$$R_1 \geq R_2 \geq \cdots R_M \geq 0$$

The remaining entries of $\mathbf{R}$ will be zero (i.e. $R_{M+1} = R_{M+2} = \cdots R_k = 0$). Practically, we will have M dominant eigenvalues and $K - M$ negligible ones whose magnitude depend on measurement noise and clutter. The number M of dominant scatterers is either given by the number of eigenvalues exceeding a certain threshold or if there is a pronounced step between two consecutive eigenvalues.

For further considerations, we set all non-dominant eigenvalues to zero so that (4.355) may also be written as

$$\begin{aligned} \hat{\underline{\mathbf{C}}}_{\text{TRO}} &\approx [\underline{\mathbf{Q}}_{\text{S}} \quad \underline{\mathbf{Q}}_{\text{N}}] \begin{bmatrix} \mathbf{R}_{\text{S}} & \mathbf{0} \\ \mathbf{0} & \mathbf{0} \end{bmatrix} [\underline{\mathbf{Q}}_{\text{S}} \quad \underline{\mathbf{Q}}_{\text{N}}]^H = \underline{\mathbf{Q}}_{\text{S}}\mathbf{R}_{\text{S}}\underline{\mathbf{Q}}_{\text{S}}^H \\ \mathbf{I} &= \begin{bmatrix} \underline{\mathbf{Q}}_{\text{S}}^H \\ \underline{\mathbf{Q}}_{\text{N}}^H \end{bmatrix} [\underline{\mathbf{Q}}_{\text{S}} \quad \underline{\mathbf{Q}}_{\text{N}}] \Rightarrow \underline{\mathbf{Q}}_{\text{S}}^H\underline{\mathbf{Q}}_{\text{N}} = \mathbf{0} \\ \underline{\mathbf{Q}}_{\text{S}} &= [\underline{\mathbf{q}}_1 \quad \underline{\mathbf{q}}_2 \quad \cdots \quad \underline{\mathbf{q}}_M] \\ \underline{\mathbf{Q}}_{\text{N}} &= [\underline{\mathbf{q}}_{M+1} \quad \underline{\mathbf{q}}_{M+2} \quad \cdots \quad \underline{\mathbf{q}}_K] \\ \mathbf{R}_{\text{S}} &= \left[\begin{bmatrix} R_1 & 0 & \cdots & 0 \\ 0 & R_2 & \cdots & 0 \\ \vdots & \vdots & \ddots & \vdots \\ 0 & 0 & \cdots & R_M \end{bmatrix}\right] \end{aligned} \tag{4.356}$$

which represents a decomposition of the measured time reversal operator into signal and noise space. The related matrices are assigned by the subscript S and N, respectively.

Since by assumption the scenario model (4.351) and (4.352), respectively, should be well adapted to the measurements (4.356), two conditions must hold

$$\begin{aligned} &\hat{\underline{\mathbf{C}}}_{\text{TRO}} = \underline{\mathbf{C}}_{\text{TRO}} = \underline{\mathbf{T}}\underline{\mathbf{\Omega}}\underline{\mathbf{T}}^H = \underline{\mathbf{G}}(\mathbf{r}, \mathbf{r}'_{\text{s}})\underline{\mathbf{\Omega}}\underline{\mathbf{G}}^H(\mathbf{r}, \mathbf{r}'_{\text{s}}) \approx \underline{\mathbf{Q}}_{\text{S}}\mathbf{R}_{\text{S}}\underline{\mathbf{Q}}_{\text{S}}^H \\ &\underline{\mathbf{Q}}_{\text{S}}^H\underline{\mathbf{Q}}_{\text{N}} = \mathbf{0} \end{aligned} \tag{4.357}$$

We will discuss their consequences for the cases of well- and non-well-resolved scatterer.

Well-Resolved Scatter The matrix $\underline{\mathbf{\Omega}} = \mathbf{\Omega}^{(wrs)}$ is real valued and diagonal as to be seen from (4.353). We decompose the transmission matrix $\underline{\mathbf{T}} = \underline{\mathbf{G}}(\mathbf{r}, \mathbf{r}'_{\text{s}}) = \underline{\mathbf{P}}\,\mathbf{D}$ in a real-valued $[M, M]$ diagonal matrix $\mathbf{D}$ and a $[K, M]$ matrix $\underline{\mathbf{P}}$ for which holds $\underline{\mathbf{P}}^H\,\underline{\mathbf{P}} = \mathbf{I}$ so that we find from (4.343)

$$\underline{\mathbf{T}}^H\underline{\mathbf{T}} = (\underline{\mathbf{P}}\,\mathbf{D})^H(\underline{\mathbf{P}}\,\mathbf{D}) = \mathbf{D}^2 \tag{4.358}$$

Hence, the entries of $\mathbf{D}$ only depend on the path attenuation $d_m^2 = \underline{\mathbf{T}}_m^H\underline{\mathbf{T}}_m$ (see also (4.353)). Insertion in (4.357) leads us to

$$\begin{aligned} &\underline{\mathbf{Q}}_{\text{S}}\mathbf{R}_{\text{S}}\underline{\mathbf{Q}}_{\text{S}}^H = \underline{\mathbf{P}}\,\mathbf{D}\,\Omega^{(\text{wrs})}\mathbf{D}\,\underline{\mathbf{P}}^H \\ &\Rightarrow \underline{\mathbf{Q}}_{\text{S}} \mathrel{\hat{=}} \underline{\mathbf{P}} \\ &\Rightarrow \mathbf{R}_{\text{S}} \mathrel{\hat{=}} \mathbf{D}\,\Omega^{(\text{wrs})}\mathbf{D} :: \quad R_\mu = \sigma_m(\underline{\mathbf{T}}_m^H\underline{\mathbf{T}}_m)^2 :: \quad m, \mu \in [1, M] \end{aligned} \tag{4.359}$$

so that we can assign the individual scatterer m to the eigenvalue R_μ of the measured time reversal operator $\hat{\underline{\mathbf{C}}}_{\text{TRO}}$. Correspondingly, the eigenvector $\underline{\mathbf{q}}_\mu$ relates to the transmission vector $\underline{\mathbf{T}}_m = \underline{\mathbf{G}}(\mathbf{r}, \mathbf{r}'_m) \propto \underline{\mathbf{q}}_\mu$ linking the antennas to scatterer m.

Since we have identified the individual transmission coefficients of the scatterer by above eigenvalue decomposition, we can perform imaging as demonstrated by (4.345)–(4.347). That is, we have to deal with

$$\underline{\Upsilon}^{(\mu)}(\mathbf{r}') = \underline{\mathbf{q}}_\mu^H\underline{\mathbf{G}}(\mathbf{r}') \tag{4.360}$$

which tends to a maximum if $\mathbf{r}'$ meets the actual position $\mathbf{r}'_m$ of scatterer m that is linked to the μth eigenvalue.

At this point, it should be mentioned that there is no fixed relation between the numbering μ of the eigenvalues and the numbering m of the scatterers. The eigenvalue decomposition is performed individually at every frequency point. Since radar cross section and path attenuation may vary with frequency, an individual scatterer may take different places within the order of eigenvalues.

Non-Well-Resolved Scatterers In the case of non-well-resolved scatters, the matrix $\underline{\mathbf{\Omega}}$ is not a diagonal one and hence we cannot uniquely relate a Green's function

$\underline{G}(\mathbf{r}_k, \mathbf{r}'_m)$ to a corresponding entry of an eigenvector so that imaging procedure (4.360) will fail. However, we can observe from (4.357) that only the signal space spanned by $\mathbf{Q}_S$ may contain appropriate Green's function $\underline{G}(\mathbf{r}_k, \mathbf{r}'_m)$ (i.e. the Green's vector can be represented by a linear combination of the eigenvectors $\underline{\mathbf{q}}_1 \cdots \underline{\mathbf{q}}_M$: $\underline{\mathbf{G}}(\mathbf{r}'_m) = \mathbf{Q}_s \underline{\mathbf{A}}^{(m)}$ where the column vector $\underline{\mathbf{A}}^{(m)} = \begin{bmatrix} \underline{a}_1^{(m)} & \underline{a}_2^{(m)} & \cdots & \underline{a}_M^{(m)} \end{bmatrix}^T$ represents a set coefficients) while the noise space $\mathbf{Q}_N$ does not. Hence, the second condition[36] in (4.357) also holds for a Green's vector $\underline{\mathbf{G}}(\mathbf{r}')$ if it meets the position $\mathbf{r}'_m$ of any scatterer. Since we replace the matrix $\mathbf{Q}_S$ by a vector, we have to express that condition in a slightly different form:

$$(\underline{\mathbf{G}}^H(\mathbf{r}'_m)\underline{\mathbf{Q}}_N)(\underline{\mathbf{G}}^H(\mathbf{r}'_m)\underline{\mathbf{Q}}_N)^H = \underline{\mathbf{G}}^H(\mathbf{r}'_m)\underline{\mathbf{Q}}_N\underline{\mathbf{Q}}_N^H\underline{\mathbf{G}}(\mathbf{r}'_m) = 0 \tag{4.361}$$

Since (4.361) is only met for scatterer positions, the imaging procedure involves either a minimum search of

$$\mathbf{r}'_m = \arg\min_{\mathbf{r}'}(\underline{\mathbf{G}}^H(\mathbf{r}')\underline{\mathbf{Q}}_N\underline{\mathbf{Q}}_N^H\underline{\mathbf{G}}(\mathbf{r}')) \tag{4.362}$$

or a maximum search

$$\mathbf{r}'_m = \arg\max_{\mathbf{r}'} \frac{1}{\underline{\mathbf{G}}^H(\mathbf{r}')\underline{\mathbf{Q}}_N\underline{\mathbf{Q}}_N^H\underline{\mathbf{G}}(\mathbf{r}')}. \tag{4.363}$$

Alternatively one can estimate the intensity image via

$$I(\mathbf{r}') = \frac{1}{\underline{\mathbf{G}}^H(\mathbf{r}')\underline{\mathbf{Q}}_N\underline{\mathbf{Q}}_N^H\underline{\mathbf{G}}(\mathbf{r}')} \tag{4.364}$$

which is similar the pseudo-spectrum of the well-known MUSIC method that was first applied for high-resolution spectral estimation [129].

The interested reader can find further discussions and examples of microwave imaging via eigenvalue decomposition (or via singular value decomposition in modified versions), for example in Refs [130–140].

4.10.3.5 Array Configurations and Remarks on UWB Radar Imaging

Many different antenna configurations and operational modes are possible. We will consider several aspects in order to have some indication for a systematic of UWB arrays:

Type of array: Here, we will distinguish between physical array (PAR – physical aperture radar or in short radar array) and synthetic array (SAR – synthetic aperture radar). A physical array means that all measurement positions are actually occupied by an antenna. In the case of a synthetic array, the measurement positions are successively taken over from one antenna, the data are recorded at every point separately and finally joint with the data captured at the other positions. The antenna displacement is either done stepwise or continuously. A combination of both array types is possible too, that is the whole (physical) array is moved

36) Note that this condition holds always, that is also for well-resolved scatterer. But in this case, it was of no profit for the imaging procedure.

over a path or surface of interest. Scanning arrays require that the scenario behaves stationary over the recoding time. The synthetic array approach may also be inverted. That is, the radar antennas are stationary but the target moves across their beams or it is rotated in front of the antennas. This is called inverse synthetic aperture radar (ISAR).

Spatial arrangement of antenna positions: The antenna positions (for both physical and synthetic arrays) may be arranged in one-, two- or three-dimensional structures. We have string configurations which may form a straight line (linear array), a circle (circular array) or an arbitrary-shaped curve. Two-dimensional shaped array structures are often planar, a surface of a sphere, a cylinder, a torus or other well-defined surfaces. But also randomly distributed antennas can be applied as long as their position and orientation is known.

Region of interest: The space region to be observed by the radar array may be differently placed. The most common option is that the region of interest is in front of a typically planar array. Furthermore, the array may be arranged at a closed (e.g. a sphere.) or partially closed surface (hemisphere, cylinder shell etc.) and the observation area is either inside or outside the array structure. Finally, the array elements may also be part of the observation scenario and they are arbitrarily distributed within the space region to be investigated.

Antenna operation: Array antennas may operate in transmit as well as in the receive mode or they may work solely as transmitter or receiver only. Correspondingly, we have several options to organize the array operation. If an antenna is simultaneously used for both – transmitting and receiving, one often talks from mono-static operation. If one has two antennas of fixed distance – one for transmitting and one for receiving – it is called bi-static measurement. In the case of short antenna distance, this operational mode is sometimes also referred to as quasi-mono-static. The multi-static mode correspondingly involves more than two antennas which may act as transmitter and/or receiver. This type of operation is also named as MiMo radar. MiMo configurations allow for several operation modes:

- One antenna is permanently transmitting, the other antennas are receiving.
- The same as before, but the transmitting antenna is switched through the array.
- The same as before, but also the transmitting antenna is able to receive. The active antenna is successively switched through the array.
- All antennas acting as receivers and transmitters at the same time and all transmitters are active. If all antennas are fed by the same but time shifted stimulus signals, the (transmission) beam direction of the antenna array may be controlled. Due to reciprocity, the control mechanism of the transmission can be considered as in the receiving case (refer to Figure 4.92) but with inverted signal flow. A possible sensor implementation for transmitter beam steering is adumbrated in Figure 3.64. If one additionally applies mutually uncorrelated sounding signals for all transmitters, one finally gets a panorama view array which can be flexibly focused purely by digital signal processing.

The technical implementation of UWB arrays may pose some challenges. If super-resolution techniques should be applied, the precision of the antenna localization or the mechanical precision of an array may be the limiting factor since sophisticated sensor electronics registers mechanical imperfections or motions down to the micrometre range (refer to Figures 4.63 and 4.64). Large MiMo arrays need elaborated antennas, electronics and synchronization networks. The incoming data amount may be huge namely if time variable scenarios should be imaged. Hence, pre-processing and data reduction should be performed in close vicinity to the radiators and the number of antenna elements should be as small as possible. UWB sounding establishes the preconditions to apply sparse arrays since no grating lobes are to be suspected (fractional bandwidth larger 100% supposed [111]). Problems to be solved refer to find appropriate array structures providing good image resolution, avoiding the reduction of sidelobe suppression by element shadowing and to combat against the reduced signal-to-noise ratio as a consequence of the low number of antenna elements.

We should also recall the conditions under which UWB radar imaging was developed in order to assess correctly the deviations between the idealized considerations above and practical measurement or application. These basic assumptions were

- **The time shape of scattered signal is not affected by the scattering object:** This is a typical 'narrow band' assumption, which correctly does only hold for specular scattering (i.e. reflection at large objects with smooth surface) under UWB conditions. Point targets perform a twofold derivation of the sounding signal. This will change the time shape of the signal but not its pulse duration or coherence time. Hence, it will not or only barely affect the image resolution. In case of resonant scattering, the signal will be stretched in time and its shape will mostly depend on aspect angle. Consequently, the different antenna signals are less coherent which finally leads to blurred images. The condition under which microwave imaging typically holds is the so-called Born approximation[37] which only allows small permittivity variations within the scattering scenario so that multiple reflections may be neglected.
- **Ideal point radiator:** Emission and reception of electric fields are referred to points. This is not correct since antennas are objects of finite dimensions. The consequences are deviations from the spherical wavefront. Furthermore, real antennas scatter electromagnetic waves which may cause multiple reflections in the test scenario affecting the imaging quality. Especially for very short-range applications antenna backscattering is an unwanted effect.
- **Idealized wave propagation:** We always supposed isotropic wave propagation within a homogenous medium. But UWB imaging is often applied to investigate opaque bodies, which involves penetration of boundaries causing reflection and

37) The Born approximation takes only the incident field as driving field of scattering at each point within the object, that is field components created by scattering at other points and reaching the considered point too are neglected. Born approximation only holds for weak dielectric contrast.

refraction. Hence, we get signals (reflection at the boundary – so-called surface clutter) which perturb imaging. These signals must be appropriately removed from the data since they often dominate the reflections of the wanted targets.

Further, the penetrated waves will change their wavefront which has to be respected correctly by designing the time shift performed by the steering vector. Wave propagation in most materials cause attenuation and dispersion leading to shape variations of the sounding signal in dependency on the propagation path length. Finally, we only considered scalar wave propagation. Actually, electromagnetic waves are vector waves which involves that electromagnetic phenomena depend on the field polarization. Hence, one can gain more information about a scenario under test by exploiting polarimetric measurements which however will extend the implementation effort of corresponding sensor array (see more on these issues in Sections 5.6 and 5.7). And last but not least, the propagation medium is often inhomogeneous due to spatial permittivity variations of the background medium, for example the soil in case of ground penetrating radar or biological tissue in case of medical microwave imaging.

4.10.4 Shape Reconstruction by Inverse Boundary Scattering

Previous imaging approaches are based on the localization of point scatterers which are more or less loosely widespread within the observation volume. Boundaries of larger objects were considered by that approach as sources of spherical elementary waves (Huygens principle), that is an interface represents an 'aggregation' of point scatterer.

We will consider scattering from a different point of view. The body of interest should have a gently curved surface which we will reconstruct by radar measurements. Under this condition, the scattered signal will have the same time shape as the incident one and it will be possible to reconstruct the body shape with high resolution by pure roundtrip time measurements.

4.10.4.1 Shape Reconstruction by Quasi-Wavefront Derivation

The basic idea is illustrated in Figure 4.104 for a two-dimensional scenario and scanning along the x-axis with a mono-static radar. Figure 4.104a explains the creation of the backscatter signals. A spherical wave emanates from a point source. It travels towards the object where it will touch the body in the specular point. This point represents the shortest distance between radiator and scattering object. Since the tangent plane related to this point is perpendicular to the travelling path, the incoming wave will be reflected immediately back to its origin where it will be registered. Hence, the measured travelling time gives us the shortest distance from the antenna to the body of interest.

By moving the point source along a known line (for illustration, we simply use a straight line), we can plot these distances in a radargram (see Figure 4.104b) as function of the scan path. From this radargram the original shape of the body has to be reconstructed.

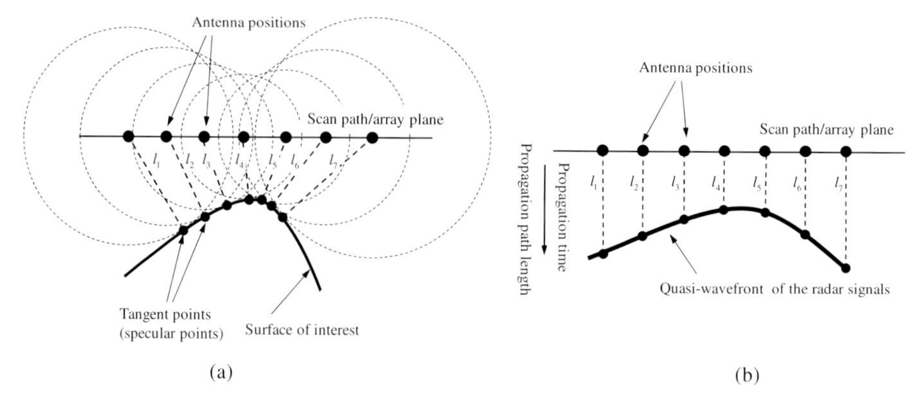

Figure 4.104 Radar measurements at a gently curved boundary. (a) Illustration of wavefronts and (b) resulting radargram.

The first reconstruction step consists in recovering the roundtrip time from the measured IRFs. This time will be translated in distance values via $r = \tau c/2$ (c is the propagation speed) as already depicted in Figure 4.104. Since wavefront detection is typically subjected to noise, it is recommended to smooth appropriately the recovered curve by low-pass filtering, curve fitting or other approaches before starting with object shape reconstruction. The curve appearing after that procedure in the radargram is called the quasi-wavefront. Practical issues of wavefront extraction are discussed in Section 6.6 and Ref. [141].

In order to develop the relation between object and radargram space, we redraw Figure 4.104 and introduce appropriate geometric relations (see Figure 4.105).

Supposing the antenna is placed at position $[x_0, 0]$, the roundtrip time measured by the radar corresponds the shortest distance r_s to the scattering object as shown in the Figure 4.105a. Let the object boundary be described in parametric or non-parametric form

$$\begin{aligned} &z_s = z(\varphi); x_s = x(\varphi) \\ &\text{or} \quad z_s = f(x_s) \end{aligned} \tag{4.365}$$

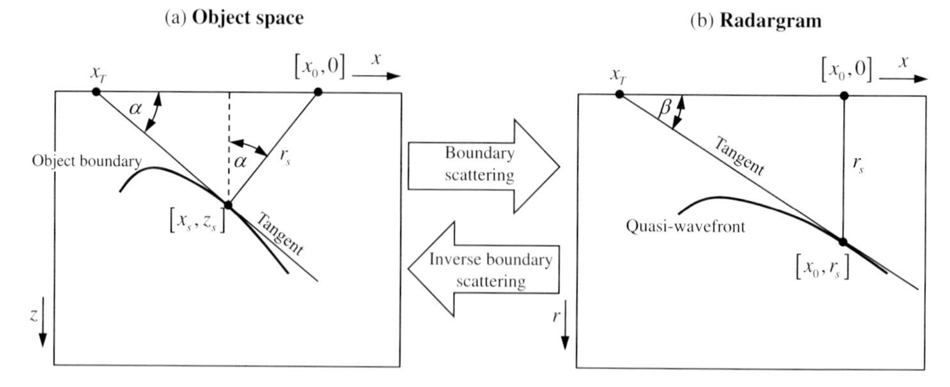

Figure 4.105 Object space $[x, z]$ versus radargram space $[x, r]$ and their mutual transformations.

then the radargram results from a simple geometric consideration (see Figure 4.105a)

$$\tan\alpha = \frac{x_0 - x_s}{z_s} \Rightarrow x_0 = x_s + f(x_s)f'(x_s)$$

$$\cos\alpha = \frac{z_2}{r_s} = \sqrt{1+\tan^2\alpha} \Rightarrow r_s = f(x_s)\sqrt{1+(f'(x_s))^2}$$

$$\text{with} \quad f'(x_s) = \left.\frac{\mathrm{d}f(x)}{\mathrm{d}x}\right|_{x=x_s} = \left.\frac{\partial z/\partial\varphi}{\partial x/\partial\varphi}\right|_{\varphi=\varphi_s} = \tan\alpha \qquad (4.366)$$

These relations are referred as boundary scattering transform. Inversely, it will also be possible to conclude from the radargram to the original shape of the object what is our actual goal of radar imaging. This procedure is correspondingly called inverse boundary scattering transform or migration. From the radar measurement, we know the developing of the wavefront $r_s = g(x_0)$ as function of the measurement position (Figure 4.105b). In order to invert the procedure, we make use of the migration relation [104]:

$$\sin\alpha = \frac{r_s}{x_0 - x_T} = \tan\beta \qquad (4.367)$$

which results from the fact that the two tangent lines drawn in both diagrams of Figure 4.105 crosses the x-axis at the same position. The common intersection point will be obvious by regarding a scattering object whose boundary would be identical with the drawn tangent in the object space. In that case, the resulting wavefront is identical with the tangent line in the radargram. Since the roundtrip time is zero if the scatterer surface crosses the x-axis, both tangent lines have the same intersection with the scanning axis (see Refs [142–144] for more details). If we again refer to Figure 4.105a and if we apply (4.367), we yield a simple set of equations which allow us to reconstruct the surface of an object based on the data given by a radargram.

$$\sin\alpha = \frac{x_0 - x_s}{r_s} = \tan\beta = g'(x_0) \Rightarrow x_s = x_0 - g(x_0)g'(x_0)$$

$$\cos\alpha = \frac{z_s}{r_s} = \sqrt{1-\sin^2\alpha} = \sqrt{1-\tan^2\beta} \Rightarrow z_s = g(x_0)\sqrt{1-(g'(x_0))} \qquad (4.368)$$

The shown examples (4.366) and (4.368) only respect two-dimensional arrangements and linear scanning. Extensions to three dimensions, an arbitrary scanning path and bi-static radar arrangements can be found in Ref. [145] and are discussed in Section 6.6. A comparative study on wavefront base reconstruction and beamforming was accomplished in Ref. [146].

So far, the inverse boundary transform is only used to determine the surface points but not the surface normal even if this would be possible. Hence, one looses information about the scattering surface.

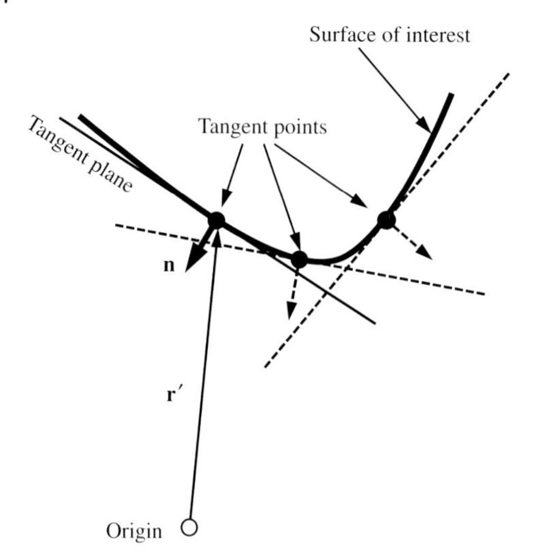

Figure 4.106 Approximation of a curved surface by a set of tangent planes.

4.10.4.2 Shape Reconstruction Based on Tangent Planes

Here, we will not further follow this way of surface reconstruction. Rather, we will generalize the method applying vector calculus which permits the use of arbitrary antenna configurations and arbitrary scanning.

The idea is to approximate the body shape by a multitude of tangent planes as it is depicted in Figure 4.106. Each of these tangential planes is given by it normal vector $\mathbf{n}$ and the position vector $\mathbf{r}'$ of the tangent point, that is the point at which the plane touches the body. Hence, we have to address the question on how to determine the position and orientation (i.e. $\mathbf{r}'$ and $\mathbf{n}$) of a flat scattering plane in space. Initially, we will consider this plane as infinite.

A plane is defined by three points which we need to determine by mono-static and bi-static radar measurements. At this, we are faced with the problem that an ultra-wideband radar measurement is only able to determine a distance but not a direction since the antennas are typically to little directive.

4.10.4.3 Planar Interface Localization by Mono-Static Measurements

For the sake of simplicity, we will use a point radiator as shown in Figure 4.107 for distance measurements. The round trip time τ of the scattered wave is

$$\tau = 2l/c \tag{4.369}$$

in which l is the distance to the unknown plane.

As obvious from Figure 4.107, the direction of l coincides with the normal unity vector $\mathbf{n}$ of the plane. If we would know $\mathbf{n}$, we could determine the point of normal incidence $\mathbf{r}'$ since the antenna position $\mathbf{r}_a$ was prior fixed:

$$\mathbf{r}' = \mathbf{r}_a - l\,\mathbf{n} \tag{4.370}$$

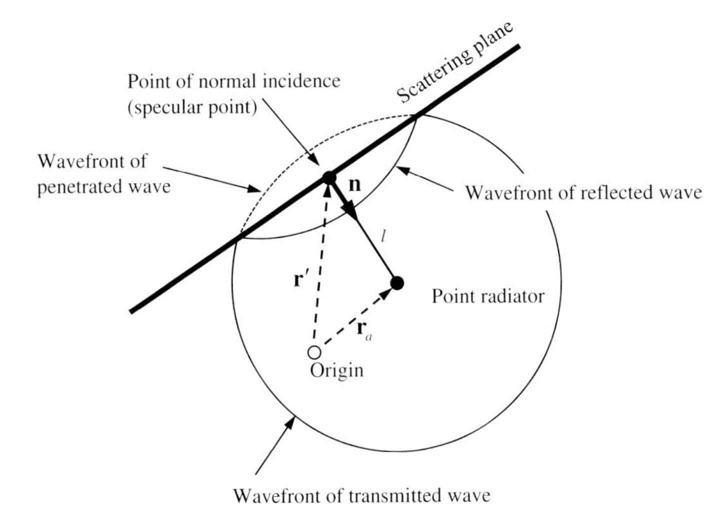

Figure 4.107 Scattering of a spherical wave at a planar surface. The radiator acts as transmitter and receiver (mono-static measurement).

Ultra-wideband antennas have typically a wide beam (we even supposed an isotropic radiator) so that the direction $\mathbf{n}$ cannot be determined by antenna beam rotation. In order to solve the problem, we perform (successively or in parallel) the same measurements from different known antenna positions. We need at least three positions in order to solve the problem for the three-dimensional case since we usually know at which side of the antenna plane the scattering surface lies. Figure 4.108 defines the corresponding quantities. The mutual position of the

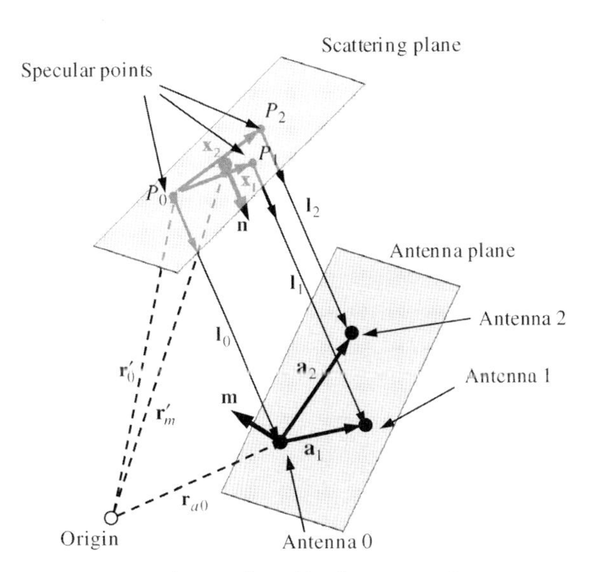

Figure 4.108 Definition of variables for mono-static measurement.

three antennas are defined by the vectors $\mathbf{a}_1$ and $\mathbf{a}_2$. They span a plane which is called antenna plane. Its orientation in space is given by the normal vector

$$\mathbf{m} = \frac{\mathbf{a}_1 \times \mathbf{a}_2}{|\mathbf{a}_1 \times \mathbf{a}_2|} \tag{4.371}$$

It is also possible to deal with a larger number of measurement positions in order to get an overdetermined system of equations. But for the sake of shortness, we restrict ourselves to the basic approach only.

Corresponding to Figure 4.107, we measure with a radar devise the shortest distance from the radiator to the plane of interest. Thus, the vectors $\mathbf{l}_i$ connecting the antenna i with the point P_i which is foremost touched by the sounding wave stands perpendicular at the scattering plane. Hence, we can write

$$\mathbf{l}_i = l_i\,\mathbf{n} \tag{4.372}$$

in which the distances l_i are known from the radar measurement. The scattering plane is spanned by the three points P_i or the two vectors $\mathbf{x}_1$ and $\mathbf{x}_2$ so that the normal vector can be expressed by

$$\mathbf{x}_2 \times \mathbf{x}_1 = \mu\mathbf{n} \tag{4.373}$$

Further we can state

$$\begin{aligned} \mathbf{x}_1 &= \mathbf{a}_1 - (\mathbf{l}_1 - \mathbf{l}_0) = \mathbf{a}_1 - \mathbf{n}(l_1 - l_0) = \mathbf{a}_1 - \Delta l_1\mathbf{n} \\ \mathbf{x}_2 &= \mathbf{a}_2 - (\mathbf{l}_2 - \mathbf{l}_0) = \mathbf{a}_2 - \mathbf{n}(l_2 - l_0) = \mathbf{a}_2 - \Delta l_2\mathbf{n} \end{aligned} \tag{4.374}$$

From (4.373), (4.374) and the unity length of the normal vector, we yield

$$\begin{aligned} \mathbf{a}_1 \times \mathbf{a}_2 &= (\Delta l_2\mathbf{a}_1 - \Delta l_1\mathbf{a}_2) \times \mathbf{n} - \mu\mathbf{n} \\ \mathbf{n}\cdot\mathbf{n} &= 1 \end{aligned} \tag{4.375}$$

(4.375) represents a set of four equations which has to be solved for μ and $\mathbf{n}$. The vector notation in (4.375) can be transformed into matrix notation (see Annex A.6 (A.164)):

$$\begin{aligned} (\mathbf{A} - \mu\mathbf{I})\mathbf{n} &= \mathbf{B} \\ \mathbf{n}^T\mathbf{n} &= 1 \end{aligned} \tag{4.376}$$

with

$$\mathbf{A} = \begin{bmatrix} 0 & \Delta l_1 a_{2z} - \Delta l_2 a_{1z} & \Delta l_2 a_{1y} - \Delta l_1 a_{2y} \\ \Delta l_2 a_{1z} - \Delta l_1 a_{2z} & 0 & \Delta l_1 a_{2x} - \Delta l_2 a_{1x} \\ \Delta l_1 a_{2y} - \Delta l_2 a_{1y} & \Delta l_2 a_{1x} - \Delta l_1 a_{2x} & 0 \end{bmatrix} = -\mathbf{A}^T$$

$$\mathbf{B} = \mathbf{a}_1 \times \mathbf{a}_2 = \begin{bmatrix} a_{1y}a_{2z} - a_{1z}a_{2y} \\ a_{1z}a_{2x} - a_{1x}a_{2z} \\ a_{1x}a_{2y} - a_{1y}a_{2x} \end{bmatrix}$$

Joining both equations of (4.376), leads to

$$\mathbf{B}^T(\mu^2\mathbf{I} - \mathbf{A}^2)^{-1}\mathbf{B} = 1 \tag{4.377}$$

which we use to determine the unknown factor μ. The matrix $(\mu^2\mathbf{I} - \mathbf{A}^2)$ is symmetric. We perform eigen decomposition for its inversion:

$$(\mu^2\mathbf{I} - \mathbf{A}^2) = \mathbf{Q\Lambda Q}^{-1}; \quad \mathbf{Q}^T = \mathbf{Q}^{-1}; \quad \mathbf{\Lambda} = \begin{bmatrix} \lambda_1 & 0 & 0 \\ 0 & \lambda_2 & 0 \\ 0 & 0 & \lambda_3 \end{bmatrix} \tag{4.378}$$

so that we yield

$$(\mu^2\mathbf{I} - \mathbf{A}^2)^{-1} = \mathbf{Q}\,\mathbf{\Lambda}^{-1}\,\mathbf{Q}^T; \quad \mathbf{\Lambda}^{-1} = \begin{bmatrix} \lambda_1^{-1} & 0 & 0 \\ 0 & \lambda_2^{-1} & 0 \\ 0 & 0 & \lambda_3^{-1} \end{bmatrix} \tag{4.379}$$

Substitution in (4.377) gives

$$\mathbf{B}^T\mathbf{Q\Lambda}^{-1}\mathbf{Q}^T\mathbf{B} = 1 \tag{4.380}$$

If we replace the matrix $\mathbf{Q}$ by the eigenvectors (column vectors) $\mathbf{Q} = [\mathbf{q}_1 \quad \mathbf{q}_2 \quad \mathbf{q}_3]$, relation (4.380) can also be expressed by

$$\mathbf{B}^T\mathbf{Q\Lambda}^{-1}\mathbf{Q}^T\mathbf{B} = \sum_{i=1}^{3} \frac{\mathbf{B}^T\mathbf{q}_i\mathbf{q}_i^T\mathbf{B}}{\lambda_i} = \sum_{i=1}^{3} \frac{(\mathbf{q}_i^T\mathbf{B})^2}{\lambda_i} = 1 \tag{4.381}$$

The calculation of the eigenvalues λ_i for an arbitrary square matrix $\mathbf{X}$ is based on the characteristic polynomial resulting from the determinate $|\mathbf{X} - \lambda\mathbf{I}| = 0$, which leads in our case to

$$\begin{aligned} &|\mathbf{A}^2 - (\mu^2 - \lambda)\mathbf{I}| = |\mathbf{A}^2 - \varepsilon\mathbf{I}| = 0; \quad \varepsilon = \mu^2 - \lambda \\ &\text{with} \quad \mathbf{X} = \mu^2\mathbf{I} - \mathbf{A}^2 \end{aligned} \tag{4.382}$$

Herein, ε_i represents the eigenvalues of the matrix $\mathbf{A}^2$. It generally holds for eigen decomposition:

$$\mathbf{X}^n = \mathbf{Q\Lambda}^n\mathbf{Q}^{-1} \tag{4.383}$$

Hence, we only need to calculate the eigenvalues ν_i of the anti-symmetric matrix $\mathbf{A}$. It is known from the eigenvalues of a real-valued anti-symmetric 3×3 matrix that

$$\begin{aligned} \nu_{1,2} &= \pm j\nu_0 \\ \nu_3 &= 0 \end{aligned} \tag{4.384}$$

Thus, (4.382) converts to

$$\varepsilon_i = \nu_i^2 = \mu^2 - \lambda_i \Rightarrow \lambda_i = \begin{cases} \mu^2 + \nu_0^2 \\ \mu^2 + \nu_0^2 \\ \mu^2 \end{cases} \tag{4.385}$$

Let finally $\mathbf{q}_1, \mathbf{q}_2$ and $\mathbf{q}_3$ be the three eigenvectors of $\mathbf{A}$, we get for (4.381)

$$\frac{(\mathbf{q}_1^T\mathbf{B})^2 + (\mathbf{q}_2^T\mathbf{B})^2}{\mu^2 + \nu_0^2} + \frac{(\mathbf{q}_3^T\mathbf{B})^2}{\mu^2} = 1 \quad (4.386)$$

$$\mathbf{B}^T\mathbf{Q}\mathbf{Q}^T\mathbf{B} = \mathbf{B}^T\mathbf{B} = \sum_{i=1}^{3}(\mathbf{q}_i^T\mathbf{B})^2$$

$$\mu^4 - \mu^2(\mathbf{B}^T\mathbf{B} - \nu_0^2) - (\nu_0\mathbf{q}_3^T\mathbf{B})^2 = 0$$

so that we finally achieve at

$$\mu^2 = \frac{1}{2}\left(\mathbf{B}^T\mathbf{B} - \nu_0^2 \pm \sqrt{\nu_0^2(\nu_0^2 - 2\mathbf{B}^T\mathbf{B} + (2\mathbf{q}_3^T\mathbf{B})^2) + (\mathbf{B}^T\mathbf{B})^2}\right) \quad (4.387)$$

Insertion of this result in (4.376) provides us the wanted normal vector $\mathbf{n}$ (i.e. the orientation) of the scattering plane and the position of the plane results from

$$\mathbf{r}' = \mathbf{r}_{a0} - l_0\mathbf{n} \quad (4.388)$$

For several reasons it could be better to place the reference point in the middle of the triangle $[P_1, P_2, P_3]$ (Figure 4.108), so that we finally get

$$\mathbf{r}'_m = \frac{1}{3}\sum_{i=0}^{2}\mathbf{r}'_i = \frac{1}{3}\sum_{i=0}^{2}(\mathbf{r}_{ai} + l_i\mathbf{n}) \quad (4.389)$$

where $\mathbf{r}_{ai}$ is the antenna positions.

4.10.4.4 Bi-Static Measurement

Now, we place two point radiators (antennas) in front of the scattering plane. One acts as transmitter, the other one as receiver (Figure 4.109). The measured travelling time of the wave between the two antennas is

$$\tau = \frac{l_1 + l_2}{c} = \frac{l_1 + l'_2}{c}; \quad l_i = |\mathbf{l}_i| \quad (4.390)$$

The length of the actual travelling path is identical with the length of the straight line which connects the transmitter with the 'receiver image'. This line intersects the scattering plane at the specular point P. Following the law of reflection the angle of incidence α and the reflection angle are identical which can be immediately seen from Figure 4.109 so that a wave coming along $\mathbf{l}_1$ will be reflected towards the focal point R:

$$\cos\alpha = \frac{-\mathbf{l}_1 \cdot \mathbf{n}}{l_1} = \frac{\mathbf{l}_2 \cdot \mathbf{n}}{l_2} \quad (4.391)$$

From a single measurement, we can only gain the overall length of the travelling path $l_p = l_1 + l_2$ which is not sufficient to determine position and orientation of the scattering plane. The only we can say is that the specular point of the unknown

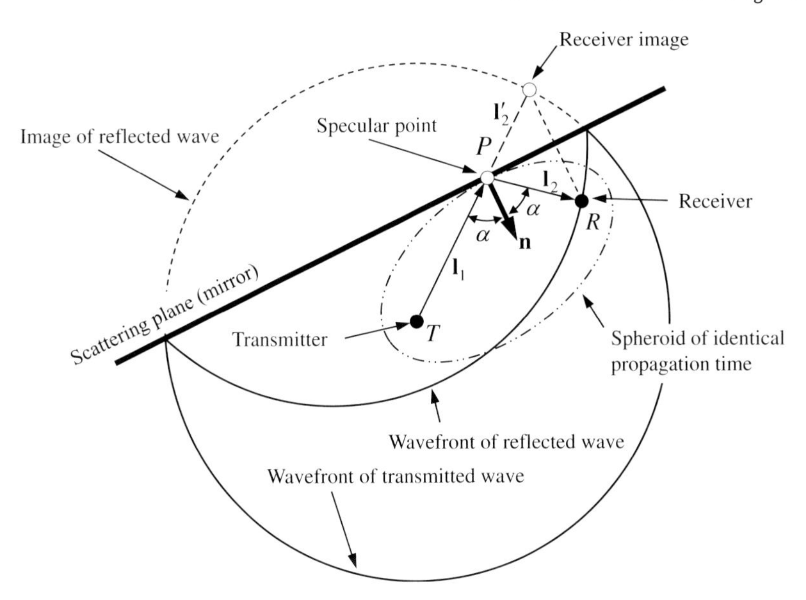

Figure 4.109 Scattering of a spherical wave at a planar surface captured by a bi-static antenna arrangement.

plane is placed anywhere at a spheroid whose focal points coincide with the antenna positions and whose axes are

$$a = \frac{1}{2} l_{\mathrm{p}}; \quad b = \sqrt{a^2 - e^2} \tag{4.392}$$

$2e$ is the distance between both antennas.

In order to solve the ambiguity, we need again three bi-static measurements from different positions. Using the bisecting vector $\mathbf{l}$ (refer to Figure 4.110), we can proceed like above in the mono-static case since $\mathbf{l}$ takes over the role of the antenna distance in Figure 4.108.

The bisecting vector $\mathbf{l} = l\,\mathbf{n}$ represents a vector which is normal to the tangent plane in point P. It intersects the antenna axis in point Q whose distance to the midpoint M is x_{q} (see Annex A.7.1).

$$\mathbf{l} = l\,n = \mathbf{r}_Q - \mathbf{r}' = (\mathbf{r}_M + x_{\mathrm{q}}\mathbf{u}) - \mathbf{r}' \tag{4.393}$$

If we need three differently placed bi-static arrangements, we get three specular points

$$\mathbf{r}'_i = \mathbf{r}_{Mi} + x_{\mathrm{q}i}\mathbf{u}_i - l_i\mathbf{n}; \quad i \in [0, 1, 2] \tag{4.394}$$

which are all placed in the wanted plane. Hence, we have the same situation as before with the mono-static arrangement. Thus applying (4.373), the conditions to be met are

$$\begin{gathered} \mu\mathbf{n} = \mathbf{x}_1 \times \mathbf{x}_2 = (\mathbf{r}'_2 - \mathbf{r}'_0) \times (\mathbf{r}'_1 - \mathbf{r}'_0) \\ \mathbf{n} \cdot \mathbf{n} = 1 \end{gathered} \tag{4.395}$$

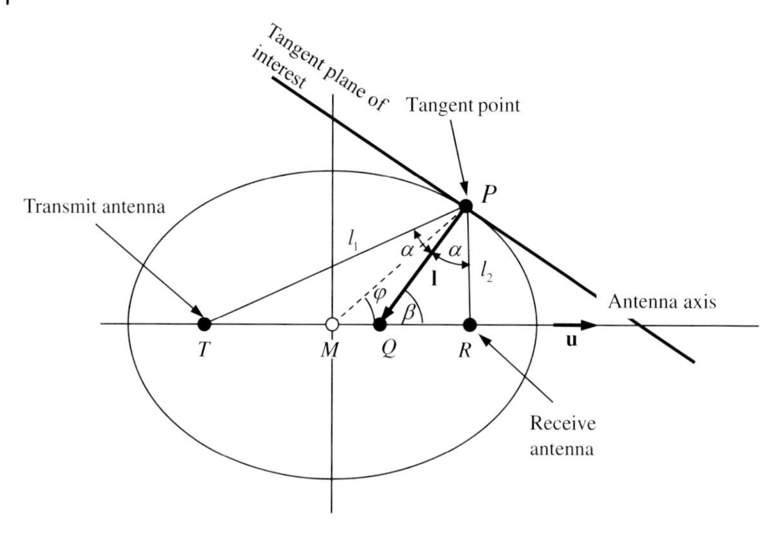

Figure 4.110 Definitions for bi-static measurement. The position vectors of the various points are assigned by $\mathbf{r}_\xi; \xi \in [T, M, Q, R]$ and $\mathbf{r}'$ for the specular point P.

This set of equations is to be solved for μ and $\mathbf{n}$. This is however more complicated as in the mono-static case since the values of x_{qi} depend on $\mathbf{r}'_i$. Closed form solutions will be found only for some symmetric array geometries.

4.10.4.5 Estimation of Reconstruction Errors

So far, we supposes idealized conditions for the reconstruction of the scattering body, that is

- **The roundtrip time of the scattered wave was supposed to be precisely known:** Random and systematic errors of time delay measurements will limit this precision.
- **The investigated object was assumed to be a flat plane:** The surface of actual bodies will have a certain curvature which requires narrowly spaced specular points in order to approach the conditions for a tangent plane.
- **The radiators are considered as ideal point sources:** In practice this will never be the case. Real antennas will add internal delay time, which may lead to erroneous time zero estimation if it is not exactly known. The measured roundtrip time relates to the radiation centre of the antenna, which must also be exactly known. Furthermore, the radiation behavior of the antenna is usually not exactly isotropic and due to its finite aperture size the captured waves do not relate to the reflection in a specular point rather in a specular spot (compare Figure 4.111). Since the antenna aperture integrates rays of different propagation length, the time shape of the receive signal is slightly different to the sounding signal which will lead to systematic deviations of time position estimation.

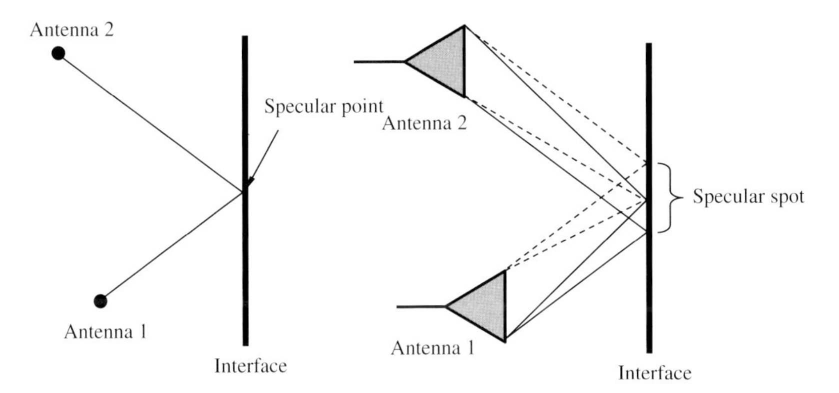

Figure 4.111 Effect of finite antenna size. The real transmit antenna emits waves over its whole aperture plane. These waves have different propagation path length to the receiving antenna. The receiving antenna integrates all waves bouncing its aperture.

- **The sounding waves were considered as scalar waves:** Actually, electromagnetic waves are vector (polarized) waves. This aspect does not influence the propagation time of the waves. But polarized waves cannot spread their energy isotropically into space, the strength of the reflection depends on their polarization direction and the angle of incident and the receive antenna may be 'blind' against the incoming wave if it is misaligned. Hence, if polarization is not adequately respected, the strength of the captured signal may be affected providing stronger noise contributions.

We will consider two of these errors.

Random Errors The stochastic errors of $\mathbf{r}'$ and $\mathbf{n}$ depend on the randomness of the propagation time measurement and the randomness of the antenna location. Without restriction on generality, we will only deal here with mono-static measurements.

If we assume identical radar channels for the measurement of l_0, l_1, l_3, the variance of all distance values will be the same $\sigma_{l_0}^2 = \sigma_{l_1}^2 = \sigma_{l_2}^2 = \sigma^2$. These errors are mutually uncorrelated. Furthermore, we will suppose that the standard deviation of the antenna positions is negligible compared to the uncertainties of the radar distance measurements. In the practical application, some attention should be paid on this simplification because the accuracy of high-resolution ultra-wideband radars may attain the μm range which has to be outperformed from the mechanical scanner at least by a factor of three. Otherwise, errors of the mechanics have to be included into the error model too.

In order to simplify the calculation, let us place the three antennas in the *xy*-plane of a Cartesian coordinate system at the edges of an isosceles triangle whose symmetry axis is parallel to the x-axis. Thus, we can write for the array vectors $\mathbf{a}_1, \mathbf{a}_2$ (Figure 4.108)

$$\mathbf{a}_1 = a\begin{bmatrix} \cos\alpha \\ \sin\alpha \\ 0 \end{bmatrix}; \quad \mathbf{a}_2 = a\begin{bmatrix} \cos\alpha \\ -\sin\alpha \\ 0 \end{bmatrix} \tag{4.396}$$

If l_0, l_1, l_3 are known from measurements, the normal vector of the scattering plane results from (4.375) and (4.387) in

$$\mathbf{n} = \begin{bmatrix} n_1 \\ n_2 \\ n_3 \end{bmatrix} = \begin{bmatrix} \dfrac{l_1 + l_2 - 2l_0}{2a\cos\alpha} \\ \dfrac{l_1 - l_2}{2a\sin\alpha} \\ \sqrt{1 - n_1^2 - n_2^2} \end{bmatrix} \tag{4.397}$$

Thus, the variance of the components of the normal vector results to

$$\boldsymbol{\sigma}_{\mathbf{n}}^2 = \begin{bmatrix} \sigma_{n_1}^2 \\ \sigma_{n_2}^2 \\ \sigma_{n_3}^2 \end{bmatrix} = \sum_{i=0}^{2}\left(\frac{\partial n}{\partial l_i}\right)\sigma_{l_i}^2 = 2\left(\frac{\sigma}{a}\right)^2 \begin{bmatrix} \dfrac{3}{\cos^2\alpha} \\ \dfrac{1}{\sin^2\alpha} \\ \dfrac{1}{1 - n_1^2 - n_2^2}\left(\dfrac{3n_1^2}{\cos^2\alpha} + \dfrac{n_2^2}{\sin^2\alpha}\right) \end{bmatrix} \tag{4.398}$$

and the variance of the position vector (4.388) is

$$\boldsymbol{\sigma}_{\mathbf{r}'}^2 = \begin{bmatrix} \sigma_{r_1'}^2 \\ \sigma_{r_2'}^2 \\ \sigma_{r_3'}^2 \end{bmatrix} = \sum_{i=0}^{2}\left(\frac{\partial \mathbf{r}'}{\partial l_i}\right)^2 \sigma_{l_i}^2$$
$$= \sigma^2 \begin{bmatrix} n_1^2 + \dfrac{4l_0 n_1}{a\cos\alpha} + \dfrac{6l_0^2}{(a\cos\alpha)^2} \\ n_2^2 \dfrac{2l_0^2}{(a\sin\alpha)^2} \\ n_3^2 + \dfrac{4n_1 l_0}{a\cos\alpha} + 2\left(\dfrac{l_0}{an_3}\right)^2\left(\dfrac{3n_1^2}{\cos^2\alpha} + \dfrac{n_2^2}{\sin^2\alpha}\right) \end{bmatrix} \tag{4.399}$$

Note that (4.398) and (4.399) suppose uncorrelated perturbations. This is usually met as long as the radar channels are working independently from each other. However, in the case of external jamming or device internal spurious signals (e.g. errors of the sampling clock) this condition is not any longer completely fulfilled.

What can we conclude form (4.398) and (4.399)?

- Clearly, the variance $\sigma^2 = \varphi^2/4c^2$ (φ^2 is the variance of pulse position estimation) of the radar range measurements should be as low as possible. As already mentioned before, it is a question of signal bandwidth, random noise, jitter and an appropriate signal model of the scattered signal.
- The distance a between the antennas should be as large as possible.
- The antenna plane $\mathbf{m} = (\mathbf{a}_1 \times \mathbf{a}_2)/|\mathbf{a}_1 \times \mathbf{a}_2|$ should be as parallel as possible to the scattering plane, that is $\mathbf{n} \approx -\mathbf{m}$. In our example this is expressed by $n_1 \approx n_2 \approx 0$. Therefore, the scan path should roughly follow the contour of the test object if possible.
- The unknown surface should be scanned from a short distance.
- The variance of the tilt angles of the plane (i.e. the direction of $\mathbf{n}$) has no spatial preference if $\sigma^2_{n_1} = \sigma^2_{n_2}$ holds. This is fulfilled for $\alpha = 30^\circ$ as it follows from (4.398). That is, the antennas should be placed at the edges of an equilateral triangle.

In summary, the random errors are for the optimum case

$$\boldsymbol{\sigma}^2_{\mathbf{n}} = \frac{8\sigma^2}{a^2}\begin{bmatrix} 1 \\ 1 \\ 0 \end{bmatrix} \quad \text{and} \quad \boldsymbol{\sigma}^2_{\mathbf{r}'} = \sigma^2 \begin{bmatrix} 8l_0^2/a^2 \\ 8l_0^2/a^2 \\ 1 \end{bmatrix} \tag{4.400}$$

(4.398) and (4.399) are also applicable to estimate the errors for the method of quasi-wavefront derivation (4.368). There, in case of 2D-scanning, we have to use derivation in x- and y-direction to span a ‘virtual antenna plane’. Translated to our notation, this corresponds to an antenna arrangement in which the antennas are placed at the edges of a rectangular triangle (i.e. $2\alpha = 90^\circ$). The length of the vectors $\mathbf{a}_1$ and $\mathbf{a}_2$ relates to the sample spacing in both scanning directions.

Errors Due to the Curvature of the Scattering Plane Our precious considerations were based on flat planes which had no limitations in their extension. But actually, we have to deal with a ‘patchwork’ in which a set of flat patches with finite dimensions has to approximate the surface of interest. In order to be able to determine $\mathbf{r}'$ and $\mathbf{n}$ of every patch, we need to shrink the size of the antennas array, that is the lengths of the vectors $\mathbf{a}_1, \mathbf{a}_2$ have an upper limit. The stronger the curvature of the surface is the shorter the antenna distance must be. But shortening the antenna distance will rise up the random errors. Hence, a reasonable compromise has to be found to balance between curvature and random error.

Figure 4.112 illustrates the problem for the 2D-case. Supposing we want to know position and inclination of the tangent line in point P. We perform a measurement at position x_0 and x_{-1}. The length of the propagation paths l_0, l_{-1} results approximately from scattering at the osculating circle, that is the lines l_0, l_{-1} are not parallel. Since (4.375) assumes parallel line, we will attain the dashed line instead of the wanted tangent line.

In order to get a better estimate of the tangent line, a third measurement at position x_1 should be made. In connection with the measurement from point x_0, it

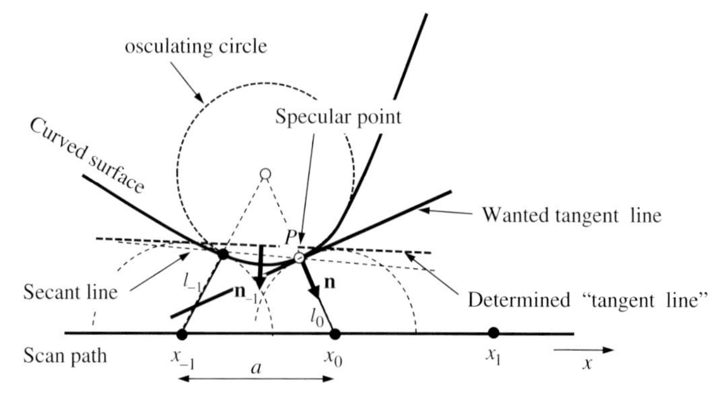

Figure 4.112 Ray geometry for curved surfaces. The surface in the specular point is approximated by the osculating circle.

provides a new and also erroneous tangent line. But now, the inclination error is opposite to the previous one, so that by taking the average one will achieve an improved estimation for the tangent line.

The ultimate solution would however be to extend (4.375) for fitting the osculating circle (osculating quadric surface in 3D-case) and to use the three measurements to estimate centre position and radius of the circle.

References

1 Lo, Y.T. and Lee, S.W. (1988) *Antenna Handbook: Theory; Applications, and Design*, Van Nostrand Reinhold Company, New York.
2 Hansen, T.B. and Yaghjian, A.D. (1999) *Plane-Wave Theory of Time-Domain Fields: Near-field Scanning Applications*, IEEE Press, New York.
3 Orfanidis, S.J. (2008) Electromagnetic waves and antennas. www.ece.rutgers.edu/~orfanidi/ewa.
4 Baum, C.E., Farr, E.G. and Frost, C.A. (1998) Transient gain of antennas related to the traditional continuous-wave (CW) definition of gain. Ultra-Wideband Short-Pulse Electromagnetics 4, 1998, pp. 109–118.
5 Lamensdorf, D. and Susman, L. (1994) Baseband-pulse-antenna techniques. *IEEE Antenn. Propag.*, **36** (1), 20–30.
6 Sörgel, W., Waldschmidt, C. and Wiesbeck, W. (2003) Quality measures for ultra wideband antennas. 48 Internationales Wissenschaftliches Kolloquium Technische Universität Ilmenau, 22–25 September 2003, Ilmenau, Germany.
7 Allen, O.E., Hill, D.A. and Ondrejka, A. R. (1993) Time-domain antenna characterizations. *IEEE T. Electromagn. C.*, **35** (3), 339–346.
8 Kunisch, J. (2007) Implications of lorentz reciprocity for ultra-wideband antennas. IEEE International Conference on Ultra-Wideband, 2007. ICUWB 2007, pp. 214–219.
9 Schantz, H.G. (2005) *The Art and Science of Ultrawideband Antennas*, Artech House, Inc., Norwood.
10 Schantz, H.G. (2004) A brief history of UWB antennas. *IEEE Aero. El. Sys. Mag.*, **19** (4), 22–26.
11 Allen, B., Dohler, M. Okon, E. *et al.* (2007) *Ultra-Wideband Antennas and Propagation for Communications, Radar and Imaging*, John Wiley & Sons, Ltd.
12 Yarovoy, A.G., Zijderveld, J.H. Schukin, A.D. *et al.* (2005) Dielectric wedge

antenna for UWB applications. IEEE International Conference on Ultra-Wideband, 2005. ICU 2005, pp. 186–189.
13 Baum, C.E. (2002) General properties of antennas. *IEEE Trans. Electromagn. C.*, **44** (1), 18–24.
14 Chew, W.C. (1995) *Waves and Fields in inhomogeneous Media*, IEEE Press, New York.
15 Ulaby, F.T. and Elachi, C. (1990) *Radar Polarimetry for Geoscience Applications*, Artech House, Inc., Norwood, MA.
16 Bowman, J.J., Senior, T.B.A. and Uslenghi, P.L.E. (1987) *Electromagnetic and Acoustic Scattering by Simple Shapes*, Hemisphere Publishing Corporation, New York.
17 Ishimaru, A. (1991) *Electromagnetic Wave Propagation, Radiation, and Scattering*, Prentice-Hall, Inc., Englewood Cliffs, NJ.
18 Astanin, L.Y. and Kostylev, A.A. (1997) *Ultrawideband Radar Measurements Analysis and Processing*, The Institution of Electrical Engineers, London, UK.
19 Herrmann, R., Sachs, J. Schilling, K. *et al.* (2008) 12-GHz bandwidth M-sequence radar for crack detection and high resolution imaging. International Conference on Ground Penetrating Radar (GPR), Birmingham, UK.
20 Tsang, L., Kong, J.A. and Ding, K.-H. (2000) *Scattering of Electromagnetic Waves: Theories and Applications*, John Wiley & Sons, Inc., New York.
21 Kong, J.A. (1990) *Electromagnetic Wave Theory*, John Wiley & Sons, Inc., New York.
22 Baum, C.E. and Kritikos, H.N. (1995) *Electromagnetic Symmetry*, Taylor Francis, Washington.
23 Tai, C.-T. (1994) *Dyadic Green Functions in Electromagnetic Theory*, IEEE Press, Piscataway, NJ.
24 Ishimaru, A. (1978) *Wave Propagation and Scattering in Random Media, Volume I: Single Scattering and Transport Theory; Volume II: Multiple Scattering, Turbulence, Rough Surfaces and Remote Sensing*, Academic Press, New York.
25 Kahn, W. and Kurss, H. (1965) Minimum-scattering antennas. *IEEE Trans. Antenn. Propag.*, **13** (5), 671–675.
26 Andersen, J.B. and Frandsen, A. (2005) Absorption efficiency of receiving antennas. *IEEE Trans. Antenn. Propag.*, **53** (9), 2843–2849.
27 Heidrich, E. and Wiesbeck, W. (1992) Reduction and minimization of antenna scattering. IEEE Antennas and Propagation Society International Symposium, 1992. AP-S. 1992 Digest. Held in Conjuction with: URSI Radio Science Meeting and Nuclear EMP Meeting, vol. 2, pp. 904–907.
28 Appel-Hansen, J. (1979) Accurate determination of gain and radiation patterns by radar cross-section measurements. *IEEE Trans. Antenn. Propag.*, **27** (5), 640–646.
29 Heidrich, E. and Wiesbeck, W. (1991) Wideband polarimetric RCS-antenna measurement. Seventh International Conference on (IEE) Antennas and Propagation, vol. 1, 1991. ICAP 91, pp. 424–427.
30 Hansen, R.C. (1989) Relationships between antennas as scatterers and as radiators. *Proc. IEEE*, **77** (5), 659–662.
31 Pozar, D. (2004) Scattered and absorbed powers in receiving antennas. *IEEE Antenn. Propag.*, **46** (1), 144–145.
32 Sato, M. and Takayama, T. (2007) A novel directional borehole radar system using optical electric field sensors. *IEEE Trans. Geosci. Remote*, **45** (8), 2529–2535.
33 Montoya, T.P. and Smith, G.S. (1996) Resistively-loaded Vee antennas for short-pulse ground penetrating radar. Antennas and Propagation Society International Symposium, vol. 3, 1996. AP-S. Digest, pp. 2068–2071.
34 Montoya, T.P. and Smith, G.S. (1999) Land mine detection using a ground-penetrating radar based on resistively loaded Vee dipoles. *IEEE Trans. Antenn. Propag.*, **47** (12), 1795–1806.
35 Wichmann, G. (2006) Non-intrusive inspection impulse radar antenna, USA.
36 Baum, C.E. (1971) On the singularity expansion method for the solution of electromagnetic interaction problems, *Interaction Notes*, Vol. Note 88.
37 Baum, C.E. (1973) *Singularity expansion of electromagnetic fields and potentials*

radiated from antennas or scattered from objects in free space, Vol. Note 179.

38 Baum, C. (1986) The singularity expansion method: background and developments. *IEEE Antenn. Propag. Soc. Newslett.*, **28** (4), 14–23.

39 Baum, C.E. (2006) Combining polarimetry with SEM in radar backscattering for target identification. The Third International Conference on Ultrawideband and Ultrashort Impulse Signals, pp. 11–14.

40 Baum, C.E., Rothwell, E.J. Chen, K.M. *et al.* (1991) The singularity expansion method and its application to target identification. *Proc. IEEE*, **79** (10), 1481–1492.

41 C Mroué, A., Heddebaut, M. Elbahhar, F. *et al.* (2012) Automatic radar target recognition of objects falling on railway tracks. *Meas. Sci. Technol.*, **23** (2), pp. 1–10.

42 Roblin, C. (2006) Ultra compressed parametric modelling of UWB antenna measurements. First European Conference on Antennas and Propagation, 2006. EuCAP 2006, pp. 1–8.

43 Roblin, C. (2008) Ultra compressed parametric modeling for symmetric or pseudo-symmetric UWB Antenna. IEEE International Conference on Ultra-Wideband, 2008. ICUWB 2008, pp. 109–112.

44 Cloude, S.R. and Pottier, E. (1996) A review of target decomposition theorems in radar polarimetry. *IEEE Trans. Geosci. Remote*, **34** (2), 498–518.

45 Immoreev, I.I. and Fedotov, P.G.S.D.V. (2002) Ultra wideband radar systems: advantages and disadvantages. IEEE Conference on Ultra Wideband Systems and Technologies, 2002. Digest of Papers, pp. 201–205.

46 Harmuth, H.F. (1981) *Nonsinusoidal Waves for Radar and Radio Communication*, Academic Press, New York.

47 Harmuth, H.F. (1984) *Antennas and Waveguides for Nonsinusoidal Waves*, Academic Press, Inc., Orlando.

48 Harmuth, H.F. and Mohamed, N.J. (1992) Large-current radiators. *IEE Proc. Microwave Antenn. Propag.*, **139** (4), 358–362.

49 Pochanin, G.P., Pochanina, I.Y. and Kholod, P.V. (2003) Radiation efficiency of the large current radiators. Electrodynamic simulation. 4th International Conference on Antenna Theory and Techniques, vol. 2, 2003, pp. 542–545.

50 Pochanin, G.P. and Kholod, P.V. (2006) LCR with a traveling wave pulse generator. The Third International Conference on Ultrawideband and Ultrashort Impulse Signals, pp. 199–202.

51 Pochanin, G.P. (2006) Large current radiators. The Third International Conference on Ultrawideband and Ultrashort Impulse Signals, pp. 77–81.

52 Balzovskii, E.V., Buyanov, Y.I. and Koshelev, V.I. (2007) An active antenna for measuring pulsed electric fields. *Russ. Phys. J.*, **50** (5), 503–508.

53 Balzovskii, E.V., Buyanov, Y.I. and Koshelev, V.I. (2010) Dual polarization receiving antenna array for recording of ultra-wideband pulses. *J. Commun. Technol. El.*, **55** (2), 172–180.

54 Heidrich, E. and Wiesbeck, W. (1989) Features of advanced polarimetric RCS-antenna measurements. Antennas and Propagation Society International Symposium, vol. 2, 1989. AP-S. Digest, pp. 1026–1029.

55 Wiesbeck, W. and Kahny, D. (1991) Single reference, three target calibration and error correction for monostatic, polarimetric free space measurements. *Proc. IEEE*, **79** (10), 1551–1558.

56 Johnk, R.T. and Ondrejka, A. (1998) *Time-Domain Calibrations of D-Dot Sensors*, NIST.

57 Andrews, J.R. (2003) *Application Note AN-14a: UWB Signal Sources, Antennas & Propagation*, Picosecond Pulse Labs.

58 Icheln, C., Vainikainen, P. and Haapala, P. (1997) Application of a GTEM cell to small antenna measurements. Antennas and Propagation Society International Symposium, vol. 1, 1997. IEEE, 1997 Digest, pp. 546–549.

59 Muterspaugh, M.W. (2003) Measurement of indoor antennas using GTEM cell.

IEEE Trans. Consum. Electr., **49** (3), 536–538.
60 Karst, J.P. and Garbe, H. (1999) Characterization of loaded TEM-waveguides using time-domain reflectometry. IEEE International Symposium on Electromagnetic Compatibility, vol. 1, 1999, pp. 127–132.
61 Thye, H., Sczyslo, S. Armbrecht, G. *et al.* (2009) Transient UWB antenna characterization in GTEM cells. IEEE International Symposium on Electromagnetic Compatibility, 2009. EMC 2009, pp. 18–23.
62 Novotny, D., Johnk, R.T. Grosvenor, C.A. *et al.* (2004) Panoramic, ultrawideband, diagnostic imaging of test volumes. International Symposium on Electromagnetic Compatibility, vol. 1, 2004. EMC 2004, pp. 25–28.
63 Cook, C.E. and Bernfeld, M. (1993) *Radar Signals: An Introduction to Theory and Application*, Artech House, Boston, London.
64 Wehner, D.R. (1995) *High Resolution Radar*, 2nd edn, Artech House, Norwood, MA.
65 Rihaczek, A.W. (1996) *Principles of High-Resolution Radar*, Artech House, Inc., Boston, London.
66 Barton, D.K. and Leonov, S.A. (1997) *Radar Technology Encyclopedia*, Artech House, Inc.
67 Ludeman, L.C. (2003) *Random Processes: Filtering, Estimation, and Detection*, Wiley-Interscience, Hoboken, NJ.
68 Haykin, S. and Steinhardt, A.O. (1992) *Adaptive Radar Detection and Estimation*, John Wiley & Sons, Inc., New York.
69 Helstrom, C.W. (1995) *Elements of Signal Detection and Estimation*, PTR Prentice Hall, Englewood Cliffs, NJ.
70 Poor, H.V. (1988) *An Introduction to Signal Detection and Estimation*, Springer-Verlag, New York.
71 Van Trees, H.L. (2001) *Radar-Sonar Signal Processing and Gaussian Signals in Noise*, John Wiley & Sons, Inc., New York.
72 Fessler, J.A. and Hero, A.O. (1994) *Space-alternating generalized EM algorithms for penalized maximum-likelihood image reconstruction*, Technical Report No. 286, University of Michigan.
73 Hantscher, S. and Diskus, C.G. (2009) Pulse-based radar imaging using a genetic optimization approach for echo separation. *IEEE Sens. J.*, **9** (3), 271–276.
74 Leib, M., Menzel, W. Schleicher, B. *et al.* (2010) Vital signs monitoring with a UWB radar based on a correlation receiver. Proceedings of the Fourth European Conference on Antennas and Propagation (EuCAP), 2010, pp. 1–5.
75 Gezici, S., Zhi, T. Giannakis, G.B. *et al.* (2005) Localization via ultra-wideband radios: a look at positioning aspects for future sensor networks. *IEEE Signal Proc. Mag.*, **22** (4), 70–84.
76 Cramér, H. (1946) *Mathematical Methods of Statistics*, Princeton University Press, Princeton, NJ.
77 Taylor, J.D. (2001) *Ultra-Wideband Radar Technology*, CRC Press, Boca Raton, FL.
78 Ballard, D.H. (1981) Generalising the Hough transform to detect arbitrary shapes. *Pattern Recogn.*, **13**, 111–122.
79 Deans, S.R. (1983) *The Radon Transform and Some of its Applications*, John Wiley & Sons, Inc.
80 Gezici, S. and Sahinoglu, Z. (2007) Theoretical limits for estimation of vital signal parameters using impulse radio UWB. IEEE International Conference on Communications, 2007. ICC '07, pp. 5751–5756.
81 Sachs, J., Zaikov, E. Helbig, M. *et al.* (2011) Trapped victim detection by pseudo-noise radar. 2011 International Conference on Wireless Technologies for Humanitarian Relief (ACWR 2011), Amritapuri, India.
82 Venkatesh, S., Anderson, C.R., Rivera, N.V. *et al.* (2005) "Implementation and analysis of respiration-rate estimation using impulse-based UWB," in Military Communications Conference, 2005. MILCOM 2005. IEEE, vol. 5, 2005, pp. 3314–3320.
83 Daniels, D.J. (2004) *Ground Penetrating Radar*, 2nd edn, Institution of Electrical Engineers, London.
84 Bonitz, F., Eidner, M. Sachs, J. *et al.* (2008) UWB-Radar Sewer Tube Crawler. 12th International Conference on

Ground Penetrating Radar Birmingham, UK.

85 Bonitz, F., Sachs, J. Herrmann, R. *et al.* (2008) Radar tube crawler for quality assurance measurements of pipe systems. European Radar Conference (EURAD), Amsterdam, The Netherlands.

86 Herrmann, R. (2011) M-sequence based ultra-wideband radar and its application to crack detection in salt mines. Faculty of Electrical Engineering and Information Technology, Ilmenau University of Technology (Germany), Ilmenau.

87 Aardal, Ø. and Hammerstad, J. (2010) Medical radar literature overview, http://rapporter.ffi.no/rapporter/2010/00958.pdf.

88 Sachs, J., Herrmann, R. Kmec, M. *et al.* (2007) Recent advances and applications of M-sequence based ultra-wideband sensors. International Conference on Ultra-Wideband, Singapore.

89 Sachs, J., Helbig, M. and Renhak, K. (2008) Unsichtbares wird sichtbar – Mit Radar den Insekten auf der Spur: Neue Möglichkeiten zum Aufspüren von Insektenaktivitäten. HOBA'08, Duisburg (Germany), pp. 45–48.

90 Zaikov, E., Sachs, J. Aftanas, M. *et al.* (2008) Detection of trapped people by UWB radar. German Microwave Conference (GeMiC 2008), Hamburg, Germany.

91 Zaikov, E. and Sachs, J. (2010) UWB radar for detection and localization of trapped people, in *Ultra Wideband* (ed. B. Lembrikov), Scivo, Rijeka, Croatia.

92 Sachs, J., Aftanas, M. Crabbe, S. *et al.* (2008) Detection and tracking of moving or trapped people hidden by obstacles using ultra-wideband pseudo-noise radar. European Radar Conference (EURAD 2008), Amsterdam, The Netherlands.

93 Nezirovic, A., Yarovoy, A.G. and Ligthart, L.P. (2010) Signal processing for improved detection of trapped victims using UWB radar. *IEEE Trans. Geosci. Remote*, **48** (4), 2005–2014.

94 Savelyev, T.G., Van Kempen, L. and Sahli, H. (2004) Deconvolution techniques, in *Ground Penetrating Radar* (ed D. Daniels), Institution of Electrical Engineers, London.

95 *MATLAB; Signal Processing Toolbox*, The Mathworks Inc, http://www.mathworks.com/help/pdf_doc/signal/signal_tb.pdf.

96 Hantscher, S., Etzlinger, B. Reisenzahn, A. *et al.* (2006) UWB radar calibration using wiener filters for spike reduction. IEEE MTT-S International Microwave Symposium Digest, 2006, pp. 1995–1998.

97 Hantscher, S., Reisenzahn, A. and Diskus, C.G. (2008) Ultra-wideband radar noise reduction for target classification. *Radar, Sonar Navigation, IET*, **2** (4), 315–322.

98 Jol, H.M. (2009) *Ground Penetrating Radar: Theory and Applications*, Elsevier.

99 Kupfer, K. (2005) *Electromagnetic Aquametry: Electromagnetic Wave Interaction with Water and Moist Substances*, Springer, Berlin.

100 Jyh-Ming, J., Hayden, L.A. and Tripathi, V.K. (1994) Time domain characterization of coupled interconnects and discontinuities. IEEE MTT-S International Microwave Symposium Digest, vol. 2, 1994, pp. 1129–1132.

101 Jyh-Ming, J., Janko, B. and Tripathi, V.K. (1996) Time-domain characterization and circuit modeling of a multilayer ceramic package. *IEEE Trans. Compon. Pack. B*, **19** (1), 48–56.

102 Aftanas, M., Rovnakova, J. Drutarovský, M. *et al.* (2008) Efficient method of TOA estimation for through wall imaging by UWB radar. International Conference on Ultra-Wideband, Hannover, Germany.

103 Grosvenor, C.A., Johnk, R.T. Baker-Jarvis, J. *et al.* (2009) Time-domain free-field measurements of the relative permittivity of building materials. *IEEE Trans. Instrum. Meas.*, **58** (7), 2275–2282.

104 Margrave, G.F. (2001) Numerical methods of exploration seismology with algorithms in MATLAB, http://www.crewes.org/ResearchLinks/FreeSoftware/EduSoftware/NMES_Margrave.pdf.

105 Huebner, C., Schlaeger, S. Becker, R. *et al.* (2005) Advanced measurement methods in time domain reflectometry for soil moisture determination, in *Electromagnetic Aquametry* (ed. K. Kupfer), Springer, Berlin, Heidelberg.

106 Schlaeger F S. (2005) A fast TDR-inversion technique for the

reconstruction of spatial soil moisture content. *Hydrolog. Earth Syst. Sci.*, **9**, 481–492.

107 Zwick, T., Haala, J. and Wiesbeck, W. (2002) A genetic algorithm for the evaluation of material parameters of compound multilayered structures. *IEEE Trans. Microwave Theory*, **50** (4), 1180–1187.

108 Bolomey, J.C. and Jofre, L. (2010) Three decades of active microwave imaging achievements, difficulties and future challenges. IEEE International Conference on Wireless Information Technology and Systems (ICWITS), 2010, pp. 1–4.

109 Schwartz, J.L. (1996) *Ultrasparse, Ultrawideband Time-Steered Arrays*, University of Pennsylvania.

110 Schwartz, J.L. and Steinberg, B.D. (1998) Ultrasparse, ultrawideband arrays. *IEEE Trans. Ultrason. Ferroelectr. Freq. Control*, **45** (2), 376–393.

111 Zhuge, X. (2010) Short-range ultra-wideband imaging with multiple-input multiple-output arrays. Faculty of Electrical Engineering, Mathematics and Computer Science, Delft University of Technology, Delft.

112 Bond, E.J., Xu, L. Hagness, S.C. *et al.* (2002) Microwave imaging via space-time beamforming for early detection of breast cancer. IEEE International Conference on Acoustics, Speech, and Signal Processing, vol. 3, 2002. Proceedings. (ICASSP '02), pp. III-2909–III-2912.

113 Sahli, H., Kempen, L.V. and Brooks, W.J. (2004) Data processing for clutter characterisation and removal, in *Ground Penetrating Radar*, 2nd edn (ed. D.J. Daniels), The Institut of Electrical Engineering, Stevenage (UK), pp. 581–610.

114 Firoozabadi F R., Miller, E.L. Rappaport, C.M. *et al.* (2007) Subsurface sensing of buried objects under a randomly rough surface using scattered electromagnetic field data. *IEEE Trans. Geosci. Remote*, **45** (1), 104–117.

115 Rappaport, C. (2004) A simple approximation of transmitted wavefront shape from point sources above lossy half spaces. IEEE International on Geoscience and Remote Sensing Symposium, 2004. IGARSS '04. Proceedings, pp. 421–424.

116 Scheers, B., Acheroy, M. and Vander Vorst, A. (2004) Migration technique based on deconvolution, in *Ground Penetrating Radar* (ed. D. Daniels), Institution of Electrical Engineers.

117 Jian F L., Stoica, P. and Zhisong, W. (2003) On robust Capon beamforming and diagonal loading. *IEEE Trans. Signal. Process.*, **51** (7), 1702–1715.

118 Stoica, P., Zhisong, W. and Jian, L. (2003) Robust Capon beamforming. *IEEE Signal Proc. Lett.*, **10** (6), 172–175.

119 Yanwei, W., Yijun, S. Jian, L. *et al.* (2005) Adaptive imaging for forward-looking ground penetrating radar. *IEEE Trans. Aero. Elec. Sys.*, **41** (3), 922–936.

120 Yao, X., Bin, G. Luzhou, X. *et al.* (2006) Multistatic adaptive microwave imaging for early breast cancer detection. *IEEE Trans. Bio.-Med. Eng.*, **53** (8), 1647–1657.

121 Klemm, M., Craddock, I.J. Leendertz, J.A. *et al.* (2009) Radar-based breast cancer detection using a hemispherical antenna array – experimental results. *IEEE Trans. Antenn. Propag.*, **57** (6), 1692–1704.

122 Sachs, J., Zetik, R. Peyerl, P. *et al.* (2005) Autonomous orientation by ultra wideband sounding. International Conference on Electromagnetics in Advanced Applications (ICEAA), Torino, Italy.

123 Zetik, R., Sachs, J. and Thoma, R. (2005) Modified cross-correlation back projection for UWB imaging: numerical examples. IEEE International Conference on Ultra-Wideband, 2005. ICU 2005, p. 5.

124 Hooi Been, L., Nguyen Thi Tuyet, N. Er-Ping, L. *et al.* (2008) Confocal microwave imaging for breast cancer detection: delay-multiply-and-sum image reconstruction algorithm. *IEEE Trans. Bio.-Med. Eng.*, **55** (6), 1697–1704.

125 Prada, C. and Fink, M. (1994) Eigenmodes of th time reversal operator: a solution to selective focusing in multiple target media. *Wave Motion*, **20**, 151–163.

126 Prada, C. and Fink, M. (1995) Selective focusing through inhomogeneous media: the DORT method. Proceedings of the IEEE Ultrasonics Symposium, vol. 2, 1995, pp. 1449–1453.

127 Fink, M. and Prada, C. (2001) Acoustic time-reversal mirrors. *Inverse Probl.*, **17**, pp. R1–R38.

128 Dinh-Quy, N. and Woon-Seng, G. (2008) Novel DORT method in non-well-resolved scatterer case. *IEEE Signal Proc. Lett.*, **15**, 705–708.

129 Marple, S.L. (1987) *Digital Spectral Analysis with Applications*, Prentice-Hall, Inc., Englewood Cliffs, NJ.

130 Devaney, A.J. (2000) Super-resolution processing of multi-static data using time reversal and MUSIC, http://www.ece.neu.edu/faculty/devaney/preprints/paper02n_00.pdf.

131 Devaney, A.J. (2005) Time reversal imaging of obscured targets from multistatic data. *IEEE Trans. Antenn. Propag.*, **53** (5), 1600–1610.

132 Dinh-Quy, N., Woon-Seng, G. and Yong-Kim, C. (2007) Detection and localization of the scatterers via DORT method. 6th International Conference on Information, Communications & Signal Processing, 2007, pp. 1–5.

133 Bellomo, L., Saillard, M. Pioch, S. *et al.* (2010) An ultrawideband time reversal-based RADAR for microwave-range imaging in cluttered media. 13th International Conference on Ground Penetrating Radar (GPR), 2010, pp. 1–5.

134 Kosmas, P., Laranjeira, S. Dixon, J.H. *et al.* (2010) Time reversal microwave breast imaging for contrast-enhanced tumor classification. Annual International Conference of the IEEE on Engineering in Medicine and Biology Society (EMBC), 2010, pp. 708–711.

135 Yavuz, M.E. and Teixeira, F.L. (2006) Full time-domain DORT for ultrawideband electromagnetic fields in dispersive, random inhomogeneous media. *IEEE Trans. Antenn. Propag.*, **54** (8), 2305–2315.

136 Yavuz, M.E. and Teixeira, F.L. (2005) Frequency dispersion compensation in time reversal techniques for UWB electromagnetic waves. *IEEE Trans. Geosci. Remote*, **2** (2), 233–237.

137 Yavuz, M.E. and Teixeira, F.L. (2008) Space-frequency ultrawideband time-reversal imaging method as applied to subsurface objects. IEEE International Symposium on Geoscience and Remote Sensing, 2008. IGARSS 2008, pp. III-12–III-15.

138 Yavuz, M.E. and Teixeira, F.L. (2008) On the sensitivity of time-reversal imaging techniques to model perturbations. *IEEE Trans. Antenn. Propag.*, **56** (3), 834–843.

139 Fouda, A.E., Teixeira, F.L. and Yavuz, M.E. (2011) Imaging and tracking of targets in clutter using differential time-reversal. Proceedings of the 5th European Conference on Antennas and Propagation (EUCAP), pp. 569–573.

140 Kerbrat, E., Prada, C. Cassereau, D. *et al.* (2000) Detection and imaging in complex media with the D.O.R.T. method. Ultrasonics Symposium, vol. 1, 2000 IEEE, pp. 779–783.

141 Hantscher, S., Etzlinger, B. Reisenzahn, A. *et al.* (2007) A wave front extraction algorithm for high-resolution pulse based radar systems. IEEE International Conference on Ultra-Wideband, 2007. ICUWB 2007, pp. 590–595.

142 Greenhalgh, S.A. and Marescot, L. (2006) Modeling and migration of 2-D georadar data: A stationary phase approach. *IEEE Trans. Geosci. Remote*, **44** (9), 2421–2429.

143 Yufryakov, B.A., Surikov, B.S. Sosulin, Y.G. *et al.* (2004) A method for interpretation of ground-penetrating radar data. *J. Commun. Technol. El.*, **49** (12), 1342–1356.

144 Sakamoto, T. and Sato, T. (2004) A fast algorithm of 3-dimensional imaging for pulse radar systems. Antennas and Propagation Society International Symposium, vol. 2, 2004. IEEE, pp. 2099–2102.

145 Helbig, M., Hein, M.A. Schwarz, U. *et al.* (2008) Preliminary investigations of chest surface identification algorithms for breast cancer detection. International Conference on Ultra-Wideband, Hannover, Germany.

146 Zhuge, X. and Yarovoy, A. (2010) Comparison between wavefront-based shape reconstruction and beamforming for UWB near-field imaging radar. European Radar Conference (EuRAD), 2010, pp. 208–211.

5
Electromagnetic Fields and Waves in Time and Frequency

5.1
Introduction

In the previous chapter, some aspects of ultra-wideband sensing were discussed on the basis of more or less intuitive comprehension founded on linear circuit theory and fundamentals of wave propagation. We disregarded the vector nature of the electromagnetic field and the numerous propagation phenomena of electromagnetic waves.

We will summarize some important interaction and propagation issues of electromagnetic fields which may be essential for ultra-wideband sensing. In the frame of this book, this can be only a short introduction. The major aim is to impart the most important propagation phenomena and their modelling in order to permit a better assessment of simplifications usually done by solving practical sensing problems. The reader is conducted to the numerously available textbooks on electromagnetic field theory for a deeper introduction to the subject. Exemplarily, following books are mentioned [1–11].

The majority of the textbooks in electromagnetism deal with the frequency domain description which simplifies mathematical derivations but sometimes this approach obscures the intuitive understanding of wave propagation phenomena. The natural way of thinking is 'time domain thinking' in which we try to comprehend the natural sequence of events. Here, we consider both, that is we use time and frequency domain description in order to show some parallelisms. Some effects are easier to understand in the frequency domain (e.g. relaxation phenomena of various substances); others are simpler to comprehend[1] in the time domain (e.g. propagation phenomena of very wideband waves).

Wave propagation phenomena which are most important for analysis of ultra-wideband sensing cover the following aspects:

- **Conversion of a guided wave to radiated waves and vice versa:** This is the task of the antennas. Usual electronics is only able to generate and measure electric

1) Better comprehension must not necessarily mean simpler to model. Calculations are often easier and more efficiently done in the frequency domain since no convolution integrals have to be solved.

Handbook of Ultra-Wideband Short-Range Sensing: Theory, Sensors, Applications, First Edition. Jürgen Sachs.

signals on a waveguide in a single point. Hence, we need an appropriate conversion between guided waves and the free field.

- **Scattering of radiated waves:** Scattering always occurs if the electric and magnetic properties within the propagation path of the wave are changed. The wave reacts to this inhomogeneity by
  - reflection
  - refraction
  - diffraction
  - mode conversion, for example generation of surface/boundary waves.
- **Beam spreading or divergence:** Waves emanating from a spatially limited source distribute their energy within a certain solid angle. Hence, they lose intensity with increasing distance from the source since they have to distribute their energy over a larger area. Sometimes this is counted as path loss even if it is not really linked to an actual loss mechanism which would lead to an energy conversion.
- **Dispersion:** Dispersion means that a physical quantity depends on the wavelength or frequency. In our case, it mainly concerns the propagation speed, the wave attenuation and the refraction and reflection index. Wave propagation in vacuum and lossless homogeneous material is dispersionless.

We can discuss the various wave phenomena occurring in ultra-wideband measurements under two opposite viewpoints. Ultra-wideband sensing is an indirect measurements principle with the goal to gain information about a scenario under test via its interaction with electromagnetic sounding waves. Hence, our first interest is to identify an interaction mechanism which offers the best opportunities to extract the wanted information. Second, we are obligated to estimate the influence of perturbing mechanisms and to reduce their impact by designing optimum experimental conditions. Since usually almost all propagation phenomena appear (more or less pronounced) in parallel, the sensing task often requires a balanced design of hardware and software components in order to meet the sensing task as best as possible.

Under this light the reader shall see the following chapter. It is intended to give not more than a short overview on wave phenomena and their impact on ultra-wideband sensing.

5.2 The Fundamental Relations of the Electromagnetic Field

The electromagnetic field is characterized by the interaction of various physical quantities. Many of them are vector fields which either represent force or flux quantities. Vector fields are composed of three components which typically depend on time and the considered point in space. The position vector of the respective points is assigned by $\mathbf{r}$. We will refer to the coordinate system as introduced in Chapter 4, Figure 4.2 and Annex A.4. For the sake of introduction we will restrict ourselves to the decomposition of the field vectors in their Cartesian or spherical components:

$$\text{Cartesian}: \quad \mathbf{E}(\mathbf{r}) = E_x(\mathbf{r})\mathbf{e}_x + E_y(\mathbf{r})\mathbf{e}_y + E_z(\mathbf{r})\mathbf{e}_z = \begin{bmatrix} E_x(\mathbf{r}) \\ E_y(\mathbf{r}) \\ E_z(\mathbf{r}) \end{bmatrix} \tag{5.1}$$

$$\text{Spherical}: \quad \mathbf{E}(\mathbf{r}) = E_\mathrm{r}(\mathbf{r})\mathbf{e}_\mathrm{r} + E_\vartheta(\mathbf{r})\mathbf{e}_\vartheta + E_\varphi(\mathbf{r})\mathbf{e}_\varphi \tag{5.2}$$

whereas Table A.2 can be used for mutual conversions between both notations, for example

$$E_y = \sin\vartheta \sin\varphi E_\mathrm{r} + \cos\vartheta \sin\varphi \, E_\vartheta + \cos\varphi E_\varphi$$

Time-dependent vector quantities are written in bold letters $\mathbf{E}(t, \mathbf{r})$ or simply $\mathbf{E}$ and underlined bold letters as $\underline{\mathbf{E}}(f, \mathbf{r})$ or $\underline{\mathbf{E}}$ refer to the frequency domain representation.

Some basic definitions and rules of vector calculus are summarized in Annex A.5.

5.2.1 Maxwell's Equations and Related Relations

In the case of ultra-wideband sensors, the stimulation of the measurement scenario and the observation of its reaction are based on electromagnetic energy. The propagation of electromagnetic waves and the distribution of electromagnetic energy in space are modelled by a set of four Maxwell's vector equations. Table 5.1 summarizes the main laws of electromagnetism in different forms of notation. They describe the mutual interaction of the involved fields ($\mathbf{E}, \mathbf{D}, \mathbf{H}, \mathbf{B}$) and their source ($\mathbf{J}, \varrho$), that is electrical charge or current. We can distinguish two kinds of fields from their physical nature. These are fields relating to a force quantity ($\mathbf{E}, \mathbf{H}$) and fields relating to a flux quantity ($\mathbf{D}, \mathbf{B}$).

Many textbooks deal also with hypothetical magnetic charge[2)] as a source of electromagnetic fields for reasons of symmetry in the equations. This would also lead to the occurrence of magnetic current extending Gauss's law for magnetism and Faraday's law. We will not deal with this theoretical model here.

Combining (5.11) and (5.7) leads to the continuity equation which states the conservation of charge, that is the flux of an electrical current, is immediately connected with a displacement of charge. Due to the vector identity $\nabla \cdot (\nabla \times \mathbf{A}) = 0$ (see Annex A.5) we achieve

$$\left.\begin{aligned} \nabla \cdot \mathbf{J} + \frac{\partial \varrho}{\partial t} = 0 \\ \nabla \cdot \mathbf{J} + u_1 * \varrho = 0 \end{aligned}\right\} \underset{\text{IFT}}{\overset{\text{FT}}{\rightleftarrows}} \nabla \cdot \underline{\mathbf{J}} + j2\pi f \underline{\varrho} = 0 \tag{5.15}$$

2) In equivalence to the unit of electric charge [A s] = [C], it would have the unit [V s] = [Wb].

Table 5.1 Maxwell's equations in terms of free charge and current.

Notation	Time domain		Frequency domain	
Faraday's law	$\nabla \times \mathbf{E} = -\dfrac{\partial \mathbf{B}}{\partial t}$ $\nabla \times \mathbf{E} = -u_1 * \mathbf{B}$	(5.3)	$\nabla \times \underline{\mathbf{E}} = -j2\pi f \underline{\mathbf{B}}$	(5.4)
	$\oint_C \mathbf{E} \cdot \mathrm{d}\mathbf{l} = -\iint_S \dfrac{\partial \mathbf{B}}{\partial t} \cdot \mathrm{d}\mathbf{S}$	(5.5)	$\oint_C \underline{\mathbf{E}} \cdot \mathrm{d}\mathbf{l} = -j2\pi f \iint_S \underline{\mathbf{B}} \cdot \mathrm{d}\mathbf{S}$	(5.6)
Maxwell–Ampère law	$\nabla \times \mathbf{H} = \mathbf{J} + \dfrac{\partial \mathbf{D}}{\partial t}$ $\nabla \times \mathbf{H} = \mathbf{J} + u_1 * \mathbf{D}$	(5.7)	$\nabla \times \underline{\mathbf{H}} = \underline{\mathbf{J}} + j2\pi f \underline{\mathbf{D}}$	(5.8)
	$\oint_C \mathbf{H} \cdot \mathrm{d}\mathbf{l} = \int_S \left(\mathbf{J} + \dfrac{\partial \mathbf{D}}{\partial t}\right) \cdot \mathrm{d}\mathbf{S}$	(5.9)	$\oint_C \underline{\mathbf{H}} \cdot \mathrm{d}\mathbf{l} = \int_S (\underline{\mathbf{J}} + j2\pi f \underline{\mathbf{D}}) \cdot \mathrm{d}\mathbf{S}$	(5.10)
Gauss's law	$\nabla \cdot \mathbf{D} = \varrho$			(5.11)
	$\oiint_S \mathbf{D} \cdot \mathrm{d}\mathbf{S} = \iiint_V \varrho \, \mathrm{d}V = Q$			(5.12)
Gauss's law for magnetism	$\nabla \cdot \mathbf{B} = 0$			(5.13)
	$\oiint_S \mathbf{B} \cdot \mathrm{d}\mathbf{S} = 0$			(5.14)

Herein are

Symbol	Description	Measuring unit
$\mathbf{E}$	Electric field strength	[V/m] = [N/C] = [N/As]
$\mathbf{D}$	Electric flux density; electric displacement field	$[As/m^2] = [C/m^2]$
$\mathbf{H}$	Magnetic field strength	[A/m] = [N/Wb] = [N/Vs]
$\mathbf{B}$	Magnetic flux density	$[Vs/m^2] = [Wb/m^2]$
$\mathbf{J}$	Flux density of free charge (free current density)	$[A/m^2]$
ϱ	Free charge density	$[C/m^3] = [As/m^3]$
Q	Charge	[As]
∇	Nabla operator (see Annex A.5 for definition)	
$\nabla\times$	Curl or rotator operator (see Annex A.5 for definition)	
$\nabla\cdot$	Divergence operator (see Annex A.5 for definition)	
$\mathbf{A}\cdot\mathbf{B}$	Scalar (inner or dot) product of the vectors **A**, **B**	
$\oint_C \mathbf{A}\cdot d\mathbf{l}$	Line integral of the vector field **A** along the contour line C enclosing the surface S. The vector field **A** typically represents a force field. **dl** is the vector of a line element	
$\iint_S \mathbf{A}\cdot d\mathbf{S}$	Surface integral of the vector field **A**. The vector field **A** typically represents a flux density. Hence, the integral gives the total flux through the considered surface S. $d\mathbf{S} = \mathbf{n}\,dS$ with $\mathbf{n}$ the normal vector of a surface element	
$\oiint_S \mathbf{A}\cdot d\mathbf{S}$	Surface integral of the vector field **A** over the surface S enclosing the volume V. The vector field **A** typically represents a flux density. Hence, the integral gives the total flux originating from (the integral has positive value) or sinking to (the integral has negative value) the considered volume V which is enclosed by the surface S	
$\iiint_V A\,dV$	Volume integral of the scalar field A	

All equations are given for the time and the frequency domain as well as in their differential and integral form. The differential forms are also given in doublet notation (see Annex A.3).

Table 5.2 Constitutive relations (linear and time invariant material supposed).

	Time domain		Frequency domain	
Ohm's law	$\mathbf{J} = \boldsymbol{\sigma} * \mathbf{E}$	(5.16)	$\underline{\mathbf{J}} = \underline{\boldsymbol{\sigma}} \cdot \underline{\mathbf{E}}$	(5.17)
Polarization	$\mathbf{D} = \boldsymbol{\varepsilon} * \mathbf{E}$	(5.18)	$\underline{\mathbf{D}} = \underline{\boldsymbol{\varepsilon}} \cdot \underline{\mathbf{E}}$	(5.19)
Magnetization	$\mathbf{B} = \boldsymbol{\mu} * \mathbf{H}$	(5.20)	$\underline{\mathbf{B}} = \underline{\boldsymbol{\mu}} \cdot \underline{\mathbf{H}}$	(5.21)

Furthermore, the force and flux fields are mutually linked by the constitutive relations describing the capability of the force fields to stimulate an appropriate flux. The commonly applied relations are summarized in Tables 5.2 and 5.3. They are dictated by the behaviour of the material exposed to the electromagnetic field. In the case of weak fields, the constitutive relations can be supposed to be linear. It should be noted that the constitutive relations can be extended to further physical quantities (e.g. mechanical tension, temperature etc.) giving access to many sensing principles for non-electric quantities.

These relations describe formally the interaction of the electromagnetic field with matter. Independently on the actually underlying mechanism of interaction, a set of three parameters is in use:

- Conductivity σ
- Permittivity ε
- Permeability μ

Typically their value and frequency behaviour depend specifically on the considered material. This gives us the opportunity to quantify and qualify substances and

Table 5.3 Material parameters.

Name	General	Isotropic (scalar)	Anisotropic (dyadic tensor)
Conductivity	$\boldsymbol{\sigma}$ $\left[\frac{\mathrm{A}}{\mathrm{V\,m}} = \frac{1}{\Omega\,\mathrm{m}} = \mathrm{S\,m^{-1}}\right]$	σ	$\boldsymbol{\sigma} = \begin{bmatrix} \sigma_{11} & \sigma_{12} & \sigma_{13} \\ \sigma_{21} & \sigma_{22} & \sigma_{23} \\ \sigma_{31} & \sigma_{32} & \sigma_{33} \end{bmatrix}$
Permittivity (electric constant)	$\boldsymbol{\varepsilon} = \boldsymbol{\varepsilon}_r \varepsilon_0$ $\left[\frac{\mathrm{A\,s}}{\mathrm{V\,m}} = \mathrm{F\,m^{-1}}\right]$ $\varepsilon_0 = 8.854\,10^{-12}\,\mathrm{F\,m^{-1}} = \frac{10^{-9}}{36\pi}\,\mathrm{F\,m^{-1}}$	$\varepsilon = \varepsilon_r \varepsilon_0$	$\boldsymbol{\varepsilon} = \begin{bmatrix} \varepsilon_{11} & \varepsilon_{12} & \varepsilon_{13} \\ \varepsilon_{21} & \varepsilon_{22} & \varepsilon_{23} \\ \varepsilon_{31} & \varepsilon_{32} & \varepsilon_{33} \end{bmatrix}$
Permeability (magnetic constant)	$\boldsymbol{\mu} = \boldsymbol{\mu}_r \mu_0$ $\left[\frac{\mathrm{V\,s}}{\mathrm{A\,m}} = \mathrm{H\,m^{-1}}\right]$ $\mu_0 = 4\pi\,10^{-7}\,\mathrm{H\,m^{-1}}$	$\mu = \mu_r \mu_0$	$\boldsymbol{\mu} = \begin{bmatrix} \mu_{11} & \mu_{12} & \mu_{13} \\ \mu_{21} & \mu_{22} & \mu_{23} \\ \mu_{31} & \mu_{32} & \mu_{33} \end{bmatrix}$

substance mixtures by electrical measurement. The electromagnetic field can interact with matter by various mechanisms showing, for example typical frequency behaviour. Representatively, we will discuss such an interaction phenomenon in Section 5.3 for aqueous substances which refers to an important class of ultra-wideband measurements.

It should be noted that the time domain notation of the constitutive relations involves a convolution of the material parameter with the field quantity as depicted in Table 5.2. This is often omitted in textbooks on electromagnetism. But, if a material parameter, for example the permittivity, is frequency dependent then their behaviour in the time domain is described by an impulse response; that is, we can write

$$\underline{\varepsilon}(f) = \varepsilon'(f) - j\varepsilon''(f) \underset{\mathrm{FT}}{\overset{\mathrm{IFT}}{\rightleftarrows}} \varepsilon(t), \text{ and}$$

hence (5.19) has to be expressed by the convolution (5.18). Herein, ε' stands for the real part of the frequency-dependent and complex-valued permittivity and ε'' stands for its imaginary part. Corresponding holds for the other constitutive relations whereas the conductivity is often joined with the imaginary part of the permittivity (see (5.59)). Since any substance must behave causal with respect to an electromagnetic stimulation, it holds for all material parameters (representatively written for the permittivity)

$$\varepsilon(t) = \begin{cases} \varepsilon(t) & \text{for} \quad t > 0 \\ \equiv 0 & \text{for} \quad t \le 0 \end{cases} \tag{5.22}$$

which means that their real and imaginary parts ε' and ε'' are mutually dependent according to the Kramers–Kronig relation (see (2.150), [12]). We will see in Section 5.4 that ε' dominates the propagation speed of an electromagnetic wave and the ε''-value relates to propagation losses. Thus, we can conclude that a medium of frequency-dependent attenuation will also have frequency-dependent propagation speed of the wave.

It is a common practice, namely by dealing with aspects of wave propagation, to replace the introduced set of material parameters by another one. In the case of an isotropic, homogenous and lossless propagation medium, usually the propagation speed c of the wave and the intrinsic impedance Z_s of the propagation medium take over the role of the material parameters

$$c = \frac{1}{\sqrt{\varepsilon\mu}} = \frac{c_0}{\sqrt{\varepsilon_r\mu_r}} \quad \text{with} \quad c_0 = \frac{1}{\sqrt{\varepsilon_0\mu_0}} \approx 2.9979\,10^8\ \mathrm{m/s} \approx 30\,\mathrm{cm/ns} \tag{5.23}$$

$$Z_s = \sqrt{\frac{\mu}{\varepsilon}} = Z_0\sqrt{\frac{\mu_r}{\varepsilon_r}} \quad \text{with } Z_0 = \sqrt{\frac{\mu_0}{\varepsilon_0}} \approx 120\pi\Omega = 376,99\ \Omega \tag{5.24}$$

Herein, c_0 and Z_0 refer to the speed of light and the intrinsic impedance of the vacuum. We will discuss later further details on that approach in connection with fundamentals of wave propagation.

5.2.2
Boundary Conditions

A scenario under test typically consists of bodies from different materials and of different size and shape. If an electromagnetic field propagates through such a scenario, it will interact with these bodies. This interaction can be split up into two parts. One refers to a volume effect which is characterized by the material parameters ε, μ, σ as introduced here, and the second results from effects at the interface between two materials. They are usually referred to as the boundary conditions. By restricting to a continuous surface, we can distinguish between two types of conditions. The one requires continuity of the normal components of flux quantities, and the second refers to the continuity of the tangential components of force quantities. Both versions are depicted in Figure 5.1 and they are expressed by (5.25) and (5.26). The analysis of wedge-shaped surface discontinuities or tips requires additionally the edge and radiation condition [8, 12].

Continuity of normal components of flux quantities:

$$\begin{aligned} \mathbf{n} \cdot (\mathbf{D}_1 - \mathbf{D}_2) &= \varrho_0 \\ \mathbf{n} \cdot (\mathbf{B}_1 - \mathbf{B}_2) &= 0 \\ \mathbf{n} \cdot (\mathbf{J}_1 - \mathbf{J}_2) &= 0 \end{aligned} \tag{5.25}$$

Continuity of the normal component of the flux quantities

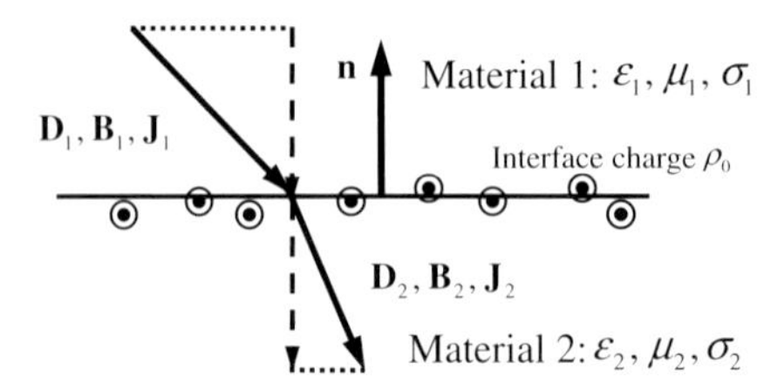

Continuity of the tangential component of the force quantities

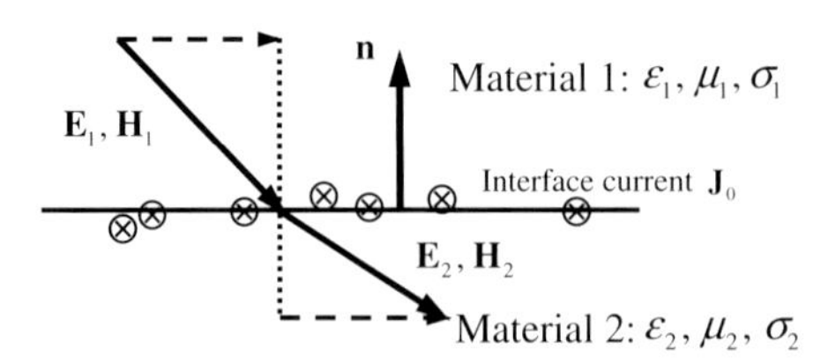

Figure 5.1 Illustration of the boundary conditions for flux and force quantities.

Continuity of tangential components of force quantities:

$$\begin{aligned} \mathbf{n} \times (\mathbf{E}_1 - \mathbf{E}_2) &= \mathbf{0} \\ \mathbf{n} \times (\mathbf{H}_1 - \mathbf{H}_2) &= \mathbf{J}_0 \end{aligned} \tag{5.26}$$

Herein, ϱ_0 and $\mathbf{J}_0$ represent the charge and the current at the interface between the two media and $\mathbf{n}$ is the unit normal vector of the considered surface. With only a few exceptions, the boundary conditions cannot be fulfilled by the incident and penetrated wave alone wherefore new waves – for example reflected wave, interface wave (Sommerfeld wave), evanescent wave – may be generated. For details, see Section 5.4.

5.2.3
Energy Flux of Electromagnetic Radiation

The stimulation of a test scenario and the feasibility to observe its reaction are always connected with an energy influx. The Poynting vector quantifies the energy flux for electromagnetic radiation. It is defined in (5.27).

$$\mathfrak{P}(t, \mathbf{r}) = \mathbf{E}(t, \mathbf{r}) \times \mathbf{H}(t, \mathbf{r}) \tag{5.27}$$

A modified version for time harmonic fields is given by

$$\underset{..}{\mathfrak{P}}(f, \mathbf{r}) = \underset{..}{\mathbf{E}}(f, \mathbf{r}) \times \underset{..}{\mathbf{H}}^*(f, \mathbf{r}) \tag{5.28}$$

Here, we referred to double-sided spectral representation of the field quantities which we assign by double points. For single-sided spectral notations, (5.28) modifies to (refer also to Section 2.2.5 (2.70) and remarks concerning (2.62))

$$\underset{.}{\mathfrak{P}}(f, \mathbf{r}) = \frac{1}{2} \underset{.}{\mathbf{E}}(f, \mathbf{r}) \times \underset{.}{\mathbf{H}}^*(f, \mathbf{r}) \tag{5.29}$$

As long as there will no difference between two- and single-sided spectral notations, we will omit the dots in what follows.

It should be noted that (5.27) represents the instantaneous power flux of an electromagnetic wave through a small surface element. Its dimension is W/m^2. The Poynting vector points in the direction of energy transport which is identical with the direction of wave propagation under isotropic propagation conditions.

Some attention has to be paid by interpreting the frequency domain representation of the Poynting vector (5.28). Its real part $\mathrm{Re}\{\underset{..}{\mathfrak{P}}(f, \mathbf{r})\}$ indicates the spectral density of the average net power flux through a surface element while its imaginary part $\mathrm{Im}\{\underset{..}{\mathfrak{P}}(f, \mathbf{r})\}$ symbolizes the reactive power transfer (idle power) through that surface. The unit of the Poynting vector in the frequency domain is given in W/(Hz m^2) for non-periodic waves and W/m^2 for waves with periodic time structure.

If the volume V contains a radiation source or sink, the net power transfer through the surface S enclosing the volume is

$$\begin{aligned} P &= \lim_{T\to\infty}\frac{1}{2T}\int_{T}^{T}\oint_{S}\mathfrak{P}(t,\mathbf{r})\cdot \mathrm{d}\mathbf{S}\,\mathrm{d}t \\ &= \int_{-\infty}^{\infty}\oint_{S}\underline{\underline{\mathfrak{P}}}(f,\mathbf{r})\cdot \mathrm{d}\mathbf{S}\,\mathrm{d}f = \int_{0}^{\infty}\oint_{S}\underline{\mathfrak{P}}(f,\mathbf{r})\cdot \mathrm{d}\mathbf{S}\,\mathrm{d}f \end{aligned} \tag{5.30}$$

5.2.4
Radiation Condition

The radiation condition implies that the energy which is radiated from the sources must scatter to infinity. It represents a useful condition for solving scattering problems. The mathematical formalism of the field equations (as summarized in Tables 5.1 and 5.2) to describe the wave scattering at an object leads to two basic solutions. In one of them, the scattered wave approaches the object, and in the second solution, it departs from the scatterer. Only the last one is causal and meets the physical reality. This fact is expressed by the radiation condition for vector fields according to Silver–Müller (5.31) [9, 11]:

$$\begin{aligned} &\lim_{r\to\infty} r\left(\mathbf{e}_0\times(\nabla\times\mathbf{X}(t,\mathbf{r}))+\frac{1}{c}\frac{\partial\mathbf{X}(t,\mathbf{r})}{\partial t}\right)=\mathbf{0} \\ &\qquad\qquad FT \ \updownarrow \ IFT \\ &\lim_{r\to\infty} r\left(\mathbf{e}_0\times(\nabla\times\underline{\mathbf{X}}(f,\mathbf{r}))+\frac{j2\pi f}{c}\underline{\mathbf{X}}(f,\mathbf{r})\right)=0 \end{aligned} \tag{5.31}$$

The vector field $\mathbf{X}$ in Eq. (5.31) stands for either the electric field $\mathbf{E}$ or magnetic field $\mathbf{H}$.

5.2.5
Lorentz Reciprocity

Concerning our topic – the characterization of a measurement scenario, the reciprocity theorem states that the locus of stimulation and the locus of observation can be exchanged without any loss of information about the transmission behaviour between the two points. That is the stimulation of the scenario by an electromagnetic radiation (caused by the current $\mathbf{J}_1$) departing from point $\mathbf{r}_1$ and capturing its impact (expressed by the electrical field $\mathbf{E}_1$) at point $\mathbf{r}_2$ leads to the same results as if the scenario would be stimulated from point $\mathbf{r}_2$ and observed in point $\mathbf{r}_1$. Figure 5.2 illustrates the statement of the reciprocity theorem, and (5.32) gives the

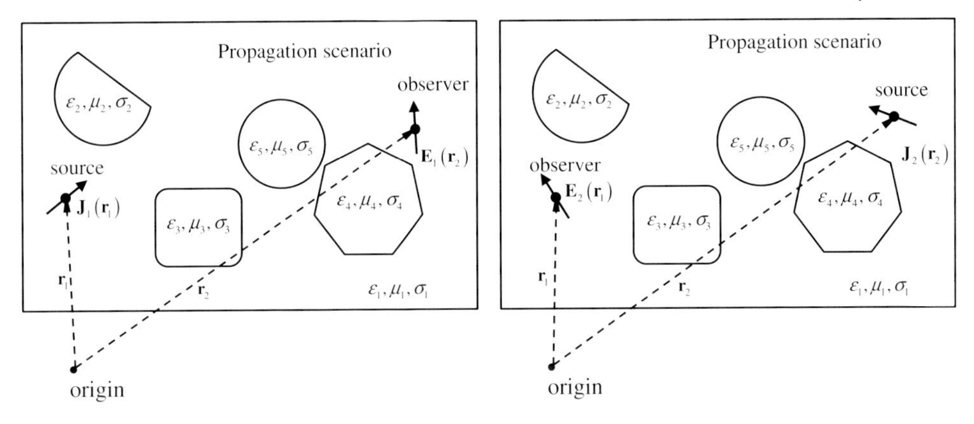

Figure 5.2 Illustration of reciprocity.

corresponding mathematical formulation.

$$
\begin{gathered}
\int_V (\underline{\mathbf{J}}_1(f, \mathbf{r}_1) \cdot \underline{\mathbf{E}}_2(f, \mathbf{r}_1) - \underline{\mathbf{E}}_1(f, \mathbf{r}_2) \cdot \underline{\mathbf{J}}_2(f, \mathbf{r}_2))\mathrm{d}V = 0 \\
\mathrm{FT} \updownarrow \mathrm{IFT} \\
\int_V (\mathbf{J}_1(t, \mathbf{r}_1) * \mathbf{E}_2(t, \mathbf{r}_1) - \mathbf{E}_1(t, \mathbf{r}_2) * \mathbf{J}_2(t, \mathbf{r}_2))\mathrm{d}V = 0
\end{gathered}
\tag{5.32}
$$

The reciprocity theorem is usually expressed by its frequency domain relations in the literature. Time domain expressions are rarely in use. In Refs [11, 13], an example can be discovered for time domain notation. However, there the time domain relation was not simply gained from the frequency domain representation by Fourier transform as it was done in (5.32). Rather the authors applied multiplication of field quantities in the time domain too instead of convolution. This approach simplifies calculations but it also obscures the physics behind the reciprocity relation since it requires non-causal advanced fields which do not appear in measurements. Annex C.1 gives further details on the time domain reciprocity relation (5.32).

The reciprocity relation can be simplified for test objects with a finite number of lumped measurement ports as depicted in Figure 5.3. We already made use of it in Sections 2.5 and 4.4. The reciprocity relation for lumped ports reads

$$\underline{I}_1(f)\underline{V}_2(f) = -\underline{I}_2(f)\underline{V}_1(f) \underset{\mathrm{IFT}}{\overset{\mathrm{FT}}{\longleftrightarrow}} I_1(t) * V_2(t) = -I_2(t) * V_1(t) \tag{5.33}$$

$$\underline{a}_1(f)\underline{b}_2(f) - \underline{a}_2(f)\underline{b}_1(f) \underset{\mathrm{IFT}}{\overset{\mathrm{FT}}{\longleftrightarrow}} a_1(t) * b_2(t) = a_2(t) * b_1(t) \tag{5.34}$$

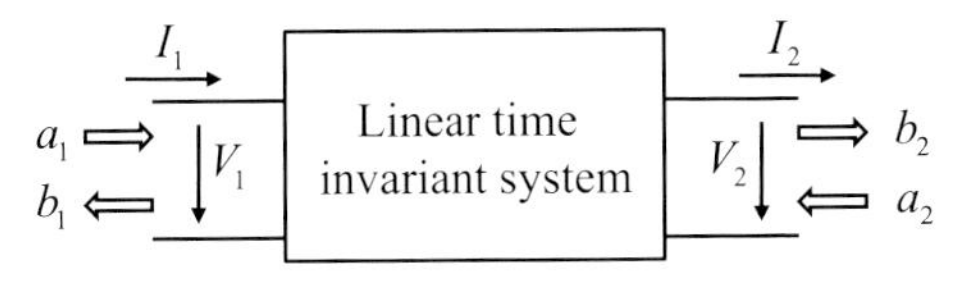

Figure 5.3 Illustration of reciprocity for systems with lumped measurement ports.

5.3
Interaction of Electromagnetic Fields with Matter

As we have seen from the previous chapter, the interaction of electromagnetic radiation with a test scenario includes geometric aspects, that is size, shape, position and orientation of involved bodies as well as the interaction with the atomic or molecular structure of substances itself. Here in this chapter, we will concentrate exclusively on substance effects by excluding geometric influences. In order to avoid any impact of unknown boundaries of the body of investigation, we will first abridge the volume of interaction to a small body section or we apply material probes of known geometry.

The interactions of matter (we will restrict ourselves to condensed matter) with electromagnetic fields are complex and manifold. The strength of this interaction depends on the material composition (including the molecular structure) and the frequency of the sounding field. Formally this is expressed by the spectral functions of $\varepsilon(f), \mu(f)$ and $\sigma(f)$. The goal of corresponding measurements is to determine these dependencies in order to quantify and qualify the material under test. Since the relative permeability of most of the materials is close to one, that is $\mu = \mu_0$, the determination of the permittivity value (also named dielectric constant) is mostly in the foreground of a measurement. Such measurements are often referred to as 'dielectric spectroscopy' or 'impedance spectroscopy' (since ε may be mapped to Z_s corresponding to (5.24)). The frequency behaviour of the permittivity will be exemplarily considered. The reader is addressed to Refs [12, 14–16] for more details. There, he can also find a comprehensive pool of publications on that topic.

It is known from the literature that measurements of dielectric parameters are spread over 19 frequency decades indicating the diversity of physical effects leading to the mutual interaction between electromagnetic field and matter. Figure 5.4

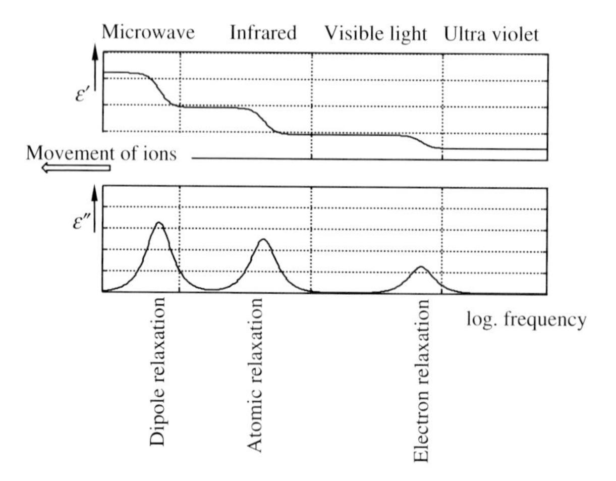

Figure 5.4 Typical frequency response of dielectric material.

schematizes some interaction phenomena and their impact on the permittivity value of the material. The frequency dependence of the permittivity is caused by the mobility, the mass and molecular forces of the involved charged particles, for example ions, dipolar molecules, atoms, nucleons, electrons and so on.

It can be observed from Figure 5.4 that most of the interaction phenomena are more or less localized at specific frequencies termed as relaxation frequency and relaxation time for its inverse.

In the case of UWB sensing, the dipole relaxation (also assigned as orientational polarizability) is of special interest since the relaxation frequency of a very important molecule – the molecule of water H_2O – is to be found in the UWB frequency range. Water plays a dominant role in biological tissues and technical processes. Hence, we will consider somewhat its behaviour in an alternating electrical field representatively for the relaxation phenomena of other substances (see also [15] for details).

The isolated water molecule is built from one oxygen and two hydrogen atoms as depicted in Figure 5.5. It possesses a permanent dipole moment $p = (1.84 \pm 0.02)\mathrm{D}$ which results from the superposition of the two dipole

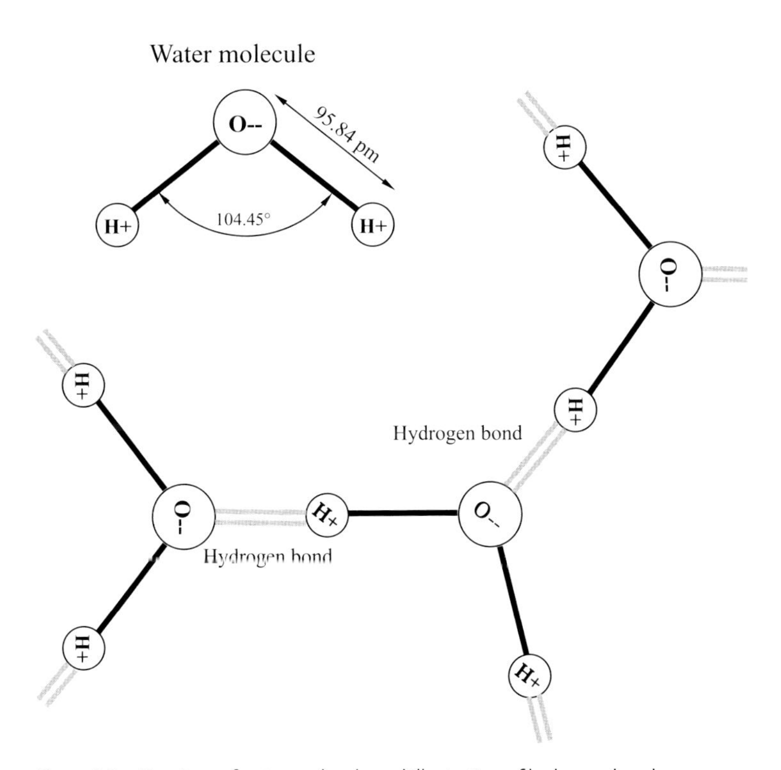

Figure 5.5 Structure of water molecule and illustration of hydrogen bonds.

moments of the H—O bonds of 1.53 D. The unit D (Debye) represents the molecular electric dipole moment ($1\,\text{D} = 3.33564 \times 10^{-30}\,\text{C m}$).

Adjacent water molecules form hydrogen bonds; thus, water molecules are finally arranged in a regular tetrahedron 'lattice'. The strength of the bond forces is given by the molecular enthalpy H_m. It constitutes $H_m = 463\,\text{kJ/mol}$ for the covalent H—O bond and it is about $H_m = 20\,\text{kJ/mol}$ only for the hydrogen bond. The thermal energy at room temperature is about $RT = 2.5\,\text{kJ/mol}$ ($R = k_B N_A = 8.314\,\text{J/(mol K)}$ is the gas constant; $k_B = 1.38 \times 10^{-23}\,\text{J/K}$ is the Boltzmann constant and $N_A = 6.022 \times 10^{23}\,\text{mol}^{-1}$ is the Avogadro constant). This is close to the energy of the hydrogen bonds. Hence, we will have a thermal motivated fluctuation/destruction of the hydrogen bonds between adjacent molecules while the covalent bonds remain stable. Most of the broken hydrogen bonds will be 'repaired' within 1 ps or less but some of the water molecules are 'missing' that opportunity so they have to 'look for a new partner'. For that purpose, these water molecules have to move or rotate. This provides electric polarization noise due to the permanent electric dipole moment of the water molecule. For water at room temperature, it takes about $\tau_r \approx 10\,\text{ps}$ in average until a new 'partner molecule' is found which promotes reorientation. The duration τ_r is called relaxation time. It is easy to accept that the process of reorientation and hence the relaxation time depends on many things, for example the number of available 'partner molecules' (depend on temperature and density), the mobility of the molecules, the bond forces, if water molecules are trapped by other substances (i.e. bounded or unbounded water) and so forth. These observations raise some hope to perform microwave measurements for the evaluation of the state of aqueous material compositions.

If we would be able to measure the polarization noise due to reorientation, we would get an auto-correlation function for this phenomenon of the form [17]

$$C_{PN,H_2O}(\tau > \tau_0) = C(\tau_0)e^{-\frac{\tau-\tau_0}{\tau_r}} \quad \text{with } \tau_0 = 2\,\text{ps} \tag{5.35}$$

Water molecules show further relaxation phenomena. Since they are running faster than the effect described above they are not of interest for UWB measurements. Hence, we disregard all physical effects which only become noticeable at a time lag τ smaller than 2 ps.

Exposing the water to an electric field, the reorientation of the molecules with free hydrogen bond will get a preferential direction along the injected field. This leads to a measurable polarization of the material. The strength of that polarization reflects in the value of relative permittivity ε_r. Water molecules have a strong dipole moment and they are only weakly trapped in a lattice. Therefore, it has one of the highest permittivity values of natural substances.

If we abruptly switch on the electric field, the reorientation cannot follow immediately the field; rather, it approaches exponentially its final value. The time constant of the exponential approach coincides with τ_r, the time constant of the auto-correlation function (5.35). The polarization and thus the permittivity value will show the same time behaviour; that is, for the permittivity step response we have

$$\varepsilon_{\text{step}}(t > t_0) = \varepsilon_\infty + (\varepsilon_{\text{DC}} - \varepsilon_\infty)\left(1 - e^{-\frac{t-t_0}{\tau_r}}\right) \tag{5.36}$$

The conversion of Eq. (5.36) into the frequency domain gives the well-known Debye-type relaxation relation [17]:

$$\underline{\varepsilon}(f) = \varepsilon'(f) - j\varepsilon''(f) = \varepsilon_\infty + \frac{\varepsilon_{\text{DC}} - \varepsilon_\infty}{1 + j2\pi f \tau_r} \tag{5.37}$$

Herein, ε_{DC} is the low-frequency permittivity value and ε_∞ represents the permittivity well above the relaxation frequency $f_r = 1/2\pi\tau_r$. Note, (5.36) and (5.37) suppose thermal equilibrium. In order to maintain the equilibrium, the electric sounding field must be restricted to reasonable strength. Its critical value is however, beyond the field strength which can be achieved by usual UWB sensors and probe geometries [18].

Typical values for the high- and low-frequency permittivity as well as the relaxation time can be taken from Table 5.4. Figure 5.6 gives an example for the frequency and temperature dependency of the water permittivity.

As already mentioned, the behaviour of water with respect to its polarizability depends on a great deal of parameters. Hence, its determination via permittivity measurements can be used to quantify the state of that substance. One example is shown in Table 5.4 and Figure 5.6 indicating the temperature influence of pure water. Other examples are given by Eqs. (5.38)–(5.40) which describe the influence of dissolved matter on the permittivity value [17].

Non-dipolar solute:

$$\underline{\varepsilon}(f) = \varepsilon_\infty + \frac{\varepsilon_{\text{DC}} - \varepsilon_\infty}{\left(1 + (j2\pi f \tau_r)^{1-h}\right)^{1-b}} \tag{5.38}$$

(5.38) is a generalization of well-known and frequently used spectra functions as the Cole–Cole relation $b = 0$, Davidson–Cole relation $h = 0$ and the already introduced Debye relation $h = b = 0$.

Table 5.4 Approximate dielectric parameters of pure water as function of temperature [17].

Temperature (°C)	$\varepsilon_{r,\text{DC}}$	$\varepsilon_{r,\infty}$	τ_r (ps)
0	88	5.7	17.7
5	86	5.8	14.5
20	80	5.7	9.4
35	75	5.3	6.5
37	74	5.3	6.3
40	73	4.6	5.8
60	67	4.2	4.0

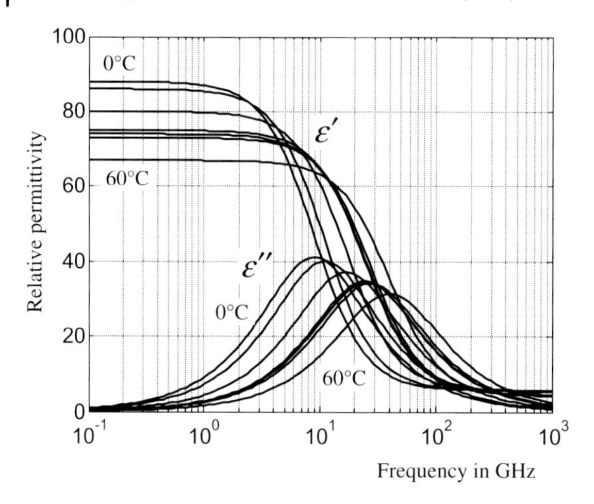

Figure 5.6 Dielectric spectrum of pure water at different temperatures (see Table 5.4).

Salt solutions lead to ionic conductivity σ. It is a usual practice to join permittivity and conductivity in the complex-valued quantity $\underline{\varepsilon}$ (see (5.59) and (5.61)), which gives us for conductive solutions:

$$\underline{\varepsilon}_{\mathrm{r}}(f) = \varepsilon_{\mathrm{r},\infty} + \frac{\varepsilon_{\mathrm{r,DC}} - \varepsilon_{\mathrm{r},\infty}}{\left(1 + (j2\pi f \tau_{\mathrm{r}})^{1-h}\right)^{1-b}} - \frac{j\sigma}{2\pi f \varepsilon_0} \tag{5.39}$$

If the solute consists of dipolar molecules too, Eq. (5.39) has to be modified to

$$\underline{\varepsilon}_{\mathrm{r}}(f) = \varepsilon_{\mathrm{r},\infty} + \frac{\varepsilon_{\mathrm{r,DC}} - \varepsilon_{\mathrm{r},\infty}}{\left(1 + (j2\pi f \tau_1)^{1-h}\right)^{1-b}} + \frac{\varepsilon_{\mathrm{r,DC}}}{1 + j2\pi f \tau_2} - \frac{j\sigma}{2\pi f \varepsilon_0} \tag{5.40}$$

We can continue arbitrarily with such kind of examples [15] in which the substance under investigation is characterized by a set of parameters. The extraction of these parameters from the measurements is usually done by appropriate estimation methods and curve-fitting procedures. If two or more substances or different bond mechanisms are involved, then the spectral behaviour of the permittivity has to be extended by additional relaxation terms.

Dielectric spectroscopy is not restricted to characterize solutions; rather, it is a general approach to investigate organic and inorganic substances [19, 20]. Here, the restriction on water was only considered as illustrative example. In order to complete the water example, we will finally refer to microwave moisture measurements which are aimed to determine the water content of solid matter or bulk. The knowledge of the moisture content is important in the field of chemistry, construction industry, food industry, medical engineering and so forth.

Moisture measurement by microwaves exploits the large difference between the high water permittivity and the usually low permittivity of the dry material. The determination of the moisture content from permittivity measurements is based on mixing rules. A comprehensive overview is given in Ref. [21]. Here, we mention only two simple examples [15].

$$V_{\mathrm{MUT}}\,\varepsilon_{\mathrm{MUT}} = V_1\varepsilon_1 + V_2\varepsilon_2 \tag{5.41}$$

$$\frac{V_{\mathrm{MUT}}}{\varepsilon_{\mathrm{MUT}}} = \frac{V_1}{\varepsilon_1} + \frac{V_2}{\varepsilon_2} \tag{5.42}$$

Herein, $\varepsilon_{\mathrm{MUT}}$ represents the measured permittivity value of the material under test, ε_i is the permittivity of the pure material components, V_i represents their partial volume and $V_{\mathrm{MUT}} = V_1 + V_2$ is the total volume. The volumetric water content results in $M_{\mathrm{V}} = V_1/V_{\mathrm{MUT}} \cdot 100\%$ if the material 1 refers to water. Often, the moisture is specified by the gravimetric water content M_{m}. It relates the mass m_1 of the included water to the total mass m of the test sample:

$$M_{\mathrm{m}} = \frac{m_1}{m} = \frac{1}{1 + (\varrho_{\mathrm{m}2}/\varrho_{\mathrm{m}1}) \cdot ((1 - M_{\mathrm{V}})/(M_{\mathrm{V}}))}, \tag{5.43}$$

whereas $\varrho_{\mathrm{m}i}$ is the mass density of the pure components.

(5.41) and (5.42) should be applied with care. They do not respect the actual molecular situation of a mixture; rather, they represent only the two extreme permittivity values which can appear by mixing two substances. (5.41), for example refers to a situation in which both components are arranged in parallel to the electrical field (comparable to the parallel circuit of capacitors) and (5.42) represents a serial arrangement (i.e. serial connection of capacitors). Both equations suppose that the molecular forces do not mutually influence the individual components. This is not always true. The reality gives typically values in between and it is very difficult to predict a reasonable model for a specific mixture. Therefore in many cases, a calibration curve is determined beforehand and the measurements are referred to a look-up table containing the moisture values.

It is also important to note that (5.41)–(5.43) are not complete if bulk material is considered. Bulk material may be available in more or less dense form depending on how it was handled before the measurement was done. We can respect this effect by adding a third partial volume connected with the permittivity ε_0 of free space simulating the air inclusions. Since we have a further unknown now, we need more independent measurement values in order to separate the different effects. Measurements over a large bandwidth may be helpful in such cases since the different substance components may behave differently with respect to frequency. Some examples of permittivity measurement with UWB sensors are shown in Section 6.3. Ref. [22] refers to a clinical application requiring online permittivity measurements.

5.4
Plane Wave Propagation

We solve Maxwell's equations for the simplest case – the propagation of planar waves – with the goal to extend our simple model for wave propagation introduced in Section 4.3. In this context, we will emphasize three aspects, namely the dispersion of waves, wave polarization and scattering at a planar interface.

5.4.1
The Electromagnetic Potentials

It is often useful to write the Maxwell's equations in alternative forms in which the original field quantities are replaced by the so-called potentials [11]. Maxwell's equations provide four field quantities all of which mutually depend. The implementation of the potentials reduces the number of involved fields and hence simplifies the solution of field equations. The mostly applied potentials are the scalar potential $\Phi(t, \mathbf{r})[\mathrm{V}]$, the vector potential $\mathbf{A}(t, \mathbf{r})[\mathrm{V\ s/m}]$ and the electric or magnetic Hertz vectors $\mathbf{\Pi}_\mathrm{e}(t, \mathbf{r})[\mathrm{V\ m}]$ or $\mathbf{\Pi}_\mathrm{m}(t, \mathbf{r})[\mathrm{A\ m}]$. The last two are also referred to as super potentials. The potentials will be defined for time domain description including frequency-dependent material parameters. That is, the constitutive relations have to be expressed by convolutions. In order to deal with a consistent notation, we will also express temporal derivations of field quantities by convolution with the unit doublet (refer to Annex A.2). We will make frequent use of the commutative law of convolution without explicitly referring to it.

The magnetic field is free of sources as to be seen from (5.13). Referring to the differential identities in Annex A.5, we can express it by a pure curl field $\mathbf{A}$ which we will call vector potential (see (A.72)):

$$\mathbf{B} = \mu * \mathbf{H} = \nabla \times \mathbf{A} \tag{5.44}$$

Insertion in (5.30) leads to

$$\nabla \times \left(\mathbf{E} + \frac{\partial \mathbf{A}}{\partial t} \right) = \nabla \times (\mathbf{E} + u_1 * \mathbf{A}) = \mathbf{0} \tag{5.45}$$

That is the expression within the brackets represents a curl-free field which can be expressed by a pure scalar field Φ applying the identity (7.76). Thus, we can write

$$-\nabla\Phi = \mathbf{E} + \frac{\partial \mathbf{A}}{\partial t} = \mathbf{E} + u_1 * \mathbf{A} \tag{5.46}$$

Insertion of (5.44) and (5.46) in (5.7) and applying the vector identity (A.74) $\nabla \times (\nabla \times \mathbf{A}) = \nabla(\nabla \cdot \mathbf{A}) - \Delta\mathbf{A}$ results in

$$\begin{aligned}\nabla \times \nabla \times \mathbf{A} = \nabla(\nabla \cdot \mathbf{A}) - \Delta\mathbf{A} &= \mu * \left(\mathbf{J}_\mathrm{s} - \sigma * \left(\nabla\Phi + \frac{\partial \mathbf{A}}{\partial t} \right) - \varepsilon * \frac{\partial}{\partial t} \left(\nabla\Phi + \frac{\partial \mathbf{A}}{\partial t} \right) \right) \\ &= \mu * (\mathbf{J}_\mathrm{s} - \sigma * (\nabla\Phi + \mathrm{u}_1 * \mathbf{A}) - \varepsilon * \mathrm{u}_1 * (\nabla\Phi + \mathrm{u}_1 * \mathbf{A}))\end{aligned} \tag{5.47}$$

Here, we divided the current flux into a source term and a part caused by the electric field $\mathbf{J} = \mathbf{J}_\mathrm{s} + \sigma * \mathbf{E}$. The vector potential $\mathbf{A}$ is not yet uniquely defined, which we achieve by additionally requiring [2]

$$\nabla \cdot \mathbf{A} = -\sigma * \mu * \Phi - \varepsilon * \mu * \frac{\partial \Phi}{\partial t} = -\sigma * \mu * \Phi - \varepsilon * \mu * u_1 * \Phi \tag{5.48}$$

This is called the Lorentz condition or Lorentz gauge. Another opportunity would be to require $\nabla \cdot \mathbf{A} = 0$ which is called the Coulomb gauge.

Applying the Lorentz gauge, (5.47) simplifies to the inhomogeneous wave equation for the vector potential:

$$\begin{aligned} \Delta \mathbf{A} - \sigma * \mu * \frac{\partial \mathbf{A}}{\partial t} - \varepsilon * \mu * \frac{\partial^2 \mathbf{A}}{\partial t^2} &= -\mu * \mathbf{J}_s \\ \Delta \mathbf{A} - \sigma * \mu * u_1 * \mathbf{A} - \varepsilon * \mu * u_2 * \mathbf{A} &= -\mu * \mathbf{J}_s \end{aligned} \tag{5.49}$$

Using finally (5.11), we also get the inhomogeneous wave equation of the scalar potential:

$$\begin{aligned} \varepsilon * \left(\Delta \Phi - \sigma * \mu * \frac{\partial \Phi}{\partial t} - \varepsilon * \mu * \frac{\partial^2 \Phi}{\partial t^2} \right) &= -\varrho \\ \varepsilon * (\Delta \Phi - \sigma * \mu * u_1 * \Phi - \varepsilon * \mu * u_2 * \Phi) &= -\varrho \end{aligned} \tag{5.50}$$

so that we succeed to reduce the numerous equations from Tables 5.1 and 5.2 to the two wave equations (5.49) and (5.50).

We achieve a further reduction to a single one by introducing the Hertz vector. The electric Hertz vector is defined by [11]

$$\mathbf{E} = \nabla \times (\nabla \times \mathbf{\Pi}_e) \tag{5.51}$$

Using (5.7), (5.16), (5.18) and supposing a source-free region, the magnetic field results to

$$\mathbf{H} = \sigma * \nabla \times \mathbf{\Pi}_e + \varepsilon * \frac{\partial}{\partial t} \nabla \times \mathbf{\Pi}_e = \sigma * \nabla \times \mathbf{\Pi}_e + \varepsilon * u_1 * \nabla \times \mathbf{\Pi}_e \tag{5.52}$$

Insertion in Faraday's law (5.3) finally yields a single equation describing the propagation of electromagnetic waves. The homogeneous wave equation in terms of the electric Hertz vector reads

$$\begin{aligned} \nabla \times (\nabla \times \mathbf{\Pi}_e) + \mu * \sigma * \frac{\partial \mathbf{\Pi}_e}{\partial t} + \mu * \varepsilon * \frac{\partial^2 \mathbf{\Pi}_e}{\partial t^2} &= \mathbf{0} \\ \nabla \times (\nabla \times \mathbf{\Pi}_e) + \mu * \sigma * u_1 * \mathbf{\Pi}_e + \mu * \varepsilon * u_2 * \mathbf{\Pi}_e &= \mathbf{0} \end{aligned} \tag{5.53}$$

and the inhomogeneous wave equation is usually written as (compare [6])

$$\begin{aligned} \varepsilon * \left(\nabla \times (\nabla \times \mathbf{\Pi}_e) + \mu * \sigma * \frac{\partial \mathbf{\Pi}_e}{\partial t} + \mu * \varepsilon * \frac{\partial^2 \mathbf{\Pi}_e}{\partial t^2} \right) &= \mathbf{p}_s \\ \varepsilon * (\nabla \times (\nabla \times \mathbf{\Pi}_e) + \mu * \sigma * u_1 * \mathbf{\Pi}_e + \mu * \varepsilon * u_2 * \mathbf{\Pi}_e) &= \mathbf{p}_s \end{aligned} \tag{5.54}$$

in which the source field $\mathbf{p}_s$ is defined by

$$\mathbf{J}_s = \frac{\partial \mathbf{p}_s}{\partial t} = u_1 * \mathbf{p}_s \quad \text{and} \quad \varrho = -\nabla \cdot \mathbf{p}_s \tag{5.55}$$

Note that the definition of $\mathbf{p}_s$ automatically fulfils the continuity relation. The vector $\mathbf{p}_s[\mathrm{As/m^2}]$ is called the electric polarization vector. It corresponds to the dipole moment per volume of the exciting source.

The same game is with the magnetic Hertz vector, which we will not repeat here.

5.4.2
Time Harmonic Plane Wave

We will solve (5.53) and (5.54) for two basic cases. Here, we deal with one of both for which we assume a propagation of a planar wave along the x-axis (i.e. $\partial/\partial y = \partial/\partial z = 0$) and we restrict ourselves to time harmonic fields in order to avoid the numerous convolutions. Hence, the Hertz vector will have one of the depicted forms

$$\underline{\mathbf{\Pi}}_{\mathrm{e}} = \begin{bmatrix} \underline{\Pi}_x \\ \underline{\Pi}_y \\ \underline{\Pi}_z \end{bmatrix} \mathrm{e}^{j2\pi(ft-px)} = \underline{\mathbf{\Pi}}_{\mathrm{e0}}\,\mathrm{e}^{j2\pi(ft-px)} = \underline{\mathbf{\Pi}}_{\mathrm{e0}}\,\mathrm{e}^{j2\pi\left(\frac{t}{t_{\mathrm{p}}}-\frac{x}{\lambda}\right)} = \underline{\mathbf{\Pi}}_{\mathrm{e0}}\,\mathrm{e}^{j(\omega t-kx)} \tag{5.56}$$

Herein, f and p represent a 'temporal' and 'spatial' frequency of the considered wave and correspondingly t_{p} and λ refer to its periodicity in time and space. Many textbooks apply angular frequencies, that is $\omega = 2\pi f = 2\pi/t_{\mathrm{p}}$ and $k = 2\pi p = 2\pi/\lambda$ where k is called wave number and λ is the wavelength.

Note that the plane wave assumption is at least theoretically never fulfilled since it requires an infinitely expanded wave which would transport an infinite quantity of energy. Furthermore, the plane wave violates the causality principle (4.18). But it represents a helpful tool to model wave propagation. In practice, one can consider any wave as planar if it is observed over a sufficiently small area at a sufficiently large distance from its source (refer to Figure 4.4). In connection with UWB short-range sensing, this approximation has to be considered with care since the distances are often quite short.

Using the differential operators defined in Annex A.5, the homogenous wave equation (5.53) simplifies to

$$(2\pi p)^2 \begin{bmatrix} 0 \\ \underline{\Pi}_y \\ \underline{\Pi}_z \end{bmatrix} + j2\pi f\mu\sigma \begin{bmatrix} \underline{\Pi}_x \\ \underline{\Pi}_y \\ \underline{\Pi}_z \end{bmatrix} - (2\pi f)^2\varepsilon\mu \begin{bmatrix} \underline{\Pi}_x \\ \underline{\Pi}_y \\ \underline{\Pi}_z \end{bmatrix} = \mathbf{0}$$

This equation is only satisfied if

$$\underline{\Pi}_x = 0 \tag{5.57}$$

and

$$(2\pi p)^2 + j2\pi f\mu\sigma - (2\pi f)^2\varepsilon\mu = 0 \tag{5.58}$$

The material parameters involved in (5.57) typically depend on frequency which leads to complex-valued quantities. In order to abridge the dispersion relation (5.58), we join conductivity and permittivity to

$$\underline{\varepsilon} = \varepsilon - j\frac{\sigma}{2\pi f} \tag{5.59}$$

Furthermore, the permeability of most materials is $\mu \approx \mu_0$, so that we can simplify (5.58) to (see Ref. [16] for complete notation with arbitrary μ)[3)]

$$\underline{p}^2(f) - f^2\underline{\varepsilon}(f)\mu_0 = 0 \tag{5.60}$$

Here, the permittivity is generalized to

$$\underline{\varepsilon} = \varepsilon' - j\varepsilon'' \tag{5.61}$$

so that ε'' includes both, the conductivity as well as relaxation phenomena of the propagation medium. Decomposing the spatial frequency in real and imaginary parts $\underline{p} = p' - jp''$, (5.60) solves in

$$\begin{aligned}
\underline{p} &= \pm f\sqrt{\underline{\varepsilon}(f)\mu_0} = \pm f\sqrt{(\varepsilon'(f) - j\varepsilon''(f))\mu_0} = \pm f\sqrt{\varepsilon'\mu_0}\sqrt{1 - j\tan\delta} \\
p' &= f\sqrt{\frac{\mu_0\varepsilon'}{2}\left(\sqrt{1 + \left(\frac{\varepsilon''}{\varepsilon'}\right)^2} + 1\right)} = f\sqrt{\frac{\mu_0\varepsilon'}{2}\left(\sqrt{1 + \tan^2\delta} + 1\right)} \\
p'' &= f\sqrt{\frac{\mu_0\varepsilon'}{2}\left(\sqrt{1 + \left(\frac{\varepsilon''}{\varepsilon'}\right)^2} - 1\right)} = f\sqrt{\frac{\mu_0\varepsilon'}{2}\left(\sqrt{1 + \tan^2\delta} - 1\right)}
\end{aligned} \tag{5.62}$$

The loss tangent of the propagation medium is defined as

$$\tan\delta = \frac{\varepsilon''}{\varepsilon'} \tag{5.63}$$

As we will see below, the real part of the spatial frequency p' and the temporal frequency f are linked by the propagation speed and the imaginary part p'' is a measure of the propagation loss. For small losses $\tan\delta \ll 1$ and applying (5.23), (5.62) can be approximated by

$$\begin{aligned}
p' &\approx \frac{f}{c}\left(1 + \frac{\tan^2\delta}{8}\right) = \frac{1}{\lambda}\left(1 + \frac{\tan^2\delta}{8}\right) \\
p'' &\approx \frac{f\tan\delta}{2c} = \frac{\tan\delta}{2\lambda}
\end{aligned} \tag{5.64}$$

From (5.62), we get two wave fields of the Hertz vector. One is propagating along the positive direction of the x-axis ($\mathbf{\Pi}_e^{\rightarrow}$) and the second one moves in the opposite direction ($\mathbf{\Pi}_e^{\leftarrow}$).

$$\underline{\mathbf{\Pi}}_e(x) = \underline{\mathbf{\Pi}}_{e0}^{\rightarrow}\,e^{j2\pi\left(ft - \underline{p}x\right)} + \underline{\mathbf{\Pi}}_{e0}^{\leftarrow}\,e^{j2\pi\left(ft + \underline{p}x\right)} \tag{5.65}$$

The components of the Hertz vector of the forward travelling wave are $\underline{\mathbf{\Pi}}_{e0}^{\rightarrow} = \begin{bmatrix} 0 & \underline{\Pi}_y^{\rightarrow} & \underline{\Pi}_z^{\rightarrow} \end{bmatrix}^T$ and the field components of the backward travelling wave are $\underline{\mathbf{\Pi}}_{e0}^{\leftarrow} = \begin{bmatrix} 0 & \underline{\Pi}_y^{\leftarrow} & \underline{\Pi}_z^{\leftarrow} \end{bmatrix}^T$. The propagation speed of the wave is about (refer also to

3) It should be noted that typically the wave number $k = 2\pi p = 2\pi/\lambda$ is considered as spatial frequency in the literature. In the same way, the frequency is often replaced by the angular frequency $\omega = 2\pi f = 2\pi/T$. The Fourier transform of waves is often denoted as *fk*-space representation which is less consistent with respect to the factor of 2π. A consistent notation should be either the ωk-space or the fp-space depending on how the Fourier transform is defined. The last one is preferred here.

(5.70) and (5.71))

$$c = \frac{\Delta x}{\Delta t} = \frac{f}{p'} \approx \frac{1}{\sqrt{\varepsilon' \mu_0}} \quad \text{if } \tan\delta \approx 0 \tag{5.66}$$

5.4.3
fp-Space Description and Dispersion Relation

The wave (5.65) is a two-dimensional function in x and t. Figure 5.7 represents an example for $f = f_0 = T_0^{-1}$ and $p' = p'_0 = \lambda_0^{-1}$ and for propagation in x-direction ($\underline{\mathbf{\Pi}}^{\leftarrow}_{e0} = \mathbf{0}$).

If we apply two-dimensional Fourier transform at (5.65) (where we restrict ourselves to the wave travelling towards the x-axis), we result in the so-called *fp*-space

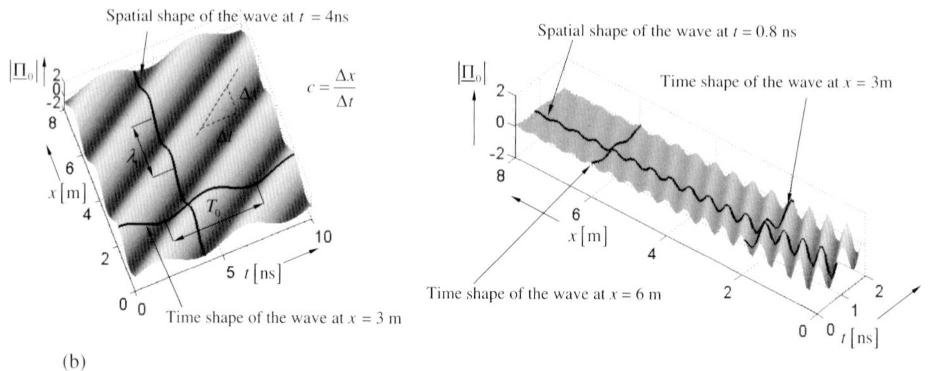

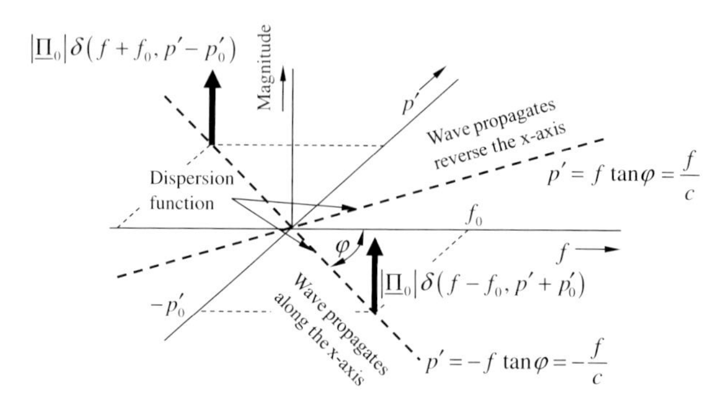

Figure 5.7 Representation of a sinusoidal plane wave in space–time coordinates (a) and *fp*-space (b). The wave propagates along the x-axis.

description of the plane wave:

$$\begin{aligned}\underline{\boldsymbol{\Pi}}_{\mathrm{e}}(f, p) &= \int \boldsymbol{\Pi}_{\mathrm{e}}(t, x))\, \mathrm{e}^{-j2\pi(ft+xp)}\,\mathrm{d}t\,\mathrm{d}x \\ &= \underline{\boldsymbol{\Pi}}_{\mathrm{e}0} \int \mathrm{e}^{-j2\pi((f-f_0)t+(p+\underline{p}_0)x)}\,\mathrm{d}t\,\mathrm{d}x \\ &= \underline{\boldsymbol{\Pi}}_{\mathrm{e}0}\delta(f \mp f_0, p \pm \underline{p}_0)\end{aligned} \tag{5.67}$$

It should be noted that propagating waves represent a special case amongst the multi-dimensional (in our case two-dimensional) functions since the function arguments f and p are not independent of each other. They are related by the dispersion relations (5.60) and (5.62). Hence, the two-dimensional Dirac function in (5.67) may only be located on a curve within the *fp*-plane which meets the dispersion relation. Figure 5.7b illustrates the *fp*-representation of a sine wave where we only respected the real part p' of the spatial frequency.

The complete representation would include still a second diagram relating to the imaginary part p''. Its physical meaning will be obvious by rewriting the plane wave (5.65) for p' and p'':

$$\mathrm{e}^{j2\pi(ft-\underline{p}x)} = \mathrm{e}^{j2\pi(ft-p'x)}\,\mathrm{e}^{-2\pi p''x}$$

The second term including the imaginary part of the spatial frequency represents a damping factor leading to a propagation loss of

$$a\left[\frac{\mathrm{dB}}{\mathrm{m}}\right] = -20\,\lg \mathrm{e}^{-2\pi p''x}\Big|_{x=1\,\mathrm{m}} = 40\,\pi\, p''\,\lg e \tag{5.68}$$

and related to the wavelength

$$a_\lambda\left[\frac{\mathrm{dB}}{\lambda}\right] = -20\,\lg \mathrm{e}^{-\frac{2\pi p''}{p'}} = 40\,\pi\,\frac{p''}{p'}\lg e \tag{5.69}$$

The dispersion relation is only real valued for lossless propagation and it is only linear for frequency-independent material parameters. While the dispersion relation depicted in Figure 5.7 is frequency independent, Figure 5.8 shows the general case in which the dispersion relation is non-linear with frequency.

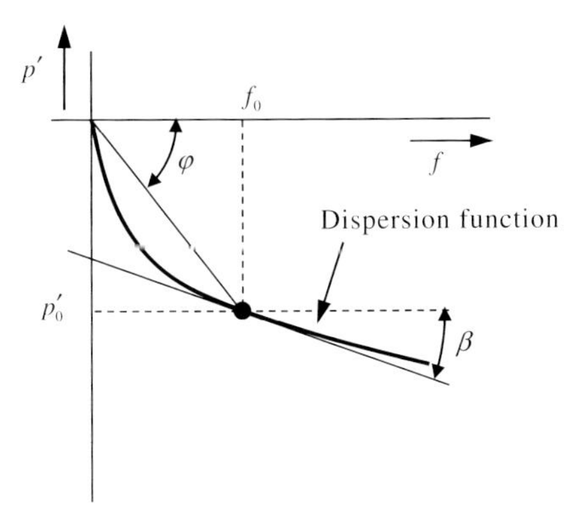

Figure 5.8 Non-linear dispersion relation and illustration of phase and group velocity.

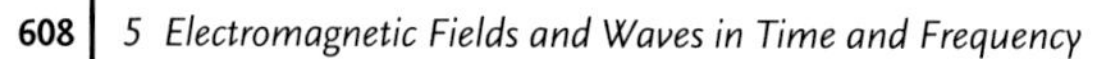

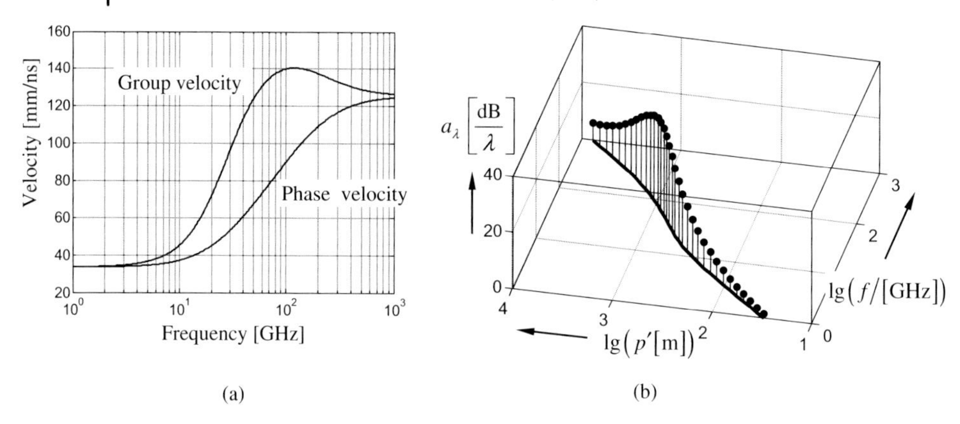

Figure 5.9 Group and phase velocity (a) and dispersion relation (b) for wave propagation in water at 20 °C. The calculations are based on the Debye relation (5.37) and the parameters are taken from Table 5.4.

From this diagram, we can define two different velocities:

$$\text{Phase velocity :} \quad \tan\varphi = \frac{1}{c} = \frac{p_0'}{f_0} \tag{5.70}$$

and

$$\text{Group velocity :} \quad \tan\beta = \frac{1}{v_g} = \left.\frac{\partial p'}{\partial f}\right|_{f=f_0} \tag{5.71}$$

For illustration, we take again water as highly dispersive media. Figure 5.9a compares group and phase velocity and the dispersion relation is shown in Figure 5.9b. Since the typical frequency band of UWB operation is located below 10 GHz, we are still at the beginning of the actual dispersion region.

5.4.4
Propagation in Arbitrary Direction

Generalizing (5.60) and (5.65) for arbitrary propagation direction $\mathbf{e}_0$, we can write for the dispersion relation and the field at (observer) position $\mathbf{r}'$:

$$\begin{aligned} &\text{Dispersion relation :} \quad && \underline{\mathbf{p}}\cdot\underline{\mathbf{p}} - f^2\,\varepsilon'\,\mu_0\,(1-j\tan\delta) = 0 \\ & && \underline{p}_x^2 + \underline{p}_y^2 + \underline{p}_z^2 - \frac{f^2}{c^2}(1-j\tan\delta) = 0 \\ & && \underline{p}_x^2 + \underline{p}_y^2 + \underline{p}_z^2 - \frac{1-j\tan\delta}{\lambda^2} = 0 \end{aligned} \tag{5.72}$$

$$\begin{aligned} &\text{Hertz vector :} \quad && \underline{\mathbf{\Pi}}_e(\mathbf{r}') = \underline{\mathbf{\Pi}}_{e0}\,e^{j2\pi\left(ft-\underline{\mathbf{p}}\cdot(\mathbf{r}'-\mathbf{r}_0)\right)} = \Pi_{e0}\,e^{j2\pi\left(ft-\underline{p}\,\Delta r\right)}\mathbf{q} \\ & && \text{with} \quad \mathbf{e}_0\cdot\mathbf{q} = 0; \quad \mathbf{q}\cdot\mathbf{q}^* = 1 \end{aligned} \tag{5.73}$$

Herein, $\mathbf{r}'$ is the position vector of the observation point, $\mathbf{r}_0$ is the reference position, that is position where the wave magnitude is known,

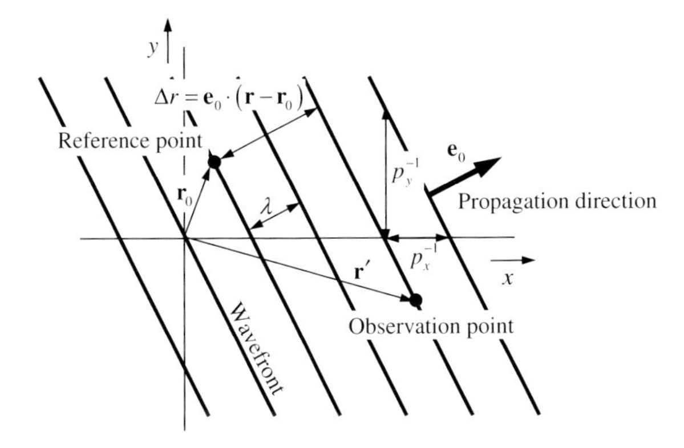

Figure 5.10 Illustration of the propagation term in (5.73).

$\underline{\mathbf{p}} = \begin{bmatrix} \underline{p}_x & \underline{p}_y & \underline{p}_z \end{bmatrix}^T$ is the vector of spatial frequencies, $\mathbf{e}_0 = \underline{\mathbf{p}}/\underline{p}$ is the unit vector pointing the propagation direction, $\underline{\mathbf{\Pi}}_{e0}$ is the vector magnitude at position $\mathbf{r}_0$, $\underline{\mathbf{q}} = \underline{\mathbf{\Pi}}_{e0}/\Pi_{e0}$ is the unitary vector (i.e. $\underline{\mathbf{q}} \cdot \underline{\mathbf{q}}^* = 1$) pointing to the wave polarization (later it will be obvious that it corresponds to the direction of the electric field) and $\Pi_{e0} = \sqrt{\underline{\mathbf{\Pi}}_{e0} \cdot \underline{\mathbf{\Pi}}_{e0}^*}$ is the magnitude of Hertz vector field.

Figure 5.10 depicts the geometric interpretation of the propagation term $(ft - \underline{\mathbf{p}} \cdot (\mathbf{r}' - \mathbf{r}_0)) = (ft - \underline{p}\, \mathbf{e}_0 \cdot (\mathbf{r}' - \mathbf{r}_0))$. It shows a snapshot of a sinusoidal wave propagating in the xy-plane (see also Figure 4.3). The scalar product $\mathbf{e}_0 \cdot (\mathbf{r}' - \mathbf{r}_0)$ represents the projection of the distance between reference and observation point onto the propagation direction.

The side condition $\mathbf{e}_0 \cdot \underline{\mathbf{q}} = 0$ in (5.73) implies the transversal nature of a propagating electromagnetic field, that is we have $\mathbf{e} = \begin{bmatrix} 1 & 0 & 0 \end{bmatrix}^T$ and $\underline{\mathbf{q}} = \begin{bmatrix} 0 & q_y & q_z \end{bmatrix}^T$; $\left(q_y^2 + q_z^2 = 1\right)$ for our example in Figure 5.7 for propagation along the x-axis.

In the generic case, the components of the polarization vector are typically complex valued (e.g. in Cartesian coordinates $\underline{\mathbf{q}} = \begin{bmatrix} q_x\, e^{j\alpha_x} & q_y\, e^{j\alpha_y} & q_z\, e^{j\alpha_z} \end{bmatrix}^T$). It is of unit length and perpendicular to the propagation direction. One speaks from a linear polarized wave if all field components are in phase $\alpha_x = \alpha_y = \alpha_z$. Otherwise the wave is elliptically polarized. Note that an initial phase α_0 of the wave can be assigned to either the magnitude Π_{e0} or the polarization vector $\underline{\mathbf{q}}$. A comprehensive overview on polarimetric issues of sine wave propagation in radar applications is to be found in Ref. [23].

We will terminate our consideration on sine wave propagation here by determining finally the electric and magnetic fields from the Hertz vector. For demonstration, we assume again propagation along the positive x-axis, that is $\mathbf{e}_0 = \begin{bmatrix} 1 & 0 & 0 \end{bmatrix}^T$. Using the complex notation of the permittivity and applying the differentiation rule of Fourier transform, (5.52) modifies to

$$\underline{\mathbf{H}} = j2\pi f \underline{\varepsilon}\, \nabla \times \underline{\mathbf{\Pi}}_e \tag{5.74}$$

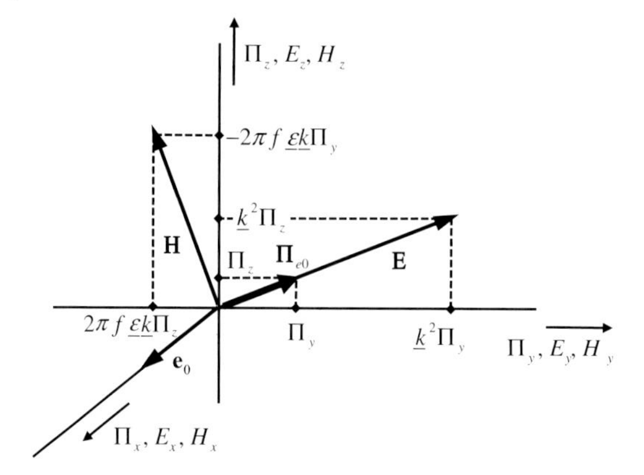

Figure 5.11 Field vectors **E**, **H** and propagation direction $\mathbf{e}_0$ of a plane wave.

Insertion of (5.65) and executing the curl operation, the magnetic field results in

$$\underline{\mathbf{H}} = (2\pi)^2 f \underline{\varepsilon}\, \underline{p} \begin{bmatrix} 0 \\ \Pi_z^{\rightarrow} \\ -\Pi_y^{\rightarrow} \end{bmatrix} e^{j2\pi(ft - \underline{p}x)} \tag{5.75}$$

The electric field is calculated via (5.51) and (A.61) of Annex A.5. It yields

$$\underline{\mathbf{E}} = (2\pi \underline{p})^2 \begin{bmatrix} 0 \\ \Pi_y^{\rightarrow} \\ \Pi_z^{\rightarrow} \end{bmatrix} e^{j2\pi(ft - \underline{p}x)} \tag{5.76}$$

Obviously, all field components are perpendicular to the propagation direction, that is $\underline{\mathbf{E}} \cdot \mathbf{e}_0 = 0$ and $\underline{\mathbf{H}} \cdot \mathbf{e}_0 = 0$. The direction of the electric field vector coincides with the direction of the Hertz vector while the magnetic field vector is perpendicularly aligned. Figure 5.11 illustrates (5.75) and (5.76). It indicates that the three vectors $[\mathbf{E}, \mathbf{H}, \mathbf{e}_0]$ are forming a right-hand system. We can join both equations indicating proportionality between the strengths of electric and magnetic fields:

$$\begin{aligned} \underline{\mathbf{E}} &= \frac{\underline{p}}{f\underline{\varepsilon}} \underline{\mathbf{H}} \times \mathbf{e}_0 = \underline{Z}_s \underline{\mathbf{H}} \times \mathbf{e}_0 \\ \mathbf{e}_0 \times \mathbf{E} &= \underline{Z}_s \underline{\mathbf{H}} \\ \underline{Z}_s \mathbf{E} \times \mathbf{H} &= \mathbf{e}_0 \end{aligned} \tag{5.77}$$

The proportionality $\underline{Z}_s$ is called intrinsic impedance of the propagation medium. Inserting (5.62) in (5.77), the intrinsic impedance is[4)]

$$\begin{aligned} \underline{Z}_s(f) &= \sqrt{\frac{\mu_0}{\underline{\varepsilon}(f)}} = \sqrt{\frac{\mu_0}{\varepsilon'(f) - j\varepsilon''(f)}} \\ &= \frac{Z_{s0}(f)}{\sqrt{1 - j\tan\delta(f)}}; \quad \text{with } Z_{s0}(f) = \sqrt{\frac{\mu_0}{\varepsilon'(f)}} \end{aligned} \tag{5.78}$$

4) Do not confuse the frequency dependent loss factor $\delta(f)$ in (5.78) with the Dirac delta of frequency.

Except for propagation in air, the intrinsic impedance is generally a complex-valued function. For small losses, we approximate (5.78) by

$$\underline{Z}_s \approx Z_{s0}\left(1 + j\frac{1}{2}\tan\delta\right) = Z_{s0}\left(1 + j\frac{\varepsilon''}{2\varepsilon'}\right); \quad \varepsilon'' \ll \varepsilon' \tag{5.79}$$

In most practical case, one usually deals with the real-valued quantity Z_{s0} omitting the phase delay between $\underline{\mathbf{E}}$ and $\underline{\mathbf{H}}$.

From Section 2.5.1, we know a relation comparable to (5.77) which relates voltage and current of a guided wave (2.200). There, we used this mutual dependency to introduce normalized waves. This we will repeat here for waves propagating freely in space [8]. We define the normalized plane wave $\mathbf{W}$ as follows:

$$\underline{\mathbf{W}} = \frac{\underline{\mathbf{E}}}{\sqrt{\underline{Z}_s}} = \sqrt{\underline{Z}_s}\underline{\mathbf{H}} \times \mathbf{e}_0 = \underline{W}_0 \mathfrak{q}\, e^{j2\pi(ft - \underline{p}\, \mathbf{e}_0 \cdot \mathbf{r})} \quad \left[\sqrt{\frac{\mathrm{W}}{\mathrm{m}^2}}\right] \tag{5.80}$$

The strength and the initial phase of the wave are given by its complex-valued amplitude $\underline{W}_0$, the polarization is defined by the vector $\mathfrak{q}$ which coincides with the direction of the electric field and $\mathbf{e}_0$ is the propagation direction.

The power flux density (Poynting vector) (5.28) of a plane wave results from (5.80) to

$$\underline{\underline{\mathfrak{P}}}(\mathbf{r}) = \underline{\underline{\mathbf{E}}}(\mathbf{r}) \times \underline{\underline{\mathbf{H}}}^*(\mathbf{r}) = \sqrt{\frac{\underline{Z}_s}{\underline{Z}_s^*}}(\underline{\underline{\mathbf{W}}}(\mathbf{r}) \cdot \underline{\underline{\mathbf{W}}}^*(\mathbf{r}))\mathbf{e}_0$$

$$= \sqrt[4]{\frac{1 + j\tan\delta}{1 - j\tan\delta}}\left|\underline{\underline{W}}_0\right|^2 e^{-4\pi p'' \Delta r}\mathbf{e}_0 = e^{j\delta/2}\, e^{-4\pi p'' \Delta r}\left|\underline{\underline{W}}_0\right|^2 \mathbf{e}_0; \quad \Delta r = \mathbf{r} \cdot \mathbf{e}_0 \tag{5.81}$$

Here we made us of the vector identity (A.52) of Annex A and $\mathbf{e}_0 \cdot \mathbf{E} = 0$. For small losses $\delta \approx \tan\delta = \varepsilon''/\varepsilon'$, (5.81) approximates to

$$\underline{\underline{\mathfrak{P}}} \approx \left(1 + j\frac{\delta}{2}\right)\left|\underline{\underline{W}}_0\right|^2 e^{-2\pi\delta\frac{\Delta r}{\lambda}}\mathbf{e}_0 \tag{5.82}$$

We observe from (5.81) and (5.82) for the simple case of plane wave propagation in isotropic media that the power transfer coincides with the propagation direction of the wave. The Poynting vector is a complex-valued quantity if propagation loss appears. Its real part $\Re\{\mathfrak{P}\}$ represents the effective power flux provided by the wave. It is reduced with increasing propagation distance due to dissipation characterized by the term $e^{-2\pi\tan\delta\frac{\Delta r}{\lambda}}$. The imaginary part $\Im\{\mathfrak{P}\}$ quantifies the idle power.

5.4.5 Time Domain Description of Wideband Plane Wave

A wideband wave occupies a large part of the dispersion curve in the *fp*-plane as demonstrated in Figure 5.12.

The related time domain function can be gained from an inverse Fourier transform of the *fp*-space description. If we assume firstly an attenuation and

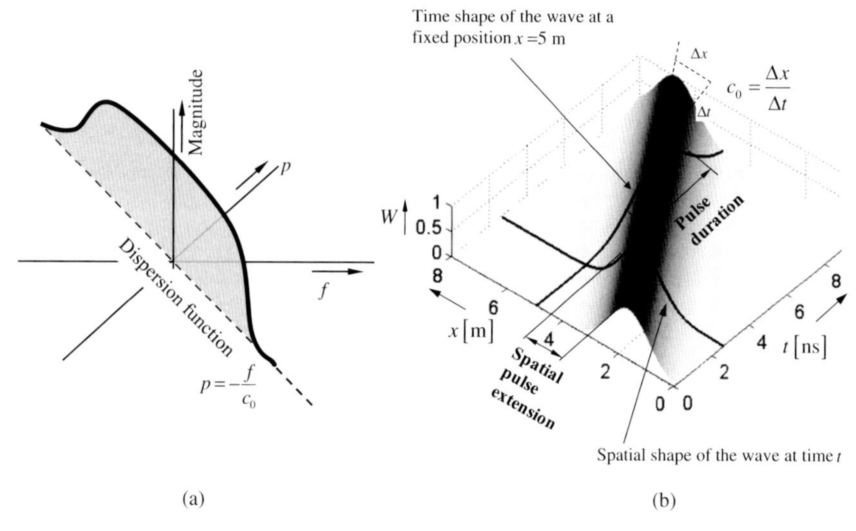

Figure 5.12 Representation of a Gaussian plan wave[5] in fp-space (a) and in space–time coordinates (b). The wave propagates along the x-axis.

dispersionless propagation as depicted in our example (i.e. the dispersion curve is a straight line), we can describe a normalized plane wave in time domain by (c_0 is the propagation speed)

$$\begin{aligned}\mathbf{W}(t,\mathbf{r}) &= W(t) * \delta\left(t - \frac{\mathbf{e}_0 \cdot \mathbf{r}}{c_0}\right) * \mathbf{q}(t) = W\left(t - \frac{\mathbf{e}_0 \cdot \mathbf{r}}{c_0}\right) * \mathbf{q}(t) \\ &\text{with} \quad \mathbf{e}_0 \cdot \mathbf{q}(t) = 0 \quad \text{and} \quad \mathbf{q}(t) * \mathbf{q}^T(-t) = \delta(t)\end{aligned} \tag{5.83}$$

which immediately follows from (5.73)

Herein, $W(t)$ represents the temporal shape of the wave magnitude which is retarded by $\Delta t = \mathbf{e}_0 \cdot \mathbf{r}/c_0$ and the unity vector $\mathbf{q}(t)$ determines the spatial orientation of the field vectors; that is, it describes the wave polarization as function of time. While simple sine waves may only have elliptical polarization (which may degenerate to a circle or a line), pulse- or arbitrary-shaped wideband waves may have arbitrary polarization. Figure 5.13 shows two simulated examples to illustrate a plane wave of linear and non-linear 'time domain' polarization. A measured example is depicted in Figure 5.26 representing the polarimetric antenna IRF of two spiral antennas. If these antennas are stimulated by a short pulse, they will provide a locally planar wave of corresponding polarization (sufficient distance to the antenna supposed). The figure clearly shows the expected rotation of the **E**-field in dependency on time.

The wave modelled by (5.83) will not change its temporal shape if it propagates through the medium. Such kind of wave propagation is very often assumed in practical applications due to its simple handling. However, some attention has to be paid since the material properties are usually frequency dependent so that phase

5) The example is only for demonstration. A propagating electric field with such time shape – that is including a DC component – does not exist.

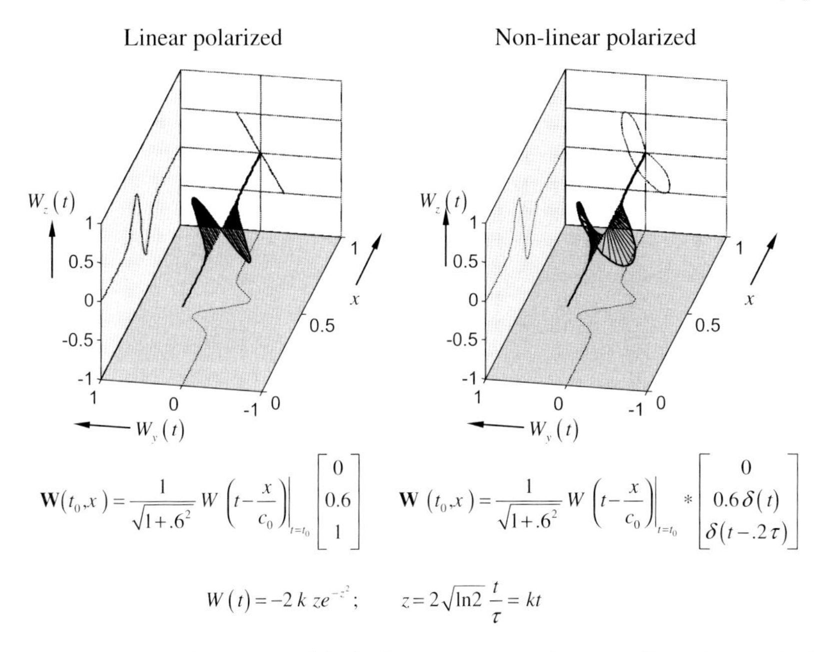

Figure 5.13 Snapshot at $t = t_0$ of the field components of a plane wave illustrating polarization in the time domain. The wave propagates along the *x*-axis. (see also the movie of wave propagation at Wiley homepage: Fig_5_13*)

velocity, intrinsic impedance and propagation loss will depend on the frequency. In such a case, the wave will increasingly be deformed by departing from its source. This may be expressed by an additional convolution term $s(t, \Delta r)$ respecting the propagation within the medium of interest:

$$\mathbf{W}(t, \mathbf{r}) = W(t) * \mathbf{q}(t) * s(t, \mathbf{e}_0 \cdot \mathbf{r}) \tag{5.84}$$

The term $s(t, \Delta r)$ includes both the propagation delay as well as the deformation of signal shape. The distance covered by the wave is expressed by $\Delta r = \mathbf{e}_0 \cdot \mathbf{r}$.

Figure 5.14 depicts the simulated propagation path IRF $s(t, \mathbf{e}_0 \cdot \mathbf{r})$ for pure water. The calculation was based on Debye relation (5.37). We can observe that for very short distances, the water relaxation is not yet as dominant so that the 'transmission' by the high-frequency permittivity ε_∞ gives steep peaks at short propagation time. However with increasing propagation distance, the relaxation phenomenon becomes more and more influential leading to the dispersion of the wave. The frequency-independent transmission via ε_∞ loses ground after a very short distance.

Basically, the two equations (5.83) and (5.84) hold for waves of pulse shape as well as for time-extended random waveforms. But the interpretation of propagation phenomena is quit complicated if time spread waveforms are involved. In order to get the intuitively better impression of pulsed wave propagation, we can proceed here in the same way as already mentioned in Section 4.3, Figures 4.6–4.8, where

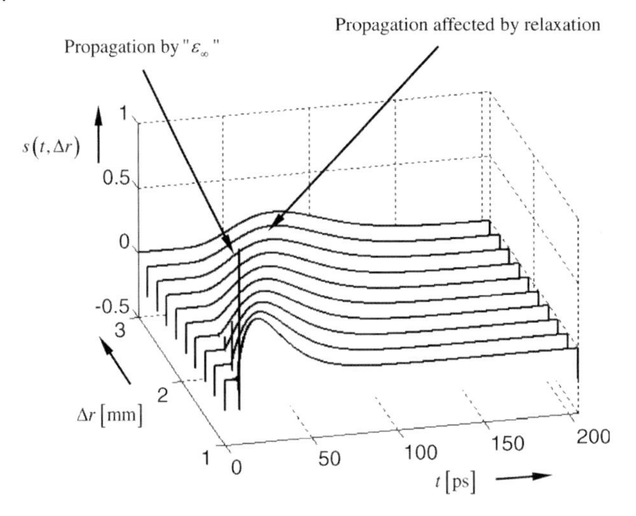

Figure 5.14 Propagation path IRF of pure water for different propagation distances.

the actual wave was correlated with an appropriate reference signal $x(t)$.

$$\begin{aligned}\mathfrak{W}(t,\mathbf{r}) &= \mathbf{W}(t,\mathbf{r}) * x(-t) = W\left(t - \frac{\mathbf{e}_0 \cdot \mathbf{r}}{c_0}\right) * \mathbf{q}(t) * x(-t) \\ \mathfrak{W}(t,\mathbf{r}) &= \mathbf{W}(t,\mathbf{r}) * x(-t) = W(t) * \mathbf{q}(t) * s(t, \mathbf{e}_0 \cdot \mathbf{r}) * x(-t)\end{aligned} \tag{5.85}$$

If $W(t)$ is of large bandwidth and $x(t)$ matches its shape, the 'virtual' vector wave $\mathfrak{W}(t,\mathbf{r})$ becomes compact in time and it can be considered as a pulse wave.

5.4.6
Scattering of a Plane Wave at a Planar Interface

We consider the situation as depicted in Figure 5.15 under simple conditions, that is the interface is perfectly planar and the two propagation media are dispersionless so that frequency and time domain expressions are equivalent since convolution with material parameters must not be applied. Both media are dielectric isolators ($\mu = \mu_0; \sigma = 0$) having the permittivity ε_1 (upper medium) and ε_2 (lower medium).

If the incident wave $\mathbf{W}_i(t,\mathbf{r})$ bounces the surface, the boundary conditions (5.26) must be satisfied. In the general case that is only possible if two new waves are generated at the boundary – a wave $\mathbf{W}_t(t,\mathbf{r})$ which penetrates the second material (the transmitted or refracted wave) and reflected wave $\mathbf{W}_r(t,\mathbf{r})$. From the observation that the boundary conditions have to be met at any time, we can deduce the reflection law and Snell's law (see Annex C.2 for details). They may be written in either of the two forms:

$$\text{Reflection law}: \quad \begin{aligned}&\mathbf{e}_r = \mathbf{e}_i - 2(\mathbf{n} \cdot \mathbf{e}_i) \cdot \mathbf{n} \\ &\alpha_i = \alpha_r = \alpha_1\end{aligned} \tag{5.86}$$

$$\text{Snell's law}: \quad \begin{aligned}&\eta_{21}\,\mathbf{e}_t = \mathbf{e}_i - \left(\sqrt{\eta_{21}^2 - (\mathbf{e}_i \cdot \mathbf{m})^2} + \mathbf{e}_i \cdot \mathbf{n}\right)\mathbf{n} \\ &\sin\alpha_i = \sin\alpha_1 = \eta_{21}\sin\alpha_t = \eta_{21}\sin\alpha_2\end{aligned} \tag{5.87}$$

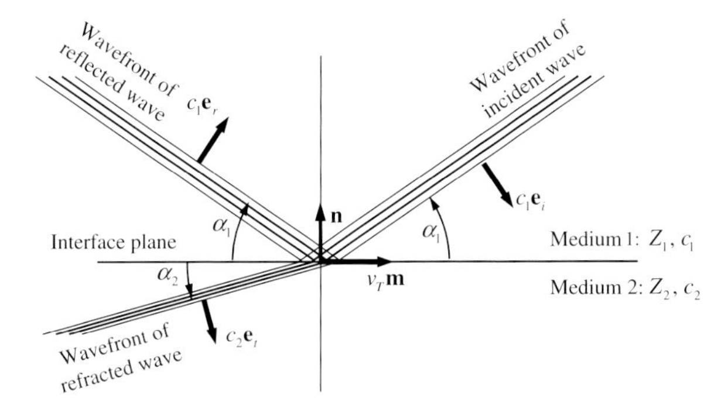

Figure 5.15 Scattering of a pulse-shaped plane wave (symbolized by contour lines) at a flat boundary.

Herein, $\mathbf{e}_\text{i}, \mathbf{e}_\text{r}, \mathbf{e}_\text{t}$ are the propagation directions of the incident, the reflected and the refracted waves. $\eta_{21} = (Z_1/Z_2) = (c_1/c_2) = \sqrt{\varepsilon_2/\varepsilon_1}$ is the refraction index.

All vectors $\mathbf{e}_\text{i}, \mathbf{e}_\text{r}, \mathbf{e}_\text{t}$ are placed in the plane of incidence, which is defined by

$$\mathbf{u} = \frac{\mathbf{n} \times \mathbf{e}_\text{i}}{|\mathbf{n} \times \mathbf{e}_\text{i}|} \tag{5.88}$$

The amplitudes of the refracted and reflected waves depend on the polarization of the incident field related to the boundary. For that purpose, we decompose all involved fields into two orthogonal components. One should be parallel to $\mathbf{u}$ (often referred as horizontal polarization) and the second one must be orthogonal to $\mathbf{u}$ and $\mathbf{e}$ (refer to (5.77) and Figure 5.11). This direction, which is different for every wave, is defined by

$$\mathbf{v}_\text{n} = \mathbf{e}_\text{n} \times \mathbf{u}; \quad n \in [i, r, t] \tag{5.89}$$

In reference to these directions (which are all perpendicular to the propagation directions), we can write for the involved normalized waves

$$\mathbf{W}_\text{n} = (\mathbf{W}_\text{n} \cdot \mathbf{u})\mathbf{u} + (\mathbf{W}_\text{n} \cdot \mathbf{v})\mathbf{v} = U_\text{n}\mathbf{u} + V_\text{n}\mathbf{v} = \begin{bmatrix} U_\text{n} \\ V_\text{n} \end{bmatrix}; \quad n \in [i, r, t] \tag{5.90}$$

wheres U, V are the field components in $\mathbf{u}, \mathbf{v}$-direction. Based on this decomposition of the fields with respect to the orientation of scattering plane, reflected and refracted wave amplitudes yield

$$\mathbf{W}_\text{r} = \mathbf{\Lambda W}_\text{i}$$
$$\begin{bmatrix} U_\text{r} \\ V_\text{r} \end{bmatrix} = \begin{bmatrix} \Lambda_\text{uu} & 0 \\ 0 & \Lambda_\text{vv} \end{bmatrix} \begin{bmatrix} U_\text{i} \\ V_\text{i} \end{bmatrix} \tag{5.91}$$

and

$$\mathbf{W}_\text{t} = \mathbf{TW}_\text{i}$$
$$\begin{bmatrix} U_\text{t} \\ V_\text{t} \end{bmatrix} = \begin{bmatrix} T_\text{uu} & 0 \\ 0 & T_\text{vv} \end{bmatrix} \begin{bmatrix} U_\text{i} \\ V_\text{i} \end{bmatrix} \tag{5.92}$$

The reflection and transmission coefficients in (5.92) are given by Fresnel's equations (see Annex C.2):

$$
\begin{aligned}
\Lambda_{uu} &= \frac{\eta_{21}\mathbf{n}\cdot\mathbf{e}_t - \mathbf{n}\cdot\mathbf{e}_i}{\mathbf{n}\cdot\mathbf{e}_r - \eta_{21}\mathbf{n}\cdot\mathbf{e}_t} = \frac{Z_2\cos\alpha_1 - Z_1\cos\alpha_2}{Z_2\cos\alpha_1 + Z_1\cos\alpha_2} \\
\Lambda_{vv} &= \frac{\eta_{21}\mathbf{n}\cdot\mathbf{e}_i - \mathbf{n}\cdot\mathbf{e}_t}{\mathbf{n}\cdot\mathbf{e}_t - \eta_{21}\mathbf{n}\cdot\mathbf{e}_r} = \frac{Z_1\cos\alpha_1 - Z_2\cos\alpha_2}{Z_1\cos\alpha_1 + Z_2\cos\alpha_2} \\
T_{uu} &= \frac{2\sqrt{\eta_{21}}\mathbf{n}\cdot\mathbf{e}_r}{\mathbf{n}\cdot\mathbf{e}_r - \eta_{21}\mathbf{n}\cdot\mathbf{e}_t} = \frac{2\sqrt{Z_1 Z_2}\cos\alpha_1}{Z_2\cos\alpha_1 + Z_1\cos\alpha_2} \\
T_{vv} &= \frac{2\sqrt{\eta_{21}}\mathbf{n}\cdot\mathbf{e}_r}{\eta_{21}\mathbf{n}\cdot\mathbf{e}_r - \mathbf{n}\cdot\mathbf{e}_t} = \frac{2\sqrt{Z_1 Z_2}\cos\alpha_1}{Z_1\cos\alpha_1 + Z_2\cos\alpha_2}
\end{aligned}
\tag{5.93}
$$

Brewster angle: If the angle of incidence α_1 is selected in such a way that the reflection coefficient becomes zero, one speaks about the Brewster angle $\alpha_1 = \alpha_B$. For purely dielectric materials, the Brewster angle can only be observed for **v**-polarized waves. It results from (5.93) to

$$\eta_{21}\mathbf{n}\cdot\mathbf{e}_i = \mathbf{n}\cdot\mathbf{e}_t \Rightarrow \tan\alpha_B = \eta_{21} \tag{5.94}$$

Angle of total reflection: Total reflection may appear if a wave propagates from a high-permittivity region into a low-permittivity region, that is if $\eta_{21} < 1$. In that case, Snell's law (5.87) provides $\alpha_2 > \alpha_1$. The critical angle of incidence $\alpha_1 = \alpha_T$ is achieved if $\alpha_2 = 90°$, that is

$$\sin\alpha_T = \eta_{21} \tag{5.95}$$

The physical reason is that the apparent speed $v_T = c_1/\sin\alpha_1$ (compare Figure 5.15) falls below the propagation speed within the second medium. The incident wave runs along the interface with this speed v_T creating the new waves according to Huygens principle. If the apparent speed falls below the propagation speed $v_T < c_2$, the wave generation in the second medium is suspended since the energy supply from the first medium cannot follow the energy evacuation in the second medium. Hence, the field magnitude in medium 2 will fall down with increasing distance from the boundary; that is, the wave is evanescent but not propagating. Since no energy transfer proceeds into medium 2, total reflection occurs in the first medium.

Since time harmonic wave propagation permits complex notation, this can be immediately seen from Snell's law. We can write (5.87) also in the form

$$
\begin{aligned}
\eta_{21}\,\mathbf{e}_t &= \mathbf{e}_i - \left(\sqrt{\eta_{21}^2 - \sin^2\alpha_1} - \cos\alpha_1\right)\mathbf{n} \\
&= \mathbf{e}_i - \left(\cos\alpha_1 - j\sqrt{\sin^2\alpha_1 - \eta_{21}^2}\right)\mathbf{n}; \quad \sin\alpha_1 \geq \eta_{21} \\
&= \sin\alpha_1\mathbf{m} - j\sqrt{\sin^2\alpha_1 - \eta_{21}^2}\,\mathbf{n}
\end{aligned}
\tag{5.96}
$$

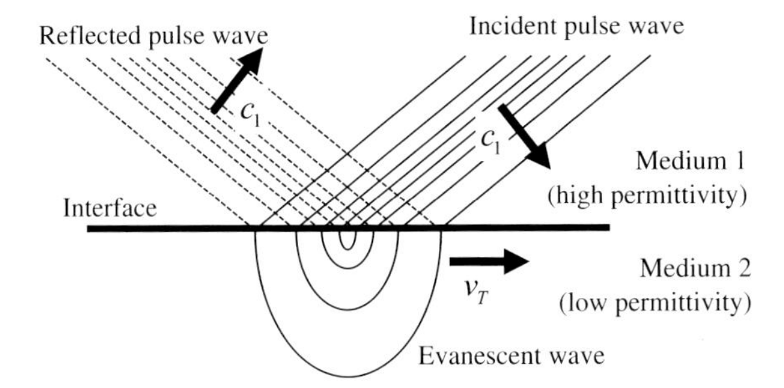

Figure 5.16 Symbolized contour plot for the total reflection of a pulse-shaped plane wave.

Hence, the propagation term of the wave in the second medium becomes complex which involves exponentially decaying amplitudes. If we insert (5.96) in a time harmonic wave of the form (5.80), we achieve

$$\begin{aligned}\underline{\mathbf{W}}_{\mathrm{t}} &= \underline{\mathbf{W}}(f,\mathbf{r}_0)\,e^{j2\pi f\left(t-\frac{\left(\sin\alpha_1\mathbf{m}-j\mathbf{n}\sqrt{\sin^2\alpha_1-\eta_{21}^2}\right)\cdot(\mathbf{r}-\mathbf{r}_0)}{\eta_{21}c_2}\right)} \\ &= \underline{\mathbf{W}}(f,\mathbf{r}_0)\,e^{j2\pi f\left(t-\frac{1}{v_T}\mathbf{m}\cdot(\mathbf{r}-\mathbf{r}_0)\right)}\,e^{-2\pi f\sqrt{\frac{1}{v_{\mathrm{T}}^2}-\frac{1}{c_2^2}}\,\mathbf{n}\cdot(\mathbf{r}-\mathbf{r}_0)}\end{aligned} \tag{5.97}$$

(5.97) represents a field which only propagates along the interface with the apparent speed v_{T}. Its amplitude drops down exponentially with increasing distance from the boundary plane. The larger the angle of incidence α_1, the stronger will be the decay rate of the field within the second medium. Figure 5.16 illustrates the total reflection in case of a pulse waveform.

Perfect metallic reflector A perfect electric conductor ($\sigma \to \infty$) forces the tangential electric field to zero at its boundary. Hence, Eqs. (C.35) and (C.38) from Annex C.2 simplify to $U_{\mathrm{i}} + U_{\mathrm{r}} = 0$ and $V_{\mathrm{i}}\mathbf{n} \times \mathbf{v}_{\mathrm{i}} + V_{\mathrm{r}}\mathbf{n} \times \mathbf{v}_{\mathrm{r}} = 0$, from which results

$$\begin{bmatrix} U_{\mathrm{r}} \\ V_{\mathrm{r}} \end{bmatrix} = \begin{bmatrix} -1 & 0 \\ 0 & 1 \end{bmatrix} \begin{bmatrix} U_{\mathrm{i}} \\ V_{\mathrm{i}} \end{bmatrix} \tag{5.98}$$

The same result we achieve by setting $Z_2 = 0$ in the Fresnel equations (5.93) which would correspond to a material of infinite permittivity.

5.5 The Hertzian Dipole

In our previous considerations of UWB radar and imaging in Chapter 4, we usually supposed a hypothetical isotropic spherical wave, which we used to stimulate the scenario under test. Such a wave will largely simplify data interpretations because propagation time, temporal shape and amplitude are independent of the radiation angle. Thus, the evaluation of target distance by round-trip time estimation or the design of the steering vectors for UWB imaging would not depend on the antenna orientation in space. We have already seen that only an electrically small antenna

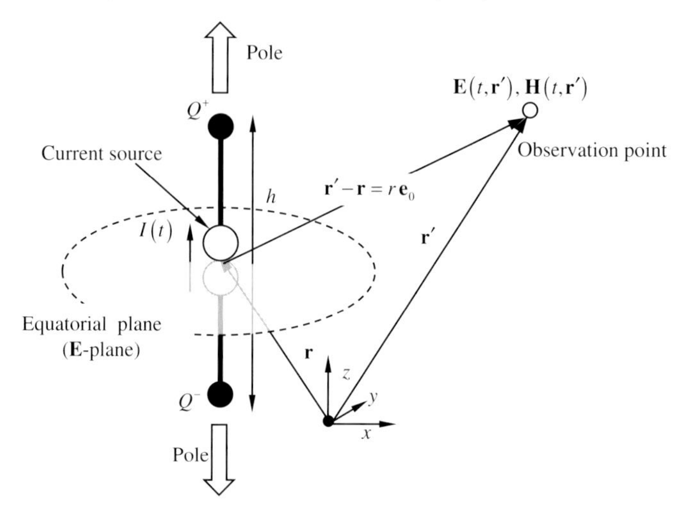

Figure 5.17 Hertzian dipole in a homogenous environment driven by a current source.

will be able to generate a sounding wave whose wavefront may well approach the surface of an expanding sphere. However, we did not yet respect the polarization of the transmitted field and its angle-dependent strength as well. This we will make up in what follows for a short radiator, namely the Hertzian dipole.

We consider a short dipole (Figure 5.17) which is driven by a pulse-shaped current[6] $I(t)$. The length of the dipole should be much shorter than the spatial extension of the current pulse $h \ll t_w c$ (t_w is the pulse duration) so that we can assume constant current across the dipole length. Our goal is to determine the electric and magnetic fields in an arbitrary observation point $\mathbf{r}'$.

The dipole consists of a short conductor whose length and orientation in space are assigned by the vector $\mathbf{h}$ (pointing in z-direction in our example). It has two small spheres at its tails which are able to accumulate the charge Q provided by the feeding current $I(t)$. We place the centre of the dipole at position $\mathbf{r}$.

5.5.1 The Dipole as Transmitter

We use again the Hertz vector for the determination of the magnetic and electric fields. For a source-free region $\nabla \cdot \mathbf{\Pi}_e = 0$ and lossless propagation medium, we can modify (5.54) to

$$\Delta \mathbf{\Pi}_e - \mu\varepsilon \frac{\partial^2 \mathbf{\Pi}_e}{\partial^2 t} = -\mathbf{p}_s \tag{5.99}$$

which has the solution

6) The same consideration can be performed for arbitrary wideband signals by referring to correlation functions of the sounding signal instead of its time functions. The pulse duration t_w has only to be replaced by the coherence time τ_{coh} of the transmitted signal.

$$\mathbf{\Pi}_{\mathrm{e}}(t,\mathbf{r}') = \frac{1}{4\pi\varepsilon}\int\frac{\mathbf{p}_{\mathrm{s}}(t-|\mathbf{r}'-\mathbf{r}|/c,\mathbf{r})}{|\mathbf{r}'-\mathbf{r}|}\mathrm{d}V \tag{5.100}$$

For a short dipole, we can approximate the polarization vector $\mathbf{p}_{\mathrm{s}}(\mathbf{r})$ by dipole moment $\mathbf{p}_0$ at position $\mathbf{r}$ since all the charge Q will be accumulated in only two small volume segments:

$$\begin{aligned}&\mathbf{p}_{\mathrm{s}} = \mathbf{p}_0\delta^{(3)}(\mathbf{r})\\&\mathbf{p}_0 \approx \mathbf{h}Q = \mathbf{h}\textstyle\int I(t)\mathrm{d}t = \mathbf{h}u_{-1}(t) * I(t)\end{aligned} \tag{5.101}$$

Insertion in (5.100) gives for the Hertz vector of a short dipole:

$$\begin{aligned}\mathbf{\Pi}_{\mathrm{e}}(t,\mathbf{r}') &= \frac{\mathbf{h}}{4\pi\varepsilon|\mathbf{r}'-\mathbf{r}|}\int I(t-|\mathbf{r}'-\mathbf{r}|/c)\mathrm{d}t\\&= \frac{\mathbf{h}}{4\pi\varepsilon|\mathbf{r}'-\mathbf{r}|}u_{-1}(t) * I\left(t-\frac{|\mathbf{r}'-\mathbf{r}|}{c}\right)\end{aligned} \tag{5.102}$$

Using doublet notation and from (5.51), (5.52) and the differential identities in Annex A.5, we yield for the magnetic and electric field (for details, see Ref. [24] Section 2.1)

$$\mathbf{H}(t,\mathbf{r}') = -\frac{1}{4\pi cr}\left(u_1(t)+\frac{c}{r}\right) * I\left(t-\frac{r}{c}\right)\mathbf{e}_0\times\mathbf{h} \tag{5.103}$$

with $r = |\mathbf{r}'-\mathbf{r}|, \mathbf{e}_0 = \frac{\mathbf{r}'-\mathbf{r}}{|\mathbf{r}'-\mathbf{r}|}$.

$$\begin{aligned}\mathbf{E}(t,\mathbf{r}') &= E_{\mathrm{T}}\mathbf{e}_0\times(\mathbf{e}_0\times\mathbf{h}) + E_{\mathrm{r}}(\mathbf{e}_0\cdot\mathbf{h})\mathbf{e}_0\\E_{\mathrm{T}} &= \frac{Z_{\mathrm{s}}}{4\pi cr}\left(u_1(t)+\frac{c}{r}+\frac{c^2}{r^2}u_{-1}(t)\right) * I\left(t-\frac{r}{c}\right)\\E_{\mathrm{R}} &= \frac{Z_{\mathrm{s}}}{2\pi cr}\left(\frac{c}{r}+\frac{c^2}{r^2}u_{-1}(t)\right) * I\left(t-\frac{r}{c}\right)\end{aligned} \tag{5.104}$$

Herein, E_{T} and E_{R}, respectively, represent a tangential and a radial field component. If the dipole coincides with the z-axis of the coordinate system (Figure 4.2 or A.1), we may write (5.103) and (5.104) also in the form

$$\begin{aligned}H_{\varphi}(t,\mathbf{r}') &= \frac{h\sin\vartheta}{4\pi cr}\left(u_1(t)+\frac{c}{r}\right) * I\left(t-\frac{r}{c}\right)\\E_{\vartheta}(t,\mathbf{r}') &= \frac{Z_{\mathrm{s}}h\sin\vartheta}{4\pi cr}\left(u_1(t)+\frac{c}{r}+\frac{c^2}{r^2}u_{-1}(t)\right) * I\left(t-\frac{r}{c}\right)\\E_{\mathrm{r}}(t,\mathbf{r}') &= \frac{Z_{\mathrm{s}}h\cos\vartheta}{2\pi cr}\left(\frac{c}{r}+\frac{c^2}{r^2}u_{-1}(t)\right) * I\left(t-\frac{r}{c}\right)\end{aligned} \tag{5.105}$$

We can observe three characteristic terms in above equations. One of them – the radiation term – depends on r^{-1}. It relates to the actually radiated fields whose time shape equals the derivation of the stimulating signal. The second term in r^{-2} resembles the Biot–Savart law for the magnetic field created by an electric current, and the third term in r^{-3} describes the electric field provided by a dipole charge.

The field components related to the last two terms are concentrated around the dipole. They will not contribute to the radiation.

The power budged of dipole radiation can be analysed with the help of the Poynting vector. Based on (5.103) and (5.104), we yield

$$\mathfrak{P}(t,\mathbf{r}') = \mathbf{E}(t,\mathbf{r}') \times \mathbf{H}(t,\mathbf{r}') = \mathfrak{P}_{\mathrm{R}}\mathbf{e}_0 + \mathfrak{P}_{\mathrm{T}}\mathbf{e}_0 \times (\mathbf{e}_0 \times \mathbf{h}) \tag{5.106}$$

with the radial and tangential components

$$\begin{aligned}
\mathfrak{P}_{\mathrm{R}} &= \frac{Z_{\mathrm{s}}}{(4\pi c r)^2} q(r,\tau)|\mathbf{e}_0 \times \mathbf{h}|^2 \\
\mathfrak{P}_{\mathrm{T}} &= \frac{2Z_{\mathrm{s}}}{(4\pi c r)^2}\left(\frac{c}{r}I(\tau) + \frac{c^2}{r^2}u_{-1}(t) * I(\tau)\right)\left(u_1(t) * I(\tau) + \frac{c}{r}I(\tau)\right)(\mathbf{e}_0 \cdot \mathbf{h})
\end{aligned}$$

wheres (applying (A.13))

$$\begin{aligned}
q(t,r) &= \left(u_1(t) * I(\tau) + \frac{c}{r}I(\tau) + \frac{c^2}{r^2}u_{-1}(t) * I(\tau)\right)\left(u_1(t) * I(\tau) + \frac{c}{r}I(\tau)\right) \\
&= \underbrace{(u_1(t) * I(\tau))^2}_{\text{provide net power transfer}} \underbrace{+2\frac{c}{r}I(\tau)(u_1(t) * I(\tau)) + \frac{c^2}{r^2}\left(u_1(t) + \frac{c^3}{r^3}\right) * (I(\tau)(u_{-1}(t) * I(\tau)))}_{\text{idle power and energy of electric field}}
\end{aligned}$$

and (applying algebraic identities of Annex A.5)

$$\begin{aligned}
(\mathbf{e}_0 \times (\mathbf{e}_0 \times \mathbf{h})) \times (\mathbf{e}_0 \times \mathbf{h}) &= -|(\mathbf{e}_0 \times \mathbf{h})|^2\mathbf{e}_0 = -\sin^2\vartheta h^2 \mathbf{e}_0 \\
(\mathbf{e}_0 \cdot \mathbf{h})\mathbf{e}_0 \times (\mathbf{e}_0 \times \mathbf{h}) &= \cos\vartheta h \mathbf{e}_0 \times (\mathbf{e}_0 \times \mathbf{h})
\end{aligned}$$

For shortening the notation, we used $\tau = t - (r/c) = t - (|\mathbf{r} - \mathbf{r}'|/c)$ in above equations.

The instantaneous net power transfer $p(t, r_0)$ through the surface of a sphere with radius r_0 is

$$\begin{aligned}
p(t,r_0) &= \oint_{S_0} \mathfrak{P}(t,\mathbf{r}') \cdot \mathrm{d}\mathbf{S} = \int_{-\pi}^{\pi}\int_{0}^{\pi} \mathfrak{P}(t,\mathbf{e}_0 r_0) \cdot \mathbf{e}_0 r_0^2 \sin\vartheta \mathrm{d}\vartheta \mathrm{d}\varphi \\
&= \int_{-\pi}^{\pi}\int_{0}^{\pi} \mathfrak{P}_{\mathrm{R}} r_0^2 \sin\vartheta \mathrm{d}\vartheta \mathrm{d}\varphi \\
&= \frac{Z_{\mathrm{s}} h^2}{(4\pi c)^2} q(t,r_0) \int_{-\pi}^{\pi} \mathrm{d}\varphi \int_{0}^{\pi} \sin^3\vartheta \mathrm{d}\vartheta \\
&= \frac{Z_{\mathrm{s}} h^2}{6\pi c^2} q(t,r_0)
\end{aligned} \tag{5.107}$$

where we find that the tangential components of the Poynting vector do not contribute to an effective power transfer.

Examining $q(t, r_0)$, we can state that only the first term provides a power quantity which propagates through the surface of a sphere of arbitrarily large diameter. The related power is solely provided by that part of the fields in (5.103) and (5.104) which we called the radiation term. The field components described by the remaining terms create only idle power. At close proximity to the dipole, the idle power crosses the surface of the sphere and returns subsequently. With increasing diameter of the sphere, this power will vanish. Thus, it will not contribute to a net power transfer. The r^{-3}-term in $q(t, r_0)$ additionally relates to the energy stored in the electric field caused by electric charging of the dipole if the feeding pulse is not of zero-mean.

Figure 5.18 shows how the electric field and power flux develop around a Hertzian dipole. A Gaussian pulse of duration $t_w = 1$ ns (spatial extension 30 cm) feeds

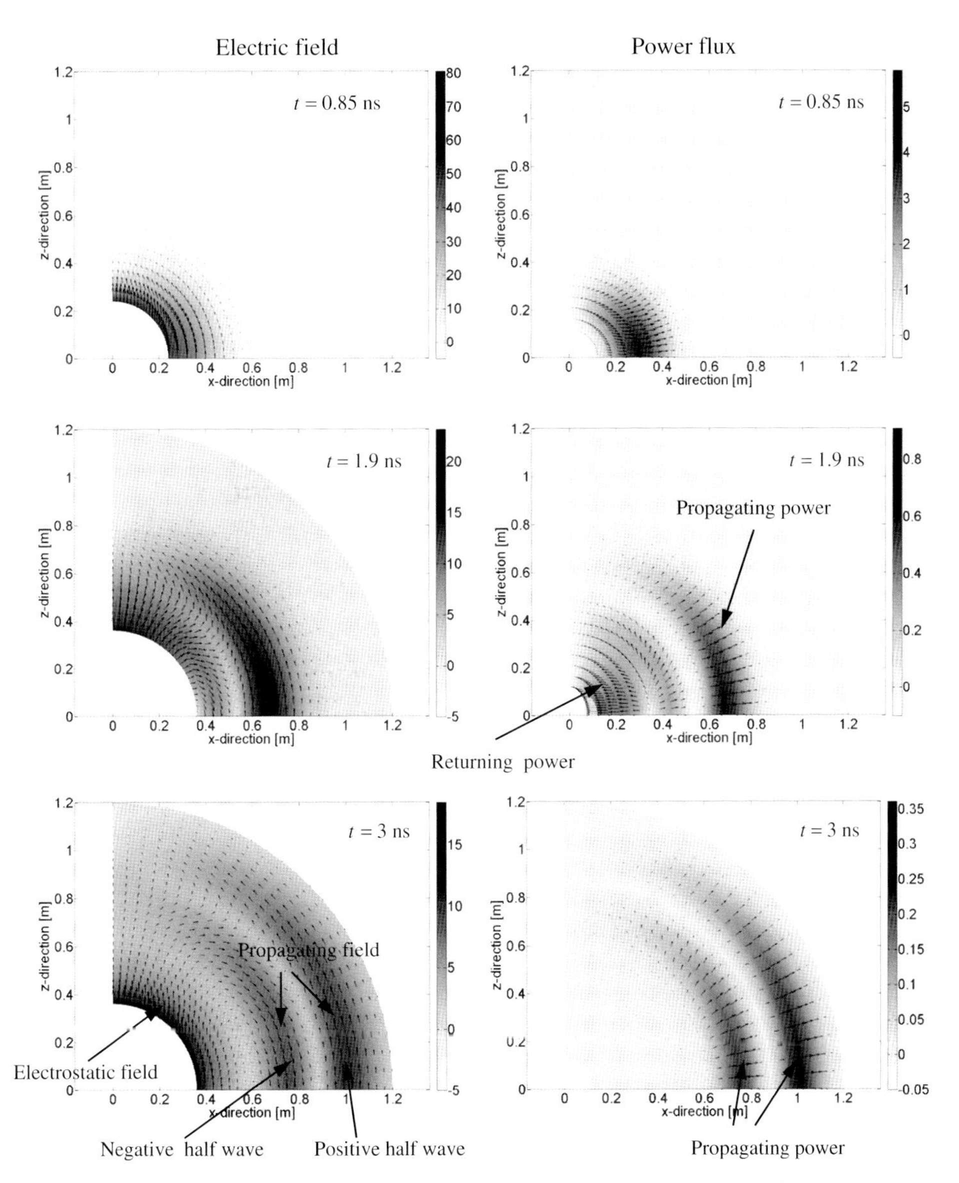

Figure 5.18 Snapshots of electric field and power flux in the vicinity of a Hertzian dipole. The near zone around the dipole is cut out due to the singularity for $\mathbf{r}' \rightarrow \mathbf{r}$. (see also movie of wave and power propagation at Wiley homepage: Fig_5_18*)

the dipole. This pulse is not of zero-mean so that it will charge the dipole causing an electrostatic field after the pulse is radiated (see snapshot at $t = 3$ ns). During the rising edge of the pulse, the field energy moves completely away from the dipole (see snapshot at $t = 0.85$ ns), while during the falling edge only a part of the energy succeeds to escape but the larger portion will return to the radiation source (see snapshot at $t = 1.9$ ns). It is also visible that the direction of the electric field reverses if the pulse passes from the rising to the falling edge.

The radiated power of a dipole only depends on the first term of $q(t, r)$. Hence, in order to generate a strong field, one is usually endeavoured to feed the dipole with a current signal of a large slew rate. Since the dipole is cumulatively charged by the feeding pulse, the current source must provide a driving voltage whose magnitude disproportionally rises above the slew rate of the current. This would pose a big challenge for the current source so that practically the operation of the Hertzian dipole will be restricted to small currents.

5.5.2
Far-Field and Normalized Dipole Wave

At a sufficiently large distance from the dipole, all field components decaying by r^{-2} or r^{-3} may be omitted; thus, we find from (5.103) and (5.104)

$$\begin{aligned} \mathbf{E}(t, \mathbf{r}') &\approx -Z_s \mathbf{e}_0 \times \mathbf{H}(t, \mathbf{r}') = Z_s \mathbf{H}(t, \mathbf{r}') \times \mathbf{e}_0 \\ &\approx \frac{Z_s}{4\pi c r} u_1(t) * I\left(t - \frac{r}{c}\right) \mathbf{e}_0 \times (\mathbf{e}_0 \times \mathbf{h}) \end{aligned} \tag{5.108}$$

Hence, the far field of the dipole is purely tangential and electric and magnetic field vectors are perpendicular to each other. The wavefront is perfectly a sphere wheres the field magnitudes decay by passing from the equatorial region to the poles.

(5.108) provides the prerequisite to normalize the dipole waves correspondingly to (5.80) which we applied for plane wave normalization:

$$\mathbf{W}(t, \mathbf{r}') = \frac{\mathbf{E}}{\sqrt{Z_s}} = \sqrt{Z_s}\, \mathbf{H} \times \mathbf{e}_0 = \frac{\sqrt{Z_s}}{4\pi c r} u_1(t) * I\left(t - \frac{r}{c}\right) \mathbf{e}_0 \times (\mathbf{e}_0 \times \mathbf{h}) \tag{5.109}$$

Since the dipole may be arbitrarily oriented in space, one often decomposes the polarization of the field with respect to a given coordinate system. If we assign the xy-plane as the horizontal plane then the field components aligned along $\mathbf{u}, \mathbf{v}$ are referred to as horizontal and vertical polarized:

$$\begin{aligned} \mathbf{u} &= \frac{\mathbf{e}_z \times \mathbf{e}_0}{|\mathbf{e}_z \times \mathbf{e}_0|} \\ \mathbf{v} &= \mathbf{e}_0 \times \mathbf{u} \end{aligned} \tag{5.110}$$

Hence in accordance with (5.90), we can write for the normalized wave generated by a short dipole of arbitrary orientation in space

$$\begin{aligned} \mathbf{W} &= (\mathbf{W}\cdot\mathbf{u})\mathbf{u} + (\mathbf{W}\cdot\mathbf{v})\mathbf{v} = U\,\mathbf{u} + V\,\mathbf{v} = \begin{bmatrix} U \\ V \end{bmatrix} = \frac{1}{\sqrt{Z_\mathrm{S}}}\begin{bmatrix} E_\varphi \\ E_\vartheta \end{bmatrix} \\ &= -\frac{\sqrt{Z_\mathrm{S}}}{4\pi c r} u_1(t) * I\left(t - \frac{r}{c}\right)\begin{bmatrix} \mathbf{h}\cdot\mathbf{u} \\ \mathbf{h}\cdot\mathbf{v} \end{bmatrix} \end{aligned} \tag{5.111}$$

since

$$\begin{aligned} (\mathbf{e}_0 \times (\mathbf{e}_0 \times \mathbf{h}))\cdot\mathbf{u} &= (\mathbf{e}_0 \times (\mathbf{e}_0 \times \mathbf{h}))\cdot\frac{\mathbf{e}_z \times \mathbf{e}_0}{|\mathbf{e}_z \times \mathbf{e}_0|} = -\frac{(\mathbf{e}_0 \times \mathbf{h})\mathbf{e}_z}{|\mathbf{e}_z \times \mathbf{e}_0|} = -\mathbf{h}\cdot\mathbf{u} \\ (\mathbf{e}_0 \times (\mathbf{e}_0 \times \mathbf{h}))\cdot\mathbf{v} &= (\mathbf{v}\times\mathbf{e}_0)\cdot(\mathbf{e}_0 \times \mathbf{h}) = -\mathbf{v}\cdot\mathbf{h} \end{aligned}$$

In order to complete notation for the normalized waves, we have to express also the feeding current $I(t)$ correspondingly, that is we have to replace it by the normalized guided wave a propagating towards the antenna. For the sake of simplicity, we will perform our discussion in the frequency domain referring to the equivalent circuit in Figure 5.19. Here, we replaced the dipole by the impedance $\underline{Z}_\mathrm{D}$ emulating the static capacity of the dipole (represents the imaginary part) and the energy loss due to radiation (related to the real part).

The normalized wave $\underline{a}$ is

$$\underline{a} = \frac{1}{2}\frac{\underline{V}_\mathrm{S}}{\sqrt{Z_0}} \tag{5.112}$$

so that we can write for the current driving the dipole

$$\underline{I} = \frac{2\sqrt{Z_0}}{Z_0 + \underline{Z}_\mathrm{D}}\underline{a} \tag{5.113}$$

if the source is perfectly matched to the waveguide. Insertion in (5.109), after having performed Fourier transformation, leads to

$$\begin{aligned} \underline{\mathbf{W}}(f,\mathbf{r}') &= \frac{\sqrt{Z_\mathrm{S}}}{4\pi c r} j 2\pi f \underline{I}(f)\,\mathrm{e}^{-j\frac{2\pi f r}{c}}\mathbf{e}_0 \times (\mathbf{e}_0 \times \mathbf{h}) \\ &= jf\frac{\sqrt{Z_\mathrm{S} Z_0}}{c(Z_0 + \underline{Z}_\mathrm{D})}\mathbf{e}_0 \times (\mathbf{e}_0 \times \mathbf{h})\frac{\underline{a}(f)\,\mathrm{e}^{-j\frac{2\pi f r}{c}}}{r} \end{aligned} \tag{5.114}$$

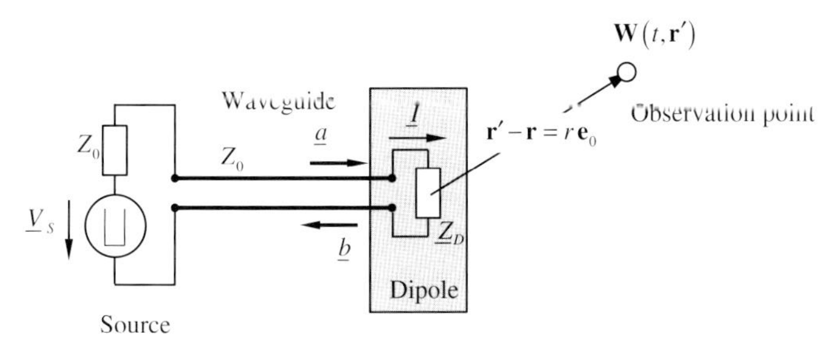

Figure 5.19 Equivalent circuit of a dipole operating in transmitter mode.

According to Section 4.4.2, we can define the polarimetric FRF of the Hertzian dipole as

$$\underline{\mathbf{T}}_{\mathrm{HD}}(f, \mathbf{e}_0) = jf \frac{\sqrt{Z_s Z_0}}{c(Z_0 + \underline{Z}_{\mathrm{D}})} \mathbf{e}_0 \times (\mathbf{e}_0 \times \mathbf{h})$$
$$\approx \frac{(j2\pi f)^2}{2\pi c} \sqrt{\frac{Z_s}{Z_0}} \tau_{\mathrm{D}}\, \mathbf{e}_0 \times (\mathbf{e}_0 \times \mathbf{h}); \quad \text{for } 2\pi f \tau_{\mathrm{D}} \ll 1;\ C_{\mathrm{D}} Z_0 = \tau_{\mathrm{D}} \tag{5.115}$$

Note, due to the transversal nature of the far field, the polarimetric antenna functions (FRF or IRF) may be expressed by two components only, for example

$$\underline{\mathbf{T}}_{\mathrm{HD}}(f, \mathbf{e}_0) = \begin{bmatrix} \underline{T}_{\mathrm{u}}(f, \mathbf{e}_0) \\ \underline{T}_{\mathrm{v}}(f, \mathbf{e}_0) \end{bmatrix} \tag{5.116}$$

A typical value for Z_0 is $50\,\Omega$ and for Z_s is about $377\,\Omega$ for propagation in air. The energy evacuation from the electric circuit by a short dipole will be very small. Therefore, the impedance $\underline{Z}_{\mathrm{D}}$ is dominated by the static capacitance C_{D} of the dipole; that is, $\underline{Z}_{\mathrm{D}} \approx (j2\pi f C_{\mathrm{D}})^{-1}$. Further, we note that this capacity will be very small so that we can assume $2\pi f C_{\mathrm{D}} Z_0 = 2\pi f \tau_{\mathrm{D}} \ll 1$ within the operational frequency band for which the dipole may be considered as short. Thus, the IRF associated to (5.115) provides a twofold derivation of the incident wave a:

$$\mathbf{T}_{\mathrm{HD}}(t, \mathbf{e}_0) \approx \frac{1}{2\pi c} \sqrt{\frac{Z_s}{Z_0}} \tau_{\mathrm{D}}\, \mathbf{e}_0 \times (\mathbf{e}_0 \times \mathbf{h}) u_2(t) * \tag{5.117}$$

One of the two derivations is caused by the radiation behaviour of the antenna and the second results from the feeding circuit. The last one may be avoided by applying high-impedance current sources to feed the antenna. This, however, would only be meaningful if the waveguide shrinks to zero length. That is the current source has to be integrated immediately into the feeding point of the antenna. Under these conditions it seems more appropriate to deal with impedance or admittance parameters than with scattering parameters to describe the propagation channel (see for comparison Section 4.6.6).

5.5.3
The Dipole as Field Sensor and Self-Reciprocity

For dipole operation in the receiver mode, we refer to an adequate equivalent circuit (Figure 5.20) as above. Now, however, the dipole acts as energy source of the circuit which is supplied from the electric field of the incident plane wave. The open source voltage of a short dipole is

$$\underline{V}_{\mathrm{D}} = \mathbf{h} \cdot \underline{\mathbf{E}} \tag{5.118}$$

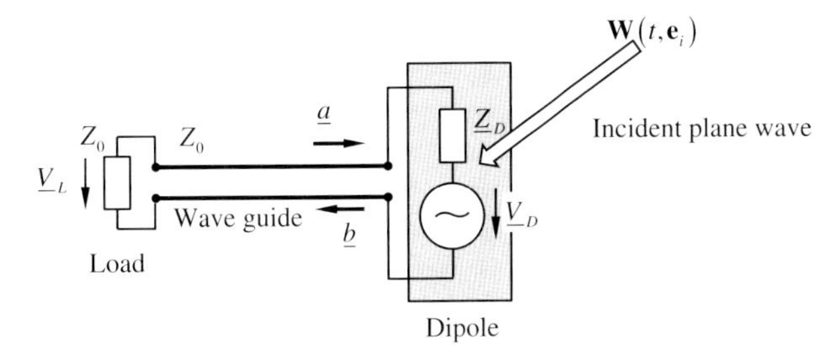

Figure 5.20 Equivalent circuit of a dipole operating in receiver mode.

where the scalar product respects the mutual orientation between the dipole and the electric field which propagates in direction $\mathbf{e}_\mathrm{i}$. If the load is perfectly matched to the waveguide, the normalized wave b results from

$$\underline{b} = \frac{\underline{V}_\mathrm{L}}{\sqrt{Z_0}} = \frac{Z_0 \mathbf{h} \cdot \mathbf{E}}{\sqrt{Z_0}(Z_0 + \underline{Z}_\mathrm{D})} = \sqrt{\frac{Z_\mathrm{s}}{Z_0}} \frac{Z_0}{(Z_0 + \underline{Z}_\mathrm{D})} \mathbf{h} \cdot \mathbf{W}(\mathbf{e}_\mathrm{i}) \tag{5.119}$$

so that by definition (4.37) the receiving FRF of the dipole is

$$\underline{\mathbf{R}}_\mathrm{HD}(f, \mathbf{e}_\mathrm{i}) = \sqrt{\frac{Z_\mathrm{s}}{Z_0}} \frac{Z_0}{(Z_0 + \underline{Z}_\mathrm{D})} \mathbf{h} \tag{5.120}$$

and under the same assumptions as for the transmitting case, the receiving IRF is approximately

$$\mathbf{R}_\mathrm{HD}(t, \mathbf{e}_\mathrm{i}) \approx \sqrt{\frac{Z_\mathrm{s}}{Z_0}} \tau_\mathrm{D} \mathbf{h} u_1(t) * \tag{5.121}$$

Comparing (5.115) with (5.119) and (5.117) with (5.121), we find the self-reciprocity of the short dipole:

$$\underline{\mathbf{T}}_\mathrm{HD}(f, \mathbf{e}_0) = -\frac{\mathbf{e}_0 \times (\mathbf{e}_0 \times \underline{\mathbf{R}}_\mathrm{HD}(f, -\mathbf{e}_0))}{j\lambda}; \quad \mathbf{e}_\mathrm{i} = -\mathbf{e}_0 \tag{5.122}$$

$$\mathbf{T}_\mathrm{HD}(t, \mathbf{e}_0) = \frac{1}{2\pi c} \mathbf{e}_0 \times (\mathbf{e}_0 \times \mathbf{R}_\mathrm{HD}(t, -\mathbf{e}_0)) u_1(t) *; \quad \mathbf{e}_\mathrm{i} = -\mathbf{e}_0 \tag{5.123}$$

These two equations may be used to extend the simplified scalar reciprocity relation (4.41) to the polarimetric case under far-field conditions.

5.5.4 Interfacial Dipole

One of the major applications of UWB sensing relates to surface-penetrating radar. This involves antennas and scatterers which are closely located to the boundary between two media. We will illustratively discuss the wave expansion in the vicinity of the interface. For deeper theoretical considerations, the reader can refer to

Refs [2, 11, 25–30]. Here, we will restrict ourselves on the basic aspect of wave propagation where we are mainly interested in the wavefront development of the involved waves since it is much more crucial for range estimation and steering vector design for UWB imaging since the precise knowledge of the wave magnitudes. For the sake of vividness, we assume a point-like source of a spherical wave. This may be a short dipole fed by a pulsed current or it may be a point object which is illuminated by a pulsed wave.

We distinguish three cases which differ by the medium in which the point source is located:

The point source is placed in the electric light medium ($\varepsilon_1 < \varepsilon_2$): Figure 5.21 shows a snapshot of symbolized wavefronts and Figure 5.23a exemplifies the development of the wave. If the primary wave hits the interface, it will be partially reflected. Except the mirroring, the reflected wave will maintain its initial spherical shape. By contrast, the penetrated wave will lose the circular wavefront. It resembles a hyperbole but correctly it is not one. Below, we will calculate its actual form. At its borders left and right, the wavefront will approach an asymptote declined by α_T, the angle of total reflection.

The point source is placed in the electric dense medium ($\varepsilon_1 < \varepsilon_2$): This case refers to Figures 5.22 and 5.23b. Now, we need to distinguish two cases. As long as the expanding wavefront touches the interface only within the borders of total reflection (see, e.g. Figure 5.22a), we have a corresponding behaviour as discussed above except the shape of the wavefront of the penetrated wave. It resembles approximately an ellipse since the wave components which first hit the surface hurry away due to the larger speed in the second medium. Its correct form will also be determined below.

If the primary wave outruns the border of total reflection (see Figure 5.22b), it cannot further create propagating waves within the second medium so that total reflection occurs. As we have seen in Section 5.4.6, this involves the creation of

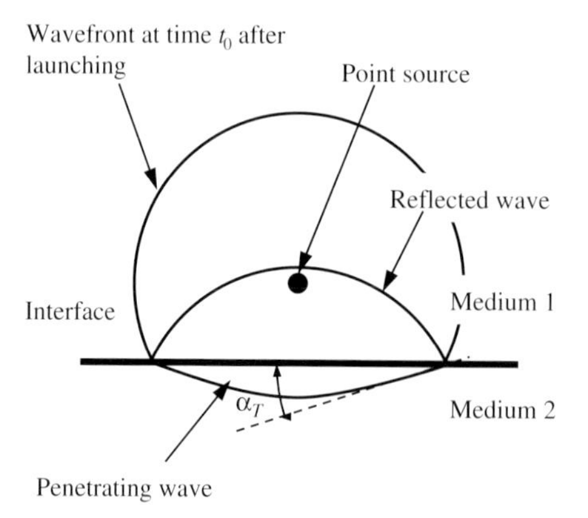

Figure 5.21 Dipole (point scatterer) in the vicinity of a planar surface if $\varepsilon_1 < \varepsilon_2$.

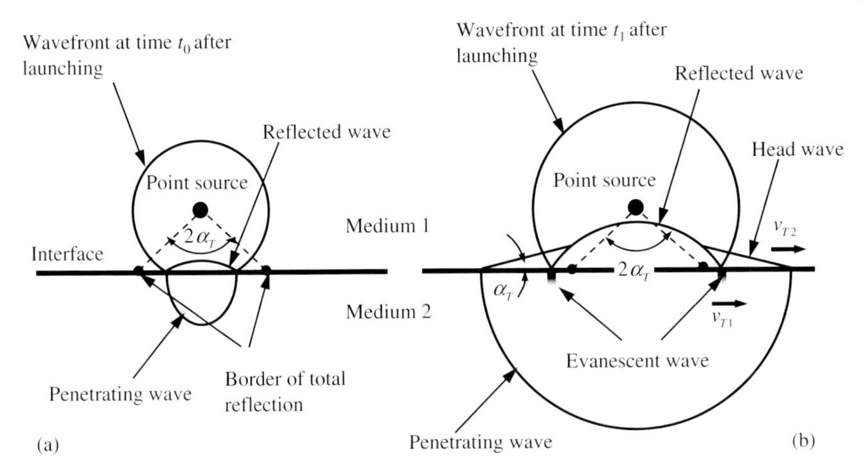

Figure 5.22 Dipole (point scatterer) in the vicinity of a planar surface if $\varepsilon_1 > \varepsilon_2$.

an evanescent wave since otherwise the boundary conditions is not fulfilled. This wave travels along the interface escorting the primary wave.

The propagation speed in the second medium is larger than that in the medium 1; therefore, the penetrated wave – created during the first phase of interaction – propagates ahead. Since it has to meet also the boundary condition, it creates a new wave in the upper medium too. Now, the apparent speed of the penetrated wave is larger than the propagation speed in the first medium, so that the new wave is able to propagate. It is known as the head wave or lateral wave. Its wavefront is declined by the angle α_T of total reflection.

We know from Figure 4.97 that the azimuth resolution of an UWB image depends on the angle ω under which the target 'sees' the imaging array. This angle will be limited here to $\omega_{max} = 2\alpha_T$ due to total reflection. But note that this would require an infinitely extended array in the second medium in order to capture all waves oozing out from the first medium within the 'angular window' of $\pm\alpha_T$.

The point source is placed at the interface: We refer to the example illustrated in Figures 5.23c and 5.24 where we supposed that the upper medium is electrically light and the lower electrically dense $\varepsilon_1 < \varepsilon_2$. Stimulating the point source, it will create a half-spherical wave in the upper medium which propagates with c_1 and a half-spherical wave in the lower medium expanding with c_2. Since these two waves must meet the boundary conditions, they create a new wave in the respective opposite medium. If the apparent speed of the 'mother' wave is higher than the propagation speed in the other medium, this wave may propagate (i.e. we may observe a head wave). In the inverse case, an evanescent wave is generated which moves along the interface with strongly decaying amplitude.

In order to deduce the correct shape of the wavefront for the penetrated wave of the two first cases, we consider Figure 5.26. We assume a point source placed at point S within the upper medium. It generates a spherical wave which partially penetrates into medium 2 and which will be partially reflected. We are interested

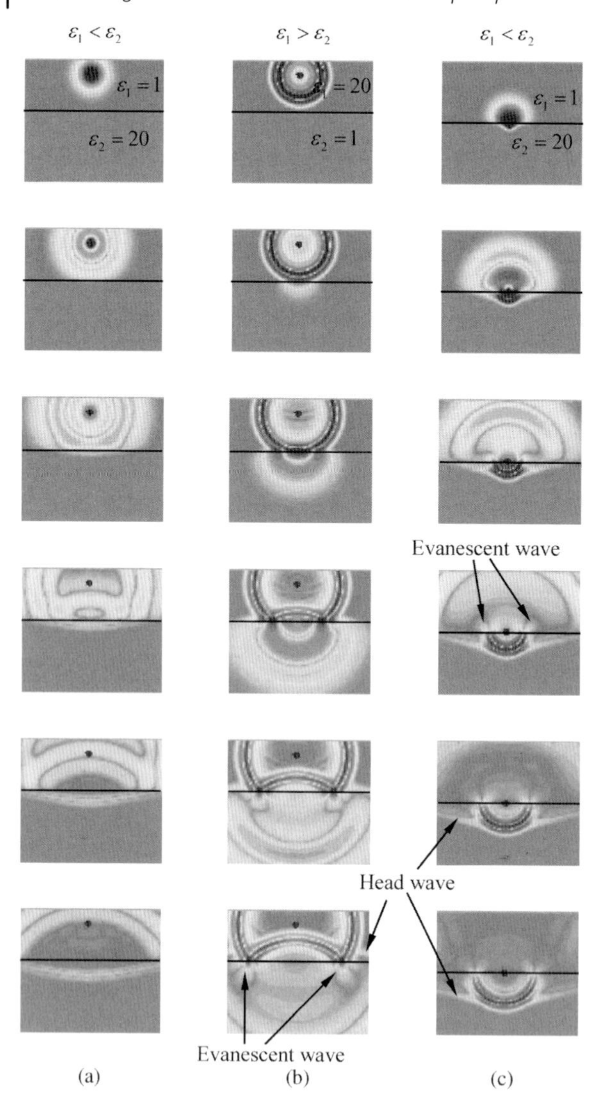

Figure 5.23 CST simulation of wave evolution in the proximity of a planar interface. (a) Dipole is placed in the low-permittivity medium; (b) dipole is placed in the high-permittivity medium; (c) dipole is placed at the interface (*Courtesy:* F. Scotto di Clemente). Colored figure at Wiley homepage: Fig_5_23)

in a simple model which describes the waveform of the refracted wave. Hence, we will exclude the reflected wave from further considerations.

To achieve a simple instruction to construct the correct wavefront in the second medium, we assume a homogenous propagation space whose properties are solely determined by the medium 2. Afterwards, we are looking for a virtual source which would provide the same field within the lower half space ($y \leq 0$) as

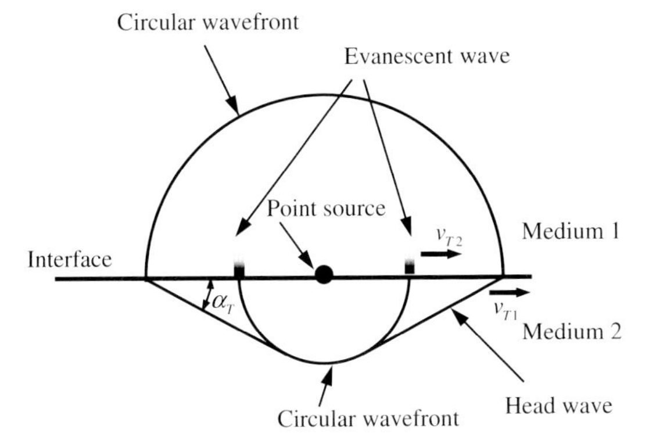

Figure 5.24 Dipole (point scatterer) at the planar interface $\varepsilon_1 < \varepsilon_2$.

it actually exists [31]. The field created by the virtual source in the upper half space is out of interest.

Figure 5.25 represents a snapshot of the front of penetrated wave at time t_0. Figure 5.25a depicts the transition from an electrical light into dense medium and Figure 5.25b deals with the inverse case. The ray which passes from the source S to point P_2 crosses the interface at point P_1. Here it alters its direction obeying Snell's law:

$$\frac{\sin\alpha_1}{\sin\alpha_2} = \frac{c_1}{c_2} = \eta_{21} \tag{5.124}$$

We can observe that the distance covered by a ray depends on its angle of incidence α_1 for our actual scenario.

In our virtual, homogeneous medium, we do not have refraction so that the origin of the considered ray must be placed at the point U. We select that point in such a way that we have straight propagation and identical propagation time from $U \to P_2$ as from $S \to P_2$ in the real scenario. Hence, we can write for the rays propagating over the time t_0

$$t_0 = \underbrace{\frac{H}{c_1} + \frac{d}{c_2} = \frac{h+d}{c_2}}_{\text{normal incidence} \quad S\to P_3} = \underbrace{\frac{l_1}{c_1} + \frac{l_2}{c_2} = \frac{l_{v1}+l_2}{c_2}}_{\text{oblique incidence} \quad S\to P_2} = \frac{l_v}{c_2} \tag{5.125}$$

From (5.125), we conclude

$$h = \frac{H}{\eta_{21}} \tag{5.126}$$

and

$$l_{v1} = \frac{l_1}{\eta_{21}} = \frac{\sqrt{H^2+x^2}}{\eta_{21}} \tag{5.127}$$

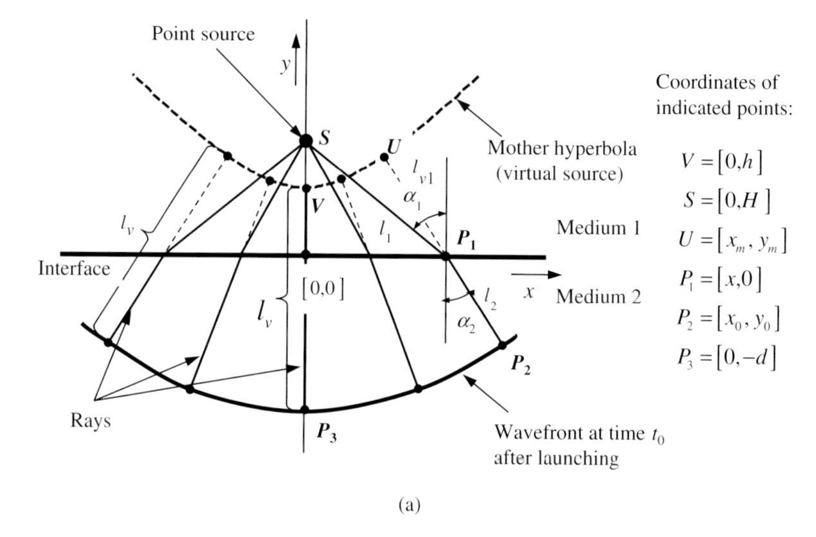

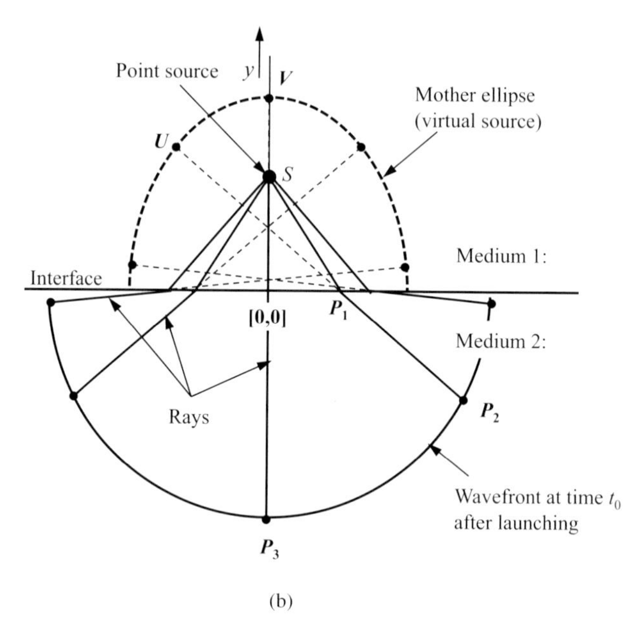

Figure 5.25 Wavefront reconstruction of a circular wave incident a planar interface. (a) $\varepsilon_1 < \varepsilon_2$ and (b) $\varepsilon_1 > \varepsilon_2$. The propagation speed in the upper medium is c_1 and in the lower it is c_2.

Furthermore, we can find from Figure 5.25 and (5.127)

$$l_{v1} \sin\alpha_2 = \frac{l_1}{\eta_{21}^2} \sin\alpha_1 = \frac{x}{\eta_{21}^2} = x - x_\mathrm{m} \tag{5.128}$$

and

$$y_{\mathrm{m}} = l_{\mathrm{v}1} \cos \alpha_2 = \frac{l_1}{\eta_{21}} \sqrt{1 - \frac{\sin^2 \alpha_1}{\eta_{21}^2}} = \frac{1}{\eta_{21}^2} \sqrt{(\eta_{21}^2 - 1)x^2 + \eta_{21}^2 H^2} \qquad (5.129)$$

so that the shape of the virtual source – also assigned as 'mother hyperbola' or 'mother ellipse' – becomes

$$y_{\mathrm{m}}^2 - \frac{x_{\mathrm{m}}^2}{\eta_{21}^2 - 1} = \frac{H^2}{\eta_{21}^2} \quad \text{or} \quad y_{\mathrm{m}} = a \cosh \varsigma; \quad x_{\mathrm{m}} = b \sinh \varsigma \quad \text{for } \eta_{21} > 1 \qquad (5.130)$$

$$y_{\mathrm{m}}^2 + \frac{x_{\mathrm{m}}^2}{1 - \eta_{21}^2} = \frac{H^2}{\eta_{21}^2} \quad \text{or} \quad y_{\mathrm{m}} = a \cos \varsigma; \quad x_{\mathrm{m}} = b \sin \varsigma \quad \text{for } \eta_{21} < 1 \qquad (5.131)^{7)}$$

ς is a real-valued parameter.

Based on this hyperbola or ellipse, we can construct the wavefront within the medium 2 for any propagation time t_x by calculating the normal vectors of the 'mother' curve and prolonging them by the factor $t_x c_2$ (see Annex A.7 (A.170))

The parameters of the 'mother' curves are

$$\text{Semi-major axis:} \quad a = \frac{H}{\eta_{21}} \qquad (5.132)$$

$$\text{Semi-minor axis:} \quad b = \begin{cases} a\sqrt{\eta_{21}^2 - 1} & \text{for} \quad \eta_{21} > 1 \\ a\sqrt{1 - \eta_{21}^2} & \text{for} \quad \eta_{21} < 1 \end{cases} \qquad (5.133)$$

$$\text{Eccentricity:} \quad \varepsilon = \eta_{21} \qquad (5.134)$$

5.6 Polarimetric Friis Formula and Radar Equation

The direct transmission between two antennas (i.e. the Friis formula) and respectively the transmission via scattering by a small object (i.e. the radar equation) have already been discussed in Sections 4.4 and 4.5. There, we supposed scalar waves. Now, we will extend it to vector waves.

As we have seen from the dipole, the emitted wave is mainly composed from tangential field components at a sufficiently large distance. This is a general observation which we may attribute to any antenna in the far field. Hence, we can establish for any antenna a polarimetric IRF or FRF which is composed of two components if we refer to polar coordinates defined by (5.110)

$$\begin{aligned} \mathbf{T}(\mathbf{e}_0) &= T_{\mathrm{u}}(\mathbf{e}_0)\mathbf{u} + T_{\mathrm{v}}(\mathbf{e}_0)\mathbf{v} = \begin{bmatrix} T_{\mathrm{u}}(\mathbf{e}_0) \\ T_{\mathrm{v}}(\mathbf{e}_0) \end{bmatrix} \\ \mathbf{R}(\mathbf{e}_{\mathrm{i}}) &= R_{\mathrm{u}}(\mathbf{e}_{\mathrm{i}})\mathbf{u} + R_{\mathrm{v}}(\mathbf{e}_{\mathrm{i}})\mathbf{v} = \begin{bmatrix} R_{\mathrm{u}}(\mathbf{e}_{\mathrm{i}}) \\ R_{\mathrm{v}}(\mathbf{e}_{\mathrm{i}}) \end{bmatrix} \end{aligned} \qquad (5.135)$$

7) Note that (5.130) and () are identical if we allow complex arguments.

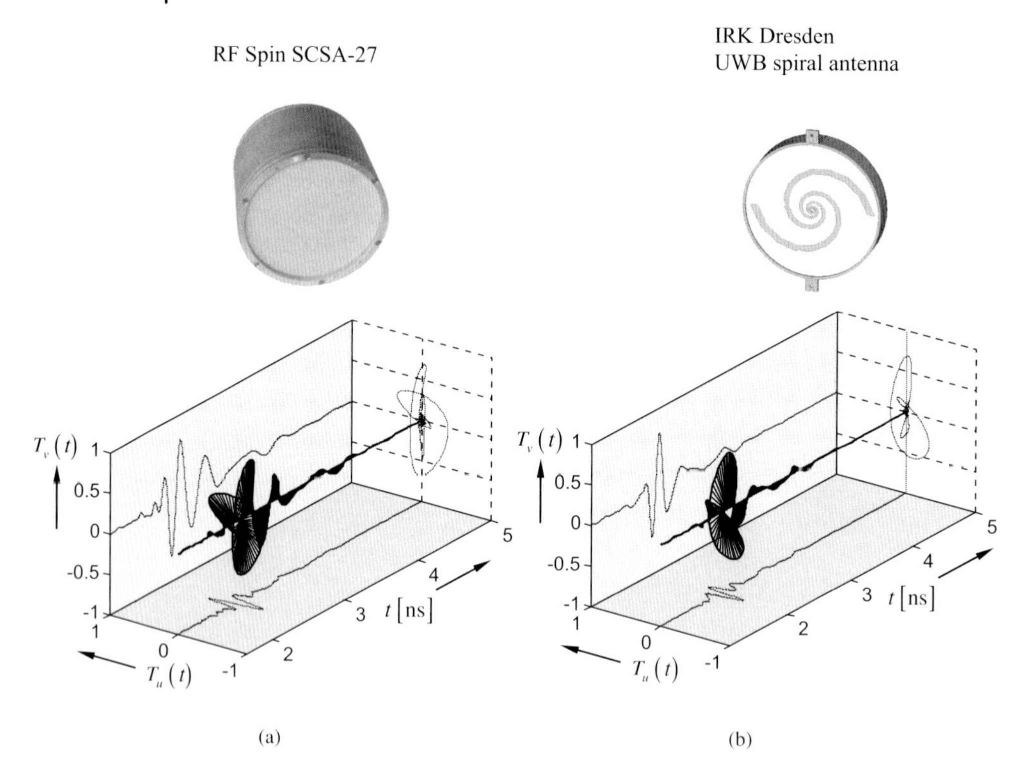

Figure 5.26 Polarimetric IRF for boresight direction ($\vartheta = 90°, \varphi = 0°$) of two spiral antennas (a: SCSA-27 from RF-spin and b: UWB spiral antenna from IRK Dresden).

If $\mathbf{T}(\mathbf{e}_0)$ or $\mathbf{R}(\mathbf{e}_\mathrm{i})$ are known for canonical antenna orientation, relation (A.46) from Annex A.4 may be used to convert the response function for arbitrary antenna orientation in space. Figure 5.26 depicts an example of the polarimetric antenna IRF for spiral antennas showing the expected variation of the **E**-field direction due to the twisted radiators. In case of a linear polarized antenna (dipole or horn antenna), both IRFs $T_\mathrm{u}(t)$, $T_\mathrm{v}(t)$ would form an arbitrary tilted plane (compare Figure 5.13 – left) according to the antenna orientation in space.

Using (A.52), the antenna reciprocity relation reads (corresponding holds for the frequency domain)

$$\begin{bmatrix} T_\mathrm{u}(\mathbf{e}_0) \\ T_\mathrm{v}(\mathbf{e}_0) \end{bmatrix} = -\frac{1}{2\pi c} \begin{bmatrix} R_\mathrm{u}(-\mathbf{e}_0) \\ R_\mathrm{v}(-\mathbf{e}_0) \end{bmatrix} * u_1(t) \tag{5.136}$$

Considering the antenna arrangement as depicted in Figure 4.14, we can now write the polarimetric Friis formula

$$\begin{aligned} b_2(t) &= \frac{1}{\Delta r} \mathbf{T}_1^T(\mathbf{e}_0, t)\big|_{P_1} * \mathbf{R}_2(-\mathbf{e}_0, t)\big|_{P_2} * a\left(t - \frac{\Delta r}{c}\right) \\ &= \frac{1}{\Delta r} \mathbf{R}_2^T(-\mathbf{e}_0, t)\big|_{P_2} * \mathbf{T}_1(\mathbf{e}_0, t)\big|_{P_1} * a\left(t - \frac{\Delta r}{c}\right) \end{aligned} \tag{5.137}$$

Δr – antenna distance
$\mathbf{T}^T * \mathbf{R} = \mathbf{R}^T * \mathbf{T} = T_u * R_u + T_v * R_v$ – dot convolution in matrix notation

Herein, the notation $\mathbf{T}(\,)|_P, \mathbf{R}(\,)|_P$ means that the antenna functions referred to a reference point and orientation symbolized by P (see Figure 4.10 in Section 4.4.1). The frequency domain notation of the complex-valued Friis formula is simply gain from the Fourier transform of (5.136).

Based on Figure 4.18, we can correspondingly extend the scalar time domain radar equation (4.60) to its polarimetric form:

$$b_2(t) = \frac{1}{\Delta r_1 \, \Delta r_2} \mathbf{T}_1^T(t, \mathbf{e}_1)\big|_{P_1} * \mathbf{\Lambda}(t, \mathbf{e}_1, -\mathbf{e}_2)|_{P_S} * \mathbf{R}_2(t, -\mathbf{e}_2)|_{P_2} * a\left(t - \frac{\Delta r_1 + \Delta r_2}{c}\right) \tag{5.138}$$

The scattering IRF $\mathbf{\Lambda}(t, \mathbf{e}_1, \mathbf{e}_2)$ of the target is now composed of four components relating to the polarization of the incident and scattered field components:

$$\mathbf{\Lambda}(t, \mathbf{e}_1, -\mathbf{e}_2) = \begin{bmatrix} \Lambda_{uu}(t, \mathbf{e}_1, -\mathbf{e}_2) & \Lambda_{uv}(t, \mathbf{e}_1, -\mathbf{e}_2) \\ \Lambda_{vu}(t, \mathbf{e}_1, -\mathbf{e}_2) & \Lambda_{vv}(t, \mathbf{e}_1, -\mathbf{e}_2) \end{bmatrix} \tag{5.139}$$

$\Lambda_{uu}, \Lambda_{vv}$ are the so-called co-polarized and $\Lambda_{uv}, \Lambda_{vu}$ are the cross-polarized components. From reciprocity of the transmission channel

$$\mathbf{T}_1^T(\mathbf{e}_1) * \mathbf{\Lambda}(\mathbf{e}_1, -\mathbf{e}_2) * \mathbf{R}_2(-\mathbf{e}_2) = \mathbf{T}_2^T(\mathbf{e}_2) * \mathbf{\Lambda}(\mathbf{e}_2, -\mathbf{e}_1) * \mathbf{R}_1(-\mathbf{e}_1)$$

we find

$$\mathbf{\Lambda}(\mathbf{e}_1, -\mathbf{e}_2) = \mathbf{\Lambda}^T(\mathbf{e}_2, -\mathbf{e}_1) \tag{5.140}$$

Referred to the scalar scattering IRF, the scattering matrix $\mathbf{\Lambda}(\mathbf{e}_1, -\mathbf{e}_2, t)$ provides additional information about the target since it is sensitive to object geometry and orientation. This may be largely of profit for target detection and imaging. Measurement examples of a mono-static scattering matrix $\mathbf{\Lambda}(\mathbf{e}_1, -\mathbf{e}_1, t)$ were shown in the previous chapter (Figures 4.24 and 4.25). Figures 5.27 and 5.28 depict additional demonstrations of polarimetric measurements for bodies of simple shape.

The figures show time slices of a radar volume gained from scanning over an area of 1 m^2. The backscattered signals are captured by a mono-static polarimetric M-sequence radar (1–13 GHz bandwidth [32]) and QRH-20 dual-polarized antenna. The antenna was moved 35 cm above the target holder. The black bold line in the radar images indicates the direction of the receiving polarization and the white bold line is the transmitter polarization. The shown data represent the captured signals which are not subject to any image processing.

The time slice in Figure 5.27 was taken when the wavefront just hits the target holder (Styrofoam block). We can observe that remarkable backscattering only appears for that polarization which is aligned with the rod, that is polarization xx (i.e. both antennas are polarized in x-direction). The residual signals of the three other antenna combinations result from small misalignments of the rod, its finite diameter and non-perfect polarization decoupling of the antenna. The tilted rod is clearly visible for all polarizations. The second example depicted in Figure 5.28

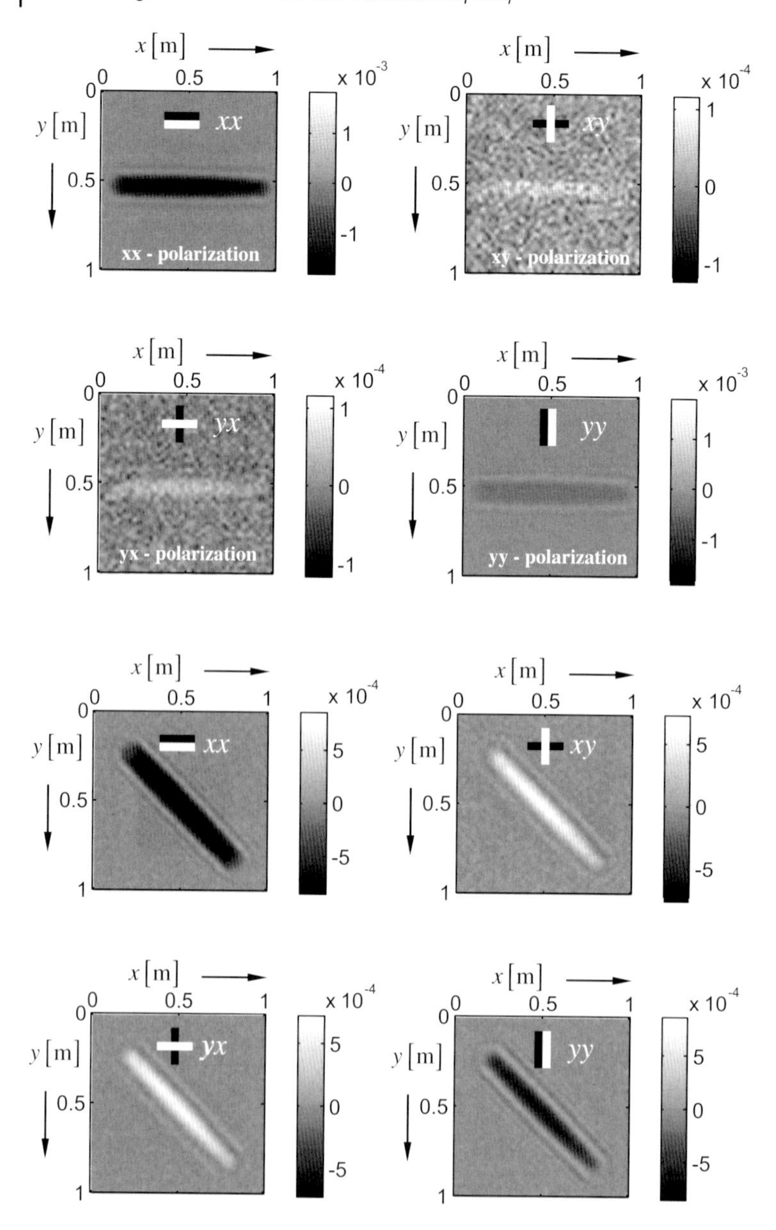

Figure 5.27 Time slice (C-scan) of a 2D-scan over an area of 1 m × 1 m showing the reflections of a metallic rod (1 m length, 6 mm diameter) oriented along the *x*-axis (above) and rotated by 45° (below). (see also movie at Wiley homepage: Fig_5_27*)

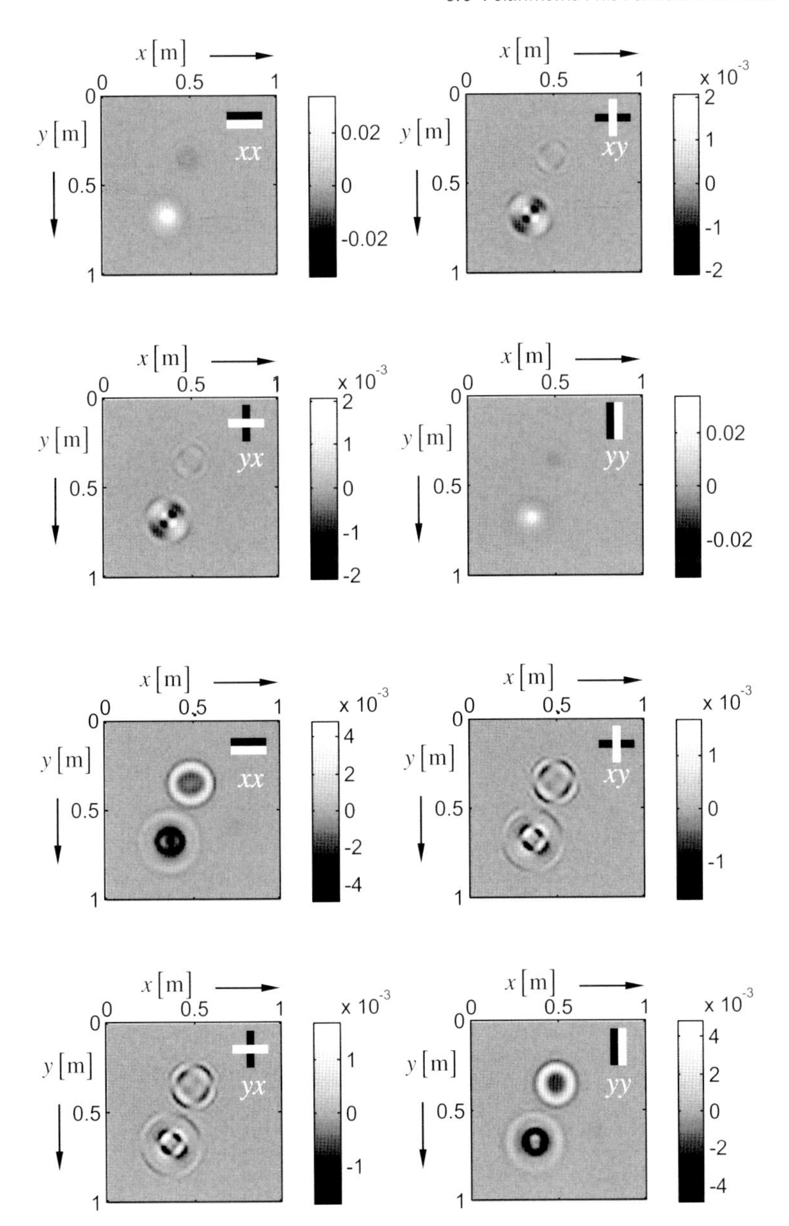

Figure 5.28 Time slices (C-scan) of a 2D-scan from a metallic cylinder (height 12 cm, diameter 10 cm; position [0.37, 0.67] m) and metallic sphere (diameter 10 cm, position [0.50, 0.35] m) taken 700 ps (above) and 560 ps (below) before the snapshots in Figure 5.27. (see for movies at Wiley homepage: Fig_5_28*. An extra movie refers to the scattering of a trihedral corner if it is illuminated at the facial plane. The trihedral corner is the same as used in Figure 4.25)

refers to scattering by a cylinder and a sphere. Here, two snapshots are shown. The upper one illustrates the backscattered fields shortly after the wavefront did bounce the highest object, that is the cylinder. The second snap shot images the fields about 140 ps later. The measurements well confirm the reciprocity relation (5.140). As we can see, the cross-polar components in Figures 5.27 and 5.28 are identical except for minor measurement errors.

The exploitation of polarimetric radar measurements is well established in remote sensing by air- or spaceborne radars. Comprehensive discussions of corresponding topics may be found in Refs [23, 33–36]. A broad survey on publications is summarized in a list of references in Ref. [36]. However, the framework of data interpretation may only be partially appropriate for our purpose here since the applied sounding signals have a relative low fractional bandwidth. Polarimetry in ultra-wideband short-range sensing is less developed so far. But recent research endeavour is increasingly directed to this field in order to improve the information yield about the targets. The following list summarizes exemplarily some research which was done in UWB polarimetry [25, 26, 37–48].

5.7
The Concept of Green's Functions and the Near-Field Radar Equation

So far, we modelled our measurement scenario always under the assumption of sufficient distance between the objects of interest and the antennas that provided the sounding signal and captured the target response. The consequences of this simplification are characteristic functions of antennas $\mathbf{T}(t, \mathbf{e}), \mathbf{R}(t, \mathbf{e})$ and targets $\mathbf{\Lambda}(t, \mathbf{e}_1, \mathbf{e}_2)$ which do not depend on the mutual distance. Further, we could omit the influence of antenna self-reflections onto the test scenario; that is, the antennas did not noticeably encroach upon the scenario.

Now, we abandon the assumption of far distances and ask for a general method on how to model the system under test under these conditions. For that purpose, we come back to an approved philosophy which we already used to characterize systems with lumped ports (Section 2.4). It provided us with the Green's functions of a black box model which we also called impulse response functions. They completely describe the input–output behaviour of a linear time invariant system. Assuming linearity and time invariance,[8] we can copy these ideas also to characterize the behaviour of distributed systems.

To illustrate our discussion, we refer to Figure 5.29 which symbolizes an arbitrary test scenario with short distances between the involved bodies. Due to the short distances between the objects and their finite size, numerous body internal and external interactions may happen. This is illustrated by the dashed lines symbolizing the propagation path of individual 'rays'. We can detect two different kinds of objects in Figure 5.29 – common scattering objects (bodies of number 2, 3, 5)

8) Time invariance is not strictly required but it would at least simplify the measurement if it can be assumed during the duration of data capturing (see also Section 2.4.2).

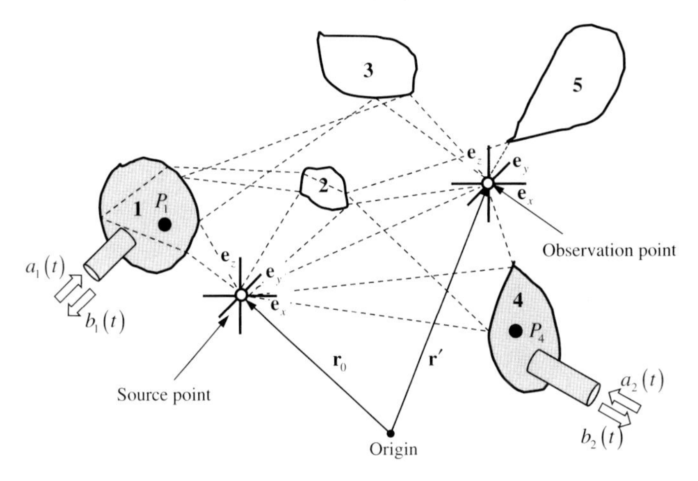

Figure 5.29 Symbolized near-field test scenario.

and scattering objects (bodies of number 1 and 4) which are also able to convert between guided and free waves (i.e. antennas). Initially, we will not make a difference between these two types of objects.

For such scenario, we can always describe the electromagnetic interactions by a linear vector equation which is subjected to related boundary conditions. Maxwell's equations would provide the theoretical base for it. We formally symbolize this vector equation by the linear operator $L\{\cdots\}$ acting on the vector field $\mathbf{\Pi}(t,\mathbf{r}')$:

$$L\{\mathbf{\Pi}(t,\mathbf{r}')\} = \mathbf{p}_s(t,\mathbf{r}) \tag{5.141}$$

Herein, $\mathbf{p}_s(t,\mathbf{r})$ (see (5.55) for definition) describes the known source of the unknown vector field $\mathbf{\Pi}(t,\mathbf{r}')$ at the observation position $\mathbf{r}'$. Exemplarily, we applied the Hertz vector to represent the field quantities. The simplest of such linear equations is the wave equation (5.54) for an unbounded isotropic media.

We presume that we know the vector field $\mathbf{\Pi}(t,\mathbf{r}') = \mathbf{G}^{(n)}(t,\tau,\mathbf{r}_0,\mathbf{r}')$ as reaction of a point source $\mathbf{p}_0^{(n)}(t,\mathbf{r}_0)$ placed at position $\mathbf{r}_0$. The source represents a dipole moment aligned along direction $\mathbf{e}_{(n)}$ launching a Dirac pulse at time τ. We may express the source term as

$$\begin{aligned}\mathbf{p}_0^{(n)}(t,\mathbf{r}_0) &= \mathbf{e}_{(n)}\delta(t-\tau,\mathbf{r}-\mathbf{r}_0) \\ &= \mathbf{e}_{(n)}\delta(t-\tau)\delta(x-x_0)\delta(y-y_0)\delta(z-z_0)\end{aligned} \tag{5.142}$$

so that the following relation holds

$$L\{\mathbf{G}^{(n)}(t,\tau,\mathbf{r}_0,\mathbf{r}')\} = \mathbf{e}_{(n)}\delta(t-\tau,\mathbf{r}-\mathbf{r}_0)$$

$$\text{with}\quad \mathbf{G}^{(n)}(t,\tau,\mathbf{r}_0,\mathbf{r}') = \begin{bmatrix} G_x^{(n)}(t,\tau,\mathbf{r}_0,\mathbf{r}') \\ G_y^{(n)}(t,\tau,\mathbf{r}_0,\mathbf{r}') \\ G_z^{(n)}(t,\tau,\mathbf{r}_0,\mathbf{r}') \end{bmatrix}$$

Since the dipole moment $\mathbf{p}_0^{(n)}(t, \mathbf{r}_0)$ can have an arbitrary direction in space, we decompose it to the orthogonal basis $[\mathbf{e}_x, \mathbf{e}_y, \mathbf{e}_z]$

$$\begin{aligned}\mathbf{p}_0^{(n)}(t, \mathbf{r}_0) &= \mathbf{e}_{(n)}\delta(t-\tau, \mathbf{r}-\mathbf{r}_0)\\ &= \left((\mathbf{e}_{(n)}\cdot\mathbf{e}_x)\mathbf{e}_x + (\mathbf{e}_{(n)}\cdot\mathbf{e}_y)\mathbf{e}_y + (\mathbf{e}_{(n)}\cdot\mathbf{e}_z)\mathbf{e}_z\right)\delta(t-\tau, \mathbf{r}-\mathbf{r}_0)\\ &= \begin{bmatrix} p_{0x}\\ p_{0y}\\ p_{0z}\end{bmatrix}\delta(t-\tau, \mathbf{r}-\mathbf{r}') \quad \text{with} \quad \sum p_{0i}^2 = 1\end{aligned}$$

and establish a set of three vector fields responding each on the sources of different alignment. We summarize these three vector fields in a matrix which is also known as the dyadic Green's function or Green's tensor:

$$\mathbf{G}(t, \tau, \mathbf{r}_0, \mathbf{r}') = \left[\mathbf{G}^{(x)}(t, \tau, \mathbf{r}_0, \mathbf{r}') \quad \mathbf{G}^{(y)}(t, \tau, \mathbf{r}_0, \mathbf{r}') \quad \mathbf{G}^{(z)}(t, \tau, \mathbf{r}_0, \mathbf{r}')\right]$$

Herein, $\mathbf{G}^{(x)}(t, \tau, \mathbf{r}_0, \mathbf{r}')$ represents the field appearing at position $\mathbf{r}'$ in reaction to a Dirac pulse launched by a point source at position $\mathbf{r}_0$ and aligned along $\mathbf{e}_x$. Corresponding holds for the other two fields.

Following the same considerations as in Section 2.4.2, we will find a space–time convolution integral for distributed systems which relates to (2.138) for black box models with lumped ports. It reads as

$$\begin{aligned}\mathbf{\Pi}(t, \mathbf{r}') &= \int \int_{V_S} \mathbf{G}(t, \tau, \mathbf{r}, \mathbf{r}')\mathbf{p}_s(t, \mathbf{r})\mathrm{d}V\,\mathrm{d}\tau; \quad \mathbf{r} \in V_S\\ &= \mathbf{G}(t, \tau, \mathbf{r}, \mathbf{r}')\bigstar\mathbf{p}_s(t, \mathbf{r})\end{aligned} \tag{5.143}$$

Herein,

$$\text{The Green's tensor}: \quad \mathbf{G}(t, \tau, \mathbf{r}, \mathbf{r}') = \begin{bmatrix} G_{xx}(t, \tau, \mathbf{r}, \mathbf{r}') & G_{xy}(t, \tau, \mathbf{r}, \mathbf{r}') & G_{xz}(t, \tau, \mathbf{r}, \mathbf{r}')\\ G_{yx}(t, \tau, \mathbf{r}, \mathbf{r}') & G_{yy}(t, \tau, \mathbf{r}, \mathbf{r}') & G_{yz}(t, \tau, \mathbf{r}, \mathbf{r}')\\ G_{zx}(t, \tau, \mathbf{r}, \mathbf{r}') & G_{zy}(t, \tau, \mathbf{r}, \mathbf{r}') & G_{zz}(t, \tau, \mathbf{r}, \mathbf{r}')\end{bmatrix}$$

$$\text{Source distribution}: \quad \mathbf{p}_s(t, \mathbf{r}) = \begin{bmatrix} p_{sx}(t, \mathbf{r})\\ p_{sy}(t, \mathbf{r})\\ p_{sz}(t, \mathbf{r})\end{bmatrix} = \begin{cases} \mathbf{p}_s(t, \mathbf{r}); & \mathbf{r} \in V_S\\ \mathbf{0}; & \mathbf{r} \notin V_S\end{cases}$$

$\mathbf{p}_s(t, \mathbf{r})$ describes the distribution of source yield within the volume V_S covered by the field source. In order to discriminate from a pure time convolution $*$, we introduced the symbol $\bigstar$ for space–time convolution. Note that the convolution (5.143) may also be converted to any other field quantities and source terms (e.g. (5.150), (5.151)).

For illustration of Green's tensor, we are coming back to Figure 5.29. In order to measure the Green's functions $\mathbf{G}(t, \tau, \mathbf{r}_0, \mathbf{r}')$ of the scenario, we place a point source at position $\mathbf{r}_0$. A Hertzian dipole can be considered as such a hypothetical field source.[9] Obviously, the dipole may take three canonical orientations. At first, we align it along $\mathbf{e}_x$ and we measure the resulting field at position $\mathbf{r}'$. This leads us to

9) Since the dipole should not influence the test scenario, it must be 'invisible' for the electric field. Hence, it must not cause any reflections. This can be achieved by driving the dipole with an ideal current source having an infinite internal impedance (see also Section 4.8.1).

three impulse response functions corresponding to the alignment (i.e. either along $\mathbf{e}_x$ or $\mathbf{e}_y$ or $\mathbf{e}_z$) of the receiving antenna, which also should be a Hertzian dipole.[10)] We arrange the three IRFs in the column vector $\mathbf{G}^{(x)}(t,\tau,\mathbf{r}_0,\mathbf{r}') = \left[G_{xx}(t,\tau,\mathbf{r}_0,\mathbf{r}') \quad G_{yx}(t,\tau,\mathbf{r}_0,\mathbf{r}') \quad G_{zx}(t,\tau,\mathbf{r}_0,\mathbf{r}')\right]^T$ and repeat the whole procedure for the other two orientations $\mathbf{e}_y$, $\mathbf{e}_z$ of the transmitting antenna. These nine IRFs cover all possible transmission paths (including object internal propagation) from position $\mathbf{r}_0$ to position $\mathbf{r}'$ as symbolized by the dashed lines in Figure 5.29. The propagation model expressed by (5.143) requires to repeat the whole procedure for any position $\mathbf{r} = \mathbf{r}_0$ of the point source and any observation point $\mathbf{r}'$. It is to be emphasized here that this applies also for the volume regions of all objects involved in the scenario in order to meet exactly the model (5.143).

Based on the physical meaning of the Green's function as explained above, we can intuitively deduce some of their basic properties:

Stability: As long as the scenario under test behaves passive, the IRFs (the related waves) must cover finite energy and they must settle down with increasing time and distance[11)] (see also radiation condition (5.31)).

$$\begin{aligned}
&\left\|G_{ij}(t,\tau,\mathbf{r},\mathbf{r}')\right\|_2 < \infty; \quad \mathbf{r} \neq \mathbf{r}' \\
&\lim_{t\to\infty} G_{ij}(t,\tau,\mathbf{r},\mathbf{r}') = 0; \quad \mathbf{r} \neq \mathbf{r}' \\
&\lim_{|\mathbf{r}-\mathbf{r}'|\to\infty} G_{ij}(t,\tau,\mathbf{r},\mathbf{r}') = 0
\end{aligned} \tag{5.144}$$

Time invariance: If the geometry and the involved materials of the test scenario do not change over time, the dynamic behaviour of the system is only a function of elapsed time after stimulation but it is not dependent on the actual time point τ of stimulation, that is

$$\mathbf{G}(t,\tau,\mathbf{r},\mathbf{r}') = \mathbf{G}(t-\tau,\mathbf{r},\mathbf{r}') \quad \overset{\tau=0}{\longrightarrow} \quad \mathbf{G}(t,\mathbf{r},\mathbf{r}') \tag{5.145}$$

Note that we mostly assumed this condition in case of a measurement; at least it was approximately supposed during the recording time of the measurement data of a full set of IRFs.

Space invariance: If the behaviour of the test scenario does not depend on the actual position of stimulation or of reception but only on their difference, the Green's functions simplify to

$$\mathbf{G}(t,\tau,\mathbf{r},\mathbf{r}') = \mathbf{G}_0(t,\tau,\mathbf{r}-\mathbf{r}') \tag{5.146}$$

It is important to note that this condition only holds for homogenous propagation medium.

Causality: Causality means that the reaction to an arbitrary stimulation may not be happen before activating the stimulation and it cannot spread faster in space

10) The voltage captured by this dipole must be recorded by a voltmeter of infinite input impedance in order to avoid perturbations of the test scenario. This guarantees that not any energy would detracted from the field and the dipole remains 'invisible' since no reflections occur.

11) Note the singularity of the Green's function if source and observation point coincides.

than the wave carries its energy. Under the condition of time and space invariance and non-dispersive propagation, this can be written as

$$\mathbf{G}_0(t,\Delta\mathbf{r}) = \begin{cases} \mathbf{G}_0(t,\Delta\mathbf{r}) & t>0 \cap \Delta r < tc \\ \equiv \mathbf{0} & t \leq 0 \cap \Delta r \geq tc \end{cases} \tag{5.147}$$

$$\mathbf{G}_0(t,\Delta\mathbf{r}) = \mathbf{G}_0(t-\tau,\mathbf{r}-\mathbf{r}') \text{ – free space Green's tensor}$$

Reciprocity: Reciprocity involves interchangeably of source and observation point (see (5.32)). From this immediately follows (refer to (5.140))

$$\mathbf{G}(t,\mathbf{r},\mathbf{r}') = \mathbf{G}^T(t,\mathbf{r}',\mathbf{r}) \tag{5.148}$$

Free space Green's function: The free space Green's tensor is space and time invariant $\mathbf{G}_0(t,\tau,\mathbf{r},\mathbf{r}') = \mathbf{G}_0(t,\Delta r)$. It simply represents the radiation of three Hertzian dipoles placed at position $\mathbf{r}$ and observed at position $\mathbf{r}'$ where one dipole is aligned along $\mathbf{e}_x$, the second along $\mathbf{e}_y$ and the third along $\mathbf{e}_z$.

Using the decomposition $\boldsymbol{\Pi}^{(n)} = \left(\boldsymbol{\Pi}^{(n)}\cdot\mathbf{e}_x\right)\mathbf{e}_x + \left(\boldsymbol{\Pi}^{(n)}\cdot\mathbf{e}_y\right)\mathbf{e}_y + \left(\boldsymbol{\Pi}^{(n)}\cdot\mathbf{e}_z\right)\mathbf{e}_z$ for the field created by a dipole moment oriented along direction $\mathbf{e}_{(n)}$, we can easily find from X the Green's tensor for the Hertz vector (i.e. the field of all three dipoles):

$$\begin{aligned}\mathbf{G}_0^{(\boldsymbol{\Pi})}(t,\Delta r) &= \frac{1}{4\pi\varepsilon\,\Delta r}u_{-1}(t)*\delta\left(t-\frac{|\mathbf{r}'-\mathbf{r}|}{c}\right)\begin{bmatrix}\mathbf{e}_x\\ \mathbf{e}_y\\ \mathbf{e}_z\end{bmatrix}\otimes\begin{bmatrix}\mathbf{e}_x\\ \mathbf{e}_y\\ \mathbf{e}_z\end{bmatrix} \\ &= \frac{1}{4\pi\varepsilon\,\Delta r}u_{-1}(t)*\delta\left(t-\frac{|\mathbf{r}'-\mathbf{r}|}{c}\right)\mathbf{I}\end{aligned} \tag{5.149}$$

The free space Green's tensor $\mathbf{G}_0^{(\mathbf{E})}, \mathbf{G}_0^{(\mathbf{H})}$ for the **E**- or **H**-field can be gained by applying (5.51) or (5.52) onto (5.149). Exemplarily, we will here give the Green's tensor of the electric field (refer also to Ref. [8]):

$$\begin{aligned} G_{0,ii}^{(\mathbf{E})}(t,\Delta\mathbf{r}) &= \frac{Z_s}{4\pi cr}\delta\left(t-\frac{r}{c}\right)*\left[\left(1-(\mathbf{e}_r\cdot\mathbf{e}_i)^2\right)u_1(t)-\frac{c}{r}\left(1+\frac{c}{r}u_{-1}(t)\right)\left(3(\mathbf{e}_r\cdot\mathbf{e}_i)^2-1\right)\right] \\ G_{0,ij}^{(\mathbf{E})}(t,\Delta\mathbf{r}) &= -\frac{Z_s}{4\pi cr}\delta\left(t-\frac{r}{c}\right)*\left[u_1(t)+\frac{3c}{r}\left(1+\frac{c}{r}u_{-1}(t)\right)\right](\mathbf{e}_r\cdot\mathbf{e}_i)(\mathbf{e}_r\cdot\mathbf{e}_j);\quad i\neq j \\ & i,j\in[x,y,z];\quad r=|\mathbf{r}-\mathbf{r}'|;\quad \mathbf{e}_r=\frac{\mathbf{r}-\mathbf{r}'}{|\mathbf{r}-\mathbf{r}'|}\end{aligned} \tag{5.150}$$

from which the electric field in the observation point $\mathbf{r}'$ may be calculated for an arbitrarily distributed current source $\mathbf{J}_s(t,\mathbf{r})$ by

$$\mathbf{E}(t,\mathbf{r}') = \int_{V_s} G_0^{(\mathbf{E})}(t,|\mathbf{r}-\mathbf{r}'|)*\mathbf{J}_s(t,\mathbf{r})\mathrm{d}V = G_0^{(\mathbf{E})}(t,|\mathbf{r}-\mathbf{r}'|)\bigstar\mathbf{J}_s(t,\mathbf{r}) \tag{5.151}$$

This equation assumes that within the source region the same propagation conditions prevail as in the rest of the propagation space. Under far-field conditions, the Green's tensor simplifies to

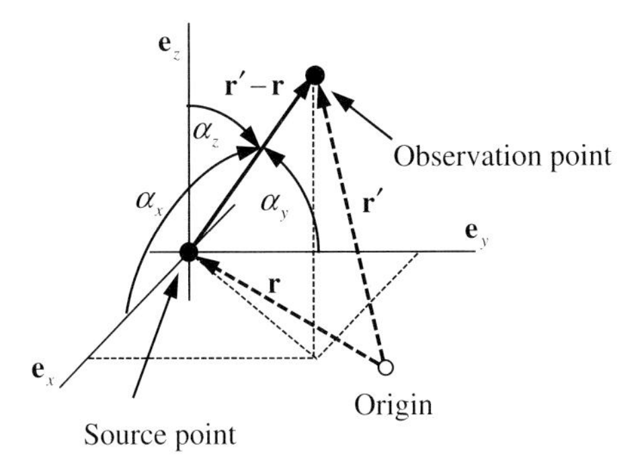

Figure 5.30 The directional angles corresponding to (5.152).

$$\mathbf{G}_0^{(\mathrm{E})}(t,\Delta\mathbf{r}) \approx \frac{Z_\mathrm{S}}{4\pi c r}\delta\left(t-\frac{r}{c}\right)\begin{bmatrix} 1-(\mathbf{e}_\mathrm{r}\cdot\mathbf{e}_x)^2 & -(\mathbf{e}_\mathrm{r}\cdot\mathbf{e}_x)(\mathbf{e}_\mathrm{r}\cdot\mathbf{e}_y) & -(\mathbf{e}_\mathrm{r}\cdot\mathbf{e}_x)(\mathbf{e}_\mathrm{r}\cdot\mathbf{e}_z) \\ -(\mathbf{e}_\mathrm{r}\cdot\mathbf{e}_x)(\mathbf{e}_\mathrm{r}\cdot\mathbf{e}_y) & 1-(\mathbf{e}_\mathrm{r}\cdot\mathbf{e}_y)^2 & -(\mathbf{e}_\mathrm{r}\cdot\mathbf{e}_y)(\mathbf{e}_\mathrm{r}\cdot\mathbf{e}_z) \\ -(\mathbf{e}_\mathrm{r}\cdot\mathbf{e}_x)(\mathbf{e}_\mathrm{r}\cdot\mathbf{e}_z) & -(\mathbf{e}_\mathrm{r}\cdot\mathbf{e}_y)(\mathbf{e}_\mathrm{r}\cdot\mathbf{e}_z) & 1-(\mathbf{e}_\mathrm{r}\cdot\mathbf{e}_y)^2 \end{bmatrix}$$
$$= \frac{Z_\mathrm{S}}{4\pi c r}\delta\left(t-\frac{r}{c}\right)\begin{bmatrix} \sin^2\alpha_x & -\cos\alpha_x\cos\alpha_y & -\cos\alpha_x\cos\alpha_z \\ -\cos\alpha_x\cos\alpha_y & \sin^2\alpha_y & -\cos\alpha_y\cos\alpha_z \\ -\cos\alpha_x\cos\alpha_z & -\cos\alpha_y\cos\alpha_z & \sin^2\alpha_z \end{bmatrix} \quad (5.152)$$

and it will be easy to show that the electric field will be purely tangential since $\mathbf{e}_\mathrm{r}\cdot\left(\mathbf{G}_0^{(\mathrm{E})}(t,\Delta\mathbf{r})\mathbf{J}_\mathrm{s}\right)=0$. The directional angles appearing in (5.152) are elucidated in Figure 5.30.

The dyadic Green's functions have large importance for theoretical investigations of the electromagnetic field. Since field theoretical investigations are not our goal, we will not continue discussing such issues here. Rather we will consider some practical aspects of their measurement and their use in date interpretation, for example UWB short-range imaging.

As argued already earlier (see Section 2.4.2), the Green's tensor represents a set of characteristic functions of the test scenario since from its knowledge the resulting field from any possible source can be calculated at any position. Consequently, it covers all possible versions on how a field can interact with the test scenario. Hence, one will be well advised to recover the Green's tensor in order to gain as much as possible information about the scenario of interest.

Within the scope of a sensing task, we can basically pose two fundamental inverse problems resulting from (5.143) – either we may be interested in the propagation scenario – described by $\mathbf{G}(t,\mathbf{r},\mathbf{r}')$ (case 1) – or we want to know more about the radiation source – modelled by $\mathbf{p}_\mathrm{s}(t,\mathbf{r})$ (case 2):

Case 1: In order to be able to determine at least a limited set of Green's functions $\mathbf{G}\left(t,\mathbf{r}_i,\mathbf{r}'_j\right)$, we need to stimulate the test scenario with known sources and to measure the resulting fields at a number of different positions. Then, we can try to extract the wanted information from the measured Green's functions if a reasonable model of the test scenario is available.

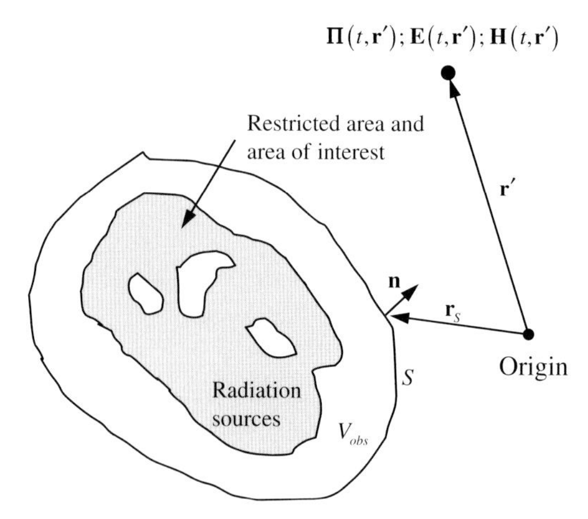

Figure 5.31 Fields outside a closed surface S. All measurement positions are located at the closed surface S.

Case 2: If we are looking for the state and location of the source, we have to measure again the field at several positions in space. From this knowledge, we can try to trace back from the measured distribution of the field to its sources supposing the propagation behaviour of the waves, that is $\mathbf{G}(t, \mathbf{r}, \mathbf{r}')$, is known.

A typical UWB sensing task covers microwave imaging which assigns to case 2 in the sense to relocate (direct or secondary) radiation sources. Hence – following to above remarks – imaging requires the knowledge of the propagation conditions which is usually not or only approximately the case in subsurface sensing. Furthermore, our typical objects of interest are not active microwave radiators so that we have to illuminate them by external sources in order to appear as 'secondary radiation sources' (i.e. scatterers). In other words, we have to perform a radar measurement which counts to case 1. Consequently, UWB short-range imaging can be considered as a mixture of the two basic inversion problems depicted by case 1 and 2.

We will summarize some aspects which will impact the solution of near-field imaging problems due to constraints imposed by practical conditions of a measurement.

Number and locations of measurement: For imaging purposes, one intends to de-convolve the source structure given by $\mathbf{p}_S(t, \mathbf{r})$ out of the measurable field[12] $\mathbf{\Pi}(t, \mathbf{r}')$ independently of the source type (active or secondary). In order to be able to perform a corresponding inversion, the space–time convolution (5.143) pretends the need of known field quantities $\mathbf{\Pi}(t, \mathbf{r}')$ within the whole space. This would require an infinite number of measurements. Fortunately, the vector diffraction theory tells us that we only need to measure the tangential components of the field (i.e. $\mathbf{n} \times (\mathbf{n} \times \mathbf{\Pi}(t, \mathbf{r}_S))$; $\mathbf{r}_S \in S$) at a surface S enclosing the volume of interest V_{obs} which covers all sources of radiation (see Figure 5.31). From the knowledge of the fields at S, we can calculate the fields at any position $\mathbf{r}'$ outside

12) We keep the Hertz vector as example for the measured quantities here, even if under real conditions it will be mostly the electric field what is measured.

of V_{obs} (the corresponding relations are given in Section 17.10 of Ref. [12]). Hence, we will not gather new information by measuring at more positions than at the surface. The access to the inner part of the observation volume is usually denied by the type of test objects (e.g. rigid bodies) or the circumstances of the measurement (e.g. remote sensing).

Practically it will most often not be possible to enclose the whole observation volume and the number of measurement positions will also be restricted. This will degrade the image resolution (see Section 4.10.3 for details). Furthermore, the measurements must not be performed simultaneously if the scenario behaves stationary. Measurement procedures dealing with sequential (by well synchronized) data capturing are assigned as synthetic aperture methods.

Near-field measurement: We are coming back to Figure 5.29 where two special objects (number 1 and 4) – namely antennas – are included. They are scattering objects as the other ones too but additionally they possess a lumped port for additional energy transfer with a waveguide. These bodies cover the volume V_1 and V_4, respectively. Object number 1 is fed by signal $a_1(t)$ which is at least partially converted into an electromagnetic wave whose source distribution is given by

$$\mathbf{p}_1(t,\mathbf{r}) = \widehat{\mathbf{T}}_1(t,\mathbf{r}) * a_1(t) \quad \mathbf{r} \in V_1 \tag{5.153}$$

Herein, $\widehat{\mathbf{T}}_1(t,\mathbf{r})$ is a vector IRF which describes how the electric feeding signal $a_1(t)$ is distributed within the antenna structure and how the antenna locally radiates at position $\mathbf{r}$. $\widehat{\mathbf{T}}_1(t,\mathbf{r})$ is fixed by the antenna geometry. The earlier introduced transmit IRF $\mathbf{T}(t,\mathbf{e}_0)$ is an integral version of $\widehat{\mathbf{T}}(t,\mathbf{r})$ under far-field conditions.

Object number 4 is able to collect wave energy and to bunch it to the signal $b_4(t)$ which can be registered by receiver electronics; that is, this body acts as receiving antenna. We use the vector IRF $\widehat{\mathbf{R}}(t,\mathbf{r})$ to specify the properties of the object structure to collect electromagnetic waves and to guide their energy to a common port. The receiving IRF $\mathbf{R}(t,\mathbf{e}_i)$ of an antenna under far-field conditions can be considered as an integral version of $\widehat{\mathbf{R}}(t,\mathbf{r})$. $\widehat{\mathbf{T}}(t,\mathbf{r})$ and $\widehat{\mathbf{R}}(t,\mathbf{r})$ of an antenna are linked via reciprocity.

For a given antenna structure, the bunching efficiency typically depends on the location where the wave is captured. If we know how an incident wave (e.g. the Hertz vector $\mathbf{\Pi}(t,\mathbf{r})$) distributes within the object 4 (receiving antenna), we can calculate the related output signal via

$$b_4(t) = \int\limits_{V_4} \widehat{\mathbf{R}}_4^T(t,\mathbf{r}) \cdot \mathbf{\Pi}(t-\tau,\mathbf{r})\mathrm{d}\tau\mathrm{d}V = \widehat{\mathbf{R}}_4^T(t,\mathbf{r})\bigstar\mathbf{\Pi}(t,\mathbf{r}); \quad \mathbf{r} \in V_4 \tag{5.154}$$

Supposing we know the Green's tensor of the scenario, (5.143), (5.153) and (5.154) may be joined which gives us the transmission from the lumped port of object 1 to the lumped port of object 4:

$$\begin{aligned} b_4(t) &= \int\limits_{V_4}\int\limits_{V_1}\int\limits_{\tau_3}\int\limits_{\tau_2}\int\limits_{\tau_1} \widehat{\mathbf{R}}_4^T(t-\tau_3,\mathbf{r}_4)\mathbf{G}(t-\tau_2,\mathbf{r}_4,\mathbf{r}_1)\widehat{\mathbf{T}}_1(t-\tau_1,\mathbf{r}_1)a_1(t)\mathrm{d}\tau_1\,\mathrm{d}\tau_2\,\mathrm{d}\tau_3\,\mathrm{d}V_4\,\mathrm{d}V_1 \\ &= \widehat{\mathbf{R}}_4^T(t,\mathbf{r}_4)\bigstar\mathbf{G}(t,\mathbf{r}_4,\mathbf{r}_1)\bigstar\widehat{\mathbf{T}}_1(t,\mathbf{r}_1) * a_1(t); \qquad \mathbf{r}_1 \in V_1, \mathbf{r}_4 \in V_4 \end{aligned} \tag{5.155}$$

This equation represents a transmission chain modelled by space–time convolution where no restrictions on distances[13] and polarization apply. It has involved a mode conversion from guided wave to free wave (expressed by $\widehat{\mathbf{T}}_1(t, \mathbf{r}')$), the spreading of the wave within the scenario (described by $\mathbf{G}(t, \mathbf{r}, \mathbf{r}')$) and the conversion back to a guided waves (modelled by $\widehat{\mathbf{R}}_4(t, \mathbf{r})$). It should be noted that the above-considered approach is not restricted to the Hertz vector $\mathbf{\Pi}(t, \mathbf{r})$; rather, it can also be applied to the **E**- or **H**-field.

We may rewrite (5.155) in the form

$$\begin{aligned} b_4(t) &= g(t) * a_1(t) \\ \text{with} \quad g(t) &= \widehat{\mathbf{R}}_4^T(t, \mathbf{r}_4) \bigstar \mathbf{G}(t, \mathbf{r}_4, \mathbf{r}_1) \bigstar \widehat{\mathbf{T}}_1(t, \mathbf{r}_1) \end{aligned} \tag{5.156}$$

Since this equation is often interpreted as filtering the signal $a_1(t)$ with a filter having the IRF $g(t)$, we can also say that (5.155) represents a filter equation in which the filter properties are implemented by the temporal and spatial behaviour of the considered scenario. The wave propagation described by Green's tensor $\mathbf{G}(t, \mathbf{r}, \mathbf{r}')$ involves the whole scenario. That is, internal wave propagations within the antennas and mutual antenna interactions are concerned too.

In the case of a measurement, the antennas are additional components of the pristine scenario under investigation. Hence, they will add new propagation paths if they are inserted, which will affect the original Green's tensor of the unperturbed scenario. Further, it is seen from (5.155) and (5.156) that the integration over the antenna volume V_1, V_4 will mask the spatial and temporal structure of the fields. Hence, we require that $\widehat{\mathbf{R}}_4(t, \mathbf{r}_4)$ and $\widehat{\mathbf{T}}_1(t, \mathbf{r}_1)$ are close to Dirac functions if we are interested to measure $\mathbf{G}(t, \mathbf{r}_4, \mathbf{r}_1)$ as complete as possible. The practical implications of this observation are that the sensing elements (i.e. the antennas) should be ideally

- as small as possible in order to avoid an integration effect due to the antenna size and
- they should be minimum scattering antennas in order to minimize their effect onto the Green's tensor of the test scenario.

Near-field Friis formula: Assuming free space propagation medium (i.e. the objects 2, 3 and 5 from Figure 5.29 are not present and the propagation within the volumes V_1, V_4 is comparable to free space[14]), we can approximate (5.155) for antennas of finite dimensions by

$$b_4(t) = \widehat{\mathbf{R}}_4^T(t, \mathbf{r}_4) \bigstar \mathbf{G}_0(t, |\mathbf{r}_4 - \mathbf{r}_1|) \bigstar \widehat{\mathbf{T}}_1(t, \mathbf{r}_1) * a_1(t) \tag{5.157}$$

This equation is the counterpart to the Friis formula(5.137) under far-field condition. Note that (5.157) respects wavefront curvature as well as the r^{-1}, r^{-2}, r^{-3} terms of the field equations while (5.137) assumes planar wavefronts and omits the higher order terms of the antenna distance $r = |\mathbf{r}_4 - \mathbf{r}_1|$. While the far-field

13) An exception may be seen for $\mathbf{r}_1 = \mathbf{r}_4$ and $V_1 = V_4$ (i.e. transmit and receiving antenna are identical) due to the singularity of the Green's functions.

14) This is a strongly simplifying assumption for real antennas which are typical metal structures. That assumption coincides with the requirement that antenna must not have structural reflections (see Section 4.6).

relations only respect tangential field components, the near-field description also takes account of the radial field components, which of course rapidly vanish if one departs from the antenna.

In addition to the already made simplifications, the relation (5.157) has to be taken with care. The antenna feeding by normalized waves implies that the antenna ports should be matched. As we have seen in Section 4.6, matched antennas reradiate partially the captured energy so that remarkable multi-path components will appear in case of short antenna distance. In order to avoid antenna reflections, the antennas have to be fed ideally with high-impedance source and load as already discussed in Section 4.6.6. The feeding port loading is however of minor interest, if the structural reflections dominate antenna backscattering.

Near-field radar equation: Including again the objects 2, 3 and 5 in our scenario and omitting the direct transmission, one can approximately separate the scenario into free space propagation (described by free space Green's functions) and 'secondary radiation sources' caused from the scatterer. For that purpose, we regard the scattering objects as a 'cloud of point' scatterer with arbitrary reflectivity and arbitrarily distributed within the scatterer volume $V_j;\ \ j \in [2, 3, 5]$. We describe reflectivity (including polarization and time dependency) and spatial distribution of the scatterer cloud belonging to object j by the function[15] $\mathbf{\Lambda}^{(j)}(t, \mathbf{r}_j)$. This function is often referred to as object or contrast function where one usually omits the dependency from the interaction time t. It is comparable with the object function $\mathfrak{O}(\mathbf{r})$ introduced by (4.312). If we further assume that the wave propagation within the volume V_j does not (only marginally) deviate from free space propagation and the point scatterers do not mutually interact, we can write for the secondary source distribution according to (5.153)

$$\mathbf{p}^{(j)}(t, \mathbf{r}_j) = \mathbf{\Lambda}^{(j)}(t, \mathbf{r}_j) * \mathbf{G}_0(t, |\mathbf{r}_j - \mathbf{r}_1|) \underset{V_1}{\bigstar} \widehat{\mathbf{T}}_1(t, \mathbf{r}_1) * a_1(t)$$
$$\mathbf{r}_1 \in V_1; \quad \mathbf{r}_j \in V_j; \quad j \in [2, 3, 5] \tag{5.158}$$

Herein, $\bigstar_{V_j}$ means integration over the volume of object j. Since we can consider $\mathbf{p}^{(j)}(t, \mathbf{r}_j)$ as source of a new wave which will illuminate the receiving antenna, the captured signal of the radar channel yields

$$b_4(t) = \widehat{\mathbf{R}}_4^T(t, \mathbf{r}_4) \underset{V_4}{\bigstar} \mathbf{G}_0(t, |\mathbf{r}_4 - \mathbf{r}_j|) \underset{V_j}{\bigstar} \mathbf{\Lambda}^{(j)}(t, \mathbf{r}_j) * \mathbf{G}_0(t, |\mathbf{r}_j - \mathbf{r}_1|) \underset{V_1}{\bigstar} \widehat{\mathbf{T}}_1(t, \mathbf{r}_1) * a_1(t) \tag{5.159}$$

This relation represents the near-field radar equation under the condition of the Born approximation which is indirectly included in our above assumptions of free space propagation within the volumes V_j and interaction free point scatterers within the 'cloud'.

15) We omit here the angular dependency of the scattering by point objects. $\mathbf{\Lambda}^{(j)}(t, \mathbf{r}_j)$ is actually a dyadic tensor.

Finally, two versions of (5.159) should still be mentioned. For the first case, the object should have a distinct surface which causes only specular scattering. Under these conditions, we can replace integration over V_j by integration over the surface $\mathbf{r}_j \in S_j$ and the time convolution may be simplified to a multiplication, that is $\mathbf{\Lambda}^{(j)}(t, \mathbf{r}_j) \Rightarrow \mathbf{\Lambda}^{(j)}(\mathbf{r}_j); \mathbf{r}_j \in S_j$.

$$b_4(t) = \widehat{\mathbf{R}}_4^T(t, \mathbf{r}_4) \underset{V_4}{\bigstar} \mathbf{G}_0(t, |\mathbf{r}_4 - \mathbf{r}_j|) \underset{r_j \in S_j}{\bigstar} \mathbf{\Lambda}^{(j)}(\mathbf{r}_j) \mathbf{G}_0(t, |\mathbf{r}_j - \mathbf{r}_1|) \underset{V_1}{\bigstar} \widehat{\mathbf{T}}_1(t, \mathbf{r}_1) * a_1(t) \tag{5.160}$$

The second case relates to scatterers with strong internal reflections. Now, the model of the 'scatterer cloud' and Born approximation cannot be applied and we have to respect all point-to-point propagations within the object. This would add another space–time convolution to (5.159)

$$b_4(t) = \widehat{\mathbf{R}}_4^T(t, \mathbf{r}_4) \underset{V_4}{\bigstar} \mathbf{G}_0(t, |\mathbf{r}_4 - \mathbf{r}_j|) \underset{\mathbf{r}_j \in V_j}{\bigstar} \mathbf{\Lambda}^{(j)}(t, \mathbf{r}_j, \mathbf{r}_i) \underset{\mathbf{r}_i \in V_j}{\bigstar} \mathbf{G}_0(t, |\mathbf{r}_i - \mathbf{r}_1|) \underset{V_1}{\bigstar} \widehat{\mathbf{T}}_1(t, \mathbf{r}_1) * a_1(t) \tag{5.161}$$

The extra space–time convolution complicates considerably the determination of the scattering behaviour of the object by inversion of (5.161). Also here it should be underlined that (5.159) and (5.161) do not include multiple reflections neither between the targets nor between targets and antennas.

In order to come to conclusions of above consideration with respect to UWB imaging, we consider again Figure 5.31. We are aimed to recover the area of interest by radar measurements as modelled by (5.159) or (5.161). For that purpose, we performed at least partially over the surface S a number of measurement which provides us with a set of measurement signals $b_4(t, \mathbf{r}_4, \mathbf{r}_1, \mathbf{q}_R, \mathbf{q}_T)$ captured from different positions $\mathbf{r}_4 \in S$ and polarization $\mathbf{q}_R \in [\mathbf{u}, \mathbf{v}]$ of the receiving antenna and from different positions $\mathbf{r}_1 \in S$ and polarization $\mathbf{q}_T \in [\mathbf{u}, \mathbf{v}]$ of the transmitting antenna. Based on these signals one is aimed at the recognition of the scattering scene which is described by the scattering tensor IRF $\mathbf{\Lambda}(\cdots)$. The simplest linkage between the scattering IRF $\mathbf{\Lambda}(\cdots)$ (object function) and the scattering scene is given under the assumptions related to (5.160) since there the object contour is immediately part of the function $\mathbf{\Lambda}(\mathbf{r}); \mathbf{r} \in S_{obj}$. In order to gain scattering scenario from the measurements, one has to de-convolve $\mathbf{\Lambda}(\mathbf{r})$ out of (5.160). This procedure would be largely simplified if one could renounce the two space–time convolution over V_1 and V_4; that is, one should select antennas of electrically small[16] size. The imaging procedures discussed in Section 4.10 represent such inversion routines, however, without respecting polarization and the r^{-2} and r^{-3} dependency of the field components.

16) Sometimes one distinguishes between electric and geometric size of antenna. Geometric small antennas have small dimensions regardless of the permittivity of the surrounding medium. Electrical small antennas are small compared to the signal extension. Hence, this characterization involves signal bandwidth and permittivity of the surrounding medium.

A further option to recover the contrast function $\mathbf{\Lambda}(\cdots)$ is given by model fitting. That is, starting from an initial guess of the scenario structure the expected receive signals $\hat{b}_4(t, \mathbf{r}_4, \mathbf{r}_1, \mathbf{q}_R, \mathbf{q}_T, \mathbf{\Lambda}(\mathbf{\Theta}))$ at the measurement positions are calculated via forward modelling. Depending on the scenario, this may be based on (5.159)–(5.161). The contrast function may be described by a set of parameters $\mathbf{\Theta}$ indicating electric or geometric properties. If the model fits the reality, calculated signals $\hat{b}_4(\cdots)$ and measured signals $b_4(\cdots)$ would coincide. Since this is generally not the case for an initial guess, one has to take the model to the reality by matching the model parameters $\mathbf{\Theta}$ which is often done via minimizing a L_2-norm cost function:

$$\mathbf{\Theta} = \arg\min_{\mathbf{\Theta}} \left\| \left(\mathbf{b} - \hat{\mathbf{b}}\right)^T \left(\mathbf{b} - \hat{\mathbf{b}}\right) \right\|_2 \qquad (5.162)$$

Herein, $\mathbf{b}$ and $\hat{\mathbf{b}}$ represent column vectors joining the measured and simulated receiving signals at all antenna positions and for all polarizations. The reader can refer to Refs [49–53] for deeper discussions on modelling and computational problems of iterative image reconstruction and resolution issues.

References

1 Balanis, C. (2005) *Antenna Theory – Analysis and Design*, John Wiley & Sons, Inc., Hoboken, NJ.

2 Chew, W.C. (1995) *Waves and Fields in inhomogeneous Media*, IEEE Press, New York.

3 Tsang, L., Kong, J.A. and Ding, K.-H. (2000) *Scattering of Electromagnetic Waves: Theories and Applications*, John Wiley & Sons, Inc., New York.

4 Tsang, L. and Kong, J.A. (2001) *Scattering of Electromagnetic Waves: Advanced Topics*, John Wiley & Sons, Inc., New York.

5 Kong, J.A. (1990) *Electromagnetic Wave Theory*, John Wiley & Sons, Inc., New York.

6 Ishimaru, A. (1991) *Electromagnetic Wave Propagation, Radiation, and Scattering*, Prentice-Hall, Inc., Englewood Cliffs, NJ.

7 Hansen, T.B. and Yaghjian, A.D. (1999) *Plane-Wave Theory of Time-Domain Fields: Near-field Scanning Applications*, IEEE Press, New York.

8 Lo, Y.T. and Lee, S.W. (1988) *Antenna Handbook: Theory; Applications, and Design*, Van Nostrand Reinhold Company, New York.

9 Tai, C.-T. (1994) *Dyadic Green Functions in Electromagnetic Theory*, IEEE Press, Piscataway, NJ.

10 Baum, C.E. and Kritikos, H.N. (1995) *Electromagnetic Symmetry*, Taylor Francis, Washington, London.

11 Bowman, J.J., Senior, T.B.A. and Uslenghi, P.L.E. (1987) *Electromagnetic and Acoustic Scattering by simple Shapes*, Hemisphere Publishing Corporation, New York.

12 Orfanidis, S.J. (2008) Electromagnetic waves and antennas, www.ece.rutgers.edu/~orfanidi/ewa.

13 Welch, W. (1960) Reciprocity theorems for electromagnetic fields whose time dependence is arbitrary. *IRE Trans. Ant. Prop.*, **8** (1), 68–73.

14 Kaatze, U. and Feldman, Y. (2006) Broadband dielectric spectrometry of liquids and biosytems. *Meas. Sci. Technol.*, **17**, R17–R35.

15 Kupfer, K. (2005) *Electromagnetic Aquametry: Electromagnetic Wave Interaction with Water and Moist Substances*, Springer, Berlin.

16 Baker-Jarvis, J., Janzic, M.D. Riddle, B.F. et al. (2005) *Measuring the Permittivity and*

Permeability of Lossy Materials: Solids, Liquids, Metals, Building Materilas, and Negativ-Index Materials, NIST.

17 Kaatze, U. (2005) *Electromagnetic Wave Interactions with Water and Aqueous Solutions, Electromagnetic Aquametry* (ed. K. Kupfer), Springer-Verlag, Berlin, Heidelberg.

18 Banachowicz, E. and Danielewicz-Ferchmin, I. (2006) Static permittivity of water in electric field higher than 10^8 V/m and pressure varying from 0.1 to 600 MPa at room temperature. *Phys. Chem. Liq.*, **44** (1), 95–105.

19 Pliquett, U. (2008) Electricity and biology. 11th International Biennial Baltic Electronics Conference, 2008. BEC 2008, pp. 11–20.

20 Pliquett, U.F. and Schoenbach, K.H. (2009) Changes in electrical impedance of biological matter due to the application of ultrashort high voltage pulses. *IEEE Trans. Contr. Syst. Technol.*, **16** (5), 1273–1279.

21 Sihvola, A.H. (1999) *Ari Sihvola – Electromagnetic Mixing Formulas and Applications*, The Institution of Electrical Engineers.

22 Nowak, K., Gross, W. Nicksch, K. *et al.* (2011) Intraoperative lung edema monitoring by microwave reflectometry. *Interact. Cardiovasc. Thorac. Surg.*, **12** (4), 540–544.

23 Ulaby, F.T. and Elachi, C. (1990) *Radar Polarimetry for Geoscience Applications*, Artech House, Inc., Norwood, MA.

24 Harmuth, H.F. (1984) *Antennas and Waveguides for Nonsinusoidal Waves*, Academic Press, Inc., Orlando.

25 van der Kruk, J., Wapenaar, C.P.A. Fokkema, J.T. *et al.* (2003) Improved three-dimensional image reconstruction technique for multi-component ground penetrating radar data. *Subsurface Sensing Technol. Appl.*, **4** (1), 61–99.

26 Streich, R. and Kruk, J.V.D. (2007) Accurate imaging of multicomponent GPR data based on exact radiation patterns. *IEEE Trans. Geosci. Remote Sensing*, **45** (1), 93–103.

27 Slob, E.C. and Fokkema, J.T. (2002) Interfacial dipoles and radiated energy. *Subsurface Sensing Technol. Appl.*, **3** (4), 347–367.

28 Banos, A. and Wesley, J.P. (1953) *The Horizontal Electric Dipole in a Conducting Half-Space*, University of California, Marine Physical Laboratory of the Scripps Institution of Oceanography, La Jolla, [1953–1954].

29 Smith, G. (1984) Directive properties of antennas for transmission into a material half-space. *IRE Trans. Ant. Prop.*, **32** (3), 232–246.

30 Engheta, N., Papas, C.H. and Elchi, C. (1982) Radiation patterns of interfacial dipole antennas. *Radio Sci.*, **17** (6), 1557–1566.

31 Rappaport, C. (2004) A simple approximation of transmitted wavefront shape from point sources above lossy half spaces. Proceedings of the 2004 IEEE International Geoscience and Remote Sensing Symposium, 2004. IGARSS '04, pp. 421–424.

32 Herrmann, R., Sachs, J. Schilling, K. *et al.* (2008) 12-GHz bandwidth M-sequence radar for crack detection and high resolution imaging. International Conference on Ground Penetrating Radar (GPR), Birmingham, UK.

33 Lee, J.-S. and Pottier, E. (2009) *Polarimetric Radar Imaging from Basics to Applications*, CRC Press, Boca Raton, London, New York.

34 Cloude, S.R. (2009) *Polarisation: Applications in Remote Sensing*, Oxford University Press, Oxford.

35 Cloude, S.R. and Pottier, E. (1996) A review of target decomposition theorems in radar polarimetry. *IEEE Trans. Geosci. Remote Sensing*, **34** (2), 498–518.

36 POLSARPRO, The Polarimetric SAR Data Processing and Educational Tool. Online tutorial, documentation and data base for polarimetry and polarimetric interferometry. http://earth.eo.esa.int/polsarpro/.

37 Chi-Chih, C., Higgins, M.B. O'Neill, K. *et al.* (2000) UWB full-polarimetric horn-fed bow-tie GPR antenna for buried unexploded ordnance (UXO) discrimination. IEEE 2000 International Geoscience and Remote Sensing Symposium, vol. 4, 2000. Proceedings. IGARSS 2000, pp. 1430–1432.

38 Chi-Chih, C., Higgins, M.B. O'Neill, K. *et al.* (2001) Ultrawide-bandwidth fully-polarimetric ground penetrating radar classification of subsurface unexploded ordnance. *IEEE Trans. Geosci. Remote Sensing*, **39** (6), 1221–1230.

39 Hansen, P., Scheff, K. and Mokole, E. (1998) Dual polarized, UWB radar measurements of the sea at 9 GHz. *Ultra-Wideband Short-Pulse Electromagnet.*, 4, 335–348.

40 Hellmann, M. and Cloude, S.R. (2001) Discrimination between low metal content mine and non-mine-targets using polarimetric ultra-wide-band radar. IEEE 2001 International Geoscience and Remote Sensing Symposium, vol. 3, 2001. IGARSS '01, pp. 1113–1115.

41 Baum, C.E. (2006) Combining polarimetry with SEM in radar backscattering for target identification. The Third International Conference on Ultrawideband and Ultrashort Impulse Signals, pp. 11–14.

42 Ibarra, C., Kegege, O. Junfei, L. *et al.* (2007) Radar polarimetry for target discrimination. Region 5 Technical Conference, 2007 IEEE, pp. 86–92.

43 Narayanan, R.M. and Kumru, C. (2005) Implementation of fully polarimetric random noise radar. *IEEE Ant. Wireless Prop. Lett.*, **4**, 125–128.

44 Herrmann, R. (2011) M-sequence based ultra-wideband radar and its application to crack detection in salt mines. Faculty of Electrical Engineering and Information Technology, Ilmenau University of Technology (Germany), Ilmenau.

45 Bonitz, F., Eidner, M. Sachs, J. *et al.* (2008) UWB-radar sewer tube crawler. 12th International Conference on Ground Penetrating Radar Birmingham, UK.

46 (2005) FCC, 05-08, Petition for Waiver of the Part 15 UWB Regulations Filed by the Multi-band OFDM Alliance Special Interest Group.

47 Sachs, J., Peyerl, P. and Alli, G. (2004) Ultra-wideband polarimetric GPR-array stimulated by pseudo random binary codes. Proceedings of the Tenth International Conference on Ground Penetrating Radar, 2004. GPR 2004, pp. 163–166.

48 van der Kruk, J., Wapenaar, C.P.A. Fokkema, J.T. *et al.* (2003) Three-dimensional imaging of multicomponent ground-penetrating radar data. *Geophysics*, **68** (4), 1241–1254.

49 Chew, W.C. (1998) Imaging and inverse problems in electromagnetics, in *Advances in Computational Electrodynamics: The Finite-Difference Time-Domain Method* (ed. A. Taflove), Artech House, Boston, London.

50 Chew, W.C., Wang, Y.M. Otto, G. *et al.* (1994) On the inverse source method of solving inverse scattering problems. *Inverse Probl.*, **10**, 547–553.

51 Tie Jun, C., Weng Cho, C. Xiao Xing, Y. *et al.* (2004) Study of resolution and super resolution in electromagnetic imaging for half-space problems. *IRE Trans. Ant. Prop.*, **52** (6), 1398–1411.

52 Zhong Qing, Z. and Qing Huo, L. (2004) Three-dimensional nonlinear image reconstruction for microwave biomedical imaging. *IEEE Trans. Bio.-Med. Eng.*, **51** (3), 544–548.

53 Persico, R., Bernini, R. and Soldovieri, F. (2005) The role of the measurement configuration in inverse scattering from buried objects under the Born approximation. *IRE Trans. Ant. Prop.*, **53** (6), 1875–1887.

6
Examples and Applications

6.1
Ultra-Wideband Sensing – The Road to New Radar and Sensor Applications

Research on ultra-wideband (UWB) topics started about 50–60 years ago driven by needs in radar and communication technology as well as to emulate EMPs of nuclear explosions. Meanwhile, ultra-wideband sensing opens new perspectives for radar applications as well as impedance spectroscopy and tomography. Small and cost-effective devices, high resolution and sensitivity, as well as low exposition by radio waves will give this radar technology access to new applications in industry, non-destructive testing (NDT), medical engineering and health care, surveillance, search and rescue and many others more.

The classical radar approaches usually employ RF-sounding signals of comparably low fractional bandwidth. These waves have a short wavelength but their envelope is spread over a large distance. Since typical dimensions of single targets are much smaller than the spatial extension of the signal envelope (more precisely: the envelope of their auto-correlation function), the captured radar signal represents a mixture of signal components from the individual scattering centres of the target.

The situation changes for ultra-wideband sensing. Here, the spatial extension of the sounding signal is in the same order as the objects to be observed. Even if narrowband and ultra-wideband radar are working on the same physical basis, they partially require different thinking on how to interpret scattering effects. Furthermore, due to the large bandwidth occupation of UWB sensors much lower power emission than in the narrowband case is allowed. Consequently, in exploring new fields of radar applications, one should mainly look for new device conceptions using weak sounding signals for short-range sensing tasks which profit from centimetre- and millimetre-resolutions as well as penetration of opaque objects. On overview on recent sensor principles was given in Chapter 3.

6.1.1
Potential of Ultra-Wideband Sensing – A Short Summary

The UWB term actually implies two aspects – a large fractional bandwidth b and a huge absolute bandwidth B. If UWB devices are used on large scale, they have to be

Handbook of Ultra-Wideband Short-Range Sensing: Theory, Sensors, Applications, First Edition. Jürgen Sachs.

bound to low emission levels in order to avoid interference with other communication systems [1]. Therefore, high-power medium and long-range radar systems will always be reserved for special (usually military) use. This chapter discusses high-resolution short-range devices and applications which deal with low radiation power (typical power <1 mW) and may be of interest for a wider audience.

Let us summarize the role of large fractional bandwidth. First, we restrict the discussion to short pulses having the duration (full width at half maximum – FWHM) of t_w and are composed from N oscillations. We can roughly assign a (centre) frequency f_m to these oscillations with $N \approx t_w f_m$. Furthermore, since bandwidth and pulse duration are related by $Bt_w \approx 1$, we yield $Nb \approx 1$ using $B = f_l - f_u$ and $f_m = (f_l + f_u)/2$. That is, wideband signals are composed from only few oscillations while narrowband signals have many of them.

We will review the consequences on the basis of the time domain radar equation. As long as the targets do not move too fast, all objects included in the radar channel (antennas and scatterers) may be considered as LTI systems. Hence, the electrodynamics of transmission, receiving and scattering can be formally described by impulse response functions (IRFs) as introduced in Chapter 2. For the sake of a short summary, we will not respect here polarization, dispersion, angular and range dependencies (refer to Chapters 4 and 5 for these issues). If we consider the bi-static radar channel as a two-port system (and correspondingly a mono-static radar as a one port), we can write for the transmission between both antennas via a number of M scatterers:

$$b_2(t) = S_{21}(t) * a_1(t) \Rightarrow b_2(t) \sim S_{21}(t) \quad \text{for } a_1(t) \approx \delta(t)$$

$$\text{with} \quad S_{21}(t) = T_1(t) * R_2(t) * \sum_{i=1}^{M} \frac{\Lambda_{21}^{(i)}\left(t - \left(r_1^{(i)} + r_2^{(i)}\right)/c\right)}{r_1^{(i)} r_2^{(i)}} \tag{6.1}$$

Herein, a_1, b_2 are the stimulus and receive signals; c is the speed of light, r_1, r_2 are the antenna to target distances, T_1, R_2 are the antenna impulse responses for transmit and receiving mode and Λ_{21} is the scatterer pulse response for incidence from antenna 1 and observation by antenna 2. Equation (6.1) assumes targets of limited size in the far field and it disregards direct antenna coupling. The responses $\Lambda_{21}^{(i)}$ of the ith target can be interpreted either as the reaction of a single body onto an incident field or the reaction of distinct scattering centres of a composed target. The scatterer responses $\Lambda_{21}^{(i)}$ contain all information about the target accessible by radar measurement. We will emphasize several features that are promoted by the large bandwidth of UWB sensors:

Target identification [2–4], Section 4.8.4: We consider a single body of complex structure. Its total scattering response $\Lambda_{21}(t)$ typically comprises a number of peaks (caused, e.g. by specular reflections) and damped oscillations (representing the eigenmodes of the target). In order to resolve these properties, the temporal width of the sounding wave must be shorter than the distances $\Delta\tau_j$ of the peaks on Λ_{21} and the sounding bandwidth should cover the eigenfrequencies

f_k of the target. $\Delta\tau_j$ and f_k are distinctive signal parameters since they relate to characteristic dimensions $D_{j,k}$ of the target, that is $c\,\Delta\tau_j \approx D_j$ and $c/2\pi f_k \approx D_k$. Hence, to achieve both – separation of specular reflections and a mix of natural frequencies – large fractional bandwidth is needed. The demands on absolute bandwidth result from the smallest dimensions to be resolved.

Detection of hidden targets and investigation of opaque structures [5, 6], Sections 4.7–4.10: On the one hand, microwave penetration in most of the substances or randomly distributed bodies (e.g. foliage, soil) is restricted to low frequencies but on the other hand, reasonable range resolution requires bandwidth. To bring both aspects together, the fractional bandwidth must be large. The absolute bandwidth is typically limited by the properties of the propagation medium.

Separation of stationary targets [7], Section 4.7.2: Scattered waves of two targets located nearby will overlap. Some knowledge about one of the targets supposed, they may be separated as long as the signals do not constitute too many oscillations. Otherwise a periodic ambiguity of the separated signals will arise. CLEAN, SAGE and MATCHING PURSUIT are numerical techniques using such approaches.

Small target detection and localization, Sections 4.8 and 4.10: Backscattering from thin layers or cracks is proportional to the first temporal derivative of the sounding wave; small volume scatterers (Rayleigh scattering) cause a second derivation. That is, high frequencies promote the detection of small defects and a large bandwidth leads to their localization.

Moving target detection, Section 4.8: Moving targets within a stationary clutter environment can be detected by weak variations in the backscattered signals. Again, large fractional and absolute bandwidths are beneficial for penetration of opaque objects, the registration of weak movements and target separation.

So far, short pulse-like signals $a(t)$ have been assumed in (6.1) for stimulation of the test scenario in order to determine its impulse response function $S_{21}(t)$. However, there are more options if we rewrite (6.1) in the forms (see Chapter 2)

$$C_{ba}(t) = S_{21}(t) * C_{aa}(t) \quad \text{with } C_{ba}(t) = b(t) * a(-t) \tag{6.2}$$

$$\text{IFT}\{\underline{\Phi}_{ba}(f) = \underline{S}_{21}(f)\Phi_{aa}(f)\} \quad \text{with } \underline{\Phi}_{ba}(f) = \text{FT}\{C_{ba}(t)\} \tag{6.3}$$

C_{aa}, C_{ba} are the auto- and cross-correlation functions of the signals and $\Phi_{aa}(f)$, $\underline{\Phi}_{ba}(f)$ are their auto- and cross-spectrums. FT{ }, IFT{ } refer to the direct or inverse Fourier transform. These two relations (6.2) and (6.3) open up several measurement principles beside the classical impulse excitation.

For UWB, $C_{aa}(t)$ should be a short pulse-like function which is met by a wideband signal of any time shape. In narrowband cases, the receiving signal $b_2(t)$ keeps the shape of $a_1(t)$ so that the required cross-correlation on the receiver side can be done via matched filtering providing optimal conditions for target detection. In UWB sensing, optimum detection or matched filtering in the classical sense can only be done if T_1, R_2, Λ_{21} are known.

6.1.2
Overview on Sensor Principles

The primary goal of an ultra-wideband sensor is to determine a single impulse response function $S_{21}(t)$ from a test scenario or a whole set $S_{ij}(t)$ of them. They represent the Green's function of the scenario (not to be confused with the free space Green's function) comprising all accessible information about the targets (see Sections 2.4 and 5.7). While an application-specific data processing has to extract the wanted information from the responses, the sensor electronics is responsible to capture them. As indicated by (6.1)–(6.3), we have several options to determine an impulse response function (see also Chapter 3).

Direct approach: The natural and so far mostly applied method to measure pulse responses is based on stimulations by sub-nanosecond pulses. The data acquisition applies stroboscopic sampling to be able to handle the large bandwidth. Impulse systems are implemented in many fashions extending from simple low-cost models to sophisticated systems with bandwidths up to 100 GHz [8]. Figure 6.1 depicts one of the first compact pulse sensing modules.

Sine wave measurement [9,10]: The transmitted sine wave is continuously (FMCW-radar) or stepwise (stepped frequency radar; network analyser) swept over the frequency band of interest. A heterodyne or homodyne receiver converts the captured signals down to low frequencies where they are digitized and subjected to the inverse Fourier transform.

Arbitrary wideband signals [11–13]: If we exploit (6.2) for the determination of the impulse response, we are free in the selection of the sounding signal. In theory, the only requirement relates to a sufficiently large bandwidth such that the auto-correlation function of the sounding signal is short compared to the target's impulse response. Apart from bandwidth, the signal can be freely selected from other points of view such as low probability of intercept, technical feasibility,

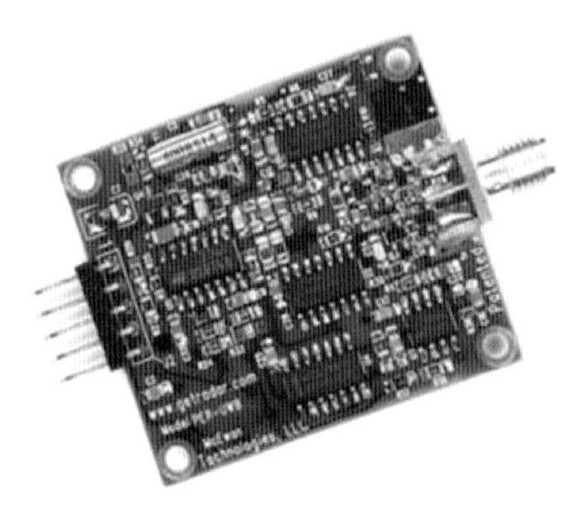

Figure 6.1 The MIR module, one of the first low-power devices for pulse measurements. (source: http://bul.ece.ubc.ca/Implementation%20Challenges%20of%20UWB%20Systems_notes1.pdf).

Figure 6.2 Integrated PN-radar (*Courtesy*: M. Kmec, Ilmenau University of Technology and Ilmsens).

achievable parameters and so on. The price to be paid for that flexibility is the need for a wideband correlation which is done either in an analogue way (e.g. by a sliding correlation) or after digitizing (for details see Sections 3.3. and 3.6). Figure 6.2 depicts an integrated RF-head of a PN-radar which is able to perform such a wideband correlation.

On a purely theoretical basis (omitting device imperfections), there is no quantitative or qualitative difference in the performance of the mentioned approaches. However, depending on the intended application, the technical implementation may lead to preferences for one over the others. Aspects such as costs, power consumption, form factor, bandwidth, noise and jitter, non-linear distortions, measurement speed and many others may be of major interest in this regard.

6.1.3 Application of Ultra-Wideband Sensing

As long as the sensors should be accessible for a larger community, they must be restricted to low-power emissions which entitles them to short-range sensing, that is up to about 100 m (Eq. (4.166) give a rough estimation). A large bandwidth combined with high jitter immunity provides high-range resolution and accuracy down to the micrometre range permitting high-resolution radar images and recording of weak movements or other target variations. Furthermore, the interaction of electromagnetic waves with matter provides the opportunity of remote material characterization via permittivity measurements.

As examples, we can find applications in following arbitrary ordered areas:

- Geology, archaeology [5]
- Non-destructive testing [14]
- Metrology [15, 16]
- Microwave imaging [17]
- Quality control [18]
- Inspection of buildings [19]
- Medical engineering [20–24]

- Search and rescue [6, 25–27]
- Localization and positioning [28]
- Ranging, collision avoidance [29]
- Law enforcement, intrusion detection [6, 29]
- Ambient-assisted living and others [30, 31]

Due to the low emissions, ultra-wideband short-range sensing will become an interesting extension of the radar approach to daily life applications. For large-scale applications, the step from the laboratory into the real world will require further system integration as well as reduction of device costs and power consumption. The processing of the captured data may become a more or less challenging issue depending on the complexity of test scenario. We should remember electromagnetic sounding is an indirect measurement approach requiring to solve ill-posed and ill-conditioned inverse problems. This may lead to unwanted cross sensitivities and ambiguities which can only be reduced by prior knowledge about the test scenario and sufficient 'orthogonal' information gained from the measurement. Mainly with respect to the last point, ultra-wideband sounding is in advantage against narrowband sensing since the larger bandwidth allows to get a broader view of the behaviour of test objects and hence to detect individual specifics of the test objects than this would be possible with narrowband excitation.

The previous Chapters 3 and 4 have shown working principles, basic features and performance figures of UWB sensors. Now, the versatility of this sensing method shall be demonstrated. For that purpose, we illustrate some classes of typical UWB sensing applications from which some aspects will be deepened by the subsequent sub-chapter. We exclude from these examples ground-penetrating radar (GPR) since it is already well documented, for example [5, 32].

One class of sensing applications refers to the *characterization of material* composite and material properties. As mentioned earlier (Section 5.3), sensing tasks involving water are well adapted to wideband microwave sensing. Related problems we can find not only in construction and food industry for determination of moisture content and quality assurance but also in medical engineering and environmental topics. Section 6.3 will deal with some aspect of food quality assessment by microwave measurement and Section 6.6 discusses the remote detection of breast cancer. Figure 6.3 anticipates some aspects of this approach by showing a contact-based example of cancer detection [33]. Both methods exploit the particular radio-frequency (RF) properties of water which is enriched in cancerous tissue. In case of the contact-based measurement, the reflection coefficient of an open-ended coaxial line will provide the water content of the tissue if an appropriate permittivity model is applied. In the shown case, the test arrangement exploits the principle shown in Figure 3.96 (top) whereas the open-ended coaxial probe is acting as reference line. Basically, the arrangement corresponding to Figure 3.95 would also be applicable. The measurement device was a ninth-order M-sequence system with 9 GHz clock rate allowing online measurements of quickly variable water concentrations.

Figure 6.4 indicates two further water-related issues. One deals with flash flood threat due to broken river embankment and the other one refers to the antipode

In vitro measurement of a human lung during NaCl-perfusion

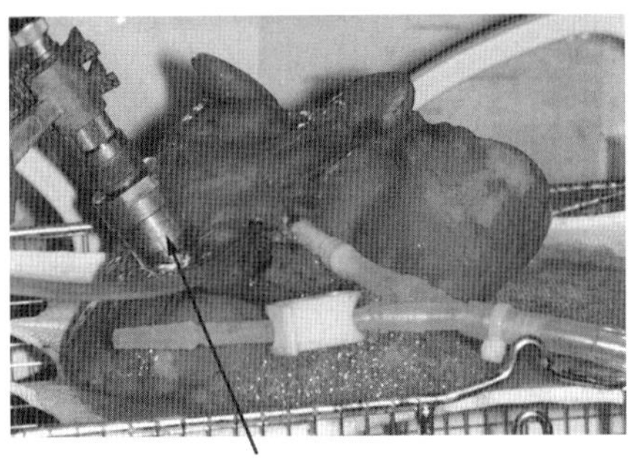

Open end coaxial probe

Variation of water content by approaching the cancer

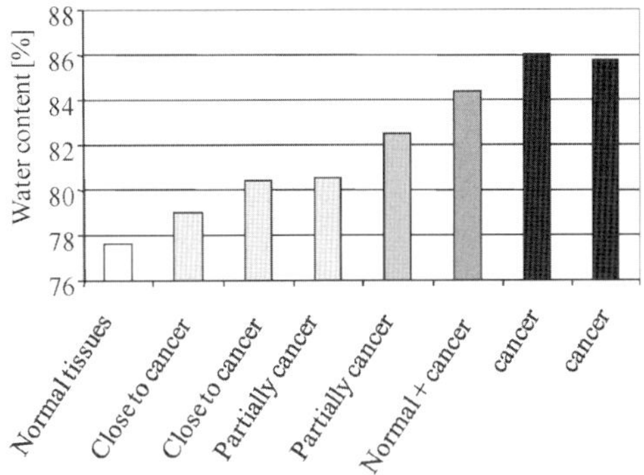

Figure 6.3 Water content determination of lung tissue via dielectric spectroscopy. If the probe approaches the cancerous area, the water content increases which is finally the reason of the measurable variations of the permittivity. (*Courtesy:* Universitätsklinikum Heidelberg, Experimentelle Chirurgie). (colored figure at Wiley homepage: Fig_6_3)

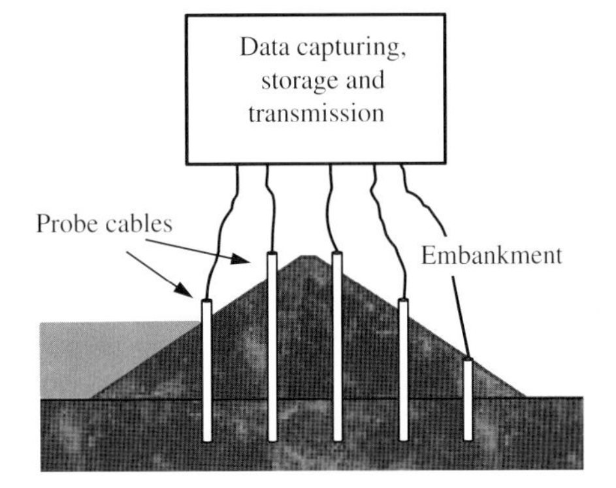

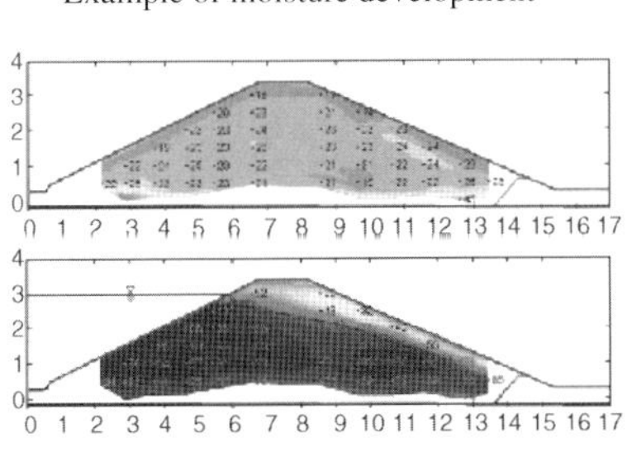

Typical sensor response for three layers of different moisture

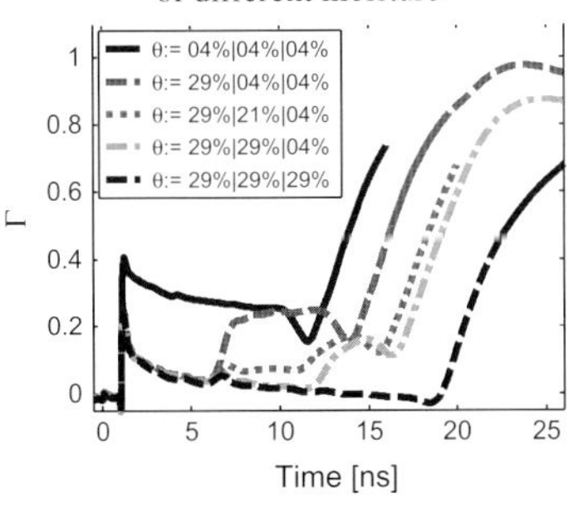

Figure 6.4 Survey of water penetration into a river embankment (*Courtesy* of MFPA Weimar). (colored figure at Wiley homepage: Fig_6_4)

namely the shortage of water resources. In case of a flood, water soaks the embankment increasing the risk to burst. In order to duly initiate countermeasures, one needs to know the development of the moisture profile. In the opposite case of water deficit, the knowledge of the actual soil moisture may of similar interest for the agriculture in order to organize reliable and efficient soil watering.

The idea is to determine it by microwave probe cables inserted into the embankment or soil as depicted in Figure 6.4 [34–38]. The basic measurement arrangement is shown in Figure 4.82 whereas the moisture distribution is considered as a layered media surround the probe cable. The moisture profile can be evaluated from the strength and the round-trip time of the reflections. The best results are gained if the probe cable may be fed from both sides.

A second class of sensing task exploits the *good penetration* of microwave in most non-metallic of object which allows to make visible obscured objects via microwave imaging. This sensor feature is already applied in ground-penetrating radar for years. It typically exploits sub-nanosecond pulses below 1 GHz of upper cut-off frequency for sounding the ground. Investigations under dry soil conditions or for shallow target detection are dealing often with signals of larger bandwidth.

Meanwhile, well-engineered devices for wall scanning are commercially available which may also be operated by less technically skilled people. The sensors are able to detect metallic and non-metallic construction elements and ducts so that their damage by drilling may be excluded.

More on these issues are to found in Section 6.6 which discusses some aspects of medical microwave imaging as well as in Section 6.4 dealing with non-destructive testing in civil engineering to visualize structural defects and foreign body inclusions. Figure 6.5 shall impart an impression about achievable resolutions of UWB microwave imaging.

A third class of sensing applications – often combined with the good signal penetration – is based on the high-*range resolution* and the sensitivity to *register motion* and *smallest target variations*. In this field, we find many different applications from which we will mention only a few. To demonstrate the sensor sensitivity to detect extremely weak motion of small targets, Figure 6.6 depicts an example of vermin detection namely Hylotrupes bajulus [38] – refer to Sections 4.7.3 and 4.8 for sensor performance analysis. Note the wood pests are only verifiably if they are moving since otherwise the reflections caused by the boundaries and inhomogeneities of the wooden beam would overwhelm the wave scattering due to the larva. Due to the high-range resolution of an UWB radar, one can limit the observation range to the actual volume of the wooden beam (region of interest – ROI) by gating out unwanted signal parts. Thus, any moving bodies outside the beam will not cause false alarms as in case of Doppler sensors since such devices do not have range sensitivity.

Body motions of a human being and movements of inner organs represent vital signs and vital activities which can be remotely captured by UWB sensor. This opens up versatile applications as intruder detection, care of helpless people, protection of dangerous areas, search and rescue, medical engineering, localization in

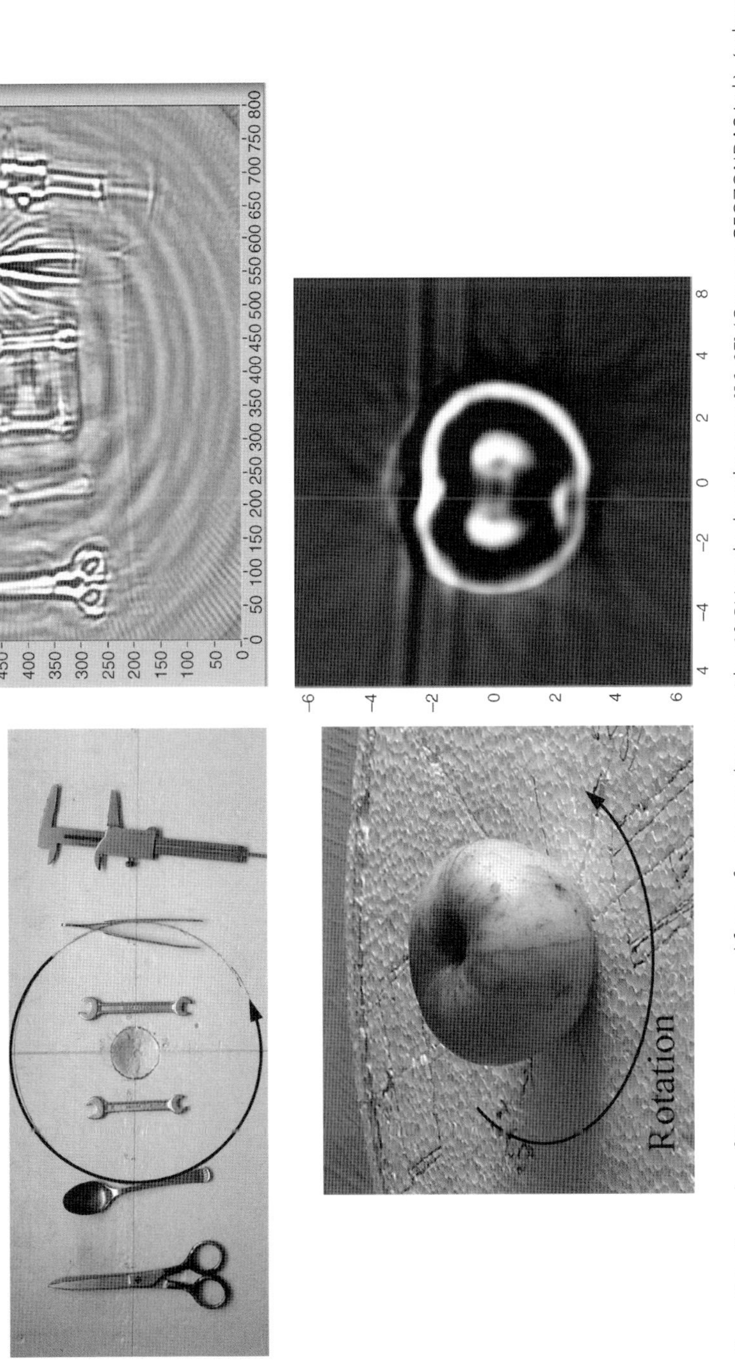

Figure 6.5 Examples of ISAR-images captured from a few meters distance with a 20 GHz pulse-based system [16, 27] (*Courtesy:* GEOZONDAS Ltd.). (colored figure at Wiley homepage: Fig_6_5)

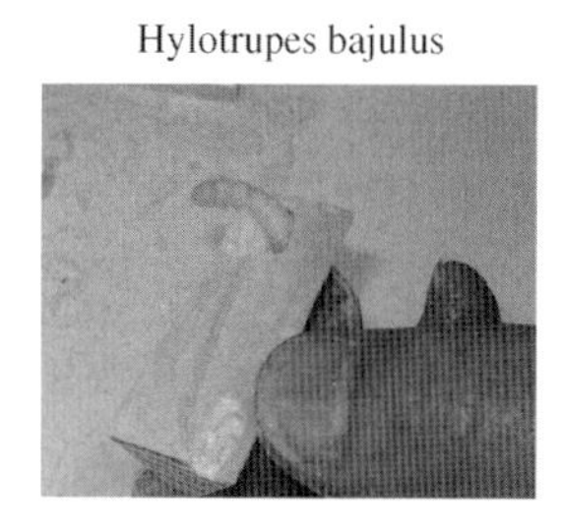

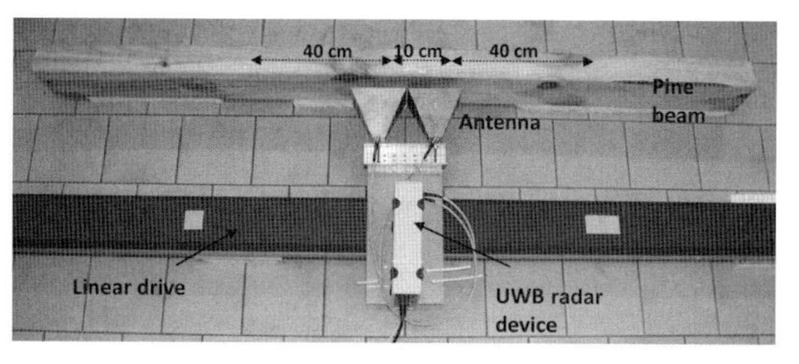

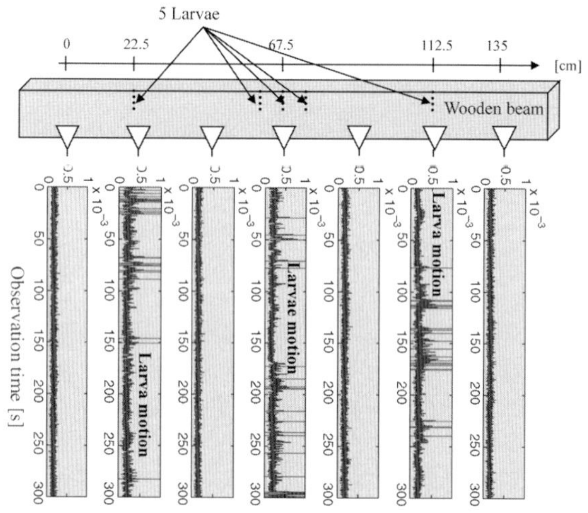

Figure 6.6 Vermin detect by registration of the feed motion of a larvae. Top left: Hylotrupes bajulus; top right: measurement arrangement. Bottom: Variations of backscatter signal within the region of interest (i.e. the beam). Vermin motion could be detected in the 2nd, 4th and 6th diagram. (colored figure at Wiley homepage: Fig6_6)

multi-path reach environment, hostage escape and many more. Figures 6.7 and 6.8 depict two illustrative examples. Respective issues are engrossed in Sections 6.5 and 6.7 which address the detection and tracking of organ motion for medical purposes and search and rescue, respectively.

Figure 6.7a demonstrates the observation of heart and respiration activity while the person is in rest. The radargram (Figure 6.7b) shows typical variations of the backscattered signals which are modulated by the deformation of the chest and inner organs. The diagram in Figure 6.7c depicts a detail of the radargram indicating the amplitude variations at the propagation time[1] of 10.9 ns. As expected, respiratory motion will cause larger backscatter modulations then the cardiac motion.

1) Time zero of the measurement data was not set to the antenna reference point. Hence, the absolute value of the propagation time cannot be taken to calculate the target distance.

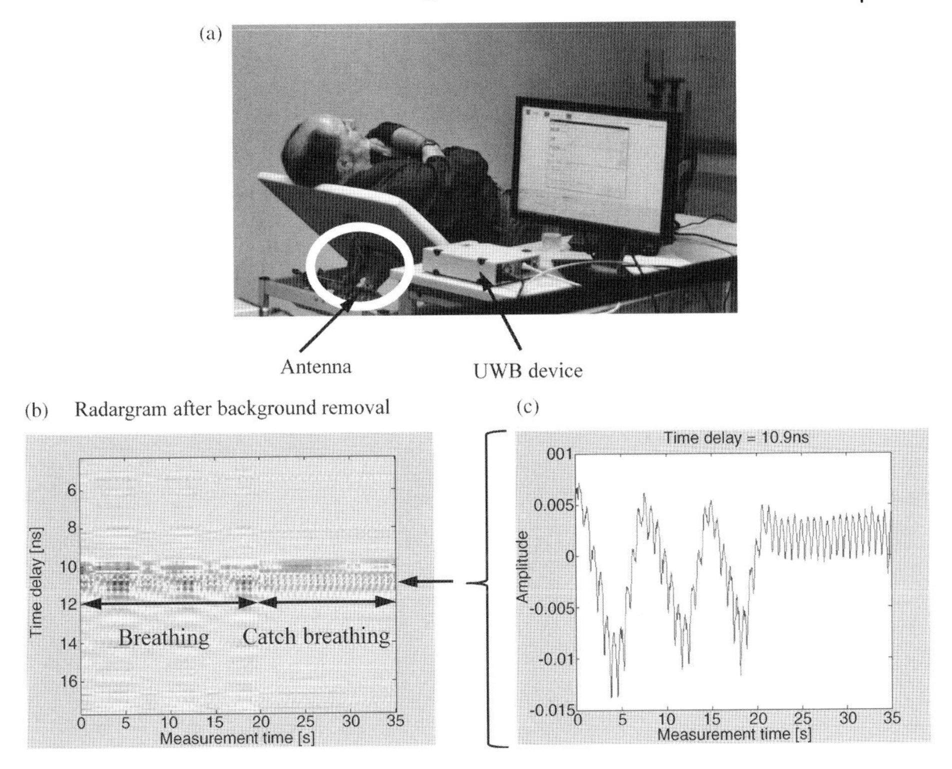

Figure 6.7 Remote vital sign monitoring of sick persons or of people in need of care (e.g. old detached people, dement people; ambient-assisted living). (colored figure at Wiley homepage: Fig_6_7)

Consequently, respiration motion tracking will be the mean of choice for the detection of hidden people trapped, for example by rubble due to an earthquake or snow of an avalanche or if they are handcuffed in case of hostage-taking and unconscious by any reason (e.g. smoke in case of fire). Figure 6.8 illustrates an example in which a motionless but alive person was detected across three levels of a building. The victim was placed at the floor of the basement and it was solely detected by its respiration motion.

Walking is a much stronger vital sign which can be detected over larger distances since the displacement of the whole body contributes to the variations of the scattering signal. Detection, tracking and qualifying the gait of walking people may be of interests not only for vital monitoring but also for other reasons like protection of restricted areas and critical infrastructure, security and emergency rescue operations (e.g. through wall radar as counter-terrorism and hostage tool), smart home and smart office purposes or home entertainment (see Sections 6.8 and 6.9 for details).

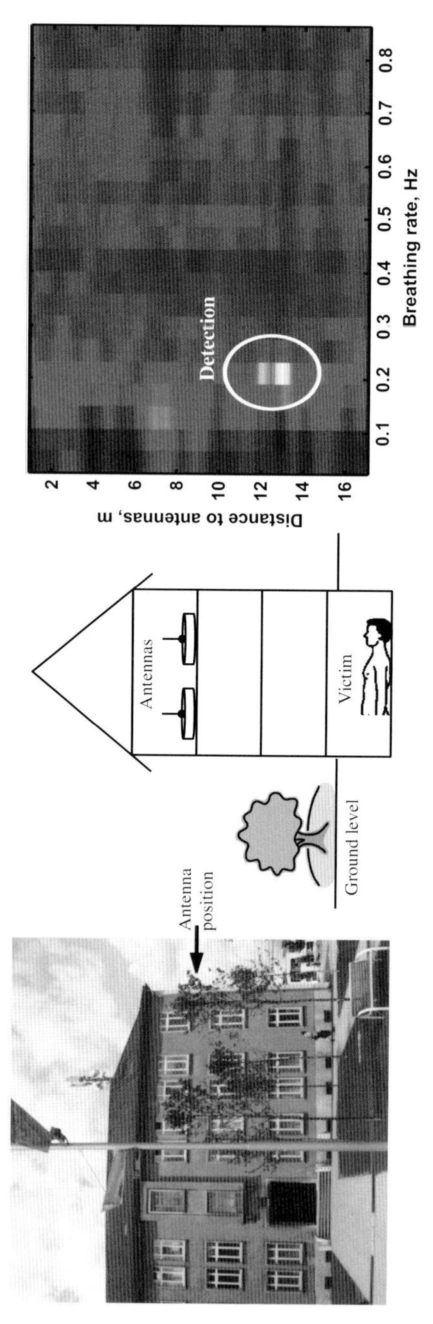

Figure 6.8 Detection of unconscious or trapped people within an empty building.

6.2 Monolithically Integration of M-Sequence-Based Sensor Head

Martin Kmec

6.2.1 Introduction

The steady growing areas of UWB application and thus potentially large consumer spotlight push the demand for new UWB sensor systems. The common requirements for these systems are high overall efficiency which correlates typically with low power consumption, flexibility in operational frequency band, architecture modularity, certain level of included intelligence, reliability, robustness and so on. On the contrary, new system should ideally share small form factor and reasonable (~ideally minimal) manufacturing costs. This trend and broad scale of various application-specific requirements result in novel UWB sensor approaches which claim usually for new high-end system architecture-specific electronic (see also Chapter 3).

For decades, miniaturization has been the key to faster performance of electronics. Therefore, ultra-high-speed monolithic integrated circuits (ICs) seem to be obligatory components for novel UWB system structures. Besides the speed performance, the ICs give a certain additional level of design freedom if compared with the classical approach, where the circuits are made using discrete, individual packaged components like resistors, capacitors, transistors and so on. During the chip design, the designer can choose single device shapes, dimensions as well as their fine placements and is able to determine their interconnections by well-defined grid structures. Thus, the mechanical dimensions of the individual IC components and related parasitic electrical effects (e.g. phase shift, attenuation) are negligible and must not be consistently considered during the design phase for the intended frequency ranges. So the chip designers have unique degree of freedom without which many state-of-the-art UWB circuits would not have been possible.

6.2.2 Technology and Design Issues

6.2.2.1 Sensor IC Technology Choice

The circuit performance efforts hand in hand to available IC engineering processes. So the realization of high-end chips is not possible without parallel efforts in material technologies, elementary circuit components and their proper modelling driven commonly by technology suppliers. Recently customer accessible chip manufacture technologies can be generally categorized into two main groups on the basis of substrate material, these are the group IV (also named 'silicon based') and group III–V (including GaAs, InP, GaN etc.) semiconductors. From engineering point of view, silicon can be cheaply grown on very large defect-free crystals yielding many low-cost ICs per wafer. The III–V group semiconductor substrates, on the other hand, are grown in smaller wafers with higher defect densities.

The locations of implementation field and performance boundaries between them are diffuse and change with time. It is apparent that the application areas depend besides technical parameters, such as frequency range, handled signal level, noise behaviour and so on, very much on costs and other non-technical aspects that are beyond the scope of this chapter.

Certainly, the technology performance is strongly affected by geometrical scaling, which defy most doom and gloom scenarios, so for example the state-of-the-art IC processes actually reach deep submicron horizontal range, for example 22 nm per ITRS in 2011 (ITRS – International Technology Roadmap for Semiconductors). Also optimizing of carrier transport properties through device materials destines the technology characteristics. So, historically considered, the III–V compound technologies have been used mainly for RF-applications due to their overall superior high-frequency performance coming from the excellent carrier mobility. Note that there have been reported III–V-based bipolar transistors (HBTs) with f_{max} (maximum oscillation frequency) of slightly over 1 THz, for example in Ref. [40]. On the other site, Si-based compound technologies, their both field-effect (e.g. CMOS) and bipolar sections, have been implemented primarily for applications up to only several GHz. Anyhow, the silicon technologies are actually making rapid inroads into high-frequency applications. Recently reported cut-off frequencies in Si-based bipolar transistors are at about 300 GHz [41–43] or up to 500 GHz under special laboratory cryogenic conditions [44] and slightly over 200 GHz for 70 nm field-effect structures – for example NMOS in Ref. [45]. Thus, the implementation frequency range from several tens of GHz up to about 100 GHz is the region in which the interplay and competition amongst IV and III–V group-based semiconductor technologies occurs. Anyway, this frequency region change with time and it is expected to move to higher allocations.

The choice of a particular semiconductor technology for a specific application must be beside frequency range of operation also discussed in many other performance points, such as predominant working domain – that is digital, mixed signal or purely analogue, as well as power- and noise-related topics, or eventually economical issues. For example, digital logic circuits require the use of a large number of switching devices (usually n-type and p-type transistors). IV group semiconductors offer the availability of both n-type and p-type devices field-effect transistors (Si can be easily doped to make both) but III–V compounds are limited in this option and hence are bounded by its applications. So silicon-based technologies are *de facto* better suited for highly integrated circuits.

Additionally, the IV group processes allow relatively simple extension of baseline Si CMOS with the high performance HBTs (e.g. SiGe) reducing the bipolar complexity by significant fraction (over 50% according to free literature, e.g. [46]) to create a complex, so-called BiCMOS technologies that outperform their CMOS peers at a nearly the same price point. This derivative of Si main CMOS process, with its bandgap engineering has made it possible to achieve low-cost high-volume production (property of Si CMOS) and provide integrated logic functionality (digital logic) with high-end analogue circuits. Consequently, this approach provides IC designers a flexibility to optimize both cost and performance of realized designs on the elementary circuit level. So for example, in the Si-based RF-mixed signal design, the bipolar transistors are preferred for circuit structures where high dynamic range is required because they exhibit higher linearity and lower $1/f$ noise compared with CMOS devices. The implementation of field-effect transistors is mainly coupled with the demand for more signal processing integrated on a single RF-chip, where the advantages of CMOS enforce.

In the purely analogue circuit world, in fact, the general tendency is to use Si-based devices for low-noise applications due to better noise behaviour and III–V-based structures (only) for high-power implementations due to higher breakdown voltages [47–50]. Anyway it should be again noted that the inherent advantage of IV group semiconductors over III–V compounds lies in a low costs and its being considered as the main stream for the broad scale for customer applications. So they evidently became central to the operations of major IC companies, with the outcome to be relatively free and easy accessible unlike the III–V technologies.

Nevertheless, the high performance systems may implement a broader mix of the most optimum semiconductor technologies and/or their particular devices that are convenient with cost and are each other supplemental in performance. These technological team up can be realized on system level (see also Figure 6.9) – flowing into multi-chip and eventually multi-technology system solutions realized on one carrier or in one small package (known as System in Package – SiP). For example, GaAs/InP for ultra-fast system I/Os with PAs/LNAs and switches, SiGe for PLLs and up/down conversion as well as baseband I/Os and CMOS for baseband DSP. An interesting alternative is the synergism on lowest chip level – flowing into single-chip integrated systems designed in one enhanced technology, where all functions of the particular application are integrated on one chip die.

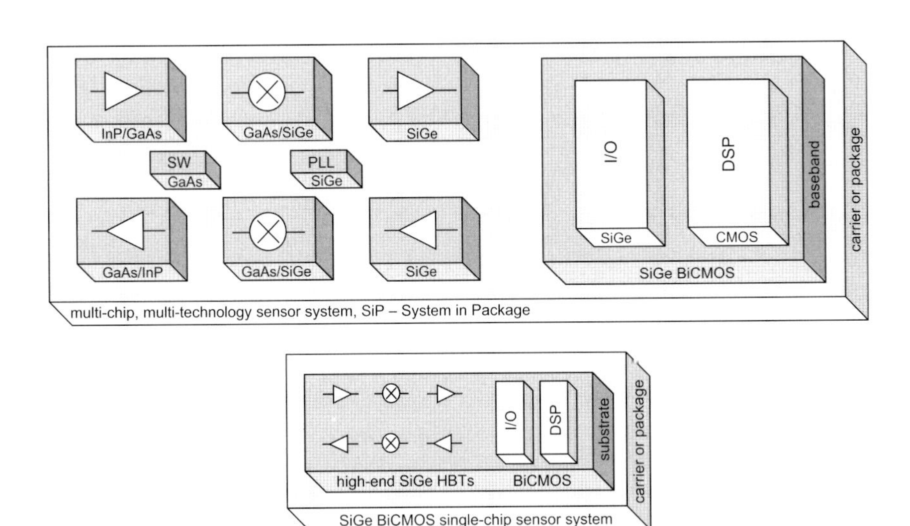

Figure 6.9 Technology team up examples.

Finally, the question of which technology family or eventually combination of more processes are most suitable for UWB system blocks is a rather complex task and strongly depends also on the implemented system architecture. Keeping in mind the UWB sensor architecture based on the M-sequence approach (see Section 3.3.4), the technical requirements for the function of new high-end integrated system components considerably depends on sensor working principle (as already discussed in Chapter 3). Remember, one of its most important benefits is in the relatively low signal levels which have to be handled by the sensor I/O circuitry. From this point of view and considered the above discussed facts, the Si semiconductor branch seems to provide the best balance between the performance, costs and future development perspective for proposed M-sequence-based UWB sensor architecture. This semiconductor branch also adds design latitude by facilitation of system partitions planning into separate functional blocks, or forward-looking in more or less proponent single-chip solutions. And especially proposed UWB sensor approach seems to be very seasonable candidate for the monolithic integration.

6.2.2.2 Design Flow

IC design flow starts usually with component parameters specification and concrete technology selection (also follow Figure 6.10). Than it continues with custom schematic phase, which briefly includes circuit entry in schematic form, simulation, adjustment the schematic and re-simulation – this is repeated until satisfied with specifications. Before an integrated circuit can be fabricated, one must specify the geometric patterns associated with the devices in it. These patterns are 'laid out' highly magnified on a computer screen. The name 'layout' refers both to the

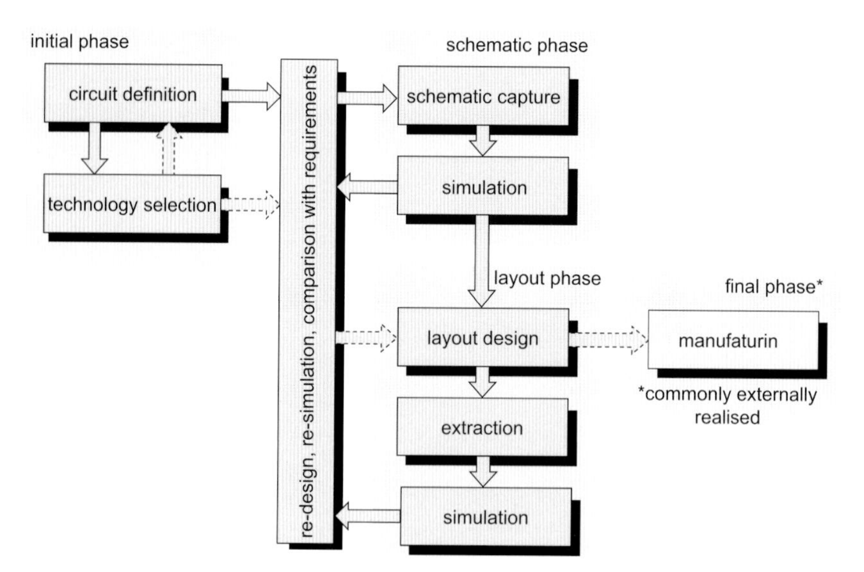

Figure 6.10 Common implemented design flow.

drawing process and to the resulting drawings. So, the layout phase is the pre-final flow step, which includes also schematic like connatural verification process.

During the verification, electrically extracted circuit networks from layout structures are again simulated and eventually re-layouted, till the electrical and engineering requirements by the particular system circuits are met. The final phase represents an IC manufacturing process. This step is usually outsourced to technology foundry in accordance to the early selected process. Here, we will not deeper penetrate into details of simulation and layout techniques as well as manufacturing issues since this goes beyond the scope of this chapter.

6.2.2.3 **Architecture-Specific Circuit Definitions**

Considering the M-sequence-based sensor architecture (see also Section 3.3.4) the fundamental requirements for customized particular ICs can be summarized in several art 'design-laws'. Thus, the desired circuits should ideally disclose

- reasonable wideband matching of the input and output ports, typically 50 Ω or (or 100 Ω for differential ports). *Note*: In the case where the electronics is placed directly in the feed point of an antenna other values may be of advantage,
- minimized power consumption and low relative jitter despite a short switching times (mainly in digital circuit blocks),
- high operational analogue bandwidth and flat gain curve over the entire band,
- good linearity (i.e. reasonable THD) and low-noise contributions by input circuitry,
- general level compatibility,
- very good reliability and robustness,
- easily testability and packageability,
- low cost and so on.

6.2.2.4 **Technology Figure-of-Merits**

With respect to obvious Si-based technology advantages and particular circuit requirements, an accessible BiCMOS process family with sufficient AC performance of its bipolar branch seems to represents a sweet point for sensor architecture bespoken IC realizations. For high-frequency AC operation, bipolar transistors are commonly assessed according to two figures-of-merit. The first is known as transition frequency f_T or unity gain cut-off. The f_T is defined as the frequency at which the common emitter short circuit AC current gain is unity [48]. It is related physically to the total delay for the minority carrier across the bipolar device from emitter to collector, τ_{ec} [49, 50]. Total transit time τ_{ec} is often related to f_T through the equation

$$f_T = \frac{1}{2\pi\tau_{ec}} \tag{6.4}$$

As written in Ref. [50], the more realistic and widely accepted parameter for a circuit environment is f_{max}, known as maximum oscillation frequency. The f_{max} characterizes the power transfer in and out of the bipolar device and is defined as the frequency at which the unilateral power gain becomes unity. Assuming the load conjugately matched to the transistor output impedance, f_{max} is related to

$$f_{max} = \sqrt{\frac{f_T}{8\pi C_{jc} R_b}} \tag{6.5}$$

where C_{jc} is base-collector junction capacitance and R_b base resistance. Equation shows that it is not sufficient to obtain a high value of f_T, but that C_{jc} and R_b must also be optimized.

Selected Technology Performance The IHP[2)] technology family provides a set of optimized npn-HBTs, ranging from high RF performance to higher breakdown voltages [43]. A middle performance bipolar sub-section, which offers npn-HBTs with peak f_{max} of around 180 GHz ($f_T = 110$ GHz) and supplemental CMOS devices (also named SG25H3), has been selected as the main engineering process for the intended IC designs. In this case, the collector–emitter breakdown voltages of the transistors are $BV_{CEO} = 2.3$ V. However, for special devices, for example for centimetre- and millimetre-wave applications, it is advantageous to use IHP's high performance branch (also named SG25H1), which provides bipolar transistor f_{max} of up to about 280 GHz, but $BV_{CEO} = 1.9$ V. Selected processes also provide four standard metal layers and one optional thick top metal, together with high dielectric stack. This option increases RF-passive component performance (e.g. device interconnections, inductors etc.), and brings a certain higher degree of designer freedom during layout phase as well as helps to minimize some chip area and reliability related problems [51].

6.2.3
Multi-Chip and Single-Chip Sensor Integration

Normally, both sensor system architects and IC designers are investigating a wide range of choices in how they implement new ideas and functions. At one end of the spectrum are the proponents of multi-chip solutions. Such approach allows primarily constant developing of new sensor system topologies or refining existing ones. However, this process is a complex and ever-evolving science. And so, in last several years of research in the field of UWB M-sequence-based sensor architectures and their application evaluations, there have been achieved diverse essential advancements with the devices based on such multi-chip structures (also see Sections 3.3.4–3.3.6).

The aim of the multi-chip approach is to be as flexible as possible, and so investigate both the performance of the developed UWB ICs within the actual circuit environment and new device concepts. For these purposes a set of architecture characteristic particular high-end UWB ICs has been specified and custom realized in given semiconductor technology (SH25H3 and SG25H1). The bespoken IC set

2) IHP GmbH – Innovations for High Performance Microelectronics; Leibniz-Institut für innovative Mikroelektronik.

includes primarily diverse versions of PN-generators, sampling circuits and integrated clock conditioning units, which all together represent so-called main M-sequence sensor chip set (referred to basic sensor principle – Section 3.3.4, Figure 3.35). A particular IC die micro-photograph examples of the main chip set are shown in Figure 6.11. From operation frequency point of view, these ICs are designed to cover the frequency band in the region from nearly DC to about 20 GHz.

While the chip internal transition time of the RF-signals is quite short (typically in the range of 15–20 ps), the chip wiring will cause some degradation due to bond pad capacitance and bond wire inductance. To reduce the impact of these parasitics, the bond wires must be as short as possible and octagon-shaped bond pads with 80 μm diameter are implemented for high-frequency I/O ports. The rectangular pads are used for power supplying, grounding and also for low-frequency control

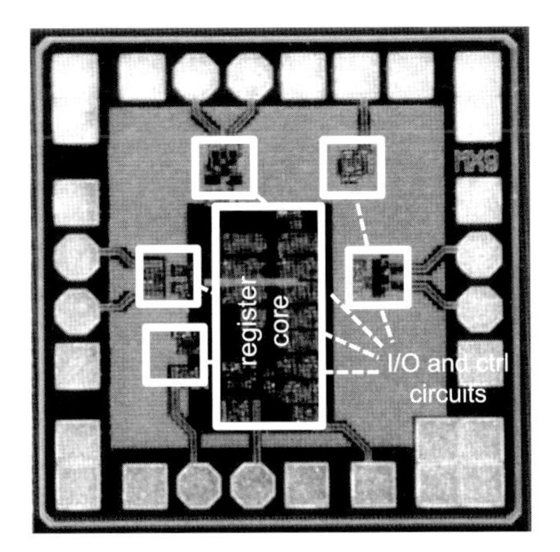

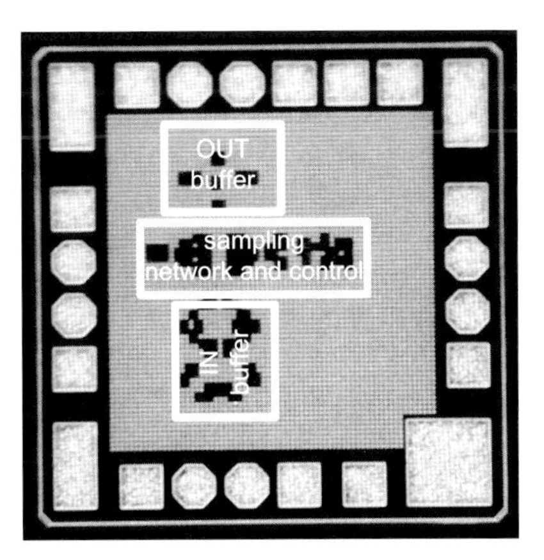

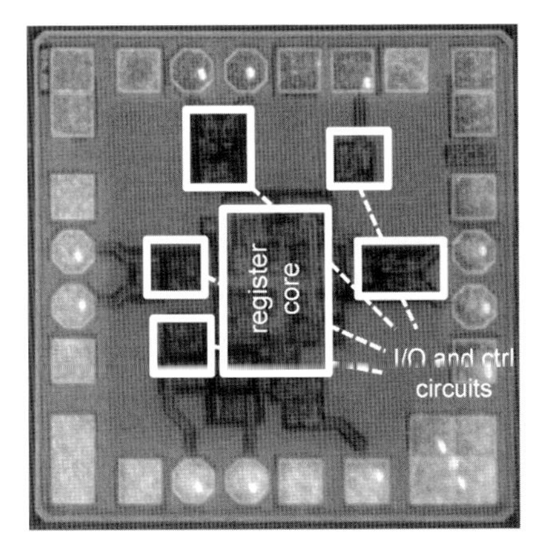

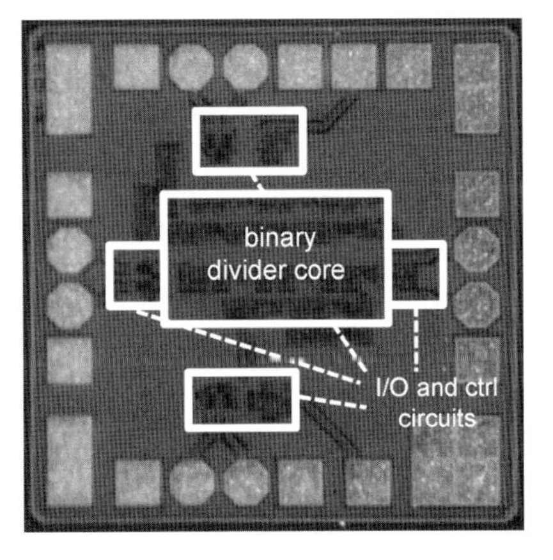

Figure 6.11 Chip die micro-photograph examples of the main M-sequence sensor IC set; each die corresponds to approximately 800 μm × 800 μm in square.

signal interconnections, if needed. Looking on Figure 6.11, it is obvious that the pad spacing contributes with a key portion to the chip area allocation. Thus, the pads are arranged to achieve a reasonable compromise between the performance, cost and to fulfil requirements for reliable contacting to the 'outside world', in this case bonding – contact realized for instance with thin (~20 μm) Ag or Al wire. The on-wafer characterization issues using concurrent both DC probes and typically RF-non-balanced ground-signal (GS) or balanced ground-signal-signal-ground (GSSG) probes must be carefully considered too. So, the pad pitch of the RF-I/O ports is adjusted to 100 μm and supply pads are placed into the chip corners.

It is understandable that the typical system architecture evolution and refining may also require some additional external components, such as front-end mixers, amplifiers (PAs or LNAs), filters and so on. This supplemental electronic devices are essential, for example, to customize the operation frequency band of the system (see also Section 3.3.6) or to adjust an average signal levels to an appropriate region, which can be sufficient processed with the receive electronic and so on. In practice, it is advantageous to integrate some of these functional devices (especially active ones) using one of the accessible semiconductor technologies, too. Figure 6.12 shows die micro-photograph examples of such supplemental ICs – realized in enhanced SG25H1 technology for applications in centimetre- and millimetre-wave range. The pad layouts and pad spacing are optimized for so-called flip-chip contacting. The connection is realized with small conductive balls (bumps) placed between face down orientated chip and periphery (package, PCB etc.)

The ability to create an optimized sensor multi-chip solution on research level is apparent, but manufacturability of such system would be much more difficult with a longer parts list and more complex assembly. Additionally, this issue is going to be very challenging and often improper neglected if the RF-UWB signals have to leave/enter the chip. The package and chip to package connection become part of

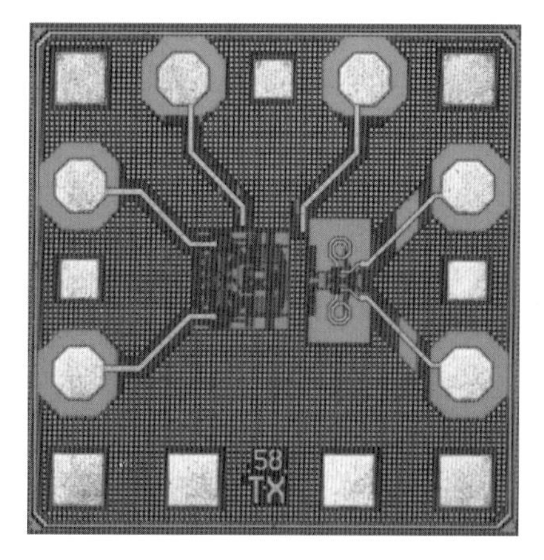

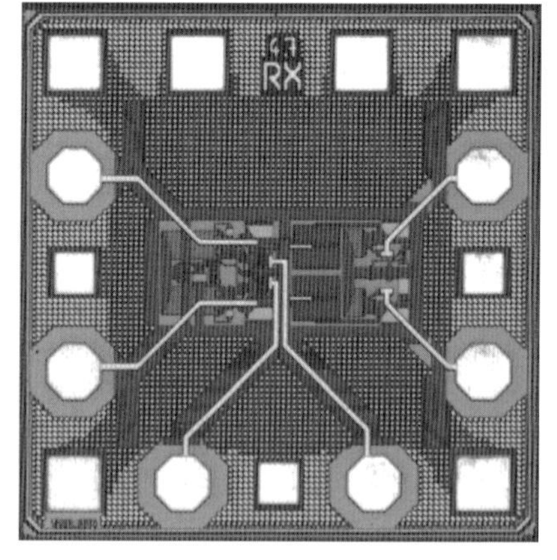

Figure 6.12 Chip die micro-photograph examples of supplemental M-sequence sensor IC set – broadband cm and mm-wave mixers.

the circuit and it is evident that they affect electrical or thermodynamical IC performance. Accessorily, the package itself is a sandwich of metal and insulator materials that convey the electrical signals to large solder bumps, which interface with the printed circuit board. This stack of materials, with their parasites brings some portion of unwanted effects, too and so on. Consequently, the IC packaging problems 'exasperate' designer. In fact, the electrical and thermodynamical problems arising from the packaged product have become a key challenge for chip and system designers the higher the application frequency is allocated. One way to overcome electrical problems is in device miniaturization, unlike heat shrink issues, which must be handled individually. An attempt to achieve the smallest die size (for example, see Figure 6.12 – 600 μm × 600 μm dies) and package footprint means that packaging must keep pace. As a result, there are several RF-packaging choices for designers – QFN (Quad Flat-Pack No Leads), BGA (Ball-Grid Array), various experimental ultra-miniature packages and so on. Figure 6.13 shows an example of an encapsulated IC in QFN as well as M-sequence-based ultra-wideband correlator 1Tx 2Rx head module assembled with multiple QFNs.

In addition, due to the efforts in PCB lithography downscaling, a direct mounting of the die on PCB-like RF-carrier has become more common. But even though, the system reliability and maintenance issues, especially in research typical small series with multiple chips on one PCB, may suffer. Figure 6.14 shows an example of an IC bonded and flip-chipped on a PCB-like carrier.

Alternatively, an intermediate solution for multi-chip approach can be achieved through so-called creative packaging. This minimizes the interconnections with PCB, so the multiple functional ICs are regularly placed into a single-chip-like module. This realization objective is to find the performance/price 'sweet spot' that

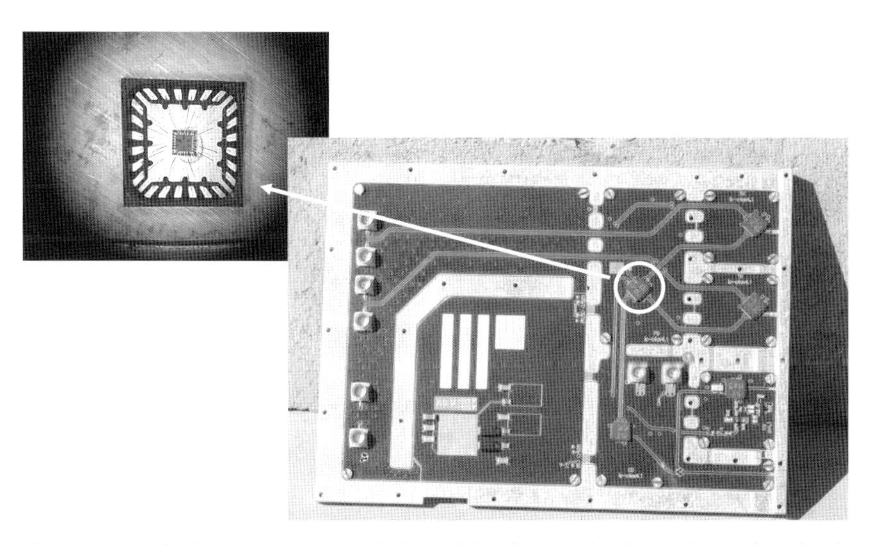

Figure 6.13 Multi-chip M-sequence-based ultra-wideband 1Tx 2Rx head module (PCB board with soldered encapsulated ICs) and micro-photograph of encapsulated IC in QFN (5 mm × 5 mm).

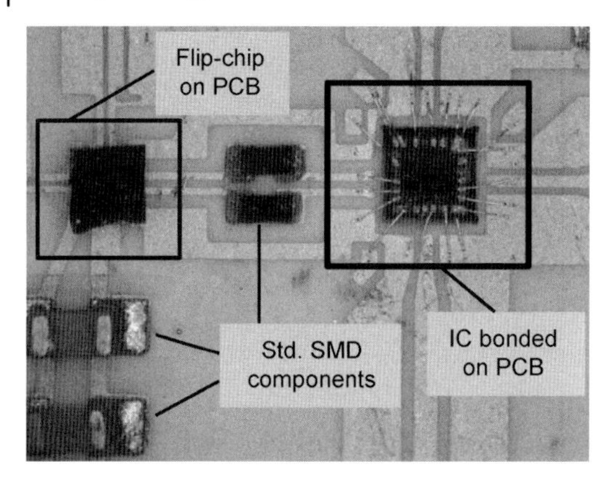

Figure 6.14 Example – IC bonded and flip-chipped on a PCB.

cannot be matched with discrete assembly approach. However, this is relative challenging task and solves the parasitic issues only limited, moreover it is strongly depended on interconnection engineering process. Therefore, a single-chip sensor solution seems to be a reasonable compromise in versatility, system complexity, power consumption and device costs with respect to single sensor devices as well as larger MiMo UWB arrays (see Section 3.3.6). This facts and gained knowledge from multi-chip approach, have motivated the first monolithic integration of the complete RF-part of the M-sequence UWB radar electronics into one silicon die. The advantages of such integration are obvious. The key benefits of system-on-chip (SoC) integration are typically in the overall system performance and reliability improvements due to reduced interconnect and package parasitics, smaller package count and the higher integration level. Furthermore, power supply requirements are reduced because fewer high-frequency signals that require usually 50 Ω interfaces are routed off-chip.

6.2.4 The UWB Single-Chip Head

6.2.4.1 Architecture and Design Philosophy

Figure 6.15 shows simplified block topology of the realized M-sequence-based sensor single chip (also named SoC). Following the picture, the chip includes fully differential transmitter and two receiver channels (also called 1Tx2Rx architecture). Suchlike architecture supports the implementation of such integrated sensor in a higher number of applications, for example in mentioned UWB arrays for high-resolution imaging (Section 3.3.6) or localizations.

The 1Tx2Rx structure is also advantageous for alone standing UWB devices where two receive channels are a priory needed, for instance in material testing and qualification applications, in which the second channel can be used as reference for device calibration purpose.

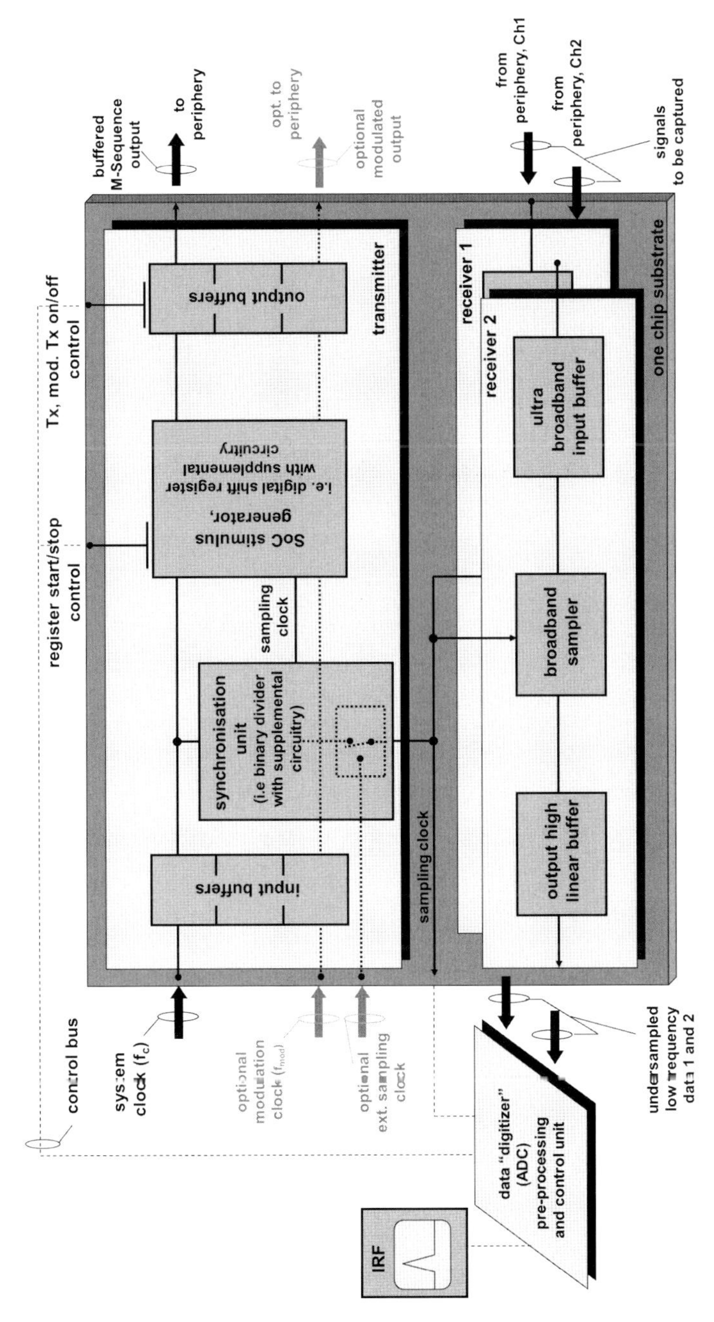

Figure 6.15 Simplified architecture of the UWB M-sequence-based sensor head single chip.

The SoC head architecture is based on the original M-sequence approach for short-range sensing, with digital pseudo-noise as sensor stimulus signal and with receiver working in under sampling mode. Thus, the transmitter is equipped with ultra-fast switching digital circuits, that is digital shift register (M-sequence generator) and sampling synchronization unit. The receiver chains are in general realized as illustrated in Figure 6.15, with ultra-wideband input buffer, sampler and output periphery driver.

It is apparent that the presented single-chip architecture envisages ultra-fast switching cells (i.e. stimulus generator and synchronization unit) with their relatively high signal swing output buffers as well as very sensitive analogue input blocks integrated on the same chip substrate. So, the undesired signal coupling or crosstalk can degrade the performance of the sensitive receive circuitry and hence of the whole system. Especially in case of analogue devices, which handle the ultra-wideband signals, the on-chip interferences can be catastrophic. For example, intermodulation/interaction of noise components with the measured signal within the frequency band of interest may cause device saturation. Therefore, the chip design has to emphasize on isolation issues of the SoC sub-systems. In general, the suppression of on-chip interactions between the functional blocks can be performed by three main ways:

- reducing of noise emissions strength,
- reducing of noise susceptibility of selected/sensitive circuits and
- reducing of noise coupling propagation

and thereby implementing separately or ideally all simultaneously combined, if possible. Thus, the first step during the SoC design phase was the definition of the noise aggressor and victims. The main source of undesired signals in our architecture is high-frequency switching noise coming from transmitter block; that is from stimulus generator and synchronization unit. Consequently, the receivers are qualified in this way as victims.

6.2.4.2 **Implemented Circuit Topology**

As one of the best-suited circuit structures which reduce the emitted switching noise is considered the balanced current steering circuit topology. The ECL circuits discussed in Section 3.2.2 belong also to this family. The basic building cell is a differential amplifier circuit, which consists of two matched transistors (in these case HBTs) whose emitters are connected into one node, which is routed to a current source as depicted in Figure 6.16. The collectors typically terminate to matched loads (e.g. resistors), whereupon these interconnections are considered as the balanced output. Matched biased bases of the amplifier transistors represent the differential input of the structure.

Implemented as switch, the emitted switching noise reduction is achieved by steering a constant current sum through "A" and "B" path by differential circuit lead-in control whereby the current switching spikes related to GND, V_{cc} or V_{ee} are avoided. Once the differential technique is reasonable implemented, all switch transistors work in active region and do not ever reach the saturation. The

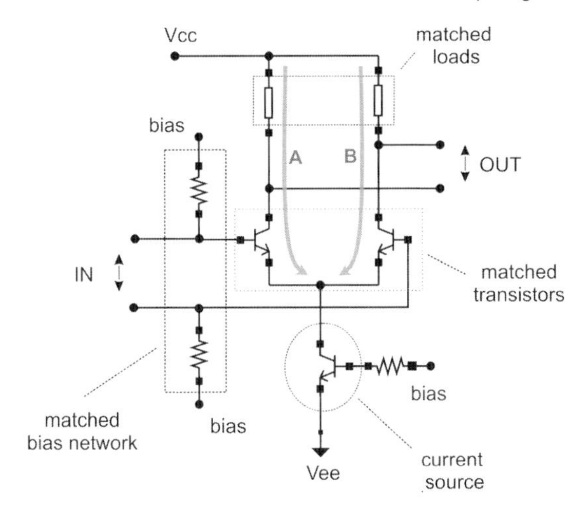

Figure 6.16 Non-saturated balanced current steering circuit.

transistors store less charge in their bases and therefore they can change their state extremely fast.

It can be also shown [52] that for a given cell power consumption the circuit should have a small output voltage swing in order to reduce propagation delay. However, there is also a lower limit on output voltage swing which is several times thermal voltage ($V_T = kt/q$) depending on the circuit concept. Below these limit the noise immunity becomes too small. Consequently, very short switching delays, by using of proper integration technology, are achievable. For example, delays of few ps/gate using IHP's 0.25 SiGe BiCMOS process have been realized [53]. A differential voltage swing of about 300 mV has found to be a reasonable compromise for the individual SoC functional blocks.

Another advantage of differential circuit architecture is the ability to control switch timing very precise. Remember that for the single-ended topology, the return currents are on an imaginary path on a reference plane. For these signals, the crossover point is either a one or a zero (which is also a subject to crosstalk or outside electromagnetic interference - EMI). Since the differential signals handled with the balanced pairs are referenced to each other, and the pair is routed as equal and opposite, it brings about a more precisely controlled crossover point. Thus, the timing jitter can be also reduced due to selected basic circuit structures. Finally, note that differential current steering structures can drive heavy capacitive loads without significant effect on switching speed which is beneficial for the on-chip signals routing, for instance for distributed or multiply connected clocks. The obvious disadvantage in the power dissipation (~constant current through the structure) and larger chip area (~more primary components needed) requirements of such balanced structures represent in case our application acceptable drawbacks.

Concerning the receiver (and its analogue nature), the additional benefit brought by using differential topology is that this acts as countermeasure to common-mode noise, that is the coupled common-mode signal is rejected by the circuit itself, and in a low-voltage system where the undistorted signal swing is limited by the operating headroom of active devices along the signal path, a differential-analogue signal enables twice the low distortion-voltage swing compared to a single-ended signal. Accessorily, the receiver differential circuit structures also maximize the power supply rejection ratio (PSRR) which is important in minimizing the effects related to power supply noise/ripple.

The degree of performance improvement by a differential device structure largely depends on the achievable circuit symmetry, that is the two signal branches should behave identically. With increasing frequency and bandwidth, respectively, this will be more and more challenging since already small imbalances becomes noticeable. Thus, we have to ask for transistors and wiring with well-matched electrical characteristics. This requires very precise custom device interconnections layouting and accurate matching of base emitter voltages as well as of collector currents of the active transistor structures, which depends on both design and technology process. For instance, NPN collector currents scale approximately with drawn emitter area, but no model can precisely predict the influence of emitter geometry on matching. It is therefore difficult to match the transistors with notwithstanding different sizes and shapes of emitters, or its disparate routing to GND or supply. Moreover, two transistors having identical dimensions and operating at equal collector currents should theoretically develop exactly the same base–emitter voltage. In practice, small differences in the emitter currents cause slightly different emitter voltages, which leads to circuit balance mismatch. Other significant sources of asymmetry are due to random fluctuations in doping during the device manufacturing process (e.g. fluctuations in base doping), than due to thermal and mechanical gradients on-chip die and so on [51]. Therefore, to minimize the impact of the discussed matching issues onto 'quality' of the balanced circuit structures, there are several common advised rules for designing and manufacturing well-matched devices. The following rules summarize the implemented principles during the design of M-sequence single chip:

- Place matched devices in close proximity.
- Keep the layout of symmetrical structures as compact as possible.
- Use geometries and shapes provided by technology vendor, that is in case of bipolar transistors use integer ratios such as 1 : 1, 2 : 1, 4 : 1, 8 : 1 and so on of basic unit.
- Place matched devices far away power devices and in low mechanical stress areas.
- Consider using circuit constructions which can transfer the burden of matching from a set of difficult 'matchable' devices (transistors) to a set of associated easy 'matchable' elements (e.g. resistors), for instance so-called emitter degeneration.

6.2.4.3 **Single-Chip Floor Plan**

Naturally, the ways of reductions of disturbance emission strength or circuitry noise susceptibility weakening are at least finite, especially in case of handling

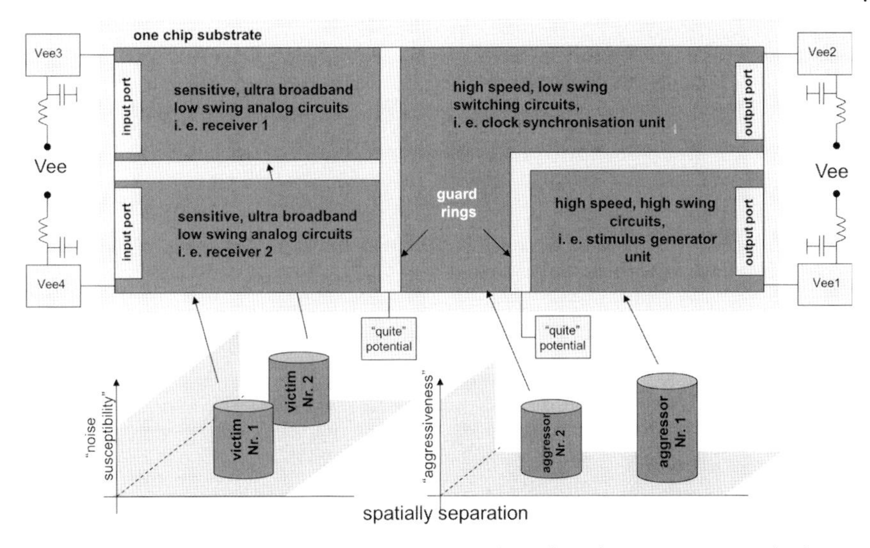

Figure 6.17 Floor plan to minimize on-chip noise coupling adapted to M-sequence single chip.

UWB signals. Thus, to avoid the unwanted interactions between the single-chip functional blocks, several on-chip crosstalk propagation reduction techniques are implemented. The commonly used method to reduce the on-chip noise coupling is carefully intended definition of the floor plan and the power domains on the chip with respect to the architecture and later application needs. Figure 6.17 shows commonly recommended methodology in the IC floor planning adapted to M-sequence single-chip architecture.

The goal is spatially separate the chip building blocks based on their abusive order, that is on their digital and analogue nature as well as the handled frequency and voltage amplitude, so the aggressor/transmitter and victim/receiver are possible widely placed. According to this, the single-chip inputs and outputs are placed relatively far away from each other, and furthermore, the worst inductive near-field coupling at the package leads can be minimized too. Additionally, each block is surrounded by a guard ring that completely enclose/separate a given cell region as shown in the figure. The guarding structure provides a low-impedance contact from the defined supply to the chip substrate. It is one of the most commonly used isolation schemes and seems to be best suited for perverting crosstalk at high operating frequencies. This is due to the voltage fixing of the substrate around the guarded device by absorbing the substrate potential fluctuations. The guard ring serves as a low-impedance path to a 'quiet' potential. Furthermore, the guarded regions should have ideally their own separated power supplies to avoid the SoC functional block interactions. Eventually, where this is not possible, that is in our case due to the requirement of maximum one off-chip power supply, a custom star power routing-based scheme has been implemented. Thus, only one supply connection is needed, what abbreviates on wafer measurements as well as packaging

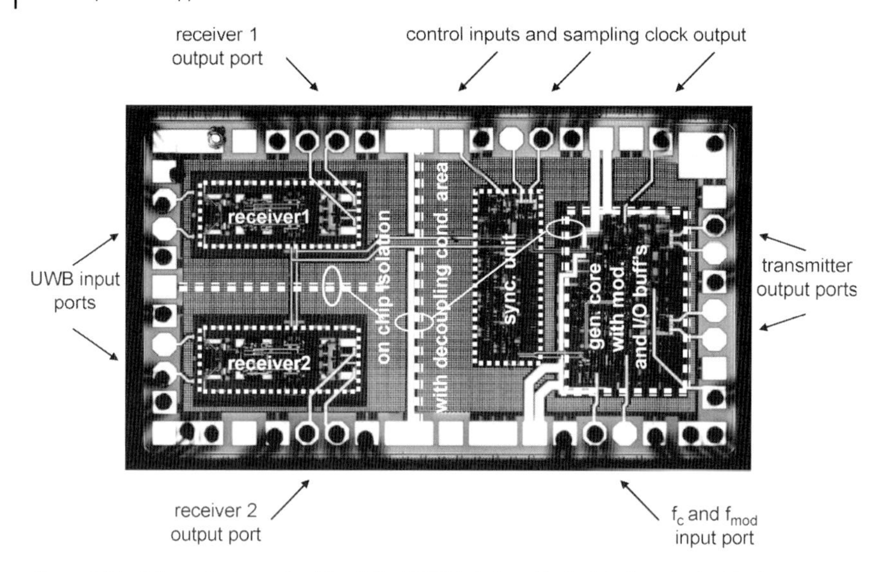

Figure 6.18 Micro-photograph of the SiGe UWB SoC die with depicted particular blocks, isolation structures and decoupling capacitor areas. Die size 2000 μm × 1000 μm.

and later chip implementations. In the star power routing scheme one supply node is established from which individual traces branch out to feed each of SoC blocks. Each trace has a finite amount of parasitic inductance and resistance which is additionally connected via large on-chip realized capacitor to the GND. Such network minimizes the influence of power supply high-frequency fluctuations. It is, however, necessary to make sure that this bypassing grid does not form a resonant circuit with the bonding wires. This can be, for example, avoided by either utilizing on-chip decoupling capacitances with high resonance frequencies or including small resistors in series with power path. Figure 6.18 shows the finalized single-chip die micro-photograph with marked particular functional blocks and decoupling capacitor blocks.

6.2.5 Particular Single-Chip Blocks

6.2.5.1 Stimulus Generator

The system stimulus generator core consists of a chain of nine low-voltage swing digital micro cells (flip-flops) with two feedback taps (see Figure 6.19) building characteristic polynomial which equals to (for more information see also Section 3.3.1)

$$P(x) = x^0 + x^4 + x^9 \tag{6.6}$$

Doing so, periodical M-sequences with 511 so-called 'signal chips' are generated and transferred via multiple-stage output buffers to SoC outside periphery.

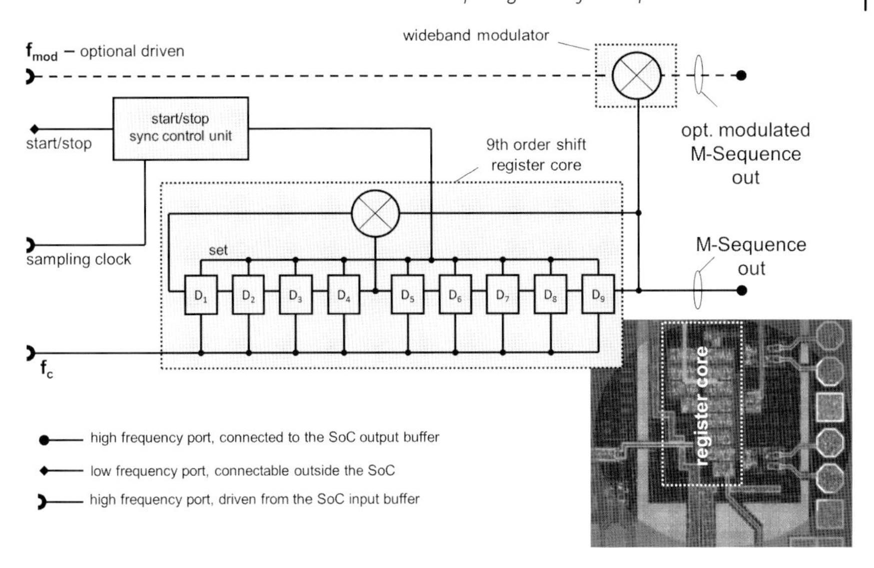

Figure 6.19 SoC generator architecture with ninth-order shift register core, modulator and generator start/stop synchronization control unit. Inset: SoC stimulus generator die part micro-photograph.

The generator core is designed to be driven with the arbitrary toggle frequency from the range inside $f_c \in \langle 0.5; 20 \rangle$ [GHz] and controlled by relative slow rising/falling TTL compatible voltage slopes (~down to 1 ns at star/stop port) coming from device control unit (see also Figure 6.15). Due to the ultra-fast switching edges inside the generator core, typically below 15 ps, it is challenging to proper define the relative starting point of the M-sequence device with the slow control voltage slopes. However, this is prerequisite for some special sensor implementations, for example in ranging applications. Hence, the SoC generator is equipped with an enhanced start/stop synchronization circuit, which allow synchronization of the stimulus to the precisely defined sampling clock. Additionally, as shown in Figure 6.19, the generator is equipped with a wideband modulator, which allows optionally increase the overall bandwidth of SoC transmit signal [54, 55] or to adapt the operational band at specific application [56] and regulation requirements (e.g. FCC or ECC rules), respectively (see also Section 3.3.6). Figure 6.20 depicts two examples of different transmit signal spectral allocations – for baseband and modulated stimulus. For illustration purpose, a clock rate of only $f_c = 1.6$ GHz and $f_{mod} = 6$ GHz has been used so that the screen of the measurement analyser was able to give a complete overview of the signal spectrum.

6.2.5.2 The Synchronization Unit

The synchronization unit has to provide stable clock signals for the signal capturing in the receiver chains (see also Section 3.3.5) and for the exact start/stop M-sequence timing inside the SoC. Moreover, outside the chip, this clock defines the sampling

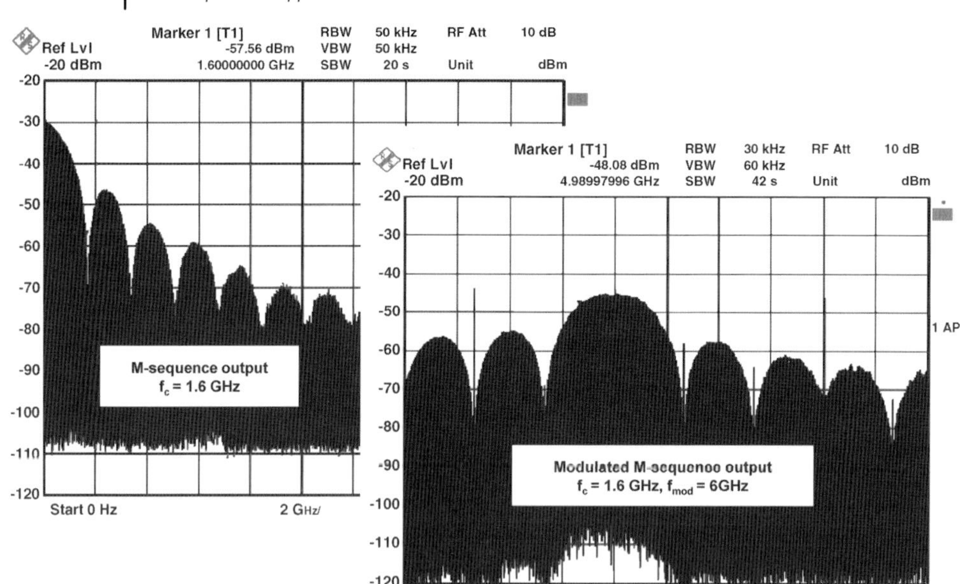

Figure 6.20 Spectrum examples of the signals at the SoC transmitter outputs for $f_c = 1.6$ GHz and $f_{mod} = 6$ GHz.

points of the ADC and is also responsible for the high-speed pre-processing time control (e.g. synchronous averaging, impulse compression etc.). Basically, the synchronization unit consists of a simple 9-stage binary counter with added periphery blocks (i.e. output buffer and supplemental control circuits), wherein the whole clock unit is toggled by the master input signal (f_c).

The synchronism error to the master clock (f_c) should not exceed several tens of femto-seconds. This is guaranteed by the introduction of the supplemental synchronization flip-flop (D_s) at the end of the divider chain (see Figure 6.21), which locks the conditioned signal to the system master clock. Also, the implementation of jitter reducing balanced flip-flop circuit topology supports ultra-precise signal edge control, too.

Conceived with various M-sequence sensor-based measurement implementations and corresponding system design variations in mind (although see Section 3.3.6), the SoC synchronization flow shall be generic enough to allow use in a broader number of different measurement system architectures. Accordingly, the time control unit is additionally equipped with optional switchable shunt sampling clock path, which allows direct clock supply from chip periphery, thereby user selectable clock rates or enhanced, architecture required, signal capturing schemes are possible.

6.2.5.3 **Transmitter I/O Buffers**

The transmitter corresponding I/O buffers are trimmed for chip matched junction to typically 50 Ω peripheries. The output circuits are designed to be primary

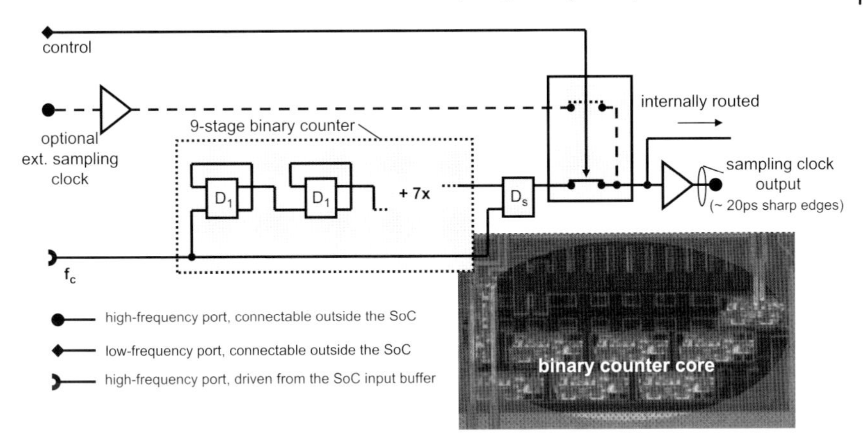

Figure 6.21 Simplified schematic of the synchronization unit architecture. Inset: SoC synchronization unit die part micro-photograph.

connected by fully differential manner, with a single-ended contacting option, for example for testing purposes without circuit performance be drastically affected. The intended 3 dB frequencies are 15 GHz for M-sequence output port and 20 GHz for the port which provides modulated output signal. Each gate offers approximately equivalent of 0 dBm of output power. The transmitter input buffers, which provide system (f_c) and modulation (f_{mod}) clock inter-chip adaptation, are active single ended to differential baluns. These are adjusted for cut-of frequencies of up to 20 GHz and added benefit of signal gain of about 18 dB each with output-related maximal balanced voltage swing of about 400 mV. So the SoC transmitter clock inputs (f_c and f_{mod}) can be driven with the non-balanced minimum signal levels below −15 dBm in the arbitrary frequency range from 0.5 up to 20 GHz. Note, the output signal of the f_c responsible balun is shared with an input of the stimulus generator and the synchronization unit, thus the output circuit of the balun is designed to drive higher capacitive loads.

6.2.5.4 **Ultra-Wideband Receivers**

The SoC receiver chains serve as ultra-wideband sampling gates, which provide under sampled received low-level signals for subsequent low-frequency digitalization and data processing, performed outside the chip (see also Section 3.2.3 as well as Section 3.3.4). In the analogue signal path, each channel includes a wideband input low-noise pre-amplifier (LNA), track-and-hold circuitry (T&H) and high linear output buffer, all arranged in chain.

The LNA is a common emitter differential amplifier with implemented miller capacitance neutralization technique using cross-connected collector base junctions and with capacitive emitter peaking to maintain desired frequency range. The LNA input is matched to 50 Ω (single ended, or 100 Ω balanced) and circuit is trimmed to provide 12 dB of balanced gain with the noise figure below 6 dB over the frequencies from 100 MHz up to 18 GHz.

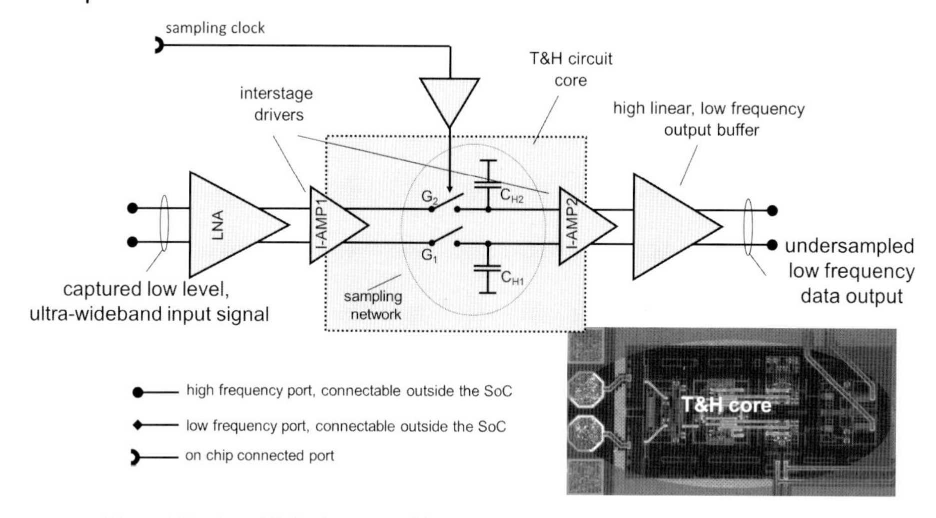

Figure 6.22 Simplified schematic of the receiver architecture. Inset: SoC receiver die part microphotograph.

The key and most challenging component of each receiver is the T&H circuit core. The implemented structure uses a differential open-loop approach. The principle is illustrated in Figure 6.22. The ultra-wideband input signal amplified by LNA is buffered via first interstage driver (*I-AMP1*) before being sampled by sampling network which roughly consists from balanced sampling gatesG_1, G_2 and storage capacitors C_{H1}, C_{H2}. The sampling network is controlled by a track-and-hold clock signal (also called sampling signal) which commands the sampling gates to either 'follow' (track) or hold the applied input ultra-wideband signal over the storage capacitors. Consequently, sampling network represents changing capacitive load in the analogue signal path as its transitions repeatedly changes between track-and-hold state. Therefore, the *I-AMP1* is designed as low output impedance unity gain driver to permit high-speed charge transfer to the sampling gates and hold capacitors without notable analogue input signal deformation. Furthermore, the driver provides additional reverse isolation whereby the impact of the T&H core to the input LNA is reduced. Finally, the processed signal over the C_{H1} and C_{H2} is buffered to the output high-linear buffer via second interstage driver (*I-AMP2*). In order to achieve the intended input frequency range (up to 18 GHz) and very high switching speed (switching aperture times lie below 10 ps), a balanced driven switched emitter followers have been chosen as one of the most usable topology for sampling gates (Figure 6.23).

Implemented differential topology is also advantageous in order to minimize T&H performance limitations formed by artefacts known as track-to-hold-step, signal droop as well as clock and signal feedthrough [57–59]. The first one is the change in T&H output voltage incurred during the transition from track mode to hold mode, caused by the charge dump from the switch into the hold capacitor.

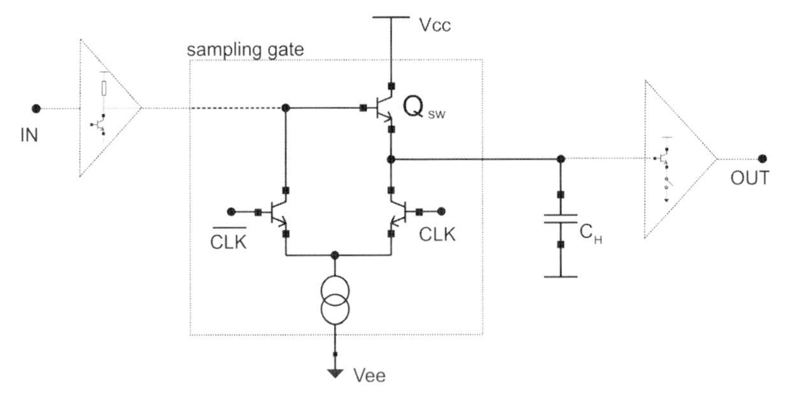

Figure 6.23 The switched emitter follower principle.

Well-designed fully differential architecture virtually eliminates the track-to-hold step if both switches (see also Figure 6.22) cause the identical charge dump. So, the differential output voltage is free from this artefact.

The signal droop (also known as decay) is the decrease in the 'held' voltage during the hold time per unit of time caused by leakages which discharge the hold capacitor. Due to the under sampling receiver working mode, the droop rate in the hold phase is one of the most essential parameters of the SoC receiver chain. Although the implemented differential topology eliminates major common-mode signal droop, that is if the both capacitors allocate the same droop the output voltage of the differential structure stays droop rate free, it is clear that real circuit components depart from the ideal response. Thus, it is advisable to combine several circuit techniques leading to lowering the droop rate. They primarily call for very low charge losses in the sampling gates, hold capacitors themselves (more or less technology given) and predominantly in the T&H core output buffer over the hold phase. The main leakages through the output buffer are minimized by implementing of high input impedance unity gain interstage circuit (*I-AMP2*). Furthermore, the tail currents of the *I-AMP2* input emitter followers are during the hold mode switched off [58]. To fully turn on/off the sampling transistor gates, the amplitudes of the control voltages must be large enough. On the other hand, the high swing gate control signals may increase unwanted amplitude errors of the captured signals, that is signal distortions caused by reason of clock presence over the hold capacitor (clock feedthrough). Due to the common-mode character of the clock feedthrough the differential architecture can compensate this phenomenon, but it has also certain limits [60, 61]. Therefore, as optimum between acceptable signal droop rate and clock feedthrough a relative high clock control differential swing of 0.6 V is selected. On behalf of this a clock conditioning buffer providing required switching levels is connected to the T&H control port. After this measure the droop rate of the SoC receiver chains is about 20%/ms relative to full scale.

The input signal hold mode feedthrough is also challenging parameter, especially if the circuit handles high-frequency broadband signals. It originates from the intrinsic B–E capacitance of the switch transistor Q_{SW} in the sampling gate. Therefore, in the hold mode the base of the Q_{SW} presents finite impedance for the input broadband signal, which afterwards provokes unwanted voltage variations over the C_H. The most used approach for hold mode feedthrough compensation in balanced T&H architectures is based on adding of feedforward capacitors (or also called compensation capacitances C_{XT}) as shown in Section 3.2.3, Figure 3.17. The charge dump of these capacitors will be of opposite sign to the charge dump of the base–emitter capacitances of the switching transistors. However, to obtain good results from this feedthrough cancellation technique, one must pay careful attention to the relevant circuit symmetries as well as the feedforward capacitors must be chosen equal to the base–emitter capacitors of Q_{SW} in the sampling gates. There are several approaches for the C_{XT} realizations. One of them is based on series-parallel construction of four diodes as described in Ref. [58]. The other method uses two identical switch pairs in main and dummy configuration (see also Section 3.2.3, Figure 3.17) as depicted in Figure 6.24. By means of monolithically integration on

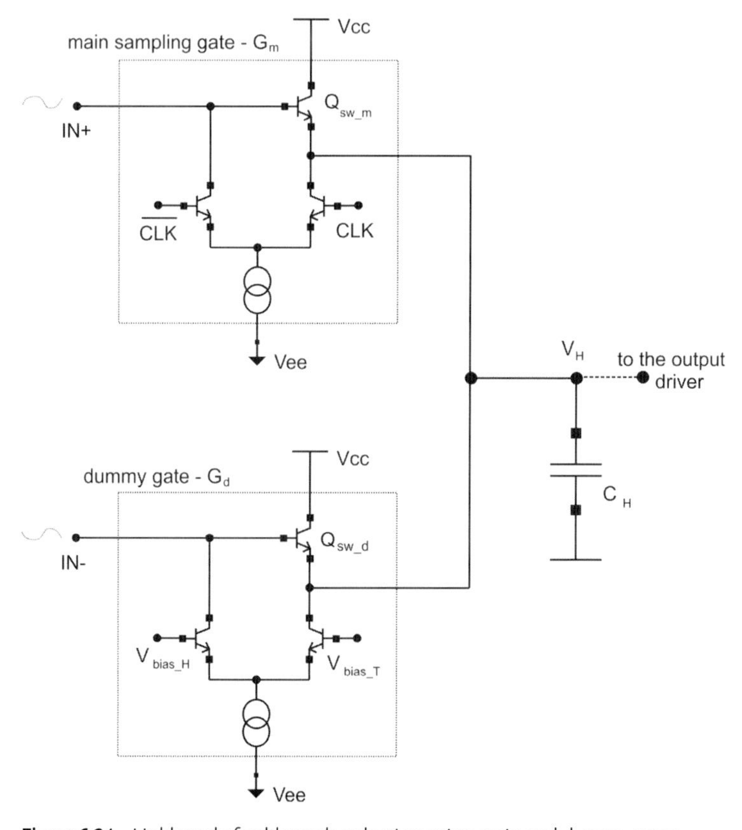

Figure 6.24 Hold mode feedthrough reduction using main and dummy gates.

one die the B–E capacitances of the Q_{sw_m} and Q_{sw_d} can be considered as nearly identical. The dummy gate (G_d) is continuously biased with $V_{bias\_H} > V_{bias\_T}$ and thereby in the hold mode. It should be noted that the input of G_d is driven with the opposite phase compared to the main switch (G_m). Thus, the output of the dummy gate is coupled to the output of the G_m and over the node V_H occurs a weighted sum of input signals IN+ and IN−. Due to this mechanism, when the main gate is in the hold mode the feedthrough is reduced. Another way around, if the G_m is in the track mode, the dummy gate remains in the hold mode and IN− signal does not significantly affect the output.

Finally, the building block after the T&H core (see also Figure 6.22) is the output high linear, low-frequency buffer. This is one stage differential amplifier with implemented emitter degeneration, 6 dB balanced gain and cut-off frequency reduced to about 4 GHz. The amplifier serves as the SoC periphery driver and it is designed to provide the output differential voltage swing of about 1 V in compression point. The restricted bandwidth between the T&H core and periphery limits the noise bandwidth of the signal applied to the periphery, which would otherwise be the full bandwidth of the input amplifier.

6.2.6 Single-Chip Test Prototypes

Generally, the performance evaluation processes of the new ultra-wideband ICs are organized in several levels. At a minimum, the first testing is performed by probing the new devices directly on the wafer using high-frequency probe station (see Figure 6.25). During this process one verifies if the realized dies work as designed

Figure 6.25 Example, on-wafer measurement using high-frequency laboratory probe station.

in fact over the full range of intended operating frequencies and input/output conditions. In doing so, the on-wafer measurements of the realized SoC show a good accordance with the expectations of the simulations. That is, the IC transmitter is able to operate at an arbitrary clock rate from the interval 0.5 to 20 GHz and the receivers can process corresponding ultra-wideband signals (see also Section 3.3.4). Furthermore, the measured mutual decoupling between the different subcomponents on die is better than 60 dB within the whole frequency band.

For higher level testing, that is in order to evaluate the performance of the realized IC also under real application orientated conditions, the SoC dies have been conventionally housed using 24 pin 5 mm × 5 mm QFN package as well as directly bonded on the enhanced circuit board. The packaged version is destined primarily for implementations in complex single- or multi-board MiMo systems (see also Section 3.3.6, Figure 3.53) whereby the enhanced MiMo device stays reasonable priced. The single-chip PCB prototype is, on the other hand, predetermined for elementary 1Tx 2Rx ultra-wideband system configurations. Such prototype board also supports broader device performance evaluation due to easy I/O accessibility via SMPs in comparison to the probe station-based practice. Beyond that, the evaluation PCB helps system level engineers to become acquainted with the SoC device and it also serves as an aid to prototype applications evaluation in the laboratory.

In both variants, the bond wires for broadband I/O ports (i.e. receiver signal inputs and transceiver outputs) are realized as short as possible in order to overcome undesirable signal losses. Nevertheless, the evaluation board as well as packaged SoC can handle signals in the frequency range from near 0.5 up to 18 GHz. This corresponds to impulses with FWHM of about 50 ps. Figure 6.26 shows both bonded SoC die in the QFN24 package and single-chip evaluation board prototype. The test board is composite multilayer carrier made from Rogers 4003C™ laminate to minimize the losses in the transmission lines caused by parasitics.

Figure 6.26 Photographs of the packaged SoC in conventional QFN24 package on the left and on the right directly bonded SoC evaluation board prototype (*Courtesy:* Ilmenau Laboratory for Microwave Measurements and Sensors – Ilmsens).

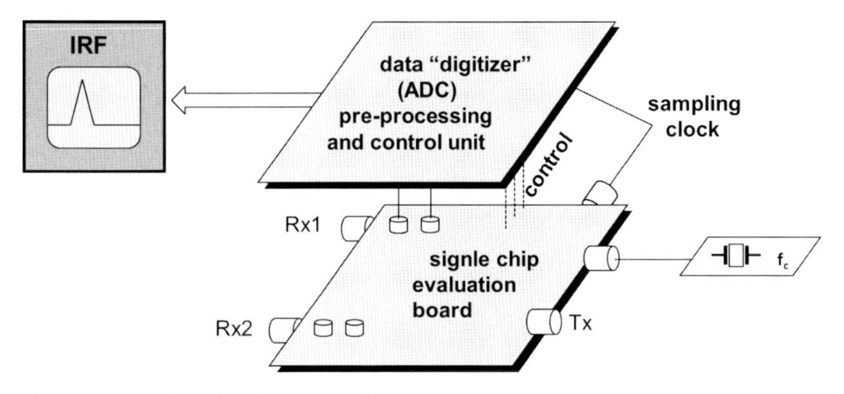

Figure 6.27 SoC board performance evaluation setup.

Figures 6.27 and 6.28 show a SoC device performance evaluation setup and an example of herewith obtained impulse response function (see also Section 3.3.2) normalized to maximum. As also depicted in the picture Figure 6.15, the evaluation board is connected with off-the-shelf ADC, pre-processing and control unit as well as stable master clock generator.

For IRF evaluation purpose, the investigated module was toggled with 9 GHz clock, meanwhile the Tx and Rx ports were non-balanced coupled with high-frequency cable and DC block. The observed impulse FWHM corresponds to about 100 ps which well coincides with the expectations for a 9 GHz master clock.

The dynamic behaviour of the SoC receiver channel is depicted in Figure 6.29. To obtain the dynamic range, the I/O ports of the evaluation board were connected via a variable attenuator block. During the measurement, the attenuation was gradually increased whilst detected IRF lost and in the meantime the estimated power on the receiver output has been recorded. The measurements were performed in single-ended mode. Thus, in case of the differential operation mode, we can expect still 3 dB better performance. The transmitter provides about 0 dBm of stimulus

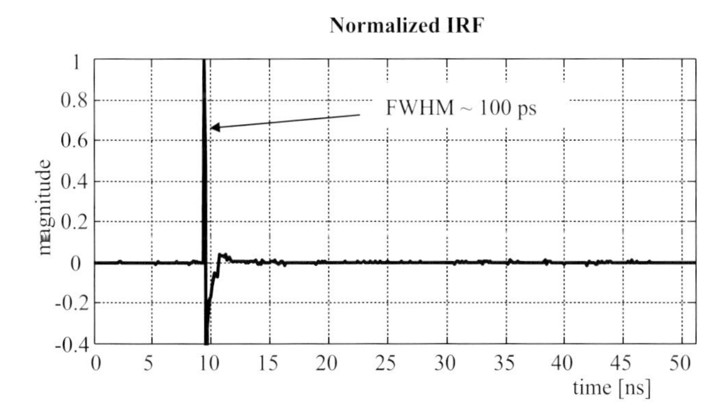

Figure 6.28 SoC device performance evaluation setup and normalized impulse response function obtained using evaluation board toggled with 9 GHz clock.

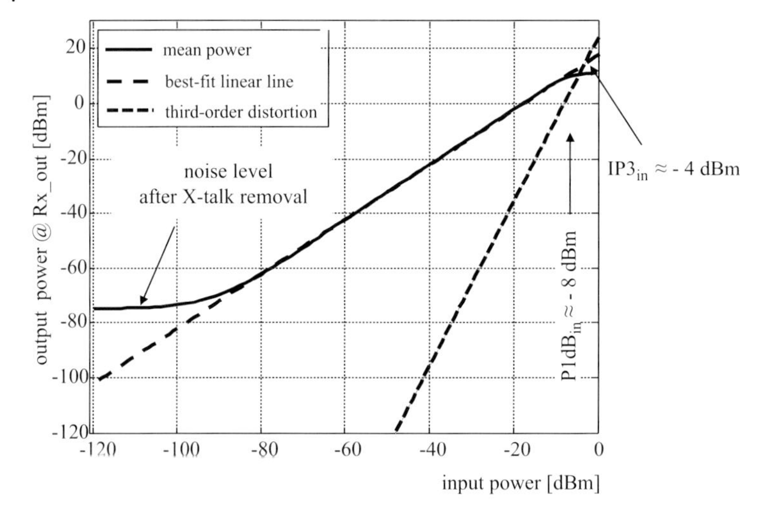

Figure 6.29 Estimated dynamic behaviour of the SoC receiver channel (single-ended mode).

power and the receiver chain exhibits about 18 dB balanced (i.e. 15 dB unbalanced) gain. The single-ended input-related 1 dB compression point ($P_{1\ \mathrm{dB,in}}$) was estimated to −8 dBm which corresponds with the expectations from the circuit simulation.

Finally, the implemented SoC offers over 80 dB dynamic range,[3)] and guarantee delay resolution below 10 fs being the only one fully integrated radar SoC head reported in the literature with these characteristics in a real-time SIMO configuration.

6.3 Dielectric UWB Microwave Spectroscopy

Frank Daschner, Michael Kent, and Reinhard Knöchel

6.3.1 Introduction

In the food industry especially many different natural materials are processed. In order to optimize the production or to verify the quality of the raw materials their properties need to be measured or defined. However, often the values to be determined are not accurately defined physical values. For example, 'quality' is more difficult to define than 'temperature'. Nevertheless when the quality of a foodstuff decreases due to a long storage time the loss of quality has an interaction with its measurable physical properties.

3) Pre-processing averaging was set to 1024 which corresponds approximately to 33 IRF/s at $f_c = 9$ GHz.

Due to the high water content of foodstuffs and other natural materials ultra-wideband dielectric spectroscopy enables the determination of many of their properties. The dielectric properties arise from polarization effects; while the electronic and ionic polarizability are in the optical and infrared frequency range the orientational polarizability is in the microwave range. Orientational polarizability occurs when the molecules have a permanent dipole moment and they try to align to the externally applied electrical field (see also Section 5.3). For low frequencies the alignment of the polar molecules is in phase with the field, hence the ability of the material to store electrical field energy is relatively high: this is the real part of the permittivity. Due to the inertia of the molecules they are not able to follow electrical fields that change more rapidly and this effect becomes greater with increasing frequency.

At the so-called relaxation frequency f_r, the average phase between the electrical field and the molecules is 90° and the losses have a maximum. The losses are expressed by the imaginary part of the permittivity. At frequencies far above f_r, the polar molecules can no longer follow the electrical field at all and the permittivity decreases to a high-frequency constant value.

The relaxation process is relatively broadband and water has its relaxation frequency at about 17 GHz depending on the temperature. The complex permittivity is a frequency-dependent function and is expressed as

$$\underline{\varepsilon}(f) = \varepsilon'(f) - j\varepsilon''(f) \tag{6.7}$$

Water has an unusually large dipole moment given by its small molecule size and dominates the dielectric properties of many natural materials because it is their main constituent. Besides water other polar molecules, for instance alcohols, are often a component of foodstuffs. Furthermore, water has a complex interaction with other, often non-polar molecules. The water molecules are rotationally hindered by these materials. This means a part of the water has a lower relaxation frequency. Additionally, many water molecules are more or less bound by substrate surfaces. This effect also reduces the permittivity and the relaxation frequency f_r of the water.

The losses of a natural material have also a strong dependence on the ionic conductivity. The ionic conductivity in foodstuffs is generally caused by dissolved salts. The contribution to the losses of such ionic conductivity depends on the frequency and is given by

$$\varepsilon''_{\sigma}(f) = \frac{\sigma}{2\pi f \varepsilon_0}, \tag{6.8}$$

where σ is the DC conductivity and ε_0 is the permittivity of free space. Obviously such ions have a much greater influence at lower frequencies while in the microwave range the polarization losses are dominant. These complex interactions between various molecules of natural materials and electric fields of varying frequency have the consequence that the more broadband the measurement of a dielectric spectrum, the more information is collected about the state of the material under test (MUT).

Often vector network analysers (VNA) are used for the dielectric spectroscopy and the measurement is performed in the frequency domain. These instruments are large, expensive and the required time for the measurements is relatively long. In order to reduce the hardware effort and the measurement time, an UWB time domain reflectometer (TDR) can also be used for the dielectric measurement of a MUT. By using the Fourier transform the dielectric spectrum can be calculated from the measured raw data.

Due to the complexity of foodstuffs it is often not possible to create adequate models in order to relate the dielectric properties to the values to be determined finally (quality, time of storage, salt content etc.). On the other hand, changes of the dielectric properties can be observed and measured. However, multivariate calibration methods enable the processing of the data and the determination of the values to be measured without the creation of a physical model.

In this chapter, the design of a TDR is presented. This instrument was used to measure the quality and storage time of several fish samples [62]. The data processing procedure is explained using the example of two selected test series. The storage time is estimated from the dielectric spectra using principal component analysis and regression (PCA/PCR). Furthermore, a calibration with artificial neural networks (ANN) is presented.

6.3.2 Time Domain Reflectometer for Dielectric Spectroscopy

In this section, a dedicated TDR suitable for the measurement of ultra-wideband dielectric spectra of foodstuff is discussed.

6.3.2.1 Probe

In the literature, many different structures for permittivity measurements have been suggested. For foodstuffs, liquids or soft materials the open-ended coaxial line has become the standard probe. With this type of probe little sample preparation is required. It consists of just a cut coaxial line with a ground plane at the aperture. At this aperture plane the wave propagation condition changes abruptly hence a reflection occurs. The absolute value of this reflection depends on the permittivity of the MUT [63, 64]. Together with the knowledge of the dimensions of the probe and a calibration measurement (in general with air, a short circuit and distilled water), the permittivity can be calculated from the measured complex scattering parameters of the probe. Many other structures could be used as probes in combination with the instrument depending on the application. However, the focus of this article is not on the probe, and the measurements using the open-ended coaxial line are just an example.

6.3.2.2 Instrument Requirements

The more broadband the measured dielectric spectrum the more information about the state and composition of the MUT is collected. For an UWB-TDR this requirement means that a generated pulse, and the rise time of a step to be generated, should be as short as possible. In the system presented here a step pulse is sent to the probe and its reflection is processed. When the influence of free and bound water as well as the salt content has to be determined a frequency range from around 200 MHz up to 10 GHz should be applied for the dielectric spectroscopy. For an instrument working in the time domain this requirement means that the rise time of the step should be $t_r < 100$ ps. The step and its duration then effectively comprise the frequency components of the required band.

The reflected pulse changes its shape depending on the wideband dielectric properties of the MUT. In order to detect details of these changes the reflected pulse should be measured with a high time resolution, for instance with $t_s = 1$ ps.

6.3.2.3 Sequential Sampling

A direct processing of such a high time resolution is actually not possible. Therefore, sequential sampling is used for the collection of the repeated waveform. This is illustrated in Figure 6.30 for a sine wave having a period of T_{max}. The sampling time is shifted by t_s from period to period. Hence, the complete sine wave can be reconstructed (see also Section 3.2.3).

This technique is only applicable for signals that are repeated exactly in each period, but for the application discussed here this requirement is fulfilled: the step signal sent to the probe is repeated with T_{max}.

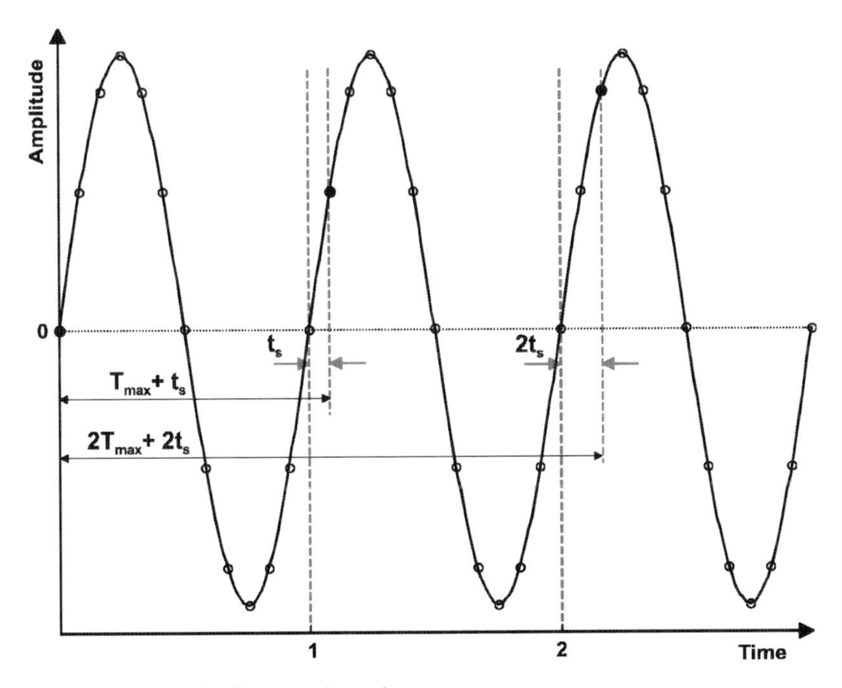

Figure 6.30 Principle of sequential sampling.

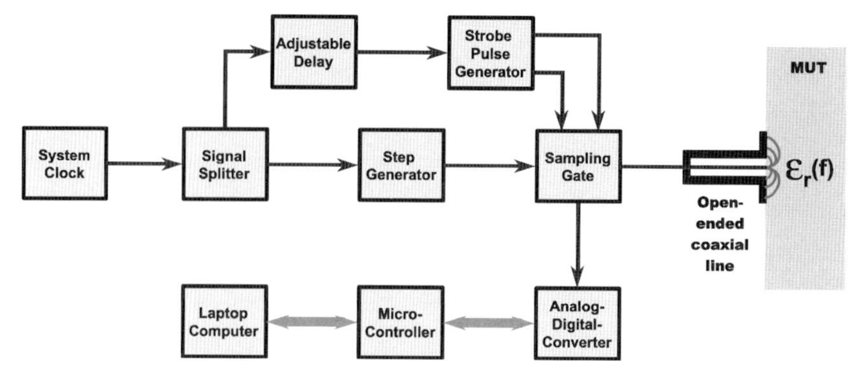

Figure 6.31 System design of a dedicated UWB-TDR.

6.3.2.4 System Design

The block diagram of the TDR is shown in Figure 6.31. All parts of the system are synchronized with the system clock. The step generator creates the step signal having a rise time of around $t_r = 100$ ps. This step passes through the sampling gate. At the open-ended coaxial line a part of the signal is reflected and travels back to the sampling gate. When the strobe pulse generator is triggered the sampling gate measures the momentary voltage and this value is held and can be read by the analogue-to-digital converter.

In order to be able to realize a sequential sampling gate as described in the previous section a precise control of the strobe pulse is required. Therefore, the triggering of the strobe pulse can be shifted from period to period by an adjustable delay. After

$$n_t = \frac{T_{\max} + t_s}{t_s} \tag{6.9}$$

repetitions of the step, the whole reflected wave is digitized and can be processed by a micro-controller and a computer. The adjustable delay can realized by switched lines with different lengths. The step generator is just a single-chip RFIC (e.g. a fast switching frequency pre-scaler).

The pulse strobe generator creates pulses to switch the diodes of the sampling gate. It is constructed as shown in Figure 6.32. First, a pulse with a half maximum

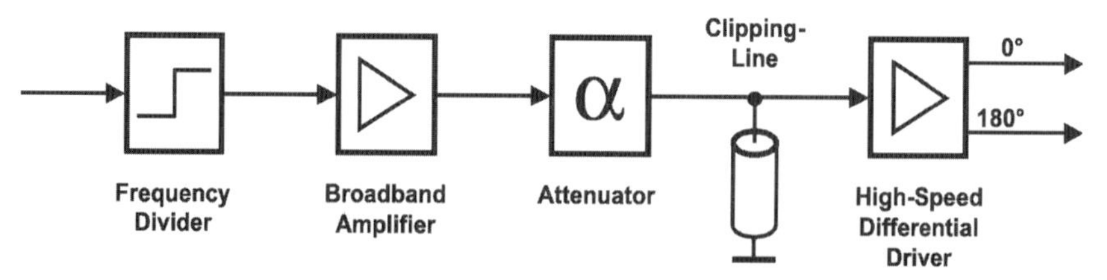

Figure 6.32 Strobe pulse generator.

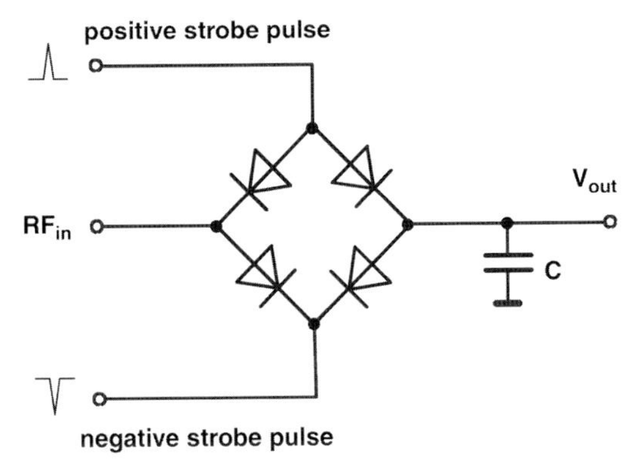

Figure 6.33 Sampling gate.

pulse width of approximately 65 ps is generated by a fast switching frequency divider. A following broadband amplifier and an attenuator adjust the amplitude and decouple the divider from the following parts of the circuit. A clipping line differentiates the signal. Hence, it is transformed from a rising edge to a pulse having a width of approximately 65 ps. The following high-speed differential driver produces a positive and negative pulse with sufficiently high voltage to switch the diodes of the sampling gate (1.5 V).

The sampling gate consists of a diode bridge as shown in Figure 6.33. When the diodes are in a non-conducting condition their RF-port is isolated from the capacitor C. The strobe pulses switch the diode bridge into a conducting state, so the instantaneous voltage of the RF_{in}-signal charges the capacitor. This value can be read by an AD-converter even when the diode bridge is again in non-conducting condition after the strobe pulses are over. Hence, the AD-converter has enough time to process the voltage of the RF_{in}-signal.

6.3.2.5 **Hardware Effort**

The overall hardware effort of the discussed design is relatively low in comparison to a broadband automatic network analyser and due to the progress in the development of integrated circuits the instrument can be constructed as a cheap and handheld device.

6.3.3 Signal Processing

For the description of the applied data processing procedures two experiments from the SEQUID project [62] are selected as examples. Cod was caught in the Baltic-Sea and some of the samples were stored on ice while the rest were frozen as fillets. The cod on ice has a shelf life of around 2 weeks due to deleterious changes arising from bacterial and enzymatic processes. In the frozen fish there is

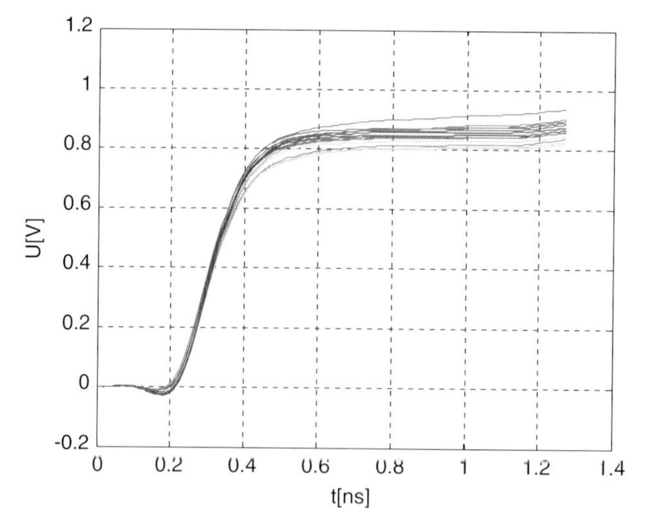

Figure 6.34 Step responses of several cod samples stored on ice.

no microbiological activity. However, only a part of the water is frozen so chemical reactions are still observable and the quality of the fish decreases, albeit over a much longer storage time (around 1 year in the experiment).

The fish samples were measured using a prototype of the instrument described in the previous section. The frozen fillets were thawed before the measurements. The diagram in Figure 6.34 shows some of the measured step responses of the cod stored on ice. It can be observed that the signal variations are relatively small. Furthermore neighbouring data points are highly correlated. A transformation of the signal to the frequency domain would not lead to data with more differentiation. This can be explained by the fact that quality losses result only in slight changes of the amount of free water in the tissue. In a dielectric spectrum such changes could not be seen clearly because the relaxation processes are relatively broadband.

Also other (partially unknown) changes happen during storage and as mentioned above a physical modelling is difficult. It can be observed that the signals change over the time but the function that describes the relation between these changes in the step response and the value to be determined is unknown. However, multivariate calibration methods enable the approximation of unknown functions, such as these relations, without physical modelling. Two multivariate approaches are demonstrated with these measurements on cod.

With principal component analysis and regression, a linear function approximation can be deduced while ANN can also approximate non-linear functions.

6.3.3.1 Principal Component Analysis and Regression

The application of multiple linear regression to the cod data for the determination of the target values (days on ice, days in freezer) is invalidated because the input data (the step responses) are too correlated. Such highly correlated data causes numerical instabilities during a necessary matrix inversion [65]. Therefore, it is advantageous to de-correlate the data and to reduce their dimension. This is the task of the principle component analysis.

The original (measured) data are linearly transformed into a set of new uncorrelated data. The new data are called principal components (PCs) and they are expressed with new coordinates. These new coordinates are related to the variance of the data set. The principal components are sorted with respect to their variance: the first PC represents the largest variance of the data, with the following PCs describing decreasing variances. The last PCs generally describe only noise and they need not be taken into account during the following steps of the data processing. This means a data reduction is also applied.

How is the calibration procedure performed with PCA/PCR? First, only a part of the measured input values should be selected for the data processing. In general, the area where the highest variation can be observed should be selected. For the application discussed here it is advantageous to select the points within the slope of the step response. The selected points are shown in Figure 6.35.

After this the data need to be separated randomly into a calibration and validation set. However, the calibration data should be as representative as possible. The measured data are arranged in a matrix where each line contains the selected points of the step response. Before the PCA is performed the data are normalized by subtracting the means of each column of the matrix and standardized by

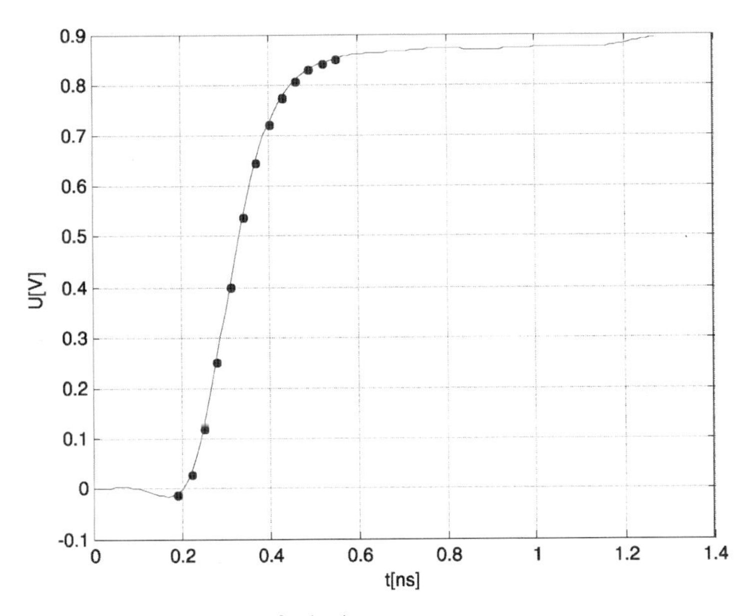

Figure 6.35 Selected points for the data processing.

dividing each column by their variances. This normalized data matrix is called $\mathbf{X}_C$ for the calibration set and $\mathbf{X}_V$ for the validation set in the following discussion. The validation data are normalized using the means and variances of the calibration data.

The principal components are calculated by a linear transformation from the normalized data:

$$\mathbf{PC}_C = \mathbf{X}_C\mathbf{L}, \tag{6.10}$$

where $\mathbf{L}$ is a matrix of so-called loadings. The loadings are calculated by an eigenvalue decomposition algorithm of the covariance matrix of $\mathbf{X}_C$ [65], which is, for example, available in the statistics toolbox of MATLAB®. The principal component regression (PCR) is a multiple linear regression applied with the principal components. Due to the fact that the PCs are uncorrelated the calculation of the coefficients is not numerically instable. The coefficients are arranged in the vector $\boldsymbol{\beta}$ and are calculated by

$$\boldsymbol{\beta} = \left(\mathbf{PC}_C^T\,\mathbf{PC}_C\right)^{-1}\mathbf{PC}_C\mathbf{t}_S^C, \tag{6.11}$$

where $\mathbf{t}_S^C$ is the vector with the storage times of the samples of the calibration group. With the determination of $\boldsymbol{\beta}$ the calibration procedure is completed and it can be evaluated using the validation data. As mentioned above it is in general sufficient to use only the first few PCs. For the cod data 6–8 PCs were used.

The first step of the validation is the calculation of the principal components using the normalized validation data and the matrix $\mathbf{L}$ as determined during the calibration procedure:

$$\mathbf{PC}_V = \mathbf{X}_V\mathbf{L}. \tag{6.12}$$

The target values (here storage times) are estimated by applying the linear regression equation with the coefficients $\boldsymbol{\beta}$ determined also during the calibration procedure:

$$\mathbf{t}_S^V = \mathbf{PC}_V\boldsymbol{\beta}. \tag{6.13}$$

The results of the application of PCA/PCR with the selected cod storage experiments are shown in Figure 6.36. The coefficient of determination is $R^2 = 0.907$ for the fresh cod stored on ice and the root mean square error of the validation group is $\mathrm{RMSE_V} = 2.77$ days. This means the error for the estimation of the storage time of a single fish sample is around 3 days. For natural materials having a high amount of variation this is a good performance. For the frozen fish one can observe a comparable accuracy: $\mathrm{RMSE_V} = 43.6$ days. This is also nearly of 10% of the value range (380 days).

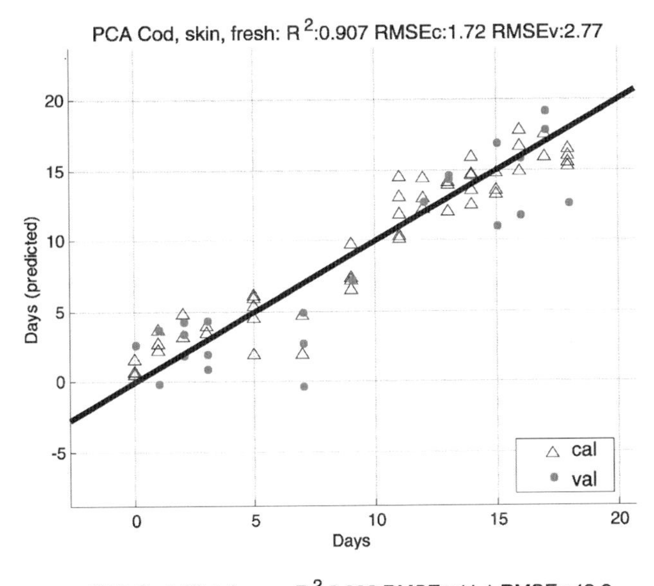

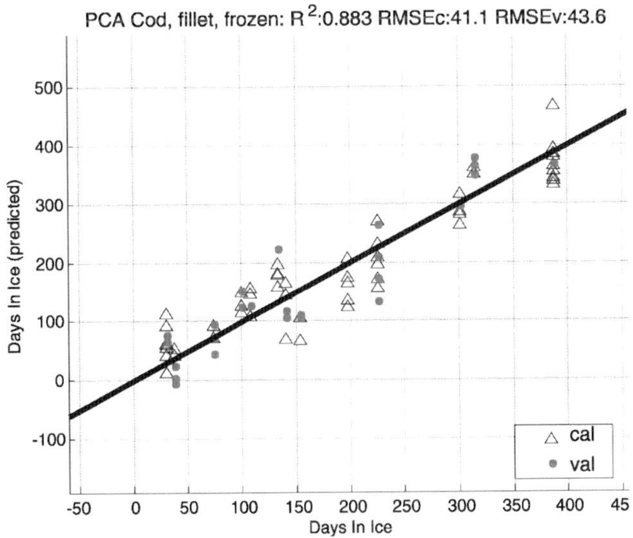

Figure 6.36 Results obtained with PCA/PCR.

6.3.3.2 **Artificial Neural Networks**

PCA/PCR is a linear calibration method and although this procedure can also be used for the approximation of non-linear functions, the residuals may be higher in comparison to methods that are better suited, for example ANN [66]. The development of this technique was inspired by the structure of nerve tissue and the high computational power available today enables its application for small instruments.

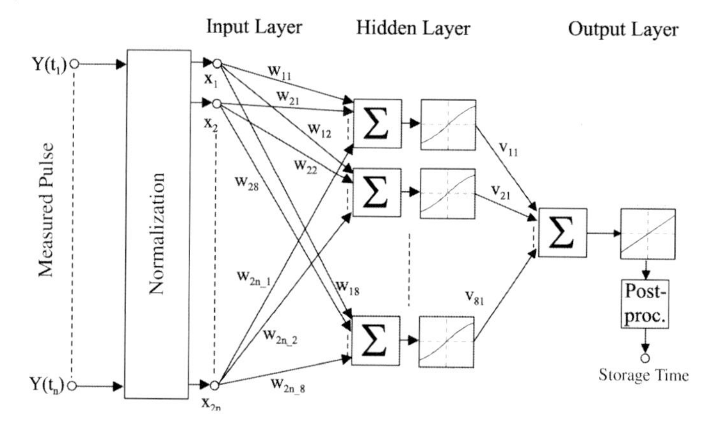

Figure 6.37 Architecture of the used artificial neural network.

The so-called architecture of the ANN used for the calibration of the cod test series is shown in Figure 6.37.

The normalized input variables are weighted and processed by the neurons of the hidden layer. The output values of the neurons of the hidden layer are weighted again and processed by the neuron of the output layer. The post-processing reverses the normalization and the storage time is determined. The activation function of the neurons in the hidden layer is the non-linear *tansig*-function. This is required for a non-linear data processing. For the neuron of the output layer the linear activation function suffices. The calibration of the ANN is called 'training' and several algorithms exist for this purpose. Training means the best weighting factors have to be found to minimize the residuals of the function approximations. For the cod data discussed here the neural network toolbox of MATLAB was used.

The results obtained with an ANN are shown in Figure 6.38. They are slightly better than the results of PCA/PCR giving $RMSE_V = 2.01$ for the fish in ice and $RMSE_V = 41.1$ for the frozen fish. However, the computational effort during calibration is higher and in general more samples are required. Whether a calibration with PCA/PCR or ANN is the best choice depends on the specific case.

6.3.4 Summary

Many natural materials have high water content. Due to the complex behaviour of water with other constituents broadband dielectric spectroscopy enables an access to much information about the composition and the state of these materials. The wider the band being analysed the more information is available for the data processing. However, a physical modelling is often difficult or impossible due to the complex interaction between the molecules. On the other

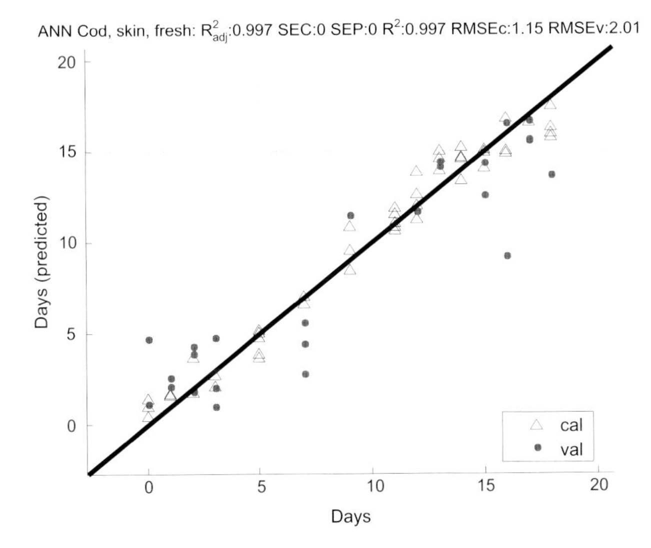

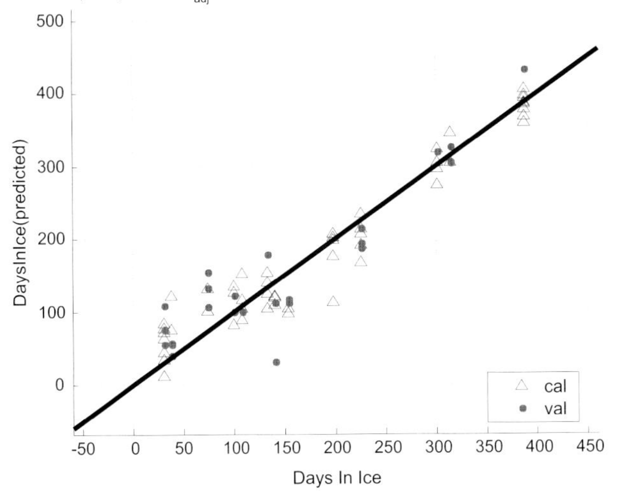

Figure 6.38 Results obtained with artificial neural networks.

hand, changes of the dielectric data are clearly observable when the material changes.

Therefore, multivariate calibration methods can be applied. Using these techniques no physical modelling is required but calibration and validation measurements need to be carried out. As examples two test series were presented and discussed in this article. Cod were both stored in ice and frozen and the dielectric spectroscopy in the time domain enabled a useful estimation of the storage time of these fish samples.

Often automatic network analysers are used for the broadband dielectric spectroscopy. However, these instruments are large and expensive. Therefore, a dedicated time domain reflectometer was suggested and tested in this work. The instrument design enables a relatively inexpensive and compact realization of a handheld instrument.

Due to the flexibility of the method many other applications with other materials and probes are imaginable. The measurements are very fast and the material under test is not changed by the measurements.

6.4
Non-Destructive Testing in Civil Engineering Using M-Sequence-Based UWB Sensors

Ralf Herrmann and Frank Bonitz

Recently, radar sensors have been considered for the inspection of men made structures: that is buildings, streets, or bridges. These applications are inspired by GPR [67–71] where microwaves are used to detect various objects underneath the ground. Low-frequency microwaves are not only able to propagate in soil or rock, but also in concrete, stoneware, and asphalt materials. Many GPR sensors can therefore be used to assess the state of various constructions, too. This also applies to UWB sensors which can help to solve tasks unsuitable for more narrowband GPR sensors. Since the microwaves do not harm the investigated objects, this field of applications belongs to NDT methodologies – more specifically non-destructive testing in civil engineering (NTDCE). One of the main tasks in NDTCE is to find disturbances like cracks, voids, or unwanted materials inside a construction. Such targets can provide information on possible security risks, immanent construction failures or breakdowns, or simply confirm the correctness of construction work.

From an application point of view, the working principle of the sensor hardware does not matter much as long as main requirements are fulfilled. There are two main aspects when choosing or designing a sensor. The useful frequency range is mainly determined by the materials to be inspected. Often the range from some 10 MHz to a few GHz can be considered. The second major aspect is clutter. There are different possible clutter sources depending on the application. Unwanted signal components can arise from objects near the measurement target or from inside the sensor (systematic errors or device clutter). For example, in most NDTCE applications the surface response is a major clutter source. Clutter removal is one of the main data processing challenges and can partly be supported by system design. Processing and interpretation of the gathered data is an equally important aspect to hardware. Radar sensors do indirect measurements and the wanted information must be extracted by solving an inverse problem.

GPR sensors are up to date dominated by wideband to ultra-wideband devices working with short pulses (pulse radar) [72–75], but M-sequence sensors are equally feasible as well, see, for example [76]. The flexibility of the M-sequence approach to cover large arbitrary frequency bands below and beyond 10 GHz and its easy extensibility make it interesting for many different tasks. Furthermore, M-sequence UWB sensors possess significant advantages in some areas, that is the superior stability of the stimulus generator and time base. Especially when small distortions or objects behind a surface are the target, system stability and calibration capabilities are major requirements to detect such weak scatterers. Another class of UWB sensors providing high stability and versatile calibration methods are VNAs. However, such devices are usually found in a laboratory environment. Contrary to this, M-sequence sensors can be highly integrated, small, and rugged. This promotes mobile deployment in the field – something that is currently the domain of classical pulse GPR sensors.

It is imaginable that M-sequence sensors will be used for ongoing supervision of constructions in the future. Small and flat devices could be attached to the surface of, for example, bridges and scan continuously for damages in the pillars over their whole lifetime. It is even possible to integrate the sensors into the buildings themselves – that is distribute them in a concrete surface. Given the possibility of mass production for tiny and flat sensor heads, device costs would be negligible compared to overall construction costs. To prove the potential of M-sequence UWB sensors in non-destructive testing, two example applications will be briefly introduced – assessment of sewer pipe embedding and inspection of the disaggregation zone in salt mine tunnels.

6.4.1
Assessment of Sewer Pipe Embedding

Efficient and long-lasting infrastructure has always been an important task for construction companies. Distribution of water and collection of sewage via underground pipe networks is a prime example. Especially in modern ever-growing urban areas, planning and maintenance of these networks is a complex and expensive task. A major problem is pipe leakage. Investigations have shown that about 80 disturbances per kilometre of pipe can be found in average in the sewer pipe system of Germany. Unlike in less-populated rural regions, chances are high that in order to repair a broken pipe, other constructions or buildings need to be touched as well. Research and experiences over the last decades have shown the main cause for leakages – disturbances in the pipes' bedding zone. When a new pipe is laid, it is usually surrounded by about 30 cm of smooth, fine-grained sand. This embedding protects the pipe by carrying and distributing all imposed forces almost equally over its whole surface thus avoiding extreme local stress to the pipe material. Examples of disturbances are larger stones, metal fragments (e.g. parts of cables or forgotten construction tools), voids filled with water or air, and regions where the bedding sand has not been compressed properly. Over many years, such objects may shift towards the pipe and damage their surface or otherwise lead to leakages due to bad force distribution. Since the causes to these problems already occur during the construction phase, they can be avoided by using NDT methods for inspection of the embedding. Optical cameras are often used to find broken pipes in the first place, but they cannot look behind the pipe into the bedding zone.

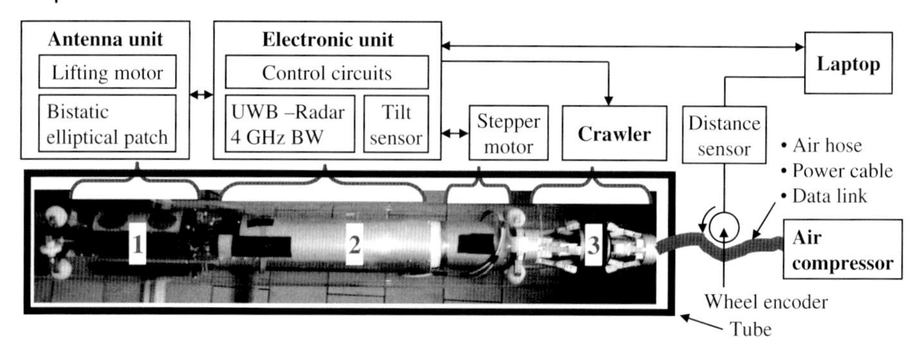

Figure 6.39 M-sequence prototype system for pipe inspection.

Sensors using microwaves have this capability. An M-sequence UWB device has been used in a prototype system to show the effectiveness of the approach.

6.4.1.1 Pipe Inspection Sensor

The sensor system for pipe inspection was intended for use with pipes of 20 cm diameter and consists of three main parts – the pipe crawler unit for movement of the device, the M-sequence sensor unit, and the antenna unit. The prototype is shown in Figure 6.39. It was built for the proof of concept and is not yet optimized with respect to miniaturization. The crawler was used to move the sensor forward in 2.5 cm steps and also featured a stepper motor for revolution of the other parts at every position in the pipe. The crawler unit is driven by a pneumatic system with the air pressure source placed outside the pipe. The whole device is attached to air- and electrical cables and can therefore be easily removed from the pipe if needed.

The middle part carried the M-sequence electronics with respective power supplies, an antenna front-end, and orientation sensors. The basic M-sequence UWB concept has been integrated with a total transmit power of 1 mW covering the frequency range from about 20 MHz to 4 GHz. Higher frequencies are significantly attenuated in soil, especially when humidity is very high. Using even higher bandwidth would consequently not lead to a much improved resolution. The actual radar electronics have become so small that they could be easily integrated into the crawler unit in future systems. The third part of the system is the antenna unit (see Figure 6.40). Two arrays of elliptical patch antennas have been placed in opposite directions. A lifting mechanism has been implemented, which moves the arrays very close to the pipe surface before scanning takes place. This way, repeatable coupling of the waves to the surface and a rather constant air gap could be ensured. Such conditions are beneficial for processing of gathered data.

6.4.1.2 Test Bed and Data Processing

To assess performance of the pipe inspection sensor, a test bed with three common types of pipes has been prepared. Materials tested were concrete pipes, stoneware pipes, and also plastic (PVC) pipes. The test bed contained the standard 30 cm bedding zone of medium-wet fine-grained sand. A number of typical disturbing

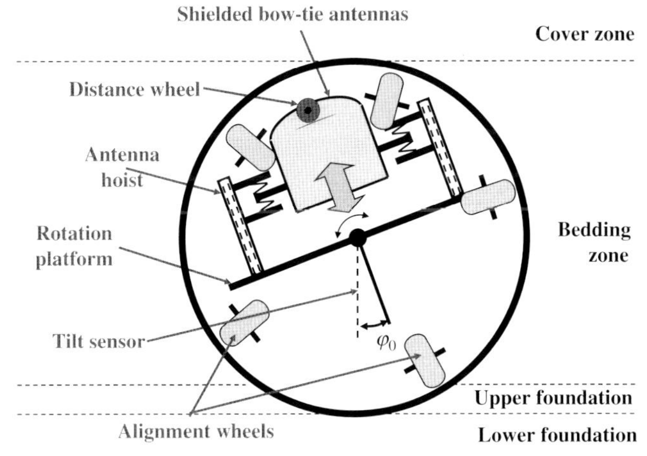

Figure 6.40 Photo and cross section of the antenna unit. A second antenna unit is mounted at the opposite side of the rotation platform. It is not shown in the drawing for the sake of clarity.

objects were placed at known locations throughout the bedding zone. Figure 6.41 shows the test bed and a map of the distortions. Many different targets were used, such as stones, pieces of wood, cables, or voids with water and air (simulated using filled plastic bottles). A number of measurement campaigns have been carried out to test different stages of sensor and antenna implementation as well as to develop data processing for visualization of distortions in the bedding zone.

The best results were achieved by placing the antennas a few millimetres away from the pipe surface to enable fast scanning while minimizing wear. The stimulus waves are therefore coupled via the air interface and hit the pipe's surface at first. Consequently, the main clutter components in received data are antenna crosstalk and the surface response. The device provides two modes of operation. In the simple mode, one pair of patch antennas is used to do a full 360° scan at each position in the pipe while the opposite antenna pair remains unused. Since the distance between antennas and surface is kept constant by the lifting mechanism, antenna crosstalk and surface response stay approximately the same throughout a scan and can be well suppressed by conventional background

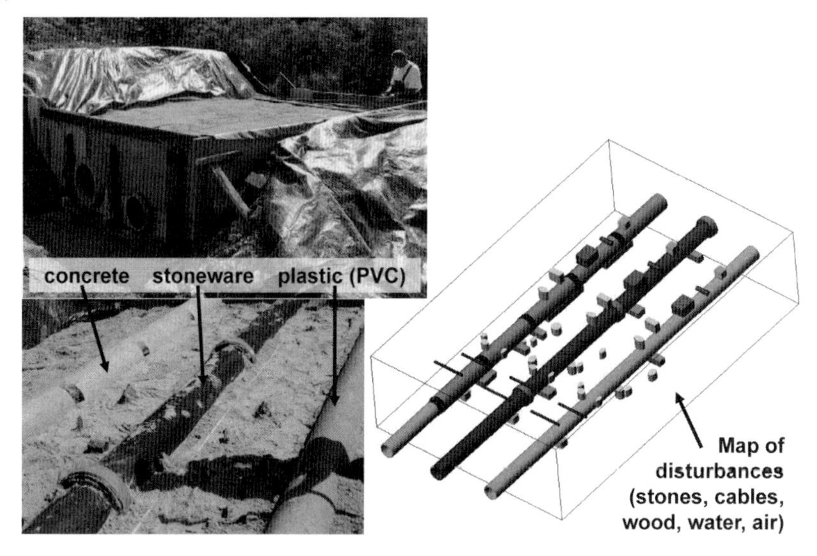

Figure 6.41 Test bed with different common types of pipes and map of disturbances.

removal, that is subtracting a reference A-scan from each profile. This operational mode is useful in situations, where surface, material or shape of the pipe are not very constant over the circumference. In the differential measurement mode, both antenna pairs are working. The transmitters are driven by the same signal and the lifting mechanism ensures a similar wave coupling to the surface on opposite sides of the pipe. The receive signals of the antennae are subtracted by a differential amplifier thus suppressing the main clutter components in hardware. This is an example of how system design can support clutter removal. A perfect cancellation of antenna crosstalk and surface response cannot be expected in reality, but the remaining clutter levels are much lower allowing for a higher sensitivity with regards to small reflections caused by bedding disturbances. Again, simple background removal techniques can suppress residual clutter in the received data. It should be pointed out that successful clutter removal highly depends on stable sensor electronics. If the device changes its behaviour too much during a scan, residual clutter level will increase and may mask weaker reflections from near the surface.

The overall data processing [77] is divided in several steps. Some data conditioning, such as matched filtering, zero time correction and calculation of correct angles using the tilt sensor information is done first. After the already mentioned surface reflection removal, band-pass filtering, interpolation and a distance-dependent gain improve the visual representation and signal-to-clutter ratio.

6.4.1.3 Measurement Example for the Bedding of a Plastic Pipe

Figure 6.42 shows a fully processed example from a plastic pipe measurement in the test bed. The gray-scale (the color scale in the colored figure which is available at

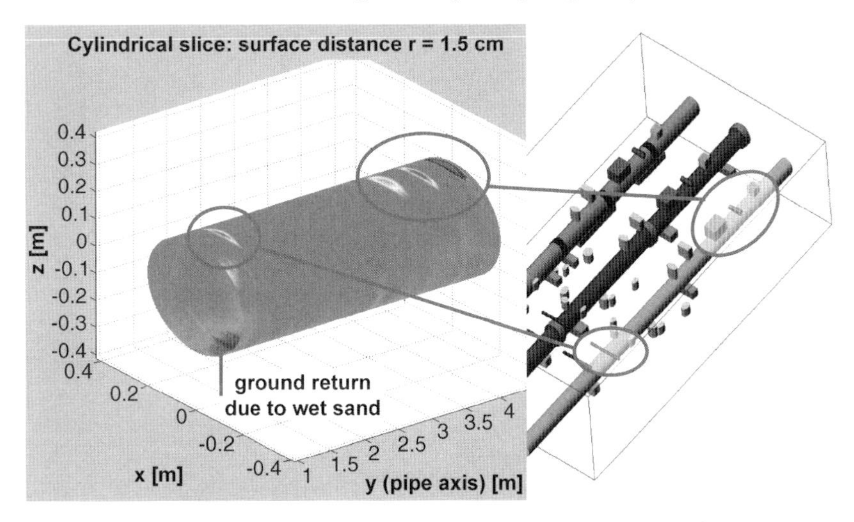

Figure 6.42 Processed result of a plastic pipe scan, slice at $r = 1.5$ cm distance from pipe. (colored figure and movie at Wiley homepage: Fig_6_42. The tube sections shown in the figure and in the movie are different)

Wiley homepage; color references are enclosed by square brackets) represents target reflectivity and the data shows a slice of the bedding zone only $r = 1.5$ cm behind the pipe's surface. 50% gray level [green] represents low reflectivity while black and white [red and blue] colours indicate high response levels. All targets placed closely above the pipe are visible in the example as indicated in the disturbance map. A cable, a piece of wood, a water inclusion, and an air void have been detected. The other targets of the test bed would become visible when the surface distance r of the shown slice is increased up to the maximum interesting range of 30 cm.

Similar results have been obtained for the other pipe materials. It is interesting to note that the amplitude and shape of the target returns are very distinct in the data set. For example, the cable (at $y \approx 1.5$ m) is imaged very sharp and with high positive amplitude (light gray and white [red and yellow colours]). The larger water inclusion (last target at $y \approx 4.5$ m) also shows a significant footprint, but this time with strong negative amplitudes (dark gray and black [blue and purple colours]). Consequently, analysis of the responses could even allow drawing conclusions on the nature of the disturbance.

Another interesting effect was observed in almost all recorded data sets. The bottom section showed generally increased reflectivity as is also evident in Figure 6.42. The reason is a difference in the water content of the embedding sand. During construction of the test bed, medium-wet sand had been used. The test bed was later protected against rain by plastic covers. Over time, humidity of the sand settled in the lower regions below the pipes. This situation could be clearly detected by the M-sequence UWB sensor.

In conclusion, the prototype system performed very well demonstrating many of the advantages of UWB sensors in common and the M-sequence approach in particular. It is also the basis for further development towards commercial use. Another interesting option would be to use such a NDTCE tool to investigate the integrity of older, that is operational, pipes.

6.4.2
Inspection of the Disaggregation Zone in Salt Mines

Salt rock has special properties [78] making salt mines candidates for long-term disposal of problematic waste [79–82]. Especially the self-healing capability in case of defects caused by geological instabilities plays a major role. Nevertheless, when a new tunnel is cut, the surface near salt rock is known to disaggregate by formation of microscopic cracks. This so-called disaggregation zone [83] usually extends 50 cm to 1 m into the rock and evolves over the lifetime of a tunnel. Due to the prospected use of salt mines for storage of waste, researchers and mine companies have been looking into closing constructions for sealing storage areas for a long time [84–87]. However, even the small disaggregation defects provide leakage paths for liquids or gases making sophisticated constructions ineffective, when the surrounding salt rock is not compact enough. To avoid such situations, an UWB sensor for non-destructive investigation of the state of salt rock disaggregation in longer tunnel sections has been developed [92, 93]. A detailed description of the device and measurement results can be found in [88].

Similar to the sewer pipe example, the purpose of inspection in salt mine tunnels is also to look behind the tunnel surface and to find disturbances in the material without damaging it. However, the nature of targets to be detected is quite different. Typical defects are very thin gaps or distributed volume distortions with sizes in the millimetre to sub-millimetre range. The distribution and alignment of such defects is generally not known a priori and may be very different from mine to mine or even among tunnels. Due to geological changes, underground force distribution, or usage-induced vibrations, many smaller distortions may eventually connect over time forming larger, macroscopic cracks. Such cracks can become a security issue and mining companies are already monitoring large cracks by using classical GPR sensors. However, these systems are unsuitable for the detection of small distortions.

A theoretical analysis of backscattering from thin gaps or volume distortions shows that their detection imposes high demands on the device properties (see also Section 4.8.3). For example, to recognize the backscattering of sub-millimetre gaps, a measurement bandwidth beyond 10 GHz must be combined with very high instantaneous dynamic range in excess of 50 dB. While the large bandwidth enables detectable returns from thin gaps in the first place, the high dynamic range is necessary to discriminate them from the surface response. This is caused by the fact that the surface reflection is again by far the strongest signal component and overlaps with the weak responses from disaggregation right behind the surface. Furthermore, device clutter must be suppressed to the same degree. Otherwise false or multiple reflections caused inside the sensor could be mistaken for disaggregation responses. To fulfil these aims, coaxial sensor calibration is mandatory. Such techniques have been known for network analysers but are not commonly found for other UWB principles. Due to the very good stability and repeatability of the M-sequence approach, coaxial calibration was successfully introduced for an M-sequence UWB sensor in this application. Feasibility of the developed prototype system has been shown by various on-site measurements in mines throughout Germany.

6.4.2.1 M-Sequence UWB Sensor for Detection of Salt Rock Disaggregation

For salt mine measurements, the basic M-sequence concept has been extended (see Figure 3.62 in Section 3.3.6.4) and the sensor was combined with an appropriate antenna scanner for acquisition along tunnel sections. The bandwidth requirement was fulfilled by shifting the stimulus spectrum to the master clock frequency (9 GHz) thus keeping everything synchronized to a single clock domain. The resulting stimulus provides 12 GHz bandwidth spanning from 1 to 13 GHz. The equivalent sampling rate of the receivers was increased to 36 GHz to allow direct acquisition of the new spectrum. Due to the unknown distribution and alignment of defects in the disaggregation zone the sensor should acquire fully polarimetric data sets. This was realized by providing two measurement ports connected to a shielded array of two horn antennas with different linear polarization. In order to measure all polarization combinations, the sensor retrieves the full set of scattering parameters (S-parameters) of the two ports, that is reflections at either port and the transmission between the two ports. An automatic calibration unit was included, too. This enables a seamless on-site calibration procedure despite the harsh environmental conditions typically found in underground mines (e.g. high ambient temperatures or the very fine-grained salt dust in the air). The device was integrated into a sealed casing and rugged to survive real-world handling. As a result, the developed prototype represents a two-port time domain network analyser with full two-port calibration for field deployment (compare Section 2.6.4 for introduction of calibration methods).

To limit the complexity of a scanner mechanism for the antenna array, test measurements were restricted to tunnels with a cylindrical shape. Such tunnels are usually cut in one step leading to a smooth surface and have often been used for transportation by a mining train. The standard diameter in German mines is 3 m. Figure 6.43 shows the whole system assembly. The antennas were mounted on a 1.5 m long arm moved along the surface by a rotational scanner. After one scan has been completed, the system was shifted along the tunnel axis in 5 cm steps. The components were mounted on a trolley or lorry and its position was recorded, too. Even though the rotational scanner was designed to place the antenna arm precisely at the aimed angles, it was impossible to ensure a constant distance between the salt rock surface and the antennas like in the pipe inspection system of Section 6.4.1. To enable fast scanning of tunnel sections, contacting the surface had to be avoided, too. Variations of the distance between antennas and salt rock in the order of a few centimetre along a single scan needed to be tolerated by data processing.

6.4.2.2 Data Processing for Detection of Disaggregation

The design of the extended M-sequence sensor combines the benefits from classical GPR sensors with those of vector network analysers. The S-parameter front-end, the calibration capability, and the polarimetric antenna array all support the

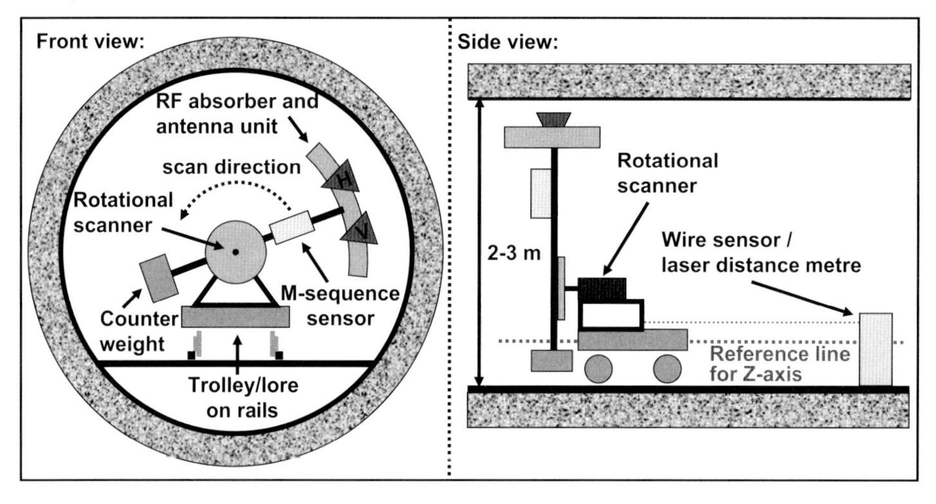

Figure 6.43 Measurement system with M-sequence sensor for inspection of salt rock disaggregation.

requirements of the measurement task in hardware. However, processing the data is a crucial part of the system as well. The realized processing flow is divided into three stages.

The first stage is devoted to data conditioning. It mainly involves the application of a full two-port calibration algorithm. There are different requirements for successful calibration. One important point is system stability. It is crucial that the electronics behave in a repeatable manner after warm-up of the device in the mine. Furthermore, the time base for sampling must be extremely stable, too. The calibration standards are a further issue. These are coaxial objects with known scattering behaviour. For field application it is necessary to select standards that do not change their scattering characteristics significantly when used in various environmental conditions. Most notably temperature dependence should be as low as possible, for example they should be mechanically short. For the salt mine system, an 8-term algorithm using an unknown transmission standard [89, 90] has been selected. This standard is merely a cable connecting the two ports, but the algorithm does not depend on its actual length. The other standards are short reflective objects. Additionally, the 8-term correction allows deriving performance figures for a quick on-site check of calibration success and quality. Device clutter could be suppressed by more than 60 dB using calibration.

The second processing stage handles clutter removal. The two main clutter components are antenna crosstalk and the surface reflection. Crosstalk stays constant during a scan and can be measured in a laboratory before going into a mine (using a calibrated sensor, of course). Its removal is straightforward similar to the pipe inspection example. However, the surface reflection must be removed using a more complex adaptive approach because the distance variations between antennas and salt rock translate into shape changes along a scan profile. Surface removal has been implemented as an optimization problem exploiting the similarity between neighbouring data vectors. The developed algorithm proved to be very effective for the rather smooth surface of cylindrical salt rock tunnels.

The final processing stage takes care of disaggregation visualization. A distance-dependent gain is applied to each data vector. This enhances responses from deeper inside the salt rock suffering from wave propagation losses. The data from individual rotational scans along a tunnel section are then combined into a single picture and represented in an ISO-surface 3D view. In this type of view, only regions showing reflection amplitudes above a tuneable threshold are enclosed and displayed in the coordinate system of the tunnel. Radar imaging techniques have not been considered because it can be predicted that individual distortions will not always be detected separately (due to limited resolution). Consequently, the averaging nature of imaging techniques might destroy traces of disaggregation in the data sets.

6.4.2.3 Example Measurement: A 3D View of Salt Rock Disaggregation in an Old Tunnel

Development of the 12 GHz bandwidth M-sequence sensor, antenna unit, and scanner have been accompanied by various measurement campaigns in salt mines to evaluate the device under real conditions. Final results have been obtained in the mine of Bernburg (Germany) in an old tunnel. Mining activities started in the early twentieth century about 500 m below ground and the investigated tunnel was cut in the 1960s. At the time of measurement, it had not been used since years and the rails were removed. The left part of Figure 6.44 shows a picture of the test site including the relative coordinate system used for scanning. As can be seen, the former completely cylindrical shape was destroyed towards the bottom leading to a disadvantageous force distribution. In the low regions cracks and distortion of the salt rock are even visible on the surface.

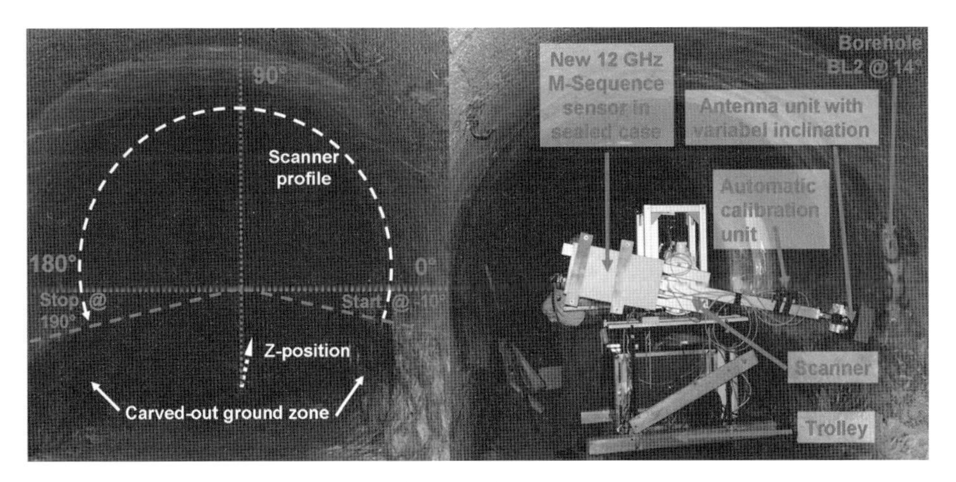

Figure 6.44 Test site and sensor system in the mine of Bernburg.

Measurements took place in two sections near old boreholes – BL1 and BL2. Each measured section was 1 m long. In some positions, metallic hooks or stubs looking through the surface were present.

The actual sensor system is in the right part of Figure 6.44. Due to the lack of rails, the scanner and M-sequence UWB sensor were mounted on a trolley, which was moved manually along the tunnel and mechanically stabilized before each scan. Acquisition of a single scan took about 3 min (in total 1 h/m). Throughout analysis of various other measurements it turned out that best results could be obtained by scanning the antennas very close to the surface. To reduce the surface reflection, the antenna array was slightly inclined by 15° with respect to the salt rock.

Figures 6.45 and 6.46 show the resulting ISO-surface view of the tunnel section around borehole BL2. The first data set represents a reflection measurement with the horizontally polarized antenna, that is it shows scattering in horizontal co-polarization HH. In the second figure, VV polarization is shown.

The figures show some interesting features of the disaggregation zone of this tunnel section. It should be emphasized that surface removal worked very well in this case and that the responses really come from the inside. There are strong distortions in the first 20 cm directly behind the surface over the whole scanned area. Furthermore, deeper cracks and stronger disaggregation are visible near the bottom as expected, where the tunnel shape is distorted. Sensitivity of the sensor drops significantly with distance from the surface. Nevertheless, especially in the ceiling region, thin distortions have been detected up to about 1 m into the top

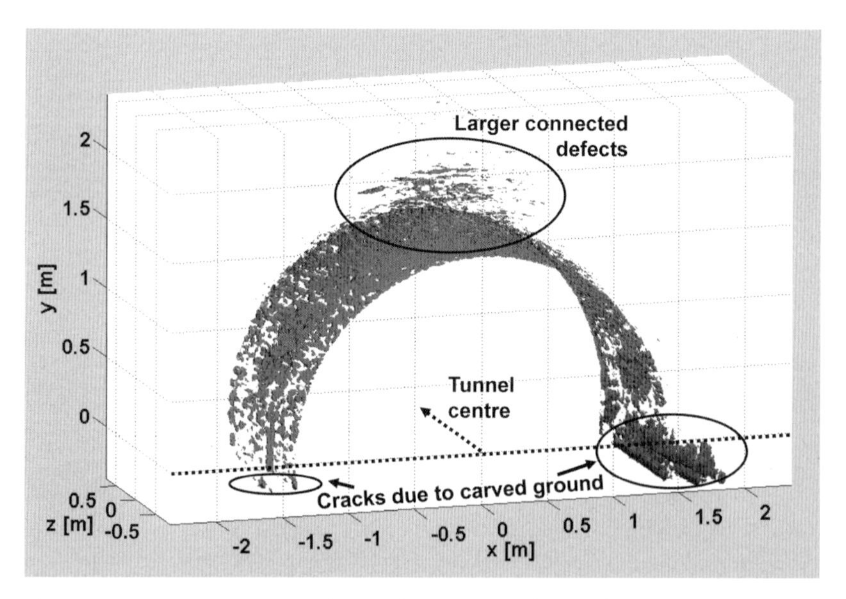

Figure 6.45 Disaggregation zone of a tunnel section around BL2, polarization HH. (movie at Wilex homepage: Fig_6_45⁺)

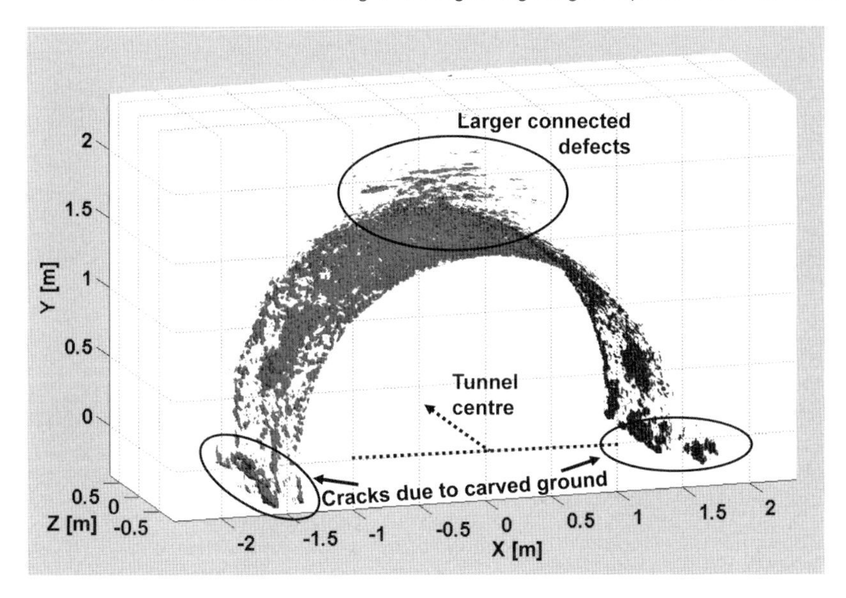

Figure 6.46 Disaggregation zone of a tunnel section around BL2, polarization VV.

rock. Figure 6.47 gives a detailed view of this region. It seems like these distortions are partly connected in a layered fashion.

Overall, the developed M-sequence UWB sensor worked very well in the mine and enabled a look into the disaggregation zone that was not possible before.

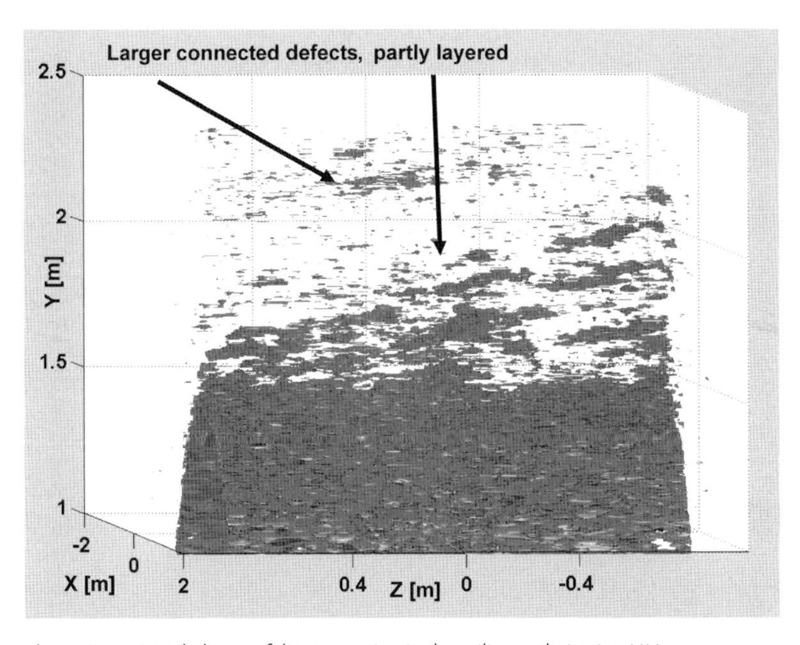

Figure 6.47 Detailed view of disaggregation in the ceiling, polarization HH.

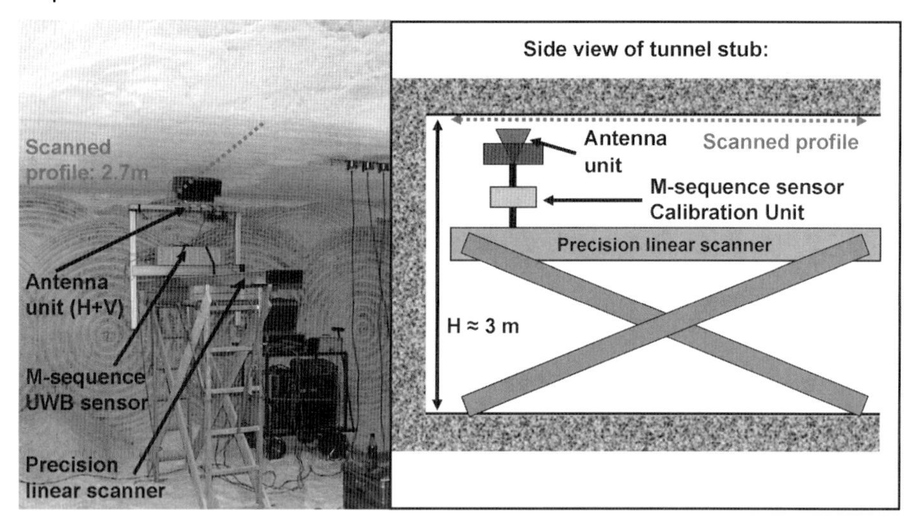

Figure 6.48 Test site and UWB sensor for subsidence analysis in Borth.

Classical destructive methods such as gas permeability tests [91] only give average information for the whole section between the necessary boreholes. Detailed information can only be obtained using modern NDT sensors such as the discussed UWB device.

6.4.2.4 Example Measurement: Subsidence Analysis in a Fresh Tunnel Stub

The feasibility of the UWB approach has been verified by a measurement of a newly cut tunnel in the mine of Borth. This mine is about 900 m below ground with very demanding environmental conditions such as ambient temperatures beyond 30 °C. Its salt rock has special properties, for example the tunnels show significant subsidence right after they have been created which means that the disaggregation zone forms very fast. We had the unique chance to do measurements in a tunnel stub of size 8 m × 3 m × 3 m starting from only 6 h after the cutting process. The stub was cut by an AW4 miner which ensures rather smooth surface conditions at the ceiling and bottom with only small grooves. Figure 6.48 shows the test site and setup. This time a linear precision scanner was used to scan a 2.7 m long track at the ceiling in the middle of the stub. The scans were repeated several times and the data sets have been collected over a 48 h period. This way, the disaggregation zone has been observed from 6 to 54 h after tunnel creation.

The tunnel height H is known to decrease after tunnel creation in Borth due to subsidence. To find out, how much settlement took place, data sets recorded at the beginning and end of observation have been compared. Pre-processing was similar to the previous example from Bernburg: calibration by 8-term correction and static removal of antenna crosstalk. Usually, the surface reflection is considered to be clutter when one is interested in the disaggregation zone. This time it was

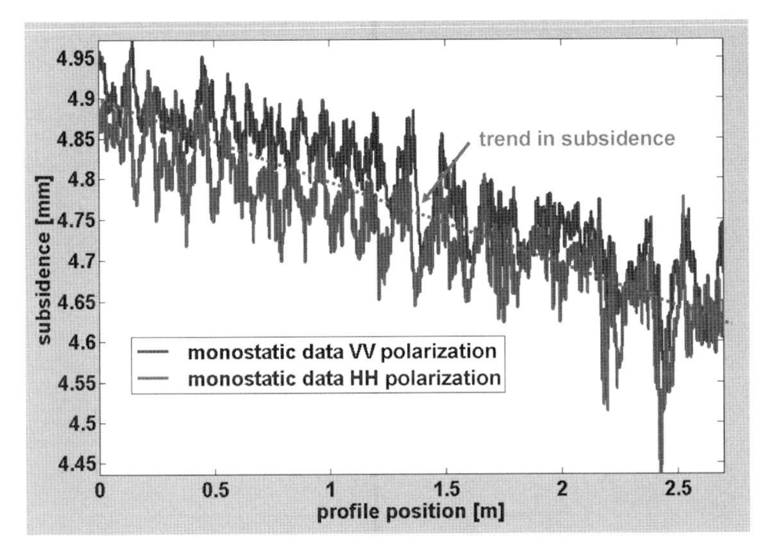

Figure 6.49 Results for subsidence analysis in Borth.

very helpful. The special surface removal algorithm used for Bernburg was modified in such a way that it matches the surface response of both data sets. The algorithm calculates the necessary delay shift and this information can be interpreted as a change in distance during the 48 h of measurements. The analysis result is shown in Figure 6.49 and it coincides very well with expectations. The overall subsidence is about 4.7 mm along the profile and there is an obvious trend. Data collected at the tunnel stub entrance (end of profile) showed a slightly smaller decrease of H than data from the inner stub. The entrance was cut several hours earlier than the inner parts so the speed of change must be slower there, that is the salt rock had already more time to settle. As can be seen in the figure, data from both co-polarizations gave similar results. This makes sense since the reflection from a smooth boundary between isotropic dielectric media should not be sensitive to polarization.

Subsidence information obtained by the M-sequence sensor is itself of use but the mining staff already knew that there would be some change. The importance of this result lies in another aspect. The developed UWB sensor prototype proved to work flawlessly and very stable despite the challenging environmental conditions in Borth. Calibration was effective and robust. The ability of the device to detect small targets and changes could be verified, too. Besides subsidence analysis, data sets from the beginning and end of observation have been compared by looking into the disaggregation zone, too [88]. Significant differences exist between those data sets which clearly shows that the disaggregation zone changed during the second day of measurements. This result verifies that disaggregation really leaves detectable traces in the backscattered data and that the sensor is applicable under real-world conditions.

Acknowledgements

For the research presented in this section we would like to acknowledge the support and help of the following organizations and partners: German Federal Ministry of Education and Research (BMBF), German Federal Ministry of Economics and Technology (BMWi), Forschungsinstitut für Tief- und Rohrleitungsbau Weimar (FITR), Fraunhofer Institut für zerstörungsfreie Prüfverfahren Dresden (IZFP), MEODAT GmbH Ilmenau, BoRaTec GmbH Weimar, Umweltsensortechnik GmbH Geschwenda (UST), Franke Maschinenbau Medingen GmbH, K + S AG Deutschland, Glückauf Sondershausen Entwicklungs- und Sicherungs-gesellschaft mbH (GSES).

6.5
UWB Cardiovascular Monitoring for Enhanced Magnetic Resonance Imaging

Olaf Kosch, Florian Thiel, Ulrich Schwarz, Francesco Scotto di Clemente, Matthias Hein, and Frank Seifert

6.5.1
Introduction

Magnetic Resonance (MR) Imaging is an active imaging technique using nuclear spins, e.g. hydrogen protons of tissue water, as endogenous magnetic field probes and three different kinds of magnetic fields to visualize the tissue distribution inside the human body. To this end, a processing transversal magnetization is excited by a static magnetic field combined with a short RF-pulse at the Larmor frequency of the spin system. Then, due to the coherent precession of the excited magnetization a macroscopic induction signal can be detected with a resonant coil or antenna yielding amplitude and phase information of the evolving magnetization. Since the precession frequency of the magnetic moments is strictly proportional to the local magnetic field, the application of additional magnetic gradient fields will modify the local Larmor frequency in a well-defined manner. By repeated RF-excitation in a proper sequence of switching magnetic gradient fields, 2D or 3D images of the spin density distribution can be reconstructed from the *magnitude and phase* of the MR signal. Furthermore, due to inevitable longitudinal and transversal relaxation of the precessing non-equilibrium magnetization, these images are affected by the relaxation properties of a given tissue type. Dependent on sequence timing, this relaxation weighting became the main source of soft tissue contrast in MRI.

Besides spin relaxation there is a multitude of other physical parameters affecting the MR signal. For instance a moving magnetic moment in a gradient field causes an additional phase of the MR signal when compared to the stationary case. Thus, MR images of moving objects can be severely deteriorated in the whole field of view, not only in regions directly affected by the motion. In particular, MR

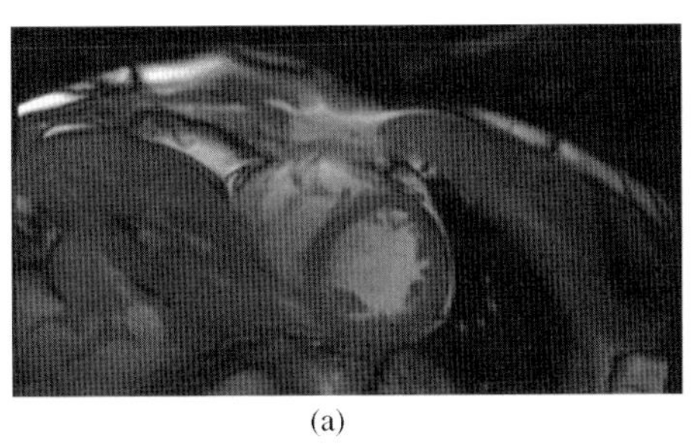
(a)

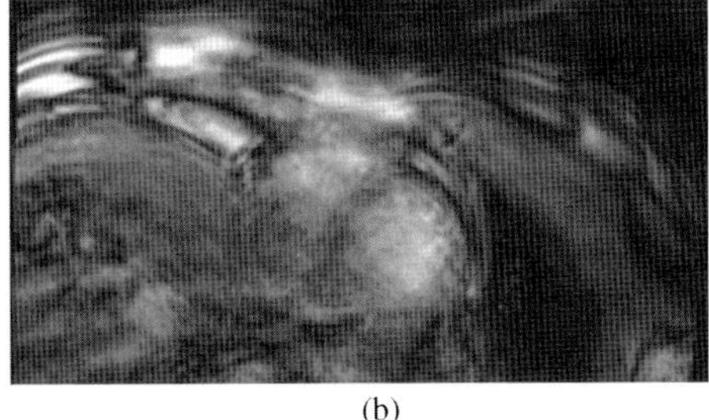
(b)

Figure 6.50 Short axis view of human heart taken from CMR at 3 Tesla. (a) Cardiac gating by pulse oximetry *and* breath hold and (b) cardiac gating only, severe image artefacts occur due to free breathing during image acquisition.

imaging of the human heart (cardiac MRI, CMR) is seriously impaired by cardiac and respiratory motion if no countermeasures are applied (Figure 6.50).

So, CMR requires proper gating with respect to both relevant motion types – cardiac and respiratory motion. To this end, state-of-the-art CMR utilizes the electrocardiogram (ECG) for cardiac gating and breath holding for freezing respiratory motion [93]. Regardless of the undisputed feasibility of state-of-the-art clinical CMR there are several unmet needs particularly when increasing the strength of the static magnetic field to improve spatial resolution in CMR [95]. First of all, at high magnetic fields the ECG is superimposed by magnetohydrodynamic effects [96]. As a result, the ECG may become completely useless at 7T CMR. Moreover, since ECG electrodes are directly attached to the patient's skin there is the risk of local RF-burns, thus compromising patient safety. Alternative approaches exist, for example pulse oximetry (PO) or acoustic cardiac triggering [97], but have their own specific limitations. And just like ECG they do not provide any information about the respiratory state. Since a cardiac patient's breath hold is often limited to 10–15 s, more complex CMR image acquisitions schemes like 3D whole heart coverage or imaging of coronaries [98] would require proper respiration gating to acquire MR data under free breathing conditions. Basically, this can be done by an intrinsic MR technique called MR navigator [99]. To this end, some extra MR excitations are used to monitor the position of the diaphragm during the respiratory cycle. Unfortunately, these extra excitations might interfere with the actual cardiac imaging sequence making this technique extremely complex and less reliable. In addition, the position of the diaphragm is only a rough measure of the position of the heart limiting the image quality and the attainable spatial resolution.

To overcome all these issues and to enhance image resolution in general, we proposed to utilize UWB radar [100, 101] as a navigation system for MRI [102–105]. The anticipated potentials of this technique are (i) to enhance patient comfort and safety and to perform a streamlined clinical workflow due to the contactless measuring principle, (ii) to monitor concurrently a variety of body movements including cardiac and respiratory motion, (iii) to relate directly the UWB signal to tissue mechanics and functional units of organs [106], (iv) to directly measure the position of evolving inner body landmarks and (v) to avoid any interferences of the navigation technique with the MR sequence.

Despite these apparent advantages the actual implementation of UWB radar as a navigation system for MRI is a challenging task. The objectives of such an implementation include (i) measures to reduce low-frequency eddy currents [107, 108] induced in UWB antenna structures by switching magnetic field gradients, (ii) optimization of UWB antenna aperture [107, 108], to fit the UWB antennas into the magnet bore of the MR scanner, (iii) optimization of MR RF-coil assemblies which are sufficient transparent for UWB radar signals, (iv) measures to reduce the crosstalk between the multi-kilowatt transmit body coil of the MR scanner and the UWB antennas (see Section 6.5.3), (v) development of sophisticated algorithms [100, 109–113] for extracting position data from *in vivo* UWB data (see Sections 6.5.2 and 6.5.4) and (vi) adaptation of MR image acquisition and reconstruction schemes [105] for proper utilization of the information from UWB radar (see Section 6.5.5).

6.5.2
Impact of Cardiac Activity on Ultra-Wideband Reflection Signals from the Human Thorax

An analysis of heart motion by numerical finite-differences time-domain (FDTD) simulations is suitable to quantify the effects on the received UWB signal and to prove the detectability of such intra-thoracic displacements. Therefore, we investigated the UWB radar illumination of a realistic human torso model incorporating the geometry of the applied antennas. A model of an adult male from the 'virtual family' [114] was utilized for that purpose (see Figure 6.51a). The human body is meshed with 150 million voxels not larger than 1 mm^3 preserving the details of the antennas, as described in Ref. [107]. A Gaussian pulse of 0.4 ns duration, generated by a current source, was applied for the excitation of the transmission antenna.

The contraction of the heart can be separated in three parts: first, the contraction of the myocardium, where the outer contour is shrinking; second, the increasing thickness of the heart wall and third, the torsion of the myocardium, which results in a further reduction of the ventricle lumen (total reduction of diameter by 25%). To carry out a detectability analysis, we simulated two static states and compared the received signals. The first state was the state of maximum dilatation of the myocardium, the end-diastolic state. The second state renders a moment in the initial contraction phase by a contraction of the outer heart contour by 2 mm and an increase of the heart wall of 3 mm and no changes at other tissue boundaries.

The waveform of the received signal Rx is shaped by the antenna's transfer characteristic, the multi-path wave propagation in the 3D space and the intra-subject reflections on tissue boundaries. The latter depends on the dielectric contrast and the angle of the wave to each boundary (Figure 6.51). The resulting Rx signal begins with the direct crosstalk between transmit and receive antenna followed by the biggest reflection on the first boundary, the air–tissue interface. Subsequently,

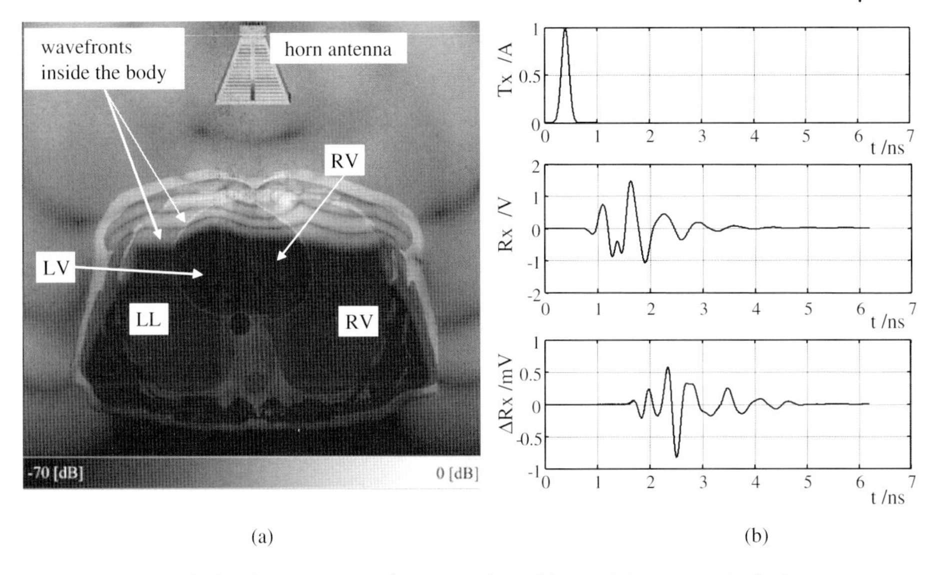

Figure 6.51 (a) Field distribution in an axial cross section extra- and intra-corporal with tissue mesh. LV/RV = left/right ventricle, LL/RV = left/right lung. (b) Transmitted Gaussian pulse, simulated output signal Rx of the receiving antenna in the heart state of end-diastolic phase of the myocardium and the difference signal ΔRx to the contracted state. (colored figure at Wiley homepage: Fig_6_51)

follow the reflections on the heart's wall and lumen. Corresponding to these timings the difference signal ΔRx of the two heart states can be observed at later propagation times. The observed signal variation lies in the range of −65 dB, indicating that with an adequate receiver the detection of such tiny myocardial displacements are indeed feasible, especially if we can assume a bigger displacement during a complete heart beat. This result is consistent with the result we gained from an analytical approach, incorporating a planar stratified model of the human thorax [109, 110].

6.5.3 Compatibility of MRI and UWB Radar

6.5.3.1 Measurements on a Stratified Human Thorax Phantom

The application of UWB systems together with MRI requires sound compatibility considerations [103, 107]. The ambient conditions inside an MR scanner are defined by three different types of fields. First, a static magnetic field of $B_{\mathrm{stat}} = 1.5 - 7\,\mathrm{T}$, generated by a superconducting coil, provides a reference orientation of the nuclear spins of the regions under inspection. Gradient magnetic fields with a slope of up to $\mathrm{d}B_{\mathrm{grad}}/\mathrm{d}t = 50\,\mathrm{T/s}$ are switched during scans to provide the required tomographic information. Furthermore, MRI is based on the resonant excitation of protons, requiring powerful RF-pulses (in the kW range) with narrow excitation bandwidth (several kHz) around the Larmor frequency (42.56 MHz/T). All experiments described below were performed in 'high-field' scanners ($B_{stat} \approx 3$ T) at 125 MHz. An UWB device, on the other hand, excites a material under test with signals offering a bandwidth of several GHz, but the applied integral power lies below $P_{rms} \sim 4$ mW. The SNR of an MR scan is not affected by the UWB signals, since the receiver bandwidth of 10–100 kHz is very low compared to the GHz bandwidth of the UWB system. Moreover, the antennas attenuate the transmitted UWB signal at 125 MHz by more than 100 dB. Comparing MR images taken from an MR head-phantom with and without UWB exposure, within measuring uncertainty, no additional noise could be observed. So, according to expectation, the MRI system was not affected by the UWB signals, as these appear as a low-power noise source to the MR system. Nonetheless special precautions must be taken to reduce eddy currents in the UWB antennas (see Section 6.5.3.2). We established a combined MRI/UWB prototype demonstrating the absence of any mutual interference between both systems, proving the feasibility of the UWB radar method to monitor respiratory and myocardial displacements in a 3T scanner on a stratified human thorax phantom [103]. The composition of the homogeneous and MR compatible phantom were adjusted to closely mimic the dielectric properties of all major organic tissue in the desired frequency range of 1–10 GHz. Additionally, we validated the physiological signatures monitored by UWB radar, utilizing reference signals provided by simultaneous MR measurements on the same subject (see Figure 6.52) [103].

6.5.3.2 Design Considerations for MR Compatible Ultra-Wideband Antennas

The design of antennas suitable for low-distortion radiation of ultra-wideband signals which are compatible with strong static magnetic fields and field gradients in MRI systems is challenging. The gradient fields induce eddy currents in the metallized sections of the antenna which, in turn, interact with the static magnetic field by exerting appreciable mechanical forces. Due to these Lorentz forces, the antenna may move or be deformed, thus degrading the UWB signal integrity. In addition, interoperability of MRI with UWB radar requires minimal interference of the strong 125 MHz RF-pulses with the detection of the tiny radar responses at frequencies above the lower cut-off frequency f_c of the antenna. For sufficient penetration of the electromagnetic UWB signals into the human body, relatively low values of $f_c \approx 1.5$ GHz are preferable, hence causing the risk of superimposing RF-power of the MRI stimulus on the UWB signal.

As described previously, for example in Ref. [107], we have focused on double-ridged horn antennas due to their wide usable bandwidth, high directivity, low side lobes and low dispersion. In order to achieve MR compatibility, the amount of metallization of the antenna had to be ultimately minimized without compromising the UWB performance beyond a tolerable level. The antenna version resulting from careful numerical electromagnetic field simulations is illustrated in Figure 6.53a. All electrically conductive parts were fabricated from thin (about 15 μm) metallized dielectrics, thick enough as compared to the skin depths at 1.5 GHz and higher frequencies, but thin enough to strongly attenuate the low-frequency (kHz) eddy

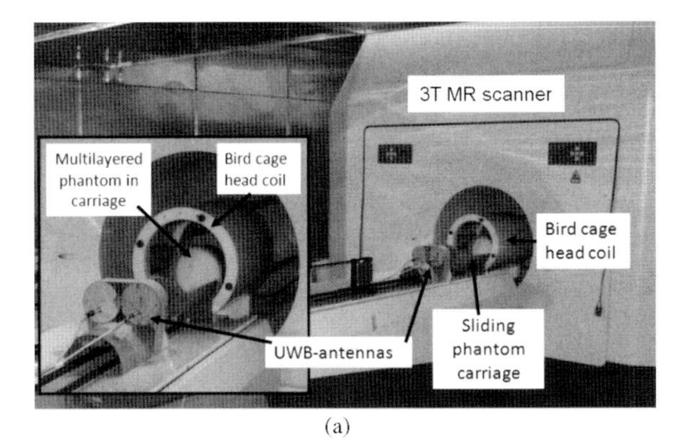

(a)

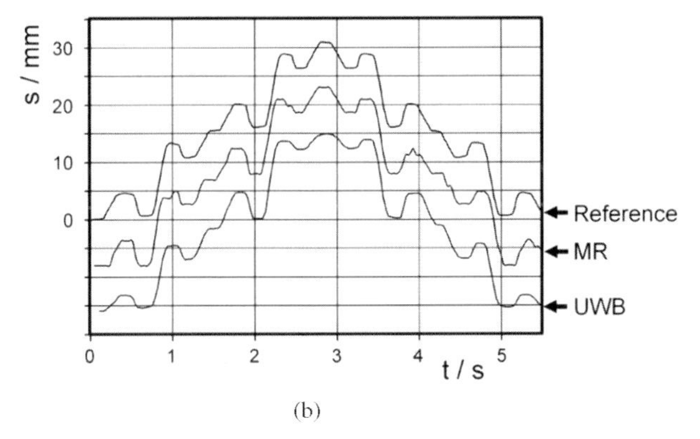

(b)

Figure 6.52 (a) Demonstrator setup inside the MR scanner (Bruker Biospin MRI). The inset shows an enlarged view of the phantom/antenna/head coil arrangement.
(b) Displacement of a stratified thorax phantom inside a MR scanner, approximating a breathing cycle with superimposed cardiac activity. Shown are the nominal displacement curve directly from the stepper motor ('reference') and the displacements detected by UWB radar and MRI. Traces are vertically offset for clarity.

currents from the magnetic gradient fields. In turn, minimizing eddy currents also reduce artefacts in the MR image accordingly. The ridges were perforated in order to reduce the cross-sectional area of contiguous metal faces until the resulting Lorentz forces became weak enough for stable operation of the UWB antenna in the MRI system. In the H-plane of the antenna, the sidewalls were reduced to bars in the aperture plane. Block capacitors were mounted at the major intersections of adjacent metallization parts.

Extensive radiation measurements were carried out, to prove the principle of operation of this antenna and to obtain quantitative information about its radiation properties, with emphasis on the expected low distortion of ultra-wideband pulses. Figure 6.53b compares the angle-dependent fidelity factor of the MR compatible double-ridged horn antenna, referring to the impulse response at bore sight, for both principle planes with the performance of a conventional solid-metal

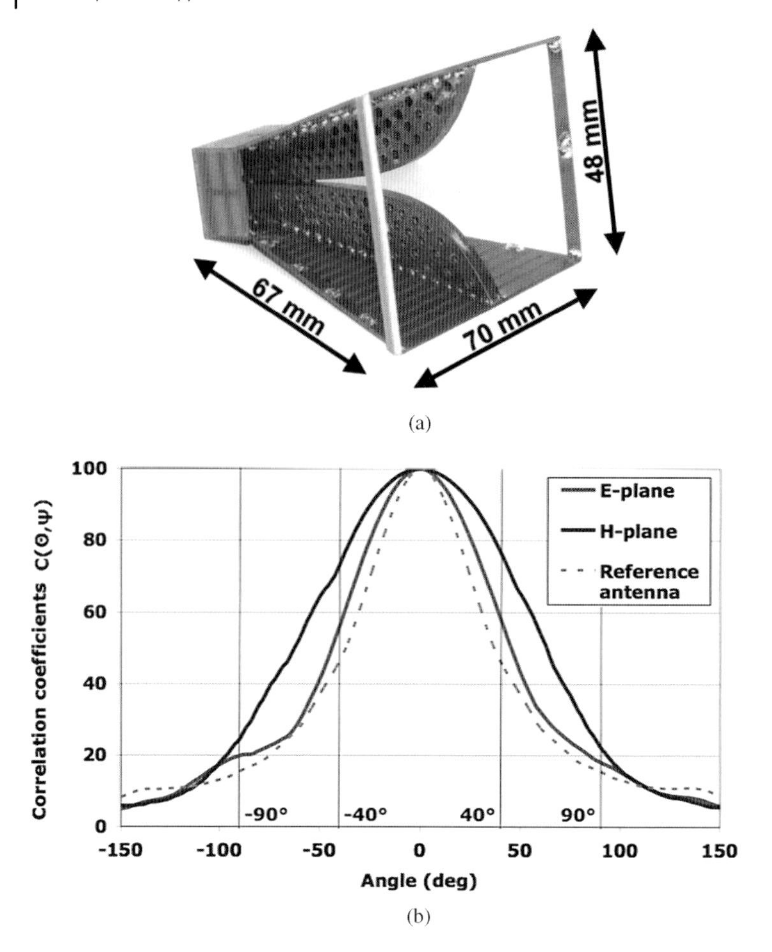

Figure 6.53 (a) MR compatible double-ridged horn antenna for a lower cut-off frequency of $f_c = 1.5$ GHz [107]. (b) Angle-dependent fidelity factor, referring to the impulse response at bore sight direction, of this antenna (solid curves) for both principal planes and of a typical conventional double-ridged horn antenna used as reference (dashed curve). All curves were derived from measured transient responses [107].

version [107]. The similarity of the results in the E-plane proves that the design modifications had little impact on the antenna performance. The differences noted for the H-plane result primarily from the removal of the sidewalls.

6.5.4 Interpretation of Physiological Signatures from UWB Signals

6.5.4.1 Simultaneous ECG/UWB Measurements

UWB and ECG were simultaneously acquired using a radar system with one transmitter (Tx) and two receiver (Rx) channels. One Rx and the Tx antenna were facing

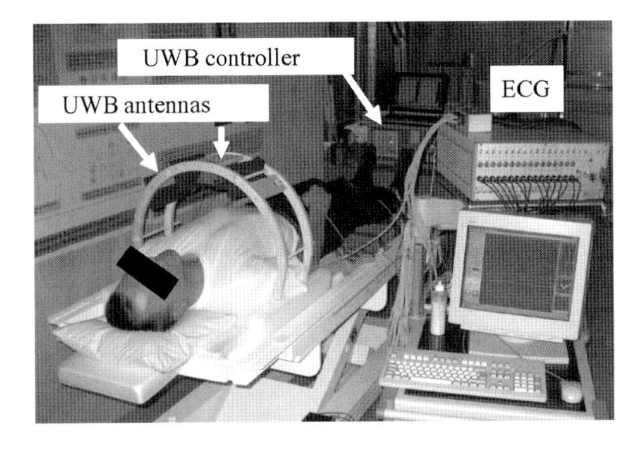

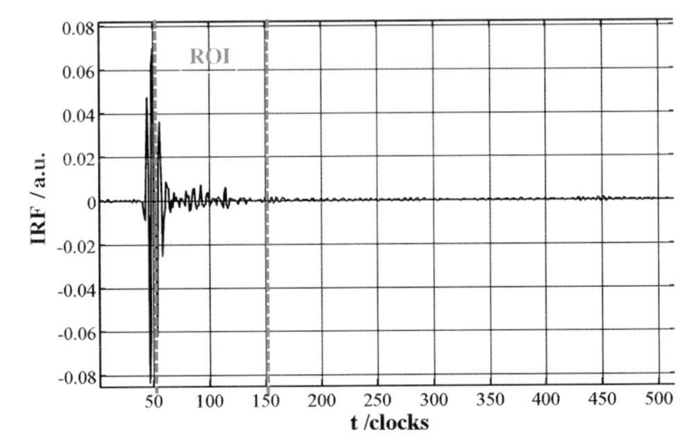

Figure 6.54 (a) Setup for simultaneously measured UWB and ECG and (b) impulse response function (IRF) with region of interest for the analysis of cardiac motion.

the antero-posterior direction, the second Rx antenna was orientated towards the left-anterior oblique direction [106] (see Figure 6.54a). The ECG was recorded with two channels. UWB and ECG data were recorded at 44.2 Hz and 8 kHz, respectively. The transmitted radar signals were generated by a pseudo-random M-sequence with a length of 511 clock signals at $f_0 = 8.95$ GHz [115]. The equivalent UWB power spectrum extends up to $f_0/2$. The lower limit is given by the cut-off frequency of the antennas at 1.5 GHz [107]. The IRF is obtained by correlating the received signal with the M-sequence [115]. In this way, we obtain IRFs with a length of 511 data points per sample.

In the IRFs of all scans, 100 time points were selected for the ROI, (see Figure 6.54b) and transformed to 100 artificial data channels. The selected interval starts after the IRF maximum, corresponding to the simulation results (see Section 6.5.2). In this manner, 200 artificial data channels are obtained from the IRFs of two real UWB channels and are available for decomposition by blind source separation (BSS).

6.5.4.2 **Appropriate Data Analysis and Resulting Multiple Sensor Approach**

The data analysis is based on the blind source separation and assumes a measured signal $x(t)$ to be a linear combination of unknown zero-mean source signals $s(t)$ with an unknown mixing matrix **A**:

$$x(t) = \mathbf{A}s(t) \quad x = (x_1, \ldots, x_m)^T \tag{6.14}$$

The original sources $s(t)$ can be estimated by the components $y(t)$ which can be calculated from the estimation of the demixing matrix $\mathbf{A}^* \approx \mathbf{A}^{-1}$:

$$y(t) = \mathbf{A}^* x(t) = \mathbf{A}^* \mathbf{A}s(t) \tag{6.15}$$

We applied a second-order time-domain algorithm (TDSEP, Temporal Decorrelation source SEParation) [112]. The components of the resulting sources are calculated via Eq. (6.15). The solution is an ill-posed problem but by extending the UWB system with a second Rx channel, we established convenient condition for the solution (see Figure 6.55b).

Automatic identification of the cardiac component was provided by a frequency-domain selection criterion because for non-pathological conditions the main spectral power density of the heart motion falls in a frequency range of 0.5–7 Hz. The algorithm searches for the highest ratio between a single narrowband signal (fundamental mode and first harmonic) within this frequency range and the maximum signal outside this range. To the cardiac component of the UWB signal, a high-order zero-phase digital band-pass filtering of 0.5–5 Hz was applied. In a similar way, respiration can be found by the BSS component with the maximum L_2-norm in the frequency range of 0.05–0.5 Hz. Prior to their comparison, the cardiac UWB and ECG signals were both re-sampled at 1 kHz to retain the more detailed information of the ECG.

To trigger on the latter signal, the usual R-peak detection was applied. In the UWB signal, we chose the points of maximum myocardial contraction during the cardiac cycle. These points appear as minima in the UWB signal (Figure 6.55a: black squares). To enhance the robustness of this detection scheme, we combined it with a gradient calculation at the trailing edge of the minima. Consistency checks on the oscillation amplitude were used to suppress double triggering.

6.5.4.3 **Physiological Interpretation**

In the comparison of cardiac UWB component and ECG, we have to consider the cardiac mechanics as well as the cardiac electric activity. Related to the R-peak, indicating the point of the myocardium's maximal electrical activity, the point of maximal mechanical contraction is delayed. More important for MRI gating, however, is the existence of a rigid relation between trigger points selected by ECG or UWB. The standard deviation of the time lag between ECG and UWB trigger events was below 20 ms and already smaller than the UWB

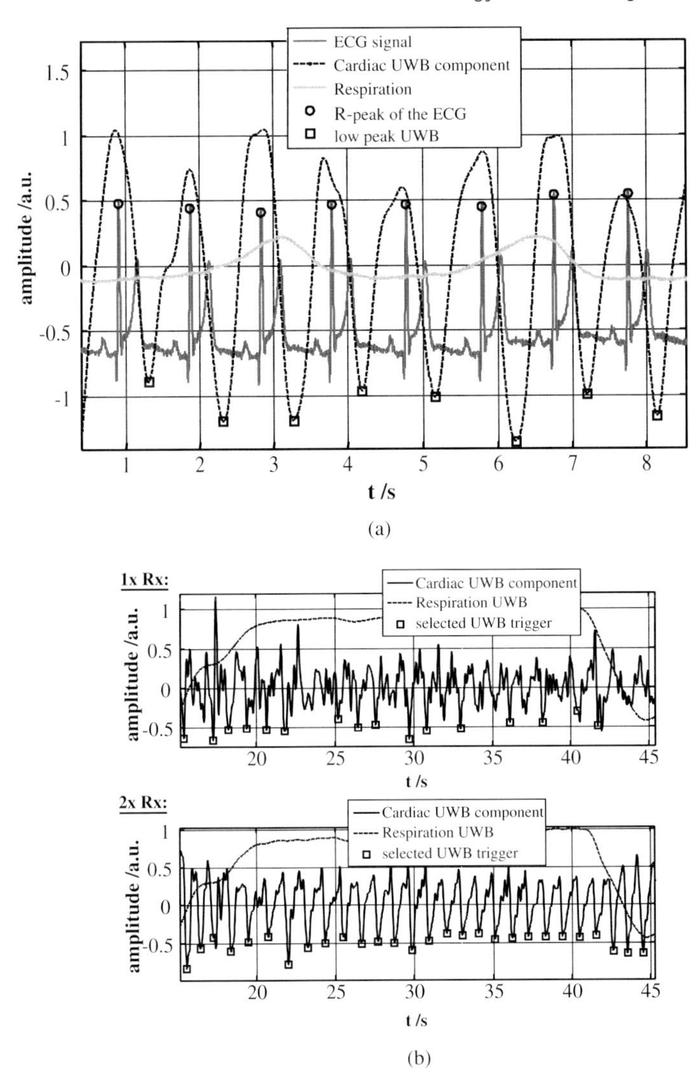

Figure 6.55 (a) Trigger events in ECG by R-peak and UWB signal by combination of low peak and slew rate calculation and (b) resulting cardiac component with trigger events from BSS with one and with two Rx-channels. (colored figure at Wiley homepage: Fig_6_55)

sampling time of 22.62 ms. This can be considered an excellent result, proving the consistency of our procedure.

We validated the proposed triggering scheme by measurements of non-uniform respiration and partial breath holding. An example for the latter is shown in Figure 6.56, where the algorithm rejected several false trigger events correctly. One real heart beat, on the other hand, at $t = 45.5$ s, was rejected, too. This particular beat exhibits quite a singular R–R-duration in the ECG, however, and we can assume a different mechanical contraction compared to the other beats. From the

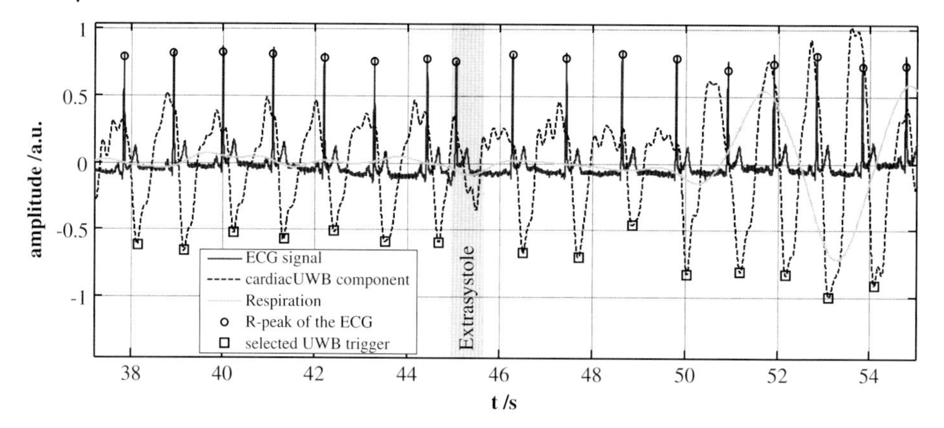

Figure 6.56 Data set with partial breath holding with UWB and ECG trigger events, the trigger event at the extrasystole at $t = 45.5$ s in the UWB component is deselected by the gradient criterion. After the breath hold (duration 30 s) at $t = 50$ s the cardiac motion amplitude was gained for the next five beats. (colored figure at Wiley homepage: Fig_6_56)

CMR point of view, it is presumably even beneficial to have it excluded as image quality suffers from such arrhythmic cycles.

6.5.5
MR Image Reconstruction Applying UWB Triggering

CMR and UWB signals were acquired simultaneously and synchronously to enable UWB triggering. The UWB antenna was mounted in the same frontal position, like before (see Figure 6.57) inside the bore of the 3T MR scanner (Magnetom Verio, Siemens, Erlangen).

To compare our approach with another trigger technique which is applicable in ultra-high fields too, PO was applied simultaneously with UWB radar. The trigger times identified by PO during a clinical MR sequence were replaced by those provided by the UWB radar. From these data, we have retrospectively reconstructed the images.

In this manner, the feasibility of CMR imaging utilizing non-contact UWB radar signals has been demonstrated. There is no significant difference in the reconstructed images of PO and UWB triggered data. An example of reconstructed images by UWB trigger is given in Figure 6.58. In contrast to established techniques like ECG or PO, the contactless UWB sensor provides cardiac and respiratory information simultaneously and thus a sequence-independent external navigator signal.

6.5.6
Outlook and Further Applications

The specific advantages of UWB sensors are high temporal and spatial resolution, penetration into object, low integral power, compatibility with established

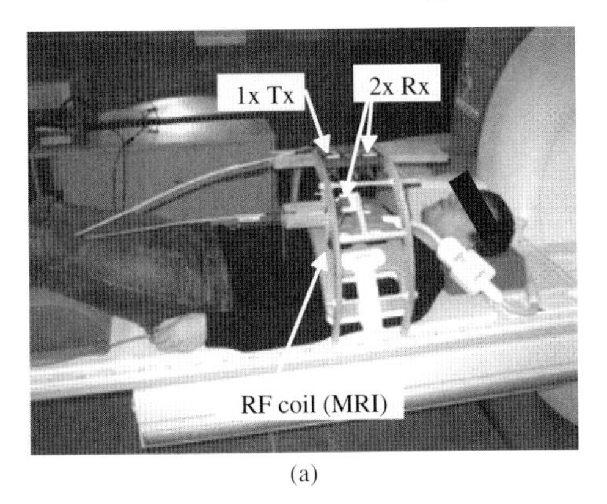

(a)

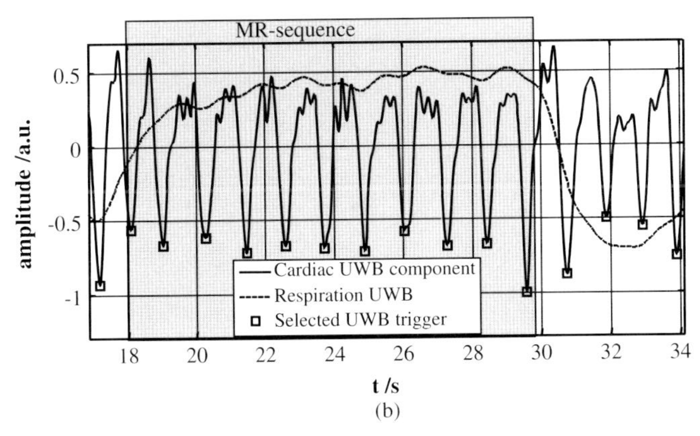

(b)

Figure 6.57 (a) UWB and CMR setup in front of the 3T Siemens Verio MR scanner and (b) cardiac UWB signal with selected trigger events.

narrowband systems and last but not least the non-contact operation. The next steps in the development of UWB sensors together with MRI are the inclusion of more components in the motion correction system, for example inclusion of UWB provided information on respiration and gating CMR for both cardiac and

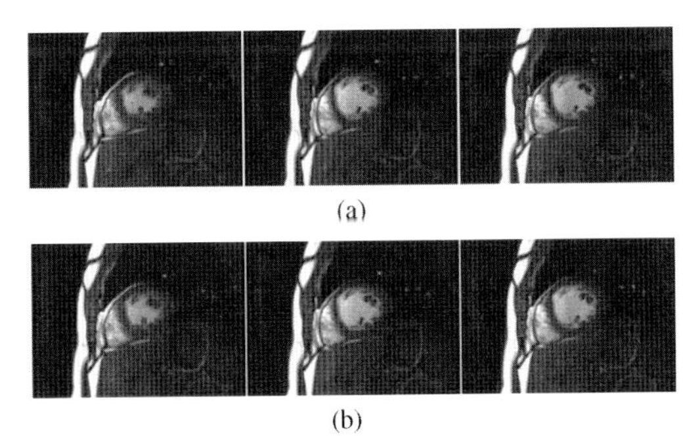

(a)

(b)

Figure 6.58 (a) Reconstructed images by PO trigger and (b) reconstructed images applying the UWB trigger technique.

breathing motions. The suitability of UWB to detect involuntary head motions and brain pulsation has also been demonstrated and could be specifically helpful for high-resolution brain imaging [104]. Further possible applications of *in vivo* UWB radar navigation systems include other imaging modalities, for example X-ray computed tomography (CT), positron emission tomography (PET) and medical ultrasonography (US). The even bigger potential for this innovative modality may even lie in therapeutic rather than diagnostic applications. UWB guided radiotherapy, particle radiotherapy or high intensity focused ultrasound (HIFU) are just a few examples. Lessons learned from all these approaches will foster medical applications of stand-alone UWB radar systems for intensive care monitoring, emergency medical aid and home-based patient care [116]. Further possible employment could be infarction detection, as ischaemic tissue exhibits a modified contraction pattern, potentially accessible by UWB radar [106]. The sensitivity to ultra-low power signals makes them suitable for human medical applications including mobile and continuous non-contact supervision of vital functions. Since no ionizing radiation is used, and due to the ultra-low specific absorption rate (SAR) applied, UWB techniques permit non-invasive, non-contact sensing with no potential risks.

Acknowledgement

This work was supported by German Research Foundation (DFG) priority program SPP1202 UKoLoS (ultraMEDIS).

6.6 UWB for Medical Microwave Breast Imaging

Marko Helbig

6.6.1 Introduction

Microwave UWB sensing and imaging represents a promising alternative for early stage screening diagnostics of breast cancer. This perspective results from advantageous properties of microwaves: sensitivity of the dielectric properties of human tissue to physiological signatures of clinical interest in this frequency range, especially water content, their non-ionizing nature (compared to X-rays) and the potential of a cost-efficient imaging technology (compared to MRI).

Numerous research groups are working in this field since the end of the 1990s. Many studies are dealing with simulations, several groups perform phantom measurements and the least number of all groups have been started first clinical measurements. The challenges which have to be overcome concerning real *in vivo* measurements are multifaceted and depend on the conditions of the measurement scenario. The developed strategies and measurement principles of microwave

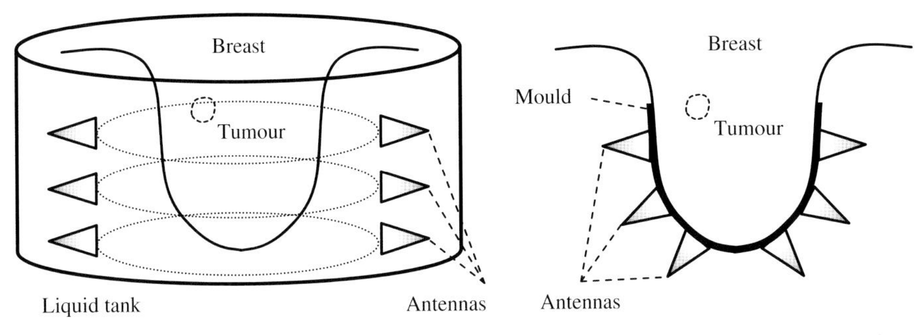

Figure 6.59 Schematization of non-contact breast imaging using a liquid contact medium (a) and contact-based breast imaging (b) in the prone examination position.

breast imaging can be classified by various characteristics: active versus passive versus heterogeneous microwave imaging systems [117]; microwave tomography (or spectroscopy) imaging [118] versus UWB radar imaging [119]; examination in prone versus supine position [117] and other differentiations more. This chapter deals exclusively with active microwave imaging based on UWB radar principle which can be applied in general in both examination positions.

Figure 6.59 shows a continuative comparison according to the contact between antennas and breast surface because this parameter has a great influence on the system configuration and the required signal processing.

6.6.1.1 Non-Contact Breast Imaging

The most significant reason for non-contact breast measurements is the largeness of the used antennas compared to the breast size. Thereby it is impossible to mount a sufficient number of antennas on the breast surface in order to achieve an adequate image quality. The displacement of the antennas from the breast increases the area, in what additional antennas can be localized. Besides that, it allows mechanical scanning where the antennas can be rotated around the breast in order to create a synthetic aperture. On the other hand, this non-contact strategy is attended by a lot of other problems and challenges.

Depending on the dielectric contrast between the medium surrounding the antennas and the breast tissue, only a fraction of the radiated signal energy will penetrate into the breast. The major part will be reflected at the breast surface and represents dynamic reducing clutter which has to be eliminated. In order to reduce the reflection coefficient several approaches use a liquid coupling medium in which the breast has to be immersed and in which the antennas can surround the breast. The same energy reduction effect appears for reflected components from inside of the breast passing the dielectric boundary in the opposite direction. Furthermore, in the opposite direction (from dielectric dense medium into a less dense medium) waves can only leave the breast below the angle of total reflection which signifies an additional reduction of the detectable signal energy outside of the breast.

The individual breast shape plays an important role in connection with these effects as well as for image processing. In Section 6.6.2, we introduce a method for breast and whole body surface reconstruction based on the reflected UWB signals.

6.6.1.2 Contact-Mode Breast Imaging

Contact-based breast imaging avoids the described disadvantages. The antennas are localized directly at the breast surface. Understandably, the applied antennas have to be small enough in order to arrange a sufficient number of antennas around the breast. The corresponding number of signal channels will be obtained by electronic scanning that means sequential feeding of all transmitter antennas with simultaneous signal acquisition of all receiver antennas. This strategy involves the problem of individual breast shapes and sizes which influences the contact pressure of the breast skin onto the antenna aperture and, thus, the signal quality [120].

However, we prefer this measurement scenario for our current investigations. We intend to weak the contact problem in the future by two or three different array sizes and an additional gentle suction of the breast into the antenna array by a slight under pressure. In Section 6.6.3, we present an experimental measuring setup where we pursue a strategy of nearly direct contact imaging in order to conjoin the advantages of contact-based imaging with the possibility of mechanical scanning.

6.6.2 Breast and Body Surface Reconstruction

6.6.2.1 Method

The benefit of the exact knowledge of the breast surface for non-contact microwave breast imaging is manifold and can improve the results significantly. The inclusion of the breast shape information is essential to calculate the wave travelling path in order to image the interior of the breast based on radar beamforming techniques. Some approaches use the surface information for initial estimations. Other non-contact measurement approaches strive to illuminate the breast from a specific distance which requires a very fast online surface identification in order to adapt the antenna position during the measurement. Furthermore, in cases of varying distances between antenna and breast an exact knowledge of the breast surface can enhance the estimation of the skin reflection component for improving the early time artefact removal. If nothing else, the region of interest for which imaging has to be processed can be curtailed based on known surface geometry in order to reduce the calculation time [121, 122].

Additionally to the significance for breast imaging, we will demonstrate the applicability of UWB microwave radar for whole body surface reconstruction which can be used in other medical microwave applications as well as in safety-relevant tasks, for example under dress weapon detection.

The boundary scattering transform (BST) represents a powerful approach for surface detection problems. BST and its inverse transform (IBST) were introduced in 2004 by Sakamoto and Sato [123] as basic algorithms for high-speed ultra-wideband imaging, called SEABED (shape estimation algorithm based on BST and extraction of directly scattered waves). Since then, this idea has been extended from mono-static 2D-imaging (see Section 4.10.4) to mono-static 3D-imaging, culminating in bi-static 3D-imaging (IBBST) [124].

The practical applicability of the original algorithm to the identification of complex-shaped surfaces is limited because of the inherent planar scanning scheme. An adequate identification of the lateral breast regions based on planar scanning over the chest would require very long scan distances. For this reason, the original approach was extended by the author towards non-planar scanning [125] and first results of breast shape identification were published in Refs [126, 127]. The objective of this section is to describe this advanced method and to discuss subsequent development steps which significantly improve the reconstruction quality.

The SEABED algorithm represents a high-speed and high-resolution microwave imaging procedure. It does not include the entire radar signal, instead it uses only wavefronts. Furthermore, changes (derivatives) of the propagation time (transmitter → object surface → receiver) depending on the antenna position during the scan process play an important role.

SEABED consists of three steps: (1) Detection of the wavefront and calculation of its derivative with respect to the coordinates of the scan plane. (2) Inverse boundary scattering transform, which yields spatially distributed points representing the surface of the object. (3) Reconstruction of the surface based on these points.

The original approach is based on a planar scan surface with the disadvantage of illuminating only one side of the object. We extended this approach to a fully three-dimensional antenna movement, including the z-direction [125] based on the idea that in cases of arbitrary non-planar scan schemes the current scan plane can be approximated by the tangential plane at each antenna position. An antenna position-dependent coordinate transform, which realizes that the antenna axis is parallel to the x-axis and the current scan plane is parallel to one plane of the coordinate system, allows the application of the IBBST for nearly arbitrary scan surfaces. Summarizing these conditions, this generalized approach is limited to scenarios where the antennas will be moved orthogonally or parallel to the antenna axis, which is fulfilled in most practical cases (Figure 6.60).

At the transformed coordinate system, the location of the specular point is described by a spheroid

$$\frac{(\bar{x}-\bar{X})^2}{D^2}+\frac{(\bar{y}-\bar{Y})^2}{D^2-d^2}+\frac{(\bar{z}-\bar{Z})^2}{D^2-d^2}=1 \tag{6.16}$$

where $\bar{x}, \bar{y}, \bar{z}$ are the coordinates of the reflective surface point (specular point), $\bar{X}, \bar{Y}, \bar{Z}$ are the coordinates of the centre between the two antennas, D is the half distance transmitter → reflection point → receiver, d is the half distance between the two antennas and the bars above the symbols mark the coordinates of the transformed coordinate system.

To conform the requirements of the IBBST (localization of the specular point based on the distance derivatives) we have to differentiate (6.16) with respect to the

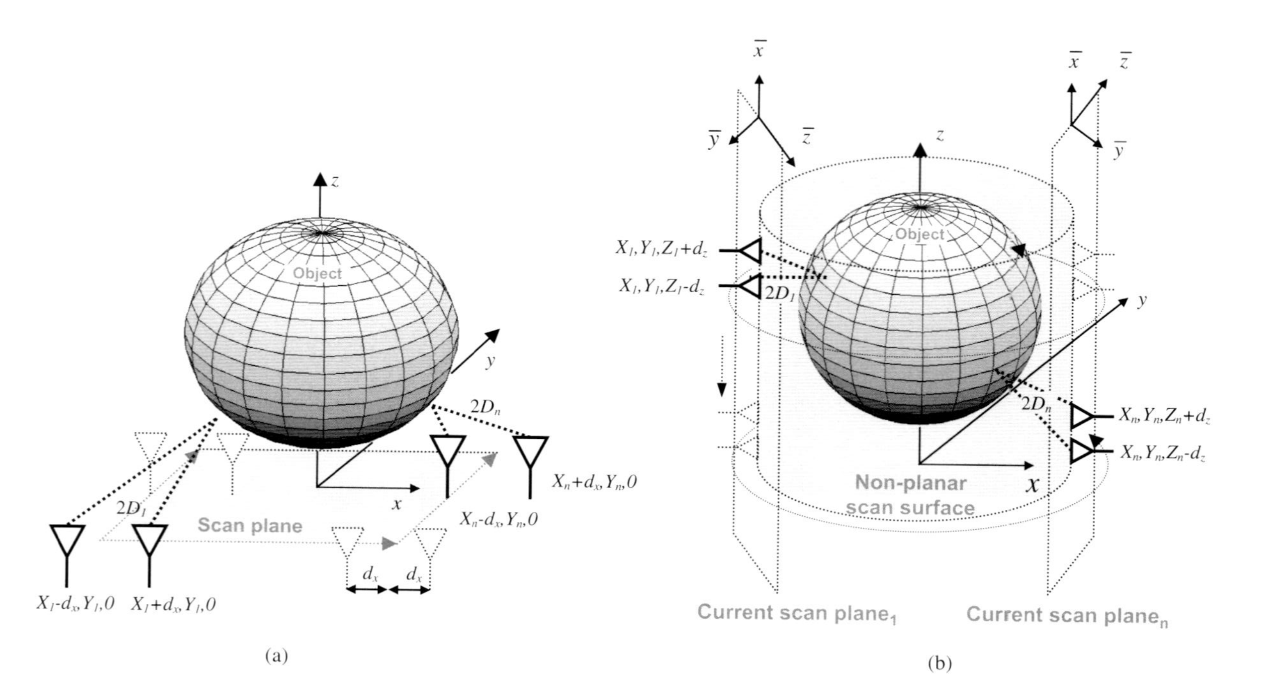

Figure 6.60 Extension of the IBBST-based surface reconstruction from common planar scanning (a) to an arbitrary scan scheme (b). The principle is exemplarily described based on a cylindrical scan.

corresponding values (directions of antenna displacement): $\partial F/\partial \bar{X} = 0$, $\partial F/\partial \bar{Y} = 0$ and $\partial F/\partial \bar{Z} = 0$, where F is simply defined by rearranging of (6.16) $F = (D^2 - d^2)(\bar{x} - \bar{X})^2 + D^2(\bar{y} - \bar{Y})^2 + D^2(\bar{z} - \bar{Z})^2 - D^2(D^2 - d^2) = 0$. In this way, we obtain equations additional to (6.16) for the determination of $\bar{x}, \bar{y}, \bar{z}$ by means of a system of equations. Its solution leads to the transform equations for the three cases, where the scan plane represents the $\bar{x}\bar{y}$-plane (6.17), $\bar{x}\bar{z}$-plane (6.18) and the $\bar{y}\bar{z}$-plane (6.19), respectively

$$\begin{aligned}
\bar{x} &= \bar{X} - \frac{2D^3 D_{\bar{X}}}{D^2 - d^2 + \sqrt{(D^2 - d^2)^2 + 4d^2 D^2 D_{\bar{X}}^2}} \\
\bar{y} &= \bar{Y} + \frac{D_{\bar{Y}}}{D^3}\left(d^2(\bar{x} - \bar{X})^2 - D^4\right) \\
\bar{z} &= \bar{Z} + \sqrt{D^2 - d^2 - (\bar{y} - \bar{Y})^2 - \frac{(D^2 - d^2)(\bar{x} - \bar{X})^2}{D^2}}
\end{aligned} \tag{6.17}$$

$$\begin{aligned}
\bar{x} &= \bar{X} - \frac{2D^3 D_{\bar{X}}}{D^2 - d^2 + \sqrt{(D^2 - d^2)^2 + 4d^2 D^2 D_{\bar{X}}^2}} \\
\bar{z} &= \bar{Z} + \frac{D_{\bar{Z}}}{D^3}\left(d^2(\bar{x} - \bar{X})^2 - D^4\right) \\
\bar{y} &= \bar{Y} + \sqrt{D^2 - d^2 - (\bar{z} - \bar{Z})^2 - \frac{(D^2 - d^2)(\bar{x} - \bar{X})^2}{D^2}}
\end{aligned} \tag{6.18}$$

$$\begin{aligned}
\bar{y} &= \bar{Y} - \frac{2D_{\bar{Y}}(D^2 - d^2)}{D + \sqrt{D^2 - 4d^2(D_{\bar{Z}}^2 + D_{\bar{Y}}^2)}} \\
\bar{z} &= \bar{Z} - \frac{2D_{\bar{Z}}(D^2 - d^2)}{D + \sqrt{D^2 - 4d^2(D_{\bar{Z}}^2 + D_{\bar{Y}}^2)}} \\
\bar{x} &= \bar{X} + \sqrt{\frac{D^2 - d^2 - (\bar{y} - \bar{Y})^2 - (\bar{z} - \bar{Z})^2}{1 - \dfrac{d^2}{D^2}}}
\end{aligned} \tag{6.19}$$

with $D_{\bar{X}} = \mathrm{d}D/\mathrm{d}\bar{X}$, $D_{\bar{Y}} = \mathrm{d}D/\mathrm{d}\bar{Y}$ and $D_{\bar{Z}} = \mathrm{d}D/\mathrm{d}\bar{Z}$ symbolizes the derivatives of the distance with respect to the denoted direction of antenna movement.

In the case of mono-static measurement with $d = 0$, (6.16) represents a sphere and all formulas will reduce to the unique simple form

$$\begin{aligned}
\bar{x} &= \bar{X} - DD_{\bar{X}} \\
\bar{y} &= \bar{Y} - DD_{\bar{Y}} \\
\bar{z} &= \bar{Z} - DD_{\bar{Z}} \\
1 &= D_{\bar{X}}^2 + D_{\bar{Y}}^2 + D_{\bar{Z}}^2
\end{aligned} \tag{6.20}$$

Finally, the calculated surface points have to be back-transformed into the real coordinate system.

The main challenge is the exact detection of the wavefronts and their proper derivative. For the purpose of wavefront detection, we use an iterative correlation-

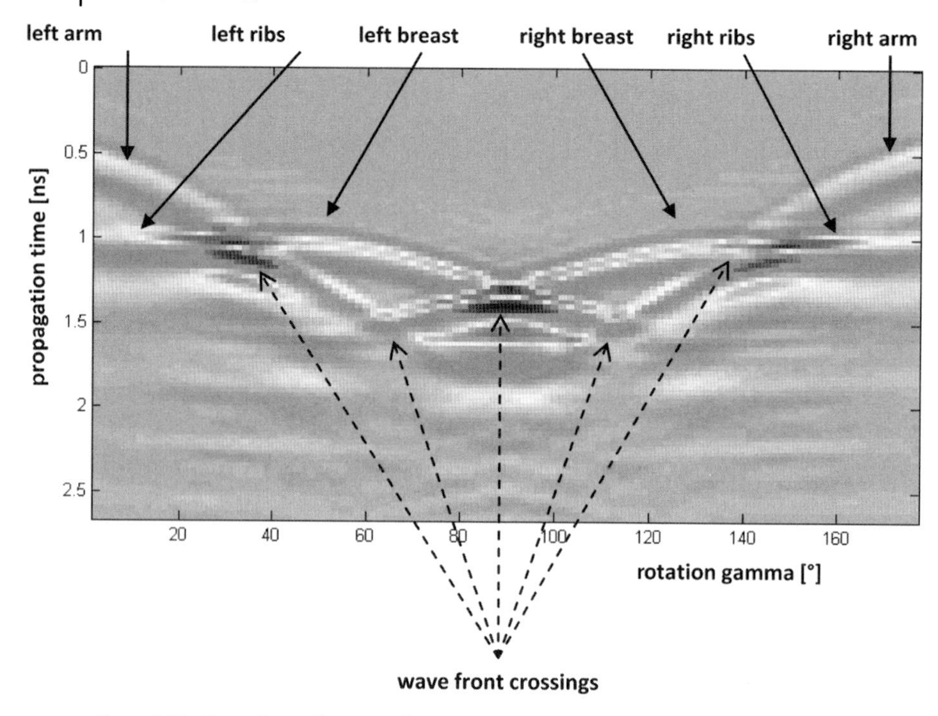

Figure 6.61 Exemplary radargram of a semi-circular scan of a female torso (see Figure 6.65) with the reflected wavefronts and their origins. Additionally, areas of crossing wavefronts are labelled. (colored figure at Wiley homepage: Fig_6_61)

based detection algorithm similar to that in Ref. [128]. The difficulties of obtaining appropriate wavefront derivatives result from the three-dimensional nature of the problem. The antennas are moved and the transmitted waves are reflected in the three-dimensional space. Especially in cases of wavefront crossing and overlapping and sparsely detected wavefronts it is very complicated to recognize, which identified wavefront at one scan position is related to which wavefront at the previous scan position and vice versa (Figure 6.61). So it may be happen that derivative values can be calculated improperly which would lead to spatially false projected surface points. In order to avoid such errors we establish thresholds of maximum feasible derivative values dependent on the antenna beam width.

6.6.2.2 Detection and Elimination of Improper Wavefronts

General Limit Values The range of values of the distance derivatives $D_{\bar{X},\bar{Y},\bar{Z}}$ is theoretically bounded between 0 and 1 depending on the slope of the reflection plane (tangent plane of the object surface at the specular point). In the case of parallelism between reflection plane and antenna axis $D_{\bar{X}} = 0$ (Figure 6.62a) and in the case of orthogonality $D_{\bar{X}} = \pm 1$ (Figure 6.62b). Thus, calculated values >1 are definitely

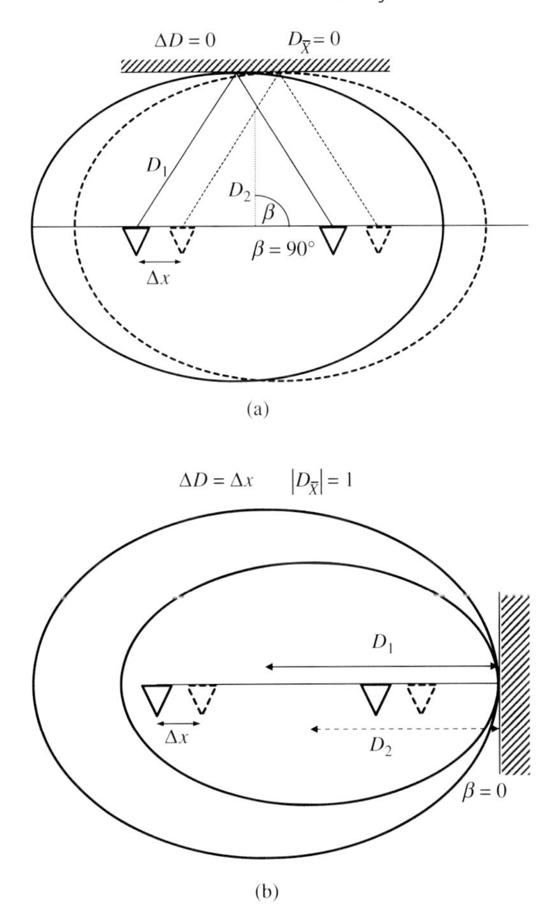

Figure 6.62 General limiting cases of derivative values.

caused by incorrect wavefront detection. Consideration these general boundaries and exclusion of wavefronts exceeding them yields a significant improvement.

Customized Plausibility Limit Values The boundary $D_{\bar{X}} = \pm 1$ assumes an antenna radiation angle of 90° and broader which is not given using directive radiators, for example horn antennas. In that case the range of plausible derivative values can further be restricted. Assuming a maximum antenna radiation angle α $(0 < \alpha \leq (\pi/2))$ and a distance between transmitter and receiver antenna of $2d$ the minimal reasonable value $D_{\min}$ can easily be defined (Figure 6.63). Wavefronts with lower D values would imply specular points which are located outside the antenna beam and, therefore, can be ignored.

$$D_{\min} = \frac{d}{\sin \alpha} \tag{6.21}$$

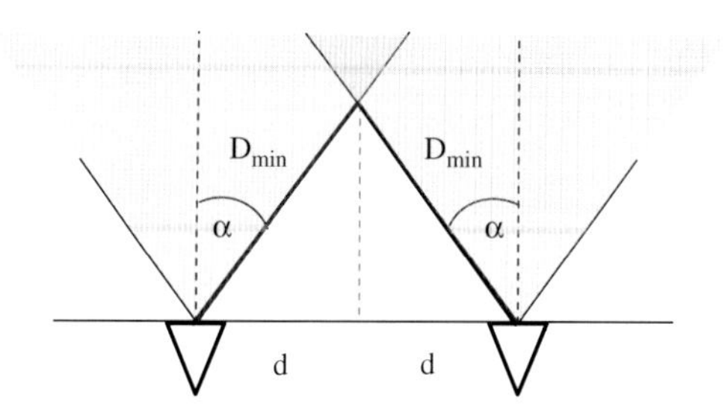

Figure 6.63 Illustration of minimal distance: values lower than D_{min} are localized outside of the illumination cone.

Furthermore, the minimal angle β representing the slope of the reflection plane can be calculated

$$\tan\beta = \frac{D\cos\alpha}{D\sin\alpha - d} \tag{6.22}$$

with the corresponding reflection angle $\gamma = \alpha + \beta - (\pi/2)$ (see Figure 6.64). The distance derivative $D_{\bar{X}}$ will be determined using measurements where the antennas are displaced $\pm\triangle x$.

$$D_{\bar{X}}(\bar{X}) = \frac{D(\bar{X} + \triangle x) - D(\bar{X} - \triangle x)}{2\triangle x} \tag{6.23}$$

We define the value $D_{\bar{x}}(\beta)$ as upper plausible boundary for the assumed antenna beam width α. For its calculation we use the perpendicular L from the reflection

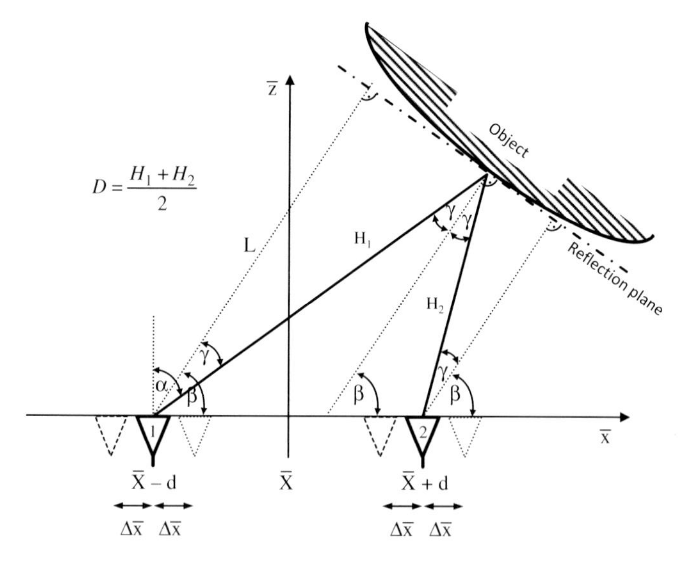

Figure 6.64 Ray geometry for derivation of the maximum plausible derivative values assuming a maximum antenna beam width α.

plane to the first antenna $L = \cos\gamma \cdot H_1 = \cos\gamma \cdot \left((D^2 - d^2)/(D - d\sin\alpha)\right)$ which leads including (6.22) to

$$D = \sqrt{L^2 - 2Ld\cos\beta + d^2} \tag{6.24}$$

The antenna displacement $\pm\triangle x$ results in a change of the perpendicular $\Delta L = \pm\cos\beta \cdot \triangle x$. In conclusion, the equation of the maximum value $D_{\bar{X}\,\max}$ depending on α, d, D can be established

$$D_{\bar{X}\,\max} = \frac{\sqrt{(L+\cos\beta\cdot\triangle x)^2 - (L+\cos\beta\cdot\triangle x)\cos\beta\cdot 2d + d^2} - \sqrt{(L-\cos\beta\cdot\triangle x)^2 - (L-\cos\beta\cdot\triangle x)\cos\beta\cdot 2d + d^2}}{2\cdot\triangle x} \tag{6.25}$$

Equation (6.25) yields $D_{\bar{X}\,\max} = \cos\beta = \sin\alpha$ for mono-static arrangements ($d = 0$) and approaches to this value in cases of $L \gg d$, respectively. That means for $D \gg d$ the threshold can be approximated by

$$D_{\bar{X}\,\max}(D \gg d) \approx \sin\alpha \tag{6.26}$$

The surface reconstruction of Figure 6.67c illustrates the improved accuracy using only wavefronts which hold $D \geq D_{\min}$ and $|D_{\bar{X}}| \leq D_{\bar{X}\,\max}(\alpha = {}^{\pi}/_{4})$ in comparison to the reconstruction based on the weaker criterion $0 \leq |D_{\bar{X}}| \leq 1$ and without verification.

6.6.2.3 **Exemplary Reconstruction Results and Influencing Factors**

For repeatable measurements we apply a female dressmaker torso which is filled with tissue-equivalent phantom material (Figure 6.65). Based on linear and rotational scanners which can move or rotate the object and/or the antennas several non-planar scan schemes can be realized in order to scan this torso efficiently. In the following, the results of two scan scenarios will be presented and discussed: identification of the breast region based on a toroidal scan and the identification of the whole body surface including detection of hidden weapons based on a cylindrical scan. The used M-sequence radar device has a bandwidth of 12 GHz [129].

Numerical problems may arise in the calculation of derivatives from discrete data (discrete time intervals; discrete antenna positions in the space) which have to be considered in setting of measurement and processing parameters. The resolutions of spatial scanning and radar signal sampling have to be harmonized carefully with each other in order to avoid derivative artefacts. The maximum possible error of the derivative according to (6.23) is $\hat{e}(D_{\bar{x}}) = (\triangle t \cdot v_0)/2\triangle x$ where $\triangle t$ is the time resolution of the wavefront detection, $\triangle x$ is the antenna displacement applied for the calculation of $D_{\bar{x}}$ and v_0 is the propagation velocity of the electromagnetic wave. Hence it will be obvious, to meet the requirement of, for example $\hat{e}(D_{\bar{x}}) \leq 0.05$ (0.05 is more than 5% relative error with respect to $D_{\bar{x}\,\max}$ for antenna beam widths $< 90°$!) with an antenna displacement such as $\triangle x = 2.5$ cm in air ($v_0 = c_0$), the wavefront detection has to be realized with an time accuracy of 8.33 ps which has

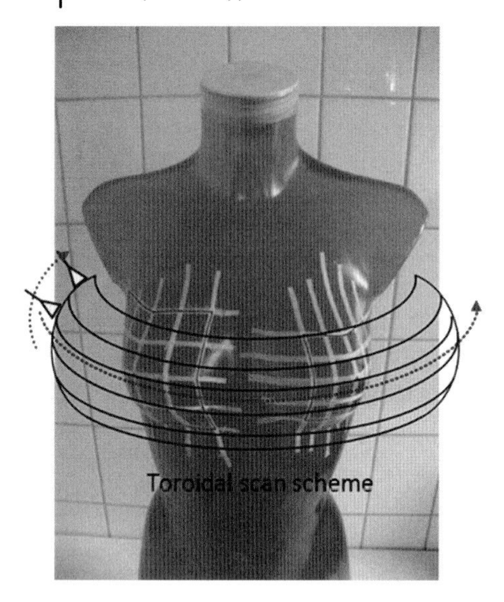

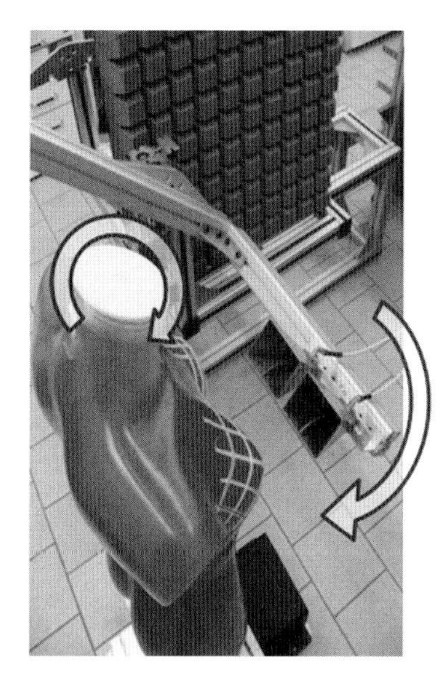

Figure 6.65 Female torso filled with phantom material mimicking the dielectric properties of human tissue and delineation of the toroidal scan scheme in order to reconstruct the chest surface. (colored figure at Wiley homepage: Fig_6_65)

to be provided by interpolation within the wavefront detection algorithm. Naturally, higher performance requirements are in need of even more precise wavefront identification. Figure 6.66 demonstrates this interrelation.

The primary objective for accurate surface reconstruction is to avoid failures and imprecisions regarding to the wavefront detection. But the supplementary detection and elimination of erroneous values represents an advisable secondary strategy. Figure 6.67 proves the applicability of the introduced containment of faultily detected wavefronts by derivative thresholds. Without any verification of the calculated derivatives numerous pretended surface points will be localized more or less far away from the real torso surface (Figure 6.67a). In a first step, the compliance of the general plausibility thresholds $0 \leq |D_{\bar{x}}| \leq 1$ unmasks extremely wrong determined derivatives, which can be seen comparing the initial left reconstruction with the middle reconstruction of Figure 6.67. Because the used horn antennas exhibit a radiation angle smaller than 45° the plausibility criterion can be further tightened applying $D_{\min}(\alpha, d) \leq |D_{\bar{x}}| \leq D_{\bar{x}\max}(\alpha, d, D)$ with $\alpha = 45°$. The improvement of the accuracy is observable (Figure 6.67c). On the other hand, the more restrictive the criterion is all the more it reduces the number of usable surface points leading to sparsely occupied areas. In order to compensate this thinning effect the resolution of the scan grid can be improved by what the number of detected wavefronts and potential surface points can be increased.

Concluding and considering all the described aspects we can demonstrate that the proposed method is a powerful tool for high-speed and high-accuracy surface and shape identification. Figure 6.68 shows the result of UWB

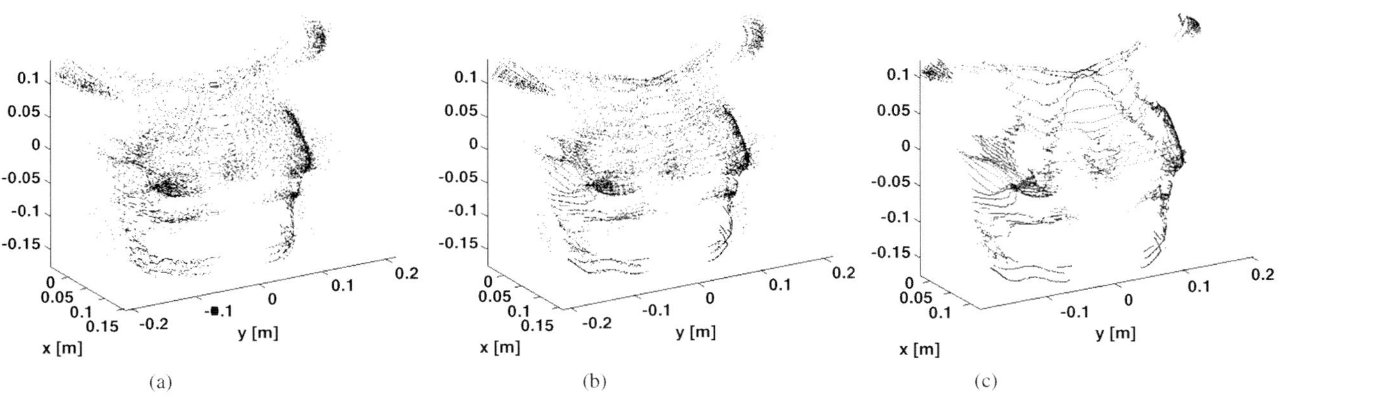

Figure 6.66 Reconstruction improvement due to increased time resolution of the wavefront detection and the respective decreasing maximum errors: $\hat{e}(D_{\dot{x}}) = 0.2$ (a), $\hat{e}(D_{\dot{x}}) = 0.05$ (b) and $\hat{e}(D_{\dot{x}}) = 0.0083$ (c). (movie at Wiley homepage: Fig_6_66*)

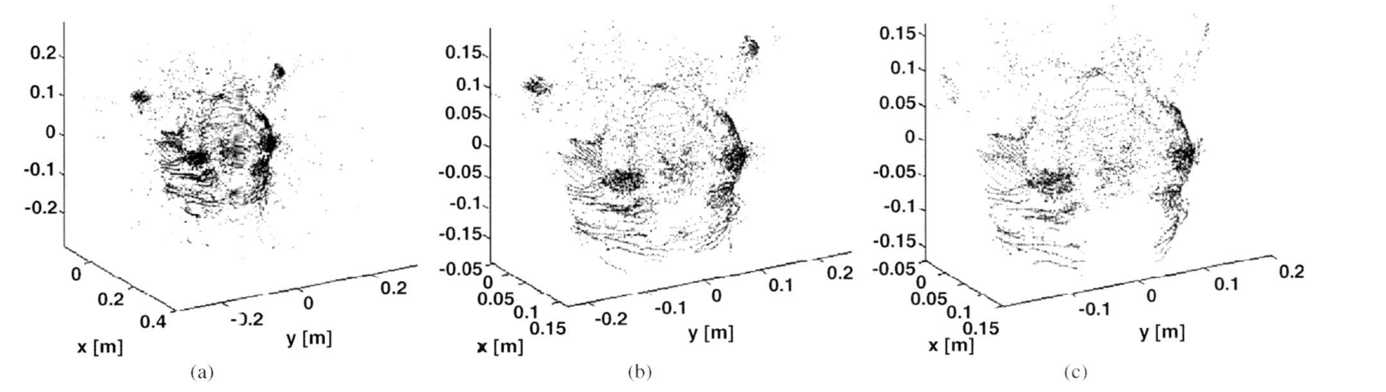

Figure 6.67 Reconstruction improvement due to the application of derivative thresholds: without verification (a), applying the general thresholds (b) and customized thresholds based on an antenna beam width of 45° (c).

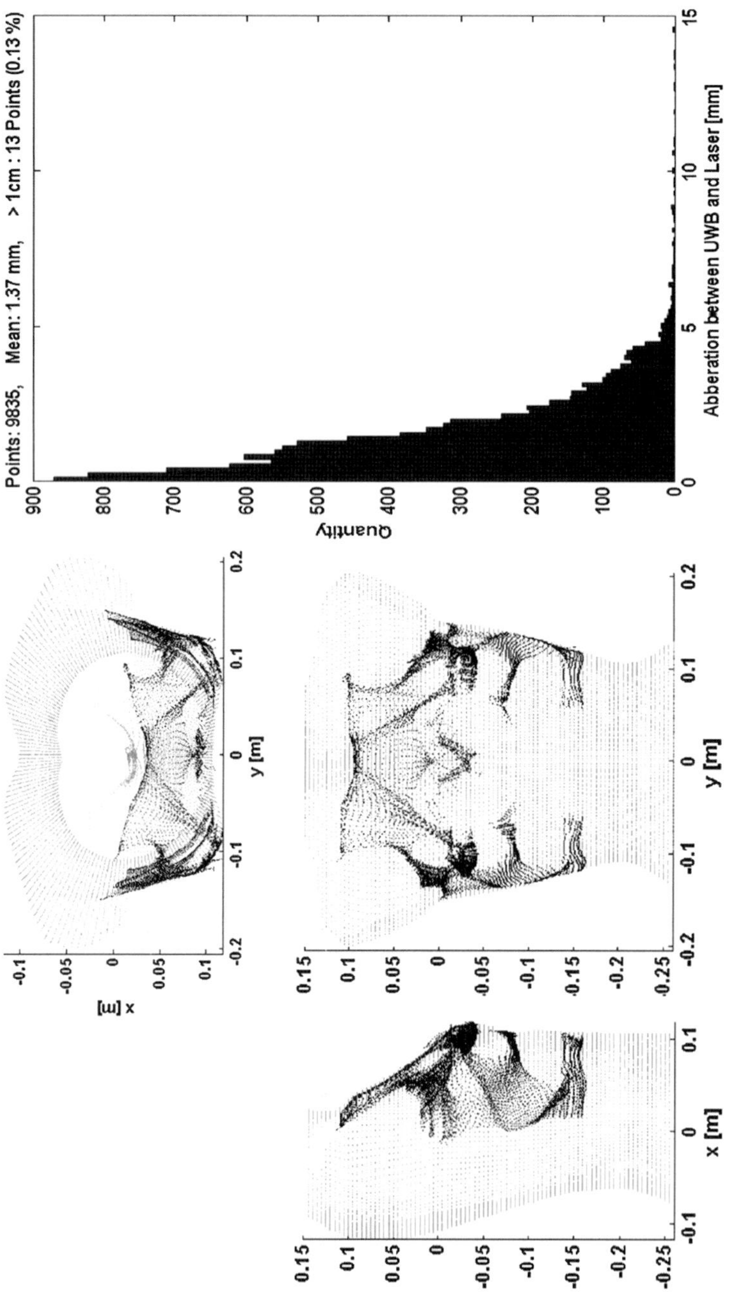

Figure 6.68 Exact UWB chest surface reconstruction and appraisal of performance values by means of a laser reference measurement: the mean aberration is lower than 1.4 mm.

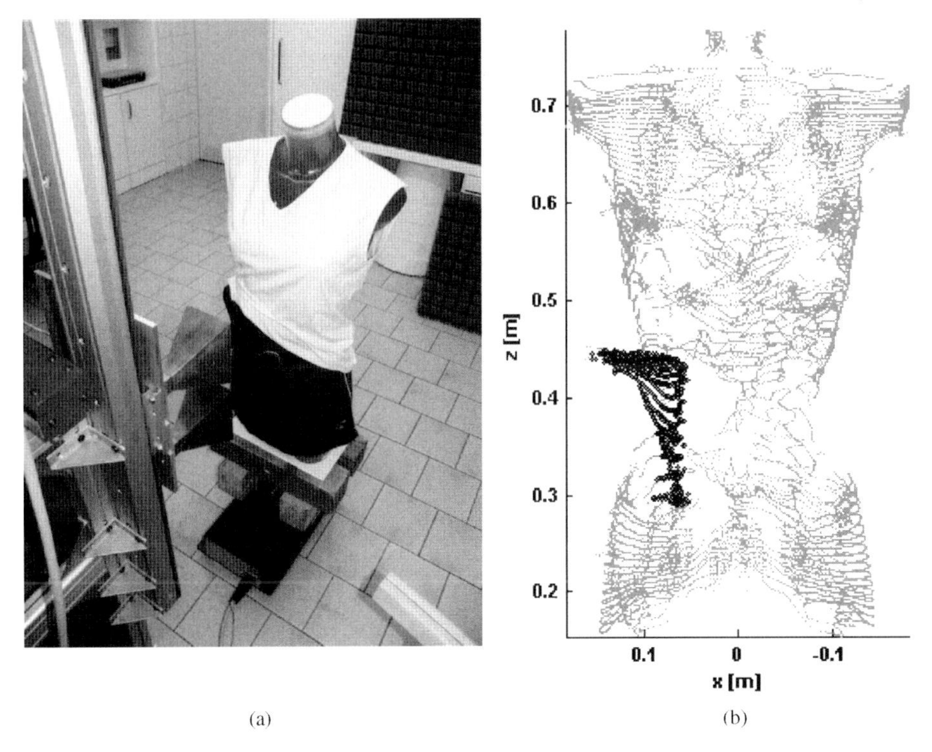

Figure 6.69 Photograph of the dressed female torso with a hidden handgun (a) and its UWB surface reconstruction based on the described algorithm (b). (movie at Wiley homepage: Fig_6_69*)

reconstruction in comparison to a laser reference measurement. In order to quantify the accuracy, the distance between each calculated UWB surface point and the laser-based detected surface is calculated. The resulting mean aberration of lower than 1.4 mm is in evidence with this appraisal. Nevertheless it is obvious that further enhancement of the wavefront detection represents a residual challenge in order to fill in increasingly the areas of sparsely distributed surface points. A genetic optimization algorithm is described in Ref. [130].

Finally, we will demonstrate the applicability of IBBST-based UWB surface reconstruction for other medical applications than breast imaging as well as for security scenarios. For the reconstruction of the whole female torso a cylindrical scan scheme is applied. In addition to the previous experiments the torso is dressed and equipped with a subjacent handgun. Figure 6.69 shows the identification result with the emphasized distinguishable weapon. The low contrast body shape representation suggests that by means of further image processing this approach allows the accentuation of only suspect regions in order to protect the privacy as far as possible.

6.6.3 Contact-Based Breast Imaging

6.6.3.1 UWB Breast Imaging in Time Domain

The both constituent parts of UWB time domain imaging are the removal of clutter (also referred to as early time artefact removal) and beamforming (also referred to as migration or back projection, see Section 4.10). Because the tumour reflections are overlapped by antenna crosstalk and skin reflection, clutter removal is a very important and critical component of signal pre-processing before beamforming can be carried out. Most clutter removal approaches assume that the clutter appears very similar in each channel and, thus, its estimation improves with increasing channel number. It must be noted that this holds only for channels with comparable clutter parameters (e.g. antenna distance Tx-Rx, radiation angle Tx-Rx). That means clutter estimation and removal has to be done separately for groups consisting of only associated channels, which accomplishes this task. In simulation works this circumstance is commonly ignored but in practical applications it has to be considered.

The simplest approach is to estimate the clutter by means of the average value. Tumour reflections are assumed to appear uncorrelated in the several channels and to be negligible in the averaged signal. Even though publications about advanced clutter removal algorithms [131] emphasize the weak points of this self-evident approach, it must be observed that it works relatively robust in cases of covering tumour response by clutter when some of the proposed alternatives are not applicable.

Most of the published image formation algorithms using time domain beamforming can be included in the following generalized formula

$$I(\mathbf{r}_0) = \sum_{\tau_\mathrm{h}=-T_\mathrm{h}/2}^{T_\mathrm{h}/2} h(\tau_\mathrm{h}, \mathbf{r}_0) \cdot \left(\sum_{n=1}^{N} \sum_{\tau_\mathrm{w}=-T_\mathrm{w}/2}^{T_\mathrm{w}/2} w_n(\tau_\mathrm{w}, \mathbf{r}_0) \cdot S_n(t + \tau_n(\mathbf{r}_0) + \tau_\mathrm{w} + \tau_\mathrm{h}) \right)^2 \tag{6.27}$$

where N is the number of channels, $S_n(t)$ is the clutter subtracted signal of channel n, $\mathbf{r}_0$ symbolizes the coordinates of the focal point (image position vector), $\tau_n(\mathbf{r}_0)$ is the focal point depended time delay of channel n and $I(\mathbf{r}_0)$ is the back scattered energy which has to be mapped over the region of interest inside the breast.

Based on two FIR filters the different extensions of the common delay-and-sum beamformer can be expressed. Path-dependent dispersion and attenuation [131, 132] can be equalized by means of $w_n(\tau_\mathrm{w}, \mathbf{r}_0)$ which can be in the simplest case only a weight coefficient. Furthermore, also other improvements can be included by convolution in time domain, for example the cross-correlated back projection algorithm [133]. $h(\tau_\mathrm{h}, \mathbf{r}_0)$ represents a smoothing window at the energy level or a scalar weight coefficient [134].

6.6.3.2 Measurement Setup Based on Small Antennas

We present an experimental measuring setup for phantom trials simulating nearly direct contact *in vivo* measurements at prone examination position. Nearly means the antennas are located very close to the breast surface in a thin (<2 mm) approximately hemispherical dielectrically matching contact layer on which the breast will be attached during the measurement. Assuming two accurately fitted bounding surfaces (e.g. glass or plastic; rotatable into each other) encase breast and contact medium, mechanical scanning without immersion the breast into a matching liquid will be possible.

For our measurements we use M-sequence radar technology [135] developed at Ilmenau University of Technology (see Section 3.3.4). The stimulation signal (M-sequences) can be generated quite simply by high-speed digital shift registers. This promotes monolithic system integration by low-cost semiconductor technologies and can be used to build very flexible and time stable (low jitter and drift) UWB sensor systems. This measurement approach distributes the signal energy equally over time, thus, the signal magnitudes remain low. The reduced voltage exposure of the medium under test prefers this technology for medical applications (spectroscopy and imaging), for example for breast cancer detection. For the described measurements we use a baseband system (bandwidth 9 GHz) containing 2 Tx and 4 Rx channels.

The efficient penetration of the electromagnetic waves into the material under test and the spatial high resolved registration of the reflected signals are crucial tasks of the antenna array design. But from our point of view in this regard efficiency is not only a matter of radiation efficiency or antenna return loss, respectively. An efficient antenna array design concerning biomedical UWB imaging purposes comprises also shape and duration of signal impulses, angle dependency of the impulse characteristics (fidelity) and physical dimensions of the antenna.

These interacting parameters are hard to accommodate to each other within one antenna design. Generally, compromise solutions have to be found considering basic conditions of scanning (mechanically or by means of an antenna array; in direct breast contact or using a matching liquid etc.), tissue properties and image processing. Here we pursue the objective of very small antenna dimension, short impulses and an application in nearly direct contact mode. Therefore, we investigate the usability of short interfacial dipoles.

The antennas are implemented on Rogers 4003® substrate (0.5 mm) using PCB technology. The dimension of the used bowties is 8 mm × 3 mm as shown in Figure 6.70.

The dipoles are differentially fed. The balanced feeding is realized by differential amplifier circuits. Eight antennas are included in this preliminary array setup distributed around a circle segment (diameter 9.5 cm) in steps of 22.5°. Four antennas acts as receiver and are permanently connected with Rx1–Rx4 of the radar device. The transmitter signal can be connected to one of four transmitter antennas by a coaxial switch matrix. Thus, 16 signal channels will be achieved without rearrangement. Their angles between the main radiation directions of Tx and Rx can be differ in the range 22.5–157.5°.

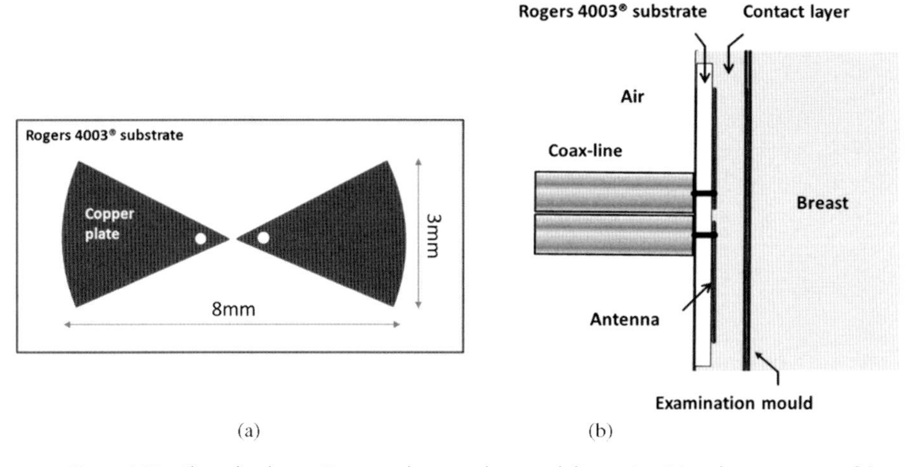

Figure 6.70 Short dipoles on Rogers substrate: shape and dimension (a) and construction of the contact layer filled with phantom material and mounted antenna inside (b).

Because this amount of signal channels is insufficient for high-resolution imaging we have to consider robust and reproducible mechanical scanning by rotating the phantom in order to achieve a sufficient number of channels. To ensure conditions for this, we use two identical plastic containers for the inclusion of the antennas. The antennas are placed in the $\sim$2 mm interspace between both casings which is filled with tissue mimicking material (40% oil phantom material). This very thin contact layer is important for signal quality. Furthermore, the low thickness of this layer and air behind (not matched material) are essential that most of the radiated energy will penetrate into the phantom and only a small part will radiate backwards. The photographs of Figure 6.71 illustrate the assembly stage before casting the contact layer and after completion with inserted phantom.

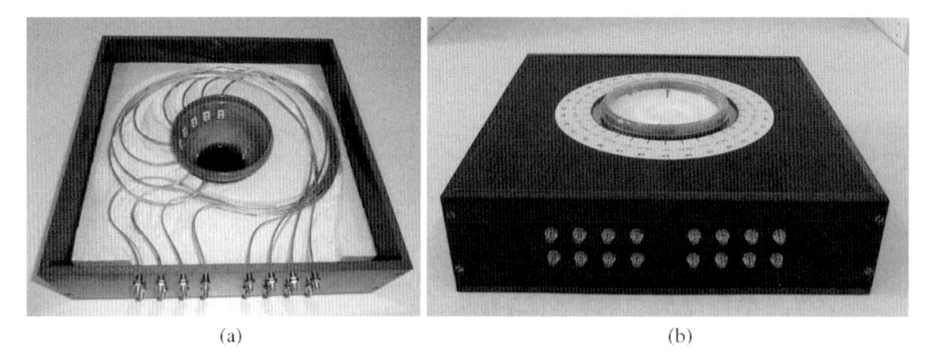

Figure 6.71 Antenna array: assembly stage before casting the contact layer. The connected and affixed differential fed antennas and the container for the outer boundary of the contact layer are still visible (a). Finished antenna array with inserted rotatable breast phantom (b).

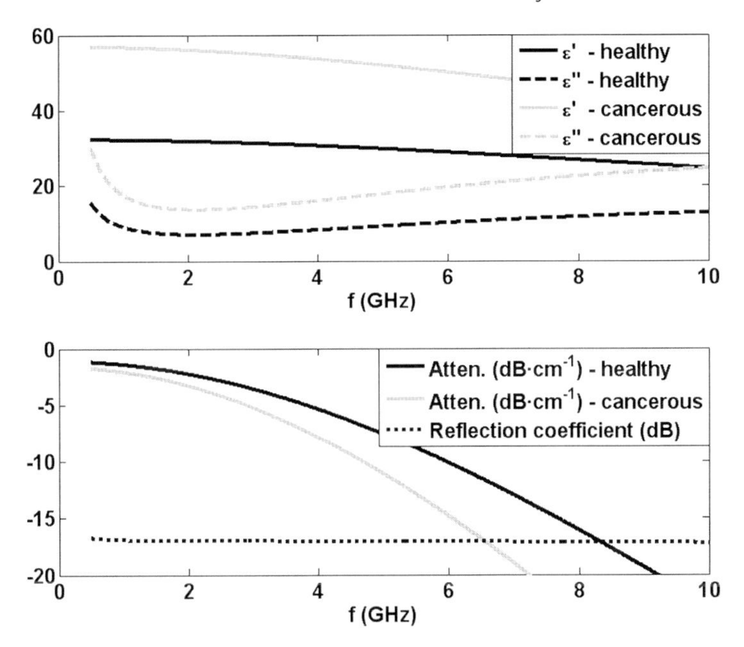

Figure 6.72 Dielectric values of the tissue mimicking phantom material: permittivity (a), transmission losses per centimetre and reflection coefficient between both (b).

6.6.3.3 Imaging Results of Phantom Trials

The phantoms are tissue mimicking oil-gelatine phantoms according to Ref. [136]. The dielectric properties can be adjusted by means of the oil content. For our measurements we used two types of material: 40% oil (57% water) content material mimics healthy tissue which approximately corresponds to group II of adipose-defined tissue (31–84% adipose tissue) [137]. The 10% oil (85.5% water) content material simulates tumour tissue. Figure 6.72 illustrates permittivity, attenuation losses and reflection coefficient between both tissues.

In order to realize an optimal contact to the antenna array the phantom material is filled in identical plastic containers (diameter 9.5 cm) as used for shaping the contact layer. The containers are hermetically sealed and stored in the fridge to avoid chemical instability of the phantom material. The phantoms have to be acclimatized at least 3 h before start of measurements.

In order to simulate antenna rotation, the phantoms will be rotated in steps of 11.25°. This results in 512 signals (16 channels × 32 rotations) which can be included into the imaging process of one phantom. Figure 6.73 shows exemplary imaging results of the described breast phantoms applying the presented measuring setup and time domain beamforming. Despite the relatively low dielectric contrast between both tissue simulations, the tumour inclusions can clearly be identified. The highest interferences (side lobes) are about 11 dB

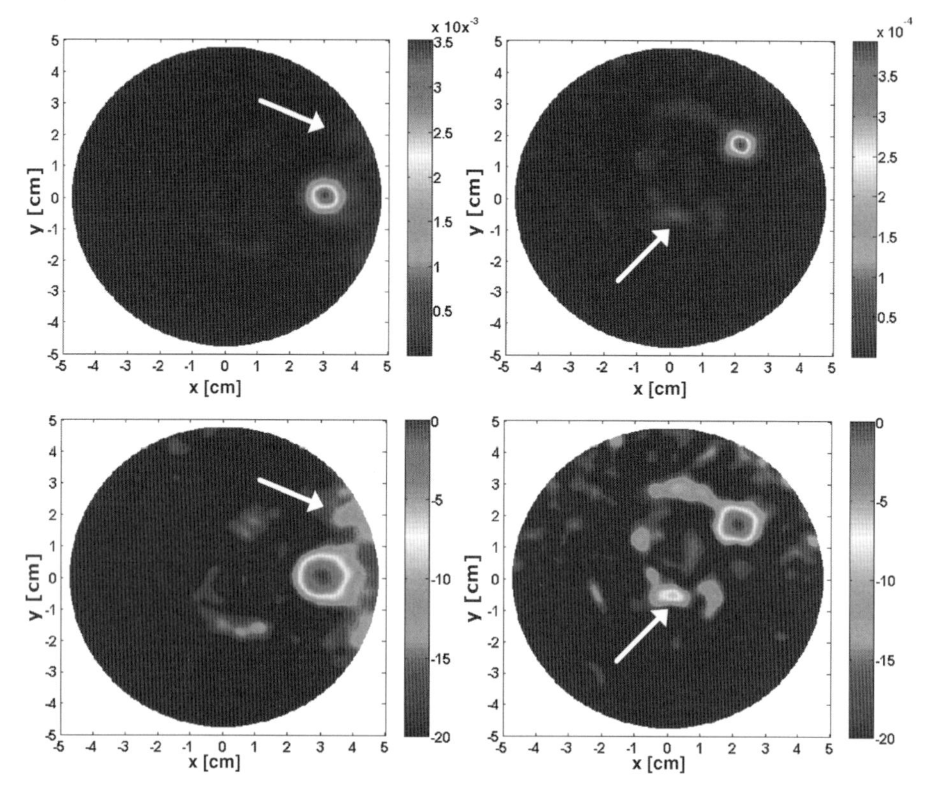

Figure 6.73 UWB images of phantom trials including a 15 mm (left) and 10 mm (right) tumour simulation with linear energy scale (above) and logarithmic scale in dB (below). The arrows mark the highest interferences in each image. (colored figure at Wiley homepage: Fig_6_73)

(15 mm tumour) and around 7 dB (10 mm tumour) lower than the tumour representation.

The results underline that short dipole antennas can profitably be applied for UWB breast imaging. The impressive identification of the used tumour simulations promises also the detection of lesser dielectric contrasts. On the other hand, it must be noted that the tumour surrounding tissue imitation is completely homogeneous which does not correspond to the reality. Therefore, our breast phantoms must be enhanced in the future towards a better approximation of the breast tissue heterogeneity.

Acknowledgement

This work was supported by German Research Foundation (DFG) priority program SPP1202 UKoLoS (ultraMEDIS).

6.7 M-Sequence Radar Sensor for Search and Rescue of Survivors Beneath Collapsed Buildings

Egor Zaikov

Like in many other applications of short-range radars the main advantage of using them in search and rescue arises from ability of electromagnetic waves to penetrate obstacles (building materials in our case) and the significant range resolution of UWB devices. The time window to rescue survivors buried under a collapsed building is restricted to only a few days after the disaster. The most critical problem is the fast and efficient detection and location of such victims in order to focus the rescue forces onto them. Following [138], the search and rescue community is still demanding technical means in order to improve the current situation which is far away from an efficient and fast search strategy.

More specifically, with relation to state of the art of search and rescue techniques, the main benefits expected from UWB radar are as follows:

- **Ability to localize trapped victims:** Techniques used by rescue services at present mostly do not provide reliable information about actual location of trapped person, although some of these techniques outperform UWB radar in terms of penetration depth in many scenarios (narrowband radars, search dogs, sledge hammers). UWB system is expected to bring the ability to localize person due to above-mentioned range resolution in the order of few decimetres.
- **Detection of unconscious victims:** Some conventional methods (like sledge hammers) rely on the response from trapped person. However, unconsciousness of person (or his/her inaptness to act in a certain way) should not be an obstacle for rescuing him or her. For this reason, we describe here a device and a procedure to detect victims by their breathing only and we will provide corresponding measurement examples and system parameters.
- **Detection of multiple persons:** In order to let rescuers evaluate the situation in disaster area, UWB radar brings also the abilities to detect and distinguish multiple victims by their breathing rate and different positions. Again, this can hardly be reached by techniques in use.
- **Detection through at least 1–2 m of rubble:** Penetration depth has to be high enough in order to make all above-mentioned points useful in practice.

First investigations of remote vital sign detection by UWB radar are going back to the time where early UWB modules became available [25]. Most of the early research in the area of breathing–detecting UWB radar was concentrated on ability to detect one person through a wall by sounding within the GHz range. Diverse hardware and algorithms taking into account periodicity of respiration are proposed for this purpose in Refs [139–141] along with measurement examples.

Importantly, the ability to detect multiple breathing persons was experimentally proven in Ref. [142]. Later, in Ref. [143], extensive experimental study of this topic with dozens of volunteers and wall/no wall measurement scenarios confirmed the ability of UWB radar to detect multiple victims.

Although most of the researches use FFT to treat the periodicity of breathing, it was also tried to extract breathing by spectral estimation via MUSIC [144]. Considerable attention to deal with the non-stationarity of breathing received recently in Ref. [145] and earlier in Ref. [146]. In Ref. [147], a tracking algorithm for varying respiration rate is proposed. The non-stationarity of respiratory activity is a major topic to extract vital signs from noisy data for rescue purposes and for health-monitoring microwave systems in general. An interesting approach is presented in Ref. [148]. It deals with the detection of slowly moving objects by moving antenna array and processing via Fourier and principal component analysis (PCA).

Here, our main goal is to report about the full cycle of data processing for breathing–detecting radar including enhancement of breathing, clutter cancellation, localization of breathing person and to test all this abilities under realistic conditions. In more details our results are presented in Ref. [149].

6.7.1
Principle and Challenges

As illustrated in Figure 6.74a, the detection of respiration motion is based on the modulation of the backscattered signal by the motion of the chest which slightly affects the round-trip time. Figures 4.71 and 4.72 depict the effect of periodic motion like breathing onto a backscattered signal and the related example discusses some issues of its detection under noisy conditions.

Under real conditions of rescue operations, the situation exhibits however much more complicated due to numerous perturbations. It shall be symbolized by the sketch in Figure 6.74b. In typical situations, the backscattered signal modulated by the respiratory motion of the victim represents the weakest component amongst all the other waves penetrating the receive antennas. The main perturbation sources and some countermeasures to suppress their influence are summarized as follows:

- **Electronic noise** will limit the sensitivity to detect weak signals. Its influence may be reduced by appropriate receiver design, high transmitter power and sophisticated signal processing. The transmitter power is limited by the saturation level of the receivers and the direct coupling between transmitter and receiving antennas. The aim of signal processing is to emphasize the wanted signal in favour of noise. The achievable processing gain depends on the duration of the observation.
- **Jamming** is caused by (typically narrowband) radio and TV stations as well as mobile devices. Their operation should largely be reduced during the search procedure. High transmitter power of the detection device suppress jamming signals and the use of pseudo-noise codes for sounding the scenario randomize the jammer so it nearly behaves like random noise.
- **Stationary clutter** is the strongest signal which is orders of magnitude larger than the reflections coming from respiratory activity. It is caused from reflections at walls, debris, construction elements and so on. Fortunately, it is time stable so it

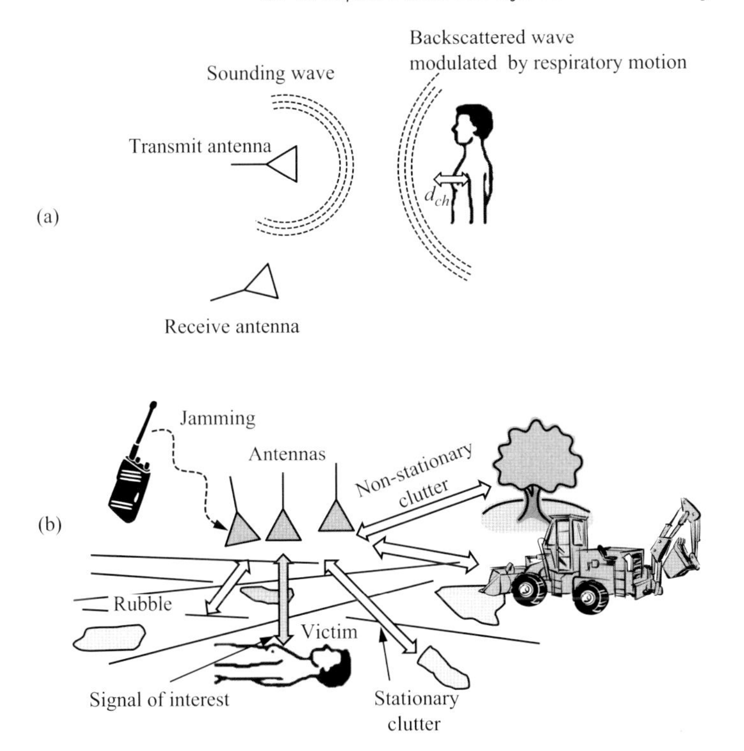

Figure 6.74 Basic principle of respiration motion detection by UWB radar (a) and illustration of perturbing signals under search and rescue conditions (b).

can be simply suppressed by high-pass filtering in observation time. However, this requires absolutely stable and rigid antennas in order to avoid any movement during the measurement. Furthermore, the radar device should be robust against jitter since otherwise the performance of stationary clutter removal will degrade (see Section 4.8.3 for related discussions). The strongest clutter component is provoked by the propagation path of the lowest attenuation which is usually the direct antenna coupling. It will limit the transmitter power because it may lead to saturation effects in the receiver. Therefore, the antennas should be designed for low direct coupling.

- **Non-stationary clutter** may arise from reflections at objects having the same range as the target or from objects which are distant. In the latter case, the clutter may be gated out as long as the unambiguity range of the radar device is sufficiently large. Hence, the unambiguity range of the radar should not be fixed by the maximum detection depth of the victim but rather by the distance of traceable clutter objects. The chance to distinguish between victim and non-stationary clutter is given by their different motion profiles. One can observe that non-stationary clutter often contains strong signal components at frequencies below the breathing rate, which can be used to cut them out. Another approach as depicted in Refs [149, 150] makes use of singular value decomposition and, in

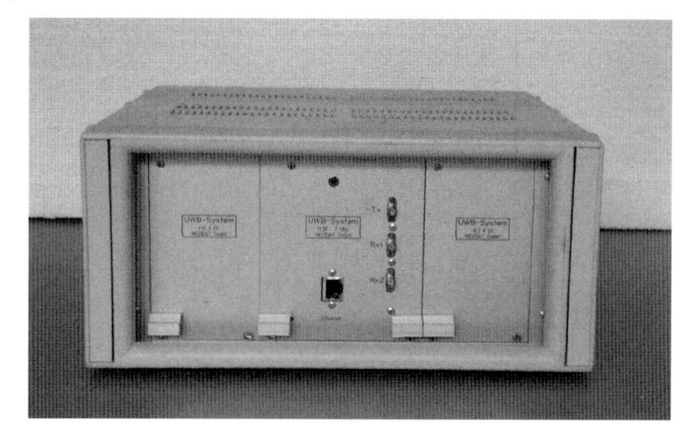

Figure 6.75 Prototype radar system (*Courtesy* of MEODAT, Ilmenau).

particular, its ability to separate the measured data set onto a number of uncorrelated sequences. Furthermore, data fusion of adjacent sensors may be used for spatial focusing by which clutter sources may also be spatially separated from the victim [149].

6.7.2
The Radar System

The core of the prototype system for rescue application reported here is an M-sequence radar with one transmitting and two receiving channels (Figure 6.75, refer to Section 3.3.4 for working principle). Built-in data processing includes impulse compression of the received signal so that the data accessible to the user represent usual impulse response function similar to the signals received by pulse radar. The captured and pre-processed data are transferred via Ethernet to a mechanical robust PC where the main processing takes place. M-sequence technology has some advantages in comparison with pulse radar, for example lower noise and lower jitter.

Additionally, we should mention some requirements to be fulfilled by radar devices if applied for breathing detecting:

- The radar device has to acquire the data fast enough in order to catch breathing. Moreover, even faster data acquisition is often needed in order to reach good separation between breathing signal and high frequent non-stationary clutter as well as to avoid aliasing in observation time. The prototype radar works stable while collecting 32 scans per second which allows us to meet both targets.
- The sensitivity of a radar device to detect weak motion improves with increasing bandwidth or operational frequency, respectively (see Section 4.7.3 for details). On the other hand, electromagnetic waves of high frequency are strongly attenuated while penetrating rubble and building material. Thus, the lower cut-off frequency of the applied rescue radar is an important issue even if it contradicts the first statement. Our experience has shown that spectral components actually

Figure 6.76 Spiral antennas with low cut-off frequency (*Courtesy:* IRK Dresden).

below 300 MHz are very important for victim detection. Frequencies as low as a few MHz can be sensed by the M-sequence radar. Hence, this reduces the choice of a meaningful frequency band to the selection of appropriate antennas.

- The choice of antennas is a compromise between retaining low frequencies, reduced cross-coupling, acceptable size and weight in order to keep the ability to deploy the system fast and easy. Planar spiral antennas seemed to be a reasonable choice from this point of view (see Figure 6.76). Opposite polarizations for transmission and reception reduces antenna crosstalk and allows us to transmit more power, increasing the depth of detection. Chosen antennas are functional in the frequency range from 0.15 to 1.1 GHz and they are about 70 cm in diameter. Evidently, the centres of antennas are always well separated during the measurement that increases our ability for localizing trapped victims.

6.7.3 Pre-Processing and Breathing Detection

First of all we should briefly explain three simple, but nevertheless important steps of data pre-processing with respect to problem under investigation:

- **Shift to time-zero** is absolutely necessary for localization of breathing person in order to compensate the length of RF cables and device internal delays. Reference for this shift can be derived either from additional calibration measurement or from the time position of the pulse related to direct wave between the antennas. Since the antenna distance can be measured, time-zero can easily be calculated.
- **Background subtraction** is used to remove antenna crosstalk and stationary reflections caused by static objects from measured data. Since we do not consider non-stationary clutter at this stage, we just apply high-pass filtering in the direction of observation time. We can do it safely, since breathing is periodic to

high extent and its lowest rate cannot be less then certain value (we used 0.1 Hz). Here we also take into account that low-frequency variation can be further used for estimating non-stationary clutter (see Section 6.7.4).

- **FIR filtering in propagation time** for SNR improvement is used to account for the actual bandwidth of the received signal which is determined by antennas and the propagation conditions of the test scenario.

In our case, the signal which is modulated by respiratory motion always appears at a fixed range and round-trip time t during the whole observation, since the affected person is usually not able to displace. That is, we can safely accumulate data in observation time T in order to improve the detection capability. If the person is not trapped and may change its position, the motion of the whole body provokes much larger but irregular effects in the radar return than it is in the case of breathing. The detection procedure will be different from what we like to consider below. See Sections 6.8 and 6.9 for detection and tracking of walking people.

In Figure 6.77, three representations of breathing signature in radar data are shown for a through-wall scenario. The time-domain representation $h(t, T)$

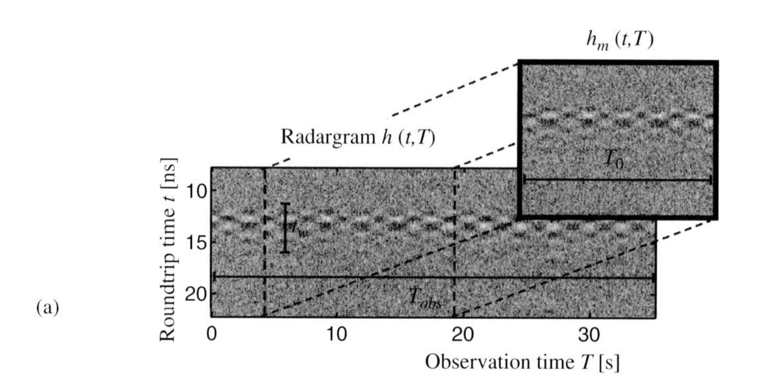

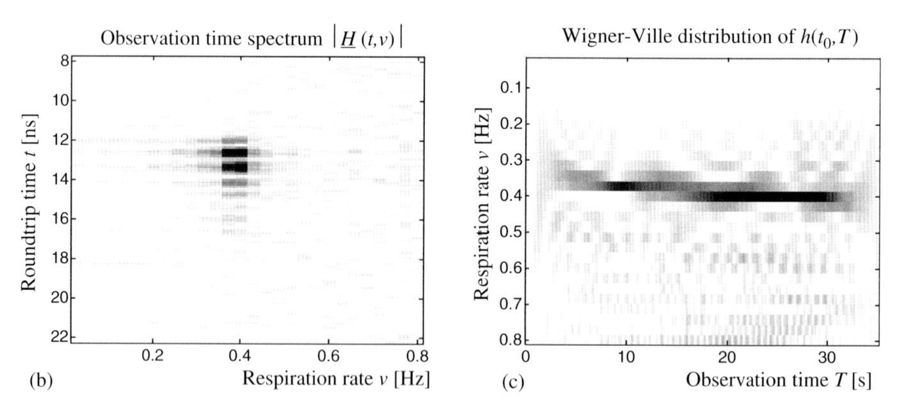

Figure 6.77 Person breathing behind a wall: radargram after background subtraction (a), observation time spectrum (b) and Wigner–Ville distribution (c) related to observation time variations at fixed round-trip time $t_0 = 13$ ns.

Table 6.1 Respiratory rate of human beings at rest in cycles per minute (http://en.wikipedia.org/wiki/Respiratory_rate).

Adult >18 years	12–20
Teenager 12–17 years	12–20
Pupil 6–12 years	18–26
Infant 3–6 years	20–30
Infant 1–3 years	23–35
Baby <1 year	30–40
Newborn baby	30–40
Infant 3–6 years	20–30

(radargram, Figure 6.77a) is the closest to what was originally measured, and one can find that respiration leaves an almost periodic and horizontal trace in the radar data. Under low SNR conditions though, which is typical for our task, this is certainly not the best signal representation for detection since the signal energy is spread over the whole observation time. After 'horizontal' Fourier transform (we call it 'observation time spectrum' $\underline{H}(t, \nu)$, Figure 6.77b), one can observe that breathing-related signal energy is collected into a hot spot (compare also Figure 4.71). In this work, such representation serves as starting point for further breathing enhancement.

The coordinates of the spot give an indications of the target distance and the respiratory rate. The example shows a rate of about 0.4 Hz (24 cycles/min). Table 6.1 summarizes typical values for human beings. The round trip time is about 13 ns. This corresponds to a target distance between about 0.8 m for propagation in concrete ($\varepsilon_r \approx 6$) and 2 m for propagation in air.

A closer look onto the periodicity of respiratory motion via Wigner–Ville distribution shows that breathing is not stationary, that is magnitude and rate of breathing usually vary with time.

Figure 6.77c depicts the joint time–frequency representation (Wigner–Ville distribution) of breathing motion gained from the horizontal data at about 13 ns. The example shows a weak decay of respiration rate during the observation time. Strong noise suppression requires long observation time. However, the fluctuation of the breathing rate will limit the performance of a simple 'horizontal' FFT over long time windows. Therefore, various time–frequency methods were applied to solve the problem [145, 146].

Figure 6.78 illustrates the problem by a simple example. The detection fails in Figure 6.78a where the Fourier transform was performed over the whole length of observation time T_{obs}. In Figure 6.78b, detection succeeds. Here, the captured data are separated into shorter (possibly overlapping) time segments $h_m(t, T)$ of duration T_0 (compare Figure 6.77) from which the spectrum was individually calculated and superimposed (Welch's method).

Obviously, the frequency grid is much thinner in the left spectrogram but we are not concerned with high-frequency resolution. In our work, we were mainly

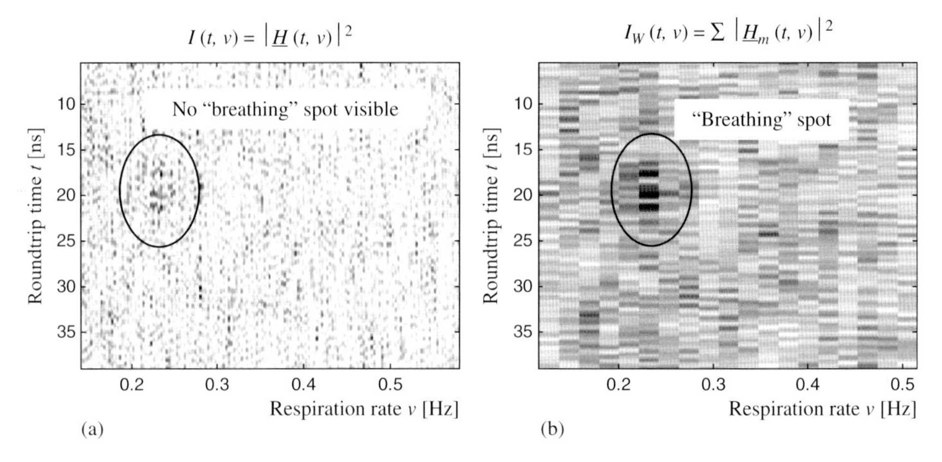

Figure 6.78 Intensity image of observation time spectrum for long (a) and short (b) time segments.

concentrated on scenarios under very poor SNR conditions, limiting ourselves to the problem of enhancing and detecting quasi-periodical motion and then localizing its source without performing frequency tracking. In case of deeply buried targets, the instantaneous amplitude of respiration is usually quite below the noise floor as illustrated in Figure 4.72 for a simulated example. Hence, the goal of further processing must be to accumulate coherently signal energy originating from the modulation by respiration motion and to enhance it against the noise. Regarding the radargram in Figure 6.77, we observe that the captured signal energy is spread within a horizontal band of width t_w (duration of impulse response) and duration T_{obs} of observation. Ideally, this energy should be compressed into a single spot in order to exceed the noise contribution.

The processing may be divided in two main steps:

- Coherent accumulation of breathing-induced signal modulation in observation time T: Here we deploy the periodicity of respiration motion.
- Reversing the time spreading of sounding signal caused from propagation effects.

6.7.3.1 **Breathing Enhancement by Its Periodicity**

This paragraph relates to the energy accumulation in observation time T. It will bring the largest effect of all processing steps since one can theoretically extend the measurement over a long duration. The energy aggregation increases linearly with observation time if one succeeds to accumulate coherently the 'breathing' energy. But due to the non-stationarity of respiration, we can achieve the theoretically best improvement only to a certain extend.

Let $h(t, T)$ be our measured radar data after pre-processing, that is pulse compression and background removal. Evidently, since we want to detect periodicals, we can compute $\underline{H}(t, \nu)$ via FFT in observation time direction (ν is the observation time frequency; refer also to (4.270)). The absolute value of $\underline{H}(t, \nu)$ is the optimal statistics for detecting a sine component of rate ν and uniformly distributed phase appearing at propagation time t. Although breathing motion does not provide an exact sinusoidal signal modulation (see Figure 4.71, (4.275), Ref. [150]), we can neglect related harmonics since they do not carry much energy.

Hence in practice, the sinusoidal modulation at a fixed frequency is a good approximation of respiratory motion for short time intervals. However – as depicted in Figure 6.77c – for longer time intervals, the breathing rate may drift away and the accumulated energy will partially spread over (observation time) frequency ν. This will hamper the detection performance for long time segments. Corresponding examples are given in Ref. [149] and Figure 6.78.

Nevertheless, (t, ν) is still the most convenient domain both for signal enhancement and for final decision about whether a person is present or not. In order to account for the non-stationarity of respiration rate, we partitioned the data in segments $h_m(t, T)$ of duration T_0 with an overlap of about $\eta \approx 50\%$ between consecutive data blocks (see Figure 6.77). Typical segment duration T_0 was about 1 min. Certainly, it is difficult to say in advance how the breathing rate of a particular person will change over time, but the indicated value was providing reasonable results in our measurements. The overall observation time results in

$$T_{\text{obs}} \approx \eta M T_0 \tag{6.28}$$

where M represents the number of segments sequentially processed. Hence, we will get M data sets $\underline{H}_m(t, \nu)$; $\nu \in [1, M]$ which are subjected to further processing in direction t.

6.7.3.2 **Signal Enhancement in Propagation Time**

As already mentioned, the signal scattered from the chest is also spread in propagation time. This gives us at least a theoretical opportunity to reverse the spreading by matched filtering in order to achieve an energy enhancement. Since this spreading is less pronounced than the energy distribution in observation time, we can of course only expect comparatively small processing gain compared with the compression in observation time.

Signal compression by matched filtering and correlation (refer to Figures 2.76 and 2.77 including corresponding considerations) is an optimal procedure for signal enhancement. However, it assumes the prior knowledge of the time shape of the signal to be compressed or to be detected.

The propagation time spreading of the radar signals is caused by the antennas and the quite chaotic wave propagation through the rubble as well as the reflections at the chest. Unfortunately, these signal deformations cannot be predicted exactly beforehand under practical constraints so that ideal and perfect matched filtering will be hardly possible. However, as shown in connection with Figure 2.77, some deviations between the signal of interest $\underline{H}_m(t, \nu)$ and a reference signal $\underline{H}_{\text{ref}}(t)$ are acceptable. Since only the round-trip time but not the time shape of the reflected wave is affected by the respiration rate, we only need a single reference function. Suppose that we know this reference signal, then we are able to determine for every individual (observation time) frequency ν and every data segment m the correlation function:

$$\begin{aligned}\underline{C}_{m,\mathrm{ref}}(\tau,\nu) &= \int\limits_{\tau-(t_w/2)}^{\tau+(t_w/2)} \underline{H}_m(t,\nu)\ \underline{H}^*_{\mathrm{ref}}(t-\tau)\mathrm{d}t \\ &= \int\limits_{-\infty}^{\infty} \underline{H}_m(t,\nu)\ \underline{H}^*_{\mathrm{ref}}(t-\tau)\mathrm{rect}\left(\frac{t-\tau}{t_w}\right)\mathrm{d}t \end{aligned} \tag{6.29}$$

where we limit the duration of integration to the approximate duration t_w of the backscattered signal. Basically, the rectangular function $\mathrm{rect}(t/t_w)$ may be replaced by any other window function $w(t/t_w)$ in order to reduce truncation effects.

Finally, the detection on target or no-target is done at the intensity map $I(\tau,\nu)$ which is determined from the superposition of the correlations functions from all segments:

$$I_C(\tau,\nu) = \sum_{m=1}^{M} \left|\underline{C}_{m,ref}(\tau,\nu)\right| \tag{6.30}$$

An approximate of the reference waveform $\underline{H}_{\mathrm{ref}}(t)$ can, for example, be gained by picking the (clear) return signal of a shallow buried victim which does not yet require signal compression in propagation time. This would allow us to 'focus' the radar to (somewhat) deeper layers in order to search for further people whose return signal is buried in noise. A second option is to collect in advance a set of reference functions for some types of rubble and typical situations of antenna-rubble contact. These functions are successively tried in (6.29). Both approaches assume that wave propagation through rubble does not strongly affect the time shape of the backscattered signal. Rather, the antenna-ground coupling will be the dominant factor.

If one does not succeed to capture an appropriate reference function, energy detection would be a further way to improve signal quality. For that purpose, the intensity map accumulates signal energy within the expected signal duration t_w by following rule:

$$I_E(\tau,\nu) = \sum_{m=1}^{M} \int\limits_{-\infty}^{\infty} \underline{H}_m(t,\nu)\ \underline{H}^*_m(t,\nu) w\left(\frac{t-\tau}{t_w}\right)\mathrm{d}t \tag{6.31}$$

(6.31) represents a simple procedure which does not require prior knowledge. The variance of the intensity values is reduced with increasing signal duration t_w. However, the method will also accumulate noise energy since the noise contributions in $\underline{H}_m(t,\nu)$ and $\underline{H}^*_m(t,\nu)$ are coherent. It provides an offset value which has to be appropriately respected in the detection procedure.

Correspondingly to (4.276) and (4.277) in Section 4.8.2, the auto-correlation function could be another mean for signal enhancement. In order to involve as less as possible noise, we limit the considered signal lengths to the window length t_w.

Since the round-trip time of the target is not known, we have to slide the window along the propagation time t. Assuming the time window is centred at time $t = t_c$, the auto-correlation function of the observation time segment m may be expressed as

$$\begin{aligned} \underline{C}_{A,m}(\tau, \nu, t_c) &= \int\limits_{t_c-(t_w/2)}^{t_c+(t_w/2)} \underline{H}_m(t,\nu)\ \underline{H}^*_m(t+\tau,\nu)\mathrm{d}t; \quad \frac{t_w}{2} \le t_c \le t_{max} - \frac{t_w}{2} \\ &= \int\limits_{-\infty}^{\infty} \underline{H}_m(t,\nu)\ \underline{H}^*_m(t+\tau,\nu)\mathrm{rect}\left(\frac{t-t_c}{t_{win}}\right)\mathrm{d}t \end{aligned} \tag{6.32}$$

t_{max} is either the duration of the sounding signal or the maximum expected propagation time to find a target. As before, we can also here replace the rectangular window $\mathrm{rect}(t/t_w)$ by another suitable window function $w(t/t_w)$. Related to (4.277), we find from (6.32) the intensity map on which the detection may be performed

$$I_A^{(p)}(t_c, \nu) = \sum_{m=1}^{M} \int \left| \underline{C}_{A,m}(\tau, \nu, t_c) \right|^p \mathrm{d}\tau \tag{6.33}$$

In the style of the L_p-norm, we applied here the pth power of the integrand where $p = 2$ seems to be a reasonable value. Figure 4.72 illustrates (6.33) for simulated data if $t_c = t_0$ coincides with the round-trip time of the target.

Finally, we can still modify (6.32) and (6.33) by respecting some specific phase relations which can be observed in the modulated signal. Let us refer to Figure 4.59 for demonstration. Supposing, we consider our radar data $h_m(t, T)$ of segment m at two different positions t_1 and t_2 in propagation time, that is $h_m(t_1, T); h_m(t_2, T)$. The (observation time) spectrum for time position t_1 may then be written as $\underline{H}_m(t_1, \nu) = H_m(t_1, \nu)e^{j\varphi_\nu}$ where φ_ν relates to the unknown phase of breathing motion. As we note from Figure 4.59, the magnitude $H_m(t_1, \nu)$ will only deviate from zero,[4] if the time position t_1 is placed at a signal edge.[5] The spectrum at time position t_2 may be expressed by the same way whereas its phase is either identical or opposite to φ_ν, that is

$$\underline{H}_m(t_2, \nu) = H_m(t_2, \nu) \begin{cases} e^{j\varphi_\nu}; & \text{if } \operatorname{sgn} \dot{h}(t,T)\Big|_{t=t_1} = \operatorname{sgn} \dot{h}(t,T)\Big|_{t=t_2} \\ e^{j(\varphi_\nu \pm \pi)}; & \text{if } \operatorname{sgn} \dot{h}(t,T)\Big|_{t=t_1} = -\operatorname{sgn} \dot{h}(t,T)\Big|_{t=t_2} \\ 0, & \text{if } \dot{h}\Big|_{t=t_2} = 0 \end{cases} \quad \text{with } \dot{h} = \frac{\mathrm{d}}{\mathrm{d}t}h \tag{6.34}$$

4) In other words, $h_m(t_1, T); t_1 = \text{const.}$ has an alternating component.

5) If the considered time position coincides with a maximum or minimum of the waveform, we have to account for distortions of the modulation signal leading to higher harmonics (see also Figure 4.71). These effects, we will disregard here for the sake of brevity.

That is, we have the same phase if the two time positions t_1 and t_2 are placed at signal edges whose slopes have the same sense and we have a phase difference by π if the points are placed at edges of opposite slobs. From this we can observe for the noise-free case that the product

$$\underline{H}_m(t,\nu)\ \underline{H}^*_m(t-\tau,\nu) = \pm H_m(t,\nu)H_m(t-\tau,\nu) \tag{6.35}$$

provides only real values. This brings us to modify (6.32) as follows:

$$C^{(p)}_{\Re,m}(\tau,\nu,t_c) = \int_{-\infty}^{\infty} \left|\Re\left(\underline{H}_m(t,\nu)\ \underline{H}^*_m(t+\tau,\nu)\right)\right|^p \mathrm{rect}\left(\frac{t-t_c}{t_{win}}\right)\mathrm{d}t; \quad \frac{t_w}{2} \le t_c \le t_{max} - \frac{t_w}{2} \tag{6.36}$$

To avoid mutual cancellation within the integrand in (6.36), we need to take the absolute value of (6.35) which is the same as the absolute value of the real part of (6.35). We applied the real part in order to suppress residual imaginary components that may be caused from erroneous signals. The power of p is again a generalization in imitation of the L_p-norm where $p = 2$ seems to be the most reasonable value. Lastly, the related intensity map yields

$$I^{(p)}_{\Re}(t_c,\nu) = \sum_{m=1}^{M} \int C^{(p)}_{\Re,m}(\tau,\nu,t_c)\mathrm{d}\tau \tag{6.37}$$

The different methods of breathing enhancement in propagation time are compared in Figure 6.79 for a real measurement. The victim was buried under dense rubble. $I_w(t,\nu)$ is depicted for reference (see Figure 6.78b for definition). We can observe some improvements with respect to the peak to noise floor ratio. However, the achievable enhancement is quite restricted as we already expected above. The quality of the intensity map $I_C(\tau,\nu)$ will largely depend on the correct choice of a reference function.

6.7.4
Non-Stationary Clutter Reduction

Most of the authors consider UWB breathing detection in the absence of interfering motion which is easy to ensure under laboratory conditions. However, in real rescue scenario clutter sources (trees wavering on the wind, people passing by, trucks) are likely to be present. Consequently, reflections of backward antenna radiation from large objects may produce stronger variations in the data then respiration of deeply buried people. Certainly, to a very considerable extent, non-stationary clutter has to be reduced by avoiding its sources at least within the first few meters around the antennas and by shielding the device which is usually not very effective for low frequencies. Non-stationary clutter cannot be completely excluded from measurement data. Hence, we also have to consider some means to suppress it algorithmically.

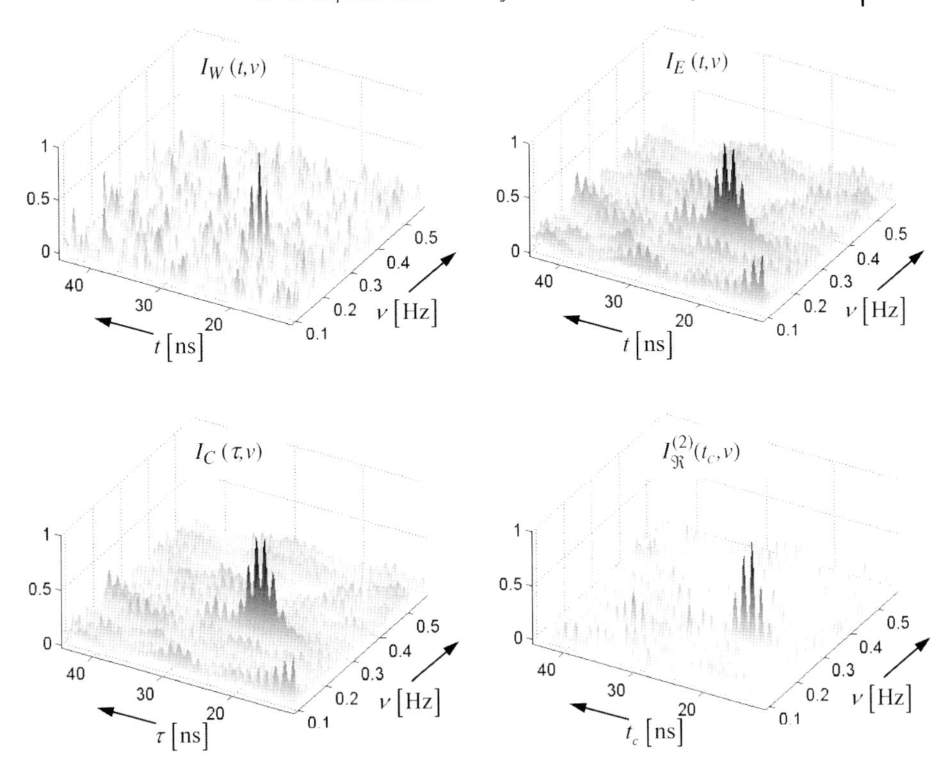

Figure 6.79 Comparison of different methods of breathing enhancement for a person, buried beneath about 1.3 m of concrete rubble. The graphs depict normalized intensity values.

From our experience, non-stationary clutter has typically lower instantaneous frequencies then breathing. Hence, it can be suppressed by filtering the captured data horizontally with a high-pass filter of appropriate cut-off frequency (typical values can be taken from Table 6.1).

A further option provides PCA which is a tool that splits the original data into a number of uncorrelated subsets. Since motions creating non-stationary clutter and breathing are typically uncorrelated, they are likely to be separated by this method. Thorough description and analysis of this approach is given in Ref. [150].

However, as we have seen in Section 4.8.2, Figure 4.71, we also have to account for non-linear effects in connection with signal modulations by round-trip time variations. In case of heavy non-stationary clutter this provokes mutually correlated components in the backscattered signal (caused from the uncorrelated sources) hampering their separation by PCA.

In order to deal with this situation, we fall back to our observation that dominant clutter perturbations are concentrated at frequencies below the breathing rate. If we filter out these signals, we can be sure that almost non-stationary clutter is included but no respiration motion. The low-pass filtered signals are taken to determine the principle components of non-stationary clutter. These components are considered as the clutter components of the whole data set which allows to remove them [149]. In detail, the process is as follows:

1) The segment $h_m(t, T)$ of captured data is low-pass filtered in T direction providing after Fourier transform $\underline{H}_{\mathrm{cl},m}(t, \nu)$. Alternatively, we separate directly the clutter from the observation time spectrum, that is $\underline{H}_{\mathrm{cl},m}(t, \nu) = \underline{H}_m(t, \nu \leq \nu_0)$ where ν_0 should be selected lower than the lowest breathing rate. $\underline{H}_{\mathrm{cl},m}(t, \nu)$ contains only information about clutter but not about breathing.
2) Principal component analysis is performed via singular value decomposition. For that purpose, the observation time spectrum is written as a matrix, that is $\underline{H}_{\mathrm{cl},m}(t, \nu) \mathrel{\hat{=}} \underline{\mathbf{H}}_{\mathrm{cl},m}$:

$$\underline{\mathbf{H}}_{\mathrm{cl},m}^{T} = \underline{\mathbf{U}}\,\mathbf{D}\,\underline{\mathbf{V}}^{T} \tag{6.38}$$

We decompose the transposed version of $\underline{\mathbf{H}}_{\mathrm{cl},m}$ to work with PCs that reflect the behaviour in observation time. Based on the distribution of the singular values in $\mathbf{D}$, we select the number M of reasonable PCs and we take the submatrix $\underline{\mathbf{U}}_M$ out of $\underline{\mathbf{U}}$ related to the M strongest singular values.
3) We 'extrapolate' the non-stationary clutter to the whole data set by

$$\underline{\mathbf{H}}_{\mathrm{cl},m} = \underline{\mathbf{U}}_M\,\underline{\mathbf{U}}_M^{T}\,\underline{\mathbf{H}}_m \tag{6.39}$$

where $\underline{\mathbf{H}}_m \mathrel{\hat{=}} \underline{H}_m(t, \nu)$ is the matrix notation of the observation time spectrum.
4) Finally, the data cleaned from non-stationary clutter yields

$$\underline{\mathbf{H}}_{0,m} = \underline{\mathbf{H}}_m - \underline{\mathbf{H}}_{\mathrm{cl},m} \tag{6.40}$$

Some examples of how algorithm performs are given in Figure 6.80. However, it should be noted that due to the big uncertainty in how clutter can look like much more strategies are possible for its removal. For example, independent component analysis (ICA) seems to be a promising tool due to maximizing 'independence' of different components when performing decomposition rather than their 'uncorrelatedness' like in PCA.

6.7.5 Localization of Breathing People

Since the antennas must have a wide beamwidth in order to guarantee a reasonable observation area, localization cannot be done by angular alignment of antennas as usual in classical narrowband radar. Rather, one has to deal with several antennas distributed over a certain area (see Figure 6.81). Based on the round-trip time measured for the different transmission paths, one can roughly estimate the target distance. For that purpose, one has to assume an average propagation speed of the rubble.

The target position is finally found by solving a geometric problem which involves the intersection of quadric curves (for a 2D-problem) or surfaces (for a 3D-problem), respectively. One knows from round-trip time measurement (often

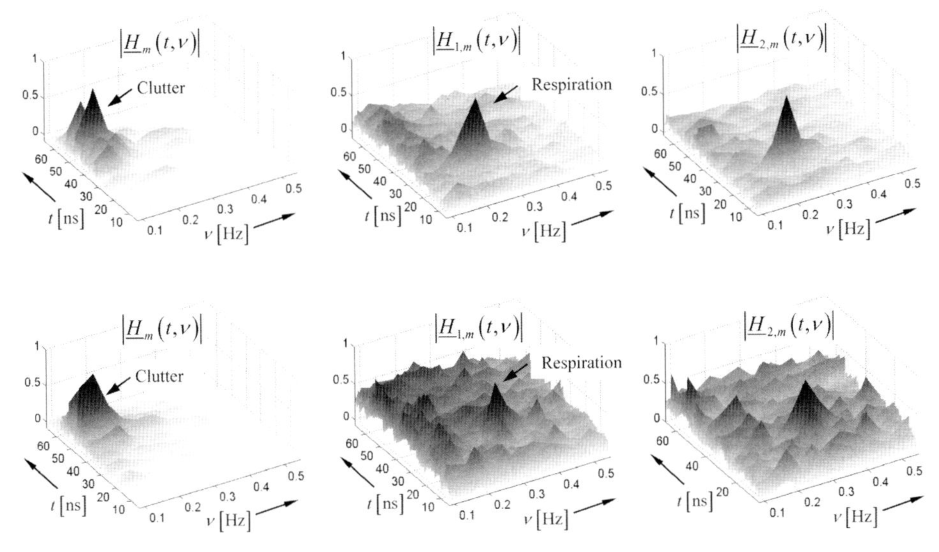

Figure 6.80 Non-stationary clutter removal for two different scenarios (top and bottom). Left: original data after Fourier transform in observation time – $\underline{H}_m(t, \nu)$. Middle: clutter reduction by PCA – $\underline{H}_{0,m}(t, \nu)$. Right: clutter reduction by high-pass filtering – $\underline{H}_{HP,m}(t, \nu)$. The example relates to a single data segment.

denoted as time of arrival – TOA) by one channel, for example Tx-Rx1 that the victim must be placed anywhere at the surface of a spheroid whose parameters are known from the measurement. Further measurements from different antenna positions, that is Tx-Rx2 and Tx-Rx3, lead to an intersection point of the spheroids which indicates the target position. Additionally, one can deal with the round-trip time difference between two different propagation paths (called as time difference of arrival – TDOA). The TDOA value provides a circular hyperboloid of potential target position which can be additionally used to find the intersection point and to reduce angular uncertainties.

The round-trip time of the victim is taken from target detection based on the intensity image as described above. If we have involved, for example one transmitter and three receiver antennas, we get three TOA values and two TDOA values.

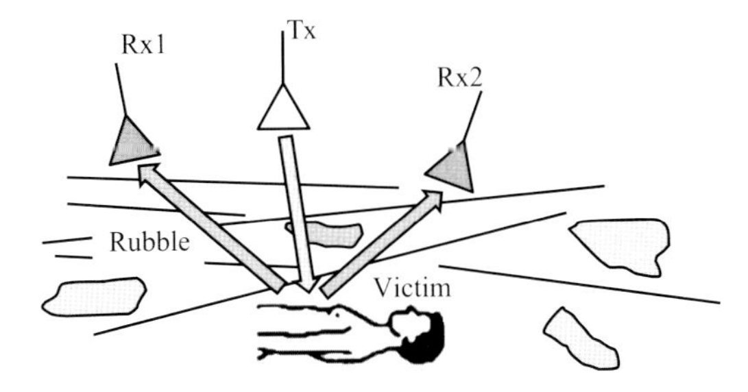

Figure 6.81 Localization principle using antennas of large beamwidth.

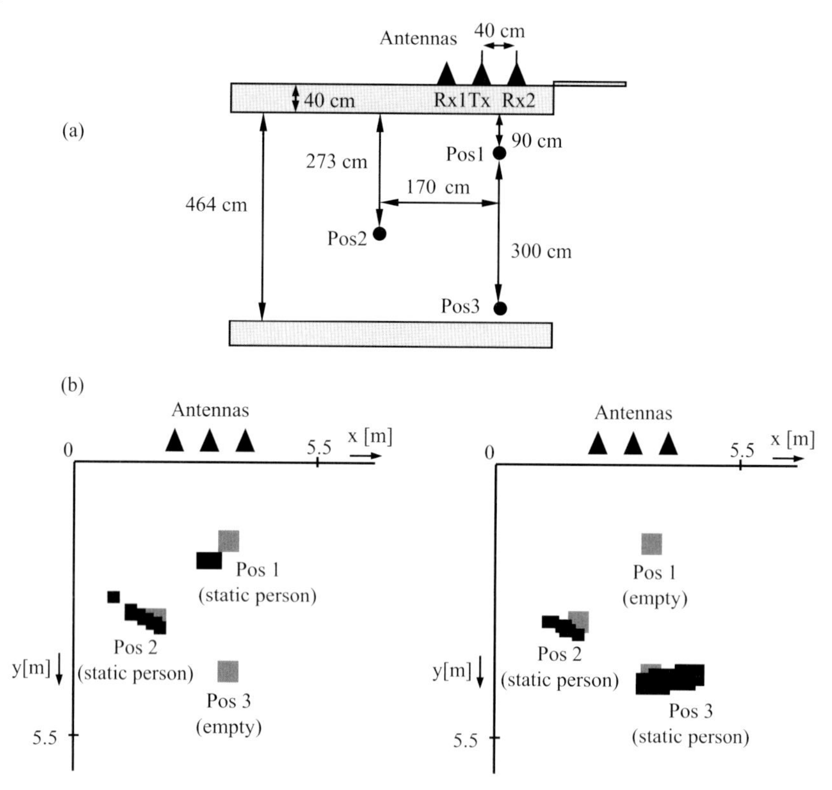

Figure 6.82 Two examples of detecting and localizing two trapped persons through a wall. (a) Measurement scenario and (b) results of the detection. The gray boxes represent the positions 1–3. The white boxes refer to detected persons.

This will lead to an overdetermined system of equations which are solved either algorithmically or by graphical means.

Two examples are shown in Figure 6.82. In order to proof the reliability of the localization procedure, we used scanning through the wall of an empty room. The applied antennas are of the same type as depicted in Figure 6.76 but they are smaller (diameter about 20 cm) and operate at higher frequencies. The distance between adjacent antennas was quite small (about 40 cm) in order to simulate ill-conditioned situations.

The measurements were carried out with two people sitting on chairs inside the room at positions indicated by the marks Pos1–Pos3. The measurements were done two times, where one volunteer changed his position from Pos1 to Pos3. Wall effects on the wave propagation are compensated.

Further results for 2d and 3d through-rubble localization of a single person buried through more than 1 m of concrete and detailed algorithm description are available in Ref. [149].

We should mention some key ideas related to the algorithm applied:

- Unlike in many other approaches, before 'fusing' TOA and TDOA images, we only find breathing rate and do not pretend for knowledge of a distance from antenna centre to the person. All information about position of a person is derived from final image, not before in order to preserve synergic effect between receiving channels. In many real-time scenarios similar method would not be applicable due to computational intensity, but in our case we can spend some seconds for computations after some minutes of collecting data.
- Any can expect that localization is a bit easier in our case in comparison with the case of walking people, since mainly lungs are moving and both antennas receive waves reflected from this small area. Besides, we accumulate data for some minutes and person is in the same place. However, in our case heterogeneity of rubble can cause serious mistakes in some scenarios and moreover, imprecise permittivity value of rubble material under real conditions can cause some error in localization.
- Unfortunately, in the case of two or more victims if they are situated at almost identical distances to antennas ghost target can appear (this happens because ellipses related to different victims cross). However, all victims in this case will be shown.

6.7.6 Conclusions and Future Work

The prototype device described in this chapter is capable of detecting a person by breathing through up to 1.3–1.5 m of dense reinforced concrete rubble and up to about 13 m of a brick building as illustrated in Section 6.1 (Figure 6.8). Appropriate software was developed that makes advantages of UWB radar useful: radar is capable of detecting multiple victims. Besides, ability to localize person by his/her respiration is demonstrated. In summary, some problems for future research are as follows:

- Although breathing detection and enhancement by UWB radar is well developed and profusely reported in the literature, reliable automatic detector that does not need decision of operator is still required.
- Both hardware and software methods are to be considered to improve clutter removal. From algorithmic point of view this problem is complex and it allows numerous solutions.
- Weak points in localization are 'ghost targets' as well as rubble heterogeneity. The problem of 'ghost targets' is not crucial though (they appear only when multiple victims are present at similar distance to antennas and multiple victims can still be detected together with additional target). Moreover, when victims have different breathing rates, this information can probably be used for separating real targets from 'ghosts'. Rubble heterogeneity can possibly be alleviated if some information about rubble structure is available (whether it is received by means of radar or any other method).

Acknowledgement

This work was supported by the European Project under the acronym RADIOTECT.

6.8
Multiple Moving Target Tracking by UWB Radar Sensor Network

Dušan Kocur, Jana Rováková, and Daniel Urdzík

6.8.1
Introduction

The word RADAR is an abbreviation for RAdio Detection And Ranging. In general, radar systems are day/night all weather sensor systems, which use modulated waveforms and antennas to transmit electromagnetic energy into a specific volume in space to search for targets. The targets within a search volume will reflect portions of this energy (echoes) back to the radar. Then the echoes are processed by the radar receiver in order to extract target information such as range, velocity, angular position and other target identifying characteristics [152]. If the fractional bandwidth of the signals emitted by the radar is greater than 0.20 or if these signals occupy 0.5 GHz or more of the spectrum, the radar is referred to as the ultra-wideband radar (UWB radar) [153, 154]. UWB technology in radar allows for a fine resolution and hence very high accuracy range, multiple target resolution, better separation between targets and clutter, rigidity to multi-path propagation (e.g. within buildings) and external electromagnetic interference [155].

With regard to these properties, short-range UWB radars are used nowadays for different applications such as subsurface sensing (e.g. ground-penetrating reconnaissance, landmine detection) [156], classification of aircrafts, automotive technologies (e.g. collision avoidance) [157], medical application (e.g. diagnosis of diseases [158]), indoor navigation, object recognition [153], material characterization and detection and positioning of human beings in a complex environment [153–155, 159].

Detection and positioning of human beings have been very interesting for military, security and emergency rescue operations. UWB radars can have many and varied applications to counter-terrorism. Reservoirs, power plants and other critical infrastructures are extremely vulnerable to terrorist attack. Here, the radar systems can be applied for monitoring of these critical environments and for the detection of unauthorized intrusion. On the other hand, several UWB radars using relatively low frequencies (e.g. typically between 100 MHz and 5 GHz) have been developed for through wall detection and tracking of moving people during security operations, through the obstacle imaging during fire, through-rubble localization of trapped people in collapsed buildings following an emergency (e.g. earthquake or explosion) or through snow detection of trapped people after an avalanche.

Detection of human beings with radars is based on their movement detection. If the position of the human being to be detected is constant, his or her vital signs such as respiratory (chest motion) and heart beating can be detected [159]. These very small movements cause changes in frequency, phase, amplitude and arrival time of reflection from a human being electromagnetic wave. Generally speaking, the changes of amplitude are negligible. Therefore, only frequency, phase and arrival time changes can be used for human being detection and localization. This approach can be applied, for example for through rubble or snow localization of trapped people. On the other hand, if the position of the human being is changing (e.g. walking persons), measuring changes in the impulse response of the environments are applied for human being detection and localization [154].

In this section, we will focus on the problem of multiple moving person detection, localization and tracking with stress to military, security and emergency rescue operations. For such applications, the handheld sensors are used by the operators (e.g. security forces) in the operation place. These devices have to operate in a stand-alone mode, so that the results of the object monitoring are provided to the operator immediately and he or she can change the sensor location in a flexible way. On the other hand, if they are applied for protection of critical infrastructures, the sensors have to be small sized and located unobtrusively in monitored region/objects. These requirements result in that these radar systems use usually only a small antenna array (e.g. one transmitting and two receiving antennas) necessary for motion detection and basic spatial positioning of the targets by trilateration methods [160] (see also Section 6.9).

For moving target localization and tracking by the handheld UWB radar, the trace estimation method described in Ref. [161] can be used with advantage. This method is the complex procedure that includes phases of signal processing such as background subtraction, weak signal enhancement, detection, TOA estimation, wall effect compensation (in the case of through wall localization), localization and tracking. The significance of the particular phases of the trace estimation method has been explained, for example in Refs [161, 162]. The mentioned trace estimation method provides excellent and low-complex performance for single target tracking scenario.

The problem of short-range detection and tracking of moving persons have been studied, for example in Refs [163–167]. However, the problem of multiple human tracking in real complex environment has been less well addressed. Our experiences received at several measurement campaigns with UWB radar systems for moving persons tracking (e.g. [168]) have shown that the single handheld radar is able to detect very often only a person moving closest to the radar antennas, whereas its ability to detect the remaining targets is meaningfully reduced. The origin of this effect can be identified as the impact of the mutual, partial or total shadowing of targets at the multiple persons tracking scenario [169]. In the case of partial shadowing, target detection can be improved by application of methods for enhancement of weak non-stationary signal components [170]. If the effect of total shadowing occurs, it is almost impossible to detect the remaining targets by the same radar.

In order to solve this problem, UWB sensor network employing proper data fusion methods can be used with advantage. This approach for multiple moving target tracking will be introduced in this chapter. The rest of the chapter is organized as follows. In Section 6.2, the key problem of multiple moving target detection by the single UWB radar consisting in the shadowing effect due to particular targets will be explained. The solution of that problem will be presented in Section 6.3, where the concept of UWB sensor network based on single handheld UWB radars networking will be given. The experimental results obtained by the UWB sensor network applied for multiple moving target tracking will be presented in Section 6.4. A discussion of the obtained results and the main conclusions of our study are given in Section 6.5.

6.8.2
Shadowing Effect

The fundamental problem of the short-range detection of moving persons by the single handheld radar consists in the reduced ability of the radar system to detect all persons at multiple moving target scenario. It has been shown in Ref. [169] that this effect is caused by time-variable mutual shadowing of the targets. Taking into account a typical frequency band employed by an UWB radar applied for the short-range detection and tracking of moving targets as well as permittivity, permeability and conductivity of a human body and clothes [171], a person reflects and absorbs the energy of electromagnetic waves emitted by the radar transmitting antenna or reflected by another object in such a way that only a negligible part of electromagnetic wave energy is transmitted through/around a human body to a region located behind him or her. This effect results in creating a bordered area behind the person where the electromagnetic waves emitted by radar or reflected by another object are received with a large attenuation. This area is referred to as the shadowed area or the dead zone.

Let us consider for the simplicity that the electromagnetic waves are transmitted by an isotropic radiator and that the principles of geometrical optics can be used for the description of their propagation. Then, if the person is represented by a simple oval in 2D space thus the effect of mutual shadowing can be illustrated by Figure 6.83. Following this figure, it can be seen that there are two basic forms of shadowing effect arising at multiple moving target tracking scenario. First, let us assume the scenario, when target A (T-A) is located in front of transmitting antenna. It results in the occurrence of the dead zone behind T-A (Figure 6.83a). Then, if target B (T-B) is located in this dead zone it can reflect only strongly attenuated electromagnetic waves and hence, it is difficult (partial shadowing) or impossible (total shadowing) to detect it. The second form of the shadowing effect manifestation is presented in Figure 6.83b. At this scenario, T-B reflects electromagnetic wave emitted by the transmitting antenna Tx without an additional attenuation. However, between T-B and the receiving antennas, T-A is located and thus it creates the dead zone, too. Because the receiving antenna is located in this dead zone, the ability of the detection of T-B is significantly reduced.

The solution of the outlined problem can be provided by the application of the UWB radar sensor network employing proper methods of the fusion of data

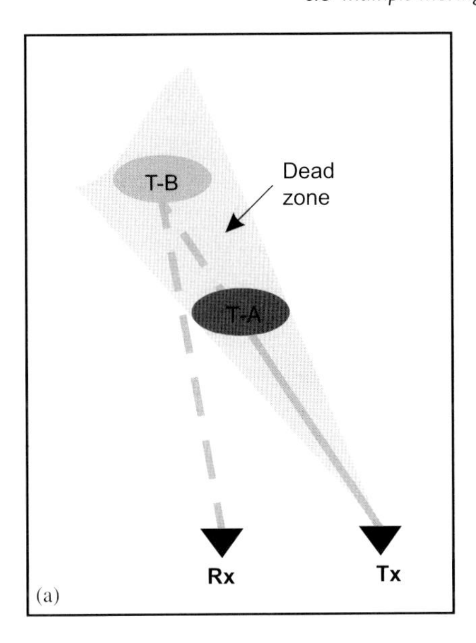

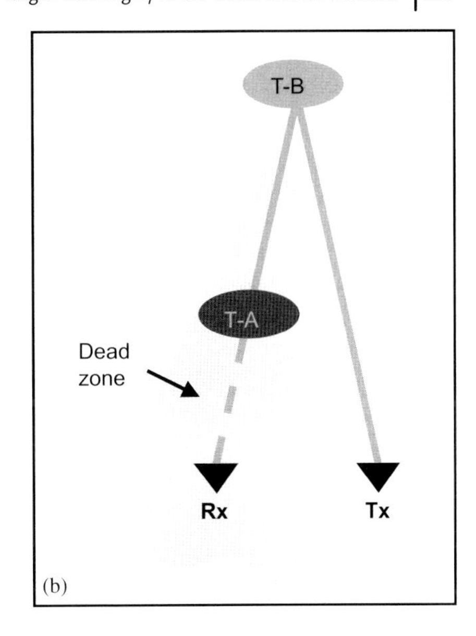

Figure 6.83 The mutual shadowing effect illustration. (consider the affinity to state space model — see section 2.4.4.4. Zeros in the input matrix **B** corresponds to case (a) and zeros in the observation matrix **C** relates to chase (b))

obtained from its particular nodes. The basic concept of such sensor network is introduced in the next sections.

6.8.3 Basic Concept of UWB Sensor Network for Short-Range Multiple Target Tracking

Recent advances in electronics and wireless communications have enabled the development of low-cost multifunctional sensor nodes that are small in size and communicate one another. These nodes consisting of sensing device, data processing unit and communicating components, leverage the idea of sensor networks. A sensor network is composed of a large number of sensor nodes that are deployed either inside the scene of interest or very close to it. The position of sensor nodes need not be engineered or predetermined. This allows random deployment also in the critical surveillance applications. On the other hand, this means also that the sensor network must possess self-organizing capabilities. Another unique feature of the sensor networks is the cooperative effort of its sensor nodes. Sensor nodes are fitted with an onboard processor. Instead of sending the raw data to the nodes responsible for the data fusion, they use their processing abilities to locally carry out necessary computations and transmit only the required and partially processed data. The above-described features ensure a wide range of applications for sensor networks. Some of the application areas are health, military and home [172]. Radar sensor networks are gaining importance in the context of target localization and

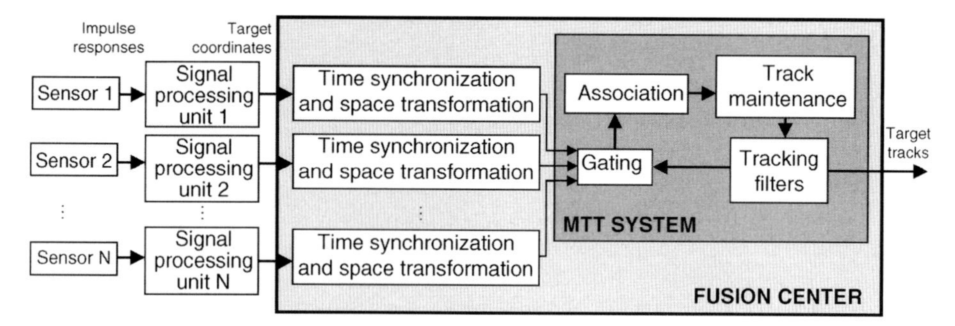

Figure 6.84 UWB sensor network architecture for multiple moving target tracking.

tracking in the complex environment. The UWB radar sensor networks applied for that purpose have been proposed, analysed and discussed, for example in Refs [172–177].

The block scheme of UWB sensor network dedicated to short-range localization and tracking of multiple persons is shown in Figure 6.84. The UWB sensor network of that kind consists of a set of UWB sensors including their signal processing units and a data fusion centre connected by WLAN. As the sensing devices, a single handheld radar system fitted with one transmitting and two receiving antennas and an onboard signal processor unit is used. This concept of the UWB sensor network allows the flexible changes of the number of the sensor network nodes. At the same time, the particular nodes can manoeuvre within a surveillance area. These two properties of the considered UWB sensor network can be very interesting for military, security and emergency rescue operations. Here, some nodes can work as a node of the sensor network or as a self-contained unit (flexible performance). On the other hand, in emergency operations it is expected that the positions of sensing devices should be changed following the requirements of the upcoming juncture.

The particular sensors have to be located in the monitored region in such a way as to cover the region of interest as good as possible. At the same time it is expected that they are able to provide some diversity with regard to target positioning. The particular sensors of UWB sensor network produce impulse responses of the environment (monitored region) through which the electromagnetic waves emitted by the sensors are transmitted. The output signals of the particular sensors are processed by signal processing unit in such a way as to estimate the coordinates of the detected targets. For that purpose, the trace estimation method without the phase of target tracking is used [161, 162]. The estimated coordinates of the particular moving targets are transmitted to the data fusion centre through WLAN by using communication part (unit) of the signal processing unit. Then, the target coordinates are processed by the fusion centre.

Data fusion is generally defined as the use of techniques that combine data from multiple sources and gather that information in order to achieve inferences, which will be more efficient and potentially more accurate than if they were achieved by means of a single source (a single sensor). In the literature, a huge number of the data fusion methods can be found (e.g. [178, 179]). In the case of the considered UWB sensor network, we have decided to apply the fusion centre with a centralized architecture consisting of a set of time synchronization and space transformation blocks and multiple target tracking (MTT) system [180, 181]. In the blocks of time synchronization and space transformation, data provided to the fusion centre are time synchronized and transformed into the common coordinate systems. The set of such data hereinafter labelled as observations represents the input data of MTT system.

The processing loop of MTT system starts when new observations are received from the time synchronization and space transformation blocks. Then, the received observations are processed at the gating block. Here, existing target tracks are updating. Gating tests determine which possible observation-to-track pairings are reasonable by attributing a cost to each pairing. The costs are calculated as the statistical distance between the predictions of the target states given by the tracking filters and the observed state coordinates received from the sensors.

The association block utilizes global nearest neighbour data association algorithm. It means that computed costs are put together in a cost matrix, which is then passed on to the assignment solver to determine the finalized pairings. The pairings are made in such a way as to ensure minimum total cost for all the pairings whereby strictly one-to-one coupling is established between observations and tracks.

The block of the track maintenance is responsible for a track management. It includes two identifiers initiating new tracks or deleting existing ones when needed. The observationless gate identifier identifies the gate where no observation falls. This indicates a probable disappearance of an already known target and hence the deletion of its track after confirmation. The new target identifier detects observations that fall outside all the gates. These observations are potential candidates for initiating new tracks after confirmation.

In the tracking filters block, the finalized observation-track pairings are passed on to the tracking filters which use them for estimating the current states of targets and predicting the next states. The linear Kalman filter is used for this block. The number of the employed filters is equal to the maximum number of the targets to be tracked.

The description of the MTT system in detail is beyond of this chapter. A reader can find it, for example in Refs [180, 181].

6.8.4 Experimental Results

The performance of the proposed approach for short-range tracking of moving persons is demonstrated by processing of real signals acquired by the UWB sensor network consisting of two sensor nodes represented by the M-sequence UWB radars [153] labelled as S1 and S2, respectively. The measurement scheme and scenario pictures are given in Figures 6.85 and 6.86, respectively.

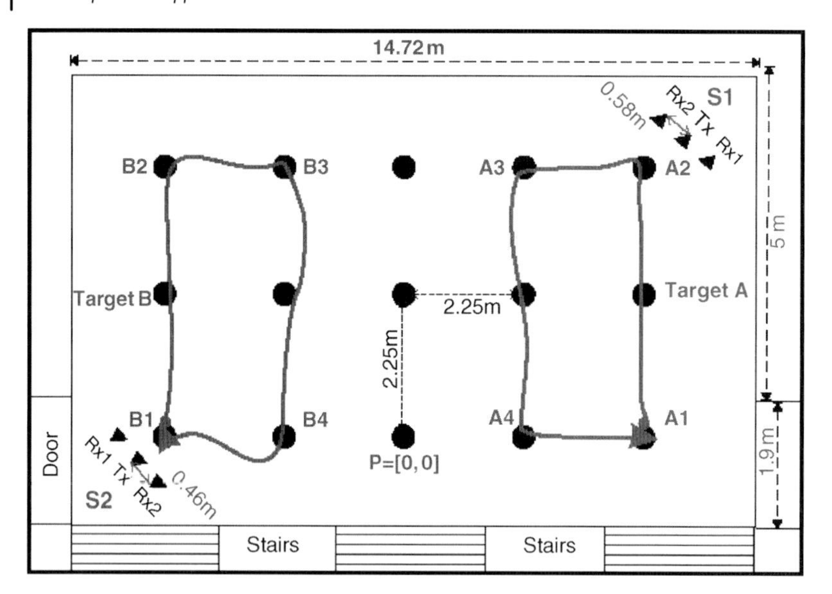

Figure 6.85 Measurement scheme.

The system clock frequency of the sensor S1 was about 4.5 GHz resulting in its operational bandwidth of about DC to 2.25 GHz (excluding the antennas). The M-sequence order emitted by radar system S1 was 9, that is the impulse response covered 511 samples regularly spread over 114 ns (refer to Section 3.3.4). It corresponds to an observation window of 114 ns leading to an unambiguous range of about 17 m. On the other hand, the system clock frequency of the sensor S2 was about 9 GHz and therefore its operational bandwidth was of about DC to 4.5 GHz (excluding the antennas). The M-sequence order emitted by sensor S2 was 9 also, that is the impulse response covered 511 samples regularly spread over 57 ns. Then, the unambiguous range of the sensor S2 was about 8.5 m. Each sensor was fitted with one transmitting (Tx) and two receiving double-ridged horn antennas (Rx1, Rx2). They were always placed in a line with Tx in the middle between Rx1 and Rx2. The pictures illustrating the measurement including photos of sensors S1 and S2 are depicted in Figure 6.86.

As the monitored area (Figure 6.86b), a school corridor in size of approximately 7 m × 15 m was used. Both sensors were located in the corridor corners whereby two persons were moving in the middle of the area in size of about 5 m × 9 m. The first target (labelled as A) was walking along the trajectory of the rectangular shape through positions A1-A2-A3-A4-A1. At the same time, the second target (B) was moving through positions B1-B2-B3-B4-B1 (Figure 6.85).

The raw radar signals acquired by each sensor were independently processed offline by the trace estimation methods. Here, the methods such as exponential averaging, CFAR detection, TOA association method (providing also de-ghosting task solution) and the direct calculation method were used for background subtraction, target detection, TOA estimation and target localization, respectively. The estimated

(a)

(b)

Figure 6.86 Measurement scenario pictures. (a) UWB radar system S1. (b) Monitored area including UWB radar system S2 (in the front) and UWB radar system S1 (at the back).

coordinates of the targets after time-synchronization and transformation into the common coordinate system are depicted for each sensor separately in Figures 6.87 and 6.88. In these figures, we can see not only the target position estimations as the results of the target localization phase but also the target tracks obtained as the target tracking phase output. Here, MTT system was applied for target tracking.

As we can see from these figures, both sensors were able to capture the biggest amount of signals reflected from the targets especially if the targets have been localized in the vicinity of the sensor antenna systems. It can be observed from Figure 6.87 that sensor S1 was able to recognize the presence of two targets in the monitored area. Here, the track of target A has been estimated well, but the track of target B could be viewed only partially. It can be explained by shadowing effect created by target A, if this target is localized between the transmitting and/or receiving antennas and target B. Sensor S2 was able to detect and localize target B

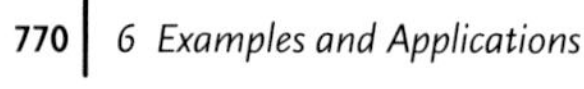

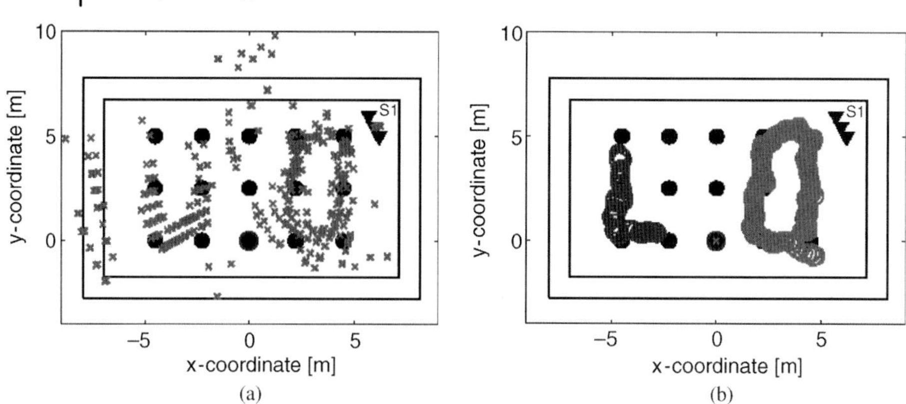

Figure 6.87 UWB radar system S1. (a) Estimated coordinates of targets entering to MTT system and (b) estimation of the final tracks of the targets.

only (Figure 6.88). It results from its small unambiguous range. If we summarize the performance of sensor systems S1 and S2, it can be concluded that neither sensor S1 nor sensor S2 were able to detect and localize the complete tracks of both the targets.

This situation was changed significantly when the fusion of data obtained from the sensors S1 and S2 were done. Here, the results provided by the data fusion centre are given in Figure 6.89. It can be observed from this figure that if the UWB sensor network is applied for the target detection and tracking by using the described data fusion method, the sensor network is able to detect, to localize and to track both targets. The estimated tracks correspond very well with the true trajectories of the targets.

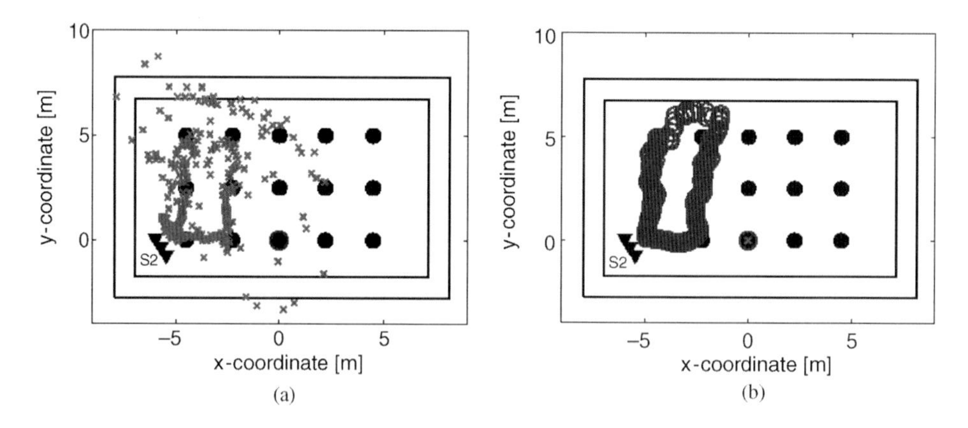

Figure 6.88 UWB radar system S2. (a) Estimated coordinates of targets entering to MTT system. (b) estimation of the final tracks of the targets.

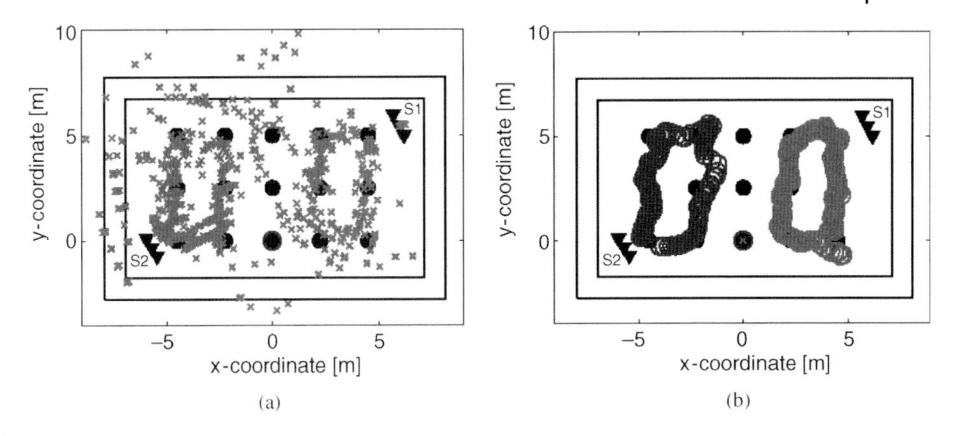

Figure 6.89 Data fusion of UWB radar systems S1 and S2. (a) Estimated coordinates of targets entering to MTT system and (b) estimation of the final tracks of the targets.

6.8.5 Conclusions

In this section, we have dealt with the problem of short-range tracking of moving persons by UWB sensor networks. First, we have analysed very shortly the problem of detection of human beings by the handheld UWB radar system. Then we have pointed out the problem of mutual shadowing arising usually at the short-range tracking of multiple moving targets. Here, we have outlined that this effect has been the key problem for that scenario because it decreases significantly the ability of the handheld UWB radar to detect and localize multiple moving targets. However, in light of applications of UWB sensor system for military, security and emergency rescue operations, the requirement for a reliable detection and localization of all moving persons is especially important. Therefore, we have proposed to apply UWB sensor network equipped with the low-complex method of data fusion with the centralized architecture as a very efficient tool for multiple moving target detection and tracking.

The performance of the proposed concept of UWB sensor network has been demonstrated by sensor network consisting of two M-sequence UWB radars. The signal processing results obtained for the analysed scenario has shown that the proposed UWB sensor network has improved remarkably the ability to detect multiple targets under shadowing effect inherence. At the same time, it has provided the better coverage of the monitored area and increased accuracy of the target positioning, too. This robust and reliable performance of the proposed UWB sensor network has been reached at the low-cost data fusion algorithm implemented in the data fusion centre. Similar results have been obtained for a number of further scenarios intent on multiple moving person tracking. Following these results it can be concluded that the described concept of UWB sensor network and signal processing procedure including data fusion method represents the promising approach for short-range detection and tracking of multiple moving persons by UWB radar systems especially for military, security and emergency rescue operations.

6.9
UWB Localization

Rudolf Zetik

UWB radio sensors promise interesting perspectives for location estimation in short-range cluttered environments. In these environments, other radio systems like GPS feature a limited performance in terms of their insufficient precision and reliability. UWB technology offers great potential for achieving high localization accuracy thanks to its huge bandwidth, which may be up to several GHz. The large bandwidth makes it possible to resolve dense multi-path, which results in a good robustness of localization algorithms. Incorporated low frequencies allow the penetration of radio waves through non-metallic materials. Such a precise and robust localization is required in many applications. We mention just two of them. Both of them represent the large field of UWB localization applications.

The first application benefits from the good penetration of UWB signals through non-metallic materials. Therefore, UWB sensors are very helpful in situations when the entering of a room or a building is considered hazardous and it is desired to inspect its interior from outside, through walls. Examples include detection, localization, tracking and monitoring of vital activities (heartbeat, breathing) of people in dangerous environments for emergency (e.g. casualties of an earthquake or an explosion), security (unauthorized intruders) or military purposes.

Another interesting application is the use of the UWB technology in a home entertainment environment. The UWB radio is one of the most promising technologies that realize communication systems with integrated localization and tracking application. UWB communication system allows to get rid of long cables by wireless data transmission. Localization capabilities offered by the UWB system allow an implementation of, for example, 'intelligent' audio algorithms. These algorithms exploit the knowledge about location of loudspeakers and instantaneous position of a listener for a perfect sound. The position estimates drive smart audio algorithms that optimize the hearing experience sensed by the user and/or direct the sound interactively according to the listener position.

6.9.1
Classification of UWB Localization Approaches

UWB localization approaches can be classified according to various criteria. Here, we provide a basic classification according to

- the output of the localization algorithm,
- the involvement of the target in its localization and
- the signal parameter exploited by the localization algorithm.

The majority of localization algorithms are based on a two-step procedure [182], which consists of a ranging step and a data fusion step. This approach provides estimated coordinates of targets. Another approach is known especially in the field of radar. It omits the ranging step and provides an image of a monitored area. The image indicates position of targets by 'hot spot' areas [183, 184].

Both approaches, two-step localization and imaging, can be subdivided according to the involvement of the target in its localization into the active and passive localization. The active localization approach presumes a target, which actively cooperates with the localization system and carries, for example a transmitter. The passive approach localizes targets just by electromagnetic waves scattered from the target. The target must not be aware about its localization. It does not have to cooperate with the localization system in any way, no tags are necessary.

Localization approaches mentioned above can be further subdivided according to the signal parameter exploited by the particular localization algorithm such as received signal strength, angle of arrival, time of arrival and time difference of arrival. The ranged-based schemes, TOA and TDOA, are proved to have the best performance due to the excellent time resolution inherently offered by UWB signals. Therefore, we concentrate just on the range-based approaches – TOA and TDOA.

6.9.1.1 Two-Step Localization versus Imaging

A two-step localization is the most common localization technique. It includes a parameter estimation step and a location estimation step. The goal of the parameter estimation is to determine a specific parameter, which is related to the position of the target such as TOA or TDOA. Multiple parameter estimates are fused together in the location estimation step, which provides coordinates of the target. There are necessary at least four parameter estimates to provide unambiguous three-dimensional coordinates of the target. These parameters have to be obtained from spatially distributed sensors or from one sensor which measures at different positions while the target is static. The parameter estimates need to be independent. Therefore, measurement positions or the sensor constellation must not create a linear array in order to obtain the unambiguous location estimate.

Another alternative way for the location estimation is based on imaging algorithms known in radar technique. Imaging performed by electromagnetic waves is well known from non-destructive testing, ground-penetrating radar, through-wall radar, medical diagnosis and so on. These methods exploit the scattering of electromagnetic waves in an unknown medium. Imaging methods involve some form of back propagation, back projection or time reversal for the image reconstruction of this medium. They predominantly rely on synthetic aperture data principles, where one sensor (or an array) creates the synthetic aperture by subsequent measurements at different positions. This approach presumes time invariance of the medium under test and restricts imaging methods to static environments. Reflections from moving objects are smeared in the focused image and gradually disappear. Since there are many analogies between conventional localization and imaging methods, the imaging algorithms can be adopted to the localization (imaging) of moving targets too. The adopted imaging algorithm provides a sequence of images – snapshots. Each snapshot is related to a certain measurement time and shows current locations of targets within the inspected environment by 'hot spots' within the image. Thus, there is no direct estimation of target's coordinates. However, the image must be interpreted by an operator or by a detection algorithm, which extracts the number and positions of targets.

This additional interpretation step is not necessarily a disadvantage of the imaging approach. In case of poor localization conditions, when some sensors cannot properly 'see' an obstructed target, the two-stage localization approach may easily fail due to incorrect parameter estimation and/or data fusion. The imaging algorithm does not apply any parameter estimation. It creates an image based on measured impulse responses and known constellation of sensors. Although, in case of the obstructed target, the obtained image is smeared and without any clear 'hot spot', it can still be correctly interpreted by a skilled operator.

6.9.1.2 **Active versus Passive Approach**

Another aspect how to divide localization systems is their classification according to the involvement of the target in its localization. If the target actively cooperates with the localization system we refer to it as the active localization. In order to cover the proactive role of the target, it has to carry a transmitter or a receiver and collaborate in measurements. In contrary, the passive approach is based on a stand-alone localization device or a localization system integrated into some existing infrastructure. It exploits wave scattering by the (unaware) target. Thus, in case of active localization systems:

- The target must cooperate by carrying some hardware. (Disadvantage)
- Wireless synchronization between transmitters and receivers is necessary. (Disadvantage)
- Signal detection is relatively easy since the localization is based on the direct wave propagation from the transmitter(s) to the receiver(s). (Advantage)

In case of passive localization systems:

- No object cooperation is necessary – it can be used as a stand-alone positioning device with wired synchronization between transmitters and receivers. (Advantage)
- Problematic signal detection, which is based on signals scattered back from the target. (Disadvantage)

By comparing features given above, it is obvious that the main advantage of the active localization systems is the easy parameter estimation (for the two-step localization), or very good 'visibility' of the direct wave, which contains information about the target (for the imaging techniques). This is due to the fact that the direct wave is the leading and the strongest signal component of the measured impulse response even if the line of sight is partially obstructed. The coverage of the active system is limited only by its signal-to-noise ratio and by the ambiguity range that is given by the length of the measured impulse response.

On the other hand, the localization of a target without the use of a tag is usually a challenging task. The receiver measures a lot of reflections or scatterings from other objects that are located within the inspected area. The passive localizer must first detect the target amongst all these echoes. If the passive system aims at the localization of, for example, people a usual criterion that differentiates them from other objects is their movement. Also calmly sitting or laying person moves due to its breathing and heart beat activity. Thanks to the superior time resolution of the UWB sensors, already such a tiny displacement of a human chest can be detected. However, a prerequisite is a large dynamic range of the UWB sensor. If the dynamic range is not high enough the weak signals are hidden by the noise or other spurious signals of the sensor and cannot be detected. In case of closely spaced antennas (stand-alone localizer) and targets at longer distances, the necessary dynamic range for the passive system may easily exceed even 50 dB values [185]. The main advantage of the passive localizer is the possibility to integrate the system into one stand-alone device, which allows the wired synchronization amongst transmitter(s) and receivers(s). This improves the stability and the precision of the passive localization.

6.9.1.3 Time of Arrival versus Time Difference of Arrival

As mentioned before, the range-based approaches, TOA and TDOA are the natural choice in case of UWB systems. Range-based approaches avoid the usage of expensive antenna arrays, which would be needed for angle of arrival methods. Since the parameter estimation is performed in the time domain, there is no need to constrict the distance of antennas to the half of the wavelength as it is known from narrowband systems. In case of range-based UWB system, the maximum antenna distance is limited by the repetition rate of the stimulation signal (relates to the unambiguous range). Antenna spacing above this limit can result in an unambiguous TOA or TDOA estimate. However, the repetition rate of an UWB system is usually in an order of at least tens of nanoseconds for short-range indoor localization applications. Therefore, the antenna spacing of up to tens or hundreds of meters is allowed in UWB systems instead of some centimetres or decimetres that is usual in narrowband systems. Thus, UWB antennas must not be collocated in an antenna array of a stand-alone localizer. They can be distributed creating an infrastructure of a sensor network as well. The antenna spacing can be selected according to the specifics of a particular application.

The TOA-based systems locate the target by measuring the absolute distance between the transmitter and the receiver (active localization), or the absolute distance amongst the transmitter, the target and the receiver (passive localization). In order to determine the absolute distance, the TOA-based localization requires a temporal synchronization between the transmitter and the receiver. The passive stand-alone localizers with collocated antennas offer the wired synchronization amongst transmitter(s) and receivers(s). However, most practical active ranging systems are unable to measure absolute distance directly. The exact time when the electromagnetic wave is transmitted is unknown at the receiver. Therefore, the ranges must be estimated indirectly. There are too common approaches:

- pseudo-range and
- round-trip time of arrival (RTOA).

The pseudo-range system requires an additional receiver. This receiver is used for the computation of the time offset. The time offset is related to the beginning of the signal transmission and it allows estimation of absolute ranges. On the other hand, the RTOA does not require an additional receiver. Instead, this approach assumes that each sensor node involved in the localization must be capable to retransmit the received signal. If the time delay caused by the signal retransmission is known it is just subtracted from the TOA estimate. The result is related to the double of the absolute range between corresponding pair of transceivers.

The TDOA-based systems locate the target by measuring the time difference of the arrival of the propagating signals between two receivers. TDOA-based localization does not require a temporal synchronization between the transmitter and the receivers. It only requires a temporal synchronization between two receivers that determine the time difference. The TDOA can be estimated by two different methods. The first method is based on the subtraction of two range (TOA) measurements from two antenna elements. The second method uses cross-correlation techniques, in which the received signal at one antenna element is correlated with the received signal at another antenna element within the array. While determining the TDOA from the TOA estimates is a feasible method, cross-correlation techniques dominate the field of TDOA estimation techniques. Many techniques have been developed that estimate TDOA with varying degrees of accuracy and robustness. These include the generalized cross-correlation and cyclostationarity-exploiting cross-correlation methods. Cyclostationarity-exploiting methods include the cyclic cross-correlation, the spectral-coherence alignment method, the band-limited spectral correlation ratio method and the cyclic Prony method. The most common cross-correlation techniques are generalized cross-correlation methods. These methods cross-correlate pre-filtered versions of the received signals at two receiving stations, then estimate the time difference of arrival as the location of the peak of the cross-correlation estimate. Pre-filtering is intended to stress frequencies at which signal-to-noise ratio is high and to attenuate the noise power before the signal is passed to the correlator.

TOA and TDOA approaches, as well as active and passive; and two-stage localization and imaging techniques can be arbitrarily combined together according to requirements of a particular application. For example, handheld through wall radar is determined for the passive localization of a target and it usually exploits TOA-based imaging. We describe selected localization approaches:

- Two-stage active localization based on TDOA estimation, which is further on referred to as active localization.
- Two-stage passive localization based on TDOA estimation, which is further on referred to as passive localization.
- Passive localization by TOA- and TDOA-based imaging, which is further on referred to as imaging.

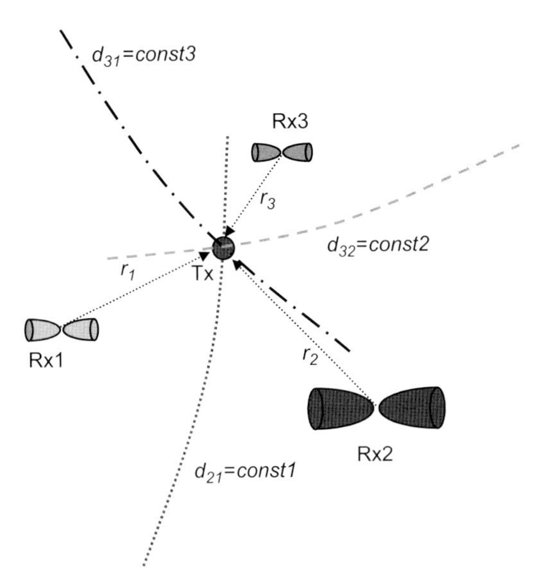

Figure 6.90 Active range difference localization.

6.9.2 Active Localization

This section introduces principles involved in the active range difference localization systems. We assume that there is one target carrying the transmitter. The target has to be localized in three dimensions by N sensor nodes receiving signals from the transmitter (see Figure 6.90). The position estimation is accomplished in two stages.

The first stage performs estimation of the TDOA of a signal at two spatially separated receivers. If the target transmits a pulse it arrives at slightly different times at these receivers. The TDOA is estimated through the use of time delay estimation techniques. For given locations of the two receivers, a whole set of the target positions would give the same TDOA estimate. The locus of possible transmitter locations $[x, y, z]$ for the two receivers situated at $[x_i, y_i, z_i]$ and $[x_j, y_j, z_j]$ is one-half of a hyperboloid or one-half of the hyperbola in the two-dimensional case (see Figure 6.90). The hyperboloid is described by

$$d_{ij} = \sqrt{(x-x_j)^2 + (y-y_j)^2 + (z-z_j)^2} - \sqrt{(x-x_i)^2 + (y-y_i)^2 + (z-z_i)^2} \tag{6.41}$$

where d_{ij} is the estimated TDOA. Multiple TDOAs determine a set of non-linear hyperbolic equations. Therefore, this localization is often referred to as a hyperbolic positioning or a multilateration. Note that the multilateration should not be confused with the trilateration or the triangulation. The trilateration uses absolute distances or TOAs to locate the target and the triangulation uses a baseline and at least two angle estimates.

The second stage of the active localization solves hyperbolic equations and produces a location estimate. Many processing algorithms, with different complexity and restrictions, have been proposed for the position estimation based on range differences. A general solution of a set of hyperbolic functions is a challenging task, which requires a lot of computational power. An example, which describes the TDOA-based localization of one target, in three dimensions and by four receivers, is given in Ref. [186]. If there are more receivers available (at least five for three dimensions), the TDOA localization problem is preferably to be described by a set of spherical equations

$$(x - x_i)^2 + (y - y_i)^2 + (z - z_i)^2 = (r_1 + d_{i1})^2 \quad i = 1 \ldots N \tag{6.42}$$

where r_1 is an additional unknown variable related to the distance between the receiver Rx1 and the transmitter. d_{i1} are estimated TDOAs between the first and the ith receiver. Note that the d_{11} is equal to zero. The summation $r_1 + d_{i1}$ represents a radius of the sphere located around the ith receiver. Without loss of generality let us assume that the origin of the coordinate system is at the position of the first receiver Rx1 ($[x_1, y_1, z_1] = [0,0,0]$). If we subtract the quadratic equation related to the first receiver from the remaining equations we obtain a set of linear equation described by

$$x_i x + y_i y + z_i z + r_1 d_{i1} = \frac{1}{2}\left(x_i^2 + y_i^2 + z_i^2 - d_{i1}{}^2\right) \quad i = 2 \ldots N \tag{6.43}$$

These equations can be rewritten into a matrix form

$$\mathbf{Ap} = \mathbf{b} \tag{6.44}$$

where

$$\mathbf{A} = \begin{bmatrix} x_2 & y_2 & z_2 & d_{21} \\ x_3 & y_3 & z_3 & d_{31} \\ \vdots & & & \vdots \\ x_N & y_N & z_N & d_{N1} \end{bmatrix}, \quad \mathbf{p} = \begin{bmatrix} x \\ y \\ z \\ r_1 \end{bmatrix}, \quad \mathbf{b} = \frac{1}{2}\begin{bmatrix} x_2^2 + y_2^2 + z_2^2 - d_{21}^2 \\ x_3^2 + y_3^2 + z_3^2 - d_{31}^2 \\ \vdots \\ x_N^2 + y_N^2 + z_N^2 - d_{N1}^2 \end{bmatrix} \tag{6.45}$$

Equation (6.44) can be solved by a modified version of the least squares approach (see Annex 7.6), which minimizes the square of the weighted Euclidian norm $\|\mathbf{A\,p} - \mathbf{b}\|^2$. Thus, the vector $\mathbf{p}$, which contains coordinates of the target, is estimated according to

$$\mathbf{p} = \arg\min_p (\mathbf{Ap} - \mathbf{b})^T \mathbf{W} (\mathbf{Ap} - \mathbf{b}) = \left(\mathbf{A}^T \mathbf{W} \mathbf{A}\right)^{-1} \mathbf{A}^T \mathbf{W} \mathbf{b} \tag{6.46}$$

where $\mathbf{W}$ is a diagonal matrix of weighing factors, which improves the localization precision of the least squares solution. The weighing matrix describes the reliability of TDOA estimates. More precise or reliable TDOA estimates are weighed by higher weights in the matrix **W**. If there is no information about TDOA reliability available, the matrix **W** is the unity matrix. In this case, all TDOA estimates are equally treated by the data fusion algorithm.

6.9.3 Passive Localization

As discussed before the passive localization of a target without a tag is a challenging task. The UWB sensor must be capable to receive and to detect the weak electromagnetic waves scattered back from a target in the presence of a strong signal such as the direct wave. According to the Friis transmission formula, the power of the direct wave increases with the decreasing Tx-Rx distance by a power of 2. According to the radar equation, the power of the wave scattered from an object decreases with the increasing distance to the antennas with the power of 4. Moreover, a target scatters only a part of the incident electromagnetic waves back to the receiver according its radar cross section. Therefore, electromagnetic waves scattered from the target and overlaid by other signals are usually almost invisible in unprocessed data. In case of moving or time-varying targets, their detection is enabled only thanks to their time variance. The raw measurements are first pre-processed by algorithms that eliminate strong time invariant signals and reveal weak time-variant signals.

6.9.3.1 Detection of Targets

The detection of a moving target is based on algorithms referred to as the 'background subtraction'. These algorithms subtract the time invariant 'background' signal from measured data. The background signal comprises especially the direct wave and waves reflected from dominant static reflectors, for example walls, furniture or metallic devices in the environment. Background subtraction can considerably enhance the dynamic range for the detection of weak time-variant signal features. Its performance strongly depends on

- the number of targets,
- their 'activity' and
- the environmental conditions.

Target's activity is related to the kind of its movement. For example, a person can walk or sit quietly. A moving person is easy to detect in comparison to the sitting or laying person. In the latter case, the person can only be detected by its respiratory or heartbeat activity. Detection of the respiratory and/or heartbeat activity is, however, a challenging task. Small movements of the human chest evoke only very weak signals. This imposes tough requirements on the dynamic of UWB sensors (see also Section 6.7).

Environmental conditions also determine the degree of complexity of the person's detection. If all objects in the person's surroundings are static then the detection of a moving person is relatively easy. In this case, the moving person can be differentiated from other static objects according to its movement. However, if the scenario is time varying with various moving objects (windows, machines), the detection of persons is much more demanding. In this case, the detection step must be followed by the target recognition.

A simple background subtraction approach starts with stacked averaging of the sequence of measured impulse responses. This way, the static background signal is estimated. In the next processing step, the background estimate is subtracted from the given sequence of impulse responses.

In the variety of realistic scenarios, estimation of the static background just by a stacked averaging is not enough due to following problems:

- The background signal can also be time-variant. This time variance is caused, for example, by undesired antenna movement or by movement of objects that are not of interest.
- The object of interest can change its state of motion. This means, after having been moving, it can also be stationary over some time interval and 'misinterpreted' by the algorithm as part of the undesired background signal.

To solve these problems, more elaborate filtering procedures are needed. An overview of more or less sophisticated background subtraction methods that are also known in video surveillance applications is given, for example in Ref. [187]. Some of them were successfully implemented also for UWB signals [188].

6.9.3.2 **Passive Localization of Targets**

The description of the passive localization principles is illustrated in Figure 6.91. One transmitter Tx illuminates the scene by electromagnetic waves. The target scatters the waves to N receivers Rx_i. We suppose that the background subtraction algorithm removes all time invariant signals from received signals. Then, the position estimation is accomplished in two stages.

The first stage performs estimation of the TOA from data processed by the background subtraction algorithm. For the given location of the transmitter and one receiver, a whole set of the target positions would give the same TOA estimate. The locus of possible target locations $[x, y, z]$ for the transmitter situated at $[x_{Tx}, y_{Tx}, z_{Tx}]$

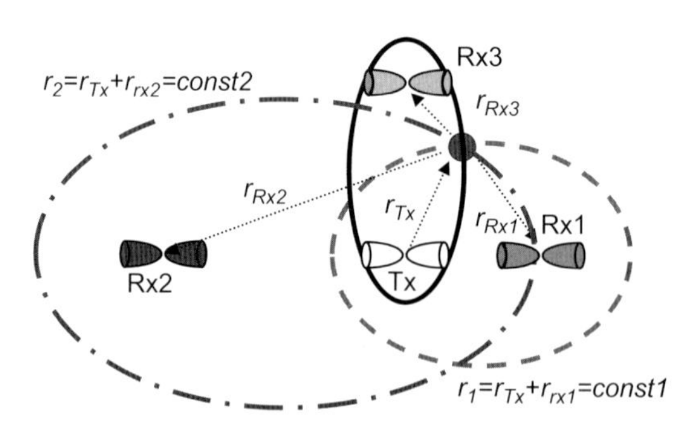

Figure 6.91 Positioning – passive approach.

and the receiver at $[x_i, y_i, z_i]$ is a spheroid or an ellipsis in the two-dimensional case (see Figure 6.91). The spheroid is described by

$$r_i = \sqrt{(x - x_{\mathrm{Tx}})^2 + (y - y_{\mathrm{Tx}})^2 + (z - z_{\mathrm{Tx}})^2} + \sqrt{(x - x_i)^2 + (y - y_i)^2 + (z - z_i)^2} \tag{6.47}$$

where r_i is the estimated TOA. Multiple TOAs determine a set of non-linear elliptic equations, which is solved in the second localization stage.

Various position estimation techniques were proposed for the TOA-based passive localization such as Taylor-series [189], spherical interpolation method [190] or least squares method [191, 192]. An overview and comparison of these methods based on measured data is given in Ref. [193].

The least squares method can be applied to solve the system of equation (6.47) similarly to the approach discussed by the active localization. If there are at least four receivers available (for three dimensions) the TOA passive localization problem can be described by a set of spherical equations

$$\begin{aligned} &(x - x_{\mathrm{Tx}})^2 + (y - y_{\mathrm{Tx}})^2 + (z - z_{\mathrm{Tx}})^2 = r_{\mathrm{Tx}}^2 \\ &(x - x_i)^2 + (y - y_i)^2 + (z - z_i)^2 = (r_i - r_{\mathrm{Tx}})^2 \quad i = 1 \ldots N \end{aligned} \tag{6.48}$$

where r_{Tx} is an additional unknown variable related to the distance between the transmitter and target. r_i are estimated TOAs of electromagnetic wave propagating from the transmitter via the target to the ith receiver. The difference $r_i - r_{\mathrm{Tx}}$ represents a radius of the sphere located around the ith receiver. If we assume that the origin of the coordinate system is at the position of the transmitter and if we subtract the quadratic equation related to the transmitter from the remaining equations we obtain a set of linear equations

$$x_i x + y_i y + z_i z - r_{Tx} r_i = \frac{1}{2}\left(x_i^2 + y_i^2 + z_i^2 - {r_i}^2\right) \quad i = 1 \ldots N \tag{6.49}$$

These equations can be rewritten into a matrix form

$$\mathbf{Ap} = \mathbf{b} \tag{6.50}$$

where

$$\mathbf{A} = \begin{bmatrix} x_1 & y_1 & z_1 & -r_1 \\ x_2 & y_2 & z_2 & -r_2 \\ \vdots & & & \vdots \\ x_N & y_N & z_N & -r_N \end{bmatrix}, \quad \mathbf{p} = \begin{bmatrix} x \\ y \\ z \\ r_{Tx} \end{bmatrix}, \quad \mathbf{b} = \frac{1}{2}\begin{bmatrix} x_1^2 + y_1^2 + z_1^2 - r_1^2 \\ x_2^2 + y_2^2 + z_2^2 - r_2^2 \\ \vdots \\ x_N^2 + y_N^2 + z_N^2 - r_N^2 \end{bmatrix} \tag{6.51}$$

Equation (6.51) can be solved by the same approach as depicted in Eq. (6.46).

6.9.3.3 Measured Example

Here, we demonstrate the passive localization by a measurement performed in the home-entertainment environment. The measurement scenario and the antenna

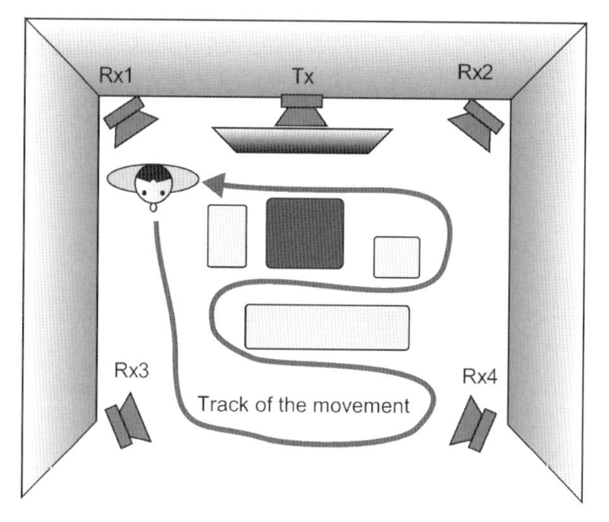

Figure 6.92 Measurement scenario.

constellation are depicted in Figure 6.92. One transmitting antenna was placed on the top of a TV. Four receiving antennas were situated at the position of satellite loudspeakers in corners of the room. One person walked along the track shown in Figure 6.92. During its movement, the M-sequence-based UWB sensors ([193], see Figure 3.57 for working principle) measured impulse responses in a real time. The measurement rate was about 25 impulse responses per second. The sensor operated in the frequency band from 3.5 to 10.5 GHz. Measured impulse responses from the receiver Rx1 are shown in Figure 6.93. The magnitude is indicated by the colour.

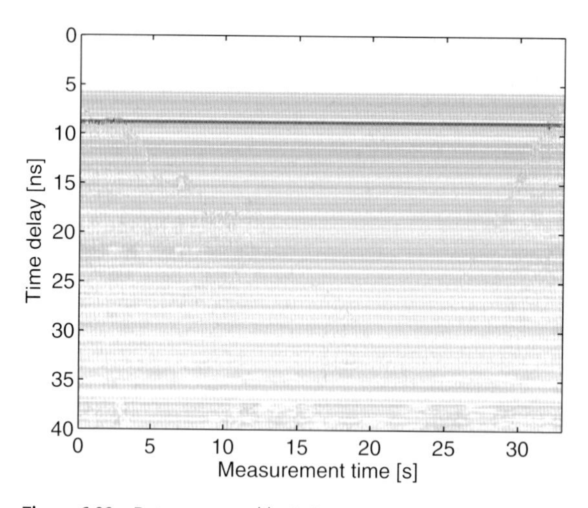

Figure 6.93 Data measured by Rx1.

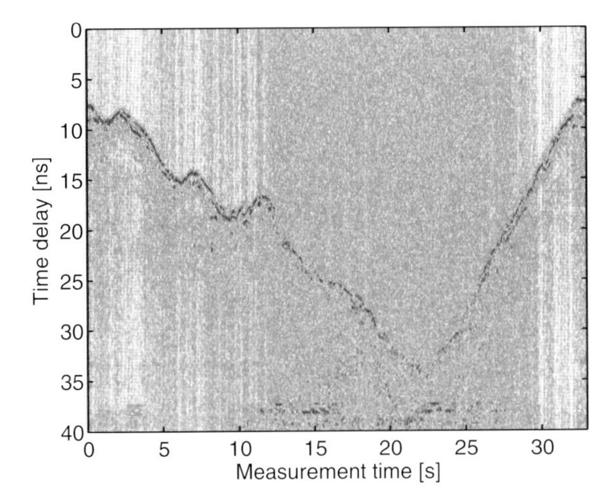

Figure 6.94 Data after background subtraction algorithm.

Due to the direct wave and strong reflections from the walls, the time-variant signal scattered by the moving person is barely visible. The time-variant signal is brought into the front by the background subtraction algorithm. Figure 6.94 shows the result of the background subtraction algorithm based on exponential averaging. Processed impulse responses are normalized in this figure. TOAs were estimated by a ranging algorithm based on the adaptive threshold [195]. The two-dimensional position of the moving person was estimated by the least squares approach described above. Estimated locations were subsequently averaged by a median filter and an exponential averaging. The group delay of both averaging filters was less than 1 s. This delay is meaningless for the application in the home-entertainment scenario, where the audio algorithms optimize or direct the sound according to the position of a listener who is usually sitting in a sofa. For more time critical applications, for example security applications, other more sophisticated tracking algorithms can be used [196]. The result of the position estimation based on signals received from all four receivers is shown in Figure 6.95. The precision of the position estimation is about 0.5 m. This precision is in the order of the body size (cross section) of the walking person and is fully sufficient for the majority of applications.

6.9.4
Imaging of Targets

Here, we describe the application of imaging algorithms for the passive localization of moving targets based on the TOA parameter (refer also to Section 4.10). The measurement constellation is the same as in the case of the two-stage passive localization of targets (see Figure 6.91) – one transmitting and N receiving antennas that

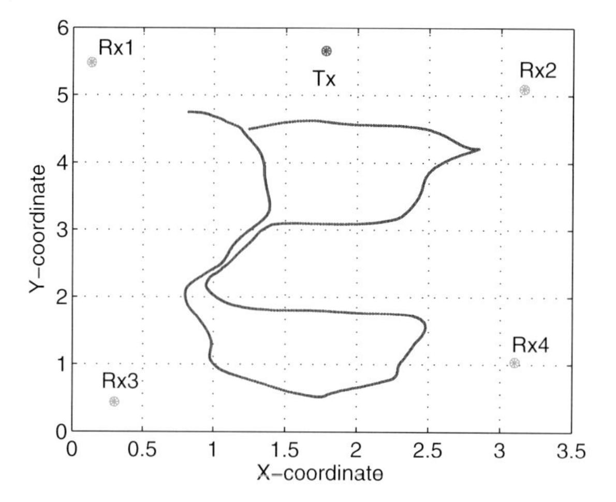

Figure 6.95 Positions of a walking person estimated by least squares approach using Rx1, Rx2, Rx3 and Rx4.

observe one moving target. We assume that the background subtraction algorithm successfully removed all time invariant components from received signals. In contrary to the two-stage passive localization, the imaging algorithm does not contain any parameter estimation step. It fuses the whole impulse responses received in parallel by a number of receivers at a certain time. The data fusion algorithm produces a two-dimensional image of the inspected area. The intensity (amplitude) of each pixel is related to the probability of the target's presence at the location given by coordinates of the pixel. Thus, the location of a target is indicated in the image by high amplitudes – hot spots.

First, the data fusion maps one-dimensional impulse response $R_n(\tau)$ onto the two-dimensional 'snapshot' according to

$$\begin{aligned} s_n(x,y) &= R_n\left(\frac{r_{\mathrm{Tx}}(x,y)+r_n(x,y)}{v}\right) \\ &= R_n\left(\frac{\sqrt{(x-x_{\mathrm{Tx}})^2+(y-y_{\mathrm{Tx}})^2}+\sqrt{(x-x_n)^2+(y-y_n)^2}}{v}\right) \end{aligned} \tag{6.52}$$

where $[x_n, y_n]$ are the coordinates of the nth receiving antenna and $[x_{\mathrm{Tx}}, y_{\mathrm{Tx}}]$ are the coordinates of the transmitting antenna, $R_n(\tau)$ is the impulse response received by the nth receiver, $r_{\mathrm{Tx}}(x, y)$ is the distance between the image's pixel at the coordinates $[x, y]$ and the transmitting antenna, $r_n(x, y)$ is the distance between the image's pixel at the coordinates $[x, y]$ and the nth receiving antenna, v stands for the velocity of electromagnetic waves. According to (6.52), an echo in the impulse response $R_n(\tau)$ scattered from a target is mapped in the snapshot $s_n(x, y)$ as an elliptical trace around the corresponding Tx-Rx antenna pair. The image of the inspected area

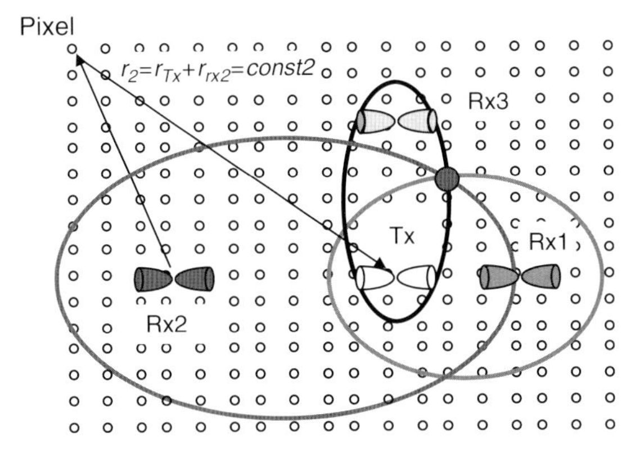

Figure 6.96 Imaging (passive localization) of moving objects.

$p(x, y)$ focuses at the objects positions (see Figure 6.96) by combining (summing) multiple snapshots

$$p(x, y) = \frac{1}{N} \sum_{n=1}^{N} s_n(x, y) \tag{6.53}$$

The more impulse responses are fused together, the more focused is the obtained image. However, the number of receivers is usually constrained by the real-time operation. All receivers must be perfectly synchronized and operate in parallel in order to observe the target at the same time and position.

Note that the same principles of the image formation are used for the creation of SAR images of static environments. Supposing time invariance of the scenario under test, the impulse responses do not have to be measured in parallel. This allows creation of large synthetic apertures and subsequently much better focusing performance of this imaging algorithm. However, this conventional SAR imaging cannot provide images of fast moving objects since they will be smeared in the focused image.

Imaging of one moving person is demonstrated by a measurement performed in a university foyer. Its size was about 15 m × 7 m. The measurement scenario and the antenna constellation are depicted in Figure 6.97. One transmitting and two receiving antennas were situated behind a brick wall. Receiving antennas were about 53 cm apart from the transmitting antenna. One person walked along the track depicted by the line in Figure 6.97 which is shaped as an eight. During its movement an UWB sensor measured impulse responses in a real time. The measurement rate was about 50 impulse responses per second. The sensor operated in the baseband from 13.5 MHz to 3.5 GHz. The low frequencies were attenuated by Horn antennas with operational frequencies starting from about 800 MHz. Measured impulse responses are shown in Figure 6.98. The magnitude is indicated by the colour.

The result of the imaging algorithm that combines TOA-based approach described above and TDOA-based approach is illustrated in Figure 6.99. First, the

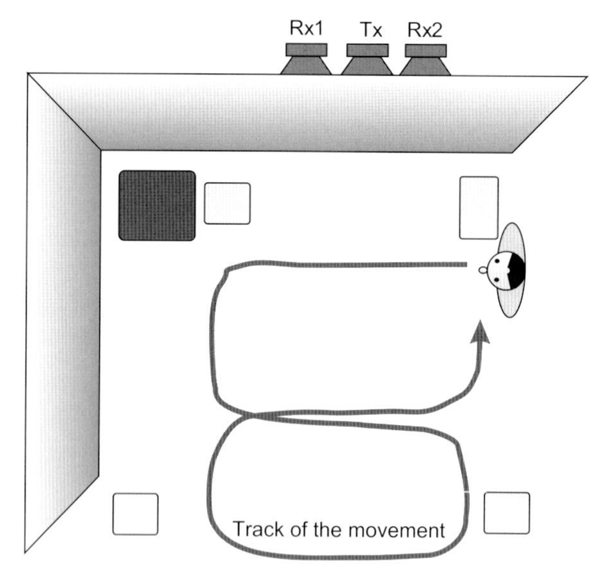

Figure 6.97 Measurement scenario.

imaging algorithm determined two snapshots according to (6.52). These two snapshots are related to envelopes of impulse responses measured at the time of 2.7 s (see Figure 6.98). The third snapshot was computed from the cross-correlation of both impulse responses $R_{xc}(\tau)$ according to

$$s_3(x, y) = R_{xc}\left(\frac{r_1(x, y) - r_2(x, y)}{v}\right) \tag{6.54}$$

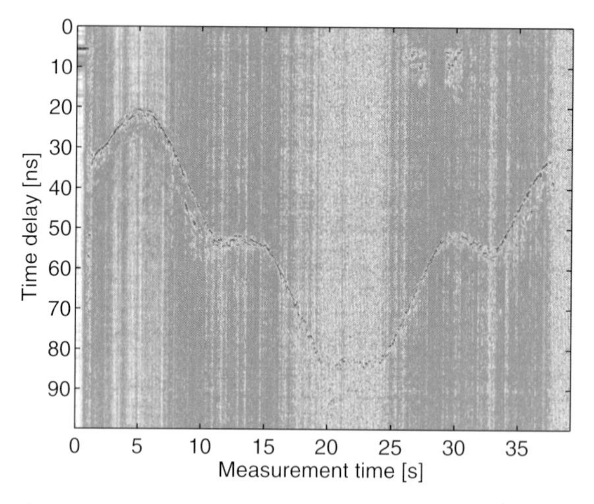

Figure 6.98 Data measured by Rx2 after background subtraction.

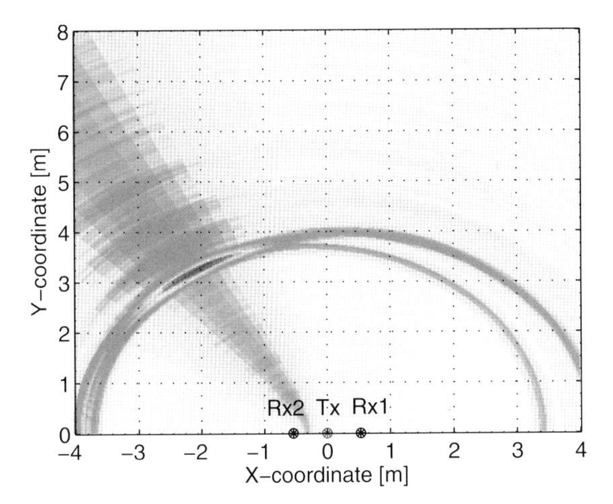

Figure 6.99 Localization of one moving person by imaging (target position at observation time $T = 2.7$ s).

where $r_1(x, y) - r_2(x, y)$ is the range difference of arrival related to the pixel of the image at the coordinates $[x, y]$. The image, which hot spot indicates the position of the moving person, was obtained by the summation of these three snapshots. Two TOA-based snapshots contribute to the final image by elliptical traces and the TDOA-based image by a hyperbolical trace. All three traces intersect at the position of a person. The performance of imaging-based localization can be improved by more sophisticated approaches [197, 198] that were successfully implemented for the imaging of environments.

6.9.5 Further Challenges

The detection and mitigation of non-line of sight (NLOS) situations seems to be the major challenge in the field of active localization. If the direct path between a sensor node and the target is obstructed, the TOA of the transmitted signal is delayed and attenuated, or even completely blocked. In the former case, the estimated TOA is delayed of the time necessary for the electromagnetic wave to path an unknown obstacle. In the latter case, the estimated TOA is falsely estimated with usually a large positive bias. Using such corrupted TOA estimates for the localization of the target may significantly degrade the positioning accuracy. Therefore, sensor nodes in NLOS situations have to be identified and their negative influence on the positioning accuracy have to be mitigated. Recently, several detection and mitigation schemes have been proposed. They include redundant reference sensor nodes, adaptive selection of nodes, sensor cooperation, soft decision localization, location tracking and others. Majority of them were developed for moving targets. The movement allows inspection of the time evolution of range estimates and their statistics. These statistical analyses are a prerequisite for the correct NLOS identification based on multiple hypothesis tests and modelling of LOS and NLOS range estimates [199]. The delay spread, running variance, confidence metrics and the SNR time evolution represent further NLOS detection algorithms that exploit the movement of the target [200]. A novel approaches, which do not require a moving target, were proposed in Refs [201, 202]. The former approach considers a joint likelihood ratio test based on the amplitude and delay statistics of the multi-path channel. The latter one is based on hypotheses tests.

Passive localization is usually applied in obstructed NLOS situations such as trough wall radar. Hence, algorithms and electronics of UWB sensors for these applications are already adopted for NLOS situations and offer large dynamic range that allows detection of weak multi-path components, for example by the background subtraction. The research field of passive localization especially concerns with localization and tracking of multiple targets. The research reported in Ref. [203] deals with through wall localization of multiple persons. It is mostly based on experiments measured in rescue facilities in Sweden. The measurement scenario is represented by closely spaced sensors with one transmitting and two receiving antennas. It was shown that the localization of multiple targets is challenging even in 'simple' environments like free room without any furniture. The reason is twofold. The first challenge is caused by the decrease of the received signal strength with distance. Due to this fact, person standing further away is very often almost invisible in the data processed by the background subtraction algorithm. The second challenge is caused by the interaction of each target with its surrounding. In scenarios with the dense multi-path the detection of multiple targets is a challenging task. An example of such a scenario is an environment containing a lot of metallic objects. A moving person 'shadows' other static objects and prohibits wave to propagate from the transmitter to, for example, a heating at the wall behind a person. If the person moves away from its position it changes the propagation conditions and the waves can again propagate to the heating and back to the receiver. Time variations occurring at the heating are misinterpreted by the background subtraction algorithm as the reflection from a time-variant target. Since the reflection form a metallic plate (heating) is much stronger than the scattering from a person where the energy is scattered into all directions, the metallic object is visible from much larger distances than a person. A background subtraction reveals so a number of strong echoes that are received after the echo evoked by the moving person. These multi-path components prohibit detection of a person that is further away. The scattering from the second person is overlaid by 'shadows' that originate from reflections on other objects in the room.

Despite of challenges described above, UWB localization is successfully applied in various applications. The large bandwidth and good temporal resolution of UWB localization systems is an essential base for an accurate positioning. Active localization reaches a precision in an order of some centimetres or in well-controlled scenarios even a fraction of a centimetre [204, 206]. The precision of the passive localization was reported to be a couple of decimetres [188, 205]. This precision is however sufficient for the majority of envisaged scenarios and is usually related to the size of targets.

References

1 Politano, C., Hirt, W., Rinaldi, N. *et al.* (2006) Regulation and standardization, in *UWB Communication Systems: A Comprehensive Overview* (eds M.G. Di Benedetto, T. Kaiser, A.F. Molischet *al.*), Hindawi Publishing Corporation, pp. 447–492.

2 Taylor, J.D. (2001) *Ultra-Wideband Radar Technology*, CRC Press, Boca Raton, FL.

3 Baum, C.E., Rothwell, E.J., Chen, K.M. *et al.* (1991) The singularity expansion method and its application to target identification. *Proc. IEEE*, **79** (10), 1481–1492.

4 Astanin, L.Y. and Kostylev, A.A. (1997) *Ultrawideband Radar Measurements Analysis and Processing*, The Institution of Electrical Engineers, London, UK.

5 Daniels, D.J. (2004) *Ground Penetrating Radar*, 2nd edn, Institution of Electrical Engineers, London.

6 Amin, M.G. (2011) Through-The-Wall Radar Imaging: CRC Press.

7 Chang, P.C., Burkholder, R.J. and Volakis, J.L. (2010) Adaptive CLEAN with target refocusing for through-wall image improvement. *IEEE Trans. Antenn. Propag.*, **58** (1), 155–162.

8 Agoston, A., Pepper, S., Norton, R. *et al.* (2003) 100 GHz through-line sampler system with sampling rates in excess of 10 Gsamples/second. IEEE MTT-S International Microwave Symposium Digest, vol. 3, 2003, pp. 1519–1521.

9 Nicolaescu, I., van Genderen, P., Van Dongen, K.W. *et al.* (2003) Stepped frequency continuous wave radar-data preprocessing. Proceedings of the 2nd International Workshop on Advanced Ground Penetrating Radar, 2003, pp. 177–182.

10 Parrini, F., Fratini, M., Pieraccini, M. *et al.* (2006) Ultra: wideband ground penetrating radar. 3rd European Radar Conference, 2006. EuRAD 2006, pp. 182–185.

11 Narayanan, R.M., Xu, X. and Henning, J. A. (2004) Radar penetration imaging using ultra-wideband (UWB) random noise waveforms. *IEE Proc.-Radar Son. Nav.*, **151** (3), 143–148.

12 Lukin, K.A. (2006) Radar design using noise/random waveforms. International Radar Symposium, 2006. IRS 2006, pp. 1–4.

13 Sachs, J., Herrmann, R., Kmec, M. *et al.* (2007) Recent advances and applications of m-sequence based ultra-wideband sensors. IEEE International Conference on Ultra-Wideband, Singapore, 2007. ICUWB 2007, pp. 50–55.

14 Sachs, J., Badstübner, A., Bonitz, F. *et al.* (2008) High resolution non-destructive testing in civil engineering by ultra-wideband pseudo-noise approaches. International Conference on Ultra-Wideband, Hannover, Germany.

15 Sachs, J., Wöckel, S., Herrmann, R. *et al.* (2007) Liquid and moisture sensing by ultra wideband pseudo noise sequences. *Meas. Sci. Technol.*, **18**, 1074–1087.

16 Levitas, B. (2007) UWB time domain measurements. The Second European Conference on Antennas and Propagation, 2007. EuCAP 2007, pp. 1–8.

17 Levitas, B. and Matuzas, J. (2005) Evaluation of UWB ISAR image resolution. European Radar Conference, 2005. EURAD 2005, pp. 89–91.

18 Daschner, F., Knoechel, R. and Kent, M. (2000) Rapid monitoring of selected food properties using microwave dielectric spectra. 2nd International Conference on Microwave and Millimeter Wave Technology, 2000, ICMMT 2000, pp. 670–673.

19 Lukin, K., Mogyla, A., Palamarchuk, V. *et al.* (2009) Monitoring of St. Sophia Cathedral interior using Ka-band ground based noise waveform SAR. European Radar Conference, 2009. EuRAD 2009, pp. 215–217.

20 Haddad, W., Chang, J., Rosenbury, T. *et al.* (2000) *Microwave Hematoma detector for the Rapid Assessment of Head Injuries*, Lawrence Livermore National Laboratory.

21 Thiel, F., Hein, M.A., Sachs, J. *et al.* (2008) Physiological signatures monitored by ultra-wideband-radar validated by magnetic resonance

imaging. International Conference on Ultra-Wideband, Hannover, Germany.

22 Helbig, M., Hein, M.A., Schwarz, U. *et al.* (2008) Preliminary investigations of chest surface identification algorithms for breast cancer detection. International Conference on Ultra-Wideband, Hannover, Germany.

23 Immoreev, I.Y. (2006) Practical application of ultra-wideband radars. The Third International Conference on Ultrawideband and Ultrashort Impulse Signals, pp. 44–49.

24 Mashal, A., Sitharaman, B., Booske, J.H. *et al.* (2009) Dielectric characterization of carbon nanotube contrast agents for microwave breast cancer detection. IEEE Antennas and Propagation Society International Symposium, 2009. APSURSI '09, pp. 1–4.

25 Haddad, W. (1997) The rubble rescue radar (RRR): a low power hand-held microwave device for the detection of trapped human personnel. Report: Lawrence Livermore National Laboratory; http://www.osti.gov/bridge/servlets/purl/571102-d7WBku/webviewable/571102.pdf

26 Zetik, R., Sachs, J. and Peyerl, P. (2007) Through-wall imaging by means of UWB radar, in *Short-Pulse Electromagnetics 7* (eds F. Sabath, E.L. Mokole, U. Schenk*et al.*), Springer, pp. 613–622.

27 Nezirovic, A., Yarovoy, A.G. and Ligthart, L.P. (2010) Signal processing for improved detection of trapped victims using UWB radar. *IEEE Trans. Geosci. Remote Sensing*, **48** (4), 2005–2014.

28 Porcino, D., Sachs, J., Zetik, R. *et al.* (2006) UWB ranging, in *UWB Communication Systems a comprehensive Overview* (eds M.G.d. Benedetto, T. Kaiser, A. Molisch *et al.*), Hindawi Publishing Corporation, pp. 411–446.

29 Fontana, R.J., Foster, L.A., Fair, B. *et al.* (2007) Recent advances in ultra wideband radar and ranging systems. IEEE International Conference on Ultra-Wideband, 2007. ICUWB 2007, pp. 19–25.

30 Ohta, K., Ono, K., Matsunami, I. *et al.* (2010) Wireless motion sensor using ultra-wideband impulse-radio. IEEE Radio and Wireless Symposium (RWS), 2010, pp. 13–16.

31 Herrmann, R., Sachs, J. and Bonitz, F. (2010) On benefits and challenges of person localization using ultra-wideband sensors. International Conference on Indoor Positioning and Indoor Navigation (IPIN), 2010, pp. 1–7.

32 Jol, H.M. (2009) *Ground Penetrating Radar: Theory and Applications,* Elsevier.

33 Nowak, K., Gross, W., Nicksch, K. *et al.* (2011) Intraoperative lung edema monitoring by microwave reflectometry. In: *Interactive cardiovascular and thoracic surgery.* Oxford : Oxford Univ. Press, ISSN 15699293, Bd. **12** (2011), 4, S. 540–544.

34 Wörsching, H., Becker, R., Schlaeger, S. *et al.* (2006) Spatial-TDR moisture measurement in a large scale levee model made of loamy soil material. TDR 2006 Purdue University, West Lafayette, USA.

35 Huebner, C., Schlaeger, S., Becker, R. *et al.* (2005) Advanced measurement methods in time domain reflectometry for soil moisture determination, in *Electromagnetic Aquametry* (ed. K. Kupfer), Springer, Berlin, Heidelberg.

36 Becker, R. (2004) *Spatial Time Domain Reflectometry for Monitoring Transient Soil Moisture Profiles,* Institut für Wasser und Gewässerentwicklung Universität, Karlsruhe (TH).

37 Schlaeger, S. (2005) A fast TDR-inversion technique for the reconstruction of spatial soil moisture content. *Hydrol. Earth Syst. Sci.*, (9), 481–492. www.copernicus.org/EGU/hess/hess/9/481/ SRef-ID: 1607-7938/hess/2005-9-481 European Geosciences Union

38 Bonitz, F., Wagner, N., Kupfer, K. *et al.* (2011) Ultra wideband measurements for spatial water content estimation in subsoil. 9th International Conference on Electromagnetic Wave Interaction with Water and Moist Substances, Kansas City, Missouri, USA.

39 Sachs, J., Helbig, M. and Renhak, K. (2008) Unsichtbares wird sichtbar – Mit Radar den Insekten auf der Spur: Neue Möglichkeiten zum Aufspüren von Insektenaktivitäten. HOBA'08, Duisburg, Germany, pp. 45–48.

40 Rodwell, M., Betser, Y., Jaganathan, S., Mathew, T., Sundararajan, P.K., Martin, S.C., Smith, R.P., Wei, Y., Urteaga, M., Scott, D. and Long, S. (2000) Submicron lateral scaling of HBTs and other vertical-transport devices: toward THz bandwidths. Proceedings of the GaAs 2000 Conference, pp. 1–4.

41 Floyd, B.A., Reynolds, S.K., Pfeiffer, U. R., Zwick, T., Beukema, T. and Gaucher, B. (2005) SiGe bipolar transceiver circuits operating at 60 GHz. *IEEE J. Solid-St. Circ.*, **40** (1), 156–167.

42 Cressler, John D. (2007) *Fabrication of SiGe HBT BiCMOS Technology*, Taylor and Francis, Boca Raton.

43 Heinemann, B., Barth, R., Knoll, D., Rucker, H., Tillack, B. and Winkler, W. (2006) High-performance BiCMOS technologies without epitaxially-buried subcollectors and deep trenches. Third International SiGe Technology and Device Meeting, 2006. ISTDM 2006, pp. 1–2.

44 Zerounian, N., Aniel, F., Barbalat, B., Chevalier, P. and Chantre, A. (2007) 500 GHz cutoff frequency SiGe HBTs. *Electron. Lett.*, **43** (14) pp.774–775.

45 Kuhn, K., Basco, R., Becher, D., Hattendorf, M., Packan, P., Post, I., Vandervoorn, P. and Young, I. (2004) A comparison of state-of-the-art NMOS and SiGe HBT devices for analog/mixed-signal/RF circuit applications. Symposium on VLSI Technology, 2004. Digest of Technical Papers, pp. 224–225.

46 Cheskis, David (2007) 130 nm SiGe BiCMOS processes optimize cost and performance www.wirelessdesignmag.com, jun.

47 Esame, O., Gurbuz, Y., Tekin, I. and Bozkurt, A. (2004) Performance comparison of state-of-the-art heterojunction bipolar devices (HBT) based on AlGaAs/GaAs, Si/SiGe and InGaAs/InP. *Microelectr. J.*, **35**, 901–908.

48 Ashburn, P. (1988) *Design and Realization of Bipolar Transistors*, John Wiley & Sons, Inc.

49 Sze, S.M. (1981) *Physics of Semiconductor Devices*, 2nd edn, John Wiley & Sons, Inc., New York.

50 Maiti, C.K. and Armstrong, G.A. (2001) *2001 Application of Silicon-Germanium Heterostructure Devices*, Institute of Physics Publishing, London.

51 Hastings, A. (2000) *The Art of Analog Layout*, Prentice Hall.

52 Fang, W. (1990) Accurate analytical delay expressions for ECL and CML circuits and their applications to optimizing high-speed bipolar circuits. *IEEE J. Solid-St. Circ.*, **25** (2), 572–583.

53 Fox, A., Heinemann, B., Barth, R., Bolze, D., Drews, J., Haak, U., Knoll, D., Kuck, B., Kurps, R., Marschmeyer, S., Richter, H.H., Rucker, H., Schley, P., Schmidt, D., Tillack, B., Weidner, G., Wipf, C., Wolansky, D. and Yamamoto, Y. (2008) SiGe HBT module with 2.5 ps gate delay. IEEE International Electron Devices Meeting, 2008. IEDM 2008, pp. 1–4.

54 Kmec, M., Herrmann, R., Sachs, J., Peyerl, P. and Rauschenbach, P. (2006) Extended approaches for integrated M-sequence based UWB sensors. 2006 IEEE AP-S International Symposium with Radio Science and AMEREM Meetings, Albuquerque, NM, USA, July 2006.

55 Kmec, M., Sachs, J., Peyerl, P., Rauschenbach, P., Thomä, R. and Zetik, R. (2005) A novel ultra-wideband real-time MIMO channel sounder architecture. XXVIIIth URSI General Assembly 2005, New Delhi, 23–29 October 2005.

56 Herrmann, R., Sachs, J., Schilling, K. and Bonitz, F. (2008) New extended M-sequence ultra wideband radar and its application to the disaggregation zone in salt rock. GPR 2008 12th International Conference on Ground Penetrating Radar, Birmingham, UK, 15–19 June.

57 Stafford, K.R., Blanchard, R.A. and Gray, P.R. (1974) A complete monolithic sample/hold amplifier. *IEEE J. Solid-St. Circ.*, **9** (6), 381–387.

58 Vorenkamp, P. and Verdaasdonk, J.P.M. (1992) Fully bipolar, 120-Msample/s 10-b track-and-hold circuit. *IEEE J. Solid-St. Circ.*, **27** (7), 988–992.

59 Karanicolas, A.N. (1997) A 2.7-V 300-MS/s track-and-hold amplifier. *IEEE J. Solid-St. Circ.*, **32** (12), 1961–1967.

60 Willingham, S.D. and Martin, K.W. (1990) Effective clock-feedthrough reduction in switched capacitor circuits. IEEE International Symposium on Circuits and Systems, vol. 4, 1–3, May 1990, pp. 2821–2824.

61 Lee, I., Tang, J. and Kim, W. (1992) An analysis of clock feedthrough noise in bipolar comparators. Proceedings of the IEEE International Symposium on Circuits and Systems, vol. 3, 10–13 May 1992. ISCAS '92, pp. 1392–1395.

62 Kent, M., Knöchel, R., Barr, U-.K., Tejada, M., Nunes, L. and Oehlenschläger, J. (eds) (2005) *SEQUID – A new Method for Measurement of the Quality of Seafood*, Shaker Verlag, ISBN 3-8322-4159-0.

63 Gajda, G. and Stuchly, S. (1983) An equivalent circuit of an open-ended coaxial line. *IEEE Trans. Instrum. Meas.*, **IM-32** (4), 506–508.

64 Grant, J.P., Clarke, R.N., Symm, G.T. and Spyrou, N.M. (1989) A critical study of the open-ended coaxial line sensor for RF and microwave complex permittivity measurement. *J. Phys. E: Instrum.*, **22**, S.757–770.

65 Martens, H. and Naes, T. (1989) *Multivariate Calibration*, John Wiley & Sons, Ltd., Chichester.

66 Patterson, D. (1998) *Artificial Neural Networks*, Prentice Hall.

67 Savelyev, T.G. and Sato, M. (2004) UWB GPR target identification for landmine detection. Proceedings of the Imaging Technology 7th International Symposium, Society of Exploration Geophysicists of Japan, 2004, pp. 124–129.

68 Sai, B. and Ligthart, L. (2000) Improved GPR data pre-processing for detection of various land mines. Proceedings of the 8th Conference on Ground Penetrating Radar, GPR, Gold Coast Australia, May 2000.

69 Nakashima, Y., Zhou, H. and Sato, M. (2001) Estimation of groundwater level by GPR in an area with multiple ambiguous reflections. *J. Appl. Geophys.*, **47**, 241–249.

70 Lualdi, M., Zanzi, L. and Sosio, G. (2006) A 3D GPR survey methodology for archaeological applications. Proceedings of the 11th International Conference on Ground Penetrating Radar, GPR, Columbus, OH, USA.

71 Porsani, J.L., Sauck, W.A. and Junior, A. O.S. (2006) GPR for mapping fractures and as a guide for the extraction of ornamental granite from a quarry: a case study from southern Brazil. *J. Appl. Geophys.*, **58** (3), 177–187.

72 MALA Geoscience Corp., Sweden (2011). Company webpage: http://www.malags.com.

73 Geophysical Survey Systems, Inc. (2011) Company webpage: http://www.geophysical.com.

74 Penetradar Corp., USA (2011). Company webpage: http://www.geophysical.com.

75 Geozondas Ltd., Lithuania (2011). Company webpage: http://www.geozondas.com.

76 Ratcliffe, J.A., Sachs, J., Cloude, S., Crisp, G.N., Sahli, H., Peyerl, P. and Pasquale, G.D. (2000) Cost effective surface penetrating radar device for humanitarian demining. Proceedings of the EUROEM.

77 Bonitz, F., Eidner, M., Sachs, J., Herrmann, R. and Solas, H. (2008) UWB-radar sewer tube crawler. Proceedings of the 12th International Conference on Ground Penetrating Radar, GPR, Birmingham, UK, June 2008.

78 Baar, C.A. (1977) *Applied Salt-Rock Mechanics 1: The In-Situ Behavior of Salt Rocks*, Elsevier/North-Holland Inc., New York, NY, 14 Bibliography USA.

79 BFS: ERAM – Endlager für radioaktive Abfälle Morsleben/Bundesamtes für Strahlenschutz. Version: December 2009. http://www.bfs.de/de/endlager/morsleben.html/morsleben_artikel.html.

80 BFS: *Einführung zum Standort Gorleben/Bundesamtes für Strahlenschutz*. Version: June 2009. http://www.bfs.de/de/endlager/gorleben/einfuehrung.html.

81 BFS: *Endlager Asse: ein Überblick/Bundesamtes für Strahlenschutz*. Version: October 2009. http://www.endlager-asse.de/cln_135/DE/5_AsseService/B_Publikationen/_node.html.

82 BFS: *Endlager Schacht Konrad: Einführung/Bundesamtes für Strahlenschutz*. Version: October 2009. http://www.endlager-konrad.de.

83 Kühnicke, H. (2002) *Entwicklung und In-situ-Test akustischer Verfahren zur zerstörungsfreien Beurteilung von Auflockerungszonen im Salinar*, FKZ: 02C0537/Fraunhofer-Institut für zerstörungsfreie Prüfverfahren Institutsteil Dresden. – BMBF research report.

84 Kecke, H.J. (2004) *Grundlagenuntersuchungen zum Dickstoffverfahren mit chem./tox. Abfällen, insbesondere MVA-Filteraschen, im Salinar*, FKZ: 02C04867/Otto-von-Guericke-Universität Magdeburg. – BMBF research report.

85 Düsterloh, U. (1996) *Modellversuche an axial gelochten Steinsalz-Großbohrkernen im Hinblick auf die Überprüfung und Erweiterung theoretischer Prognosemodelle zum Sicherheitsnachweis von Untertagedeponien*, FKZ: 02C00922/Technische Universität Clausthal – Institut für Bergbau. – BMBF research report.

86 Sitz, P. (2002) *Entwicklung eines Grundkonzepts für langzeitstabile Streckenverschlussbauwerke im Salinar - Bau und Test eines Versuchsverschlussbauwerkes unter realen Bedingungen*, FKZ: 02C0547/Technische Universität Bergakademie Freiberg. – BMBF research report.

87 Kudla, W. and Gruner, M. (2006) Streckenverschlüsse: Konzepte und Materialien. Proceedings of the 7th Annual Review for BMBF/BMWi Research Projects on Toxic Waste Disposal, Projektträger Forschungszentrum Karlsruhe WTE, Karlsruhe, Germany, May 2006.

88 Herrmann, R. (2011) M-sequence based ultra-wideband radar and its application to crack detection in salt mines, URN: urn:nbn:de:gbv:ilm1-2011000344, Ilmenau University of Technology.

89 Rytting, D.K. (1996) Network analyzer error models and calibration methods. *ARFTG/NIST: Notes of the Short Course On RF & Microwave Measurement for Wireless Application*, Vol. 1, 1996.

90 Ferrero, A. and Pisani, U. (1992) Two-port network analyzer calibration using an unknown THRU. *IEEE Microwave Guided. Wave Lett.*, **2** (12), 505–507.

91 Häfner, F. (2001) *In-situ-Ermittlung von Strömungskennwerten natürlicher Salzgesteine in Auflockerungszonen gegenüber Gas und Salzlösungen unter den gegebenen Spannungsbedingungen im Gebirge*, FKZ: 02C0527/Technische Universität Bergakademie Freiberg. – BMBF research report.

92 Herrmann, R., Sachs, J. and Peyerl, P. (2006) System evaluation of an M-sequence ultra wide-band radar for crack detection in salt rock. Proceedings of the 11th International Conference on Ground Penetrating Radar, GPR, Columbus, OH, USA.

93 Herrmann, R., Sachs, J., Schilling, K. and Bonitz, F. (2008) New extended M-sequence ultra wideband radar and its application to the disaggregation zone in salt rock. Proceedings of the 12th International Conference on Ground Penetrating Radar, GPR, Birmingham, UK, June 2008.

94 Haacke, E.M., Li, D. and Kaushikkar, S. (1995) Cardiac MR imaging: principles and techniques. *Top. Magn. Reson. Imaging*, **7**, 200–217.

95 Dieringer, M.A., Renz, W., Lindel, T., Seifert, F., Frauenrath, T., von Knobelsdorff-Brenkenhoff, F., Waiczies, H., Hoffmann, W., Rieger, J., Pfeiffer, H., Ittermann, B., Schulz-Menger, J. and Niendorf, T. (2011) Design and application of a four-channel transmit/receive surface coil for functional cardiac imaging at 7T. *J. Magn. Reson. Imaging*, **33** (3), 736–741.

96 Nijm, G.M., Swiryn, S., Larson, A.C. and Sahakian, A.V. (2006) Characterization of the magnetohydrodynamic effect as a signal from the surface electrocardiogram during cardiac magnetic resonance imaging, in *Computers in Cardiology Valencia*, Spain, pp. 269–272, ISBN 978-1-4244-2532-7.

97 Frauenrath, T., Hezel, F., Renz, W., de Geyer, T., Dieringer, M., von Knobelsdorff-Brenkenhoff, F., Prothmann, M., Schulz-Menger, J. and Niendorf, T. (2010) Acoustic cardiac triggering: a practical solution for synchronization and gating of cardiovascular magnetic resonance at 7 Tesla. *J. Cardiovasc. Magn. Reson.*, **12** (1), 67.

98 van Geuns, R.J., Wielopolski, P.A., de Bruin, H.G. *et al.* (2000) MR coronary angiography with breath-hold targeted volumes: preliminary clinical results. *Radiology*, **217**, 270–277.

99 Hinks, R.S. (1988) Monitored echo gating (MEGA) for the reduction of motion artifacts (abstr). *Magn. Reson. Imaging*, **6** (Suppl 1): 48.

100 James, C.J. and Hesse, C.W. (2005) Independent component analysis for biomedical signals. *Physiol. Meas.*, **26**, R15–R39.

101 Semenov, S.Y., Bulyshev, A.E., Posukh, V.G., Sizov, Y.E., Williams, T.C. and Souvorov, A.E. (2003) Microwave tomography for detection/imaging of myocardial infarction. I. Excised Canine Hearts. *Ann. Biomed. Eng.*, **31**, 262–270. doi: 10.1114/1.1553452.

102 Thiel, F., Helbig, M., Schwarz, U., Geyer, C., Hein, M., Sachs, J., Hilger, I. and Seifert, F. (2009) Implementation of ultra-wideband sensors for biomedical applications. *Frequenz, J. RF/Microwave-Eng. Photon. Commun.*, **63** (9–10), 221–224, ISSN 0016-1136.

103 Thiel, F., Hein, M., Sachs, J., Schwarz, U. and Seifert, F. (2009) Combining magnetic resonance imaging and ultra-wideband radar: a new concept for multimodal biomedical imaging. *Rev. Sci. Instrum.*, **80** (1), 014302, ISSN 0034-6748, Melville, NY, American Institute of Physics (AIP), 10 pages.

104 Thiel, F., Kosch, O. and Seifert, F. (2010) Ultra-wideband sensors for improved magnetic resonance imaging, cardiovascular monitoring and tumour diagnostics, sensors. Special Issue *Sens. Biomechan. Biomed.*, **10** (12), 10778–10802. doi: 10.3390/s101210778, ISSN 1424-8220, 25 pages.

105 Kosch, O., Thiel, F., Ittermann, B. and Seifert, F. (2011) Non-contact cardiac gating with ultra-wideband radar sensors for high field MRI. Proceedings of the International Society for Magnetic Resonance in Medicine **19**, (ISMRM), Montreal, Canada, ISSN 1545-4428, p. 1804.

106 Thiel, F., Kreiseler, D. and Seifert, F. (2009) Non-contact detection of myocardium's mechanical activity by ultra-wideband RF-radar and interpretation applying electrocardiography. *Rev. Sci. Instrum.*, **80** (11), 114302, 0034-6748, American Institute of Physics (AIP), Melville, NY, 12 pages.

107 Schwarz, U., Thiel, F., Seifert, F., Stephan, R. and Hein, M. (2010) Ultra-wideband antennas for magnetic resonance imaging navigator techniques. *IEEE Trans. Antenn. Propag.*, **58** (6), 2107–2112.

108 Hein, M.A., Geyer, C., Helbig, M., Hilger, I., Sachs, J., Schwarz, U., Seifert, F., Stephan, R. and Thiel, F. (2009) Antennas for ultra-wideband medical sensor systems. *EuCAP*, 1868–1872, ISBN 978-1-4244-4753-4.

109 Thiel, F. and Seifert, F. (2009) Non-invasive probing of the human body with electromagnetic pulses: modelling of the signal path. *J. Appl. Phys.*, **105** (4), 044904, ISSN 0021-8979, American Institute of Physics (AIP), Melville, NY, 8 pages.

110 Thiel, F. and Seifert, F. (2010) Physiological signatures reconstructed from a dynamic human model exposed to ultra-wideband microwave signals. *Frequenz, J. RF/Microwave-Eng. Photon. Commun.*, **64** (3–4), 34–41, ISSN 0016-1136.

111 Hunt, S., Roseiro, A. and Siegel, M. (1988) Signal processing for a non-invasive microwave heart rate estimator. Proceedings of the Annual International Conference of the IEEE Engineering in Medicine and Biology Society, vol. 1, 4–7 November 1988, p. 158.

112 Ziehe, A. and Müller, K.R. (1998) TDSEP – an efficient algorithm for blind separation using time structure.

Proceedings of the International Conference on Artificial Neural Networks (ICANN'98), Skövde, Sweden, pp. 675–680.

113 Kosch, O., Thiel, F., Yan, D.D. and Seifert, F. (2010) Discrimination of respirative and cardiac displacements from ultra-wideband radar data. International Biosignal Processing Conference (Biosignal 2010), Berlin, Germany, 14–16 July.

114 Christ, A. *et al.* (2010) The virtual family – development of surface-based anatomical models of two adults and two children for dosimetric simulations. *Phys. Med. Biol.*, **55**, N23–N38.

115 Sachs, J., Peyerl, P., Wöckel, S. *et al.* (2007) Liquid and moisture sensing by ultra-wideband pseudo-noise sequence signals. *Meas. Sci. Technol.*, **18**, 1074–1088.

116 Muehlsteff, J., Thijs, J., Pinter, R., Morren, G. and Muesch, G. (2007) A handheld device for simultaneous detection of electrical and mechanical cardio-vascular activities with synchronized ECG and CW-Doppler Radar. IEEE International Conference on Engineering in Medicine and Biology Society (EMBS), 22–26 August 2007, pp. 5758–5761.

117 Fear, E.C., Hagness, S.C., Meaney, P.M. *et al.* (2002) Enhancing breast tumor detection with near-field imaging. *IEEE Micro Mag.*, **3** (1), 48–56.

118 Meaney, P.M., Fanning, M.W., Li, D. *et al.* (2000) A clinical prototype for active microwave imaging of the breast. *IEEE Trans. Microwave Theory*, **48** (11), 1841–1853.

119 Klemm, M., Craddock, I.J., Leendertz, J.A. *et al.* (2009) Radar-based breast cancer detection using a hemispherical antenna array – experimental results. *IEEE Trans. Antenn. Propag.*, **57** (6), 1692–1704.

120 Klemm, M., Craddock, I.J., Leendertz, J. A. *et al.* (2010) Clinical trials of a UWB imaging radar for breast cancer. Proceedings of the Fourth European Conference on Antennas and Propagation (EuCAP), 2010, pp. 1–4.

121 Williams, T.C., Bourqui, J., Cameron, T. R. *et al.* (2011) Laser surface estimation for microwave breast imaging systems. *IEEE Trans. Bio.-Med. Eng.*, **58** (5), 1193–1199.

122 Winters, D.W., Shea, J.D., Kosmas, P. *et al.* (2009) Three-dimensional microwave breast imaging: dispersive dielectric properties estimation using patient-specific basis functions. *IEEE Trans. Med. Imaging*, **28** (7), 969–981.

123 Sakamoto, T. and Sato, T. (2004) A target shape estimation algorithm for pulse radar systems based on boundary scattering transform. *IEICE Trans. Commun.*, **E87b** (5), 1357–1365.

124 Kidera, S., Kani, Y., Sakamoto, T. *et al.* (2008) Fast and accurate 3-D imaging algorithm with linear array antennas for UWB pulse radars. *IEICE Trans. Commun.*, **E91b** (8), 2683–2691.

125 Helbig, M., Hein, M.A., Schwarz, U. *et al.* (2008) Preliminary investigations of chest surface identification algorithms for breast cancer detection. IEEE International Conference on Ultra-Wideband, 2008. ICUWB 2008, pp. 195–198.

126 Helbig, M., Geyer, C., Hein, M. *et al.* (2009) A breast surface estimation algorithm for UWB microwave imaging, in *IFMBE Proceedings 4th European Conference of the International Federation for Medical and Biological Engineering* (eds J. Sloten, P. Verdonck, M. Nyssen *et al.*), Springer, Berlin, Heidelberg, pp. 760–763.

127 Helbig, M., Geyer, C., Hein, M. *et al.* (2009) Improved breast surface identification for UWB microwave imaging, in *IFMBE Proceedings World Congress on Medical Physics and Biomedical Engineering, 7–12 September 2009, Munich, Germany* (eds O., Dössel and W.C. Schlegel), Springer, Berlin, Heidelberg, pp. 853–856.

128 Hantscher, S., Etzlinger, B., Reisenzahn, A. *et al.* (2007) A wave front extraction algorithm for high-resolution pulse based radar systems. IEEE International Conference on Ultra-Wideband, 2007. ICUWB 2007, pp. 590–595.

129 Sachs, J., Herrmann, R., Kmec, M. *et al.* (2007) Recent advances and applications of M-sequence based ultra-wideband sensors. IEEE International Conference on Ultra-Wideband, 2007. ICUWB 2007, pp. 50–55.

130 Salman, R. and Willms, I. (2011) Super-resolution object recognition approach for complex edged objects by UWB radar, in *Object Recognition* (ed. T.P. Cao), InTech.

131 Bond, E.J., Li, X., Hagness, S.C. *et al.* (2003) Microwave imaging via space-time beamforming for early detection of breast cancer. *IEEE Trans. Antenn. Propag.*, **51** (8), 1690–1705.

132 Xie, Y., Guo, B., Xu, L. *et al.* (2006) Multistatic adaptive microwave imaging for early breast cancer detection. *IEEE Trans. Bio.-Med. Eng.*, **53** (8), 1647–1657.

133 Zetik, R., Sachs, J. and Thoma, R. (2005) Modified cross-correlation back projection for UWB imaging: numerical examples. IEEE International Conference on Ultra-Wideband, 2005. ICU 2005, pp. 650–654.

134 Klemm, M., Craddock, I.J., Leendertz, J. A. *et al.* (2008) Improved delay-and-sum beamforming algorithm for breast cancer detection. *Int. J. Antenn. Propag.*, **2008**, p. 9.

135 Sachs, J. (2004) M-Sequence radar, in *Ground Penetrating Radar, IEE Radar, Sonar, Navigation and Avionics Series 15* (ed. D.J. Daniel), The Institution of Electrical Engineers, London, UK, p. 225–237.

136 Lazebnik, M., Madsen, E.L., Frank, G.R. *et al.* (2005) Tissue-mimicking phantom materials for narrowband and ultrawideband microwave applications. *Phys. Med. Biol.*, **50** (18), 4245–4258.

137 Lazebnik, M., Popovic, D., McCartney, L. *et al.* (2007) A large-scale study of the ultrawideband microwave dielectric properties of normal, benign and malignant breast tissues obtained from cancer surgeries. *Phys. Med. Biol.*, **52** (20), 6093–6115.

138 Bäckström, C.-J. and Christoffersson, N. (2006) *Urban Search and Rescue – An Evaluation of Technical Search Equipment and Methods*, Department of Fire Safety Engineering, Lund University, Sweden, Lund.

139 Chernyak, V. (2006) Signal processing in multisite UWB radar devices for searching survivors in rubble. 3rd European Radar Conference, 2006. EuRAD 2006, pp. 190–193.

140 Ossberger, G., Buchegger, T., Schimback, E. *et al.* (2004) Non-invasive respiratory movement detection and monitoring of hidden humans using ultra wideband pulse radar. International Workshop on Ultra Wideband Systems, 2004. Joint with Conference on Ultrawideband Systems and Technologies. Joint UWBST & IWUWBS, pp. 395–399.

141 Levitas, B. and Matuzas, J. (2006) UWB radar for human being detection behind the wall. International Radar Symposium, 2006. IRS 2006, pp. 1–3.

142 Levitas, B., Matuzas, J. and Drozdov, M. (2008) Detection and separation of several human beings behind the wall with UWB Radar. International Radar Symposium, 2008, pp. 1–4.

143 Zhang, Y., Jing, X., Jiao, T. *et al.* (2010) Detecting and identifying two stationary-human-targets: a technique based on bioradar. First International Conference on Pervasive Computing Signal Processing and Applications (PCSPA), 2010, pp. 981–985.

144 Wei Chong, Y. and Wang Da, Q. (2010) On the signal processing in the life-detection radar using an FMCW waveform. Third International Symposium on Information Processing (ISIP), 2010, pp. 213–216.

145 Loschonsky, M., Feige, C. Rogall, O. *et al.* (2009) Detection technology for trapped and buried people. IEEE MTT-S International Microwave Workshop on Wireless Sensing, Local Positioning, and RFID, 2009. IMWS 2009, pp. 1–6.

146 Narayanan, R.M. (2008) Through wall radar imaging using UWB noise waveforms. IEEE International Conference on Acoustics, Speech and Signal Processing, 2008. ICASSP 2008, pp. 5185–5188.

147 Lin, Y., Jianchao, Y. and Yap-Peng, T. (2010) Respiratory rate estimation via simultaneously tracking and

segmentation. IEEE Computer Society Conference on Computer Vision and Pattern Recognition Workshops (CVPRW), 2010, pp. 170–177.

148 Lidicky, L. (2008) Fourier array processing for buried victims detection using ultra wide band radar with uncalibrated sensors. IEEE International Geoscience and Remote Sensing Symposium, 2008. IGARSS 2008, pp. II-831–II-834.

149 Zaikov, E. and Sachs, J. (2010) UWB radar for detection and localization of trapped people, in *Ultra Wideband* (ed. B. Lembrikov), Scivo, Rijeka, Croatia.

150 Nezirovic, A., Yarovoy, A.G. and Ligthart, L.P. (2010) Signal processing for improved detection of trapped victims using UWB radar. *IEEE Trans. Geosci. Remote Sensing*, **48** (4), 2005–2014.

151 Sachs, J., Zaikov, E., Helbig, M. *et al.* (2011) Trapped victim detection by pseudo-noise radar. 2011 International Conference on Wireless Technologies for Humanitarian Relief (ACWR 2011), Amritapuri, India.

152 Mafhaza, B.R. and Elsherbeni, A.Z. (2004) *MATLAB Simulation for Radar System Design*, Chapman & Hall, CRC Press, LLC.

153 Zetik, R., Sachs, J. and Thomä, R.S. (2007) UWB short-range radar sensing. *IEEE Instrum. Meas. Mag.*, **10** (2), 39–45.

154 Withington, P., Fluhler, H. and Nag, S. (2003) Enhancing homeland security with advanced UWB sensors. *IEEE Micro Mag.*, **4** (3), 51–58.

155 Yarovoy, A.G., Zhuge, X., Savelyev, T.G. and Ligthart, L.P. (2007) Comparison of UWB technologies for human being detection with radar. Proceedings of the EuMA 2007, Munich, Germany, pp. 295–298.

156 Daniels, D.J. (2004) *Ground Penetrating Radar, IEE Radar Sonar and Navigation Series 15*, 2nd edn, The Institution of Engineering and Technology, London, UK.

157 Rohling, H., Höß, A. and Lübbert, U. (2002) Multistatic radar principles for automotive radarnet applications. International Radar Symposium (IRS), Bonn, Germany, pp. 181–185.

158 Staderini, E. (2002) UWB radars in medicine. *IEEE Aero. Electron. Syst. Mag.*, **17** (1), 13–18.

159 Yarovoy, A.G., Ligthart, L.P., Matuzas, J. and Levitas, B. (2008) UWB radar for human being detection. *IEEE Aero. Electron. Syst. Mag.*, **23** (5), 36–40.

160 Rovňáková, J. and Kocur, D. (2011) Data fusion from UWB radar network: preliminary experimental results. The 21th International Conference Radioelektronika, Brno, Czech Republic, pp. 353–356.

161 Rovňáková, J. and Kocur, D. (2010) UWB radar signal processing for through wall tracking of multiple moving targets. Proceedings of the 7th European Radar Conference (EuRAD), Paris, France, pp. 372–375.

162 Kocur, D., Rovňáková, J. and Şvecová, M. (2009) Through wall tracking of moving targets by M-sequence UWB radar, in *Computational Intelligence in Engineering*, Springer's Book Series "Studies in Computational Intelligence" (eds I.J. Rudas, J. Fodor and J. Kacprzyk), Springer, Berlin, Heidelberg, pp. 394–364.

163 Chang, S., Sharan, R., Wolf, M., Mitsumoto, N. and Burdick, J.W. (2010) People tracking with UWB radar using a multiple-hypothesis tracking of clusters (MHTC) method. *Int. J. Soc. Robot.*, **2**, 3–18.

164 Gurbuz, S.Z., Melvin, W.L. and Williams, D.B. (2007) Comparison of radar based human detection techniques. Proceedings of the 41st Asilomar Conference on Signals, Systems and Computers, pp. 2199–2203.

165 Gauthier, S.S. and Chamma, W. (2004) Surveillance through concrete walls. Proceedings of the SPIE-C3I Technologies for Homeland Security and Homeland Defense III, pp. 597–608.

166 Nag, S., Barnes, M.A., Payment, T. and Holladay, G. (2002) Ultrawideband through-wall radar for detecting the motion of people in real time. Proceedings of the SPIE-Radar Sensor

Technology and Data Visualization, vol. 4744, pp. 48–57.
167 Ram, S.S. and Ling, H. (2008) Through-wall tracking of human movers using joint Doppler and array processing. *IEEE Geosci. Remote Sensing Lett.*, **5** (3), 537–541.
168 Rovňáková, J. (2010) *Complete Signal Processing for through Wall Tracking of Moving Targets*, LAP LAMBERT Academic Publishing, Germany.
169 Kocur, D., Rovňáková, J. and Urdzík, D. (2011) Experimental analyses of mutual shadowing effect for multiple target tracking by UWB radar. 34th International Conference on Telecommunications and Signal Processing : pp. 302–306, 18–20 Aug. 2011, Budapest, Hungary, pp. 302–306, doi: 10.1109/TSP.2011.6043721.
170 Rovňáková, J. and Kocur, D. (2010) Weak signal enhancement in radar signal processing. Proceedings of the Radioelektronika, Brno, Czech Republic, pp. 147–150.
171 Andreuccetti, D., Fossi, R. and Petrucci, C. (1997–2010) *Calculation of the dielectric properties of human body tissues in the frequency range 10 Hz-100 GHz*, Italian National Research Council, Institute for Applied Physics, Florence, Italy. Version: May 2010, http://niremf.ifac.cnr.it/tissprop/.
172 Akyildiz, I.F., Su, W., Sankarasubramaniam, Y. and Cayirci, E. (2002) A survey on sensor networks. *IEEE Commun. Mag.*, **40** (8), 102–114.
173 Bartoletti, S., Conti, A. and Giorgetti, A. (2010) Analysis of UWB radar sensor networks. Proceedings of the IEEE ICC, pp. 1–6.
174 Shingu, G., Takizawa, K. and Ikegami, T. (2008) Human body detection using MIMO-UWB radar sensor network in an indoor environment. Proceedings of the International Conference on Parallel and Distributed Computing, Applications and Technologies (PDCAT), pp. 437–442.
175 Hunt, A.R. (2005) Image formation through walls using a distributed radar sensor network. *Proc. SPIE*, **5778**, 169. doi: 10.1117/12.602655.
176 Zetik, R., Jovanoska, S. and Thoma, R. (2011) Simple method for localisation of multiple tag-free targets using UWB sensor network. Ultra-Wideband (ICUWB), 2011 IEEE International Conference on, pp. 268–272, 14–16 Sept. 2011, Bologna, Italy doi: 10.1109/ICUWB.2011.6058843.
177 Chiani, M., Giorgetti, A., Mazzotti, M., Minutolo, R. and Paolini, E. (2009) Target detection metrics and tracking for UWB radar sensor networks. Proceedings of the IEEE International Conference on Ultra Wideband (ICUWB), pp. 469–474.
178 Hall, D.L. and Llinas, J. (2001) *Handbook of Multisensor Data Fusion*, CRC Press.
179 Ng, G.W. (2003) *Intelligent Systems: Fusion, Tracking and Control*, Research Studies Press, Baldock.
180 Blackman, S.S. and Popoli, R. (1993) *Design and Analysis of Modern Tracking Systems*, Artech House Publishers.
181 Khan, J., Niar, S., Menhaj, A. and Elhillali, Y. (2008) Multiple target tracking system design for driver assistance application. Conference on Design and Architectures for Signal and Image Processing (DASIP 08), Belgium.
182 Gezici, S., Tian, Z., Giannakis, G.B., Kobayashi, H., Molisch, A.F., Poor, H.V. and Sahinoglu, Z. (2005) Localization via ultrawideband radios. *IEEE Signal Process Mag.*, **22** (4), 70–84.
183 Kocur, D., Gamec, J., Švecová, M., Gamcová, M. and Rovňáková, J. (2010) Imaging method: a strong tool for moving target tracking by a multistatic UWB radar system. The 8th International Symposium on Applied Machine Intelligence and Informatics, Herľany, Slovakia, 28–30 January 2010.
184 Yan, H., Shen, G., Hirsch, O., Zetik, R. and Thomä, R. (2010) A UWB active sensor localization method based on time-difference back-projection. 2010 IEEE International Conference on Ultra-Wideband, Nanjing, China, 20–23 September 2010.
185 Zetik, R., Shen, G. and Thomä, R. (2010) Evaluation of requirements for UWB localization systems in home-

entertainment applications. International Conference on Indoor Positioning and Indoor Navigation, Zürich, Switzerland, September 2010.

186 Bucher, Ralph and Misra, D. (2002) A synthesizable VHDL model of the exact solution for three-dimensional hyperbolic positioning system. *VLSI Des.*, **15** (2), 507–520.

187 Piccardi, M. (2004) Background subtraction techniques: a review. IEEE International Conference on Systems, Man and Cybernetics, vol. 4, 10–13 October 2004, pp. 3099–3104.

188 Zetik, R., Crabbe, S., Krajnak, J., Peyerl, P., Sachs, J. and Thomä, R. (2006) Detection and localization of persons behind obstacles using M-sequence through-the-wall radar. SPIE Defense and Security Symposium, Orlando, FL, USA, 17–21 April 2006.

189 Foy, W.H. (1976) Position-location solutions by Taylor-series estimation. *IEEE Trans. Aero. Electron. Syst.*, **AES-12** (2), 187–194.

190 Smith, J.O. and Abel, J.S. (1987) The spherical interpolation method of source localization. *IEEE J. Oceanic Eng.*, **OE-12** (1), 246–252.

191 Yu, K., Montillet, J.-P., Rabbachin, A., Cheong, P. and Oppermann, I. (2006) UWB location and tracking for wireless embedded networks. *Signal Proc.*, **86** (9), 2153–2171.

192 Cheung, K.W., So, H.C., Ma, W.-K. and Chan, Y.T. (2004) Least squares algorithms for time-of-arrival-based mobile location. *IEEE Trans. Signal Process.*, **52** (4), 1121–1128.

193 Svecova, M., Kocur, D., Zetik, R. and Rovnakova, J. (2010) Target localization by a multistatic UWB radar. 20th International Conference Radioelektronika, Brno, Czech Republic, 19–21 April 2010.

194 Zetik, R., Sachs, J. and Thomä, R. (2007) UWB short range radar sensing. *IEEE Instrum. Meas. Mag.*, **9** (1), pp. 39–45.

195 Dardari, D., Chong, C.-C. and Win, M.Z. (2006) Analysis of threshold-based TOA estimator in UWB channels. Proceedings of the European Signal Processing Conference (EUSIPCO), Florence, Italy, September 2006.

196 Zetik, R., Hirsch, O. and Thomä, R. (2009) Kalman filter based tracking of moving persons using UWB sensors. IEEE MTT-S International Microwave Workshop on Wireless Sensing, Local Positioning, and RFID, Cavtat, Croatia, 24–25 September 2009.

197 Zetik, R., Sachs, J. and Thomä, R. (2005) Modified cross-correlation back projection for UWB imaging: numerical examples. IEEE International Conference on Ultra-Wideband, Zürich, Switzerland, September 2005.

198 Zetik, R. and Thomä, R.S. (2008) Monostatic imaging of small objects in UWB sensor networks. Invited paper to the International Conference on Ultra-Wideband (ICUWB 2008), Germany.

199 Borras, J., Hatrack, P. and Mandayam, N. B. (1998) Decision theoretic framework for NLOS identification. Proceedings of the 48th IEEE Vehicular Technology Conference (VTC '98), vol. 2, Ottawa, Canada, May 1998, pp. 1583–1587.

200 Schroeder, J., Galler, S., Kyamakya, K. and Jobmann, K. (2007) NLOS detection algorithms for Ultra-Wideband localization. IEEE Workshop on Positioning Navigation and Communication, Hannover, 2007, pp. 159–166.

201 Guvenc, I., Chong, C.-C., Watanabe, F. and Inamura, H. (2008) NLOS identification and weighted least-squares localization for UWB systems using multipath channel statistics. *EURASIP J. Adv. Sig. Process.*, **2008**, Article ID 271984.

202 Shen, G., Zetik, R., Hirsch, O. and Thomä, R.S. (2010) Range-based localization for UWB sensor networks in realistic environments. *EURASIP J. Wirel. Commun. Netw.*, **2010**, 9. Article ID 476598. doi: 10.1155/2010/476598.

203 Rovnakova, J. (2009) Complete signal processing for through wall target tracking by M-sequence UWB radar system. Ph.D. dissertation. Department of Electronics and Multimedia Communications, Faculty of Electrical

Engineering and Informatics, Technical University of Kosice, Slovak Republic.

204 Zetik, R., Sachs, J. and Thomä, R. (2005) Imaging of propagation environment by UWB channel sounding. XXVIIIth General Assembly of URSI, New Delhi, India, 23–29 October 2005.

205 Aftanas, M., Rovňáková, J., Rišková, M., Kocur, D. and Drutarovský, M. (2007) An analysis of 2D target positioning accuracy for M-sequence UWB radar system under ideal conditions. Proceedings of the 17th International Conference Radioelektronika, Brno, Czech Republic, 24–25 April 2007, pp. 189–194.

206 Tuchler, M., Schwarz, V. and Huber, A. (2005) Location accuracy of an UWB localization system in a multi-path environment. IEEE International Conference on Ultra-Wideband, pp. 414–419.

Appendix

Symbols and Abbreviations

Symbols

Symbol	Assignment
a	Propagation loss
a	Radius of a sphere
$a(t),\ \underline{a}(f)$	Incident normalized wave in time and frequency domain
$a_n,\ b_m$	Coefficients of a linear ordinary differential equation
$\mathbf{A}$	Chain parameters of a network
$A(f,\mathbf{e})$	Antenna aperture
$\mathbf{A}(t,\mathbf{r})$	Vector potential
$\mathbf{A},\ \mathbf{B},\ \mathbf{C},\ \mathbf{D}$	State space matrices
$\mathbf{B}$	Magnetic flux density
$b(t),\ \underline{b}(f)$	Emanated normalized wave in time and frequency domain
$b_{\mathrm{f}},\ b_{\mathrm{rb}},\ b_{\mathrm{fr}}$	Fractional bandwidth; relative bandwidth; frequency ratio
$B_3,\ B_6,\ B_{10}$	Absolute bandwidth
B_{eff}	Effective bandwidth
B_{n}	Bandwidth of a noise signal
B_{N}	Effective noise bandwidth of a filter
B_{rect}	Rectangular bandwidth
B_{win}	Bandwidth of a window function
c	Speed of light
C	Capacity
C	Contour curve enclosing an area; perimeter line of an area
C_{B}	Bit rate [bits/s]

$C_{xx}(\tau)$, $C_{yx}(\tau)$	Auto- and cross-correlation function
CF	Crest factor; coupling factor
coh (f)	Coherence function
cov(τ)	Covariance function
CP1 dB	1 dB compression point
d	Distance
d_c	Duty cycle
D	Dynamic range; directivity
D_{cl}	Clutter-free dynamic range
D_{max}	Maximum signal-to-noise ratio of a receiver
D_{opt}	Spurious-free dynamic range
D_R	Dynamic range of the receiver circuit
D_{sys}	System performance
DR	Decay rate
$\mathbf{D}$	Electric flux density
e	Error
e_q	Quantization error
$\mathbf{e}$	Unit vector assigning a coordinate system or a propagation direction
$[\mathbf{e}_x \quad \mathbf{e}_y \quad \mathbf{e}_z]$	Unit vectors defining a global coordinate system in Cartesian coordinates
$[\mathbf{e}_r \quad \mathbf{e}_\vartheta \quad \mathbf{e}_\varphi]$	Unit vectors defining a global coordinate system in spherical coordinates
$[\mathbf{e}_1 \quad \mathbf{e}_2 \quad \mathbf{e}_3]$	Unit vectors defining a local coordinate system in Cartesian coordinates
$\mathbf{e}_0$, $\mathbf{e}_T$, $\mathbf{e}_R$, $\mathbf{e}_i$, $\mathbf{e}_s$, $\mathbf{e}_B$	Propagation direction
$\mathbf{E}$, E	Electric field (vector and magnitude)
EIRP	Effective or equivalent isotropically radiated power
ENOB	Effective number of bits
$\mathfrak{E}$	Energy of a signal or a wave
$\mathfrak{E}_{acc}$	Accumulated energy
f_d	Doppler frequency; Doppler shift
$f_{eq} = 1/\Delta t_{eq}$	Equivalent sampling rate
f_l, f_u, f_m	Lower, upper and centre frequency
$f_s = 1/\Delta t_s$	Sampling rate; Nyquist rate
F	Noise factor
FI_{yx}	Fidelity of two signal
FoM	Figure-of-merit
FSR	Full-scale range
g	Gain [dB]
g_M	Conversion gain of a multiplier
$g(t)$; $\underline{G}(f)$; $\underline{G}(s)$	Impulse response function (IRF); frequency response function (FRF), transfer function
$G(f, \mathbf{e})$	Antenna gain

$\mathbf{G}(t, \mathbf{r}, \mathbf{r}')$	Green's tensor
h	Planck's constant (h= . . . eVs) $h = 6.62 \times 10^{-34}$ J s = 4.13×10^{-15} eV s
$h(t)$	Step response
h, $\mathbf{h}$	Length and orientation of a Hertzian Dipole
H_0	Data amount [bit]; data quantity
$\mathbf{H}$	Hadamard matrix
$\mathbf{H}$, H	Magnetic field (vector and amplitude)
$i = \sqrt{-1}$	Imaginary unit
i, j, k, n	Sample numbers
$I(t)$, $\underline{I}(f)$	Current in time and frequency domain
$I^{\rightarrow}$; $I^{\leftarrow}$	Current wave travelling towards right or left
I; $I(t)$	In-phase signal
ICP2, ICP3	Intercept points
IL	Insertion loss
IS	Isolation
$\mathbf{I}$	Identity matrix
$J(\boldsymbol{\theta})$	Cost function
$\mathbf{J}$	Current density
k	Coupling factor
k	Boltzmann's constant $k = 1.38 \times 10^{-23}$ W s/K
$\mathbf{k} = k\,\mathbf{e}_0, \quad k = 2\pi/\lambda$	Wave vector; wave number
K, L, k, l	Number of input- and output channels
l	Length of a cable or an object
L	Inductivity
$L_{\text{cut-off}}$	Cut-off level
L_{n}	Noise level [dBm]
L_{T}	Transmitter level [dBm]
L_{SL}	Side-lobe level
L_{ϕ}	Phase noise level
m_{R}	Measurement rate
$m(t)$	M-sequence
$\mathbf{M}_{\text{circ}}$	Circulant M-sequence matrix
$n(t)$	Random noise
$\mathbf{n}$	Unit vector of surface normal
N	Number; number of bits
$N_{\square}$	Number of samples per scan; number of samples of a signal
NF	Noise figure
$\mathfrak{O}(\mathbf{r})$	Object function
p	Number of synchronous averaging/stacking
$p(t)$	Instantaneous power
$p_x(x)$, $p_{\underset{\sim}{x}}(x)$	Probability density function for signal $x(t)$ and the random variable $\underset{\sim}{\mathbf{x}}$

$\mathbf{p} = p\,\mathbf{e}_0$	Spatial frequency
$\mathbf{p}_s(t, \mathbf{r})$	Electric polarization vector
P	Average active power
P_V	Power dissipation per volume
PAPR	peak-to-average power ratio
PSLR	Peak side-lobe ratio
$P_x(\xi)$	Probability function
$P_{A,x}(\xi)$	Amplitude probability
$\overline{\mathfrak{P}}$	Magnitude of average power flux density
$\underline{\mathfrak{P}}(f)$	Frequency domain Poynting vector
$\mathfrak{P}(t)$	Time domain Poynting vector
q	Quantization interval
$q(t), \underline{Q}(s)$	State of a system in time or Laplace domain
$Q;\ Q(t)$	Quadrature-phase signal
$\mathfrak{q}$	Unit vector of wave polarization
$r = \|\mathbf{r}\| = \|\mathbf{r}_2 - \mathbf{r}_1\|$	Distance between two points
r_D	Differential resistance of a non-linear device
$r_{ua},\ r_{max}$	Unambiguous range
$\mathbf{r}$	Position vector
$\mathbf{r}'$	Position of an observation point
$R,\ R_0$	Ohmic resistor; reference resistor
$R(t, \mathbf{e});\ \mathbf{R}(t, \mathbf{e})$	Scalar and vector antenna impulse response for the receive mode (far field assumption)
RJR	Rise-time-jitter ratio
RL	Return loss
s	Doppler scaling factor
$s = \sigma + j2\pi f$	Complex frequency
$S,\ d\mathbf{S}$	Surface; surface enclosing a volume; surface element
$\mathbf{S}$	Scattering parameters/matrix of a network
SACF	Slope-amplitude coherence function
SF	Shape factor
SNR	Signal-to-noise ratio
SR	Slew rate
t	time, interaction time, short time
t_{eff}	Effective pulse width
t_d	Delay time
t_D	Duration of a time spread signal
t_f	Fall time
t_I	Integration time
t_{ot}	Time on target
t_P	Period duration
t_r	Rise time

t_{rect}	Rectangular pulse width
t_R	Reverberation time
t_s	Storage time
t_{sett}	Settling time
t_w	Pulse width; pulse duration
t_{win}	Duration of a window function; duration of a time function
Δt_c	Period duration of clock signal
Δt_{eq}	Equivalent sampling interval
Δt_s	Sampling interval (related to interaction time t)
T	Time, observation time, slow time
T	Temperature [K] $T_0 = 290$ K
T_{obs}	Observation time
T_R	Recording time (Duration to capture all data samples for one response function)
$T(t, \mathbf{e})$; $\mathbf{T}(t, \mathbf{e})$	Scalar and vector antenna impulse response for the transmit mode (far field assumption)
T_0, T_S, T_{DUT}	Noise temperatures
ΔT	Sampling interval (related to observation time T)
$\mathbf{T}$	Transformation matrix; rotation matrix
$\mathbf{T}$	Transmission parameters of a network
$\mathbf{T}_H^{(\mathbf{u})}$	Householder matrix; mirroring a vector at a plane of unit vector $\mathbf{u}$
TB	Time-bandwidth product
$u(t) = u_{-1}(t)$	Step function; Heaviside function
$u_n(t)*$	doublet
$\mathbf{u}$	Unit vector (used for field polarization)
v_r	Target speed in range direction
$\mathbf{v}$	Unit vector (used for field polarization)
$V(t)$, $\underline{V}(f)$	Voltage in time and frequency domain
$V^{\rightarrow}$; $V^{\leftarrow}$	Voltage wave travelling towards right or left
V, V_p	Pulse amplitude
V_{cc}	Supply voltage
V_{RF}	RF-voltage; input voltage
V, V_{obs}, dV	Volume; observation volume; region of interest; volume element
$w(t)$, $W(f)$	Time or frequency domain window
W	Photon energy

$W(t,\mathbf{r})$; $\underline{W}(f,\mathbf{r})$ $\mathbf{W}(t,\mathbf{r})$; $\underline{\mathbf{W}}(f,\mathbf{r})$	Normalized scalar and vector wave in time and frequency domain
$\mathfrak{W}(t,\mathbf{r})$; $\mathfrak{W}(t,\mathbf{r})$	"correlated" normalized scalar and vector wave
W_{P}	Planer wave
W_{SP}	Spherical wave
W_{Ω}	Normalized source yield of a spherical wave
x, y, z	Space coordinates
$x(t)$, $\underline{X}(f)$, $y(t)$, $\underline{Y}(f)$, $z(t)$, $\underline{Z}(f)$	Arbitrary signals in time or in frequency domain
$\mathbf{Y}$	Admittance parameters of a network
Z_0	Characteristic impedance; intrinsic impedance of free space
Z_{s}	Intrinsic impedance of an arbitrary substance
$\mathbf{Z}$	Impedance parameters of a network
$\alpha, \beta, \varphi, \theta, \vartheta, \omega$	angle
α_p	Moment of *p*th order
β_p	Central moment of *p*th order
$\chi_{xx}(\tau, s)$	Wideband ambiguity function
δ_{A}	Range accuracy
δ_{D}	Minimum detectable displacement
δ_{f}	Spectral resolution
δ_{r}	Range resolution
δ_{V}	Voxel size
$\delta(n,m) = \begin{cases} 1 & m = n \\ 0 & m \neq n \end{cases}$	Kronecker delta
$\delta(t) = u_0(t)$	Dirac-pulse ($u_0(t)$ – doublet notation)
$\Delta_{\Delta t_{\mathrm{s}}}$	Dirac comb
$\Delta\tau_{\mathrm{j}}(t)$	Random jitter
$\Delta\Omega_{\mathrm{B}}$	Antenna beamwidth
ε, ε', ε''	Permittivity (real and imaginary part)
ϕ, φ	Phase
$\Phi(f)$	Power spectrum, power spectral density
Φ_{n0}	Thermal noise power spectral density
$\Phi(t, r)$	Scalar potential
$\varphi_{\mathrm{i}}(t)$	Instantaneous phase
φ_{j}^2	Jitter variance
φ	Azimuth angle
η	Efficiency; refraction index

η_S	Efficiency of a sampling gate
η_m	Mixer efficiency
η_r; η_{sr}; η_{cr}; η_{mr}	Receiver efficiency; sampling receiver efficiency; correlation receiver efficiency; digital M-sequence receiver efficiency
κ	Condition number
λ	Wavelength
λ_{coh}	Coherence length
$\Lambda(t, \mathbf{e}_1, \mathbf{e}_2)$; $\mathbf{\Lambda}(t, \mathbf{e}_1, \mathbf{e}_2$ $\underline{\Lambda}(f, \mathbf{e}_1, \mathbf{e}_2)$; $\underline{\mathbf{\Lambda}}(f, \mathbf{e}_1, \mathbf{e}_2)$	Scalar and matrix scattering IRF respectively FRF; scatter length
μ	Permeability
	μ Expected value
$\mathbf{\Pi}(t, \mathbf{r})$	Hertz vector
$\Theta(f, \mathbf{e})$	Antenna radiation pattern
$\boldsymbol{\theta} = [a_n, b_m]^T$	Parameter vector
ϑ	Inclination angle
$\boldsymbol{\vartheta}$	Rotation matrix
σ	Conductivity
	σ Standard deviation
σ	Radar cross section
σ^2	Variance
τ	Time lag; time constant
τ_a	Aperture time of a sampling gate
τ_{coh}	Coherence time; coherence width
$\tau_g = \tau_2$	Pulse position – energetic centre of gravity
τ_H	Time constant of the hold phase
Υ	Point spread function
Ω	Solid angle
Ξ	Structural efficiency
Ψ_n	Spectral density of noise
$\Psi_{xx}(f)$, $\underline{\Psi}_{yx}(f)$	Auto- or cross-spectrum

Notations

Notation	Description
$\mathbf{a}$, $\mathbf{A}$, $\mathbf{b}$, $\mathbf{B}$	Vector or matrix (bold letters)
$\det(\mathbf{A})$	Determinate of matrix $\mathbf{A}$
$E\{\underset{\sim}{x}\}$	Expected value of the random variable $\underset{\sim}{x}$
$FT\{\ \}$; $IFT\{\ \}$	Fourier- and inverse Fourier transform
$L\{y(t)\}$	(Arbitrary) linear operation subjecting the signal $y(t)$
$HT\{\ \}$	Hilbert transform
$N(\mu, \sigma^2)$	Normal distribution of variance σ^2 and expected value μ
$U(x_1, x_2)$	Uniform distribution between the boundaries x_1 and x_2
$\Re\{\underline{X}\}$; $\Im\{\underline{X}\}$	Real and imaginary part of $\underline{X}$
$\mathrm{var}\{\underset{\sim}{x}\}$	Variance of the random variable $\underset{\sim}{x}$
x_{rms}, y_{rms}	rms value; effective value
$x(t)$	Analogue time signal
$x[n\Delta t_s] = x[n]$ $\mathbf{x}[N] = [x[1]\ x[2] \cdots x[n] \cdots x[N]]^T$	Sampled signal and sampled signal expressed as vector
$\underset{\sim}{x}(t) = [\underset{\sim}{x}_1(t)\ \underset{\sim}{x}_2(t) \cdots \underset{\sim}{x}_N(t)]$	Random processes
$\mathbf{x}(t) = [x_1(t)\quad x_2(t)\quad x_3(t)\quad \cdots]$	Multi-port signal in vector notation
$\underline{X}^*$	Conjugate complex of a complex variable
$\mathbf{x}^T$	Transpose of a vector or matrix
$\underline{\mathbf{X}}^{\mathrm{H}} = (\underline{\mathbf{X}}^*)^T$	Conjugate transpose (Hermitian-, self-adjoint matrix)
$\mathbf{X}^+$	Pseudo-inverse (Moore Penrose inverse) of a matrix
$\Vert x(t) \Vert_p$	p-norm of the signal $x(t)$
$\dot{x}(t) = \frac{\partial x(t)}{\partial t} = u_1(t) * x(t)$	Time derivation
$\overline{x(t)}$	Temporal mean value
$\underset{\cdot}{B}$, $\underset{\cdot}{\Psi}$, $\underset{\cdot\cdot}{B}$, $\underset{\cdot\cdot}{\Psi}$	One- and two-sided spectral quantities

$\cap$; $\cup$	Conjunction; disjunction
$\mathbf{C} = \mathbf{A}\,\mathbf{B};\, c_{nm} = \sum_{k=1}^{K} a_{nk}\, b_{km}$	Matrix product
$\mathbf{C} = \mathbf{A} \circ \mathbf{B};\, c_{nm} = a_{nm}\, b_{nm}$	Hadamard or Schur product (entry wise product)
$c = \mathbf{a}\,.\mathbf{b} = \sum_{n=1}^{N} a_n b_n$	vector scalar product; dot or inner product (vector notation)
$c = \mathbf{a}^T\mathbf{b} = \mathbf{b}^T\mathbf{a} = \sum_{n=1}^{N} a_n b_n$	vector scalar product; dot or inner product (matrix notation)
$c = \mathbf{a} \otimes \mathbf{b} = [\mathbf{a}b_1 \quad \mathbf{a}b_2 \ \ldots \ \mathbf{a}b_N]$	outer or dyadic product of two vectors (vector notation)
$\mathbf{c} = \mathbf{a}\mathbf{b}^T$	outer or dyadic product of two vectors (matrix notation)
$\mathbf{a} \times \mathbf{b}$	Cross or exterior product
$a(t) * b(t) = \int a(\tau)\, b(t-\tau)\mathrm{d}\tau$	Convolution
$\mathbf{a}(t) \boxtimes \mathbf{b}(t) = \int \mathbf{a}(\tau) \times \mathbf{b}(t-\tau)\mathrm{d}\tau$	Exterior convolution
$\mathbf{a}(t) * \mathbf{b}(t) = \int \mathbf{a}(\tau) \cdot \mathbf{b}(t-\tau)\mathrm{d}\tau$	Dot convolution (vector notation)
$\mathbf{a}^T(t) * \mathbf{b}(t) = \int \mathbf{a}^T(\tau)\, \mathbf{b}(t-\tau)\mathrm{d}\tau$	Dot convolution (matrix notation)
$\mathbf{a}(t) \circledast \mathbf{b}(t) = \int \mathbf{a}(\tau) \circ \mathbf{b}(t-\tau)\mathrm{d}\tau$	Hadamard or Schur convolution
∇	Nabla operator
$\Delta = \nabla \cdot \nabla = \nabla^2$	Laplace operator
∇a	Gradient of the scalar field a
$\nabla \cdot \mathbf{a}$	Divergence of the vector field $\mathbf{a}$
$\nabla \times \mathbf{a}$	Rotation (curl) of vector field $\mathbf{a}$
$\mathbb{N}$	Natural number excluding zero
$\mathbb{N}_0$	Natural number including zero
$\mathbb{Z}$	Integer number
$\mathbb{C}$	Complex numbers

Structure of Multi-Dimensional Data

2D-data: Left: Observation of a scenario at a fixed position over a given time T. Right: Scanning of a scenario over a given path x (synthetic array measurements) and measurement of a scenario at a multitude of positions x (physical array). It holds corresponding for multi-dimensional data capturing

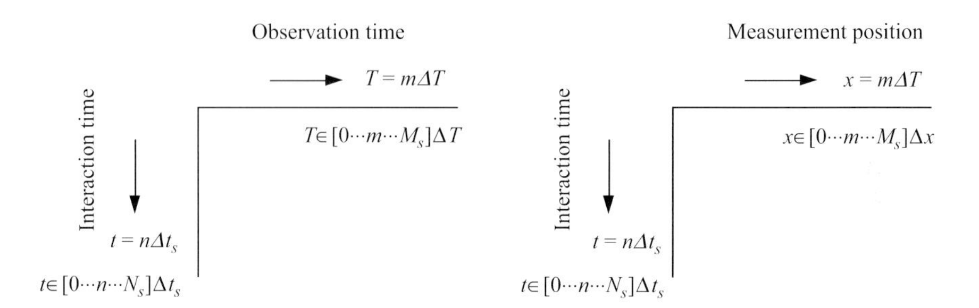

3D-data, radar volume: It is gained either from an antenna array or from two-dimensional scanning.

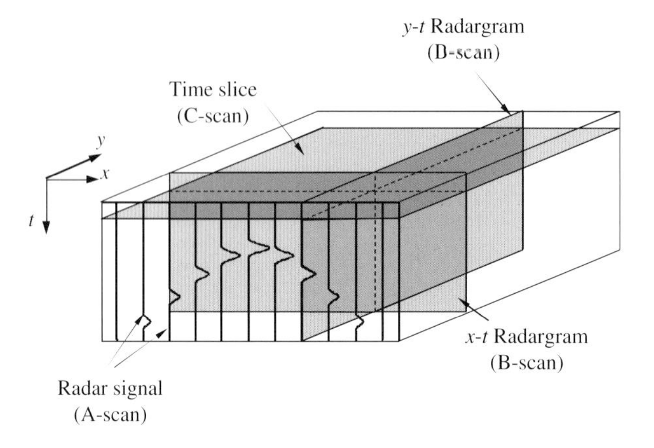

Abbreviations

Abbreviation	Assignment
AAL	Ambient-assisted living (assistance of handicapped people)
AC	Alternating current
ACF	Auto-correlation function
ADC	Analogue-to-digital converter
ANN	Artificial neural network
ARMA	Auto-regressive moving average model
AUT	Antenna under test
AWGN	Additive white Gaussian noise
BGA	Ball-grid array (IC housing)
BiCMOS	Bipolar-CMOS semiconductor technology

BSS	Blind source separation
BST	Boundary scattering transform
CCF	Cross-correlation function
CFAR	Constant false alarm rate
CFR	Code of Federal Regulations
CMOS	Complementary metal–oxide semiconductor
CMR	Cardiac magnetic resonance imaging
CW	Continuous wave
DAC	Digital-to-analogue converter
DARPA	US Defence Advanced Research Projects Agency
DC	Direct current
DDS	Direct digital synthesiser
DLL	Delay locked loop
DORT	Diagonalization de l'Operateur de Retournement Temporel
DSP	Digital signal processor
DUT	Device under test
ECC	Electronic Communication Committee
ECG	electrocardiogram
ECL	Emitter coupled logic
EIRP	Effective isotropically radiated power
EMI	Electromagnetic Interference
EMP	Electromagnetic pulse
FCC	Federal Communications Commission
FDHM	Full duration at half maximum
FFT	Fast Fourier transform
FHT	Fast Hadamard transform
FIR	Finite impulse response (filter)
FMCW	Frequency modulated continuous wave
FRF	Frequency response function
FSK	Frequency shift keying
FSR	Full-scale range
FWHM	Full wave at half maximum
GaAs/InP	Gallium arsenide/indium phosphide
GPR	Ground-penetrating radar
GTEM	Gigahertz transverse electromagnetic
HBT	Hetero-junction bipolar transistor
IBBST	Inverse bi-static boundary scattering transform
IBST	Inverse boundary scattering transform
IC	Integrated circuit
IF	Intermediate frequency
IIR	Infinite impulse response (filter)
IPCP	Inter-period correlation processing
IRF	Impulse response function
ISAR	Inverse synthetic aperture radar
ITRS	International technology roadmap for semiconductors

LCR	Large current radiator
LNA	Low noise amplifier
LO	Local oscillator
LOS	Line of sight
LSB	Least significant bit
LTI	Linear time-invariant
MIC	Ministry of Internal Affairs and Communications
MiMo	Multiple input and multiple output
MLE	Maximum likelihood estimation
MR	Magnetic resonance
MRI	Magnetic resonance imaging
MSB	Most significant bit
MSCW	M-sequence-modulated continuous wave
MTI	Moving target indication
MTT	Multiple target tracking
MUSIC	Multiple signal classification
MUT	Material under test
NDTCE	Non-destructive testing in civil engineering
NLOS	Non-line of sight
PA	Power amplifier
PAR	Physical aperture radar
PC	Principal component
PCA	Principal component analysis
PCB	Printed circuit board
PCR	Principal component regression
PDF	Probability density function
PLL	Phase-locked loop
PN	Pseudo-noise
PSD	Power spectral density
PSLR	Peak-side-lobe ratio
PSSR	Power supply rejection ratio
QFN	Quad flat-pack no leads (IC housing)
RAM	Random access memory
RAM	Radar absorbing material
RCS	Radar cross section
ROI	Region of interest
RTOA	Roundtrip time of arrival
SAGE	space-alternating generalized expectation-maximization
SAR	Synthetic aperture radar
SEABED	Shape estimation algorithm based on BST and extraction of directly scattered waves
SFCW	Stepped frequency continuous wave
SiGe	Silicon–germanium
SiP	System in package
SiSo	Single input–single output

SMP	Sub-miniature push-on connector
SoC	System on chip
SPR	Surface-penetrating radar
TDOA	Time difference of arrival
TDNA	Time domain network analyser
TDR	Time domain reflectometry
TDSEP,	Temporal de-correlation source SEParation
TDT	Time domain transmission
THD	Total harmonic distortion
TOA	Time of arrival
TWR	Through wall radar
UWB	Ultra-wideband
VCDL	Voltage-controlled delay line
VCO	Voltage-controlled oscillator
VNA	Vector network analyser
VANA	vector automatic network analyser
YIG	Yttrium iron garnet (ferrimagnetic[1] material); $Y_3Fe_2(FeO_4)_3$

1) Do not confuse with ferromagnetic.

Index

Handbook of Ultra-Wideband Short-Range Sensing: Theory, Sensors, Applications, First Edition. Jürgen Sachs.